Proceedings of the 1st International Conference on Water Energy Food and Sustainability (ICoWEFS 2021)

João Rafael da Costa Sanches Galvão ·
Paulo Sérgio Duque de Brito ·
Filipe dos Santos Neves ·
Flávio Gabriel da Silva Craveiro ·
Henrique de Amorim Almeida ·
Joel Oliveira Correia Vasco ·
Luís Miguel Pires Neves ·
Ricardo de Jesus Gomes ·
Sandra de Jesus Martins Mourato ·
Vânia Sofia Santos Ribeiro
Editors

Proceedings of the 1st International Conference on Water Energy Food and Sustainability (ICoWEFS 2021)

Springer

Editors
João Rafael da Costa Sanches Galvão
School of Technology and Management
Polytechnic of Leiria
Leiria, Portugal

Paulo Sérgio Duque de Brito
School of Technology and Management
Polytechnic Institute of Portalegre
Portalegre, Portugal

Filipe dos Santos Neves
School of Technology and Management
Polytechnic of Leiria
Leiria, Portugal

Flávio Gabriel da Silva Craveiro
School of Technology and Management
Polytechnic of Leiria
Leiria, Portugal

Henrique de Amorim Almeida
School of Technology and Management
Polytechnic of Leiria
Leiria, Portugal

Joel Oliveira Correia Vasco
School of Technology and Management
Polytechnic of Leiria
Leiria, Portugal

Luís Miguel Pires Neves
School of Technology and Management
Polytechnic of Leiria
Leiria, Portugal

Ricardo de Jesus Gomes
School of Technology and Management
Polytechnic of Leiria
Leiria, Portugal

Sandra de Jesus Martins Mourato
School of Technology and Management
Polytechnic of Leiria
Leiria, Portugal

Vânia Sofia Santos Ribeiro
School of Health Sciences
Polytechnic of Leiria
Leiria, Portugal

ISBN 978-3-030-75314-6 ISBN 978-3-030-75315-3 (eBook)
https://doi.org/10.1007/978-3-030-75315-3

This Springer imprint is published by the registered company Springer Nature Switzerland AG
The registered company address is: Gewerbestrasse 11, 6330 Cham, Switzerland

Preface

The **International Conference on Water Energy Food and Sustainability – ICoWEFS 2021**, taking place in Leiria/Portugal (May 10–12), Portugal, aims to be a major forum to foster innovation and exchange knowledge in the water-energy-food nexus, embracing the sustainable development goals (SDGs) of the United Nations, bringing together leading academics, researchers and industrial experts.

A climate-neutral continent by 2050 will drive technological, economic and societal transformations towards circular economies using more green and clean technologies and the decarbonisation of energy-intensive industries.

Innovation regarding water, energy, agri-food, bioeconomy, natural resources and the environment will speed up the transition towards sustainability, promoting water and food security in the world.

The forest and wood pine tree from the Leiria region welcomed us, and this material was the basis in the manufacture of the ships used in the Portuguese Navigations in the fifteenth and sixteenth centuries. Now in the twenty-first century, this region has several wind and biogas power plants supported by a solid industrial cluster.

The conference will be a networking and collaboration among participants to advance the knowledge and identify major trends in the fields mentioned above, even in an online format according to the health rules.

We are grateful to the authors from 33 countries with their contribution of 98 papers accepted to be presented at ICoWEFS 2021 and published by Springer Nature, to the directors and staff of the School of Technology and Management of the Leiria and Portalegre Polytechnics for their support, to the research centres and sponsoring companies, to the members of the scientific committee and external reviewers, keynote speakers, and, finally, to the members of the organisation, who with redoubled efforts during a pandemic time managed to carry out this conference.

We hope that we can meet again at the next ICoWEFS.

João Galvão
Paulo Brito
Conference Chairs

ICoWEFS
2021

Organisation

Conference Co-chairs

João Rafael da Costa Sanches Galvão
Polytechnic of Leiria, Portugal

Paulo Sérgio Duque de Brito
Polytechnic of Portalegre, Portugal

Organising Committee

Filipe dos Santos Neves

Flávio Gabriel da Silva Craveiro

Henrique de Amorim Almeida

Joel Oliveira Correia Vasco

Luís Miguel Pires Neves

Ricardo de Jesus Gomes

Sandra de Jesus Martins Mourato

Vânia Sofia Santos Ribeiro

Scientific Committee

Abel Rodrigues	INIAV/Research Institute, Portugal
Ahmad Pourmovahed	Kettering University, Michigan, USA
Alfeu Sá Marques	Coimbra University, Portugal
Ana Cristina Rodrigues	Leiria Polytechnic, Portugal
Ana Gomes	ESB-UCP/Catholic University, Portugal
Ana Vale	School of Science IT Sligo, Ireland
Annukka Pakarinen	HAMK University, Finland
Antonio Macías	Extremadura University, Spain
Carla Ferreira	ESAC/Coimbra Polytechnic, Portugal
Carlos Borrego	Aveiro University, Portugal
Carlos Henggeler	INESCC/Coimbra University, Portugal
Eduardo Álvarez	Oviedo University, Spain
Eliseu Monteiro	Coimbra University, Portugal
Fausto Freire	Coimbra University, Portugal
Fernando Lidon	Nova University, Portugal
Hassan El Bari	Ibn Tofail University, Kenitra, Morocco
Helena Bártolo	Leiria Polytechnic, Portugal
Jerónimo Cortés	CICTE/Extremadura, Research Centre Spain
João Pin	FCT/Nova Lisbon University, Portugal
João Sampaio	Polytechnic of Cávado and Ave, Portugal
Joaquim Sousa	Coimbra Polytechnic, Portugal
John Cosgrove	Limerick Institute of Technology, Ireland
Jorge MGP Isidoro	Algarve University, Portugal
José Lobo	Extremadura University, Spain
Kirsi Knuuttila	JAMK University, Finland
Liliana Vitorino	Leiria Polytechnic, Portugal

Lisiane Librelotto	Universidade Federal de Santa Catarina, Florianópolis, Brazil
Lucélia Hoehne	Universidade do Vale do Taquari, UNIVATES, Brazil
Luís Loures	C3i/Portalegre Polytechnic, Portugal
Luís Silva	Porto Polytechnic, Portugal
Luís Tarelho	Aveiro University, Portugal
Luísa Scmidt	ICS/Lisbon University, Portugal
Madalena Moreira	Évora University, Portugal
Margarida Gonçalves	Nova University, Portugal
María Hernández	Extremadura University, Spain
Marika Tossavainen	HAMK University/Finland
Marina Antunes	ISCAL/Lisbon Polytechnic, Portugal
Marritta Kymalainen	HAMK University, Finland
Milica Kasanin-Grubin	University of Belgrade, Serbia
Nicolau Almeida	Portalegre Polytechnic, Portugal
Noel Perera	Birmingham City University, UK
Nuno Simões	Coimbra University, Portugal
Pedro Mucharreira	ISCE/Lisbon University, Portugal
Ramón Sanguino	Extremadura University, Spain
Raquel Antunes	Leiria Polytechnic, Portugal
Ricardo Chacartegui	Sevilha University, Spain
Ricardo Simões	Polytechnic of Cávado and Ave, Portugal
Rodrigo Moruzzi	UNESP/São Paulo State University, Brazil
Rui Ganhão	Leiria Polytechnic, Portugal
Sérgio Leitão	UTAD/Trás-os-Montes e Alto Douro University, Portugal
Sónia Pereira	Leiria Polytechnic, Portugal
Susana Neto	IST/Lisbon University, Portugal
Teresa Mouga	Polytechnic of Leiria, Portugal
Valentim Realinho	VALORIZA/Portalegre Polytechnic, Portugal
Zahra Kalantari	Stockholm University, Sweden

Keynote Speakers

Smart Bioeconomy to Enable Sustainability

Annukka Pakarinen

HAMK – Häme University of Applied Sciences, Finland

Dr Annukka Pakarinen holds a PhD degree in Environmental Solutions in Agriculture and a Master's degree in Chemical Engineering. She is a director of HAMK Bio Research Unit. She has expertise in utilising and treating (physical, chemical, enzymatic, etc.) plant or waste-based raw materials in biofuel production. As a research unit director, she closely follows the field of bioeconomy throughout the chain from primary production to value-added refining and smart solutions. She has also been a founder of a start-up company.

The Conditional Framework of Regional Biobased Innovations; Insights and Experiences from Regional Hubs in H2020 Bloom Project

Remco Kranendonk

Wageningen University and Research, The Netherlands

Remco Kranendonk is a senior researcher at Wageningen Environmental Research, working on regional development and bioeconomy regional clusters. Advices and facilitates regional innovation processes and has a broad experience in connecting science and regional networks of practitioners. He participated in the FP7 project BERST as WP leader in setting up a community of practice. He is the program leader of the Climate Smart Agriculture Program in the Netherlands within Climate KICs. He was the organiser of the TCI 2016 Global Cluster Conference and responsible for WUR Strategy on Regional Partnerships. He is active in the H2020 Bloom project, focussing on raising awareness of biobased perspectives within the civil society; he is responsible for extending the regional triple helix networks to a quadruple helix interplay.

Ecodesign and Sustainable Packaging: Challenges, Trends and Perspectives

Ana L. Pires

MARE/FCT NOVA University of Lisbon, Portugal

Ana Pires received a PhD in Environmental Engineering from the Faculty of Science and Technology of the NOVA University of Lisbon (FCT NOVA). She is a researcher and invited professor at FCT NOVA in solid waste management and circular economy and a author of several articles and communications, with a book awarded by the International Solid Waste Association. Her research areas focus on developing sustainable eco-design methodologies, reuse-based circular economic models and sustainability metrics for hierarchical waste management. She is a co-founder of the waste@NOVA network of researchers dedicated to innovation in waste treatment and management technologies.

All life on Earth depends upon healthy Oceans—Choose your lifeline

Howard Dryden

GOES Foundation, University of Edinburgh, Scotland UK

Howard is an internationally acclaimed marine biologist, now takes the helm as Chief Scientific Officer of GOES Foundation (Global Oceanic Environmental Survey). As an academic and successful businessman, Howard's 30+ year career has been focussed on one vision—to make the aquatic environment safer and cleaner for humans and all the other organisms on which our life depends. Along with his team of marine biologists and an experienced globally connected group of policymakers, communication experts and management consultants are the driving force behind GOES. Being aware that all life on Earth depends on a healthy oceanic ecosystem and that climate change is all about carbon dioxide (but that most of the effort by Governments has gone into reducing emissions) and that it is known that elimination of aquatic environmental pollution from toxic-for-ever chemicals such as PCBs and oxybenzone is primordial to avoid ocean acidification and a complete collapse of the entire marine environment, Howards' focus is about restoring oceanic ecosystems to sequester carbon dioxide.

Stating that if we do not eliminate pollution within ten years, then we lose the marine ecosystem and biodiversity in 25 years, along with the food supply for 2 billion people, the GOES Foundation is now the primary focus of Dr. Howard Dryden.

Biochemical Biorefineries for Advanced Biofuels

Francisco Gírio

LNEG-National Laboratory of Energy and Geology, Portugal

Francisco Gírio, PhD, is a principal researcher and the head of Bioenergy Department of LNEG-National Laboratory of Energy and Geology, Lisbon, Portugal. In his research career, he participated as a researcher in more than 47 research projects on lignocellulose bioconversion, being the coordinator of 14. He is currently a member of the Management Board of Bioenergy Joint Program of the European Energy Research Alliance (EERA-Bioenergy), member of the Steering Committee of the European Technology and Innovation Platform on Bioenergy (ETIP-Bioenergy), member of the Scientific Council and Board of Trustees of IMDEA Energy (Spain) and President of SIADEB—Iberoamerican Society for the Development of Biorefineries. He did publish over more than 100 peer-review papers and more than 200 papers and lectures at international conferences.

The Future Challenges of Water Management as a Finite Resource

Alberto C. Neto

Be Water (PT) CEO

Since 2017, Alberto Carvalho Neto is the chief executive officer at BEWG (PT), a company working on water, energy, waste sectors and environment projects, improving the circular economy. Leader in a group with more than 400 workers in Portugal, he is also involved in the challenging international development of the group, essentially in Portuguese speaking countries, North Africa and Europe, through the group's water expertisein the last three decades in Portugal. The challenges in the environment for the following years are relevant and strategic for the future of humankind, and companies such as BEWG will have a role to play in the management and preservation of limited natural resources.

Global Electric Power Sector: Engaging with Environment Issues

Saifur Rahman

Joseph Loring Professor and Director Virginia Tech Advanced Research Institute, USA

Professor Saifur Rahman is the founding director of the Advanced Research Institute (https://www.ari.vt.edu) at Virginia Tech, USA, where he is the Joseph R. Loring professor of electrical and computer engineering. He directs the Centre for Energy and the Global Environment (https://www.ceage.vt.edu). He is Life Fellow of the IEEE and IEEE Millennium Medal winner and was the President of the IEEE Power and Energy Society (PES) in 2018 and 2019. He was the vice president for IEEE publications and a member of the IEEE Board of Directors in 2006. He was a member of the IEEE Society's governing board on the Social Implications of Technology (SSIT) from 2016 to 2018. He was the founding editor-in-chief of the IEEE Electrification Magazine and the IEEE Transactions on Sustainable Energy. He served as a chair of the USA National Science Foundation Advisor Committee for International Science and Engineering from 2010 to 2013. He is a member at the large of the IEEE-USA Energy Policy Committee. He is a distinguished lecturer for the IEEE PES and has lectured on smart grid, energy-efficient buildings, renewable energy, demand response, distributed generation and critical infrastructure protection topics in over 30 countries on all six continents. He has published over 140 journal papers and has made over four hundred conferences and invited presentations.

In recognition of his expertise and contributions to energy efficiency research, Dr Rahman was appointed to the State of Virginia Governor's Executive Committee on Energy Efficiency in 2015. He is the only member of the academia of this 12-member committee. He was also a member of the Virginia Energy Efficiency board of governors Council from 2015 to 2018. He received his PhD in electrical engineering from Virginia Tech in 1978. His MS degree is from the Stony Brook University and has BScEE degree from the Bangladesh University of Engineeringand Technology. His industry and government experience includes work with the Tokyo Electric Power Company in Japan, the Brookhaven National Laboratory in New York, Duke Energy in North Carolina and consultancy for the World Bank, the United Nations, US Agency for the International Development and the Asian Development Bank.

Conference Sponsors

The conference co-chairs and organising committee wish to acknowledge the support and sponsorship given in the organisation of the ICoWEFS 2021—International Conference on Water Energy Food and Sustainability, held at the School of Technology and Management of the Polytechnic of Leiria, Portugal:

Official Sponsors

POLITÉCNICO DE LEIRIA

POLITÉCNICO DE PORTALEGRE

ICoWEFS 2021 is Organised in Cooperation with

inesc coimbra

VALORIZA

Event Sponsors and Exhibitors

Contents

Water

AGRIFOOD

Effects of COVID-19 on Food Security, Safety and the Supply Chain

Haley Fox, Jennyfer Piña, Miranda Reinhart, and Ahmad Pourmovahed(✉)

Mechanical Engineering Department, Kettering University, Flint, MI 48504, USA
{fox4147,pina8706,rein9130,apourmov}@kettering.edu

Abstract. COVID-19 has heightened all factors leading to food insecurity globally. Currently, governments and food charity organizations are working to alleviate the outbreak's adverse effects on food security. To combat the heightened food insecurity, the focus needs to be on safely reopening workplaces for the purpose of increasing household incomes, expanding the food supply distribution network, and providing food for children who typically receive subsidized lunches at school. Concern for food safety has risen during the pandemic due to the hypothesis that the disease stems from a similar strain of coronavirus found in bats. Regulating safe food is difficult due to stay-at-home orders but is more important now than ever. Production, accessibility, and sustainability are taken more seriously because of COVID-19 to ensure there is enough food available while keeping society healthy and safe. COVID-19's effects on the food supply chain vary drastically within developed and developing countries. Overall, in more developed countries, the food supply chain remained resilient to delays, understaffing, and demand switches thanks to increased automation within the supply chain and innovations within the "consumer" stage. Due to the dependence on labor-intensive methods in developing countries, the supply chain has been detrimentally affected at all levels. Through different integrations of technology, developing countries will rely less on labor and be better prepared to withstand similar situations in the future. This paper reports the results of the research done to investigate various ways in which community and government assistance can counteract the effects of COVID-19 on food.

Keywords: Food security · Food safety · Food supply chain · COVID-19

1 Food Security

1.1 Current Stance

Food insecurity occurs when a person or household has insufficient means to regularly provide enough food to meet its needs. Before March of 2020, 135 million people were acutely food insecure, 690 million people were chronically food insecure [1], and 2 billion people did not have regular access to safe and sufficient food [2]. Places like Africa, Yemen and Southeast Asia are the most at-risk areas for food insecurity. With

J. R. da Costa Sanches Galvão et al. (Eds.): *Proceedings of the 1st International Conference on Water Energy Food and Sustainability (ICoWEFS 2021)*, pp. 3–11, 2021.
https://doi.org/10.1007/978-3-030-75315-3_1

extreme weather patterns, political conflicts and high poverty rates, the food supply is low, and many people cannot afford the available food. Before the COVID-19 crisis began these areas already had above-average levels of malnourished or underfed populations.

1.2 Direct Impact from COVID-19

COVID-19 swept across the globe, disrupted the usual working conditions, and had a massive impact on global food security. By the end of 2020, the number of people acutely food insecure was expected to double to 265 million due to COVID-19 [3]. Stay-at-home orders, quarantines, and forced (often unpaid) sick time, have caused businesses to close resulting in laying off, furloughing, or firing many of their employees. Of all the industries affected by COVID-19, none have been as disrupted as the service industry. With in-person dining restricted, servers who largely rely on tips for their pay have been hit especially hard. Leisure and hospitality industries have also seen the highest increase in food insecurity since the start of COVID-19. Workers in these industries experienced above-average rates of food insecurity before the pandemic, 16–17% in the United States. As seen in Fig. 1, the service industry has the highest percentage of workers qualifying as food insecure at around 17.5% while employing 18% of the workforce [4].

Though the service industry has experienced the worst of the pandemic's economic effects, nearly every industry is facing hardship. Estimates [4] show that if unemployment continues to rise, food insecurity rates will increase 1–5.2% in the United States with 3.3 to 17.1 million people affected (Fig. 2). Other countries around the world are facing even more severe unemployment rates and poverty level. As seen in Fig. 2, even a one percent hike in the unemployment rate will result in over three million people falling into food insecurity in the United States alone. Nearly every country is following the same trend.

The pandemic is also negatively affecting certain age groups more than others. Children around the world are the most at-risk age group for food insecurity. Children in low-income families who qualify for reduced or free lunch at school, are no longer able to attend regular classes in COVID-19's online schooling systems to get their daily meal. This has caused these students to face the challenges of a typical summer break malnutrition since March of 2020. Single-mother families are facing an even tougher increase of 170 percent in food insecurity cases for their children under the age of 12 [5]. Along with help from several cities and school districts, the United States Department of Agriculture (USDA) also made an order allowing schools to extend summer meal programs to continue to provide free meals to children but this order was terminated on December 31, 2020 [6].

1.3 Other Factors Affecting Food Insecurity in 2020

Other factors are currently spreading famine and food insecurity around the world. A locust infestation worse than anything anyone has seen in a generation is swarming in the Middle East and the horn of Africa. Locust destruction of crops and grazeland is projected to leave 5 million people in Africa facing the starvation level of food insecurity [7]. Conflict, economic shocks, health shocks, and weather extremes are just a few more

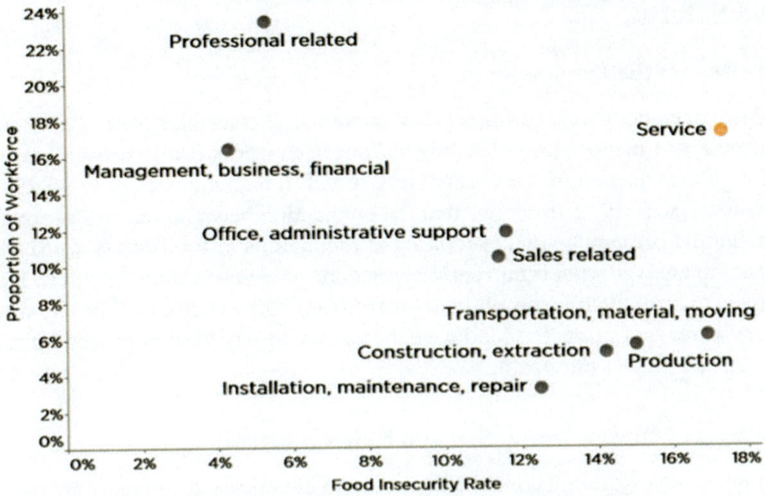

Fig. 1. Food insecurity rate by industry in the USA [4]

	Scenarios		
	A	B	C
Unemployment rate increase (% pts)	1.1	4.5	7.6
Poverty rate increase (% pts)	1.5	2.6	4.8
Food insecurity rate increase (% pts)	1.0	3.0	5.2
Increase to number of food-insecure individuals	3.3 million	9.9 million	17.1 million

Fig. 2. Estimated increase in food insecurity during the pandemic based on possible increases in unemployment rate as the pandemic continues in the United States [4]

ways that have been contributing to food insecurity within nations long before this pandemic [8].

The current actions of governments have not been enough to face the global challenge of food security, as seen by the near doubling of the world population qualifying as food insecure. Outside of the benefits of food programs like SNAP in the USA which are assisting in the effort to get meals to students while in school, overall food security systems need to be improved and expanded. When the first step to beating the virus is cooperation and mutual support so that everyone can remain safe, communities, schools, governments, and global organizations need to contribute their part to that effort meeting the needs of those who are struggling to provide the basic necessities to themselves and their families.

2 Food Safety

2.1 The Root of the Pandemic

Studies have been done suggesting a possible source of transmission was consumption of bats from a wet market, however, this has not been scientifically proven. According to [9], the gene sequence of the coronavirus found in bats matches only 96.2% to the coronavirus COVID-19. This shows that the connection between the two viruses is not absolute, but the percentage makes looking at regulations of food safety worthwhile. It is necessary to analyze what is currently being done about food safety and how it should be improved to limit the transfer of diseases from animals to humans. The situation also raises our awareness of how the food we eat on a day-to-day basis is processed and how this process can impact our health.

2.2 Methods of Disease Prevention and New Standards

Most countries have different codes in place to regulate safe food and these are monitored during production, imports, sales, and other transactions. For instance, the Food and Drug Administration (FDA) is responsible for food safety in the United States, and the entirety of the European Union has acted by utilizing the European Food Safety Authority (EFSA) [10]. These organizations do their best when it comes to trade to make sure that food going across borders is safe for consumption. Because of COVID-19, some countries have restricted trade from entire regions or made the inspection process more intense. ISO 22000 is one newly established model to achieve food safety with mandatory implementation of *Hazard Analysis and Critical Control Points* (HACCP) to processing plants in several countries all over the world [10]. This helps achieve good practices in food safety no matter where you live and should help keep food safe during global pandemics.

For companies to sell or produce food and meet the regulations provided in these countries, inspection processes must be implemented. However, because of COVID-19, more inspections are being done from home. Food inspectors are learning to do their job remotely which can bring up certain issues. In an interview, Gudrun Gallhoff, Minister Counsellor for health and food safety, Delegation of the European Union to China, speaks about the new risks of process modifications and ingredient substitutions. She says that "Many of us are doing inspections via audio and you have to adapt your processes accordingly" [11].

As seen from COVID-19, sometimes regulations are not enough to make sure safe food is being consumed. There are ways of getting around regulations when it comes to the people in charge of producing and selling food. Due to this mentality, it ultimately comes down to the consumer to take appropriate safety measures. To avoid potential risk of infection at the consumer end, different countries are taking extra precautions when it comes to food safety. For example, some European restaurants have decided to stop serving raw steaks because, "Although foodborne transmission of COVID-19 has not been reported, avoiding raw and undercooked foods of animal origin (meat, eggs, milk products) will reduce exposure to all viruses and other foodborne pathogens" [12]. In meat, COVID-19 and many other pathogens can be killed if the internal temperature reaches 70 °C.

2.3 Food Safety in the World Post-COVID-19

Paying attention to food safety in this new world is more important than ever. Not only does food need to be looked at more closely to make sure everything is clean and safe to eat, but with a growing population, this needs to be achieved at an even larger scale. "The increasing demands for proteins, the increasing population, as well as the depletion of resources lead researchers to investigate more sustainable and safer food sources to feed the world and meet markets' needs" [13]. Accessing food by going to grocery stores and markets can also be a concern, but with social distancing protocols and plastic panels, getting food is a little safer. Finding different sources of nutrition is essential in order to make sure food is accessible to the growing population and also safe to consume. Inventive ways of producing alternative sources of food such as lab-grown meat are on the rise.

3 The Food Supply Chain

3.1 What is the Food Supply Chain?

The food supply chain (FSC) is the process in which food goes from farmer to consumers. This includes farming, production, processing, distribution, and consumption. COVID-19 has managed to impose various obstructions and difficulties through each link of the food supply chain. In developed countries, the food supply chain has been rather resilient compared to that of developing countries due to the difference in dependence on labor-intensive methods.

3.2 How Has the Food Supply Chain Been Affected in Developed Countries?

Farm production is the first phase within the FSC. This stage consists of planting seeds, maintaining crops and animals, and harvesting. COVID-19 has affected this stage of the food supply chain through labor and source displacement. Many farmers within the United States depend on immigrant seasonal workers for labor-intensive produce like fruits and vegetables. Thus, 53% of all U.S. farmworkers were born outside the country and most of these immigrants enter the U.S. with H-2A Visas [14]. With travel regulations set in place to control the spread of COVID-19, in-person interviews for the visa have been delayed causing many farms to be temporarily short-staffed. There was also a shortage of workers due to immigrant workers sharing housing which makes the spread of the virus harder to control [15]. Sourcing of seeds and fertilizers was also an issue due to the pandemic.

Processing plants, the next phase of the FSC may look large from the outside but only use small areas for their workers during the packaging process. Due to the confined working environment, processing plants were required to reduce staff or shut down completely. The drop in demand from the plants caused a fall of about 40% in U.S. pig slaughter [16]. This, unfortunately, forced farmers to euthanize their animals to prevent overcrowding and disease.

Lastly, distribution includes all methods in which food is moved around the world such as transportation by boats, trains, planes, or trucks. The main issues that affected this phase were the reduction of overall staff, increased safety regulations and checks during a border crossing. In European countries, the food supply chain is heavily dependent on imports and exports. This means that food is often moved between countries with trucks and the lack of staff can lengthen the time to reach stores and unload food.

Developed countries have been very quick to adjust to the drastic switch in consumer demand. During the first few months of the pandemic, the demand for retail food spiked as people panicked and stockpiled. Since then, the demand for retail food has remained higher than last year as shown in Fig. 3.

Fig. 3. Food demand index for the United States, same period last year = 100 [17]

3.3 Initiatives in Developed Countries

Many developed countries experienced severe damage caused by COVID-19 with many small businesses losing income and closing. Most restaurants began offering options for their customers to eat at home. Options included takeout, delivery, and even at-home meal kits. Grocery stores began offering similar services and reserving certain hours for their most at-risk customers, like senior citizens. Farmers were also able to ship produce that required very limited processing, directly to customers. In New Zealand, new initiatives allowed farmers to donate produce to food banks in desperate need. With government support in the U.S., the requirement for in-person interviews was dropped to expedite the H-2A Visas for immigrants entering to work on farms.

3.4 The Food Supply Chain in Developing Countries

The most prevalent type, transitional food supply chains (Fig. 4) rely heavily on labor-intensive methods for the production, processing, and distribution stages. This means that the main method for the consumers to obtain food is through small and medium-sized enterprises (SMEs). These are more commonly known as wet markets or street food markets. They have independent sellers who are not held to any governmental standards which leaves the consumers to decide if the food is safe. These SMEs are generally in densely populated areas and are usually very busy as they are often the only place to purchase food. This presents an obvious problem with COVID-19 regulations requiring social distancing.

	Traditional FSC	Transitional FSC	Modern FSC
Approximate prevalence in Africa & South Asia as share of food economy	10%	70%	20%
Approximate prevalence in Southeast Asia & Latin America	5%	50%	45%
Main enterprise type	Home microenterprise	SMEs, wetmarkets	Supermarkets, large processors
Length	Short, local	Long, rural-urban	Long, rural-urban, international
Use of Arrangements	No contracts, no standards	No contracts, public standards	Emerging contracts, private standards
Technology	Labor-intensive	Labor-intensive	Capital-intensive

Fig. 4. Three stages of the food supply chain and their prevalence in the food economy of developing countries [18]

Food distribution by air is often carried out through commercial passenger airplanes and has experienced negative effects in developing countries. In Kenya, up to 70% of the population depends on income from exports of fresh produce [18]. With the COVID-19 regulations, reduced airplane capacity, and doubled export costs, Kenya reported that they were only able to export 40% of their total capacity at the beginning of the pandemic due to bans on passenger flights [19]. Enforcement of regulations caused an increase in transit time resulting in the loss of some perishable goods.

3.5 Methods for Helping Developing Countries

The main reason the food supply chains in most developing countries are not as resilient as those in developed countries is their dependence on labor-intensive methods [20]. To lessen the gap between developed and developing countries, implementation of technology in the food supply chain is imperative. Government or private funding will be needed for building automated processing plants, sanitation plants, industrial sowing and harvesting tools, and proper refrigeration and storing facilities. A shorter-term method that could be implemented more immediately is governments using cash-for-work schemes that would employ citizens for the distribution of emergency food rations.

4 Concluding Remarks

In the year 2020, COVID-19 engulfed the world quickly and devastated many parts of the globe through physical illness and economic ruin. In most developed countries, sizable economic stimulus programs have helped many businesses cope with the financial crisis although some businesses, in particular many restaurants, may not be able to recover. Take-out options and grocery deliveries have become commonplace in many parts of the world. Many governments have instituted new immigration and food inspection policies to better protect food safety and availability. Children who typically receive subsidized lunches at school have been particularly impacted by the outbreak.

Many developing countries, especially those heavily depending on food exports have been adversely affected by the disruptions in the food supply chain. Sustained assistance from governments and food charity organizations is crucial in alleviating the adverse effects of COVID-19 on food security. The focus needs to be on safely reopening schools and businesses through widespread vaccination. The public is understandably concerned about food safety more now than ever. By implementing automation and various new technologies – such as vertical farming - developing countries can rely less on labor and be better prepared to combat the effects of similar outbreaks in the future.

References

1. Food Security and COVID-19 (worldbank.org). https://www.worldbank.org/en/topic/agriculture/brief/food-security-and-covid-19. Accessed 1 Mar 2021
2. United Nations World Food Programme (WFP) - WFP.org. https://www.wfp.org/. Accessed 1 Mar 2021
3. The coronavirus could double severe hunger around the world to 265 million people without 'swift action,' a UN report warns | Business Insider India. https://www.businessinsider.in/international/news/the-coronavirus-could-double-severe-hunger-around-the-world-to-265-million-people-without-swift-action-a-un-report-warns/articleshow/75281526.cms#:~:text=Severe%20food%20insecurity%20across%20the%20globe%20could%20double,,by%20the%20end%20of%202020,%20the%20UN%20says
4. Food Insecurity and Poverty in the US - Feeding America. https://www.feedingamerica.org/sites/default/files/2020-04/Brief_Impact%20of%20Covid%20on%20Food%20Insecurity%204.22%20%28002%29.pdf. Accessed 1 Mar 2021
5. The COVID-19 crisis has already left too many children hungry in America (brookings.edu). https://www.brookings.edu/blog/up-front/2020/05/06/the-COVID-19-crisis-has-already-left-too-many-children-hungry-in-america/. Accessed 1 Mar 2021
6. USDA Extends Free Meals for Kids through December 31, 2020, Food and Nutrition Service | U.S. Department of Agriculture Webpage. https://www.fns.usda.gov/news-item/usda-035520. Accessed 1 Mar 2021
7. Millions in Africa face starvation as huge locust swarms descend | World Economic Forum (weforum.org). https://www.weforum.org/agenda/2020/06/locusts-africa-hunger-famine-covid-19/. Accessed 1 Mar 2021
8. Ghosh, I.: "Hunger Pandemic: The COVID-19 Effect on Global Food Insecurity," Visual Capitalist. https://www.visualcapitalist.com/COVID-19-global-food-insecurity/
9. Rizou, M., Galanakis, I.M., Aldawoud, T.M.S., Galanakis, C.M.: Safety of foods, food supply chain and environment within the COVID-19 pandemic. Trends Food Sci. Technol. **102**, 293–299 (2020). https://doi.org/10.1016/j.tifs.2020.06.008. https://www.ncbi.nlm.nih.gov/pmc/articles/PMC7295520/. Accessed 1 Mar 2021
10. A Country-by-Country Look at Regulations and Best Practices in the Global Cold Chain (food-safety.com). https://www.food-safety.com/articles/4618-a-country-by-country-look-at-regulations-and-best-practices-in-the-global-cold-chain. Accessed 1 Mar 2021

11. Whitworth, J.: Experts share regional and global COVID-19 insight. Food Safety News, 1 August 2020. https://www.foodsafetynews.com/2020/08/experts-share-regional-and-global-COVID-19-insight/. Accessed 1 Mar 2021
12. Food Safety in the time of COVID-19, Food and Agriculture Organization of the United Nations. https://www.fao.org/documents/card/en/c/ca8623en/. Accessed 1 Mar 2021
13. Galanakis, C.M.: The Food Systems in the Era of the Coronavirus (COVID-19) Pandemic Crisis. Foods | Free Full-Text. https://www.mdpi.com/2304-8158/9/4/523/htm. Accessed 1 Mar 2021
14. Willingham, Z., Mathema, S.: Protecting Farmworkers from Coronavirus and Securing the Food Supply. Center for American Progress, 23 April 2020. https://www.americanprogress.org/issues/economy/reports/2020/04/23/483488/protecting-farmworkers-coronavirus-securing-food-supply. Accessed 1 Mar 2021
15. Coronavirus US: Every Worker at One Farm Has Covid on Harvest Eve - Bloomberg. https://www.bloomberg.com/news/articles/2020-05-29/every-single-worker-has-covid-at-one-u-s-farm-on-eve-of-harvest. Accessed 1 Mar 2021
16. Kallingal, M.: Even as grocery stores limit meat sales, US farmers may have to euthanize 10 million pigs, CNN, 16 May 2020. https://edition.cnn.com/2020/05/15/us/hog-farmers-euthanize-coronavirus-trnd/index.html. Accessed 1 Mar 2021
17. Food Supply Chains and COVID-19: Impacts and Policy Lessons, Organisation for Economic Co-operation and Development, 2 June 2020. https://www.oecd.org/coronavirus/policy-responses/food-supply-chains-and-covid-19-impacts-and-policy-lessons-71b57aea. Accessed 1 Mar 2021
18. Reardon, T., Bellemare, M.F., Zilberman, D.: How COVID-19 may disrupt food supply chains in developing countries. International Food Policy Research Institute (IFPRI), 2 April 2020. https://www.ifpri.org/blog/how-covid-19-may-disrupt-food-supply-chains-developing-countries. Accessed 1 Mar 2021
19. Organization for Economic Co-operation and Development. OECD Scheme for the Application of International Standards for Fruit and Vegetables, 11 May 2020. https://www.oecd.org/agriculture/fruit-vegetables/oecd-COVID-19-impact-on-fruit-and-vegetables-trade.pdf. Accessed 1 Mar 2021
20. COVID-19 and resilience innovations in food supply chains | IFPRI : International Food Policy Research Institute. https://www.ifpri.org/publication/covid-19-and-resilience-innovations-food-supply-chains. Accessed 1 Mar 2021

Understanding the Behavior of the Low-Income Population When Purchasing Fruits and Vegetables. Case Study of Cochabamba (Cercado) - Bolivia

Agatha Clarice da Silva-Ovando[1,2](✉) and Carla Andrea Ocampo-Terceros[2]

[1] Universidad de la Sabana, Chía, Colombia
agathadaov@unisabana.edu.co, agathadasilva@upb.edu
[2] Universidad Privada Boliviana UPB, Cochabamba, Bolivia

Abstract. A region is considered food secure when all of its four pillars are met: accessibility, affordability, safety and utilization. Unfortunately, developing countries such as Bolivia presents a very high food insecurity index, due its high levels of poverty. The paper aims to discover relevant factors that people with an average income for the region, located in peripherical areas, considers at the moment of acquire their family basket, especially fruits and vegetables, and how their environment affects their purchase decision regarding fresh and healthy food. The study was performed on Cochabamba – Bolivia during the time of the pandemic crisis of 2020.

Keywords: Food safety · Food security pillars · Food insecurity

1 Introduction

According to the Food and Agriculture Organization (FAO), food security is possible when people have physical, social and economic access to safe, healthy and enough food steadily, in order to satisfy their nutritional requirements and food preferences. The Food and Agriculture Organization concluded that food insecurity in Latin America increased by 4.5 million making a total of 42.5 million people fighting against hunger by 2018. From a long-term perspective the region (Latin America and the Caribbean) has managed to reduce the undernourished people by 20 million compared to 2000. However, 2018 has been the fourth consecutive year where the hunger level has increased and continued to grow during 2020 [1, 2].

In line with the above, the estimations made consider only the regular issues attacking this chain, and ignoring further disruptions that may appear to hinder the chain, such as the one caused by the novel COVID-19 pandemic. Food insecurity is a reality on any food chain of every country, but has gained more weight due to the pandemic situation.

FAO has dedicated several years on making estimations of the Global Food Security Index (GFSI), which considers the core issues of affordability, availability and quality across a set of 113 countries, inside Latin America; Bolivia has the fourth place with the lowest GFSI (see Fig. 1) [3].

J. R. da Costa Sanches Galvão et al. (Eds.): *Proceedings of the 1st International Conference on Water Energy Food and Sustainability (ICoWEFS 2021)*, pp. 12–19, 2021.
https://doi.org/10.1007/978-3-030-75315-3_2

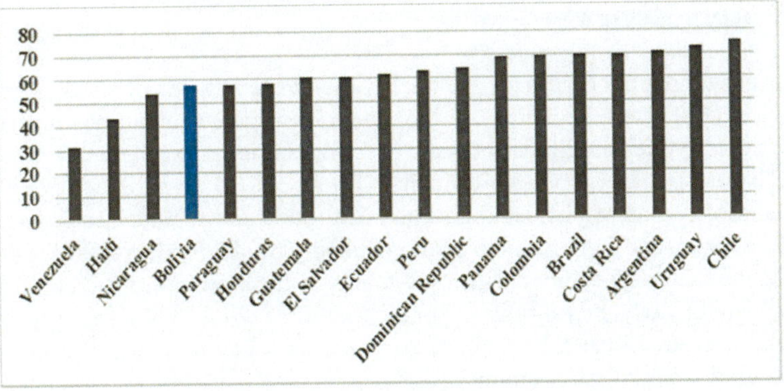

Fig. 1. Global food security index

In Bolivia, 60% of the urban population has an income of $726.92 or less per month, and in the city of Cochabamba, the 28.35% of the total population belongs to that percentage. This group have expenses around $130.84 to $232.61 per month to acquire a basic family basket. This means that around 25% of the family income is intended for food purchases. For this reason, it is relevant to understand the triggers and preferences of this part of the population when purchasing fresh food.

However, affordability is only one of the pillars for food security. Accessibility is another pillar that must be tackled when analyzing the conditions presented in the environment that may influence the consumer in the moment of purchase.

For that reason, within the city of Cochabamba, certain areas were selected according its distance to the city center, and its low presence of fresh food retailers, such as nanostores, local markets, and supermarkets. The selected areas were distributed on the zones presented below (see Fig. 2). With this, it is expected to be able to narrow the gap between good nutrition and low resources [4–6].

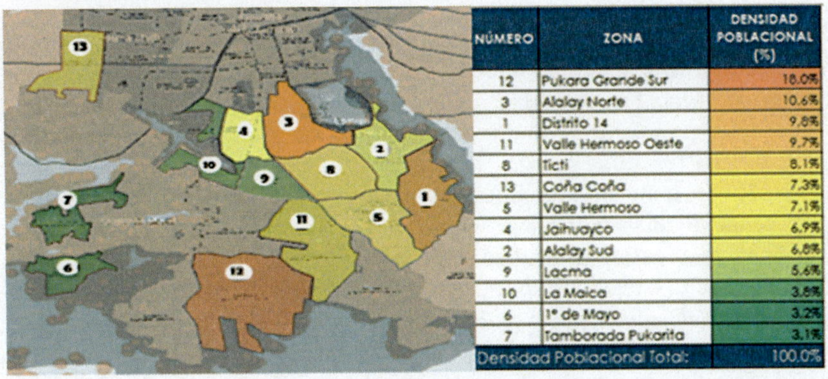

NÚMERO	ZONA	DENSIDAD POBLACIONAL (%)
12	Pukara Grande Sur	18.0%
3	Alalay Norte	10.6%
1	Distrito 14	9.8%
11	Valle Hermoso Oeste	9.7%
8	Ticti	8.1%
13	Coña Coña	7.3%
5	Valle Hermoso	7.1%
4	Jaihuayco	6.9%
2	Alalay Sud	6.8%
9	Lacma	5.6%
10	La Maica	3.8%
6	1° de Mayo	3.2%
7	Tamborada Pukarita	3.1%
Densidad Poblacional Total:		100.0%

Fig. 2. Target population density

2 Literature Review

In order to ensure regular access to high quality food to people, the Food and Agriculture Organization (FAO) has implemented four main dimensions, also known as the four pillars, to quantify the conditions and state of food security in each country and specific regions. The mentioned variables are Availability Access, Utilization and Stability [1].

Availability means to physical existence of food in the region; access means that people have opportunity to access to economic and physical food; stability is about the balance of the other three dimensions, and finally, utilization in terms of absorption of nutrients from the body. During this paper the first three will be taking into account, but it is important to emphasize the importance of each one of them to measure and evaluate the global food index of every country [2].

The global Food Security Index measures the overall index scores of affordability, availability, quality and safety of food in the region. This index according to Maxwell, Watkins, Wheeler and Collins, is a costly and complicated exercise. The same is a critic point to determine that any region is suffering, is going to suffer a severe impact, or in the worse scenario, a need of food aid interventions. To understand the variability of the food security index; the most recent publications of FAO during December of 2020, showed that due the pandemic events, the already fragile conditions of most of the countries of Latin America has raised sharply [3, 4].

Bolivia is a clear example of a frail region that have been heavily affected by the recent disruptions. This country adopted a strict lockdown since March to June of 2020, after months of conflicts due political issues. As effect, the supply chain of food experienced the severe consequences of empirical work; specifically, when talking about resilience on trade of fruits and vegetables, among other kind of food.

Similarly, cases have occurred around other countries in Latin America, and even when there are a lot of variables to analyze with Sheffi on his book: The Power of Resilience, where exposes different study cases of countries and enterprises that have achieved resilience when disruptions have occurred, affirms that the key of any organization, state or even a country to impulse the resilience - the ability to bounce back from large scale disruptions, is to understand the key client [5, 6].

Consequently, due that the global security index is related with the poverty level, arises a need to understand the relevant factors that people with an average income for the region, located in peripherical areas, considers at the moment of acquire their family basket, especially fruits and vegetables, and how their environment affects their purchase decision regarding fresh and healthy food. Since the analysis of people behavior or preferences are quite different from one to another, there will be a better scope if the range of people to analyze is homogeneous and the method chosen allows to stablish the functional relationship or mathematical equation that relates the variables, as well as the strength of that relationship [7].

To conclude, in line with the mentioned above, the multiple linear regression model is the one that fits with the mentioned needs. Chao, Zhao, Lawrence and Nylander French, explain the importance of using multiple linear regression analysis to quantify the relative importance of predictors on fields like public health, where the number of possible variables to consider are great, leading to a large and robust model. On the same direction Maradiaga, Rodríguez, Lozano and Vallejo aims the advantages to any professional of any field to the good use of this tool that is often little used [7, 8].

3 Methodology

To achieve the expected results for this study, the following methodology have been proposed.

The first approach to the study was to determine which variables and factors would be considered as influent. For that, some key aspects and definitions were considered and a literature review was carried.

In the first stage was the potential zones identification, where, an initial survey was released to different sectors of the city of Cochabamba. This, to determine more specific values for the output variables when designing and releasing the second study set of surveys and interviews, and to identify and select the most relevant region to focus the study.

After analyzing the first results, statistical analyses were performed, which led to the selection of type key variables of study to characterize the group, such as "age, gender, number of members per family, etc."

After the first approach to the group to be studied, a second release of the survey was made in the selected regions of the city of Cochabamba-Cercado, so we could validate and characterize the profile of the consumer and their preferences regarding fresh food.

A final regression study was carried with the final results of the fully released survey. These results provided a deep insight of the key variables that must be considered in the study.

After the first approach to the target group of the study, the information gathered was complemented with extended interviews to active actors in this supply chain. This allowed a deeper understanding of how the functioning of the supply chain may affect the consumer's decisions.

Finally, the results were analyzed and discussed.

4 Analysis and Experimental Results

On this segment, a brief evaluation on the surveys results and the linear regression used to understand the behavior of people with low income in the region of Cochabamba – Cercado is explained.

4.1 Surveys Analysis

To obtain the correct results, two polls were launched, the first one had the purpose of finding the areas where people with $ 726,92 were located. Then the second questionnaire was more detailed, with the main objective of characterizing this group.

It is recommended that before applying the model, we sought to locate the areas with the major percentage of people with low income. In the case of Cochabamba-Cercado, this percange can reach up to 28.35%, and due to the proliferation and mix of different range on income that can be found on the region there is not a concrete zone where people with low income but enough resources to acquire a family basket can be found.

During the procedure of understanding the characteristics of the population with an income less than Bs. 5000 ~ $726,92 the following results were found. First, these group are formed by 4 or 5 family members, a big part of them are composed by at least two elderly people and the youngest mostly tend to have modest jobs in the public sector. The most frequent jobs are workers of the public or private sector, then professionals and technicians and finally service staff.

Second, during the performance of the surveys, some problems concerning the people with low income were found. For example; they walk long distances to acquire fruits and vegetables, a big part of them cannot afford a place near to the city or even to some traditional big market, classic of some countries in Sudamerica. Elsewhere people with less than $726,92 per month affirm there were not any mobile markets by their zone during the quarantine restrictions, and even when some mobile markets go by the zone the fruits and vegetables arrive in bad conditions, which means that even if there is any access to healthy food, the safety or quality of the produce is not guaranteed in those regions.

4.2 Multiple Linear Regression Model – Experimental Results

As explained before, the group of study are people with relatively low income, but with the capacity to afford at least their family basket. Initially, based on the first poll launched, it was found that there is an important relation between the monthly household income and the weekly budget for the purchase of Fruits and Vegetables (F&V) (see Fig. 3).

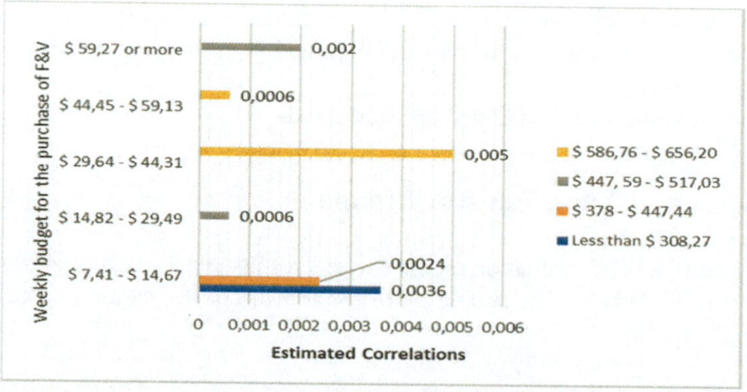

Fig. 3. Test of statical significance of estimated correlations

The analysis showed that people with low income does not necessarily indicate that they cannot afford fruits and vegetables, especially in South America, where the price of fruits and vegetables tends to be lower compared to other countries. Allocating a low budget for the purchase of fruits and vegetables tends to be more related to a lack of knowledge about the importance of good nutrition in people of Cochabamba – Cercado.

Taking in count the results of the polls released on Cochabamba, most of the people eat what they think is a balance and healthy food, their weekly diet is often based on fruits and vegetables of the type B and C, that is to say of the tomato, onion, carrot and lemon; according to the interviews to professionals and academic investigations, an adequate nutrition is based most on the consumption of fruits and vegetables of the type A and B, and in less quantity fruits & vegetables of the type C, also is important to point that, at least in Bolivia, most of the fruits & vegetables that are located in the group A tent to have a low cost due that most of them are well known herbs [9]. Is important to clarify that fruits and vegetables are classified in three types, (Type A, B and C). Each one divided by the quantity of carbs that contains, being those of the type A the ones that contain the least amount of carbohydrates and sugars.

Further, to understand the relevant elements that these people with low-income bear in mind to the moment of acquire their fruits and vegetables (F&V), the analysis was made as a support in the process of the making of the model, to understand if there is any relation between the monthly household income and the weekly budget for the purchase of F&V. The model below, was carried out with STATGRAPHICS CENTURION XV. II (Table 1).

$$\textbf{Purchase of F\&V} = 0{,}37 + 0{,}23 * A - 0{,}28 * B + 0{,}23 * C + 0{,}26 * D - 0{,}25 * E \quad (1)$$

Table 1. Regression variable's meaning

Letter	Description
A	Closeness
B	Hygiene
C	Security
D	Home Service
E	Variety and Assortment of products

The R-square is equal to 21.75% and the p-value of the model is less than 0.05 shows the existence of a statistically significant relationship between the variables with a confidence level greater than 95% (See Fig. 4), below the p-value for every relevant factor is presented.

To conclude, the variables that leads the customer of an income least than Bs. 5000 ~ $726,92 per month to acquire a family basket of fruits & vegetables are essentially the closeness, hygiene, security; in terms of reliability, home service and finally the variety and assortment of products. Concerning the low percentage of the R-square, this value is still a stage study, any discipline that attempts to predict human behavior, such as psychology, typically has an R-squared value less than 50%, this due human being tends to be more difficult to predict than; for example, physical process [10].

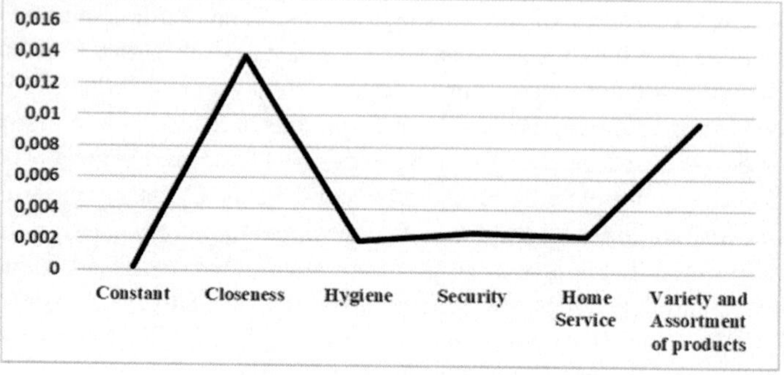

Fig. 4. P-Value of each element

5 Conclusions

The proposed method provides an analysis to the results found on the investigation, and it is possible to replicate the same to other groups and regions. It is important to clarify that the coefficient of determination may increase considering other variables, in this particular case, only those factors related to the user's preferences and way of purchasing products were considered.

In order of a better understanding on the preferences of people with low income, the objective is to collaborate to some regions, especially on Sudamerica, to set real targets in the short and long term to accomplish the Sustainable Development Goals Scheduled in FAO's 2030 Agenda.

The model initially considered many other factors like price, discounts, etc.; but the ones found relevant by the model were closeness, hygiene, security, home service, variety and assortment of products. Hygiene is based on the way the products are prepared before they are sold and security refers to the environment conditions at the moment of purchase.

Closeness and Security are somehow variables that can be inferred due the population of study and the type of products analyzed, however variables like variety and assortment of products are not. Taking into account the pillars mentioned at the beginning of the paper, this means people of low income in Cochabamba-Bolivia, have some fruits and vegetables within reach but not necessarily all of them. Finally, another intriguing variable found was home service, considering the population analyzed and the factors around them, thinking on home service is clearly a surprise, and will be interesting to discuss on further studies.

From the obtained results, some interesting new avenues for future research are presented, for example, consider other variables that are also important on this type of population or region, such as the number of children or elderly people on each family. Also, we encourage researchers to keep paying attention on this type of analysis in a humanitarian context.

References

1. Boliko, M.C.: FAO and the situation of food security and nutrition in the world. J. Nutr. Sci. Vitaminol. **65** (2019)
2. An Introduction to the Basic Concepts of Food Security. FAO. Security Programme, Rome
3. Maxwell, D., Watkins, B., Wheeler, R., Collins, G.: The Coping Strategies Index: a tool for rapidly measuring food security and the impact of food aid programmes in emergencies. CARE Eastern and Central Africa Regional Management Unit and the World Food Programme Vulnerability Assessment and Mapping Unit, Nairobi (2003)
4. Food and Agriculture Organization of the United Nations. FAO Food Price Index rises sharply. FAO-Newsroom. Sustainable Development Goals (2020)
5. Sheffi, Y.: The Power of Resilience: How the best Companies Manage the Unexpedted. MIT Press, Cambridge (2017)
6. Morales, R.: The resilience of informal workers to COVID19 and the difficulties of trade. Sociol. Int. J. **4** (2020). https://doi.org/10.15406/sij.2020.04.00232
7. Cardona Madariaga, D.F., González Rodríguez, J.L., Rivera Lozano, M., Cárdenas Vallejo, E.H.: Application of linear regression on the problem of poverty. Universidad del Rosario, Rosario (2014)
8. Chao, Y.C.E., Zhao, Y., Kupper, L.L., Nylander-French, L.A.: Quantifying the relative importance of predictors in multiple linear regression analyses for public health studies. J. Occup. Environ. Hyg. **5**(8) (2008). https://doi.org/10.1080/15459620802225481
9. Hernández, A., Borrel, M.A., Cotilla, L.L., Gonzáles, I.M.: Healthy food. Cuban Nurs. Mag. **20** (2004). ISSN 1561-2961
10. Rights, J.D., Sterba, S.K.: Quantifying explained variance in multilevel models: an integrative framework for defining R-squared measured. Psychol. Methods **24**(3), 300–309 (2019)

Gendered Video Production for Livestock Farmers in Senegal

Eija Laitinen(✉)

Häme University of Applied Sciences, P.O. Box 230, 13101 Hämeenlinna, Finland
eija.laitinen@hamk.fi

Abstract. In developing countries, women are the backbone of rural economies. Overlooking the needs of female farmers and their limited access to relevant information are among the key components hampering agricultural productivity development and poverty reduction. ICT based-knowledge transfer has transformative potential to enhance female farmers' productivity, to build adaptive capacity and enable access to critical information. However, this can be successful only if women's special needs are recognized. In the FoodAfrica project framework, the rationale for a gendered approach to video-mediated knowledge transfer in arid and semi-arid lands (ASAL), where the roles of women and men in agriculture are strongly gendered, was developed.

Keywords: Agricultural extension · Women farmers · Gender · ICT · Video

1 Introduction

The role of women in agriculture is fundamental. In Sub-Saharan Africa, women are about 50% of the agricultural labor force [1], some sources even claim that women's and girls' representation in the agricultural labor force is 80% [2]. If the yields of women's farms would be increased by 20–30%, it is estimated that the agricultural output overall would increase by 2.5–4% [1]. Rural women are critical stakeholders in water and energy management and environmental conservation. Almost two-thirds of fresh water is used in agriculture. For the sustainability of food production and scarce water resources, female farmers need access to relevant information.

Agricultural Knowledge and Information Systems (AKIS) and Agricultural Innovation Systems (AIS) are in place to help male and female farmers to make informed decision to improve their agricultural practices and livelihoods. High-quality agricultural extension means relevant, understandable, and reliable information [3].

In recent years, information, and communication technologies (ICT) have raised considerable attention as prominent tools to address the challenges of AKIS/AIS. The increased number of mobile phones and constantly improving access to internet have huge potential to provide quality extension services in remote rural areas and address the gender challenges of knowledge transfer. ICT-based knowledge transfer, including videos, has the transformative potential to enhance women farmers' access to critical

J. R. da Costa Sanches Galvão et al. (Eds.): *Proceedings of the 1st International Conference on Water Energy Food and Sustainability (ICoWEFS 2021)*, pp. 20–27, 2021.
https://doi.org/10.1007/978-3-030-75315-3_3

information and to increase agricultural productivity and the sustainable use of scarce natural resources. However, this can be successful only if, by extension, women's special needs are recognized.

In their historical overview on gender in agricultural extension, Manfre et al. [4] present the stumbling steps of extensions systems from the 1970s to achieve a gender-equitable knowledge transfer system. Today, the persistent gender gap still exists and challenges to women farmers' access to information, knowledge and advice is still restricted [e.g. 1, 5, 6].

The impacts of climate change increase the urgency to provide relevant and timely information to facilitate the adaption to changing production conditions. According to Aryal, Maraseni & Cockfield [7], grazing-based livestock, which is a complex production system compared to other agricultural production systems, suffers from the impacts of climate change more than other production systems. In pastoralism, livelihood is based on the direct utilization of natural resources as livestock is moved to utilize the seasonal availability of grazing resources. According to FAO [8], pastoralism is the main livelihood of an estimated 268 million people. Pastoralists living in harsh ecological conditions of arid and semi-arid lands (ASALs) of Sub-Saharan Africa are at a higher risk and vulnerable to extreme climate change impacts compared to those living in less fragile ecosystems.

The multidimensional nature of climate change requires innovative strategies that strengthen the resilience of vulnerable communities to first respond to the changes effectively and then adapt to such changes [9, 10]. The use of ICT/technology can play a mediating role and produce timely, targeted, and relevant information. A gendered approach to ICT-based knowledge transfer would be especially critical to helping women in pastoralist communities develop resilience and mitigation strategies for climate change.

2 Potential of ICT in Rural Resilience Building

2.1 Gender Gap in ICT-Based Knowledge Transfer

The Intergovernmental Panel on Climate Change [11] defines resilience as the ability of a social or ecological system to absorb disturbances while retaining the same basic structure and means of functioning, the capacity for self-organization, and the capacity to adapt to stress and change. Ospina & Heeks [12] describe resilience as a complex system with three primary foundations for resilience and six secondary enablers. The fundamentals of resilience are robustness, self-organization and learning and the enablers are redundancy, rapidity, scale, diversity, flexibility, and equality. The stronger the attributes are, the more resilient the ecosystem is. ICT-based capacity building can be applied to all of them. The resilience attributes assist in the design of strategic and effective ICT interventions to build or boost resilience.

In agricultural extension, ICT has enabled new solutions for knowledge and skills development. ICT-based solutions have reshaped the models of generating, storing and sharing information. Among others, ICT has paved the way to accessing downscaled weather forecasts, improved agricultural practices, increased access to reliable water sources and the promotion of a revolving microcredit system for women's self-help groups. Social media, discussion groups and various networks can all be parts of extension ecosystem. The success stories of ICT in agriculture in information service provision in agriculture and rural development are many [e.g. 13] and ICT's potential for climate change adaptation is high [14]. However, only a few of the ICT-based services have been able to scale up to large farmer numbers [15].

Further, ICT has substantial potential in the gender mainstreaming of agricultural information and innovation services, but this does not happen without systematically addressing the needs of girls and women. The challenges do not only link to the content of the service provision but may relate to the access and capacity of women to use the ICT-based solutions. Despite the progress in use and access to ICT, the persistent gender gap remains [16]. Over half (54%) of the women in low- and middle-income countries use mobile internet, but still women are 20% less likely to use internet than men. In Sub-Saharan Africa, the gender gap of internet usage (37%) has remained consistent. In his review article, Arjjumend [17] concludes that until now, only cell phone usage has gained footing in pastoralist communities, while the use of other technologies still requires input and training.

2.2 Women Use ICT Differently Than Men

In addition to women farmers' ability to access to information, there are a number of other barriers and obstacles to consider in use of ICT for women farmers' service provision and empowerment. Illiteracy as well as technical or digital illiteracy can strongly limit women's possibilities to even use text messaging. Because of lower ownership of mobile phones, women would need to use cyber cafés or tele-centers, but they may not feel comfortable doing so or may live in a place where women visiting premises frequented by men is seen as inappropriate. Women's traditional role as caretaker in homes may be seen as a more relevant duty for women than to engage in ICT. Women's participation in household decision-making can also be limited. Due to heavy workload and domestic responsibilities, women may have very limited time to engage with ICT. Also, the cost of acquiring facilities or using them, may hamper women's' ICT usage [18].

Female farmers' use of information and method of sharing information differs from the patterns and channels of male farmers. Also, regarding the use of digi-based information sources, women differ from men. Female farmers may have fewer networks than male farmers, their networks may be closer to home, but they develop stronger relationships with value chain actors and are active consumers of radio, TV and mobile phones. Women require more detailed information of required product/practice, but armed with details, they are more likely than men to use the new know-how [19, 20].

Videos can be powerful tools for empowering women. The Zossou, Van Mele, Vodouhe & Wanvoeke [21] research showed that through a farmer-to-farmer video on parboiling technology of rice, the members of a cooperative of female rice farmers were able to gain multiple benefits. They did not only learn the new technology, but the videos also enhanced women's entrepreneurial mindset, which then encouraged them to develop activities for profit. The videos also strengthened and built social cohesion by encouraging women to organize themselves, and then the new groups were able to demand and access more training workshops. Lecoutere, Spielman & Van Campenhout

[22] also point out the multiple impacts of videos on women's empowerment, including women's increased participation in agricultural decision making.

3 The FoodAfrica Project Framework for Extension Videos

The FoodAfrica project, funded by the Ministry for Foreign Affairs of Finland worked to find solutions to West and East African food security. Strengthening the rights of the most vulnerable and promoting gender equality was at the core of all FoodAfrica activities. Häme University of Applied Sciences' (HAMK) contribution to the project was to work on innovative extension and knowledge transfer methodologies.

One of the FoodAfrica work packages focused on finding the right gene combination of crossbreed cows in ASALs in Senegal with the objective to increase milk production and thus improve the livelihoods and diets of pastoralists. Gender roles in pastoral societies are strong [23]. Women are more vulnerable and their adaptive capacity to climate change is less than that of men [24]. This article describes the rationale of HAMK's gendered approach to video-mediated knowledge transfer among the beneficiaries of the FoodAfrica work package in Senegal. During the project work, the HAMK team first internalized the project outcomes and based on this, prepared a plan to produce the video. This article uses data gathered during the interviews of the local Food Africa team, regional and local organizations that use the project's research findings, and the local communities who are ultimately the final end-users of the new knowledge. Some interviews were individual discussions, some were group discussions and based on the findings, more interviews were organized to cross-check and clarify the findings. The first set of interviews was done when refining the manuscripts, but during the shooting process, complementing interviews were conducted as needed. Also, the filmed raw material was watched with the end-users and their feedback was used as guidance. Literature findings and the experiences of the video production teams have been utilized as well.

4 Gender in HAMK FoodAfrica Video Production

4.1 Gendered Information Needs and Value Propositions

Gender roles define men's and women's work in agricultural production, processing and marketing, even if they work in the same value chain. In the videos, it is important to be able to meet the needs and different motivations of the users to adapt the new technology. Therefore, it is critical to understand the specific informational needs and motivational drivers of both women and men. In a pastoralist family, women are responsible for feeding children, and thus would need to understand the nutritional value of milk. In Senegal, men are responsible for feeding the animals. Through discussions and observations, the HAMK video team gained understanding of the roles and responsibilities of each gender, and based on this, it was possible to plan the details of the messages for both genders. A thorough analysis of each gender's roles and responsibilities opens up the information needs of women and men.

Collecting background data through discussions and gender analysis takes time and may bring out findings that lead to changes in plans. A report prepared by Olson, Rubin & Wangui [25] found that there are institutional challenges in finding the time, resources, and expertise to design and implement gender-responsive climate change programs. Data collection on gender demands time, and in the hectic process of producing extension materials, it is easy to proceed based on assumptions. However, this time spent on gender is well justified because it lays the foundation for effectiveness and sustainability of the extension materials.

Farmers are very capable of finding their own solutions [26]. Farmers are good at recognizing their problems and they have the necessary competencies to find localized solutions when they understand the reasons behind the new way of doing things [27]. Farmers are innovative and by nature are willing to explore. A participatory approach to development calls for a healthy respect of the knowledge of the beneficiaries. This is crucial because all knowledge and solutions are context-bound. Further, even with good intentions, the professionals often end up getting an incorrect understanding and not recognizing the fundamental facts that frame the reality of the community, and as a result, the solutions imposed are often doomed to fail. These statements are very much supported by the Ospina & Heeks [12] RABIT model, which recognizes self-organization and learning as part of the fundamental component of resilience building.

Ownership of knowledge is the most powerful tool in learning [28] and change. When the farmers, men or women understand, the reasons and need for the change, they can start shaping their own localized solutions which fit into the available resources as well as to the local environmental context. For this reason, it is important to carefully think how ready solutions the videos need to give; the reply to the question why might be more important than detailed instructions on how to apply the new technology. In Senegal, the metis cows, which can bring about benefits for both women and men through increased production of milk and meat, require changes in the production system; to produce more, the crossbreeds require more water and fodder. Men and women need to find sustenance solutions to provide the animals to be able to benefit from their increased production capacity.

4.2 Female Role Models for Empowerment

In Senegal, female extension workers are well respected. This made it easy to give a strong role to women in the videos. Agricultural extension is often very biased towards men and men are used as agents of informational resources. Role models are seen to act as examples, they can inspire, and they can build trust in one's personal capability. The findings on the role-model effect on women acting as the information messengers of agricultural extension videos are mixed [22] but the HAMK video team put a strong focus on gender to encourage women to feel ownership and opportunities to contribute. For example, a female veterinary expert was a very well-received informant.

Videos can encourage women's groups toward technology innovations and the adoption of new practices [21]. The socio-constructivist learning theory of Vygovsky [29] allows for understanding of the impact of peers on learning. The socially constructed learning process is based on situational collaboration and communication in which shared meanings are created. This means that when sharing and working together, humans can produce more than individuals alone.

5 Conclusions

Gender is the very key in improving agricultural productivity and enhancement of sustainable agricultural practices. The key principles of the gendered approach to extension videos include understanding the different value propositions of female and male farmers, recognition of innovativeness of both females and males, careful selection of the people featured in the videos to build trust in the message and the use of role modeling for women's empowerment. Despite the persistent gender gap in female farmers' access to internet, lower ownership of smartphones, and illiteracy as well as technical or digital illiteracy, the videos can be a powerful tool in helping rural women contribute to the sustainability of agricultural practices, including climate change mitigation. The necessity of climate change resilience augments the knowledge intensiveness of agriculture. Decision making becomes more and more complex, and for informed decisions, both female and male farmers need accurate and easily accessible information.

Videos can be powerful tools in empowering women, and they can give a voice to rural women. There is evidence in research that women gain not only the substance of the information in the videos, but also the videos support their group building and encourage them to take more active roles in their communities. Recognizing the opportunities to address gender issues through videos is crucial because the role of women in agriculture is so vital. To achieve sustainable results through extension, it must be kept in mind that a mere understanding of agriculture is not enough to guarantee the right approach from a humanistic or educational point of view. Also, male and female farmers' motivations may include improved income, but motivational triggers may be deeper in the life values and norms.

Videos help extensionists visualize new possibilities in a very controlled manner. This minimizes the risks in the training sessions by guaranteeing the proper understanding. Success in communication is based on everyone utilizing the same context for the words used. It is challenging to use words based on a reality that everyone interprets based on words from their own context and experiences. Even in the same sector, the words can have different meanings and therefore images are more reliable because interpreting them in the same way is more probable than interpreting the words. The other fact linked to videos being based on images is that one of the fundamental principles of human learning is that hearing is not as effective as seeing during the process of learning.

Further, videos can also change the role of extensions. Trust between extension agents and farmers is crucial for the motivation of both parties and a win-win situation where the farmer feels personal gain and the extensionist feels the experience of success is the foundation of successful cooperation. In all communication and in the process of transformation, the impact comes through trust. People will change their practices if they trust new information sources. In many countries, extension services need more trustworthiness in the eyes of beneficiaries and technology-based tools can contribute to this process of trust building.

Through videos, more people can be reached because, despite the challenges, more and more male and female farmers have access to internet, in addition to the extension services being able to operate with various actors to screen the videos in the farming community or simply encourage community members to use their laptops with long battery life. Women, if given access to ICT and web-based extension, are at least as interested as men in utilizing these tools.

References

1. FAO: The state of food and agriculture 2010–2011. FAO, Rome (2011)
2. GFRAS: Gender Equality in Rural Advisory Services. https://www.agrilinks.org/sites/default/files/resource/files/GFRAS%202012_11%20Concept%20Note%20-%20Gender%20Equality%20in%20RAS.pdf. Accessed 25 Jan 2021
3. Mur, R., Oonk, L., Bitzer, V.: Quality of content in agricultural extension. https://www.kit.nl/wp-content/uploads/2018/08/Quality-of-content-in-agricultural-extension.pdf. Accessed 25 Jan 2021
4. Manfre, C., Rubin, D., Allen, A., Summerfield, G., Colverson, K., Akeredolu, M.: Reducing the gender gap in agricultural extension and advisory services. https://meas.illinois.edu/wp-content/uploads/2015/04/Manfre-et-al-2013-Gender-and-Extension-MEAS-Discussion-Paper.pdf. Accessed 25 Jan 2021
5. FAO, IFAD, WFP: The State of Food Insecurity in the World 2015. Meeting the 2015 international hunger targets: taking stock of uneven progress. FAO, Rome (2015)
6. World Bank, FAO, IFAD: Gender and Agriculture Sourcebook. World Bank, Washington D.C. (2009)
7. Aryal, S., Maraseni, T.N., Cockfield, G.: Climate change and indigenous people: perceptions of transhumant herders and implications to the transhumance systems in the Himalayas. Geol. Geosci. **3**(4), 1–5 (2014)
8. FAO: Pastoralism in Africa's drylands. FAO, Rome (2018)
9. Ickowicz, A., Ancey, V., Corniaux, C., Duteurtre, G., Poccard-Chappuis, R., Touré, I., Vall, E., Wane, A.: Crop–livestock production systems in the Sahel – increasing resilience for adaptation to climate change and preserving food security. In: Proceeding of FAO/OECD Workshop on Building Resilience for Adaptation to Climate Change in the Agriculture Sector, pp. 261–294. FAO-OCDE, Rome (2012)
10. Herrero, M., Addison, J., Dedelian, C., Carbine, E., Havlík, P., Henderson, B., van de Steek, J., Thornton, P.K.: Climate change and pastoralism: impacts, consequences and adaptation. Revue scientifique et technique (International Office of Epizootics) **35**(2), 417–433 (2016)
11. IPCC: Climate Change 2007 Impacts, Adaptation and Vulnerability. Contribution of Working Group II to the Fourth Assessment Report of the IPCC. Cambridge University Press, Cambridge and New York (2007)
12. Ospina, A.V., Heeks, R.: Benchmarking, Visualizing and Strengthening ICTs' Impact on Agricultural Livelihoods' Resilience: A Ugandan Case Study. https://www.niccd.org/wp-content/uploads/2017/11/RABITUgandaShortCaseStudye-Resilience.pdf. Accessed 25 Jan 2021
13. FAO: Success stories on information and communication technologies for rural development. RAP Publication 2015/02. FAO Regional Office for Asia and the Pacific, Bangkok (2015)
14. ITU: Resilient Pathways: The adaptation of the ICT sector to climate change. https://www.itu.int/en/ITU-T/climatechange/Documents/Publications/Resilient_Pathways-E.PDF. Accessed 25 Jan 2021
15. Barber, J., Mangnus, E., Bitzer, V.: Harnessing ICT for agricultural extension. https://www.kit.nl/wp-content/uploads/2018/08/Harnessing-ICT-for-agricultural-extension.pdf. Accessed 25 Jan 2021

16. GSMA: The Mobile Gender Gap Report 2020. https://www.gsma.com/mobilefordevelopment/wp-content/uploads/2020/05/GSMA-The-Mobile-Gender-Gap-Report-2020.pdf. Accessed 25 Jan 2021
17. Arjjumend, H.: Review of pastoralists' resilience and adaptation to climate change: can technology help pastoralists mitigate the risks? Int. J. Bio-resource Stress Manag. **9**(1), 118–128 (2018)
18. FAO: Gender and ICT. FAO, Rome (2018)
19. Manfre, C., Nordehn, C.: Exploring the promise of information and communication technologies for women farmers in Kenya. https://pdf.usaid.gov/pdf_docs/PA00KVFN.pdf. Accessed 25 Jan 2021
20. Sebstad, J., Manfre, C.: Behavior change Perspectives on Gender and Value Chain Development: A Framework for Analysis and Implementation. https://www.usaid.gov/sites/default/files/documents/1862/gender_and_value_chain.pdf. Accessed 25 Jan 2021
21. Zossou, E., Van Mele, P., Vodouhe, S.D., Wanvoeke, J.: Women groups formed in response to public video screenings on rice processing in Benin. Int. J. Agric. Sustain. **8**(4), 270–277 (2010)
22. Lecoutere, E., Spielman, D., Van Campenhout, B.: Women's empowerment, agricultural extension, and digitalization: disentangling information and role model effects in rural Uganda. IFPRI Discussion Paper 1889. International Food Policy Research, Washington, D.C. (2019)
23. FAO: Pastoralism in the new millennium. FAO, Rome (2001)
24. Balehey, S., Tesfay, G., Balehegn, M.: Traditional gender inequalities limit pastoral women's opportunities for adaptation to climate change: evidence from the Afar pastoralists of Ethiopia. Pastoralism: Res. Policy Pract. **8**(23) (2018)
25. Olson, J., Rubin, D., Wangui, E.: Gender, Agriculture, and Climate Change: An Analysis for USAID/East Africa. USAID, Washington D.C. (2010)
26. Bentley, J.W., Van Mele, P., Acheampong, G.K.: Experimental by nature: rice farmers in Ghana. Hum. Organ. **69**(2), 129–137 (2010)
27. Zossou, E., Van Mele, P., Vodouhe, S.D., Wanvoeke, J.: The power of video to trigger innovation: rice processing in central Benin. Int. J. Agric. Sustain. **7**(2), 119–129 (2009)
28. Biggs, J., Tang, C.: Teaching for Quality Learning at University. Society for research into Higher Education & Open University Press, Glasgow (2011)
29. Vygotsky, L.: Mind in Society. Harvard University Press, London (1978)

How Sustainable is the Millennials' Diet? Reflections on a Qualitative Study

María Suárez-Gómez[1](✉) and Rosalina Pisco Costa[2]

[1] José Joaquim Fernandes Hospital, R. Dr. António Fernando Covas Lima, 7801-849 Beja, Portugal
[2] Universidade de Évora, Largo dos Colegiais 2, 7004-516 Évora, Portugal
rosalina@uevora.pt

Abstract. The world population is growing at a rapid pace. Economic advance is increasing the global demand for food and diversified diets. Agricultural production contributes to the pollution of the environment. Given the limitation of water and land resources, a dietary change is necessary to contribute to food security and ensure the care of the planet. Sustainable and healthy eating, among which the Mediterranean diet is identified, could be the answer. Leaving the parental home to live independently during young adulthood is likely to influence food choice behavior towards the development of unhealthy and unsustainable food choice patterns. This paper draws on a small scale, mixed-methods qualitative study made with young Portuguese and Spanish millennials in 2019 to question the eating practices and perceptions when experiencing transition to adulthood. Based on food diaries and semi-structured interviews, results show that often participants choose the Mediterranean diet. Millennials express interest in continuing to prepare the homemade and traditional dishes they ate before becoming independent, arguing that they are accustomed to flavors, identifying traditional food with healthy food and valuing the affective memories with their family of origin. Overall, millennials' food practices transversally claim for a water-energy-food nexus, embracing SDGs.

Keywords: Family · SDGs · Sustainable food · Socialization · Millennials

1 Introduction

Transition into adulthood appears to be a critical stage in the development of unhealthy food choice patterns by young people [1]. Particularly, it is likely to influence food choice behavior [2]. When living independently, do young adults cook at home, and if so, what kind of dishes do they cook? Are they incorporating some of the so-called traditional dishes in their daily life? This paper departs from such questions to explore senses of family food among millennials, specifically homemade traditional dishes. This topic is as interesting as millennials are presented as a generation where technology and social media are strongly replacing family rituals and traditions, which used to be passed from generation to generation through tight socialization norms and values [3].

J. R. da Costa Sanches Galvão et al. (Eds.): *Proceedings of the 1st International Conference on Water Energy Food and Sustainability (ICoWEFS 2021)*, pp. 28–37, 2021.
https://doi.org/10.1007/978-3-030-75315-3_4

Sociology has long been devoted to the study of generations as a way of looking at social change [4]. While their parents and grandparents lived all these changes as adults, the millennials' socialization was mediated by technology and social media [5, 6]. In fact, this generation accuses some of the characteristics often pointed out to "digital natives", a label created by Prensky [7]. Millennials employ technology and social media for academic, work and personal life, develop daily tasks simultaneously in digital environments (multitasking) [8] and tend to be both producers and consumers of content in social media [9].

At present, the millennials are entering in the adult life, here understood in its traditional meaning of living independently, away from the parents' house [10]. Some studies have come to say that the millennials tend to live longer with their parents, delaying their entry into adulthood as usually associated with the milestones of marrying or having children, which is planned only from the age of 30 onwards [11–13]. Portugal and Spain are characterized by having a strong familiastic tradition, especially when compared to other countries in Europe [14].

Previous research found contradictory trends among millennials [15]. While some studies advocate that the millennials belong to a generation that eats away from home and presents bad food habits [12]; others show that this is a generation which bets on cooking and eating at home [16].

Currently, environment is being threatened by scarcity of resources. Which is the role of millennials in this issue? Do they follow a sustainable diet? Are they having friendly habits with environment?

Following a survey conducted by Gastromedia, millennials prefer products that provide health benefits, such as gluten and caffeine free, high in protein, organic, fortified and controlled portions [17]. Millennials are also demanding for more healthful and fresh ingredients, with a greater intake of cereals, grains, fruit and fruit juices, and dairy [18]. They also stand out to demand to pay a fair price in relation to quality [19].

According to the marketeers [12], millennials can be defined as demanding consumers regarding food; they are impatient and original in their tastes [20], and calling a family member to ask for a prescription for information on cooking or food issues is often replaced by a quick search on the Internet [21].

This article focuses particularly in the so called homemade traditional food. This comprises dishes that are considered part of culture and social identity, including recipes that go from generation to generation during the socialization process [22]. In as much as the socialization of these young people was strongly influenced by technology, although always mediated by the family [23], it is worth understanding the extent to which millennials incorporate the gastronomic culture into their daily life.

The primary aim of this paper is to unveil food practices among millennials when leaving the house of the family of origin in the context of the transition to adulthood. More specifically, the objectives comprise to: (1) contrast and compare food practices between the past (at the parents' home) and the present life of the millennials (while leaving independently); (2) understand the place of homemade traditional food in the daily food practices of the millennials (specific dishes, arguments put forward to justify the inclusion (or not) of such dishes in the actual diet and meanings attached to it); (3)

analyze if food practices and diet chosen by millennials are perceived as respecting the environment and being sustainable.

2 Methods

2.1 Research Design

A qualitative, small-case research project [24] was designed aiming to understand latent meanings behind food practices among millennials when they leave the family of origin in the context of the transition to adulthood. The use of a creative and mixed approach [25], based on the combination of data collected through food diaries and semi-structured interviews, was purposely designed as a way to meet in-depth understandings and additional qualitative insight of data [26].

The study used data collected between November 2018 and February 2019. A total of 22 individuals from the millennial generation, at the time of transition to adult life, participated in this study. The selection of cases was determined by theoretical arguments rather than by mathematical or statistical assumptions, as befits quantitative oriented studies [27]. As such, a multiple cases sample, by contrast and depth [28], was developed. Thus, individuals were recruited according to the following inclusion criteria: young people up to 35 years old, both Portuguese and Spanish, with a high level of education (bachelor degree, minimum), experiencing transition into independent life, and living in Portugal. Both Portugal and Spain are well known for a late age in leaving parental home [29], what means that 35 years of age will serve as a milestone for such an event.

Individuals were recruited according to the personal and professional networks of the researchers, and after through a snowball procedure [25]. As the researchers had different occupations and nationalities, the network of participants recruitment was sufficiently broad to avoid entropy, one of the most common criticisms addressed to this sampling process [30].

2.2 Procedures

Data collection was developed in two phases, sequentially. Firstly, individuals were asked to take part of the study and to complete a food diary for at least one week. This study took advantage of new technologies, namely the usage of smart phones and its applications, for the purposes of data collection regarding food practices [31, 32]. Specifically, individuals were asked to share through the mobile phone the photos of their lunch and dinner dishes, conjointly with a small description and to use WhatsApp Messenger for sharing it, being this media useful nowadays to share information, ideas and emotions [33, 34].

Secondly, participants were invited to engage in a semi-structured, comprehensive interview [26]. After a brief set of questions aiming to get socio-demographic data, the interview was structured in three main sections. In section one, the interviewee was asked about the photographs that come closest to a regular meal. Section two was centered in food practices in the transition to adult life and about differences when comparing their own food practices before leaving the parental home. The third section comprised a set of questions devoted to the daily food practices.

The interviews were carried in Spanish and Portuguese, digitally recorded, and transcribed into the native language. In all phases of data collection, individuals were previously informed about the procedures; their participation was voluntary and unpaid; and oral consent was obtained. Broadly, researchers adhered to the International Sociological Association ethical code [35] and specifically to the Deontological Code of the Portuguese Sociological Association [36].

2.3 Data Analysis

Data was explored through a qualitative content analysis for the photos [37] and interview transcripts [38]. The qualitative analysis was performed with NVIVO12® software (QSR, International) [39].

Coding followed a mixed procedure, according to which data segments, either text or image, were associated with meaning coming from the literature and other, including in vivo categorization [40]. Translation, grammatical and style adaptations into English were done only for the purposes of the present paper.

3 Results

A total of 22 individuals, five Portuguese men, five Spanish men, seven Portuguese women and five Spanish women took part in the study. Ages range from 20 to 35 years old, which comprised individuals born across 15 years, from 1983 to 1998. All the participants completed a bachelor's degree, and at the moment they were mostly living with a partner (boy/girlfriend, 6; partner and children, 3), solo living (7), co-workers (5) and friends (1).

3.1 Then and Now: Continuities and Changes During the Transition into Adulthood

When comparing what the participants in the study eat now with their food practices before leaving their parents' home, the Portuguese millennials highlight the greater presence of "traditional food" then:

"My parents cooked typical Alentejo food, like boiled beans, beans, soups…" [E07, woman, Portuguese, 31 years old, nurse, lives with her husband and two children].

Often, they also noted that they used to eat more food cooked in the oven, which they associate with a "healthy diet".

"When I lived in the village, my grandmother used to cook, and I ate very well, pan food, in the oven." [E04, woman, Portuguese, 31 years old, doctor, lives with a partner].

Regardless, Spanish young repeatedly mention the presence of fried food at their parents' home and associate it with unhealthy food.

"In my parents' house, I ate a lot of meat and fried fish." [E11, woman, Spanish, 33, psychologist, lives with a partner].

When asked about their actual practices, Portuguese participants refer that they continue to make and eat the dishes that they ate with their family of origin before becoming independent. Examples of this are bean stew and grilled codfish.

"I continue to eat food very similar to my parents, which is traditional Portuguese, like seafood rice, and typical food from the Alentejo region, bean stew, several stews, Portuguese stew with grain..." [E19, man, Portuguese, 21 years old, sociologist, solo living].

Participants often refer to the introduction of new elements in their habits since they left the house with their parents, namely pasta and vegetarian food. Nonetheless, they kept underlining the will of adapting their diet, incorporating those "new" elements with aspects of the diet that they followed when they lived in the family of origin.

"We now eat vegetarian food, and we adapt dishes from Alentejo by replacing animal protein with vegetable protein, and we really like it." [E07, woman, Portuguese, 31, nurse, lives with her husband and two children].

As for the Spanish participants actual food practices, they mentioned more often the will of not including in their daily meals some dishes "inherited" from their family of origin, which are sometime considered to be "unhealthy". Alternatively, they mentioned the frequent consumption of pastas, assuming that sometimes they made use of the precooked ones when they do not have enough time to cook.

"I prefer to save time, not have to cook and spend more money having to look for a good place to eat, because although there are more fast-food places, these do not give you a good quality." [E05, man, Spanish, 24 years old, teacher, lives with friends].

Facing the pressure of time, the Spanish participants often refer to salads as an easy and healthy alternative to traditional "heavy" food.

"I really like cooked vegetables and salads [...] In the markets I buy fruits, and vegetables, and in supermarkets the pasta, vegetables and other products, where they also come cheaper, but I always look for a good value for money." [E14, woman, Spanish, 33 years old, teacher, lives with a co-worker].

3.2 Homemade Traditional Food Matters

Among the criteria behind their current food practices, the millennials who were interviewed for this study expressed a general interest in continuing to elaborate the homemade traditional dishes that they ate before they became independent. Regardless their nationality, the main arguments used to prepare some of those meals were the fact that they want to enjoy the food and flavors they are accustomed to, and because traditional food is often perceived as healthy food.

"We like the traditional Portuguese food, because of the flavors. What is homemade "does not taste like plastic" [E08, woman, Portuguese, 32 years old, secretary, lives with her husband and daughter].

Often, the interviewees pointed that they intend to pursue a healthy and balanced diet, reason why they keep cooking some of the dishes they learnt with their families, as they are included in the broader label of "the Mediterranean diet", which is identified as a good choice and practice.

"I like Spanish food and [that] follow[s] the Mediterranean diet" [E14, woman, Spanish, 33, teacher, lives with a co-worker].

4 Discussion

These results confirm the leaving of parental home as an important milestone in the individuals life course as far as food practices are concerned. As Marquis [41] advocates, recognizing some food as familiar is important in decision making. The Portuguese millennials highlight the greater presence of homemade traditional food then, what implies a certain set of ingredients (e.g., beans, vegetables), but also a specific way of cooking it (e.g. stews, soups, use of the pan or of the oven, etc.). From here results a kind of food considered to be "traditional", made of "typical dishes", which is recognized as being "healthy".

Healthy food is in vogue. In recent years, several trends have been created regarding healthy eating among millennials, which advocate that artificial or "ultra-processed" food is harmful to health, leading millennials to choose natural products, also called "real food". Regarding ingredients, millennials tend to consume more fruit and vegetables, grains, nuts and seeds, which allow them to connect with the history of the product, a fundamental characteristic to gain the trust of this demanding generation [42].

The Spanish interviewees also remember the stews and soups as frequent meals for their families of origin yet repeatedly mention the presence of fried food (e.g. both meat and fried fish). Fried food at the parents' home is strongly perceived as "unhealthy" by the Spanish millennials. Differently, the millennials interviewed for this study refer to the importance of maintaining a healthy diet, which already was reported as a trend in foregoing studies [18], and their preferences go to fresh and sustainable products, as previous studies also show [43]. Thus, millennials can, for instance, make a vegetarian dish using the basis of a Portuguese bean stew or incorporate more vegetables and salads when eating dishes considered unhealthier, such as pasta.

Millennials make decisions that they think benefit animal welfare, reducing its consumption, becoming vegan, or partially reducing it, being vegetarian or flexitarian. Likewise, they avoid buying products "ultra-processed", in order to avoid products that can be harmful, not only to their health, but also to the environment [44]. Accordingly, when justifying the choice for such dishes, millennials often point to the preference for fresh, barely seasoned food, products of easy preparation, with a good aesthetical presentation on the plate and being in good value for money. A sustainable product generally has a higher price than a normal product. However, previous studies point that 37% of millennials would be willing to pay up to 10% more if the product is sustainable [45]. Additionally, millennials prioritize buying at farmers markets, directly from the producer, in small stores, acquiring local products or even growing their own fruit and vegetables.

Traditional Portuguese and Spanish cuisine are under the broad label of the Mediterranean cuisine and millennials who were interviewed for this study are aware of this fact. Specifically, they point to the fact that the Mediterranean diet is healthy and balanced, based upon the preparation of fresh, barely seasoned food [46]. In this sense, millennials compare homemade with fast food, showing preference for the first as the last is often associated with food that "tastes like plastic". As millennials are aware of the pollution caused by using plastic, they try not to buy packed products. In fact, according to the data from the Center for Culinary Development, 70% of millennials say they do not buy bottled water due to its high environmental impact [45]. As the Mediterranean cuisine is also a promotor of a cuisine of proximity [46], by using fresh and seasoned products, millennials transversally claim to the quality of the products, their origin and environmentally sustainable impact [47].

Food provides comfort by bringing together affective memories, which connect millennials with the people they love, places of their childhood and adolescence time, and specific moments or occasions considered to be happy or special [48]. This way, food can be perceived as a protective factor [49]. It is a rather interesting remark, as traditional food is usually immersed in family secrets [50], passing from generation to generation. Being given attention by the millennials' generation, it is not at risk of being lost.

5 Conclusions

This study found that transition into adulthood continues to be an important moment where individuals have to make, on their own, decisions about what to eat. Moreover, it shows that among the millennials there is a strong willingness in continuing to eat some of the dishes they used to eat at their parents' home. Thus, millennials maintain the greater presence of "traditional food" in their actual food practices and identify "traditional food" with "healthy and balance diet". At the same time, it is possible to conclude that millennials consider "unhealthy food" the presence of fried food in the diet.

Millennials better their opinion about food practices of the family of origin when they become independent, when they realize the value of eating a healthy diet. Additionally, millennials seem to adapt their diet, in a way, to eliminate unhealthy products they consumed in the family of origin, such as fried dishes, and incorporate those "new" elements that they consider to be healthier or more respectful to the environment. For this reason, millennials tend to consume more fruit and vegetables, grains, nuts, and seeds. This also implies a specific way of cooking it (e.g. stews, soups, use of the pan or the oven, etc.). However, millennials admit that sometimes they make use of precooked or cooked pasta when they do not have enough time to cook, but they are aware that it is an unhealthy practice. Participants often refer to salads as an easy and healthy alternative to traditional "heavy" food.

By conclusions, we can point that millennials show preference for fresh, barely seasoned food, products of easy preparation, with a good aesthetical presentation on the plate and being in good value for money. On the other hand, more millennials make decisions that they think benefit animal welfare, reducing its consumption, becoming vegan, or partially reducing it, being vegetarian or flexitarian. Traditional recipes are adapted and personalized according to individual preferences and values that generally shape the millennials diet and consumption practices, namely, ideas of health, sustainability, aesthetics, and creativity. Combining the use of fresh, local, and seasoned products with an easy preparation, typical regional cuisine is being appropriated, rather than refused, by the millennials. Most of the Spanish and Portuguese millennials on this study identify the Mediterranean diet as their diet of choice because it meets the criteria they choose for their food, because it is a healthy, sustainable diet and culturally engaged. In short, even at a time of strong pervasiveness of technologies in the daily life, understanding cooking practices among the millennials cannot be done without looking at the cultural and family contexts of both origin and current belonging.

References

1. Beasley, L., Hackett, A., Maxwell, S.: The dietary and health behaviour of young people aged 18–25 years living independently or in the family home in Liverpool UK. Int. J. Consum. Stud. **28**(4), 355–363 (2004). https://doi.org/10.1111/j.1470-6431.2004.00394.x
2. Sogari, G., Vélez-Argumedo, C., Gómez, M.I., Mora, C.: College students and eating habits: a study using an ecological model for healthy behavior. Nutrients **10**(12), 1823 (2018). https://doi.org/10.3390/nu10121823
3. Haddon, L.: Social media and youth. In: Mansell, R., Hwa Ang, P. (eds.) The International Encyclopedia of Digital Communication and Society, pp. 1–9 (2015)
4. Mannheim, K.: Essays on the Sociology of Knowledge. Oxford University Press, Oxford (1952)
5. Zemke, R., Raines, C., Filipczak, B.: Generations at work: managing the clash of Boomers, Gen Xers, and Gen Yers in the workplace. AMACOM, New York (2013)
6. Mastrolia, S., Willits, S.: Millennials: what do we really know about them? Adv. Account. Educ. Teach. Curriculum Innov. **14**, 45–72 (2013). https://doi.org/10.1108/S1085-4622(2013)0000014009
7. Prensky, M.: Digital natives, digital inmigrants. Horizon **9**(5), 1–6 (2001). https://doi.org/10.1108/10748120110424816
8. Piscitelli, A.: Nativos digitales. Santillana, Buenos Aires (2009)
9. Urresti, M.: Ciberculturas juveniles: vida cotidiana, subjetividad y pertenencia entre los jóvenes ante el impacto de las nuevas tecnologías de la comunicación y la información. In: Urresti, M. (ed.) Ciberculturas juveniles. Buenos Aires: La Crujía, pp. 13–66 (2008)
10. Furlong, A., Cartmel, F.: Young People and Social Change: Individualization and Risk in Late Modernity, 2nd edn. Open University Press, Buckingham (2007)
11. Stein, J.: Millennials: The Me Me Me generation. Time Magazine, May 20. https://time.com/247/Millennials-the-me-me-me-generation. Accessed 18 Feb 2019
12. Nielsen: Reporte de estudio global "Estilos de vida generacionales". https://www.nielsen.com/content/dam/nielsenglobal/latam/docs/reports/2016/EstilosdeVidaGeneracionales.pdf. Accessed 18 Feb 2019
13. Mínguez, A.M.: The youth emancipation in Spain: a socio-demographic analysis. Int. J. Adolesc. Youth **23**(4), 496–510 (2018). https://doi.org/10.1080/02673843.2018.1438299
14. Torres, A.: Igualdade de género ao longo da vida: Portugal no contexto europeu. Lisboa: Fundação Francisco Manuel dos Santos (2018)
15. Brillat-Savarin, J.A.: A Quantitative Study to the Underlying Motives for Food Choices: And How Do These Motives Contribute to Healthy Food Choices Among Dutch Young Adults? Communication Studies (MSc) Essay. School of Behavioural Science: University of Twente. https://www.semanticscholar.org/paper/A-Quantitative-Study-to-the-Underlying-Motives-for-Brillat-Savarin/ed4f4cdce801ccfc97601b244f3ac91e8c8c0b69. Accessed 18 Feb 2019
16. Gastromedia: Estudio de Hábitos digitales de los españoles en alimentación. https://docplayer.es/docview/77/75525470/#file=/storage/77/75525470/75525470.pdf. Accessed 18 Feb 2019
17. Rumiati, R.I., Foroni, F.: We are what we eat: how food is represented in our mind/brain. Psychon. Bull. Rev. **23**(4), 1043 (2016). https://doi.org/10.3758/s13423-015-0908-2
18. Kuhns, A., Saksena, M.: How millennial food purchase decisions compare to previous generations. In: 2016 Annual Meeting of the Agricultural and Applied Economics Association, Boston, Massachusetts, 31 July–2 August (2016). https://ideas.repec.org/p/ags/aaea16/235907.html. Accessed 18 Feb 2019

19. Todd, J.E.: Changes in Eating Patterns and Diet Quality Among Working-Age Adults, 2005–10, Economic Research Report 262214. United States Department of Agriculture, Economic Research Service. https://ageconsearch.umn.edu/record/262214/files/41995_err161.pdf. Accessed 18 Feb 2019
20. Doub, A.E., Levin, A., Heath, C.E., LeVangie, K.: Mobile appetite: consumer attitudes towards and use of mobile technology in the context of eating behaviour. J. Direct Data Digit. Mark. Pract. **17**(2), 114–129 (2015). https://doi.org/10.1057/dddmp.2015.44
21. García-González, Á., Achón, M., Alonso-Aperte, E., Varela-Moreiras, G.: Identifying factors related to food agency: cooking habits in the spanish adult population—across-sectional study. Nutrients **10**(2), 217 (2018). https://doi.org/10.3390/nu10020217
22. Mura Paroche, M., Caton, S.J., Vereijken, C., Weenen, H., Houston-Price, C.: How infants and young children learn about food: a systematic review. Front. Psychol. **8**, 1046 (2018). https://doi.org/10.3389/fpsyg.2017.01046
23. Fingerman, K.L.: Millennials and their parents: implications of the new young adulthood for midlife adults. Innov. Aging **1**(3), 1–16 (2017). https://doi.org/10.1093/geroni/igx026
24. Denscombe, M.: The Good Research Guide for Small-Scale Social Research Projects. Open University Press, Buckingham (1998)
25. Creswell, J.: Research Design. Qualitative, Quantitative, and Mixed Methods Approaches, 5th edn. Sage Publications, Los Angeles (2018)
26. Mason, J.: Qualitative Researching, 2nd edn. Sage Publications, London (2002)
27. Denzin, N., Lincoln, Y.: The Sage Handbook of Qualitative Research, 5th edn. Sage Publications, Los Angeles (2018)
28. Pires, Á.: Échantillonnage et recherche qualitative: Essai théorique et méthodologique. In: Poupart, J.D., Deslauriers, J.P., Groulx, L.H., Laperriere, A., Mayer, R., Pires, A. (eds.) Enjeux Épistémologiques et Méthodologiques, pp. 113–167. Gaëtan Morin, Montreal (1997)
29. Cinalli, M., Giugni, M.: New challenges for the welfare state: the emergence of youth unemployment regimes in Europe? Int. J. Soc. Welfare **22**, 290–299 (2013). https://doi.org/10.1111/ijsw.12016
30. Bryman, A.: Social Research Methods, 3rd edn. Oxford University Press, Oxford (2008)
31. Chatzitheochari, S., Fisher, K., Gilbert, E., Calderwood, L., Huskinson, T., Cleary, A., Gershuny, J.: Using new technologies for time diary data collection: instrument design and data quality findings from a mixed-mode pilot survey. Soc. Indic. Res. **137**(1), 379–390 (2018). https://doi.org/10.1007/s11205-017-1569-5
32. Garcia, B., Welford, J., Smith, B.: Using a smartphone 'app' in qualitative research: the good, the bad and the ugly. Qual. Res. **16**(5), 508–525 (2016). https://doi.org/10.1177/1468794115593335
33. Bhatt, A., Arshad, M.: Impact of WhatsApp on youth: a sociological study. IRA-Int. J. Manag. Soc. Sci. **4**(2), 376–386 (2016). https://doi.org/10.21013/jmss.v4.n2.p7
34. DESI: Digital Economy and Society Index Report 2019. Brussel: European Comission (2019). https://ec.europa.eu/digital-single-market/en/desi. Accessed 18 Feb 2019
35. ISA: Code of Ethics of the International Sociological Association, Madrid: ISA. https://www.isa-sociology.org/about/isa_code_of_ethics.htm. Accessed 18 Feb 2019
36. APS: Código Deontológico. Lisboa: Associação Portuguesa de Sociologia. https://aps.pt/pt/codigo-deontologico/. Accessed 20 June 2019
37. Rose, G.: Visual Methodologies: An Introduction to the Interpretation of Visual Materials. Sage, London (2001)
38. Miles, M.B., Huberman, M., Saldaña, J.: Qualitative Data Analysis: A Methods Sourcebook, 3rd edn. Sage, Los Angeles (2014)
39. Bazeley, P., Jackson, K.: Qualitative Data Analysis with NVivo, 2nd edn. Sage Publications, Los Ángeles (2013)

40. Krippendorff, K.: Content Analysis: An Introduction to Its Methodology, 4th edn. Sage Publications, Los Angeles (2019)
41. Marquis, M.: Exploring convenience orientation as a food motivation for college students living in residence halls. Int. J. Consum. Stud. **29**(1), 55–63 (2005). https://doi.org/10.1111/j.1470-6431.2005.00375.x
42. Pina, P.: Tendencias alimentarias de 2016: el auge de los alimentos funcionales. Recuperado de. https://www.thinkwithgoogle.com/intl/es-es/canales-de-publicidad/busqueda/tendencias-alimentarias-de-2016/
43. Achón, M., Serrano, M., García-González, Á., Alonso-Aperte, E., Varela-Moreiras, G.: Present food shopping habits in the spanish adult population: a cross-sectional study. Nutrients **9**(5), 508 (2017). https://doi.org/10.3390/nu9050508
44. Sharma, B., Harker, M., Harker, D., Reinhard, K.: Youth transition to university in Germany and Australia: an empirical investigation of healthy eating behaviour. J. Youth Stud. **13**(3), 353–367 (2010). https://doi.org/10.1080/13676260903447510
45. Mintel: Tendencias globales de alimentación y bebidas 2017. Recuperado de. https://es.scribd.com/document/344980428/Mintel-Tendencias-de-Alimentacion-y-Bebidas-2017
46. Gregório, M.J., Rodrigues, A.M., Graça, P., Sousa, R.D., Dias, S.S., Branco, J.C., Canhão, H.: Food insecurity is associated with low adherence to the mediterranean diet and adverse health conditions in portuguese adults. Front. Public Health **6**, 38 (2018). https://doi.org/10.3389/fpubh.2018.00038
47. Ristino, L.: Back to the new: millennials and the sustainable food movement. Vermont J. Environ. Law **15**, 1–20 (2013). https://vjel.vermontlaw.edu/publications/back-new-millennials-sustainable-foodmovement/
48. Piqueras-Fiszman, B., Jaeger, S.R.: The incidental influence of memories of past eating occasions on consumers' emotional responses to food and food-related behaviors. Front. Psychol. **7**, 943 (2016). https://doi.org/10.3389/fpsyg.2016.00943
49. Hoffman, R., Gerber, M.: Food processing and the mediterranean diet. Nutrients **7**(9), 7925–7964 (2015). https://doi.org/10.3390/nu7095371
50. Smart, C.: Families, secrets and memories. Sociology **45**(4), 539–553 (2011). https://doi.org/10.1177/0038038511406585

What Do 3rd Level Students Consider a Sustainable Diet? A Qualitative Study

Diarmuid Loftus and Maria Dermiki(✉)

Department of Health and Nutritional Sciences, Institute of Technology, Sligo, Ash Lane, Sligo F91 YW50, Ireland
Dermiki.Maria@itsligo.ie

Abstract. The world population is increasing along with the need for food. However, the way in which we grow and consume food has environmental, economic, and social implications, which are the pillars upon which the concept of sustainability is built. Third-level students will become the generation that will be working and starting families soon, therefore the aim of the current study is to explore the views of Irish third-level students regarding sustainable diets. For this reason, semi-structured interviews were used to collect data on the understanding of sustainability, sustainable food and diets of students ($n = 7$) from different courses (Environmental, Human Nutrition and Marketing). Moreover, the barriers they faced when trying to follow a sustainable diet were explored. Thematic analysis of the qualitative data showed that the perception of sustainability depended partly on the course they were studying. Sustainable diet consisted of foods which are local, healthy, without packaging, and producing minimum waste when consumed. The barriers they faced, regardless of their studies, were linked to the lack of information on the packaging in relation to sustainability and when living independently versus living with their parents. For the future, it is suggested to incorporate sustainability and sustainable diets in the learning of students across courses. Moreover, it is suggested to explore the views on sustainable diets of adults from different age groups.

Keywords: Sustainable diets · Third level students · Barriers · Thematic analysis

1 Introduction

There is a growing concern that the current production and consumption behavior pose a significant threat to the environment and to the health of humans [1]. Therefore, there is a need for a more sustainable way of living. Sustainability refers to the situation where humans and natural systems survive and thrive in harmony [2]. The central concept of 'sustainability' is built upon three interconnected pillars: environmental, economic and social well-being [1–3], which also provide the foundations on which the seventeen Sustainable Development Goals (SDG's) of the United Nations are based.

With the world population projected to increase from 7.7 billion, to 9.7 in 2050, producing enough food that is safe, accessible, affordable and nutritious represents a

J. R. da Costa Sanches Galvão et al. (Eds.): *Proceedings of the 1st International Conference on Water Energy Food and Sustainability (ICoWEFS 2021)*, pp. 38–45, 2021.
https://doi.org/10.1007/978-3-030-75315-3_5

major challenge of the 21st century [4, 5]. Climate change, environmental degradation, and increased competition for natural resources are adding pressure to the global food system [6], while changes in global dietary patterns, high in saturated fat, sugar, and salt, have contributed to the increased prevalence of non-communicable disease and greenhouse gas emissions [7, 8]. On the other hand, malnutrition is prevalent in many low-income countries [9], highlighting the need for food security as stated in SDG 2 [10, 11]. According to the EAT Lancet report [12], a unified and widespread response involving multiple sectors and stakeholders is necessary to change the current trends of food production and its impact on the environment and health [12]. Ireland has been making significant strides in promoting sustainable food production through the Origin Green programme [13]. This is a multi-disciplinary programme involving various stakeholders including the government, the private sector, members of the food supply chain, and retailers, in order to ensure that food and drink are sustainably produced, from farm to fork.

According to the Food and Agriculture Organization of the United Nations, sustainable diets have low impact, are respectful of the biodiversity and ecosystems, economically fair and affordable, safe and healthy while optimising human and natural resources [14]. Despite the urgent need towards sustainable diets, there are still many gaps in our understanding of what these may consist of [7]. Furthermore, there is a limited literature pertaining to the views of third-level students and their understanding of a sustainable diet. This cohort, often dubbed 'post-millennials', or 'Gen-Z', have been observed to possess a considerable amount of knowledge in regard to sustainability compared to previous generations, and studies show that post-millennials are content to pay more money for foods that they perceive to be healthier and more eco-friendly [15]. While this observation was published in a U.S study [15], to what extent this observation translates to post-millennial, third-level students in Ireland remains unclear. It is expected that students attending specific courses might be more exposed to the term "sustainability". Educational institutes across Ireland are working towards promoting and enhancing sustainability on campus, by aiming to obtain the Green Campus Flag. Participation in this scheme allows students to understand the importance of developing sustainable lifestyles regardless of the course they are attending and following a sustainable diet is part of a sustainable lifestyle [7].

The aim of the present study was to explore how third-level students studying in different courses perceive the term 'sustainable diet', and to identify perceived barriers and facilitators of adherence to it.

2 Methodology

2.1 Rationale

In order to understand the perception of sustainable diets of third-level students a qualitative study design with semi-structured interviews was used.

2.2 Participants

A convenience sample was employed for the purpose of the study; specifically, third-level students, both undergraduate and postgraduate, who were studying in the Institute of Technology Sligo in Ireland. A total of 7 individuals (4 males, 3 females) participated in the study, from three different courses (Human Nutrition (HN), Environmental (ES) and Marketing and Sales (MS)). This was in order to eliminate bias that might arise from including participants only from one area of study, and to investigate if there was an effect of the area of study on the perception of sustainability. The age range of the participants was from 18 to 29 years old.

2.3 Data Collection

The study was approved by the Ethical Review Board of the Department of Health and Nutritional Sciences of the Institute of Technology Sligo. One-to-one, semi-structured interviews were conducted online using Microsoft Teams. Informed consent was obtained from each participant prior to the collection of data. Data was collected by asking open-ended questions in relation to understanding of the term 'sustainability', such as: "*Have you heard of the term 'sustainability' before? Can you describe it*?", in relation to sustainability and food: *"When you think of food and sustainability, what comes to your mind"*, in relation to sustainable diets asking them to complete the sentence "*sustainable diets are ...* ". They were also asked to give characteristics of sustainable and unsustainable foods, and to discuss about their current dietary practices at home vs on campus, in order to explore perceived facilitators and barriers to adhering to a sustainable diet. Moreover, the effect of labelling in communicating the sustainability of the food products was explored by asking them "*what sort of information is there on the labels that determines the sustainability of a food product?*". The questions were first piloted by the two authors before the data collection. The interviews were recorded and transcribed verbatim, all information was anonymized, and the recordings were deleted after transcribing.

2.4 Data Analysis

Thematic analysis was used for the identification, analysis, and reporting of patterns of meaning both within, and between participant transcripts [16]. Thematic analysis was conducted in six phases, as described by Braun and Clarke [16]. The interviews were recorded, transcribed verbatim and anonymized. Transcripts were read by both authors to avoid bias and ensure validity.The transcribed data was then transferred to a Microsoft Excel worksheet. Each individual cell of the Microsoft Excel worksheet was then assigned to a code that was sorted by color [17]. This phase was completed in a systematic fashion across the complete data set. The codes were gathered and sorted into themes. Themes were generated using both an inductive, and deductive approach [18] based on the generated data, and on existing literature respectively. The themes were then reviewed and refined. After review and refinement of the themes, it was identified how each theme contributed to the broader scope of sustainability.

3 Results and Discussion

3.1 Perception of Sustainability and Sustainable Diets

Themes relating to how third-level students perceived the term 'sustainable diets' can be seen in Fig. 1. These included the environment, health, lifestyle, ethical and financial considerations.

Terms related to the environment such as renewable energy and environmental protection were the most cited dimensions within this theme and were mentioned by most students regardless of the area of studies. For example, sustainability was associated with renewable energy, and energy conservation.

"*Sustainable energy, wind farms, solar energy… Things that can constantly be used, without being used up*" (Participant 1, ES).

"*Sustainability means things can continue forever or indefinitely*" (Participant 2, HN).

In some cases, the word 'sustainability' carried different meanings for different participants, depending on the course they were studying. For example, a participant from Marketing studies had only heard of the term sustainability within the context of businesses.

"I have heard it in terms of a lot of businesses, but I haven't heard of it in terms of diet" (Participant 5, MS).

When asked what came to mind in relation to sustainable foods, the main theme was again the environment. Participants reported that sustainable foods were foods which when produced did not harm the environment, did not produce carbon emissions, were produced with minimal natural resources and energy, could be produced continuously, local, and grown organically.

"*I guess, local, lack of packaging. In terms of carbon emissions, there are no emissions, so instead of buying apples shipped from another country, you should buy apples that are grown nearby*" (Participant 6, HN).

"*Organic options come to mind*" (Participant 7, HN).

Conversely, participants reported unsustainable foods as foods that required a lot of energy to produce, with carbon emissions as a by-product. Imported goods were also understood to contribute to increased carbon emissions. Local and Irish products were favored by participants not only due to personal lifestyle choice, but also due to concern for the environment. The use of pesticides on crops was also reported as being unsustainable.

The other common theme in relation to sustainable foods and diets was the aspect of health. Participants understood sustainable foods to offer longevity and protection, within the context of both the environment, and health as also seen by other researchers [19]. Specifically, the students with nutritional background would mention the health effects of sustainable foods, considering they promote health and well-being. At the same time, other participants reported that sustainable diets consisted of fresh, unprocessed foods. On the other hand, convenience foods such as take-away meals were considered unsustainable since they do not lead to satiety when consumed.

"Sustainable foods are less harsh on the body." (Participant 2, HN).

"Organic, non-processed foods. I think of home-cooked foods being sustainable too. When it comes to unsustainable foods, definitely processed, takeaways, and frozen foods come to mind. They're full of preservatives" (Participant 4, MS).

"Foods like McDonalds where you think you've eaten a lot but you're not full" (Participant 5, MS).

Under the theme "lifestyle" was the subtheme of dietary patterns, such as veganism and vegetarianism, since the meat industry was considered unsustainable by some participants. On the other hand, one participant considered meat to be sustainable, due to the satiating effect of animal protein. In the same line, of satiety, one participant expressed that calorie restrictive diets were unsustainable and were considered unsustainable.

Another sub-theme under the main theme of lifestyle, was the one related to motivation and values. Specifically, it was mentioned that sustainable diets are trendy, i.e. popular and socially desirable. Knowing that social influences can affect food choices [20], this finding could be used to promote sustainable eating to this generation.

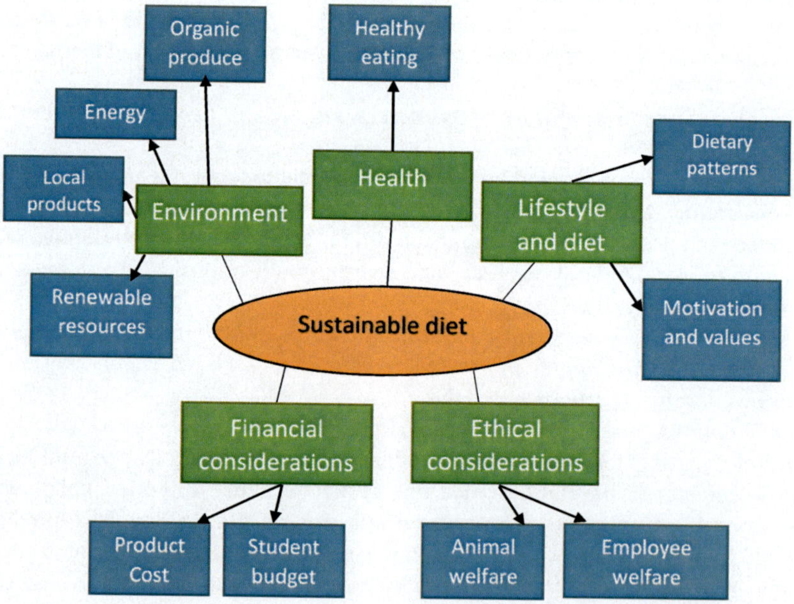

Fig. 1. Main themes and subthemes generated from the qualitative data and describing the perceptions of third-level students from IT Sligo on sustainability and sustainable diets.

Ethical considerations were the least discussed dimension of sustainable diets. One participant reported that sustainable food production considered animal welfare, and the welfare of those who worked in the food production industry, while another referred to equity and fair trade.

In recent years, the most pertinent topics surrounding sustainability have been focused on the protection of the natural environment, while ethical considerations such as socioeconomic equality has received less attention [21]. The findings from this study reflect on the current study, whereby environmental protection was often cited by the participants, while ethical considerations were cited less frequently. Lack of participant awareness with regards to ethical aspects of sustainability could be reflected by lack of awareness of the three pillars sustainability. Each participant would consider the aspects of sustainability mainly linked to the knowledge gained from their course. Based on this finding, it could be hypothesized that the course that the student is studying may influence their perception of sustainability.

3.2 Barriers and Facilitators Towards a Sustainable Diet

In order to explore the barriers and facilitators towards adhering to a sustainable diet, participants were asked about their dietary habits at home compared to when they are on campus. Participants' diet on campus was different from what they would eat at home, which is in agreement with the findings of a study on the effects of living arrangements on the dietary habits of undergraduate university students [22]. Specifically, when living on campus due to lack of time students would consume processed foods due to convenience. On the other hand, others were willing to spend more time and money to choose the sustainable option. Previous research has shown that food choices are affected by the socio-economic background [23], and despite the fact that the current study did not collect information on participants' financial status, this is suggested for future studies.

While in most cases life on campus was associated with unhealthy and unsustainable eating habits, for others staying at home would be a barrier to a sustainable diet. In most cases when students are living with their parents, they are not responsible for cooking and shopping, but they follow their family's eating patterns, which might be far from being sustainable.

"I would definitely consider my diet more varied in student accommodation. I can ask my mum to take my lifestyle into consideration, but I don't like to inconvenience her....I would consider myself to buy products with less plastic and I buy loose fruit whereas my mum doesn't".

When asked how easy participants thought it would be to adhere to a sustainable diet, many believed that it would be difficult due to factors including cost, lack of education, awareness and motivation and lack of accessibility. Living in Ireland with a long-standing tradition in dairy and beef production might be a reason why some participants felt that their choices were restricted. These findings are in agreement with the results of a qualitative study conducted [24] among British consumers, where dairy and beef production has a significant impact on the economy as well.

Labelling is a means of communicating information to the consumers in relation to the attributes of the food they are buying. Therefore, research has shown that it can affect buying behavior and food choices of consumers [25]. In the current study, every participant reported that they examined the label when buying food. The nutritional information provided on the front-of-pack labelling was the most reported aspect of the label that caught the eye of the participant. The participants' consumer behavior seemed to be influenced by the composition of the foods that they were purchasing. However, labelling information was not clear in communicating sustainability to the participants. Some participants were unsure of the meaning of the Origin Green logo found on Irish products, but considered local and organic products sustainable, while little attention was paid to the ethical considerations of sustainability.

The challenges when communication sustainable practices through food labels has been shown in other studies as well [21]. There is a need for better education at all levels of society in relation to the information presented on the labels, not only because of the impact on public health nutrition, but also on the information related to sustainability. Currently there are working groups at European level investigating how food labels can communicate sustainable practices to consumers [26, 27].

4 Conclusions

Despite the small number of participants, this qualitative study provided rich data which helped to understand how third-level students perceive the term "sustainable diets", and it identified the perceived barriers and facilitators of adhering to a sustainable diet. The understanding is partly affected by the course they are studying, but overall, they associate sustainable diets with environmental protection, and minimal generation of waste, such as plastic. They also associate sustainable diets with healthy, fresh, and locally sourced, non-imported produce. While students may be open to the concept of following sustainable diets, they may be deterred to following such a diet due to the perceived cost and limited student budgets, or due to family influence. They may be able to develop a greater understanding of what sustainable diets are through information and available resources; however, whether they utilize this information to follow sustainable diets could be considered a matter of personal values and motivation. Despite the small number of participants, the findings of this study are further validated by similar studies conducted in other countries. It is suggested to expand this study to participants of different ages, varying socio-economic background and education level.

References

1. Sustainability Research. www.epa.ie. Accessed 20 Jan 2021
2. Behrens, P., Bosker, T., Ehrardt, D.: Food and Sustainability. Oxford University Press (2020)
3. Purvis, B., Mao, Y., Robinson, D.: Three pillars of sustainability: in search of conceptual origins. Sustain. Sci. **14**(3), 681–695 (2019)
4. Leridon, H.: World population outlook: explosion or implosion? Popul. Soc. **1**, 1–4 (2020)
5. Fraser, E.D.G.: The challenge of feeding a diverse and growing population. Physiol. Behav. **221**, 112908 (2020)
6. Godfray, H.C.J., Beddington, J.R., Crute, I.R., Haddad, L., Lawrence, D., Muir, J.F., et al.: Food security: the challenge of feeding 9 billion people. Science **327**(5967), 812–818 (2010)
7. Johnston, J.L., Fanzo, J.C., Cogill, B.: Understanding sustainable diets: a descriptive analysis of the determinants and processes that influence diets and their impact on health, food security, and environmental sustainability. Adv. Nutr. **5**(4), 418–429 (2014)
8. Conrad, Z., Niles, M.T., Neher, D.A., Roy, E.D., Tichenor, N.E., Jahns, L.: Relationship between food waste, diet quality, and environmental sustainability. PloS One **13**(4), e0195405 (2018)
9. Popkin, B.M., Corvalan, C., Grummer-Strawn, L.M.: Dynamics of the double burden of malnutrition and the changing nutrition reality. Lancet **395**(10217), 65–74 (2020)

10. Mc Carthy, U., Uysal, I., Badia-Melis, R., Mercier, S., O'Donnell, C., Ktenioudaki, A.: Global food security – Issues, challenges and technological solutions. Trends Food Sci. Technol. **77**, 11–20 (2020)
11. Fanzo, J.: Healthy and sustainable diets and food systems: the key to achieving sustainable development goal 2? Food Ethics **4**(2), 159–174 (2019)
12. Willett, W., Rockström, J., Loken, B., Springmann, M., Lang, T., Vermeulen, S., et al.: Food in the anthropocene: the EAT–Lancet commission on healthy diets from sustainable food systems. Lancet **393**(10170), 447–492 (2019)
13. Origin Green. https://www.origingreen.ie/. Accessed 31 Jan 2021
14. Burlingame, B., Dernini, S.: Sustainable diets and biodiversity directions and solutions for policy, research and action: FAO Headquarters, Rome (2012)
15. Su, C.-H.J., Tsai, C.-H.K., Chen, M.-H., Lv, W.Q.: US sustainable food market generation Z consumer segments. Sustainability **11**(13), 3607 (2019)
16. Braun, V., Clarke, V.: Using thematic analysis in psychology. Qual. Res. Psychol. **3**(2), 77–101 (2006)
17. Bree, R.T., Gallagher, G.: Using Microsoft Excel to code and thematically analyse qualitative data: a simple, cost-effective approach. Ireland J. High. Educ. **8**(2) (2016)
18. Attride-Stirling, J.: Thematic networks: an analytic tool for qualitative research. Qual. Res. **1**(3), 385–405 (2001)
19. Pieniak, Z., Żakowska-Biemans, S., Kostyra, E., et al.: Sustainable healthy eating behavior of young adults: towards a novel methodological approach. BMC Public Health **16**, 577 (2016)
20. Salazar, H.A., Oerlemans, L., van Stroe-Biezen, S.: Social influence on sustainable consumption: evidence from a behavioural experiment. Int. J. Consum. Stud. **37**, 172–180 (2013)
21. Hanss, D., Böhm, G.: Sustainability seen from the perspective of consumers. Int. J. Consum. Stud. **36**, 678–687 (2012)
22. Papadaki, A., Hondros, G.A., Scott, J., Kapsokefalou, M.: Eating habits of University students living at, or away from home in Greece. Appetite **49**(1), 169–176 (2007)
23. Köster, E.P.: Diversity in the determinants of food choice: a psychological perspective. Food Qual. Prefer. **20**(2), 70–82 (2009)
24. Chambers, S., Lobb, A., Butler, L., Harvey, K., Bruce, T.: Local, national and imported foods: a qualitative study. Appetite **49**(1), 208–213 (2007)
25. Cecchini, M., Warin, L.: Impact of food labelling systems on food choices and eating behaviours: a systematic review and meta-analysis of randomized studies. Obes. Rev. **17**, 201–210 (2016)
26. Storcksdieck Genannt Bonsmann, S., et al.: Front-of-pack nutrition labelling schemes: a comprehensive review, EUR 29811 EN, Publications Office of the European Union, Luxembourg (2020)
27. A European Green Deal. https://ec.europa.eu/info/strategy/priorities-2019-2024/european-green-deal_en. Accessed 31 Feb 2021

Integrating the Sustainable Development Goals with the Water-Energy-Food Nexus: A Model for Agro-Industrial Companies

Fernando Caixeta[1,2](✉), Pedro Saraiva[3,4], and Fausto Freire[5]

[1] Federal Institute of Education, Science and Technology of Triângulo Mineiro, Campus Uberlândia, Uberlândia, Brazil
fernandocaixeta@iftm.edu.br
[2] University of Coimbra, Coimbra, Portugal
[3] Department of Chemical Engineering, CIEPQPF, University of Coimbra, Coimbra, Portugal
pas@eq.uc.pt
[4] NOVA IMS, NOVA University of Lisbon, Lisbon, Portugal
[5] University of Coimbra, ADAI, Department of Mechanical Engineering, Rua Luís Reis Santos, Pólo II, 3030-788 Coimbra, Portugal
fausto.freire@dem.uc.pt

Abstract. The Sustainable Development Goals (SDGs) express the ambition to achieve targets by 2030 for the Planet, as defined by the United Nations (UN). Under a similar perspective, the Water-Energy-Food (WEF) nexus approach highlights that these three elements are intrinsically interconnected. Consequently, they should be evaluated and managed in an integrated way. The agro-industrial sector makes extensive use of the most demanded natural resources, and plays a major role in both frameworks. Therefore, proposing an integrated view for this sector, dealing with both the UN SDGs and the WEF nexus is relevant and promising, but challenging. This short note provides a proposition on the connections between the SDGs and WEF nexus for the agro-industrial sector, based upon available literature. Finally, the overall expected benefits are to impact the decreasing of the water used volume (ensuring its availability, and quality), the energy efficiency-enhancing (stimulating the replacement by clean energy sources), and to avoid food wastage and contamination during the process.

Keywords: Agro-industrial · Sustainability · UN SDGs and WEF nexus

1 Introduction

Sustainability is becoming increasingly important and a major concern, due namely to the depletion of natural resources, as well as raising concerns regarding the disparity of wealth and lack of corporate social responsibility [1]. Policy makers, thus, must implement environmental policies able to support sustainable alternatives to the existing practices [2]. Having these concerns in mind, the General Assembly of the United Nations

J. R. da Costa Sanches Galvão et al. (Eds.): *Proceedings of the 1st International Conference on Water Energy Food and Sustainability (ICoWEFS 2021)*, pp. 46–53, 2021.
https://doi.org/10.1007/978-3-030-75315-3_6

launched the 2030 Agenda with its Sustainable Development Goals (SDGs). This Agenda is a plan of action for people, planet and prosperity. It seeks to strengthen universal peace in larger freedom. Moreover, one of the main goals is to use the interlinkages and integrated nature of the SDGs. In response to increasing concerns about the long-term sustainability of human societies, the SDGs include 17 goals and 169 targets [3].

The SDGs provide an evidence-based framework for national, regional and global sustainable development planning and programming over a 15-year period, until 2030 [3]. Thus, countries need to take into account concerns about natural resources management and some of the SDGs major highlights are related with water, energy and food. Demand for these resources is increasing [4]. Their availability may indicate poverty decline and sustainable expansion [5]. Enhancing their security separately requires increasing efficiency. To reduce trade-off needs, one must also try to build synergies and improve public and private governance.

In 2011, the WEF nexus approach was incorporated into international discussions on sustainable development [6, 7]. According to [6], WEF nexus implies that water availability, energy and food security are inextricably linked. Consequently, actions in any one area have impacts in one or both of the others [5]. Focusing on the promotion of inseparable links between the use of resources to provide basic rights to food, water and energy security, there is a 'nexus thinking'. This concept highlights the inseparable linkages between the use of resources and the universal human rights to water, energy, and food security [7].

Among the 17 SDGs, there are at least three specifically dedicated to this nexus approach: water security (SDG 6 – Ensure availability and sustainable management of water and sanitization for all); energy security (SDG 7 – Ensure access to affordable, reliable, sustainable and modern energy for all); food security (SDG 2- End Hunger, achieve food security and improve nutrition and promote sustainable agriculture) [3].

Some researchers proposed new models to address the relationships between SDGs for renewable energy [8], and natural resources [9], among others. However, not many attempts have been reported in connecting in more detail the SDGs and the WEF nexus. Therefore, the main aim of this short note is to integrate the Sustainable Development Goals with the Water-Energy-Food nexus, with a special focus regarding the agro-industrial sector.

1.1 The Agro-Industrial Sector

The agro-industrial sector is very energy intensive with important environmental impacts resulting from this sector and its companies. For reducing the GHG emissions in industrial sectors alternative energy (solar and wind energy for instance) and water resources (sustainable water production), should be employed here [10].

Water pollution owing to agro-industrial activities is a crucial problem to human beings and the ecosystem environment [11]. Approximately 85% of global water consumption is used for irrigation, and the demand is still increasing [12]. In order to avoid wastage, adequate water treatment and reuse through non-pollutant methodologies is suggested [13].

The Food scenario is not different. One out of nine people in the world go hungry every day according to The State of Food Insecurity section in the World Report [5].

Annually one third of food produced for human consumption is wasted on the food production chain [14]. Thus, it is a big challenge for all agro-industrial chains to produce food while avoiding food wastage.

1.2 Why Integrate WEF Nexus and SDGs for the Agro-Industrial Sector?

Due to the situation mentioned above, the literature about the WEF nexus expresses great ambitions to achieve policy coherence and overcome the unintended consequences of uncoordinated policies between different sectors [15]. Indeed, interdisciplinary research is essential for effective management of WEF nexus systems. While the various science disciplines have long histories of working independently in components of the WEF nexus, future research should integrate physical, agro ecological, and social sciences with economics [16], and the agro-industrial sector seems to be quite appropriate for testing and doing so. Nevertheless, different measures for overcoming issues related to the WEF nexus do vary by adopting diverse perspectives, which steer in different directions [15]. The majority of the matrixes available concerning WEF nexus perspectives do correspond to governmental, sectorial, cross national or national analysis [4, 6, 17, 18].

Additionally, at present, there is no universally recognized methodology for conducting nexus analysis [18], and few efforts have been made to integrate SDGs and the WEF nexus. The integration of other sustainable approaches is, therefore, recommended, namely with regards to the UN Sustainable Development Goals. Although there are some advances in the SDGs implementation, key gaps remain in terms of the assessment of interlinkages, trade-offs and synergies between targets [19].

Additionally, there is limited SDGs research about how exactly national interests, public expectations, technical and economic feasibility, and environmental performance can be accounted for at the country, local or company level [8]. Therefore, the integration between SDGs and WEF nexus can also be expected to lead to more sustainable actions in agro-industrial companies, and thus cover some of the gaps identified above.

2 Integration

Based on previous studies [20–22], a literature review was performed underscoring the synergies between the mentioned approaches (SDGs and the WEF nexus) with agro-industrial characteristics. Thus, the results were expressed in this article in the introduction text. Based on that, the selected criteria were first designated due to SDGs' list proposition [3].

The results are shown as follows:

- SGD 2 – End Hunger
  2.1 – End Hunger and ensure access by all people;
  2.2 – End all forms of malnutrition;
  2.3 – Double the agriculture productivity to provide more food production;
  2.4 – Ensure Sustainable Food production of small-scale food producers;
  2.5 – Maintain the genetic diversity of the fauna and flora;
  2.6 – Increase investment in rural infrastructure;
  2.7 – Adopt measures to ensure the proper functioning food.

- SDG 3 – Good health and well-being
  3.9 – Reduce the number of deaths and illnesses from pollution and contamination, including the ones coming from misuse of pesticides and fertilizers.
- SDG 6 – Clean water and Sanitization
  6.1 – Universal and equitable access to water for all;
  6.3 – Improve water quality;
  6.4 – Water-use efficiency across all sectors;
  6.5 – Implement integrated water resources management.
- SDG 7- Affordable and clean energy
  7.1 – Ensure universal access to energy services;
  7.3 – Double the rate of energy efficiency.
- SDG 8 – Decent work and economic growth
  8.2 – Higher levels of economic productivity;
  8.4 – Global resource efficiency in consumption and production.
- SDG 9 – Industry innovation and infrastructure
  9.2 – Promote inclusive and sustainable industrialization;
  9.4 – Upgrade sustainable infrastructure and retrofit industries.
- SDG 12 – Sustainable consumption and production
  12.1 – Implement framework of programs in Sustainable Consumption and Production;
  12.2 – Achieve the sustainable management and efficient use of natural resources;
  12.3 – Reduce Food losses;
  12.4 – Achieve the environmentally sound management;
  12.5 – Reduce waste generation;
  12.6 – Adopt sustainable practices.
- SDG 14 – Life below water
  14.4 – Regulate harvesting and end overfishing, an important substitute to wild harvesting.
- SDG 15 – Life on land
  15.1 – Ensure terrestrial and inland freshwater ecosystems;
  15.2 – Promote the implementation of sustainable management;
  15.3 – Restore degraded land and soil.

With this set of criteria, the aforementioned topics were classified under the three main topics, that correspond to the pillars of the WEF nexus: Water, Energy and Food. The results obtained indicate that the overall criteria correspond to the following percentages, by WEF domain: Water (23%); Energy (14%) and Food (63%). Then, at a more detailed level, topics were cascaded into WEF sub-topics. For Food, the sub-topics considered were: Food security and Improved Nutrition; Promote Sustainable Agro-industrial Sector; and Food Safety. For Water, we considered: Water security and safety; and Water resource efficient management. Finally, for Energy: Energy Security and efficiency; and Use of renewable energy.

In the end, criteria were allocated into these sub-topics, leading to the final correspondences between the SDGs and WEF nexus summarized in Table 1.

Table 1. Topic, Sub-topic and Criteria for the integration of SDGs with WEF nexus for agro-industrial companies.

WEF nexus Topic	WEF nexus Sub-Topic	SDGs Criteria
Food	Food Security and Improved Nutrition	Ensure access of food for all people
		Improve the food nutrition
	Promote Sustainable Agro-industrial sector	Increase small-scale food producers' productivity
		Sustainable Production
		Crop Diversity
		Increase investments in Infrastructure
		Build Measures to ensure the food sustainable Production
		Food Productivity
		Reduce Food Losses
		Sustainable Fishing Production
		Sustainable Forest
		Sustainable Land and soil use
	Food Safety	Reduce Food contamination
		Achieve the environmentally sound management of chemicals and all wastes
Water	Water security and safety	Water security
		Water safety
	Water resource efficient Management	Water-use efficiency
		Water resources management
		Sustainable Water Use
Energy	Energy Security and efficiency	Energy Security
		Energy efficiency
	Use of Renewable Energy	Increase substantially the share of renewable energy

3 Discussion

As noticed, the SDGs and the WEF nexus must be integrated. Both of them aim to implement sustainable actions into different systems. However, it is a challenge to integrate both systems, that were built for reflection in a macro perspective (governments, regions, river basins, among others), and adapt them to a given specific more micro context, in this case, the agro-industrial sector and its companies.

The aforementioned sector has responsibility for the increasing damage found in the environmental systems, including water, energy and also food production. It may therefore benefit, the same also being the case for its companies, from aspects that were highlighted in this integrated view. This may include actions that ensure food security for the population and food sustainability important concerns, such as empowering small producers, reduce food losses or sustainable production. But may also refer to food safety actions (providing food free of contaminants). In the water perspective, it is inevitable to secure its availability, quality, and also appropriate management. Finally, regarding energy issues, an effort is made to point out the production control and efficiency, and the use of renewable sources.

In this context, case studies are suggested to validate the relationships found between the SDGs and the WEF nexus. Such case studies may cover a variety of agro-industrial companies, such as forests, fisheries, crop production, organizations that use soil, and even the use of chemicals. Subsequently, based on the selected items presented in this research, evaluation criteria should be proposed, such as the Linkert scale, a check-list, among other score methodologies. Given that, the agro-industrial production site should be inspected. Consequently, with the results, suggestions should be given on how to perform better in a sustainability vision. Thus, aspects for instance the use of renewable energy, water re-circulation, and the use of optimization tools can appear as the mentioned solution. Therefore, those aspects enable the proposed approach to target the challenge to enable information to the consumers on how the industrial transformation can be managed.

4 Conclusions

From the analysis conducted and presented here, one can conclude that:

It is possible to develop and propose integrations of the Sustainable Development Goals and the Water-energy-food nexus for agro-industrial companies, as was exemplified.

The suggested approach can be used by different agro-industrial companies in order to expand their sustainable actions and priorities in an integrated way.

As for possible additional work, the practical testing and implementation of such a framework in agro-industrial companies from different countries is suggested.

Acknowledgements. Fernando Caixeta acknowledges financial support from the Brazilian National Council for Scientific and Technological Development - CNPq.

References

1. Govindan, K., Khodaverdi, R., Jafarian, A.: A fuzzy multi criteria approach for measuring sustainability performance of a supplier based on triple bottom line approach. J. Clean. Prod. **47**, 345–354 (2013). https://doi.org/10.1016/j.jclepro.2012.04.014
2. Bilgili, F., Koçak, E., Bulut, Ü., Kuşkaya, S.: Can biomass energy be an efficient policy tool for sustainable development? Renew. Sustain. Energy Rev. **71**, 830–845 (2017). https://doi.org/10.1016/j.rser.2016.12.109

3. UN: financing the 2030 agenda for sustainable development. Gov. Through Goals Sustain. Dev. Goals as Gov. Innov. **16301**, 259–273 (2015). https://doi.org/10.7551/mitpress/9780262035620.003.0011
4. Rasul, G.: Food, water, and energy security in South Asia: a nexus perspective from the Hindu Kush Himalayan region☆. Environ. Sci. Policy. **39**, 35–48 (2014). https://doi.org/10.1016/j.envsci.2014.01.010
5. FAO, IFAD, WFP: Strengthening the enabling environment for food security and nutrition (2014)
6. Hoff, H.: Understanding the Nexus. Background paper for the Bonn2011 Nexus Conference: stock. Environ. Inst. 1–52 (2011)
7. World Economic Forum: Water security: the water-food-energy-climate nexus: the World Economic Forum water initiative (2011). https://doi.org/10.5860/choice.49-1001
8. Büyüközkan, G., Karabulut, Y., Mukul, E.: A novel renewable energy selection model for United Nations' sustainable development goals. Energy **165**, 290–302 (2018). https://doi.org/10.1016/j.energy.2018.08.215
9. Merino-Saum, A., Baldi, M.G., Gunderson, I., Oberle, B.: Articulating natural resources and sustainable development goals through green economy indicators: a systematic analysis. Resour. Conserv. Recycl. **139**, 90–103 (2018). https://doi.org/10.1016/j.resconrec.2018.07.007
10. Pashangpour, R., Faghihi, F., Soleymani, S.: Optimized scheduling for electric lift trucks in a sugarcane agro-industry based on thermal, biomass and solar resources. Int. J. Environ. Sci. Technol. **15**, 2349–2358 (2018). https://doi.org/10.1007/s13762-017-1535-4
11. Chen, Y., Wang, F., Duan, L., Yang, H., Gao, J.: Tetracycline adsorption onto rice husk ash, an agricultural waste: its kinetic and thermodynamic studies. J. Mol. Liq. **222**, 487–494 (2016). https://doi.org/10.1016/j.molliq.2016.07.090
12. Zou, Y., Duan, X., Xue, Z., Mingju, E., Sun, M., Lu, X., Jiang, M., Yu, X.: Water use conflict between wetland and agriculture. J. Environ. Manag. **224**, 140–146 (2018). https://doi.org/10.1016/j.jenvman.2018.07.052
13. Mo, J., Yang, Q., Zhang, N., Zhang, W., Zheng, Y., Zhang, Z.: A review on agro-industrial waste (AIW) derived adsorbents for water and wastewater treatment. J. Environ. Manag. **227**, 395–405 (2018). https://doi.org/10.1016/j.jenvman.2018.08.069
14. Gustavsson, J., Cederberg, C., Sonesson, U., Emanuelsson, A.: The methodology of the FAO study: global Food Losses and Food Waste - extent, causes and prevention - FAO, 2011 (2013)
15. Weitz, N., Strambo, C., Kemp-Benedict, E., Nilsson, M.: Closing the governance gaps in the water-energy-food nexus: insights from integrative governance. Glob. Environ. Chang. **45**, 165–173 (2017). https://doi.org/10.1016/j.gloenvcha.2017.06.006
16. Scanlon, B.R., Ruddell, B.L., Reed, P.M., Hook, R.I., Zheng, C., Tidwell, V.C., Siebert, S.: The food-energy-water nexus: transforming science for society. Water Resour. Res. **53**, 3550–3556 (2017). https://doi.org/10.1002/2017WR020889
17. Kaddoura, S., El Khatib, S.: Review of water-energy-food Nexus tools to improve the Nexus modelling approach for integrated policy making. Environ. Sci. Policy **77**, 114–121 (2017). https://doi.org/10.1016/j.envsci.2017.07.007
18. Al-Ansari, T.: The development of the energy, water and food nexus systems model. 313 (2016)
19. Allen, C., Metternicht, G., Wiedmann, T.: Initial progress in implementing the Sustainable Development Goals (SDGs): a review of evidence from countries. Sustain. Sci. **13**, 1453–1467 (2018). https://doi.org/10.1007/s11625-018-0572-3

20. McElwee, P., Calvin, K., Campbell, D., Cherubini, F., Grassi, G., Korotkov, V., Le Hoang, A., Lwasa, S., Nkem, J., Nkonya, E., Saigusa, N., Soussana, J.F., Taboada, M.A., Manning, F., Nampanzira, D., Smith, P.: The impact of interventions in the global land and agri-food sectors on Nature's Contributions to People and the UN Sustainable Development Goals. Glob. Chang. Biol. **26**, 4691–4721 (2020). https://doi.org/10.1111/gcb.15219
21. Siegel, K.M., Bastos Lima, M.G.: When international sustainability frameworks encounter domestic politics: the sustainable development goals and agri-food governance in South America. World Dev. **135**, 105053 (2020). https://doi.org/10.1016/j.worlddev.2020.105053
22. Serhan, H., Yannou-Lebris, G.: The engineering of food with sustainable development goals: policies, curriculums, business models, and practices. Int. J. Sustain. Eng. 1–14 (2020). https://doi.org/10.1080/19397038.2020.1722765

Island Housing and Peri-Urban Agriculture in the City of Luanda

Filomena do Espírito Santo Carvalho[1](✉), Miguel Amado[2], Manuel Duarte Pinheiro[3], and António Gameiro[1]

[1] Department of Architecture, Faculty of Engineering, University Agostinho Neto, Talatona, Angola
{filomena.carvalho,antonio.gameiro}@feuan.ao
[2] Instituto Superior Técnico, University of Lisbon, Lisbon, Portugal
miguelpamado@tecnico.ulisboa.pt
[3] Department of Civil Engineering, Architecture and Georesources, University of Lisbon, Lisbon, Portugal
manuel.pinheiro@tecnico.ulisboa.pt

Abstract. Luanda presents a growing scenario of island housing, with its neighbourhoods comprising a significant percentage of the precarious income population living in a confined regime, in backyards behind houses and buildings making shared use of spaces for everyday use such as the kitchen and bathroom. This scenario is ignored by official institutions and has been aggravating the general picture of existing urban poverty and the lack of access and availability of quality and quantity food for this population. The purpose of this article is to contribute to the diagnosis and analysis of the phenomenon of this "new dwelling" in the city of Luanda, which we call new island housing. It is viewed from the perspective of comfort in housing and the importance of the practice of peri-urban agriculture considered as residual rural practice existing in many of the city's neighbourhoods and in the lives of the residents of this island-dwelling and these neighbourhoods. It is intended to understand how to integrate this reality into the panorama of housing supply in the city of Luanda in order to discuss quality housing, healthy food, decent work with a fair salary and a sustainable environment in the neighbourhoods, all within a framework of sustained economic development and social development of families.

Keywords: Island Housing (IH) · Peri-urban agriculture · Housing comfort · Sustainable development

1 Introduction

Urbanization is an irreversible global phenomenon. Estimating that Luanda will have 12 million people in 2030, what is the approach to housing, considering that, according to the United Nations, quality urbanization generates economic growth and development?

J. R. da Costa Sanches Galvão et al. (Eds.): *Proceedings of the 1st International Conference on Water Energy Food and Sustainability (ICoWEFS 2021)*, pp. 54–64, 2021.
https://doi.org/10.1007/978-3-030-75315-3_7

The gap between urbanization and the dynamics of the city of Luanda's economy shows a contradictory scenario in economic activities. It is marked *by an urban development of exclusion in various dimensions of the city, with the presence of migration whose growth pressures resources, infrastructures, the leadership of the city, in a scenario of fast-growing cities with untapped economies,* resulting in accentuated imbalances between salaries and the cost of living. These social and economic imbalances, reflected in the search for housing, lead a specific fringe of the population with the low financial capacity to build and rent spaces on the outskirts of the city of Luanda or to opt for spaces in areas of the consolidated city and in the surrounding peripheral areas. They live in house-spaces built in the backyards of houses and buildings or in the corridor spaces between this building [1].

A very common scenario is the cultivation of land with the practice of agriculture in apparently vacant free areas. This activity, which can be seen in various neighbourhoods in the city, is practised by migrants from the south.

1.1 New Island Housing in Luanda

This article aims to study thermal comfort and evaluate the perception of the residents' thermal environment in the island housing. This is achieved with the application of questionnaires related to heat exchanges between the body and the environment based on the variables of the environment and the variables of the residents (metabolism and clothes), adjusted to the place and influenced by the uses and customs [2].

Island housing appears similarly to working-class neighbourhoods as a result of the demographic increase in cities with the agricultural crisis and the development of industrialization. It concerns accommodation that emerges in long and narrow corridors next to empty land at the back of the bourgeois houses and in their backyards, a type of basic, social housing, with unsanitary problems, poor ventilation, poor construction quality and intensive use of the soil, registered in several countries such as Portugal, England and Brazil at the end of the 19thcentury.

Amaral (1968, 47,115) refers to the lower city of Luanda in the year 1770 as ... "*the lower city, with commercial port activity, with its warehouses and backyards for slaves*"... "*in Coqueiros, Nazaré, Carmo*". *The backyards of the houses in Coqueiros, Bungo and Ingombotas in Luanda were used to house people...*" [3].

Morphologically, the Island Housing (IH), a form of popular and social housing common in Luanda since the year 2000, appear as the nucleus of housing built in the backyards of houses and buildings (Fig. 1) or in spaces of narrow corridors existing between houses. It consists of small houses at ground level, usually with one or two small rooms, with kitchen and toilet facilities located in the outside yard for common use by different tenants (Fig. 2, Fig. 3). These are spaces with dense occupancy and maximized utilization, with overload in the use of the local network of water, energy and sewage infrastructure. With its back to the city, the yard or corridor (Fig. 4, Fig. 5), it is the common distribution space and the only access to the outside. The small size of the house and the common use of facilities provide a certain social coexistence among the residents and promotes promiscuity.

The new self-produced urbanity (constructed by the inhabitants) in Luanda can be refered to three scales of urban analysis that constitute the city's built space in the

Fig. 1. Common Backyard (Island housing-Cassendaneighbourhood, 2021, credit: W.E. Santo)

Fig. 2. Entrance to a house (Prenda neighbourhood, 2021, credit: C.Soares)

Fig. 3. Common kitchen (Prenda neighbourhood, 2021, credit: C.Soares)

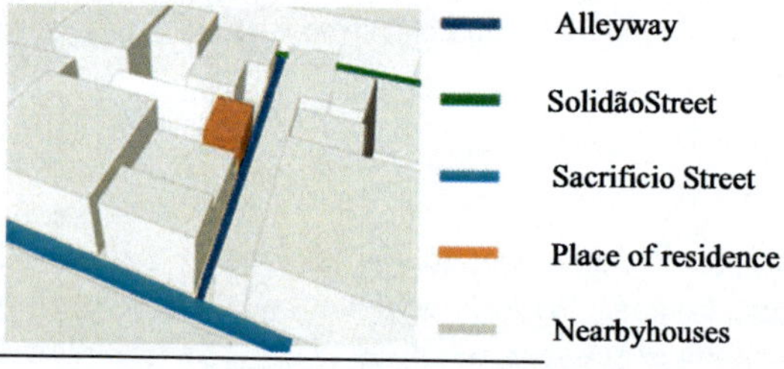

Fig. 4. Implantation of island housing (Estalagem neighbourhood, Viana, 2021 credit: S. Cajolo)

Fig. 5. Island housing corridor (Estalagem neighbourhood, Viana, 2021 credit: S. Cajolo)

city centre, the muceques and the residential condominiums, centralities and others, perceived in the current context as found solutions for urban transformation and production, as an "*unpredictable city*" [4]. The new housing islands are developing in the city centre and slums (*muceques*) in new peripheral neighbourhoods. The National Urbanism and Housing Program of Angola (NUHP, 2002) represents the main strategic instrument of the Housing Promotion Policy, based on the Housing Promotion Act (Law nº 3 3/07, of September 3, 2015) and includes three fundamental axes: fostering housing (with social housing, directed self-construction, affordable rent housing), development of new centralities, and urban requalification. With the housing island built in a format adjusted to the promoter's availability, could the potential residents' economic capacity be considered a contribution, if informal, to one of the fundamental axes, which is offering living spaces? The relationship between the population and income is an important fact in view of the percentage of the population in need of assistance (Fig. 6). Island housing and peri-urban agriculture are growing in some places in the city (see Fig 7).

1.2 Other Islands Cases

The existence of the islands in the 19thcentury, with their location in narrow corridors or in the back voids of bourgeois houses, and precarious conditions were the common characteristics of these popular housing nuclei in different latitudes such as Portugal (Porto), the United Kingdom and Brazil. The city of Porto registered the existence of

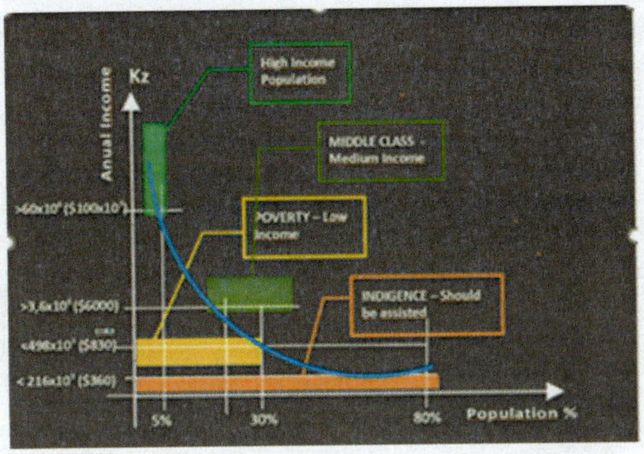

Fig. 6. Relation between population and income (those at the bottom of the graphic should be assisted). (Gameiro, DW 2017, adjusted by FESanto)

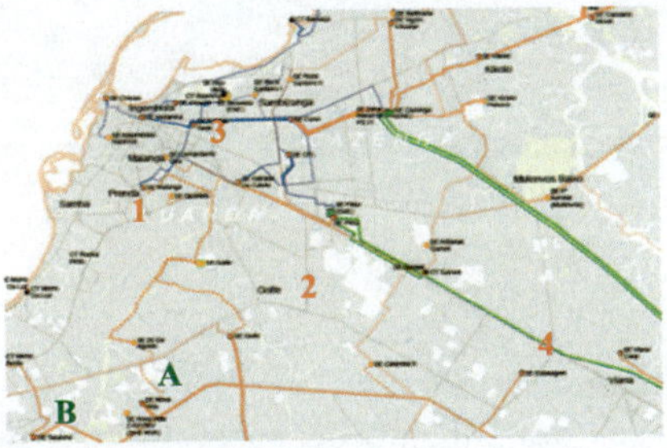

New island housing	
1	Neighbourhood Prenda
2	Neighbourhood Golfe
3	Neighbourhood Cassenda
4	Neighbourhood Estalagem
	Peri-urban agriculture
A	Neighbourhood Nova Vida
B	Neighbourhood Talatona

Fig. 7. Location of four new areas of island housing in Luanda and neighbourhood with periurban agriculture. (Map of Metropolitan Master Plan of Luanda (MMPL 2030). Consider only the new island housing.

small houses with only one floor, built with precarious material, located in long, narrow corridors facing a single facade or in empty grounds behind the bourgeois houses with low real estate value. They housed the rural population that arrived in the city and were later transformed into the poor working classes' shelter. It was a type of unhealthy, overoccupied, unventilated dwelling [5, 6]. The United Kingdom presented several samples of these nuclei in London and other English cities. Rio de Janeiro in Brazil was home for the called tenements. Designated as rooming houses, located close to the workplace, they were a low-income collective residence in a small two-storey building with common spaces for sharing activities. They were built at the back of the old homes of the middle class, with only a single access to the street through a corridor [6].

1.3 Peri-Urban Agriculture

According to the World Food Program, food security is linked to the population's right to regular and permanent access to food in sufficient quantity and quality.With poverty reaching the weakest social groups with low financial capacity, they are weak and without the ability to have adequate food, constituting one of the causes of food insecurity. The National Statistics Institute (NSI,2020) reports that 63% of the population of Angola lives in an urban context where 40% of the country's poor life. The demographic increase in Luanda, reflected in the supply of food, has motivated the use of free and unoccupied areas, normally close to water courses, in some districts of Luanda for the cultivation of land. In the Talatona neighbourhood, floriculture is common practice with residual production; in the Nova Vida neighbourhood, subsistence crops such as cassava, potato, okra, kale, *gindungo* (*Capsicum frutescens*) and others are grown (see Fig. 8, Fig. 9, Fig. 10). This activity is carried out by migrants with an agricultural vocation and is practised as an opportunity for survival; undervalued by local residents and ignored by institutions, it is often a source of supply of products in informal neighbourhood markets.

Fig. 8. Floriculture (1)

Fig. 9. Diverse plantation (1)

Fig. 10. Cassava plantation (2) Talatonaneighbourhood (2), Nova Vida neighbourhood (2020)

The 2015–2016 Multiple Indicator Cluster Health Survey states that the national incidence of poverty is estimated at 54.0%, i.e., more than 5 out of every 10 people in Angola are multidimensionally poor. The Multidimensional Poverty Index-Angola (MPI-A, INE, 2020) consists of four dimensions: a) health; b) education; c) quality of life; and d) employment – and sixteen indicators that express the different deprivations faced by people living in poverty, such as access to electricity, characterization of housing (types of wall, floor and ceiling materials), access to appropriate water, access to sanitation. The approach is to identify the level of importance of the practice of peri-urban agriculture in the lives of these people in order to understand the level of integration of this situation and to discuss healthy eating and a sustainable environment in the neighbourhoods within a framework of sustainable economic and social development of families.

2 Methodology

2.1 Material

The study evaluates thermal comfort in housing and the perception of the sensation of thermal comfort reported by residents of four "new housing islands". It is organized in the neighbourhoods of Prenda, Golfe, Cassenda and Estalagem in Luanda and is undertaken with research on climate adaptation and preferences regarding the thermal sensation of residents. The analysis of the practice of peri-urban agriculture in the Nova Vida and Talatona neighbourhoods in Luanda is made with the comparison of the data on the quality of life of the population in Luanda and the survey data from the study areas [1].

2.2 Method, Collection and Sampling

The research on housing comfort carried out in Luanda in four housing islands in the warm and rainy season was applied to six families, of which four households were in the Prenda neighbourhood, one in the Golfe neighbourhood and one in the Estalagem-Viana neighbourhood. The survey phase involved a literature review of dissertations, theses and scientific articles related to thermal comfort. For the characterization of the housing, a survey was carried out of the constructive solutions, the building materials' characteristics, the spatial organization and size. The local climate information supports the analysis.

In order to obtain the perceptual responses of the residents, the research survey comfort votes and thermal sensations of the residents about the thermal environment inside the houses, with the application of questionnaires. These were related to heat exchanges between the body and the environment based on the variables of the environment and the variables of the residents (metabolism and clothes), adjusted to the place and influenced by uses and customs. The comfort scales used to collect information on the residents' thermal sensation and preferences are based on the scale of thermal sensations and preferences (Table 1) [5].

Table 1. Thermal sensation scale Source: ISO7730, 2005

Thermal sensation value *(PMV)*		Thermal Preference	
Scale	Description	Scale	Description
+3	Very warm	+3	Very, very warm
+2	Slightly warm	+2	Very warm
+1	Hot	+1	A little warmer
0	Neutral	0	Neither
−1	Slightly cold	−1	A little colder
−2	Cold	−2	Very cold
−3	Very cold	−3	Very, verycold

In the analyse of the residents' thermal perception of the environment, other issues were included, such as the indication of three important points for comfort in the home, suggestions for improvement for comfort conditions, the importance of the backyard, the reason for living in the island housing and expectations regarding housing. The study of peri-urban agriculture involved the collection of data with the Multidimensional Poverty Index in Angola. Visits were made to the places where agricultural activity is carried out; research into motivation was conducted based on a conversation with farmers (normally women) to find out what leads them to undertake this practice and, if possible, to understand how and at what level the practice of peri-urban agriculture can contribute to the reduction of urban poverty and be considered as a support for the food subsistence of the most disadvantaged population.

3 Results and Discussion

The results are presented in two groups. The first reports on the comfort in indoor housing with the analysis of the quality of life used a set of indicators that can lead the residents of the new island housing to deprivations of human life [1] and the analysis of the thermal comfort to understand the residents'perception of the thermal environment of housing on the islands. The second is the understanding of peri-urban agriculture in Luanda.

3.1 Quality of Life and Comfort in Indoor Housing

Housing as the centre of social, emotional and sometimes economic life is a human right and not a commodity (United Nations Organization), and the relationship that is established with one's environment depends on how residents see, feel and can use their physical space. One of the most important aspects of human comfort is thermal comfort, which provides satisfaction with the surrounding environment. Thermal comfort is affected by heat conduction, convection, radiation and evaporative heat loss [7], and the main constraint on thermal comfort is the combination of temperature and humidity. In order to achieve such thermal comfort in a hot, humid city environment such as Luanda, the objective of housing is to provide protection against heat from the sun, sufficient ventilation and cooling with adjustable air quality. On the island, the dwellings are built with sheet metal on the roof, cement brick in the walls, few window openings and the absence of false ceilings (Fig. 11, 12, 13).

Fig. 11. House cover – zinc sheet (Estalagem neighbourhood, Luanda (2020), credit: S. Cajolo)

Fig. 12. Cement brick wall (Estalagem neighbourhood, Luanda (2020), credit: S. Cajolo)

Fig. 13. Absence of false ceiling (Estalagem neighbourhood, Luanda (2020), credit: S. Cajolo)

The households have deprivations [1] such as inadequate housing related to the quality, performance and durability of construction materials, shown in the coverage with sheet metal, absence of false ceilings, inappropriate material in the walls in relation to thermal issues and no access to piped water or sanitation (Tables 2, 3, 4, 5, 6 and 7).

Table 2. Roofs of houses. 100% of the houses are covered in roof metal. The walls are made of cement bricks.

Sheet metal and cement wall =100%

Table 3. Typology of the Housing Island. 67% of island housing is organized around a **non**-linear yard and 33% in a corridor space.

Island housing –**non**-linear yard (67%)	Island housing in a corridor (33%)

Table 4. House size: 50% of houses have (01) one-bedroom, and 50% have (02) two

One bedroom = 50%	Two bedrooms =50%

Table 5. Location of WC: 50% of houses do not have a WC inside the house

WC inside the house = 50%	WC outside the house = 50%

Table 6. Water supply situation: 67% of the houses do not have running water at home

No running water at home = 67%	With running water at home = 33%

Table 7. Electricity situation. 100% of the houses have electricity

Electricity = 100%

100% of the residents are tenants. The backyard is the opportunity for ventilation. Sanitation is limited to the common kitchen and toilet facilities located in the outside yard. Living on the island is not what is desirable but it is a necessity. A survey was carried out to observe and record the perception of the existing thermal environment conditions [2, 7] to raking the satisfaction of the residents and understand the dynamics of comfort and the perception and eventual thermal adaptation of the residents in the indoor environment of the dwellings. Due to the complexity and subjectivity of thermal comfort, it was considered important to know the meaning of comfort for island residents of different ages, gender, and working activity (Tables 8, 9, 10 and 11).

Table 8. Evaluation of the thermal environment based on the ***thermal sensation*** of the residents

How do you consider your home?						
Very hot	Hot	Slightly hot	Not hot, not cold	Slightly cold	Cold	Very cold
Responses: Very hot 100%						

Table 9. Evaluation of the thermal environment based on the thermal sensation of the residents

How do you consider your home environment?				
Very comfortable	Comfortable	Neutral	Uncomfortable	Very uncomfortable
Very uncomfortable 74%				Neutral 16%

The experiences of the place relate to the response to what they claim to feel during the survey period. The way space is seen and felt, their experience of the conditions and the way it relates to the environment, based on expectations, motivation and preference [8].

Table 10. Evaluation of the thermal sensation of the inhabitants during the hot season (season at the time of the questionnaire). The residents feel the houses are "very hot".

Thermal sensation. As a resident, how you generally feel in the hot season?						
Very hot	Hot	Slightly hot	Not hot, not cold	Slightly cold	Cold	Very cold
Very hot 73%			Responses	Slightly hot 27%		

Table 11. Assessment of the thermal preference of residents according to the time of year

Thermal preference. How would you like the environment and temperature to be?					
Much hotter	Hotter	Slightly hotter	Slightly colder	Colder	Much colder
Colder 50%		Slightly colder 33%		Hotter 17%	

The thermal preference at the time of the interview and the sensation they would like to feel at the time of the interview were analyzed with the residents. It appears that most of the residents perceived the microclimate of the environment as very hot and claim to prefer to stay in the outside yard where there is natural ventilation, as there are few possibilities to interact with the elements of the house.

Important Elements for Comfort. In a range of elements such as temperature, ventilation, natural light, artificial light, relationship with the outside, organization of the house, the colour of the walls and privacy, among others, the participants were requested to indicate three elements in order of priority in terms of importance for their comfort. The responses indicated temperature as the first priority; the second was natural ventilation and natural light, and the third was privacy.

Freedom of Control over the Thermal Environment

Some studies show the importance of residents' freedom of control in the environment they use, assuming that there is greater satisfaction with the thermal conditions when it is possible to interact with the environment. Within the scope of this work, issues related to the opportunities offered by housing that allow residents to control some mechanism to change thermal conditions were analyzed [9, 10].100% of the residents presented suggestions to change the thermal environment at that moment; however, there is no possibility (conditions or mechanism) to interact with the environment to change the thermal conditions (Table 12).

Table 12. Assessment of the thermal sensation of residents according to the season

Open windows 50%	Install air conditioning 50%

3.2 Peri-Urban Agriculture

No relation or predisposition of the island's inhabitants to the activity of peri-urban agriculture was analyzed. This practice is carried out by migrants from the southern provinces of Angola; they live mostly in communities in informal occupation neighbourhoods very far from the city in precarious housing and move daily to the cultivated areas in the city due to the availability of soil for cultivation in a favourable and profitable sales environment.

4 Final Considerations

This survey of comfort in housing indicates that all residents feel "uncomfortable" (74%), evidenced by the indication of temperature, natural ventilation and privacy as residents'priorities for comfort in the house. Most residents (100%) say they do not like their houses as they are small, hot, uncomfortable, and would like the environment to be slightly colder (83%) in the hot season and to make changes to improve indoor thermal conditions, such as opening the windows or changing materials. The willingness to interact with the devices of the house (100%) may indicate some level of adaptive intention to seek thermal comfort in the interior environment. However, there is some conformism in view of the impossibility of change, a situation justified by some tolerance based on the thermal history of the residents and their economic capacity. The strategy of using the backyard can be considered as an adaptive opportunity in the hot season. They claim to be residing in the new housing islands out of necessity and that if given the opportunity, they would move to a better, larger and more comfortable space. They emphasized the advantage of the good location of the islands, the proximity of services and the socialization that the city allows. These considerations may constitute a premise for the study of adaptive thermal comfort in housing and indicators for redefining social housing policy and programmes. These are objective indicators to rethink the NUHP's popular social housing strategy with the appreciation of the region's climate and the real economic capacity of future residents. **Peri-urban agriculture:** this practice seems to present itself as the basis of subsistence for a certain layer of the population, an alternative to alleviate the consequent food insecurity in the face of the level of multidimensional poverty. Its importance and the management challenges highlight the need to institutionalize a localized agri-food system (AFS) as a management tool to integrate this productive activity on the development agenda. This is justified by the cultivation of vegetable and ornamental plants in the neighbourhoods of Luanda.

References

1. Multidimensional Poverty Index-Angola. National Statistic Institute, 2020MPA (NSI2020)
2. Nicol, J.F., Humphreys, M., Roaf, S.: Adaptative Thermal Comfort: Principles and Practice. Oxford Brookes University (2012)
3. Amaral, I.: Luanda: Estudo de Geografia Urbana. Memórias da Junta de Investigação Ultramarina, Lisboa (1968)
4. Paulo, A.: Buatica: Gestão Sócio-Espacial e Ambiental dos Bairros Autoproduzidos consolidados ou em consolidação em Luanda. Relatório para grau mestrado em Gestão Ambiental. Instituto Politécnico de Coimbra. Escola Superior Agrária de Coimbra
5. Almeida, B.G.: Ilhas do lado de lá. Reflexões sobre a habitação popular e social e a sua integração na cidade do Porto. Mestrado Integrado, FAUP (2015)
6. Pereira, G.M.: As Ilhas no percurso das famílias trabalhadoras do Porto em finais do século XIX. Capítulo de livro,Família, espaço e património, 2011, p.477–493. Faculdade de Letras.Universidade do Porto (2013). https://hdl.handle.net/10216/63550
7. Frota, A.B., Schiffer, S.R.: Manual de Conforto Térmico, 5° edição, Studio Nobel (2001)
8. ASHRAE: ASHRAE Handbook: Heating, Ventilation, and Air-Conditioning (2004)
9. De Dear, R., Kim, J., Parkinson, T.: Residential adaptive comfort in a humid subtropical climate-Sydney Australia. Energy Build. **158**, 1296–1305 (2018)
10. Matias, L., Almeida, S., Rebelo, M., Santos, C.: Perception and thermal adaptation in indoor environments. LNEC, Report 302/2009 ES/LNEC, I&D Sustainable building (2009)

Greenhouse Gas Emissions from Subsistence Dairy Livestock in Rural Livelihoods in the Northern Andes of Ecuador

Jhenny Cayambe[1(✉)], Marco Heredia-R[2], Luis Valencia[3], Bolier Torres[4], Carlos G. H. Díaz-Ambrona[2], and Theofilos Toulkeridis[5,6]

[1] School of Agricultural and Environmental Sciences, Pontificia Universidad Católica del Ecuador Sede Ibarra (PUCESI), 100112 Imbabura, Ecuador
jmcayambe@pucesi.edu.ec

[2] AgSystems, Ceigram, itdUPM, Centro de Innovación en Tecnología para el Desarrollo, Universidad Politécnica de Madrid (UPM), 28040 Madrid, Spain

[3] Gobierno Autónomo Descentralizado de Montúfar, Carchi, Ecuador

[4] Universidad Estatal Amazónica (UEA), 160101 Pastaza, Ecuador

[5] Universidad de las Fuerzas Armadas ESPE, Sangolquí, Ecuador

[6] Universidad de Especialidades Turísticas, Quito, Ecuador

Abstract. Greenhouse gas emissions (GHG) from livestock raise the increase of temperature worldwide. They contribute 14.5% of global emissions while enteric fermentation is the largest contributor (39.1%). The main objectives of the current study were: a) to determine the socio-productive characteristics at the household/farm level and b) to estimate the GHG emissions of dairy farms. This study was conducted in the Andes of northern Ecuador. Sociodemographic information and characteristics of the farms were obtained through field surveys. In order to estimate GHG emissions, the Cool Farm Tool program (version 0.11.36) was used. The probabilistic technique of cluster sampling was performed, with the support of a key informant, while the survey lasted some 45 min each. Hereby, we surveyed 73 mestizo people, being 54.79% male and 45.21% female, with an average experience in milk production of 23 years, distributed in 20 homes in 15 communities. The average values have been of pastures with 3.1 ha in 5.07 ha of farm, with a herd of seven animals, a milk production of 33 l/day and a price of 0.37 dollars/liter. The estimate of GHG emissions indicated that 81% of livestock emissions derived from enteric digestion. Methane from enteric digestion is higher for some species (ruminants) than others, demonstrating that changes in diet potentially improve such scenario. Most of the emissions originated from the management of pasture residues, occupying 82% of the total. These may be compensated by directly incorporating such residues or its compost and using it in soil.

Keywords: Climate change · Greenhouse gas emissions · Livestock · Pasture residues · Ecuador

J. R. da Costa Sanches Galvão et al. (Eds.): *Proceedings of the 1st International Conference on Water Energy Food and Sustainability (ICoWEFS 2021)*, pp. 65–74, 2021.
https://doi.org/10.1007/978-3-030-75315-3_8

1 Introduction

Most dairy farmers manage production costs, the productivity and animal welfare [1]. As expected, the search for higher yields and therefore income is paramount, leaving aside about the impact on the environment. Thus, several investigations have been developed in order to analyze the environmental implications of intensifying productivity [2, 3], demonstrating that the stronger the intensification, the better results may be achieved in mitigation measures of gases [4–6]. However, in protected areas of biodiversity, the intensification of a livestock system should not be pursued, but rather the proper management of the animals, achieving the yield that meets the needs of rural families and in turn not affecting the biodiversity in protected areas [7].

Cool Farm Tool software becomes a decision maker in the quest to reduce greenhouse gas emissions. Although this software has been applied in different countries and not only in livestock, it is fundamental to understand the response of each livestock system according to the location and type of management [8–10].

Estimates and proposals for emission mitigation measures have been made in intensive mixed systems (milk and potato production) in the Andes of northern Ecuador [11]. However, the emissions of the productive systems located in the protected area under study have not yet been quantified. Based on the aforementioned, in the current study the following objectives were proposed: a) to determine the socio-productive characteristics at the household/farm level and b) to quantify the greenhouse gas (GHG) emissions of farms dedicated to livestock. Dairy farmers that have not integrated modern production technologies, considered as subsistence systems, the conclusions of this research may serve to small milk producers as a valuable tool for the implementation of changes in production in order to reduce emissions of GHG in protected areas.

2 Materials and Methods

2.1 Geographic Location

The study area is located in the northern Ecuadorian Andes, within the Area of Conservation and Sustainable Use (ACUS), created in 2016 (Fig. 1), with the objective of protecting water sources, moorland and forests belonging to the ecosystem Upper Montane Evergreen Forest of the northern part of the Eastern Andean Cordillera (BsAn01) [12]. It comprises an area of some 17,560 ha, while the protected area belongs to the hydrographic basin of the Mira River [13]. ACUS adjoins the El Ángel Ecological Reserve, with the La Bonita Municipal Conservation Ecological Area and is superimposed in the province of Carchi. This constitutes a sample of dairy cattle farms in the Andean region [11].

2.2 Data Collection and Sampling System

The current research was performed in 19 dairy cattle farms. The data were collected through visits to the site and interviews with the heads of households of the mestizo peasants, with a duration of 45 min, between the months of January and February 2020

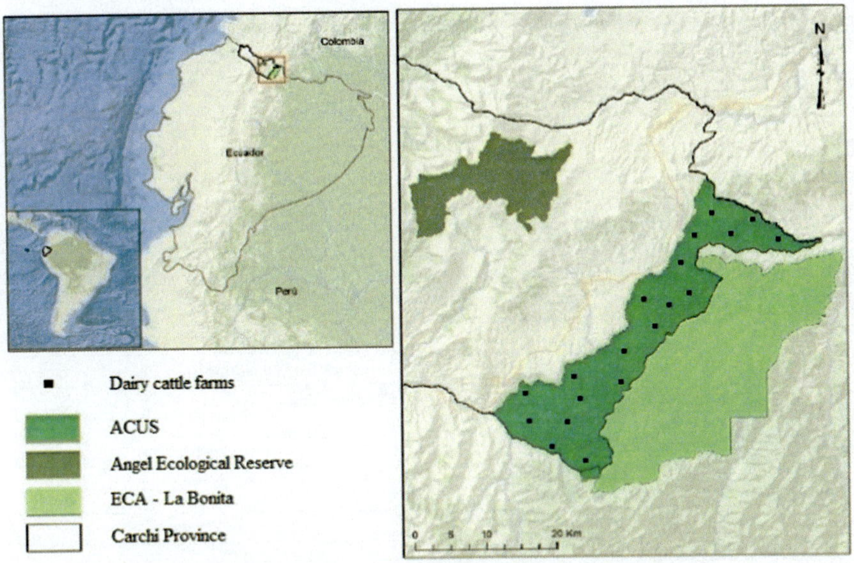

Fig. 1. Study area: conservation and sustainable use area (ACUS)

(Fig. 1). The used non-probabilistic sampling technique, was for convenience following the criteria: i) producers maintain subsistence systems ii) milk production in the area is a source of economic income, iii) the selected places are representative for the region, considering that the surface of the dairy farms is less than 20 ha., and finally also ethnicity. The questionnaire was adapted from the Poverty and Environment Network (PEN) template [14]. The descriptive statistics calculated were discrete and continuous quantitative multidimensional variables, using the SPSS version 22 program [15].

2.3 Socio-productive Characterization at the Household Level and Dairy Farms

In order to obtain the needed information we studied the 1) demographic characteristics, the population structure and its distribution by gender, age, household size, 2) characteristics of the productive systems such as: areas (ha), pastures (ha) and crops (ha), 3) characteristics of milk production such as: total herd, months of milk production, milk production per day and milk price (average in dollars per liter).

2.4 Quantification of Greenhouse Gas Emissions

The methodology for quantifying GHG emissions in the selected livestock systems consisted of entering the data from each survey in the Cool Farm Tool software (Table 1) in order to obtain emissions data for each dairy livestock farm. The software, through a series of data (inputs), quantifies the emissions (outputs) of each livestock system studied, based on two criteria: per unit of product (tons) and per area used (hectares) [16] (Table 1).

Table 1. Inputs of the Cool Farm Tool software for the quantification of emissions from livestock systems.

Inputs	Outputs
General information	Pasture emissions
Location of the farm	Direct and indirect field N_2O
Surface used	Pasture residue management
Climate of the area	
Pasture management	Livestock
Soil type	Livestock enteric emissions
Fertilizers used	Livestock manure management emissions
Use of pesticides	Livestock feed emissions
Pasture residue management	CH_4 emissions
	N_2O emissions
Livestock management	CO_2 emissions
Cattle breed and weight	
Number of young animals	
Number of productive animals	
Number of non-productive animals	
Characteristics of the food	
Amount of milk produced per day per farm	
Milk fat/protein content	
Grazing form	
Manure management	

The survey collected information related to pasture and livestock management. Simultaneously, additional information was obtained from face-to-face interviews with a group of experts from the National Autonomous Institute of Agricultural Research (INIAP) which conducted studies in the area. The information referring to the characteristics of the soil and climate of the selected places was collected through secondary sources. With the information generated on emissions, critical points were defined, from which action plans were developed in order to improve the management of livestock systems in the study area and, in turn, to reduce GHG emissions.

3 Results and Discussion

Based on the results obtained, two sections are presented: 1) the socio-productive characteristics and 2) the quantification of greenhouse gas emissions from dairy farms.

3.1 Socio-productive Characteristics at the Household Level and of Dairy Farms

The mestizo peasants involved in the study were a total of about 73 people, being 33 men (45.21%) and 40 women (54.79%). The average age of the head of the household is 33 years (SD: 20.78; max 84 years and min 1 year), 12 years less for the dairy farmers in the province of Chimborazo [17]. Of the total involved, 39 (53.42%) mestizo peasants are married, 2 are separated (2.74%), there is 1 divorced and a widower (1.37%), 22 are single (30.14%) and 8 children (10.96%). Regarding the level of education, only 3 peasants (4.11%) attended university without completing it, 26 completed secondary level (35.62%), 39 completed primary education (53.42%) and 6,85% have not received any type of formal instruction, this determines that 98.1% of mestizo peasants do not have a high-level technical training, for which it is essential to develop a capacity-building program [18–22]. The average experience of heads of households in milk production is 22 years (SD: 11.41; max: 50; min: 7), of spouses 23 years (max: 50; min 7), of children 6.5 years (max 8; min: 3) and parents/in-laws have 45 years of experience.

Regarding the characteristics of dairy production systems, the average pasture area per production system is 3.1 ha (SD: 3.8; max 18 ha and min 1 ha), the average crop area is 1.97 ha (SD: 1.5; max 6 ha and min 0.4 ha), and the average of the farms is 5.07 (ha), (SD: 1.4; max 19 ha and min 1.4 ha). It was evidenced that most dairy farms do not have facilities or adequate infrastructure, which is a limitation, since it is important to provide the animals with comfort [23]. In several investigations, it is evidenced that, with the improvement of the comfort and well-being of the cows, the productivity and profitability of the herd are increased [24–26].

Regarding the characteristics of milk production, there are seven average animals per herd (SD: 2.5; max: 10, min: 2), while the peasants of the San Rafael community (Province of Chimborazo) have 6 average animals per herd [17]. The average months of milk production are 6.7 (SD: 1.49; max: 11, min: 4), while the liters of production per day between the farms are 33 L on average (SD: 19.36; max: 60, min: 6), which depends on climatic variables and herd and pasture management [27]. The price per liter of milk is $ 0.37 (avg) US dollars (SD: 0.03; max: 0.44, min: 0.30).

3.2 Greenhouse Gas Emissions in Production Systems

Subsistence dairy farming developed in the ACUS presents emission values in terms of surface equivalent to 10.91 tonCO_2e ha^{-1} (Table 1). These values are similar to the emissions of the intensive mixed systems in three study sites according to the research of Cayambe et al. [11], who reported between 10 and 11 ton CO_2e ha^{-1}. Additionally, low annual farm system emissions per finished product are observed, equivalent to 364,689 kg CO_2e.ton^{-1}. These results are much lower than presented in the study by González-Quintero et al. [9] in the Colombian Andes, being between 2100 and 4200 kg CO_2e.ton^{-1}. However, as mentioned by Figueiredo et al. [28] in other GHG quantification studies, the comparison of emissions between studies is complicated due to the diversity in the limits of the analysis .

Annual farm system emissions per ton of milk in Ecuador's subsistence dairy farming systems (Table 2) represent less than 80% of Colombia's emissions in dual-purpose cattle systems (milk and meat). There are several reasons that explain these differences in emissions, such as improved pastures [4], higher fertilizer application rates and livestock feeding [29].

In our study, the direct and indirect emissions field N_2O are equivalent to 149.3 kgCO_2e ha^{-1}, which is 52% less than that presented in Colombia [29], whose emissions

Table 2. GHG emissions from subsistence dairy farming systems in rural livelihoods in Ecuador.

Outputs	Average	Unit	SD	CV (%)
Annual farm system emissions by area	10915,5	$kgCO_2e\ ha^{-1}$	3581,69	33%
Emissions Annual farm system per finished product	364,68	$kgCO_2e\ ton^{-1}$	117,99	32%
Direct and indirect emissions field N_2O	149,3	$kgCO_2e\ ha^{-1}$	0,16	3%
Emissions from crop residue management	192,20	$kgCO_2e\ ton^{-1}$	0,00	0%
Livestock enteric emissions	133,31	$kgCO_2e\ ton^{-1}$	93,52	70%
Livestock manure management	29,22	$kgCO_2e\ ton^{-1}$	24,77	85%
Livestock feed	2,88	$kgCO_2e\ ton^{-1}$	2,50	87%
CH_4 emissions	288,80	$kg\ CH_4\ ton^{-1}$	132,99	0,46
N_2O emissions	5,09	$kg\ N_2O\ ton^{-1}$	3,09	0,60
CO_2 emissions	152,80	$kg\ CO_2\ ton^{-1}$	93,06	0,60

CV: Coefficient of variation; SD: Standard deviation

amount to 306.94 $kgCO_2e\ ha^{-1}$. This may be due to the different feeding of dual-purpose cattle versus subsistence dairy farming.

Most of the pasture emissions originated from the management of pasture residues, as producers abandon the residues in the field [10]. This is able to be compensated by directly incorporating these residues or composting them and applying them to the soil or to fertilize other crops on farms. No emissions from fertilizers were observed, due to the almost zero application of these.

The estimation of GHG emissions yielded that 80.96% of livestock emissions derived from enteric digestion. The largest component of these emissions corresponds to methane emissions equivalent to 288.80 $kgCH_4\ ton^{-1}$, followed by 152.80 kg $CO_2\ ton^{-1}$ and, finally, 5.09 kg $N_2O\ ton^{-1}$. This is comparable to many other studies where enteric fermentation is responsible for the highest percentage of emissions. [3, 4, 30].

Emissions by livestock manure management were 29.22 kg $CO_2e.ton^{-1}$. This is because producers do not disperse feces. In the study conducted by Cayambe et al. [11], one of the measures recommended in the Livestock systems was the dispersion of feces, since it could reduce GHG emissions, entering the soil as compost and reducing the amount of gases to the environment.

Of the total emissions from enteric fermentation (CH_4, N_2O and CO_2), methane represents the gas that most influences this study. Such scenario occurs in most subsistence livestock systems in the Andes, due to the absence of management of excreta. Diet changes based on feed management options is the main mitigation proposal identified, as indicated by [31].

Intensive farms in Brazil [32] generate bioenergy by converting manure waste into energy, thus translating costs into profits, making the process more efficient and respectful of the environment. However, this practice as well as smart farming [33–38], would be very difficult to implement in the ACUS, due to the lack of policies on biofuel. Given this, the incorporation of feces into the soil would also become a practice in order to reduce emissions.

4 Conclusions

The characteristics of the household and the level of education of the household heads can influence the management of the herd and pastures, which is important to formulate programs and policies for the sustainable management of dairy farms, considering that they are in a conservation area.

Estimation of GHG emissions from subsistence farms in the conservation area yielded that 81% of livestock emissions emerged from enteric digestion, of which, 21% correspond to emissions from manure management, since no management measures are performed. Methane from enteric digestion is the gas in the highest proportion according to the study, followed by nitrous oxide emissions from excreta deposited on pastures. Alternatives for feed management, offering balanced amounts of feed according to the needs of the cattle according to age or production phase could improve this scenario both to increase milk production and to reduce emissions in subsistence farms in the conservation area.

It is advisable to expand the research with a greater number of cattle farms in the conservation area, including semi-intensive and intensive farms, which would be a priority for future research.

References

1. Von Keyserlingk, M.A., Weary, D.M.: A 100-year review: animal welfare in the Journal of Dairy Science—the first 100 years. J. Dairy Sci. **100**(12), 10432–10444 (2017)
2. Alemu, A.W., Janzen, H., Little, S., Hao, X., Thompson, D.J., Baron, V., Iwaasa, A., Beauchemin, K.A., Kröbel, R.: Assessment of grazing management on farm greenhouse gas intensity of beef production systems in the Canadian Prairies using life cycle assessment. Agric. Syst. **158**, 1–13 (2017). https://doi.org/10.1016/j.agsy.2017.08.003
3. Mazzetto, A.M., Bishop, G., Styles, D., Arndt, C., Brook, R., Chadwick, D.: Comparing the environmental efficiency of milk and beef production through life cycle assessment of interconnected cattle systems. J. Cleaner Prod. **277**, 124108 (2020). https://doi.org/10.1016/j.jclepro.2020.124108
4. Bogaerts, M., Cirhigiri, L., Robinson, I., Rodkin, M., Hajjar, R., Junior, C.C., Newton, P.: Climate change mitigation through intensified pasture management: estimating greenhouse gas emissions on cattle farms in the Brazilian Amazon. J. Cleaner Prod. **162**, 1539–1550 (2017). https://doi.org/10.1016/j.jclepro.2017.06.130
5. Ahmed, M., Ahmad, S., Waldrip, H.M., Ramin, M., Raza, M.A.: Whole farm modeling: a systems approach to understanding and managing livestock for greenhouse gas mitigation, economic viability and environmental quality. Anim. Manure: Prod. Characteristics Environ. Concerns Manag. **67**, 345–371 (2020). https://doi.org/10.2134/asaspecpub67.c25
6. Toulkeridis, T., Tamayo, E., Simón, D., Merizalde, M.J., Reyes, D.F., Viera, M., Heredia, M.: Cambio Climático según los académicos ecuatorianos - Percepciones versus hechos. La Granja: Revista de Ciencias de la Vida **31**(1), 21–46 (2020). https://doi.org/10.17163/lgr.n31.2020.02

7. Barreto-Álvarez, D.E., Heredia-Rengifo, M.G., Padilla-Almeida, O., Toulkeridis, T.: Multi-temporal evaluation of the recent land use change in Santa Cruz Island, Galapagos, Ecuador. In: Rodriguez Morales, G., Fonseca, C.E.R., Salgado, J.P., Pérez-Gosende, P., Orellana Cordero, M., Berrezueta, S. (eds.) Information and Communication Technologies, TICEC 2020. Communications in Computer and Information Science, vol. 1307. Springer, Cham (2020). https://doi.org/10.1007/978-3-030-62833-8_38
8. Egas, D., Ponsá, S., Colon, J.: CalcPEFDairy: a Product Environmental Footprint compliant tool for a tailored assessment of raw milk and dairy products. J. Environ. Manag. **260**, 110049 (2020). https://doi.org/10.1016/j.agsy.2017.08.003
9. González-Quintero, R., Kristensen, T., Sánchez-Pinzón, M., Bolívar-Vergara, D., Chirinda, N., Arango, J., Pantevez, H., Barahona-Rosales, R., Trydeman, M.: Carbon footprint, non-renewable energy and land use of dual-purpose cattle systems in Colombia using a life cycle assessment approach. Livestock Sci. 104330 (2020)
10. Gilardino, A., Quispe, I., Pacheco, M., Bartl, K.: Comparison of different methods for consideration of multifunctionality of Peruvian dairy cattle in Life Cycle Assessment. Livestock Sci. **240**, 104151 (2020). https://doi.org/10.1016/j.livsci.2020.104151
11. Cayambe, J., Iglesias, A.: The cost of mitigating greenhouse gas emissions in farms in Central Andes of Ecuador. Spanish J. Agric. Res. **18**(1), 1–17 (2020)
12. Santiana, J., Báez, S., Guevara, J.: Sistema de Clasificación de los Ecosistemas del Ecuador Continental. Ministerio del Ambiente del Ecuador (2013)
13. GADPC: Creación del Área de Conservación y Uso Sustentable de la Cordillera Oriental del Carchi. Gobierno Autónomo Descentralizado de la Provincia del Carchi/CONDESAN-Proyecto EcoAndes. Tulcán (2016)
14. Cavendish, W.: How do forests support, insure and improve the livelihoods of the rural poor? A research note. Center for International Forestry Research. Bogor, Indonesia (2003)
15. Aroca, P.R., García, C.L., López, J.J.G.: Estadística descriptiva e inferencial. Revista el auge de la estadística en el siglo XX, 22, 165–176 (2009)
16. Haverkort, A., Hillier, J.: Cool farm tool–potato: model description and performance of four production systems. Potato Res. **54**(4), 355–369 (2011)
17. León Alvear, V., Torres, B., Luna, M., Torres, A., Ramírez, P., Andrade-Yucailla, V., Muñoz-Rengifo, J.C., Heredia-R, M.: Percepción sobre cambio climático en cuatro comunidades orientadas a la ganadería bovina en la zona central de los Andes Ecuatorianos. Livestock Res. Rural Dev. **32**, Article #165 (2020)
18. Heredia, M., Bravo, C., Torres, B., Alemán, R.: Innovación para el fortalecimiento de capacidades sobre sostenibilidad de los recursos naturales en poblaciones indígenas y mestizas – colonas: Reserva de Biosfera Yasuní. Iberian J. Inf. Syst. Technol. RISTI, No. E25, 01/2020 (2020). https://www.risti.xyz/issues/ristie25.pdf. ISSN: 1646–9895
19. Heredia-R, M., Falconí, V., H-Silva, J., Amores, K., Endara, C.A., F-Ausay, K.: Technological innovation for the sustainability of knowledge and natural resources: case of the Choco Andino Biosphere reserve. In: Botto-Tobar, M., Zambrano Vizuete, M., Díaz Cadena, A. (eds.) Innovation and Research. CI3 2020. Advances in Intelligent Systems and Computing, vol 1277. Springer, Cham (2021). https://doi.org/10.1007/978-3-030-60467-7_38
20. Heredia-R, M., Falconí, A.K., Barreto, D., Amores, K., Silva, J.H., Torres, B.: Conductas sustentables sobre el marco de evaluación SAFA–FAO: Un aporte para poblaciones rurales vulnerables de la Amazonía. Revista Ibérica Sistemas Tecnologias Informação **33**, 312–326 (2020)
21. Heredia-R, M., Torres, B., Guerrero, E., Gallardo, D., Núñez, M., Alemán, R., Cayambe, J., Díaz-Ambrona, C.G.H.: Evaluación de la sostenibilidad de sistemas productivos en la franja de diversidad y vida: Reserva De Biosfera Yasuní, Amazonía. Revista Científica Axioma (22), 5–11 (2020)

22. Heredia, M., Torres, B., Alemán, R., Bravo, C., Díaz-Ambrona, C.: SAFA and GeoGebra allies to evaluate natural and cultural sustainability: Yasuni biosphere reserve. Int. J. Adv. Sci. Eng. Inf. Technol. **10**(2), 880–887 (2020)
23. Losada-Espinosa, N., Miranda-De la Lama, G.C., Estévez-Moreno, L.X.: Stockpeople and animal welfare: compatibilities, contradictions, and unresolved ethical dilemmas. J. Agric. Environ. Ethics **33**, 71–92 (2020). https://doi.org/10.1007/s10806-019-09813-z
24. Robichaud, M.V., Rushen, J., de Passillé, A.M., Vasseur, E., Orsel, K., Pellerin, D.: Associations between on-farm animal welfare indicators and productivity and profitability on Canadian dairies: I. On freestall farms. J. Dairy Sci. **102**(5), 4341–4351 (2019)
25. Wang, Y., Wang, D., Wang, J., Li, K., Heng, C., Jiang, L., Cai, C., Zhan, X.: Effects of different stocking densities on tracheal barrier function and its metabolic changes in finishing broilers. Poultry Sci. **99**(12), 6307–6316 (2020)
26. De Vries, A., Marcondes, M.I.: Overview of factors affecting productive lifespan of dairy cows. Animal **14**(S1), s155–s164 (2020)
27. Mylostyvyi, R., Chernenko, O.: Correlations between environmental factors and milk production of Holstein cows. Data **4**(3), 103 (2019)
28. Figueiredo, E.B., Jayasundara, S., de Oliveira Bordonal, R., Berchielli, T.T., Reis, R.A., Wagner-Riddle, C., La Scala Jr., N.: Greenhouse gas balance and carbon footprint of beef cattle in three contrasting pasture-management systems in Brazil. J. Cleaner Prod. **142**, 420–431 (2020). https://doi.org/10.1016/j.jclepro.2016.03.132
29. Kinley, R.D., Martinez-Fernandez, G., Matthews, M.K., de Nys, R., Magnusson, M., Tomkins, N.W.: Mitigating the carbon footprint and improving productivity of ruminant livestock agriculture using a red seaweed. J. Cleaner Prod. **259**, 120836 (2020). https://doi.org/10.1016/j.jclepro.2020.120836
30. Escribano, M., Elghannam, A., Mesias, F.J.: Dairy sheep farms in semi-arid rangelands: a carbon footprint dilemma between intensification and land-based grazing. Land Use Policy **95**, 104600 (2020). https://doi.org/10.1016/j.landusepol.2020.104600
31. Wilkes, A., Wassie, S., Odhong, C., Fraval, S., Dijk, S.: Variation in the carbon footprint of milk production on smallholder dairy farms in central Kenya. J. Cleaner Prod. **265**, 121780 (2020). https://doi.org/10.1016/j.jclepro.2020.121780
32. Herrera, A.M., Esteves, E.M., Morgado, C.R., Esteves, V.P.: Carbon footprint analysis of bioenergy production from cattle manure in the Brazilian central-west. BioEnergy Res. 1–12 (2020)
33. Cayambe, J., Diaz-Ambrona, C.G., Torres, B., Heredia-R.M.: Decision support systems for the Imbabura Geopark: Ecuadorian andes. In: International Conference on Information Technology & Systems, pp. 310–320. Springer, Cham (2021)
34. Cayambe, J., Iglesias, A., Garcia de Jalon, S., Chuquillanqui, C., Riga, P.: Economic evaluation of strategies for mitigating greenhouse gas emissions in potato production systems. ITEA **111**(2), 154–173 (2015)
35. Moncayo Cevallos, L.N., Andrade Suárez, B.I., Sinde González, I., Maiguashca Guzmán, J.A., Rivadeneira García, J.L., Leiva González, C.A., Yépez Campoverde, J.A., Toulkeridis, T.: A NDVI analysis contrasting different spectrum data methodologies applied in pasture crops previous grazing – a case study from Ecuador. In: 2018 5th International Conference on eDemocracy and eGovernment, ICEDEG 2018, 8372318, pp. 126–135 (2018)
36. Cañarte Ruiz, D.A., Mora Villacís, M.G., Kirby, E., Maiguashca Guzmán, J.A., Toulkeridis, T.: Correlation of NDVI obtained by different methodologies of spectral data collection in a commercial crop of quinoa (Chenopodium quinoa) in central Ecuador. In: 2020 7th International Conference on eDemocracy and eGovernment, ICEDEG 2020, pp. 207–214 (2020)

37. Mora Villacís, M.G., Cañarte Ruiz, D.A., Kirby, E., Maiguashca Guzmán, J.A., Toulkeridis, T.: Index relationship of vegetation with the development of a quinoa crop (Chenopodium quinoa) in its first phenological stages in central ecuador based on GIS techniques. In: 2020 7th International Conference on eDemocracy and eGovernment, ICEDEG 2020, pp. 190–199 (2020)
38. Viera-Torres, M., Sinde-González, I., Gil-Docampo, M., Bravo, V., Toulkeridis, T.: Generation of the base line in the early detection of bud rot and the red ring disease in oil palms by geospatial technologies. Remote Sens. **12**(19), 3229 (2020)

Sustainability and Rural Development Through the Production of Organic Cocoa in Calceta – Ecuador: Economic Analysis and Assessment of the Implications on the Living Standards of Farmers

Marlon Arturo Cedeño Álava[1](✉), David Giband[2], and Luz Cecilia García Cruzatty[1]

[1] Universidad Técnica de Manabí, POB. 130108, Portoviejo, Ecuador
marlon.cedeno@utm.edu.ec
[2] Université de Perpignan Via Domitia, POB. 66100, Portoviejo, France

Abstract. According to development theory, in the 20th century, the concept of sustainability emerged with three important variables, stability, resilience and adaptability; giving them an economic approach based on productivity, efficiency and effectiveness without neglecting the perceptions of equality and common welfare. Sustainable organic agriculture is a production system that largely avoids or excludes the use of synthetically compounded fertilizers, pesticides, growth regulators, and food additives. Cocoa (Theobroma cacao) is a small sector of all agricultural production in the world. In Latin America, between 70% and 100% of cocoa exports correspond to fine or aroma cocoa, making this region the first producer of this variety at an international level according to the International Cocoa Organization. This paper looks for evaluating the economic sustainability of cocoa production, both in organic and conventional agriculture. An economic analysis was carried out based on a mathematical programming model that projected the maximizations of production and income in Fortaleza del Valle Corporation in Calceta Ecuador. The results show a greater economic sustainability of organic farming compared to conventional agriculture. The higher profitability of organic agriculture was due to the lower requirement of labor and a greater appreciation of the market for organic products that awarded a price higher than conventional prices. Furthermore, the higher profitability of organic agriculture and the use of environmentally friendly inputs in the production process make the farms competitive. Additionally, the environmental results in terms of reducing carbon emissions suggest that the organic system is more effective in mitigating climate change.

Keywords: Organic agriculture · Conventional agriculture · Cost-benefit analysis · Mathematical programming · Cost effectiveness

J. R. da Costa Sanches Galvão et al. (Eds.): *Proceedings of the 1st International Conference on Water Energy Food and Sustainability (ICoWEFS 2021)*, pp. 75–82, 2021.
https://doi.org/10.1007/978-3-030-75315-3_9

1 Introduction

Sustainability is synonym of stability, adaptability, and resilience[1]. This approach has a huge scope of action including the agroecological production which in certain way is getting the attention of developing countries due to the possibility of managing natural resources in rural areas in a more effective way. The definition for organic farming then will be:

"It is a production system which avoids the use of synthetically compounded fertilizers, pesticides, growth regulators, and livestock feed additives. To the maximum extent feasible, organic farming systems rely upon crop rotations, crop residues, animal manures, legumes, green manures, off-farm organic wastes, mechanical cultivation, mineral-bearing rocks, and aspects of biological pest control to maintain soil productivity and tilth, to supply plant nutrients, and to control insects, weeds, and other pests." [1].

Cocoa (*Theobroma cacao*) is a small sector of the whole agricultural production in the world. In Latin America between 70% and 100% of the cocoa exports corresponds to the fine flavour cocoa, making this region the first producer of this variety at an international level according to the International Cocoa Organization ICCO.

Since the 70's farmers have had to use pesticides and synthetic material for controlling diseases in cocoa farms [2]. However, cocoa production is in jeopardy due to many factors such as price fluctuations, supply deficits or surpluses, production issues, soil erosion, fungal diseases, farmers too old and at the same time living at extreme poverty or under the poverty threshold, and so forth.

If we add that for the last decades, the climate change has come up with increasing rainfall, water scarceness or floods instead, and high temperatures, the options for developing countries started to worsen; mainly for those who focused its main activities in agriculture. And in fact, agriculture is one of the problems that can lead the planet to these events, but it can also be a solution to better off the environmental conditions through changes in global food systems that meet the demand of a growing population [3].

For that reason, current cocoa producers are thinking of moving towards a new scenario i.e., the organic cocoa farming which at the end of the day is more sustainable and will provide a better future production aimed at international markets according to several researchers. Indeed, due to globalization, exporters are seeing a new opportunity to boost their business by adopting this new way of production, as it gives innovation and creativeness when offering fresh fruits with new production methods and at the same time concerned of the environment and health [4].

This research proposal is going to be involved within the area of sustainable development that will show the benefits of implementing organic agriculture in a rural community

[1] An agroecosystem is considered sustainable when it produces, in a state of stable equilibrium, a specific combination of goods and services which satisfy a set of productive goals, without degrading its resources (stability). The level of sustainability will depend on the ability to face and recover quickly form disturbances (resilience); as well as finding new alternative states of stable equilibrium without compromising its productivity and reproducibility. (adaptability) (Masera 2000).

called Fortaleza del Valle. The main reason of this study is to make farmers aware of the economic benefits they can obtain by using this technique. In Ecuador, the share of farms that produces organic food is extremely low and that is why studies like this should be carried on. According to INEC[2] less than 10% of rural areas are producing organically.

2 Current Situation

Ecuador has four main provinces where cocoa is produced, Guayas Manabí, Los Ríos, and Sucumbíos. There are several points that were analyzed to choose the Corporation Fortaleza del Valle in Calceta – Manabí as the research area of the project. First, there is a Cooperation Plan[3] between Universidad Técnica de Manabí and Fortaleza del Valle to promote and develop academic, technological, and technical programs that enhance fields such as research, innovation, and community service to guarantee a sustainable development for the region and the institutions.

Second, considering the results of the Survey of Surface and Continuous Agricultural Production with a Spanish acronym ESPAC carried out in 2018, Manabí is the province with the largest area aimed for agriculture, 45.1% (777.088 hectares). The main agricultural products are banana, sugar cane, African palm, rice, corn, and potato but all of them are produced in the conventional way, i.e., with pesticides and chemicals [5].

Despite Ecuadorian fine aroma cocoa is not on the list of leading products in the province, the importance of conducting a research about this topic comes up as because more than 62% of the worldwide production comes from it, becoming the first exporter in the globe [6].

Manabí is the third province with the highest production of fine aroma cocoa (organic and non-organic) with an output of 33.487 metric tons compared to the 221.130 metric tons of the country. Fortaleza del Valle is an Association of organic cocoa producers located in Calceta, a city established in the district of Bolívar which is mainly dedicated to agriculture, livestock production and handcrafted products with a population of 40.735 habitants [7].

Fortaleza del Valle is formed by four Agricultural Associations: Valle del Carrizal, La Fortaleza, Quiroga, and Rio Grande from Canuto, which belong mainly to the districts of Bolívar, Tosagua, Chone, Junín and Portoviejo in the Province de Manabí, and gather up small and medium producers whose objective is to promote the production and marketing of Ecuadorian organic fine aroma cocoa with organic and fair trade certification, promoting the social, economic and environmental development of its members and families [8].

Fortaleza del Valle, has been participating in the Fair-Trade system since 2005, through the certification issued by the independent commission of FLO-Cert. In recent years, generic Fair-Trade standards have been evaluated within small producer organizations, with the main priority of allowing producers to choose their own development path,

---

2 Instituto Nacional de Estadísticas y Censos.

3 https://www.utm.edu.ec/vinculacion/convenios/2019/file/559-76-convenio-marco-de-cooperacion-interinstitucional-entre-la-universidad-tecnica-de-manabi-y-la-coorporacion-fortaleza-del-valle.

to adjust more closely to the reality of producers and to simplify and reduce compliance costs.

At domestic level, Fortaleza del Valle is very well positioned, serving as a benchmark for other organizations dedicated to the export of cocoa beans. Furthermore, the Association is member of the Corporation of Producing Organizations of Domestic Fine Aroma Cacao of Ecuador "CONCACAO" and holds the vice-presidency of this trade union (Fig. 1).

Fig. 1. Drying beds for cocoa and fermentation area at Fortaleza del Valle Corporation in Calceta, Ecuador

3 Methodology

As the main purpose of this research is to assess the economic sustainability of organic cocoa farming, a series of analyses need to be conducted to determine if switching from conventional towards organic production is convenient. There are many researchers who propose different models to evaluate the impact of this new trend. The assessment should not only focus on economic welfare; it must propose a scientific evaluation of the environment impact, people's health improvement, and the social concerns of the community [9].

To assess the economic sustainability of cocoa organic farming in Fortaleza del Valle, the methodology proposed might use several aspects of the model proposed by [10] with several adjustments; the application of General Algebraic Modelling (GAM) will search to maximize the use of limited resources in an area, thus, thanks to a production function that satisfies several constraints, we will be able to know the values that minimize and maximize the economic profits. First, we need to define the economic part and the GAM system to be applied; a general view of this can be seen in the model proposed by [11] as follows:

$$Max\,Z = \sum_{i=1}^{n} \mathrm{C}_i X_i$$

$$St = \sum_{i=1}^{n} A_{ji} X_i \leq B_j\ all\ j = 1 \ldots .m$$

And $X_i \geq 0,\ for\ all\ i = 1 \ldots .m$

In their research, Binta & Barbier apply this model to the production of several crops, such as cabbage, eggplant, green pepper, onion, and tomato; they compute prices according to the different seasons and they used three types of labor: permanent, temporary, and seasonal workers. For our case, a modification has been made in several aspects of their original formula that will remain as:

$$Max\,Z = \sum_{j}^{n} Y_j P_j - \sum_{f=1}^{D} \sum_{j=1}^{n} P_{fj} Q_{fj} - \sum_{m=1}^{H} \sum_{j=1}^{n} P_{mj} Q_{mj} - \sum_{p=1}^{V} \sum_{j=1}^{n} P_{pj} Q_{pj} -$$

$$\sum_{s=1}^{B} \sum_{j=1}^{n} P_{sj} Q_{sj} - \left(\sum_{l=1}^{R} \sum_{j=1}^{n} W_{lj} N_j D_{lj} \right) - c_j W$$

The main components of this formula are: Income, costs of fertilizer, manure, pesticides, labor, and the cost of water, all the variables represents whether an input or an output but the point is that in the formula, they take into consideration permanent, seasonal and temporary workers; and as we will only work with cocoa production, we established in advance and clarify how Fortaleza del Valle hire workers for their production and what are the other vectors of income as they will produce different products aside the cocoa trees to provide shadow. To obtain all this data, semi-structure questionnaires and interviews will be applied before using the linear programming approach that will expose the position of the farms. With the use of a random sampling technique, we will obtain the sample which will be independent and identically distributed.

Concerning the cost of production, [12] mentioned the production of cacao using conventional farming in Ecuador, this study can be used to elaborate the constraints for the economic analysis. Those constraints will be:

Labor constraint:

$$\sum_{t}^{E} \sum_{j=1}^{n} L_{tj} X_j L_j \leq L$$

Land constraint:

$$\sum_{j=1}^{n} L_j X_j \leq L$$

Water constraint:

$$\sum_{j=1}^{n} W_j X_j \leq W$$

Capital constraint:

$$\sum_{j=1}^{n} K_j X_j \leq K$$

Fertilizer constraint

$$\sum_{j=1}^{n} Q_{fj} X_j \leq Q_{fr}$$

Carbon constraint

$$\sum_{j=1}^{n} Q_{fj} C_j X_j \leq Q_{cr}$$

Non-negativity

$$X_j \geq 0; j = l \ldots$$

We can anticipate certain problems during these analyses as most of the farmers have financial problems in developing countries due to the terms of trade that are the reason of having low prices in relation with the costs of inputs. But organic farming has a market where the consumers are willing to pay more for a product that better off their health and take care of the environment.

4 Results

The main findings after conducting 60 surveys indicated that the mean age of farmers was around 69 years. The gender equality analysis explained that the participation of women is increasing, in all cases not only the wife of the owner but also their daughters are directly involved in the process of production and harvest of organic cocoa which indicates that the whole family are the main source of labor.

The average farm size under exploitation was about 1.1 ha. Therefore, we use 1 ha as an upper bound for land constraint in the simulation analysis between organic and conventional. The last one presents a greater use of land as they work in an intensive basis which requires more resources compared to organic farms.

The gross margin of organic cocoa and conventional favored the first one, even though, the traditional system is more intensive and generates higher yields per hectare, the premium price of 48 USD (organic) above the 33 USD (conventional) per quintal marked the difference at the moment of computing the loss and profit balance of farmers. The costs of cultivation were mainly compounded of labour, fertilizers, manure, pesticides, organic pesticides, irrigation, etc.).

5 Conclusions

At the end of the research, we could notice that cocoa organic farming is more lucrative for farmers of Fortaleza del Valle, compared to the conventional cocoa farmers. This is explained mainly for the premium price that they received for holding a Fair Trade and Green certification which is around 12 dollars above the normal price of conventional cocoa.

The cost of production also showed a downward tendency and that is mainly due because they work as a cooperative so they can use the economies of scale to produce their biological pesticides and fertilizers, thus, farmers can have access to cheaper inputs and in this way obtain a competitive advantage over their rivals.

Despite the fine flavour organic cocoa is a premium product overseas, within the country there is no market for this kind of products which creates a certain type of risk if for any external shock the international sales suffer a drop or even worse bilateral trading ends. The government should promote the creation of a domestic market that provides an alternative for farmers.

Obtaining information about ecological research in cocoa was extraordinarily complex as there is not a lot of research on this topic in Ecuador. Universities along with the private sector must look for agreement of cooperation that boost organic agriculture indicating the multiples impacts farmers and society can gain from it.

Although this research focused on the maximization of income and production when assessing costs, we could perceive that the carbon print left by organic producers is lower as they use environmentally friendly inputs in their processes. The diffusion of this kind of work should be socialized among farmers so they can be aware of the implications of moving towards organic.

Finally, we must indicate that the simulation model could not be tested yet as it was impossible to obtain data from two regions (Quiroga and Rio Grande) belonging to the corporation. The COVID19 and the strong winter made extremely difficult to reach those farms. It is expected that at the end of this year the production function will be modeled obtaining thus an indicator of the optimum values to maximize production and income.

References

1. Usda, U.: Report and Recommendations on Organic Farming. Washington (1980)
2. Batman, R.: Pesticide Use in Cocoa: A Guide for Training Administrative and Research Staff. ICCO, London (2010)
3. Middelbergm, S.L.: Sustainable agriculture: a review of challenges facing the South African agricultural sector. J. Hum. Ecol. **42**, 163–163 (2013)
4. Tudisca, S., Di Trapani, A.M., Sgroi, F., Testa, R.: The case of Sicilian durum wheat. Org. Farming Econ. Sustain. **15**, 93–96 (2014)
5. Instituto Nacional de Estadísticas y Censos: Instituto Nacional de Estadísticas y Censos. Encuesta de Superficie y Producción Agropecuaria Continua 2018. https://www.ecuadorencifras.gob.ec/documentos/webinec/Estadisticas_agropecuarias/espac/espac2018/Presentacion%20de%20principales%20resultados.pdf. Accessed 02 July 2020
6. Ballesteros, H.: ECUADOR: Programa de empleo productivo y. Iniciativa Latinoamericana del Cacao, p. 5 (2018)
7. Instituto Nacional de Estadísticas y Censos. Instituto Nacional de Estadísticas y Censos. Encuesta de Superficie y Producción Agropecuaria Continua 2018. https://www.ecuadorencifras.gob.ec/documentos/webinec/Estadisticas_agropecuarias/espac/espac2018/Presentacion%20de%20principales%20resultados.pdf. Accessed 20 July 2020
8. Corporación Fortaleza del Valle: Corporación Fortaleza del Valle . Corporación Fortaleza del Valle. https://fortalezadelvalle.org/. Accessed 07 Apr 2020
9. Nemes, N.: Comparative analysis of organic and non-organic farming systems: a critical assessment of farm profitability. Food Agric. (2009)

10. Binta, A., Barbier, B.: Economic and environmental performances of organic farming system compared to conventional farming system: a case farm model to simulate the horticultural sector of the Niayes Region in Senegal. J Hortic. **2**(4), 1 (2015). https://doi.org/10.4172/2376-0354.1000152
11. Hazell, P.B., Norton, R.D.: Mathematical Programing for Economic Analysis in Agriculture. Macmillan Publishing Company, New York (1986)
12. Chavez , G., Olaya, R., Maza , J.: Costo de producción de Cacao clonal ccn-51 en la parroquia Bellamaria, Ecuador. Universidad y Sociedad **10**(4), 179–185 (2018). https://rus.ucf.edu.cu/index

Cultivation of *Gracilaria gracilis* in an Aquaculture System at Mondego River (Portugal) Estuary Adjacent Terrain

Ana Carolina Inácio[1], Tiago Morais[2,3], João Cotas[3], Leonel Pereira[3], and Kiril Bahcevandziev[1](✉)

[1] Research Centre for Natural Resources Environment and Society (CERNAS), Institute of Applied Research (IIA), Agricultural College of Coimbra (ESAC/IPC), 3045-601 Coimbra, Portugal
kiril@esac.pt

[2] Lusalgae, Lda, IIncubadora de Empresas da Figueira da Foz, Rua das Acácias N° 40-A, 3090-380 Figueira da Foz, Portugal

[3] Department of Life Sciences, University of Coimbra, 3001-456 Coimbra, Portugal

Abstract. Seaweeds are one of the most sustainable and feasible natural sources to the imminent food shortage, due to their nutritional profile. Seaweed aquaculture is an important tool for seaweed food safety and maintain the nutritional value. The specie *Gracilaria gracilis* (Rhodophyta) is an edible seaweed, which can contribute to a balanced diet. The objective of this work was to evaluate the nutritional profile of *G. gracilis* in semi-controlled aquaculture (SCA), located in the estuarine zone of the Mondego River (Figueira da Foz, Portugal) compared to specimens harvested in the natural environment (SW). *G. gracilis* that presented good growth rate in the aquaculture system (4.70%/day), and its biomass, analyzed by Weende method, showed that SCA had different nutritional profile than SW. A fresh weight of cultivated *G. gracilis* showed some differences in moisture (SCA = 85.49% ± 0.02; SW = 86.04% ± 0.01), ash (SCA = 3.96% ± 0.01; SW = 4,10% ± 0.01), fat (SCA = 0.01% ± 0.00; SW = 0.01% ± 0.00), fiber (SCA = 0.79% ± 0.01; SW = 0,68% ± 0.00), protein (SCA = 2.07% ± 0.01; SW = 2.54% ± 0.01) and carbohydrate (SCA = 7.69% ± 0.02; SW = 6.64% ± 0.01). These results support the importance of *G. gracilis* aquaculture cultivation, maintaining the seaweed nutritional quality, when compared to the wild specimens obtained from the sea. Thus, promoting an eco-sustainable way of cultivating seaweeds for nutritious food purposes and even promoting the global food safety.

Keywords: Food safety · Nutritious food · Seaweed aquaculture cultivation · *Gracilaria gracilis* · Sustainability

1 Introduction

The ocean is a promising food resource due to the comprehensive nutritional profile and source of bioactive molecules, the seaweeds [1]. Seaweeds have been used worldwide

J. R. da Costa Sanches Galvão et al. (Eds.): *Proceedings of the 1st International Conference on Water Energy Food and Sustainability (ICoWEFS 2021)*, pp. 83–92, 2021.
https://doi.org/10.1007/978-3-030-75315-3_10

for several food and non-food applications for thousands of years. In recent decades, they have received much attention due to their nutritional and therapeutic benefits [2, 3]. Seaweeds will be a viable alternative for achieving a more sustainable diet in the future, they can contribute to improving/correcting nutritional deficiencies and consequently promote health [1, 4]. In this sense the seaweed aquaculture, which is constantly increasing, plays an important role due to the fact that it involves human intervention both at the economic level, in increasing production, and at the level of restocking, species protection and has contributed to human nutrition and food security [1, 4]. The nutrients and light are essential environmental factors that can determine the yield of biomass and the productivity of an aquaculture system [5]. Seaweeds are a rich source of nutrients and metabolize interesting bioactive molecules. The nutritional components, including the main nutrients (such as proteins and minerals) and micronutrients (lipids and vitamins) provide high nutritional value to humans [6]. These bioactive molecules vary, depending on the species, seasonal variations, geographical location and environmental conditions, such as temperature and light intensity, as well as the concentration of nutrients in the water [6–11].

Gracilaria gracilis is an edible red macroalgae belonging to the Rhodophyta phylum. Like other red algae, because of its agar content, it is also included in the agarophytes. It is distributed worldwide and lives in temperate seas, at a depth, between 0 and 20 m. It is found in protected areas, on rocks and stones, and require the presence of sand to develop. This seaweed is red with greenish shades, cartilaginous consistency and irregular ramifications, which can reach up to 50 to 60 cm in length [12]. *G. gracilis* has a remarkable economic relevance for its content in agar and extractable compounds and is used as food or feed supplement for humans and several species of animals. The rapid growth rate, ease of vegetative reproduction, and other attributes of this algae promoted its initiation into aquaculture [13].

This study aimed to understand if there are differences in the nutritional composition between *G. gracilis* from aquaculture system (SCA) and *G. gracilis* harvested directly from its natural habitat (SW), furthermore comparing with the results presented in the bibliography and understanding if algae from aquaculture has the potential to integrate human food chain, contributing for global food safety.

2 Materials and Methods

2.1 Seaweed Collection

The specimens of *Gracilaria gracilis* were collected at the rocky shore in the Praia da Tamargueira in Buarcos (Figueira da Foz, Portugal) (40°10′18.6"N, 8°53′44.4"W), Portugal. Sampling was conducted in July 2020 from the locations with well-established *G. gracilis*, without epiphytes or degradation visible at eyesight. Once collected, the samples were stored in plastic bags, and transported to the laboratory in a cool box. In the laboratory, the *G. gracilis* specimens were washed with filtered seawater to remove sands, contaminants, and other unlikely organisms. After, the samples were separated, half of the specimens were weighted and placed in aquaculture systems (SCA), while the remaining biomass was rapidly washed with distilled water, cleaned with paper, and stored at −20 °C for further analysis (SW).

2.2 Seaweed Cultivation in Semi-controlled Aquaculture (SCA)

For seaweed cultivation, we recurred to the prototype designed to the MENU - Marine macroalgaE: alternative recipes for a daily Nutritional diet project. To evaluate the growth potential of *G. gracilis* was used one cultivation tank from Lusalgae, Lda (Figueira da Foz), to growth enough biomass to do the nutritional profile of the cultivated *G. gracilis* (SCA) and analyze the seaweed growth rate. The SCA was cultivated in a tank exposed to direct sunlight with aeration during the day (on during the daylight, off during the night). The medium used for cultivation was estuarine seawater (29–35 ‰) of the Mondego River estuary, in Figueira da Foz, Coimbra, Portugal, mechanically filtered and without additional nutrients. The cultivation tank had a capacity of 1,000 L, containing 800 L of filtered estuarine water. The cultivation started from an initial biomass of 1050 g in a single tank. Approximately 75% of the water volume in the tank was renewed three times a week and after three weeks the entire biomass was retired to further analysis. The seaweed cultivation was established on the Lusalgae cultivation standardized method to the red seaweeds [14].

2.2.1 Specific Growth Rate

The specific growth rate (SGR) was calculated based on the equation below (1), according to Pérez-Mayorga, et al. [15] and regularly used as a standard for determining the growth rate of algae.

$$SGR\left(\frac{\%}{day}\right) = \frac{\ln|m_2| - \ln|m_1|}{t_{final} - t_{intial}} \times 100 \quad (1)$$

Where:

m_1 represents the initial wet mass in grams;
m_2 represents the final wet mass in grams;
t_ (initial) the initial day on which the culture started;
t_ (final) the final day on which the culture ended.

2.3 Seaweed Nutrient Profile

2.3.1 Preparation of Seaweed Samples

The samples (SCA and SW) were dried on an oven (Memmert, Germany) at 65 °C during and ground for the determination of moisture, ash, fat, fiber, protein, and carbohydrates.

2.3.2 Biochemical Analysis

For the nutritional profile, the algae cultivated in semi-controlled aquaculture (SCA) and specimens harvested in the natural environment (SW) were compared. The analyses were carried out following the AOAC International standards, formerly known as the Weende scheme [16].

Moisture
The moisture content was determined according to the AOAC method 930.04 (1997) [16], with some changes.

Ash
The ash content was determined according to the AOAC method 930.05 (1997) [16].

Fat
The fat content was determined according to the AOAC method 930.09 (1997) [16].

Dietary Fiber
The dietary fiber content was determined according to the AOAC method 930.10 (1997) [16].

Protein
The protein content was determined according to the AOAC method 930.04 (1997) [16].

Carbohydrates
Non-nitrogenous extractives were defined as the difference for 100 of the remaining constituents (moisture, fat, protein, dietary fiber and ash) and corresponded approximately to carbohydrates [17].

Fatty Acids
The fatty acids content was determined according to Portuguese Standard, NP EN ISO 5508 (1996). The process was divided into two steps: sample preparation and chromatographic analysis.

3 Results

3.1 Seaweed Cultivation and Specific Growth Rate (SGR)

The main objective of this culture was to obtain enough fresh material to be able to carry out the analysis of the nutritional composition, which implied the observation of the growth of algae biomass. During the aquaculture cultivation it was found that the biomass grew for 15 days (Table 1), but after, approximately, 20 days of cultivation, the seaweed maintained the biomass without any growth. *G. gracilis* showed a good growth rate in the aquaculture system (4.70%/day) [18].

Table 1. *Gracilaria gracilis* growth rate during the cultivation assay.

Date	Weight (Kg)	SGR (%)
26/05/2020	1050	–
09/06/2020	2027	4.70
16/06/2020	1750	−1.05

The abiotic parameters were collected by MAREFOZ Laboratory near the aquaculture location, due to the pandemic situation, only have the parameters values (Table 2) from June 2020 (in other reports, the value of the 6 of June is similar to end of May 2020). The average air temperature during aquaculture cultivation was around 18 °C–19 °C. *G. gracilis* has an optimum growth temperature in the range 19–27 °C [19], with the values presented indicates that the algae was not constantly at the ideal temperature for its growth. Regarding the pH, during cultivation it is crucial that it is kept stable [19]. The algae under study is able to withstand a wide range of salinity, it can grow with very low salinity values [18, 19]. Conductivity results from the relationship between the content of dissolved mineral salts in water and the resistance it offers to the passage of electric current, the high values obtained allow us to quickly and globally evaluate the degree of mineralization [20].

Table 2. Physico-chemical parameters from adjacent location of the cultivation assay.

Day	pH	Water Temperature (°C)	Salinity (PSU)	Conductivity (uS/cm)
08/06/2020	8.45	19.8	35.10	51238
19/06/2020	8.06	17.6	30.79	39570

3.2 Seaweed Nutrient Profile

The analysis of the nutritional composition of *G. gracilis* aquaculture (SCA) and *G. gracilis* harvested from the natural habitat (SW) (Table 3), showed that they have a nutritional profile with some significant differences, namely in moisture, ash, fiber, protein and carbohydrate content. SW had higher moisture, ash, and protein content than SCA. The fat, fiber and carbohydrate content were higher in SCA than in SW.

Comparing the results of fatty acids content in *G. gracilis* SCA and SW (Table 4), it can be seen that there are no significant differences. The total percentage of saturated fatty acids was 30.56 and 30.98%, respectively, with stearic acid (C18:0) as the main saturated fatty acid. The total percentage of PUFAs, were 69.44 and 69.02%, respectively, with linolenic acid (C18:3) being more abundant.

4 Discussion

The growth of the *G. gracilis* in the aquaculture system indicated that this seaweed should be collected in short periods, approximately each 10–15 days, due to the algae being unable to regenerate after exceeding the culture limit. The growth rate was lower than the one obtained by Bastos [19], where he obtained an specific growth rate of 8.78% / day in the first 14 days of cultivation. However, the cultivation method can have impact in these results, the assay was in an open cultivation and the Bastos cultivation was a Lab cultivation. However, with the same species and in an identical geographic region.

Table 3. Nutritional composition of *Gracilaria gracilis* specimens harvested in aquaculture (SCA) and the natural environment (SW), dry matter and green. The extraction yield results are expressed in mean ± standard deviation ($n = 3$)[a, b]. The same letters indicate no significant differences at the p-value < 0.05 level.

	Gracilaria gracilis semi-controlled aquaculture (SCA)		*Gracilaria gracilis* specimens harvested in the natural environment (SW)	
	Dry matter	Green matter	Dry matter	Green matter
Moisture (g/100 g)	-	85,49 ± 0.02[b]	-	86.04 ± 0.01[a]
Ash (g/100 g)	27.31 ± 0.05[b]	3.96 ± 0.01	29.34 ± 0.05[a]	4.10 ± 0.01
Fat (g/100 g)	0.05 ± 0.00[a]	0.01 ± 0.00	0.04 ± 0.00[a]	0.01 ± 0.00
Fiber (g/100 g)	5.44 ± 0.05[a]	0.79 ± 0.01	4.89 ± 0.03[b]	0.68 ± 0.00
Protein (g/100 g)	14.24 ± 0.06[b]	2.07 ± 0.01	18.18 ± 0.07[a]	2.54 ± 0.01
Carbohydrate (g/100 g)	52.97 ± 0.07[a]	7.69 ± 0.02	47.55 ± 0.02[b]	6.64 ± 0.01

Table 4. Fatty acids composition of *Gracilaria gracilis* specimens cultivated in aquaculture (SCA) and harvested in the natural environment (SW) specimens.

Fatty accids	*Gracilaria gracilis* SCA	*Gracilaria gracilis* SW
Palmitic acid (C16:0) (%)	12.22	9.23
Stearic acid (C18:0) (%)	12.90	15.47
Oleic acid (C18:1) (%)	25.44	25.88
Linoleic acid (C18:2) (%)	10.50	10.73
Linolenic acid (C18:3) (%)	33.08	32.83
Araquic acid (C20:0) (%)	5.86	5.86
Saturated fatty acids (%)	30.98	30.56
PUFAs (%)	69.02	69.44

Regarding the nutritional profile of the algae (SCA and SW) under study, although there are different environments, *G. gracilis* in the aquaculture maintains the nutritional quality. The data obtained relatively to the nutritional composition of both *G. gracilis* SCA and SW are different from those found from the literature (Table 5). These values are concerned only with e moisture, ash, fat, fiber, protein, and carbohydrates. The protein and carbohydrate values of the SCA and SW samples (dry matter) from our study are in the range of the values mentioned in the literature. However, the ash and fiber content of our sample are higher than those mentioned. This difference, namely in humidity, ash, fat and fiber, can be result from the environmental conditions of the collection site (sea or cultivation) and / or abiotic (salinity, temperature, maturity, season, nutrients) and biotic factors (herbivory, plant development), and the methods applied to determine each parameter according to the literature.

Table 5. Nutritional composition of *Gracilaria gracilis*, in g/100 g dry matter. Used to compare the composition of *Gracilaria gracilis* studied with other studies found in the literature.

Parameters	Dry matter composition (g/100g)	References
Moisture	7.99 –19.04	[11, 21–23]
Ash	6.78–24.8	
Fat	Low	
Dietary fiber	0.19–0.60	
Protein	5.83–20.2	
Carbohydrates	9.52–68.13	

The fatty acid contents obtained in this study are similar to other studies, already carried out, in this seaweed. However, the results of this study have higher values when compared with results from the other works [22, 23].

It is necessary to have a balance between the consumption of different types of fatty acids, with a reduced intake of saturated fatty acids and trans fatty acids being important and a higher consumption of monounsaturated acids and PUFAs in order to prevent cardiovascular diseases, osteoarthritis and diabetes [24]. In our assay, this balance was observed, both samples showed that the unsaturated fatty acids were more than twice the saturated ones.

This balance between saturated and unsaturated fatty acids is important, namely, between hypocholesterolemic (h) and hypercholesterolemic (H) fatty acids, as LDL-cholesterol, bad cholesterol, respectively, increase. Hypocholesterolemic fatty acids are C18:1 cis-9, C18:2 cis-9,12, C18:3 cis-9,12,15 acids and polyunsaturated fatty acids of the $n-3$ and $n-6$ families. And hypercholesterolemic fatty acids are C12:0, C14:0 and C16:0 acids [24]. The observed h:H ratio was 5.35% in the SCA *G. gracilis* and 7.52% in the SW *G. gracilis*, indicating that the proportion of hypocholesterolemic acids is higher than that of hypercholesterolemic, which have health benefits.

For any food to be introduced into human food, food safety must be taken into account, so one of the rules is to check the patterns of nutritional needs through the energy that the food provides [25]. The samples (SCA and SW) were analyzed again in relation to their nutritional composition, but for green matter, considering the water present, to determine the energy that each sample provides. The samples showed a moisture content (SCA = 4.25% ± 0.10; SW = 6.84% ± 0.03), ash (SCA = 26.14% ± 0.07; SW = 27.33% ± 0.04), fat (SCA = 0. 04% ± 0.00; SW = 0.04% ± 0.00), dietary fiber (SCA = 5.21% ± 0.05; SW = 4.56% ± 0.03), protein (SCA = 13.64% ± 0.05; SW = 16.93% ± 0.06), carbohydrates (SCA = 50.71% ± 0.14; SW = 44.30% ± 0.04) and consequently SCA provides 258 kcal/100 g and SW 245 kcal/100 g.

In European Union countries, Nutrient Reference Values (NRV) (Regulation EU 1169/2011) are declared for a healthy daily diet for the total 8400 kJ/2000 kcal that must be ingested per day by an adult to maintain a healthy state. However, the Standing Committee on Scientific Assessment of Dietary References of the Food and Nutrition Council of the US National Academy of Sciences (NAS), with the participation of Health Canada, has established the daily nutritional needs for each age group, sex and pregnancy or lactation status. In this case, the Dietary Reference Intake (DRI) was defined, considering the Recommended Food Allowance (RDA), the Estimated Average Requirement (EAR), the Adequate Intake (AI) and the Acceptable Macronutrient Distribution Ranges (AMDR). The RDA is the average level of daily intake sufficient to meet the nutritional needs of almost all (97–98%) healthy individuals in a group (Regulation EU 1169/2011). Still, according to NAS, AMDR is the range of macronutrients that must be consumed into the diet to ensure a reduced risk of chronic disease and to provide essential micronutrients such as vitamins and minerals. For an adult, the recommended protein intake is 50 g NRV, 56 g RDA and AMDR is 10–35% energy. For carbohydrates it is 260 g NRV, 130 g RDA and AMDR is 45–65% energy. For total fibers, AMDR is 38–30% energy, total lipids 20–35% energy, omega-6 5–10% energy and omega-3 0.6–0.1% energy [25]. Based on the energy that SCA *G. gracilis* and the recommended intake indicates that it has the potential to be used as a dietary supplement, presents low-calorie content due to its low-fat content and has high nutritional value in terms of protein and carbohydrates. Moreover, the range of macronutrients that must be consumed in human diet to ensure a reduced risk of chronic diseases and provide essential micronutrients is very important. The results obtained are between the values recommended [25], which is very important and SCA *G. gracilis* can be considered a good alternative as nutritious food source for human consumption.

5 Conclusion

The aquaculture cultivation of *G. gracilis* presented good seaweed growth with promising nutritional values for human diet. Consequently, the cultivation of this alga promotes eco-sustainability and food safety. From the aquaculture cultivation (outdoor) it is concluded that the collection of *G. gracilis* must be in short periods due to the algae having a rapid growth behavior, producing more seaweed biomass. In addition, the factors that influence and affect the cultivation of this alga must be considered, namely, light, temperature, salinity, nutrients and water movement in the next steps of this evolution of the seaweed cultivation and should be continuously optimized.

However, assessing the seaweed nutritional parameters and nutritional composition, it is concluded that *G. gracilis* has promising values for the human diet. Moreover, the results obtained indicate that this alga is promising source of unsaturated fatty acids.

In general, the sample that showed the best nutritional composition was the wild collected sample, just having less carbohydrates and less fiber, it is the one that has the most balance between fatty acids and that has the highest protein content. In this case, seaweed can play an important role as a nutritious food when produced and consumed within safety standards, if produced in controlled systems as SCA.

Acknowledgments. Tiago Morais and Lusalgae, Lda. Thanks to the project MENU - Marine Macroalgae: Alternative recipes for a daily nutritional diet (FA_05_2017_011) which co-financed this research, funded by the Blue Fund under Public Notice No. 5 - Blue Biotechnology. João Cotas thanks to the European Regional Development Fund through the Interreg Atlantic Area Program, under the project NASPA (EAPA_451/2016).

References

1. Mahadevan, K.: Seaweeds: A sustainable food source. In: Tiwari, B.K., Troy, D.J. (eds.) Seaweed Sustainability: Food and Non-food Application, pp. 347–364. Academic Press, Cambridge (2015)
2. Tiwari, B.K., Troy, D.J.: Seaweed Sustainability: Food and Non-food Applications, pp. 1–6. Academic Press, Cambridge (2015)
3. Batool, A., Menaa, F.: Concentration and purification of seaweed components by chromatography methods. In: Torres, M.D., Kraan, S., Dominguez, H. (eds.) Advances in Green Chemistry, pp. 315–370. Sustainable Seaweed Technologies, Elsevier, Amsterdam (2020)
4. Pérez-Lloréns, J.L., Mouritsen, O.G., Rhatigan, P., Cornish, M.L., Critchley, A.T.: Seaweeds in mythology, folklore, poetry, and life. J. Appl. Phycol. **32**, 3157-3182. (2020)
5. Harrison, P.J., Hurd, C.L.: Nutrient physiology of seaweeds: application of concepts to aquaculture. Cah. Biol. Mar. **42**(1), 71–82 (2001)
6. Peng, Y., et al.: Chemical composition of seaweeds. In: Tiwari, B.K., Troy, D.J. (eds.) Seaweed Sustainability: Food and Non-food Application, pp. 79–124. Academic Press, Cambridge, , USA (2015)
7. Goñi, I., Gudiel-Urbano, M., Saura-Calixto, F.: In vitro determination of digestible and unavailable protein in edible seaweeds. J. Sci. Food Agric. **82**(15), 1850–1854 (2002)
8. Mišurcová, L.: Isolation and Chemical Properties of Molecules Derived from Seaweeds Chemical Composition of Seaweeds. In: Kim, S.-K. (ed.) Handbook of Marine Macroalgae, pp. 171–192. Wiley, Chichester (2011)
9. Rajapakse, N., Kim, S.K.: Nutritional and digestive health benefits of seaweed. In: Kim, S.-K. (ed.) Advances in Food and Nutrition Research, pp 17–28. Academic Press, Cambridge (2011)
10. Rajauria, G., Cornish, L., Ometto, F., Msuja, F.E., Villa, R.: Identification and selection of algae for food, feed, and fuel applications. In: Tiwari, B.K., Troy, D.J. (eds.) Seaweed Sustainability: Food and Non-Food Application, pp 315–345. Academic Press, Cambridge (2015)
11. Padam, B.S., Chye, F.Y.: Seaweed components, properties, and applications. In: Torres, M.D., Kraan, S., Dominguez, H. (eds.) Advances in Green Chemistry, pp. 22–87. Sustainable Seaweed Technologies, Elsevier, Amsterdam (2020)
12. Pereira, L.: As Algas Marinhas e Respectivas Utilidades. Monografias **913**, 1–9 (2008)
13. Gioele, C., et al.: *Gracilaria gracilis*, Source of Agar: A Short Review. Curr. Org. Chem. **21**(5), 380–386 (2017)
14. Araujo, G.S., Cotas, J., Morais, T., Leandro, A., García-Poza, S., Gonçalves, A.M.M., Pereira, L.: *Calliblepharis jubata* Cultivation potential—a comparative study between controlled and semi-controlled aquaculture. Appl. Sci. **10**(21), 7553 (2020)
15. Pérez-Mayorga, D.M., Ladah, L.B., Zertuche-González, J.A., Leichter, J.J., Filonov, A.E., Lavín, M.F.: Nitrogen uptake and growth by the opportunistic macroalga *Ulva lactuca* (Linnaeus) during the internal tide. J. Exp. Mar. Biol. Ecol. **406**(1–2), 108–115 (2011)
16. AOAC: Official Methods of Analysis of AOAC international, 16th edn, vol. 1. AOAC International, Gaithersburg, USA (1997)

17. FAO: Methods of food analysis. Food energy methods Analysis Conversion factors, FAO, Rome (2003)
18. Ministro, M.: Otimização do sistema de cultivo de macroalgas. Bacheralate Curricular Internship Report, Instituto Politécnico de Coimbra (2020)
19. Cabral Bastos, G.F.P.: Cultivation of the wild seaweed Gracilaria *gracilis* under laboratory scale: The Effect of Light Intensity and Nutrients on Growth, Pigment and Total Soluble Protein Content. MSc Thesis, Instituto Politécnico de Leiria (2019)
20. Pacheco, D.: Cultivo de Microalgas em Águas Residuais : Avaliação de Crescimento e da Capacidade de Remoção de Nutrientes. Bacheralate Curricular Internship Report, Instituto Politécnico de Leiria (2018)
21. Rioux, L., Turgeon, S.L.: Seaweed carbohydrates. In: Tiwari, B.K., Troy, D.J. (eds.) Seaweed Sustainability: Food and Non-Food Application, pp. 141–192. Academic Press, Cambridge , USA (2015)
22. Rasyid, A., Ardiansyah, A., Pangestuti, R.: Nutrient Composition of Dried Seaweed *Gracilaria gracilis*. ILMU KELAUTAN Indones. J. Mar. Sci. **24**(1), 1–6 (2019)
23. Rodrigues, D., et al.: Chemical composition of red, brown and green macroalgae from Buarcos bay in Central West Coast of Portugal. Food Chem. **183**, 197–207 (2015)
24. Paiva, L.S.: Desenvolvimento de tecnologias de extração e de quantificação dos principais componentes nutricionais de macroalgas do litoral dos Açores tendo em vista o seu aproveitamento como suplemento alimentar. MSc Thesis, Universidade dos Açores (2014)
25. Leandro, A., Pacheco, D., Cotas, J., Marques, J.C., Pereira, L., Gonçalves, A.M.M.: Seaweed's bioactive candidate compounds to food industry and global food security. Life **10**(8), 140 (2020)

The Challenges of Water Saving in Rice Irrigation: Field Assessment of Alternate Wetting and Drying Flooding and Drip Irrigation Techniques in the Lis Valley, Portugal

José M. Gonçalves[1(✉)], Manuel Nunes[1], António Jordão[2], Susana Ferreira[1], Rui Eugénio[3], Javier Bigeriego[4], Isabel Duarte[1,5], Paula Amador[1,5], Olga Filipe[1,5], Henrique Damásio[3], and Kiril Bahcevandziev[1,5]

[1] Instituto Politécnico de Coimbra, Escola Superior Agrária, Coimbra, Portugal
jmmg@esac.pt

[2] Direção Regional de Agricultura e Pescas Do Centro, Coimbra, Portugal

[3] Associação de Regantes e Beneficiários do Vale do Lis, Leiria, Portugal

[4] Universidad de Extremadura, Badajoz, Spain

[5] IIA - Institute of Applied Research, CERNAS - Research Centre for Natural Resources, Environment and Society, Coimbra, Portugal

Abstract. The rice irrigation by continuous flooding is highly water demanding, in comparison with most methods applied in irrigation of other crops. This is due to a significant deep percolation, and the need for surface drainage of water from the basin. Currently, rice irrigation in basins with precision land leveling requires much less water than it was used in the past; nevertheless, there are still some recognized problems, like water scarcity, environmental impacts on water quality and agroecosystems and the methane emissions to the atmosphere. Saving water in rice production becomes a priority, to safeguard its economic and environmental sustainability. This study was elaborated, under the MEDWATERICE project, aiming to experiment the alternative wetting and drying flooding (AWD), using rice cultivar "Ariete" and the drip irrigation techniques applied in three rice cultivars, "Ariete", "Teti", and "Crono". The field experiment was installed in the Lis Valley, during the 2020 campaign, on farmer's fields. The AWD results showed that there is a potential of saving 10% of irrigation water, with no significant yield impacts, allowing additional 28 days with non-flooded soil. The precise land levelling is a priority to reduce the water level above soil surface. The drip irrigation essay, innovative in this area, uncover problems with soil lateral wetting and fertilizers leaching in a light soil, explaining yield losses. However, the maximum plot yield, makes glimpse a potential good performance. In the next season experiment adjustments of the distance between row crops, position of drip lines and the fertigation plan, will allow to prove this potential.

Keywords: Rice irrigation · Oryza Sativa L. · Water saving · AWD · MEDWATERICE · www.medwaterice.org · Lis Valley · Portugal

J. R. da Costa Sanches Galvão et al. (Eds.): *Proceedings of the 1st International Conference on Water Energy Food and Sustainability (ICoWEFS 2021)*, pp. 93–102, 2021.
https://doi.org/10.1007/978-3-030-75315-3_11

1 Introduction

Rice crop has an important economic and social value in several countries, like Portugal, being traditionally cultivated under continuous flooding (CF) irrigation. This practice is highly water demanding, in comparison with most methods applied in irrigation of other crops, due to a significant deep percolation, and the need for surface drainage of water from the basin. There main problems related with the CF refers to water scarcity, environmental impacts on water quality and agroecosystems and the soil methane emissions to the atmosphere. Although the common use of laser land leveling have allowed a significant reduction of water use, there are still common problems of water management to deal with, to face the climate global changes and the raising of social emergent consensus. The issue of rice irrigation water saving is challenging and intensively studied, demonstrating several irrigation management systems, such as, zero-grade fields, alternate wetting and drying, multiple-inlet, furrow, pivot irrigation and drip irrigation [1–3].

Alternate wetting and drying (AWD) consists of intermittent flooding, through a sequence of irrigation cycles with flooding for a certain period, followed by interruption of supply, until the soil dries up, creating conditions of non-saturation in the surface layer [4]. In this way, the volume of water applied to irrigation is reduced, compared to CF, as well as the reduction of greenhouse gas emissions [5] and lower arsenic accumulation in the grain [6], due to soil aeration conditions during non-saturation periods. AWD has been successfully used in several countries, as India [7], Philippines [8], and Mi-South of USA [9].

Drip irrigation is a more recent solution, taking advantage of the great potential of these systems in terms of automation and fertigation and water savings, allowing the expansion of rice cultivation to highland areas, with promising results [10–12].

This paper presents the preliminary results of research carried out on Lis Valley, Portugal, sponsored by the project MEDWATERICE (PRIMA-Sect. 2-2018; www.medwaterice.org) [13]. This project aims to explore the sustainability of innovative rice irrigation methods and technologies in the Mediterranean basin, in order to reduce rice water use and environmental impacts, and to extend rice cultivation outside of traditional paddy areas to meet the growing demand. This study, based on field experiments, had the objective to assess the actual practices of rice irrigation by CF, considered as the 'reference' irrigation method, and test two types of irrigation strategies, considered as alternative to water saving rice irrigation, the AWD and the surface drip method.

2 Material and Methods

2.1 Study Site

The Lis Valley Irrigation District is a public irrigation district managed by a Water Users' Association (WUA), located in the Coastal Center of Portugal (coordinates, 39°51′22.1"N 8°50′56.1"W), under Mediterranean Temperate climate, belonging to the counties of Leiria and Marinha Grande (Fig. 1a). The total area is about 2000 ha, cultivated with forage corn, forage grass, vegetables, fruits and rice. The hydraulic infrastructures have the objectives of perimeter drainage defense through slope collectors and valley ditches, the irrigation water supply through a canal conveyance system and the field drainage based on a ditch network. Water is supplied by an open-channel conveyance network from weirs insalled along the Lis river and tributaries, and by pumping from drainage ditches [14]. The soils are mainly alluvial with high agricultural quality, some are poorly drained, with waterlogging and salinization risks, particularly on the downstream areas. The structure of the on-farm parcel property is characterized mainly by small parcels [15].

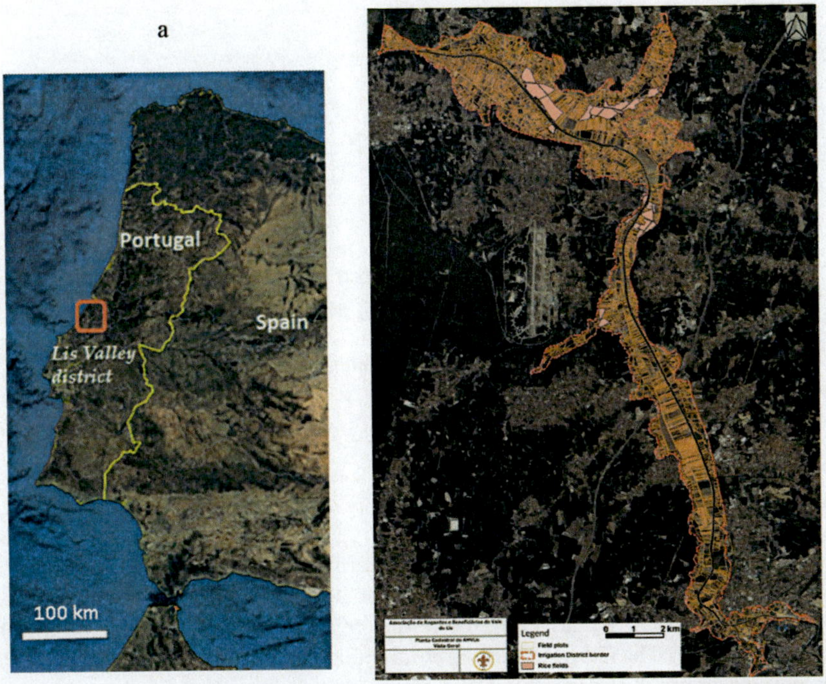

Fig. 1. (a) Location of the Lis Valley Irrigation District in Portugal (red square) (source: Google Maps, https://maps.google.pt); (b) Location of rice fields (rose color) in Lis Valley Irrigation District (source: Lis Valley Irrigation District Water Users' Association).

Rice is cultivated in traditional paddies, with ca. 10 cm of ponding depth, and an irrigation frequency varying from daily to a few days. Water in rice plays a main role in a temperature regulation and weed control. This crop is traditionally cultivated on lower soils with heavy texture, drainage problems, and a shallow groundwater table, totalizing an area of about 140 ha (Fig. 1b). A water is supplied from irrigation canals, or in some fields, by farmer's pumping from ditches. The water shortage is an endemic problem during summer months in Lis Valley, being the major constraint to rice crop sustainability and expansion. Water-saving practices on rice irrigation are welcome to mitigate a problem with water scarcity. Other constraints refer to water distribution conditions, surface water organic pollution, soil salinization, pumping energy for irrigation and economic sustainability of rice production. The essays took place in farmer fields, in a close relationship between the research team and the farmer. This proximity to the production conditions facilitated the connection and involvement with stakeholders and the dissemination activity. Meteorological observations were carried out with automatic stations installed near the experiments, being the air daily temperatures presented in Fig. 2. Daily reference evapotranspiration was calculated by Penman-Monteith method, being applied crop coefficients of 1.25 for flooding, and between 1.0 and 1.1 for dry periods and drip irrigation.

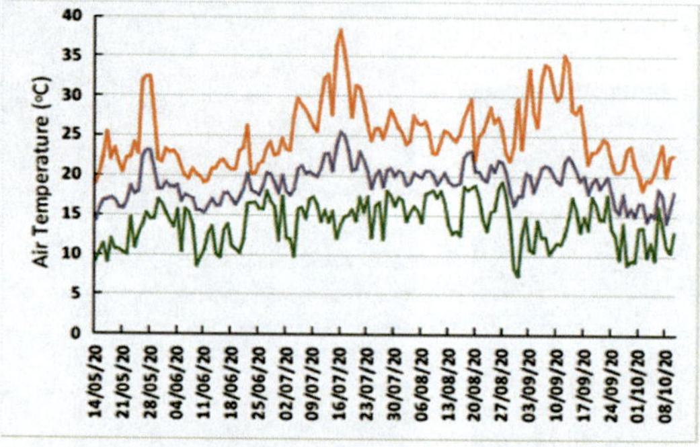

Fig. 2. Minimum (green), average (blue), and maximum (red) daily air temperatures during 2020 rice crop season.

2.2 Flooding Irrigation

The CF system essay was carried on a private farm, using two plots: i) area of 3.0 ha (215 m × 140 m), using traditional practices of flood irrigation full managed by the farmer, Lis Valley representative, used like the 'reference' irrigation method; ii) other one, with area of 0.20 ha (20 m × 100 m), contiguous to the traditional one with identical agronomic practices, applying the AWD irrigation practice. The soil has clay-loamy, neutral (pH = 7.4), with a deep profile, and with a shallow groundwater table level, between 75 cm to 85 cm below the soil surface. The agronomic practices in both plots used were soil preparation, including the ploughing and harrowing, wet sowing of the rice cultivar "Ariete", carried out on 14th May, and the fertilizer doses of 90 kg N/ha, and herbicide treatments. The monitoring system installed followed Gonçalves et al. [14]. The evaluation of crop yield, considered the sampling of aerial part at harvest time, was done in several points of the plots, with a unit sampling area of 0.5 m^2.

The AWD essay applied the following procedure based on Bouman et al. [16], with the adjustments to local agronomic practices: i) The initial flooding allowing the wet sowing, like the traditional practice (follows an initial drying event to favour the emergence); ii) Shallow ponding during the vegetative phase, considering the drying periods required for herbicide application, usually twice; iii) AWD technique applied after the flower initiation, until last irrigation; the target was a flood water depth not higher than 5–7 cm; the irrigation schedule considered was an interval between 10 to 14 days of irrigation events; ensuring that the water level should not fall to 15 cm below the soil surface, measured in a water tube; particular attention during the flowering period because it is very sensitive to water stress; iv) The last irrigation should be about 20 days before the harvest.

2.3 Drip Irrigation

The drip irrigation experiment carried out on a plot of a private farmer that usually produces horticultural crops, sown on 20th May. The soil has a light texture, very good internal drainage, acid (pH = 5.3), 1.5% of organic matter, and with a groundwater level about 3–4 m depth below soil surface. The plot had an effective area of 240 m^2 (12 m width × 25 m length), was divided into 8 strips, with two drip lines per each strip, having each strip 6 rows of plants. Three cultivars, of Portuguese rice paddies, were used: "Ariete", "Teti", and" Crono" in each third of all strips, to assess their performance to the drip irrigation condition. Spacing between rows was 20 cm, and the target plant spacing within row was 5–10 cm. The weed control was by two applications of chemical herbicide control, with the active substance Bentazona, complemented with thinning.

A water was supplied using a submersible electric pump (Hidrobex, model Vetax-1000, 1 kW) and automatically controlled. The irrigation system comprised a sand-filter, complemented with two plastic mesh-filters, a fertigation injector, a water counter and two pressure gauges; a manifold of PE 50 mm, and PE drip lines of 16.2 mm diameter, brand NETAFIM, model Thyphon Plus 16150, non-regulated, dripper spacing of 0.30 m, with a flow of 1.00 L/h with pressure of 1.0 bar; the field installation had 16 drip lines with a length of 25 m, spaced of 0.60 m, working at a pressure of 1.0 bar, with a total discharge of 1.33 m^3/h. The monitoring system installed in this plot evaluated the following aspects: water use, soil moisture, crop development, yield quantity and quality. The evaluation of crop yield was considered the sampling of aerial part at harvest time, for each strip of 100 cm linear at row crop close to drip line, and identical sample on the row far from the drip line, to assess the medium and maximum yield. This procedure was applied to each of the eight strips, for every rice variety.

3 Results and Discussion

3.1 Flooding irrigation

The analysis of the water level above soil surface of flooding irrigation plots during crop season (Fig. 3) shows that the traditional CF considers four dry periods during the vegetative phase, and that after 16th July, at the end of flowering (64 DAS), the AWD technique have been applied, in five wet-dry cycles, until the final period before the harvest (Fig. 4). These cycles corresponded to a period between 12 and 14 days, with irrigation depths between 92 and 109 mm, and 5 or 6 days, with dry soil, per cycle. The irrigation water was 1292 mm and 1169 mm, for CF and AWD, respectively (Table 1), resulting in a water saving of 9.4%, corresponding to 122 mm, and a decrease of 5.6% of yield (Table 2), although the production differences were not significant (5.6%). The water productivity with CF and AWD was 0.436 kg/m^3 and 0.452 kg/m^3, respectively.

The results obtained show that the traditional practice of CF can save water if the precise levelling of the soil is improved, which in turn allows to decrease the water depth,

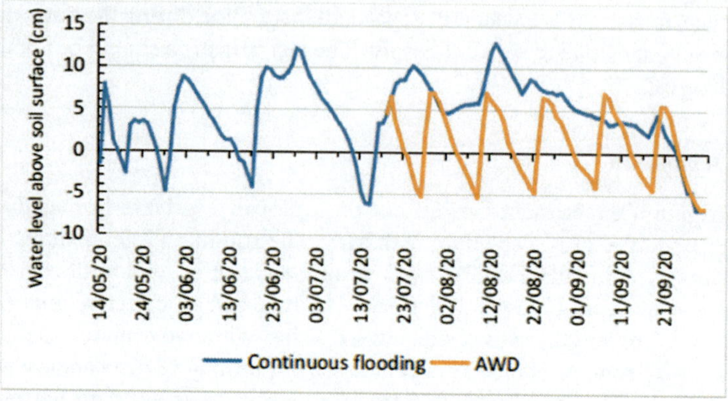

Fig. 3. Water level above soil surface (cm) of continuous flooding irrigation and alternate wetting and drying (AWD) plots, during 2020 rice crop season.

Fig. 4. Rice crop at maturation phase, during a dry period of AWD plot, with a water tube (arrow).

with the consequent decrease of the irrigation depth and percolation. However, there are some practical aspects that justify a higher value of this water level, related to the greater ease of controlling weeds and protecting the crop from the adverse effects of the strong wind that appeared during the early rice plant developing phases. The AWD practice, although it translates into a water saving of 9.4% of the total volume of the CF practice, allowed additional 28 days with non-flooded (dry) soil.

The issues to be studied in 2021 crop season experiment are: (i) continue with the CF and AWD trials; (ii) improve precise land levelling to try to reduce water level above soil surface, not to exceed 5 cm; (ii) increase the dry soil periods, elevating the critical water table threshold, closer to the 15 cm.

Table 1. Flooding irrigation plots water use (mm).

Water use (mm)	Complete crop season period		Period after flower initiation[c]	
	CF[a]	AWD[b]	CF	AWD
Evapotranspiration	674.7	656.3	347.4	329.0
Deep Percolation	587.8	488.7	330.2	249.3
Precipitation	81.8	81.8	68.4	68.4
Irrigation	1291.5	1169.3	639.2	517.0
Surface Drainage	154.8	118.2	36.5	0

[a]CF - Continuous flooding; [b]AWD - alternate wetting and drying; [c]period 16July-9October

Table 2. Irrigation trials, yield, and water productivity (rice cultivar "Ariete").

Irrigation trial	Y[c] (t/ha)	WP[d] (kg/m^3)	WG[e] (g)	SY[f] (t/ha)
CF[a]	5.993 ± 1.264	0.436	31.98 ± 1.943	4.122 ± 1.029
AWD[2]	5.659 ± 0.290	0.452	30.84 ± 0.168	3.655 ± 0.305

[a]CF - Continuous (traditional) flooding; [b]AWD - alternate wetting and drying flooding; [c]Y- Rice grain yield at 14% humidity (t/ha); [d]WP - Water Productivity = grain yield at 14% humidity/(irrigation + precipitation) (m^3/ha); [e]WG - Weight of 1000 grains at 14% humidity (g); [f]SY - Straw yield, dry matter (t/ha)

3.2 Drip Irrigation

This trial was a first experiment of aerobic drip irrigated rice (Fig. 5) applied in the Lis Region. The soil preparation included the usual practice of ploughing and harrowing. The dry sowing was manual, carried out on 20th May (Table 3).

Due to the dry and hot weather that was felt in those days (Fig. 2), the irrigation after sowing was by sprinkler solid-set, 30 min per day, during 7 days: 20th to 26th May, with a depth of 5 mm/day, to favor the seed germination and plant emergence. The drip system started from 27th May, with daily events, being the irrigation time regulated on the automatic controller. The higher daily irrigation depths were applied during the flowering and milky seed phases (19th August to 17th September), close to 14 mm/day. Two problems have occurred during vegetative and reproductive phases. First, the crop showed water stress signals on leaves, particularly in plants located on the rows far from the drip line. To overcome this problem, an over irrigation practice was applied (total of irrigation of 1114 mm, much higher than crop evapotranspiration (ETc) of 601 mm), to favour the lateral wetting, with a partial success. The second problem resulted of over irrigation on a very permeable soil, that led to an excessive deep percolation (total of 623 mm), that resulted to a significant fertilizer leaching, close to the drip line, justifying the symptoms of chlorosis on the plant leaves. Regarding the drip irrigation test, problems observed were related to the soil texture, and it was found that the lateral spacing of 20 cm between the drip line and the furthest row of plants was excessive, especially in spots with lighter soil.

Fig. 5. Drip irrigation experiment, at harvest, showing the drip lines (orange arrows).

Table 3. Crop development phases, fertigation, and irrigation depths of drip experiment.

Phases	Vegetative	Panicle D[a]	Flowering	Milky	Maturation	Final	Total
Start date	20 May	28 Jul	19 Aug	30 Aug	18 Sep	2 Oct	18 Oct
Start DAS[b]	0	69	91	103	121	135	150
N (kg/ha)	175	25					200
K_2O (kg/ha)	39	20	27	15			101
P_2O_5 (kg/ha)	25		25				50
DI[c] (mm/d)	6.1	10.3	14.5	14.0	5.0	0	
I[d] (mm)	415	216	160	238	85	0	1114
Etc[e]	320.7	89.0	51.5	72.2	36.1	32.0	601.5
P[f]	10.4	1.4	17.0	0.4	49.4	12.8	91.4
DP[g]	104.7	128.4	125.5	166.2	98.3	0	623.1

[a]Panicle D. - Panicle Differentiation; [b]DAS - Days After Sowing, at 20th May; [c]DI - Daily Irrigation depth, average value, mm/day; [d]I - Phase Irrigation depth, mm; [e]ETc - crop evapotranspiration (mm); [f]P - precipitation (mm); [g]DP

These problems led to high water consumption and very significant production losses (Table 4). However, in the three cultivars ("Ariete", "Teti", and "Crono"), control areas with high production were identified and yield assessed.

Table 4. Drip irrigation trial yield and water productivity, in three rice cultivars.

Rice cultivar	Ymed[a] (t/ha)	Ymax[b] (t/ha)	WPmed[c] (kg/m^3)	WPmax[c] (kg/m^3)	WG[d] (g)	SY[e] (t/ha)
Ariete	4.218 ± 1.646	7.138 ± 2.544	0.379	0.641	25.475 ± 0.883	10.96 ± 2.845
Teti	3.306 ± 1.850	6.341 ± 2.401	0.297	0.569	23.729 ± 0.969	8.483 ± 3.363
Crono	2.545 ± 2.408	8.129 ± 2.432	0.228	0.730	22.273 ± 4.564	8.290 ± 6.147

[a]Ymed - Medium plot rice grain yield at 14% humidity (t/ha); [b]Ymax - Maximum plot rice grain yield at 14% humidity (t/ha); [c]WP - Water Productivity = grain yield at 14% humidity (kg/ha)/(irrigation + precipitation) (m^3/ha); [d]WG - Medium Weight of 1000 grains at 14% humidity (g); [e]SY - Medium Straw yield, dry matter (t/ha)

4 Final Considerations

The experimental results obtained in the 2020 season concerning rice irrigation water saving techniques show that this issue is complex and challenging. Regarding the AWD technique, the results seem to be in accordance with several published studies [4, 7, 8], showing that there is a relative potential for saving water. The 10% of irrigating water was saved, with a negative impact of 5.6% on the production. AWD allowed a 28 days period of dry soil, expecting lower levels of greenhouse gas emissions and the arsenic content in the rice grain, in the future studies. It is considered a priority to improve the precise land levelling to optimize the water level above the soil surface, aiming water saving. On the other hand, the need to carry out frequent and planned irrigation events in the AWD period (after mid-July), makes the inflow control devices more demanding, making place for its automation.

Regarding the drip irrigation test, in 2021 the crop row distance and the drip lines position will be adjusted, reducing the spacing, to try to achieve more yield uniformity and decrease the water use.

Acknowledgments. Project MEDWATERICE: Towards a sustainable water use in Mediterranean rice-based agro-ecosystems. The project is financed in the context of the PRIMA Programme (PRIMA-Section-2018; Topic: 1.1.3: Irrigation technologies and practices). (MEDWATERICE-PRIMA-0005–2018, www.medwaterice.org).

References

1. Datta, A., Ullah, H., Ferdous, Z.: Water management in rice. In: Chauhan, B.S. (eds.) Rice Production Worldwide, pp. 255–277. Springer Int. Pub. AG (2017)
2. Steven, G., Nelson, J., Heiser, J.: Guidelines for producing rice using furrow irrigation. University of Missouri, Columbia (2020). www.extension.missouri.edu
3. Vories, E., Stevens, W.G., Rhine, M., Straatmann, Z.: Investigating irrigation scheduling for rice using variable rate irrigation. Agric. Water Manag. **179**, 314–323 (2017)
4. Tuong, T.P., Bouman, B.A.M.: Rice production in water-scarce environments. In: Kijne, J.W., Barker, R., Molden, D. (eds.) Water Productivity: Limits and Opportunities for Improvement. CABI Pub., Oxon, pp. 53–67 (2003)

5. Runkle, B.R.K., Suvoscarev, K., Reba, M.L., Reavis, C.W., Smith, S.F., Chiu, Y.L., Fong, B.: Methane emission reductions from the alternate wetting and drying of rice fields detected using the eddy covariance method. Environ. Sci. Technol. **53**, 671–681 (2019)
6. Linquist, B., Anders, M.M., Adviento-Borbe, M.A.A., Chaney, R., Da Roda, E.F.F., Van Kessel, C.: Reducing greenhouse gas emissions, water use, and grain arsenic levels in rice systems. Glob. Change Biol. **21**, 407–417 (2014)
7. Jalindar, M.K., Rao, V.P., Ramulu, V., Kumar, K.A., Devi, M.U.: Effectiveness of field water tube for standardization of alternate wetting and drying (AWD) method of water management in lowland rice (Oryza Sativa L.). Irrig. Drain. **68**, 679–689 (2019)
8. Lampayan, R.M., Rejesus, R.M., Singleton, G.R., Bouman, B.A.M.: Adoption and economics of alternate wetting and drying management for irrigated lowland rice. Field Crops Res. **170**, 95–108 (2015)
9. Reba, M.L., Massey, J.H.: Surface irrigation in the Lower Mississippi River Basin: trends and innovations. In Trans. of the ASABE; St. Joseph, MI, USA, vol. 63, pp. 1305–1314 (2020)
10. Parthasarathi, T., Mohandass, S., Senthivel, S., Vered, E.: Effect of drip irrigation systems on yield of aerobic rice. Environ. Ecol. **31**(4A), 1826–1929 (2013)
11. Arbat, G., Cufí, S., Duran-Ros, M., Pinsach, J., Puig-Bargués, J., Pujol, J., Ramírez de Cartagena, F.: Modeling approaches for determining dripline depth and irrigation frequency of subsurface drip irrigated rice on different soil textures. Water, **12**, 1724 (2020)
12. McClung, A.M., Rohila, J.S., Henry, C.G., Lorence, A.: Response of U.S. rice cultivars grown under non-flooded irrigation management. Agronomy, **10**, 55 (2020)
13. MEDWATERICE. https://www.medwaterice.org/. Accessed 31 Jan 2021
14. Gonçalves, J.M., Ferreira, S., Nunes, M., Eugénio, R., Amador, A., Filipe, O., Duarte, I.M., Teixeira, M., Vasconcelos, T., Oliveira, F., Gonçalves, M., Damásio, H.: Developing irrigation management at district scale based on water monitoring: study on Lis valley, Portugal. AgriEngineering **2**, 78–95 (2020)
15. Ferreira, S., Oliveira, F., Silva, F.G., Teixeira, M., Gonçalves, M., Eugénio, R., Damásio, H., Gonçalves, J.M.: Assessment of factors constraining organic farming expansion in Lis valley, Portugal. AgriEngineering **3**, 111–127 (2020)
16. Bouman, B.A.M., Humphreys, E., Tuong, T.P., Barker, R.: Rice and water. Adv. Agron. **92**, 187–237 (2007)

Application of Remote Sensing Based Tools for Improving Agricultural Management in Paddy Fields in the Lower Mondego Region

Romeu G. Jorge[1,2](✉), Isabel P. de Lima[1,2], and João L. M. P. de Lima[1,2]

[1] Department of Civil Engineering, University of Coimbra, Coimbra, Portugal
romeu.jorge@uc.pt

[2] MARE - Marine and Environmental Sciences Centre, Coimbra, Portugal

Abstract. Rice, which is strategic for food security in some countries, is traditionally grown under continuous flooding, thus requiring much more irrigation water than non-ponded crops. Fresh water shortage issues and steadily increasing competition for water (human consumption, agriculture) make the accurate and up-to-date assessment of the spatial distribution of rice cultivated area and crop condition (health, water requirement) key information for stakeholders including policy makers, rice farmers and consumers.

In recent years, remote sensing data, such as images obtained from satellites and Unmanned Aerial Systems (UAS), are offering attractive routes to acquire field data in a fast and easy way. Those data are applied to extract e.g., Vegetation Indices (VI) such as the Normalized Difference Vegetation Index (NDVI) or the Green Normalized Difference Vegetation Index (GNDVI).

This work is embraced by the international project MEDWATERICE that focus on improving the sustainable use of water in Mediterranean rice agro-ecosystems. It concentrates on the Lower Mondego region (Portugal) and aims at the increased understanding of the usefulness of remote sensing tools for rice agriculture, for the local conditions. This includes investigating: 1) NDVI trend during the rice cropping season, and its spatial variability, using Sentinel-2A data; 2) Comparison between VI estimated from Sentinel-2A and UAS data. Preliminary results show that Sentinel-2A data are potential useful tools for monitoring paddy fields, but the full understanding of the relation between the signals obtained from Sentinel-2A and UAS data still requires the analysis of longer paddy fields monitoring campaigns.

Keywords: Water management · Precision agriculture · Rice farming · Vegetation index · NDVI · MEDWATERICE

1 Introduction

Rice cultivation has a long tradition in Portugal and is of great importance to local economies. Portugal produces annually about 180 million kilograms of rice in the dedicated downstream rice cultivation areas of the Tagus, Sado and Mondego River basins

J. R. da Costa Sanches Galvão et al. (Eds.): *Proceedings of the 1st International Conference on Water Energy Food and Sustainability (ICoWEFS 2021)*, pp. 103–109, 2021.
https://doi.org/10.1007/978-3-030-75315-3_12

[1]. These areas have a Mediterranean climate influenced by the proximity of the Atlantic Ocean. Portugal is the fourth rice producer of the European Union (EU), accounting for about 6% of the EU rice production [2].

The rice irrigation practice by continuous flooding is often identified as contributing largely to low agricultural water use efficiency, an issue that is often studied [3–5] and discussed in water users and managers' forums. In fact, concerns on water are boosted by the increasing water demand and number and type of users (e.g., human consumption, agriculture, industry), and the vulnerability of Portugal to water stress that is predicted by climate change and hydrological models.

Remote sensing is capable of providing timely and reliable information for various purposes related to rice agricultural systems, in particular hydrologic assessment and modelling (e.g., monitoring of water use and productivity in agricultural fields, and relationship between crop yield and hydrological processes) [6]. The sensors' data allow to calculate Vegetation Indices – VI (e.g., Normalized Difference Vegetation Index – NDVI) using several spectral bands that are sensitive to plant biomass and vigour. Multi-temporal data acquired during the cropping seasons can provide information about the crop response to adopted agronomic practices.

Satellite remote sensing is increasingly being used as a complementary source of information to *in situ* environmental monitoring networks [7]. The satellites MODIS and Landsat have been widely used [8–10], but Sentinel-2, especially due to its finer resolution, constitutes nowadays a major asset for this application [11, 12]. However, satellite data resolutions are still too coarse to address plot scale conditions, therefore Unmanned Aerial Systems (UAS) offer a useful solution to obtain higher resolution data. For surveying relatively small areas (tens of hectares), the use of lightweight UAS can be a good alternative [13–15], especially due to the data's finer resolution (better than $0.20 \times 0.20\ m^2$ pixels), and their versatility and lower economic costs than other remote sensing techniques.

This paper presents preliminary results of an experimental study focusing on the Lower Mondego Valley rice production region, aiming at contributing to improve the estimation of rice water consumption using remote sensing based products. It uses satellite Sentinel-2A and UAS images: 1) to determine the temporal profile of NDVI during the rice cropping season and its spatial variability; and 2) to better understand the relation between the signals estimated from Sentinel-2A and UAS data.

2 Material and Methods

The study uses data from 2020, collected for an agricultural farm (Quinta do Canal; coordinates: 40° 7′ 0.16" N, 8° 47′ 47.34" W) located in the Lower Mondego Valley, near the city of Figueira da Foz, Central Portugal (Fig. 1). The total area of Quinta do Canal is about 335 ha that is predominantly dedicated to rice cultivation (332 ha). Based on information provided by the Local Farmers' Association (Associação de Beneficiários da Obra de Fomento Hidroagrícola do Baixo Mondego), Quinta do Canal is composed of 71 rice fields. The rice produced has long grain, of the variety Ariete, sub-species *Oryza sativa* ssp. *japonica*; commercially, it is known as "Carolino" rice. Irrigation is by flooding. In 2020 the rice growing season lasted 151 days: sowing date was May 10 and harvest date was October 10.

This study investigates two of the rice fields in Quinta do Canal, identified as plot 1 (3.15 ha) and plot 2 (1.6 ha), which are characterized using the Quantum GIS (QGIS) open-source software. The location of these plots is shown in Fig. 1.

Fig. 1. Location the two pilot rice irrigation fields (plots 1 and 2) in "Quinta do Canal" and in relation to the Mondego river (Google Maps, 2021).

Available satellite Sentinel-2 images of the study area, obtained from March 1 to October 31, 2020, were downloaded from the Copernicus open-access website (https://scihub.copernicus.eu). Level-2A images at 10×10 m^2 full resolution, already atmospheric corrected, were selected. Only images showing a cloud cover lower than 8% were retained. Sentinel-2A optical instruments sample in 13 spectral bands that allow to calculate not only vegetation vigour indices but also those of water stress and chlorophyll. Vegetation absorbs solar radiation in different bands (i.e., in different frequency ranges and wavelengths), and emits a different percentage of it. The percentage of refracted radiation in specific bands varies with plant health and water stress.

The NDVI was calculated by the equation: NDVI = (NIR-RED)/(NIR + RED), where NIR represents the near infrared band reflectance and RED is the red band reflectance. NIR is represented by spectral band 8 and RED is represented by spectral band 4. This index is directly related to the photosynthetic capacity and hence energy absorption of plant canopies. Healthy vegetation absorbs blue- and red-light energy to fuel photosynthesis and create chlorophyll; thus, a plant with more chlorophyll will reflect more near-infrared energy than an unhealthy plant. The value range of the NDVI is -1 to 1: i) higher positive NDVI values indicate peak growth rates of vegetation whereas lower positive values may reveal the existence of unhealthy plants; ii) values close to zero generally indicate dead plants or inorganic objects; iii) negative values, approaching -1, correspond to water. NDVI average values were computed for each studied plot and for all images using QGIS software.

A UAS flight was made on July 9, 2020, with a DJI Matrice 600 drone equipped with a camera X5 for RGB and a Micasense RED Edge –M multi-spectral sensor. The images obtained during the UAS flight, between 2:30 pm and 4:00 pm, without wind and in clear sky conditions, had a spatial resolution of $0.074 \times 0.074\ m^2$. The UAS was autonomous; its operation respected a pre-set flight plan. The UAS flight images data, including orthorectification and elaboration of VI (e.g., NDVI) maps, were performed using Pix4D software (Pix4D S.A., Prilly, Switzerland).

3 Results and Discussion

3.1 Temporal Variation of the Rice Plots' NDVI Estimated from Sentinel-2A Data

The temporal variation of the NDVI average values estimated from Sentinel-2A data during the 2020 rice growing season, for plots 1 and 2, is shown in Fig. 2.

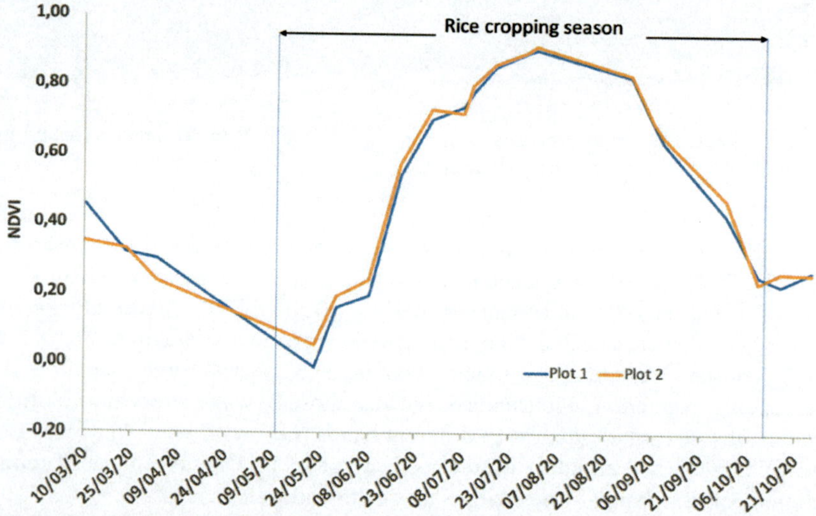

Fig. 2. NDVI average values calculated for irrigated agricultural plots 1 and 2 (Quinta do Canal, Fig. 1) using satellite Sentinel-2A imagery obtained in 2020. The period plotted include the rice cropping season, that lasted 151 days in 2020: sowing date was May 10 and harvest date was October 10.

Overall, the variation in NDVI shown in Fig. 2 behaves as expected. The NDVI profile highlights the different conditions of the field plots, over time, namely the different rice cropping phases: initial growth, crop development, reproduction, and maturity phases. There are also differences between the two field plots, explained by differences in field conditions and crop responses to the applied water and agronomic practices. The lowest NDVI values, around zero, observed in the initial phase, are related to the flooding of the fields, which is first established close to rice sowing time. NDVI values attain maximum magnitude, of approximately 0.90, in August (the peak greenness of rice that corresponds to the heading stage). Towards the end of the cultivation period, the NDVI values decrease until the plant reaches full maturation and is harvested, in the beginning of October.

The temporal profile of NDVI extracted from Sentinel-2 imagery for the rice crop in Quinta do Canal plots are alike those reported by previous studies in the Mediterranean region for this type of crop [16–18]. For the Lower Mondego agricultural region, the specific conditions' signature depicted by the studied vegetation index is a useful tool to better characterize, at the local scale, rice crop responses and, thus, water requirements, and to support water management decisions.

3.2 Comparison Between Sentinel-2A and UAS Based NDVI Estimates

For rice plots 1 and 2, NDVI values obtained from satellite Sentinel-2A and UAS data during the rice reproductive phase were compared; the data were extracted using the zonal statistic plugin in QGIS software. For the satellite data obtained on July 8 and the UAS data obtained on July 9, comparison of the corresponding NDVI estimates for plots 1 and 2 show that the plots' NDVI average values estimated from UAS data (0.87 and 0.88, respectively) are higher than the values estimated based on the Sentinel-2A data (0.74 and 0.72, respectively). However, because there is a one-day lag between the satellite and UAS records and in view of the irregularity of the satellite NDVI curves (Fig. 2) around those record dates, it is expected that the difference between the two platforms' NDVI estimates might be smaller. It is also expected that data from different sensors could lead to slightly different NDVI values.

The corresponding NDVI maps were obtained for each plot from both Sentinel-2A and UAS platforms' images, using QGIS software. Figure 3 shows these maps for plot 1. Places with healthy vegetation are seen in the green areas of the images. In a field, where a crop tends to be at the same phenological stage, areas with a significantly lower than average NDVI value reveal problems in vegetative development. The two maps in Fig. 3 illustrate that the quality of the mapping is, as expected, affected by the different resolution of the two products: the number of analysed pixels for plot 1 is 321 for the Sentinel-2 images and $\approx 6.26 \times 10^6$ for the UAS images. The NDVI map obtained from the UAS platform, is closer to the original conditions in the field; ditches, roads and tracks are easily identified in these maps.

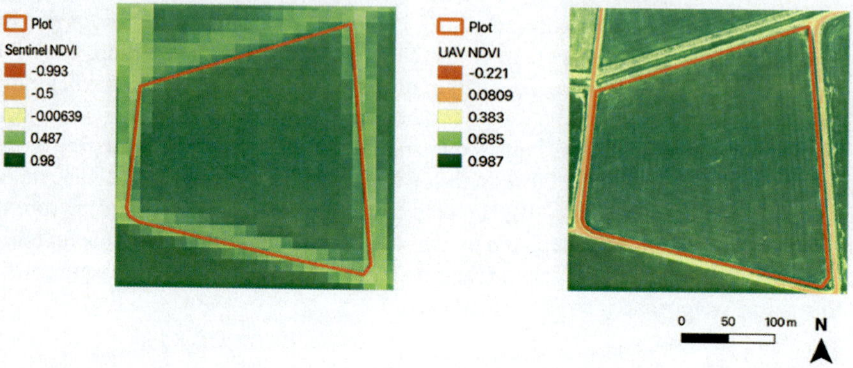

Fig. 3. NDVI map calculated for rice plot 1 (Quinta do Canal), using data from Sentinel-2A (left) and UAS (right) imagery. The data were obtained in 2020, during the rice reproductive phase.

4 Conclusions

This study is focused on characterizing, for the rice production conditions in the Lower Mondego region, the variations in time of rice NDVI estimates based on remote sensing products provided by satellite Sentinel-2A. It also compares the values of this vegetation index estimated from Sentinel-2A and UAS data. Preliminary results of this study provide increased insight into the usefulness of remote sensing tools to monitoring rice fields, which is perceived as being a valuable asset for agriculture. The interpretation of the absolute value of the NDVI is highly informative for rice cultivation, as it allows the immediate recognition of the areas of the rice farm or field that have problems. This type of data allows to obtain information of important temporal and spatial resolutions that, combined with other water management tools, have also the potential to routinely assist to better estimating rice water requirements and, therefore, to support better informed water management decisions.

Acknowledgments. The research was conducted under the umbrella of project MEDWATERICE - Towards a sustainable water use in Mediterranean rice-based agro-ecosystems, financed by the PRIMA Programme (PRIMA-Section-2018; Topic: 1.1.3: Irrigation technologies and practices) through the Portuguese Foundation for Science and Technology (PRIMA/0006/2018), and Action Cost CA16219 HARMONIOUS - Harmonization of UAS techniques for agricultural and natural ecosystems monitoring.

References

1. INE: Main crops yield (kg/ha) by geographic localization (agrarian region) and specie. Annual. Instituto Nacional de Estatística, I.P., Lisboa, Portugal (2019)
2. USDA: Portuguese rice imports pick up as production declines. USDA Gain Report, p. 13. United States Department of Agriculture, Washington, DC, USA (2017)
3. Howell, T.: Enhancing water use efficiency in irrigated agriculture. Agron. J. **93**, 281–189 (2001)

4. Chaves, M.M., Oliveira, M.M.: Mechanisms underlying plant resilience to water deficits: prospects for water-saving agriculture. J. Exp. Botany **55**(407), 2365–2384 (2004)
5. Gonçalves, J.M., de Lima, I.P.: Improving water productivity in irrigated agriculture: challenges from climate change and new water resources paradigms. In: Alves, F., Leal Filho, W., Azeiteiro, U. (eds.) Theory and Practice of Climate Adaptation, pp. 347–361. Springer, Cham (2018)
6. Kuenzer, C., Knauer, K.: Remote sensing of rice crop areas. Int. J. Remote Sens. **34**(6), 2101–2139 (2013)
7. Sheffield, J., Wood, E.F., Pan, M., Beck, H., Coccia, G., Serrat-Capdevila, A., Verbist, K.: Satellite remote sensing for water resources management: potential for supporting sustainable development in data-poor regions. Water Resour. Res. **54**, 1–35 (2018)
8. Oguro, Y., Suga, Y., Takeuchi, S., Ogawa, H., Tsuchiya, K.: Monitoring of a rice field using Landsat-5 TM and Landsat-7 ETM+ data. Adv. Space Res. **32**(11), 2223–2228 (2003). ISSN 0273-1177
9. Coll, C., Galve, J.M., Sanchez, J.M., Caselles, V.: Validation of Landsat-7/ETM+ thermal-band calibration and atmospheric correction with ground-based measurements. IEEE Trans. Geosci. Remote Sens. **48**(1), 547–555 (2010). https://doi.org/10.1109/TGRS.2009.2024934
10. Dong, J., Xiao, X., Kou, W., Qin, Y., Zhang, G., Li, L., Jin, C., Zhou, Y., Wang, J., Biradar, C., Liu, J., Moore, B.: Tracking the dynamics of paddy rice planting area in 1986–2010 through time series Landsat images and phenology-based algorithms. Remote Sens. Environ. **160**, 99–113 (2015)
11. Segarra, J., Buchaillot, M.L., Araus, J.L., Kefauver, S.C.: Remote sensing for precision agriculture: Sentinel-2 improved features and applications. Agronomy **10**, 641 (2020)
12. Belgiu, M., Drăgu, L.: Random forest in remote sensing: A review of applications and future directions. ISPRS J. Photogramm. Remote Sens. **114**, 24–31 (2016)
13. Duan, B., Fang, S., Zhu, R., Wu, X., Wang, S., Gong, Y., Peng, Y.: Remote estimation of rice yield with unmanned aerial vehicle (UAV) data and spectral mixture analysis. Front. Plant Sci. **10**, 204 (2019). https://doi.org/10.3389/fpls.2019.00204
14. Wan, L., Cen, H., Zhu, J., Zhang, J., Zhu, Y., Sun, D., Du, X., Zhai, L., Weng, H., Li, Y., Li, X., Bao, Y., Shou, J., He, Y.: Grain yield prediction of rice using multi-temporal UAV-based RGB and multispectral images and model transfer – a case study of small farmlands in the South of China. Agric. Forest Meteorol. **291**, 108096 (2020). ISSN 0168-1923
15. Chosa, T., Miyagawa, K., Tamura, S., Yamazaki, K., Iiyoshi, S., Furuhata, M., Motobayashi, K.: Monitoring rice growth over a production region using an unmanned aerial vehicle: preliminary trial for establishing a regional rice strain. IFAC Proc. Volumes **43**(26), 178–183 (2010)
16. Rolim, J., Navarro, A., Vilar, P., Saraiva, C., Catalao, J.: Crop data retrieval using earth observation data to support agricultural water management. Engenharia Agrícola **39**(3), 380–390 (2019)
17. Corvino, G., Lessio, A., Borgogno-Mondino, E.: Monitoring rice crops in Piemonte (Italy): towards an operational service based on free satellite data. In: IGARSS 2018 - 2018 IEEE International Geoscience and Remote Sensing Symposium, Valencia, pp. 9070–9073 (2018). https://doi.org/10.1109/IGARSS.2018.8518634
18. Sarvia, F., Xausa, E., Petris, S.D., Cantamessa, G., Borgogno- Mondino, E.: A possible role of Copernicus Sentinel-2 data to support common agricultural policy controls in agriculture. Agronomy **11**, 110 (2021). https://doi.org/10.3390/agronomy11010110

Risks Perceptions on Rice Production: *Baixo Mondego*, Portugal

Maria de Fátima Oliveira[1(✉)], Sérgio Oliveira[2], António Russo[3], Kiril Bahcevandziev[1], Ana Bela Lopes[1], José M. Gonçalves[2], António Dinis Ferreira[1], and Pedro Arnaldo Reis[4]

[1] Agriculture School, Research Centre for Natural Resources, Environment and Society, Polytechnic Institute of Coimbra, 3045-601 Coimbra, Portugal
{foliveira,Kiril,anabela,aferreira}@esac.pt
[2] Agriculture School, Polytechnic Institute of Coimbra, 3045-601 Coimbra, Portugal
jmmg@esac.pt
[3] Lower Mondego Water Users Association, Montemor-o-Velho, Portugal
[4] National Institute of Agrarian and Veterinary Research, Oeiras, Portugal
pedro.reis@iniav.pt

Abstract. This paper aims to analyse the rice production system in the lower Mondego Valley, in Portugal, in order to understand the main concerns of farmers and risk perception. A field survey was conducted and rice production systems as well as marketing systems were analysed. A study on the worries was also made and a correlational attempt was done, considering different types of risk. The results show a production system strongly oriented by agricultural policies and agri-environmental measures. The perceived risks mainly arise from rice production costs, prices and their relative instability which results from the uncertainty about the evolution of worldwide demand and production, and from the effects of global changes. However, the choices made by farmers, who possess sound knowledge of agricultural policies and marketing systems, are not completely correlated with their risk perceptions.

Keywords: Rice production · Risk · Worries · Lower Mondego Valley

1 Introduction

The agricultural business always bears risks associated with the nature of its production, since it is prone to climatic, soil and biotic factors. Duong *et al.* [1] carried out a systematic review of the literature and concluded that research on risk perception, risk management in agriculture has increased since 1985. This research found out that market risk is the most significant risk followed by biosecurity, which highlights the gaps between the risks mentioned by farmers and the research on socioeconomic factors that explain the perception of risks. Some papers discuss both risks and adaptation strategies (Crane *et al.* [2], Ahsan and Roth [3] and Harwood *et al.* [4]).

J. R. da Costa Sanches Galvão et al. (Eds.): *Proceedings of the 1st International Conference on Water Energy Food and Sustainability (ICoWEFS 2021)*, pp. 110–119, 2021.
https://doi.org/10.1007/978-3-030-75315-3_13

Girdžiūtė [5] and Komarek *et al.* [6] examine risks in agriculture and point out that research work focuses on one risk, particularly on production risk, and that there are a limited number of studies exploring the various different sources of risk, also showing the importance of risk assessment methods and the importance of studying the interrelationship and interaction between risks. The results show that different types of risk are relevant at different levels and that it is necessary to interconnect risk so as to understand risk relevance to farmers. Farmers face many risk causes, from natural risks to economic ones. Waage and Mumford [7] define biosecurity as "the protection of countries against alien pests (insects, vertebrates, etc.) and diseases".

The existence of other risks, such as climate risk, financial risk, governing restraints and new technologies, should not be ignored, though. Crane *et al.* [2] explain risk as the chance of damage or a negative outcome linked to an action and uncertainty falls within these definitions. Harwood *et al.* [4] reveal that risk affects individual welfare and is related to loss. Risk perception influences farmers' behaviour and future business decisions such as the continuity in the market. Risk perception and behaviour are linked with the resilience of the agricultural sector itself. According to Keil *et al.* [8] loss, level is the most important factor in shaping risk perception, and there is a significant relationship between risk perception and decision-making. Sjöberg [9] says that cultural biases are not major factors in risk perception and the variability within the public in a country is probably due to factors such as trust, beliefs, and human concepts.

Risk perception encompasses the mental processing of information and the skill mechanisms used by people to deal with uncertain events. In addition to definition issues and risk perception, risk categorization is another issue dealt with by literature presenting different risk categories according to the sector and within agriculture. The categorization varies according to the subject focused on by research. Crane *et al.* [2] and Harwood *et al.* [4] identified production, marketing, economics, legal matters and human resources related issues as the main five sources of risk. According to Hardaker *et al.* [10] business risks are those affecting farm business performance such as production, market, institutional and personal risks and financial risks (firm's financing).

The aim of this paper is to provide empirical analyses of how farmers grow rice in the Mondego Valley in Portugal, how they prioritize their concerns, how their concerns are related to risk perception, and how these concerns interact with rice production systems and sociodemographic characteristics. The focus was not to directly address the concept of risk, but rather to concentrate on worries or concerns. The word "worries" was used instead of "risk" in OECD [11] as stated by Winsen *et al.* [12] (p. 44) "in order to avoid influencing the interviewee". The word "concerns" was also applied in Sjoberg's work [9].

Rice was introduced in southern Iberian Peninsula by the Arabs in the eighth century. Thus, rice growing was a key factor in the economic development of originally very underprivileged areas contributing to the emergence of social and cultural traditions and modelling these regions reputation up to the present-day. Rice plays a role of great social and economic importance in Portugal. We must bear in mind that rice cultivation uses areas that could not or could hardly be allocated to other productions, thus playing an important role in the conservation of these specific ecosystems. Due to cultural issues, rice is of utmost importance in the Portuguese diet. Rice producers are entitled to the Basic Payment Scheme of the Common Agricultural Policy whose payments are allocated to land rights.

2 Material and Methods

Field work was carried out in the *Baixo Mondego* Valley, located in the *Beira Litoral* region. A survey was conducted to grasp the full range of farming enterprises typology. This methodology provided indicators to address present farm conditions and represents current agronomic managing practices as noted by Dantsis *et al.* [13] (p. 261). The survey was personally conducted on the farms. The questionnaire was divided into four groups of questions: (i) the farmers' socioeconomic profile; (ii) agricultural cultural practices and their production inputs; (iii) production costs; (iv) production mode according to its environmental sustainability, mainly related to fertilizers and plant protection products application. The questions presented were closed, single, or multiple choice one with a maximum of two possible choices.

To assess the degree of fulfilment in relation to certification instruments, policy support and marketing channels, we held an analysis of the degree of satisfaction. The five points of Likert scale were used Vagias [14] to perform the satisfaction analysis. In our analysis, we applied the categorization of Crane et al. [2], which considers marketing risk (price and market channel), production risk (production costs and yield risks) and legal and financial risks. When considering nonparametric correlations, the Spearman coefficient was applied, using the SPSS 25 software and STATA15.

The 34 validated surveys were carried out in June 2019 in the central region of the *Vale do Mondego*. The 34 questionnaires represent 20% of rice farmers of the study area. The interviewed farmers were selected at random due to contact access obstacles and farmer's limited time availability. The number of small farmers is too large to perform an interview. Small famers duplicate big farmers' behaviour both at structural and commercial levels. Due to the high number of very small agricultural properties and their heterogeneity in terms of cultivated area and to the importance of medium or large sized agricultural properties as decision makers in the rice chain and in production systems, the cultivated area was the basis to decide on the size of the sample. If we apply for the number of hectares of rice cultivated area, and for an acceptable error of 5%, the sample size will be 328 hectares (ha). The sample size we questioned is displayed in Table 1, and the surveyed farm holders own 764 ha (42% of the rice area in the *Baixo* Mondego valley), that is, the sample is larger than the minimum required. We understand that the size sample is small if we consider the number of farmers but if we consider the analysed hectares as sample size, we can consider the sample to be representative.

Table 1. Sample categories and number of producer's holders: production areas

Hectares	Farmers survey		Rice farmer		Representative sample
	Number	%	Number	%	%
≤6.30	5	14.7	103	59.2	4.9
≥6.31 and ≤22.60	16	47.1	52	9.2	30.8
≥22,61 and ≤38.80	5	23.5	8	4.6	62.5
≥38.81	5	14.7	11	6.3	45.5
Total	34	100	174	100	19.5

3 Results

3.1 Sample Characterization

The study conducted with rice farmers shows that 82% of respondents are full-time farmers, 3% are part-time farmers and 15% are agricultural pensioners. About 32% of respondents: are between the ages of 20 and 40; 27% are older than 60 years. Approximately 62% of the interviewees had nine years of schooling and 27% of the respondents had university level education. When compared to national results, these figures are higher in terms of both farmers' age and educational level. The average literacy level in Portugal shows that 71.4% farmers have elementary education and only 6% attended university. In terms of age group, about 55% Portuguese farmers are over 65 years [15]. All farmers apply integrated production systems.

A weak correlation was identified (significant at 5% and 1% level of significance), with the importance of agricultural activity in the professional life of the farmer (full-time; part-time, retirees) and the farmer's level of education and age, respectively. In the case of young rice producers, agriculture is a main activity when compared to other age groups. At the same time, young producers have higher academic degrees. We found no correlation between farmers characteristics and the size of the farm they manage. The farmers' education level is positively related to the educational level of the turnover generation. Carolino rice variety, Ariete, (Oryza sativa L. subspecies Japonica), represents approximately 77% of the area rice cultivated area. About 56% farmers only produce one rice variety, 19% grow two varieties and 6% produce grow more than two rice varieties. Carolino Ariete is grown by about 91%, followed by Euro and Opale variety. Variety diversification can reduce production risk.

3.2 Fixed Assets and Costs

In agriculture, fixed assets refer to farmers' technical equipment. The fixed assets to total assets ratio are the main factor distinguishing farms. Farmers' competitiveness depends on the use of technical capacity. An analysis of a farmer's ability to invest in equipment and his dependence on equipment from others can explain and help understanding the farmer´s notion or perception of risk in relation to long-term production or investment. All respondents own their tractors (88% respondents own three or four) which are used with different types of needed for rice cultivation equipment. Most farmers own their equipment. Only two farmers rent rice dryers and only one has peeling and bleaching equipment. Land ownership may be crucial for risk perception. Not knowing if you can access land in the future may be a concern, because some land is rented. All respondents have their own land but about 47% use leased land.

When controlling weeds, plagues and diseases farmers are subject to integrated production rules regarding the quantity and quality of herbicides and pesticides applied. All farmers apply herbicide one to four times. Regarding others pesticide use (insecticides and fungicides), the number of applications is less frequent: 5.9% do not apply, 50% apply once and 44% apply twice.

Production costs are the most important variable on farm profitability, and they can affect risk perception. Production risk involves both the quantity and the quality of the output, as noted by Chartier *et al.* [16]. Income variability depends on the correlation between price and income risks, the farm's cost (fixed and variable costs), particularly the variable costs that constitute the activity's crop account and bring up to date the farmer´s efficiency and even their ability to support themselves and risk awareness brought by their activity to family life. Climate, biosecurity, technology and yields issues are in the production risks category (Duong *et al.* [1]).Respondents were asked about rice production and processing costs in the last three years. Taxes, interest, amortization of fixed assets and certification costs were not included. It is worth noting that water costs include a fixed component (conservation fees) and a variable component of water use, that is, exploitation fees (0.00262 €/m^3; one hectare uses about 16,390 m^3 of water). Table 2, shows that five variables represent 70% rice production costs. To understand the relationship between costs and production costs the Ordinary Least Squares (*OLS*) was performed. The presence of multicollinearity, heteroscedasticity and residual autocorrelation tests (*Durbin-Watson)* was searched for. The heteroscedasticity test was performed by analysing regression residues in SPSS25 and later confirmed by regression performed by STATA15. According to Table 2, we can safely conclude that there is no collinearity and the absence of multicollinearity since Tolerance (Tol.) is higher than 0.1 and the Variance Inflation Factor (VIF) is less than 10.

The production and area variables introduced and improved the value of the Durbin-Watson statistic (DW). According to DW test, we can reject autocorrelation in residuals. The model is fit and explains the independent variable. Production is negatively correlated with costs, that is, the higher the production, the lower the costs and the larger the cultivated area the higher the costs. The rental cost variable is significant and has the highest partial regression coefficient. Labour costs represent 17% of the total cost, although, according to regression, this variable does not help explaining production costs. It is interesting to note that the weight of each component of total costs is not directly related to the weight of this component in explaining rice production costs.

It is important to emphasize that despite the questions about total costs of rice production and the costs *per* component in rice production have been carried out in different parts of the questionnaire, there were no significant differences between total costs calculated by component and total production costs indicated by the interviewees. This means that farmers have a good perception of both costs per component and the activity cost.

Table 2. OLS regression and weight of costs components in costs rice production

OLS Regression (exogenous variable: Costs for Rice production; Euros/hectare). Endogenous Variables. Costs in Euros/hectare; The variables Surface are in hectares and Production are in tons						Costs (Euros/ha)			
Explanory Variables	Beta	*t*-test	*p-value*	Tol.	VIF	Costs	Average	*Std* Desv	Weight
Intercept	6857	1.22	.24			**Total**	1731,3	*299,7*	100%
Costs Labour	.23	1.65	.11	.18	5.60	Labour	287.0	*116.1*	17%
Costs Herbicides	.23	**2.59**	**.02**	.43	2.32	Herbicides	305.9	*54.8*	18%
Costs Fungicides/insecticide	.12	1.39	.18	.46	2.16	Fungicides/ insecticide	111,4	*52.7*	6%
Costs Seeds	.20	**2.49**	**.02**	.50	1.98	Seed	191,1	*24.3*	11%
Costs Machine & fuel	.20	**2.13**	**.05**	.39	2.56	Machines & fuel	225.0	*43.1*	13%
Costs Fertilizers	19	**2.21**	**.04**	.46	2.16	Fertilizers	209.9	*55.8*	12%
Costs Land Rent	.59	**7.84**	**.00**	.58	1.71	Rent	188.4	*193.0*	11%
Costs Insurance	.01	0.07	.95	.44	2.29	Insurance	2.7	*1.2*	0%
Costs Water/Irrigation	-.10	−1.20	.24	.48	2.09	Water/Irrigation	88.1	*0.3*	5%
Costs Dryer Rice	.19	**2.25**	**.04**	.45	2.25	Dryer Rice	117.6	*46.5*	7%
Costs Peeler/Bleaching	.34	**2.85**	**.01**	.24	4.15	Peeler/Bleaching	4.4	*25.5*	0%
Surface	-.43	**-3.32**	**.00**	.20	5.07				
Production	. 45	**4.22**	**.00**	.30	3.39				
Adj R2 = .889		**F(13., 20)** = 21.50				**Breusch-Pagan chi2(1)** = .00			
DURBIN-WATSON = 1.890		**Prob > F** = .00				**Prob > chi2** = .98			

We checked whether there is a correlation between cost components and education level, age group and activity relevance to farmers. There is a negative relation, significant at the 0.05 level between herbicide costs and farmer relevance to activity (Spearman's coefficient: = − .525; *p*-value = 0.001). Full-time farmers have lower herbicide costs, and the time spent by farmers on farm activity explains 28% herbicide costs. No other correlation was found.

3.3 Channel Marketing

Market risk is considered in literature as the most significant risk and is related to market process, price volatility and the marketing channel efficiency. Market risks can also arise from the loss of market access due to the loss of one element of the value chain, *i.e.* either the cooperative or the industry that is able to turn rice into a finished product.

To analyse market behaviour, respondents were asked about the market channel used to sell their rice, the reason for their choice and their degree of satisfaction. With respect to processing capacity, only one interviewee has the ability to transform and sell under his own brand. All other respondents depend on either the cooperative or the paddy rice processing industry. Considering the market channels, one of the farmers sells 80% of its yield to the retail sector, a single case, only possible because he has his own brand. The remaining 20% are sold to the industry. About 20 rice farmers (59% of the respondents) sell all their harvest straight to manufacturing and 11 (32%) of the interviewers sell their entire crop to the Cooperative of *Montemor-o-Velho*. Only three producers use two market channels simultaneously. The degree of satisfaction is higher towards the industry then to the cooperative and we observe that they are satisfied (average value: 2,93) with the market channel used.

These results match the results found by Székely and Pálinkás [17] who state that selling agricultural products through contracts with industry or cooperatives is less risky due to contractual factors. Selling products individually is probably the most profitable way to market them, especially when there is greater competition and the farmer has no bargaining power. Out of the 34 farmers interviewed, 32 respondents (representing 94% of the sample) answered a question concerning the reason underlying their market channel choice, in which each respondent could choose two options. As a result of this procedure, a set of 55 responses was obtained. About 38% respondents (in 32 respondents) chose the market channel because of "Flow ease production" and 34% because they have no processing capacity (in 32 respondents).

Regarding quality certification standards applied by the farmer, all respondents provided the same answer: they only have the certification for integrated production. "Arroz Carolino do Baixo Mondego" was certified as a product with "Protected Geographical Indication" in 2015 but none of the interviewed producers claimed to have PGI certification. All respondents answered the question on the reason for choosing the certification system: about 59% of farmers (in 34 farmers) chose this type of certification due to the "ease production flow", and 44% due to the "guarantee production flow". Ease flow and ensuring flow production are key factors for choosing the market channel and influenced the quality certification choice.

3.4 Main Concerns

The producers' concerns analysis provides an important explanation for the degree of risk perception. Farmers were presented a list of "worries" and asked to point out the degree of "worry" raised by issues. The degree of worry was presented on a 1 to 5 points Likert scale (Table 3). There were 33 responses to all items, except for the question "Flow production" which received 32 responses. Respondents indicated seven most important worries (values above the average: 3.21).

The most important concern was the weeds issue, followed by seed costs, which is easily understandable. Because production is made in an integrated production mode, the application and the type of herbicides used by farmers must comply with the regulation and they must apply certified seeds. Five out of the seven concerns are about the costs of paddy rice produced; market and environmental concerns, mainly climate changes, complete the list of issues. We found a positive correlation (at 0.05 level of significant) between education level and water issues, ($\rho = 0.375$), a negative correlation (at 0.05 level of significant) between class age level and water issues ($\rho = -0.412$), and a negative correlation (at 0.05 level of significant) between class age level and soil salinity ($\rho = -0{,}351$). Several correlations between production area and different worries were found (at 0.05 level of significant) such as water issues ($\rho = 0.388$); soil salinity ($\rho = 0.395$); seed quality ($\rho = -0.414$). We found (at 0.01 level of significant) a negative correlation between rice price producer and production area ($\rho = -0.520$).

This finding is interesting to be explored in future works because it goes against the economic theory of the supply model. Small farmers are more rice price sensitive.

Table 3. Degree of concerns by worries or issues

Degree	1. No worry		2. Little worry		3. Worry		4. Pretty worried		5. Very worried		Average	*Std, Desv* Value
Issues	N°	%	N°	%	N°	%	N°	%	N°	%		
Weeds	0	0.0	0	0.0	0	0.0	4	12.1	29	87.9	4.88	0.33
Seeds cost	0	0.0	1	0.0	2	6.1	9	27.3	22	66.7	4.61	0.61
Climate	0	0.0	0	0.0	3	9.1	11	33.3	19	57.6	4.48	0.67
Plagues and Diseases	1	3.0	0	0.0	2	6.1	11	33.3	19	57.6	4.42	0.87
Rice price producer	0	0.0	3	9.1	4	12.1	9	27.3	17	51.5	4.21	0.99
Fertilization	0	0.0	5	15.6	13	40.6	10	31.3	4	12.5	3.41	0.91
Machine and Fuel costs	0	0.0	5	15.2	17	51.5	7	21.2	4	12.1	3.30	0.88
Rice dryer	1	3.0	10	30.3	14	42.4	6	18.2	2	6.1	2.94	0.93
Labor availability	4	12.1	13	39.4	7	21.2	3	9.1	6	18.2	2.82	1.31
Labor cost	2	6.1	19	57.6	5	15.2	4	12.1	3	9.1	2.61	1.09
Machine availability	10	30.3	13	39.4	2	6.1	6	18.2	2	6.1	2.30	1.26
Seeds quality	0	0.0	1	3.0	3	9.1	10	30.3	19	57.6	2.30	0.79
Water	14	42.4	5	15.2	12	36.4	2	6.1	0	0.0	2.06	1.03
Flow production	8	24.2	21	63.6	1	3.0	1	3.0	2	6.1	2.03	0.98
Soil salinity	23	69.7	3	9.1	0	0.0	4	12.1	3	9.1	1.82	1.42

4 Discussion

Business diversification is a strategy used by farmers as risk protection. Mishra and El-Osta [18] suggest that diversification and farm size may be negatively correlated according to economic theory, but the present study provides evidence that farms receiving government payments are more diversified than others.

Diversification is the most widely used risk avoidance instrument and is based on risk spreading. The motivation for diversification is based on the principle that when one activity generates low revenue, other activities may be profitable. The multiple activity agricultural enterprise significantly reduces the possibility for local natural disasters to have a simultaneous negative impact on all activities and allows diverse income source on and off the farm. From an environmental point of view, diversifying agricultural production on a farm or agricultural region is one of the main tactics adopted by farmers to restrain the long-term changes induced by environmental changes.

About 50% respondents said rice farming is their main income source (≥50% of farm income) and for 24% it represents 100% of income. About 47% respondents have two and 29% have at least three agricultural activities, including rice, and 18% have non-agricultural income. According to the responses, all farmers surveyed received support from the agricultural policies, namely the Rural Development Program (RDP) over the past decade. However, there are three support policies where support requests were not submitted: support policies for organic farming; for diversification and finally for commercialization and processing. Farmers were satisfied or more than satisfied with the progress of the measures on their farms, and all respondents stated that they want this production system in the near future, at least during the next couple of years. In this question, respondents could justify their answer by choosing at most two main motives to maintain the agricultural activity. The financial reasons and family income balance represented 24% of the responses, respectively, but 22% stated that agricultural activity would only continue due of the nonexistence of profitable alternatives.

Regarding rice crop, only one farmer replied he/she does not intend to keep producing rice due to the crop's absence of economic sustainability. Approximately 89% of farmer want to maintain rice production, and 45% said they preserve this activity since there are no suitable agronomic systems available for their land, given the current agronomic and technological systems' availability for farmers in the region. Around 24.2% of the sampled producers consider that the culture knowledge is an important reason to keep rice crop. About of 21% of the farmers in this work believe that the economic viability justifies the option to continue cultivating rice.

Rice farmers are reasonably satisfied with the functioning of this farming system. The fact that they use marginal lands that cannot be used for other cultural systems is an important factor. Agri-environmental policies support their production decisions. The lack of development of an own brands is highlighted, but this fact can be justified by the high level of satisfaction they have with the industry and also with the cooperatives that supports it. The diversification of the crops makes it possible to reduce the economic risk in the most complex risks particularly in places with high pest incidence.

5 Conclusions

The work carried out an analysis of the rice production and marketing systems in the Mondego valley. This system is supported by RDP measures of which farmers are aware, benefit from and are satisfied with. The concern with production costs is clear, as well as the concern about rice price. Production risk is important and is related to the price of rice. Market risk is observed by the producer despite the reduction in volatility and price growth in the last five years. The choice of market channels has to do with rice flow output guarantees, but this is not a relevant concern. In terms of market risk, there is a difference between what farmers consider a price risk and the safest choice. Farmers are aware of the problems and try to find risk mitigation strategies, diversifying the activity, choosing the channel, controlling production costs and applying specific policies for Portuguese as well as agri-environmental measures.

Climate change issues and how these changes may affect crop production, such as rice that uses sensitive ecosystems, are important matters to be considered in European governance policies and in agricultural policy discussions in Europe.

The perception of farmers' risk in crops is vital to ensure food sovereignty and to guarantee national and European food self-sufficiency, namely in cereals. Food sovereignty and self-sufficiency are fundamental pillars in the implementation of agricultural policies for an environmentally and economically sustainable development of agri-food production.

Funding. This research was conducted within Project MEDWATERICE, PRIMA/0005/2018. This work is financed by national funds through FCT - Foundation for Science and Technology, I.P., within the scope of the project «Centre for Natural Resources, Environment and Society-CERNAS: UIDB/00681/2020».

References

1. Duong, T., Brewer, T., Luck, J., Zander, K.: A global review of farmers' perceptions of agricultural risks and risk management strategies. Agriculture **9**(1), 10 (2019)
2. Crane, L., Gantz, G., Isaacs, S., Jose, D., Sharp, R.: Introduction to Risk Management: Understanding Agricultural Risks. Extension Risk Management Education and Risk Management Agency (2013)
3. Ahsan, D., Roth, E.: Farmers' perceived risks and risk management strategies in an emerging mussel aquaculture industry in Denmark. Mar. Resour. Econ. **25**(3), 309–323 (2010)
4. Harwood, J., Heifner, R., Coble, K., Perry, J., Somwaru, A.: Managing risk in farming: concepts, research, and analysis. Agricultural Economic Report. No, 774 (1999)
5. Girdžiūtė, L.: Risks in agriculture and opportunities of their integrated evaluation. Procedia –Soc. Behav. Sci. **62**, 783–790 (2012)
6. Komarek, A., De Pinto, A., Smith, V.: A review of types of risks in agriculture: what we know and what we need to know. Agric. Syst. **178**, 102738 (2020)
7. Waage J., Mumford, J.: Agricultural biosecurity. Philos. Trans. R Soc. Lond. B Biol. Sci. **363**(1492), 863–876 (2008)
8. Keil, M., Wallace, L., Turk, D., Dixon-Randall, G., Nulden, U.: An investigation of risk perception and risk propensity on the decision to continue a software development project. J. Syst. Softw. **53**, 145–157 (2000)
9. Sjöberg, L.: World views, political attitudes and risk perception. Risk-Health, Saf. Environ. **9**, 137–152 (1998)
10. Hardaker, J., Huirne, R., Anderson, J.: Coping with risk in agriculture, Cab International, Wallingford, UK (1997)
11. OECD: Risks that Matter: Main Findings from the 2018 OECD Risks that Matter Survey (2019). https://www.oecd.org/social/risks-that-matter.htm
12. Winsen, W., Mey, Y., Ludwig, L., Passel, S., Vancauteren, M., Wauters, E.: Cognitive mapping: a method to elucidate and present farmers' risk perception. Agric. Syst. **122**, 42–52 (2013)
13. Dantsis, T., Douma, C., Giourga, C., Loumou, A., Polychronaki, E.: A methodological approach to assess and compare the sustainability level of agricultural plant production systems. Ecol. Ind. **10**, 256–263 (2010)
14. Vagias, W.: Likert-type scale response anchors, Clemson International Institute for Management. Clemson University (2006)
15. INE: Farm Structure Survey 2016, INE, I,P (2017)
16. Chartier, O., Cronin, E., Jongeneel, A., van Asseldonk, M., Bergevoet, M., Cordier, J., Bocci, M.: Study on risk management in EU agriculture: Final report. European Commission (2017)
17. Székely, C., Pálinkás, P.: Agricultural risk management in the european union and in the USA. Studies in Agricultural Economics, No. 109, pp. 55–72 (2009)
18. Mishra, A., El-Osta, H.: Risk management through enterprise diversification: a farm-level analysis. American Agricultural Economics Association, 28–31 July 2002

Urban Hydroponic Herb and Spice Garden

J. Miller, Z. Spinning, O. Swain, M. Sanchez, and A. Pourmovahed(✉)

Kettering University, Flint, MI 48504, USA
apourmov@kettering.edu

Abstract. Hydroponics and vertical farming are unconventional methods that forgo the typical large growing area, in favor of faster and more space efficient growing methods. Hydroponics, also known as soilless agriculture, is a method of farming that replaces standard soil with a nutrient rich solution in which the roots of plants are submerged. This method boasts much higher plant yield as the nutrient solution can be formulated specifically for each plant type. Vertical farming is the use of vertical elevation in farming allowing for a larger surface area of growth in confined spaces. Plants are grown on walls or in stacked rows like library books. These two methods are gaining increased popularity due to the need for faster and more space efficient farming methods to feed the rapidly growing world population. Everyday improvements in technology are making these formerly expensive methods cheaper and more efficient as plants can be monitored to ensure they stay in ideal conditions on an ongoing basis. The goal of this ***student project*** was to design, build, and test a self-sustaining hydroponic unit for the *urban gardener*. This unit will allow people to grow their own vegetables and herbs within the city. The hydroponic system was constructed and used to test the practicality of a small-scale gardening unit for home use. The system presented in this article can be used to grow the most expensive spice in the world - saffron - but the eleven-week time frame of this project was a limiting factor in testing the potential of this system to grow a bulb flower. The total cost of this project was USD 217, 45% of which was the cost of plexiglass.

Keywords: Urban farming · Hydroponics · Saffron

1 Introduction

Farming has been a crucial part of the survival of human beings for thousands of years. According to [1], 80% of the world's food comes from family farms. With a rapidly rising world population, an increase in urbanization and industrialization and a pandemic sweeping across the world, the ability to grow some food at home has become paramount.

Hydroponic farming does not require any soil. Instead, plant roots are immersed in a liquid nutrient solution and allowed to grow to maturity with little supervision. In a hydroponic system, nutrients can be tailored to specific plants. Hydroponic units can be used in any home to grow a variety of small plants such as leafy greens, herbs, and spices. Since the same water is circulated through the system with no run-off, these systems are highly water-efficient and particularly suitable in areas where the soil is

J. R. da Costa Sanches Galvão et al. (Eds.): *Proceedings of the 1st International Conference on Water Energy Food and Sustainability (ICoWEFS 2021)*, pp. 120–126, 2021.
https://doi.org/10.1007/978-3-030-75315-3_14

poor. Hydroponics will not completely remedy food shortages, but it can help to alleviate some of the stress in the food supply system. Hydroponics can bring farming into the city and facilitate "*Urban Farming* ". Under greenhouse conditions, fruits, vegetables, edible mushrooms, and algae can be produced year-round in building complexes. Having much greater control over the plants encountering fertilizers, pesticides and herbicides, hydroponics can make the path to becoming *certified organic* simpler. The *United States Department of Agriculture*, USDA specifies that to be certified an organic producer, "Any land used to produce raw organic commodities must not have had prohibited substances applied to it for the past three years" [2]. Hydroponic systems offer a great remedy to this as they do not "use the land" as a medium for plant growth.

While hydroponics is a relatively new farming method, *Vertical Farming* is an ancient method used by farmers in East Asia and South America dating back many centuries. With an ever-growing population, inefficiency of current farming methods, and exposure to pests, disease and chemicals plaguing the traditional farming methods, the introduction of hydroponics and vertical farming are the new waves of modern agriculture. Traditional and current farming methods have many disadvantages not faced by hydroponics and vertical farming. The use of natural soil means that a vast land area must be used to produce large amounts of food. Plants must be placed a certain distance apart to prevent the spreading roots from competing to share the nutrients. The large land area utilized is also associated with a steep number of labor hours, and the use of heavy machinery. Along with the many hours of run time on the heavy machinery comes the strong adverse impact on the environment caused by emissions and pollution. Finally, the use of natural bodies of land means that the crops are subject to pests and disease. Farmers counter these measures with harsh chemicals that keep pests and diseases away, but those chemicals are harsh on the environment and have side effects, like runoff into nearby bodies of water. Traditional farming methods are not plausible in urban areas since space is limited. However, hydroponics and vertical farming methods are perfect for those desiring to grow crops in urban areas.

One major advantage of hydroponic farming is its much higher crop yield per unit area. The compact nature of hydroponic farming coupled with a significant reduction in crop loss due to environmental factors, allow hydroponic farms to harvest more crops and for longer duration. According to [3], hydroponic lettuce production in Yuma, Arizona produced up to 11 times higher yield per unit area but required 82 times more energy compared to conventionally produced lettuce.

Growing plants in space for food is another important application of hydroponics. The *International Space Station* (ISS) requires periodic resupply missions to transport scientific equipment as well as food, water, and oxygen. These resources are needed to keep humans alive in space and take valuable space in these vessels. The cost to send food to the ISS is estimated at about $10,000 a pound [4]. Growing food in space using conventional methods and soil is nearly impossible but highly feasible with hydroponics. The principles can be applied to further human exploration and building settlements on other celestial bodies. Remaining in space for a prolonged duration will require a sustainable food supply. Having fresh food in space is beneficial for many reasons. First and foremost, growing food will allow for prolonged human habitation almost anywhere with the facilities to grow crops. This will allow humanity to explore further and further away from the Earth. Growing plants will also have positive psychological benefits on the humans growing them [5]. Early studies suggest that simple acts of taking care of plants while they grow is a stress reducing task for the humans involved while they are in the very stressful environment of space. All these benefits can be applied to both space exploration and settlement on other planets, moons, or asteroids. Without vertical farming and hydroponics, it will be impossible to set up a human colony somewhere like Mars or even the moon as they would require constant resupply missions that would have to be on time, every time or else the colonists would starve. Vertical farming is the best way to produce food for a new colony on another celestial body. This is because vertical farming is very compact, and space will be at a premium for a long time in this new colony. Most colonies will likely be underground to protect the settlers from solar radiation and provide some natural walls to build upon. This will work well with vertical farming as it can be put into a small space and still feed the colony. A small room full of vegetation could also help colonists as a place to decompress. As humans continue to stretch the bounds of space travel and with an ever-increasing world population and urban living, hydroponic farming may become a focal point of food production.

2 The Demonstration Unit

The hydroponic unit designed is intended for the health conscious who may not have access to adequate land for growing their own crops. The unit will allow them to control the way some of their food is grown from the comfort of their own home. Since most people living in urban areas do not have access to a lot of space, the unit was designed to fit in most windows. This unit will allow apartment dwellers to control the inputs to their plants such as pesticides, herbicides, and fertilizers. This will give them a level of comfort knowing exactly what went into their food.

The system will also be environmentally beneficial because it allows the production of vegetables, herbs, and spices with very little carbon footprint. A typical farm uses machinery for preparing the land, planting the seeds, irrigating, and harvesting the crops while relying on fuel as well as plastics and other materials used in packaging of the utilized resources. A well-designed hydroponic home garden requires very little maintenance and leaves a much smaller carbon footprint. As the plants are grown close to where they are consumed, no packaging or transportation will be required for the harvest. Once the system is primed, only a small amount of water is required to make up for evaporation and the water consumed by the plants.

The hydroponic unit is environmentally friendly since it conserves water and other resources. The tested proof-of-concept display system was designed and built for indoor testing. It uses a grow light and a pump powered by a standard 120-V outlet. An actual window unit will be outfitted with an exterior material like that used to construct most greenhouses. It will trap the energy from the sun as well as an optional solar panel and battery to power the water pump.

3 Saffron, a Potential Hydroponic Crop

Saffron is the most expensive and a highly prized and sought-after spice in the world. Because of its distinctive flavor, it is often added to many dishes particularly seafood, chicken, and rice. Saffron has been used for thousands of years as a culinary spice, a dye, perfume and most importantly medicine for therapeutic treatments as an antidepressant. Saffron is produced by handpicking and drying the delicate stigmas of "saffron crocus" flowers (Fig. 1).

Fig. 1. Saffron crocus flower [6]

Most of the world's saffron is grown and harvested in northeastern Iran. Other countries growing saffron include Spain, India, Greece, Azerbaijan, and Morocco. New Zealand, Italy, and the United States have also produced this spice on a small scale. Producing saffron is not a simple task. Hundreds of delicate flowers must be harvested early in the morning to produce a few grams of saffron. This includes the handpicking of the flowers, manual separation of the red stigmas followed by the drying process. According to [7], only 6 kg of saffron are typically harvested from every hectare by picking about 900,000 flowers by hand. The petals which make up much of the waste, can be processed through steam hydrodistillation to extract saffron essential oil and perfume. The main factors influencing the flowering of saffron include temperature, humidity during storage, and bulb size and weight [8]. The labor-intensive process of producing saffron has made it very costly. This provides an opportunity to cultivate this flower in a more controlled and cost-effective way.

4 Demonstration Unit Specifications

The system was designed to take water from a 5-gallon bucket, pump it through the 2-inch-diameter PVC pipe, and then return it into the bucket (Figs. 2 and 3). The bucket is large enough to make this system self-sustaining for up to 3 weeks. The user can test the quality of the water to ensure that it is at the proper PH and that there are ample nutrients in the water to supply the plants. The system also includes a grow light, biodegradable cups, and a timer with a switch to turn the system on and off. The timer turns the pump on for about 3–5 min 4 times a day. This will help reduce the system's carbon footprint while keeping the water moving for optimum plant growth. It will also allow nutrients added in the bucket to get to the plants in the system.

One initial design flaw was using the same size fittings for both the inlet and the outlet of the system. This was a cause for more water coming into the system than what could flow out which made the system nearly overflow through the 20 holes in the PVC. The solution to this was to make the outlet diameter much larger than the inlet diameter. In addition, the vertical height of the outlet fitting on the PVC is critical. The bottom of the fitting sets the height for the water within the PVC. If the height of water is too low, the water will not get to the roots of the plants; if it is too high, the water will spill out the 20 holes. Further models will require testing to optimize the vertical position of the fittings on the PVC. Initially, the PVC system was going to be at a slight downward angle towards the bucket, this proved to be an issue when the system was built because the holes closest to the bucket overflowed. Because of this, the PVC was placed flat on the frame of the window unit.

Fig. 2. Proof of concept display

Figure 4 shows the intended latching system for the unit. Although the concept model does not have such a latch, the CAD model shows how the latch will be installed in a window. The latch will automatically spring out from the bottom of the unit when pushed through the opening in the window. When the system needs to be removed, the latch can be reached through the bottom of the hydroponic unit for safe removal.

Fig. 3. Close up of the hydroponic system

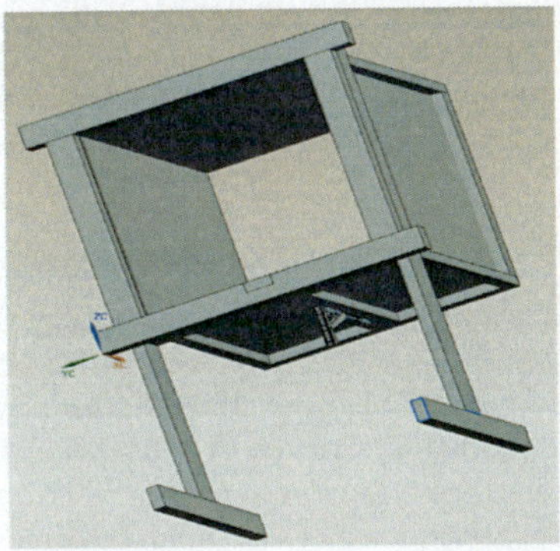

Fig. 4. 3-D CAD model depicting the latching system

An environmentally friendly, yet significantly lighter material will need to be used for the frame of the next model. Wood works well for display purposes but is quite heavy and would make installation in a window difficult.

The size of a window-mounted hydroponic system will depend on the application. While the display model was designed for some of the largest windows, scaling of this model would allow for use in smaller windows. It is estimated that this system can produce 15 kg of radishes per year while using 1 L of water and 0.4 kWh of energy per day. The total cost of this project was USD 217, 45% of which was the cost of plexiglass.

5 Conclusions

The unit built was successful in showing that growing radishes hydroponically in a short period of time is plausible. A substantial number of people live in cramped quarters lacking usable land to grow plants. The designed unit can assist those who wish to grow their own leafy greens, herbs and spices at home but do not have the available land or the time to tend to plants. A window-mounted design will allow apartment dwellers with access to a window to grow some of their own food. The design, as built, is not usable for most windows. However, it can be scaled to fit in any size window. An actual window unit also needs to be well insulated or it can lead to high heating or air conditioning bills for the house. Also, if the users were living in an area without much sunlight, like Seattle, the window mount could prove to be ineffective and the plants may not thrive. A hydroponic unit can also be used to liven up an office. The unit built and tested can yield about 15 kg of radishes per year, a significant harvest from a window sized hydroponic system. Herbs such as basil, mint and tarragon and many others have proven through the centuries that they can provide astonishing flavor along with therapeutics benefits. They are excellent candidates as plants for a hydroponic urban garden. These herbs are often pricey and can easily be grown in a window- mounted hydroponic garden all around the world. Traditional farming methods have proven tried and true for millennia and are likely not going to be phased out any time soon. However, as technology advances and the world population continue to increase, the idea of large-scale hydroponic farming operations becomes more of an ideal solution to aid the global food supply. Higher cost and the increased energy consumption of indoor farming can be offset by the ease of environmental control and the overall yield that hydroponic farming can achieve.

References

1. Arsenault, C.: Family farms produce 80 percent of world's food, speculators seek land. Thomson Reuters Foundation, 16 Oct. 2014. Accessed 18 Jan 2021
2. Becoming a Certified Operation, USDA Agricultural Marketing Service (usda.gov)
3. Barbosa, G., Gadelha, F., Kublik, N., Proctor, A., Reichelm, L., Weissinger, E., Wohlleb, G., Halden, R.: Comparison of land, water, and energy requirements of lettuce grown using hydroponic vs. conventional agricultural methods. Int. J. Environ. Res. Public Health **12**, 6879–91 (2015). https://doi.org/10.3390/ijerph120606879
4. Hirsch, J.: Space farming. The final frontier. Modern Farmer, 10 Sep. 2013. Accessed 22 Jan 2021
5. Nasa astronauts take first bites of lettuce grown in space: 'Tastes like arugula', | Space | The Guardian. Accessed 22 Jan 2021
6. Wikipedia, Saffron – Wikipedia. Accessed 22 Jan 2021
7. Souret, F.F., Weathers, P.J.: The growth of saffron (Crocus sativus L.) in aeroponics and hydroponics. J. Herbs Spices Med. Plants, **7**(3), 25–35 (2000). https://doi.org/10.1300/J044v07n03_04
8. Kordestani, M.: Hydroponic saffron cultivation and the effects of soil salinity, 4 Nov 2019. Hydroponic Saffron Cultivation and the Effects of Soil Salinity | by Milan Kordestani | Medium. Accessed 22 Jan 2021

Sand-Dune Plants from the Atlantic Coast of the Iberian Peninsula: Features and Applications

Tiago Parracho[1,2,3,4], Daniela C. Vaz[1,2], Paula Veríssimo[3], and Vânia Ribeiro[2,4](✉)

[1] Coimbra Chemistry Centre, University of Coimbra, 3004-535 Coimbra, Portugal
[2] CiTechCare, Centre for Innovative Care and Health Technology, School of Health Sciences, Polytechnic Institute of Leiria, 2411-901 Leiria, Portugal
vania.ribeiro@ipleiria.pt
[3] Centre for Neurosciences and Cell Biology, University of Coimbra, 3004-504 Coimbra, Portugal
[4] GeoBioTec, GeoBioSciences, GeoTechnologies and GeoEngineering, Earth Sciences Department, Universidade NOVA de Lisboa, 2829-516 Caparica, Portugal

Abstract. Plants with one or more consumable parts are considered edible. Although many plants have been classified as edible (about 27 thousand species), few are used as food. Nonetheless, to overcome food scarcity and excessive dependence on the same plant species, humans have always consumed wild plants, either through direct intake, or as spices, condiments, or oils. Thus, edible wild plants are part of a cultural and genetic heritage assigned to different geographical areas, as well as important sources of essential oils, antioxidants, vitamins, minerals, and special flavours. Therefore, edible wild plants have been the subject of a growing interest, not only due to their nutritional and medicinal value, but also as a way of diversifying eating habits and of promoting biodiversity and ecological sustainability.

The Atlantic coast of the Iberian Peninsula is characterized by rocky cliffs, sandy dunes, and maritime pine forests, where several halophyte wild plants flourish and thrive. Different endemic species such as *Corema album* ("Camarinha"), *Crithmum maritimum* ("Funcho-do-mar"), *Eryngium maritimum* ("Cardo-marítimo"), *Helichrysum italicum* ("Perpétua-das-areias") and *Otanthus maritimus* ("Cordeiros-da-praia") prosper in these sand-dune environments and are known to present important bioactive compounds, which also show relevant antioxidant, antimicrobial, and anti-inflammatory properties. Here, we review on the several special features and characteristics presented by this group of autochthonous plants, that show great potential, both in terms of agri-food applications, as well as in terms of cosmetics and other biotechnological uses.

Keywords: Edible wild plants · Sand-dune plants · Nutraceutical value · Review

J. R. da Costa Sanches Galvão et al. (Eds.): *Proceedings of the 1st International Conference on Water Energy Food and Sustainability (ICoWEFS 2021)*, pp. 127–136, 2021.
https://doi.org/10.1007/978-3-030-75315-3_15

1 Introduction

A sustainable use of edible wild plants can be a way overcoming food scarcity and varying the human diet, as well as of obtaining bioactive compounds, with nutritional and medicinal interest. The dune systems of the west coast of the Iberian Peninsula serve as habitat for several endemic species, (some of them with various edible parts, such as leaves, flowers, and/or fruits), that have been consumed by local inhabitants since remote ages but are still poorly described. Therefore, its crucial to carry out the proper characterization of these plants, in order to explore new applications and value their rational use and subsistence. Here, we describe some of the features presented by five sand-dune autochthonous plants of the Atlantic coast of the Iberian Peninsula: *Corema album (C. album), Crithmum maritimum (C. maritimum), Eryngium maritimum (E. maritimum), Helichrysum italicum (H. italicum)* and *Otanthus maritimus* (O. *maritimus)* that hold great potential for agri-food, biotechnological, cosmetic, and pharmaceutical applications (Table 1).

Table 1. Sand-dune plants features and biotechnological applications.

Species	Special features	Applications
C. album	Rich in minerals, essential oils, antioxidants (phenolic acids)	Agri-food, cosmetics, medicines
C. maritimum	Rich in minerals, essential oils, antioxidants, antimicrobial, insecticide, diuretic, digestive	Agri-food, aromatherapy, cosmetics, medicines
E. maritimum	Rich in essential oils, antioxidants, antimicrobial, anti-inflammatory	Cosmetics, medicines
H. italicum	Rich in essential oils, antioxidants, anti-inflammatory, antimicrobial, insecticide	Agri-food, aromatherapy, cosmetics, medicines
O. maritimus	Rich in essential oils, antioxidants, antimicrobial anti-inflammatory	Cosmetics, medicines

Fig. 1. Sand-dune plants from the Atlantic coast of the Iberian Peninsula: a) *Corema album*, b) *Crithmum maritimum L.*, c) *Eryngium maritimum*, d) *Helichrysum italicum* and e) *Otanthus maritimus*.

1.1 Corema Album

Corema album (*C. album*) is an edible plant (Fig. 1a), due to its fruit, the "Camarinha", which is a round pulpy berry characterized by a pleasant acid and fresh flavour [1–3]. The *Corema album* species is a perennial shrub of the *Ericaceae* family, found along the Atlantic coast of the Iberian Peninsula, at rocky cliffs, costal dunes, and their proximal maritime pine forests [2, 3]. The designation *Corema* (in Greek = Korema) which means "broom", has been attributed to this plant genus due to its numerous branches overly filled with leaves [2, 4]. *C. album* includes the subspecies: *C. album azoricum* originally from Azores, Portugal [5] and *C. album ssp album*, found at the Atlantic coast of the Iberian Peninsula [2–4]. *C. album* shrubs seed and grow specially in granular soils (70% sand) with at least 30% of clay, up to 2.6% of organic matter and a pH between 5.6 and 7.8 [3, 4] The fruit appears in the summer or early autumn, from July to October [5, 6].

The *Corema album* fruit is small and spherical, varies in diameter between 5 and 8 mm, presents a consistent and uniform epidermis and has been used as food supply in the Iberian Peninsula since Palaeolithic ages [1]. Fruit seeds have a thick and woody endocarp [3, 5] and are rich in omega-3 and omega-6 polyunsaturated fatty acids (PUFAs), such as, alpha-linolenic (C18:3 n-3) and linoleic (C18:2 n-6), respectively [7, 8]. Like other similar fruits, the *C. album* berry presents a high content in minerals and antioxidants, such as polyphenols [8–10], especially phenolic acids (e.g. chlorogenic acid, neochlorogenic acid, caffeic acid, p-hydroxybenzoic acid, p-coumaric acid and ferulic acid), flavonoids (e.g. myricetin 3-O-hexoside, quercetin 3-O-hexoside, rutin, quercetin 3-O-pentoside and Kaempferol 3-O-hexoside) and anthocyanins (e.g. delphinidin 3-O-hexoside, cyanidin 3-O-glucoside and cyanidin 3-Opentoside) [8, 11]. Thus, this fruit pulp has shown to present several nutraceutical properties such as, to protect against diseases related to oxidative stress, prevent urinary infections, reduce the risk of cardiovascular disease, provide neuroprotection and contribute to the prevention of cancer [7, 9, 10, 12]. Hence, given its composition, these berries can either be consumed as fresh or undergo processing, to produce lemonades, jams, appetizers, and liqueurs with acid flavours [1, 7, 9, 11, 13]. Seed, leaf, and fruit extracts can also be biotechnologically explored, to take advantage of the health beneficial properties associated to the PUFAs, polyphenols, essential oils, and minerals present in the plant.

1.2 Crithmum Maritimum

Crithmum maritimum, also known as "sea fennel" or "rock samphire" is an aromatic halophyte plant (Fig. 1b) of the *Apiaceae* family, 50 cm in height, with woody stems and succulent leaves [14]. *C. maritimum* can be found at rocky and sandy environments by the sea, especially in Azores, Madeira, the Canary Islands and coast of the Mediterranean and Black Seas [14]. The plant is known by the local communities for its medicinal properties, and for its use in cosmetics and in gastronomy [15–18].

Despite some edaphoclimatic differences, from a nutritional point of view, the plant can be a source of minerals (sodium, potassium, and calcium) and fibres [14–17] Additionally, *C. maritimum* flowers are rich in antioxidants and essential oils, specially γ-terpinene, limonene, sabinene and dillapiole; while seeds are rich in oleic, linoleic, and palmitic acids [15–19]. The beneficial properties of the plant are associated to anti-scorbutic, antibacterial (especially Gram (+)), carminative, diuretic, digestive, purgative and vermifuge effects [15, 16, 19, 20].

Regarding its applications, the plant is being used in industry for cosmetics manufacturing, due to its essential oils and antioxidants [15–20], and as a natural repellent, given its insecticide power [19]. Due to its antimicrobial effect, the plant can also be used as a natural food additive [16, 21] in the formulation of new mixtures of spices and herbs, since it can be a natural substitute of chloride sodium, given its salty flavour [18, 21]. Thus, the plant is widely used in the preparation of salads, herbal infusions, sauces, soups, and side dishes, to provide nutrients, colour and aroma.

1.3 Eryngium Maritimum

Eryngium maritimum L. (sea holly) is a rare perennial subshrub (Fig. 1c) of the Apiaceae family, usually flowering and fruiting in the second year of its growth [22, 23]. *E. maritimum* prospers on the coastal dunes of the Iberian Peninsula, Baltic Sea, Mediterranean basin, and Black Sea, and is a species considered endangered in some countries [22–24]. It normally grows in yellow dunes, where salinity and nitrogen availability are lower and where sand engulfment is stronger. Fruits are densely covered with spines and crowned by a persistent and spiny calyx [22, 23].

E. maritimum was traditionally used as diuretic, anti-inflammatory, and antinociceptive [22–25]. The plant is rich in essential oils, mainly hydrocarbon sesquiterpenes and oxygenated sesquiterpenes [26–29]. Germacrene D was found to be the dominant constituent in the fruits; while hexadecanoic acid, menthol, and menthone were the compounds most present in the roots; and 2,3,4-trimethyl benzaldehyde and germacrene D the most present in the leaves [23, 26]. Other features such as antimicrobial, antioxidant, anti-fibrotic, hepatoprotective and anti-inflammatory properties have been associated to the plant [25, 30–32]. The antibacterial activity has been confirmed against several bacteria [29–31], namely, Gram (+) bacteria: *Listeria monocytogenes, Propionibacterium acnes, Streptococcus bovis, Streptococcus pyogenes, Streptococcus dysgalactiae, Streptococcus pneumoniae*, and Gram (−) bacteria: *Echerichia coli, Pseudomonas aeruginosa, Pseudomonas fluorescens* and *Salmonella arizonae*, which has been associated to the sesquiterpenes present in the plant [29].

Plant extracts have also showed promising anticancer properties, as judged by the experiments conducted on human cell models [26, 33]. Thus, *E. maritimum* is considered a potential candidate to use as food, cosmetic or pharmaceutical ingredient. The plant has already been described in terms of glycerophospholipid composition [34], but additional analysis will have to be carried out, to completely characterize the plant. Nonetheless, *E. maritimum* shows great potential to be used in the preparation of cosmetics and medicines, due to the presence of high amounts of essential oils and antioxidants.

1.4 Helichrysum Italicum

Helichrysum italicum (*H. italicum*) is an aromatic shrub (Fig. 1d) of the Asteraceae family, 30 to 70 cm tall, blossoming yellow flowers with a strong curry-like aroma [35, 36]. The species is divided into six subspecies being the *H. italicum Franco picardii* the one mainly found at the Atlantic coast of the Iberian Peninsula [36–38]. This xerophytic plant, also known as "perpetual-of-the-sand", "perpetual-of-Italy", "Immortelle" and "everlasting" is adapted to dry, sandy and stony soils [35–38].

The beneficial properties of *H. italicum* leaves and flowers, that have been used in traditional medicine, relate to its antimicrobial properties, namely, antibacterial, antifungal, and antiviral [39–42]. Other health benefits are associated to antioxidant [35, 43–46], anti-inflammatory and antiallergic [35, 43], antispasmodic [47], antidiabetic [48] and diuretic properties [47]. Although the composition of the plant extracts may vary according to different extraction procedures, the bioactive compounds present in *H. italicum* are mostly phenolic compounds and terpenes [38]. Phenolic compounds include flavonoids, floroglucinois, acetophenones, chloroglucinols, tremetones, coumarins, coumarates, phenolic acids and esters [35, 38, 44, 45]. Terpenes include monoterpenes (αpinene, limonene, nerol, neryl acetate and neryl propylate) and sesquiterpenes (α-selinene, βselinene, ϒ-curcumene, trans-β-karyophylene and eudesm- 5-en-11-ol) [35, 38, 44, 45].

H. italicum essential oils have been used in aromatherapy and cosmetics as an anti-aging agent, as well as in the treatment of burns and erythema [43]. *H. italicum* extracts have demonstrated beneficial effects as a glycemic control agent [48] along with its antimicrobial properties (e.g., against *Staphylococcus aureus* [41], *Candida albicans* and *Acinetobacter baumanii* [39]). Thus, *H. italicum* leaves and flowers can be used in food industry, as a natural food additive and food preservative [49, 50], that also shows important natural insecticide properties, that can be used in agro-industry. Due to its characteristic curry aroma, flowers can also be used to season olives, salads, and meals and to flavor sweets, bakery products and soft drinks [35, 37, 45, 49, 50].

1.5 Otanthus Maritimus

Otanthus maritimus (L.) Hoffmans. & Link is an aromatic perennial herb (Fig. 1e) of the Asteraceae family, 15 to 30 cm tall, growing in maritime sands of the Atlantic coast of the Iberian Peninsula and Mediterranean basin [51–54]. Flowers of *O. maritimus* present a high content in monoterpene compounds, such as oxygenated monoterpenes, yomogi alcohol, camphor, artemisyl acetate and artemisia alcohol [51]. Roots of *O. maritimus* contain fatty acid amides, acetylene derivatives and sesquiterpenes hydrocarbons [51–53]. The plant has been traditionally used in decoctions as tonics, in the treatment of toothache, asthmatic bronchitis, and dysentery, due to its anti-inflammatory properties [51–53].

Essential oils obtained from *O. maritimus* flowers have shown antimicrobial properties. Strong antibacterial activities were detected both against Gram (+) (e.g.: *Staphylococcus aureus*) and Gram (−) bacterial strains (e.g.: *Campylobacter jejuni*, *Enterobacter cloacae*, *Pseudomonas aeruginosa*, *Salmonella typhi*) [52, 55]. Antifungal properties were described against pathogenic yeast (*Candida albicans*) and filamentous phytopathogenic fungi (*Botrytis cinerea* and *Rhyzoctonia solani*) [52]. Additionally, piperidinyl amides obtained from the aerial parts of the plant have also showed insecticidal and larvicidal activity [53, 54], explaining why dry specimens have been used traditionally as decorative elements, as well as a way of repelling insects from household areas. Also, essential oils extracted from the aerial parts of the plant (chrysanthenone isomer 1 and verbenyl acetate) have been able to produce cytotoxic effects against cancer cells, by inducing apoptosis in 2D and 3D models of cancer cell lines: colorectal cancer RKO cells and breast cancer MCF7 cells [26]. Thus, due to its anti-inflammatory, antimicrobial and antiproliferative properties, the aerial parts of *O. maritimus* may have important applications in the production of new cosmetics and pharmaceuticals. In addition, the insecticidal properties present in the plant, can also be applied to the production of natural repellents, environmentally friendly.

2 Conclusions

Sand-dune wild plants have been used by our ancestors as a way of overcoming food scarcity and need for therapies. However, due to the growing supply of food products and medicinal treatments, the benefits of these plants have been forgotten over the years. Nevertheless, given the beneficial properties of these species, that adapt to different climates and present relevant features, such as, a high composition in minerals, antioxidants, and essential oils, that are associated to antimicrobial (*i.e.*, antibacterial, antifungal, and antiviral) and anti-inflammatory properties, this coastline autochthonous biodiversity ought to be valued and well preserved. Additionally, since these plant materials can either be used as fresh or under low carbon and low-cost processing, this can be a sector for investment and future opportunities for eco-friendly and biodynamic farming. In addition to their promising medicinal (aromatherapy, cosmetics, pharmaceuticals) applications, these plants can be a source of nutrients and bioactive compounds, with possible agro-industrial and culinary uses. Various parts of the plants, plant extracts or compounds can be valued and used as aromatic herbs, food additives, food preservatives, organic fertilizers or incorporated in functional food packaging. In this sense, the relevance and need for further studies, mainly with the objective of properly characterize these plant systems and evaluate their safety for human use, is paramount. Moreover, these wild plants of the sand-dune systems of the Iberian Peninsula provide a reservoir of genetic variation, admirably adapted to the various challenges of climate change. Hence, the relevance of this study is also in line with the objectives of nature protection, ecosystem conservation and sustainable development of the Biodiversity Strategy for 2030.

References

1. Martínez-Varea, C., Ferrer-Gallego, P., Raigón, M., Badal, E., Ferrando-Pardo, I., Laguna, E., Real, C., Roman, D., Villaverde, V.: *Corema album* archaeobotanical remains in western Mediterranean basin. Assessing fruit consumption during Upper Palaeolithic in Cova de les Cendres (Alicante, Spain). Quatern. Sci. Rev. **207**, 1–12 (2019)
2. Álvarez-Cansino, L., Zunzunegui, M., Barradas, M., Esquivias, M.: Gender-specific costs of reproduction on vegetative growth and physiological performance in the dioecious shrub *Corema album*. Ann. Bot. **106**(6), 989–998 (2010)
3. Oliveira, P., Dale, A.: *Corema album* (L.) D. Don, the white crowberry - a new crop. J. Berry Res. **2**(3), 123–133 (2012)
4. Santos, M., Oliveira, C., Valdiviesso, T., Oliveira, P.: Effects of pretreatments on *Corema album* (L.) D. Don (subsp. *album*) seeds' germination. J. Berry Res. **4**(4), 183–192 (2014)

5. Calviño-Cancela, M.: Spatial patterns of seed dispersal and seedling recruitment in *Corema album* (*Empetraceae*): the importance of unspecialized dispersers for regeneration. J. Ecol. **90**, 775–784 (2002)
6. Zunzunegui, M., Díaz-Barradas, M.C., Jáuregui, J., Rodríguez, H., Álvarez-Cansino, L.: Season-dependent and independent responses of Mediterranean scrub to light conditions. Plant Physiol. Biochem. **102**, 80–91 (2006)
7. Martin, D., Marques, J., Amado, A., Barroca, M., Moreira da Silva, A., Carvalho, L., Marques, M.: Shedding light into the health-beneficial properties of *Corema album*—A vibrational spectroscopy study. J. Raman Spectrosc. **51**, 313–322 (2020)
8. Brito, C., Bertotti, T., Primitivo, M., Neves; M., Pires, C., Cruz, P., Martins, P., Rodrigues, A., Moreno, M., Brito, R., Campos, M., Vaz; D., Pessoa, M., Lidon; F., Reboredo, F., Ribeiro, V.: *Corema album* spp: edible wild crowberries with a high content in minerals and organic acids. Food Chem. **345**, 128732 (2021)
9. Andrade, S., Guiné, R., Gonçalves, F.: Evaluation of phenolic compounds, antioxidant activity and bioaccessibility in white crowberry (*Corema album*). J. Food Measur. Charact. **11**(4), 1936–1946 (2017)
10. León-González, A.J., Mateos, R., Ramos, S., Martín, M.Á., Sarriá, B., Martín-Cordero, C., López-Lázaro, M., Bravo, L., Goya, L.: Chemo-protective activity and characterization of phenolic extracts from *Corema album*. Food Res. Int. **49**(2), 728–738 (2012)
11. León-González, A.J., Truchado, P., Tomás-Barberán, F.A., López-Lázaro, M., Barradas, M., Martín-Cordero, C.: Phenolic acids, flavonols and anthocyanins in *Corema album* (L.) D. Don berries. J. Food Compos. Anal. **29**(1), 58–63 (2013)
12. Macedo, D., Tavares, L., McDougall, G., Miranda, H., Stewart, D., Ferreira, R., Tenreiro, S., Outeiro, T., Santos, C.: (Poly)phenols protect from α-synuclein toxicity by reducing oxidative stress and promoting autophagy. Human Mol. Genet. **24**(6), 1717–1732 (2015)
13. Barroca, M.J., Silva, A.M.: Chapter 7 - From folklore to the nutraceutical world: the *Corema album potential*. In: Cockle, C. (ed.) Gastronomy and Food Science, p 119–135. Elsevier Inc., The Netherlands (2021). ISBN: 978-0-12-820057-5
14. Arafeh, R.M.: Molecular phylogeography of the European coastal plants *Crithmum maritimum* L., *Halimione portulacoides* (L.) Aellen, *Salsola kali* L. and *Calystegia soldanella* (L.) R. Br. Johannes Gutenberg-Universität, Mainz (2005)
15. Nabet, N., Boudries, H., Chougui, N., Loupassaki, S., Souagui, S., Burló, F., Hernández, F., Carbonell-Barrachina, A., Madani, K., Larbat, R.: Biological activities and secondary compound composition from *Crithmum maritimum* aerial parts. Int. J. Food Prop. **20**(8), 1843–1855 (2017)
16. Senatore, F., Napolitano, F., Ozcan, M.: Composition and antibacterial activity of the essential oil from *Crithmum maritimum* L. (Apiaceae) growing wild in Turkey. Flavour Fragr. J. **15**(3), 186–189 (2000)
17. Pateira, L., Nogueira, T., Antunes, A., Venâncio, F., Tavares, R., Capelo, J.: Two chemotypes of *Crithmum maritimum* L. from Portugal. Flavour Fragr. J. **14**(5), 333–343 (1999)
18. Petropoulos, S., Karkanis, A., Martins, N., Ferreira, I.: Edible halophytes of the Mediterranean basin: potential candidates for novel food products. Trends Food Sci. Technol. **74**, 69–84 (2018)
19. Tsoukatou, M., Tsitsimpikou, C., Vagias, C., Roussis, V.: Chemical Intra-Mediterranean variation and insecticidal activity of *Crithmum maritimum*. Zeitschrift Für Naturforschung C **56**(3–4), 211–215 (2001)
20. Mekinić, I., Blažević, I., Mudnić, I., Burčul, F., Grga, M., Skroza, D., Jerčić, I., Ljubenkov, I., Boban, M., Miloš, M., Katalinić, V.: Sea fennel (*Crithmum maritimum* L.): phytochemical profile, antioxidative, cholinesterase inhibitory and vasodilatory activity. J. Food Sci. Technol. **53**(7), 3104–3112 (2016)

21. Romojaro, A., Botella, M., Obón, C., Pretel, M.: Nutritional and antioxidant properties of wild edible plants and their use as potential ingredients in the modern diet. Int. J. Food Sci. Nutr. **64**(8), 944–952 (2013)
22. Savchenko, K., Carris, L., Castlebury, L., Heluta, V., Wasser, S., Nevo, E.: Revision of *Entyloma* (Entylomatales, Exobasidiomycetes) on *Eryngium*. Mycologia **106**(4), 797–810 (2014)
23. Clausing, G., Vickers, K., Kadereit, J.: Historical biogeography in a linear system: genetic variation of sea rocket (*Cakile maritima*) and sea holly (*Eryngium maritimum*) along European coasts. Mol. Ecol. **9**(11), 1823–33 (2000)
24. Kikowska, M., Kalemba, D., Dlugaszewska, J., Thiem, B.: Chemical composition of essential oils from rare and endangered species- Eryngium maritimum L. and E. alpinum L. Plants **9**(4), 417 (2020)
25. Küpeli, E., Kartal, M., Aslan, S., Yesilada, E.: Comparative evaluation of the anti-inflammatory and antinociceptive activity of Turkish *Eryngium* species. J. Ethnopharmacol. **107**(1), 32–37 (2006)
26. Beeby, E., Magalhães, M., Poças, J., Collins, T., Lemos, M., Barros, L., Ferreira, I., Cabral, C., Pires, I.: Secondary metabolites (essential oils) from sand-dune plants induce cytotoxic effects in cancer cells. J. Ethnopharmacol. **258**, 112803 (2020)
27. Lajnef, H., Ferioli, F., Pasini, F., Politowicz, J., Khaldi, A., D'Antuono, F., Caboni, M., Nasri, N.: Chemical composition and antioxidant activity of the volatile fraction extracted from air-dried fruits of Tunisian *Eryngium maritimum* L. ecotypes. J. Sci. Food Agric. **98**(2), 635–643 (2018)
28. Kowalczyk, M., Masullo, M., Thiem, B., Piacente, S., Stochmal, A., Oleszek, W.: Three new triterpene saponins from roots of *Eryngium planum*. Nat. Prod. Res. **28**(9), 653–60 (2014)
29. Darriet, F., Bendahou, M., Desjobert, J., Costa, J., Muselli, A.: Bicyclo [4.4.0] decane oxygenated sesquiterpenes from *Eryngium maritimum* essential oil. Planta Medica **78**(4), 386–389 (2012)
30. Traversier, M., Gaslonde, T., Lecso, M., Michel, S., Delannay, E.: Comparison of extraction methods for chemical composition, antibacterial, depigmenting and antioxidant activities of *Eryngium maritimum*. Int. J. Cosmet. Sci. **42**(2), 127–135 (2020)
31. Meot-Duros, L., Floch, G., Magné, C.: Radical scavenging, antioxidant and antimicrobial activities of halophytic species. J. Ethnopharmacol. **116**(2), 258–62 (2008)
32. Lisciani, R., Fattorusso, E., Surano, V., Cozzolino, S., Giannattasio, M., Sorrentino, L.: Anti-inflammatory activity of Eryngium maritimum L. rhizome extracts in intact rats. J. Ethnopharmacol. **12**(3), 263–270 (1984)
33. Yurdakök, B., Baydan, E.: Cytotoxic effects of Eryngium kotschyi and Eryngium maritimum on Hep2, HepG2, Vero and U138 MG cell lines. Pharm. Biol. **51**(12), 1579–85 (2013)
34. Zitouni, M., Wewer, V., Dörmann, P., Abdelly, C., Youssef, N.: Quadrupole time-of-flight mass spectrometry analysis of glycerophospholipid molecular species in the two halophyte seed oils: *Eryngium maritimum* and *Cakile maritima*. Food Chem. **213**, 319–328 (2016)
35. Viegas, D., Palmeira-de-Oliveira, A., Salgueiro, L., Martinez-de-Oliveira, J., Palmeira-de-Oliveira, R.: *Helichrysum italicum*: from traditional use to scientific data. J. Ethnopharmacol. **151**(1), 54–65 (2014)
36. Hassan, M., Pacurar, A., Gaspar, A., Vicente, O., Boscaiu, M.: Growth and reproductive success under saline conditions of three Plantago species with different levels of stress tolerance. Notulae Botanicae, Horti Agrobotanici, Cluj-Napoca **42**(1), 180–186 (2014)
37. Romano, A., Gonçalves, S.: Plantas silvestres comestíveis do Algarve. Faro: Universidade do Algarve (2015). ISBN 978-989-8472-64-9

38. Pereira, C.G., Barreira, L., Bijttebier, S., Pieters, L., Neves, V., Rodrigues, M.J., Rivas, R., Varela, J., Custodio, L.: Chemical profiling of infusions and decoctions of *Helichrysum italicum subsp. picardii* by UHPLC-PDA-MS and *in vitro* biological activities comparatively with green tea (*Camellia sinensis*) and rooibos tisane (*Aspalathus linearis*). J. Pharm. Biomed. Anal. **145**, 593–603 (2017)
39. Oliva, A., Garzoli, S., Sabatino, M., Tadić, V., Costantini, S., Ragno, R., Božović, M.: Chemical composition and antimicrobial activity of essential oil of *Helichrysum italicum* (Roth) G. Don fil. (Asteraceae) from Montenegro. Nat. Prod. Res. **34**(3), 445–448 (2020)
40. Djihane, B., Wafa, N., Elkhamssa, S., Pedro, D., Maria, A., Mihoub, Z.: Chemical constituents of *Helichrysum italicum* (Roth) G. Don essential oil and their antimicrobial activity against Gram-positive and Gram-negative bacteria, filamentous fungi and *Candida albicans*. Saudi Pharm. J. **25**(5), 780–787 (2017)
41. D'Abrosca, B., Buommino, E., Caputo, P., Scognamiglio, M., Chambery, A., Donnarumma, G., Fiorentino, A.: Phytochemical study of *Helichrysum italicum* (Roth) G. Don: Spectroscopic elucidation of unusual amino-phlorogucinols and antimicrobial assessment of secondary metabolites from medium-polar extract. Phytochemistry **132**, 86–94 (2016)
42. Nostro, A., Bisignano, G., Cannatelli, M., Crisafi, G., Germano, M., Alonzo, V.: Effects of *Helichrysum italicum* extract on growth and enzymatic activity of *Staphylococcus aureus*. Int. J. Antimicrob. Agents **17**(6), 517–520 (2001)
43. Fraternale, D., Flamini, G., Ascrizzi, R.: *In Vitro* Anticollagenase and Antielastase activities of essential oil of *Helichrysum italicum* subsp. *italicum* (Roth) G. Don. J. Med. Food **22**(10), 1041–1046 (2019)
44. Maksimovic, S., Tadic, V., Skala, D., Zizovic, I.: Separation of phytochemicals from *Helichrysum italicum*: An analysis of different isolation techniques and biological activity of prepared extracts. Phytochemistry **138**, 9–28 (2017)
45. Mari, A., Napolitano, A., Masullo, M., Pizza, C., Piacente, S.: Identification and quantitative determination of the polar constituents in *Helichrysum italicum* flowers and derived food supplements. J. Pharm. Biomed. Anal. **96**, 249–255 (2014)
46. Melito, S., Petretto, G., Podani, J., Foddai, M., Maldini, M., Chessa, M., Pintore, G.: Altitude and climate influence *Helichrysum italicum* subsp. *microphyllum* essential oils composition. Ind. Crops Prod. **80**, 242–250 (2016)
47. Rigano, D., Formisano, C., Senatore, F., Piacente, S., Pagano, E., Capasso, R., Borrelii, F., Izzo, A.: Intestinal antispasmodic effects of *Helichrysum italicum* (Roth) Don ssp. *italicum* and chemical identification of the active ingredients. J. Ethnopharmacol. **150**(3), 901–906 (2013)
48. Garza, A., Etxeberria, U., Lostao, M., Román, B., Barrenetxe, J., Martínez, J., Milagro, F.: *Helichrysum* and grapefruit extracts inhibit carbohydrate digestion and absorption, improving postprandial glucose levels and hyperinsulinemia in rats. J. Agric. Food Chem. **61**(49), 12012–12019 (2013)
49. Karača, S., Trifković, K., Bušić, A., Đorđević, V., Belščak-Cvitanović, A., Cebin, A.V., Bugarski, B., Komes, D.: The functional potential of immortelle (*Helichrysum italicum*) based edible films reinforced with proteins and hydrogel particles. LWT **99**, 387–395 (2019)
50. Kim, Y.S., Ahn, E.S., Shin, D.H.: Extension of shelf life by treatment with allyl isothiocyanate in combination with acetic acid on cooked rice. J. Food Sci. **67**(1), 274–279 (2002)
51. Ruiu, S., Anzani, N., Orrù, A., Floris, C., Caboni, P., Maccioni, E., Distinto, S., Alcaro, S., Cottiglia, F.: N-Alkyl dien- and trienamides from the roots of *Otanthus maritimus* with binding affinity for opioid and cannabinoid receptors. Bioorg. Med. Chem. **21**(22), 7074–82 (2013)
52. Basile, A., Rigano, D., Sorbo, S., Conte, B., Rosselli, S., Bruno, M., Senatore, F.: Antibacterial and antifungal activities of *Otanthus maritimus* (L.) Hoffmanns. & Link essential oil from Sicily. Nat. Prod. Res. **27**(17), 1548–55 (2013).

53. Christodoulopoulou, L., Tsoukatou, M., Tziveleka, L., Vagias, C., Petrakis, P., Roussis, V.: Piperidinyl amides with insecticidal activity from the maritime plant *Otanthus maritimus*. J. Agric. Food Chem. **53**(5), 1435–9 (2005)
54. Pavela, R.: Larvicidal effects of various Euro-Asiatic plants against *Culex quinquefasciatus* Say larvae (Diptera: Culicidae). Parasitol. Res. **102**(3), 555–9 (2008)
55. Muselli, A., Rossi, P., Desjobert, J., Bernardini, A., Berti, L., Costa, J.: Chemical composition and antibacterial activity of *Otanthus maritimus* (L.) Hoffmans. & Link essential oil from Corsica. Flavour Fragr. J. **22**, 217–223 (2007)

Comparison of Chemical Parameters in Zinc Biofortified Flours of *Triticum aestivum* L.: Development of a Functional Food

Inês Carmo Luís[1,2(✉)], Ana Coelho Marques[1,2], Cláudia Campos Pessoa[1,2], Diana Daccak[1,2], Ana Rita F. Coelho[1,2], Fernando C. Lidon[1,2], Manuel Patanita[2,3], José Dôres[3], Manuela Simões[1,2], Ana Sofia Almeida[2,4], Ana Sofia Bagulho[2,4], José Moreira[2,4], Maria Fernanda Pessoa[1,2], Maria Manuela Silva[2,5], Fernando Henrique Reboredo[1,2], Paulo Legoinha[1,2], Paula Scotti Campos[2,6], Isabel P. Pais[2,6], and José C. Ramalho[2,7]

1 Earth Sciences Department, Faculdade de Ciências e Tecnologia, Universidade Nova de Lisboa, Campus da Caparica, 2829-516 Caparica, Portugal
idc.rodrigues@campus.fct.unl.pt

2 GeoBioTec Research Center, Faculdade de Ciências e Tecnologia, Universidade Nova de Lisboa, Campus da Caparica, 2829-516 Caparica, Portugal

3 Escola Superior Agrária, Instituto Politécnico de Beja, R. Pedro Soares S/N, 7800-295 Beja, Portugal

4 Instituto Nacional de Investigação Agrária e Veterinária, I. P. (INIAV), Estrada de Gil Vaz 6, 7351-901 Elvas, Portugal

5 ESEAG-COFAC, Avenida do Campo Grande 376, 1749-024 Lisboa, Portugal

6 Instituto Nacional de Investigação Agrária e Veterinária, I. P. (INIAV), Avenida da República, Quinta do Marquês, 2780-157 Oeiras, Portugal

7 PlantStress & Biodiversity Lab, Centro de Estudos Florestais (CEF), Instituto Superior Agronomia (ISA), Universidade de Lisboa (ULisboa), Quinta do Marquês, Av. República, 2784-505 Oeiras, Portugal

Abstract. In the next 30 years a sharp increase in the world population is expected, which determines the need of an increasing food production, linked to a high nutritional pattern, to ensure human health. In this context, as bread wheat is one of the most produced cereals worldwide, its mineral enrichment through agronomic biofortification is of great interest. Under this framework, considering that zinc deficiency triggers some pathologies in the human population, this study presents an agronomic workflow for Zn enrichment of bread wheat and aimed to evaluate the technological quality of the related flours (whole and refined bread wheat flours), to create an innovative product, with added value and capable of supplying nutritional deficiencies in zinc. Tests were carried out after three foliar applications of Zn-EDTA, in three different concentrations, in Paiva and Roxo varieties of *Triticum aestivum* L. Therefore, an analytical determination of ash and moisture contents, Zn, Ca and K on bread wheat flours and scanning colorimetry were carried out. Whole bread wheat flours submitted to the workflow of Zn enrichment revealed, relatively to refined flours, higher contents of ash, Zn, Ca and K. Nevertheless, in the scanning colorimetric analysis, two transmittance peaks were recorded at

J. R. da Costa Sanches Galvão et al. (Eds.): *Proceedings of the 1st International Conference on Water Energy Food and Sustainability (ICoWEFS 2021)*, pp. 137–146, 2021.
https://doi.org/10.1007/978-3-030-75315-3_16

550 and 650 nm in the regions of the visible spectrum, in which the whole flour presented lower values. The obtained data is discussed, being concluded that whole wheat flours resulted in flours nutritionally more interesting, allowing the design of innovative and functional foods produced from Zn enriched flours with added value.

Keywords: Agronomic biofortification · Bread wheat flours · Functional food · Zinc

1 Introduction

An increasing demand for production of staple foods stands, as there are projections that point to a growth in the world population. In 2019 human population reached about 7.7 billion, being expected to reach around 8.5, 9.7 and 10.9 billion of people in 2030, 2050 and 2100, respectively [1]. Yet, WHO declared, on 11th of March of 2020, COVID-19 as a pandemic, in which this virus has already spread to more than 200 countries [2]. There is still no data regarding births being affected by this virus, nevertheless, the projections of population growth might be affected [3].

Independently of this pandemic crisis, it is estimated that world wheat will reach, in 2020/2021, a production of 761.7 million tons and a utilization of 757.6 million tons, in which the world cereal will reach 2741.7 and 2744.4 million tons of production and utilization, respectively [4].The consumption of whole wheat has a myriad of benefits for human health as it contains phytochemicals (flavonoids, carotenoids, polyphenols, phenolic acids and others), dietary fibers, vitamins and minerals. These components help in reducing obesity, cancer, cardiovascular disease, and type II diabetes. What differs the whole wheat flour form from refined wheat flour is the fact that the first one presents the outer layer of the kernel where those components mentioned above are more concentrated in [5, 6]. On the other hand, zinc (Zn) deficiencies trigger some human health problems linked to loss of brain functions, weakening of the immune system and some troubles related to physical growth. Nevertheless, this mineral has a fundamental role at regulatory, functional, and structural levels in the human body and interacts with a diversity of proteins and enzymes [7, 8]. Therefore, biofortification of Zn can be a strategy to enhance this mineral content or even to make it available in the edible parts of staple crops, during the plant growth [8–11]. This strategy has been implemented in a large amount of staple foods namely, maize, rice, bread wheat, bean, potato, pear, spinach, broccoli, mango and others [12–18].

Bearing in mind the importance of increasing the Zn levels in bread wheat, this study aims to compare and asses the chemical composition and technological quality of whole and refined bread wheat flours biofortified in Zn, in order to create a functional food with added value and also capable of diminish Zn nutritional deficiencies.

2 Materials and Methods

2.1 Experimental Field

Two Portuguese varieties of *Triticum aestivum* L., namely Roxo and Paiva, were cultivated in Beja, Portugal (at 37°57′27.59"N; 8°08′05.50"W). Sown took place at 30

December of 2018 with a rate of 350 seeds/m^2 and harvest occurred on 12 July of 2019. Climate during plant cycle was characterized by an average of maximum temperature of 22 °C and an average of minimum of 11 °C. During this period, the maximum and minimum averages of air humidity ranged between 69% and 11%, respectively, and the total rainfall accumulation was about 5.43 mm (with a daily maximum of 1.85 mm). The agronomic biofortification of wheat experiment comprised zinc foliar application on three different moments, which occurred during April and May. Zn-EDTA was applied with three different concentrations, distributed between three moments (0 – control (P0Tr and R0Tr), 6.3 (P1Tr and R1Tr) and 12.60 (P2Tr and R2Tr) kg.ha^{-1}) and with 46% urea. Before sowing, the field was fertilized with 50 kg Zn.ha^{-1} and with classical NPK fertilizer. The field was sown in a randomized block design with four repetitions, where the field presented 24 plots, with an area of 12 m^2 (10 m × 1.2 m), comprising 0.4 m between plots and 3 m between repetition. The field harvest was conducted using a plot harvester combine (Hege), recovering all the produced grain. After harvest, all the grains from the different plots were stored in bags for vacuum sealing and placed in the freezer at −20 °C, separating the different plots.

2.2 Analysis of Whole and Refined Bread Wheat Flours

The whole wheat flour was obtained by milling grains in the Disintegrator MPD-102 Biobase Biodustry Co., Ltd (Shandong) mill. On the other hand, the grains were milled by Chopin CD1 mill (Group Tripette & Renaud, France) to obtain refined flour. Subsequently, both whole and refined flours were stored in sealed bags in the freezer at − 20 °C.

The ash content (%) in the flours (whole and refined) was determined according to [15, 19]. Five grams of each sample, in triplicate, were weighted and placed in a muffle, for 2 h, at 900 °C, until total incineration of the sample. Thereafter, the samples were placed in a desiccation until it reached room temperature, weighted, and the ash content determined.

Determination of the moisture content (%) for the flours (whole and refined) was carried out according to [20, 21]. About five grams of each sample was weighted, in triplicate, and the samples were placed in the oven for about 90 min at 130 °C, and then left to cool in a desiccation until it reached room temperature. Then, the samples were weighted, and the moisture content was determined.

The colorimetric parameters were determined in the flour samples in triplicate, using an Agrosta scanning spectrophotometric colorimeter, which covers practically the entire visible region of the electromagnetic spectrum (shows measurements at wavelengths 450, 500, 550, 570, 600 and 650 nm). The light was provided by a white LED light [12, 22]. The main purpose of the colorimetric analysis was to understand if this analysis can be used to differentiate whole and refined flours, as well as, to distinguish varieties. Moreover, it is of major importance to verify if Zn biofortification affects color parameters.

The zinc content in the whole and refined wheat flours were determined, using an XRF analyzer (model XL3t 950 He GOLDD+) with helium atmosphere [15, 23]. Measurements were carried out in triplicates for each sample of whole and refined wheat flours, with a radiation emission of 180 s. The software NITON Data Transfer (XL 3t-36653) was used for data analysis.

2.3 Statistical Analysis

Data were statistically analyzed using software R (R version 3.6.3.). Statistical analysis included One-Way ANOVA ($p \leq 0.05$) to assess differences between samples with different concentrations of the same treatment. A Tukey's test for mean comparison was performed, based on the results, considering a 95% confidence level. The four biological replicates in the field were analyzed in triplicate.

3 Results

Whole wheat flour presented about four times higher ash contents than refined wheat flour (Fig. 1). There were no significant differences between both varieties, regardless the type of flour for the ash content. There was no direct influence between the increase of the Zn concentration applied and the ash content.

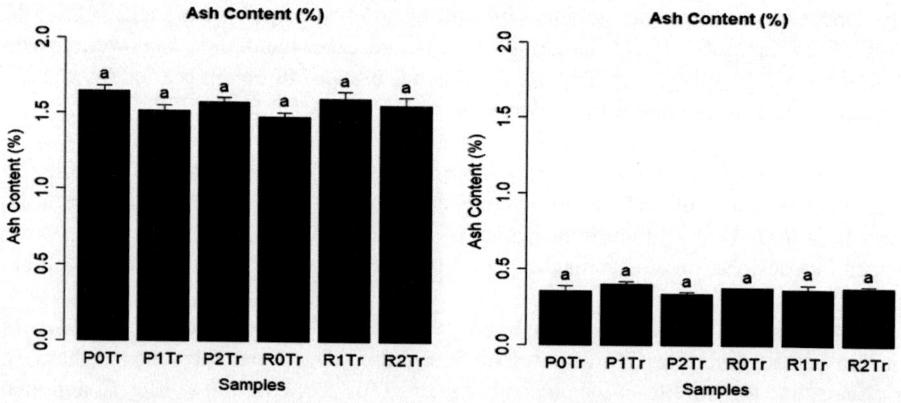

Fig. 1. Ash Content (%) of whole wheat flour (on the left) and of refined wheat flour (on the right) of *Triticum aestivum* L., varieties Roxo and Paiva (n = 3). With the foliar application of Zn-EDTA (Tr): P corresponds to Paiva and R to Roxo; 0 = control; 1 and 2 correspond respectively, to initial and upper concentration. Letter a indicate no significant differences among treatments for both varieties ($p \leq 0.05$).

Relatively to moisture content, despite the significant differences found among treatments for the same type of wheat flour, the results showed only minor deviations for the whole wheat flour (Table 1). Regarding whole wheat flour, in both varieties the higher values correspond to the higher application of the fertilizer. However, in refined wheat flour R1Tr has the higher values and P0Tr and P2Tr did not diverge.

Table 1. Moisture Content (%) of whole wheat flour and of refined wheat flour of *Triticum aestivum* L., varieties Roxo and Paiva (n = 3). With the foliar application of Zn-EDTA (Tr): P corresponds to Paiva and R to Roxo; 0 = control; 1 and 2 correspond respectively, to initial and upper concentration. Letters a, b, c indicates significant differences among treatments for both varieties ($p \leq 0.05$).

Samples	Whole wheat flour	Refined wheat flour
P0Tr	$8{,}61 \pm 0{,}0348^{b}$	$9{,}05 \pm 0{,}1804^{b}$
P1Tr	$8{,}21 \pm 0{,}0686^{c}$	$5{,}58 \pm 1{,}69^{c}$
P2Tr	$9{,}07 \pm 0{,}0407^{a}$	$8{,}58 \pm 0{,}126^{bc}$
R0Tr	$8{,}82 \pm 0{,}0632^{ab}$	$10{,}5 \pm 0{,}0958^{ab}$
R1Tr	$8{,}75 \pm 0{,}01101^{b}$	$13{,}7 \pm 0{,}0377^{a}$
R2Tr	$8{,}88 \pm 0{,}0996^{ab}$	$8{,}69 \pm 0{,}0922^{bc}$

In general, the whole flour presented lower values than refined wheat flour, in all bands (from 450 to 650 nm) (Fig. 2). Differences were found essentially, in the peaks at 550 (green region) and 650 nm (red region), between both different flours. In the whole wheat flour, these two peaks remained slightly lower than refined wheat flour. Nevertheless, the 550 nm peak is lower than 650 nm for the whole wheat flour, whereas refined wheat flour is just the opposite. Moreover, Roxo variety showed slightly higher values for both flours. There is no direct influence between the increase of the Zn concentration applied and this analysis.

Whole wheat flour presented higher values for the three minerals (Ca, K and Zn), specially for Zn (Fig. 3). Zn has the higher values for both types of flours, followed by K and Ca. For Ca and K there were no significant differences among treatments in both flours. However, Ca and K tended to show higher values when applying the higher concentration of Zn for whole and refined wheat flours (in particularly with Paiva). Relatively to Zn content in both flours, as the applied fertilizer concentration increases, the greater the Zn content in the sample, with Roxo variety presenting higher values.

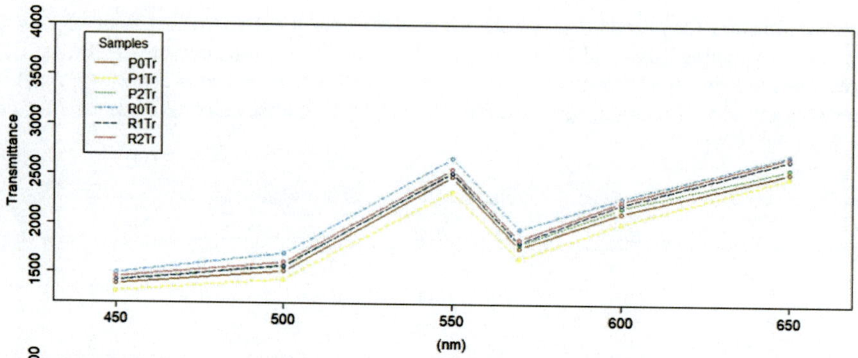

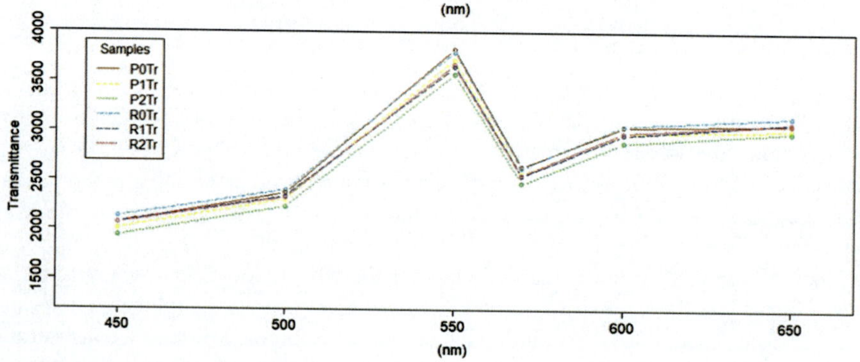

Fig. 2. Colorimetric transmittance, in the visible spectrum region between 450 and 650 nm, of whole wheat flour (above, with n = 12) and of refined wheat flour (below, with n = 9) of *Triticum aestivum* L., varieties Roxo and Paiva. With the foliar application of Zn-EDTA (Tr): P corresponds to Paiva and R to Roxo; 0 = control; 1 and 2 correspond respectively, to initial and upper concentration.

4 Discussion

The Ca, K and Zn contents of both types of flour suggest that industrial processing of wheat flour transform it in a product less nutritionally complete, as it removes the layers where these minerals preferably accumulate [24, 25] (Fig. 3). The higher values of ash content may indicate that the samples have higher inorganic content (*i.e.* minerals), which was observed in our study, as the whole wheat flour presented higher values of mineral contents and of ash (Fig. 1 and 3). Nevertheless, it was not possible to verify a direct correlation between the increase of Zn foliar application with the ash content as previously found in some other studies [26]. Probably this is due to the interactions between minerals and even with the fact that both genotypes (Paiva and Roxo) have different physiological responses [27–29]. Additionally, some studies report that Zn has an antagonist interaction with Ca and a synergetic interaction with K [27, 28, 30], however, in this study Ca and K suggested both a synergetic interaction with Zn, even though it was slight.

Some studies revealed that moisture content can be affected by variations in temperature or even inappropriate storage conditions of flours [31] (Table 1). It was found

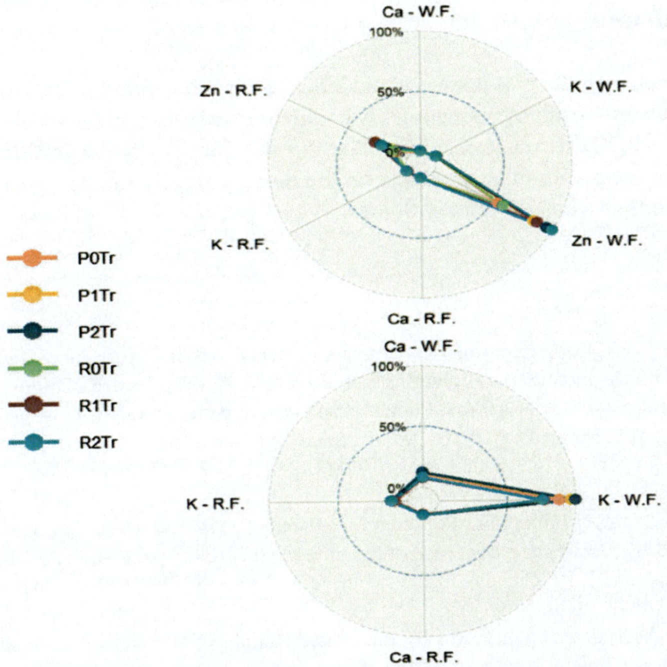

Fig. 3. Average Ca, K and Zn contents (above) of whole wheat flour (n = 12) and of refined wheat flour (n = 9) of *Triticum aestivum* L., varieties Roxo and Paiva. The lower part of the figure represents only Ca and K contents. With the foliar application of Zn-EDTA (Tr): P corresponds to Paiva and R to Roxo; 0 = control; 1 and 2 correspond respectively, to initial and upper concentration; W.F. and R.F. correspond to Whole Flour and Refined Flour, respectively.

that when storing wheat flour in paper bags, at 20 °C, over the months (about 30 months of study), the moisture content decreased [31]. Even though the differences between samples are not significant, they might be related to the fact that the milling of wheat grains of all the samples (whole and refined wheat flours) did not happen at the same time, it was done gradually. Moreover, the storage conditions may have contributed to the different values. It is important that the moisture content of wheat flours remains low so that it does not allow the development of microorganisms and the flours might have an extended shelf life [32].

It can be observed that the refined flour showed higher values in all bands, as well as the Roxo variety (regardless of the type of flour). This fact might be related to the fact that Roxo and refined flour was visible to the human eye as a lighter color. In addition, the upper peak in the 550 nm band refers to a transition in the green region. The industrial processing of the wheat flour favors the lighter color in the flours, due to the removal of the outer layers of the grain [24], being more visible in the Paiva variety than in Roxo, which does not alter its color tone so considerably.

5 Conclusions

To sum up, between the two flours, whole wheat flour biofortified in Zn is a functional product which will result in a nutritionally healthier choice, as it presents higher mineral content and ash, in detriment of refined flour. In this framework, biofortified flour will be able to compete with other products on the market, mostly due to innovation of Zn enrichment, either whole or refined flours.

References

1. United Nations, Department of Economics and Social Affairs, Population Division. World Population Prospects 2019: Highlights (ST/ESA/SER.A/423). United Nations, Department of Economics and Social Affairs, Population Division: New York, NY, USA (2019)
2. Chakraborty, I., Maity, P.: COVID-19 outbreak: Migration, effects on society, global environment and prevention. Sci. Total Environ. **728**(138882), 1–7 (2020). https://doi.org/10.1016/j.scitotenv.2020.138882
3. Micelli, E., Cito, G., Cocci, A., Polloni, G., Russo, G.I., Minervini, A., Carini, M., Natali, A., Coccia, M.E.: Desire for parenthood at the time of COVID-19 pandemic: An insight into the Italian situation. J. Psychosom. Obstet. Gynaecol. **41**(3), 183–190 (2020). https://doi.org/10.1080/0167482X.2020.1759545
4. Food and Agriculture Organization of the United Nations (FAO). https://www.fao.org/worldfoodsituation/csdb/en/. Accessed 28 Jan 2021
5. Zhang, Y., Pei, F., Fang, Y., Li, P., Zhao, Y., Shen, F., Zou, Y., Hu, Q.: Comparison of concentration and health risks of 9 *Fusarium mycotoxins* in commercial whole wheat flour and refined wheat flour by multi-IAC-HPLC. Food Chem. **275**, 763–769 (2019). https://doi.org/10.1016/j.foodchem.2018.09.127
6. Tian, W., Chen, G., Tilley, M., Li, Y.: Changes in phenolic profiles and antioxidant activities during the whole wheat bread-making process. Food Chem **345**(128851), 1 (2021). https://doi.org/10.1016/j.foodchem.2020.128851
7. Ciccolini, V., Pellegrino, E., Coccina, A., Fiaschi, A.I., Cerretani, D., Sgherri, C., Quartacci, M.F., Ercoli, L.: Biofortification wih iron and zinc improves nutritional and nutraceutical properties of common wheat flour and bread. J. Agric. Food Chem. **65**, 5443–5452 (2017). https://doi.org/10.1021/acs.jafc.7b01176
8. Cakmak, I., Kutman, U.B.: Agronomic biofortification of cereals with zinc: a review. Eur. J. Soil Sci. **69**, 172–180 (2018). https://doi.org/10.1111/ejss.12437
9. Bouis, H.E., Saltzman, A.: Improving nutrition through biofortification: a review of evidence from HarvestPlus, 2003 through 2016. Glob. Food Sec. **12**, 49–58 (2017). https://doi.org/10.1016/j.gfs.2017.01.009
10. Saini, D.K., Devi, P., Kaushik, P.: Advances in genomic interventions for wheat biofortification: a review. Agronomy **10**(62), 1–28 (2020). https://doi.org/10.3390/agronomy10010062
11. Izydorczyk, G., Ligas, B., Mikula, K., Witek-Krowiak, A., Moustakas, K., Chojnacka, K.: Biofortification of edible plants with selenium and iodine – a systematic literature review. Sci. Total Environ. **754**(141983), 1–15 (2021). https://doi.org/10.1016/j.scitotenv.2020.141983

12. Marques, A.C., Lidon, F.C., Coelho, A.R.F., Pessoa, C.C., Luís, I.C., Soctti-Campos, P., Simões, M., Almeida, A.S., Legoinha, P., Pessoa, M.F., Galhano, C., Guerra, M.A.M., Leitão, R.G., Ramalho, J.C., Semedo, J.M.N., Bagulho, A., Moreira, J., Rodrigues, A.P., Marques, P., Silva, C., Ribeiro-Barros, A., Silva, M.J., Silva, M.M., Oliveira, K., Ferreira, D., Pais, I.P., Reboredo, F.H.: Quantification and tissue localization of selenium in rice (*Oryza sativa* L., Poaceae) grains: a perspective of agronomic biofortification. Plants **9**(12), 1670 (2020). https://doi.org/10.3390/plants9121670
13. Buturi, C.V., Mauro, R.P., Fogliano, V., Leonardi, C., Giuffrida, F.: Mineral biofortification of vegetables as a tool to improve human diet. Foods **10**(223), 1–23 (2021). https://doi.org/10.3390/foods10020223
14. Coelho, A.R.F., Lidon, F.C., Pessoa, C.C., Marques, A.C., Luís, I.C., Caleiro, J., Simões, M., Kullberg, J., Legoinha, P., Brito, M., Guerra, M., Leitão, R.G., Galhano, C., Scotti-Campos, P., Semedo, J.N., Silva, M.M., Pais, I.P., Silva, M.J., Rodrigues, A.P., Pessoa, M.F., Ramalho, J.C., Reboredo, F.H.: Can foliar pulverization with $CaCl_2$ and $Ca(NO_3)_2$ trigger Ca enrichment in *Solanum tuberosum* L. tubers? Plants **10**(2), 245 (2021). https://doi.org/10.3390/plants10020245
15. Luís, I.C., Lidon, F.C., Pessoa, C.C., Marques, A.C., Coelho, A.R.F., Simões, M., Patanita, M., Dôres, J., Ramalho, J.C., Silva, M.M., Almeida, A.S., Pais, I.P., Pessoa, M.F., Reboredo, F.H., Legoinha, P., Guerra, M., Leitão, R.G., Campos, P.S.: Zinc enrichment in two contrasting genotypes of *Triticum aestivum* L. grains: interactions between edaphic condition and foliar fertilizers. Plants **10**(2), 204 (2021). https://doi.org/10.3390/plants10020204
16. Motoya, M., Guardia, G., Recio, J., Castellano-Hinojosa, A., Ginés, C., Bedmar, E.J., Álvarez, J.M., Vallejo, A.: Zinc-nitrogen co-fertilization influences N_2O emissions and microbial communities in an irrigated maize field. Geoderma **383**(114735), 1–12 (2021). https://doi.org/10.1016/j.geoderma.2020.114735
17. Pessoa, C.C., Lidon, F.C., Coelho, A.R.F., Caleiro, J.C., Marques, A.C., Luís, I.C., Kullberg, J.C., Legoinha, P., Brito, M.G., Ramalho, J.C., Guerra, M.A.M., Leitão, R.G., Simões, M., Campos, P.S., Semedo, J.M.N., Silva, M.M., Pais, I.P., Leal, N., Alvarenga, N., Gonçalves, E.M., Silva, M.J., Rodrigues, A.P., Abreu, M., Pessoa, M.F., Reboredo, F.H.: Calcium biofortification of Rocha pears, tissues accumulation and physicochemical implications in fresh and heat-treated fruits. Sci. Hortic. **277**(109834), 1–11 (2021). https://doi.org/10.1016/j.scienta.2020.109834
18. Vaiknoras, K., Larochelle, C.: The impact of iron-biofortified bean adoption on bean productivity, consumption, purchases and sales. World Dev **139**(105260), 1–17 (2021). https://doi.org/10.1016/j.worlddev.2020.105260
19. NP 519: Cereais e Derivados. Determinação do Teor de Cinza a 900 °C. Processo Corrente. Instituto Português da Qualidade, Portugal, p. 5 (1993)
20. NP 516: Cereais e Derivados. Determinação do Teor de Água. Método Prático de Referência. Instituto Português da Qualidade: Portugal, p. 5 (2000)
21. Lidon, F.C., Daccak, D., Scotti-Campos, P., Silva, M.M., Bagulho, A.S., Pais, I., Galhano, C., Ramalho, J.C., Moreira, J., Pessoa, M.F., Reboredo, F.H.: An integrated chemical and technological approach for assessing portuguese wheat flours quality and lengthening bread shelf-life. Emir. J. Food Agric. **31**(11), 884–894 (2019). https://doi.org/10.9755/ejfa.2019.v31.i11.2037
22. Ramalho, J.C., Pais, I.P., Leitão, A.E., Guerra, M., Reboredo, F.H., Máguas, C.M., Carvalho, M.L., Scotti-Campos, P., Ribeiro-Barros, A.I., Lidon, F.C., DaMatta, F.M.: Can elevated air $[CO_2]$ conditions mitigate the predicted warming impact on the quality of coffee bean? Front. Plant Sci. **9**(287), 1–14 (2018). https://doi.org/10.3389/fpls.2018.00287

23. Pelica, J., Barbosa, S., Reboredo, F., Lidon, F., Pessoa, F., Calvão, T.: The paradigm of high concentration of metals of natural or anthropogenic origin in soils – The case of Neves-Corvo mine area (Southern Portugal). J. Geochem. Explor. **186**, 12–23 (2018). https://doi.org/10.1016/j.gexplo.2017.11.021
24. Ramos, I., Pataco, I.M., Mourinho, M.P., Lidon, F., Reboredo, F., Pessoa, M.F., Carvalho, M.L., Santos, J.P., Guerra, M.: Elemental mapping of biofortified wheat grains using micro X-ray fluorescence. Spectrochim. Acta Part B At **120**, 30–36 (2016). https://doi.org/10.1016/j.sab.2016.03.014
25. Liu, J., Yu, L.L., Wu, Y.: Bioactive components and health beneficial properties of whole wheat foods. J. Agric. Food Chem. **68**, 12904–12915 (2020). https://doi.org/10.1021/acs.jafc.0c00705
26. Gomez-Coronado, F., Poblaciones, M.J., Almeida, A.S., Cakmak, I.: Zinc (Zn) concentration of bread wheat grown under Mediterrranean conditions as affected by genotype and soil/foliar Zn application. Plant Soil **401**, 331–346 (2016). https://doi.org/10.1007/s11104-015-2758-0
27. Loneragan, J.F., Webb, M.J.: Interactions between zinc and other nutrients affecting the growth of plants. In: Robson, A.D. (eds.) Zinc in Soils and Plants: Developments in Plant and Soil Sciences. Springer, vol. 55, pp. 119–134. Dordrecht, Netherlands (1993). https://doi.org/10.1007/978-94-011-0878-2_9
28. Prasad, R., Shivay, Y.S., Kumar, D.: Interactions of zinc with other nutrients in soils and plants—a review. Indian J. Fertil. **12**, 16–26 (2016)
29. Rengel, Z.: Carbonic anhydrase activity in leaves of wheat genotypes differing in Zn efficiency. J. Plant Physiol. **147**, 251–256 (1995). https://doi.org/10.1016/S0176-1617(11)81513-0
30. Pendias, A.K.: Trace Elements in Soils and Plants, 4th ed. CRC Press, Boca Raton (2011). https://doi.org/10.1201/b10158.
31. Lancelot, E., Fontaine, J., Grua-Priol, J., Le-Bail, A.: Effect of long-term storage conditions on wheat flour and bread baking properties. Food Chem. **346**(128902), 1–8 (2021). https://doi.org/10.1016/j.foodchem.2021.128902
32. Nasir, M., Butt, M.S., Anjum, F.M., Sharif, K., Minhas, R.: Effect of moisture on the shelf life of wheat flour. Int. J. Agri. Biol. **5**(4), 458–459 (2003)

Calcium Biofortification in *Solanum tuberosum* L. cv.Agria: A Technical Workflow

Ana Rita F. Coelho[1,2](✉), Ana Coelho Marques[1,2], Cláudia Campos Pessoa[1,2], Inês Carmo Luís[1,2], Diana Daccak[1,2], Manuela Simões[1,2], Fernando Henrique Reboredo[1,2], Maria Pessoa[1,2], Maria Manuela Silva[2,3], Paulo Legoinha[1,2], Mauro Guerra[4], José C. Ramalho[2,5], Paula Scotti Campos[2,6], Isabel P. Pais[2,6], and Fernando C. Lidon[1,2]

[1] Earth Sciences Department, Faculdade de Ciências e Tecnologia, Universidade Nova de Lisboa, Caparica, Portugal
arf.coelho@campus.fct.unl.pt

[2] GeoBioTec Research Center, Faculdade de Ciências e Tecnologia, Universidade Nova de Lisboa, Caparica, Portugal

[3] ESEAG – Grupo Universidade Lusófona, Lisboa, Portugal

[4] LIBPhys, Physics Department, Faculdade de Ciências e Tecnologia, Universidade Nova de Lisboa, Campus da Caparica, 2829-516 Caparica, Portugal

[5] PlantStress and Biodiversity Lab, Centro de Estudos Florestais, Instituto Superior Agronomia, Universidade de Lisboa, Oeiras, Portugal

[6] INIAV, Instituto Nacional de Investigação Agrária e Veterinária, Oeiras, Portugal

Abstract. In the human body, Ca is required in high quantities, being crucial for bone synthesis, specific metabolic features, and tooth mineralization. However, its deficiency can trigger osteoporosis and rickets development. To avoid Ca deficiency, agronomic biofortification is a strategy that can improve this nutrient enhancement in food crops providing, in the long term, its intake and physiological use in the human body. In this context, being *Solanum tuberosum* L. one of the most important staple food crops worldwide and a primary source in human diet, this study aimed to develop a technical workflow for Ca biofortification of cv. Agria and assess some related quality parameters. The technical workflow was applied during the production cycle, through four foliar sprays with solutions containing calcium chloride (3 and 6 $kg.ha^{-1}$) and calcium nitrate (0.5, 2 and 4 $kg.ha^{-1}$). The accumulation kinetics of Ca in the shoot was carried out and, at harvest, the average index of Ca enrichment in the tubers varied between 5% and 40%. Naturally enriched tubers with Ca showed minor changes in dry weight matter content and colorimetric parameters. It was concluded that potato tubers naturally fortified with Ca can be used as raw material for further processing into functional food products, thus adding product value.

Keywords: Calcium accumulation · Calcium biofortification · Harvest and post-harvest · *Solanum tuberosum* L

J. R. da Costa Sanches Galvão et al. (Eds.): *Proceedings of the 1st International Conference on Water Energy Food and Sustainability (ICoWEFS 2021)*, pp. 147–154, 2021.
https://doi.org/10.1007/978-3-030-75315-3_17

1 Introduction

Potato (*Solanum tuberosum* L.) is the 3rd staple food crop most consumed worldwide (after rice and wheat) [1] and is the 4th most cultivated after rice, wheat, and maize [2, 3]. Due to the major role that potato has in human diet all over the world, different studies of mineral enrichments in potato tubers has been carried out, namely with selenium [4–6], zinc [7–9], iron [9] and zinc and manganese [10]. Among them, agronomic biofortification is a fast, reliable, and cost-effective way to increase different minerals content in the edible part of the plants [11]. As such, foliar application is considered one of the most common methods of biofortification [12]. Accordingly, some studies with potatoes have showed a higher Ca content through foliar spraying [13–15].

Calcium is an essential mineral nutrient for plants, being required as Ca^{2+} [16]. Acts as a signal transduction agent [17], has an important central role in stress responses [16] and provide integrity and stability to cell walls [17]. Also plays a determining role as a cofactor by enzymes that are involved in the catabolism of ATP and phospholipids [18]. Regarding Ca function in human body, has an important role in a wide range of functions, namely in bone development, skeletal mineralization [19], blood clotting, and in the fluid balance within cells [20]. However, Ca deficiency can trigger some pathologies, as such osteoporosis [21] and rickets [22]. In this context, to minimize the deficiency of Ca in the human population, this study aimed to develop a workflow for Ca biofortification of *Solanum tuberosum* L. cv. Agria. This variety was chosen because of the wide range of industrial processing, namely dehydrated products [23].

2 Materials and Methods

2.1 Experimental Field

The experimental potato field, located in the Western of Portugal (39° 16’ 38,816” N; 9° 15’ 9,128”O), was used to growth cv. Agria (*Solanum tuberosum* L.). During the agricultural period, from 4th May (planting date) to 4th September of 2018 (harvest date), air temperatures reached an average daily of 23 °C and 15 °C (with maximum and minimum values of 41 °C and 6 °C, respectively). Regarding rainfall, the average was 0.41 mm, with daily maximum of 18.03 mm and an accumulation of 60.4 mm. After the beginning of tuberization, four foliar spraying (with 8–10 days interval) with $CaCl_2$ (3 and 6 $kg.ha^{-1}$) or $Ca(NO_3)_2$ (0.5, 2 and 4 $kg.ha^{-1}$) were performed. Control plants were not sprayed at any time with $Ca(NO_3)_2$ or $CaCl_2$. Each treatment was performed in quadruplicate (compass, 60–80 cm), in a plot 20 × 20 m.

2.2 Calcium Content in Potato Tubers

After harvest, considering the four replications in the experimental field, eight randomized tubers with similar sizes, were washed, dried at 60 °C until constant weight, and grounded in an agate mortar. Thereafter, an acid digestion procedure was performed with a mixture of HNO_3- $HClO_4$ (4:1), according to [24, 25]. After filtration, Ca content was measured by atomic absorption spectrophotometry, using a model Perkin Elmer AAnalyst 200, and the absorbency in mg / L was determined with a coupled AA WinLab software.

2.3 Quality Parameters

Dry weight was measured considering four randomized tubers per treatment. Colorimetric parameters, using fixed wavelength, followed [26].

Brightness (L) and chromaticity parameters (a* and b* coordinates) were obtained with a Minolta CR 400 colorimeter (Minolta Corp., Ramsey, NJ, USA) coupled to a sample vessel (CR-A504). Using the illuminant D_{65}, the system of the Commission Internationale d'Éclaire (CIE) was applied. The parameter L represents the brightness of the sample, indicating the variation of the tonality between dark and light, with a range between 0 (black) and 100 (white). Parameters a* and b*, indicate color variations between red (+60) and green (−60), and between yellow (+60) to blue (−60), respectively. The approximation of these coordinates to the null value translates neutral colors like white, gray and black. Chroma is the relationship between the values of a* and b*, where the real color of the analyzed object is obtained. Hue is the angle formed between a* and b*, indicating the saturation of the object's color. To calculate Chroma (C), Eq. (1) was used and, to calculate Hue-Angle (H), Eq. (2). Measurements were carried out in quadrupled in the pulp of fresh tubers at harvest.

$$C* = \sqrt{a*^2 + b*^2} \tag{1}$$

$$H* = \mathrm{arctg}\frac{b*}{a*} \tag{2}$$

2.4 Statistical Analysis

Data were statistically analyzed using a One-Way ANOVA to assess differences among treatments in cv. Agria, followed by a Tukey's for mean comparison. A 95% confidence level was adopted for all tests.

3 Results

After harvest, Ca accumulation in the whole tubers of cv. Agria was assessed (Fig. 1). Regarding the control, the content of Ca was significantly higher in all treatments (except with $Ca(NO_3)_2$ 0.5 kg.ha^{-1}). The biofortification indexes ranged between 5% and 40%, with maximum levels obtained with the $CaCl_2$ 6 kg.ha^{-1} and $Ca(NO_3)_2$ 4 kg.ha^{-1} treatments.

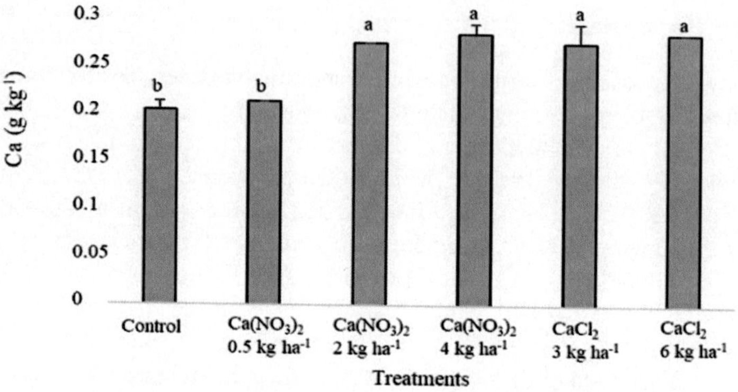

Fig. 1. Mean values ± S.E. (n = 4) of Ca content in tubers of *Solanum tuberosum* L., cv. Agria, at harvest. Letters a and b indicates significant differences, between treatments ($P \leq 0.05$). Foliar spray was carried out with three concentrations of $Ca(NO_3)_2$ (0.5, 2 and 4 $kg.ha^{-1}$) and two concentrations of $CaCl_2$ (3 and 6 $kg.ha^{-1}$). Control was not sprayed.

In the biofortified tubers a decrease was found (Fig. 2), which was significant the in $Ca(NO_3)_2$ treatments.

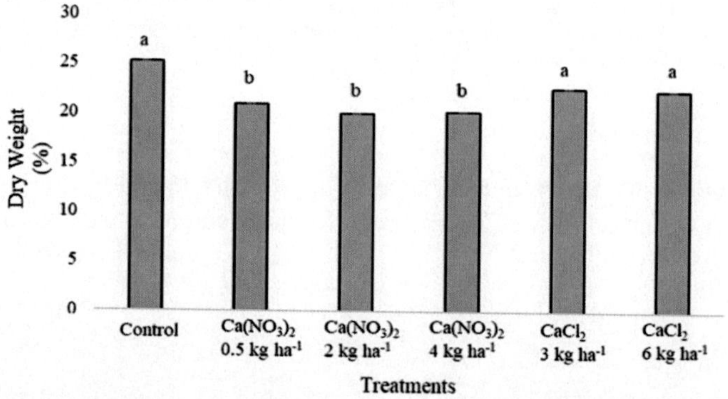

Fig. 2. Mean values ± S.E. (n = 4) of dry weight in tubers of *Solanum tuberosum* L., cv. Agria, at harvest. Letters a and b indicates significant differences, between treatments ($P \leq 0.05$). Foliar spray was carried out with three concentrations of $Ca(NO_3)_2$ (0.5, 2 and 4 $kg.ha^{-1}$) and two concentrations of $CaCl_2$ (3 and 6 $kg.ha^{-1}$). Control was not sprayed.

The colorimetric parameters in the fresh tubers of cv. Agria was assessed (Table 1) and it was found that regardless of the biofortification indexes obtained, Hue parameter did not vary significantly. However, L (brightness/luminosity) and Chroma parameter (saturation) did varied significantly. In fact, $Ca(NO_3)_2$ treatments at 4 $kg.ha^{-1}$ and 0.5 $kg.ha^{-1}$ showed a lighter color. Regarding Chroma, the $Ca(NO_3)_2$ treatment with 0.5 $kg.ha^{-1}$ showed a more intense color.

Table 1. Mean values ± S.E. (n = 4) of colorimetric parameters (L, Chroma and Hue) in fresh tubers of *Solanum tuberosum* L., cv. Agria, at harvest. Letter a indicates no significant differences, of each parameter, between treatments ($P \leq 0.05$). Foliar spray was carried out with three concentrations of $Ca(NO_3)_2$ (0.5, 2 and 4 $kg.ha^{-1}$) and two concentrations of $CaCl_2$ (3 and 6 $kg.ha^{-1}$). Control was not sprayed.

Treatments	L	Chroma	Hue
Control	55.67bc ± 0.04	20.71bc ± 0.24	98.00a ± 0.40
$Ca(NO_3)_2$ (0.5 $kg.ha^{-1}$)	59.15a ± 0.31	23.49a ± 0.38	97.42a ± 0.44
$Ca(NO_3)_2$ (2 $kg.ha^{-1}$)	57.43ab ± 0.59	21.89abc ± 0.30	97.75a ± 0.18
$Ca(NO_3)_2$ (4 $kg.ha^{-1}$)	59.54a ± 0.36	22.49ab ± 0.33	98.45a ± 0.14
$CaCl_2$ (3 $kg.ha^{-1}$)	58.84a ± 0.50	22.15abc ± 0.26	97.67a ± 0.62
$CaCl_2$ (6 $kg.ha^{-1}$)	54.86c ± 0.93	20.69c ± 0.63	97.50a ± 0.30

4 Discussion

The accumulation of minerals in potatoes, are mainly obtained via the phloem (from above-ground redistribution) [27, 28]. Yet, considering Ca, its delivery depends of the xylem [29], being less mobile in the phloem [28]. In the other hand, different studies showed that thought foliar pulverizations with Ca, its levels increased in potatoes [13–15], suggestion that thought foliar spraying Ca is redistributed by phloem. As seen in Fig. 1, Ca content increased with foliar spraying with $Ca(NO_3)_2$ and $CaCl_2$. Additionally, some studies showed that increasing the Ca levels, using $Ca(NO_3)_2$ or $CaCl_2$, (namely with potatoes, apples, mangos) triggered beneficial effects in quality parameters [30]. In this context, dry matter content is one of the characteristics more relevant for industrial processing and is being used as a criterion for the classification of potato tubers. Also, for industrial use, potatoes must have the dry matter content higher than 20% (which is the case of all the treatments, except with $Ca(NO_3)_2$ at 2 $kg.ha^{-1}$, being the dry matter content exactly 20%), as higher dry matter content decreases the absorption of fat during the frying process [31]. In this context, Ca-biofortification process with $Ca(NO_3)_2$ did interfered in the dry matter content (Fig. 2), but regarding this variety, the values of dry matter obtained in this study, remained higher relatively to the values obtained by [32] and similar to other authors [33, 34].

Color, is also a very important parameter in potato tubers, affecting the acceptability of consumers [35]. Nevertheless, regarding L parameter, data obtained in this study (Table 1) remained lower relatively to other studies [34, 36, 37]. Besides, Chroma showed lower values and the parameter Hue showed values that agree with those obtained by [38]. As such, there were minor effects in Ca biofortified tubers.

5 Conclusion

S. tuberosum cv. Agria showed an increase in Ca content in all the treatments (*i.e.*, through foliar spraying with $Ca(NO_3)_2$ and $CaCl_2$). In fact, the highest concentrations applied of $Ca(NO_3)_2$ (4 kg.ha^{-1}) and $CaCl_2$ (6 kg.ha^{-1}) showed the highest Ca content, and a similar efficacy. Considering the technical workflow used for Ca biofortification of potato tubers, our data suggest that foliar application of Ca redistributed by phloem complements the xylem mass flow of this essential nutrient. However, Ca biofortified tubers showed a lower dry matter content and trigger some minor changes in the colorimetric parameters (L and Chroma).

References

1. CIP—International Potato Center. Potato Facts and Figures. https://cipotato.org/crops/potato/potato-facts-and-figures/. Accessed 08 Jan 2021
2. FAO (Food and Agriculture Organization). Potato world: Production and Consumption—International Year of the Potato (2008). https://www.fao.org/potato-2008/en/world/. Accessed 08 Jan 2021
3. Muthoni, J., Mbiyu, M., Nyamongo, D.: A review of potato seed systems and germplasm conservation in Kenya. J. Agric. Food Inf. **11**, 157–167 (2010). https://doi.org/10.1080/10496501003680565
4. Poggi, V., Arcioni, A., Filippini, P., Pifferi, P.: Foliar application of selenite and selenate to potato (*Solanum tuberosum*): effect of a ligand agent on selenium content of tubers. J. Agr. Food Chem. **48**, 4749–4751 (2000). https://doi.org/10.1021/jf000368f
5. Oliveira, V., Faquin, V., Andrade, F., Carneiro, J., Júnior, E., Souza, K., Pereira, J., Guilherme, L.: Physiological and physicochemical responses of potato to selenium biofortification in tropical soil. Potato Res. **62**, 315–331 (2019). https://doi.org/10.1007/s11540-019-9413-8
6. Zhang, H., Zhao, Z., Zhang, X., Zhang, W., Huang, L., Zhang, Z., Yuan, L., Liu, X.: Effects of foliar application of selenate and selenite at different growth stages on Selenium accumulation and speciation in potato (*Solanum tuberosum* L.). Food Chem. **286**, 550–556 (2019). https://doi.org/10.1016/j.foodchem.2019.01.185
7. Mousavi, S., Galavi, M., Ahmadvand, G.: Effect of zinc and manganese foliar application on yield, quality and enrichment on potato (*Solanum tuberosum* L.). Asian J. Plant Sci. **6**, 1256–1260 (2007). https://doi.org/10.3923/ajps.2007.1256.1260
8. White, P., Thompson, J., Wright, G., Rasmussen, S.: Biofortifying Scottish potatoes with zinc. Plant Soil **411**, 151–165 (2016). https://doi.org/10.1007/s11104-016-2903-4
9. Kromann, P., Valverde, F., Alvarado, S., Vélez, R., Pisuña, J., Potosí, B., Taipe, A., Caballero, D., Cabezas, A., Devaux, A.: Can Andean potatoes be agronomically biofortified with iron and zinc fertilizers? Plant Soil **411**(1–2), 121–138 (2017). https://doi.org/10.1007/s11104-016-3065-0
10. Mousavi, S.R., Galavi, M., Ahmadvand, G.: Effect of zinc and manganese foliar application on yield, quality and enrichment on potato (*Solanum tuberosum* L.). Asian J. Plant Sci. **6**(8) 1256–1260 (2007). https://doi.org/10.3923/ajps.2007.1256.1260
11. Bouis, H., Saltzman, A.: Improving nutrition through biofortification: a review of evidence from HarvestPlus, 2003 through 2016. Glob. Food Sec. **12**, 49–58 (2017). https://doi.org/10.1016/j.gfs.2017.01.009
12. Alshaal, T., El-Ramady, H.: Foliar application: from plant nutrition to biofortification. EBSS **1**, 71–83 (2017). https://doi.org/10.21608/jenvbs.2017.1089.1006
13. El-Hadidi, E.M., El-Dissoky, R.A., Abdelhafez, A.A.H.: Foliar calcium and magnesium application effect on potato crop grown in clay loam soils. JSSAE **8**(1), 1–8 (2017). https://doi.org/10.21608/JSSAE.2017.37074517

14. El-Zohiri, S.S.M., Asfour, H.E.: Effects of foliar sprays of potassium, magnesium and calcium on yield, quality and storageability of potato. In: Proceedings of the Fifth International Congress of Sustain Agriculture Development, pp. 57–71 (2009). Faculty of Agriculture, Fayoum University, 21–23 December
15. Seifu, Y.W., Deneke, S.: Effect of calcium chloride and calcium nitrate on potato (*Solanum tuberosum* L.) growth and yield. J. 521 Hortic. **4**(3), 1–4 (2017)
16. Hocking, B., Tyerman, S., Burton, R., Gilliham, M.: Fruit calcium: transport and physiology. Front. Plant Sci. **7**, 569 (2016). https://doi.org/10.3389/fpls.2016.00569
17. Wei, S., Qin, G., Zhang, H., Tao, S., Wu, J., Wang, S., Zhang, S.: Calcium treatments promote the aroma volatiles emission of pear (*Pyrus ussuriensis* 'Nanguoli') fruit during post-harvest ripening process. Sci. Hortic-Amst. **215**, 102–111 (2017). https://doi.org/10.1016/j.scienta.2016.12.008
18. Taiz, L., Zeiger, E.: Plant Physiology. 3rd ed, p. 665. Sinauer Associates, Inc. Sunderland (2002)
19. Peacock, M.: Calcium metabolism in health and disease. Clin. J. Am. Soc. Nephrol. **5**(Suppl. 1), S23–S30 (2010). https://doi.org/10.2215/cjn.05910809
20. Pravina, P., Sayaji, D., Avinash, M.: Calcium and its role in human body. Int. J. Res. Pharm. Biomed. Sci **4**(2), 659–668 (2013)
21. NIH (National Institutes of Health). https://pubchem.ncbi.nlm.nih.gov/compound/calcium. Accessed 12 Jan 2021
22. IOM—Institute of Medicine. Dietary Reference Intakes for Calcium and Vitamin D. The National Academies Press, Washington, DC, USA (2011). ISBN 978-0-309-16395-8
23. AHDB—Agriculture and Horticulture Development Board. https://varieties.ahdb.org.uk/varieties/view/Agria. Accessed 13 Jan 2021
24. Carrondo, M., Reboredo, F., Ganho, R., Santos Oliveira, J.F.: Heavy metal analysis of sediments in Tejo estuary, Portugal, using a rapid flameless atomic absorption procedure. Talanta **31**, 561–564 (1984). https://doi.org/10.1016/0039-9140(84)80141-1
25. Reboredo, F.H.S., Ribeiro, C.A.G.: Vertical distribution of Al, Cu, Fe and Zn in soil salt marshes of the Sado estuary. Portugal. Int. J. Environ. Stud. **23**, 249–253 (1984). https://doi.org/10.1080/00207238408710160
26. Ramalho, J.C., Pais, I.P., Leitão, A.E., Guerra, M., Reboredo, F.H., Máguas, C.M., Carvalho, M.L., Scotti- Campos, P., Ribeiro-Barros, A.I., Lidon, F.J.C., et al.: Can elevated air [CO_2] conditions mitigate the predicted warming impact on the quality of coffee bean? Front. Plant Sci. **9**, 287 (2018). https://doi.org/10.3389/fpls.2018.00287
27. Palta, J.P.: Role of calcium in plant responses to stresses: linking basic research to the solution of practical problems. HortScience **31**(1), 51–57 (1996)
28. Subramanian, N.K., White, P.J., Broadley, M.R., Ramsay, G.: The three-dimensional distribution of minerals in potato tubers. Ann Bot. **107**, 681–691 (2011). https://doi.org/10.1093/aob/mcr009
29. Weinl, S., Held, K., Schlücking, K., Steinhorst, L., Kuhlgert, S., Hippler, M., Kudla, J.: A plastid protein crucial for Ca^{2+}-regulated stomatal responses. New Phytol. **179**, 675–686 (2008). https://doi.org/10.1111/j.1469-8137.2008.02492.x
30. Dayod, M., Tyerman, S., Leigh, R., Gilliham, M.: Calcium storage in plants and the implications for calcium biofortification. Protoplasma **247**, 215–231 (2010). https://doi.org/10.1007/s00709-010-0182-0
31. Braun, H., Fontes, P., Finger, F., Busato, C., Cecon, P.: Carboidratos e matéria seca de tubérculos de cultivares de batata influenciados por doses de nitrogênio. Cienc Agrotec **34**, 285–293 (2010). https://doi.org/10.1590/s1413-55370542010000200003
32. Romano, A., D'Amelia, V., Gallo, V., Palomba, S., Carputo, D., Masi, P.: Relationships between composition, microstructure and cooking performances of six potato cultivars. Food Res. Int. **114**, 10–19 (2018). https://doi.org/10.1016/j.foodres.2018.07.033

33. Arvanitoyannis, I., Mavromatis, A., Vaitsi, O., Korkovelos, A., Golia, E.: Effect of genotype and geographical origin on potato properties (physical and sensory) for authenticity purposes. J. Agric. Sci. **4**, 1916–1960 (2012). https://doi.org/10.5539/jas.v4n4p63
34. Bordoloi, A., Kaur, L., Singh, J.: Parenchyma cell microstructure and textural characteristics of raw and cooked potatoes. Food Chem. **133**(4), 1092–1100 (2012)
35. Xiao, Q., Bai, X., He, Y.: Rapid screen of the color and water content of fresh-cut potato tuber slices using hyperspectral imaging coupled with multivariate analysis. Foods **9**(1), 94 (2020). https://doi.org/10.3390/foods9010094
36. Mesías, M., Holgado, F., Márquez-Ruiz, G., Morales, F.J.: Impact of the characteristics of fresh potatoes available in-retail on exposure to acrylamide: case study for French fries. Food Control **73**, 1407–1414 (2017). https://doi.org/10.1016/j.foodcont.2016.11.005
37. Cabezas-Serrano, A.B., Amodio, M.L., Cornacchia, R., Rinaldi, R., Colelli, G.: Suitability of five different potato cultivars (*Solanum tuberosum* L.) to be processed as fresh-cut products. Postharvest Biol. Technol. **53**, 138–144 (2009). https://doi.org/10.1016/j.postharvbio.2009.03.009
38. Yang, Y., Achaerandio, I., Pujolà, M.: Classification of potato cultivars to establish their processing aptitude. J. Sci. Food Agric. **96**(2), 413–421 (2016). https://doi.org/10.1002/jsfa.7104

Agronomic Biofortification in Se of *Oryza sativa* L.: Food Quality Control for Baby Food Products

Ana Coelho Marques[1,2](✉), Cláudia Campos Pessoa[1,2], Ana Rita F. Coelho[1,2], Diana Daccak[1,2], Inês Carmo Luís[1,2], Paula Scotti Campos[2,3], Manuela Simões[1,2], Ana Sofia Almeida[2,4], Maria Fernanda Pessoa[1,2], Fernando Reboredo[1,2], Mauro Guerra[5], Roberta G. Leitão[5], Carlos Galhano[1,2], José C. Ramalho[2,6], Paula Marques[7], Ana Bagulho[4], José Moreira[4], Maria Manuela Silva[2,8], Paulo Legoinha[1,2], Isabel Pais[2,3], and Fernando C. Lidon[1,2]

[1] Earth Sciences Department, Faculdade de Ciências e Tecnologia, Universidade Nova de Lisboa, Campus da Caparica, 2829-516 Caparica, Portugal
amc.marques@campus.fct.unl.pt

[2] GeoBioTec Research Center, Faculdade de Ciências e Tecnologia, Universidade Nova de Lisboa, Campus da Caparica, 2829-516 Caparica, Portugal

[3] Instituto Nacional de Investigação Agrária e Veterinária, I. P. (INIAV), Avenida da República, Quinta do Marquês, 2780-157 Oeiras, Portugal

[4] Instituto Nacional de Investigação Agrária e Veterinária, I. P. (INIAV), Estrada de Gil Vaz 6, 7351-901 Elvas, Portugal

[5] LIBPhys, Physics Department, Faculdade de Ciências E Tecnologia, Universidade Nova de Lisboa, Campus da Caparica, 2829-516 Caparica, Portugal

[6] PlantStress and Biodiversity Lab, Centro de Estudos Florestais, Instituto Superior Agronomia (ISA), Universidade de Lisboa (ULisboa), Quinta do Marquês, Av. República, 2784-505 Oeiras and Tapada da Ajuda, 1349-017 Lisboa, Portugal

[7] Centro Operativo e Tecnológico do Arroz COTArroz, 2120-014 Salvaterra de Magos, Portugal

[8] ESEAG-COAF, Avenida do Campo Grande 376, 1749-024 Lisboa, Portugal

Abstract. Deficiency in selenium (Se) intake has been associated with the occurrence of various diseases. Rice, being the staple food for many populations, it has enormous potential to reduce this micronutrient deficit. The growing demand for efficient, bioeconomic and sustainable strategies to increase Se contents in rice is therefore justified, as it is our the study of the technical and nutritional implications inherent in its production and industrial processing for baby food products. Thus, the natural rice Se enrichment constitute an added value of its food chain and constituted the aims of our study. The technical workflow was implemented, using one commercial variety of rice (Ceres) and one advanced line of INIAV National Rice Genetic Improvement Program (OP 1505). Five Se concentrations (0, 25, 50, 75 and 100g Se.ha^{-1}) were tested in the forms of sodium selenate and selenite by foliar application. After grain harvesting, the application of Se showed 2.0–8.4 and 1.3–12.6 fold increases in OP 1505 and Ceres, respectively. Selenium did not reveal significant changes in the size of the biofortified grains. The color of the whole and refined flours showed two peaks at 550 nm and 650 nm, with the refined flour having the highest value. Total protein increased in both cultivars

J. R. da Costa Sanches Galvão et al. (Eds.): *Proceedings of the 1st International Conference on Water Energy Food and Sustainability (ICoWEFS 2021)*, pp. 155–163, 2021.
https://doi.org/10.1007/978-3-030-75315-3_18

when selenite was applied mainly in the Ceres cultivar. In a context of innovation, Se biofortification of both rice cultivars can be used for development of functional food products, namely baby food products.

Keywords: Rice cultivar · Sodium selenate · Sodium selenite · Selenium biofortification

1 Introduction

In humans Selenium (Se) is essential, having an antioxidative activity [1, 2]. In fact, the biological functions of Se include the maintenance of the body's defense systems against infection, the protection of tissues against oxidative stress and the modulation of growth and development. Severe deficiency can result in Keshan or Kaschin-Beck disease [2]. This deficiency affects approximately 15% people worldwide [3], being babies further reported to be at risk due to the low levels of this micronutrient in human milk and infant formula, which are their only or main food source during early life [4].

A relevant amount of studies showed positive effects of Se application, improving physiological and morphological parameters in plants. The mobilization of Se in plants occurs in the form of sodium selenate or sodium selenite [5]. For food fortification purposes, sodium salts are generally considered to be most suitable source of Se. The absorption of sodium selenite is approximately 50% [2]. Also, some studies indicates that foliar spraying of selenite is more effective than selenate in rice culture [6]Sodium selenate has lower solubility in water and more stable than selenite, especially in the presence of copper and iron [2]. However, several studies have found that selenate has similar chemical properties to sulphate, sharing metabolic pathways in plants. Thus, it is the main chemical form of Se absorbed by plants [7].

Considering that rice (*Oryza sativa* L.) is a predominant food in more than 30 countries [8], further being the second largest cereal crop in the world, biofortification is one of the safest ways to increment Se content in rice grains, as it can be assimilated by plants under organic forms bioavailable to humans [9]. This strategy can help populations with limited access to markets and health systems for reducing this micronutrient deficiency [10, 11]. In this context, this study aimed at developing an agronomic workflow for Se biofortification by foliar spraying of sodium selenate and sodium selenite in two cultivars (*Oryza sativa* L.).

2 Materials and Methods

2.1 Experimental Fields

The experimental rice field was carried out in the experimental station of the Rice Technological Center (COTArroz) located in Ribatejo - Portugal (39° 02′ 21.8′'N; 8° 44′ 22.8′'W), to grow two cultivars (Ceres and OP 1505). The agriculture period went from 30 May (planting date) to 2 November of 2018 (harvest date), for both cultivars.

The cultivar OP 1505 is a new advanced rice (*Oryza sativa* L.) line of the national Breeding Program carried out by the Instituto Nacional de Investigação Agrária e Veterinária (INIAV, Elvas, Portugal) and Ceres cultivar were used as test systems. The analysis was carried out in paddy, brown and white rice grains and additionally to the whole and refined flours. In both cultivars, the agronomic Se biofortification was carried out through foliar spraying with sodium selenate (Na_2SeO_4) and sodium selenite (Na_2SeO_3) (at 25, 50, 75 and 100g Se.ha^{-1}), and presented three distinct phases: end of booting, anthesis and milky grain stage. Regarding the experimental design in the field, the biofortification was performed in 80 plots (2 cultivars × 2 forms selenium × 5 concentrations × 4 replicates), each plot with 9.6 m^2.

2.2 Analysis of Grain Selenium Content

After harvest, considering control and biofortified grains (from each plot), Se was quantified in rice grains using an Energy Dispersive X-Ray Fluorescence (μ-EDXRF system, M4 Tornado™, Bruker, Germany) and following [12].

2.3 Thousand Grain Weight, Colorimetric and Crude Protein Content

The thousand grain weight was carried out for each cultivar and treatments, in triplicate. After that, random grains were hulled and whitened, according to [13]. Colorimetric parameters were determined in quadruplicate in rice grains, according to [14]. Following [15], crude protein was quantified using the Kjedhal method.

2.4 Statistical Analysis

Data were statistically analyzed using a One-Way ANOVA ($P \leq 0.05$), to assess differences among treatments, followed by a Tukey's for mean comparison. A 95% confidence level were adopted for all tests.

3 Results

After harvest, average yields (in kg.ha^{-1}) were quantified, being 5499 and 5322 for Ceres and 4669 and 4927 for OP 1505 (after foliar spraying of selenate and selenite, respectively). Additionally, the Se accumulation in brown grains was assessed (Fig. 1), showing an increase of Se through foliar spraying with sodium selenate and sodium selenite. In fact, with selenate spraying, the content of Se ranged between 4.61–12.6 and 1.95–7.07 fold, regarding to the control in cultivar Ceres and cultivar OP 1505, respectively. Still, with selenite spraying, biofortified rice showed an increase in Se content ranging between 1.33–7.69, and 1.75–8.40 fold, for Ceres and OP 1505, respectively.

Independently of the Se content in rice, grain weight did not vary with foliar spraying with Se fertilizers, in both forms (Fig. 2). Yet, relatively to OP 1505, Ceres cultivar always presents higher values in the paddy, brown and white grains.

The colorimetric analysis of rice at harvest, did not showed significant differences between treatments with both fertilizers' forms (except in white flour of OP 1505 cultivar). However, selenite and selenate application showed two highest values at 550 and 650 nm, which corresponds to yellow and red color, respectively (Fig. 3).

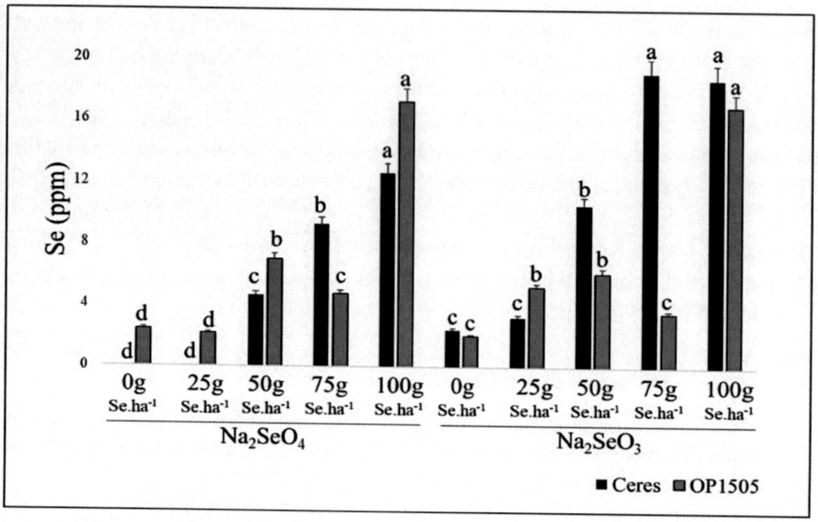

Fig. 1. Average values (ppm) ± S.E. ($n = 4$) of Se in the brown grains of *Oryza sativa* L., cultivars Ceres and OP 1505, at harvest. Letters *a, b, c,* and *d* indicate significant differences between treatments (Na_2SeO_4 and Na_2SeO_3) for each cultivar (single factor ANOVA test, $P \leq 0.05$).

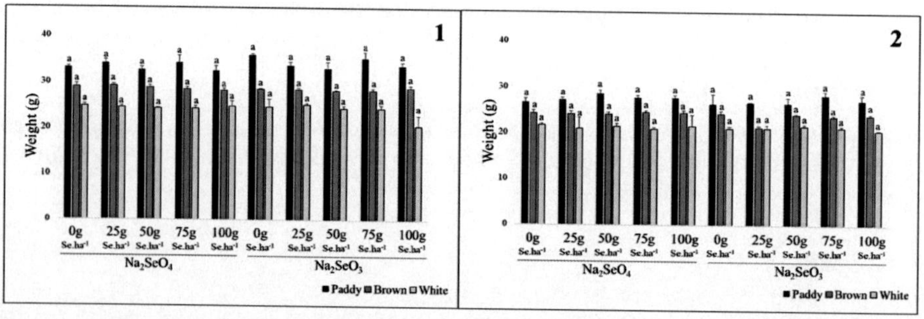

Fig. 2. Average (g) ± S.E. ($n = 4$) of 1000-grain weight of cultivar Ceres (1) and OP 1505 (2), at harvest. Letter *a* revealed the absence of significant differences among treatments (Na_2SeO_4 and Na_2SeO_3) for each cultivar (single factor ANOVA test, $P \leq 0.05$).

Protein content of rice was assessed in both cultivars, after biofortification process with selenate and selenite and did not show significant variations between the control and different treatments(Fig. 4). Yet, selenite application in whole flour of Ceres cultivar, showed a maximum value with treatment 100g $Se.ha^{-1}$ (6.72%). In refined flour, compared to control, the highest value,, were obtained with 50g of selenite (5.75%). In OP 1505 cultivar, the maximum protein values was achieved in 75g $Se.ha^{-1}$ of selenite (6.82%) and 50g $Se.ha^{-1}$ of selenate (5.83%) treatments, in whole and refined flour, respectively.

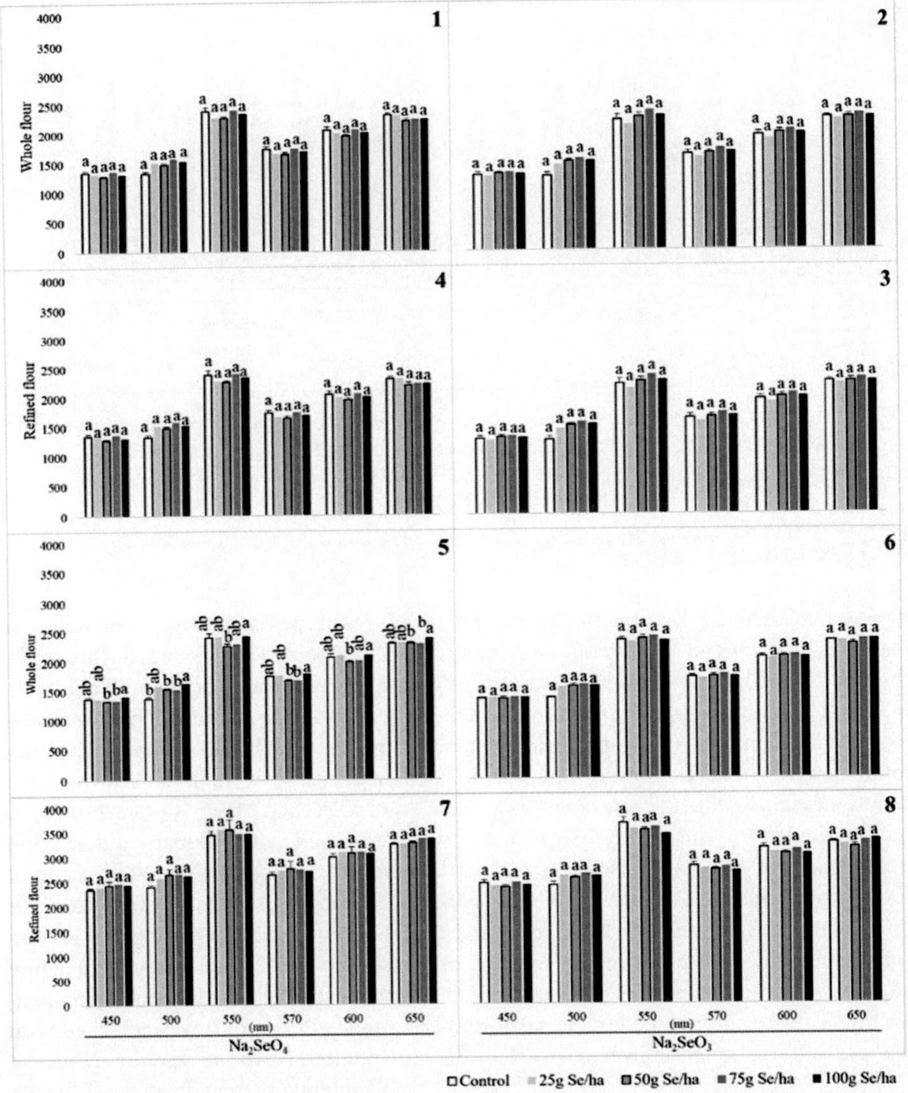

Fig. 3. Average colorimetric transmittance parameters ($n = 4$) ± S.E., in the visible spectra (450–650 nm band), of whole and refined flours of cultivars Ceres (1–4) and OP 1505 (5–8), at harvest. Letters *a* and *b* indicate significant differences among treatments (Na2SeO4 and Na2SeO3) for each cultivar (single factor ANOVA test, $P \leq 0.05$).

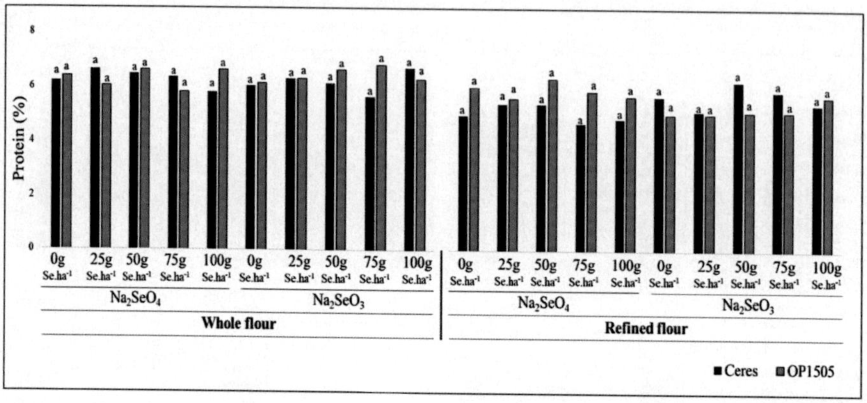

Fig. 4. Mean content of total protein ($n = 4$) in whole and refined rice flours, at harvest. Letter *a* revealed the absence of significant differences among treatments (Na_2SeO_4 and Na_2SeO_3) for each cultivar (single factor ANOVA test, $P \leq 0.05$).

4 Discussion

In order to increase Se content in rice grains of genotypes with high accumulation potential, it is crucial to develop new fertilization strategies [16]. As such, this study showed that selenite and selenate forms promoted biofortification in Se for both cultivars. In fact, through foliar spraying of Se, enabled the accumulation of this mineral in brown grains, and thereafter in the produced flours. Also, was possible to verify that, as the foliar spraying concentration of Se increase, also the increase of Se in the brown grains occurred(except with the treatment 75g $Se.ha^{-1}$ of OP 1505) (Fig. 1). Nevertheless, for both cultivars, the highest Se contents were observed with selenate and selenite treatments (100g $Se.ha^{-1}$). Sodium selenate is currently used to fortify a wide range of foods worldwide. In Finland and United States of America, is added to fertilizers for increasing the Se content of milk, meat and cereals grown on soils having low Se and sodium selenate is used to fortify infant foods, respectively [2, 17]. Actually, selenite has been reported to be more efficiently in Se biofortification than selenate [18]. In fact, for the Ceres genotype, the highest biofortification index (12.6 fold) was achieved with the selenate treatment. However, in OP 1505 was found to be in the selenite treatment (8.4 fold). Although, this behavior is probably due to selenite being a form that is easily absorbed and very mobile in plants [19]. Also, other study reported that foliar spraying of sodium selenite is more effective than sodium selenate in rice biofortification, being in agreement with our data [6].

Selenite fortification of table salt (15mg.kg^{-1}) has significantly reduced the prevalence of Keshan disease in regions of China where Se deficiency is endemic [20].

The grain weight did not reveal significant changes (Fig. 2), which indicated that the biofortification workflow implemented did not affected productivity. Besides, the absence of substantial variations in color parameters of whole and refined flours among treatments pointed that there were no symptoms of toxicity (Fig. 3). In OP 1505, whole flour showed differences in selenate treatment, which is probably related with industrial processes. Accordingly, it seems that external factors such as post-harvest processes (in this case dehusking) can influence the whole color and must be considered [21, 22].

The average of total protein content in flours varied between 5.65–6.82% and 4.33–5.83% in whole and refined flour, respectively (Fig. 4), increasing the quality of the flours. According to the data obtained, suggest that the nutritional value was not affected, mainly due to the interactions between Se and N [23, 24]. In fact, in both cultivars, the highest protein content was obtained with the selenite treatment in whole and refined flour, respectively, suggesting the presence of a higher content of amino acid, leading to an increase of Maillard reactions, creating brown compounds. The brown compounds can contribute to coloring of cookies surface [25]. Nevertheless, it is important to note that some studies showed that high protein levels can interfere to rice flour pasta viscosity [26, 27]. Indeed, for products that aim higher pasta viscosity, flours with lower protein content will have greater applicability.

5 Conclusions

Foliar spraying with selenium forms (sodium selenate and sodium selenite) at the tested concentrations (25, 50, 75, 100 g Se.ha^{-1}) did not showed symptoms of toxicity in both cultivars, allowing and a greater increase of Se in rice grains. The average of Se biofortification index ranged between 2.0–8.4 and 1.3–12.6 fold increases in OP 1505 and Ceres, respectively. Grain weight and total protein did not varied significantly in both cultivars. It is concluded that the develop workflow for Se biofortification in Ceres and OP 1505 can be used to increase this chemical element in the crop.

References

1. Zembala, M., Filek, M., Walas, S., Mrowiec, H., Kornaś, A., Miszalski, Z., Hartikainen, H.: Effect of selenium on macro- and microelement distribution and physiological parameters of rape and wheat seedlings exposed to cadmium stress. Plant Soil **329**, 457–468 (2010). https://doi.org/10.1007/s11104-009-0171-2
2. Allen, L., de Benoist, B., Dary, O., Hurrell, R.: Guidelines on food fortification with micronutrients. World Health Organization and Food and Agriculture Organization of the United Nations, Geneva (2006). ISBN: 9241594012
3. White, P.J., Broadley, M.R.: Biofortification of crops with seven mineral elements often lacking in human diets – iron, zinc, copper, calcium, magnesium, selenium and iodine. New Phytologist **182**, 49–84 (2009). https://doi.org/10.1111/j.1469-8137.2008.02738.x
4. Litov, R.E., Combs Jr., G.F.: Selenium in pediatric nutrition. Pediatrics **87**(3), 339–51 (1991). PMID: 2000274
5. White, P.J.: Selenium metabolism in plants. BBA- Gener. Sub. **1862**(11), 2333–2342 (2018). https://doi.org/10.1016/j.bbagen.2018.05.006
6. Li, H.F., McGrath, S.P., Zhao, F.J.: Selenium uptake, translocation and speciation in wheat supplied with selenate or selenite. New Phytologist **178**(1), 92–102 (2008). https://doi.org/10.1111/j.1469-8137.2007.02343.x
7. Silva, V.M., Tavanti, R.R., Gratão, P.L., Alcock, T.D., dos Reis, A.R.: Selenate and selenite affect photosynthetic pigments and ROS scavenging through distinct mechanisms in cowpea (*Vigna unguiculata* (L.) *walp*) plants. Ecotoxicol. Environ. Saf. **201**, 110777 (2020). https://doi.org/10.1016/j.ecoenv.2020.110777

8. Lucca, P., Poletti, S., Sautter, C.: Genetic engineering approaches to enrich rice with iron and vitamin A. Physiologia Plantarum **126**(3), 291–303 (2006). https://doi.org/10.1111/j.1399-3054.2006.00609.x
9. Hartikainen, H.: Biogeochemistry of selenium and its impact on food chain quality and human health. J. Trace Elem. Med. Biol. **18**(4), 309–318 (2005). https://doi.org/10.1016/j.jtemb.2005.02.009
10. Hu, Q., Chen, L., Xu, J., Zhang, Y., Pan, G.: Determination of selenium concentration in rice and the effect of foliar application of Se-enriched fertiliser or sodium selenite on the selenium content of rice. J. Sci. Food Agric. **82**(8), 869–872 (2002). https://doi.org/10.1002/jsfa.1115
11. Bouis, H.E., Hotz, C., McClafferty, B., Meenakshi, J.V., Pfeiffer, W.H.: Biofortification: a new tool to reduce micronutrient malnutrition. Food Nutr. Bull. **32**(1), 31–40 (2011). https://doi.org/10.1177/15648265110321S105
12. Pessoa, C.C., Lidon, F.C., Coelho, A.R.F, Caleiro J.C., Marques, A.C., Luís, I.C., Kullberg, J.C., Legoinha, P., Brito, M.G, Ramalho, J.C., Guerra, M., Leitao, R., Simoes, M., Campos, P.S., Semedo, J.M.N., Silva, M.M, Pais, I.P., Leal, N., Alvarenga, N., Gonçalves, E.M., Silva, M.J, Rodrigues, A.P., Abreu, M., Pessoa, M.F., Reboredo, F.H.: Calcium biofortification of Rocha pears, tissues accumulation and physicochemical implications in fresh and heat-treated fruits. Scientia Horticulturae **277** (2021). https://doi.org/10.1016/j.scienta.2020.109834
13. Lidon, F., Oliveira, K., Ribeiro, M., Pelica, J., Pataco, I., Ramalho, J., Leitão, A., Almeida, A., Campos, P., Ribeiro-Barros, A., Pais, I.P., Silva, M.M., Pessoa, M.F., Reboredo, F.H.: Selenium biofortification of rice grains and implications on macronutrients quality. J. Cereal Sci. **81**, 22–29 (2018). https://doi.org/10.1016/j.jcs.2018.03.010
14. Coelho, A., Pessoa, C., Marques, A.C., Luís, I., Daccak, D., Silva, M.M., Simões, M., Reboredo F., Pessoa, M., Legoinha, P., Ramalho J., Scotti-Campos P., Pais, I., Lidon, F.: Natural Mineral Enrichment in *Solanum tuberosum* L. cv. Agria: Accumulation of Ca and Interaction with Other Nutrients by XRF Analysis. PCR. Plants MDPI. https://doi.org/10.3390/IECPS2020-08724
15. Lidon, F.C., Daccak, D., Scotti-Campos, P., Silva, M.M., Bagulho, A.S., Pais, I., Galhano, C., Ramalho, J.C., Moreira, J., Pessoa, M.F., Reboredo, F.H.: An integrated chemical and technological approach for assessing Portuguese wheat flours quality and lengthening bread shelf-life. Emirates J. Food Agric. **31**(11), 884–894 (2019). https://doi.org/10.9755/ejfa.2019.v31.i11.2037
16. Sharma, S., Kaur, N., Kaur, S., Nayyar, H.: Selenium as a nutrient in biostimulation and biofortification of cereals. Indian J. Plants Physiol. **22**, 1–15 (2016). https://doi.org/10.1007/s40502-016-0249-9
17. Aro, A., Alfthan, G., Varo, P.: Effects of supplementation of fertilizers on human selenium status in Finland. Analyst **120**(3), 841–843 (1995). https://doi.org/10.1039/an9952000841
18. Deng, X., Liu, K., Li, M., Zhang, W., Zhao, X., Zhao, Z., Liu, X.: Difference of selenium uptake and distribution in the plant and selenium form in the grains of rice with foliar spray of selenite or selenate at different stages. Field Crops Res. **211**, 165–171 (2017). https://doi.org/10.1016/j.fcr.2017.06.008
19. Lyons, G.H., Genc, Y., Stangoulis, J.C., Palmer, L.T., Graham, R.D.: Selenium distribution in wheat grain, and the effect of postharvest processing on wheat selenium content. Biol. Trace Elem. Res. **103**, 155–168 (2005). https://doi.org/10.1385/BTER:103:2:155
20. Cheng, Y.Y., Qian, P.C.: The effect of selenium-fortified table salt in the prevention of Keshan disease on a population of 1.05 million. Biomed. Environ. Sci. **3**, 422–428 (1990). PMID: 2096847
21. Dillahunty, A.L., Siebenmorgen, T.J., Mauromoustakos, A.: Effect of temperature, exposure duration, and moisture content on color and viscosity of rice. Cereal Chem. **78**, 559–563 (2001). https://doi.org/10.1094/cchem.2001.78.5.559

22. Lanning, S., Siebenmorgen, T.: Effects of preharvest nighttime air temperatures on whiteness of head rice. Cereal Chem. **90**, 218–222 (2013). https://doi.org/10.1094/CCHEM-07-12-0082-R
23. Seregina, I.I., Niloskaya, N.T., Ostapenko, N.O.: The role of selenium in the formation of the grain yield in spring wheat. J. Cereal Sci. **1**, 44–50 (2001). https://doi.org/10.1016/j.jcs.2018.01.004
24. Reis, H.P.G., de Queiroz, J.P.B., Furlani, E., Jr., Santos, E.F., Silva, V.M., Moraes, M.F., Putti, F.F., Dos Reis, A.R.: Agronomic biofortification of upland rice with selenium and nitrogen and its relation to grain quality. J. Cereal Sci. **79**, 508–515 (2018). https://doi.org/10.1016/j.jcs.2018.01.004
25. Mancebo, C., Rodriguez, P., Gomez, M.: Assessing rice flour-starch- protein mixtures to produce gluten free sugar-snap cookies. Food Sci. Technol. **67**, 127–132 (2016). https://doi.org/10.1016/j.lwt.2015.11.045
26. Puncha-Arnon, S., Uttapap, D.: Rice starch *vs.* rice flour: differences in their properties when modified by heat–moisture treatment. Carbohydr. Polym. **91**, 85–91 (2013). https://doi.org/10.1016/j.carbpol.2012.08.006
27. Mangueze, A., Pessoa, M.F., Silva, M.J., Ndayiragije, A., Magaia, H., Cossa, V., Reboredo, F., Carvalho, M.L., Santos, J.P., Guerra, M., Ribeiro-Barros, A.I., Lidon, F.C., Ramalho, J.C.: Simultaneous zinc and selenium biofortification in rice accumulation, localization and implications on the overall mineral content of the flour. J. Cereal Sci. **82**, 34–41 (2018). https://doi.org/10.1016/j.jcs.2018.05.005

Agronomic Biofortification with Fe and Zn in Organic Tomatoes (*Lycopersicum esculentum* L.)

Ana Rita F. Coelho[1,2(✉)], Cláudia Campos Pessoa[1,2], Ana Coelho Marques[1,2], Inês Carmo Luís[1,2], Diana Daccak[1,2], Manuela Simões[1,2], Fernando Henrique Reboredo[1,2], Maria Pessoa[1,2], Maria Manuela Silva[2,3], Paulo Legoinha[1,2], Carlos Galhano[1,2], José C. Ramalho[2,4], Paula Scotti Campos[2,5], Isabel P. Pais[2,5], and Fernando C. Lidon[1,2]

[1] Earth Sciences Department, Faculdade de Ciências e Tecnologia, Universidade Nova de Lisboa, Caparica, Portugal
arf.coelho@campus.fct.unl.pt
[2] GeoBioTec Research Center, Faculdade de Ciências e Tecnologia, Universidade Nova de Lisboa, Caparica, Portugal
[3] ESEAG – Grupo Universidade Lusófona, Lisboa, Portugal
[4] PlantStress and Biodiversity Lab, Centro de Estudos Florestais, Instituto Superior Agronomia, Universidade de Lisboa, Oeiras, Portugal
[5] INIAV, Instituto Nacional de Investigação Agrária e Veterinária, Oeiras, Portugal

Abstract. The human metabolism requires Fe to support growth and the immune system. Iron further plays an essential role in the synthesis of oxygen transport proteins (namely, hemoglobin and myoglobin), while its deficiency can lead to anemia. Zinc also plays major physiological roles, working in multiple basic biochemical functions in the cells of the organism, with its deficiency triggering multiple systems, namely gastrointestinal, central nervous, immune, and reproductive. Considering the physiological impact of these both micronutrients, this study aimed to develop a technological workflow for biofortification of tomato (*Lycopersicum esculentum* L.) fruits (variety beef heart) in Fe and Zn, following an organic production mode. The experimental field was selected considering the edaphoclimatic parameters and the protocols for tomato growth according to the organic production mode. Four foliar applications were carried out during the production cycle, with a mix of two products (Zitrilon – 15%) at a concentration of 0.40 and 1.20 $kg.ha^{-1}$ and Maxiblend with two concentrations (1 and 4 $kg.ha^{-1}$). At harvest, Fe content showed a 1.8–2.5 fold increase, whereas Zn content showed increase of 34%. Morphologic parameters did not vary significantly in biofortified tomatoes. It was concluded that direct consumption of tomato (variety beef heart) production, through the organic mode, can mitigate the nutritional zinc and iron deficiency, supporting the maintenance of public health.

Keywords: Biofortification · Iron · *Lycopersicum esculentum* L. · Organic tomato production · Zinc

J. R. da Costa Sanches Galvão et al. (Eds.): *Proceedings of the 1st International Conference on Water Energy Food and Sustainability (ICoWEFS 2021)*, pp. 164–169, 2021.
https://doi.org/10.1007/978-3-030-75315-3_19

1 Introduction

The human body requires iron (Fe), being involved in several biological processes [1] and essential for blood production [2]. In fact, most of our body's iron is needed for the synthesis of oxygen transport proteins, mainly hemoglobin and myoglobin [2, 3]. However, Fe deficiency is very common, leading to anemia [3] and major health consequences, namely, in physical and cognitive development [4]. Besides Fe, zinc (Zn) is also an essential micronutrient for humans [5], playing major physiological roles [6], standing out its need in the immune system (having a major role in prevention and recovery from infectious diseases) [5]. As such, Zn deficiency can trigger problems in the immune system. However, it can also affect the gastrointestinal, central nervous and reproductive system [6]. Unfortunately, Fe and Zn deficiencies are a current global problem [5].

Considering that in human diet, nutrients come ultimately through plants [7], agronomic biofortification can be used as a strategy to increase the content of different minerals in the edible part of plants, through foliar fertilization. Besides, foliar fertilization is widely used in horticultural crops, providing essential nutrients through the leaves, being a faster way compared to soil applications [8]. Despite all nutrient's translocation from the roots to the growing portions of the plants [9], both Fe and Zn have low kinetic mobility in soils [10] and in plants [9, 10]. Yet, some studies indicate that Zn is translocated though xylem and phloem tissues [11, 12]. Also, Fe behavior in plants varies regarding the specie. In fact, in tomato plants, Fe has low mobility namely in the phloem [13, 14]. Additionally, due to soil interactions that reduce Fe uptake, foliar applications are most effective [9, 15]. Nevertheless, foliar application of Zn has also been identified as the most efficient method for increasing its content in the edible part of plants [16].

Considering the importance and the physiological impact of Fe and Zn, and being tomato one of the most popular and consumed horticultural crop worldwide [17, 18], this study aimed to develop a technological workflow for biofortification of tomato (*Lycopersicum esculentum* L.) fruits (variety beef heart – also known as Coeur de Boeuf) in Fe and Zn, following an organic production mode.

2 Materials and Methods

2.1 Experimental Field

The experimental tomato-growing field (in a plot of 15 × 60 m), located in the Western of Portugal (39° 41' 48,517" N; 8°35' 45,524"O), was used to growth beef heart variety (*Lycopersicum esculentum* L.) following an organic production mode. During the agricultural period, from 27th June (planting date) to 12th October of 2018 (harvest date), air temperatures reached an average daily of 27.5 and 11.8 °C (with maximum and minimum values varying between 0 and 41 °C, respectively). Foliar application with Fe and Zn was carried out with two treatments during the production cycle, with four foliar sprays (with 10–16 days interval). Treatments were carried out with a mix of two products (Zitrilon- 15% and Maxiblend), in which treatment 1 (T1) corresponds to a mix of 0.40 $kg.ha^{-1}$ Zitrilon (15%) and 1 $kg.ha^{-1}$ Maxiblend and treatment 2 (T2) corresponds to a mix of 1.20 $kg.ha^{-1}$ Ziltrilon (15%) and 4 $kg.ha^{-1}$ Maxiblend. Control plants were not sprayed at any time with Fe and Zn. Each treatment was performed in quadruplicate.

2.2 Iron and Zinc Content in Tomatoes

At harvest, considering the four replications in the experimental field, eight randomized tomatoes with similar sizes, were washed, dried at 60 °C until constant weight and grounded in an agate mortar. Thereafter, an acid digestion procedure was performed with a mixture of HNO_3- $HClO_4$ (4:1) according to [19, 20]. After filtration, Fe and Zn contents were measured by atomic absorption spectrophotometry, using a model Perkin Elmer AAnalyst 200, and the absorbency in mg.L^{-1} was determined with a coupled AA WinLab software.

2.3 Quality Parameters

Height, diameter, and dry weight was measure in four randomized tomatoes per treatment.

2.4 Statistical Analysis

Data were statistically analyzed using a One-Way ANOVA to assess differences among treatments in beef heart variety, followed by a Tukey's for mean comparison. A 95% confidence level was adopted for all tests.

3 Results

At harvest, Fe and Zn content was assessed (Fig. 1 and 2) in beef heart variety. Relatively to the control, biofortified tomatoes showed an increase in Fe content (Fig. 1) ranging between 1.8–2.5 fold, with T2 showed a significantly higher content of Fe.

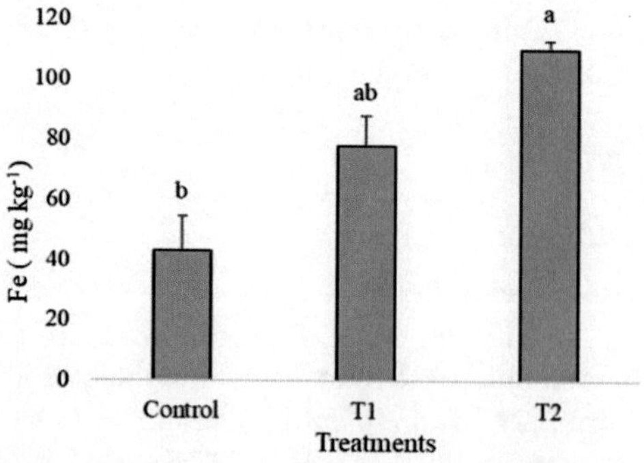

Fig. 1. Mean values ± S.E. (n = 4) of Fe content in tomatoes of *Lycopersicum esculentum* L., beef heart variety, at harvest. Letters a and b indicates significant differences, between treatments ($P \leq 0.05$). Foliar spray was carried out with two concentrations (T1 and T2). Control was not sprayed.

Regarding Zn content, only T2 showed a higher content compared to the control (Fig. 2), with an increase of 34%.

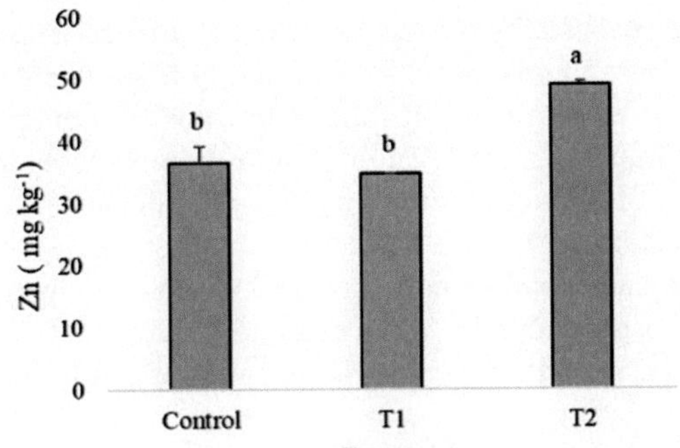

Fig. 2. Mean values ± S.E. (n = 4) of Zn content in tomatoes of *Lycopersicum esculentum* L., beef heart variety, at harvest. Letters a and b indicates significant differences, between treatments ($P \leq 0.05$). Foliar spray was carried out with two concentrations (T1 and T2). Control was not sprayed.

Independently of higher contents of Fe and Zn, height, and diameter of biofortified tomatoes did not varied significantly (Table 1). However, the average height of biofortified tomatoes decreased with the increase of the fertilizer's concentrations applied (Control > T1 > T2).

Table 1. Mean values ± S.E. (n = 4) of height (cm) and diameter (cm) in tomatoes of *Lycopersicum esculentum* L., beef heart variety, at harvest. Letters a and b indicates significant differences, between treatments ($P \leq 0.05$). Foliar spray was carried out with two concentrations (T1 and T2). Control was not sprayed.

Treatments	Height (cm)	Diameter (cm)
Control	10.17a ± 1.36	7.43a ± 0.68
T1	9.77a ± 0.66	7.90a ± 0.15
T2	8.60a ± 0.60	7.73a ± 0.73

4 Discussion

The accumulation of minerals in tomatoes grown according to the organic production mode, is not consensual. Kelly et al. [21] found that Zn and Fe displayed higher contents compared to conventional growth. However, Ordóñez-Santos et al. [22], reported that Zn content was dependent of the variety. Additionally, with tomato varieties 'Lladó' and 'Antillas', these authors [22] additionally found that Fe remained lower. Nevertheless, in this context, through our workflow for tomato biofortification with Fe and Zn, it was possible to enhance these micronutrients contents in beef heart variety (Fig. 1, 2). As so, despite the low mobility of Fe [13], our data suggest that through foliar spraying, of

Fe and Zn are also redistributed by the phloem [11, 12]. In fact, relatively to the control, Fe contents increased after spraying with increasingly higher contents of Fe fertilizer (Fig. 1), confirming the effectiveness of foliar applications with this micronutrient [9, 15]. Regarding the highest treatment (T2), Zn content also increased in tomatoes (Fig. 2), pointing that, at this fertilizer's concentrations, foliar applications with this nutrient can also be an efficient method [16].

When considering shape, biofortified tomatoes presented a slight decrease of height and an increase in diameter (Table 1). Yet, these small differences may be related to this variety's characteristics, namely presenting large and irregular fruits, heart shaped with heavy ribbing [23].

5 Conclusion

Through foliar spraying with Fe and Zn fertilizers, at concentrations reported in this study, the tomato variety "beef heart" can be enriched, through the organic production mode, with these nutrients. In fact, the maximum content obtained of Zn and Fe was achieved, without triggering the threshold of toxicity, at T2 (the highest treatment applied) and with no relevant changes in height and diameter thus, remaining in accordance to consumers expectations. Considering this technical workflow, agronomic biofortification of tomato variety "beef heart", further indicates that foliar applications with Fe and Zn fertilizers are also redistributed by phloem, complementing these nutrients uptake by roots and translocation through the xylem. Additionally, Fe and Zn can be applied to increase both nutrients in tomatoes meant for fresh consumption, while presenting a health benefit.

References

1. Swelm, R.P.V., Wetzels, J.F., Swinkels, D.W.: The multifaceted role of iron in renal health and disease. Nat. Rev. Nephrol. **16**(2), 77–98 (2020)
2. University of California San Francisco Health. Hemoglobin and Functions of Iron. https://www.ucsfhealth.org/education/hemoglobin-and-functions-of-iron. Accessed 10 Jan 2021
3. Abbaspour, N., Hurrell, R., Kelishadi, R.: Review on iron and its importance for human health. J. Res. Med. Sci.: Off. J. Isfahan Univ. Med. Sci. **19**(2), 164 (2014)
4. Ciccolini, V., Pellegrino, E., Coccina, A., Fiaschi, A.I., Cerretani, D., Sgherri, C., Quartacci, M.F., Ercoli, L.: Biofortification with iron and zinc improves nutritional and nutraceutical properties of common wheat flour and bread. J. Agric. Food Chem. **65**(27), 5443–5452 (2017)
5. Li, H., Lian, C., Zhang, Z., Shi, X., Zhang, Y.: Agro-biofortification of iron and zinc in edible portion of crops for the global south. Adv. Plants Agric. Res. **6**, 52–54 (2017)
6. Roohani, N., Hurrell, R., Kelishadi, R., Schulin, R.: Zinc and its importance for human health: an integrative review. J. Res. Med. Sci.: Off. J. Isfahan Univ. Med. Sci. **18**(2), 144 (2013)
7. Díaz-Gómez, J., Twyman, R.M., Zhu, C., Farré, G., Serrano, J.C., Portero-Otin, M., Muñoz, P., Sandmann, G., Capell, T., Christou, P.: Biofortification of crops with nutrients: Factors affecting utilization and storage. Curr. Opin. Biotechnol. **44**, 115–123 (2017)
8. Alshaal, T., El-Ramady, H.: Foliar application: From plant nutrition to biofortification. Environ. Biodivers. Soil Secur. **1**, 71–83 (2017)

9. Pagani, A., Sawyer, J.E., Mallarino, A.: Site-specific nutrient management: for nutrient management planning to improve crop production, environmental quality, and economic return. Extension and Outreach Publications 116 (2013)
10. Cornell University, Northeast Region Certified Crop Adviser (NRCCA). Nutrient Management. https://nrcca.cals.cornell.edu/soilFertilityCA/CA1/CA1_print.html. Accessed 13 Jan 2021
11. Pearson, J.N., Rengel, Z., Jenner, C.F., Graham, R.D.: Transport of zinc and manganese to developing wheat grains. Physiologia Plantarum **95**(3), 449–455 (1995)
12. Haslett, B.S., Reid, R.J., Rengel, Z.: Zinc mobility in wheat: uptake and distribution of zinc applied to leaves or roots. Ann. Bot. **87**(3), 379–386 (2001)
13. Guzmán, M., Valenzuela, J.L., Sánchez, A., Romero, L.: A method for diagnosing the status of horticultural crops – 2. Micronutrients. Phyton, Int. J. Exp. Bot. **51**, 43–56 (1990)
14. Marschner, H.: Mineral Nutrition of Higher Plants, 3rd edn. Academic Press, London (2012)
15. Sakya, A.T.: Foliar iron application on growth and yield of tomato. In: IOP Conference Series: Earth and Environmental Science, vol. 250, no. 1, p. 012001 (2019)
16. Zhang, Y., Shi, R., Rezaul, K.M., Zhang, F., Zou, C.: Iron and zinc concentrations in grain and flour of winter wheat as affected by foliar application. J. Agric. Food Chem. **58**(23), 12268–12274 (2010)
17. Sainju, U.M., Dris, R., Singh, B.: Mineral nutrition of tomato. Food Agric. Environ. **1**(2), 176–183 (2003)
18. Dorai, M., Papadopoulos, A., Gosselin, A.: Influence of electric conductivity management on greenhouse tomato yield and fruit quality. Agronomie **21**(4), 367–383 (2001)
19. Carrondo, M., Reboredo, F., Ganho, R., Santos Oliveira, J.F.: Heavy metal analysis of sediments in Tejo estuary, Portugal, using a rapid flameless atomic absorption procedure. Talanta **31**, 561–564 (1984)
20. Fernando, H., Reboredo, C., Ribeiro, G.: Vertical distribution of al, cu, fe and zn in the soil salt marshes of the Sado estuary, Portugal. Int. J. Environ. Stud. **23**(3–4), 249–253 (1984)
21. Kelly, S.D., Bateman, A.S.: Comparison of mineral concentrations in commercially grown organic and conventional crops – tomatoes (*Lycopersicon esculentum*) and lettuces (*Lactuca sativa*). Food Chem. **119**, 738–745 (2010)
22. Ordóñez-Santos, L.E., Vázquez-Odériz, M.L., Romero-Rodríguez, M.A.: Micronutrient contents in organic and conventional tomatoes (*Solanum lycopersicum* L.). Int. J. Food Sci. Technol. **46**(8), 1561–1568 (2011)
23. Specialty Produce. https://specialtyproduce.com/produce/Cow_Heart_Tomatoes_15444.php. Accessed 22 Jan 2021

Influence of Zinc Fertilization for Physical and Chemical Parameters and Sensory Properties of Grapes

Diana Daccak[1,2](✉), Cláudia Campos Pessoa[1,2], Ana Rita F. Coelho[1,2], Inês Carmo Luís[1,2], Ana Coelho Marques[1,2], Manuela Simões[1,2], Fernando Reboredo[1,2], Maria Fernanda Pessoa[1,2], Maria Manuela Silva[2,3], Carlos Galhano[1,2], Paula Scotti Campos[2,4], Isabel Pais[2,4], Nuno Alvarenga[2,4], Elsa Gonçalves[2,4], José C. Ramalho[2,5], and Fernando C. Lidon[1,2]

[1] Earth Sciences Department, Faculdade de Ciências e Tecnologia, Universidade Nova de Lisboa, Caparica, Portugal
d.daccak@campus.fct.unl.pt
[2] GeoBioTec Research Center, Faculdade de Ciências e Tecnologia, Universidade Nova de Lisboa, Caparica, Portugal
[3] ESEAG/Grupo Universidade Lusófona, Lisboa, Portugal
[4] INIAV - Instituto Nacional de Investigação Agrária e Veterinária, Quinta do Marquês, 2780-157 Oeiras, Portugal
[5] PlantStress and Biodiversity Lab, Centro de Estudos Florestais, Instituto Superior Agronomia, Universidade de Lisboa, Oeiras, Portugal

Abstract. As nowadays there is a growing concern about micronutrients deficiency in human food products, agronomic biofortification is being considered an enrichment potential technological option. In this context, to surpass limitations of Zn intake, ensuring proper level of this element for growth and maintenance of human health, this study aimed to develop a technological workflow for Zn enrichment of grapes (Syrah variety of *Vitis vinifera*) and assess the related quality. A foliar application of zinc oxide (ZnO) or zinc sulfate ($ZnSO_4$) (respectively, at a concentration of 30% and 60% - 450g ha^{-1} and 900g ha^{-1}) was carried out. At harvest, Zn enrichment in grapes revealed a 1.6 fold increase relatively to control. In Zn biofortified grapes, the analysis of the color parameters, through the Cielab system, a low luminosity, low contribution of green and greater input of blue (as indicated by L, a and b parameters) was found. Concerning to firmness, penetration force and hardness of the pulp, no significant differences were found between the control and treated samples, which additionally pointed that fruit texture was not affected. Sensory analysis of the grapes further pointed a positive response to biofortified grapes. It was concluded that biofortification of grapes can mitigate the deficiency of zinc without negative effects on quality.

Keywords: Syrah grapes · Sensory analysis · Zinc biofortification

J. R. da Costa Sanches Galvão et al. (Eds.): *Proceedings of the 1st International Conference on Water Energy Food and Sustainability (ICoWEFS 2021)*, pp. 170–177, 2021.
https://doi.org/10.1007/978-3-030-75315-3_20

1 Introduction

After the global efforts to combat the hunger in the last decades, this problem is on the rise again. Besides hunger, micronutrient deficient diets caused by lower levels of micronutrients in food, or their availability for absorption by the intestinal mucosa, reaches about 3 billion people worldwide [1].

Dietary deficiency of micronutrients leads to physiological and structural anomalies that may be surpassed throw its consumption [2]. Once micronutrients intake is necessary for the normal organism functions, different approaches are being explored, namely agronomic biofortification, consisting in enhancing mineral content in edible parts of food crops and their bioavailability throw agronomic practices [1, 3]. Yet, the efficiency of biofortification depends on the type of fertilizer and application method [4], and some aspects must be considered, namely bioavailability, side effects and technological aspects [5]. Bioavailability of micronutrients further depends on the type of fertilizer, as the form of nutrients and the interactions between them, can affect plant yields and nutrient use efficiencies [4]. Relatively to the application method, foliar spraying allows plants to assimilate nutrients with high efficiency, without depending upon root-to-shoot translocation [6], and avoid immobilization in the soil [4]. Nevertheless, foliar fertilization is a faster and more cost-effective way of agronomic biofortification [7].

Zn deficiencies are one of the most common and devastating health problems [1], resulting in retarded growth, anorexia, and hypogeusia in children, pregnancy problems and several chronic diseases in adults [8]. In the human body, Zn develops a large number of functions related to the immune system, sensory function, neurobehavioral development, reproductive health, growth, and physical development [9]. In this context many evidences have emerged that shows that Zn biofortification increases the yield and nutritional quality of crops, being the existent research mostly done in numerous cereals (maize, sorghum, barley, wheat) and dicotyledonous (soybean, safflower, pea, common bean, canola, common vetch), showing increases in yields and grain Zn concentrations [4].

Grapes are suitable fruits for Zn biofortification, being one of the most economically important fruit-tree crops in the world [6, 10]. Accordingly, as Zn enrichment in this fruit might have a high potential for mitigation of human malnutrition [6], this study aimed to develop a workflow for grapes (Syrah variety of *Vitis vinifera*) biofortification with this nutrient and assess some of the related quality parameters.

2 Materials and Methods

2.1 Experimental Field

An irrigated Syrah vineyard located in Palmela, Portugal (38° 35′23.629″N; 8° 51′ 46.208″ W), was used for Zn biofortification. The workflow was performed between 16th of June and 25th September in 2018, through foliar application with zinc sulfate ($ZnSO_4$) and zinc oxide (ZnO), at concentrations of 0% (control), 30% (SZn30 and OZn30, respectively) and 60% (SZn60 and OZn60, respectively), equivalent to 0, 450 and 900 g ha^{-1} of the respective fertilizers. Harvest was carried out on 11th October of 2018.

2.2 Physicochemical Parameters of Grapes

Density (kg m^{-3}), dry weight (%) and colorimetric CIElab analyses were measured considering three randomized grapes per treatment. For the colorimetric parameters, it was used a Minolta CR 300 colorimeter (Minolta Corp., Ramsey, NJ, USA), with D_{65} as illuminant, coupled to a sample vessel (CR-A504). Brightness (L), can vary between 0 (black) and 100 (white), indicating the variation of the tonality between dark and light. Chromaticity parameters (a* and b* coordinates) express variations between green (−60) and red (+60), or between blue (−60) and yellow (+60), respectively. Colors as white, black or gray correspond to values near the coordinates (null values) [11].

Physical characterization was carried out in the control and treated samples with 60% of ZnSO4 and ZnO (SZn60 and OZn60, respectively), using texturometer "Texture Analyzer Model TAHDiR" (Stable Micro Systems, Go-dalming, UK), with a 5 kg load cell" Interchangeable Low Force Load Cells Model LC/25". The test was carried out through penetration in the pulp of the grapes (10 per sample), with a 10 mm diameter aluminum cylindrical probe, at room temperature. Programmed test conditions were: test speed 1 mm/s and probe penetration depth 10 mm. Thus, the force vs. time graphs were obtained, where the parameters Firmness (N), Penetration work (N * s) and Pulp hardness (N) were calculated.

Sensory analyses were performed in the control and SZn60 and OZn60 treated grapes, considering four randomized grapes and a semi-trained panel of provers ($n = 28$) in individual cabins, according to NP 4258:1993 (ISO 8589:1988).

2.3 Quantification of Zn in Syrah Grapes

Grapes were cut, dried (at 60 °C, until constant weight), grounded and processed into pellets [12], being Zn contents determined at harvest, using a XRF analyzer (model XL3t 950 He GOLDD+), under He atmosphere.

2.4 Statistical Analyses

Data were statistically analyzed using a One-Way ANOVA ($p \leq 0, 05$), to access differences. Using the results, a Tukey's for mean comparison was performed (95% confidence level).

3 Results

3.1 Density and Dry Weight

Density of the grapes ranged between 870.59–1323.10 kg m^{-3}, both concentrations of $ZnSO_4$ showing the highest values relatively to the remaining treatments (significant relatively to the control) (Table 1). Moreover, the dry weight did not vary significantly among treatments, ranging between 18.09–20.86% (Table 1).

Table 1. Average content ± S.E. ($n = 3$) of density and dry weight in fruits at harvest of *Vitis vinifera* L., variety Syrah. Letters a, b and c indicates significant differences among treatments ($P \leq 0.05$).

Treatments	Density ($kg\ m^{-3}$) mean	SE	Dry weight (%) mean	SE
Control	870.59c	±0.29	18.84a	±2.13
OZn30	1229.56ab	±69.94	19.55a	±1.91
OZn60	1013.96bc	±9.75	18.09a	±2.57
SZn30	1323.10a	±76.90	19.88a	±0.28
SZn60	1263.79a	±18.51	20.86a	±1.27

3.2 Colorimetric Analyses

Through colorimetric analyses it was observed that the parameter L and b* did not have significative differences among treatments (Table 2), being values for L more closely to 0, which shows low luminosity. As b* values are negative, a more contribution of blue was found. Moreover, parameter a* showed significantly lower values in the control relatively to the remaining treatments (Table 2), therefore showing a greater contribution of red (whereas the treated samples revealed a higher contribution of green).

Table 2. Average content ± S.E. ($n = 3$) of colorimetric parameters Cielab, using the illuminant D_{65} in fruits at harvest of *Vitis vinifera* L., variety Syrah. Letters a and b indicates significant differences among treatments ($P \leq 0.05$).

Treatments	L Mean	SE	a*Mean	SE	b*Mean	SE
Control	12.037a	±0.124	0.417a	±0.020	−0.653a	±0.078
OZn30	16.193a	±1.137	−0.287b	±0.188	−1.370a	±0.614
OZn60	15.847a	±1.320	−0.457b	±0.129	−1.283a	±0.495
SZn30	15.133a	±2.058	−0.383b	±0.143	−1.360a	±0.492
SZn60	14.453a	±2.482	−0.563b	±0.160	−1.127a	±0.388

3.3 Physical and Sensory Characterization

At harvest, the firmness, penetration force and pulp hardness did not revealed significative differences (Table 3), with values ranging between 3.55–4.03, 10.58–11.24 and 0.22–0.23, respectively.

Table 3. At harvest, average content ± S.E. (n = 3) of grapes firmness, penetration force and pulp hardness of *Vitis vinifera* L., variety Syrah. Letters a and b indicates significative differences among treatments ($P \leq 0.05$).

Samples	Firmness [N]	Penetration force [N*s]	Pulp hardness [N]
Syrah control	3.84 ± 0.33a	11.24 ± 2.08a	0.23 ± 0.05a
Syrah OZn60	4.03 ± 0.78a	11.14 ± 2.55a	0.22 ± 0.04a
Syrah SZn60	3.55 ± 0.67a	10.58 ± 1.85a	0.22 ± 0.04a

Concerning to the sensory analysis, significant differences were not found for the appearance, color, texture, aroma and global appreciation (Table 4).

Table 4. At harvest, average content ± S.E. (n = 3) of some sensory parameters of grapes (*Vitis vinifera* L., variety Syrah). Letters a and b indicates significative differences among treatments ($P \leq 0.05$). The scale varied between 1 and 5.

Samples	Appearance	Color	Texture	Aroma	Global Appreciation
Syrah control	2.9 ± 0.8a	3.4 ± 0.8a	3.2 ± 0.8a	4.0 ± 0.9a	3.6 ± 0.8a
Syrah OZn60	3.0 ± 0.8a	3.2 ± 0.8a	3.3 ± 1.0a	3.6 ± 0.7a	3.4 ± 0.6a
Syrah SZn60	3.2 ± 0.9a	3.2 ± 0.9a	3.1 ± 0.9a	3.6 ± 0.8a	3.3 ± 0.9a

3.4 Quantification of Zn in Grapes

Zinc contents in the grapes varied between 13.460–21.400 ppm with treatments OZn60, SZn30 and SZn60 displaying higher values and significant differences relatively to control (Table 5).

Table 5. At harvest, average content ± S.E. (n = 3) in grapes of *Vitis vinifera* L. variety Syrah. Letters a and b indicates significant differences of Zn content among treatments ($P \leq 0.05$).

Treatments	Zn Mean (ppm)
Control	13.460 ± 0.876b
OZn30	18.843 ± 1.799ab
OZn60	21.400 ± 0.892a
SZn30	19.287 ± 1.487a
SZn60	19.580 ± 0.800a

4 Discussion

Since the beginning of the International HarvestPlus Programme (www.harvestplus.org) and its sub-project HarvestZinc (www.harvestzinc.org), the agronomic biofortification with Zn has become a global topic of interest. In this framework it was found that foliar spraying with Zn is more effective than soil application on wheat, rice and maize [13]. Indeed, foliar application can triggered a more efficient capture and allocation of nutrients [4]. In this study, foliar application was performed with $ZnSO_4$ and ZnO, resulting in a maximum increase (1.6 fold) in Zn concentration with treatment OZn60. Nevertheless, the selection of foliar fertilizers must consider the price, water solubility and the taste in foods. ZnO it is the cheapest but has low water solubility, yet in previous studies it was noted that absorption of zinc from cereal products fortified with ZnO resembled to those fortified with $ZnSO_4$ [14]. Therefore, this data suggests that under distinct edaphoclimatic conditions and using different genotypes might require different fertilizers.

In previous studies, density was related to higher quality, with values greater than 1088 $kg.m^{-3}$ being considered advantageous at nutraceutical and sensory levels [15]. Considering our data (Table 1), all Zn treated grapes presented higher densities, being more interesting these characteristics for consumers. Besides, the water content for red grapes usually rounds a value of 78.9% [16], which matches the values experimentally, varying the values between 79.14–81.91% in Syrah grapes (Table 1).

Color parameters of Syrah grapes showed (Table 2) a low luminosity (*i.e.*, values near 0), which is a characteristic of red grapes varieties. In the control, parameter a* had a greater contribution of the axis corresponding to red, but in general the values ranged between −0.563 and 0.417, being close to 0 corresponds to neutral colors as white, black or gray [11]. This effect might be linked to the climatic (light, temperature and water) and soil conditions of the growing local, but the viticulture practices have a direct impact on grape maturation and phenolic composition [17]. Besides, depending on cultivar, the skin color of red grapes is further related to anthocyanin biosynthesis and accumulation in the cells [17, 18]. The results obtained for a* could also be explained by the degree of bunch exposure, affecting the composition of anthocyanins.

The grapes size, appearance, texture and sensory attributes have an important role in acceptance of consumers [19]. Relatively to firmness, soft berries are less appreciated, being negative for commercialization [20]. Besides commercialization, texture is also essential to producers because fruit resistance during transportation depends on it [21]. At harvest, the results obtained for firmness did not present significant differences among treatments (Table 3) and after sensory analyses it was found that Zn biofortification do not interfere negatively at a sensorial level (Table 4). Nevertheless, it must be pointed that these sensory attributes depend on some aspects like grape genetic backgrounds, cultivation conditions, post-harvesting operations, and packaging treatments [19].

5 Conclusions

In Syrah grapes, the workflow for agronomic biofortification with Zn displayed a 1.6 fold increase of this nutrient. The effectiveness of both fertilizers was somewhat similar.

Negative alterations in quality, which is important for commercialization, could not be found, yet in Zn treated grapes density increased, which is advantageous at a nutraceutical and sensory level.

References

1. Freire, B., Pereira, R., Lange, C., Batista, B.: Biofortification of crop plants: a practical solution to tackle elemental deficiency. In: Sustainable Solutions for Elemental Deficiency and Excess in Crop Plants, pp. 135–182. Springer, Singapore (2020)
2. Zand, N., Christides, T., Loughrill, E.: Dietary intake of minerals. In: Handbook of Mineral Elements in Food, p. 23. Wiley, New York (2015)
3. Luís, I.C., Pessoa, C., Marques, A., Daccak, D., Coelho, A., Lidon, F., Patanita, M., Silva, M., Almeida, A., Bagulho, A., Moreira, J., Ramalho, J., Ribeiro, A., Rodrigues, A., Pessoa, M., Simões, M., Reboredo, F., Galhano, C., Legoinha, P., Leal, N., Scotti-Campos, P., Pais, I., Guerra, M., Leitão, R., Dores, J.: Tissue accumulation and quantification of Zn in biofortified triticum aestivum grains—interactions with Mn, Fe, Cu, Ca, K, P and S. In: The 1st International Electronic Conference on Plant Science session Plant Nutrition, Plant Physiology and Metabolism (2020)
4. Valença, A.W., Blake, A., Brouwer, I.D., Giller, K.E.: Agronomic biofortification of crops to fight hidden hunger in sub-Saharan Africa. Glob. Food Secur. **12**, 8–14 (2017)
5. Alegría-Toran, A., Barberá-Sáez, R., Cilla-Tatay, A.: Bioavailability of minerals in foods. In: Handbook of Mineral Elements in Food, p. 48. Wiley, New York
6. Zhao, H., Xie, X., Read, P., Loseke, B., Gamet, S., Li, W., Xu, C.: Biofortification with selenium and lithium improves nutraceutical properties of major winery grapes in the Midwestern United States. Int. J. Food Sci. Technol. **56**, 825–837 (2020)
7. Coelho, A., Pessoa, C., Marques, A., Luís, I., Daccak, D., Silva, M.M., Simões, M., Reboredo, F., Pessoa, M., Guerra, M., Leitão, R., Galhano, C., Legoinha, P., Leal, N., Ramalho, J., Ribeiro, A., Rodrigues, A., Scotti-Campos, P., Pais, I., Lidon, F.: Natural mineral enrichment in Solanum tuberosum L. cv. Agria: accumulation of ca and interaction with other nutrients by XRF analysis. In: The 1st International Electronic Conference on Plant Science session Plant Nutrition, Plant Physiology and Metabolism (2020)
8. Liu, D., Liu, Y., Zhang, W., Chen, X., Zou, C.: Agronomic approach of zinc biofortification can increase zinc bioavailability in wheat flour and thereby reduce zinc deficiency in humans. Nutrients **9**, 465 (2017)
9. Rugeles-Reyes, M., Filho, A.B.C., Aguilar, M.A.L., Silva, P.H.S.: Foliar application of Zinc in the agronomic biofortification of arugula. Food Sci. Technol. **39**(4), 1011–1017 (2019)
10. Jiang, J., Fan, X., Zhang, Y., Tang, X., Li, X., Liu, C., Zhang, Z.: Construction of a high-density genetic map and mapping of firmness in grapes (Vitis vinifera L.) based on whole-genome resequencing. Int. J. Mol. Sci.**21**, 797 (2020)
11. Marques, A., Luís, I., Coelho, A., Pessoa, C., Daccak, D., Scotti-Campos, P., Simões, M., Almeida, A., Pessoa, M., Reboredo, F., Guerra, M., Leitão, R., Galhano, C., Ramalho, J., Marques, P., Bagulho, A., Moreira, J., Ribeiro, A., Rodrigues, A., Silva, M., Legoinha, P., Leal, N., Pais, I., Lidon, F.: Rice (Oryza sativa L.) biofortification with selenium: enrichment index and interactions among nutrients. In: The 1st International Electronic Conference on Plant Science session Plant Nutrition, Plant Physiology and Metabolism (2020)

12. Pessoa, C.C., Lidon, F.C., Coelho, A.R.F., Caleiro, J.C., Marques, A.C., Luís, I.C., Kullberg, J.C., Legoinha, P., Brito, M.G., Ramalho, J.C., Guerra, M.A.M., Leitão, R.G., Simões, M., Campos, P.S., Semedo, J.M.N., Silva, M.M., Pais, I.P., Leal, N., Alvarenga, N., Gonçalves, E.M., Silva, M.J., Rodrigues, A.P., Abreu, M., Pessoa, M.F., Reboredo, F.H.: Calcium biofortification of Rocha pears, tissues accumulation and physicochemical implications in fresh and heat-treated fruits. Scientia Horticulturae **277** (2021)
13. Cakmak, I., Kutman, U.B.: Agronomic biofor-tification of cereals with zinc: a review. Eur. J. Soil Sci. **69**(1), 172–180 (2017)
14. Allen, L., Benoist, B., Dary, O., Hurrell, R.: Zinc, folate and other B vitamins, vitamin C, vitamin D, calcium, selenium and fluoride. In: Guidelines on Food Fortification with Micronutrients, p.124. World Health Organization, Switzerland (2006)
15. Rolle, L., Torchio, S.G., Segade, S.R.: Berry density and size as factors related to the physicochemical characteristics of Muscat Hamburg table grapes (Vitis vinifera L.). Food Chem. **173**, 105–113 (2015)
16. PortFIR – Portal de Informação Alimentar. https://portfir.insa.pt/foodcomp/food?19124. Accessed 15 Jan 2021
17. Blancquaert, E.H., Oberholster, A., Ricardo-da-Silva, J.M.R., Deloire, A.J.: Effects of abiotic factors on phenolic compounds in the grape berry– a review. S. Afr. J. Enol. Viticult. **40**(1), 92–105 (2019)
18. Olivares, D., Contreras, C., Muñoz, V., Rivera, S., González-Aguero, M., Retamales, J., Defilippi, B.G.: Relationship among color development, anthocyanin and pigment-related gene expression in 'Crimson Seedless' grapes treated with abscisic acid and sucrose. Plant Physiol. Biochem. **115**, 286–297 (2017)
19. Qian, X., Sun, L., Xiao-Qing, X., Zhu, Bao-Qing., Hai-Ying, X.: Differential expression of VvLOXA diversifies C6 volatile profiles in some Vitis vinifera table grape cultivars. Int. J. Mol. Sci. **18**(12), 2705 (2017)
20. Balic, I., Ejsmentewicz, T., Sanhueza, D., Silva, C., Peredo, T., Olmedo, P., Barros, M., Verdonk, J.C., Paredes, R., Meneses, C., Prieto, H., Orellana, A., Defilippi, B.G., Campos-Vargas, R.: Biochemical and physiological study of the firmness of table grape berries. Postharvest Biol. Technol. **93**, 15–23 (2014)
21. Vargas, A.M., Fajardo, C., Borrego, J., Andrés, M.T., Ibáñez, J.: Polymorphisms in VvPel associate with variation in berry texture and bunch size in the grapevine. Aust. J. Grape Wine Res. **19**(2), 193–207 (2013)

A Technological Workflow for Ca Enrichment in Rocha Pears: Implication in Quality

Cláudia Campos Pessoa[1,2(✉)], Ana Rita F. Coelho[1,2], Inês Carmo Luís[1,2], Ana Coelho Marques[1,2], Diana Daccak[1,2], Manuela Simões[1,2], Fernando Reboredo[1,2], Maria Manuela Silva[2,3], Maria Fernanda Pessoa[1,2], Carlos Galhano[1,2], Paula Scotti Campos[2,4], Isabel Pais[2,4], José C. Ramalho[2,5], and Fernando C. Lidon[1,2]

[1] Earth Sciences Department, Faculdade de Ciências e Tecnologia, Universidade Nova de Lisboa, Caparica, Portugal
c.pessoa@campus.fct.unl.pt
[2] GeoBioTec Research Center, Faculdade de Ciências e Tecnologia, Universidade Nova de Lisboa, Caparica, Portugal
[3] ESEAG/Grupo Universidade Lusófona, Lisboa, Portugal
[4] INIAV, Instituto Nacional de Investigação Agrária e Veterinária, Oeiras, Portugal
[5] PlantStress & Biodiversity Lab, Centro de Estudos Florestais, Instituto Superior Agronomia, Universidade de Lisboa, Oeiras, Portugal

Abstract. In the next 30 years, scarcity of food resources for a growing population will be linked to specific nutritional requirements for keeping public human health. In this framework, target minerals increase in the edible part of plants, through foliar spraying, can minimize the development of some pathologies, but to guarantee consumers acceptance and industry processing, quality must be monitored. In this context, considering that a lack of Ca can promote osteoporosis, this study aimed to develop a workflow for Ca enrichment of Rocha pears and assess its nutrient contents. Rocha pears were submitted to seven foliar sprayings with $CaCl_2$ and $Ca(NO_3)_2$ at different concentrations (Calcium nitrate at 0.1, 0.3, 0.6 $kg.ha^{-1}$; Calcium chloride – 0.4, 0.8, 1.6 $kg.ha^{-1}$) and the implications on Ca accumulation, as well as some quality parameters upon storage for 3 months with and without a bath in a solution with 1.3% $CaCl_2$ determined. During fruits growth, Ca accumulation in the shoot was measured at four different periods, and at harvest. It was found that increases of Ca contents in Rocha pears varied between 12.8–21.1%. After 3 months storage, some minor changes occurred in the quality parameters, namely: slight decrease in L parameter and dry weight, but bath with 1.3% $CaCl_2$ did not reveal negative impacts in the fruits. It was concluded that, under the designed workflow, Ca enrichment of Rocha pears can be achieved, while post-harvest emersion with 1.3% $CaCl_2$ did not compromise its quality.

Keywords: Agronomic biofortification · Calcium · Post-harvest application · *Pyrus communis* L. · Quality parameters

J. R. da Costa Sanches Galvão et al. (Eds.): *Proceedings of the 1st International Conference on Water Energy Food and Sustainability (ICoWEFS 2021)*, pp. 178–186, 2021.
https://doi.org/10.1007/978-3-030-75315-3_21

1 Introduction

Rocha pears (*Pyrus communis* L.) are a typical Portuguese fruit produced mainly in the West region, occupying 11 000 ha, and presenting an average annual production of 173000 tons, where 60% is exported [1]. This variety's resistance and storability are recognized, and under controlled atmosphere conditions, Rocha pears can be stored up to 10 months, thus having few losses and waste [1, 2].

In the agroindustry, monitorization of quality parameters at harvest, and consequent storage, are vital when considering the perishable properties of fruits. These factors are important to assure a chance to further processing into new processed foods with added value, or commercialization to consumers. Furthermore, the food sector faces the need to provide food to an exponential growing population [3], while needing to maintain specific nutritional requirements to ensure public health.

Pathologies such as osteoporosis, rickets, infantile bone deformity and hypertension are linked to low intakes of calcium [4–6], an important mineral that performs both structural and signaling function in the body [4, 7]. Regarding this matter, fruits and vegetables are an alternative source of Ca to dairy products, especially when considering some food intolerances, and they also contain higher Ca levels than cereals [5].

The increase of a target mineral in the edible parts of plants, can be achieved through agronomic biofortification, by means of either foliar sprays or soil fertilizers [8]. In two apple varieties, Ca biofortification (with $Ca(NO_3)_2$ and $CaCl_2$) promoted increases of Ca in fruits, without affecting quality parameters [9, 10]. For pears, although some leave damages were observed for higher concentrations, pre-harvest applications of Ca between 10–25 $kg.ha^{-1}$, resulted in Ca accumulation in fruits [11, 12]. Post-harvest benefits were further reported, for two different pear varieties, namely decreases in developments of hard end disorder [13] and positive changes in organoleptic properties (such as aroma) [14].

Post-harvest application of Ca delays fruit ripening. This is especially important for climacteric fruits such as pears, which are prone to quality deterioration, firmness loss, and disease development (such as internal browning disorders that affects the inner part of fruits), while in storage, due to remaining physiologically active [15, 16]. Papaya are climacteric fruits like pears, and immersion in 2.5% $CaCl_2$ for 15 min delayed firmness loss, reduced ethylene production, and increased total soluble solids [16].

This study aimed to assess Ca assimilation in leaves and consequent enrichment of Rocha pear fruits. Quality parameters were then monitored at harvest and after 3 months in storage, to assess changes in fruits after a Ca post-harvest bath.

2 Materials and Methods

2.1 Technological Workflow

An orchard located in the West region of Portugal (39° 29′ 52.641″N; 9° 1′ 19.604″W) was used as a test system, and from 11th May to 23th August 2018, seven foliar sprays were manually applied, with 15 days between each application.

For the first two foliar sprays, $CaCl_2$ (0.4, 0.8 and 1.6 $kg.ha^{-1}$) was used. At the same time, $Ca(NO_3)_2$ (0.1, 0.3 and 0.6 $kg.ha^{-1}$) was also applied in another group of trees, located in the same orchard. Treatments were applied in three tree rows, with a tree row in between each to avoid cross contaminations. All $CaCl_2$ treatments were applied in the same row, in a total of 4 trees per concentration, keeping one tree in between each concentration to avoid contaminations. The same layout was used for treatments with $Ca(NO_3)_2$. The control trees were monitored in a separate row from the ones with treatments. No foliar spray of neither $Ca(NO_3)_2$ or $CaCl_2$ were applied in the control. For the 3rd foliar spray, $CaCl_2$ (4 kg.ha^{-1}) was used in all 24 trees, and the remaining 4 sprays were done using $CaCl_2$ (8 kg.ha^{-1}) in the same number of trees. Harvest was performed on the 10th of September 2018. Pears from all treatments with exception of the control, were then separated into two groups, and one group was bathed in a solution with 1.3% $CaCl_2$. All fruits were kept in storage (in temperatures varying between -0.5 °C and 1 °C, and 95% humidity) and analyzed at 9th January 2019.

2.2 Ca Monitorization in Leaves and Assessment in Fruits

In randomized leaves, previously cleaned with deionized water and dried (at 60 °C, until constant weight), grounded and processed into pellets, Ca contents were determined four times during pre-harvest pulverizations (8th, 15th June and 10th, 20th July 2018), with a XRF analyzer (model XL3t 950 He GOLDD +) under He atmosphere [17].

At harvest, fruits were firstly cut and dried (until constant weight, at 60 °C), and an acid digestion procedure was performed with a mixture of HNO_3-HCl (4:1), according to [18, 19]. After filtration, Ca content was quantified by means of atomic absorption spectrophotometry (using a model *Perkin Elmer AAnalyst 200*), and the absorbency in mg/L was determined with a coupled *AA WinLab software.*

2.3 Quality Parameters

At harvest, and after 3 months of storage, dry weight (%), total soluble solids (°Brix) and colorimetric parameters (CieLab scale) were measured in bathed and unbathed fruits.

For dry weight (%) and colorimetric parameters, four randomized fruits per treatment were considered. Total soluble solids (°Brix) were measured in fruit juice, using a digital refractometer Atago (Atago, Tokyo, Japan). Colorimetric parameters (L - Brightness varying between 0 (black) and 100 (white); a* and b* coordinates - chromaticity parameters varying between −60 (green or blue) and +60 (red or yellow) were measured in the pulp, using a Minolta CR 300 colorimeter (Minolta Corp., Ramsey, NJ, USA), with D_{65} as illuminant, coupled to a sample vessel (CR-A504) [20].

2.4 Statistical Analysis

Statistical analysis, using an One-Way and a Two-Way ANOVA ($P \leq 0.05$), was performed to assess differences between treatments (a to d), and between different time analysis (r to t) and a Tukey's for mean comparison was done, considering a 95% confidence level.

3 Results

Leaves Ca content (Fig. 1) after the first two applications of $CaCl_2$ and $Ca(NO_3)_2$, was significantly higher than the control in NCa0.3, CCa0.8 and CCa1.6. After the 3rd spray (only with $CaCl_2$ 4 $kg.ha^{-1}$), all treatments were significantly higher than the control. For the 4th spray, the concentration doubled ($CaCl_2$ 8 $kg.ha^{-1}$), an after one application, treatments NCa0.3 and 0.4CCa were significantly higher than the control, while NCa0.6 was significantly lower. A second application with $CaCl_2$ 8 $kg.ha^{-1}$, did not reveal significant differences, however all Ca contents of sprayed trees were higher than control.

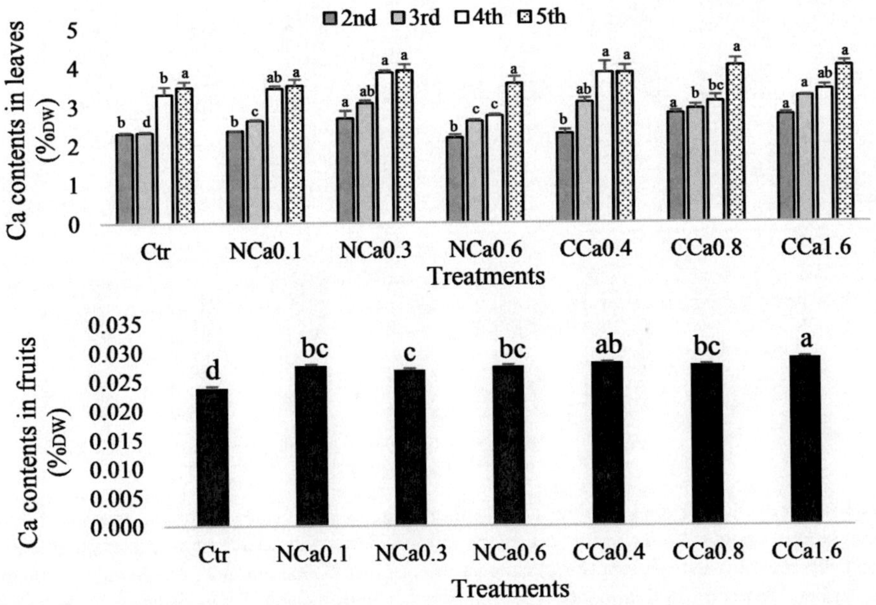

Fig. 1. Mean values of Ca contents ± SE (n = 3) in leaves from *Pyrus communis* L., variety Rocha, after the 2nd (8-6-2018), 3rd (15-6-2018), 4th and 5th leaves spray (at 10-7-2018 and 20-7-2018 respectively), and in pear fruits and at harvest (10-09-2018). Letters a, b, c, d indicate significant differences, between the treatments (statistical analysis using the single factor ANOVA test, $P \leq 0.05$). Ctr = Control; NCa0.1, NCa0.3, NCa0.6 correspond, respectively, to the initial pulverization of 0.1 $kg.ha^{-1}$, 0.3 $kg.ha^{-1}$ and 0.6 $kg.ha^{-1}$ $Ca(NO_3)_2$; CCa0.4, CCa0.8, and CCa1.6, correspond respectively to the initial pulverization of $CaCl_2$ 0.4.$kg.ha^{-1}$, 0.8.$kg.ha^{-1}$ and 1.6 $kg.ha^{-1}$.

Calcium contents in all sprayed fruits at harvest (Fig. 1) increased, being all significantly different from the control. Furthermore, treatments initially pulverized with $CaCl_2$ achieved higher values than treatments with $Ca(NO_3)_2$, with CCa1.6 presenting the highest.

At harvest, total soluble solids of the fruits (Fig. 2) varied between 9.9–13.2, with all treatments being significantly different from the control (corresponding to the highest value). After 3 months of storage (Fig. 2), values varied between 9.9–12.0, with NCa0.1 and CCa0.8 being the only two treatments significantly lower than the control. For the same period of storage, fruits with Ca post-harvest bath varied between 10.1–12.8, with NCa0.1 and NCa0.6 being significantly inferior to the control. Only treatments NCa0.1 and CCa1.6 did not present significant differences between harvest and storage fruits.

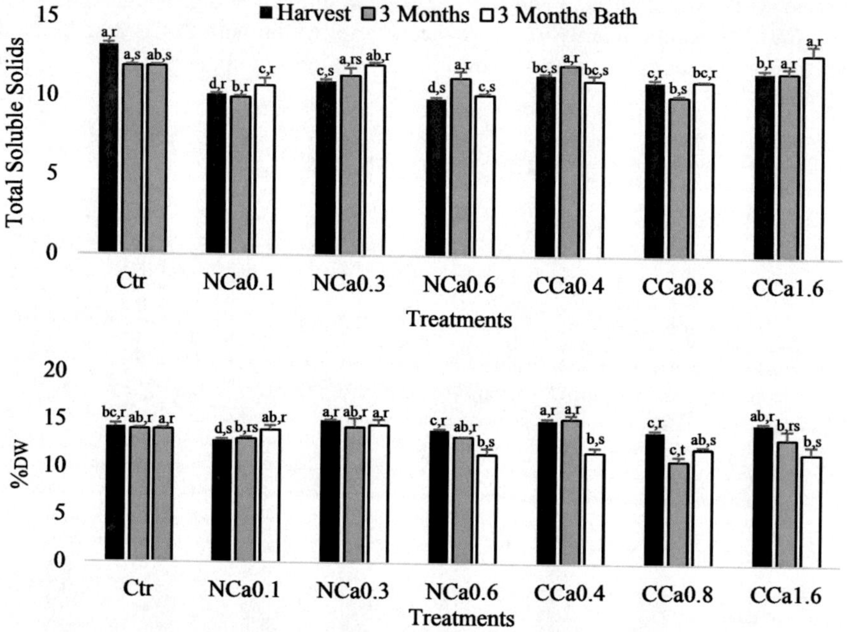

Fig. 2. Mean values of total soluble solids and dry weight ± SE (n = 4) in pear fruits of *Pyrus communis* L., variety Rocha, at harvest (black, 10-09-2018), and after 3 months storage (09-01-2019), with (white) and without (grey) Ca post-harvest bath. Letters a, b, c, d indicate significant differences, between the treatments, and letters r, s, t indicate significant differences, between harvest and stored fruits (statistical analysis using the single factor ANOVA test, $P \leq 0.05$). Ctr = Control; NCa0.1, NCa0.3, NCa0.6 correspond, respectively, to the initial pulverization of 0.1 $kg.ha^{-1}$, 0.3 $kg.ha^{-1}$ and 0.6.$kg.ha^{-1}$ $Ca(NO_3)_2$; CCa0.4, CCa0.8, and CCa1.6, correspond respectively to the initial pulverization of $CaCl_2$ 0.4.$kg.ha^{-1}$, 0.8 $kg.ha^{-1}$ and 1.6 $kg.ha^{-1}$.

At harvest, dry weight (Fig. 2) ranged from 12.8–15.1%, with NCa0.1 being significantly lower than the control, while NCa0.3 and CCa0.4 were significantly higher. In stored fruits, the value for this parameter ranged between 10.9 to 15.2%, with CCa0.8 being the only treatment significantly lower than the control. For fruits with post-harvest bath, values varied between 11.5–14.5% with NCa0.6, CCa0.4 and CCa1.6 being significantly lower than the control. Regarding stored fruits, significant differences prevailed mainly between Ca post-harvest bathed pears and fruits from harvest.

Regarding colorimetric parameters (Fig. 3), among treatments, significant differences were only observed for a*, in pears after 3 months of storage with post-harvest bath, with the control being significantly higher than NCa0.1, NCa0.6 and CCa0.8. At harvest, values ranged between 73.1–76.0, −2.2 – −1.0 and 10.0–12.4 for L, a* and b*, respectively. After 3 months of storage, values ranged from 70.1 to 72.5, −0.9 to −0.7 and 10.4 – 11.0, while fruits with Ca post-harvest bath varied between 70.0–72.5, −1.2 – −0.8 and 10.7–11.1 for L, a* and b*, respectively. When comparing storage time, significant differences were mainly observed for L parameter, with values from fruits at harvest being significantly higher than stored fruits.

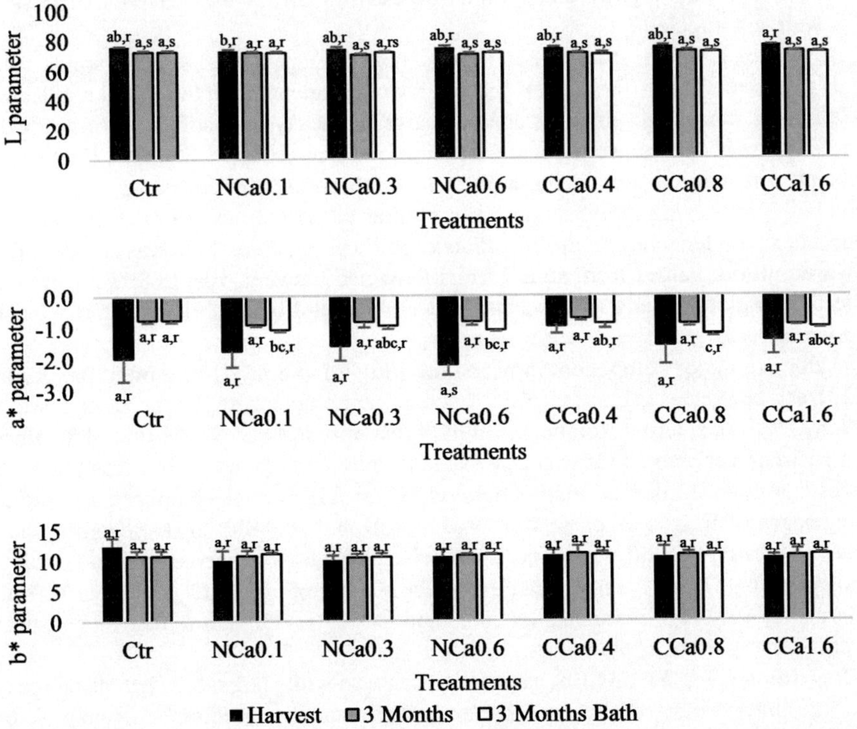

Fig. 3. Mean values ± SE (n = 4) of colorimetric analysis in pear fruits of *Pyrus communis* L., variety Rocha, at harvest (black, 10-09-2018), and after 3 months storage (09-01-2019), with (white) and without (grey) Ca post-harvest bath. Letters a, b, c indicate significant differences, between the treatments, and letters r, s indicate significant differences, between harvest and stored fruits (statistical analysis using the single factor ANOVA test, $P \leq 0.05$). Ctr = Control; NCa0.1, NCa0.3, NCa0.6 correspond, respectively, to the initial pulverization of 0.1 kg.ha^{-1}, 0.3 kg.ha^{-1} and 0.6 kg.ha^{-1} $Ca(NO_3)_2$; CCa0.4, CCa0.8, and CCa1.6, correspond respectively to the initial pulverization of $CaCl_2$ 0.4 kg.ha^{-1}, 0.8 kg.ha^{-1} and 1.6 kg.ha^{-1}.

4 Discussion

The increase of Ca content in leaves (Fig. 1) during the pre-harvest phase relates to the foliar sprays with Ca. Smaller concentrations were used in the beginning to avoid any physiological disruption on fruit development, and since no toxicity was detected, a progressive increase in the fertilizer's concentration was performed. The absence of significant differences in the last measurement, suggests a Ca mobilization to other tissues other than leaves. Ultimately, an increase in Ca was observed in fruits at harvest (Fig. 1). Our results are in accordance with another study [12], where six pre-harvest sprays with $CaCl_2$ were performed, using 2–5 $kg.ha^{-1}$ for the first five sprays, and using between 10 to 25 $kg.ha^{-1}$ for the last one. It further reported damage to leaves like defoliation, with the highest concentration presenting an intense leave burn, however fruits were not affected.

At harvest (Fig. 2), Rocha pear fruits should present a total soluble solids value ranging between 11–13 °Brix [21], and only two treatments with the initial application of $Ca(NO_3)_2$, presented values outside this range. Similarly, total soluble solids of 'Conference' pears at harvest were not affected after pre-harvest applications of $CaCl_2$ [12]. By being climacteric fruits, increase of total soluble solids is expected [15]. In general, fruits with pre-harvest sprays stored for 3 months either maintained or increased their value, but a clear tendency regarding influence of Ca post-harvest bath was not identified. However, all our values from stored fruits remained between 10−13.8 °Brix, a range indicated by another study as being an interval of variability used to characterize Rocha pear variety [15].

Water is a major component in pears, reaching up to 85.5g/100 g of edible portion [1, 22], and our values at harvest (Fig. 2) were in accordance with it, presenting losses of 84.9–87.2%. Considering storage, the highest loss only reached 89.1% (CCa0.8 without a Ca post-harvest bath). This was possible since while in storage, water loss from fruit tissues is minimized due to relative humidity control in storage chambers, helping the maintenance of fruits weight, texture, and overall marketability [15]. Regarding post-harvest bath, although differences occurred between fruits from harvest and fruits with Ca post-harvest bath (mainly those treated from the beginning with $CaCl_2$), comparing fruits at 3 months, only two treatments (NCa0.6 and CCa0.4) with bath remained significant lower relatively to the fruits without bath.

Regarding color (Fig. 3), Rocha pears presented a white pulp [1, 21], and changes in color are unwanted by consumers and the transformation food sector, being an indicator of disease development (such as internal browning) or fruits senescence. Our values reflect a high brightness (L parameter) and predominance of yellow (b* parameter), with minor contributions of green (a* parameter), being in accordance with other studies [15]. These results revealed the absence of visible internal diseases and senescence signals. Minor changes in a* parameter between treatments is not alarming since they were not significantly different from harvest. The slight decrease in brightness was not, concerning since the previous authors, considered the brightness/lightness (70.4–87.2) and b* coordinate (6.8–17.7) of pulps to be fingerprint markers, for Rocha pear variety, with our values falling within their declared intervals [15].

5 Conclusion

The biofortification workflow presented in this study, led to an increased Ca content in Rocha pear fruits. Furthermore, at harvest and in storage pears, quality parameters were not majorly affected by pre or post-harvest Ca application, reporting only minor effects in dry weight (after storage). Overall, agronomic biofortification resulted in the production of fruits with added value, and this variety storability was not negatively impacted, allowing its marketability to consumers and possibly further processing to attain products with functional properties.

References

1. Associação Nacional de Produtores de Pera Rocha. https://perarocha.pt/. Accessed 12 Jan 2021
2. Deuchande, T., Larrigaudière, C., Guterres, U., Carvalho, S.M.P., Vasconcelos, M.W.: Biochemical markers to predict internal browning disorders in Rocha pear during storage under high CO_2. J. Sci. Food Agric. **97**, 3603–3612 (2017). https://doi.org/10.1002/jsfa.8219
3. Food and Agriculture Organization of the United Nations - How to Feed the World in 2050. https://www.fao.org/fileadmin/templates/wsfs/docs/expert_paper/How_to_Feed_the_World_in_2050.pdf. Accessed 12 Jan 2021
4. EFSA Panel on NDA: Scientific opinion on dietary reference values for calcium. Efsa J. **13**, 1–82 (2015). https://doi.org/10.2903/j.efsa.2015.4101
5. Dayod, M., Tyerman, S.D., Leigh, R.A., Gilliham, M.: Calcium storage in plants and the implications for calcium biofortification. Protoplasma **247**, 215–231 (2010). https://doi.org/10.1007/s00709-010-0182-0
6. Sharma, D., Jamra, G., Singh, U.M., Sood, S., Kumar, A.: Calcium biofortification: three pronged molecular approaches for dissecting complex trait of calcium nutrition in Finger Millet (*Eleusine coracana*) for devising strategies of enrichment of food crops. Front. Plant Sci. **7**, 2028 (2017). https://doi.org/10.3389/fpls.2016.02028
7. Programa Nacional para a Promoção da Alimentação Saudável – Cálcio. https://alimentacaosaudavel.dgs.pt/nutriente/calcio/. Accessed 12 Jan 2021
8. Singh, U., Praharaj, C., Chaturvedi, S., Bohra, A.: Biofortification: introduction, approaches, limitations, and challenges. In: Singh, U., Praharaj, C., Singh, S., Singh, N. (eds.) Biofortification of Food Crops, pp. 3–18. Springer, New Delhi, Heidelberg (2016). https://doi.org/10.1007/978-81-322-2716-8
9. Lidon, F., Ribeiro, V., Reboredo, F., Pessoa, M., Santos, M., Ramos, P., Sánchez, C.: Calcium biofortification of apples: interaction with Macronutrients. In: COST Action Project FA 0905 Mineral-Improved Crop Production for Healthy Food and Feed, Final Conference Agronomic, Molecular Genetics and Human Nutrition Approaches for Improving the Nutritional Quality and Safety of Food Crops, pp. 102–103. Sabanci University, Turkey (2014)
10. Reboredo, F., Sánchez, C., Santos, M., Ramos, P., Rodrigues, C., Ribeiro, V., Pessoa, M., Lidon, F.: Calcium biofortification of apples: implications on fruit quality parameters. In: COST Action Project FA 0905 Mineral-Improved Crop Production for Healthy Food and Feed, Final Conference Agronomic, Molecular Genetics and Human Nutrition Approaches for Improving the Nutritional Quality and Safety of Food Crops, pp. 52–53. Sabanci University, Turkey (2014)
11. Gąstoł, M., Domagała-Świątkiewicz, I.: Mineral composition of "conference" pears as affected by different foliar sprays. Pol. J. of Environ. Stud. **18**, 739–742 (2009)
12. Wójcik, P., Skorupińska, A., Filipczak, J.: Impacts of preharvest fall sprays of calcium chloride at high rates on quality and "conference" pear storability. Sci. Hortic-Amsterdam **168**, 51–57 (2014). https://doi.org/10.1016/j.scienta.2014.01.017
13. Wang, Y., Zhang, X., Wang, Y., Yang, S., Qu, H.: The changes of intracellular calcium concentration and distribution in the hard end pear (*Pyrus pyrifolia* cv. "Whangkeumbae") fruit. Cell Calcium **71**, 15–23 (2018). https://doi.org/10.1016/j.ceca.2017.11.002

14. Wei, S., Qin, G., Zhang, H., Tao, S., Wu, J., Wang, S., Zhang, S.: Calcium treatments promote the aroma volatiles emission of pear (*Pyrus ussuriensis* "Nanguoli") fruit during post-harvest ripening process. Sci. Hortic-Amsterdam **215**, 102–111 (2017). https://doi.org/10.1016/j.scienta.2016.12.008
15. Pedro, S.I., Coelho, E., Peres, F., Machado, A., Rodrigues, A., Wessel, D.F., Coimbra, M.A., Anjos, O.: Physicochemical fingerprint of "Pera Rocha do Oeste". A PDO Pear Native from Portugal. Foods **9**, 1209 (2020). https://doi.org/10.3390/foods9091209
16. Gao, Q., Tan, Q., Song, Z., Chen, W., Li, X., Zhu, X.: Calcium chloride postharvest treatment delays the ripening and softening of papaya fruit. J. Food Process. Pres. **44**(8), 1–12 (2020). https://doi.org/10.1111/jfpp.14604
17. Pelica, J., Barbosa, S., Lidon, F., Pessoa, M.F., Reboredo, F., Calvão, T.: The paradigm of high concentration of metals of natural or anthropogenic origin in soils – the case of Neves-Corvo mine area (Southern Portugal). J. Geochem. Explor. **186**, 12–23 (2018). https://doi.org/10.1016/j.gexplo.2017.11.021
18. Carrondo, M., Reboredo, F., Ganho, R., Santos Oliveira, J.F.: Heavy metal analysis of sediments in Tejo estuary, Portugal, using a rapid flameless atomic absorption procedure. Talanta **31**, 561–564 (1984)
19. Reboredo, F.H.S., Ribeiro, C.A.G.: Vertical distribution of Al, Cu, Fe and Zn in the soil salt marshes of the Sado estuary, Portugal. Int. J. Environ. Stud. **23**, 249–253 (1984). https://doi.org/10.1080/00207238408710160
20. Ramalho, J.C., Pais, I.P., Leitão, A.E., Guerra, M., Reboredo, F.H., Máguas, C.M., Carvalho, M.L., Scotti-Campos, P., Ribeiro-Barros, A.I., Lidon, F.J.C., DaMatta, F.M.: Can elevated air [CO_2] conditions mitigate the predicted warming impact on the quality of coffee bean? Front. Plant Sci. **9**, 287 (2018). https://doi.org/10.3389/fpls.2018.00287
21. PrimoFruta – Producing norms for Pera Rocha do Oeste. https://primofruta.pt/en/rocha-pear. Accessed 13 Jan 2021
22. PortFIR (Portal de Informação Alimentar) – Food Composition Pear. https://portfir.insa.pt/foodcomp/food?19540. Accessed 13 Jan 2021

A Deep Learning-Based Fruit Quality Assessment System

S. Harini[1], Parijat Deshpande[2](✉), Jayita Dutta[2], and Beena Rai[2]

[1] PSG College of Technology, Coimbatore, India
[2] CTO, TCS Research, Pune, India
parijat.deshpande@tcs.com

Abstract. Fresh and ideally ripe fruits contribute to nutrition, dietary diversity, and consequently to consumer health and are therefore imperative to ensure their quality. Fruit quality check is conventionally conducted by visual inspection performed by humans, which is time-consuming and may also become inconsistent. Cost-effective and accurate sorting can be achieved by automated sorting. We have proposed a deep learning-based technique that detects the quality of fruits with an accuracy of 89.5% . This process has two major aspects, first is the fruit recognition task, subsequently followed by detecting the quality of the fruit, which is a good/bad classification of the recognized fruit. We have curated a database of the 12 most commonly used fruits in India. However, to operate for fruits outside our database, we have employed a transfer learning approach. To facilitate this, we have used the Siamese network with triplet loss, which finds a similarity between the fruit and gives us the closest match. For each fruit, based on the availability of data, we have developed models by using a deep learning approach and machine learning models on extracting image-based features. This technique presents an automatic approach for the detection of spoilage in fruits efficiently.

Keywords: Fruit quality · Deep learning · Transfer learning

1 Introduction

Fruits and vegetables are an important supplement to the human diet as they provide the essential minerals, vitamins, and fibre required for maintaining health. There are different varieties of fruits available and there are many ways to prepare, cook, and serve them. Apart from consuming raw fruits, nowadays, there is an increasing demand for processed food such as jams, jellies, fruit beverages, pickles, etc. in the market. Fruits are highly perishable and are highly susceptible to bruising, senescence, decay, and rotting. The quality of a product depends on the quality of fruits being used in the preparation of that product. This in turn affects the sales of the product in the market. In an agricultural country, all the pre and post-harvest removal of spoiled fruits are done manually with help of labor. The manual process is very time-consuming, less efficient. This may cause a delay in the supply chain. Any delay in the supply chain or disruption

J. R. da Costa Sanches Galvão et al. (Eds.): *Proceedings of the 1st International Conference on Water Energy Food and Sustainability (ICoWEFS 2021)*, pp. 187–192, 2021.
https://doi.org/10.1007/978-3-030-75315-3_22

in storage conditions can seriously impair the product's quality. Food safety is a major issue [1], with a large number of people around the world suffering from illness due to the consumption of contaminated food products. An estimated 600 million – almost 1 in 10 people in the world fall ill after eating contaminated food [2]. Early detection of food spoilage helps industries to reduce wastages as well as delivering good quality products to their customers. This article gives an approach for the detection of spoilage in fruits using Artificial Intelligence (AI) techniques and an application that runs on the Amazon cloud.

On page 16, 1st paragraph in El-Bendery et al. [3], multi-class classification approaches investigating and classifying stages of ripeness of tomato are used. A study Lal et al. [4] automatically detected and classified the maturity level of apple fruit. The purpose of the study in Saadl et al. [5] is to detect the ripeness of bananas and classify it into three categories such as ripe, unripe, and overripe. On page 2, the 2nd paragraph in Castro et al. [6] aims to classify the Cape gooseberry fruit into different ripeness levels. [7] have used the fuzzy logic algorithm to determine the ripeness of bananas. On page 3, the 1st paragraph in Ashok et al. [8] has used a probabilistic neural network approach for the segmentation of defective apples from the non-defective ones. Kamble et al. [9] has used a faster Convolutional Neural Networks (CNN) model to identify the ripening stages in Apple, Banana, and Mango. In all the previously published work, the detection of the quality of fruits is done for a limited number of fruits and also for fruits only in the database. But in this paper, we propose techniques, using which we can detect the quality of many fruits that are not in our database as well.

2 Materials and Methods

We have used 12 different fruits in this work namely apple, banana, guava, jackfruit, lemon, mango, orange, papaya, pear, pineapple, pomegranate, and strawberry with two classes good and bad. Data for fruit recognition was obtained from the dataset [10]. Data for goodness detection of fruits are obtained from different sources. For apple, banana, and orange we have used the dataset from [11] which has about 1000 images in each class of apple, banana, and orange respectively. For the remaining other fruits, data was curated via web scraping following due protocol to avoid copyrighted images. Web scraping is implemented using the Selenium package, which is available in Python. Following this, we had curated limited data of 50 images in each class of guava, jackfruit, lemon, mango, papaya, pear, pineapple, pomegranate, and strawberry respectively. Thus, we have two types of data, one belonging to the 'adequate data category' and the other one to the 'limited data category'.

The proposed work entails a two-stage approach, fruit recognition, and bi-classification between good and bad. As an initial step, there is a need to pre-process the input image. Following this is to recognize the fruit, which can be accurately done using conventional learning algorithms. But, since we need to extend the recognition task for other fruits that are not in our database, we have implemented one-shot learning using Siamese Neural Network (SNN) [12] with triplet loss [13].

Traditionally, a neural network learns to predict multiple classes. This poses a problem when we need to add/remove new classes to the data. In this case, we have to update the neural network and retrain it on the whole dataset. Also, deep neural networks need a large volume of data to train on. SNN, on the other hand, learns a similarity function. Thus, we can train it to see how similar the two images are. This enables us to classify new classes of data without training the network again. To analyze the similarity, we have transformed the images into embeddings. So, the mathematical distance between the embeddings of two similar images is small and for two different images it is large (Fig. 1, Fig. 2).

Fig. 1. Embeddings of fruit 'Durian' not used in our database. The red box denotes the closest similarity to Durian fruit, which is jackfruit.

Fig. 2. Embeddings of a strawberry bunch being similar to a strawberry. The red box denotes the closest similarity to the fruit

Followed by fruit recognition, we have used separate approaches for "adequate data category" and "limited data category". For the "adequate data category", we have used deep CNN, which has 3 convolutional blocks with the Rectified Linear Unit (ReLU) activation function using binary cross-entropy as a loss function. We have used Adam as an optimizer since it performed better than other optimizers like Stochastic Gradient Descent (SGD) for our task. For the "limited data category", we have developed individual models for each of the 9 fruits. For each model, we have extracted color and texture features from the respective fruit images. For color-based features, we have calculated the mean and standard deviation for each of Red Green Blue (RGB), Hue Saturation Value (HSV), and L*A*B* color spaces [6]. For texture-based features, we have used the Grey Level Coherence Matrix (GLCM) [14] to extract features like entropy, contrast, etc. [15]. Since we have extracted a lot of features, dimensionality reduction is required. To facilitate that, Linear Discriminant Analysis (LDA) is used [16]. Further, we have used these features with machine learning algorithms like K- Nearest Neighbors, Logistic Regression (LR), Support Vector Machines (SVM) with Voting Classifier for final predictions.

3 Results and Discussions

The solution is implemented in 2-stages namely, Fruit recognition and Fruit quality assessment. Fruit recognition is achieved using SNN, and the Receiver Operator Characteristics (ROC) curve with Area Under the Curve (AUC) value along with specificity value (Fig. 3) is chosen as the performance metric.

For fruit quality detection, we present the results for 3 learning categories which are 'adequate data, limited data, and unknown data.

1. For the fruits belonging to the 'adequate data category', we report an accuracy of 97% using CNN. E.g. Estimation of quality of Bananas.
2. For the fruits belonging to the 'limited data category', an average accuracy of 85.61% is achieved for the combination model with majority voting, in which the performance of the individual fruit model is reported in Table 1. E.g. Estimation of quality of Kesar Mango which has limited training data.
3. For the fruits which are not in our database i.e. unknown data, we have used transfer learning [17] with the closest similar fruit as recognized by SNN. We have tested it on 240 random fruit images from the Google image database, which had both good and spoiled fruits, from which we have achieved an accuracy of 81.5%. E.g. Estimation of the quality of Durian which is not in the training database but similar Jackfruit data is available.

Table 1. Accuracy values for each fruit's model based on the voting classifier

Fruit	Accuracy
Apple	98.00%
Banana	95.50%
Orange	97.50%
Guava	90.91%
Jackfruit	81.82%
Lemon	81.80%
Mango	91.66%
Papaya	90.90%
Pear	85.70%
Pineapple	75.00%
Pomegranate	81.82%
Strawberry	90.92%
Unknown data	81.50%

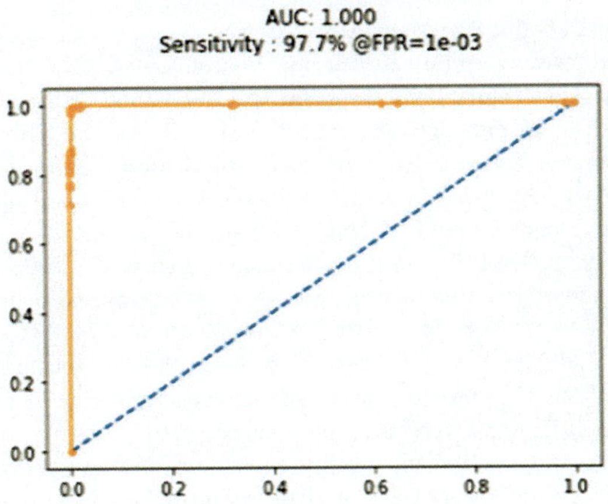

Fig. 3. ROC curve for SNN

In the future, with the availability of additional data, we aim to detect the quality of many different food items like vegetables, meats, seafood, etc. This technique has immense potential and can be implemented in an E-commerce/grocery portal for assessing the quality of their food items to establish the authenticity of rejects. This can dispel the need for operator-based manual intervention for quality assessment of relatively low-cost items such as fresh produce.

4 Conclusions

Ensuring safe and adequate food to all is one of the Sustainable Development Goals in the United Nations Development Programme (UNDP's) charter [18] and therefore a major concern for society. We propose a deep learning-based automated food quality rapid check system to ensure food safety and avoid the need for manual intervention. In this paper, we have presented a solution for automated fruit quality checks based on AI techniques. Such a system will ensure quality fruit supply to the respective food processing industry and therefore corresponding healthy and safe food products.

Acknowledgements. This work is done as a part of an internship project at Tata Research and Development and Design Centre (TRDDC), Pune, India. This project is funded by TCSL. All the authors declare that they have no conflict of interest.

References

1. FAO Homepage. https://www.fao.org/home/en/. Accessed 16 Jan 2021
2. WHO Homepage - Food Safety Fact sheets. https://www.who.int/en/news-room/fact-sheets/detail/food-safety/. Accessed 05 Jan 2021
3. El-Bendary, N., El Hariri, E., Hassanien, A.E., Badr, A.: Using machine learning techniques for evaluating tomato ripeness. Expert Syst. Appl. **42**(4), 1892–1905 (2015)

4. Lal, S., Behera, S.K., Sethy, P.K., Rath, A.K.: Identification and counting of mature apple fruit based on BP feed forward neural network. In: Proceedings of 2017 3rd IEEE International Conference on Sensing, Signal Processing and Security, ICSSS 2017, pp. 361–368 (2017)
5. Saadl, H., Ismaie, A.P., Othmanl, N., Jusohl, M.H., Naim, N.F., Ahmad, N.A.: Recognizing the ripeness of bananas using artificial neural network based on histogram approach. In: ICSIPA09 - 2009 IEEE International Conference on Signal and Image Processing Applications, Conference Proceedings, pp. 536–541 (2009)
6. Castro, W., Oblitas, J., De-La-Torre, M., Cotrina, C., Bazan, K., Avila-George, H.: Classification of cape gooseberry fruit according to its level of ripeness using machine learning techniques and different color spaces. IEEE Access **7**, 27389–27400 (2019)
7. Kulkarni, M.N., Mudhalwadkar, R.P.: Data driven modelling for banana ripeness assessment. In: Proceedings of the 2017 International Conference on Intelligent Computing and Control Systems, ICICCS 2017, vol. 2018-Janua, pp. 662–665 (2017)
8. Ashok, V., Vinod, D.S.: Automatic quality evaluation of fruits using probabilistic neural network approach. In: Proceedings of 2014 International Conference on Contemporary Computing and Informatics, IC3I 2014, pp. 308–311 (2014)
9. Kamble, P.R., Marathe, R.S., Jha, S.K., Sunil, S., Katti, J.V., Chinchwad, P.: Development of an effective system to identify fruit ripening stage for apple, banana and mango. **29**(12), 2766–2772 (2020)
10. Kaggle. https://www.kaggle.com/moltean/fruits/. Accessed 16 Dec 2020
11. Kaggle. https://www.kaggle.com/sriramr/fruits-fresh-and-rotten-for-classification/. Accessed 16 Dec 2020
12. Schroff, F., Kalenichenko, D., Philbin, J.: FaceNet: a unified embedding for face recognition and clustering. In: Proceedings of the IEEE Computer Society Conference on Computer Vision and Pattern Recognition, pp. 815–823 (2015)
13. Hermans, A., Beyer, L., Leibe, B.: In defense of the triplet loss for person re-identification (2017)
14. Based, C., Retrieval, I.: Chapter 5 color and texture based image, pp 219–278 (2019)
15. Mohanaiah, P., Sathyanarayana, P., Gurukumar, L.: Image texture feature extraction using GLCM approach. Int. J. Sci. Res. Publ. **3**(5), 1–5 (2013)
16. Song, F., Mei, D., Li, H.: Feature selection based on linear discriminant analysis. In: Proceedings - 2010 International Conference on Intelligent System Design and Engineering Application, ISDEA 2010, vol. 1, pp. 746–749 (2010)
17. Torrey, L.: Transfer learning. In: Soria, E., Martin, J., Magdalena, R., Martinez, M., Serrano, A. (eds.) Handbook of Research on Machine Learning Applications, pp. 657–665. IGI Global (2014)
18. UNDP Webpage. https://www.undp.org/content/undp/en/home/sustainable-development-goals.html. Accessed 20 Nov 2020

Storage Procedures Influence the Antioxidant Capacity of Bee Pollen

Telma Orvalho[1], Daniela C. Vaz[2,3], Vânia Ribeiro[2,4], and Maria Jorge Campos[5(✉)]

[1] Inov'Linea - TagusValley, Parque Tecnológico do Vale do Tejo, 2200-062 Abrantes, Portugal
[2] CiTechCare, Centre for Innovative Care and Health Technology, School of Health Sciences, Polytechnic of Leiria, 2411-901 Leiria, Portugal
[3] Coimbra Chemistry Centre, University of Coimbra, 3004-535 Coimbra, Portugal
[4] GeoBioTec, GeoBioSciences, GeoTechnologies and GeoEngineering, Earth Sciences Department, Universidade NOVA de Lisboa, 2829-516 Caparica, Portugal
[5] MARE-IPLeiria, Marine and Environmental Sciences Centre, School of Tourism and Marine Technology, Polytechnic of Leiria, 2520-630 Peniche, Portugal
mcampos@ipleiria.pt

Abstract. Bee products, including bee pollen, are known for their therapeutic and nutritional properties due to the presence of several nutrients (essential lipids and amino acids, fat-soluble and water-soluble vitamins and bioelements) as well as natural antioxidants, such as flavonoids (e.g.: kaempferol, quercetin, and isorhamnetin), phenolic acids (chlorogenic acid) and catechins. Therefore, given its health beneficial properties, bee pollen is considered a valuable dietary supplement to the human diet, being currently marketed, and used as a food supplement. Nonetheless, the chemical and antioxidant properties of pollen can be altered by several factors, such as improper storage, type of sample handling and processing and sample ageing. Thus, we have evaluated the influence of different processing and conservation methods (freeze-drying, drying, and freezing) in the amount of total phenolics and antioxidant capacity present in 1-year samples, submitted to the different storing procedures. Total phenolic content (TPC) was estimated spectrophotometrically by using the *Folin-Ciocalteu* method and nuclear magnetic resonance spectroscopy (^{1}H NMR); while the antioxidant capacity (AC) was determined by the ABTS (2,2′-azino-bis (3-ethylbenzothiazoline-6-sulfonic acid)) and DPPH (2,2-Diphenyl-1-picrylhydrazyl) methods, that measure radical scavenging activities. Different solvent extraction procedures (ethanol and acetone based) were also applied to the several samples prior to analysis. The results showed that the antioxidant activity and total phenolics reduce considerably after one year of storage, especially when bee-pollen is submitted to freezing procedures.

Keywords: Bee-pollen · Antioxidant capacity · Total polyphenols · Conservation procedures

J. R. da Costa Sanches Galvão et al. (Eds.): *Proceedings of the 1st International Conference on Water Energy Food and Sustainability (ICoWEFS 2021)*, pp. 193–200, 2021.
https://doi.org/10.1007/978-3-030-75315-3_23

1 Introduction

Bee pollen is a natural api-material formed by agglutination of bee-collected floral pollen, nectar, and bee saliva (rich in enzymes, such as amylase and catalase) in the form of granules or pellets of 1.4–4 mm in size [1–4]. Pollen grains may also differ in shape (round, cylindrical, bell-shaped, triangular, or thorny) and colour (from bright yellow to black) [3, 4]. Bee pollen is rich in a wide variety of primary and secondary metabolites with important properties and bioactivities, such as antioxidant, anti-inflammatory, anti-carcinogenic, antibacterial, antifungal, hepatoprotective, anti-atherosclerotic, hypolipidemic and detoxifying, among others [1–3, 5–9]. Although with variable composition, depending on the season, ecological habitat, biogeographic region and floral origin, certain components are always present in bee pollen, *i.e.*, carbohydrates (generally the most abundant), protein and essential amino acids, fatty acids, crude fibre, minerals, and vitamins [1, 2, 10, 11]. Bee-pollen mineral composition includes calcium, magnesium, iron, zinc, copper, and potassium; while its composition in vitamins includes both water- and fat-soluble vitamins, like tocopherol, niacin, thiamine, biotin, folic acid, ascorbic acid and provitamin A (β-carotene) [1, 2, 10, 11].

Bee pollen is also a source of natural antioxidants capable of counteracting the effects of oxidative stress [5, 13]. Oxidative stress results from the increase in the concentration of reactive oxygen species (ROS), that contribute to cell membrane, protein and/or DNA damage, induce inflammation and are associated to several degenerative diseases [14]. In general, antioxidant compounds possess a phenolic character, with the ability to scavenge these free radicals. Thus, the antioxidant effect of bee pollen is both attributed to the activity of antioxidant enzymes, as well as to the presence of secondary plant metabolites, such as phenolic compounds, carotenoids, vitamin C, vitamin E, and glutathione [1–3, 5, 7, 9, 11, 13]. Bee-pollen is rich in polyphenols, mainly phenolic acids, and flavonoids, such as caffeic acid, catechins, kaempferol, quercetin, and isorhamnetin [5, 7, 9, 12]. Phenolic compounds are amongst the most studied antioxidants due to their biological effects and therapeutic properties, namely, their effect as metal chelators and free radical scavengers or as modulators of enzyme activity and protectors of signal transduction pathways [15]. Thus, given its medicinal and nutritional value, this bee product is considered a perfect whole health-promoting food and a valuable dietary supplement [10, 11].

Bee pollen should be consumed fresh, soon after harvesting, to preserve its composition and special properties. Nonetheless, in order to reach more consumers and markets this api-material must be preserved and stored. Therefore, storage conditions are of major importance to retain the product's biological activity. While reducing sugars, proteins and vitamins do not tolerate high temperatures; many other bioactive compounds can suffer oxidation. Thus, low processing preserving methods should be applied. The objective of this study was to analyse the total phenolic content (TPC) and antioxidant capacity (AC) present in three bee pollen samples submitted to three preserving procedures, namely, freezing, freeze-drying and oven drying, after 1-year of storage.

2 Materials and Methods

2.1 Sample Preparation

Fresh bee pollen was harvested from the hive at A-dos-Francos (Caldas da Rainha, Portugal) 39°19′27″N 9°1′10″W; 119 m a.s.L.. and submitted to three preservation procedures, namely, freezing at −20 °C, freeze-drying for 120 h, 0.140 mbar, −89 °C (Labconco freezone 6 plus) and oven drying for 32 h, 40 °C, according to the scheme depicted in Fig. 1. Freeze dried and oven dried samples were vacuum packed and stored at room temperature, away from light, for 1 year (Fig. 2). Fresh commercial bee pollen was also used for comparison.

Bee pollen extracts in ethanol (ethanol: water, 70:30, v/v) and acetone (acetone: water, 70:30, v/v) were prepared by manually crushing in a mortar 1.0 g of sample (Metller Toledo AE200-S), diluted in 5 ml of solvent, submitted to magnetic stirring for 1 h, 9000 rpm, at room temperature, protected from light (JP-Selecta magnetic stirrer). Extracts were then centrifuged at 3000 rpm, 20 °C, for 5 min (Heraeus Labofuge 200). Supernatants were filtered through Whatman filters (pore size ~10 mm). Pollen extracts were further diluted in the respective solvents for the *Folin-Ciocalteu,* ABTS and DPPH assays. NMR experiments used deuterated solvents.

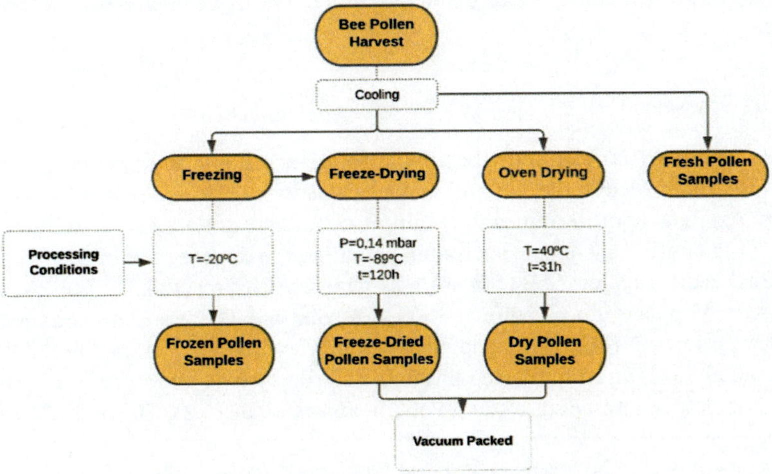

Fig. 1. Processing scheme of the bee pollen samples.

2.2 *Folin-Ciocalteu* Assay

Total phenolic content (TPC) was determined by the *Folin-Ciocalteau* assay [16, 17]. Bee-pollen extracts in ethanol (ethanol: water, 70:30, v/v) and acetone (acetone: water, 70:30, v/v) were tested for phenolic compounds by the *Folin-Ciocalteau* assay, as

Fig. 2. Samples of bee-pollen submitted to (from left to right) freezing, freeze-drying and oven drying.

described and adapted by Brito et al. [16, 17]. Phenolic compounds react with the *Folin-Ciocalteu* reagent to produce phosphomolybdate and phosphotungstate resulting in a blue colour, detected spectrophotometrically at 750 nm.

Small aliquots of the bee pollen extracts and standard solutions of gallic acid (0.03, 0.05, 0.10, 0.25 and 0.50 g/L) were mixed with the *Folin-Ciocalteau* reagent. Sodium carbonate Na_2CO_3 (70.0 g/L) and Milli-Q ultrapure water were added to the mixture and let to incubate for 60 min, at 20 °C. Absorbances were measured at 750 nm (Varian Cary 50 Scan UV Visible spectrophotometer). Control samples used ethanol (ethanol: water, 70:30, v/v) and acetone (acetone: water, 70:30, v/v) diluted in the *Folin-Ciocalteau* reagent. Total phenolic content (TPC) was expressed as grams of gallic acid equivalents (g GAE)/100 g, by making use of the calibration curve. All samples were analysed in triplicate.

2.3 ABTS Assay

The ABTS (2,2′-Azino-bis(3-ethylben zothiazoline-6-sulfonic acid) diammonium salt) assay was used to determine the antioxidant capacity of the bee-pollen samples [18, 19]. The ABTS assay determines the ability to eliminate antioxidant activity, reacting with a strong antioxidant agent (potassium persulfate) in the presence of the ABTS salt. The ABTS radical cation (ABTS● +) was generated by reaction of 7.0 mM ABTS with 2.45 mM potassium persulfate ($K_2O_2S_8$) solution. The assay was conducted as described previously [18], with adaptations [19]. The mixture was held in the dark, at room temperature, for 16 h, filtered through a sterile syringe and filter (0.22 μm) and diluted with Milli-Q ultrapure water, to obtain an absorbance of 0.700 ± 0.020 units, at 734 nm.

Samples of bee pollen extracts and standards were added to the ABTS solution. All tubes and cuvettes were properly stoppered and sealed for mixing and to avoid contact with atmospheric oxygen. After a 6 min-reaction period, sample absorbances were read at 734 nm (Varian Cary 50 Scan UV-Vis spectrophotometer). Trolox was used as standard for the calibration curve (0.03, 0.05, 0.10, 0.25, 0.50 g/L). Control samples used ethanol (ethanol: water, 70:30, v/v) and acetone (acetone: water, 70:30, v/v) diluted in the ABTS solution. The results report on antioxidant activity expressed in grams of TEAC (Trolox Equivalent Antioxidant Capacity) per 100 g of sample (g TEAC)/100 g), by making use of the calibration curve. All samples were analysed in triplicate.

2.4 DPPH Assay

The DPPH (●) (2,2-diphenyl-1 picrylhydrazyl) assay was also used to determine the antioxidant capacity of the bee-pollen samples [19, 20]. Different *in vitro* assays can help to better characterize the redox active properties present in the samples. The DPPH radical exhibits a strong purple colour that changes to yellow after reaction with antioxidant species. The assay was conducted as described previously [20], with adaptations [19]. Stock solutions of 0.099 mM DPPH in ethanol (ethanol: water, 70:30, v/v) and acetone (acetone: water, 70:30, v/v), were prepared to obtain absorbance values in the range of 0.500 ± 0.040, at 515 nm.

Samples (bee pollen extracts and standards) were added to the DPPH solution in stoppered glass tubes, vortexed and let to rest in the dark, at room temperature, for 30 min. Absorbance was measured at 515 nm (Varian Cary 50 Scan UV Visible spectrophotometer). Control samples used ethanol (ethanol: water, 70:30, v/v) and acetone (acetone: water, 70:30, v/v) diluted in the DPPH solution. The results report on antioxidant activity expressed in grams of TEAC (Trolox Equivalent Antioxidant Capacity) per 100 g of sample (g TEAC)/100 g), by making use of the calibration curve. All samples were analysed in triplicate.

2.5 NMR Experiments

NMR solutions were prepared in deuterated solvents ethanol-d6 70% (v/v) and acetone-d6 70% (v/v). ^{1}H NMR spectra were obtained on a Bruker Avance III 400 spectrometer operating at 400.13 MHz (1H), at 25 °C. 1D data acquisition was performed with 16k complex points, 512 scans, a spectral width of 6402 Hz and a recycle delay of 2 s. Final data was processed and analyzed with Topspin v4.0 (Bruker) and MestReNova 9.1 (Mestrelab). Chemical shifts were reported in ppm.

3 Results and Discussion

3.1 Total Phenolic Content (TPC)

Quantification of total phenolic compounds (TPC) was conducted by the *Folin-Ciocalteu* method. Bee pollen extracts in ethanol 70% (v/v) and acetone 70% (v/v) of fresh bee-pollen and of the samples submitted to the three preservation procedures (freezing, freeze-drying and oven drying) were analysed. Different extraction methods were used, since aqueous, ethanolic and acetonic extracts are known to have significantly higher contents of bioactive elements, when compared to natural bee pollen [11]. Results are presented in terms of the final product obtained and with the variations in moisture content characteristic of the processing method used, with the aim of marketing the product [2]. Results are expressed in Table 1 in grams of gallic acid equivalents per 100 g of sample (g GAE/100 g).

Total phenolic contents vary between 0.43 ± 0.02 and 0.82 ± 0.12 g GAE/100 g and are within the range of previously reported values [5, 7, 13] stating for the presence of polyphenolic compounds, such as phenolic acids (hydroxybenzoic acids and hydroxycinnamic acids) and flavonoids (flavones, flavonols, flavanones). Extracts in ethanol

Table 1. Total phenolic content (TPC) of extracts of bee pollen.

Bee pollen	TPC g GA/100 g (ethanol 70%)	TPC g GA/100 g (acetone 70%)
Oven dried	0.74 ± 0.02	0.66 ± 0.01
Freeze-dried	0.82 ± 0.12	0.59 ± 0.03
Frozen	0.53 ± 0.12	0.43 ± 0.02
Fresh	0.72 ± 0.13	0.71 ± 0.01

70% (v/v) exhibited a higher content in phenolics than extracts in acetone 70% (v/v). Furthermore, freeze-dried and oven dried pollen samples preserved most of their phenolic content, when compared to fresh pollen; while the frozen sample showed a reduction to about 74% in the ethanolic extracts and to 61% in the acetonic extracts.

Proton NMR experiments were also conducted to identify major groups of phenols present in the samples. The aromatic region (6.0 to 8.0 ppm) of the 1D ^{1}H NMR spectra of the bee pollen in ethanol-d6 70% (v/v) and acetone-d6 70% (v/v) state for the presence of phenolic compounds, especially in the ethanolic extracts of the freeze-dried and oven dried samples, corroborating the data obtained by the *Folin-Ciocalteu* assay.

3.2 Antioxidant Capacity (AC)

The overall intrinsic reducing capability of the extracts in ethanol 70% (v/v) and in acetone 70% (v/v) of commercial bee-pollen and of the samples submitted to the three preservation procedures (freezing, freeze-drying and oven drying) were evaluated via the ABTS and DPPH assays. Different *in vitro* assays can help to better characterize the redox active properties present in the samples. Results are shown in Table 2.

Table 2. Trolox equivalent antioxidant capacity (TEAC) of extracts of bee pollen.

Bee pollen	g TEAC/100 g (ethanol 70%)		g TEAC/100 g (acetone 70%)	
	ABTS	DPPH	ABTS	DPPH
Oven dried	0.24 ± 0.01	0.19 ± 0.02	0.56 ± 0.07	0.28 ± 0.09
Freeze-dried	0.27 ± 0.01	0.19 ± 0.01	0.43 ± 0.02	0.25 ± 0.01
Frozen	0.20 ± 0.01	0.07 ± 0.01	0.23 ± 0.01	0.14 ± 0.12
Fresh	0.58 ± 0.11	0.48 ± 0.04	1.59 ± 0.01	1.74 ± 0.01

The antioxidant capacity determined by the power of reduction of the ABTS radical expressed in g TEAC/100 g of sample (Trolox Equivalent of antioxidant capacity) varied between 0.20 ± 0.01 and 0.27 ± 0.01 g TEAC/100 g for the extracts in ethanol 70% (v/v) and between 0.23 ± 0.01 and 0.56 ± 0.07 g TEAC/100 g for the extracts in acetone 70% (v/v). Likewise, the DPPH radical scavenging assay showed a similar profile, with values varying between 0.07 ± 0.01 and 0.19 ± 0.04 g TEAC/100 g for the extracts in ethanol 70% (v/v) and between 0.23 ± 0.01 and 0.28 ± 0.09 g TEAC/ 100 g for the extracts in acetone 70% (v/v).

The results show that acetone is more efficient than ethanol in extracting antioxidant compounds from bee pollen; according to both ABTS and DPPH assays. Nonetheless, even if all samples submitted to the different storing procedures showed a reduction in antioxidants, relatively to fresh commercial pollen; the dried and freeze-dried samples presented a higher antioxidant content, when compared to the frozen samples. Extracts in ethanol 70% (v/v) of the dried and freeze-dried samples showed a reduction to around 45% by the ABTS method and to 40% by the DPPH; while the frozen sample showed an even more pronounced decrease to 34% and 16% by the ABTS and DPPH methods, respectively. Similarly, extracts in acetone 70% (v/v) of the dried and freeze-dried samples showed a reduction to around 30% by the ABTS method, and to 16% by the DPPH, whereas the frozen sample showed a decrease to 14% and 8% of the initial amount of antioxidants, by the ABTS and DPPH methods, respectively. Thus, samples submitted to drying procedures, such as freeze-drying and oven drying are better preserving and maintaining the antioxidant capacity (AC) of the pollen, given the ability of dry pollen pellets to resist decay, due to their tough outer coat composed by cellulose and sporopollenin [4, 6].

4 Conclusions

Honeybee products, such as bee pollen, are recognized by their medicinal and health-promoting properties. Bee pollen is a source of nutrients and natural antioxidants, such as flavonoids and phenolic acids, and can be used as a food supplement. Therefore, to reach more consumers and wider markets, it is paramount to characterize and preserve the chemical composition of bee pollen, in order to expand its shelf life.

Our results showed that 1-year samples of bee pollen are better preserved through drying procedures than through simple freezing. Freeze-drying and oven-drying preserved most of the phenolic content of the samples, while maintaining nearly half of the antioxidant capacity present in fresh samples. Thus, other processes like oven drying, freeze-drying and conventional drying, followed by vacuum packing, must be considered instead of freezing, when storing bee pollen, given their preserving capabilities and more eco-friendly features.

References

1. Denisow, B., Denisow-Pietrzyk, M.: Biological and therapeutic properties of bee pollen: a review. J. Sci. Food Agric. **96**(13), 4303–9 (2016)
2. Komosinska-Vassev, K., Olczyk, P., Kaźmierczak, J., Mencner, L., Olczyk, K.: Bee pollen: chemical composition and therapeutic application (2015)
3. Mărgăoan, R., Stranţ, M., Varadi, A., Topal, E., Yücel, B., Cornea-Cipcigan, M., Campos, M.G., Vodnar, D.C.: Bee collected pollen and bee bread: bioactive constituents and health benefits. Antioxidants **8**(12), 568 (2019)
4. Salazar-González, C., Rodríguez-Pulido, F., Terrab, A., Díaz-Moreno, C., Fuenmayor, C., Heredia, F.: Analysis of multifloral bee pollen pellets by advanced digital imaging applied to functional food ingredients. Plant Foods Hum. Nutr. **73**, 328–335 (2018)

5. Belina-Aldemita, D., Schreine, M., D'Amico, S.: Characterization of phenolic compounds and antioxidative potential of pot-pollen produced by stingless bees (*Tetragonula biroi* Friese) from the Philippines. J. Food Biochem. **44**(1), e13102 (2020)
6. Didaras, N., Karatasou, K., Dimitriou, T., Amoutzias, G., Mossialos, D.: Antimicrobial activity of bee-collected pollen and beebread: state of the art and future perspectives. Antibiotics **9**(11), 811 (2020)
7. Kocot, J., Kiełczykowska, M., Luchowska-Kocot, D., Kurzepa, J., Musik, I.: Antioxidant potential of propolis, bee pollen, and royal jelly: possible medical application. Oxidative Med. Cell. Longevity, Article ID 7074209 (2018)
8. Kurek-Górecka, A., Górecki, M., Rzepecka-Stojko, A., Balwierz, R., Stojko, J.: Bee products in dermatology and skin care. Molecules **25**, 556 (2020)
9. Jannesar, M., Shoushtari, M., Majd, A., Pourpak, Z.: Bee pollen flavonoids as a therapeutic agent in allergic and immunological disorders. Iran. J. Allergy Asthma Immunol. **16**(3), 171–182 (2017)
10. Kostić, A., Milinčić, D., Barać, M., Shariati, M., Tešić, Z., Pešić, M.: The application of pollen as a functional food and feed ingredient-the present and perspectives. Biomolecules **10**(1), 84 (2020)
11. Kroyer, G., Hegedus, N.: Evaluation of bioactive properties of pollen extracts as functional dietary food supplement. Innov. Food Sci. Emerg. Technol. **2**, 171–174 (2001)
12. Campos, M.G., Bogdanov, S., de Almeida-Muradian, L.B., Szczesna, T., Mancebo, Y., Frigerio, C., Ferreira, F.: Pollen composition and standardisation of analytical methods. J. Apicult. Res. **47**, 154–161 (2008)
13. Pérez-Pérez, E., Vit, P., Rivas, E., Sciortino, R., Sosa, A., Tejada, D., Rodríguez-Malaver, A.: Antioxidant activity of four-color fractions of bee pollen from Mérida, Venezuela. Arch. Latinoamericanos Nutr. **62**(4), 375–80 (2012)
14. Schieber, M., Chandel, N.: ROS function in redox signaling and oxidative stress. Curr. Biol. **24**, R453–R462 (2014)
15. Valko, M., Leibfritz, D., Moncol, J., Cronin, M., Mazur, M., Telser, J.: Free radicals and antioxidants in normal physiological functions and human disease. Int. J. Biochem. Cell Biol. **39**, 44–84 (2007)
16. Singleton, V.L., Rossi, J.A.: Colorimetry of total phenolics with phosphomolybdic-phosphotungstic acid reagents. Am. J. Enol. Viticult. **16**(3), 144–158 (1965)
17. Brito, C., Bertotti, T., Primitivo, M., Neves, M., Pires, C., Cruz, P., Martins, P., Rodrigues, A., Moreno, M., Brito, R., Campos, M., Vaz, D., Pessoa, M., Lidon; F., Reboredo, F., Ribeiro, V.: *Corema album* spp: edible wild crowberries with a high content in minerals and organic acids. Food Chem. **345**, 128732 (2021)
18. Re, R., Pellegrini, N., Proteggente, A., Pannala, A., Yang, M., Rice-Evans, C.: Antioxidant activity applying an improved ABTS radical cation decolorization assay. Free Radical Biol. Med. **26**, 1231–1237 (1999)
19. Ribeiro, V., Leitão, A., Ramalho, J., Lidon, F.: Chemical characterization and antioxidant properties of a new coffee blend with cocoa, coffee silverskin and green coffee minimally processed. Food Res. Int. **61**, 39–47 (2014)
20. Brand-Williams, W., Cuvelier, M., Berset, C.: Use of a free radical method to evaluate antioxidant activity. LWT - Food Sci. Technol. **28**, 25–30 (1995)
21. Rzepecka-Stojko, A., Stojko, J., Kurek-Górecka, A., Górecki, M., Kabała-Dzik, A., Kubina, R., Moździerz, A., Buszman, E.: Polyphenols from bee pollen: structure, absorption metabolism and biological activity. Molecules **20**, 21732–21749 (2015)

Antioxidant Capacity Evaluation of Porphyran Extracted from *Porphyra Dioica*: Incorporation in Films for Food Packaging

Marco Teles[1], Pedro Adão[2], Raul Bernardino[1,3], Clélia Afonso[1], and Susana Bernardino[1](✉)

[1] MARE, School of Tourism and Maritime Technology, Polytechnic of Leiria, Leiria, Portugal
susana.bernardino@ipleiria.pt
[2] MARE, Polytechnic of Leiria, Leiria, Portugal
[3] Laboratory of Separation and Reaction Engineering-Laboratory of Catalysis and Materials (LSRE-LCM), School of Tourism and Maritime Technology (ESTM), Polytechnic of Leiria, 2411-901 Leiria, Portugal

Abstract. Increasing the shelf life of food is a constant challenge in the food industry. In recent years, edible coatings and biodegradable films have been considered one of the most promising technologies to achieve this goal, ensuring microbiological safety and food protection from the influence of external factors. The potential assessment of algae polysaccharides as a substitute for petroleum-based plastics has been the focus of many scientific efforts. Films derived from alginate, agar and carrageenan have been widely studied, due to their gelling properties and film-forming ability, which are already well established, however, there is scarce information regarding films derived from porphyran, a sulphated galactan which is produced by the red macroalgae belonging to the *Porphyra* taxonomic genus. In the present study porphyran was extracted using a Soxhlet system and water as a solvent, following precipitation and washing with isopropanol with a yield of 26.63 ± 0.29% for the semi-refined porphyran, with 67.74 ± 4.13% of D-galactose and protein not detectable by *p*-Benzoquinone assay. The antioxidant capacity was evaluated through different methods: the DPPH and ABTS radical scavenging activity, 6.49 ± 1.2% and 13.83 ± 1.5%, respectively, the HPSA scavenging activity, from a ΔAbs of 0.089 ± 0.004 to 0.069 ± 0.008, the ferric reducing/antioxidant power, FRAP, 0.4 ± 0.003 μg mL^{-1} equivalents of ascorbic acid and the total polyphenols content, QTP, 1.2 ± 0.006 μg mL^{-1} equivalents of gallic acid.

Keywords: Active packaging · Biodegradable plastic · Edible films/Coatings · Porphyran · *Porphyra dioica*

1 Introduction

Plastic is an anthropogenic compound, which has physical and mechanical properties that makes it a material widely used today, especially in the food industry, in product packaging [1].

J. R. da Costa Sanches Galvão et al. (Eds.): *Proceedings of the 1st International Conference on Water Energy Food and Sustainability (ICoWEFS 2021)*, pp. 201–209, 2021.
https://doi.org/10.1007/978-3-030-75315-3_24

However, after its use, when it is discarded in the environment, and because it is virtually non-biodegradable, plastic residues interact negatively with organisms, through entanglement, hindering locomotion and physical development of the organism [2], by ingestion, by its toxicity and effects of filling and obstruction of the digestive and respiratory systems [3], which contaminates the upper trophic levels [4], and by skin contact, which can cause diseases, namely eczema [5]. As such, plastics represent one of the biggest global environmental problems [1].

Due to the fact that plastics have such adverse environmental effects and at the same time their use is so extensive, it is necessary to develop alternative materials with physical and mechanical properties similar to plastic, capable of being environmentally degraded, within a reasonable time scale and that during this process do not cause environmental damage [6]. Biodegradable plastics have all these characteristics, being generally formed by natural or synthetic polymers [7]. Sulphated polysaccharides are examples of these polymers, being carbohydrate structures that can be extracted from the cell wall of various types of algae and other organisms [8, 9]. These compounds can be included as a basis in the formulation of films and coatings, making them ideal substitute for plastics and coatings applied in the food industry [9, 10]. Edible films and coatings ensure that there is no accidental ingestion of macro and microparticles from the packaging, as is often the case with plastic packaging, which leads to the accumulation of these substances in the body, which can lead to health problems [10, 11]. To be a good substitute for conventional plastic packaging, these films need to establish a physical barrier between food products and the outside, which protected them from external influences and conferred a certain resistance to mechanical forces [12, 13].

The genus *Porphyra*, to which the species *Porphyra dioica* belongs, is widely used in traditional Japanese cuisine where it is known as *nori* [14]. The group of sulfated polysaccharides present in the cell wall of these algae is called porphyran, which, although little explored, has high potential in several areas, mainly due to its antioxidant activity [15]. Porphyran is the sulfated polysaccharide that forms the cell wall of algae of this genus, composed of units of disaccharides, that consist of (1,3-β) D-galactose residues that alternate with (1,4-α) 3,6-anhydrous-α-L-galactose [16], which, due to the interactions between their chains and the underlying chains, form complex matrices with the ability to adhere to surfaces and retain water due to their adaptable sulfate groups, allowing the development edible films and aqueous coatings [17]. The formation of these matrices and bioactivities depends on the molecular weight, glycosidic remission, degree of sulfation and sulfate groups of the porphyran molecules, parameters that can differ [18].

In addition to the referred great environmental benefits, these can have active groups with properties that allow extended food products shelf life, such as antioxidants, and these can be intrinsic to the material used in the composition of the films and coatings, that then became active packaging [19].

This project aims to develop a fast, low-cost, low environmental impact and high yield method to obtain porphyran from *Porphyra dioica* and the formulation of biodegradable films and coatings that can be applied to food products.

2 Materials and Methods

2.1 Extraction

In three portions of absorbent paper, 15 g of powdered *Porphyra dioica* were distributed, which were rolled into cartridges and placed in the Soxhlet extractor with 250 mL of distilled water in the collector. The extraction solvent was heated to boiling point with a heating mantle set at 200 °C. The extract obtained in the first hour was discarded and two subsequent cycles of seven hours were performed, with each fraction recovered separately.

The recovery took place by evaporation on a rotary evaporator, followed by precipitation with isopropanol in two steps, the precipitate was then crushed and vacuum filtered, using a quantitative paper filter, and washed with isopropanol. The recovered mass was kept in an oven at 60 °C for 24–48h.

2.2 *p*-Benzoquinone Assay

This method was adapted from Amin and El-Didamony (2003) [20]. The p-Benzoquinone (p-BQ) solution (Sigma-Aldrich, St. Louis, Missouri, USA) at 1 mg mL^{-1} of initial concentration was prepared for each analysis in phosphate buffer (PB) at pH 9.2, 0.08 M anhydrous monobasic sodium (Amresco, Solon, Ohio, USA). The external standard was performed with L-alanine (PanReac AppliChem, Castellar del Vallès, Barcelona, Spain) 0.030 mg mL^{-1} at 0.005 mg mL^{-1} in PB at pH 9.2. The samples were prepared in PB at pH 9.2 at a concentration of 0.6 $mgmL^{-1}$. The 2.5 M HCl solution (VWR Chemical, Radnor, Pennsylvania, USA) was prepared in 96% ethanol (Aga, Loures, Portugal). The reaction medium was composed of 2 ml of L-alanine solution or sample and 1 ml of p-BQ, immediately after its constitution, it was kept in a bath at 50 °C for 15 min. After that period, 3 mL of 2.5 M HCl was added. The absorbance spectrum was measured between 400 and 700 nm.

2.3 Phenol-Sulfuric Acid Method

This method was adapted from Rao and Pattabiraman (1989) [21]. The external standard was D-glucose (VWR, Radnor, Pennsylvania, USA) with concentrations from 0.0450 mg mL^{-1} to 0.0075 $mg.mL^{-1}$ in distilled water, and the samples at 0.6 mg mL^{-1}. The reaction medium consisted of 0.6 ml of phenol at 10 mg mL^{-1}, phenol crystals (Fisher Scientific, Waltham, Massachusetts, USA), 0.6 ml of standard or sample and 1.8 ml of sulfuric acid 95% (VWR, Radnor, Pennsylvania, USA) composed the reaction medium. After the end of the exothermic reaction, 5 mL of distilled water was added, and the absorbance spectrum was recorded between 400 and 700 nm.

2.4 QTP

In the quantification of the total phenols, the Folin – Ciocalteu (F-C) reagent method (AppliChem, Darmstadt, Darmstadt, Germany) was used based in Fu et al. (2010) [22]. The analysis was performed by adding 500 μL of the sample, to 2 mg mL^{-1} in distilled water, and 2500 μL of diluted F-C reagent, 1:10 in distilled water. After 4 min, 2000 μL of sodium carbonate (Scharlau, Sentmenat, Barcelona, Spain) saturated at 75 mg mL^{-1} was added, the reaction was carried out for 2 h at room temperature and its analysis was carried out by reading the absorbance at 760 nm. Calibration was performed using gallic acid (Merck KGaA, Darmstadt, Darmstadt, Germany) with concentrations 0.006 mg mL^{-1} to 0.001 mg mL^{-1}.

2.5 HPSA

NaFeEDTA, composed of ethylenediamine tetra-acetic acid disodium dihydrate salt (VWR Chemicals, Radnor, Pennsylvania, USA) and Iron (III) chloride (Labkem, Barcelona, Spain), 0.251g in 250 mL of filtered methyl red solution. A filtered solution of methyl red, 12 mg of methyl red with 0.2 ml of acetic acid (VWR Chemicals, Radnor, Pennsylvania, USA) in 500 ml of distilled water was used. The method is based on Bektaşoğlu et al. (2008) with changes [23]. In quartz cuvettes 1 ml of sample was placed at 0.6 mg mL^{-1} and 1.8 ml of NaFeEDTA/acetic acid/methyl red, 0.2 ml of 30% hydrogen peroxide (Chem-Lab, Zedelgem, Belgium) was added and immediately after reading the absorbance for 16 cycles with an interval of 1 min, between 650 to 400 nm. The change in absorbance after 15 min was determined using the equation:

$$\Delta\text{Abs} = Abs_{Sample\ t0} - Abs_{Sample\ t15} \tag{1}$$

2.6 DPPH

The method is based on Miceli et al. (2018) with changes [24]. To solutions of 2,2-diphenyl-1-picryl-hydrazil (DPPH) (Sigma-Aldrich, St. Louis, Missouri, USA) at 0.1 mg mL^{-1} in acetonitrile (Chem-Lab, Zedelgem, West Flanders, Belgium) In ascorbic acid concentrations (Vilassar de Dalt, Barcelona, Spain) from 0.006 mg mL-1 to 0.001 mg mL^{-1} in TF at pH 5.5 and the samples at 0.6 mg.mL^{-1} in PB at pH 5.5. Calibration and samples corresponded to a volume in the reaction medium of 1 ml to which 2 ml of DPPH and 0.5 ml of acetonitrile and 0.5 ml of PB were added at pH 5.5, in the scoring at room temperature, after 30 min, the absorbance at 517 nm, the %inhibition was determined using the equation:

$$\%Inhibition = \frac{Abs517_{Control} - Abs517_{Sample}}{Abs517_{Control}} \times 100 \tag{2}$$

2.7 FRAP

The standard, ascorbic acid (Vilassar de Dalt, Barcelona, Spain) was used at concentrations from 2.60 μg mL^{-1} to 0.65 μg mL^{-1}. This method was adapted from Berker et al. (2007) [25]. The samples at 0.6 mg mL^{-1} were dissolved in distilled water. In the reaction medium, 2 ml of standard solution and 1 ml of Phen (Phen = 1,10-phenanthroline) reagent, $[Fe\ (Phen)_3]^{3+}$, 1,10-phenanthroline monohydrate ($C_{12}H_8N_2 \cdot H_2O$) (Sigma-Aldrich, St. Louis, Missouri, USA) at 3.3 mM were kept for 30 min in the dark at room temperature. After that period, 6 mL of distilled water was added and the absorbance was measured at 510 nm. The reaction medium of the samples consisted of 1 ml of sample, 1.5 ml of distilled water and 1 ml of Phen reagent, then, after 30 min, 6 ml of distilled water was added, and the analysis took place under the same conditions.

2.8 ABTS

The ABTS reagent consisted of 48 mg of 2,2′-azino-bis (3-ethylbenzothiazoline-6-sulfonic acid) ($(NH_4)_2C_{18}H_{16}N_4O_6S_4$) (Alfa Aesar, Haverhill, Massachusetts, USA) and 12.8 mg potassium peroxydisulfate ($K_2S_2O_8$) (Alfa Aesar, Haverhill, Massachusetts, USA) was dissolved in 250 mL of distilled water before applying and the solution was reserved for 16 h at room temperature in the dark. Calibration was performed from concentrations of 6.12 $\mu g\ mL^{-1}$ to 1.02 $\mu g\ mL^{-1}$. In the case of the sample, it was prepared at 0.6 mg mL^{-1} in distilled water.

The method based on Pellegrini et al. (1999) with changes [26]. The reaction medium consisted of 2 ml of ABTS reagent, 2 ml of the sample or ascorbic acid and 2 ml of distilled water, the reaction took place in 1 min and its absorbance was measured at 734 nm, the % inhibition was determined using the equation:

$$\%Inhibition = \frac{Abs734_{Control} - Abs734_{Sample}}{Abs734_{Control}} \times 100 \tag{3}$$

3 Results and Discussion

The extraction was performed using hot water as solvent. However, due to the use of the Soxhlet device, and as shown in Table 1, we obtained a semi-refined porphyran with undetectable protein contamination, as measured through the *p*-Benzoquinone assay. The level of D-galactose, the main monomer of porphyran, of 67.74%, a value similar to that obtained with other commonly applied methods used in the extraction of porphyran (around 70%) [27], as well as an extraction yield of 26.63%, quite high when compared to the yields reported in the literature, which range between 5–18% dry weight [28, 29].

However, the porphyran obtained probably had some contamination due to the presence of other components of the cell wall that naturally associate with it, due to its to its complexity and heterogeneity [30]. Among them may be phenols as observed in the QTP, test with values in gallic acid equivalents of 1.2 $\mu g\ mL^{-1}$, however, as this test is based on the reduction of the active component, it does not exclusively detect phenolic compounds [31], and interference from other active agents may occur, such as porphyran itself.

This extraction method stands out for using non-toxic and inedible components, for its high rate of extraction and for being relatively fast.

The antioxidation tests were carried out in order to address different modes of action. The main differences are between the primary antioxidant actions, assessed by HPSA [32], and secondary actions, by the DPPH, ABTS and FRAP methods [33].

The primary antioxidant activity is the uptake of antioxidant agents before their interaction with the compounds [34]. Sulfated polysaccharides such as carrageenan are

Table 1. Extraction yield in percentage of dry mass, protein composition (p-BQ), concentration of D-galactose in percentage from external standard and phenol composition (QTP) in gallic acid equivalents, by the developed extraction method Porf P2.

	Extraction yield (%)	*p*-BQ	[D-Gal] (%)	QTP (Gallic acid Eq. μg/mL)
Porf P2	26.63 ± 0.29	*ND	67.74 ± 4.13	1.2 ± 0.006

*ND – Not detected. Values represent the mean ± standard error of the mean (SEM) of three independent experiments carried out in triplicate.

recognized for this potential [35], due to the presence of hydroxyl functional groups, thus performing antioxidant activities very similar to those of secondary alcohols [32]. As shown in Table 2, the extract had a high primary antioxidant potential, exhibiting a smaller change in absorbance after 15 min when compared to the control reaction in the HPSA assay (0.089 ± 0.004 for the control compared to 0.069 ± 0.008 for Porf P2). The smaller DAbs observed for Porf P2 was associated with the increased suppression of the oxidation of the dye induced by hydrogen peroxide.

The secondary antioxidant activity is based on other mechanisms, mainly in the reduction of oxidizing agents making them inactive [36]. Table 2 highlights the ABTS value, 13.83%, although the DPPH method, 6.49%, is based on the same principle. However, due to the fact that the reagent in ABTS method reacts much faster [33], and porphyran reacts slower [37], that influences differences in detection and for this reason, in the presence of faster reagents, the perception of an unequal inhibition rate, this being the main reason for carrying out several methods in the detection of this type of activities [38]. These tests, as well as the FRAP, with a result of 0.4 μg equivalent of ascorbic acid.mL^{-1}, showed a high activity, not expected for this compound, since it is not normally attributed secondary mechanisms of antioxidation, [37]. This phenomenon can be attributed, as already mentioned, to the likely presence of phenols from the alga cell wall [39].

Table 2. Antioxidant activity of the extracts obtained by the developed extraction method, Porf P2.

	DPPH (%Inhibition)	ABTS (%Inhibition)	FRAP (Ascorbic acid Eq μg/mL)	HPSA (ΔAbs)
Porf P2	6.49 ± 1.2%	13.83 ± 1.5%	0.4 ± 0.003	0.069 ± 0.008
				0.089 ± 0.004[a]

Values represent the mean ± standard error of the mean (SEM) of three independent experiments carried out in triplicate.

[a] Control reaction carried out with distilled water instead of porphyran sample.

4 Conclusion

The use of a low-cost, non-polluting extraction method, with a high yield of 26.63 $\pm$ 0.29%, as well as extraction from edible products such as Porf P2, allows access to compounds that can be integrated in packaging films, allowing the formulation of an accessible and high-quality final material. In addition to the high ecological value generated, due to porphyran extracts being non-toxic and biodegradable, their antioxidant properties (both primary, HPSA with ΔAbs of 0.069 $\pm$ 0.008 vs. 0.089 $\pm$ 0.004 obtained for the control, and secondary, DPPH 6.49%, ABTS 13.83% and FRAP 0,4 μg of ascorbic acid equivalent. mL^{-1}), guarantee higher shelf-life to food products, ensuring, simultaneously, better food safety and reduced environmental impact.

References

1. PlasticsEurope, PlasticsEurope Market Research Group (PEMRG)/Consultic Marketing & Industrieberatung GmbH: Plastics – the Facts 2017 (2017)
2. Mucientes, G., Queiroz, N.: Presence of plastic debris and retained fishing hooks in oceanic sharks. Mar. Pollut. Bull. **143**, 6–11 (2019). https://doi.org/10.1016/j.marpolbul.2019.04.028
3. Bejgarn, S., MacLeod, M., Bogdal, C., Breitholtz, M.: Toxicity of leachate from weathering plastics: an exploratory screening study with Nitocra spinipes. Chemosphere **132**, 114–119 (2015). https://doi.org/10.1016/j.chemosphere.2015.03.010
4. Van Cauwenberghe, L., Janssen, C.R.: Microplastics in bivalves cultured for human consumption. Environ. Pollut. **193**, 65–70 (2014). https://doi.org/10.1016/j.envpol.2014.06.010
5. Wright, S.L., Kelly, F.J.: Plastic and human health: a micro issue? Environ. Sci. Technol. **51**, 6634–6647 (2017). https://doi.org/10.1021/acs.est.7b00423
6. Iwata, T.: Biodegradable and bio-based polymers: future prospects of eco-friendly plastics. Angew. Chemie Int. Ed. **54**, 3210–3215 (2015). https://doi.org/10.1002/anie.201410770
7. Shen, M., Song, B., Zeng, G., Zhang, Y., Huang, W., Wen, X., Tang, W.: Are biodegradable plastics a promising solution to solve the global plastic pollution? Environ. Pollut. **263**, 114469 (2020). https://doi.org/10.1016/j.envpol.2020.114469
8. Fonseca, R., Mourão, P.: Fucosylated chondroitin sulfate as a new oral antithrombotic agent. Thromb. Haemost. **96**, 822–829 (2006). https://doi.org/10.1160/TH06-06-0304
9. Cunha, L., Grenha, A.: Sulfated seaweed polysaccharides as multifunctional materials in drug delivery applications. Mar. Drugs **14**, 42 (2016). https://doi.org/10.3390/md14030042
10. Galus, S., Kadzińska, J.: Food applications of emulsion-based edible films and coatings. Trends Food Sci. Technol. **45**, 273–283 (2015). https://doi.org/10.1016/j.tifs.2015.07.011
11. Arvanitoyannis, I.S., Bosnea, L.: Migration of substances from food packaging materials to foods. Crit. Rev. Food Sci. Nutr. **44**, 63–76 (2004). https://doi.org/10.1080/10408690490424621
12. Ozen, B.F., Floros, J.D.: Effects of emerging food processing techniques on the packaging materials. Trends Food Sci. Technol. **12**, 60–67 (2001). https://doi.org/10.1016/S0924-2244(01)00053-X
13. Marsh, K., Bugusu, B.: Food packaging-roles, materials, and environmental issues. J. Food Sci. **72**, 39–55 (2007). https://doi.org/10.1111/j.1750-3841.2007.00301.x
14. Blouin, N.A., Brodie, J.A., Grossman, A.C., Xu, P., Brawley, S.H.: Porphyra: a marine crop shaped by stress. Trends Plant Sci. **16**, 29–37 (2011). https://doi.org/10.1016/j.tplants.2010.10.004

15. Venkatraman, K.L., Mehta, A.: Health benefits and pharmacological effects of porphyra species. Plant Foods Hum. Nutr. **74**, 10–17 (2019). https://doi.org/10.1007/s11130-018-0707-9
16. Zhang, Q., Li, N., Liu, X., Zhao, Z., Li, Z., Xu, Z.: The structure of a sulfated galactan from Porphyra haitanensis and its in vivo antioxidant activity. Carbohydr. Res. **339**, 105–111 (2004). https://doi.org/10.1016/j.carres.2003.09.015
17. Han, J.H., Gennadios, A.: Edible films and coatings: a review. In: Innovations in Food Packaging, pp. 239–262. Elsevier (2005)
18. Raposo, M., De Morais, R., de Morais, A.B.: Bioactivity and applications of sulphated polysaccharides from marine microalgae. Mar. Drugs **11**, 233–252 (2013). https://doi.org/10.3390/md11010233
19. Han, J.-W., Ruiz-Garcia, L., Qian, J.-P., Yang, X.-T.: Food packaging: a comprehensive review and future trends. Compr. Rev. Food Sci. Food Saf. **17**, 860–877 (2018). https://doi.org/10.1111/1541-4337.12343
20. Amin, A.S., El-Didamony, A.M.: Colorimetric determination of benzocaine, lignocaine and procaine hydrochlorides in pure form and in pharmaceutical formulations using p-benzoquinone. Anal. Sci. **19**, 1457–1459 (2003). https://doi.org/10.2116/analsci.19.1457
21. Rao, P., Pattabiraman, T.N.: Reevaluation of the phenol-sulfuric acid reaction for the estimation of hexoses and pentoses. Anal. Biochem. **181**, 18–22 (1989). https://doi.org/10.1016/0003-2697(89)90387-4
22. Fu, L., Xu, B.-T., Xu, X.-R., Qin, X.-S., Gan, R.-Y., Li, H.-B.: Antioxidant capacities and total phenolic contents of 56 wild fruits from South China. Molecules **15**, 8602–8617 (2010). https://doi.org/10.3390/molecules15128602
23. Bektaşoğlu, B., Özyürek, M., Güçlü, K., Apak, R.: Hydroxyl radical detection with a salicylate probe using modified CUPRAC spectrophotometry and HPLC. Talanta. **77**, 90–97 (2008). https://doi.org/10.1016/j.talanta.2008.05.043
24. Miceli, M., Roma, E., Rosa, P., Feroci, M., Loreto, M., Tofani, D., Gasperi, T.: Synthesis of Benzofuran-2-one derivatives and evaluation of their antioxidant capacity by comparing DPPH assay and cyclic voltammetry. Molecules **23**, 710 (2018). https://doi.org/10.3390/molecules23040710
25. Berker, K.I., Güçlü, K., Tor, I., Apak, R.: Comparative evaluation of Fe(III) reducing power-based antioxidant capacity assays in the presence of phenanthroline, batho-phenanthroline, tripyridyltriazine (FRAP), and ferricyanide reagents. Talanta **72**, 1157–1165 (2007). https://doi.org/10.1016/j.talanta.2007.01.019
26. Re, R., Pellegrini, N., Proteggente, A., Pannala, A., Yang, M., Rice-Evans, C.: Antioxidant activity applying an improved ABTS radical cation decolorization assay. Free Radic. Biol. Med. **26**, 1231–1237 (1999). https://doi.org/10.1016/S0891-5849(98)00315-3
27. Bojko, L., de Jonge, G., Lima, D., Lopes, L.C., Viana, A.G., Garcia, J.R., Pessôa, C.A., Wohnrath, K., Inaba, J.: Porphyran-capped silver nanoparticles as a promising antibacterial agent and electrode modifier for 5-fluorouracil electroanalysis. Carbohydr. Res. **498**, 108193 (2020). https://doi.org/10.1016/j.carres.2020.108193
28. Chen, Y., Xue, Y.: Optimization of microwave assisted extraction, chemical characterization and antitumor activities of polysaccharides from porphyra haitanensis. Carbohydr. Polym. **206**, 179–186 (2019). https://doi.org/10.1016/j.carbpol.2018.10.093
29. He, D., Wu, S., Yan, L., Zuo, J., Cheng, Y., Wang, H., Liu, J., Zhang, X., Wu, M., Choi, J., Tong, H.: Antitumor bioactivity of porphyran extracted from Pyropia yezoensis Chonsoo2 on human cancer cell lines. J. Sci. Food Agric. **99**, 6722–6730 (2019). https://doi.org/10.1002/jsfa.9954
30. Rahman, M.: An overview of the medical applications of marine skeletal matrix proteins. Mar. Drugs. **14**, 167–176 (2016). https://doi.org/10.3390/md14090167

31. Ainsworth, E.A., Gillespie, K.M.: Estimation of total phenolic content and other oxidation substrates in plant tissues using Folin-Ciocalteu reagent. Nat. Protoc. **2**, 875–877 (2007). https://doi.org/10.1038/nprot.2007.102
32. Rong, S., Sun, Y.: Degradation of TAIC by water falling film dielectric barrier discharge – influence of radical scavengers. J. Hazard. Mater. **287**, 317–324 (2015). https://doi.org/10.1016/j.jhazmat.2015.02.003
33. Dudonné, S., Vitrac, X., Coutière, P., Woillez, M., Mérillon, J.-M.: Comparative study of antioxidant properties and total phenolic content of 30 plant extracts of industrial interest using DPPH, ABTS, FRAP, SOD, and ORAC assays. J. Agric. Food Chem. **57**, 1768–1774 (2009). https://doi.org/10.1021/jf803011r
34. Galano, A., Alvarez-Idaboy, J.R.: A computational methodology for accurate predictions of rate constants in solution: application to the assessment of primary antioxidant activity. J. Comput. Chem. **34**, 2430–2445 (2013). https://doi.org/10.1002/jcc.23409
35. Zhu, M., Ge, L., Lyu, Y., Zi, Y., Li, X., Li, D., Mu, C.: Preparation, characterization and antibacterial activity of oxidized κ-carrageenan. Carbohydr. Polym. **174**, 1051–1058 (2017). https://doi.org/10.1016/j.carbpol.2017.07.029
36. Pisoschi, A.M., Pop, A.: The role of antioxidants in the chemistry of oxidative stress: a review. Eur. J. Med. Chem. **97**, 55–74 (2015). https://doi.org/10.1016/j.ejmech.2015.04.040
37. Zhang, Z., Zhang, Q., Wang, J., Song, H., Zhang, H., Niu, X.: Chemical modification and influence of function groups on the in vitro-antioxidant activities of porphyran from Porphyra haitanensis. Carbohydr. Polym. **79**, 290–295 (2010). https://doi.org/10.1016/j.carbpol.2009.08.004
38. Antolovich, M., Prenzler, P.D., Patsalides, E., McDonald, S., Robards, K.: Methods for testing antioxidant activity. Analyst **127**, 183–198 (2002). https://doi.org/10.1039/b009171p
39. Kazłowska, K., Hsu, T., Hou, C.-C., Yang, W.-C., Tsai, G.-J.: Anti-inflammatory properties of phenolic compounds and crude extract from Porphyra dentata. J. Ethnopharmacol. **128**, 123–130 (2010). https://doi.org/10.1016/j.jep.2009.12.037

Eco-Design and Sustainable Packaging: Challenges, Trends and Perspectives in the Food Sector

Ana Pires[1,2](✉)

[1] CENTIMFE, Zona Industrial, Rua de Espanha, Lote 8, 2430-028 Marinha Grande, Portugal
ana.pires@centimfe.com
[2] MARE, Campus de Caparica, 2829-516 Caparica, Portugal

Abstract. Food packaging is important to the modern society, to ensure healthy and safe distribution of food to the consumers. Also, packaging is the way that food producers can communicate with the consumer, being an important marketing tool. However, packaging, mostly plastic packaging, when wrongly managed, has been identified as one of the biggest environmental problems facing today, including the marine plastic litter problem. This study intends to present an overview of the existing methodologies to design sustainable packaging, which could prevent the environmental problems from packaging life cycle. Most methodologies are based on life cycle assessment (LCA) or combine life cycle assessment with indicators from sustainability spectrum. Nevertheless, the drawbacks of life cycle assessment are related with the incomplete inclusion of environmental impacts related to littering, as well with the missing indicators to measure packaging circularity. In the future, the food packaging must clarify the dichotomy food-packaging study strategy, to quantify the impacts from plastic litter in the short term and to relate circularity as an impact category, where the management of end-of-life phase must be mandatory.

Keywords: Food packaging · Life cycle assessment · Circularity indicators

1 Introduction

To define sustainable packaging is important to understand the sustainable concept first. The Brundtland report defined sustainability as a concept that promotes a type of development whereby the needs of the present generation will not compromise the ability of future generations to meet their own needs, focusing on economic, environmental, and social interdependence [1]. With the 2030 United National Agenda for a Sustainable Development, 17 Sustainable Development Goals (SDGs) have been established, with targets for 2030, helping to the implementation of the sustainability concept [2]. To be sustainable, a packaging should include applicable SDG into the life cycle of packaging.

Packaging, in particular food packaging, can contribute to SDG for several reasons: food protection, by avoiding its deterioration and preventing food waste generation;

J. R. da Costa Sanches Galvão et al. (Eds.): *Proceedings of the 1st International Conference on Water Energy Food and Sustainability (ICoWEFS 2021)*, pp. 210–215, 2021.
https://doi.org/10.1007/978-3-030-75315-3_25

improving food transportation, through the efficient packaging and respecting logistics constraints, leading to a reduction of greenhouse gas emissions. However, without a proper design, packaging can have severe drawbacks to the environment: plastic lightweight packaging can easily get out of technological cycle and reach marine environment, with impacts to marine organism of all dimensions (microorganisms like plankton to cetaceans and mammals, including humans, at the top of the food chain); non-recyclable packaging is a problem for waste managers, which cannot complete the technological cycle and re-introduce packaging material again in the economy, not promoting its circularity, neither reducing the extraction of virgin raw materials (renewable and non-renewable). Such drawbacks are addressed by consumers and national authorities, pressuring the food business to develop and use more eco-friendly packaging. To respond to this demand, businesses have tried to develop and implement assessment tools which could be capable of delivering the eco-designed packaging. The objective of the present study is to understand the existing eco-design assessment tools applied for packaging and how they are capable to face future challenges.

2 Current Packaging and Packaging Waste Policy in EU

European Union (EU) strategies to face packaging and packaging waste issues have a long background. The first attempt to minimize packaging waste impacts were the Packaging and Packaging Waste Directive 1994/62/EC and its amendments [3], where recycling rates were introduced to impose some circularity at the management of such waste streams. The Extended Producer Responsibility (EPR) was also presented, meaning that the packaging producers and importers bear the financial (or financial and organizational) responsibility of the management of the packaging when it becomes a waste. The financial support is made through a fee, which should encourage producers and importers to improve the packaging design and packaging components in an eco-friendly way [4]. Although the intention, there is no evidence of the effect of EPR fee in the promotion of eco-design [5, 6].

More recently, the amendment of 1994/62/EC Directive will have consequences on the packaging supply chain, due to higher recycling rates that impose a redesign of packaging and the preparation of infrastructure to conduct the packaging recycling. Concerning end-of-life, Directive 2008/98/EC [7] imposes a waste hierarchy management of waste streams as the preferable from an environmental point of view. However, the waste hierarchy principle has been challenged by life cycle assessment, which may give different ordering on waste management in particular exceptions like wastepaper [8]. The waste hierarchy principle is now embraced by the Circular Economy Action Plan [9], where packaging should be reduced, as well packaging waste, increase recyclability of packaging and increase its safety to be used in contact with consumer goods, as well reduce packaging complexity. Also, the establishment of the tax applicable to non-recycled plastic packaging of 0.80 €/kg at European Union will be a significant booster to the eco-design of plastic packaging [9].

3 Sustainable Packaging Assessment Tools

The review of sustainable packaging assessment tools was searched at Google Scholar articles, mostly. The timeframe starts in 2015 until nowadays and is presented in Table 1. Other assessment tools were obtained from Google search engine.

The search for sustainable packaging assessment tools devoted to answering three main questions: is the principle of promoting eco-design based on economic, environmental or technical features; what is the methodology used to promote eco-design (KPI, LCA, other) and which environmental impacts are included in the assessment. In this short review it is noticed that EPR schemes in Europe are investing in eco-design tools based on LCA, when mostly academic research tools are devoted to test other methodologies, also including LCA.

Table 1. Examples of packaging assessment tools.

Authors	Eco-design promotion	Methodology	Impacts
Pires et al. [10]	Modulation of EPR of packaging	LCA and MCDM	Environmental and economic
Huang and Ma [11]	Choose packaging material	Cluster including LCA and MCDM	Environmental
Pires and Martinho [12]	Waste hierarchy principle with CE of waste streams	Indicator	Environmental
RecyClass [13]	Flexible plastic packaging design for recycling	Decision tree	Technical, resources use (recyclability)
Civancik-Uslu et al. [14]	Eco-design of cosmetic products	LCA	Environmental
Foschi et al. [15]	Packaging eco-design	LCA with eco-design principles	Environmental
EcoImpact [16]	Packaging eco-design	LCA	Environmental
Instant LCA Packaging [17]	Packaging eco-design	LCA	Environmental
PackageSmart [18]	Packaging eco-design	LCA	Environmental
e-DEA eliPack [19]	Packaging eco-design	LCA	Environmental
Eco Tool CONAI [20]	Packaging eco-design	LCA (streamlined)	Environmental
Pack4Ecodesign [21]	Packaging eco-design	LCA (streamlined)	Environmental
De Koeijer et al. [22]	Product-packaging eco-design	Circularity indicators	

4 Future Perspective on Assessment Tools for Packaging

Until now, eco-design tools could assess its sustainability mostly focused on life cycle assessment and the environmental impact indicators resulting from such exercise. LCA has been capable of reducing the amount of raw materials used in packaging, justify the proliferation of plastic packaging due to the reduction of weight which would contribute to a reduction of the packaging transportation (with and without the product). The proliferation of plastic packaging led to an environmental problem of plastic litter in marine environment, where LCA is not yet capable of measuring and including the assessment. Efforts have been made to develop environmental impact models to be included in LCA, but it is expected that more years will be needed to have a robust impact assessment tool. Sustainable packaging eco-design tools must be aware of this problem and work to solve it.

The circularity issue is also of great relevance in sustainable eco-design of packaging. The recycling rates of packaging waste increase and the need to incorporate recycled material has imposed a different shift on how packaging must be conceived. Circularity indicators must be included in future eco-design tools.

The development of eco-design tools has the advantage to be easily accessed by non-experts. However, the need to harmonize the rules on how to calculate the LCA or to conceive those tools at a European level requires the adoption of the same rules, especially when those tools have LCA. The adoption of Product Environmental Footprint Category Rules (PEFCR), included in the Product Environmental Footprint (PEF) initiative by the European Commission, would be of great relevance to the improvement of eco-design tools for packaging at European level.

Nevertheless, sustainable packaging must be in accordance with SDGs. A holistic perspective of what should be the future development of packaging is presented in Fig. 1. All SDGs must influence researchers to develop better assessment tools. The Industry, governments and consumers must transpose their needs to the researchers to also improve those assessment tools. At the end, the assessment tools must have the life cycle thinking in mind, to ensure that the impacts will not migrate from one life stage to another, ensuring a full sustainability.

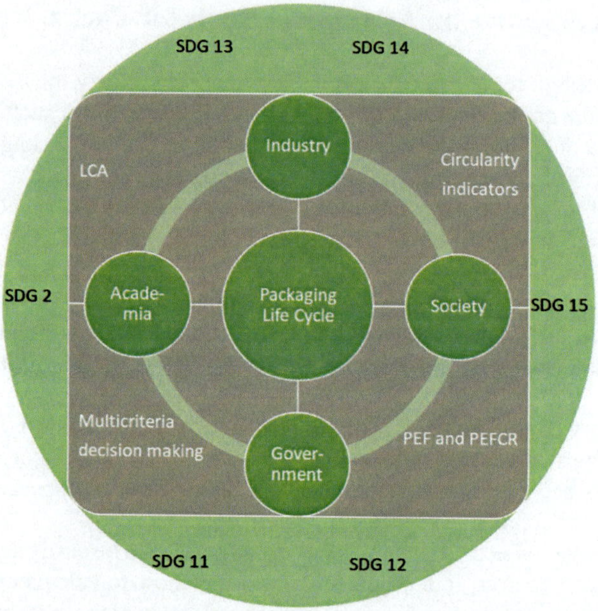

Fig. 1. Holistic approach to the sustainable packaging framework.

5 Conclusions

The future for the sustainable packaging involves a coordination between LCA and indicators, to ensure that environmental impacts not yet calculated by the LCA can be brought into the eco-design, namely impacts related to packaging circularity. Also, the need to include the impacts of the dichotomy product-packaging has also to be more explored.

Nevertheless, the stakeholders involved in the entire life cycle of the packaging must be brought into the definition of sustainable packaging, which is influenced by recycling technology available at regional and national scales, consumption patterns and recyclable patterns of population, and existing recycling schemes, at least. There are high differences between EU countries concerning packaging waste management, which may dictate different sustainable packaging concepts, which can be another drawback for the dissemination of harmonized rules of eco-design operating in the EU.

Acknowledgments. The authors acknowledge the financial support of Sociedade Ponto Verde, Portugal, through the research project Pack4Sustain and the strategic project UIDB/04292/2020 granted to MARE – Marine and Environmental Sciences Centre.

References

1. Martinho, G., Pires, A., Portela, G., Fonseca, M.: Factors affecting consumers' choices concerning sustainable packaging during product purchase and recycling. Resour. Conserv. Recycl. **103**, 58–68 (2015)

2. UN homepage. https://www.un.org/. Accessed 31 Jan 2021
3. Eur-Lex Homepage. https://eur-lex.europa.eu/legal-content/EN/TXT/?uri=CELEX%3A01994L0062-20180704. Accessed 31 Jan 2021
4. Pires, A., Martinho, G., Rodrigues, S., Gomes, M.I.: Sustainable Solid Waste Collection and Management. Springer International Publishing, Switzerland (2019)
5. Watkins, E., Hogg, D., Mitsios, A., Mudgal, S., Neubauer, A., Reisinger, H., Troeltzsch, J., van Acoleyen, M.: Use of economic instruments and waste management performances – final report. Report for the European Commission (DG ENV) (2012)
6. Dempsey, M., van Rossem, C., Lifset, R., Linnell, J., Gregory, J., Atasu, A., Perrym J., Sverkman, A., van Wassenhove, L.N., Therkelsen, M., Sundberg, V., Mayers, C.K., Kalimo, H.: Individual producer responsibility: a review of practical approaches to implementing individual producer responsibility for the WEEE Directive. INSEAD working paper no. 2010/71/TOM/INSEAD Social Innovation Centre (2010)
7. European Parliament and Council. Directive 2008/98/EC on waste and repealing certain Directives. Off. J. Eur. Union **L312**, 3–30 (2008)
8. Merrild, H., Damgaard, A., Christensen, T.H.: Life cycle assessment of waste paper management: the importance of technology data and system boundaries in assessing recycling and incineration. Resour. Conserv. Recycl. **52**, 1391–1398 (2008)
9. Circular Economy Action Plan EUR-Lex - 52020DC0098 - EN - EUR-Lex (europa.eu)
10. Pires, A., Martinho, G., Ribeiro, R., Mota, M., Teixeira, L.: Extended producer responsibility: a differential fee model for promoting sustainable packaging. J. Clean. Prod. **108A**, 343–353 (2015)
11. Huang, C.-C., Ma, H.-W.: A multidimensional environmental evaluation of packaging materials. Sci. Total Environ. **324**, 161–172 (2004)
12. Pires, A., Martinho, G.: Waste hierarchy index for circular economy in waste management. Waste Manage. **95**, 298–305 (2019)
13. RecyClass Homepage. https://recyclass.eu/. Accessed 31 Jan 2021
14. Civancik-Uslu, D., Puig, R., Voigt, S., Walter, D., Fullana-i-Palmer, P.: Improving the production chain with LCA and eco-design: application to cosmetic packaging. Resour. Conserv. Recycl. **151**, 104475 (2019)
15. Foschi, E., Zanni, S., Bonoli, A.: Combining eco-design and LCA as decision-making process to prevent plastics in packaging application. Sustainability **12**, 9378 (2020)
16. Trayak Homepage. https://trayak.com/solutions/packaging-solutions. Accessed 31 Jan 2021
17. RDC Environment Homepage. https://packaging.instantlca.com/. Accessed 31 Jan 2021
18. Earthshift Global Homepage. https://www.earthshiftglobal.com/software/packagesmart. Accessed 31 Jan 2021
19. Elipso Homepage. https://www.elipso.org/en/eco-design/. Accessed 31 Jan 2021
20. CONAI Homepage. https://www.ecotoolconai.org/. Accessed 31 Jan 2021
21. Pack4Ecodesign Homepage. pack4ecodesign.org/index_en.html. Accessed 31 Jan 2021
22. De Koeijer, B., Borgman, I., Henseler, J., ten Klooster, R., de Lange, J.: Embedding a sustainability focus in packaging development processes. In: Kishita, Y., Matsumoto, M., Inoue, M., Fukushige, S. (eds.) EcoDesign and Sustainability, vol. II, pp. 49.60. Springer, Singapore (2020)

Sustainable Use of Seaweeds from S. Martinho do Porto, Portugal – Past, Present, and Future Perspective

Mário Cavaco[1], Adriana Duarte[2], Susana Bernardino[1], Clélia Afonso[1], and Teresa Mouga[1](✉)

[1] MARE, School of Tourism and Maritime Technology, Polytechnic of Leiria, Leiria, Portugal
mougat@ipleiria.pt
[2] ESTM, Polytechnic of Leiria, Leiria, Portugal

Abstract. Seaweeds have long been used by populations on the coast of Portugal as fertilizers or food during periods of scarcity. Both uses have decreased dramatically since the introduction of chemical fertilizers and horticulture food. Even so, in the past decades, globalization has brought new interest to the consumption of these organisms, raising interest on algae as a healthy and tasty food. Much scientific research is underway on the properties of seaweeds, and it is proven that past uses are well-founded.

S. Martinho do Porto, in the Centre of Portugal is the only location where the commercial harvesting of seaweeds still takes place. *Gelidium corneum* is harvested because this species produces high-quality agar, a much sought-after phycocolloid, which is used in the food industry as gelling or thickening agent. However, the species shows other properties that have been disregarded by the seaweeds' industry. There are also some invasive species, namely *Asparagopsis armata*, which is causing considerable ecological problems to local ecosystems.

Thus, the Mar2020 project Limo do Cais was approved to study the sustainable use the algal biomass from S. Martinho do Porto taking into account its industrial potential. In this way, the properties of *G. corneum* are being studied, to increase its commercial value. As to *A. armata*, it shows important antimicrobial properties that are being studied aiming the use of the algal biomass in the pharmaceutical industry.

Thus, past uses of seaweeds will be discussed, as well as new potential uses that scientific research has uncovered.

Keywords: Seaweeds · Sustainable use · Bioactivities · Applications

1 Introduction

Seaweeds have long been harvested by coastal populations of Portugal for various purposes, such as fertilizers, feed, or food during periods of scarcity. This is common to many European maritime countries, such as Spain, the French Brittany, Ireland, Scotland, Norway, Denmark and Iceland, and the Azores and Cape Verde [1].

J. R. da Costa Sanches Galvão et al. (Eds.): *Proceedings of the 1st International Conference on Water Energy Food and Sustainability (ICoWEFS 2021)*, pp. 216–226, 2021.
https://doi.org/10.1007/978-3-030-75315-3_26

These uses have decreased dramatically in Portugal since the introduction of chemical fertilizers and horticulture food, which occurred mainly after the second half of the 20th century.

Yet, during the second half of the 20th century Portugal met a new interest in harvesting seaweeds for the extraction of agar, used in the food industry. Again, this activity declined during the 90s. Even so, in the past decades globalization has brought new interest to the consumption of seaweeds, arousing a new interest on algae as a healthy and tasty food. Much research is underway on the properties of seaweeds, and it has proven that past uses are well-founded: bio stimulants, biofertilizers, in food, and feed. New discoveries have also been made in the field of medicine, cosmetics, bioremediation, biofuels, or as bioindicators, confirming the importance of these marine resources.

S. Martinho do Porto, in the Centre of Portugal is the main location where the commercial harvesting of seaweeds still takes place. The red seaweed *Gelidium corneum* is harvested due to its high agar quality. However, the species shows other properties that are being disregarded by the seaweed's industry.

Also, among the many seaweed found in the Portuguese coast there are some invasive species that periodically cause harmful blooms. Among these *Asparagopsis armata* is causing considerable ecological problems to local ecosystems.

Thus, the Mar2020 project Limo do Cais aimed to study the sustainable use of this algal biomass taking into account its industrial potential.

This paper discusses the past and present uses of seaweeds from the Centre of Portugal, and well as the state-of-the art on *G. corneum* and *A. armata* potential uses.

2 Traditional Uses of the Seaweeds from the Centre of Portugal

2.1 The Use of Seaweeds as Fertilizers

The activity of seaweed harvesting was an old and common activity all along the Portuguese coast being more markedly north of the river Douro. Different mixtures of seaweeds were harvested depending on the location. In the North it was called s*argaço* and was a mixture of brown, red, and green algae of the species *Fucus* sp. (bodelha), *Pelvetia canaliculata* (botelho bravo), *Laminaria hyperborea* (folha de Maio), *Saccorhiza polyschides* (fitas), *Himanthalia elongata* (cordas), *Codium* sp. (chorão), *Palmaria palmata* (botelho-comprido) and *Chondrus crispus* (musgo da Irlanda).

In the Ria de Aveiro the mixture of seaweeds was called *moliço* and it was constituted by *Ulva* sp. (alface do mar), *Rhizoclonium* sp., *Gracilaria* sp. (cabelo de velha) and the marine plants *Zostera, Ruppia* and *Potamogeton* [1–3]. In the Centre of Portugal this harvest has started more recently and in a more sporadic way, but it took place near Peniche, e.g., on the beaches of Baleal, Consolação and S. Martinho do Porto, in Ericeira, on the Alentejo (Sines) and the Algarve coast (Fig. 1). Here the seaweeds harvested were called *limo* or *golfo* being likewise made up of a mixture of various red, brown, and green algae such as *Saccorhiza, Laminaria, Fucus, Codium, Palmaria, Gelidium,* and *Chondrus* [1, 4] that were washed to the seashore on days of rough tides. These seaweeds were used mainly as fertilizers in agricultural fields.

Seaweed harvesting is an old activity in Portugal, widely documented and regulated. The first royal charter dates to 1308, signed by king D. Dinis. Along with other legal documents it is also worth mentioning the decree of queen D. Maria I of 1798, which sets up a period of defence, as well as other decrees limiting the harvesting to certain periods of the year aiming at the recovery and the protection of natural populations. In the 20th century a regulation from 1909 was published to regulate the catch of seaweeds and remained in force until 1964 [1, 2].

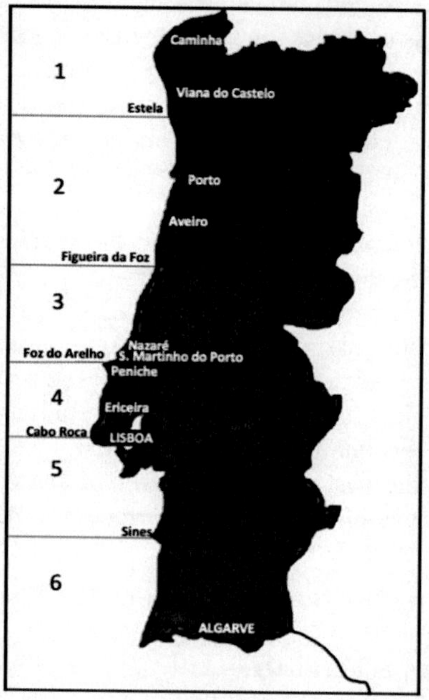

Fig. 1. Map of continental Portugal showing the main littoral areas referred in the text, and the territorial division into harvesting zones: the zone 1 – north, zone 2 centre – until Figueira da Foz; zone 3 – until Foz do Arelho and including S. Martinho do Porto, zone 4 – until Roca cape, zone 5 – until Sines Cape, and zone 6 – southern coast, including Algarve (as defined by Decree-Law n. 504/80, of 20th of October).

The high number of legal documentation proves the importance of this activity for the coastal populations. Initially developed by coastal farmers to be used as fertilizer, the activity of seaweed harvesting came to be an independent activity aiming the sale of the seaweeds because there was a greater demand for seaweeds to be used in fertilization [1]. To this end people developed their own tools, especially nets and long rakes for collecting the seaweeds from the beach or from a boat.

In the Centre of Portugal, the *encinho* (or *gravanço*) as it was called in Peniche (Fig. 2a), was used to collect unattached seaweeds on the beach, during the low tide [1]. Other common instrument used in Peniche (Baleal) is the *caramoeiro*, to harvest seaweeds suspended in the seawater (Fig. 2b). The *limo* was then dried on the sand of transported to the agriculture fields of all the western region of Portugal, namely vineyards and vegetable gardens, and other crops [1]. This *limo* is an excellent natural fertilizer, being rich in nitrogen, phosphoric acid, and potassium, the three main nutrients required by plants to thrive [4].

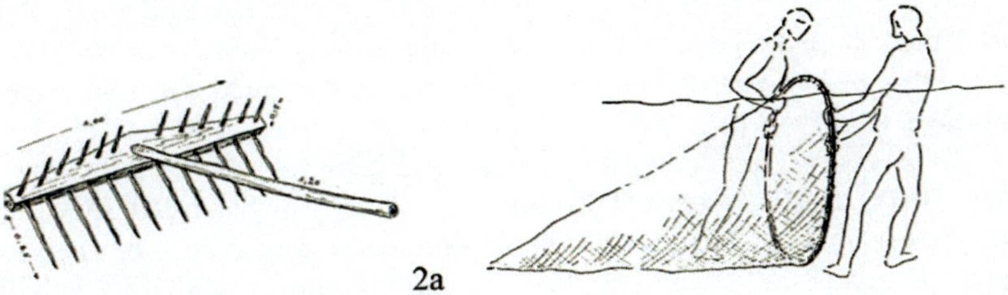
2a 2b

Fig. 2. Instruments used in Centre of Portugal to collect seaweeds: a - *Encinho* or *gravanço*; b – *Caramoeiro*. Adapted from [1].

Although this activity has decreased dramatically since the introduction of chemical fertilizers, it is still residually practiced in the coastal area of northern Portugal [3].

Regarding the use of seaweeds as feed, there are some records of consumption by domestical animals, namely by pigs, which prefer *Fucus* sp. [1] and of feed basic industrial production in Peniche [5].

The use of seaweeds as food by the littoral populations is registered on the Azores islands, but not on continental Portugal [6]. Yet, the oral tradition in the Centre littoral of Portugal (Peniche) refers to the consumption of seaweeds in periods of famine especially until the World War II.

2.2 The Harvest of Agarophytes

The word “agar” derives from the Malaysian name meaning "red alga" and refers to the members of a class of polysaccharides that occur in the intercellular material of many species of red algae. Because red algae grow in aquatic environment, they have developed a gelatinous, hydrophilic, and flexible polymer which has been used by man for over 350 years as a gelling agent in many food applications [7].

All agar-producing species - agarophytes - are red algae, most of which are harvested in the wild. The most harvested is *Gelidium*, currently being harvested in Morocco, followed by Spain and Portugal where the species *Gelidium corneum* (formerly known as *G. sesquipedale*) is the most harvested species. Other species of *Gelidium, Pterocladiella, Gracilaria, Ahnfeltia,* and *Gelidiella,* are also harvested in France, Chile, Japan, South Korea, and South Africa to produce agar [8].

The shortage of Asian agar in Europe that arose during World War II has restarted the interest in harvesting seaweeds in Portugal. Thus, in 1944 around 30 tons of dry *Gelidium* were collected and in 1947 the first industrial agar unit was installed [2].

Since the 60s of the 20th century the harvest started to be carried out by divers, and this recent activity was regulated by a law published in 1967 and reprinted in 1980 [2]. This law defined six harvesting zones in continental Portugal (Fig.1) nine in the Azores’ archipelago (one for each island) and one in Madeira archipelago, to promote the rational exploitation and to support programs to preserve the resources. Currently the seaweed harvest is regulated by the order n.10089/2019 which defines the security measures necessary, the number of divers, and so on.

By 1971 Portugal was the second world agar producer with six factories running and a production capacity of 1620 tons of agar [9]. But most factories closed at the beginning of the 90s of the 20th century due to a shortage in international demand and to a decrease in the availability of biomass, and currently only one agar producing Iberian company runs in Portugal (Iberagar) [3].

Presently harvest occurs during the summer and autumn, between July 15th and November 15th [1]. In 2017 almost 1000 tons (dry weight) were harvested, showing the new interest these agarophytes represent in the Portuguese economy, with three zones being explored: zone 3 in S. Martinho do Porto, zone 4 which restarted its activity in 2015, but with modest figures, and Azores region where *G. microdon* and *Pterocladiella capillacea* are currently being harvested (Fig. 3).

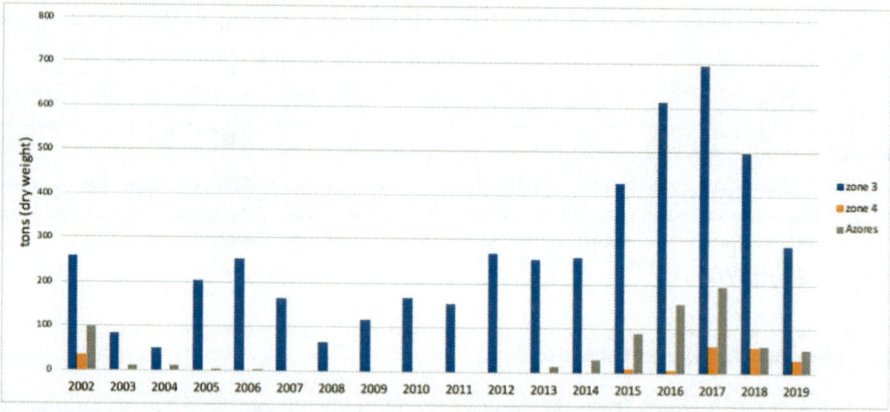

Fig. 3. Annuals seaweed harvest zone 3 (S. Martinho do Porto), zone 4 (Peniche) and Azores islands [1]. Data from [10]

Each boat can harvest up to 4 tons a day [3, 9] and twenty-two boats are presently licensed to harvest seaweeds in Peniche, Sesimbra and S. Martinho do Porto.

It is important to notice that a tendency of decrease is noticeable for 2018 and 2019, which may be due to overharvesting, as it occurred previously in the 90s of the 20th century [2]. Previously the overexploitation of the resource led to the ending of the activity in most of the littoral, namely in Peniche (zone 4) in the early 90s, Sesimbra (zone 5) around 1996, and finally Azenha do Mar (zone 6) in 2004 [3], which raises new concerns on the sustainability of the resource.

3 New Uses of the Seaweeds from the Centre of Portugal

3.1 *Gelidium corneum* (Hudson) J.V. Lamouroux 1813

Gelidium corneum may be called Atlantic agar (English), Tengusa, Tokorotengusa; (Japanese), Cabelo-de-cão, Febra, Francelha, Limo-fino, Limo-preto, (Portuguese), or

Ocle, Caloca (Spanish) [6]. It is widely distributed in the Atlantic and Indian Oceans as well as on the coasts of Japan, China, Indonesia, and Myanmar [11].

It is a perennial Rhodophyta, always attached to rocks, that inhabits the coastal area at shallow depths, in cold, light exposed agitated waters. Fronds are large and robust, deep red and shiny, with a rigid and cartilaginous texture. The fronds are cylindrical at the base but becoming flattened at the top, branching in the upper half, through secondary and tertiary usually opposite branches (Fig. 4a) [6].

It is the most important species for the extraction of agar, producing the highest quality sea gelling agent. In Europe the species is not commonly used as food, but it is edible and is eaten in the Hawaiian Islands [6].

Besides these classical uses the species has also been used in the treatment of ulcers and stomach problems [12]. Also, there are positive effects on the intestine microbiota promoted by the effects of probiotics from *Gelidium* and other algal species [13]. Finally, the fact that agar is not digested by the human organism determines that it can be considered a fibre. These soluble polysaccharides, as well as the algae extracts, and bio gels, present antioxidant, antiviral, anti-proliferative, anti-inflammatory properties, phyto stimulant, and bioremediatory effects (Table 1).

Table 1. List of the main bioactivities of *Gelidium* sp.

Taxa	Fraction	Activity	Reference
Gelidium corneum	Edible biofilms Methanolic extracts Dichloromethane-methanol extracts	Antimicrobial Antifungal	[14] [15] [16]
	Agar-based aerogel Ethanol/aqueous extracts Chloroform extracts Aqueous extracts of the mycosporine-like amino acids	Antioxidant	[17] [18] [19] [20]
	Ethanolic extracts	Anti-inflammatory	[21]
	Aqueous extracts Dichloromethane-methanol extracts	Anti-proliferative	[21] [22]
	Oligosaccharides	Phyto stimulant	[23]
	Non-living biomass	Bioremediation	[24]
Gelidium foliaceum	Polysaccharides	Antioxidant	[25]
Gelidium amansii	Ethyl acetate, acetone, methanol extracts	Antioxidant	[26]
	Phosphate-buffered saline, methanol extracts	Antiproliferative	[27]
Gelidium latifolium	Ethanol and dichloromethane fraction	Antimicrobial Antifouling	[28]
Gelidium pulchellum Gelidium spinulosum	Aqueous extracts	Antiviral	[29]

The seaweed *G. corneum* therefore may be used in the pharmaceutical industry due to its ability to treat obesity, hypercholesterolemia, diabetes, and intestine cancer [7], in the food industry as packaging due to the ability to produce stable gels and their antimicrobial and antioxidant activities [17], as well as in cosmetic industry, as photo protectant and antioxidant [30]. Besides, the phyto stimulant effect also opens the possibility of its use in the field of agriculture as a substitute for chemical fertilizers [23]. Finally, the *Gelidium* biomass consist of hard rhizoidal filaments whose fibres can also be used in the manufacture of paper [31].

As such, from a circular economy perspective and given the bioactivities evidenced by the above-mentioned studies, many of them showed in recent years, the by-products of *G. corneum* biomass used in agar extraction have immense potential to be used in other industries.

Fig. 4. Red seaweeds mentioned. a.: *Gelidium corneum*; b: *Asparagopsis armata.*

3.2 *Asparagopsis armata* Harvey 1855

The species is known as harpoon grass, which arises from the similarity of its thorny branches in the form of a harpoon or hook [32]. It is originated from South Australia and New Zealand [33] and is now an invader in Europe where it is widely distributed across the Atlantic, causing considerable ecological and economic problems. It was introduced from the Mediterranean Sea and was first detected in Italy (island of Elba) in 1880, then in Algeria in 1923 and in Cherbour and Biarritz in 1925 [32].

It has two very distinct morphological forms, a larger gametophyte (Fig. 4b) and a smaller, darker, filamentous tetrasporophyte, that forms small spherical tufts [34].

All the red macroalgae species of the Bonnemaisoniaceae family, such as *A. armata*, are well known as halogenated sources of compounds with strong antibacterial, antiviral, and anti-cancer activity (Table 2).

As such they are being investigated as potential source of compounds that can be used in the cosmetic and pharmaceutical industries [35–37], more recently as its ability to reduce methane emissions from cattle [38]. Despite the presence of the halogenated compounds, both species *A. armata* and *A. taxiformis* are also considered edible species [6].

The chemical properties of this seaweed are related to its invasive character. This property, along with the ability to use the hocked filaments to attach to diverse substrate, are the key to the success of this species as an invader. Therefore, harvesting this biomass

Table 2. List of the main bioactivities of *Asparagopsis* sp.

Taxa	Fraction	Activity	Reference
Asparagopsis armata	Crude, ethanolic, methanolic extracts	Antimicrobial	[39, 40]
	Aqueous extracts	Antiviral	[29]
	Crude extracts and fractions	Antifouling	[40]
	Raw biomass	Methanogenesis	[38, 41]
	Dichlorometane extracts	Anti-tumour	[35, 42]
Asparagopsis taxiformis	Methanol, chloroform, petroleum, ether ethyl acetate extracts	Anti-tumour Antioxidant	[43]
	Cinnamaldehyde, methanol extracts	Antifouling	[44]
	Raw biomass	Methanogenesis	[45]

decreases its abundance in nature where it is a nuisance, and it can be converted into an industrial product, due to the potential it has for the feed, cosmetics, and pharmaceutical industries, as well as anti-fouling coatings.

4 Conclusions

Macroalgae are a remarkably interesting source of nutrients and bioactive compounds that can be used in various industries. From the resources perspective, sustainability must be considered and therefore the need to cultivate native species must be addressed ensuring sufficient biomass for the industry and protecting wild seaweeds' populations.

It should also be noted that invasive species are so because they show their inherent chemical compounds which can be used in biotechnology as bioactive compounds in a range of industries. In this way, threat is converted into an opportunity, creating value from biological residues.

Finally, it is also important to ensure that all the seaweed's biomass should be used, from a biorefinery perspective, that is, successive extractions can be made for different purposes, with the final biomass also being used as a bioproduct. This process adds value to biomass and ensures full use of the resource, without producing waste.

Acknowledgments. This research was funded by Fundação para a Ciência e Tecnologia (FCT), through the strategic project UIDB/04292/2020 granted to MARE, and the project Operational Programme MAR2020 through the project MAR-04.03.01-FEAMP-0101 – Limo do Cais.

References

1. Veiga de Oliveira, E., Galhano, F. Pereira, B.: Actividades agro-marítimas em Portugal. Etnográfica Press, Lisboa (2020)

2. Melo, R.A.: Exploração dos recursos Algológicos em Portugal. In: Martins-Loução, M.A. (ed.), Fragm. em Ecol. FCUL-Livraria Esc. Ed., pp. 45–65 (2002)
3. Gaspar, R., Pereira, L., Sousa-Pinto, I.: The seaweed resources of Portugal. Bot. Mar. **62**, 499–525 (2019)
4. Pereira, L., Cotas, J.: Seaweeds as plant fertilizer, agricultural biostimulants and animal fodder. In: Pereira, L., Bahcevandziev, K., Joshi, N.H. (eds.) Seaweeds as Plant Fertilizer, Agricultural Biostimulants and Animal Fodder, pp. 1–22. CRC Press (2019)
5. Engenheiro, F.: O 'limo" e o seu aproveitamento ao longo dos anos na jurisdição marítima de Peniche. Cabo Carvoeiro Memórias (2006). https://cabo-carvoeiro-historico.blogspot.com/2006/06/o-limo-e-o-seu-aproveitamento-ao-longo.html. Accessed 02 Jan 2021
6. Pereira, L.: Edible Seaweeds of the World. CRC Press (2016)
7. Cardoso, S., et al.: Bioproducts from Seaweeds: a review with special focus on the Iberian Peninsula. Curr. Org. Chem. **18**, 896–917 (2014)
8. Williams, P.: Food Polysaccharides and Their Applications. Taylor & Francis Group, Boca Raton (2006)
9. Santos, R., Duarte, P.: Marine plant harvest in Portugal. J. Appl. Phycol. **3**, 11–18 (1991)
10. DGRM: Publicações Recursos da Pesca - Estatísticas da Pesca 2000–2019 (2021). https://www.dgrm.mm.gov.pt/web/guest/recursos-da-pesca-dgrm. Accessed 28 Jan 2021
11. Guiry, D.: AlgaeBase. In: Guiry, M.D., Guiry, G.M.: AlgaeBase. World-wide electronic publication. National University of Ireland, Galway (2021)
12. Almeida, A.R.G.S.: Guia de Campo das Algas do Intertidal da Praia da Vigia, Master's thesis. University of Algarve, Faro (2007)
13. de Jesus Raposo, M., De Morais, A.M.B., De Morais, R.M.S.C.: Emergent sources of prebiotics: seaweeds and microalgae. Mar. Drugs **14**, 27 (2017)
14. Kanmani, P., Rhim, J.W.: Antimicrobial and physical-mechanical properties of agar-based films incorporated with grapefruit seed extract. Carbohydr. Polym. **102**, 708–716 (2014)
15. Lakhdar, F., Sidi, O.A., Ben, M.: Antimicrobial effect of two marine algae *Gelidium sesquipedale* and *Laminaria ochroleuca* collected from the coast of El Jadida-Morocco. J. Bio. Innov **5**, 16–23 (2016)
16. Sidi, O.A., Ben, M.: Preparation of an antifungal soap from *Gelidium sesquipedale* waste. Bothalia J. **44**, 9–16 (2014)
17. de Oliveira, J.P., et al.: Development of food packaging bioactive aerogels through the valorization of *Gelidium sesquipedale* seaweed. Food Hydrocoll. **89**, 337–350 (2019)
18. Vega, J., Bonomi-Barufi, J., Gómez-Pinchetti, J.L., Figueroa, F.L.: Cyanobacteria and red macroalgae as potential sources of antioxidants and UV radiation-absorbing compounds for cosmeceutical applications. Mar. Drugs **18**, 659 (2020)
19. Metidji, H., et al.: In vitro screening of secondary metabolites and evaluation of antioxidant, antimicrobial and cytotoxic properties of *Gelidium sesquipedale* Thuret et Bornet red seaweed from Algeria. J. Mater. Environ. Sci. **6**, 3184–3196 (2015)
20. De La Coba, F., Aguilera, J., Figueroa, F.L., De Gálvez, M.V., Herrera, E.: Antioxidant activity of mycosporine-like amino acids isolated from three red macroalgae and one marine lichen. J. Appl. Phycol. **21**, 161–169 (2009)
21. Matos, J., et al.: Commercial Red Seaweed in Portugal (*Gelidium sesquipedale* and *Pterocladiella capillacea*, Florideophyceae): going beyond a single-purpose product approach by valorizing bioactivity. Thalassas **36**, 213–224 (2020)
22. Grozdanic, N., et al.: In vitro evaluation of antioxidant and antitumoral activities of marine Algae *Gelidium sesquipedale* and *Fucus spiralis*. Eur. J. Cancer **48**, S26 (2012)
23. Lebbar, S.: Valorisation biologique de co-produits de l'extraction de l'agar issu du *Gelidium sesquipedale*. Limoges. https://www.theses.fr (2018)

24. Trioui, L., et al.: Study of chromium biosorption by four Atlantic Algae: Gelidium sesquipedale, Gelidium corneum, *Corallina oficinalis*, and *Ulva lactuca.* Am. J. Eng. Res. **6**(6), 103–108 (2017)
25. Rengasamy, K.R.R., et al.: Phenolic profiles, antioxidant capacity, and acetylcholinesterase inhibitory activity of eight South African seaweeds. J. Appl. Phycol. **27**, 1599–1605 (2015)
26. Yan, X., Nagata, T., Fan, X.: Antioxidative activities in some common seaweeds. Plant Foods Hum. Nutr. **52**, 253–262 (1998)
27. Chen, Y.-H., Tu, C.-J., Wu, H.-T.: Growth-inhibitory effects of the red alga *Gelidium amansii* on cultured cells. Biol. Pharm. Bull. **27**, 180–184 (2004)
28. Hellio, C., De La Broise, D., Dufossé, L., Le Gal, Y., Bourgougnon, N.: Inhibition of marine bacteria by extracts of macroalgae: potential use for environmentally friendly antifouling paints. Mar. Environ. Res. **52**, 231–247 (2001)
29. Rhimou, B., Hassane, R., Nathalie, B.: Antiviral activity of the extracts of Rhodophyceae from Morocco. Afr. J. Biotechnol. **9**, 7968–7975 (2010)
30. Pandey, A., et al.: Mycosporine-Like Amino Acids (MAAs) profile of two marine red Macroalgae, *Gelidium sp.* and *Ceramium* sp. Int. J. Appl. Sci. Biotechnol. **5**, 12–21 (2017)
31. Seo, Y.-B., Lee, Y.-W., Lee, C.-H., You, H.-C.: Red algae and their use in papermaking (2009)
32. Ní Chualáin, F., Maggs, C.A., Saunders, G.W., Guiry, M.D.: The invasive genus *Asparagopsis* (Bonnemaisoniaceae, Rhodophyta): molecular systematics, morphology, and ecophysiology of Falkenbergia isolates. J. Phycol. **40**, 1112–1126 (2004)
33. Bonin, D.R., Hawkes, M.W.: Systematics and life histories of New Zealand Bonnemaisoniaceae (*Bonnemaisoniales, Rhodophyta*); I. The genus Asparagopsis. New Zeal. J. Bot. **25**, 577–590 (1987)
34. Blanco, A., Lemos, M.F.L., Pereira, L., Gaspar, R., Mouga, T., Neto, J.M., Troncoso, J.S., Olabarria, C.: Mapping invasive macroalgae in the western Iberian Peninsula: a methodological guide. In: Olabarria, C., Blanco, A., Troncoso, J.S. (eds.). Servizo de Publicacións da Universidade de Vigo, Vigo, España (2020)
35. Pinteus, S., Silva, J., Alves, C., Horta, A., Fino, N., Rodrigues, A.I., Mendes, S., Pedrosa, R.: Cytoprotective effect of seaweeds with high antioxidant activity from the Peniche coast (Portugal). Food Chem. **218**, 591–599 (2017)
36. Kolanjinathan, K., Ganesh, P., Saranraj, P.: Pharmacological importance of seaweeds: a review. World J. Fish Mar. Sci. **6**, 1–15 (2014)
37. Pinteus, S., et al.: Marine invasive macroalgae: turning a real threat into a major opportunity - the biotechnological potential of *Sargassum muticum* and *Asparagopsis armata.* Algal Res. **34**, 217–234 (2018)
38. Vijn, S., et al.: Key considerations for the use of seaweed to reduce enteric methane emissions from cattle. Front. Vet. Sci. **7**, 597430 (2020)
39. Salvador, N., Garreta, A.G., Lavelli, L., Ribera, M.A.: Antimicrobial activity of Iberian macroalgae. Sci. Mar. **71**, 101–113 (2007)
40. Pinteus, S., et al.: Marine invasive species for high-value products' exploration – Unveiling the antimicrobial potential of *Asparagopsis armata* against human pathogens. Algal Res. **52**, 102091 (2020)
41. Roque, B.M., Salwen, J.K., Kinley, R., Kebreab, E.: Inclusion of *Asparagopsis armata* in lactating dairy cows' diet reduces enteric methane emission by over 50 percent. J. Clean. Prod. **234**, 132–138 (2019)
42. Gutiérrez-Rodríguez, A.G., Juárez-Portilla, C., Olivares-Bañuelos, T., Zepeda, R.C.: Anticancer activity of seaweeds. Drug Discov. Today **23**, 434–447 (2018)
43. Neethu, P., Suthindhiran, K., Jayasri, M.: Antioxidant and antiproliferative activity of *Asparagopsis taxiformis*. Pharmacogn. Res. **9**, 238–246 (2017)

44. Jha, B., Kavita, K., Westphal, J., Hartmann, A., Schmitt-Kopplin, P.: Quorum sensing inhibition by *Asparagopsis taxiformis*, a marine macro alga: separation of the compound that interrupts bacterial communication. Mar. Drugs **11**, 253–265 (2013)
45. Roque, B.M., Duarte, T.L.: Red seaweed (*Asparagopsis taxiformis*) supplementation reduces enteric methane by over 80 percent in beef steers. bioRxiv 2020.07.15.204958 (2020)

Biochemical Composition of Six Native Seaweeds from Buarcos Bay, Central West Coast of Portugal

Ana Marta Mendes Gonçalves[1,2](✉), Sara García-Poza[1], João Cotas[1], João Carlos Marques[1], and Leonel Pereira[1]

[1] Marine and Environmental Sciences Centre, Department of Life Sciences, University of Coimbra, 3001-456 Coimbra, Portugal
amgoncalves@uc.pt

[2] Department of Biology and CESAM, University of Aveiro, 3810-193 Aveiro, Portugal

Abstract. Seaweeds have a great variety of compounds with different properties and benefits to human health. Marine macroalgae provide a high nutritional value along with low caloric value, poor in fat, and with the presence of polysaccharides that behave as fibers with no calories. This is not widespread, but some macroalgae and by-products are used in various applications, inclusively in food products. Under the project MENU - Marine Macroalgae: Alternative recipes for a daily nutritional diet, six native seaweed species were collected in Praia da Tamargueira, Buarcos, Figueira da Foz, Portugal: *Ulva* spp. (green seaweed), *Chondrus crispus, Gracilaria gracilis, Mastocarpus stellatus, Porphyra umbilicalis* (red seaweeds) and *Bifurcaria bifurcata* (brown seaweed) to further biochemical characterization. Polysaccharides' monomeric composition was determined after sulphuric acid hydrolysis, derivatization to alditol acetates, then analyzed by gas chromatography with flame ionization detector (GC-FID), whereas protein content was quantified following colorimetric Bradford method. Results showed red seaweeds presenting the highest polysaccharide profile followed by the green macroalgae and then the brown seaweed species. In the case of protein content, *Porphyra umbilicalis* was the seaweed with the highest content, followed by *Ulva* spp., *Gracilaria gracilis, Chondrus crispus* and *Mastocarpus stellatus*. The brown seaweed *Bifurcaria bifurcata* presented the lowest protein content.

Keywords: Nutritional content · Polysaccharides · Protein content · Marine macroalgae

1 Introduction

There are diverse countries that have nutritional supplements based in seaweeds, in a daily basis from ancient times, this occurs mainly in the Southeast Asiatic countries (China, Japan and Korea) [1]. This seaweed consumption is due to the seaweed high nutritional value (good source of carbohydrates, proteins, lipids, vitamins, minerals) [2–4]. However, the seaweed nutritional value can vary from species to species and even

J. R. da Costa Sanches Galvão et al. (Eds.): *Proceedings of the 1st International Conference on Water Energy Food and Sustainability (ICoWEFS 2021)*, pp. 227–236, 2021.
https://doi.org/10.1007/978-3-030-75315-3_27

between specimens from the same species with different geolocation, temporal season, or from ecological parameters (biotic and abiotic factors) [4–6]. For this, there is a necessity to analyze the seaweed nutritional values before considering available and acceptable for human consumption, as nutritious food [4]. This step of seaweed characterization is an essential key to observe if it is possible to take advantage of a specific seaweed species as a food source, with the mindset of the global food security [4], which the seaweed need to have values approximated to the nutritional values recommended by the World Health Organization (WHO) and Health Organizations [4]. Because a nutritious food is assimilated as a food supplement a low intake of this food can satisfy the human daily intake of caloric energy and essential nutrients [4]. The seaweed is already considered a food supplement in the Japanese cuisine, where the normal seaweed intake in a normal adult is 4 g up to 7 g of seaweed dry weight daily. On the contrary, in the western countries there is a lack of information about this topic in the native seaweeds. However, this food source is gaining more interest, due to be a natural source and even, if its cultivation do not compete for terrestrial arable land, being considered a natural sustainable food source if the food safety is guaranteed [7, 8].

There are nutrients which are essential to the human daily intake, such as proteins and carbohydrates that play an important role in human metabolism and health [4, 9].

The carbohydrates in seaweeds are commonly dietary fibers, such alginate, fucoidan, laminarin, carrageenan, agar, porphyran or ulvan [4, 9]. However, these carbohydrates are not the only carbohydrates in seaweeds, where also, can appear glucose, galactose, mannose, rhamnose or xylose as simple carbohydrates. These simple molecules are important energy source for the human organism. These fast-absorbed carbohydrates are similar to a fuel source for normal cells function, without this intake the cell consumes proteins until the intake or the cell death. This energy source (carbohydrates) inhibits the protein to be metabolized as energy source in human cells [4, 7, 10]. However, these carbohydrates content varies greatly between species [11].

The protein intake is very important due to the various and essential functions in the human organism. They are considered the essential block units for the well function of muscles, bones, cartilage, skin, blood, being mainly precursors of macromolecules, such as enzymes, antibodies and hormones [12]. Unlike, the carbohydrates, the organism does not retain protein content, so there is a need to a daily intake of protein content food, to satisfy the organism requirement to repair and build muscle tissue, mainly [4].

In this study, a characterization of carbohydrates profile and protein content was performed in six native seaweeds from the west coast of Portugal to assess the seaweeds nutritional value and their potential to be inserted as natural food source for human consumption (based in the food global safety).

2 Materials and Methods

2.1 Seaweed Collection

Under the project MENU—Marine macroalgaE—alternative recipes for a daily Nutritional diet project (Project Reference: FA_05_2017_011), a biochemical characterization was carried out on six different marine macroalgae (green, red and brown) that are already collected and known in coastal communities: *Ulva* spp. (Fig. 1), *Chondrus crispus* (Fig. 1), *Gracilaria gracilis* (Fig. 2), *Mastocarpus stellatus* (Fig. 2), *Porphyra umbilicalis* (Fig. 3) and *Bifurcaria bifurcata* (Fig. 3). Specimens of these six different macroalgae were collected during the winter of 2020 from Praia da Tamargueira in Buarcos, located in Figueira da Foz city, Portugal. Samples were harvested by hand from well-established patches and without epiphytes or degradation visible at eyesight and later the samples were transported in cold boxes to the laboratory. After that, samples were washed with filtered seawater with the objective to remove sands and contaminants. Afterwards, the biomass was washed with distilled water, cleaned with paper and stored at −80 °C until further biomass utilization.

Fig. 1. *Ulva* spp. and *Chondrus crispus*, respectively.

Fig. 2. *Gracilaria gracilis* and *Mastocarpus stellatus*, respectively.

Fig. 3. *Porphyra umbilicalis* and *Bifurcaria bifurcata*, respectively.

2.2 Seaweed Biochemical Profile

2.2.1 Samples Preparation

To carry out the biochemical characterization, the biomass collected from the field was prepared for analysis. All the samples were duly identified, weighed and placed in aluminum containers. The samples were dried in the oven at a constant temperature of 60 °C for 48 h. Lastly, the samples were crushed in a commercial mill (Taurus aromatic, Oliana, Spain) and the dry biomass was stored, protected from light so that they did not undergo oxidation, until the analysis proceeded.

2.2.2 Carbohydrates Analysis

Carbohydrate analysis from algal biomass comprised the quantification of polysaccharides (analysed in the form of monosaccharides) that were subjected to hydrolysis followed by reduction and acetylation, as described by Coimbra et al. [13]. These analyzes were carried out starting from differently processed biomasses (fresh and dry) and with different proportions of biomass and volume of reagents used during the extraction. The internal standard 2-deoxyglucose was incorporated to the sugar quantification. The samples were centrifuged and resuspended in acetone for injection into the GC-FID. The samples were passed through a Thermo Scientific Trace 1310 chromatography equipment (Waltham, MA, USA) defined with a flame ionization detector (GC-FID). A TG-WAXMS A (30 m length, 0.32 mm i.d., 0.25 μm film thickness) GC column was utilized and the oven was set at an initial temperature of 180 °C, following a linear increase of temperature of 5 °C min^{-1} until reach a final temperature of 230 °C, keeping this temperature for 12 min. The carrier gas was helium at a flow rate of 2.5 ml min^{-1}. Monosaccharides were identified by comparing retention time with standards. The quantification of sugars was carried out by comparing the chromatographic sugar peaks with the peaks obtained for the internal standard used (2-deoxyglucose).

2.2.3 Protein Content

The total protein concentration of the seaweeds was determined by the spectrophotometric (wavelength 595 nm) method of Bradford [14] adapted to microplate. This colorimetric method depends in linkage of the dye Coomassie Blue G-250 to proteins, where the protein concentration is estimated by a comparative analysis to the Bovine Serum albumin (BSA) standard solution.

2.3 Statistical Data Analyses

For monosaccharides and protein data, the statistical analysis was performed with the software Sigma Plot v.14. It was employed an ANOVA analysis to assess statistically differences. While, Tukey test was used after the rejection of the ANOVA null hypothesis. Differences were considered significant when the p value was less than 0.05.

3 Results

3.1 Carbohydrates Analysis

In the polysaccharide analysis performed on the seaweeds six different polysaccharides were identified (Table 1), namely rhamnose, arabinose, xylose, fucose, mannose, galactose and glucose, although the content of galactose and glucose was provided jointly. Glucose and galactose (Glc+ Gal) content was, in general, the most abundant residue in all the seaweeds analyzed, achieving the highest content in *Chondrus crispus* (737.55 μgg^{-1}) and the lowest quantity in *B. bifurcata* (28.89 μgg^{-1}). The red and brown seaweeds do not present rhamnose and arabinose, unlike the green seaweed (*Ulva* spp.). In addition, the fucose residues only appear in *B. bifurcata* (26.66 μgg^{-1}) as expected for the presence of fucose units in the fucoidan. Referring to the xylose concentration, *Ulva* spp. present the highest content (100.76 μgg^{-1}) whereas *B. bifurcata* showed the lowest amount (1.68 μgg^{-1}). The highest content of mannose was observed in *P. umbilicalis* (161.39 μgg^{-1}) which differs greatly from the other seaweeds, showing *Ulva* sp. the lowest content (10.29 μgg^{-1}). These results showed that *B. bifurcata* showed the lowest content in polysaccharide whereas the red seaweeds (*C. crispus, P. umbilicalis* and *M. stellatus*) have the highest polysaccharide concentration.

Table 1. Carbohydrates profile of polysaccharides in the studied seaweeds species. The results are expressed in μgg^{-1} of seaweed dried weight. Statistically significant differences found in the content of a residue among the studied species are expressed by letters.

Polysaccharide	*Bifurcaria bifurcata*	*Chondrus crispus*	*Gracilaria gracilis*	*Mastocarpus stellatus*	*Porphyra umbilicalis*	*Ulva spp.*
Rhamnose	nd	nd	nd	nd	nd	143.19
Arabinose	nd	nd	nd	nd	nd	18.82
Xylose	1.68^{b}	$14.35^{a,b}$	$4.05^{a,b}$	$17.23^{a,b}$	$88.18^{a,b}$	100.76^{a}
Fucose	26.66	nd	nd	nd	nd	nd
Mannose	$26.71^{a,b}$	$18.01^{a,b}$	$13.63^{a,b}$	$11.17^{a,b}$	161.39^{a}	10.29^{b}
Glucose + Galactose	28.89^{b}	737.55^{a}	$290.06^{a,b}$	$523.96^{a,b}$	$588.12^{a,b}$	$306.85^{a,b}$

Nd–non detectable.

3.2 Protein Analysis

Total protein content (Table 2) was determined using less than 100 mg of wet biomass for the six species. Protein concentration was calculated according to the standard protein curve of bovine serum albumin. Although these results are expressed in mgg^{-1} of wet biomass, and most of the studies in literature analyses are made in dry matter, those are in agreement, as *Porphyra umbilicalis* (105.40 mgg^{-1}) presents the highest levels of protein, and, generally, brown seaweeds (in this case, *Bifurcaria bifurcata* showed a protein concentration of 1.28 mgg^{-1}) have less protein content [15].

Table 2. Total protein concentration, in mg per gram of wet seaweed biomass. (Mean ± SEM; n = 8 replicates. one-way ANOVA - a, b: there are no statistically significant differences of protein concentration within species with the same *letter*).

Species	*Bifurcaria bifurcata*	*Chondrus crispus*	*Gracilaria gracilis*	*Mastocarpus stellatus*	*Porphyra umbilicalis*	*Ulva spp.*
Protein concentration (mgg^{-1})	1.28 ± 0.04^{a}	3.75 ± 0.07^{a}	$15.40 \pm 2.20^{a,b}$	3.59 ± 1.49^{a}	105.40 ± 6.78	30.86 ± 8.80^{b}

4 Discussion

This study highlights the highest amount of polysaccharides' monomeric composition of the red seaweeds *Chondrus crispus, Gracilaria gracilis, Mastocarpus stellatus, Porphyra umbilicalis* followed by the green seaweed *Ulva* spp. and at the least the brown seaweed *Bifurcaria bifurcata* here studied. In the seaweed polysaccharides' monomeric analysis in the targeted species, there it is a general lack of studies with a full characterization. *B. bifurcata* diverges from the results obtained by Gómez-Ordoñez et al. [16], mainly in the fucose and mannose content, where the results obtained are very high in our sample. However the results obtained, in *C. crispus*, are similar to the study of Kulshreshtha et al. [17], nevertheless, the difference ratio between Glucose + Galactose and mannose is higher in our study than in the results obtained in the authors' study. In *M. stellatus*, the monosaccharide ratio is similar to the work of Gómez-Ordoñez et al. [18], thus, there is not a great difference from different collection locations (Center of Portugal and the North of Spain). The monosaccharide fraction of *Ulva* spp. is very different from the analysis of the Yaich et al. [19], where the seaweed collection occurred in the Mediterranean sea, the monosaccharide differ totally in the monosaccharide and total concentration in the seaweed biomass. In *Gracilaria* sp. and *Porphyra* sp. there is a general lack of the characterization of polysaccharide fractioning. This result in polysaccharide fraction demonstrates that the seaweeds analyzed have interesting monosaccharides concentration, mainly in the Glucose + Galactose, to be applied in human food as possible energy source [4].

In the case of protein content, *Porphyra umbilicalis* was the seaweed with the highest content, followed by *Ulva* spp., *Gracilaria gracilis, Chondrus crispus* and *Mastocarpus stellatus*. The brown seaweed *Bifurcaria bifurcata* presented the lowest protein content. All the seaweeds analyzed stayed between the range of the data from the literature to the red, green and brown seaweed (due to lack of literature in the species analyzed), where the brown seaweeds have low content, the green seaweeds have intermediate and the red seaweeds have the higher content of protein, mostly *Gracilaria* sp. and *Porphyra* sp. have the highest protein content [4, 11, 20].

These results demonstrate that seaweeds have a wide range of polysaccharide and protein concentration not depending directly from the phylogenetic group. But between species demonstrating that each seaweed has different nutritional value. And so, different maximum limit of seaweed intake daily needs to be based in the monosaccharide concentration, but more importantly, in the protein concentration which it was demonstrated higher concentration in the seaweeds analyzed [4]. If the seaweed intake surpass the limit the nutrients will have adverse effect in the human health [9].

These macromolecules like polysaccharides, lipids and proteins are needed to ingest daily, in large amounts supplying energy [21], being associated with crucial physiological functions in the cells. Seaweeds are known to be a nutrient rich dietary source in proteins, dietary fibers, vitamins, minerals, with low caloric composition that varies according to several factors such as geographic location, season, environmental factors, part of the seaweed harvested, etc. [4]. Indeed, in calm waters there is a lower production of polysaccharides. Seaweeds have a wide application in several industrial sectors like cosmetic and care products, agriculture, animal feed, human food and pharmaceutical due to their bioactivities making this marine resource a promising natural product with multiple applications, increasing the interest in these organisms. For instance, *Porphyra* spp. and *Gracilaria* spp. are intensively cultivated, as other algae such as *Eucheuma* spp., *Kappaphycus alvarezii*, for the hydrocolloid and food industry [22]. The red macroalgae polysaccharides, agar and carrageenan, are widely studied and applied in food and pharmaceutical industries as multifunctional ingredients like homogenizers, stabilizer and emulsifiers [23, 24]. On the other hand, the level of sulfation of these molecules, agar, carrageenan and porphyrin may change the bioactivity of the polysaccharides [25].

Seaweeds and seaweeds-derived products like extracts from derived sulfated polysaccharides showed a high potential like in crops production, plant germination and absorption, animals' efficiency immune response, methane reduction, skin and hair care [4]. Furthermore it is known the benefits in human health like blood pressure [26] decreasing the risk of ischemic heart disease [27], and even related with a decreased incidence of depression symptoms [28].

Seaweeds synthetizes a high amount and variety of proteins with red macroalgae producing a high amount of protein that around 35% to 47%, higher than terrestrial plants (average between 30–40% dry weight). Indeed, red seaweeds present a huge amount of essential amino acids, showing a ratio of EAA/NEAA higher than brown seaweeds [29, 30]. Pepsin, which digests from *Neopyropia yezoensis* is also related to decrease of cholesterol levels, inhibition of calcium precipitation, reduction of blood sugar, calcium precipitation inhibition, decrease of cholesterol levels, antioxidant, and improvement of hepatic function activities [31].

Information in literature corroborate our data by highlighting the great nutritional value of seaweeds and correlating their composition with the bioactivity properties of each group – red, brown and green marine macroalgae - and thus the benefits we can take from them. As previously referred, seaweed composition is dependent of several factors varying their composition according to those factors.

5 Conclusion

Marine macroalgae present a great potential with several applications in several daily life products, some of them unknown to some citizens. Seaweeds are promising natural marine resources due to their bioactivities, some of them not yet explored, and thus need to be investigated in a sustainable way, to improve human well-being. The usage of natural products instead of synthetic compounds are indeed a better way to a healthier life, avoiding harmful effects. The growing interest in their potentialities will conduct to a further research in this area and the increase of knowledge and the discovery of bioactive compounds with specific and unique characteristics that can improve the life of human population with the development of novel products.

Acknowledgments. This work is financed by national funds through FCT – Foundation for Science and Technology, I.P., within the scope of the projects UIDB/ 04292/2020 – MARE – Marine and Environmental Sciences Centre and UIDP/50017/2020 + UIDB/50017/2020 (by FCT/MTCES) granted to CESAM – Centre for Environmental and Marine Studies, and financed by project MENU (Marine Macroalgae: Alternative recipes for a daily nutritional diet) (FA_05_2017_011) funded by the Blue Fund under Public Notice No. 5—Blue Biotechnology. Ana M. M. Gonçalves acknowledges University of Coimbra for the contract IT057–18-7253. Sara García-Poza thanks to the project MENU - Marine Macroalgae: Alternative recipes for a daily nutritional diet (FA_05_2017_011). João Cotas thanks to the European Regional Development Fund through the Interreg Atlantic Area Program, under the project NASPA (EAPA_451/2016).

References

1. Dawes, C.J.: Marine Botany. Wiley, New York (1995)
2. Arasaki, S., Arasaki, T.: Vegetables from the Sea. Japan Publishing Inc., Tokyo (1983)
3. Wong, K.H., Cheung, P.C.K.: Nutritional evaluation of some subtropical red and green seaweeds. Part I - proximate composition, amino acid profiles and some physico-chemical properties. Food Chem. **71**, 475–482 (2000). https://doi.org/10.1016/S0308-8146(00)00175-8
4. Leandro, A., Pacheco, D., Cotas, J., Marques, J.C., Pereira, L., Gonçalves, A.M.M.: Seaweed's bioactive candidate compounds to food industry and global food security. Life **10**, 140 (2020). https://doi.org/10.3390/life10080140
5. Zepeda, E., Freile-Pelegrín, Y., Robledo, D.: Nutraceutical assessment of *Solieria filiformis* and *Gracilaria cornea* (Rhodophyta) under light quality modulation in culture. J. Appl. Phycol. **32**, 2363–2373 (2020). https://doi.org/10.1007/s10811-019-02023-0
6. Pacheco, D., Araújo, G.S., Cotas, J., Gaspar, R., Neto, J.M., Pereira, L.: Invasive seaweeds in the Iberian Peninsula: a contribution for food supply. Mar. Drugs **18**, 560 (2020). https://doi.org/10.3390/md18110560
7. Godfray, H.C.J., Beddington, J.R., Crute, I.R., Haddad, L., Lawrence, D., Muir, J.F., Pretty, J., Robinson, S., Thomas, S.M., Toulmin, C.: Food security: the challenge of feeding 9 billion people. Science. **327**, 812–818 (2010). https://doi.org/10.1126/science.1185383
8. Bleakley, S., Hayes, M.: Algal proteins: extraction, application, and challenges concerning production. Foods **6**, 33 (2017). https://doi.org/10.3390/foods6050033
9. Cherry, P., O'Hara, C., Magee, P.J., McSorley, E.M., Allsopp, P.J.: Risks and benefits of consuming edible seaweeds. Nutr. Rev. **77**, 307–329 (2019). https://doi.org/10.1093/nutrit/nuy066
10. Rioux, L., Turgeon, S.L.: Seaweed carbohydrates. In: Tiwari, B.K., Troy, D.J. (eds.) Seaweed Sustainability: Food and Non-Food Applications, pp. 141–192. Academic Press (2015). https://doi.org/10.1016/B978-0-12-418697-2/00007-6.
11. Pereira, L.: A review of the nutrient composition of selected edible seaweeds. In: Pomin, V.H. (ed.) Seaweed: Ecology, Nutrient Composition and Medicinal Uses, pp. 15–47. Nova Science Publishers, Inc. (2011)
12. Salehi, B., Sharifi-Rad, J., Seca, A.M.L., Pinto, D.C.G.A., Michalak, I., Trincone, A., Mishra, A.P., Nigam, M., Zam, W., Martins, N.: Current trends on seaweeds: looking at chemical composition, phytopharmacology, and cosmetic applications. Molecules **24**, 4182 (2019). https://doi.org/10.3390/molecules24224182

13. Coimbra, M.A., Waldron, K.W., Selvendran, R.R.: Isolation and characterisation of cell wall polymers from olive pulp (Olea europaea L.). Carbohydr. Res. **252**, 245–262 (1994). https://doi.org/10.1016/0008-6215(94)90019-1
14. Bradford, M.: A Rapid and sensitive method for the quantitation of microgram quantities of protein utilizing the principle of protein-dye binding. Anal. Biochem. **72**, 248–254 (1976). https://doi.org/10.1006/abio.1976.9999
15. Fleurence, J.: Seaweed proteins. Trends Food Sci. Technol. **10**, 25–28 (1999). https://doi.org/10.1016/S0924-2244(99)00015-1
16. Gómez-Ordóñez, E., Jiménez-Escrig, A., Rupérez, P.: Dietary fibre and physicochemical properties of several edible seaweeds from the northwestern Spanish coast. Food Res. Int. **43**, 2289–2294 (2010). https://doi.org/10.1016/j.foodres.2010.08.005
17. Kulshreshtha, G., Burlot, A.-S., Marty, C., Critchley, A., Hafting, J., Bedoux, G., Bourgougnon, N., Prithiviraj, B.: Enzyme-assisted extraction of bioactive material from chondrus crispus and codium fragile and its effect on herpes simplex virus (HSV-1). Mar. Drugs **13**, 558–580 (2015). https://doi.org/10.3390/md13010558
18. Gómez-Ordóñez, E., Jiménez-Escrig, A., Rupérez, P.: Bioactivity of sulfated polysaccharides from the edible red seaweed Mastocarpus stellatus. Bioact. Carbohydrates Diet. Fibre **3**, 29–40 (2014). https://doi.org/10.1016/j.bcdf.2014.01.002
19. Yaich, H., Garna, H., Besbes, S., Paquot, M., Blecker, C., Attia, H.: Chemical composition and functional properties of Ulva lactuca seaweed collected in Tunisia. Food Chem. **128**, 895–901 (2011). https://doi.org/10.1016/j.foodchem.2011.03.114
20. Morais, T., Inácio, A., Coutinho, T., Ministro, M., Cotas, J., Pereira, L., Bahcevandziev, K.: Seaweed potential in the animal feed: a review. J. Mar. Sci. Eng. **8**, 559 (2020). https://doi.org/10.3390/jmse8080559
21. Institute of Medicine of The National Academies: Dietary Reference Intakes. National Academies Press, Washington, D.C. (2006). https://doi.org/10.17226/11537.
22. Ferdouse, F., Løvstad Holdt, S., Smith, R., Murúa, P., Yang, Z.: The global status of seaweed production, trade and utilization. FAO Globefish Res. Program. **124**, 120 (2018)
23. Michalak, I., Chojnacka, K.: Algae as production systems of bioactive compounds (2015). https://doi.org/10.1002/elsc.201400191
24. FAO: The State of the World Fisheries and Aquaculture - Meeting the sustainable development goals, Rome (2018). https://doi.org/10.1093/japr/3.1.101
25. Cardozo, K.H.M., Guaratini, T., Barros, M.P., Falcão, V.R., Tonon, A.P., Lopes, N.P., Campos, S., Torres, M.A., Souza, A.O., Colepicolo, P., Pinto, E.: Metabolites from algae with economical impact. Comp. Biochem. Physiol. - C Toxicol. Pharmacol. **146**, 60–78 (2007). https://doi.org/10.1016/j.cbpc.2006.05.007
26. Wada, K., Nakamura, K., Tamai, Y., Tsuji, M., Sahashi, Y., Watanabe, K., Ohtsuchi, S., Yamamoto, K., Ando, K., Nagata, C.: Seaweed intake and blood pressure levels in healthy pre-school Japanese children. Nutr. J. **10**, 83 (2011). https://doi.org/10.1186/1475-2891-10-83
27. Murai, U., Yamagishi, K., Sata, M., Kokubo, Y., Saito, I., Yatsuya, H., Ishihara, J., Inoue, M., Sawada, N., Iso, H., Tsugane, S.: Seaweed intake and risk of cardiovascular disease: the Japan public health center–based prospective (JPHC) study. Am. J. Clin. Nutr. **110**, 1449–1455 (2019). https://doi.org/10.1093/ajcn/nqz231
28. Guo, F., Huang, C., Cui, Y., Momma, H., Niu, K., Nagatomi, R.: Dietary seaweed intake and depressive symptoms in Japanese adults: a prospective cohort study. Nutr. J. **18**, 58 (2019). https://doi.org/10.1186/s12937-019-0486-7
29. Vieira, E.F., Soares, C., Machado, S., Correia, M., Ramalhosa, M.J., Oliva-teles, M.T., Paula Carvalho, A., Domingues, V.F., Antunes, F., Oliveira, T.A.C., Morais, S., Delerue-Matos, C.: Seaweeds from the Portuguese coast as a source of proteinaceous material: total and free amino acid composition profile. Food Chem. **269**, 264–275 (2018). https://doi.org/10.1016/j.foodchem.2018.06.145

30. Abirami, R.G., Kowsalya, S.: Phytochemical screening, microbial load and antimicrobial activity of underexploited seaweeds. Int. Res. J. Microbiol. **3**, 328–332 (2012)
31. Harnedy, P.A., FitzGerald, R.J.: Bioactive proteins, peptides, and amino acids from macroalgae. J. Phycol. **47**, 218–232 (2011). https://doi.org/10.1111/j.1529-8817.2011.00969.x

Sustainable Treatments for Wastewater Deriving from the Coffee Processing

Subramanian Shanthakumar[1] and Fulvia Chiampo[2(✉)]

[1] Department of Environmental and Water Resources Engineering, School of Civil Engineering, Vellore Institute of Technology (VIT), Vellore 632014, India
shanthakumar.s@vit.ac.in

[2] Department of Applied Science and Technology, Politecnico di Torino, Corso Duca degli Abruzzi 24, 10129 Turin, Italy
fulvia.chiampo@polito.it

Abstract. A large quantity of water is used in the production process of the coffee industry, with a specific production of wastewater up to 20 L/kg of coffee berries. Considering that the current world production of berries is around 10 million metric tons, a huge volume of wastewater is generated, characterized by suspended solids, and high concentration of soluble organic matter, Biological Oxygen Demand (BOD), and Chemical Oxygen Demand (COD). It is a great threat to the soil, surface, and groundwater environment if this wastewater is discharged without treatment. Hence, the treatment of coffee processing wastewater (CPWW) is essential through innovative, sustainable, and eco-friendly techniques. Although conventional biological treatment methods proved to be very efficient, economic constraints and challenges to operational strategies are the limitations. Recently, studies started to test Advanced Oxidation Processes to treat CPWW. Among them, Fenton and solar photo-Fenton process already demonstrated good efficacy to remove recalcitrant molecules. The current presentation reports the findings achieved when these two processes are applied to treat CPWW for COD and color removal. The two sets of runs were carried out with the same operative conditions. The results evidenced similar performances for COD and color removal, except in one case where a better removal was achieved with the solar photo-Fenton process. This pushes further tests to clarify some features and optimize the operative conditions since the solar-driven processes can be efficiently exploited in the countries where the solar irradiance is high, exactly where the coffee berries processing is usually carried out. In this way, the sustainability of wastewater treatment will be enhanced using sunlight.

Keywords: Coffee processing · Wastewater · Fenton process · Solar Fenton process · COD · Color

1 Introduction

Water is the essential requirement for almost all industries and the wastewater generated from industrial processes leads to a serious environmental concern when it is discharged

J. R. da Costa Sanches Galvão et al. (Eds.): *Proceedings of the 1st International Conference on Water Energy Food and Sustainability (ICoWEFS 2021)*, pp. 237–244, 2021.
https://doi.org/10.1007/978-3-030-75315-3_28

without proper treatment. The pollution due to the discharge of industrial wastewater ultimately affects the ecosystem which is an integrated system that includes water, land, and air.

Globally, coffee is one of the most commercialized commodities and is the second-largest traded commodity after petroleum. Coffee belongs to the genus *Coffea* of the *Rubiaceae* family. Coffee Arabica and Coffee Robusta are the two main varieties of coffee grown in India, where the area under cultivation is about 0.418 million hectares with Arabica (50%) and Robusta (50%) [1].

A large quantity of water is being used in the coffee industry during coffee processing, and hence a high volume of wastewater is generated. Although the coffee industries operate only during the coffee picking season (October-December), the wastewater generated is large in volume and is strongly acidic with high Biochemical Oxygen Demand (BOD) and Chemical Oxygen Demand (COD). This is a threat to the quality of surface and groundwater, and also the land contamination. When discharged into the natural water body the highly acidic coffee processing wastewater causes damage to the aquatic life due to the decrease in the dissolved oxygen concentration in the water [2].

Even though the conventional methods of wastewater treatment yield good results, they are not successful in implementation because of various limitations like cost, operation efficiency, etc. As the coffee processing is seasonal, its wastewater is not generated throughout the year, and hence the treatment unit may not be feasible. Therefore, it becomes essential to have suitable techniques to overcome the limitations caused by the biological treatment units.

Advanced Oxidation Processes (AOPs) are the innovative wastewater treatment technologies that have been successful in degrading most of the organic compounds present in the polluted water especially when they are recalcitrant to the biological removal.

Photocatalytic degradation is part of these processes and it is considered as an effective system for the mineralization of many organics through the generation of radicals such as •OH and $O_2^{-•}$, reducing considerably the organic load of effluents [3]. The principle involves the initial absorption of a photon having energy equal to or greater than the bandgap energy of the semiconductor, leading to the formation of electrons and holes.

The Fenton process involves reactions of hydrogen peroxide (H_2O_2) with iron salts to form active oxygen species with high oxidizing power. The Fenton process has been found effective, less time-consuming, and cost-effective in treating industrial wastewaters.

In recent years, AOPs have been widely used because of their capacity to degrade a huge number of recalcitrant substances in liquid and gaseous systems using procedures with relatively low cost [4]. The main mechanism of AOPs is the generation of highly reactive free radicals. Hydroxyl radicals are effective in destroying organic chemicals because they are reactive electrophiles and react rapidly with nearly all electron-rich organic compounds. Most of AOPs use a combination of strong oxidizing agents (e.g. H_2O_2, O_3) with catalysts (e.g. transition metal ions) and irradiation (e.g. ultraviolet, visible). Among different available AOPs producing hydroxyl radicals, titanium dioxide/UV light process, hydrogen peroxide/UV light process and Fenton's reactions seem to be promising technologies for wastewater treatment [5].

The photocatalytic degradation process has been in discussion for the past few decades since the photochemical splitting of water into hydrogen and oxygen with TiO_2 was discovered [6].

Zayas et al. [7] studied the removal of the natural organic matter present in coffee processing wastewater through coagulation-flocculation and advanced oxidation processes. The effectiveness of the removal of natural organic matter using commercial flocculants and UV/H_2O_2, UV/O_3, and UV/H_2O_2/O_3 processes was determined under acidic conditions. It was reported that a reduction in COD of 67% was achieved when the coffee wastewater was treated by chemical coagulation-flocculation with lime and coagulant T-1. Further, the processing of coffee wastewater by coagulation-flocculation in combination with UV/H_2O_2 gave a COD reduction of 86%.

Satori and Kawase [8] studied degradation of model dark brown colored coffee effluent using zinc oxide (ZnO) as photocatalyst by varying ZnO dosage from 0 to 4000 mg/L, coffee loading from 0 to 90 mg/L, and intensity of UV light having the radiation peak at 352 nm. Almost complete decolorization was achieved after 180 min for the initial coffee concentration of 50 mg/L with a ZnO dosage of 3000 mg/L and three UV lamps. The dissolved oxygen (DO) largely affected the photo decolorization process. The study indicated that ZnO photocatalyst has the potential for the treatment of coffee processing wastewaters.

The Fenton process is proved to an efficient treatment for various types of industrial wastewaters such as chemical, pharmaceutical, textile, paper and pulp, cosmetics, cork processing wastewater [9].

Researchers have proved that Fenton oxidation is an efficient treatment for the removal of color and hence it is used for the decolorization of dyes [10, 11]. It is most effective as a pretreatment tool when COD is greater than 500 mg/L. The biological wastewater treatment may be improved with the addition of a Fenton pretreatment step. It is capable to nearly complete the mineralization of organic components to CO_2 and H_2O. However, it is a less time-consuming process. The main advantage of the Fenton process is its simplicity. No energy input is necessary to activate hydrogen peroxide. The Fenton reagents are readily available, easy to store and handle, safe, and require a relatively shorter reaction time [12, 13]. However, this process has disadvantages such as the high cost of hydrogen peroxide, and the homogeneous catalyst that is added as iron salt cannot be retained in the process which further requires separation to prevent additional water pollution [9].

Barbusinski and Filipek [14] studied the Fenton process for the treatment of pesticides, achieving a degradation of 97–100%. Dantas et al. [15] evaluated the efficiency of the Fenton and photo-Fenton process for the treatment of wastewater from the leather industry. According to the study, 70% of COD reduction was obtained. Park et al. [16] investigated the feasibility of Fenton oxidation for the removal of color and the non-biodegradable organics from wastewater produced in animal breeding facilities. It was found that the removal efficiencies reached more than 90% and around 50% for color and COD, respectively were obtained. Fenton oxidation has proved to be a suitable pretreatment for extremely polluted pharmaceutical wastewater [17].

Kumar et al. [18] conducted the treatment of coffee wastewater using Fenton's oxidation process. The coffee pulping wastewater used for the study was characterized by low pH with a high concentration of COD, ammonia nitrogen, nitrate-nitrogen, and phosphorus, with a BOD/COD ratio varying from 0.037–0.050. The effect of different operating conditions, namely Fenton dosage, mixing speed, pH, reaction time was evaluated. The results obtained a COD removal of 84.4% and percentage removal of other pollutants such as ammonia nitrogen, nitrate-nitrogen, and phosphorous was found to be 90.8, 57.5, and 80% respectively.

Tokumura et al. [19] studied the decolorization of dark brown colored coffee effluent by the solar photo-Fenton process. Effects of accumulated solar light energy and dosage of Fenton reagents (iron and hydrogen peroxide) on the color removal were examined. With increasing Fe dosage, the rate of the decolorization increased but the enhancement was not pronounced beyond 10 mg/L. Although the addition of H_2O_2 increased the decolorization rate up to around 1000 mg/L of H_2O_2, further addition of H_2O_2 could not enhance the color removal. At excess dosages of Fenton reagents, the color removal did not improve due to their scavenging of hydroxyl radicals. It was found that the pseudo-first-order decolorization kinetic constant based on the accumulated solar energy is a sole parameter unifying solar photo-Fenton decolorization processes under different weather conditions. The results suggested that abundant solar energy driving decolorization of coffee effluent by the photo-Fenton reaction is highly efficient.

In this study, the Fenton and the solar photo-Fenton process was applied to real coffee processing wastewater, to compare the performance in terms of COD and color removal, when the same operative conditions are adopted. For the solar photo-Fenton process, the sunlight irradiance at Vellore latitude was used.

Previous experimental runs to study the UV-photocatalytic degradation of coffee processing wastewater using TiO_2 as catalyst demonstrated the feasibility and efficacy of photolysis processes [20], and this stimulated the trial to test the Fenton process, too.

The preliminary test sets gave similar results, except with one dosage of reactants where the solar photo-Fenton process showed higher removal for COD and color. This is encouraging and challenges to go on the study, to clarify the lacking details, and optimize the process.

2 Materials and Methods

2.1 Properties of the Wastewater

The wastewater used for this study was collected from an Indian coffee processing company. The physicochemical properties are reported in Table 1.

Table 1. Characteristics of the tested coffee processing wastewater.

Parameter	Value
pH	4.4
Colour Unit (CU)	1023
Electrical Conductivity (mS/cm)	1718
Total Dissolved solids (mg/L)	2110
Total Suspended Solids (mg/L)	4223
Total Solids (mg/L)	6333
Chemical Oxygen Demand (mg/L)	28800
Biochemical Oxygen Demand (mg/L)	14000
Phosphate (mg/L)	10
Nitrate (mg/L)	23

2.2 Reagents

The Fenton reagents consisted of H_2O_2 and Fe^{2+} from $FeSO_4 \cdot 7H_2O$.

For all the experiments, single distilled water was used. HCl and NaOH were used to have pH = 3.5.

2.3 Experimental Set-Up and Methodology

The experimental runs were carried out with a volume of wastewater equal to 0.5 L contained in a 1 L beaker, placed on a magnetic stirrer.

The Fenton reagents dosage was varied, keeping constant the molar ratio $Fe/H_2O_2 = 0.2$, respectively: $Fe^{2+} = 50$ mg/L and $H_2O_2 = 150$ mg/L, $Fe^{2+} = 100$ mg/L and $H_2O_2 = 300$ mg/L, $Fe^{2+} = 200$ mg/L and $H_2O_2 = 600$ mg/L.

Two sets of runs were done to study:

1. The Fenton process;
2. The solar Fenton process.

In these last runs, the effect of the sunlight was studied. During the test period, the average sunlight irradiance was found to be 878 W/m^2, measured by a laboratory luxmeter at Vellore Institute of Technology (12.97″ N, 79.16″ E).

Each test lasted 45 min.

2.4 Analytical Methods

COD was analyzed according to the standard method prescribed by the American Public Health Association (APHA) [21], making use of a mineralizer.

Color removal was measured as a decrease in absorbance at 465 nm by a UV-VIS spectrophotometer. The removal of COD and color was calculated using these Equations:

$$\text{COD removal \%} = \frac{C_i - C_f}{C_i} \times 100$$

$$\text{Color removal \%} = \frac{Cl_i - Cl_f}{Cl_i} \times 100$$

where: C_i = initial COD (mg/L), C_f = final COD (mg/L), Cl_i = initial color (color unit), Cl_f = final color (color unit).

3 Results

3.1 Wastewater Treatment by the Fenton Process

The first set of experimental runs was done with the Fenton process, testing different dosages of reagents at molar ratio $Fe/H_2O_2 = 0.2$.

Table 2 shows the results. For both the target parameters, COD and color, the highest removal was achieved with the dosage of 100 mg/L of Fe^{2+} and 300 mg/L of H_2O_2.

Table 2. COD and color removal with the Fenton process.

Dosage of Fe^{2+} (mg/L)	Dosage of H_2O_2 (mg/L)	COD removal (%)	Color removal (%)
50	150	69	65
100	300	80	71
200	600	74	65

3.2 Wastewater Treatment by the Solar Photo-Fenton Process

The same experimental runs were carried out exploiting the sunlight, to realize the solar photo-Fenton process. The operative conditions were the same used for the Fenton process.

Table 3 shows the results for COD and color removal. The best performance was achieved for the dosage of 100 mg/L Fe^{2+} and 300 mg/L of H_2O_2, with 86% removal of COD and 73% removal of color.

With the other dosages, the results are similar to the ones obtained with the Fenton process. The data are encouraging, even if an in-depth analysis is compulsory to clarify some features not evidenced in these first tests.

These first data are the starting point for the process optimization aiming to define the effects of Fe^{2+} and H_2O_2 dosage on the objective variables COD and color removal.

Table 3. COD and color removal with the solar Fenton process.

Dosage of Fe^{2+} (mg/L)	Dosage of H_2O_2 (mg/L)	COD removal (%)	Color removal (%)
50	150	59	63
100	300	86	73
200	600	72	66

About the solar Fenton process, it is expected that the reactants dosage can be reduced, even if to a limited extent, to achieve the same removal results. This will entail a lower cost and at the same time a lower environmental impact of the wastewater treatment due to the lower use of chemicals.

4 Conclusions

The preliminary experimental runs to test and compare the performance of the Fenton and the solar photo-Fenton process to remove COD and color from coffee processing wastewater gave optimistic findings.

The comparison evidenced similar performances, except with one dosage, which showed a better result for the solar photo-Fenton process.

This encourages to continue the study, to clarify the features that can influence the process and optimize the operative conditions. Moreover, sunlight exploitation is a way to improve the sustainability of the coffee processing wastewater, considering that this treatment occurs in the countries where usually the coffee is harvested and processed.

References

1. USDA. Foreign Agricultural Service, Coffee Annual, India (2020). https://www.fas.usda.gov/data/india-coffee-annual. Accessed 22 Jan 2021
2. Yemane, D.T., Beyene, A.H., Wassie, T.A., Gebresilassie, A.T.: Effect of coffee processing plant effluent on the physicochemical properties of receiving water bodies. Am. J. Environ. Protect. **4**(2), 83–90 (2015)
3. Vigo, F., Cagliari, M.: Photocatalytic oxidation applied to olive mill wastewaters treatment. Riv. Ital. Sostanze Grasse **76**, 345–353 (2010)
4. Hoffmann, M.R., Martin, S.T., Choi, W.Y., Bahnemann, D.W.: Environmental applications of semiconductor photocatalysis. Chem. Rev. **95**(1), 69–96 (2009)
5. Stasinakis, A.S.: Use of selected advanced oxidation processes (AOPs) for wastewater treatment - a mini review. Glob. Nest J. **10**(3), 378–385 (2008)
6. Fujishima, A., Honda, K.: Electrochemical photolysis of water at a semiconductor electrode. Nature **238**, 37–38 (1972)
7. Zayas, P.T., Geissler, G., Hernandez, F.: Chemical oxygen demand reduction in coffee wastewater through chemical flocculation and advanced oxidation processes. J. Environ. Sci. **19**, 300–305 (2007)

8. Satori, H., Kawase, Y.: Decolorization of dark brown colored coffee effluent using zinc oxide particles: the role of dissolved oxygen in degradation of colored compounds. J. Environ. Manage. **139**, 172–179 (2014)
9. Bautista, P., Mohedano, A.F., Casas, J.A., Zazo, J.A., Rodriguez, J.J.: An overview of the application of Fenton oxidation to industrial wastewaters treatment. J. Chem. Technol. Biotechnol. **83**, 1323–1338 (2008)
10. Lucas, M.A., Peres, J.A.: Decolorization of the azo dye reactive black 5 by Fenton and photo-Fenton oxidation. Dyes Pigments **71**, 236–244 (2006)
11. Muruganandham, M., Swaminathan, M.: Photocatalytic decolourisation and degradation of reactive orange 4 by TiO_2-UV process. Dyes Pigments **68**, 133–142 (2006)
12. Cuerda-Correa, E.M., Alexandre-Franco, M.F., Fernández-González, C.: Advanced oxidation processes for the removal of antibiotics from water. An overview. Water **12**, 102 (2020)
13. Pignatello, J.J., Oliveros, E., MacKay, A.: Advanced oxidation processes for organic contaminant destruction based on the Fenton reaction and related chemistry. Crit. Rev. Environ. Sci. Technol. **36**(1), 1–84 (2006)
14. Barbusinski, K., Filipek, K.: Use of Fenton's reagent for removal of pesticides from industrial wastewater. Pol. J. Environ. Stud. **10**(4), 207–212 (2001)
15. Dantas, T.L.P., Mendonca, V.P., Jose, H.J., Rodrigues, A.E., Moreira, R.F.P.M.: Treatment of textile wastewater by heterogeneous Fenton process using a new composite Fe_2O_3/carbon. Chem. Eng. **118**, 77–82 (2006)
16. Park, J.H., Shin, D.S., Lee, J.K.: Treatment of high-strength animal wastewater using photo-assisted Fenton oxidation coupled to photocatalytic technology. Water **11**, 1553 (2019)
17. Hofl, C., Sigl, G., Specht, O., Wurdack, I., Wabner, D.: Oxidative degradation of AOX and COD by different advanced oxidation processes: a comparative study with two samples of a pharmaceutical wastewater. Water Sci. Technol. **35**, 257–264 (1997)
18. Kumar, M.B., Ulavi, S.U., Ramesh, H.S., Asha, G., Pallavi, R.: Pretreatment of coffee pulping wastewater by Fenton's reagent. Indian J. Chem. Technol. **19**, 213–217 (2012)
19. Tokumura, M., Znad, H.T., Kawase, Y.: Decolorization of dark brown colored coffee effluent by solar photo-Fenton reaction: effect of solar light dose on decolorization kinetics. Water Res. **42**, 4665–4673 (2008)
20. Sujatha, G., Shanthakumar, S., Chiampo, F.: UV Light-irradiated photocatalytic degradation of coffee processing wastewater using TiO_2 as a catalyst. Environments **7**(6), 47 (2020)
21. APHA: Standard methods for the examination of water and wastewater. 22nd edn. American Public Health Association, Washington (2012)

Check for updates

Integrated Treatment of Pig Production Wastewaters Using Pre-treatment with Biomass Ash and Bioremediation by Microalgae

Catarina Viegas[1](✉) and Margarida Gonçalves[1,2]

[1] Mechanical Engineering and Resource Sustainability Center (MEtRICs), NOVA School of Science and Technology, Lisbon, Portugal
cv.sousa@campus.fct.unl.pt

[2] Centro de Investigação Para a Valorização de Recursos Endógenos (VALORIZA), Instituto Politécnico de Portalegre, Portalegre, Portugal

Abstract. Animal production leads to effluents with high loads of macro and micronutrients, and therefore with a huge potential of water bodies eutrophication. Conventional wastewater treatments are expensive, energy-consuming, release greenhouse gases (GHG), and produce a residual sludge. The use of microalgae for wastewater treatment allows recovery of nutrients (N, P, COD), minimize GHG emissions, and can significantly reduce costs relatively to conventional treatments. Microalgae have been used in the bioremediation of various effluents, such as sewage, manure, brewery, dairy, urban, among others.

In this work, piggery effluents were remediated by combining a physico-chemical pre-treatment with biomass ash and bioremediation with microalgae (*Chlorella vulgaris*, *Chlorella protothecoides* and *Tetradesmus obliquus*). The mixture of piggery effluent with biomass ash was stirred and fractionated by decantation to yield a liquid fraction and a solid precipitate. The fortification of the liquid fraction with olive-oil mill wastewater was also evaluated. Microalgae grown in the pre-treated effluent, in semi-continuous mode reached productivities of 258 and 237 mg $L^{-1}day^{-1}$ for *C. vulgaris* and *T. obliquus*, respectively. Both microalgae reached nutrient removal efficiencies of 100, 100, 90, and 100% for N, P, COD, and BOD_5, respectively. The microalgae composition was evaluated in terms of protein, sugar, lipid, and ash contents.

The produced microalgae biomass and the solid precipitate were tested as biostimulants for the germination of wheat and watercress seeds with positive results. In particular, fortification with *C. vulgaris* biomass produced an increase of 86% in the germination index of watercress seeds.

Keywords: Bioremediation · Wastewater treatment · Piggery effluents · Biomass ash · Olive-oil mill wastewater · Fertilization

1 Introduction

Urban, agro-industrial and animal production wastewaters have been treated conventionally, by complex and costly technologies to avoid pollution of water streams, air and

J. R. da Costa Sanches Galvão et al. (Eds.): *Proceedings of the 1st International Conference on Water Energy Food and Sustainability (ICoWEFS 2021)*, pp. 245–257, 2021.
https://doi.org/10.1007/978-3-030-75315-3_29

soils. Water contamination with pathogens and excess nutrients, is an environmental problem since it can lead to the eutrophication of fresh and marine ecosystems [1].

In 2018 the meat production sector in the European Community (EU28) amounted to 172,076.32 million EUR [2]. In that year, EU produced 23.85 million tonnes of pork, and Portugal around 361,510 tonnes [3]. The pig sector has annual emissions of 668 million tonnes CO_2-eq., 27% of which corresponds to emissions from manure management. The total manure and feed emissions in industrial pig production systems was 5.2 kg CO_2-eq/kg carcass weight [4]. In Portugal it is estimated that around 12 million cubic meters of manure are produced annually [3]. The pig farm effluents are mostly composed by the natural production of faeces and urine and the wastewater from facility cleaning. Normally these effluents are sent to stabilization ponds and a fraction may be valorised by composting or anaerobic digestion. The stabilization ponds requires the existence of several ponds for the treatment process to take place over time, so the effluent could be transferred to the next pond as it matures [5]. Anaerobic digestion implies tight control of the operating parameters and a high dilution ratio of the effluents, so that microorganisms are not inhibited, being a limited solution for processing large volumes of piggery effluent [6].

The use of microalgae reduces to half the energy consumption of conventional treatment, allowing the recovery up to 90% of the nutrients present into wastewater, with the advantage of production a valuable biomass for animal feeding and agriculture uses, and/or feedstock for biofuel production, thus enhancing the productivity and sustainability of the whole process [7, 8].

Olive-oil mill wastewater (OMW) is a wastewater produced in large quantities in olive oil producing countries like Portugal. This residue is originated by the olive-oil extraction process and generates 1 to 1.6 m^3 of wastewater per tonne of olive fruit processed [9]. It constitutes a significant environmental problem mainly due to its high chemical oxygen demand (COD) and difficult biodegradability related to its antibacterial activity induced by the high quantity of polyphenols and tannins. In 2018 olive oil production in Portugal was 135,000 tonnes, resulting in a production of 725,000 to 1,161,000 m^3 of OMW [10]. The olive-oil wastewaters has been previously tested for the growth of the microalga *Arthrospira platensis*, however, OMWs were used after a sodium hypochlorite treatment (NaClO) which significantly decreases the concentration of phenols and turbidity of the medium but also raises environmental problems [9].

In this work, the treatment and valorisation of piggery wastewater was evaluated using a precipitation step in the presence of forestry biomass ash, followed by bioremediation with microalgae. Incorporating the ash into the piggery effluent is a way to enrich the culture medium in mineral components. The addition of OMW was also tested as an approach to the co-treatment of these two effluents. Following the effluent treatment, the microalgae and the precipitate obtained in the pre-treatment were analysed for their composition, and both were tested as biostimulants in germination tests of watercress and wheat seeds. The advantage of using biomass ash in the precipitation step is to value a residue from biomass combustion and avoid using specific chemicals to achieve the same purpose. The semi-continuous mode proposed for the bioremediation tests, aims to ensure a high production of algal biomass with a low concentration of nutrients in the last reactor, in order to improve the treatment efficiency.

2 Material and Methods

2.1 Culture Medium

The piggery effluent consisting essentially of pig urine from suckling sows came from Raporal, from a farm near Setúbal, Portugal and was stored at 4 °C. This effluent was pre-treated with biomass ash in a proportion of 120 g per litre under agitation for 48 h. The bottom ash sample used was supplied by Prélis Cerâmicas Lda. and is a by-product of the combustion of forestry biomass with a small percentage of polymeric residues of ceramic furnaces. The olive-oil mill wastewater (OMW) came from Beira Baixa - Portugal and was provided by Herdade da Tapada da Tojeira, an organic olive oil producer.

The trials were performed testing two conditions: piggery effluents pre-treated with ash (P + A) and piggery effluents pre-treated with ash plus 2% of olive-oil mill wastewater (P + A + O). The microalgae were also grown in a synthetic culture medium – BG-11 [11] that served as control (C).

2.2 Microorganisms and Culture Conditions

The tested microalgae were: *Chlorella vulgaris* (INETI 58, LNEG_UBB, Portugal) (Cv), *Chlorella protothecoides* (UTEX 25, USA) (Cp) and *Tetradesmus* (formerly known as *Scenedesmus*) *obliquus* (ACOI 204/07, Portugal) (To). The trials were conducted in 1L Erlenmeyer flasks, agitated by an air flow of 15.2 L L^{-1}culture h^{-1}. The microalgae grew at room temperature (22 °C ± 2 °C), under artificial lighting (±200 μE m^{-2} s^{-1}) with cycles of 12h light/12h dark. The inoculations were performed using approximately 16 mL of inoculum (2% of the final volume), in order to have an initial optical density (at 540 nm) between 0.2 and 0.4 [1].

2.3 Trials

Four sets of trials were performed: batch mode tests with the three microalgae and control (C – synthetic medium), piggery effluents pre-treated with ash (P + A), piggery effluents pre-treated with ash plus 2% of olive-oil mill wastewater (P + A + O) (1st set of trials), batch mode tests with *C. vulgaris* and *T. obliquus* with piggery effluent + ash (P + A) (2nd set of trials), semi-continuous mode tests with *C. vulgaris* and *T. obliquus* with periodic transfer of 50mL of (P + A) (3rd set of trials) and with periodic transfer of 100 mL of (P + A) (4th set of trials). The batch mode trials lasted for 12 days and semi- continuous for 20 days. The 2nd, 3rd and 4th trials were performed with *C. vulgaris* and *T. obliquus*, using a series of three reactors with 1L for each microalga (Cv-1, Cv-2, Cv-3 and To-1, To-2, To-3). In the 2nd trial the 3 reactors of each microalga served as replicates. In 3rd and 4th trials each reactor received every other day *x* mL of effluent from the previous reactor, and the first reactor received *x* mL of piggery effluent, according to Fig. 1. In the 3rd trial the *x* was 50 mL and in the 4th trial the *x* was 100 mL.

In the 3rd and 4th trials (semi-continuous mode) reactors 2 and 3 were also supplemented with aqueous $NaNO_3$ to a final concentration of 20 mgN L^{-1} in the culture medium, every week [12] and with aqueous KH_2PO_4 to a final concentration of 10 mgP L^{-1}.

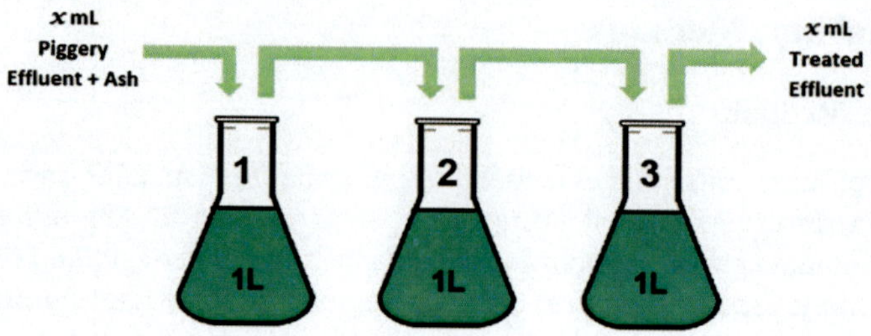

Fig. 1. Scheme of the transfer process for the 3rd and 4th trials. In the 3rd trial *x* is 50 mL and in the 4th trial *x* is 100 mL.

2.4 Microalgae Growth Evaluation

During the trials, samples were collected every other day to analyse total N, P, COD, BOD_5 and total solids content and total ash content in the medium according to the Standard Methods for Water and Wastewater [13]. Total phenolics were measured by the Folin-Ciocalteu method adapted by Singleton et al. [14]. Samples were also taken every week to evaluate biomass dry weight by filtering the samples through a Whatman GF/C 47mm. In the 1st and 4th trials after the decrease of N, P, and COD below the discharge limits (15, 10 and 150 mg L^{-1}, respectively) [15], the culture was harvested by centrifugation at 7000 rpm for 5 min. The supernatant was evaluated for all the same parameters as in the beginning, and the biomass was dried at 45 °C for 48 h. For the 2nd and 3rd trials at the end of the experiment part of the culture was used for analysis and the rest was used for the following trial.

2.5 Microalgae Biomass Characterization

The algal biomass was grounded using a Retsch ball mill - model MM400 for 4 min at a speed of 25 s^{-1} and characterized for total protein, total carbohydrate, and total lipid contents. Total nitrogen was quantified by the modified Kjeldahl method [16]. Total protein was calculated by multiplying total nitrogen content by the conventional conversion factor of 6.25 [17]. Carbohydrate content was determined by quantitative acid hydrolysis according to Miranda et al. [18] followed by the phenol-sulfuric method [19]. Lipid content was determined after Soxhlet extraction with n-hexane for 6 h.

2.6 Precipitate Analysis

The proximate and ultimate compositions of the precipitate was evaluated using standard methods. Moisture (M), ash content (Ash) and volatile matter (VM) were determined gravimetrically according to the ASTM methods 949-88, 830-87, and 897-88, respectively. Fixed carbon (FC) was established by difference. Ultimate analysis was performed using an elemental analyser (Thermo Finnigan – CE Instruments Model Flash EA 112 CHNS series). Oxygen content was achieved by difference. The mineral composition was evaluated through ICP-AES (Inductively Coupled Plasma – Atomic Emission Spectrometer, Ultima). Conductivity and pH were determined using a conductivity meter (Mettler Toledo) and a pH meter (Crison MicropH), respectively.

2.7 Seed Germination Tests

The germination tests of *Triticum aestivum* (wheat) and *Nasturtium officinale* (watercress) were done by the method proposed by Zucconi et al. [20], using *C. vulgaris* and *T. obliquus* biomass produced in piggery wastewater, at fortification levels of 0.2 and 0.5 g L^{-1}. Seed germination tests using precipitate extracts were also done following a method adapted from Monteiro and colleagues [21]. The dried precipitate sample was sieved (< 2 mm) and mixed with distilled water at 60 °C in the proportions of 0 (control), 5, 10, 20, and 40% and subsequently stirred with a magnetic stirrer for 3 h. This procedure was also applied to the ash used in the pre-treatment, at 5 and 10%. The aqueous extracts were then filtered through filter paper (Whatman 2) and used in the germination tests at volume of 3 mL/Petri dish. The seeds to be tested (50 watercress and 50 wheat seeds) were distributed over the Petri dishes lined with sterile absorbent paper, with 3 replicates per treatment. The Petri dishes were sealed with parafilm and placed in an incubator at 28 °C, without lighting for 5 days. The germination index was determined by the Eq. (1):

$$Germination\,index\,(\%) = \frac{G \times W}{G_C \times W_C} \times 100 \qquad (1)$$

where G is the number of germinated seeds and W is the seedling weight. Gc and Wc are the same parameters, but in the control (distilled water).

2.8 Statistical Analyses

Trials were performed in duplicate for microalgae growth and in triplicate for germination tests and for all the analysis, data were reported as mean ± standard deviation (SD). The parameters such as productivities, biomass composition and germination index were compared using analysis of variance with one-way ANOVA, by IBM SPSS statistical 23 software. The mean values were compared using the Tukey HSD test and correlation was considered statistically significant when $p < 0.05$.

3 Results and Discussion

3.1 Microalgal Productivity and Remediation Rates

Conducting a physico-chemical pre-treatment with forestry biomass ash in the agro-industrial effluents leads to an almost sterilization of the effluent by increasing the pH to 12.5, as well as the precipitation of several particles in suspension allowing a better light penetration and its use as culture medium, namely for microalgae.

The (P + A), (P + A + O) and the control medium were evaluated and had the characteristics presented in Table 1. Biomass ash composition is showed in Table 2.

The 1st and the 2nd trials ran for 12 days because after this time the discharge limits for total N, P, and COD were reached for (P + A). The 1st trial demonstrate that the three microalgae were able to grow in the tested effluents, achieving average productivities of 18.9, 30.9, and 16.1 mg L^{-1} d^{-1} for Cv, Cp, and To, respectively in (P + A + O), and 15.8, 11.4, and 19.2 mg L^{-1} d^{-1} for Cv, Cp, and To, respectively, in (P + A) (Fig. 2).

Table 1. Characterization of the Control and the Effluents used as culture medium. Piggery effluents pre-treated with ash (P + A), piggery effluent pre-treated with ash and olive-oil mill wastewater (P + A + O).

	Total nitrogen	Total phosphorus	COD	BOD5	Phenols	Total solids content	Total ash content	Initial optical density
	$mgN\,L^{-1}$	$mgP\,L^{-1}$	$mgO_2\,L^{-1}$	$mgO_2\,L^{-1}$	$mg\,L^{-1}$	$g\,L^{-1}$	$g\,L^{-1}$	(540 nm)
Control	173.1 ± 0.2	284.5 ± 7.6	24.3 ± 1.1	0.5 ± 0.0	2.1 ± 0.2	4.6 ± 0.3	3.2 ± 0.1	0.052
(P + A)	913.7 ± 6.9	6.2 ± 1.4	1171.4 ± 40.4	800.0 ± 11.3	17.2 ± 0.3	27.7 ± 0.3	24.0 ± 0.4	0.766
(P + A + O)	913.1 ± 11.5	6.8 ± 0.3	2100.5 ± 14.2	970.0 ± 35.3	87.2 ± 4.0	29.9 ± 0.1	24.0 ± 0.2	1.067

Table 2. Main chemical composition of biomass ash (wt. %).

Parameter	Value	Parameter	Value
pH	13.02 (± 0.03)	SO_3	0.915 (± 0.01)
CaO	65.9 (± 0.2)	P_2O_5	0.765 (± 0.01)
Cl	11.5 (± 0.04)	Na_2O	0.556 (± 0.01)
SiO_2	6.59 (± 0.06)	MnO	0.17 (± 0.008)
Al_2O_3	3.97 (± 0.04)	BaO	0.163 (± 0.002)
MgO	3.16 (± 0.02)	ZnO	0.0883 (± 0.001)
TiO_2	2.52 (± 0.06)	SrO	0.0855 (± 0.001)
Fe_2O_3	2.28 (± 0.03)	Cr_2O_3	0.0679 (± 0.006)
K_2O	1.180 (± 0.030)	CuO	0.0582 (± 0.003)

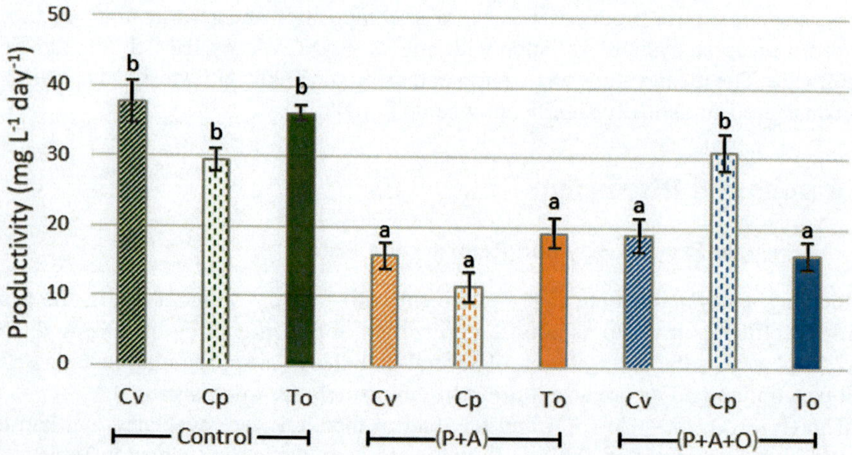

Fig. 2. Average productivity for the 1st trail in 12 days (mean ± SD, n = 3). (Control; (P + A) – piggery effluent + ash; (P + A + O) - piggery effluent + ash + olive oil mill wastewater; Cv – *Chlorella vulgaris*; Cp – *Chlorella protothecoides*; To – *Tetradesmus obliquus*)

The addition of a reduced quantity of olive-oil mill wastewater (OMW) led to an increase in the productivity of microalga *C. protothecoides*, what may reflect a positive effect from a higher abundance of nutrients, expressed by the higher values of COD and BOD_5 (Table 1). The remaining microalgae were not sensible to those differences and did not present significant differences between (P + A) and (P + A + O).

For 2nd, 3rd, and 4th trials, the two algae with the best growth and remediation performance were selected, for which reason *C. vulgaris* and *T. obliquus* were chosen.

In the semi-continuous mode, the 3rd and the 4th trials last for 20 days. The highest productivity was reached in the 3rd trial with 258.2 ± 7.1 and 236.7 ± 8.9 mg L^{-1} day^{-1} for *C. vulgaris* and *T. obliquus* respectively (Fig. 3).

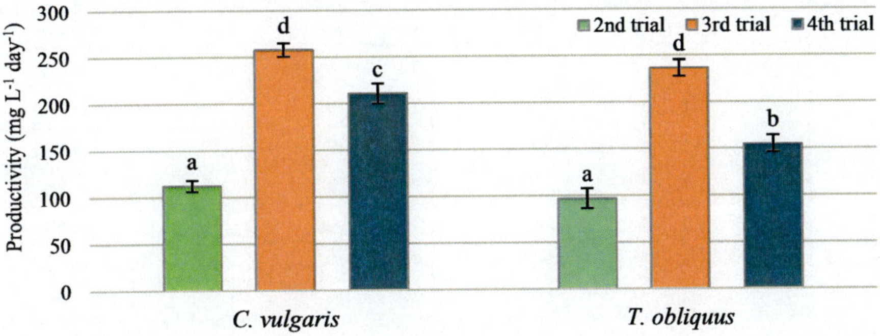

Fig. 3. Average productivity of the 2nd (12 days), 3rd (20 days) and 4th (20 days) trials for *Chlorella vulgaris* and *Tetradesmus obliquus* (mean ± SD, n = 3).

The growth of the microalgae led to an effective removal of nitrogen and phosphorus and other components contributing to COD (Table 3).

The removal efficiency was superior to 82% for N and more than 77% for P in the batch mode. The three tested microalgae have a high potential to remove the N, P, and total solids. Regarding COD removal, microalgae in (P + A) was shown to be effective in reducing COD. In the case of (P + A + O) the microalgae were able to significantly reduce N and P, but the COD was not reduced to levels that would allow its discharge. The raw effluent (P + A + O) had a higher COD (2100 mgO_2 L^{-1}) and the values attained were 511.3 and 574.9 mgO_2 L^{-1}, for To and Cv, respectively. In the 3rd trial the remediation to allow the effluent release was reached after 8 days of transfers. In the 4th trial the COD remediation never achieved the required levels for discharge (150 mgO_2 L^{-1}) [15].

Another study with piggery effluents reached biomass yields of 0.53 g L^{-1} and 0.49 g L^{-1} for *T. obliquus* and *C. vulgaris*, respectively, significantly higher than those achieved in the present study. However the starting effluent had only 1/20 of the total nitrogen of the current study, and the removal rate for nitrogen and phosphorus was only 49% and 18% for *C. vulgaris* and 58% and 24% for *T. obliquus*, respectively [22].

Table 3. Remediation rates for the microalgae (Cv – *Chlorella vulgaris*, Cp – *Chlorella protothecoides*, To – *Tetradesmus obliquus*) in the four trials (Control and Piggery effluents).

			Total nitrogen (%)	Total phosphorus (%)	COD (%)	BOD_5 (%)	Ash (%)	Phenols (%)	Optical density (%)
1st trial	Control	Cv	99	34	–	–	27	100	51
		Cp	99	43	–	–	26	100	34
		To	100	42	–	–	29	100	4
	(P + A)	Cv	91	77	88	99	61	65	95
		Cp	93	100	82	100	59	59	98
		To	89	100	89	99	61	33	99
	(P + A + O)	Cv	98	91	73	98	58	45	90
		Cp	99	100	76	99	56	40	90
		To	100	86	76	99	57	45	90
2nd trial		Cv	100	100	90	100	61	70	96
		To	100	100	91	100	61	61	99
3rd trial		Cv	99	19*	89	99	67	68	97
		To	99	35*	91	99	65	59	99
4th trial		Cv	83	3	40	98	60	63	91
		To	82	27	39	96	59	53	94

3.2 Biomass Composition

The characterization of the algal biomass was made in terms of protein, sugar, lipid, and ash contents (Fig. 4). Protein is the most abundant content, independently the microalgae species or culture conditions. There was a tendency of an ash increase from the 1st to 4th trial. The microalga with the highest protein content was *T. obliquus* (51.3%) in the (P + A), the one that had the highest sugar content was *C. protothecoides* (51.8%) in (P + A + O) and the one with the highest lipid content was again *T. obliquus* (16.0%) in (P + A + O).

3.3 Application of Microalgal Biomass as Biostimulant

Germination tests were used to directly test the impact of microalgal biomass on seed development. The potential biostimulant activity was assessed determining the germination index (GI) of the control with distilled water (100%), and with microalgae biomass of *C. vulgaris* and *T. obliquus* obtained in the 4th trial. If the GI was higher than 100% it was considered that there was a biostimulating activity (Fig. 5).

Both microalgae had a positive effect on seed germination, between 16% and 86%. The biostimulant effect was more evident in watercress seeds, especially for Cv - 0.2 g/L with an increment of 86%. Cv - 0.5 g/L had similar effect on watercress and wheat with

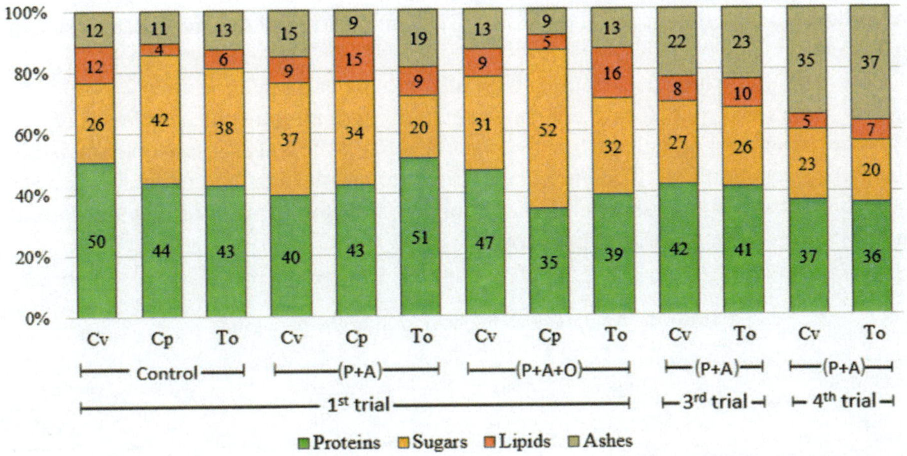

Fig. 4. Characterisation of the biomass (% DW) for the four trials (mean, n = 3). (Control – synthetic medium; (P + A) – piggery effluent + ash; (P + A + O) - piggery effluent + ash + olive oil mill wastewater; Cv – *C. vulgaris*; Cp – *C. protothecoides*; To – *T. obliquus*)

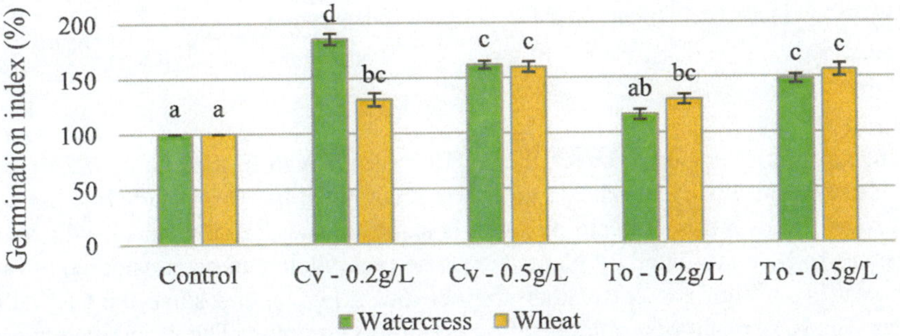

Fig. 5. Germination index of wheat and watercress seeds, using the two microalgae cultures (Cv-*Chlorella vulgaris* and To- *Tetradesmus obliquus*), at two concentrations (0.2 and 0.5 g/L).

an increment around 62%. Other authors have also obtained promising results of a 40% increase in the germination of watercress seeds, using algae biomass of *T. obliquus*, obtained in the treatment of brewery effluents [23]. The results obtained in the present study showed that the use of algal biomass grown in piggery effluent has the capability to be used as a biostimulant in seed germination, to a more sustainable agriculture practice. Both microalgae biomass had important macro and micronutrients for plant nutrition, namely calcium, potassium, magnesium, and sodium which could be, besides phytohormones and amino acids, the responsible for the good results.

3.4 Precipitate Composition and Germination Tests

The precipitate obtained after the pre-treatment of the piggery effluent with biomass ash is mainly composed of ash, but it also contains some precipitated solids that were removed from the effluent by precipitation. The composition of the precipitate is shown in Table 4.

The precipitate had an extremely high alkaline pH and low moisture (27.5%). The organic matter was quite low (5.2%) however, most agricultural soils have levels of 1 to 2%. The presence of organic matter generates benefits to the soil as it conserves moisture and soil aggregation [24]. The presence of calcium, magnesium, and potassium cations are important as they are essential elements for plant development and are usually present in low concentrations in Portuguese soils.

Table 4. Analytical characterization of the precipitate.

Parameter	Units	Value	Parameter	Units	Value
Moisture	%	27.5 ± 0.8	C/N		69.8
Ash content (Ad%)	%	81.2 ± 0.3	C/P		44.5
Volatile Matter	%	17.0 ± 0.9	Total nitrogen	$g\ Kg^{-1}$	1.5
Organic matter	%	5.24 ± 07	Total phosphorus	$g\ Kg^{-1}$	5.2
pH		12.3	Calcium	$g\ Kg^{-1}$	233.2
Electrical conductivity	$dS\ m^{-1}$	6.2	Magnesium	$g\ Kg^{-1}$	20.5
Bulk density	m/v	1.29	Potassium	$g\ Kg^{-1}$	2.3

The C/N ratio is used to check the stability of nitrogen in the compost. Composts with low C/N ratios (<15) indicate that their decomposition can provide high amounts of nitrogen, high ratios indicate nitrogen is stable in the compost and it will be less accessible to be assimilated by plants [25]. The precipitate can be considered a source of P, as it has a high content of this nutrient (5.2 g kg^{-1}), and since the C/P ratio is not very high, phosphorus mineralization is not too complex. The precipitate had low concentrations of micronutrients and heavy metals therefore its incorporation in the soil would not exceed the legal limits allowed for sludge applications [26].

The germination index of watercress and wheat seeds for precipitate extracts, for biomass ash extracts and for control is shown in Fig. 6. The results obtained demonstrate that the precipitate presented some toxicity for the watercress seeds, since the germination index showed a decline close to 40% compared to the control. However, in the case of wheat, there was an increase in the germination rate of 16.2% for incorporations of 5% of precipitate. Basically, both precipitate or ash, have a negative influence on watercress and wheat seed germination. Nevertheless, its incorporation on the soil could have a different behaviour. The precipitate had a higher concentration of minerals than the ash, namely aluminium, boron, barium, copper, iron, manganese, sodium, titanium, and zinc.

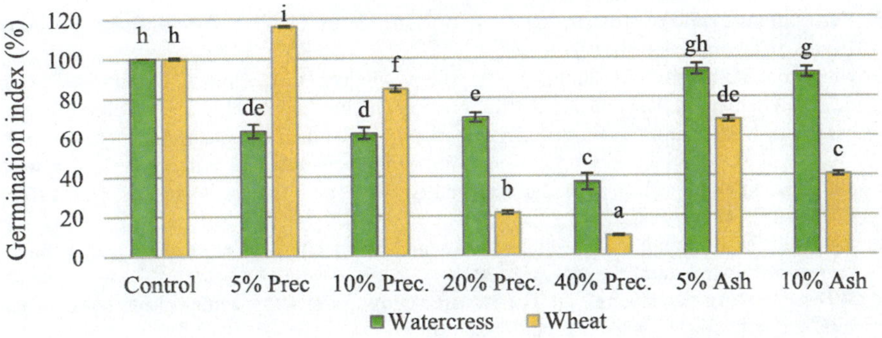

Fig. 6. Germination index in watercress and wheat seeds (mean ± SD, n = 3) for control and different aqueous extract of precipitate and biomass ash (Prec. - Precipitate aqueous extract; Ash - Ash aqueous extract)

4 Conclusions

Piggery effluent pre-treated with forestry ash is suitable to the growth of the tested microalgae (*Chlorella vulgaris*, *Chlorella protothecoides* and *Tetradesmus obliquus*) and the incorporation of 2% olive-oil wastewaters did not have a negative effect on microalgae growth. These microalgae achieved excellent rates of bioremediation for total nitrogen, total phosphorus, and COD. The produced microalgae showed high protein contents what suggests its possible use as a supplement for pig feed, in a circular economy strategy. However, further testing would be necessary to validate the incorporation of microalgae in the pig diet in order to select appropriate doses and ensure animal safety conditions. Also, the biomass carbohydrate components could be used to produce hydrogen (by dark fermentation) or ethanol (by alcoholic fermentation). The precipitate obtained after the pre-treatment may serve as fertilizer, after appropriate formulation. The treated effluent can be used as a water resource to irrigate crops or washing operations of animal production facilities. The microalgae biomass could be used as a biostimulant for seed germination with good results in the case of wheat and watercress, thus promoting the production of animal feed.

Acknowledgments. This work has been supported by FCT – Fundação para a Ciência e Tecnologia within the PhD Grant SFRH/BD/131178/2017 and R&D Units Project Scope: UIDP/04077/2020. The authors would like to thank Dr. Luísa Gouveia for the suggestions, insights and helpful discussions.

References

1. Gouveia, L., et al.: Microalgae biomass production using wastewater: treatment and costs. Scale-up considerations. Algal Res. **16**, 167–176 (2016)
2. Eurostat: Animal output - basic and producer prices. European Statistics (2018). https://ec.europa.eu/eurostat/tgm/. Accessed 15 Jan 2020

3. Eurostat: Agriculture , forestry and fishery statistics in the EU, 2019 editi. Luxembourg: European Commission (2019)
4. MacLeod, M., Gerber, P., Mottet, A., Tempio, G., Falcucci, A., Opio, C., Vellinga, T., Henderson, Greenhouse gas emissions from pig and chicken supply chains - a global life cycle assessment. Food and Agriculture Organization of the United Nations (FAO), Rome (2013)
5. Ho, L., et al.: Spatial and temporal variations of greenhouse gas emissions from a waste stabilization pond: effects of sludge distribution and accumulation. Water Res. **193**, 116858 (2021)
6. Siddique, N.I., Wahid, Z.A.: Achievements and perspectives of anaerobic co-digestion: a review. J. Clean. Prod. **194**, 359–371 (2018)
7. Wang, R., Peng, B., Huang, K.: The research progress of CO_2 sequestration by algal biofertilizer in China. J. CO2 Util. **11**, 67–70 (2015)
8. Viegas, C., Gouveia, L., Gonçalves, M.: Aquaculture wastewater treatment through microalgal. Biomass potential applications on animal feed, agriculture, and energy. J. Environ. Manage. **286**, 112187 (2021)
9. Markou, G., Chatzipavlidis, I., Georgakakis, D.: Cultivation of Arthrospira (Spirulina) platensis in olive-oil mill wastewater treated with sodium hypochlorite. Bioresour. Technol. **112**, 234–241 (2012)
10. Eurostat: Market situation in the olive oil and table olives sectors (2020)
11. UTEX: Algal Culture Medium Recipes. UTEX Culture Collection of Algae (2020). https://utex.org/pages/algal-culture-media-recipes. Accessed 19 Aug 2020
12. Ansari, F.A., Singh, P., Guldhe, A., Bux, F.: Microalgal cultivation using aquaculture wastewater: integrated biomass generation and nutrient remediation. Algal Res. **21**, 169–177 (2017)
13. Rice, E.W., Baird, R.B., Eaton, A.D. (eds.): Standard Methods for the Examination of Water and Wastewater. In: 23th ed. American Public Health Association, American Water Works Association, Water Environment Federation, Washington, DC (2017)
14. Singleton, V., Orthofer, R., Lamuela-Raventós, R.M.: Analysis of total phenols and other oxidation substrates and antioxidants by means of Folin-Ciocalteu Reagent. Methods Enzymol. **299**, 152–178 (1999)
15. Portuguese Ministry of the Environment: Decree-Law No. 236/98, 1998, of 1 August Establishing Water Quality Standards. Diário República - Série I**176**, 3676–3722 (1998)
16. AOAC (Association of Official Analytical Chemists), Official Methods of Analysis of AOAC International, 18th editi. Association of Official Analytical Chemists (2006)
17. Jones, D.B.: Factors for converting percentages of nitrogen in food and feed into percentages of proteins. United States Dep. Agric. **Circular**(183), 1–22 (1931)
18. Miranda, J.R., Passarinho, P.C., Gouveia, L.: Pre-treatment optimization of Scenedesmus obliquus microalga for bioethanol production. Bioresour. Technol. **104**, 342–348 (2012)
19. Dubois, M., Gilles, K.A., Hamilton, J.K., Rebers, P.A., Smith, F.: Colorimetric method for determination of sugars and related substances. Anal. Chem. **28**(3), 350–356 (1956)
20. Zucconi, F., Forte, M., Monaco, A., De Bertoldi, M.: Biological evaluation of compost maturity. Biocycle **22**(4), 27–29 (1981)
21. Monteiro, M.C., Carneiro, J., Águas, P., Batista, M., Cipriano, V., Sebastião, P.: Validação de um Composto como Fertilizante para Efeitos de Legalização de Uso Agrícola. IPCB. ESA (2011)
22. Abou-Shanab, R.A.I., Ji, M., Kim, H., Paeng, K., Jeon, B.: Microalgal species growing on piggery wastewater as a valuable candidate for nutrient removal and biodiesel production. J. Environ. Manage. **115**, 257–264 (2013)
23. Navarro-López, E., Ruíz-Nieto, A., Ferreira, A., Gabriel Acién, F., Gouveia, L.: Biostimulant potential of Scenedesmus obliquus grown in brewery wastewater. Molecules **25**(3), 1–16 (2020)

24. USDA: Inherent Factors Affecting Soil Organic Matter (2010)
25. Deng, X., et al.: Spatial and temporal trends of soil total nitrogen and C/N ratio for croplands of East China. Geoderma **361**, 114035 (2020)
26. Portuguese Ministry of the Environment: Decree-Law No. 276/2009, 2 October, Soil protection in the agricultural use of sewage sludge. Diário da República, pp. 7154–7165 (2009)

Current and Future Trends of 3D Food Printing

Henrique Almeida[1,2(✉)], Joel Vasco[1], and Vânia Ribeiro[3]

[1] School of Technology and Management, Polytechnic of Leiria, Leiria, Portugal
henrique.almeida@ipleiria.pt
[2] Computer Science and Communication Research Centre, Polytechnic of Leiria, Leiria, Portugal
[3] Center for Innovative Care and Health Technology, ciTechCare, School of Health Sciences, Polytechnic Institute of Leiria, Leiria, Portugal

Abstract. 3D food printing is the process of manufacturing food products using a variety of additive manufacturing technologies. Similar to the industrial systems, the food is also produced in a layer-by-layer fashion. Most commonly, food grade syringes hold the printing material, which is then deposited through a food grade nozzle.

The selection of additives considering the desired rheological and sensory properties of food products is decisive for the success of 3D food printing, as well aspects related with shelf life and increased functionality of the food. The food can be customized in shape, colour, texture, flavour or nutrition, which makes it very useful for designer gourmet meals. This paper intends to provide a current state-of-the-art of 3D food printers and which will be the future trends. Not only the technological issues will be discussed, but also the nutritional, food safety and sustainability of raw materials issues will be taken into account for a better understanding of this particular industry.

Keywords: 3D printing · Food printing · 3D food printing

1 Introduction

3D food printing is the process of manufacturing food products using a variety of additive manufacturing technologies. Similar to the industrial systems, the food is also produced in a layer-by-layer fashion. Most commonly, food grade syringes hold the printing material, which is then deposited through a food grade nozzle. The food can be customized in shape, colour, texture, flavour or nutrition, which makes it very useful for designer gourmet meals.

In comparison to 3D printing, 3D food printing presents other similarities besides the layer-by-layer fabrication and the freedom of design. The system requires a substrate, which in this case is an edible substrate, but with similar material properties to be considered, such as rheological properties. Pre-processing may be required such as ingredient mixing for 3D food printing, similar to functional gradient material 3D printing. Regarding the printing processes, while 3D printing has seven categories of printing

J. R. da Costa Sanches Galvão et al. (Eds.): *Proceedings of the 1st International Conference on Water Energy Food and Sustainability (ICoWEFS 2021)*, pp. 258–267, 2021.
https://doi.org/10.1007/978-3-030-75315-3_30

processes, 3D food printing is mainly limited to one, namely extrusion based. Both 3D food printing and 3D printing have post-processing, but for 3D food printing, the post-processing procedures are cooking, oven drying or freeze drying. In terms of attributes to evaluate the produced parts, while 3D printing considers other attribute that aren't sensory besides appeal and textures, 3D food printing considers only sensory attributes, namely appeal, texture, aroma and taste.

3D food printing offers a wide range of potential benefits. It can be healthy and good for the environment because helps to convert alternative ingredients such as proteins from algae, beet leaves, or insects into tasty products. It also opens the door to food customization allowing individual needs and preferences in the meals.

Figure 1 illustrates some examples of 3D printed food in terms of design and concept.

Fig. 1. Illustrations of 3D printed dishes from the Byflow (top row) and Foodini (bottom row) website galleries.

The advantages of 3D food printing are:

- Personalized designed meals: Besides controlling the diversity and amount of nutrients and calories per meal, 3D food printing allows for precision. This issue is extremely important for hospitals where restricted diets are required, and this technology offers the potential of easy patient customization.
- Solution for problems related to allergies and cross-contamination.
- Freedom to design novel and complex visual structures and textures with the aid of CAD software.
- Unconventional food consumption: When you process nutritious plants and protein-rich insects or small animals in a semi-liquid state, these 3D printed foods can be

presented in a more attractive way and thus incentivize their consumption. In the future, even synthetic food might be 3D printed.
- Easy reproducibility: Sharing recipes could be as simple as transferring a digital file over the internet, namely the ingredients or the food design. All that would be required is the same raw materials, printing settings, and compatible printing equipment.
- Food waste reduction and Reductions in energy and transportation needs.
- Requires no or very less human interaction.

On the other hand, the disadvantages are:

- The processing time of the meals might be higher to the conventional cooking due to the fact that some 3D food printing technologies require an additional preparation time for the meals such as bakery. The edibles used in these machines either require pre-cooking or pre-processing to achieve the consistency required.
- The cost of equipment and consumables (that must be purchased directly from the manufacturers) also presents barriers as well as the time spent on training.
- The expected reproducibility and reliability for these machines depend heavily on the proper preparation of those materials.
- Limited printing accuracy and surface finishes.
- Due to the processing times, 3D food printing lacks scalability.
- Food safety and printer sanitation concerns.
- Consumers' perception.

The applications of 3D food printing isn't just the freedom of design for the sophisticated meal, but as mentioned before, personalized designed meals (children, athletes and military) by controlling the diversity and amount of nutrients and calories per meal, along with solving allergies and cross-contamination issues. The freedom of design may also meet the children's curiosity and influence them to other nutritional needs. 3D food printing may also be used for people facing mastication problems. Space food is another application of 3D food printing allowing to reconstitute food with micronutrient materials due to the long shelf-life and nutritional needs of long-term missions.

This paper intends to provide a current state-of-the-art of 3D food printers and which will be the future trends. Not only the technological issues will be discussed, but also the nutritional, food safety and sustainability of raw materials issues will be considered for a better understanding of this particular growing industry.

2 Food Grade Additives

The use of food grade additives can be decisive in 3D printing of food, mainly in improving the printing capacity, in the rheological and organoleptic properties of the final products.

The incorporation of food grade additives in 3D printing formulations has proven to be determinant in printing efficiency, providing the desired viscosity for the different layers [1, 2]. Some hydrocolloids from algae such as agar and alginate have shown a stabilising effect on mashed potatoes and at the same time a variation in viscosity that increase or decrease, considering the concentrations of the additive used (0,5 and 1%) [2].

Some widely studied foods such as hydrogel, dough and chocolate are discussed from the point of view of printability and potential benefits for the consumer.

2.1 Hydrogels

Gelatine (protein derived from the irreversible breakdown of collagen fibrous structure) is added into food material to improve food viscosity and enhance food printability. Concentrations between 0.5% and 4% in formulations are indicated for a good food impression, however the higher the concentration of gelatine, the firmer the food structure [1]. Recently Kuo et al. (2021) studied formulations with different percentages and ratios of hybrid gelatine and alginate hydrogels to verify the best texture profile in 3D printing and the shape retention capacity of hybrid hydrogels. The authors concluded that the 7% concentration of the hybrid gelatine/algae with the 1/2 gelatine/algae ratio showed the best formulation with respect to hardness and adhesiveness. In this study, the lyophilization of the printed bioscaffold was chosen to extend the useful life and improve the mechanical properties of the manufactured structure, the moisture content and the water activity were thus significantly reduced. Another aspect considered relevant concerns the porous structure obtained after lyophilization of the product which has the potential for encapsulation of bioactive compounds [3]. Another study by Dick et al. 2020 on the feasibility of incorporating hydrocolloids xanthan gum and guar gum into cooked pork for the formulation of 3D foods for dysphagia patients found that the combined addition of 0.36% with varying proportions of the 2 hydrocolloids (0.5:0.5, 0.7:0.3, and 0.3:0.7) has the potential to be incorporated into follow-on foods within the framework of the International Dysphagia Diet Standardization Initiative (IDDSI). This categorisation is since these hydrocolloid samples exhibit a visible enlarged network with heterogeneously sized cavities, which suggests additional water retention and higher viscosity than the control sample after post-processing [4].

2.2 Dough

Dough from wheat flour is widely used in 3D formulations due to its viscoelastic properties, mainly due to the presence of gluten in contact with water [1]. Instability of the form can be optimised through the formulation (amount of butter, eggs and sugar added to the dough may affect its stability) and the inclusion of food additives. It should be noted that too much butter is introduced into the dough increases the extrusion speed during printing, however it does not allow the stability of the structure, the egg yolk influences the height of the structure construction [5] and the sugar (sucrose) when added in a percentage of 6.6 produces an improved intact form [6].

2.3 Chocolate

Chocolate represents a complex material, its properties (ingredients such as cocoa butter, soy lecithin) and material properties (rheology, melting temperature and solidification rate) have a direct impact on the extrusion and solidification of chocolate during the 3D printing process [7]. According to Mantihal et al. the temperature of the nozzle before deposition should be 32 °C to extrude the melting state of the sample, since the flow behaviour curves indicate that the chocolate melting starts between 28 °C and 30 °C. Incorporating magnesium stearate into the forming chocolate helps to lubricate the material and increases the flow efficiency during deposition [8]. Extrusion, speed and nozzle height are determinants for chocolate deposition and 3D formation [7]. Future trends aim to improve chocolate printing, invest in more efficient printers (equip with the cooling system) [1], and improve formulations with potentially functional ingredients that modify desserts (without sugar and hydrocolloids incorporation) and transform them into nutritional supplements.

3 Intellectual Property of 3D Food Printing

The main purpose of intellectual property law is to encourage the development of of products. The intellectual property law gives people and businesses property rights to the information and intellectual products that they create. This policy provides economic incentives for creation allowing the inventors to profit from the information and intellectual products that they created. These economic incentives are expected to stimulate innovation and contribute to the technological progress. In other words, it can also be considered as an innovation matrix. The higher number of patents of a specific domain, represents the societies' interest. And if one considers the number of patents as function of years, this will provide an indication of the growing or declining interest of inventors for that specific domain.

From a patent search in Espacenet with the keywords "Food 3D Printing", 107 results were obtained. The first patents appeared in 2011 and since then the number of patents per year has increased with a particular boom of patents in 2016.

After a careful analysis of the existing patents, excluding those that aren't available in the English language and those that don't fall under the correct definition of 3D printing (building a part by adding material in a layer-by-layer fashion), the following table presents the most recent and relevant patents regarding 3D food printing (Table 1).

4 Comercial 3D Food Printing Systems

Conventional food processing methods based on chemical and physical changes may not be directly adaptable for 3D printing. Food properties may vary significantly as well as its processing parameters. For a successful implementation, food 3D printers must embed knowledge upon preliminary processes to enhance flavour, to provide dietary food and shape [9].

Commercially, several are the 3D food printers available on the market and are already classified into two categories, namely professional printers and consumer printers. Below, some of the systems will be discussed.

In the category of professional 3D food printers, you may find three printers, namely [10] (Fig. 2):

Table 1. Listing of the most relevant and recent patents on 3D Food Printing.

Publication number	Title	Earliest priority	Type of food	Post-processing
CN111700298A	Extruding device for 3D food printing	2020-06-24	Suitable for several types of dough	May require baking
CN211458754U	Ice cream 3D printing equipment and food 3D printing equipment	2020-01-17	Suitable for ice cream	May require freezing
CN111248474A	Food microwave 3D printing equipment, system and printing method	2020-01-17	Suitable for heated food slurry such as surimi, minced meat and dough using microwave energy	Not specified/required
CN111067121A	Heating integrated 3D food printing machine and printing method thereof	2019-12-09	Suitable for deposition of heated chocolate and dough	Not specified/required
CN211091699U	Internal heating type 3D printing food heating spray head	2019-09-30	Suitable for continuous deposition of heated chocolate	Not specified/required
CN211091838U	Internal heating type food 3D printing and extruding device	2019-09-30	Suitable for continuous deposition of heated chocolate	Not specified/required
CN110326807A	Molding and reducing spray nozzle for 3D printing of food	2019-06-27	Suitable for heated food deposition	Not specified/required
CN210130341U	3D printing spray head for plastic food	2019-02-27	Suitable for several types of dough	May require baking
EP3698647A1	Food composition for 3D printing, process for producing a food product by 3D printing and food product obtained thereby	2019-02-21	Suitable for several types of dough	May requires baking
CN209694918U	Novel food 3D printing garland machine	2019-01-15	Suitable for several types of dough in a single deposition operation	May requires baking
KR20200071861A	3D Intelligent 3D food printing apparatus	2018-12-05	Suitable for space foods, combat foods, and customized health foods according to patient characteristics	Not specified/required
CN109363221A	Microwave-assisted 3D food printing device and method	2018-11-21	Suitable for food deposition upon a pre-established recipe	Microwave heating included on the patent
CN109198650A	Food 3D printing material containing rich dietary fiber	2018-11-09	Preparation of dietary fiber-rich food for 3D printing	May requires baking
CN209219224U	Food 3D printing device	2018-10-24	Suitable for the deposition of several types of dough or chocolate	May require baking

(*continued*)

Table 1. (*continued*)

Publication number	Title	Earliest priority	Type of food	Post-processing
CN209017876U	Food printer based on 3D printing	2018-09-12	Suitable for several types of dough	May require baking
CN109170875A	Enzymatic 3D printing food material and preparation method thereof	2018-08-27	The enzymatic 3D printing food material is prepared from the following raw materials in percentage by mass: 10–20% of enzymes, 30–45% of pregelatinized starch, 2–4% of xylitol, 2–4% of maltodextrin, 0.5–3% of L-arabinose, 0.5-3% of dietary fibers, 0.3–0.8% of soyabean lecithin, 0.05–0.2% of essence and the balance of water	Not specified/required
CN110856553A	Multifunctional food 3D printing system	2018-08-23	The 3D printer is developed for binder jetting of food in powder form, using a suitable binder to aggregate the food	May require baking
CN109049675A	3D food model printing device and printing machine with same	2018-08-08	Suitable for heated food deposition	Not specified/required
KR101971093B1 KR20190001948A	3D Food 3D Printing System and Food printing method using it	2018-07-11	Suitable for food deposition in liquid, solid and powder form, based on a cloud to enable the user to produce its own food without the need of the 3D modelling capacity	Not specified/required
CN109090616A	Accurate 3D printing method for high-protein semifluid instant food	2018-06-14	Suitable for 3D print for high-protein semifluid instant food	Not specified/required
CN109042788A CN109042788B	3D printing system for food	2018-06-13	Suitable for heated food slurry such as minced meat, dough or others using an integrated oven to bake	Not specified/required
CN109007945A CN109007945B	Cooperative type precision nutritious food 3D printing system and method	2018-06-12	Integrated manufacturing cell for several types of AM processes suitable for food and several material forms	Not specified/required
CN208909117U	3D food and drug printing equipment with drying structure	2018-03-28	Suitable for personalized cereal food directly without using molds, such as pasta and artificial rice including drug delivery and alarm system to detect faults	Not specified/required

- byFlow Focus: This printer is a material-extrusion 3D printer used by professional chefs around the world. It's Wi-Fi controllable and uses reusable cartridges (syringes) that can be filled with edible pastes.
- Foodini: This printer is one of the most common 3D food printers. It is capable of processing almost any ingredient, from jellies to tasty burgers, using up to five different cartridges simultaneously.
- MMuse: This chocolate 3D desktop printer is robust and unique since it is capable of processing chocolate chips up to 4 mm in diameter and melts them for extrusion.

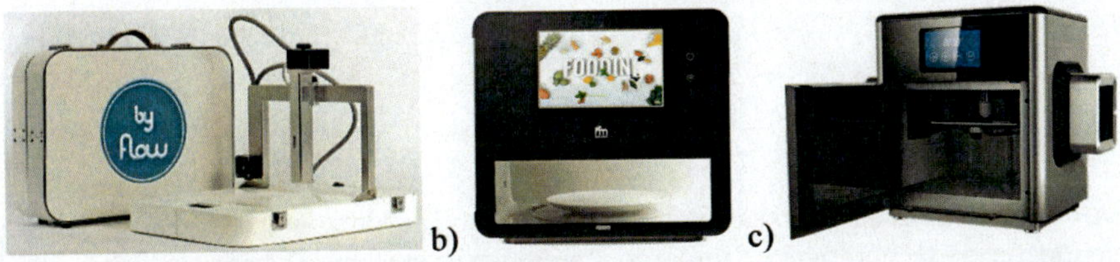

a) b) c)

Fig. 2. Illustrations of professional 3D food printer: a) byFlow Focus, b) Foodini and c) MMuse.

Regarding the consumer 3D food printers for household applications, you may find the following systems [10] (Fig. 2):

- Mycusini: This printer is a chocolate 3D printer and it displays great design and is compact enough for a kitchen countertop, while its internal battery allows off-grid printing for up to two hours. The stainless-steel cartridges are easy to fill and clean, but the refills must be purchased directly from the manufacturer.
- PancakeBot: In spite of this printer being characterized as a 3D printer, it is actually a 2D printer because it is only capable of printing pancakes. This printer includes a heated surface so the batter can be cooked immediately after printing the desired design of pancake.
- Procusini: This printer is capable of handling small catering events. The cartridges can be heated up to 60 °C to process thicker materials like marzipan or fondant for cake decoration. Similar to the Mycusini printer, the refills must be purchased directly from the manufacturer (Fig. 3).

Other 3D printing manufacturers have realized the potential market of 3D food printers and have started working on adjustments and modifications for 3D food printing. The first to take these steps was the F@bHome printers, but now also the WASP and the ZMORPH printers are moving into this market.

Another aspect that improves and increases approachability to 3D food printing is that the manufacturers have created recipe repositories where verified designs can be easily downloaded and reproduced. Once one is familiarized with the printer and how it works, the printers are also supplied with proprietary software tools that supports "food" designing. These systems are practical and user-friendly programs that limit the tools required for simple 3D food printing projects. For more complex designs, similar to additive manufacturing, advanced knowledge of CAD modelling tools is required.

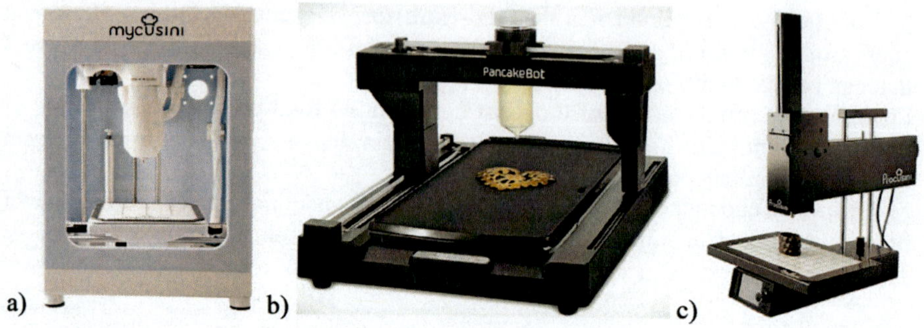

Fig. 3. Illustrations of professional 3D food printer: a) Mycusini, b) PancakeBot and c) Procusini.

5 Conclusions

3D food printing is the process of manufacturing food products using a variety of additive manufacturing technologies. Similar to the industrial systems, the food is also produced in a layer-by-layer fashion. Most commonly, food grade syringes hold the printing material, which is then deposited through a food grade nozzle. The food can be customized in shape, colour, texture, flavour or nutrition, which makes it very useful for designer gourmet meals.

The selection of additives considering the desired rheological and sensory properties of food products is decisive for the success of 3D food printing, as well aspects related with shelf life and increased functionality of the food.

The organoleptic characteristics of 3D foods and the validation of their preference through hedonic scales are yet to be explored. The magnificent textures that these prototypes promise to present still lack structurally optimised formulations, in which the food additives to be incorporated are selected, at the same time as the aim is to produce food of high nutritional quality.

3D food printing is a real and existing technology with everyday applications. One may still not be at the point that one places a few pills in the microwave and after a few seconds, a roasted chicken with vegetables is then removed from the microwave, as seen in one of the scenes from the movie "5th Element" or the Replicator, the Star Trek food manufacturer. Putting aside science fiction, this technology provides novel designs and gastronomic solutions for gourmet meals with novel potentials to cooking chiefs.

References

1. Mantihal, S., Kobun, R., Lee, B.-B.: 3D food printing of as the new way of preparing food: a review. Int. J. Gastron. Food Sci. **22**, 100260 (2020)
2. Dankar, I., et al.: 3D printing technology: the new era for food customization and elaboration. Trends Food Sci. Technol. **75**, 231–242 (2018)
3. Kuo, C.-C., et al.: An integrated manufacturing strategy to fabricate delivery system using gelatin/alginate hybrid hydrogels: 3D printing and freeze-drying. Food Hydrocoll. **111**, 106262 (2021)

4. Dick, A., et al.: Feasibility study of hydrocolloid incorporated 3D printed pork as dysphagia food. Food Hydrocoll. **107**, 105940 (2020)
5. Lipton, J., et al.: Multi-material food printing with complex internal structure suitable for conventional post-processing. In: Solid Freeform Fabrication Symposium. Citeseer (2010)
6. Yang, F., et al.: Physical properties of 3D printed baking dough as affected by different compositions. Innov. Food Sci. Emerg. Technol. **49**, 202–210 (2018)
7. Hao, L., et al.: Chapter 8 - material, process and business development for 3D chocolate printing. In: Godoi, F.C., et al. (eds.) Fundamentals of 3D Food Printing and Applications, pp. 207–255. Academic Press (2019)
8. Mantihal, S., et al.: Optimization of chocolate 3D printing by correlating thermal and flow properties with 3D structure modeling. Innov. Food Sci. Emerg. Technol. **44**, 21–29 (2017)
9. Sun, J., Peng, Z., Yan, L., Fuh, J., Hong, G.: 3D food printing an innovative way of mass customization in food fabrication. Int. J. Bioprint. **1**(1), 27–38 (2015)
10. All3DP: 3D Printed Food: All You Need to Know in 2021. https://all3dp.com/2/3d-printed-food-3d-printing-food/. Accessed March 2021

Energy

Comparative Performance and Efficiency of EURO VI Heavy-Duty Engines Fueled by Biodiesel, HVO and Diesel

Luis Serrano[1,2](✉), Nuno Pereira[1], and Paulo Matos de Carvalho[1,2]

[1] School of Technology and Management, Polytechnic Institute of Leiria, Leiria, Portugal
luis.serrano@ipleiria.pt
[2] ADAI –LAETA, Coimbra, Portugal

Abstract. In the last years, road mobility has been changing driven by environmental, energetic, and economy concerns, leading to a growing research and development of new anti-emissions technologies, better fuel management systems and the emergence of new fuels. These concerns encourage to consider the use of biofuels in engines. This work aims to understand how the engine with actual evolved technologies are affected by the use of biofuels. This paper will focus on evaluating several biofuels blends on a 2019 performance diesel EURO VI heavy-duty truck, as to review some injection parameters and fuels properties effects. It was also reviewed the respective fuel properties, standards, and emissions expectations. With the objective of preparing a fuel consumption measurement and a performance analysis, an experimental procedure was developed based on the homologation cycles and processes of heavy-duty truck on a chassis bench roll, testing the following samples: B7, B15, B100 and a HVO15. The powertrain performance results reveal that the fuel which presented higher power values was the HVO15, followed by B0, B7, B15 and B100. On fuel consumption, the smaller results were obtained for B7 and B15, without reasonable differences, followed by B100 and then HVO15, at which, in some measured engine regimes, reveal extremely high fuel consumption. To evaluate the fuel economy performance of neat biodiesel compared to fossil diesel on a real road use, three buses on their daily utilization were analyzed. In this experiment, the biodiesel buses had an increase on fuel consumption about 4,5%.

Keywords: Biodiesel · HVO · Biofuels · Engines performance · Fuel consumption · CO_2 emissions

1 Introduction

1.1 Contextualization

One of the sectors with the highest energy consumption in the world is the transportation. Considering the data from the International Energy Agency (IEA), in 2018 this sector alone accounted for about 29% of global energy consumption [1]. Even in Europe,

J. R. da Costa Sanches Galvão et al. (Eds.): *Proceedings of the 1st International Conference on Water Energy Food and Sustainability (ICoWEFS 2021)*, pp. 271–279, 2021.
https://doi.org/10.1007/978-3-030-75315-3_31

where there has been a very demanding attitude towards the implementation of measures to promote an effective reduction in fossil fuel energy consumption in mobility and transport, this sector still represented, in the same year, about a quarter of greenhouse gas emissions (GHG). In fact, there is still a significant growth trend for this type of emissions in transport [2].

The information revealed in that same IEA report, transport sector was the one that showed the highest growth rate in energy consumption in the recent year 2018 [1]. Even with the advantages that come from the current technological developments in mobility and transport, the existence of an implemented and safe infrastructure network, the high energy density presented by petroleum fuels and also the economic and financial system associated with oil, what actually happens is that the energy needs of this sector have been suffering successive growth.

Given the evidence and the impacts of this relevant trend, the urgent need to find measures to minimize the dependence of transport on oil consumption is as complex as it is absolutely essential. According to the available data [3]. the transport sector is expected to account for about 97% of the increase in global primary oil consumption between 2007 and 2030. If nothing is changed in the current context of this sector, the implications in terms of global warming could become very serious and irreversible, with a very significant increase in the temperature values of the earth's atmosphere by the year 2050. The same report also reveals that these temperature values would make life on our planet practically unsustainable.

1.2 The Use of Biofuels in Heavy Duty Engines

It is a fact that, in the last years, road mobility has been changing driven by environmental, energetic and economy concerns, leading to a growing research and development of new anti-emissions technologies, better fuel management systems and the emergence of new fuels [4] . These concerns lead to the need of using biofuels in engines. Biofuels, derived from organic sources, can be sustainable, produced locally, more environment friendly in their life cycle and present similar properties as the fossil fuels used [5–7]. Thus, the use of biodiesel in heavy duty engines appears as a possible and viable solution to reduce the GHG emissions from these vehicles and to have a lower dependence from fossil fuels. To accomplish this task, the European Commission defined the need to increase the use of biofuels in the transport sector obtained from residues and advanced technologies, with the development of advanced biofuels, like HVO (Hydrotreated Vegetable Oils). For this purpose, the use of B15, B100 and HVO15 was chosen. Currently, the available fuel in the stations has 7% of biodiesel (B7). It is expected that this percentage will grow, and that some transport companies will start using higher percentages of biodiesel. There was also the need to include the use of HVO15 (15% of HVO in diesel) to see the effects of this new and advanced fuel on performance and consumption.

While Biodiesel blends have a superior viscosity and density, oxygen content and an inferior calorific power. [8–17] HVO have similar viscosity, lower density, no oxygen, and present a lower calorific power compared with diesel on a volumetric basis [13, 18]. About emissions, they typically become reduced for both biofuels, except for the NOx, especially on biodiesel [6, 12, 19–22]. On HVO, the increase of NOx tends to be smaller and it is not always verified [13, 18, 23–25].

2 Experimental Apparatus

This work aims to understand how a Diesel engine with state of the art tail pipe emission control technologies is affected by the use of biofuels.

To accomplish this goal, a 2020 DAF heavy truck, model CF 430, equipped with an MX-13 315 engine was used. This vehicle is Euro 6 compliant. The engine has a brake power of 428 cv (315 kW). The brake torque is 2300 N.m ranging from 900 rpm to 1125 rpm. The experimental apparatus setup is presented in Fig. 1

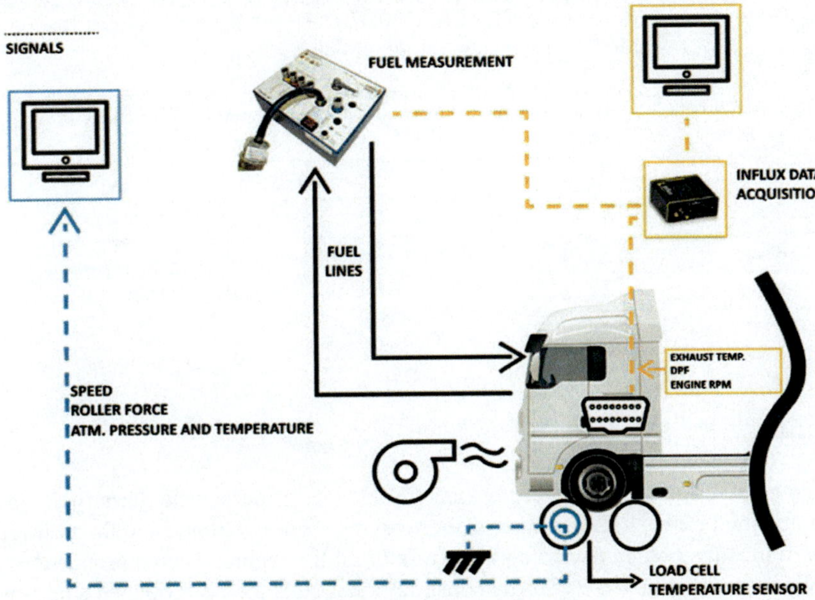

Fig. 1. Experimental apparatus

The chassis dynamometer tests were performed on a DimTech DF2T model. The vehicle's fuel tank was not used to store the test fuels. Instead, an external deposit equipped with rapid and dry fuel hose connections was used. Each time there was a need to perform a fuel change, the management of both suction fuel line from the external tank and the return one to the truck's own fuel tank, allowed to insure the complete removal of all the previous fuel from the supply line and preventing any test fuel contamination.

A real time Kistler-Datron DFL3x5 fuel measurement system connected to an Influx data acquisition was used.

OBD data was also connected to the Influx data card. Monitored parameters included: Engine liquid coolant temperature; Fuel temperature; Engine lubricant temperature; Intercooler air temperature; Engine thermostat operating mode.

With the objective of having a comparative analysis about the fuel consumption measurement and a performance analysis considering the different fuels, an experimental procedure was developed considering the homologation cycles and processes of heavy-duty trucks, which are the WHSC (World Harmonized Stationary Cycle) and the WHTC (World Harmonized Transient Cycle) [26]. For this study, it was chosen to adapt the WHSC procedure to a heavy-duty 2019 diesel truck on a chassis dynamometer equipment, testing the following samples: 3 different blends of biodiesel in diesel B7, B15, B100 (7%, 15% and 100%) and HVO15 (advanced fuel with 15% diluted in fossil fuel).

The most relevant properties of each of the fuels are presented in the Table 1. All of these properties were measured, with the exception of HVO15 ones. In this case, they were defined based in the bibliography [27].

Table 1. Most relevant properties of the fuels

Fuels	Density at 15°C kg/m^3	Heating value MJ/kg	Viscosity at 40°C mm^2/s	Cetane Number	Biofuel content %
B7	835,6	42,62	3	52,5	7,9
B15	836,2	41,87	3	53,7	15,7
B100	882,3	37,21	4,6	56,2	97
HVO15	825,4	43,18	3	53,9	15,0

3 Results

The results obtained in the tests were presented below considering the mass fuel consumption obtained for each fuel in each of the working regimes defined and the average of the several results obtained considering a mix of all the regimes with a proportional and equivalent predominance. After the consumption results the performance data was also presented for each one of the fuels considered in the present study.

3.1 The Fuel Consumption

As it can be observed in the results presented in Fig. 2, for the 10 engine operation points defined, the B7 is the fuel that presents a lower mass fuel consumption, but it is not so notorious for all the points. In fact, that conclusion is correct for the lower and higher engine rotation points, but for engine rotation from 1000 rpm to 1250 rpm, that is not so obvious. Considering the use of B100 and HVO15, it becomes clear that there is an increase in fuel consumption, however there are some particular operation points where that increase becomes extremely higher (almost 20%), but for other regimes, the differences is not so notorious either for B100 as for HVO15.

When the average of all the tests were made, considering that each of the operation points contribute in an equality manner for an overall mass fuel consumption, it becomes obvious that the lower consumption correspond to B7, with a small increase of 2% for B15 and higher differences for B100 and HVO15, respectively 9% and 8%. This is represented in Fig. 3.

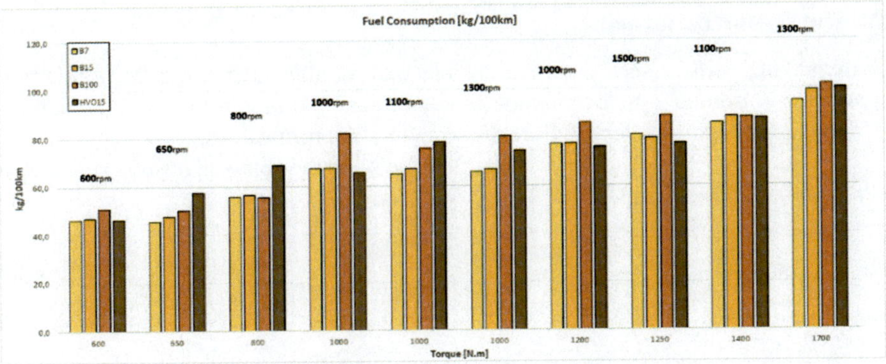

Fig. 2. Mass fuel consumption for the fuels in 10 different engine operation points

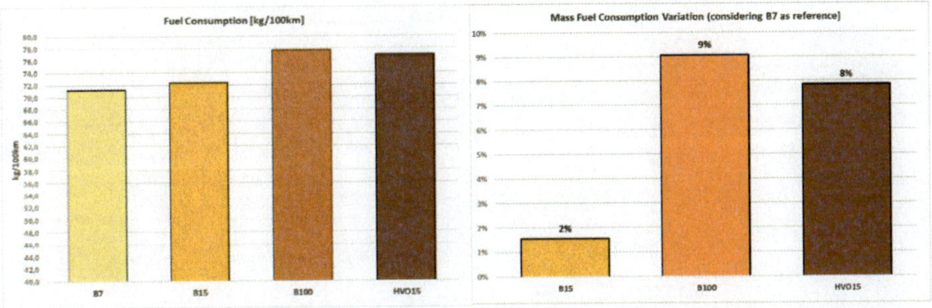

Fig. 3. Average of mass fuel consumption for the fuels, considering B7 as the reference

From the analysis of the Figs. 2 and 3 it can be expressed that the use of B15 does not penalize the fuel consumption, because the variations are within the accepted uncertainty of the tests (2%). Another factor that contributes for this is that the expected operation of the vehicle can mislead this variation. In fact if the use of the vehicle in road can be characterized by middle engine operation regimes, this difference can be ringed or even represented by a small decrease. Face to that, it is only possible to state that the use of B15 does not introduce significative changes in the fuel consumption of the vehicle. By the contrary, it becomes clearer that the use of B100 and HVO15 penalize the fuel consumption. However, it is a fact that HVO15 also is highly penalized by 3–4 operation points where the fuel consumption presents significative values, but for most of the operation points considered there are no significative differences between HVO15 and B7 or B15. This requires a deeper study in chassis dynamometer and road tests to acquire the most common operation points in a real vehicle use.

It is not obvious the possible gains in efficiency obtained from the HVO higher cetane number and higher energy availability associated with the properties of this advanced fuel.

3.2 The Engine Performance

For the engine performance data, the vehicle was running in pushing the accelerator pedal to the bottom and the dynamometer was imposing torque for each of the defined 200rmp engine rotation interval, from 800 rpm to 2000 rpm values.

The tests were repeated once for each of the fuels and the obtained results are revealed in the graphic of Fig. 4 and in the Table 2.

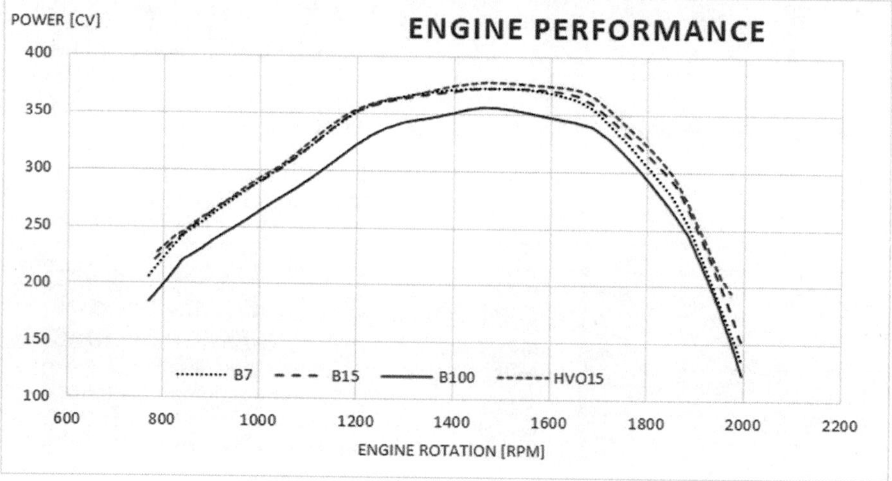

Fig. 4. Engine performance results for the 4 fuels tested.

Contrary to the fuel consumption results, the higher values of the engine power were for the HVO15, with an increase of 1,5% compared to the values for B7. The differences are very small and inside the uncertainty of the experimental data, but it becomes clear that the use of B15 or HVO15 does not introduce visible differences or any king of penalty in the performance of the engine. It is also clear that the use of B100 correspond to a significative decrease in the available power of the engine. Even so, the difference is not as higher as it was expected once the heating value of B100 introduces a higher difference considering B7 than the 4,5% penalization in observed Power.

Table 2. Results for the maximum power obtained for each of the fuels.

Engine Power max. [cv]			
B7	B15	B100	HVO15
372,8	372,1	355,9	378,3
Variation to B7			
0,0%	-0,2%	-4,5%	1,5%

3.3 Real Data of Heavy-Duty Vehicles

To compare the effects of using B100 in a heavy-duty vehicle under real road conditions, 6 similar buses traveling in a Portuguese city were analyzed. This experimental test, conducted with PRIO (fuel supplier) and Carris (transport owner), considers the same route traveled by 3 buses fueled with B100 and another 3 fueled with commercial diesel (B7). This experience was accomplished for 8 months and correspond to almost 200000 km for all the vehicles. During this period, the drivers did not know that this experience was being done, so they were not aware if they were using a different fuel.

The first conclusion about this experience is that there were no problems in the operationalization of the fleet and no difference in maintenance procedures and costs. During the period of the experience, there were no noticeable vehicles malfunction.

The results reveal a 4,5% increase in the fuel consumption of B100 fueled-vehicles comparing the results to the vehicles that were using B7. Even considering this increase in the fuel consumption, it reveals an increase in the efficiency, due to the lower energy offered by B100, corresponding to 10% lower energy content per liter. This means that in real conditions, the expected CO_2 emissions penalization associated with the tank to wheel cycle is not as critical as it expected and in a full life cycle analysis, the biodiesel introduces a significative reduction in these emissions. By one side owing to the renewable origin of the biodiesel and by the other side offering a double environmental gain, once the biofuel was produced from used cooking oil, which obtains useful energy from a harmful residue.

4 Conclusions

It was observed that biologic fuels (B7, B15, B100 and HVO) can be used in a heavy Diesel truck, Euro 6 emission compliant, without the need of further changes to the fuel lines or injection system. In this condition, volumetric fuel consumption results showed that these fuels present similar values or slightly higher ones, depending on the fuel, and when compared with B7 (commercially available Diesel in Portugal). One of the biggest advantages of the use of biodiesel is the decrease in the amount of CO_2 emissions and the replacement of fossil fuels by renewable energy sources.

There is a potential to further increase the biofuels performance, if injection timing and pressure are tuned for each one of them. Some of these fuels have a lower Air-Fuel ratio, different cetane numbers and different droplet paths than B7. The engine injection parameters are tuned using other fuel's characteristics, different from the fuels considered in this work.

Even considering that the use of B15 or HVO15 introduce an increase in overall fuel consumption of 2 and 8% respectively, the overall CO_2 emissions can be reduced, due to the renewable character of this fuels. In fact, all the tested biofuels provide a reduction in CO_2 emissions, even B100. There were no noticeable differences between smallest blends, B15, B7 and HVO15, which present similar engine power, with the best result revealed by HVO15 with a small increase of 1,5% relatively to B7. The use of B100 cause a reduction of 4,5% in available power.

There were no problems detected in the vehicle, but for a more complete and adequate analysis, some more tests should be considered, namely using the vehicle on road and detect the preferable operation regimes to make a proper design of a better representativeness of the vehicle operation in real conditions.

A first experiment was made considering a real use of heavy buses fueled with B100 for 8 months in a Portuguese city and the perspectives were very good, without detection of breakdowns or increased problems in the maintenance and operation of the fleet. The results obtained were surprisingly positive, with an increase of only 4.5% in fuel consumption compared to B7. However, this study needs to be continued, considering other fleets and other blends, and also obtaining information on engine speeds for a possible cross-analysis of the results with the chassis dynamometer data.

Acknowledgements. The authors acknowledge Nuno Pires for the experimental procedure development and implementation and NTA | Novas Tecnologias Auto Lda, for the accessibility of their chassis dynamometer and installations.

Authors also acknowledge Prio Energy for all the support given to this work and Sintética, Lda, for having provide the HVO.

References

1. IEA international Energy Agency -Statistics report - World Energy Balances, Overview, July 2020
2. European Environmental Agency - Transport and Environment Reporting Mechanism (TERM), Transport: increasing oil consumption and greenhouse gas emissions hamper EU progress towards environment and climate objectives, February 2020.Author, F., Author, S., Author, T.: Book title. 2nd edn. Publisher, Location (1999)
3. IEA international Energy Agency - Transport Energy Efficiency, Implementation of IEA Recommendations since 2009 and next steps, Information Paper OECD/IEA, September 2010
4. The New York Times - LN France Plans to End Sales of Gas and Diesel Cars by 2040 – Accessed 08 Feb 2021. https://www.nytimes.com/2017/07/06/business/energy-environment/france-cars-ban-gas-diesel.html
5. Nuno Pereira: Análise sobre a utilização de biodiesel em veículos rodoviários equipados com tecnologias avançadas de controlo de emissões (2020)
6. Serrano, M.V.: Análise comparativa do desempenho de motores de combustão interna quando utilizam biocombustíveis (2012)
7. Kegl, B.: NOx and particulate matter (PM) emissions reduction potential by biodiesel usage. Energy Fuels **21**, 3310–3316 (2007). https://doi.org/10.1021/ef7002393
8. European Standards: EN 590 - Automotive fuels - Diesel - Requirements and test methods. 1–12 (2009)
9. Product PD, Standards A, CFPP I, Anon. Sulphur Free Diesel BS EN 590 : 2013. 10:1–18 (2013)
10. European Standards, Motor vehicle fuels Fatty acid methyl ester (FAME) for diesel engines Requirements , test methods and threshold values DIN EN 14214 Threshold values. 14538
11. Product PD, Standards A, Cfpp I: Product specification sheet fatty acid methylesters (FAME). 5–6 (2019)
12. Demirbas, A.: Biodiesel, A Realistic Fuel Alternative For Diesel Engines, 2008th edn. Springer, 20 March 2008. ISBN-13: 978–1846289941
13. Neste Corporation: Neste Renewable Diesel Handbook. Neste **1–33**, (2015). https://doi.org/10.1016/S0997-7538(02)01247-0

14. Bacha, J., Freel, J., Gibbs, A., et al.: Diesel fuels technical review. Chevron Glob Mark **1–116**, (2007). https://doi.org/10.1063/1.3575169
15. Canakci, M., Sanli, H.: Biodiesel production from various feedstocks and their effects on the fuel properties. J. Ind. Microbiol. Biotechnol. **35**, 431–441 (2008). https://doi.org/10.1007/s10295-008-0337-6
16. Organisation Internationale des Constructeurs d'Automobiles (OICA), Biodiesel Guidelines - Worldwide Biofuels Harmonization, From the worldwide fuel chapter committee, march 2009
17. Arbeitsgemeinschaft Qualitätsmanagement Biodiesel e.V. Biodiesel Analytics: Important Parameters and their Meaning Issue 6/2018
18. Dimitriadis, A., Seljak, T., Vihar, R., et al.: Improving PM-NOx trade-off with paraffinic fuels: a study towards diesel engine optimization with HVO. Fuel **265**, (2020). https://doi.org/10.1016/j.fuel.2019.116921
19. Guarieiro, N., Guarieiro, A.: Vehicle Emissions: What Will Change with Use of Biofuel?, Biofuels - Economy, Environment and Sustainability, Zhen Fang, IntechOpen, 23rd January 2013. https://doi.org/10.5772/52513
20. Kaya, T., Kutlar, O.A., Taskiran, O.O.: Evaluation of the effects of biodiesel on emissions and performance by comparing the results of the new European drive cycle and worldwide harmonized light vehicles test cycle. Energies **11**, (2018). https://doi.org/10.3390/en11102814
21. Pullen, J., Saeed, K.: Factors affecting biodiesel engine performance and exhaust emissions - part I: review. Energy **72**, 1–16 (2014). https://doi.org/10.1016/j.energy.2014.04.015
22. McCormick, R.L., Alleman, T.L., Yanowitz, J., et al.: Exhaust emissions. Biodiesel Handb. Second Ed. **253–298**, (2010). https://doi.org/10.1016/B978-1-893997-62-2.50012-7
23. Dobrzyńska, E., Szewczyńska, M., Pośniak, M., et al.: Exhaust emissions from diesel engines fueled by different blends with the addition of nanomodifiers and hydrotreated vegetable oil HVO. Environ. Pollut. **259**, (2020). https://doi.org/10.1016/j.envpol.2019.113772
24. Bortel, I., Vávra, J., Takáts, M.: Effect of HVO fuel mixtures on emissions and performance of a passenger car size diesel engine. Renew. Energy **140**, 680–691 (2019). https://doi.org/10.1016/j.renene.2019.03.067
25. Omari, A., Pischinger, S., Bhardwaj, O.P., et al.: Improving engine efficiency and emission reduction potential of HVO by fuel-specific engine calibration in modern passenger car diesel applications. SAE Int. J. Fuels Lubr. **10**, (2017). https://doi.org/10.4271/2017-01-2295
26. United Nations: Global technical regulation No. 4 Test procedure for compression-ignition (CI) engines and positive-ignition (PI.) engines fuelled with natural gas (NG) or liquefied petroleum gas (LPG) with regard to the emission of pollutants (2015)
27. Aatola, H., Larmi, M., Sarjovaara, T., Mikkonen, S.: Hydrotreated Vegetable Oil (HVO) as a Renewable Diesel Fuel: Trade-off between NO x , Particulate Emission, and Fuel Consumption of a Heavy Duty Engine, SAE 2008–01–2500

Study of a Mid-Scale Prototype of Biogas in a Pig Farm Using Grape Waste in Anaerobic Co-digestion

A. I. Parralejo(✉), L. Royano, M. A. Serrano, J. Cabanillas, and J. González

Department of Field and Energy Crops: Biogas, Biomass and Bioproducts, Centre for Scientific and Technological Research of Extremadura (CICYTEX), Consejeria of Economy, Science and Digital Agende, Junta de Extremadura, Guadajira, 06187 Badajoz, Spain
ana.parralejoa@juntaex.es

Abstract. In Spain there are large rural areas where the primary sector is an important source of economic income and emissions generated must be taken into account because they are quite hazardous, especially emissions produced by manure and slurry. Anaerobic digestion is applied to process this waste, and sometimes mixed with other feedstocks, which in this case were carried out with grape waste (anaerobic co-digestion). Results obtained in anaerobic co-digestion of pig manure with grape waste were 50% higher than anaerobic mono-digestion of pig manure, referred to methane yield (207 NL/kg VS versus 138 NL/kg VS, respectively) employing an Organic Load Rate (OLR) of 1.27 g VS/L_D d and 1.36 g VS/L_D d, respectively. The feasibility analysis from a big pig farm was developed to use the biogas obtained to produce energy and the digestate characterized to evaluate its quality.

Keywords: Methane yield · Payback time · Grape waste

1 Introduction

Strategic guides or protocols have been fixed by different countries in the world to reduce the greenhouse emissions in 2021 to 2030. Concretely, according PEIEC (Extremeño Integrate Plan of Energy and Climate) [1] in the region where the analysis has been developed, an important objective is to reduce 10% the greenhouse gas emissions. Main sectors generating these kinds of gases are mentioned in the Plan as agriculture and cattle raising. In the two last sectors an appropriate technology to solve the identified problem is the anaerobic digestion, a process developed without oxygen where waste is valorised and two products are obtained, biogas and biofertilizer. Methane component of biogas is responsible to increase the heating value in a way that can be applied as biofuel. Biofertilizer is considered the digested material from anaerobic digestion, called digestate, and its content in nitrogen (N), potassium (K) and phosphorous (P) are available for use as biofertilizer along with its organic matter [2]. This technology is improved when manure or slurry is mixtured with different substrates such as agroindustrial waste,

J. R. da Costa Sanches Galvão et al. (Eds.): *Proceedings of the 1st International Conference on Water Energy Food and Sustainability (ICoWEFS 2021)*, pp. 280–290, 2021.
https://doi.org/10.1007/978-3-030-75315-3_32

crops residues, food waste, used cooked oils [3–5]. In this case the substrate used has been the grape waste, problematic waste produced in the pressing of the grape to obtain wine. Previous results showed successful methane yield values when the mechanical pretreatment was manually carried out (increase of 50% in methane yield was obtained when a complete mechanical pretreatment was carried out in semicontinuous regime opposite to simple mechanical pretreatment of winery waste) [6]. Based on the information available in previous research, the main objective of this study is the development of anaerobic digestion with pig manure and winery waste in a prototype installed in a pig farm with an innovation mechanical pretreatment, and a feasibility analysis for this pig farm taking advantage of the results obtained in the experiences carried out.

2 Materials and Methods

2.1 Raw Materials

Anaerobic digestion process in semi-continuous regime will be carried out in a pig farm to achieve results "in situ" with the pig manure itself generated. This farm is located to Alburquerque, town from Badajoz, in Extremadura (Spain). The grape waste employed in anaerobic co-digestion was previously crushed and seeds ground manually, after. The processed material was subjected to a silage process for three months before its use. Grape waste was colleted to winery industry in Almendralejo, a town dedicated to wine manufacture, in Extremadura.

2.2 Anaerobic Digestion Prototype and Assays Design

A prototype designed and manufactured to treat manures in co-digestion with other substrates have been installed in a pig farm to demostrate its effectiveness (Fig. 1). The idea has emerged as a demonstration action within a cross-border European project between the region of Alentejo (Portugal) and Southwestern of Spain (Extremadura), IDERCEXA project (Research, Development and Renewable Energies for the improvement of the business clusters in Centro, Extremadura and Alentejo). The mixture to introduce into the digester can be pumped at once. Solid raw material can be crushed by a second time through a shredder at the feeding pipeline of the biodigester just before the pump. The work volume employed has been 2650 L and the biodigester has been fed two times a week. Additions amounts for each assay have been 146 kg/d of pig manure in mono-digestion assay (corresponding to an ORL average of 1.36 g VS/L_D d) and 100 kg/d of pig manure and 1 kg/d of grape waste (corresponding to 1.27 g VS/L_D d). Assays were evaluated 43 days each one.

Fig. 1. Prototype employed in experiments carried out.

2.3 Analytical Methods

The characterization of the substrates was carried out according to the standard methods [7]. The Total Solids (TS) content was determined by drying the sample to constant weight in an oven (JP Selecta Digitheat, USA) at 105 °C for 48 h (2540 B method). The content of Volatile Solids (VS) was obtained by drying at 550 °C for 2 h in a muffle oven (Hobersal 12PR300CCH, Spain) in an inert atmosphere (2540 E method). The pH and redox potential values were measured through their corresponding electrodes with a pH meter (Crison Basic 20, Spain); the alkalinity of the medium according to method 2320; Chemical Oxygen Demand (COD) according to method 410.4 [8]; ammonia nitrogen (N-NH4) by volumetric titration according to the E4500-NH3 B method and Total Volatile Fatty Acids (VFA) according to Buchauer's volumetric method [9]. The initial C/N ratio in the substrates was determined using a True-Spec CHN Leco 4084 elementary analyzer (USA), according to the UNE-EN 16948 standard for biomass analysis C, N, H [10]. The composition of the biogas and the volume generated were automatically monitored on-site throughout the experiments with an Awite System of Analysis Process series 9 analyzer (Bioenergie GmbH, Germany). The gas meter (Ritter model MGC-1 V3.2 PMMA, Germany) was used to individually measure the biogas produced, which was stored in Tedlar bags. The volume of dry gas was corrected at standard conditions (0 °C, 101,325 kPa). The elements analyzed in the digestate obtained were detected by spectroscopy using an ICP-OES Varian 715 ES (Australia).

2.4 Feasibility Study Estimate

A feasibility study of biogas plant needs initial values to calculate the Payback Time (PBT) (expression 1), one of the most important parameters to know if this plant is economically feasible. This parameter is obtained when the Annual Total Revenue (ATR) is calculated, included Renewable Source Replaced (RSR) to actual energy of the farm, and digestate sale (expression 2). The digestate produced in the process can be considered as income to the study because the price of it is estimated in 6 €/t.

$$PBT = TI/ATR \tag{1}$$

$$ATR = \{Electricity\,price * RSRt\ (or\ RSRe) + digestate\ sell\ income\} \tag{2}$$

$$TI = (Biodigester\,price + electricity\,generator\,price + operating\ and\ maintenance\,price) \tag{3}$$

Total Inversion (TI) can be obtained according to Eq. 3, allowing for the biodigester, the electrical and heat generator, and the operating and maintenance costs, as it is reflected in Table 1. To know the biodigester price, it is necessary to know its volume (V_B) (calculated through the Eq. 4).

$$V_B = (Total\ treated\ waste\ dairy * HRT) \tag{4}$$

The RSR to actual energy of the farm has been obtained according to Eq. 5 considering its own energy consumption of the biogas plant, the methane calorific value and the thermal and electrical efficiency of the electrical and heat generator.

$$RSR = (Methane\ calorific\ value * Methane\ yield * Vsconcentration * Total\,feed\ amount)/100 \tag{5}$$

$$RSR_t = \{0.5 * RSR * (1 - 0.68)\}$$

$$RSR_e = \{0.35 * RSR * (1 - 0.05)\}$$

In Table 1 the necessary values to carried out a feasibility study of anaerobic mono-digestion and co-digestion of pig manure, and pig manure with grape waste are shown, respectively.

Table 1. Values employed to develop the feasibility study of anaerobic mono-digestion and co-digestion.

	Mono-digestion	Co-digestion
Feed amount, kg/d	40000	33000
HRT, days	18	22
VS content in the feed, %	0.80	1.20
Methane yield, N m^3/kg VS	0.138	0.207
Biodigester prototype cost, €	400–600 €/m^3 [11]	
Operation and maintenance costs, c€/kWh_e	4 c€/kWh_e [12]	
Operation time, h/year	7500 h/year	
Energy consumption, %	68% and 5% electrical and thermal produced energy, respectively [13]	
Methane calorific value, kWh/N m^3 CH_4	9.96 [14]	
Thermal efficiency, %	50% [14]	
Electrical efficiency, %	35% [14]	
Electricity sell price, €/kWh	0.13 €/kWh [15]	

3 Results and Discussion

3.1 Comparative Study of Variable Assays

Assays carried out in prototype of anaerobic digestion were evaluated in the process analyzing samples taken two times a week. In the Table 2 a characterization of the two feedstocks used in this work are presented.

Further, the evolution for some parameters analyzed have been represented in Figs. 2 and 3. Average values shown in the figures indicate a stability of the process: ammonia nitrogen parameter for anaerobic digestion process is found in values below inhibitory range (1500–2000 ppm [16]). VFA values shown in the Table 2 are not too much high. When overloading is happened in some investigation, VFA is increased and pH in the digestion medium is decreased, thus the methane production is affected [17]. In this case the OLR employed have been appropriate to carry out the mixed feedstock. The observed trend in the represented variables (see Figs. 2 and 3) is very similar for two assays, except alkalinity. In this case, values lightly higher are presented in co-digestion assay than mono-digestion assay. Grape waste used in co-digestion assay has low value in pH, probably the alkalinity of the medium provided by the pig manure, has experimented a small increase to buffer the low value of grape waste. On the other hand, VFA parameter shows higher values as assays progresses, however values reached are far away to inhibit the anaerobic digestion process (4000 ppm, [16]).

Methane yield obtained increased by -50% in co-digestion assay in opposition to mono-digestion assay (Table 3). Pellera and Gidarakos [17] achieved a value of methane

Table 2. Waste samples characterization.

	Pig manure	Grape waste
pH	6.82	3.23
Redox potential, mV	−382	107
Chemical Oxygen Demand (COD), mgO_2/L	18400	368571
Ammonia nitrogen, mg/L	1600	286
Total Solid (TS), %	1.56	45.9
Volatile Solid (VS), %	51.49	91.76
Alkalinity, mg $CaCO_3$/L	5986	-
Volatile Fatty Acid, mg/L	686	-
C/N ratio	2.33/0.13	21.20/0.57

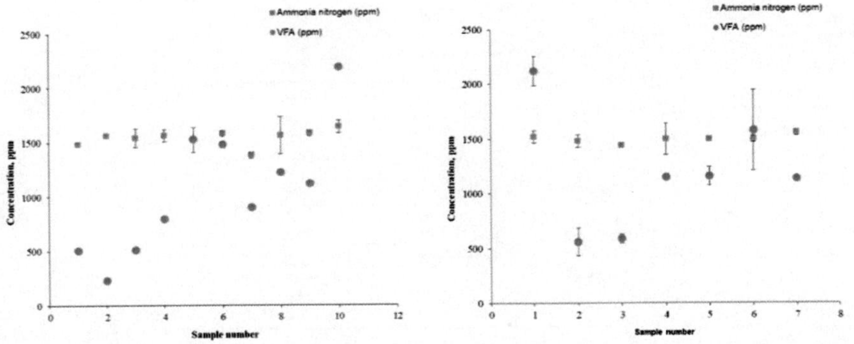

Fig.1. Ammonia nitrogen and VFA evolution for anaerobic mono-digestion (right) and co-digestion (left) assays carried out in prototype.

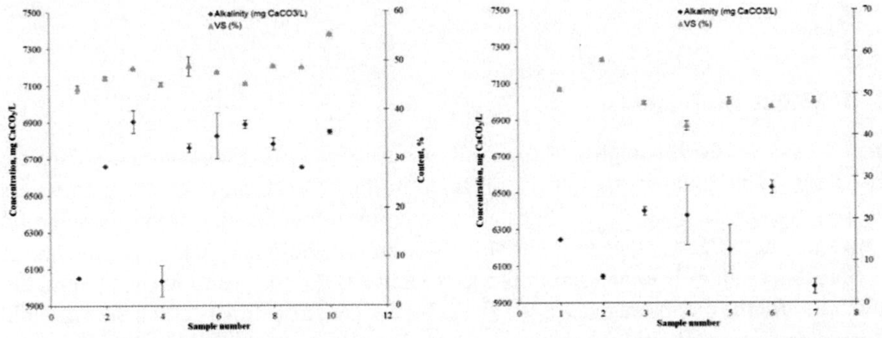

Fig. 2. Alkalinity and VS evolution for anaerobic mono-digestion (right) and co-digestion (left) assays carried out in prototype.

yield of 347 NL/kg VS, higher than 207 N L/kg VS obtained in this work. Nevertheless, winery waste and manure employed had a carbon content extremely higher than mixture used in this work (45% versus 2.40%). However, previous studies developed in the same initial conditions obtained methane yield lower (177 N L/kg VS [6]).

Table 3. Productions and yields obtained in assays.

	Mono-digestion assay	Co-digestion assay
Average methane concentration in biogas, %	51	59
Methane yield, N L/kg VS	138	207

Methane yield obtained in assays carried out, experiment a decrease in the methane production at the end of the mono-digestion assay (Fig. 3). It was due to mechanical problems with the prototype.

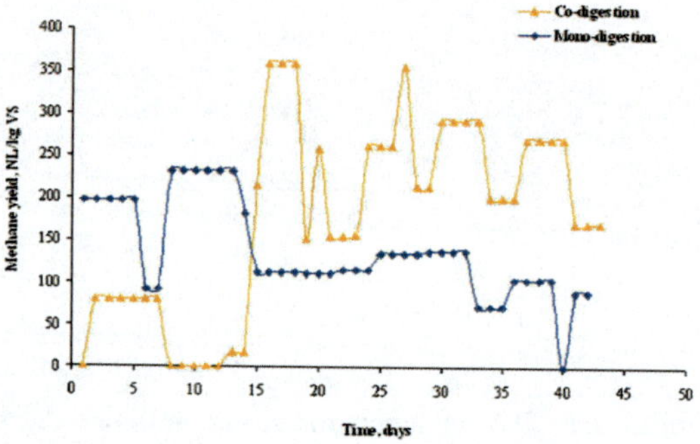

Fig. 3. Methane yield evolution of mono-digestion and co-digestion assays carried out in the prototype.

3.2 Digestate Evaluation

When assays finished, samples of digestate were taken and a chemical analysis were carried out. Table 4 shows the results obtained in the characterization. More complete nutrients content is obtained in digestate to employ as biofertilizer than commercial fertilizers. The NPK content in the digestate is one of the main quality parameter. In practice, this content is expressed in oxide content for P (P_2O_5) and K (K_2O), and the availability for the plants are not 100%. The Table 5 shows an estimate about the available nutrients content by plants in digestate compared to commercial fertilizer. Nevertheless, the digestate should have a low heavy metal content (Cu, Cr and Ni) and pH values close to neutrality. Concretely, the normative (RD 999/2017) [20] for digestates classify in three groups the fertilizer according to the heavy metal contents. In these assays digestates obtained belong to A fertilizer (fertilizer with the lowest heavy metal content).

Table 4. Digestate characterization in mono-digestion and co-digestion assays.

	Mono-digestion assay	Co-digestion assay
pH	7.54	8.34
Redox potential, mV	−43	−184
Chemical Oxygen Demand (COD), mgO_2/L	6800	9200
Ammonia nitrogen, mg/L	1460	1600
Total Solids (TS), %	0.68	0.89
Volatile Solids (VS), %	48.27	50.30
Alkalinity, mg $CaCO_3/L$	5987	6515
Volatile Fatty Acid, mg/L	1137	2155
C/N ratio	2.36/0.12	2.42/0.07

Table 5. Digestate evaluation for fertilizer use.

	Mono-digestion assay			Co-digestion assay		
	N	P_2O_5	K_2O	N	P_2O_5	K_2O
Total nutrients in digestate, %	0.12	0.010	0.13	0.07	0.07	0.19
Nutrients available by plants in digestate*, %	0.07	0.006	0.12	0.04	0.04	0.18
Nutrients contributed to plants for the digestate equivalent to comercial fertilizer**, kg/ha	21	1.80	36	12	12	54
	Cu, ppp	Cr, ppp	Ni,ppm	Cu, ppm	Cr, ppm	Ni, ppm
Heavy metal content, ppm	2.08	0.75	0.52	7.31	3.19	2.33
Fertilizer kind	A			A		

* Available nutrients for plants: 55%, 64% y 92% for N, P_2O_5 and K_2O, respectively[19]
** Digestate dose: 30 m^3/ha [19]. (assays developed with 2650 L work volume in the biodigester, when RHT is gone 0.09 ha could be fertilized).

3.3 Economic Analysis

A farm with 2500 pigs is considered to carried out the analysis, generating 40 m^3/d of pig manure. For co-digestion, 330 kg/d of grape waste has been employed. The grape waste production is considered are located near the biogas plant (<50 km). Results obtained referred to V_B, TI, TR and RSR is shown in the Table 6. A value of V_B similar has been pretended to achieve an easier comparative study between mono-digestion and co-digestion assays. To calculate the TI the price for biodigester has been considered 450 €/m^3, for energetic generator a price of 500000 € were attributed, and the digestate sell has been fixed to 6 €/t generated. RSR have been associated to electrical and thermal energy to employ in the power generator. As it can be shown in Table 6 the PBT for anaerobic mono-digestion and co-digestion doesn´t exhibit any difference. The digestate sell effect in the economic analysis is very important to get an economically feasible biogas plant.

Table 6. Results obtained in an economic feasibility of biogas plant to reach the PBT.

	Mono-digestion	Co-digestion
V_B, m^3	720	726
TI, €	824183	826788
RSR, kWh/year	137448	213432
RSR_e, kWh/year	45702	70966
RSR_t, kWh/year	21992	34149
TR, €/year	83800	88665
PBT_t, years	9.84	9.32

4 Conclusions

A complete study about anaerobic digestion have been developed in a farm performing a comparative study about mono-digestion and co-digestion processes with pig manure and, pig manure with grape waste, respectively. Results obtained for each assay have achieved an increase of 60% in methane yield in co-digestion anaerobic digestion opposite to mono-digestion process (207 NL/kg VS versus 138 NL/kg VS, respectively) in semi-continuous regime. A feasibility analysis in a farm with 2500 pig manures has been carried out and a PBT very similar have been obtained in mono-digestion and co-digestion (9.84 years and 9.32 years for mono-digestion and co-digestion assays, respectively). A theoretical valorization of digestate have been developed and both digestates obtained have been classificated according to RD 99/2017 as A fertilizers.

Acknowledgements. This research work has been carried out as part of the work developed by CICYTEX associated with the IDERCEXA project (Research, Development and Renewable Energies for the improvement of the business clusters in Centro, Extremadura and Alentejo) belonging to the INTERREG-POCTEP program, Programs of Cross-Border Cooperation Spain and Portugal, financed through FEDER Funds.

References

1. Junta de Extremadura: Consejería para la Transición Ecológica y Sostenibilidad. Plan Extremeño Integrado de Energía y Clima (PEIEC) (2020)
2. Sherrard, A., Persson, D.N., Fogelmark, J., Shen X., Björkman, M., Haaker, A.: Feedstock and pre-treatment, Novel biogas plants, Upgrading and post-treatment, biogas market. Bioenergy International. Digital Biogas Special 1 (2020)
3. Rubio, J.A., Romero, L.I., Wilkie, A.C., García-Morales, J.L.: Mesophilic anaerobic co-digestion of olive-mill waste with cattle manure: effects of mixture ratio. Front. Sustain. Food Syst. **3**, 9–11 (2019)
4. Vavouraki, A.I., Zakoura, M.V., Dareioti, M.A., Kornaros, M.: Biodegradation of Polyphenolic Compounds from Olive Mill Wastewaters (OMW) during two-stage anaerobic co-digestion of agro-industrial mixtures. Waste Biomass Valorization **11**, 5783–5791 (2020)
5. Kabouris, J.C., Tezel, U., Spyros, G., Pavlostathis, G., Engelmann, M., Dulaney, J., Gillette, R.A., Todd, A.C.: Methane recovery from the anaerobic co-digestion of municipal sludge and FOG. Bioresour. Technol. **100**, 3701–3705 (2009)
6. Parralejo, A.I., Royano, L., Acevedo, M., Cabanillas, J., González, J.: Study of the influences of pre-treatment in the grape waste biogas generation. Mod. Environ. Sci. Eng. **5**, 207–214 (2019)
7. APHA: Standard Methods for Examination of Water and Wastewater, 22 edn. American Public Health Association Washington, DC (2012)
8. Methods for Chemical Analysis of Water and Wastes. EPA-600/4-79-020. U.S. Environmental Protection Agency. Environmental Monitoring and Support Laboratory (1983)
9. Buchauer, K.: Titrations verfahren in der Abwasserund Schlam-manalytik zur Bestimmung von flüchtigen organischen Säuren. Das Gas- und Wasserfach (gfw). Wasser Abwasser, 138 (6), 313–320 (1997)
10. Norma UNE-EN 16948. Biocombustibles sólidos, Determinación de contenido total de C, H y N. Método instrumental (2015)
11. Agencia Andaluza de la Energía. Consejería de Economía Innovación y Ciencia. Junta de Andalucía. Estudio básico del biogás (2011)
12. Casson, A., Muriastra, Y., Obidzinski, K. Large scale platations, bioenergy developments and land use change in Indonesia. In: CIFOR Working Paper (2014)
13. Ministerio de Medio Ambiente y Medio Rural y Marino. Recursos Agrícolas y ganaderos. El sector del biogás agroindustrial en España, Madrid (2010)
14. Mas, M.: Feasibility Study of Anaerobic Digestion of Olive Mill Waste for Energy Production. Case Study Applied to Two Olive Oil Extraction Industries of Spain. Universitat Politècnica de Catalunya, Barcelonatech (2017)
15. Spain, R.D.: 9/2013 of 12 July on regulating electricity generation as a special regime. BOE of 167 (2013)
16. Zubayeda, Z., Stelios, G., Tim, H.M., Maazuza, Z.O.: Semi-continuous anaerobic co-digestion of chicken litter with agricultural and food wastes: a case study on the effect of carbon/nitrogen ratio, substrates mixing ratio and organic loading. Bioresour Technol. **270**, 245–254 (2018)
17. Pellera, F.-M., Gidarakos, E.: Anaerobic digestion of solid agroindustrial waste in semi-continuous mode: evaluation of mono-digestion and co-digestion systems. Waste Manage. **68**, 103–119 (2017)

18. Drosg, B. Process monitoring in biogas plant. International Energy Agency (IEA) Bioenergy. Task 37 – Energy from Biogas and Landfill Gas (2013)
19. Agriculture & Horticulture Development Board (AHDB). Nutrient Management Guide (RB209), Section 2: Organic materials, 35 (2017)
20. Real Decreto 999/2017: de 24 de Noviembre, por el que se modifica el Real Decreto 506/2013, de 28 de Junio, sobre productos fertilizantes

Enhancement of Biogas Production from Swine Wastewater Using Electrochemical Pretreatment and Anaerobic Co-digestion with Rice Straw

G. Lourinho(✉) and P. S. D. Brito

VALORIZA - Research Centre for Endogenous Resource Valorization, Campus Politécnico, 10, 7300-555 Portalegre, Portugal
goncalo.lourinho@ipportalegre.pt

Abstract. Co-digestion and pretreatment approaches were tested using batch tests for the enhancement of anaerobic digestion (AD) of swine wastewater (SW). AD was evaluated in mesophilic conditions to understand the effects of the addition of rice straw (RS) to the medium. The results showed that an increase in the SW/RS ratio led to higher cumulative methane production until 30 wt.% RS. The cumulative methane productions for SW/RS 70:30 and 50:50 were 903.66 $\pm$ 111.0 mL, and 625.55 $\pm$ 55.1 mL, respectively, which represented an increase of 41.3% and 17.8% compared with the control (525.78 $\pm$ 18.77 mL). Kinetic modeling translated well to methane production in the different treatments. The hydrolysis constant and maximum methane production were higher in the 70:30 SW/RS ratio, while the addition of higher quantities of RS increased the duration of the lag-phase. Electrochemical pretreatment was performed to increase the bioavailability of suspended particulate matter in SW at 10 mA cm^{-2} and two temperatures (30 °C and 45 °C). Final methane production increased by 8% and 19% in the two pretreatments, respectively, when compared to the control. The pretreatment at 10 mA cm^{-2} and 45 °C showed the best performance but delayed the time for stable methane production. Kinetic modeling confirmed these results as the parameters obtained for the second pretreatment group were better than the control. These findings support the potential of both enhancement approaches to enhance energy recovery via AD in swine farms.

Keywords: Anaerobic digestion · Swine wastewater · Biogas

1 Introduction

AD enhancement is expected to attract much attention in the future as a way of raising process reliability and improving performance. Enhancing the methane yields of current biogas plants could be a key strategy to increase the economic feasibility in farm-scale plants as the real driving force behind waste-to-energy systems is the revenue from electricity sales [1]. There are two primary ways of improving the digestion process: enhancing hydrolysis or methanogenesis [2]. The first concept functions through

J. R. da Costa Sanches Galvão et al. (Eds.): *Proceedings of the 1st International Conference on Water Energy Food and Sustainability (ICoWEFS 2021)*, pp. 291–300, 2021.
https://doi.org/10.1007/978-3-030-75315-3_33

the acceleration of the hydrolysis step where hydrolytic-fermentative bacteria assist in the structural breakdown of complex biopolymers present in the digesting media. This enzymatic hydrolysis of particulate into dissolved organic matter is hindered in complex substrates and can be promoted with effective pretreatment steps. For example, physical pretreatment (including thermal, mechanical, microwave [3]) has been explored, as well as chemical pretreatment (acid, alkali, electrochemical [4, 5]) and biological treatment. The second approach to enhance methane production in AD systems works in response to macronutrient balance inside the reactors by co-digesting two different substrates simultaneously. Co-digestion of two complementary substrates has been proven to favor process stability and feedstock reliability, as well as synergisms and dilution of toxic compounds [6]. Balancing the disrupted quantities of carbon and nitrogen inside a reactor can also prevent ammonia inhibition, as high ammonia concentrations affect methanogenic microorganisms. This inhibition is a known practical problem in AD that often results in poor methane yields, especially in the treatment of animal wastes such as SW. Thus, pretreatment and co-digestion strategies are critical in enhancing process performance and compose two of the most relevant topics in AD research.

AD studies have shown that an electrochemical pretreatment allows the rupture of cell structure, releases intracellular substances, and results in increased microbial access to biopolymers, including proteins and polysaccharides [5, 7]. Applying current before AD also promotes the conversion of these organic polymers into low molecular weight products which can be easily metabolized by acidogenic bacteria. Based on these principles, electrochemical methods have been explored as a pretreatment approach. However, previous studies have been mainly focused on substrates other than swine wastes with positive results in AD performance. For example, Veluchamy et al. [7, 8] adopted this electrochemical approach to enhance the digestion of pulp and paper sludge and waste activated sludge. They reported a 13.8% and 9.85% enhancement in methane production, respectively. Concerning swine-derived wastes, Safavi et al. [9] applied pulsed electric fields with different intensities and found that methane production and Chemical Oxygen Demand (COD) removal increased up to 58% and 44% at optimal conditions, respectively. Although other pretreatment approaches have been applied to SW, the effects of several pretreatments differ considerably. More information is thus important to fully explore pretreatment methods and their practical application.

On the other hand, the co-digestion of SW with other substrates has been more widely studied. Previous research evaluating the effects of this enhancement approach includes co-substrates such as lignocellulosic crop residues [10–13]. Among these substrates, agricultural wastes are the most common substrates to co-digest with SW within a perspective of nutrient balance and process stability. Wu et al. [12] tested corn stalks, oat straw, and wheat straw to enhance the methane productivity of swine wastes. The results showed that all crop residues significantly increased methane volume at all C/N ratios. Corn stalks, for instance, increased the daily maximum biogas volume 11 times as compared to the control. Jiménez et al. [10] also reported higher methane yields from the co-digestion of swine manure, RS, and clay materials when compared with the digestion of swine manure alone. This increased performance implies that RS could be a promising feedstock as co-substrate for the AD of swine wastes.

This work presents the results of (1) the addition of RS into the digester environment during the AD of SW and (2) the application of an electrochemical treatment before the digestion of SW. The effects of both enhancement approaches on the performance of AD were assessed in terms of cumulative methane volume and daily methane production in batch assays. Two kinetic models were also used to assess methane production, link kinetic parameters with methane volume, and help determine the practical application of both enhancement technologies.

2 Materials and Methods

2.1 Characteristics of Substrate and Inoculum

SW was obtained from a large-scale farm located in Alentejo, Portugal. The farm operates in a closed cycle in which pigs are raised in separated breeding units. On-farm wastewater management is based on a sequential process comprising a solid-liquid separation and anaerobic lagoons. Samples were collected from the growing fattening phase before biological treatment. Inoculum was obtained from the Wastewater Treatment Plant (WWTP) of Castelo Branco, Portugal. The plant has an anaerobic digester for the treatment of domestic wastewater and operates under a mesophilic regimen. Air-dried RS was collected from an agroindustrial facility located in Extremadura, Spain. The chemical characteristics of the substrates used in this study are presented in Table 1.

Table 1. Basic characteristics of SW, RS, and inoculum. Results are the average ± standard deviation of triplicate measurements. N.D.: not determined.

Parameters	Swine wastewater	Rice straw	Inoculum
pH	8.50 ± 0.00	N.D.	6.55 ± 0.05
Conductivity/mS cm^{-1}	38.0 ± 5.00	N.D.	6.05 ± 0.28
TS/%	9.28 ± 0.07	92.1 ± 0.08	0.93 ± 0.05
S/%TS	67.3 ± 0.43	76.8 ± 0.35	65.0 ± 0.44
TDS/g L^{-1}	24.5 ± 2.50	N.D.	2.72 ± 0.01
Total COD/g L^{-1}	23.3 ± 0.13	N.D.	1.64 ± 0.01
Soluble COD/g L^{-1}	7.02 ± 0.05	N.D.	N.D.
VFA/mg L^{-1}	1287 ± 240	N.D.	N.D.
TAN/mg L^{-1}	3982 ± 48.0	N.D.	N.D.
FAN/mg L^{-1}	137.9 ± 1.65	N.D.	N.D.
C/N ratio	8.10 ± 0.30	27.9 ± 0.41	N.D.

2.2 Batch Co-digestion

Co-digestion experiments were carried out under batch conditions in serum bottles of 1000 mL (Schott Duran, Germany). Each bottle was filled with 60 mL of inoculum and appropriate amounts of digesting substrates, flushed with an inert gas (argon), sealed and closed with 5 mm thick silicone discs and plastic screw caps (Schott Duran, Germany). Three groups containing different substrate combination ratios (SW/ RS) were set up: the T1 group (SW/ RS 70:30), the T2 group (SW/RS 50:50), and the T0 control group (SW/ RS 100:0) as a wt.% of the total feedstock measured. The initial Total Solids (TS) concentration was between 7 and 12%. The temperature was controlled by a water bath and kept at a constant mesophilic temperature (37 ± 1 °C). Agitation was performed manually during the incubation period at regular times. Cumulative biogas and methane production were measured indirectly by mass loss (gravimetric method) using the biogas package in R software as detailed by Hafner et al. [14]. Gas measurements were standardized to a temperature and pressure of 0 °C and 1 atm, respectively.

2.3 Electrochemical Pretreatment

The electrochemical system consisted of a glass cell of 1000 mL with the electrodes connected from the top and a Reference 600 potentiostat (Gamry Instruments, Warminster, PA) controlled by the associated Gamry Framework software (version 6.3). Electrochemical pretreatment was carried out for 30 min at two temperatures (T1 and T2 groups, 30 °C and 45 °C, respectively) in galvanostatic conditions (10 mA cm^{-2}) using a mesh plate Ti/MMO anode and an aluminum rod cathode. The Disintegration degree (DD) of SW was used to evaluate the effectiveness of pretreatment and calculated as described in Yu et al. [4]. After the pretreatment, batch AD assays were conducted to both unpretreated SW (T0 control group) and pretreated SW for comparison purposes (T1 and T2 groups). The initial Volatile Solids (VS) ratio of substrate to inoculum was kept at 1:3 for all experimental setups. Methane volume was measured as described earlier.

2.4 Kinetic Modeling

The kinetics of methane production was characterized during co-digestion and pretreatment tests by applying a first-order kinetic model, as well as a Gompertz model. The first-order model allows evaluating the digestion suitability by estimating a hydrolysis constant, k, that describes the velocity of substrate degradation and methane production. On the other hand, the fitting of the Gompertz model to the cumulative methane production curves allows the estimation of four parameters and constants with biochemical importance: the maximum methane volume, A (mL), the maximum rate of methane production, μ (mL d^{-1}), and the length of the lag phase, λ (d^{-1}). Model implementation and parameter estimation were performed with Origin software using the Levenberg-Marquardt algorithm.

2.5 Statistical Interpretation

The results of batch co-digestion tests were tested for significant differences in the methane production with a nested Analysis of Variance (ANOVA) followed by a posthoc analysis (Dunnett's test) using the decision diagram suggested by Hofmann et al. [15]. Methane production datasets with and without electrochemical pretreatment were analyzed with a paired t-test. The significance level for all tests was chosen at 0.05. Data analysis was carried out using R software and all values correspond to the mean of two independent replicates (n = 2) ± standard deviation.

2.6 Analytical Methods

TS, VS, pH and conductivity were measured following standard methods [16]. Total Ammonia Nitrogen (TAN) concentration was measured by steam distillation followed by titration also according to APHA standard methods. Free Ammonia Nitrogen (FAN) was calculated based on the ionization equation at the digestion temperature. Total Volatile Fatty Acids (VFA) were determined using the Kapp four-point titration method. Total COD was analyzed according to the closed reflux colorimetric method (5220D). Appropriate amounts of SW sample, digestion solution containing potassium dichromate and sulfuric acid reagent were introduced into digestion vials and kept at 150 °C for 2 h in a COD digestion unit (Aqualytic AL32). After cooling to room temperature, absorbance was measured for all samples using a UV-vis spectrophotometer at 600 nm (Turner Model 690). Soluble COD was analyzed using the same procedure after the sample's centrifugation at 10 000 rpm for 10 min. Biogas composition was determined by a portable gas analyzer (GasData GFM406) that allows measuring the volume percentages of CO_2 and CH_4 in the gas mixture.

3 Results and Discussion

3.1 Batch Co-digestion

The effects of the SW/RS ratio on the total biogas and methane production are illustrated in Fig. 1. From the results, two main findings can be highlighted: (1) an increase in the SW/ RS ratio leads to higher cumulative biogas and methane production until 30% RS and (2) the cumulative methane production was significantly different for the T1 group but not for T2 group at the 95% confidence level. However, the maximum biogas and methane yields obtained for T1 and T2 groups (SW/ RS ratio of 70:30 and 50:50) were always higher than the control. This distribution of the measured methane volume is illustrated in the boxplots of Fig. 1A. The cumulative biogas and methane productions for SW/RS 70:30 and 50:50 were 1559.99 ± 219.6 and 903.66 ± 111.0 mL, and 1236.48 ± 45.99 and 625.55 ± 55.1 mL, respectively (Fig. 1B). In terms of methane production, these results represented an increase of 41.3% and 17.8% compared with single SW digestion (525.78 ± 18.77 mL). These findings showed that the co-digestion of SW and RS enhanced AD performance at the studied mixing ratios, as well as waste utilization. Others in the literature such as Wang et al. [13] and Wu et al. [12] also acknowledged the value of co-digesting swine wastes with agricultural by-products to increase methane yield. Moreover, these results confirmed those of Mao et al. [11] who reported that a swine manure/corn straw ratio of 70:30 resulted in the best cumulative methane production. This improved performance is related to an improved C/N ratio and balanced nutrients that provide a more suitable environment for microbial communities to develop.

The daily biogas and methane production by the co-digestion of SW and RS was also calculated under different mixing ratios (Fig. 1C). In the reactors of the T1 group,

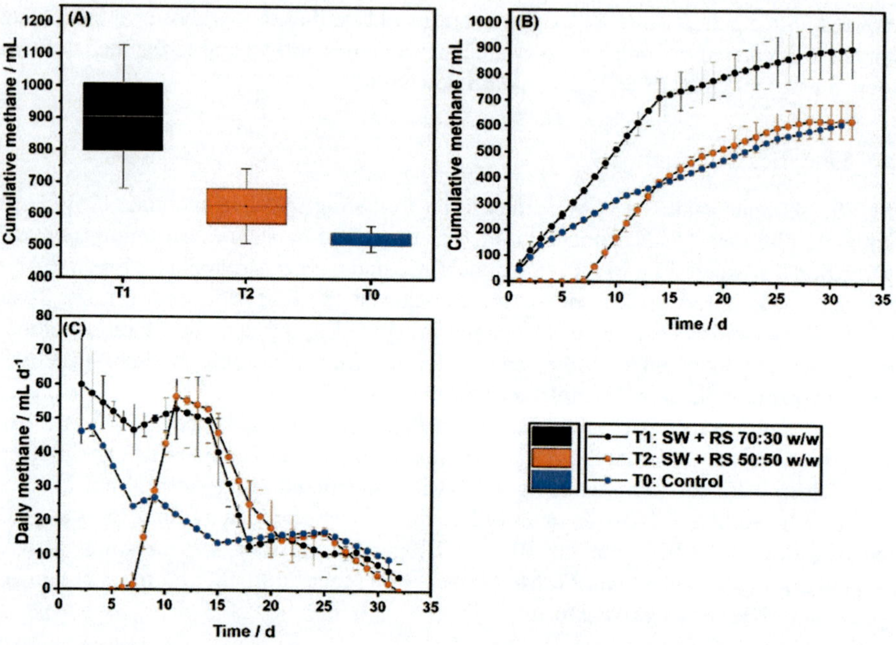

Fig. 1. Boxplots of methane production (A), Cumulative methane production (B), and Daily methane production (C) for each treatment group and control.

the peak biogas and methane yields were obtained at day 2 (154.77 and 59.93 mL d^{-1}, respectively) and remained stable until day 14. In the T2 group, a prolonged lag phase was observed when compared with the other two groups. Rapid biogas and methane production occurred only on day 7 with a peak at day 11 (88.19 and 56.53 mL d^{-1}). From day 7 to day 32, the production rates were similar between the T1 and T2 groups. The digestion of the T0 control group (100:0) produced biogas and methane earlier than group T2 but had only one peak at day 3 (103.71 and 46.99 mL d-1) before decreasing steadily, especially after day 9. These results indicated that the addition of RS could significantly delay the production of biogas because lignocellulosic materials are known to be more difficult to hydrolyze particularly when in the solid form [17].

Reaction kinetics was also studied to better characterize the degradation behavior of the substrates and enable the translation of the experimental results for continuous operation. For the first-order model, the resulting hydrolysis constant, k, increased for the T1 group in comparison to the untreated sample (T0 group). The k value for the T1 group (0.070 d^{-1}) was 1.3 times higher than that of the T0 group (0.052 d^{-1}). These hydrolysis rates were consistent with other co-digestion studies that used swine wastes. For example, Zhang et al. [18] reported k values in the range of 0.04–0.07 d^{-1} for pig manure and sewage sludge mixtures. In this model, k describes the velocities of degradation and methane production; thus, the addition of 30% RS to SW results in faster degradation and higher methane production during batch operation. On the other hand, the model failed to converge to a solution for the T2 group. As for the

Gompertz equation, the results indicated that the model can be used to describe the methane production of the tested substrate mixtures with R^2 coefficients ranging from 0.9845 to 0.9955. The maximum methane production, *A*, increased with the addition of RS until 50 wt.% while the lag-phase, λ showed the opposite behavior. In fact, *A* was the lowest (618.92 mL) and λ was the longest (7.12 days) for the T2 group. The co-digestion of SW with RS at a mixing ratio of 50 wt.% had a shorter time of effective biogas duration than other treatments. This performance indicated that the addition of higher amounts of slowly biodegradable organics such as RS made the process slower at the beginning by increasing the hydrolysis phase. For the other treatments, the calculated λ was found to be less than one day for the T1 group, while for the T0 group methane production started almost immediately after incubation. Indeed, the duration of the lag phase has been linked to the buffer capacity of co-digestion systems and the time needed for bacterial acclimation [11].

3.2 Electrochemical Pretreatment

Electrochemical pretreatment was performed to increase the bioavailability of particulate matter and thus enhance AD performance. As a result of the pretreatment process, organic matter concentration was expected to increase in the aqueous phase. Indeed, the soluble COD and the DD increased for both the T1 and T2 pretreatment groups when compared to the control. This increase could be due to the breakdown of the complex structure of polymeric substances, which released proteins and polysaccharides into the soluble phase [4]. Figure 2A shows the cumulative methane production of pretreated SW. The final methane production was higher in the T1 pretreatment group followed by the T2 group, and lower in T0 control. The most significant improvement in AD performance occurred with the T1 group digester, which had a 19% increase in the final methane production volume which was considered significant statistically at the 95% confidence level. The T2 pretreatment resulted in a final cumulative methane production volume (mL) higher by 8% when compared to the control. A paired t-test verified that this difference was also significant ($p < 0.05$). The daily biogas production after different pretreatment groups is shown in Fig. 2B and different behaviors can be identified between groups. For the T1 pretreatment group, there was little methane production in the beginning but methane started to accumulate after 15 days. Thus, the T1 pretreatment increased total methane production but delayed the time for a stable production rate probably due to the need for microorganism acclimation to the temperature conditions. For the T2 pretreatment and T0 control groups, cumulative methane production quickly reached a peak after 3 days and followed a similar pattern until day 11 ($p = 0.629$) with a second peak around day 10 for T0 control and day 11 for T2 pretreatment. After this second peak, T2 pretreatment showed superior performance in daily methane production, which was deemed significant at a 95% confidence level. These results indicated a positive correlation between methane production volume and SW disintegration. Thus, the AD performance of pretreated SW benefited from the increase of the degradable fraction of the substrate. In the literature, electrochemical pretreatment has been reported to increase methane production volumes by 13.8 - 63.4% in the AD of waste activated sludge [19], and pulp and paper sludge [7].

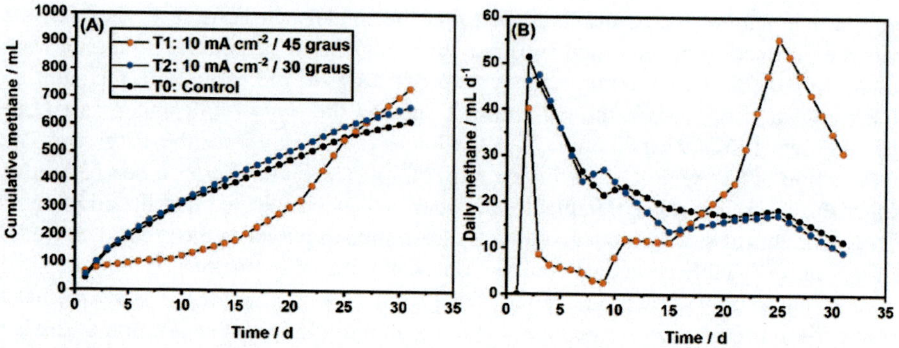

Fig. 2. Cumulative methane production (A) and Daily methane production (B) for each pretreatment group and control.

The results of cumulative methane production after pretreatment tests were also described by the first-order and Gompertz models for kinetic analysis, except for the T1 pretreatment group which did not follow the typical sigmoidal or exponential shape of production curves. For this reason, in this case, the common model structures used to simulate methane production gave unreliable results. The first-order model fitted well to the accumulated methane in T2 pretreatment and T0 control. The correlation coefficient R^2 was 0.993 for both tests. Parameter A was enhanced in the T2 pretreatment group and thus the pretreatment could enhance cumulative methane production. As for the Gompertz model, the lag phase time, λ, was very short for both groups. The final cumulative methane also improved after pretreatment but the maximum daily production, μ, did not change significantly. R^2 ranged between 0.9845 and 0.992. The times for 90% of total methane production and the effective biogas production duration were about 24 days.

4 Conclusion

Two AD enhancements techniques (co-digestion of SW with RS and electrochemical pretreatment) were successfully performed under batch conditions. In terms of co-digestion, the addition of RS improved process stability and methane yield by balancing the C/N ratio of the digestion media. Thus, a linear relationship was found between methane volume and the introduction of RS. The best results occurred in the T1 treatment group (70:30 SW/RS ratio) after 32 days of digestion time. The performance increase amounted to 41.3% in terms of methane production. The first-order and Gompertz models fitted the experimental results well and may provide valuable knowledge for the treatment of SW in continuous processes. The maximum methane production increased with the addition of 30 wt.% RS and decreased for 50 wt.% RS while the lag-phase showed the opposite behavior. Therefore, the T2 treatment group (50:50 ratio) could not achieve high methane yields due to the long lag-period. The electrochemical pretreatment alone did not have a great impact on the methane production volume during the AD of SW. Still, a linear relationship was observed between methane production and SW solubilization.

Both the T1 and T2 pretreatment groups yielded a higher final methane volume when compared to a control. For the T1 pretreatment, a long lag-time was observed due to the need for microbial acclimation. Concerning practical application, the performance of co-digestion by the addition of 30 wt.% RS was superior to the pretreatment tests. This co-digestion allows the utilization of co-substrates that might be unsuitable for agronomic valorization. Moreover, pretreatment methods are not highly recommended mainly due to higher energy consumption. In summary, this study offers useful practical insights into the biogas production from SW and provides information for the establishment of a cost-effective treatment process in real conditions.

Acknowledgments. The authors would like to thank the Portuguese Foundation for Science and Technology (FCT, Portugal) for Ph.D. grant no. SFRH/BDE/111878/2015 (G. Lourinho).

References

1. Pavan, P., Bolzonella, D., Battistoni, E., Cecchi, F.: Anaerobic co-digestion of sludge with other organic wastes in small wastewater treatment plants: an economic considerations evaluation. Water Sci. Technol. **56**, 45–53 (2007)
2. Li, Y., Chen, Y., Wu, J.: Enhancement of methane production in anaerobic digestion process: a review. Appl. Energy **240**, 120–137 (2019)
3. Wu, J., Hu, Y., Wang, S., Cao, Z., Li, H., Fu, X.-M., Wang, K., Zuo, J.: Effects of thermal treatment on high solid anaerobic digestion of swine manure: enhancement assessment and kinetic analysis. Waste Manag. **62**, 69–75 (2017)
4. Yu, T., Deng, Y., Liu, H., Yang, C., Wu, B., Zeng, G., Lu, L., Nishimura, F.: Effect of alkaline microwaving pretreatment on anaerobic digestion and biogas production of swine manure. Sci. Rep. **7**, 1668 (2017)
5. Zhen, G., Lu, X., Li, Y.Y., Zhao, Y.: Combined electrical-alkali pretreatment to increase the anaerobic hydrolysis rate of waste activated sludge during anaerobic digestion. Appl. Energy **128**, 93–102 (2014)
6. Astals, S., Nolla-Ardèvol, V., Mata-Alvarez, J.: Thermophilic co-digestion of pig manure and crude glycerol: Process performance and digestate stability. J. Biotechnol. **166**, 97–104 (2013)
7. Veluchamy, C., Raju, V.W., Kalamdhad, A.S.: Electrohydrolysis pretreatment for enhanced methane production from lignocellulose waste pulp and paper mill sludge and its kinetics. Bioresour. Technol. **252**, 52–58 (2018)
8. Veluchamy, C., Raju, V.W., Kalamdhad, A.S.: Prerequisite – an electrohydrolysis pretreatment for anaerobic digestion of lignocellulose waste material. Bioresour. Technol. **235**, 274–280 (2017)
9. Safavi, S.M., Unnthorsson, R.: Enhanced methane production from pig slurry with pulsed electric field pre-treatment. Environ. Technol. (UK) **39**, 479–489 (2018)
10. Jiménez, J., Guardia-Puebla, Y., Cisneros-Ortiz, M.E., Morgan-Sagastume, J.M., Guerra, G., Noyola, A.: Optimization of the specific methanogenic activity during the anaerobic co-digestion of pig manure and rice straw, using industrial clay residues as inorganic additive. Chem. Eng. J. **259**, 703–714 (2015)
11. Mao, C., Zhang, T., Wang, X., Feng, Y., Ren, G., Yang, G.: Process performance and methane production optimizing of anaerobic co-digestion of swine manure and corn straw. Sci. Rep. **7**, 9379 (2017)

12. Wu, X., Yao, W., Zhu, J., Miller, C.: Biogas and CH4 productivity by co-digesting swine manure with three crop residues as an external carbon source. Bioresour. Technol. **101**, 4042–4047 (2010)
13. Wang, G., Gavala, H.N., Skiadas, I.V., Ahring, B.K.: Wet explosion of wheat straw and codigestion with swine manure: effect on the methane productivity. Waste Manag. **29**, 2830–2835 (2009)
14. Hafner, S.D., Koch, K., Carrere, H., Astals, S., Weinrich, S., Rennuit, C.: Software for biogas research: tools for measurement and prediction of methane production. SoftwareX. **7**, 205–210 (2018)
15. Hofmann, J., Peltri, G., Sträuber, H., Müller, L., Schumacher, B., Müller, U., Liebetrau, J.: Statistical interpretation of semi-continuous anaerobic digestion experiments on the laboratory scale. Chem. Eng. Technol. **39**, 643–651 (2016)
16. APHA/AWWA/WEF: Standard Methods for the Examination of Water and Wastewater. Stand. Methods. 541 (2012)
17. Saady, N.M.C., Massé, D.I.: Starting-up low temperature dry anaerobic digestion of cow feces and wheat straw. Renew. Energy. **88**, 439–444 (2016)
18. Zhang, W., Lang, Q., Wu, S., Li, W., Bah, H., Dong, R.: Anaerobic digestion characteristics of pig manures depending on various growth stages and initial substrate concentrations in a scaled pig farm in Southern China. Bioresour. Technol. **156**, 63–69 (2014)
19. Yu, B., Xu, J., Yuan, H., Lou, Z., Lin, J., Zhu, N.: Enhancement of anaerobic digestion of waste activated sludge by electrochemical pretreatment. Fuel **130**, 279–285 (2014)

Effect of Combined Alkali-Thermal Pretreatment on Methane Potential from BMP of Date Palm Empty Fruit Bunch

Nabila Lahboubi[1](✉), Ikram Naim[2], Sanae Habchi[2], Azzouz Essamri[1], and Hassan El Bari[2]

[1] Laboratory of Organic Chemistry, Catalysis and Environment, Faculty of Sciences Kenitra, University Ibn Tofail, Kenitra, Morocco
nabila.lahboubi@uit.ac.ma

[2] Laboratory of Electronic Systems, Information Processing, Mechanics and Energetics, Faculty of Sciences Kenitra, University Ibn Tofail, Kenitra, Morocco

Abstract. This study focuses on evaluating the effect of combined alkali-thermal pretreatment on the methane potential of the date palm empty fruit bunch (EFB). For this purpose, the Biochemical Methane Potential (BMP) tests were conducted with substrate:inoculum ratio of 2, the alkali pretreatment was tested with 6% (w/w) (ratio weight of NaOH/weight of VS) and with the thermal condition at 50 °C for 10 min. The anaerobic digestion experiments were conducted at mesophilic conditions (37 °C). The combined alkali-thermal pretreatment result (144 NmL/gVS) was better than the alkali pretreatment one (132 NmL/gVS), knowing that the methane potential for the control test was only 118.5 NmL/gVS. Furthermore, the kinetic study was applied for all experimental results to explore the best fitting and process parameters (Modified Gompertz, Logistic function Transference function and First-order function). The Transference function corresponds to the higher μ and R^2, which allows concluding that this model is the more adequate to the kinetic modeling fitting.

Keywords: Combined alkali-thermal pretreatment · Date palm empty fruit bunch · Biochemical methane potential · Anaerobic digestion · Kinetic modeling

1 Introduction

Date fruits (Phoenix dactylifera L.) contains an essential nutrients for human nutrition, with high rates of carbohydrates, dietary fiber and minerals [1]. Date palm fruit represents proximately 82.25% of the total world production in the Arab countries (FAOSTAT, 2009). Agriculture waste that are produced from one date palm tree were about 20 kg of dry leaves yearly [2]. These agricultural wastes are in general sent to landfill and used by rural household for home indoor heating. The palm waste can be also burned at random way causing several environmental problems [3].

J. R. da Costa Sanches Galvão et al. (Eds.): *Proceedings of the 1st International Conference on Water Energy Food and Sustainability (ICoWEFS 2021)*, pp. 301–310, 2021.
https://doi.org/10.1007/978-3-030-75315-3_34

Anaerobic digestion (AD) is one of methods that treat organic waste. It consists of degrading organic matter contains in waste to produce two main products: biogas and digestate [4]. AD is described in four phases of degradation: Hydrolysis, Acidogenesis, Acetogenesis and Methanogenesis. The biogas is considered as a source of clean and renewable energy, which consist mainly of methane and carbone dioxide [5]. The content of methane in biogas depends on waste characterization and process technology. The digestate is a by-product of AD, and rich of organic matters that did not fully degrade in AD process. It is used in agriculture as a fertilizer. The temperature condition for AD is either mesophilic (37 °C) or thermophilic (55 °C) to provide optimal working process [6].

Biochemical methane potential (BMP) is most used test that determine the methane potential and biodegradability from AD. The design of full-scale AD plants could be used from the BMPs of the substrates and their specific organic loads, which can define the size of the digesters and possibilities of exploiting the produced biogas [6]. The methane potential of the substrate is expressed as the cumulative methane production per mass of volatile solids added or chemical oxygen demand (COD) added.

Palm waste is considered as a substrate of lignocellulosic nature due to its cellulose (30.2 ± 1.2%) and hemicellulose (14.8 ± 1.2%) content [7]. The lignocellulosic substrates could be used as biomass for energy conversion using AD. A previous study of producing methane from EFB of date palm using a fed-batch AD under mesophilic conditions results a methane yield of 118.5 NmL/gVS with an organic loading rate of 0.27 gVS/L.d [3].

Ensuring the proper functioning of the AD, the stability of the process and the potential improvement of biogas production are still the main objectives of pretreatment studies. So identifying suitable pretreatment techniques appropriate for Palm waste as lignocellulosic biomass allows an efficient degradation of the biomass in order to eliminate lignin and ensures carbohydrate accessibility to enzymes [8]. Basically, the types of pretreatments for lignocellulosic materials are numerous, but not all pretreatment methods can be used to treat agriculture waste that are produced from date palm tree [9].

In particular, the literature considers alkaline pretreatment as one of the most powerful methods to solubilize lignin and hemicelluloses for this type of substrate [10]. And to further develop and improve methane yield, this study focuses the evaluation of the effect of combined alkali thermal pretreatment at low temperature as an important strategy with advantages for both environmental conservation and renewable energy production. The use of a high concentration NaOH pretreatment (usually with 6–20% NaOH) at relatively low temperatures (< 100 °C) can be very effective in reducing the crystallinity of cellulose, thus improving the biological conversion of lignocelluloses [11].

The kinetic modeling is used recently more in literature to describe the methane production from AD [3, 12, 13]. The bacterial growth rate was predicted using Gompertz equation, which contained mathematical parameters that did not have biological meaning [14]. The kinetic equations have being modified from many authors in order to make the mathematical parameters have a biological meaning. The Gompertz equation is modified in order to switch the mathematical to biological parameters, where new parameters have defined: the maximum value reached, the maximum specific growth rate and the lag time [15].

The objective of this study is to explore the methane potential using combined alkali-thermal pretreatment on date palm waste: empty fruit bunch. The alkali pretreatment used in this study was sodium hydroxide (NaOH), with a ratio weight of NaOH/weight of VS of 6% (w/w) and the thermal pretreatment was conducted at 50 °C for 10 min. The kinetic study was applied to all experimental results to describe adequate fitting. The four kinetic models are: Modified Gompertz, Logistic function Transference function and First-order function.

2 Material and Methods

2.1 Substrate Preparation

The substrate was collected from Valley Ziz, province Tinghir, Morocco [3]. The Total Solid (TS) was determined by drying the sample for 24 h at 105°C. The ashes were determined by burning the dried sample at 550°C for 2 h. The Volatile Solid (VS) was calculated by the difference between TS and ash. The moisture was determined according to the APHA standard method [16]. The results of the substrate characterization are shown in Table 1.

Table 1. Physico-chemical characterization of the inoculum and substrate

Parameters	Inoculum	Raw substrate	Pretreated NaOH ratio 6% (w/w)
pH	8.03	-	8.6
TS (g/kg)	23.63	867.23	170.06
VS* (g/kg)	8.86	691.39	24.22
TS (%)	2.35	86.74	16.98
VS (%TS)	37.50	91.91	83.89

For alkali pretreatment, 6% NaOH (w/w) concentration were added to substrate:inoculum ratio (SIR) 2 for AD experiment. The samples were mixed to get good homogenization conditions using magnetic stirring for 10 min, and then a sample was incubated for 10 min at 50 °C to evaluate the effect of combined alkali-thermal pretreatment. The prepared substrates were fed directly into the digester to avoid having other effect. The digesters were numbered using the Table 2.

2.2 Inoculum

The inoculum was provided from a full scale digester of the vinasse industry, working under mesophilic conditions (37 °C). The inoculum was stored at 4 °C until use. The pH of inoculum was 8.03 and the TS and VS content were 2.35 and 37.5%, respectively. The inoculum used was chosen based on its high methanogenic activity resulted from the laboratory teams [17–19].

Table 2. Digesters numbering

Digesters	SIR	NaOH concentration	Incubation time
D1	2	-	-
D2	2	6% NaOH	-
D3	2	6% NaOH	10 min (at 50°C)

2.3 Experimental Set-Up and Method

Several 1 L Duran bottles (2) were used as batch AD digesters, each with a working a volume of 0.8 L (Fig. 1). Each digester was sealed with a rubber bung with two outlets, first one was a closed port for sampling/feeding (5) and the other (6) was connected to a bubbler (7) containing NaOH solution. The biogas produced passed to NaOH solution (6N) to separate the CO_2 from the biogas. The methane produced was then measured using water displacement method. The bottles were shacked manually twice a day to facility the contact between bacteria and substrate [20]. The produced methane volume was standardized under ambient conditions [21]. At the beginning of the experiment, all digesters were filled with 0.8 L of inoculum, and flushed with N_2 to ensure anaerobic conditions. All digesters were placed in a water-bath (1) controlled at an operating temperature of 37 °C (4). All experiments were run at triplicate.

The methane potential produced under experiment conditions is shown in the Eq. 1 [22] which is calculated by the volume of methane produced divided by the quantity of VS added (mL/gVS):

$$PCH_4\ (mL/gVS) = \frac{VCH_4(mL)}{added\ VS(gVS)} \tag{1.}$$

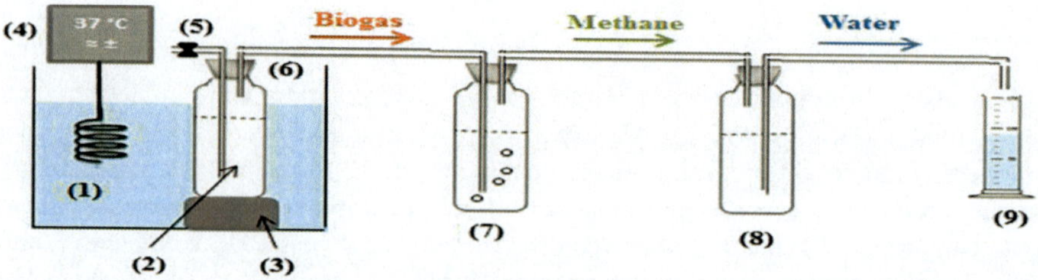

Fig. 1. Experimental BMP set-up. 1: water bath; 2: digester; 3: support; 4: thermo controller; 5: sample inlet/outlet; 6: biogas outlet; 7: NaOH solution; 8: gasmeter; 9: graduated tube

2.4 Kinetic Models

In this study, four kinetic equations were applied to fit the experimental data: modified Gompertz, Transference function, Logistic function and First-order. These models are used to fit the experimental data and determine the main kinetics parameters. The theoretical data were obtained by adjusting the experimental methane production (NmL/gVS) for each experiment. The modified Gompertz, Logistic function and Transference function are described using Eqs. (2), (3) and (4), respectively [22]. These models could estimate the methane production rate and lag phase. The First-order model (5) was calculated by the relationship between the concentration of VS and methane production; however the maximum biological activity rate could not be predicted [22].

$$P(t) = A * \exp\left(-\exp\left(\frac{\mu e}{A}(\lambda - t) + 1\right)\right) \tag{2}$$

$$P(t) = \frac{A}{1 + \exp\left(\frac{4\mu}{A}(\lambda - t) + 2\right)} \tag{3}$$

$$P(t) = A*\left[1 - \exp\left(\frac{\mu*(\lambda - t)}{A}\right)\right] \tag{4}$$

$$P(t) = A * (1 - \exp(-K_0 * t)) \tag{5}$$

Where (P) represents the simulated cumulative methane production (NmL/gVS), (A) represents the simulated maximum cumulative methane production (NmL/gVS), (μ) represents the methane production rate (NmL/gVS.d), (λ) represents the lag phase time (d), (K_0) represents the apparent hydrolysis rate coefficient (d^{-1}) and (t) represents the digestion time (d).

3 Results and Discussions

3.1 pH

The process stability was controlled by the acidity of the digester with the pH parameters. The pH was controlled before the start of the experiment and after, for each digester. At the beginning of the experiment, the pH results were between 7.5 and 8.02 for all digesters. After AD experiment, the pH results were: 7.6, 7.45 and 7.85 (D1, D2 and D3; respectively). The pH were between 6.5 and 8.5, which are around the neutrality [3, 5]. As results, the experiment process was working under the standard conditions.

3.2 Effect of Pretreatment on Methane Potential

Figure 2 and Table 3 show the results of combined alkali thermal pretreatment on improving methane potential. After alkali pretreatment combined with heat treatment, the improvement in methane production in the substrate ranged from 11.4 to 21.52%. Under alkaline pretreatment conditions (6% NaOH (w/w)) incubated for 10 min at 50°C, the maximum BMP was 144 NmL/gVS improved over alkaline pretreatment (132 NmL/gVS) as well as for the control (118.5 NmL/gVS) and this indicates that the improvement in methane production was due to the improved hydrolysis rate of the substrate. Moreover, in another study of producing methane from an oil palm empty fruit bunch, alkali pretreatment was effective with an improvement of 83% after pretreatment with 8% (w/v) NaOH for 60 min [23].

Table 3. Effect of pretreatment on the methane potential and its improvement

Digester	Methane potential (NmL/gVS)	Improvement (%)
D1	118.5	-
D2	132	11.4
D3	144	21.52

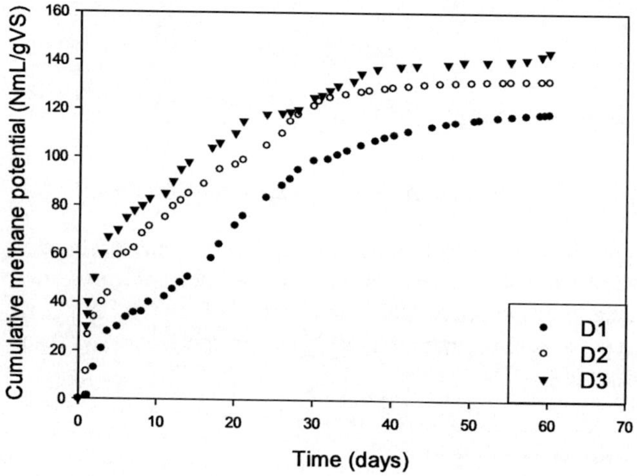

Fig. 2. Effect of alkali pretreatment on methane potential

3.3 Kinetic Study Results

The experimental data were fitted with Modified Gompertz, Eq. (2), Logistic function, Eq. (3), Transference function, Eq. (4) and First-order function, Eq. (5). The kinetic study results of these models are represented in the Table 4 and Fig. 3, whereas Fig. 4 shows the fitting accuracy of the models for all digester. The Fig. 4 has shows that all the models had best fitting in terms of visual seeing. We notice that the maximum methane potential reached in 25 days in D3 fast than the D1 and D2 which is reached in 35 days. The proposed models had shown the stability (R^2 > 90%) and error low than 15%. The lag phase of the all simulated digester is low (negligible) which show that the anaerobic bacteria consume quickly the substrate. The methane production rate increases by increasing the NaOH pretreatment condition. The parameter K_0 (apparent hydrolysis rate coefficient) increases also in simultaneous with μ from D1 to D3. The proposed improvement pretreatment had a positive effect on the methane production rate. The higher μ and R^2 are corresponding to transference function, which allow concluding that this model is the best in kinetic modeling fitting. In other kinetic study, the transference function was found also suitable for fitting experimental data [3].

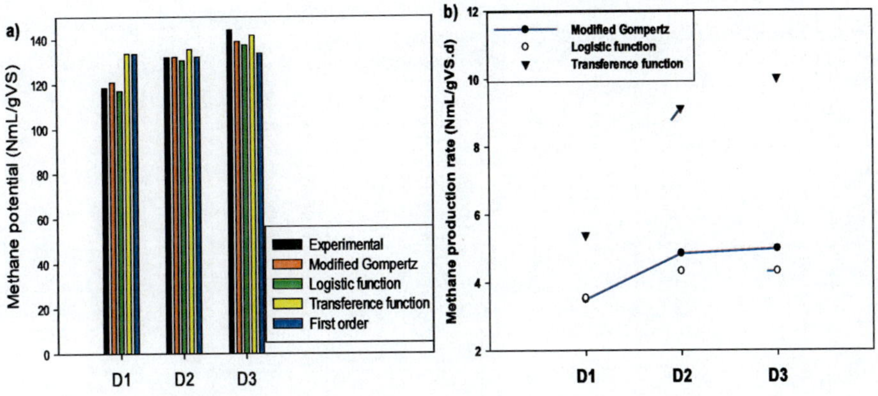

Fig. 3. Experimental and kinetic simulated results of all digesters: a) methane potential and b) methane production rate

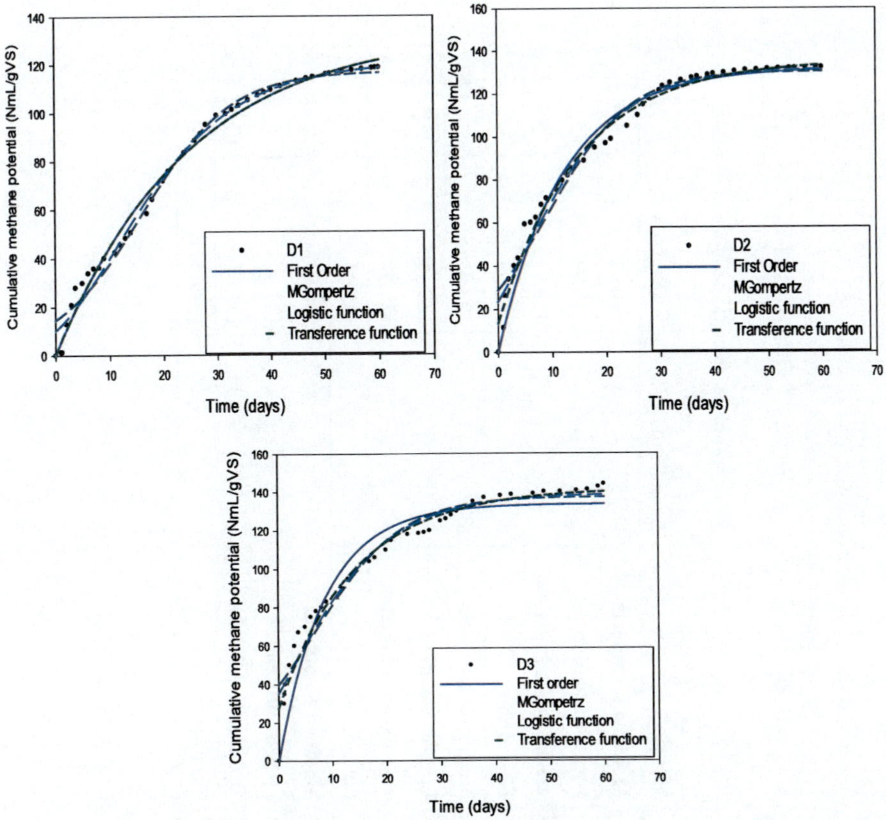

Fig. 4. Kinetic results and fitting accuracy of the models for all digesters.

Table 4. Kinetic parameters results of each digester

Digester	Models	P_{CH4} (NmL/gVS)	A (NmL/gVS)	μ (NmL/gVS.d)	λ (d)	$K_0(d^{-1})$	R^2	Error (%)
D1	**Modified Gompertz**	118.5	120.94	3.5	0	-	0.9895	2.05
	Logistic function		116.99	3.54	0	-	0.9851	1.27
	Transference function		133.63	5.39	0	-	0.9904	12.76
	First-order		133.53	-	-	0.04	0.9904	12.68
D2	**Modified Gompertz**	132	132.03	4.84	0	-	0.9747	0
	Logistic function		130.45	4.32	0	-	0.9642	1.17
	Transference function		135.52	9.12	0	-	0.9864	2.66
	First-order		132.03	-	-	0.08	0.9756	0
D3	**Modified Gompertz**	144	138.99	4.98	0	-	0.9565	3.47
	Logistic function		137.38	4.32	0	-	0.9445	4.59
	Transference function		141.67	10.02	0	-	0.9711	1.61
	First-order		133.66	-	-	0.11	0.9203	7.18

4 Conclusion

In this study, the empty fruit bunch substrate was prepared under different conditions: alkali and combined alkali-thermal pretreatment. The sodium hydroxide (NaOH) was used as alkali pretreatment, with a ratio weight of NaOH/weight of VS of 6% (w/w) and the thermal condition was at 50 °C for 10 min. The AD experiment process was working under the standard conditions. The maximum methane potential was found with the combined alkali thermal pretreatment with a yield of 144 NmL/gVS, while the control test was only 118.5 NmL/gVS, which correspond to 21.5% increasing. The kinetic study was applied to all experimental results to describe adequate fitting and kinetic parameters. The four kinetic models are: Modified Gompertz, Logistic function, Transference function and First-order function. The higher μ and R^2 are corresponding to Transference function, which allow concluding that this model is the best in kinetic modeling fitting. Moreover, the results obtained from the alkali pretreatment suggest to taking into consideration the complexity of the substrate while the evaluation of the AD process that should be carefully conducted. This study provides useful information for the alkali pretreatment on EFB biomass for producing methane in the contribution of the development society and a perspective of clean energy production in the future.

References

1. Lattieff, F.A.: A study of biogas production from date palm fruit wastes (2016)
2. Chandrasekaran, M., Bahkali, A.H.: Valorization of date palm (Phoenix dactylifera) fruit processing by-products and wastes using bioprocess technology – Review (2013)
3. Lahboubi, N., Kerrou, O., Karouach, F., Bakraoui, M., Schüch, A., Schmedemann, K., Stinner, W., El Bari, H., Essamri, A.: Methane production from mesophilic fed-batch anaerobic digestion of empty fruit bunch of palm tree. Biomass Convers. Biorefinery. (2020). https://doi.org/10.1007/s13399-020-00864-1
4. Hamraoui, K., Gil, A., El Bari, H., Siles, J.A., Chica, A.F., Martín, M.A.: Evaluation of hydrothermal pretreatment for biological treatment of lignocellulosic feedstock (pepper plant and eggplant). Waste Manag. **102**, 76–84 (2020). https://doi.org/10.1016/j.wasman.2019.10.020
5. Karouach, F., Bakraoui, M., El Gnaoui, Y., Lahboubi, N., El Bari, H.: Effect of combined mechanical–ultrasonic pretreatment on mesophilic anaerobic digestion of household organic waste fraction in Morocco. In: Energy Reports, pp. 310–314 (2020). https://doi.org/10.1016/j.egyr.2019.11.081
6. Filer, J., Ding, H.H., Chang, S.: Biochemical methane potential (BMP) assay method for anaerobic digestion research. Water **11**, 921 (2019). https://doi.org/10.3390/w11050921
7. Galiwango, E., Rahman, N.S.A., Al-Marzouqi, A.H., Abu-Omar, M.M., Khaleel, A.A.: Klason Method: An Effective Method for Isolation of Lignin Fractions from Date Palm Biomass Waste (2018)
8. Issah, A.-A., Kabera, T., Kemausuor, F.: Biogas optimisation processes and effluent quality: a review. Biomass Bioenergy. **133**, 105449 (2020). https://doi.org/10.1016/j.biombioe.2019.105449
9. Aziz, M.M.A., Kassim, K.A., ElSergany, M., Anuar, S., Jorat, M.E., Yaacob, H., Ahsan, A., Imteaz, M.A., Arifuzzaman: Recent advances on palm oil mill effluent (POME) pretreatment and anaerobic reactor for sustainable biogas production. Renew. Sustain. Energy Rev. **119**, 109603 (2020). https://doi.org/10.1016/j.rser.2019.109603

10. Millati, R., Wikandari, R., Ariyanto, T., Putri, R.U., Taherzadeh, M.J.: Pretreatment technologies for anaerobic digestion of lignocelluloses and toxic feedstocks. Bioresour. Technol. **304**, 122998 (2020). https://doi.org/10.1016/j.biortech.2020.122998
11. Mirahmadi, K., Kabir, M.M., Jeihanipour, A., Karimi, K., Taherzadeh, M.J.: Alkaline Pretreatment of Spruce and Birch to Improve Bioethanol and Biogas Production 11 (2010)
12. Bakraoui, M., El Gnaoui, Y., Lahboubi, N., Karouach, F., El Bari, H.: Kinetic study and experimental productions of methane production from UASB reactor treating wastewater from recycled pulp and paper for the continuous test. Biomass Bioenergy. 105604 (2020)
13. El Gnaoui, Y., Sounni, F., Bakraoui, M., Karouach, F., Benlemlih, M., Barz, M., El Bari, H.: Anaerobic co-digestion assessment of olive mill wastewater and food waste: effect of mixture ratio on methane production and process stability. J. Environ. Chem. Eng. **8**, 103874 (2020). https://doi.org/10.1016/j.jece.2020.103874
14. Syaichurrozi, I., Budiyono, Sumardiono, S.: Predicting kinetic model of biogas production and biodegradability organic materials: biogas production from vinasse at variation of COD/N ratio. Bioresour. Technol. **149**, 390–397 (2013) https://doi.org/10.1016/j.biortech.2013.09.088
15. Zwietering, M.H., Jongenburger, I., Rombouts, F.M.: Modeling of the Bacterial Growth Curve (1990)
16. APHA: Standard Methods for examination of water and wastewater. Washington, DC. (1989)
17. Bakraoui, M., Karouach, F., Ouhammou, B., Lahboubi, N., Gnaoui, Y.E., Aggour, M., Bari, H.E.: Kinetics study of methane production from anaerobic digestion of sludge and wastewater recycled pulp and paper. Mater. Sci. Eng. 13 (2020)
18. El Gnaoui, Y., Karouach, F., Bakraoui, M., Barz, M., El Bari, H.: Mesophilic anaerobic digestion of food waste: effect of thermal pretreatment on improvement of anaerobic digestion process. Energy Rep. **6**, 417–422 (2020). https://doi.org/10.1016/j.egyr.2019.11.096
19. Joute, Y., El Bari, H., Belhadj, S., Karouach, F., Gradi, Y., Stelte, W., Bjerre, A.B.: Semi-continuous anaerobic co-digestion of cow manure and banana waste: effects of mixture ratio. Appl. Ecol. Environ. Res. **14**, 337–349 (2016). https://doi.org/10.15666/aeer/1402_337349
20. Angelidaki, I., Alves, M., Bolzonella, D., Borzacconi, L., Campos, J.L., Guwy, A.J., Kalyuzhnyi, S., Jenicek, P., van Lier, J.B.: Defining the biomethane potential (BMP) of solid organic wastes and energy crops: a proposed protocol for batch assays. Water Sci. Technol. **59**, 927–934 (2009). https://doi.org/10.2166/wst.2009.040
21. Bakraoui, M., Karouach, F., Ouhammou, B., Aggour, M., Essamri, A., El Bari, H.: Kinetics study of the methane production from experimental recycled pulp and paper sludge by CSTR technology. J. Mater. Cycles Waste Manag. (2019). https://doi.org/10.1007/s10163-019-00894-6
22. Blasius, J.P., Contrera, R.C., Maintinguer, S.I., Alves de Castro, M.C.A.: Effects of temperature, proportion and organic loading rate on the performance of anaerobic digestion of food waste. Biotechnol. Rep. **27**, e00503 (2020). https://doi.org/10.1016/j.btre.2020.e00503
23. Nieves, D.C., Karimi, K., Horváth, I.S.: Improvement of biogas production from oil palm empty fruit bunches (OPEFB). Ind. Crops Prod. **34**, 1097–1101 (2011). https://doi.org/10.1016/j.indcrop.2011.03.022

Size Screening and Washing to Improve Pellet Quality Solid Biofuel from Plum Tree Pruning Biomass

Luis Royano Barroso[1](✉), Ana Parralejo Alcobendas[1], Manuel Acevedo Serrano[1], Juan Cabanillas Patilla[1], Juan Félix González González[2], and Jerónimo González Cortés[1]

[1] Centre for Scientific and Technological Research of Extremadura (CICYTEX), Consejería of Economía, Ciencia and Agenda Digital. Junta de Extremadura, 06187 Badajoz, Guadajira, Spain
luis.royano@juntaex.es

[2] Escuela de Ingenierías Industriales, Universidad de Extremadura. Campus Universitario, Av. de Elvas, s/n, 06006 Badajoz, Spain

Abstract. Decarbonisation is at the core of energy and climate policy in the European Union. The European Green Deal reasserts the European Commission´s commitment to tackling climate and environmental challenges to reach a carbon-neutral society by 2050. Biomass is part of our low-carbon future. The use of biomass from pruning fruit trees to obtain biofuel pellets can be an interesting alternative in Southern Europe, generating added value in rural areas and favoring the development of the circular economy and green economy. Sustainable use of biomass replaces fossil fuels. Size fractionation and washing proceed to demineralization of the plum wood residues with high efficiencies. These biomass pretreatments reduce the ash content in the wood biomass with bark or debarked in water treatment and increase and reduce the ash content in different size fractions. The ash content has been studied in different treatments for plum biomass (with bark, without bark and fractionated at 3.15 and 8 mm). In turn, a washing treatment was carried out at room temperature with different times (5, 30, 60 and 180 min), together with a washing at a temperature of 45 ºC for the 8 mm fractionation.

The results obtained have confirmed that the debarking, fractionation and 30 min washing treatments are the most efficient for reducing the ash content of plum biomass.

Keywords: Wood washing · Size separation · Pellet quality · Reduce ash

1 Introduction

Decarbonisation is at the core of energy and climate policy in the European Union. The European Green Deal reasserts the European Commission's commitment to tackling climate and environmental challenges to reach a carbon-neutral society by 2050. Biomass is part of our low-carbon future. Many biomass technologies have already made

J. R. da Costa Sanches Galvão et al. (Eds.): *Proceedings of the 1st International Conference on Water Energy Food and Sustainability (ICoWEFS 2021)*, pp. 311–317, 2021.
https://doi.org/10.1007/978-3-030-75315-3_35

significant progress towards commercial applications. There are major opportunities to enhance the role of biomass and bioenergy as part of a circular bioeconomy. The growing global demand for fossil products and fuels has contributed to the increasing unsustainability of non-renewable resources. Biomass is now being given an important role in the blueprints to lessen climate change, and represents an important source of renewable and sustainable fuel for energy and bioproducts production. These biomass resources can be converted into different forms of fuels through chemical and mechanical treatments. Size fractionation can separate fractions with high and low ash contents. Biomass low ash fractions could dedicate to obtain high quality pellets. The water washing of biomass also allows the decrease of ash content, a process after which it must be dried and the profitability of said process evaluated. Size fractionation and water washing of plum tree pruning biomass can improve biofuel characteristics and it is possible to obtain high quality pellets.

Biomass needs to be processed and handled in an efficient manner. However, biomass resources are difficult to handle, transport, store and utilize in their original form due to their higher moisture or ash content, low bulk density, and irregular shape and sizes [1]. Pelletization is a suitable and established technique for upgrading raw biomass to consistent and high quality feedstock with uniform shape or size for making biofuel and bioproducts in order to meet specifications of the conversion facilities, and safe and economical handling and storage [2].

Crop residues are considered to be potential fuels for power generation. The fuel quality of herbaceous species was much lower compared to wood due to higher ash content and high percentage of N, Si, K and Na [3]. Researchers have explored various types of technologies to reduce ash content and other inorganic constituents in biomass prior to the utilization in the conversion processes [4]. The methods utilized can be classified as mechanical (physical) and chemical pretreatment techniques. Mechanical pretreatment techniques include leaching by water, acid or chelating agent. In this study, size fractionation and le3aching by water pretreatments are used with plum tree pruning biomass.

Bioenergy production represents one of the opportunities for agriculture and rural development, providing alternative options for using agriculture, forestry and wood processing residues [5]. The large amount of biomass obtained from pruning trees could be used as a source of energy [6]. The use of biomass from pruning fruit trees to obtain biofuel pellets can be an interesting alternative in Southern Europe, generating added value in rural areas and favouring the development of the green and circular economy. Nowadays, residues are often directly burned or mulched in the field. Pome and stone fruit trees have to be cleared when they reach a senescence stage and fruit production is no longer economic due to yield decline or change of cultivar, The senescence stage is reached after 12–15 years, although longer periods are possible. Pruning and clearance wood's main advantage is that these residues can be used without interfering with fruit production, because it is already part of the food production chain [7].

The washing of biomass allows the decrease in its environmental impact during its combustion in a pellet stove although the wood species plays an important role [8]. Sieving for size classification yielded fractions elevated in ash content with the smaller size fractions (<1.9 mm) containing the highest concentrations of ash (2.3–6.0 wt%) [9].

Sustainable use of biomass replaces fossil fuels. Size fractionation and washing proceed to demineralization of the plum wood residues with high efficiencies. These biomass pretreatments reduce the ash content in the wood biomass with bark or debarked in water treatment and increase and reduce the ash content in different size fractions. Three treatments have been studied in the biomass, separation into different fractions, washing, and both fractionation and washing treatment.

This study aims at studying the leaching at room temperature of plum pruning biomass. Wood chips were washed using distilled deionized water. Biofuel analyses were done on all washed biomass obtained while the non-treated raw material was also analyzed in order to calculate the change in the ash content. Separation step consisted in a dry separation. A sieve shaker (Haver UWL 400, Germany) with various-sized sieves was used to remove the majority of biomass debris and group the collections in a series of sizes ranges.

2 Materials and Methods

2.1 Biomass

The plum trees pruning biomass material was taken from a biomass logistic centre in Badajoz-Extremadura, Southern Spain. A representative sample of branches of 6 cm diameter was selected in this study. After cutting, some branches were debarked. The two fractions of plum biomass were chipped. The moisture content of the wood chips was 14%.

2.2 Experimental Procedures

All experiments of wood chips without bark and wood chips with bark samples were carried out with three replicates. The two fractions (with and without barked), wood chip samples of 200 g were batch-leached at room temperature (22 °C) with distilled deionized water (10 l kg^{-1} leaching ratio). Leaching extractions were carried out for different periods: 5 min, 30 min, 60 min and 3 h. Debarked fraction was separated in different sizes (8 mm, 3.15 mm) with a sieve shaker applying vibration for 15 min.

2.3 Statistical Analysis

The results obtained were analyzed using univariate analysis of variance for a bifactorial ANOVA, by which, the differences among biofuel types were tested. The statistical treatment was carried out with spss 20.0 program.

3 Results and Discussion

The ash content of bark was higher than that of raw material with bark and debarked raw material; these results are similar to the results of previous studies [9, 10]. Because of their higher ash content, bark material was placed in a poorer class than woodchips from raw material with bark and debarked raw material.

Washing or leaching proceeds to demineralization of the wood with very high efficiencies of some minerals as potassium, sodium, calcium, sulfur and phosphor [11, 12]. Schmidt *et al.* found ash mean reduction rate in washed biomass of 25%, 32% and 38%, for fir, beech and oak respectively. Water leaching reduced the ash content of debarked and delimbed Douglas fir by 40% and Miscanthus biomass by 39% [13]. The ash and mineral concentrations reduction rate of grasslands biomass were 37.5% for ash, 23% for N, 38.3% for S, 83.5% for K, 32% for Mg and 92% for Cl [14]. The mean ash content of plum feedstock in the present study was 2.39% with bark and 0.95% debarked. Water leaching treatments reduced the ash content in the wood chips with bark between 12% and 28%, and between 5% and 43% for wood chips debarked (Table 1). Longer treatment of 2 days or treatments with more than a period of washing didn´t show a clear tendency of reduction even an increase in ash was observed that it could be due to the loss of solubilized biomass.

Table 1. Average ash content (%). Plum biomass residues with different pretreatments: size fractionation and washing.

	No washing	5' washing	30' washing	60' washing	3h washing
With bark no screening	2.39	1.73	1.76	1.91	2.09
Debarked no screening	0.95	0.94	0.92	0.91	0.92
3.15 mm Fraction	0.75	0.73	0.70	0.70	0.74
8 mm Fraction	0.71	0.72	0.61	0.79	0.75
8 mm Fraction 45 °C washing	0.71	0.61	0.69	0.58	0.52

Plum woodchips debarked of 3.15 mm and 8 mm fractions allow to obtain woodchips and pellets with A1 quality requirements, while plum woodchips debarked complies only with A2 quality threshold (Fig. 1).

Table 2 shows how there are significant differences in the ash content for all plum samples, except for 3.15 and 8 mm fractions between them. It is important to highlight that the best result in terms of reducing the ash content is achieved with the fractionation treatment at 8 mm with washing at 45 °C, obtaining a biomass of the highest quality (A1) according to the ISO 17225 standard.

In Table 3 it can be seen how there are significant differences in the ash content between all the wash treatments. It is important to note that the best average results are achieved with a 30-min biomass wash.

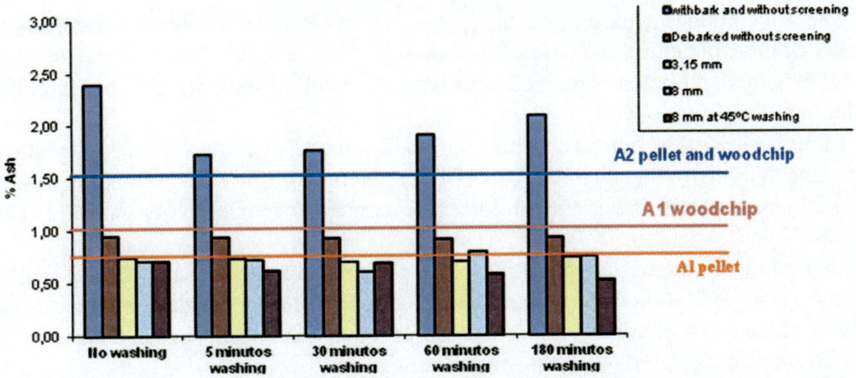

Fig. 1. Influence of the size fractionation and washing treatment duration in the variation of ash content and threshold for the different qualities of the ISO 17225 standard for solid biofuel.

Table 2. Result of statistical analysis of ash content (%)Plum biomass residues with different pretreatments: size fractionation.

	% Ash mean and statistical significance
With bark no screening	1,97 a
Debarked no screening	0,93 b
3.15 mm Fraction	0,73 c
8 mm Fraction	0,71 c
8 mm Fraction 45 °C washing	0.62 d

Table 3. Result of statistical analysis of ash content (%)Plum biomass residues with different pretreatments: washing

	% Ash mean and statistical significance
No washing	1,20 a
5 min washing	1,03 d
30 min washing	1,00 e
60 min washing	1,07 c
180 min washing	1,12 b

4 Conclusions

The results obtained in the quality parameters for woodchips from plum tree pruning biomass show that it is possible to obtain high grade woodchips and pellets in Southwestern Spain conditions with size fractionation and water washing pretreatments.

The ash content increased with the presence of bark in the woodchips, affecting the quality of the woodchips and pellets as biofuel.

Woodchips and pellets obtained with biomass without bark are able to reach high pellet qualities (A1).

Liquid water extraction on several samples of plum tree pruning biomass was shown to be able to improve fuel characteristics from reduction of ash. Size separation pretreatment allows to obtain fractions with lower ash content for high quality woodchips and pellets.

Regarding the pretreatment of the plum pruning biomass, it can be said that the debarking of the biomass and its screening to obtain different fractions reduce the percentage of ashes. In this study, it has been observed that the results in ash content do not vary with the sieves used (3.15 mm and 8 mm).

All the washing treatments carried out in this study have considerably reduced the percentage of ashes in the plum pruning biomass, the 30-min treatment being the most effective.

Acknowledgment. The authors thank the financial support of the MESOCEX project (CICYTEX - Junta de Extremadura – co-financed with FEDER funds) and IDERCEXA POCTEP Project.

JUNTA DE EXTREMADURA
Consejería de Economía, Ciencia y Agenda Digital

CICYTEX
CENTRO DE INVESTIGACIONES CIENTÍFICAS Y TECNOLÓGICAS DE EXTREMADURA

IDERCEXA

Interreg
España - Portugal

Unión Europea
Fondo Europeo
de Desarrollo Regional
"Una manera de hacer Europa"

References

1. Kaliyan, N., Morey, R.V.: Factors affecting strength and durability of densified biomass products. Biomass Bioenergy **33**(3), 337–359 (2009)
2. Zhijia, L., Xing'e, L., Benhna, F., Zehui, J., Zhiyong, C., Yan, Y.: The properties of pellets from mixing bamboo and rice straw. Renewable Energy **55**, 1–5 (2013)
3. Olanders, B., Steenari, B.M.: Characterization of ashes from wood and straw. Biomass Bioenergy **8**, 105–115 (1995)
4. Clarke, S., Preto, F.: Biomass burn characteristics. Food and Rural Affairs. Ontario. Canada, Ministry of Agriculture (2011)

5. Scarlat, N., Blukdea, V., Dallemand, J.F.: Assessment of the availability of agricultural and forest residues for bioenergy production in Romania. Biomass Bioenergy **35**, 1995–2005 (2011)
6. Pérez-Arévalo, J.J., Velázquez-Martí, B.: Evaluation of pruning residues of Ficus benjamina as a primary biofuel material. Biomass Bioenergy **108**, 217–223 (2018)
7. Winzer, F., Kraska, T., Elsenberger, C., Kötter, T., Pude, R.: Biomass from fruit trees for combined energy and food production. Biomass Bioenergy **107**, 279–286 (2017)
8. Smidt, G., Trouve, G., Leyssens, G., Schönnenbeck, C., Genevray, P., Cazier, F., Dewaele, D., Vandelbicke, C., Faivre, E., Danance, Y., Le Dreff-Lorimier, C.: Wood washing: influence on gaseous and particulate emissions during wood combustion in a domestic pellet stove. Fuel Process. Technol. **174**, 104–117 (2018)
9. Filbakk, T., Raida, J., Nurmi, J., Hoibo, O.: The effect of bark content on quality parameters of Scots pine (*Pinus sylvestris* L.) pellets. Biomass Bioenergy **35**, 3342–3349 (2011)
10. Lerma-Arce, V., Oliver-Villanueva, J.V., Segura- Orenga, G.: Influence of raw material composition of Mediterranean pinewood on pellet quality. Biomass Bioenergy **99**, 90–96 (2017)
11. Deng, I., Zhang, T., Che, D.: Effect of water washing on fuel properties, pyrolisis and combustion characteristics, and ash fusibility of biomass. Fuel Process. Technol. **106**, 712–720 (2013)
12. Eom, I.Y., Kim, K.H., Kim, J.Y., Lee, S.M., Yeo, H.M., Choi, I.G., Choi, J.W.: Characterization of primary thermal degradation features of lignocellulosic biomass after removal of inorganic metals by diverse solvents. Bioresour. Technol. **102**, 3437–3444 (2011)
13. Yu, C., Thy, P., Wang, L., Anderson, S.N., VanderFheynst, J.S., Upadyaya, S.K., Jenkins, B.M.: Influence of leaching pretreatment on fuel properties of biomass. Fuel Process. Technol. **128**, 43–53 (2014)
14. Lacey, A.J., Aston, J.E., Delezene-Briggs, K.: Biomass fractionation methods for ash removal: sieving, anatomical, and air classification. In: 37th Symposium on Biotechnology for Fuels and Chemicals. San Diego, CA. USA (2015)

Hydrogen Peroxide Pretreatment of Lignocellulosic Biomass (Pepper Plant and Eggplant) for Anaerobic Digestion

K. Hamraoui[1](✉), J. A. Siles[2], A. F. Chica[2], M. Ángeles Martín Santos[2], and H. El Bari[1]

[1] Renewable Energy and Environment Laboratory, Faculty of Sciences Ibn Tofail, Kénitra, Morocco

[2] Department of Inorganic Chemistry and Chemical Engineering, University of Cordoba, Campus Universitario de Rabanales, Edificio Marie Curie (C-3). Ctra.N-IV, km 396.14071, Cordoba, Spain

Abstract. Since many years the anaerobic digestion (AD) process has been confirmed as an effective technique for converting organic material into methane-rich-biogas. Nonetheless, the complex lignocellulosic biomass (LB) structure impedes the ability of microorganisms to degrade and convert these compounds into biogas through AD. In this context, a pretreatment step is necessary to enhance LB degradability in order to achieve higher biogas production rate and yield. H2O2 of LB (such as pepper plants and eggplant) is a promising approach to increasing the production of biogas in AD among the different pretreatment methods.

This study reports the effects of hydrogen peroxide pretreatment (H2O2) on the C/N ratio and volatile fatty acids (VFA) of pepper plant and eggplant.

The optimal concentration of H2O2 for the use of pretreated pepper plant and eggplant plant was 3% and 1%,respectively reaching 709,5 ± 3,5g N/kg dry matter and 509,3 ± 2,93g N/kg dry matter.

The results showed that the concentration of soluble total nitrogen (STN) of LB increased from 46,0g ± 0,98 g N/kg dry matter to 61,5 ± 1,41 g N/kg dry matter of pepper plant at 3% (H_2O_2) and from 45,5 ± 0,48g N/kg to 60,7 ± 1,11 g N/kg dry sample of eggplant. Finally, the maximum VFA concentration was reached at 2% of H_2O_2 in pepper plant (160% higher than the value observed in the raw material) and at 3% of eggplant (238% VFA higher than the original substrate).

Keywords: Anaerobic digestion · Lignocellulosic biomass · Hydrogen peroxide pretreatment

1 Introduction

Increasing energy demand and issues related to non-renewable energy resources have led to numerous searches for alternative sources of energy [1]. Among energy sources,

J. R. da Costa Sanches Galvão et al. (Eds.): *Proceedings of the 1st International Conference on Water Energy Food and Sustainability (ICoWEFS 2021)*, pp. 318–324, 2021.
https://doi.org/10.1007/978-3-030-75315-3_36

anaerobic digestion of organic matter has been documented over the past decades as an appropriate method of processing residual organic matter that produces biogas which can be used directly or converted to heat and electricity for local energy needs [2].

Lignocellulosic biomass (LB), one of the most important renewable resources available nowadays, is considered to be an alternative energy source, accounting for approximately 10% of the total energy obtained from biomass. However, the valorization process is limited by its complex composition and structure [3], and anaerobic digestion efficiency of LB is generally low [4]. Therefore, pretreatment of the lignocellulosic structure is an essential step in biomass conversion processes [5]. This step changes the lignocellulosic structure and increases the accessibility of enzymes to cellulose during hydrolysis.

To create an efficient, economic and sustainable chain of conversion of biomass, the choice of the best pretreatment technology and operational conditions is crucial.

Pretreatment technologies of LB, including physical, chemical, physical-chemical and biological methods, have been investigated and developed with the aim of modifying and/or removing hemicellulose and lignin, improving the access of enzymes to the cellulose structure and, as a result, increasing the availability of fermentable sugar after enzymatic hydrolysis. The main focus has been paid to chemical pretreatment because, in general, it is easier than the other processes.

The pretreatment methods in anaerobic digestion are physical, chemical and biological pretreatments [6, 7, 8, and 9]. Physical processes include mechanical (mechanical jet, high pressure homogenization, or mechanical ball mill), thermal and ultrasonic pretreatments. Chemical pretreatments include alkali processing, organosolv process, wet oxidation, ozonolysis pretreatment and acid hydrolysis), while biological processed are based in the use of microorganisms. If physical, chemical or biological pretreatment does not demonstrate effective results, combined pretreatments (physical-chemical, chemical-biological, etc.) might be recommended.

A well-known reagent in the paper and cellulose industry, hydrogen peroxide (H_2O_2) is used as a bleaching agent. Specifically, one of the best methods for improving hydrolysis and sugar release from various lignocellulosic substrates is alkaline H2O2 pretreatment. Pretreatment with hydrogen peroxide has been shown to significantly improve the digestibility of biomass as this not only eliminates lignin selectively, but also deconstructs the cell walls [10]. The goal of the pretreatment stage is to adjust the structure of complex (usually cellulosic) materials with decreasing degree of polymerization, weakening lignin bonds with carbohydrates and -particle surface area. Pretreatment affects surface area, crystallinity, the physical distribution of lignin in the biomass matrix, the degree of polymerization, pore volume, and biomass particle size [11]. The correlation of all these parameters with digestibility varies and depends on the type of biomass.

The major component of LB is cellulose and hemicellulose. Cellulose fibers are stabilized laterally by hydrogen bonds established between hydroxyl groups and create linear chains. Cellulose chains are stiffened by hydrogen bonds and a crystalline structure is aggregated [12]. The second most common polymer is hemicellulose. It is a complex carbohydrate structure with different polymer forms, e.g. the second most popular in nature are polysaccharides [13]. In addition to cellulose, hemicellulose consists of pentoses and/or hexoses and glucose combinations.

After cellulose, lignin is the second most abundant organic compound on earth. It is responsible for cross-linking between the polysaccharides (cellulose and hemicellulose) into a rigid woody structure. Lignin is a large and complex hydrophobic and aromatic macromolecule with highly variable chemical structure [11].

The composition and quantity of such components in LB vary depending on the plant species, tissues, and maturity. To optimize the use of lignocellulosic feedstocks, efficient fractionation must be performed prior to any kind of bioconversion by pretreatment. Cellulose and hemicellulose may then be further hydrolyzed and used, while residual lignin and/or its degradation products might be further processed through other conversion procedures. The composition of each type of feedstock is highly influenced by the appropriate fractionation of lignocellulose, not only because of the very diverse proportions of the three main biopolymers, but also because of the presence of other compounds.

A pretreatment of biomass (derived from eggplant and pepper processing) with hydrogen peroxide is presented in this study. The effect of pretreatment on the soluble carbon/soluble nitrogen ratio and the solubilization of volatile fatty acids (VFAs) associated with the use of such procedure is discussed.

2 Materials and Methods

2.1 Lignocellulosic Biomass: Chemical Composition

The substrate (pepper plant and eggplant) was collected from a farm located in Córdoba (Spain). Cutting and ground to particle size of 1-2mm and stored dry at room temperature (~25 °C). The values obtained according to the standard method (APHA, 2005) [14] are shown in Table 1. It is worthy to note the volatile solid (VS) for pepper plant and eggplant was 87% and 89% (dry basis), respectively. The C/N ratio of pepper plant and eggplant was 32 and 44, respectively.

Table 1. Physical-chemical characterization of the agricultural substrates (dry basis).

Parameter	Substrate	
	Eggplant	Pepper plant
TS(g/kg)	983 ± 10	977 ± 10
FS(g/kg)	114 ± 5	173 ± 10
VS(g/kg)	869 ± 10	805 ± 15
TOC (mg C/kg)	121 ± 1	108 ± 1
TSN (mg N/kg)	22 ± 1	16 ± 1
TKN(mg N/kg)	15,817 ± 110	18,357 ± 69
TP(mg P/kg)	326 ± 1	378 ± 1
TCOD(g O_2/kg)	1095 ± 50	905 ± 50
C/N	32	44

2.2 Hydrogen Peroxide Pretreatment of Lignocellulosic Biomass: Experimental Set-Up and Procedure

Pretreatment of hydrogen peroxide was conducted in Erlenmeyer flasks (250 mL) at different concentration of H_2O_2 (0%, 1%, 2% and 3%). In pretreatment with H_2O_2, 4 g of substrate (dry weight) were diluted fitting at pH 11.5 in 100mL of 0%, 1%, 2% and 3% of H2O2. Pretreatment was performed at room temperature (~25 °C) for 1 h in an orbital shaker at 150 rpm.

The values of soluble organic carbon (TOC, g/kg dry sample), soluble total nitrogen (STN (g/kg dry sample), volatile acid (VFA, mg acetic acid/L extracted fraction) were also determined according to the standard method (APHA, 2005).

3 Results and Discussion

3.1 H_2O_2 Pretreatment

A hydrogen peroxide pretreatment process was developed to enhance the solubilization of organic matter. With increasing concentration of hydrogen peroxide, lignin, cellulose, and hemicellulose of pepper plant and eggplant were significantly released. The higher concentration of H_2O_2 resulted in higher concentrations of cellulose, hemicellulose, and lignin for the same time of pretreatment (1 h).

The variation insoluble TOC with the concentration of H_2O_2 is shown in Fig. 1. As can be seen, the concentration of soluble TOC increased with the addition of hydrogen peroxide up to 3% (H_2O_2) with pepper plant, and 1% with eggplant and then stabilized at higher concentrations, suggesting that the optimum concentration of H2O2 pretreatment might be set at those values, under the study conditions.

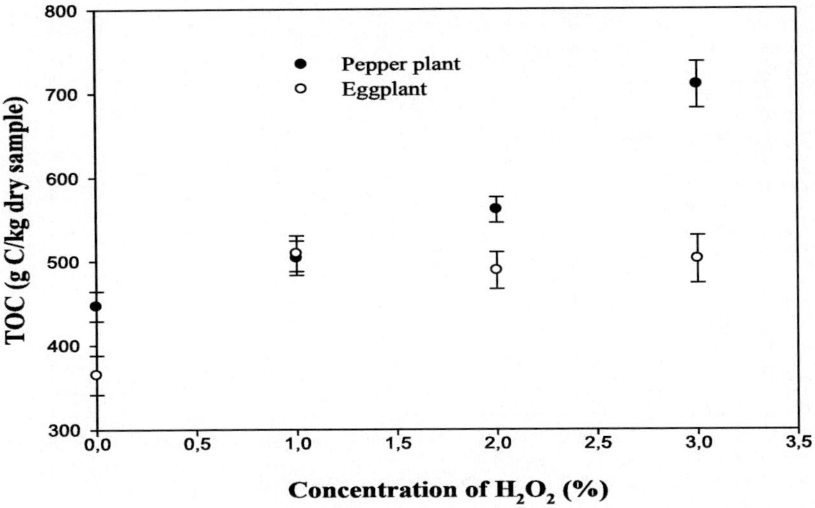

Fig. 1. Evolution of TOC as a function of hydrogen peroxide concentration.

As can be seen in Fig. 2, the STN increased also with the concentration of H_2O_2. The highest values of such a variable reached 61,54 ± 1,25 g N/kg dry sample of pepper plant at 3% H_2O_2, and 60,72 g N/kg dry sample of eggplant at 1% H_2O_2.

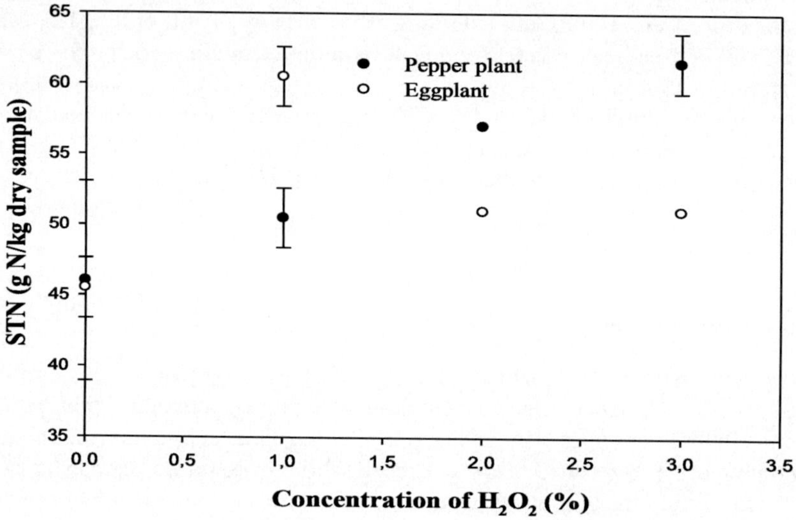

Fig. 2. Soluble total nitrogen evolution as a function of hydrogen peroxide concentration.

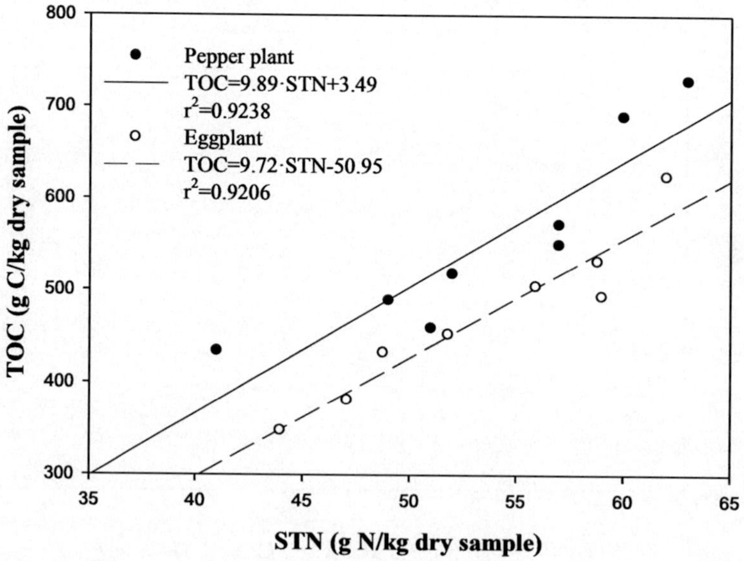

Fig. 3. Variation of the concentration of TOC with STN from pretreated substrates.

Figure 3 shows the relationship between the concentration of soluble TOC and STN for pepper plant and eggplant after pretreatment with hydrogen peroxide at all the concentrations evaluated. As can be seen, in both substrates the vale pairs TOC-STN can be fitted to a straight line, which indicates the simultaneous solubilization of both compounds during the pretreatment.

Another important variable to be monitored during the pretreatment of LB is the concentration of VFAs, which in general are advisable for subsequent biomethanization processes. Figure 4 shows the variation in the concentration of VFAs from pretreated pepper plant and eggplant throughout the experimentation. The concentration of VFAs showed clear increase compared to untreated substrate, after 1 h of pretreatment. The highest concentration of VFAs was reached for 2% H_2O_2 in pepper plant and 3% H2O2 in eggplant. Specifically, the concentration of VFAs from pepper plant and eggplant increased by 160% and 238%, respectively.

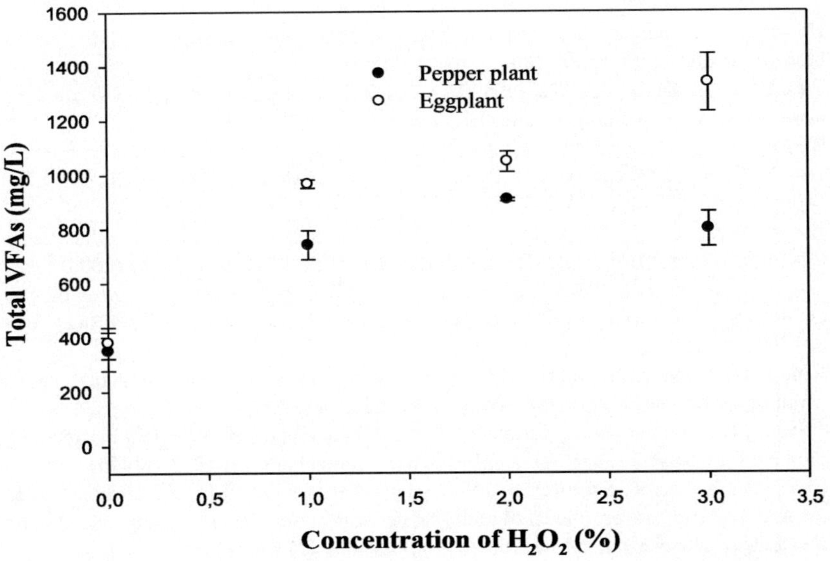

Fig. 4. Variation in the concentration of VFAs with hydrogen peroxide pretreatment.

4 Conclusions

Lignocellulosic biomass is a highly potential feedstock of anaerobic digestion to produce methane, but the complexity of biomass structure (contains 10–25% lignin) might inhibit anaerobic digestion process.

A lab-scale hydrogen peroxide pretreatment showed that soluble total carbon and soluble total nitrogen were solubilized proportionally. Moreover, pretreatment was optimal at 3 and 1% of H_2O_2 in the degradation of pepper plant and eggplant, respectively, increasing the concentration of VFAs about 160% and 238% from pepper plant and eggplant, respectively.

Finally, the results showed that pretreatment with hydrogen peroxide improve the availability of carbon and nitrogen, also increasing the formation of VFA, which could improve subsequent biomethanization of such substrates.

This research provides useful information for the hydrogen peroxide pretreatment on lignocellulosic biomass for producing methane in the contribution of the development of society.

References

1. Isci, A., Demirer, G.N.: Biogas production potential from cotton wastes. Renewable Energy **32**, 750–757 (2007)
2. Westerman, P.W., Bicudo, J.R.: Management considerations for organic waste use in agriculture. Bioresour. Technol. **96**, 215–221 (2005)
3. Hendriks, A., Zeeman, G.: Pretreatments to enhance the digestibility of lignocellulosic biomass. Bioresour. Technol. **100**(1), 10–18 (2009)
4. Wei, Y., Li, X., Yu, L., Zou, D., Yuan, H.: Mesophilic anaerobic co-digestion of cattle manure and corn stover with biological and chemical pretreatment. Bioresour. Technol. **198**, 431–436 (2015)
5. Sun, Y., Cheng, J.: Hydrolysis of lignocellulosic materials for ethanol production: a review. Bioresource Technology **83**, 1–11 (2002)
6. Liu, C., Wyman, C.E.: The effect of flow rate of compressed hot water on xylan, lignin and total mass removal from corn stover. Ind. Eng. Chem. Res. **42**, 5409–5416 (2003)
7. Lin, Y.Q., Wang, D.H., Wu, S.Q., Wang, C.M.: Alkali pretreatment enhances biogas production in the anaerobic digestion of pulp and paper sludge. J. Haze. Mat. **170**(1), 366–373 (2009)
8. Kim, K.H., Hong, J.: Supercritical CO_2 pretreatment of lignocellulose enhances enzymatic cellulose hydrolysis. Bioresource Technol. **77**, 139–144 (2001)
9. Teixeira, L.C., Linden, J.C., Schroeder, H.A.: Alkaline and peracetic acid pretreatments of biomass for ethanol production. Appl. Biochem. Biotechnol. **77**, 19–34 (1999)
10. Banerjee, G., Car, S., Scott-Craig, J.S., Hodge, D.B., Walton, J.D.: Alkaline peroxide pretreatment of corn stover: effects of biomass, peroxide, and enzyme loading and composition on yields of glucose and xylose. Biotechnol. Biofuels **4**, 1 (2011)
11. Saha, B.C.: Hemicellulose bioconversion. J. Ind. Microbiol. Biot. **30**, 279–291
12. Pu, Y., Zhang, D., Singh, P.M., Ragauskas, A.J.: The new forestry biofuels sector. Biofuels Bioprod. Bioref. **2**, 58–73 (2003)
13. Yang, L., Fuqing, X., Ge, X., Li, Y.: Challenges and strategies for solid-state anaerobic digestion of lignocellulosic biomass (2015)
14. APHA: Standard methods for the examination of water and wastewater, APHA (2005)

Effect of Temperature on the Gasification of Olive Prune Pellets

Carla Barros[1], Eveline Santos[1], Ana Assis[1], Carline Ribeiro[1], Ligia Gomes[1], Vitor Matos[1], Victor Arruda[2], Roberta Panizio[1(✉)], Luís Calado[1], Pedro Romano[1], Luiz Rodrigues[1], and Paulo Brito[1]

[1] VALORIZA - Research Center for Endogenous Resource Valorization, Polytechnic Institute of Portalegre, Campus Politécnico 10, 7300-555 Portalegre, Portugal
pbrito@ipportalegre.pt

[2] School of Engineering, Guaratinguetá, Department of Energy, Laboratory of Optimization of Energy Systems (LOSE), Institute of Bioenergy Research (IPBEN - UNESP) - Associated Laboratory of Guaratinguetá, São Paulo State University (UNESP), São Paulo, Brazil

Abstract. The present work studied the thermal gasification of olive pruning. Olive pruning from agricultural processes went through a pre-treatment of grinding and pelletizing in order to standardize the raw material. The tests of the experimental process were carried out with a fixed bed gasifier, downdraft type, with the raw material of olive pruning pellets and the oxidizing agent atmospheric air. The main idea of the present work was to study the influence of the gasification temperature, on the quality of the synthesis gas produced, in a temperature range between 650 to 800 °C. The pre-treatment of grinding, with a 6 mm sieve, transformed the olive pruning into particles smaller than 0.85 mm and pelletizing on pellets with a diameter of 6–16 mm. This raw material was studied in terms of elementary analysis, thermogravimetric analysis and calorific value. During the performed tests, several samples of synthesis gas were taken for further characterization through gas chromatography. The results of the tests carried out allow to demonstrate that relatively lower temperatures favor the production of gas and the calorific value, mainly due to the capitalization in light hydrocarbons. The temperature with the highest lower calorific value was on the range of 650 °C, with 5.35 $MJ.kg^{-1}$, with a gas flow of 26.3 $m^3.h^{-1}$.The thermal gasification process is a viable alternative for the energetic recovery of lignocellulosic waste, namely olive pruning pellets.

Keywords: Gasification · Olive waste · Energy recovery

1 Introduction

Currently, biomass has been arousing a growing interest as an alternative source, from the energy point of view, to current energy sources. Biomass is the most available energy source among all sources and the most sustainable, where its use, would allow the

J. R. da Costa Sanches Galvão et al. (Eds.): *Proceedings of the 1st International Conference on Water Energy Food and Sustainability (ICoWEFS 2021)*, pp. 325–334, 2021.
https://doi.org/10.1007/978-3-030-75315-3_37

diversification of energy sources, which can be used as a local source and thus contribute to decentralization and local economy, and reduction pollutant emissions into the atmosphere.

The olive tree, with the scientific name of *Olea europea*, is a species disseminated throughout the world and, more particularly, in southern Europe (around 64% of the world total production), where it occupies a prominent place in the production of olive oil and also table olives [1, 2]. Portugal also has a dominant position worldwide, being the fourth most representative country in the production of olive oil, where the olive tree represents over 50% of the cultivated area in the country, with more than 366 thousand hectares. According to the National Statistics Institute of Portugal (INE), in 2020 olive manufacture touched a historical peak of more than 940 thousand tons, the highest since 1941 [3].

Derived from this activity, large quantities of leftovers are produced annually, from olive pruning and some by-products originating from the processing of olives. Ordinarily, these agricultural leftovers, which are characterized by pruning of logs and leaves, purchased to produce charcoal, although the vast majority, are burned in the field without any use of energy. Consequently, the treatment of residues associated with the production of olive oil is extremely important and needs to be resolved. One of the alternatives for the treatment of residues, may be the direct production of energy, which can be an interesting alternative for the sector, that has high energy costs [4].

For the use of olive pruning in the energy recovery process, it is necessary to carry out a pre-treatment process. Such a process is necessary due to the physical and chemical characteristics of the biomass. The tests aimed at two important processes for the energy recovery of this type of biomass, the reduction of the particle size and subsequent pelletization. Both mechanical processes where the first process serves as support for the second. The pelletizing process consists of increasing the density of the biomass to reduce the costs associated with handling, transport/storage.

The production of biomass pellets, so that it has a high quality, depends mainly on the raw material composition, moisture, particle size and equipment specifications. The way the biomass is distributed, and its size are dependent on the milling process. The quality of the grinding is related to the size of the sieves and the speed of rotation of the motor [5]. Studies carried out tests to verify the effects of the particle size produced on the sieves and their influence on the pelletizing process. They were able to observe that, as the size of the biomass particles reduced, the density of the pellet increases [6, 7]. It is noted that, although the particle size during the grinding process is of great importance for the quality of the produced pellets, moisture and temperature are factors that must be considered when it is intended to make quality and resistant pellets [6, 8].

Due to a better particle distribution, both the density and the hardness of the pellet improve with the decrease of the particle after grinding and the production capacity of the equipment increases when using biomass with a dimension of less than 0.8 mm.

In this context, the process of thermal gasification appears. Gasification is a thermochemical process that converts organic material, with low moisture content (biomass, industrial, agro-industrial waste, etc.) into a combustible gas fraction, a biochar and a liquid fraction, known as tar, through partial oxidation of the organic material, in a temperature range of 600 to 900 °C. Another advantage of this process is the small size and high performance that units of this type have. Generally, the thermal gasification process consists of several processes inside the reactor, such as drying, pyrolysis, oxidation, and reduction. As mentioned, gasification products are characterized by synthesis gas, chars, and tars. The solid fraction is formed through the devolatilization of the raw material, and the liquid fraction is formed through the condensation of phenols, light hydrocarbons. The gasification process with atmospheric air, as an oxidizing agent, has as main objective the production of synthesis gas with a calorific value between 3.5 and 7 $MJ.m^{-3}$. Generally, this synthesis gas is composed of CH_4, C_xH_y, H2, CO, CO N_2 [9].

The performance of the gasification process is dependent on many factors, mainly the characteristics of the raw material, the type of gasifier, the type of oxidizing agent, operating conditions, such as the temperature and flow rate of the oxidizing agent that enters the process, calculated through efficiency ratio (ER) [10].

The reactions present in the gasification process are complex and are sometimes competing reactions. In the initial phase of the process, the raw material undergoes devolatilization, where it forms small hydrocarbons, carbon monoxide and hydrogen. In the next stage, partial oxidation, the gases produced during devolatilization may undergo oxidation, forming carbon dioxide and water.

In the literature it is possible to find some studies of olive gasification, which study the influence of temperature on the synthesis gas. One study report that the concentration of CO tends to increase with increasing temperature, with this increase becoming more manifest above 850 °C. The author also mentions that the concentration of CH_4 decreases, as the temperature increases and, consequently, the opposite occurs with the concentration of H2, which increases as the temperature increases. The concentration of CO_2 tends to decrease as the process stabilizes [11]. Another author reports that the production of synthesis gas increased with the gasification temperature. With the increase in the gasification temperature, the concentrations of C_xH_y, CH_4 and CO_2 tends to decrease. An increase in the efficiency ratio, caused a decrease in the production of synthesis gas and, at the same time, led to a decrease in the low heating value (LHV) of the gas and a decrease in the production of tars [12].

The present study intends to demonstrate the evolution of the syngas characteristics during the gasification of olive can pellets, a residue of the agro-industrial, in a fixed bed reactor, downdraft.

2 Materials and Methods

The residual biomass obtained from olive pruning is essentially composed of small trunks, branches and leaves. The composition of olive pruning can vary according to location, season, variety, etc. Currently, the mechanism for collecting olive pruning waste, although not very usual, is only used for the manufacture of charcoal (pyrolysis), and the use of this type of waste for other types of energy recovery is not common. The raw material was already crushed and ready to undergo the grinding process.

The raw material underwent a grinding process in an industrial mill, with a capacity of 30 ton h^{-1}, and a power of 100 kW. The unit performs the crushing and grinding of biomass to diameters less than 0.40 mm. The ground olive pruning was sent to the densification process, using a KAHL 14–175 pelletizer with 3 kW of power. The tests were carried out in late summer, where the ambient temperatures were between 31 and 35 °C, with an average relative humidity of 45%. The pelletizer die was flat and the dimensions of the die press channel were 6 mm in diameter and 15 mm thick.

2.1 Characterization of Biomass

The inorganic, carbonaceous, volatile and moisture content were analyzed using gravimetry. The thermogravimetric profile, considering the inflection points of the mass derivative, were determinate using a PerkinElmer thermogravimetric scale, model STA 6000, with an inert atmosphere (nitrogen of 20 mL min^{-1}) and a temperature ramp of 20 °C min^{-1}.

To determine the high heating value of the raw material, an IKA C 2000 calorimeter was used, where the samples were subjected to complete combustion, in an adiabatic environment.

For the determination of carbon, hydrogen, nitrogen, sulfur, and oxygen present in the raw material, an elemental analyzer model ThermoFisher Scientific Flash 2000 CHNS-O was used. Of note, oxygen was determined by difference.

2.2 Gasification Test

The gasification tests were carried out on an AllPowerLabs PP20 Power Pallets gasifier. This equipment has a power of 15 kW and is the combination of a fixed bed reactor, downdraft type, and an internal combustion engine connected to a generator. The equipment consists of a silo that has the purpose of storing and drying the raw material, through the recirculation of the gases produced during the process. The oxidizing agent is preheated through the reactor walls, before entering the process, to favor its performance. There is a deposit at the bottom of the reactor, which serves as a collector of the unconverted material (chars), and then there is a cyclone filter, in order to retain the finer particles that were "dragged" with the synthesis gas. The cleaning of the syngas still undergoes one more step, in the filter composed of several biomass granulometries, and where much of the produced tars are retained. The synthesis gas produced and cleaned, is subsequently burned in an internal combustion engine 3.0, which produces mechanical energy for the coupled generator.

2.3 Characterization of Syngas

During the gasification tests, several samples of synthesis gas were retained, in suitable bags, for later analysis in gas chromatography. The samples were taken at the outlet of the reactor, with the aid of a vacuum pump.

The synthesis gas analysis were performed in a Varian 450-GC Gas Chromatograph with TCD detector (used for identification and quantification of gaseous constituents CO, CO2, H2, CH4 and light hydrocarbons present). The injection of the samples took place through the help of a peristaltic pump, where the gas subsequently enters a combination of columns that allows to separate and identify all the elements present.

3 Results and Discussions

It was noted that after the milling process the raw material was not homogeneous, thus containing particles with a maximum diameter of 2.8 mm, with a greater amount of material (48.16%) in the 0.85 mm sieve. The homogeneity of the particles is important for determining the quality of the pellets, as it is the one that determines the fragility of the pellets as it alters the connections between the particulars and their interaction.

The raw material after going through the pelletizing process, as mentioned, was tested for its mechanical durability. For durability tests, the UNI EN 15210–1 standard was used. The durability is checked according to the number of pellets that are retained in the sieve after the rotation process. The tests carried out on 2.8 mm sieves used to collect the pellets after the rotation process at 50 rpm for 15 min. For the tests, a mass of 500 g of pelleted biomass was used.

The quality of a pellet is determined by its durability, which in turn is influenced by factors such as humidity, particle size and chemical composition. The resistance of the pellets produced (87%) during the tests is slightly lower than the normalized value (EN 17225–6 and the normalized value is 96%), but with an improved selection process (for example, pelletizing more fines) and an optimization in the pelletizing process (for example pelletizing with lower moisture content), the quality of the pellets can be improved without major additional costs.

3.1 Characterization of Biomass

All characteristics of biomass, elementary analysis, thermogravimetric analysis and higher calorific value are shown in Table 1.

Table 1. Characterization of the raw material.

Parameters	Olive
C	47.94
H	6.41
N	0.62
S	44.98
O	<0.05
HHV (MJ/kg)	19.13
Moisture (%)	4.84
Volatiles (%)	78.03
Fixed Carbon (%)	16.23
Ashes (%)	0.9

3.2 Gasification Test

Table 2 presents the results obtained for the samples collected during the gasification tests.

Table 2. Characterization of syngas.

		Sample			
Parameters	Uni.	1	2	3	4
CO_2	%	11.74	10.56	9.16	8.52
C_2H_4	%	0.66	0.57	0.17	0.47
C_2H_6	%	0.18	0.15	0.04	0.09
C_2H_2	%	0.02	0.02	0.04	0.02
H2S	%	0.00	0.05	0.05	0.05
N_2	%	52.02	52.95	54.58	55.01
CH_4	%	2.61	2.24	1.09	2.04
CO	%	19.47	19.77	20.92	19.91
H2	%	13.30	13.91	14.19	13.56
LHV syngas	MJ/m^3	5.35	5.26	4.72	5.07
T_{oxi}	°C	648	707	751	802
T_{red}	°C	402	446	487	509
P_{comb}	KPa	−25	−21	−19	−11
P_{React}	KPa	−32	−31	−27	−18
P_{filt}	KPa	−41	−38	−37	−29
V_{ar}	m^3/h	17.40	16.20	15.70	12.40
T_{ar}	°C	25.70	26.90	26.50	26.70
V_{tars}	m^3/h	0.000131			
Q_{chars}	kg/h	0.0546			
ER	-	0.36	0.34	0.32	0.26
LHV_{comb}	MJ/kg	17.88			
η_{syngas}	%	93.55			
V_{syngas}	m^3/h	26.13	23.87	22.29	17.61
Time the test	h	4			

3.3 Effect of Temperature on Syngas Composition

One of the main parameters of a thermal gasification process is the temperature, as it directly affects the cracking reactions of the process. During the tests carried out it was possible to observe a trend in the behavior of the syngas composition with respect to temperature variations. The results obtained regarding the percentage of syngas composition regarding the content of H2, CO, CH_4, CO_2, N_2 and light hydrocarbons during the test are shown in Table 2. It is possible to observe that temperatures of 650 °C, 700 °C were used, 750 °C and 800 °C to be able to compare the gases produced.

The highest rate of gas production was verified in the first tests, where more oxidizing agent is needed to increase the gasification temperature. Over the course of the test, and the inertia and stabilization of the equipment, less oxidizing agent will be needed to maintain the gasification temperature, producing less volume of syngas.

With an increase in temperature from 648 °C (Fig. 1) to 751 °C (Fig. 2) there was a volumetric increase in relation to concentrations CO and H_2 in the syngas was observed and a decrease in the concentration of CH_4 was also observed. In the 802 °C test (Fig. 4), the reverse was observed, with an increase in the volumetric concentration of CH_4 and a decrease in the concentration of CO and H2 (Fig. 3).

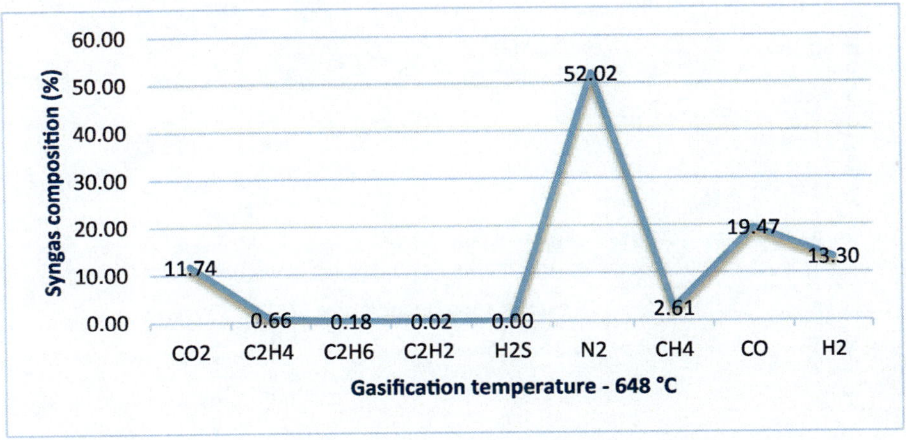

Fig. 1. Analysis of syngas at 648 °C.

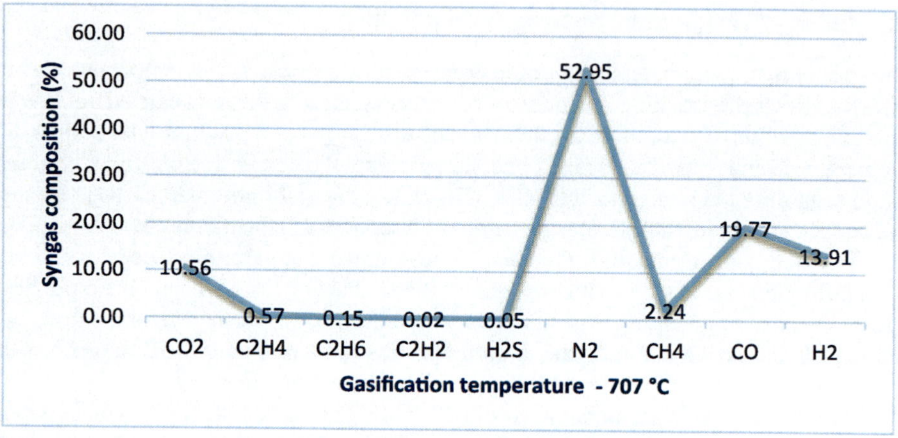

Fig. 2. Analysis of syngas at 707 °C.

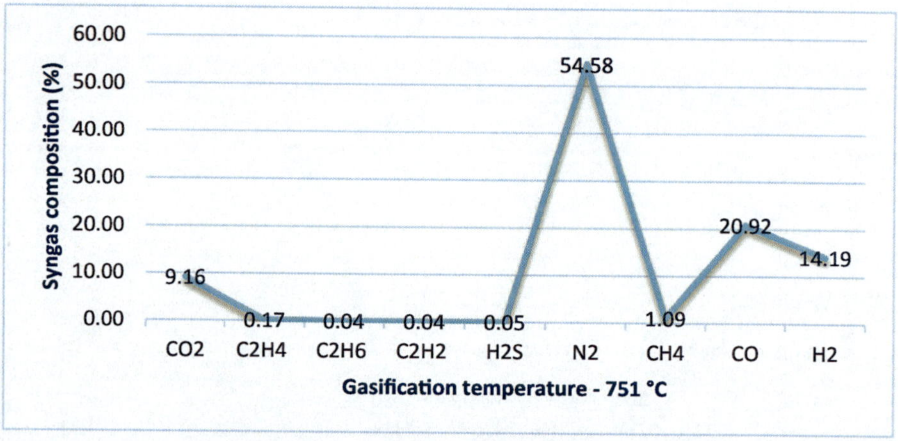

Fig. 3. Analysis of syngas at 751 °C.

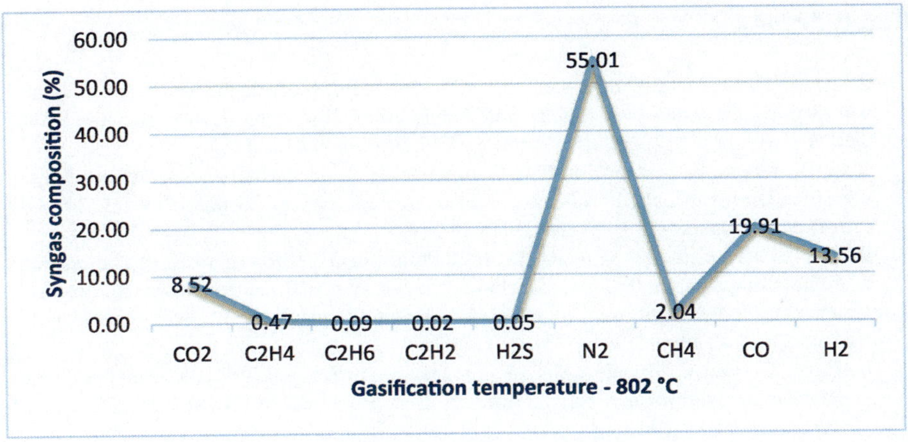

Fig. 4. Analysis of syngas at 802 °C.

4 Conclusion

The present work aimed to energetically valorize an agro-industrial residue, olive pruning. For this, it was necessary to perform a pre-treatment to the raw material, grinding and pelletizing and later to gasify the material. During the tests on the fixed bed gasifier where air was used as an oxidizing agent, it was possible to evaluate the synergistic effect which the increase in temperature has on the quality of syngas.

It was observed that, at 800 °C there was an important change regarding the composition of the syngas, with a reduction in the content of CO and H_2 in the syngas. As the temperature rises, it is possible to notice that the number of light hydrocarbons tends to decrease. This change may be associated with a change in the reaction mechanism, and methane production may be the most important reaction.

With the increase in temperature it was possible to verify that not all performance parameters have improved. The best operating conditions presented were in the temperature range of 650 and 750 °C. In this temperature range, PCI values of 5.35 and 5.26 MJ/m^3 were obtained, many derived from the presence of light hydrocarbons in the composition of the synthesis gas.

The tests presented that the use of olive pruning can obtain a rich syngas is possible and that the performance of the gasification process is optimized when operating conditions with a longer time than experimental are used.

Acknowledgments. The authors are grateful for the support given by the Foundation for Science and Technology, I.P. (FCT), through the "Summer with Science" Program, within the Summer School "Technologies for a waste biorefinery", held at VALORIZA - Research Center for the Valorization of Endogenous Resources - Polytechnic Institute of Portalegre (05064). The authors are grateful to VALORIZA - Research Center for Endogenous Resource Valorization, Polytechnic Institute of Portalegre. The authors also would like to thank the Coordination for the Improvement of Higher Education Personnel (CAPES) and the Portuguese Foundation for Science and Technology (FCT) for the funding provided through project number 88881.156267/201701.

References

1. Zribi, M., Lajili, M., Escudero-sanz, F.J.: Gasification of biofuels blended from olive mill solid wastes and pine sawdust under diff erent carbon dioxide/nitrogen atmospheres. Fuel **282**(March), 118822 (2020). https://doi.org/10.1016/j.fuel.2020.118822
2. Vera, D., Jurado, F., Margaritis, N.K., Grammelis, P.: Experimental and economic study of a gasification plant fuelled with olive industry wastes. Energy Sustain. Dev. **23**, 247–257 (2014). https://doi.org/10.1016/j.esd.2014.09.011
3. INE, "Previsões Agrícolas 31 de janeiro 2020 Produção de azeitona para azeite com máximo histórico ultrapassa as 940 mil toneladas," pp. 1–5 (2020). https://www.ine.pt/xportal/xmain?xpid=INE&xpgid=ine_destaques&DESTAQUESdest_boui=399592136&DESTAQUESmodo=2&xlang=pt.
4. Lajili, M., Guizani, C., Escudero Sanz, F.J., Jeguirim, M.: Fast pyrolysis and steam gasification of pellets prepared from olive oil mill residues. Energy **150**, 61–68 (2018). https://doi.org/10.1016/j.energy.2018.02.135
5. Pradhan, P., Mahajani, S.M., Arora, A.: Pilot scale production of fuel pellets from waste biomass leaves: effect of milling size on pelletization process and pellet quality. Fuel **285**(August 2020), 119145 (2021). https://doi.org/10.1016/j.fuel.2020.119145
6. Shaw, M.D., Karunakaran, C., Tabil, L.G.: Physicochemical characteristics of densified untreated and steam exploded poplar wood and wheat straw grinds. Biosyst. Eng. **103**(2), 198–207 (2009). https://doi.org/10.1016/j.biosystemseng.2009.02.012
7. Carone, M.T., Pantaleo, A., Pellerano, A.: Influence of process parameters and biomass characteristics on the durability of pellets from the pruning residues of Olea. Biomass Bioenergy **35**(1), 402–410 (2010). https://doi.org/10.1016/j.biombioe.2010.08.052
8. Stelte, W., Holm, J.K., Sanadi, A.R., Barsberg, S., Ahrenfeldt, J., Henriksen, U.B.: Fuel pellets from biomass: the importance of the pelletizing pressure and its dependency on the processing conditions. Fuel **90**(11), 3285–3290 (2011). https://doi.org/10.1016/j.fuel.2011.05.011
9. Basu, P.: Biomass Gasification and Pyrolysis: Pratical Design and Theory (2010)
10. Carmo-Calado, L., Hermoso-Orzáez, M.J., Mota-Panizio, R., Guilherme-Garcia, B., Brito, P.: Co-combustion of waste tires and plastic-rubber wastes with biomass technical and environmental analysis. Sustainability **12**(3) (2020). https://doi.org/10.3390/su12031036
11. González, J.F., Román, S., Bragado, D., Calderón, M.: Investigation on the reactions influencing biomass air and air/steam gasification for hydrogen production, vol. 9 (2008). https://doi.org/10.1016/j.fuproc.2008.01.011
12. Skoulou, V., Zabaniotou, A., Stavropoulos, G., Sakelaropoulos, G.: Syngas production from olive tree cuttings and olive kernels in a downdraft fixed-bed gasifier. Int. J. Hydrogen Energy **33**(4), 1185–1194 (2008). https://doi.org/10.1016/j.ijhydene.2007.12.051

Factors Affecting Landowners' Decision to Invest in Photovoltaics on Their Farmland: The Case of Landowners in a Greek Rural Area

Evangelia Karasmanaki(✉), Despoina Sfiri, and Georgios Tsantopoulos

Department of Forestry and Management of the Environment and Natural Resources, Democritus University of Thrace, Pantazidou 193, 68200 Orestiada, Greece
evagkara2@fmenr.duth.gr

Abstract. Investments in renewable energy have taken a center-stage discussion among policymakers and several incentives are now provided to landowners to encourage them to install photovoltaics on their land. However, such investments remain below their potential and thus insights into what affects potential investors' decision are needed. This study focuses on landowners who are willing to install photovoltaics on their farmland and analyzes the effect of various factors on the amount of money they would pay in order to install photovoltaics on their land. According to our analysis, landowners' personal view on the profitability of the investment, the degree to which they agree with the installation of solar parks in a location visible from their residence and the hectares of the dry and irrigated land they own, explain the amount of money the landowners would invest in photovoltaics. The findings of the present study may be particularly useful to policymakers and may help them prepare practical and attractive propositions for potential investors thereby improving the existing investment environment for photovoltaics on farmland.

Keywords: Solar panels on farmland · Investment behavior · Factors affecting investment decision

1 Introduction

In the European Union (EU), various types of subsidies have been adopted to promote investments in renewable energy sources (RES). Regarding solar energy, the Feed-in-System is a popular subsidy policy that encompasses two types of contracts, the Feed-in-Tariff (FIT) and the Feed-in-Premium (FIP). These schemes have led to an impressive increase in private, small-scale investments in photovoltaics while solar installations have become a common sight throughout EU. At the same time, the public in EU has a clear preference for solar energy in comparison to other renewable types [1–4].

As individuals can participate in the diffusion of renewable energy through investments, a considerable number of research works has sought to detect the factors that affect the investment decision of potential investors [5–7]. For example, Hai et al. [8]

J. R. da Costa Sanches Galvão et al. (Eds.): *Proceedings of the 1st International Conference on Water Energy Food and Sustainability (ICoWEFS 2021)*, pp. 335–343, 2021.
https://doi.org/10.1007/978-3-030-75315-3_38

observed that although Finnish interviewees had a positive attitude to solar energy, they did not purchase or invest in solar energy due to their concerns about the limited availability of sunlight in their country, investment costs, expenses of grid connection and the absence of cost-effective mechanisms for selling produced power. Kastner and Matthies [9] explored the effect of value orientations on investment decision among German houseowners and found that the intention to invest, the possibility of saving energy costs along with the duration of guarantee period and reliable recommendations had a positive effect while investment costs had an adverse impact on investment decisions.

Other factors that have a discernible effect on renewable investments involve environmental values, trust in the technology of the systems, former investment experience, age, place of residence, education level, preference for comfort, social class, and house ownership [5–7, 10–12]. Regarding environmental values, different studies have concluded that the environmental values of individuals often drive them to invest in renewables. For instance, Swedish respondents, who reported to be acutely concerned about the environment and its problems, were motivated to install renewable systems and emphasized the need to think about the environment and lead their lives in a way that would be 'as ecologically sustainably as possible' [13]. The positive effect of environmental values was further confirmed in studies performed in the US [14] and also in Austria, Italy, and Germany [12, 15]. It is also significant to note that environmental concerns emerged as an important investment factor in countries which present relatively low installed RES capacity such as Tunisia, Egypt, Lebanon, Jordan [16] suggesting that environmental values and concerns can induce investments even in countries which have probably a less favorable investment environment.

Moreover, the decision to invest is often reliant on the perceived profitability of the investment, which if deemed substantial, the investment proceeds [17]. Bergek et al. [18] noted that investments in renewables are made whenever an opportunity is recognized and when the value of pursuing it, is high enough. In addition, access to financial resources, which are often necessary for investments, is another influential factor [17].

While environmental values and perceived profitability have a positive effect on investments, low levels of trust in the technology of renewable systems and limited understanding of how these systems work can affect adversely an individual's decision to invest [7, 12]. On the other hand, if an adequate understanding of the systems has been attained, the sight of photovoltaic systems on the rooftops of neighboring houses can spark interest and can induce investments [11]. This has been observed in California where the installation of photovoltaic panels in the neighborhood of potential investors led many to proceed to the installation of the same systems on their houses [19]. Furthermore, the age of potential investors has an ambiguous influence on investment decisions with some studies showing that younger individuals are more likely to invest [7, 16] and other indicating that older people are more likely to adopt them [10].

Another factor that plays a notable role in investments is the place of residence with research showing that rural dwellers are more positive to investments than urban dwellers [12, 20]. Moreover, investment decisions are affected by social norms [5] and the tendency to keep the existing level of comfort and status quo [12]. As for the latter, behavioral economics analysis showed that individuals tend to make social comparisons, follow others' behavior, and align with social norms; that is, explicit and implicit rules as well as behavioral expectations [21]. Hence, individuals are expected to make investments if people in their close environment have invested or approve such investments.

In the relevant literature, most previous studies have explored individuals' willingness to invest in renewable energy [5–7, 22]. However, there is a scarcity of studies focusing explicitly on landowners who are already willing to make investments. Given that in many EU countries incentives are provided to landowners to encourage installations on their land, it is interesting to analyze what affects the decision of this particular category of investors. Hence, the aim of the present study is to identify the factors that affect the decision to invest in photovoltaics among landowners in the Municipal Unit of Didymoteicho, in Northern Greece.

2 Methodology

The findings presented in this paper are part of a larger research which was conducted from June 2020 to September 2020 and analyzed landowners' views on investments in photovoltaic systems in the Municipal Unit of Didymoteicho, which is located in the Prefecture of Evros, in northern Greece. The Municipal Unit of Didymoteicho covers an area of 354.134 km^2.

For the purposes of this research, a questionnaire was developed which involved nominal, ordinal and ratio variables. To make sure that the questionnaire would be able to give both accurate and coherent results, a pilot study was conducted on a limited scale and few minor changes were made.

As for the sampling method, simple random sampling was followed because this method is easy to perform and requires very little knowledge about the population under study. In total, 235 landowners participated in the study and among these respondents, 205 were willing to invest in photovoltaics. In this paper, we analyze the responses of these 205 willing landowners using the Statistical Package for the Social Sciences (SPSS). More specifically, descriptive statistics and categorical regression were performed. Categorical regression extends the principles of classical linear regression and logarithmic analysis. By means of scaling, it assigns values to each category of variables so that they are suitable for regression and reflect the characteristics of the original variables. Categorical regression scales nominal, ordinal and numerical variables optimally and quantifies their categories, so that the squared correlation between the quantified dependent variable and the linear combination of the quantified independent variables is maximized. The interpretations are related to the transformed variables, but they also relate to the original variables, due to the relationship that exists between the original variables and the transformed ones [23].

3 Results

3.1 Landowners' Characteristics

Respondents were first asked to state their demographic characteristics as well as the exact number of hectares of dry and irrigated land they owned. In terms of gender (Q1), male respondents (50.2%) outnumbered slightly their female counterparts. Regarding their age (Q2) and family status (Q3), the majority was older than 41 years (69.7%) and married (68.8%). As for their main profession (Q4), considerable shares of respondents were farmers (42.9%), private employees (14.1%) and public employees (11.2%). Regarding respondents' education level (Q5), 29.8% were high school graduates, 14.1% were university graduates, 11.2% were middle school graduates, 9.8% were vocational school graduates and another 9.8% were primary school graduates. Moreover, 44.4% stated that their annual family income (Q6) ranged between 10.001 and 30.000 per year. Finally, the respondents owned 2272.7 "Hectares of dry land" (Q7) and 2306.7 "Hectares of irrigated land" (Q8).

3.2 Views on Topics Related to Photovoltaics

Respondents' "Personal view on the profitability of the investment" (Q9) was first examined. As shown in Table 1, 42.9% perceived that investing in photovoltaics would be moderately profitable and 37.5% perceived that it would be very and extremely profitable. Only 1% of respondents thought that investments in photovoltaics would be no profitable at all.

Table 1. Frequency and percentages regarding landowners' views on the profitability of photovoltaics on their farmland.

	Frequency	Percentage
Not at all	2	1.0
Slightly	38	18.5
Moderately	88	42.9
Very	63	30.7
Extremely	14	6.8

Next, respondents were asked whether they agreed with the installation of solar parks in a location which would make the solar park visible from their residence (Q10). As it can be seen in Table 2, respondents were divided as 43.4% neither agreed nor disagreed with such a possibility. In addition, 32.7% agreed and 23.9% disagreed.

Table 2. Frequency and percentages concerning respondents' level of agreement with the installation of photovoltaic parks near their residence.

	Frequency	Percentage
Strongly disagree	11	5.4
Disagree	38	18.5
Neither agree nor disagree	89	43.4
Agree	55	26.8
Strongly agree	12	5.9

Then, the "Amount of money landowners would invest in photovoltaics" (Q11) was examined. As it can be seen in Table 3, a significant share of respondents (17.2%) would invest an amount of money ranging between 2.000 and 5.000 €, a considerable share (15.9%) would invest between 5.000–10.000 € and an appreciable share (17.6%) would invest between 10.000–20.000 €. However, only 8.6% of respondents would invest more than 20.000 €.

Table 3. Frequency and percentages regarding the amount of money landowners would invest in photovoltaics

	Frequency	Percentage
Not willing to invest	26	11.2
Less than 500€	16	6.9
500–1000€	24	10.3
1000–2000€	29	12.4
2000–5000€	40	17.2
5000–10000€	37	15.9
10000–20000€	41	17.6
Over 20000€	20	8.6

Categorical regression was performed in order to analyze what affects the amount of money landowners would invest in photovoltaic systems on their land. In this analysis, Q11 "Amount of money landowners would invest in photovoltaics" served as the dependent variable. As for independent variables, it was sought to select those variables which account for the problem to a higher degree.

According to our analysis, the correlation matrices of the independent variables showed that in Pratt's measures of relative importance measures and tolerance there was a multicollinearity problem. More specifically, certain independent variables exhibited high correlation while Pratt's relative importance measures had high negative values but low tolerance values. As these independent variables rendered the categorical regression model unstable, they had to be removed from the subsample. To remove these variables, the value of F statistic, which determines whether the removal of an independent variable from the model reduces significantly the predictive abilities of the subsample, were considered. It is important to note that these independent variables were not removed altogether, but only one independent variable was removed each time based on the value of the F statistic for each independent variable. Of all the tests conducted in this analysis, here we present only the results of Pratt's most important relative importance coefficients, standardized beta coefficients, transformation plots and the relevant interpretations.

Several tests were conducted to detect the variables that best explain the dependent variable "Amount of money landowners would invest in photovoltaics" (Q11). It was concluded that the independent variables which explain it best are: "Gender" (Q1), "Age" (Q2), "Family status" (Q3), "Main profession" (Q4), "Education level" (Q5),

"Annual family income" (Q6), "Hectares dry land" (Q7), "Hectares of irrigated land" (Q8), "Personal view on the profitability of the investment" (Q9), "Agreement with the installation of solar parks in a location visible from residence" (Q10).

Hence, categorical regression gave coefficient value of multiple determination $R^2 = 0.433$, indicating that 43.3% of the dependent variable's variance was accounted for by the transformed values of the independent variables in the regression equation. Moreover, the analysis of the variance (ANOVA) gave value $F = 5.002$, which corresponds to a zero level of statistical importance, showing that the subsample of the categorical regression was adapted well to transformed data (Table 4).

Table 4. Categorical regression results regarding the factors affecting the amount of money respondents would invest in photovoltaic systems on their farmland

	Beta	Std error	Df	Importance	F	Sig
Gender (Q1)	0.005	0.044	1	0.000	0.014	0.904
Age (Q2)	−0.186	0.115	4	0.064	2.598	0.038
Family status (Q3)	0.085	0.066	3	0.017	1.626	0.185
Main profession (Q4)	0.206	0.070	8	0.076	8.756	0.000
Education level (Q5)	−0.016	0.158	2	0.001	0.010	0.990
Annual family income (Q6)	−0.001	0.142	1	0.000	0.000	0.996
Hectares of dry land (Q7)	0.182	0.076	1	0.124	5.732	0.018
Hectares of irrigated land (Q8)	0.148	0.091	1	0.113	2.663	0.104
Personal view on the profitability of the investment (Q9)	0.408	0.085	4	0.453	23.115	0.000
Agreement with the installation of solar parks in a location visible from residence (Q10)	0.218	0.119	2	0.152	3.327	0.038

Based on the standardized regression coefficients, the "Amount of money landowners would invest in photovoltaics" (Q11) was mostly affected by the following variables: "Personal view on the profitability of the investment" (Q9), "Agreement with the installation of solar parks in a location visible from residence" (Q10) and "Main profession" (Q4). Moreover, the F values of each independent variable show that the removal of variables with high F values make the subsample weak whereas the removal of the variable "Gender" (Q1) exerts only a negligible effect on the predictability of the subsample. In addition, independent variables' measures of relative importance indicate that the following independent variables make the biggest contribution to the dependent variable: "Personal view on the profitability of the investment" (Q9) (45.3%), "Agreement with the installation of solar parks in a location visible from residence" (Q10) (15.2%), "Hectares of irrigated land" (Q8) (12.4%) and "Hectares of dry land" (Q7) (11.3%). In aggregate, these variables account for 84.1% of independent variables' total importance. Finally, the transformation plots of dependent and independent variables show that as the value of each independent variable increases, so does value of the dependent variable "Amount of money landowners would invest in photovoltaics" (Fig. 1).

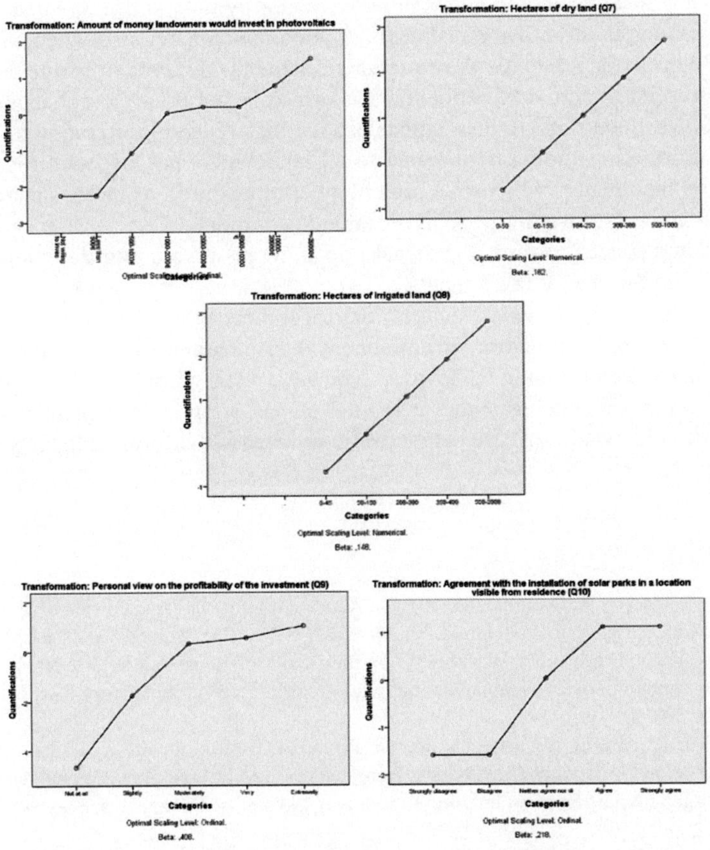

Fig. 1. Transformation plots of dependent and independent variables

4 Conclusions

Investments in photovoltaics made by landowners can contribute significantly to the achievement of renewable targets and promote the diffusion of renewables. This study provides new insights into the decision-making of potential investors in photovoltaics as it explored the factors affecting the amount of money landowners would invest in photovoltaics.

Our analysis showed that landowners' view on the profitability of the investment, their agreement with the installation of solar parks in a location visible to their residence as well as the amount of hectares they own, is what mostly affects the amount of money they invest in photovoltaics. It was also indicated that as these variables increase, so does the investment amount. That is, positive attitudes to the visual impact of solar installations can also increase the amount of money invested in photovoltaics. Moreover, the more hectares of dry and irrigated land landowners own, the more money they invest.

These new insights should be taken into account by policymakers when designing strategies aiming at attracting investments in photovoltaics but also when considering ways to turn investment intentions into actual investments. In order to ensure that willing investors proceed to actual investments, it is critically important to enhance the profitability of investments and to disseminate profitability-related information to landowners. This information should be directed to all landowners but a special focus should be placed on landowners who own a high number of hectares, as there is evidence that these potential investors will invest higher amounts of money. Policymakers should also be aware that attitudes towards the visual impact of photovoltaic installations may be associated with the size of investments.

Finally, this study provides context for researchers who want to further analyze the influence of various factors on investment decisions and to detect those that can encourage investments among landowners who are overall willing to invest but hesitate. Moreover, it can be concluded that it is vital to pay attention to potential investors' mindset as if landowners are certain that the investment will be profitable, they are likely to make investments.

References

1. Roddis, P., Carver, S., Dallimer, M., Ziv, G.: Accounting for taste? Analysing diverging public support for energy sources in Great Britain. Energy Res. Soc. Sci. **56**, 101226 (2019)
2. Sposato, R.G., Hampl, N.: Worldviews as predictors of wind and solar energy support in Austria: bridging social acceptance and risk perception research. Energy Res. Soc. Sci. **42**, 237–246 (2018)
3. Koirala, B.P., Araghi, Y., Kroesen, M., Ghorbani, A., Hakvoort, R.A., Herder, P.M.: Trust, awareness, and independence: insights from a socio-psychological factor analysis of citizen knowledge and participation in community energy systems. Energy Res. Soc. Sci. **38**, 33–40 (2018)
4. Flacke, J., De Boer, C.: An interactive planning support tool for addressing social acceptance of renewable energy projects in the Netherlands. ISPRS Int. J. Geo-information **6**(10), 313 (2017)
5. Gamel, J., Menrad, K., Decker, T.: Which factors influence retail investors' attitudes towards investments in renewable energies? Sustain. Prod. Consumption **12**, 90–103 (2017)
6. Curtin, J., McInerney, C., Gallachóir, B.Ó., Salm, S.: Energizing local communities—what motivates Irish citizens to invest in distributed renewables? Energy Res. Soc. Sci. **48**, 177–188 (2019)
7. Masini, A., Menichetti, E.: Investment decisions in the renewable energy sector: an analysis of non-financial drivers. Technol. Forecasting Soc. Change **80**, 510–524 (2013)
8. Hai, M.A., Moula, M.M.E., Seppälä, U.: Results of intention-behaviour gap for solar energy in regular residential buildings in Finland. Int. J. Sustain. Built Environ. **6**(2), 317–329 (2017)
9. Kastner, I., Matthies, E.: Investments in renewable energies by German households: a matter of economics, social influences and ecological concern? Energy Res. Soc. Sci. **17**, 1–9 (2016)
10. Claudy, M.C., Michelsen, C., O'Driscoll, A., Mullen, M.R.: Consumer awareness in the adoption of microgeneration technologies. Renewable Sustain. Energy Rev. **14**, 2154–2160 (2010)

11. Parkins, J.R., Rollins, C., Anders, S., Comeau, L.: Predicting intention to adopt solar technology in Canada: the role of knowledge, public engagement, and visibility. Energy Policy **114**, 114–122 (2018)
12. Michelsen, C.C., Madlener, R.: Switching from fossil fuel to renewables in residential heating systems: an empirical study of homeowners' decisions in Germany. Energy Policy **89**, 95–105 (2016)
13. Palm, J., Tengvard, M.: Motives for and barriers to household adoption of small-scale production of electricity: examples from Sweden. Sustain.: Sci. Practice Policy **7**, 6–15 (2011)
14. Zhai, P., Williams, E.D.: Analyzing consumer acceptance of photovoltaics (PV) using fuzzy logic model. Renew. Energy **41**, 350–357 (2012)
15. Braito, M., Flint, C., Muhar, A., Penker, M., Vogel, S.: Individual and collective socio-psychological patterns of photovoltaic investment under diverging policy regimes of Austria and Italy. Energy Policy **109**, 141–153 (2017)
16. Strazzera, E., Statzu, V.: Fostering photovoltaic technologies in Mediterranean cities: Consumers' demand and social acceptance. Renew. Energy **102**, 361–371 (2017)
17. Ek, K., Persson, L., Johansson, M., Waldo, Å.: Location of Swedish wind power—Random or not? A quantitative analysis of differences in installed wind power capacity across Swedish municipalities. Energy Policy **58**, 135–141 (2013)
18. Bergek, A., Mignon, I., Sundberg, G.: Who invests in renewable electricity production? Empirical evidence and suggestions for further research. Energy Policy **56**, 568–581 (2013)
19. Carlisle, J.E., Solan, D., Kane, S.L., Joe, J.: Utility-scale solar and public attitudes toward siting: a critical examination of proximity. Land Use Policy **58**, 491–501 (2016)
20. Tabi, A., Wüstenhagen, R.: Keep it local and fish-friendly: social acceptance of hydropower projects in Switzerland. Renew. Sustain. Energy Rev. **68**, 763–773 (2017)
21. Leete, S., Xu, J., Wheeler, D.: Investment barriers and incentives for marine renewable energy in the UK: an analysis of investor preferences. Energy Policy **60**, 866–875 (2013)
22. Tsantopoulos, G., Arabatzis, G., Tampakis, S.: Public attitudes towards photovoltaic developments: case study from Greece. Energy Policy **71**, 94–106 (2014)
23. Siardos, G.: Multivariable Statistical. Part II. Siardos, Thessaloniki (2000)

Factors and Barriers Towards Investments in Photovoltaics Among Landowners: A Case Study from Greece

Evangelia Karasmanaki(✉), Despoina Sfiri, and Georgios Tsantopoulos

Department of Forestry and Management of the Environment and Natural Resources, Democritus University of Thrace, Pantazidou 193, 68200 Orestiada, Greece
evagkara2@fmenr.duth.gr

Abstract. Renewable energy production is undergoing an unprecedented expansion across Europe but investments are still required in order to transition to a sustainable energy future. A deeper understanding of the factors affecting landowners' decision to invest in photovoltaics and the barriers preventing such investments is critically important for policymakers and energy developers. Hence, the aim of this study is to examine the economic and environmental factors affecting the decision to invest in photovoltaics as well as to investigate the barriers inhibiting such investments among landowners in a Greek rural area who are willing to invest in photovoltaics. Results obtained via structured questionnaires show that the investment decision is highly and positively affected by economic advantages. In specific, earning a stable income, reduced electricity costs and low taxation on investment are the most important economic factors. The respondents are also driven by their desire to contribute to flora and fauna conservation as well as to the reduction of air pollution which is caused by conventional energy production. However, high interest rates on bank loans for renewable energy investments along with high initial capital requirement prevent respondents from making such investments. The results presented in this paper not only assist a more thorough understanding of the factors and barriers affecting investments but also consist a basis for improving the relevant policies to facilitate the implementation of investments in renewable energy.

Keywords: Willingness to invest · Small-scale photovoltaic installations · Landowners' perceptions

1 Introduction

The impressive technical advances in solar photovoltaic systems along with supporting policies have rendered solar energy a realistic option for decarbonizing energy systems and achieving ambitious energy targets. Moreover, the prices of photovoltaic modules have decreased and the cost of owning a photovoltaic system has fallen [1]. At the same time, PV systems acquired by end-users can improve the visibility of renewable systems and allow individuals to become part of the "energy turnaround" [2].

J. R. da Costa Sanches Galvão et al. (Eds.): *Proceedings of the 1st International Conference on Water Energy Food and Sustainability (ICoWEFS 2021)*, pp. 344–351, 2021.
https://doi.org/10.1007/978-3-030-75315-3_39

At the end of 2019, solar installed capacity reached 580.1 GW, and off-grid PV reached 3.4 GW at global level [3] suggesting that the solar sector is experiencing a huge growth. Europe reached 138.2 GW of installed solar power, with 129.8 GW having being installed across the European Union. During the same period, in Greece solar installed capacity was 2.763 MW whereas in 2010 the same rate was as low as 202 MW [3]. This remarkable increase can, to some degree, be ascribed to the country's compliance with the European legislative framework which promoted energy generation from renewable and mandated member-states to meet specific energy targets (with the enforcement of Law 2009/28/EC and Law 3851/2010).

According to analysts, there is a strong tendency to view energy transition as being technology-driven, however, technology is effective only if it is accompanied by changes in individuals' behavior, perceptions and norms [4]. Therefore, insights into individuals' behavior and thoughts are required [5]. When it comes to investments in renewable systems, researchers have been interested in understanding the barriers underlying investment decisions. Most importantly, reduced economic means, such as lack of savings and inaccessible loans, was found to be the most significant barrier [6]. Another reported barrier was that potential investors are often insufficiently informed or misinformed on the real costs and economic characteristics of investments in renewables [6, 7]. Most of the existing studies focus on citizens and house owners, whereas there is a paucity of research when it comes to landowners. Moreover, most of these works measure investment willingness and detect the factors affecting it, while very few studies examine the perceptions of individuals who are already willing to invest in renewable energy. In addition, most willingness-to-invest studies examine investments in wind energy and residential photovoltaics but neglect photovoltaics on lots of land. Hence, the aim of this paper is to examine the economic and environmental factors affecting the decision to invest in photovoltaics as well as to investigate the barriers inhibiting such investments among landowners in a Greek rural area who are willing to invest in photovoltaics.

2 Methodology

The population under study was landowners in the Municipal Unit of Didymoteicho, which is located in the Prefecture of Evros, Greece. According to the last national population census conducted in 2011, the study area has 9.263 residents and, in terms of agricultural production, it is considered an important rural area. The findings presented in this paper are part of a larger research which was conducted from June 2020 to September 2020 and analyzed landowners' views on investments in photovoltaic systems on their land. However, in this paper we focused explicitly on landowners who are willing to invest in photovoltaics on their land.

To achieve the research objectives, a questionnaire was designed. Variables were measured at nominal and ordinal scales. To make sure that all items in the questionnaire would be fully understood by respondents, a pilot study was conducted on a limited scale and few minor changes were made in the phrasing and formulation of few items.

As for the sampling method of the study, the respondents were selected by simple random sampling. Simple random sampling was considered the most appropriate method because it is quite easy to perform and requires very little knowledge about the population under study. In total, 235 landowners participated in the study and among these respondents, 205 were willing to invest in photovoltaics. In this paper, we analyze the responses of these 205 willing landowners using descriptive statistics and the non-parametric Friedman test.

3 Results

3.1 Respondents' Demographic Characteristics

Data on respondents' demographic characteristics were collected. There were more male than female respondents in the sample (50.2% vs 49.8%) and most were aged over 41 (69.7%) and married (68.8%). A high share of respondents (42.9%) stated that farming was their main profession. The percentages of private (14.1%) and public employees (11.2%) were also considerable. Moreover, 29.8% were high school graduates and 11.2% were middle school graduates. Finally, a high share of respondents (44.4%) earned between 10.001 and 30.000 per year.

3.2 Respondents' Views on Economic Factors Affecting Their Decision to Invest in Photovoltaics

Respondents then evaluated a set of economic factors which would affect their decision to invest in solar photovoltaics. As shown in Table 1, about seven out of ten respondents agreed and strongly agreed that reduced electricity expenses, increased income due to the sale of solar energy and the acquisition of a stable income affect their decision to install photovoltaics on their farmland. Although all economic factors were highly rated by respondents, respondents seemed to be quite confused about certain economic factors and, in specific, they neither agreed nor disagreed with moderate investment cost, short depreciation time and protection against energy price fluctuations due to geopolitical crises.

Table 1. Percentages regarding economic factors affecting respondents' investment decision.

	Strongly disagree	Disagree	Neither agree nor disagree	Agree	Strongly agree
High subsidies	0	1.5	36.1	48.8	13.7
Increased income due to the sale of produced energy	1	1	28.8	48.8	20.5
Reduced electricity bill	0	2.9	25.9	44.9	26.3
Low taxation	0.5	3.9	32.7	32.7	30.2
Acquisition of stable income	0.5	2.9	24.9	45.4	26.3
Capital investment	0	3.9	36.1	42.0	18.0
Moderate investment cost	0	6.3	37.1	39.0	17.6
Short depreciation time of the investment	1.5	6.8	36.1	41.0	14.6
Protection against energy price fluctuations due to geopolitical crises	0.5	4.9	36.1	36.1	22.4

Statistical analysis was then carried out by the non-parametric Friedman test. As presented in Table 2, the "Acquisition of a stable income" ranked first (mean rank 5.51), whereas "Short depreciation time of the investment" ranked last (mean rank 4.37).

Table 2. The application of the non-parametric test for ranking economic factors affecting respondents' investment decision.

Variables	Mean rank
High subsidies	4.85
Increased income due to the sale of produced energy	5.22
Reduced electricity bill	5.47
Low taxation	5.29
Acquisition of stable income	5.51
Capital investment	4.74
Moderate investment cost	4.64
Short depreciation time of the investment	4.37
Protection against energy price fluctuations due to geopolitical crises	4.90
$N = 205$ Chi-Square = 52.136 $df = 8$ $p < 0.001$	

3.3 Respondents' Views on Environmental Factors Affecting Their Decision to Invest in Photovoltaics

Next, respondents were asked to assess various environmental factors which would affect their decision to invest in photovoltaics. According to Table 3, the strong majority of respondents (by 83.4%) agreed or strongly agreed that investments in photovoltaics can contribute to the reduction of air pollution. Moreover, 81.9% stated that investments can halt the depletion of natural resources.

Table 3. Respondents' degree of agreement with various environmental factors affecting investment decision.

	Strongly disagree	Disagree	Neither agree nor disagree	Agree	Strongly agree
Contribution to the reduction of air pollution	1.0	3.9	11.7	52.2	31.2
Contribution to the reduction of natural resources	0	1.5	16.6	46.8	35.1
Protection of flora	0.5	0	17.6	43.9	38.0
Protection of fauna	0.5	0	16.1	47.3	36.1

Then, to detect statistical differences among respondents' responses regarding the factors affecting the decision to install photovoltaics on their farmland, statistical analysis was conducted using the non-parametric Friedman test. According to the analysis

(Table 4), "Protection of flora" was ranked first (mean rank 2.55), followed by "Protection of fauna" (mean rank 2.54) whereas the "Contribution to the reduction of air pollution" ranked last (mean rank 2.39).

Table 4. Hierarchy of factors affecting the decision to install photovoltaics on farmland

Variables	Mean rank
Contribution to the reduction of air pollution	2.39
Contribution to the reduction of natural resources	2.52
Protection of flora	2.55
Protection of fauna	2.54
N = 205 Chi-Square = 6.936 df = 3 p < 0.001	

3.4 Respondents' Views on the Barriers Preventing Them from Making Investments in Photovoltaics

Then, respondents were asked to state the degree to which various barriers prevent them from investing in photovoltaic systems. Results in Table 5 show that 81.4% of landowners stated that high interest rates of bank loans which are available for investments in renewables prevent them from making actual investments in photovoltaics. High taxes on investments in photovoltaics as well as the complexity of loan application process were reported as other significant barrier by 76.6% and 72.7% of respondents, respectively.

Table 5. Percentages regarding the degree that various barriers inhibit investments in photovoltaics

	Strongly disagree	Disagree	Neither agree nor disagree	Agree	Strongly agree
High initial capital	2.0	2.9	15.1	35.1	41.0
Lack of subsidies	1.0	2.0	21.5	34.6	35.1
High taxation	0	1.0	17.1	39.5	37.1
The instability of the energy institutional framework	0	2.0	22.0	37.1	35.1
Bureaucratic license process	0.5	2.9	22.4	30.7	39.0
The unreliability of the electricity distribution network	1.5	2.9	24.9	34.1	28.8
The lack of experienced technical staff for the installation of the photovoltaics on farmland	2.9	4.4	23.9	34.6	25.4
The lack of experienced technical staff for the maintanance of the photovoltaics on farmland	1.5	5.9	26.8	36.6	22.9
Lack of trust in the technological quality and efficiency of photovoltaic systems	2.0	5.9	33.2	30.7	22.4
Operational problems due to weather	0.5	5.9	31.7	32.2	22.9

(continued)

Table 5. (*continued*)

	Strongly disagree	Disagree	Neither agree nor disagree	Agree	Strongly agree
Difficulty in obtaining relevant information	2.9	5.4	28.3	33.7	25.4
Complex loan application	1.0	3.9	20.5	36.1	36.6
High interest rates of loans for investments in photovoltaics	0.5	1.5	15.1	38.5	42.9
Limited competitiveness of renewable energy due to the lower production costs of conventional energy sources	2.0	3.4	24.4	35.6	25.4

The barriers were then analyzed using the non-parametric Friedman Test. According to the results of this analysis (Table 6), "High interest rates of loans for investments in photovoltaics" (mean rank 8.32), "High taxation" (mean rank 8.26) and "High initial capital" (mean rank 8.15) received the highest rankings. In contrast to these variables, "Lack of trust in the technological quality and efficiency of photovoltaic systems" (mean rank 6.46) was ranked last.

Table 6. The application of the non-parametric test for ranking the barriers preventing landowners from investing in photovoltaic systems

Variables	Mean rank
High initial capital	8.15
Lack of subsidies	7.91
High taxation	8.26
The instability of the energy institutional framework	7.71
Bureaucratic license process	7.86
The unreliability of the electricity distribution network	7.56
The lack of experienced technical staff for the installation of the photovoltaics on farmland	7.37
The lack of experienced technical staff for the maintanance of the photovoltaics on farmland	6.88
Lack of trust in the technological quality and efficiency of photovoltaic systems	6.46
Operational problems due to weather	6.86
Difficulty in obtaining relevant information	6.57
Complex loan application	7.53
High interest rates of loans for investments in photovoltaics	8.32
Limited competitiveness of renewable energy due to the lower production costs of conventional energy sources	7.55
N = 205 Chi-Square = 76.083 df = 13 p < 0.001	

4 Discussion

Renewable energy sources can potentially play a dynamic role in decreasing carbon emissions from electricity production. Nevertheless, substantial investments made in renewable energy systems such as solar photovoltaics are required in order to realize this potential. In this regard, future policymaking needs to incorporate an understanding of how landowners, who are willing to invest, perceive investments and what affects their decision-making.

The analysis presented in this paper has brought some interesting findings to light. In terms of the economic characteristics of investments in photovoltaics installed on farmland, our analysis revealed that the acquisition of stable income, revenues from selling the produced energy and lower energy bills are the economic factors that have a major effect on the decision to invest in photovoltaics. However, the respondents appeared to be somewhat confused about some other factors. In specific, they were confused about the depreciation time of investments in photovoltaics, the height of investment costs as well as whether investments in photovoltaics protect from energy price fluctuations which result from geopolitical crises. This confusion suggests that respondents may lack relevant information which, in turn, may make them hesitant to proceed to actual investments. Therefore, energy developers and those responsible for disseminating information on the economic characteristics of investments in photovoltaics ought to ensure that potential investors obtain adequate information on these important matters. However, the provision of information should not be neglected because the findings of Ebers Broughel and Hampl [6] as well as Tsantopoulos et al. [7] have indicated that investments are unlikely to proceed when potential investors are insufficiently informed or misinformed about the economic characteristics and real costs of investments in renewables.

Moreover, our study has confirmed previous findings which shown that environmental awareness and concerns about the environment play a role in individuals' decision to pay for on invest in environmentally-benign energy technologies [2, 7]. In specific, a surprisingly high percentage of respondents in our study stated that investments in photovoltaics can contribute to the reduction of air pollution and also halt the depletion of natural resources which are used for electricity production.

Special attention should also be given to landowners' perceived investment barriers because these landowners have not invested so far despite their willingness. In essence, the perceived barriers can reveal what explicitly prevents investment willingness to turn into actual investment. For the strong majority of respondents, high interest rates of bank loans which are available for investments in renewable energies along with the complexity of loan applications prevented them from making investments. This finding is in line with the results of a study conducted by Ebers Broughel and Hampl [6] where many respondents reported that inaccessible loans hindered their investment. Hence, these findings bring to surface the need to provide bank loans with favorable conditions to those interested in making small-scale investments but also to landowners who express their interest in installing photovoltaics on their land. Nevertheless, there are more barriers that need to be alleviated in order to promote investments. In specific, the lack of subsidies, the bureaucracy of licensing process as well as the high initial capital cost were regarded by respondents as significant barriers to investments. These barriers need to be addressed in the future design of policies which aim at increasing investments by landowners in photovoltaics. Although the creation of a more favorable investment environment is particularly challenging in a period of global economic slowdown, efforts should not be neglected because the deployment of renewable energy can help avert the most devastating possible effects of our changing climate.

5 Conclusions

This paper explored the factors and barriers that affect the decision to invest in photovoltaic systems among agricultural landowners who are already willing to invest. Our findings can help policymakers and other actors involved in renewable investments understand why potential investors do not proceed to actual investments and can guide them as to the steps they can take to reduce existing legal and market barriers and encourage investments.

Acquiring a stable income and increasing their existing income through selling the produced energy emerged as the most important factors that affect positively landowners' decision to invest. In addition, reduction in electricity bills due to installing photovoltaic systems was another significant economic factor with a positive effect on investment decisions. As for environmental factors, landowners were positively affected by their wish to protect flora and fauna as well as to contribute to the issue of the depletion of natural resources which is caused by conventional energy production. However, certain barriers prevent respondents from proceeding to actual investments. In specific, the high initial capital which is required for investments along with high interest rates on loans granted for renewable investments and high taxation on these investments are the barriers that prevent landowners from installing photovoltaics on their land.

In order to overcome these barriers, policymakers and other actors involved in renewable investments ought to reduce the interest rates of loans granted for investments in photovoltaic systems while easier access of potential investors to loans should be provided to interested agricultural landowners. At the same time, governments should establish measures to reduce the capital cost for investments in photovoltaic systems. Finally, since landowners perceive that photovoltaic installations on their land can offer them new or additional income and reduce electricity bills, future strategies aiming at promoting investments should highlight these benefits in order to attract more investors.

References

1. Goodrich, A., James, T., Woodhouse, M.: Residential, commercial, and utility-scale photovoltaic (PV) system prices in the United States: current drivers and cost-reduction opportunities (No. NREL/TP-6A20–53347). National Renewable Energy Lab (NREL). United States (2012)
2. Baur, L., Uriona, M.: Diffusion of photovoltaic technology in Germany: a sustainable success or an illusion driven by guaranteed feed-in tariffs? Energy **150**, 289–298 (2018)
3. International Renewable Energy Agency. https://www.irena.org/solar. Accessed 15 Jan 2021
4. Karasmanaki, E., Tsantopoulos, G.: Exploring future scientists' awareness about and attitudes towards renewable energy sources. Energy Policy **131**, 111–119 (2019)
5. Schot, J., Kanger, L., Verbong, G.: The roles of users in shaping transitions to new energy systems. Nat. Energy **1**(5), 1–9 (2016)
6. Ebers Broughel, A., Hampl, N.: Community financing of renewable energy projects in Austria and Switzerland: profiles of potential investors. Energy Policy **123**, 722–736 (2018)
7. Tsantopoulos, G., Arabatzis, G., Tampakis, S.: Public attitudes towards photovoltaic developments: case study from Greece. Energy Policy **71**, 94–106 (2014)

Anaerobic Digestion of Cork Boiling Wastewater – Effect of Agitation

Vitor de Oliveira Matos[1], Luís do Carmo-Calado[1], Victor Arruda Ferraz de Campos[2], Pedro Romano[1], Paulo Sérgio Duque de Brito[1], Luiz Rodrigues[1], and Roberta Mota-Panizio[1](✉)

[1] VALORIZA - Research Center for Endogenous Resource Valorization, Polytechnic Institute of Portalegre, Campus Politécnico 10, 7300-555 Portalegre, Portugal
pbrito@ipportalegre.pt

[2] School of Engineering, Guaratinguetá, Department of Energy, Laboratory of Optimization of Energy Systems (LOSE), Institute of Bioenergy Research (IPBEN - UNESP) - Associated Laboratory of Guaratinguetá, São Paulo State University (UNESP), São Paulo, Brazil

Abstract. The main objective of the work was to perform the energy recovery of cork boiling wastewater (CBW) through anaerobic digestion process. The experiments were performed in reactors of work volume of 6 L and different substrate and co-substrate ratios, with temperature control and mechanical agitation, for obtaining greater biogas and biomethane amounts. The operational temperature was mesophilic (38 °C) with the agitation of 15 rpm being carried out for five minutes three times a day. The reactors were fed with different ratios of Inoculum (IN), CBW and cow manure (CM). Two different mixtures were made, namely IN: CBW and IN:CBW/CM, for ratios of 1:2 and 1:2/1 respectively, with agitation being performed 3 times a day and without any agitation.. For tests IN:CBW no continuous agitation and with agitation there was no significant biogas production. For the IN:CBW/CM mixture with continuous agitation in the ratio 1:2/1, there was no biogas production due to the high conductivity and excess production of volatile fatty acids. For IN mixture: CBW/CM no agitation had an accumulated production of 10 L of biogas in 33 days. With the tests carried out, it can be concluded that the process of anaerobic digestion of CBW can be a viable alternative for energy recovery.

Keywords: Cork boiling wastewater · Anaerobic digestion · Biogas · Cow Manure

1 Introduction

Anaerobic digestion (AD) is one of the most efficient and ecological technologies in the treatment of wastewater with a high organic load, treating not only organic fraction, but also making it possible to reach an important source of renewable energy due to the

J. R. da Costa Sanches Galvão et al. (Eds.): *Proceedings of the 1st International Conference on Water Energy Food and Sustainability (ICoWEFS 2021)*, pp. 352–360, 2021.
https://doi.org/10.1007/978-3-030-75315-3_40

concomitant production of biomethane and, in addition, of organic fertilizer for plants [1].

One of these residues that still constitutes a huge environmental problem, but with a huge potential for transformation into energy and chemicals is cork boiling water (CBW). Portugal is the largest producer of cork in the world, reaching 52% of the cork extracted in the world, with 60% of this national extraction being transformed into cork products by national companies. The cork transformation process involves large amounts of water, about 400 L per ton of processed cork, and the current production of cork is around 100 mil tons per year, only in Portugal [2, 3]. This situation is aggravated by the fact that the sector works in discontinuous operation and many companies are small, so the treatment of effluents resulting from the cooking of cork has been thrown in natural environments, without any previous treatment, over time., since the beginning of the industry in Portugal around the 1960s [2].

Over many decades, this effluent was discharged directly into water due to its difficulty in biological treatment, violating the legal emission limits of Decree-Law 236/98 [4, 5].

Currently, several methods have been studied to treat these effluents, among them, physical, physicochemical and biological.

Physical methods such as nanofiltration membrane/modules using hydraulic permeability to filter and purify the water from organic and heavy compounds reaching COD, BOD_5, color and TOC rejection values remained well above the 90% value [6], although the permeate characterization showed that it cannot be directly discharged to the environment as it does not fulfil the values of the Portuguese discharge legislation.

Physicochemical method involving Fenton oxidation and coagulation/flocculation for the removal of chemical oxygen demand (COD), total polyphenols (TP), and aromatic compounds (AC) from cork manufacturing process wastewater reach COD reduction from 22% to 85%, TP from 4% to 98% and aromatic compounds from 22% to 85% [7].

Treatment of cork boiling wastewater using chemical oxidation and biodegradation using minor amounts of oxidative reagents and culture enrichments composed of cork boiling wastewater autochthonous microbial biota, could stand the stressful conditions generated by the chemical oxidation reagents. Such combination of methodologies seems a good alternative for cork boiling wastewater treatment, because it promotes extensive TOC reduction in a short period of time. Additionally, it may be used during the cooling period, allowing the release of a pre-treated effluent [8].

Anaerobic digestion (AD) is a complex microbial process that occurs naturally in the absence of oxygen, in which it involves a series of metabolic reactions for the decomposition of organic substances in biogas and organic fertilizer [1]. It is the only biological waste recovery technology capable of producing biogas, an average mixture of 55–75% biomethane (CH_4) and 25–45% carbon dioxide (CO_2) with a calorific value of 21–24 MJ /m^3 [9], this wide range of average values was due to the fact that the vast array of organic compounds constituting biomass, in addition to the many physical-chemical and microbial parameters necessary for the resolution of this biological process [1, 10].

2 Materials and Methods

The digestion system was composed of four benchtop reactors with a capacity of 8 L, as described in Fig. 1. Each digester consists of a temperature sensor (1), which allows for temperature control in the digesting media; an inlet valve (2), which allows for feeding and digestate sampling; a collection tube and exhaust valve (3), which connect the digester to a biogas collector; a graduated cylinder (4), which acts as a water displacement system and is filled with acidified water to avoid CO_2 dissolution; and a gas analysis valve (5) at the top of the biogas collector, which allows for biogas sampling for further analysis. The digestion experiments were performed under discontinuous mode with a working volume of 6 L.

The tests were performed in order to be able to compare the influence of continuous agitation with periodic agitation. Two different mixtures were performed in duplicate where the conditions of continuous and no continuos agitation (3 times a day, for 5 min) were placed. For the IN:CBW mixture the mixing ratio was 1:2, for the IN:CBW/CM mixture, the ratio used was 1:2/1.

Fig. 1. Schematic representation of anaerobic digester and biogas collection system.

2.1 Characterization of Biomass

The substrates and inoculum were characterized based on selected physicochemical and biochemical parameters relevant for AD. Total solids (TS), and volatile solids (VS) were determined gravimetrically. The ultimate analysis was performed using a CHNS elemental analyzer (TrueSpec CHN Levo, 4084 series). The pH and redox potential were determined with a pH meter (Basic 20, Crison). Alkalinity was determined by titration according to Standard Methods, Section 2320. VFA was determined using the Kapp titration method. Chemical oxygen demand (COD) was determined by the standard dichromate method using an Aqualytic AL32 thermoreactor along with a spectrophotometer (Aqualytic COD Vario-PC compact). Ammonia nitrogen (NH4-N) analysis was performed photometrically using test kits (NANOCOLOR® Ammonium 200, PF12 photometer, Macherey–Nagel). Temperature, pH, and biogas production were monitored daily. TS, VS, and C/N ratio were monitored regularly, usually each week. Biogas volume was measured via water displacement and immediately analyzed by a portable gas analyzer (GasData GFM406), which allows for measuring the volume percentages of CO_2, CH_4, O_2, H2S, and CO in the mixture.

3 Results and Discussions

Table 1 describes the characterizations carried out on the substrate, inoculum and co-substrates used in the AD process. CBW is the only substrate that has an acid character, with a pH of 4.6. The I and the other co-substrates used have a neutrality, helping the CBW to reach the optimum pH (between 6.5 and 7.5) so that the AD process occurs in a favorable way. CBW waters are rich in phenolic-polyphenolic compounds, which characterizes them as an acid effluent, with an acid pH between 4.5 and 5.5 [4, 8, 11]. The pH of the CM was not possible to perform the measurement, however according to Kozan and Avarenga the pH varies between 6.5 and 7.8 [12].

The CBW substrate used has a COD of 7060.0 mg/L. According to the literature, the CBW effluent presents COD with very wide values, varying between 2,600 and 12600 mg/L [8, 13]. These values are dependent on the quality of the boards, the number of times the water is reused, the location of the cork oak forest, among other factors that can change the characteristics of the raw material.

According to Pintor et al. and Gonçalves et al. the conductivity of CBWs varies between 1,200 and 1,500 μS/cm [14, 15]. Also according to Gonçalves et al. the total solids of the effluents are in the order of 5200 mg/L [15].

Ponce-Robles et al. used in its tests the effluents from the association plant of the cork industries of San Vicente de Alcântara. Despite being from different lots, the pH (5.0) and conductivity (1684 μS/cm) values are relatively close. The authors characterized the amount of polyphenols present in the effluent, being in the order of 455 mg/L. Although the characterization of the total polyphenols in the effluent used for AD is not done, due to the similarities it is possible to assume that the values are close.

To correct the pH in the digesters a NaOH solution was added until approximately pH 7 was obtained, conductivity and dissolved solids give an idea of the amount of complex organic matter and to some degree inorganic matter in each digester and confirm that the substrate increases the conductivity parameters and solids in the preparation. The COD values, as expected from the initial analysis of the raw materials, are well above the values determined by the legislation to be sent to the bodies of water.

Digestor IN:CBW - no continuous agitation and with agitation, shows an increase in dissolved solids, which indicates an initial degradation of biological macromolecules, but the DA process in the initial faces stagnated or only a small percentage of these solids managed to reach the methanogenesis phase. In the digester IN:CBW/CM - no continuous agitation and with agitation, the decreases in these dissolved solids mean higher amounts of microbiological flora, consequently their greater and better digestate biodegradation.

There was a small decrease in COD values in the IN:CBW – no continuous agitation, with agitation digesters and in the IN:CBW/CM - with continuous agitation digester,

Table 1. Characterization of the raw materials.

Parameters	Unit	CBW	IN	CM
pH	-	4.6	6.7	-
Dissolved solids	mg/L	740	2840	-
Conductivity	μS/cm	1480	5770	-
C	(%)	44.17	35.08	36.56
H	(%)	6.03	6.06	5.91
N	(%)	3.7	7.52	3.44
S	(%)	0	0	0
O	(%)	29.96	11.42	27.13
Ratio C/N	-	12.06	4.66	10.63
COD	mg/L	7060	1639.5	84475
Total solids	mg/L TS	8375	14525	187250
Volatile solids	mg/L VS	6875	9500	136600

considering that a small part of the organic compounds were consumed or oxidized, however this parameter is not directly proportional to the biogas production. The opposite occurred in the IN:CBW/CM – no continuous agitation digester. In which the COD value after 33 days was significantly higher than the initial one and this one has the best biogas results. The increase in the concentration of COD in this digester is due, in particular, to the CBWs that after many days begin to biodegrade, increasing and accumulating to the production of volatile fatty acids (VFA) of the co-substrate during the phase of acidogenesis which can constitute up to 95% of carbon total organic [16]. For an effective decrease in the COD value in a digester in optimized production more than 33 days will be necessary, the reduction in the COD value continues until the end of the methanization leading to a reduction in the volume of biogas produced [17].

The total biogas production in digester IN:CBW/CM – no continuous agitation was 10 L, as shown in Fig. 2. Accumulated biomethane production, in which after 33 days biomethane production was still taking place in the IN:CBW/CM no continuou agitation digester. It had an accumulated production of approximately 7 L, Fig. 3.

3.1 Effect of Agitation

In general, the digestate is mixed by mechanical agitation, pumping and forced circulation by pressure of the gas produced. Mixing is a major challenge in achieving high AD performance, particularly in dry AD, since a thick paste requires more energy to homogenize the raw material. The comparison of the three mixing methods indicates

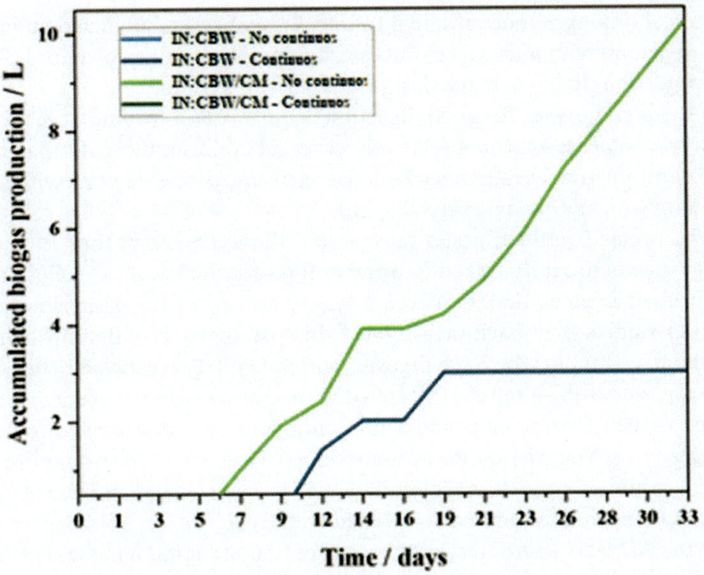

Fig. 2. Accumulated biogas production.

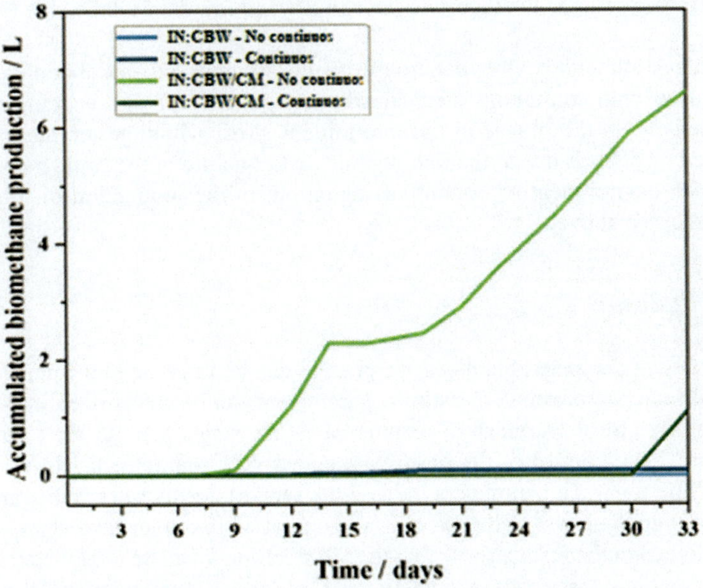

Fig. 3. Accumulated biomethane production.

that mechanical mixing is more efficient followed by gas circulation and pumped circulation. Co-digestion with manure exhibits strong non-Newtonian behavior [18], and the fluid rheology strongly impacts the design of a mixing system.

One of the prerequisites for good digestion is the uniform feeding of the digester, so that microorganisms are kept in a relatively constant concentration of organic solids at all times. Therefore, the digester must be fed at the same time every day with a balanced diet of the same quantity and quality [19, 20].

Since the bacteria in the digester have a very limited reach in food, it is necessary that the digestate is properly mixed to provide food for the bacteria. A light mixture is found to improve fermentation; however, a violent stirring of the paste slows digestion.

In several studies they have investigated this parameter at different intensities, as Hoffmann et al. (2008) in which the digester without agitation showed better CH_4 yield than in the digester with agitation, 100 rpm demonstrating to be a violent agitation [21].

Also the mode of continuous and semi-continuous agitation proved to be relevant in the efficiency of AD, where digesters continuously agitated according to McMahon et al. (2001), exhibited persistent accumulation of propanoic acid and less abundance of methanogens compared to minimally agitated studies [22].

During the AD tests with digesters without continuous agitation, it was observed that the production of CH4 reached a percentage in the biogas of 50%. While in the digesters that the stirring was continuous they did not produce practically biogas. Focusing only on the agitation parameter, these results validate that low rotations of continuous mechanical agitation are favorable to the digestate poor in SV and low% of total solids (very moist digestion).

In co-digestion with bovine manure the results are reversed, not obtaining biogas in the biodigester with continuous mechanical stirring (15 rpm) and reaching more than 71% of methane in the biogas in the intermittent stirring biodigester, also at 15 rpm when turned on. This is due to the rich organic matter nature of the cattle manure which results, in an environment of continuous agitation, in the production of inhibitors as already discussed above.

4 Conclusion

The efficiency of the anaerobic digestion process can be improved by controlling some vital operational parameters. Operational parameters can be categorized into basic and key parameters. Basic parameters include pH, C/N ratio, particle size, temperature and agitation. The control of the basic parameters will guarantee the ideal microbial activity. While the main parameters include the type of co-digestion substrate, type of pre-treatment applied, wet or dry process and operation in one or several stages.

In co-digestion the results show that the CBW effluent can be treated and recovered in DA. Without any pretreatment in CBW and CM (only a small initial pH correction in the digesters) a potential of 0.65 m^3 CH_4 kg/VS_{add} was obtained.

The agitation proved to be essential for the optimization of biogas production with different behaviors depending on the mixture and amount of volatile and total solids in each digester. However, a low speed agitation (15 rpm used) is key to a good operation in all types of parameters or mixtures. With rotation of continuous mechanical agitation being favorable to the digestate poor in VS and TS and with intermittent mechanical agitation beneficial to digestate with a greater amount of VS and TS, such as co-digestion with CM.

Acknowledgments. The authors are grateful for the support given by the Foundation for Science and Technology, I.P. (FCT), through the "Summer with Science" Program, within the Summer School "Technologies for a waste biorefinery", held at VALORIZA - Research Center for the Valorization of Endogenous Resources - Polytechnic Institute of Portalegre (05064). The authors are grateful to VALORIZA - Research Center for Endogenous Resource Valorization, Polytechnic Institute of Portalegre. The authors also would like to thank the Coordination for the Improvement of Higher Education Personnel (CAPES) and the Portuguese Foundation for Science and Technology (FCT) for the funding provided through project number 88881.156267/201701.

References

1. Kumar, A., Samadder, S.R.: Performance evaluation of anaerobic digestion technology for energy recovery from organic fraction of municipal solid waste: a review. Energy **197**, 117253 (2020). https://doi.org/10.1016/j.energy.2020.117253
2. Silva, S.P., Sabino, M.A., Fernandas, E.M., Correlo, V.M., Boesel, L.F., Reis, R.L.: Cork: properties, capabilities and applications. Int. Mater. Rev. **50**(6), 345–365 (2005). https://doi.org/10.1179/174328005X41168
3. APCOR, "Associação Portuguesa da Cortiça" (2020)
4. Guedes, A.M.F.M., Madeira, L.M.P., Boaventura, R.A.R., Costa, C.A.V.: Fenton oxidation of cork cooking wastewater - overall kinetic analysis. Water Res. **37**(13), 3061–3069 (2003). https://doi.org/10.1016/S0043-1354(03)00178-7
5. Ministério do Ambiente, *Decreto-Lei n.º236/98*. Decreto-Lei, pp. 3676–3722 (1998)
6. Oliveira, J., Nunes, M., Santos, P., Cantinho, P., Minhalma, M.: Cork processing wastewater treatment/valorisation by nanofiltration. Desalin. Water Treat. **11**(1–3), 224–228 (2009). https://doi.org/10.5004/dwt.2009.805
7. Beltrán De Heredia, J.B., Domínguez, J.R., López, R.: Treatment of cork process wastewater by a successive chemical-physical method. J. Agric. Food Chem. **52**(14), 4501–4507 (2004). https://doi.org/10.1021/jf035289t
8. Dias-Machado, M., Madeira, L.M., Nogales, B., Nunes, O.C., Manaia, C.M.: Treatment of cork boiling wastewater using chemical oxidation and biodegradation. Chemosphere **64**(3), 455–461 (2006). https://doi.org/10.1016/j.chemosphere.2005.11.067
9. Bond, T., Templeton, M.R.: History and future of domestic biogas plants in the developing world. Energy Sustain. Dev. **15**(4), 347–354 (2011). https://doi.org/10.1016/j.esd.2011.09.003
10. Náthia-Neves, G., Berni, M., Dragone, G., Mussatto, S.I., Forster-Carneiro, T.: Anaerobic digestion process: technological aspects and recent developments. Int. J. Environ. Sci. Technol. **15**(9), 2033–2046 (2018). https://doi.org/10.1007/s13762-018-1682-2
11. Silva, C.A., Madeira, L.M., Boaventura, R.A., Costa, C.A.: Photo-oxidation of cork manufacturing wastewater, vol. 55, pp. 19–26 (2004). https://doi.org/10.1016/j.chemosphere.2003.11.018
12. Konzen, E.A., Alvarenga, R.C.: Adubação Orgânica, *Embrapa* (2019)
13. Picado, A., Silva, L., Anselmo, A.M.: Ecotoxicological evaluation of cork-boiling wastewaters, vol. 66, pp. 384–390 (2007). https://doi.org/10.1016/j.ecoenv.2006.02.013

14. Pintor, A.M.A., Vilar, V.J.P., Boaventura, R.A.R.: Decontamination of cork wastewaters by solar-photo-Fenton process using cork bleaching wastewater as H2O2 source. Sol. Energy **85**, 579–587 (2011). https://doi.org/10.1016/j.solener.2011.01.003
15. Gonçalves, M.R., Gil, L., Marques, I.P.: Cork boiling wastewater management by anerobic digestion. In: 4th International Conference on Engineering for Waste and Biomass Valorisation, pp. 6–11 (2012)
16. Harmsen, J.: Identification of organic compounds in leachate from a waste tip. Water Res. **17**(6), 699–705 (1983). https://doi.org/10.1016/0043-1354(83)90239-7
17. Bouabdesselam, H., Khelafi, M., Ahmed, T.: Energetic Valorization of Adrar City Step Sludge by Batch Solar Digester, no. February 2017 (2016)
18. Wu, B.: CFD simulation of mixing for high-solids anaerobic digestion. Biotechnol. Bioeng. **109**(8), 2116–2126 (2012). https://doi.org/10.1002/bit.24482
19. Giuliano, A., Bolzonella, D., Pavan, P., Cavinato, C., Cecchi, F.: Co-digestion of livestock effluents, energy crops and agro-waste: feeding and process optimization in mesophilic and thermophilic conditions. Bioresour. Technol. **128**, 612–618 (2013). https://doi.org/10.1016/j.biortech.2012.11.002
20. Wang, X., Yang, G., Feng, Y., Ren, G., Han, X.: Optimizing feeding composition and carbon–nitrogen ratios for improved methane yield during anaerobic co-digestion of dairy, chicken manure and wheat straw. Bioresour. Technol. **120**, 78–83 (2012). https://doi.org/10.1016/j.biortech.2012.06.058
21. Hoffmann, R.A., Garcia, M.L., Veskivar, M., Karim, K., Al-Dahhan, M.H., Angenent, L.T.: Effect of shear on performance and microbial ecology of continuously stirred anaerobic digesters treating animal manure. Biotechnol. Bioeng. **100**(1), 38–48 (2008)
22. McMahon, K.D., Stroot, P.G., Mackie, R.I., Raskin, L.: Anaerobic codigestion of municipal solid waste and biosolids under various mixing conditions—II: microbial population dynamics. Water Res. **35**(7), 1817–1827 (2001)

Analysis of Specific Energy Consumption of Wastewater Treatment Plants in the North of Portugal

Bruno J. Cardoso[1(✉)], Eugénio Rodrigues[1], Adélio R. Gaspar[1], and Álvaro Gomes[2]

[1] ADAI, Department of Mechanical Engineering, University of Coimbra, Rua Luís Reis Santos, Pólo II, 3030-788 Coimbra, Portugal
bruno.cardoso@adai.pt

[2] INESC Coimbra, Department of Electrical and Computer Engineering, University of Coimbra, Rua Sílvio Lima, Pólo II, 3030-290 Coimbra, Portugal

Abstract. Wastewater treatment plants (WWTPs) are becoming more energy intensive due to a growing urban population, an increment in the contaminant load to be treated, and a stringent regulatory and environmental protection standard for effluent quality and water reuse. As the energy demand in the water sector will continue to grow in the forthcoming decades, benchmark the energy performance of WWTPs is essential to improve their energy efficiency. Typically, the specific energy consumption of WWTPs is expressed through two common key performance indicators (KPIs): energy consumption per cubic meter of treated wastewater (kWh/m^3) and per unit of chemical oxygen demand removed (kWh/kg COD rem). The first KPI (kWh/m^3) is the most applied due to its reliability and accuracy as the volume of wastewater is measured continuously whereas, for example, COD concentrations are determined discontinuously. In this paper, these KPIs are used to compare the specific energy consumption of fifteen WWTPs located in the north of Portugal; these values are also compared against similar facilities and types of treatment found in literature. Despite the use of specific energy consumption indicators allows a simplified, yet quick, assessment of energy performance of WWTPs, the results show that is possible to identify some worse performances. In addition, results show that the joint utilization of several KPIs allows more complete and reliable conclusions.

Keywords: Specific energy consumption · Wastewater treatment plant · Energy performance · Benchmarking · Key performance indicators

1 Introduction

The water sector in 2014 had a global energy consumption of 120 Mtoe and typically represents the main contributor to municipal energy use, accounting for about 44% of the energy costs of municipalities [1, 2]. According to International Energy Agency,

J. R. da Costa Sanches Galvão et al. (Eds.): *Proceedings of the 1st International Conference on Water Energy Food and Sustainability (ICoWEFS 2021)*, pp. 361–369, 2021.
https://doi.org/10.1007/978-3-030-75315-3_41

the largest share of the water sector's electricity consumption in developed countries belongs to the wastewater treatment.

Wastewater treatment plants (WWTPs) are important for global development by ensuring adequate sanitation and purifying wastewater to control water pollution, thereby minimizing environmental and human health impacts [3, 4]. The diversity, quantity and complexity of the processes involved in the wastewater treatment means that WWTPs consume substantial amounts of energy [4, 5], typically representing the second largest portion of the operating costs [1, 6].

As the number of people with access to improved sanitation conditions has increased in recent decades, the number of WWTPs operating globally and the associated energy consumption have also increased considerably [7]. Furthermore, energy consumption is expected to maintain a growth trend as wastewater treatment processes are also expected to become even more energy intensive due to population growth and the corresponding increase in the contaminant load to be treated, as well as increasing requirements of regulatory and environmental protection standards for effluent quality and water reuse [5]. This increase in energy consumption will be accompanied by increases in greenhouse gases emissions due to the use of fossil fuels in power plants and in operating costs [8, 9].

Therefore, it is essential to improve the energy efficiency of WWTPs and, according to the literature [7, 10], and the benchmarking efficiency of WWTPs is a suitable approach for identifying inefficiencies and improving the energy efficiency of these plants. Within the benchmarking methodologies applied in the water sector, the normalization approach is generally the method most used by plant operators and all other stakeholders due to the simplicity of implementation and interpretation [11] as well as ease of calculation with normally available data [3]. This approach consists of employing simple key performance indicators (KPIs) and single input-output ratios to calculate the specific energy consumption, such as for example reporting the energy consumption in WWTPs referred to the volume of treated wastewater (kWh/m^3) and/or to the Chemical Oxygen Demand (COD) removed (kWh/kg COD removed) [12].

These simple indicators are important because they allow a quick and easy comparison with the values of other studies and countries available in the literature. Typically, the first KPI (kWh/m^3) is preferred for assessing the performance of WWTPs due to the accuracy and reliability of data since the volume of treated wastewater is easy to measure and is usually measured continuously while COD and other concentrations are determined discontinuously despite their significant daily and seasonal variations, namely once or twice a week or month or even less frequently in the case of small plants [6].

In this study data from fifteen WWTPs in the north of Portugal were analyzed to compare their energy performance. For this purpose, the specific energy consumption of the WWTPs was evaluated by calculating two KPIs that relate energy consumption with (a) the volume of treated wastewater or (b) Chemical Oxygen Demand (COD) removed.

2 Framework and Methodology

Data from fifteen WWTPs located in the north of Portugal were used to perform an evaluation of their specific energy consumption. All the WWTPs of the sample apply an activated sludge (AS) treatment configuration to allow a comparison of plants with a similar layout as possible.

The WWTPs have very different annual volumes of treated wastewater, ranging from nearly 65 × 10^3 m^3 to about 2541 × 10^3 m^3 (Fig. 1). In fact, the sample may be divided into three main groups, namely plants that treat annually a volume (a) below 500 × 10^3 m^3 (5 WWTPs); (b) above 500 × 10^3 m^3 and below 1500 × 10^3 m^3 (6 WWTPs); and (c) above 1500 × 10^3 m^3 (4 WWTPs).

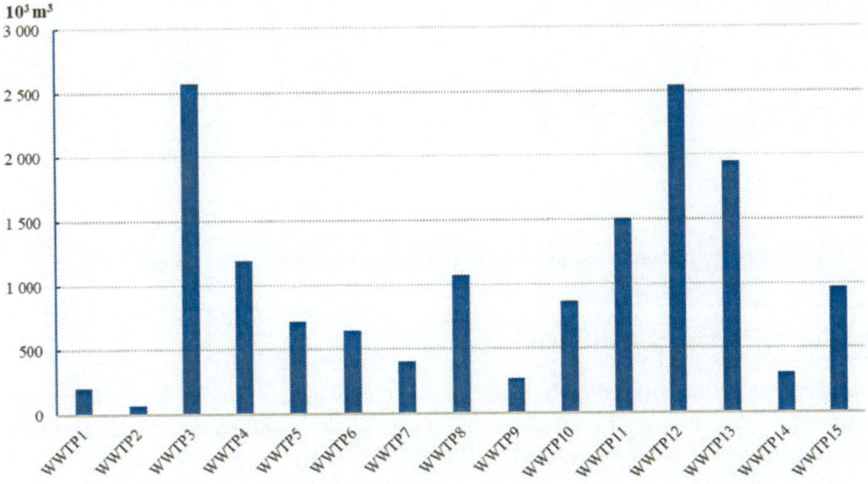

Fig. 1. Volume of treated wastewater (10^3 m^3) of the sampled WWTPs.

Similarly, annual energy consumption also shows significant variations between WWTPs. However, these differences are distinct from those for the annual volume of treated wastewater, since, for example, by observing the annual energy consumption values (Fig. 2) it is possible to perceive the existence of a plant (WWTP12) with a much higher value than the others.

Regarding energy consumption, the sample can be divided into three groups, namely WWTPs with an annual energy consumption (a) below 500 MWh (6 WWTPs); (b) above 500 MWh and below 1000 MWh (6 WWTPs); and (c) above 1000 MWh (3 WWTPs).

Finally, by converting annual energy consumption into tonnes of oil equivalent (toe) we can see that only one (WWTP12) is covered by the threshold of the SGCIE[1] (500 toe) which establishes the obligation to carry out energy audits. This situation may show that, unlike WWTP12, the remaining WWTPs may not usually be subject to an analysis of their energy performance.

[1] *Sistema de Gestão dos Consumos Intensivos de Energia* (SGCIE) stipulates an obligation to carry out energy audits, with viable measures implemented through rationalization plans, on facilities with an annual consumption of 500 toe or more in the year preceding the analysis.

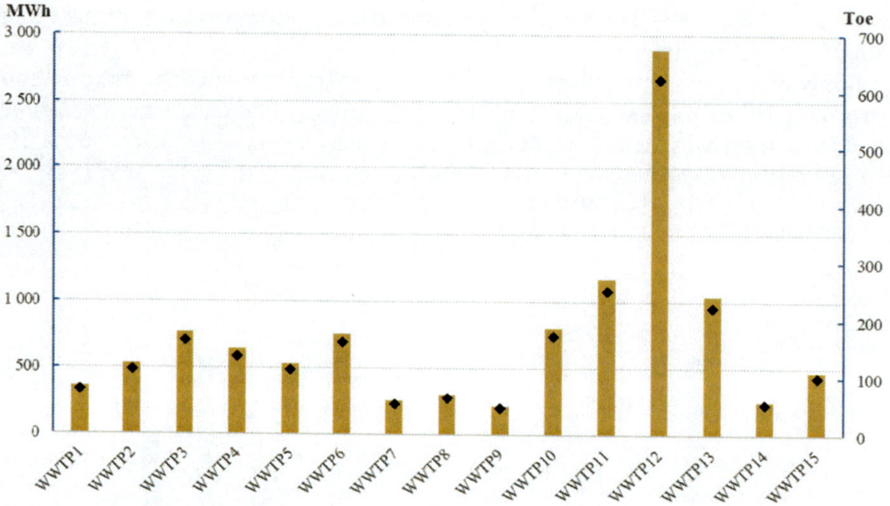

Fig. 2. Energy consumption in MWh (bars) and in toe (dots) of the sampled WWTPs

2.1 Key Performance Indicators

After data collection, the following step was to calculate the two KPIs, namely energy consumption per volume of treated wastewater (kWh/m^3) and energy consumption per Chemical Oxygen Demand removed (kWh/kg COD rem):

$$KPI_1 = \text{Energy consumption (kWh)/Treated wastewater}(m^3) \quad (1)$$

$$KPI_2 = \text{Energy consumption (kWh)/COD removed (kg COD rem)} \quad (2)$$

Annual data were used to calculate these two ratios as KPIs typically refer to a reference period of one year.

3 Results and Discussion

Since it is the indicator preferably used in the sector, the KPI_1, that relates energy consumption to the volume of treated wastewater, was the first to be calculated for the 15 WWTPs (Fig. 3). The representation of these values shows that two plants have higher values compared to the others, namely WWTP1 and WWTP2. In fact, the latter plant (WWTP2) has an extremely higher value, i.e., the specific energy consumption value of this plant is 8,11 kWh/m^3 while for other plants the values range between 0,28 kWh/m^3 and 1,79 kWh/m^3. This alone demonstrates the need to analyze in more detail the causes for this plant to present such a disparate value of specific energy consumption and such a lower energy performance. On the contrary, WWTP3 and WWTP8 have much lower specific energy consumption values (kWh/m^3) than the other WWTPs in this study, namely 0,30 kWh/m^3 and 0,28 kWh/m^3 for WWTP3 and WWTP8, respectively. These

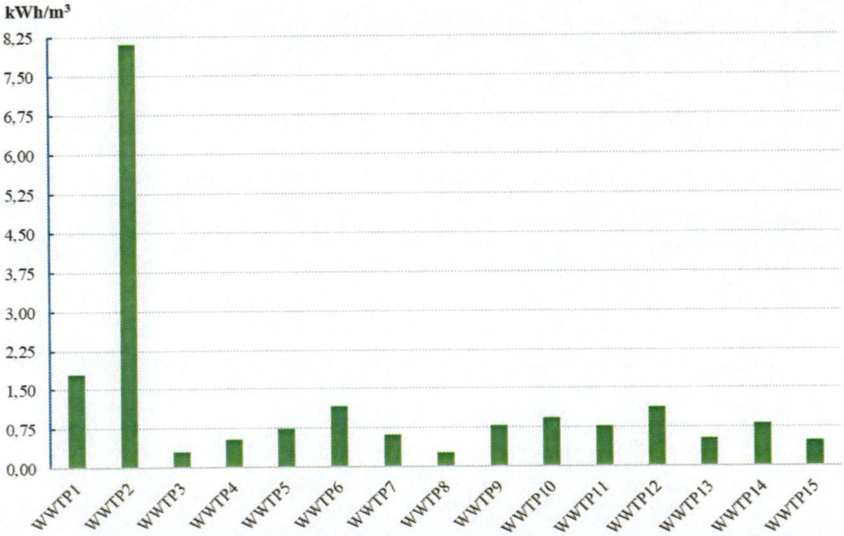

Fig. 3. Specific energy consumption (kWh/m^3) of the sampled WWTPs.

figures may show that the two plants have a significantly better energy performance than other plants.

By representing the specific energy consumption (kWh/m^3) of the different plants according to the annual volume of treated wastewater (Fig. 4), it can be concluded that the worst performance of WWTP1 and WWTP2 may partly be due to their smaller size. In fact, the representation of the trendline obtained from the various points shows that for this sample, similarly to what is widely shown in the literature, the specific energy consumption tends to decrease as the size (annual volume of treated wastewater increases. That is, typically smaller WWTPs, such as WWTP1 and WWTP2, have worse energy performance because, unlike larger WWTPs, they do not benefit from economies of scale or more stable conditions in the treatment process, which makes them more prone to variations leading to further intensive energy consumption [13].

This representation also shows that considering its size (annual volume of treated wastewater), WWTP12 has a specific energy consumption higher than expected because the verified value is not only much higher than what the trendline shows, but it also is much higher than that of the other plant with a similar annual volume of treated wastewater (approximately 2500 × 10^3 m^3). This situation may also be indicative of lower-than-expected energy performance and/or of the existence of possible improvements.

Finally, a literature review was carried out to find reported values for similar WWTPs and compare them with the sample. For consistency, only studies analyzing WWTPs in Portugal were selected. From this survey, it was found that the WWTPs had specific energy consumption values between 0,34 kWh/m^3 and 1,29 kWh/m^3 [14], or 0,4 kWh/m^3 and 1,0 kWh/m^3 [15], or even 0,34 kWh/m^3 and 1,27 kWh/m^3 [6]. When comparing the values (Fig. 5) most plants fall within the range of values found in literature, with

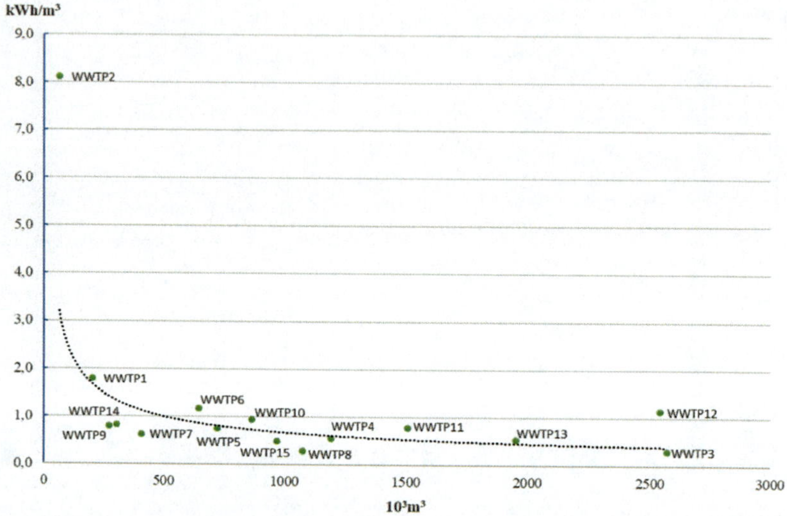

Fig. 4. Specific energy consumption as a function of the annual volume of treated wastewater.

only two plants having lower values (WWTP3 and WWTP8) and other two facilities presenting much higher values (WWTP1 and WWTP2).

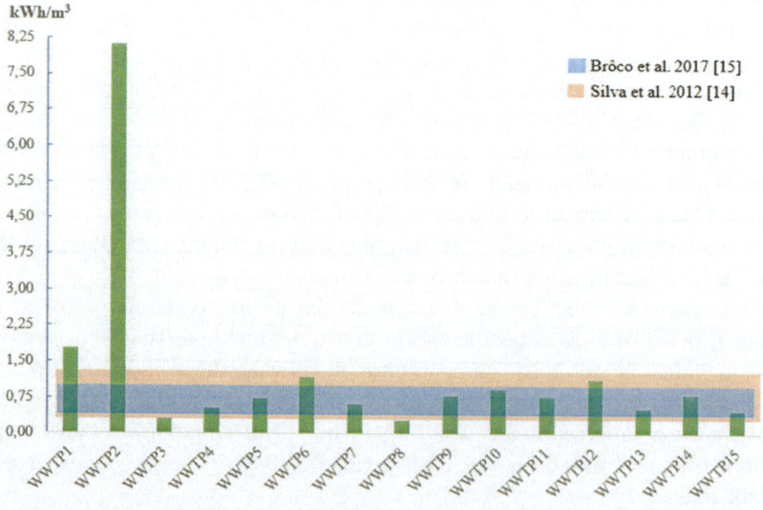

Fig. 5. Comparison of the sample and literature specific energy consumption values (kWh/m^3).

According to the literature, the treatment process is not only dependent on the amount of wastewater to be treated, but also on the pollutant load, nutrients, suspended solids or even the treatment technology applied, among others [11]. In fact, some studies state that the mass removed of COD and Biochemical Oxygen Demand (BOD) are typically more relevant than the volume of treated wastewater for some assessment criteria since the removal of organic matter and nutrients have a large contribution to the energy consumption of WWTPs. It is therefore advantageous to report the energy consumption per unit of pollutant removed.

Hence, the second KPI linking the annual energy consumption to the annual quantity of COD removed (KPI2) has been calculated (Fig. 6). For this indicator, the worst energy performance was found in WWTP1, WWTP6 and WWTP12. This shows that although WWTP2 has a much higher value of energy consumption per volume of treated wastewater, the value obtained by this plant for this KPI is lower than for six other WWTPs and is in line with a large part of the remaining sample. Thus, besides its size, another of the reasons for the lower energy performance of this plant may be the influent saturated with pollutants. Regarding this KPI, the highest values are presented by WWTP1, WWTP6 and WWTP12, thus showing a worse performance since also for the other KPI (kWh/m^3) their values are among the highest.

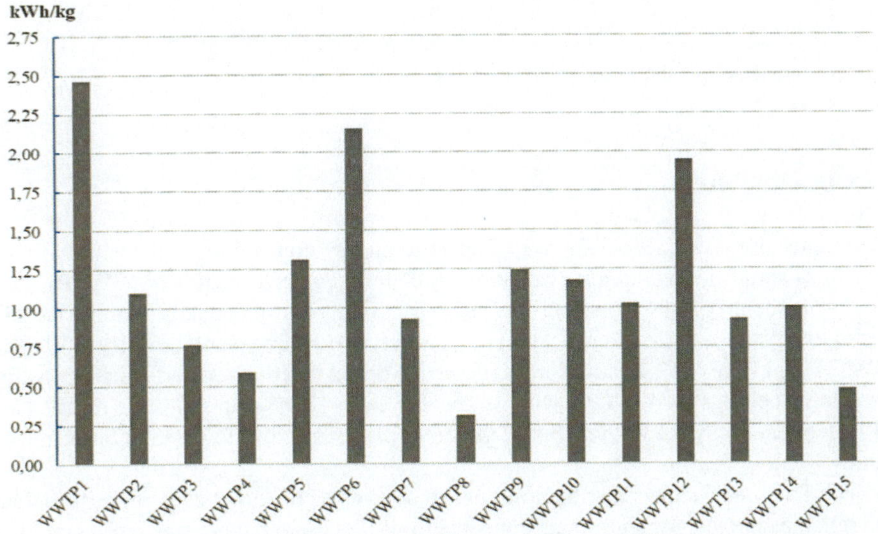

Fig. 6. Specific energy consumption (kWh/kg COD rem) of the sampled WWTPs.

By comparing the sample values with the range of values found in the literature for WWTPs in Portugal (Fig. 7) it is clear that not only WWTP1, WWTP6 and WWTP12 have the worse specific energy consumption values of the sample, but also the values shown are higher than the range of values found in the literature for these plants in Portugal. In view of this, it should be noted that these plants must be subject to a detailed analysis to assess the possible causes of this worse energy performance. On the contrary, and similarly to the other KPI, WWTP8 has a lower specific energy consumption value than the range of values found in the literature.

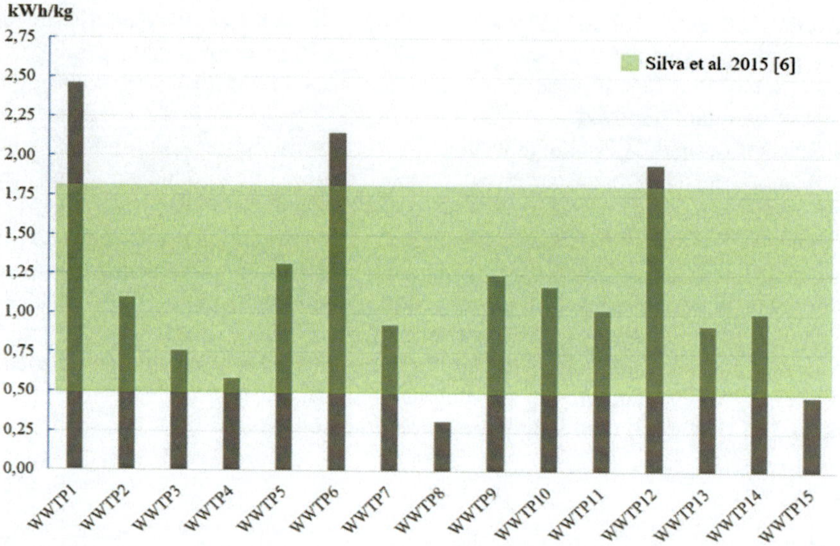

Fig. 7. Comparison of the sample and literature specific energy consumption values (kWh/kg COD rem).

4 Conclusions

The results show that while the use of specific energy consumption indicators allows only for a simplified yet quick evaluation of the energy performance of WWTPs, it is possible to identify some worse performances accurately and adequately. In this case this methodology allowed the identification of the worst performing plants, namely WWTP1, WWTP6 and WWTP12. In addition, it allowed to detect that the best performance within the sample belongs to WWTP8, with its specific energy consumption value for the two KPIs being not only lower than that of the other plants, but also lower than the range of values found in the literature.

This study also shows that the joint use of several KPIs allows even more complete and reliable conclusions since it accounts the impact of more factors, e.g., pollutant load removed, on the energy consumption of this type of plants. In fact, this joint utilization led to the conclusion that, unlike the WWTP1, although the WWTP2 presented a rather high value of energy consumption per volume of treated wastewater, its energy performance was not as low as it seemed because by relating the energy consumption to the quantity of COD removed it was found that after all the value presented was in line with the other plants and was in the range of values found in literature.

Acknowledgements. The presented work is framed under the Energy for Sustainability Initiative of the University of Coimbra (UC).

Funding:. Bruno J. Cardoso acknowledges the support provided by the Portuguese Foundation for Science and Technology (FCT) under Doctoral grant PD/BD/142812/2018.

References

1. Copeland, C., Carter, N.T.: Energy-Water Nexus: The Water Sector's Energy Use (2017)
2. IEA: Water Energy Nexus - Excerpt from the World Energy Outlook 2016. Int. Energy Agency **60** (2016). https://doi.org/10.1021/es903811p
3. Di Fraia, S., Massarotti, N., Vanoli, L.: A novel energy assessment of urban wastewater treatment plants. Energy Convers. Manag. **163**, 304–313 (2018). https://doi.org/10.1016/j.enconman.2018.02.058
4. Molinos-Senante, M., Sala-Garrido, R., Iftimi, A.: Energy intensity modeling for wastewater treatment technologies. Sci. Total Environ. **630**, 1565–1572 (2018). https://doi.org/10.1016/j.scitotenv.2018.02.327
5. Hernández-Sancho, F., Molinos-Senante, M., Sala-Garrido, R.: Energy efficiency in Spanish wastewater treatment plants: a non-radial DEA approach. Sci. Total Environ. **409**, 2693–2699 (2011). https://doi.org/10.1016/j.scitotenv.2011.04.018
6. Silva, C., Rosa, M.J.: Energy performance indicators of wastewater treatment: a field study with 17 Portuguese plants. Water Sci. Technol. **72**, 510–519 (2015). https://doi.org/10.2166/wst.2015.189
7. Gómez, T., Gémar, G., Molinos-Senante, M., Sala-Garrido, R., Caballero, R.: Assessing the efficiency of wastewater treatment plants: a double-bootstrap approach. J. Clean. Prod. **164**, 315–324 (2017). https://doi.org/10.1016/j.jclepro.2017.06.198
8. Hao, X., Liu, R., Huang, X.: Evaluation of the potential for operating carbon neutral WWTPs in China. Water Res. **87**, 424–431 (2015). https://doi.org/10.1016/j.watres.2015.05.050
9. Molinos-Senante, M., Hernández-Sancho, F., Mocholí-Arce, M., Sala-Garrido, R.: Economic and environmental performance of wastewater treatment plants: potential reductions in greenhouse gases emissions. Resour. Energy Econ. **38**, 125–140 (2014). https://doi.org/10.1016/j.reseneeco.2014.07.001
10. Hernández-Chover, V., Bellver-Domingo, Á., Hernández-Sancho, F.: Efficiency of wastewater treatment facilities: the influence of scale economies. J. Environ. Manag. **228**, 77–84 (2018). https://doi.org/10.1016/j.jenvman.2018.09.014
11. Longo, S., D'Antoni, B.M., Bongards, M., Chaparro, A., Cronrath, A., Fatone, F., Lema, J.M., Mauricio-Iglesias, M., Soares, A., Hospido, A.: Monitoring and diagnosis of energy consumption in wastewater treatment plants. A state of the art and proposals for improvement. Appl. Energy **179**, 1251–1268 (2016). https://doi.org/10.1016/j.apenergy.2016.07.043
12. Foladori, P., Vaccari, M., Vitali, F.: Energy audit in small wastewater treatment plants: methodology, energy consumption indicators, and lessons learned. Water Sci. Technol. **72**, 1007–1015 (2015). https://doi.org/10.2166/wst.2015.306
13. Christoforidou, P., Bariamis, G., Iosifidou, M., Nikolaidou, E., Samaras, P.: Energy benchmarking and optimization of wastewater treatment plants in Greece. Environ. Sci. Proc. **2**, 36 (2020). https://doi.org/10.3390/environsciproc2020002036
14. Silva, C., Ramalho, P., Quadros, S., Alegre, H., Rosa, M.J.: Results of "PASt21" - the Portuguese initiative for performance assessment of water and wastewater treatment plants. Water Sci. Technol. Water Supply **12**, 372–386 (2012). https://doi.org/10.2166/ws.2012.004
15. Brôco, N., Carvalho, M.: Management tool to assess, benchmark and support energy efficiency actions in more than 800 WWTP. Water Pract. Technol. **12**, 745–749 (2017). https://doi.org/10.2166/wpt.2017.079

Energy-Efficient Dewatering Technique for Recycled Films

Joel C. Vasco[1,2(✉)], Nelson Silva[3], Fernando Lagoa[3], Carlos Capela[1,4], Lizete Heleno[1], Marcelo Gaspar[1,5], and Nelson Oliveira[1,6]

[1] School of Technology and Management, Polytechnic of Leiria, Leiria, Portugal
joel.vasco@ipleiria.pt

[2] Institute for Polymers and Composites, University of Minho, Guimarães, Portugal

[3] Plasmaq – Recycling Systems, Lda., Leiria, Portugal

[4] Centre for Mechanical Engineering, Materials and Processes, University of Coimbra, Coimbra, Portugal

[5] Centre for Rapid and Sustainable Product Design, Polytechnic of Leiria, Marinha Grande, Portugal

[6] Laboratory of Separation and Reaction Engineering-Laboratory of Catalysis and Materials, Leiria, Portugal

Abstract. Plastic films used in products packaging represent a significant amount of plastic waste that requires to be recirculated, despite the advantages that these films provide such as longer shelf-life for food products or performance enhancement for many other products. The recycling process for post-consumer plastic waste often requires washing and dewatering procedures to be carried out before the material can be reprocessed and transformed into a new product. The dewatering procedure, in particular, is an intensive energy-consuming process due to the need to remove moisture from the material. Centrifugation is typically used for dewatering, many times supported by air dryers, resulting on a significant high energy consumption and a larger foot-print to install this equipment.

This work describes a novel approach for dewatering shredded films that have been washed to remove contaminants from post-consumer use or post-industrial.

The dewatering system is based on a compression stage in which the material is being pressed to remove the largest amount of water, forming a cake that is dismantled afterwards for transportation to the following recycling step. Consequently, this dewatering approach is fully comparable with the pre-existent ones, providing a throughput of 1,5tonnes/hour with a significantly reduced energy consumption, economic advantages and a quite smaller footprint.

Keywords: Dewatering efficiency · Energy savings · Footprint

1 Introduction

Packaging represents almost 40% of the total plastic demands in Europe and surrounding countries, being also the largest source of plastic waste in Europe (63%), followed by

J. R. da Costa Sanches Galvão et al. (Eds.): *Proceedings of the 1st International Conference on Water Energy Food and Sustainability (ICoWEFS 2021)*, pp. 370–381, 2021.
https://doi.org/10.1007/978-3-030-75315-3_42

construction (6%), electronics (5%), among other activity sectors (D'Amato et al. 2019). These numbers show that packaging films have a significant share in food packaging (or secondary packaging), thus generating plastic waste that needs to be addressed in terms of material circularity (Singh et al. 2017).

The recycling process for shredded packaging films often requires washing and dewatering procedures to be carried out before proceeding to the downstream recycling steps. These procedures are mandatory in the cases of post-consumer use and/or mishandled post-industrial use to remove contaminants which represents an important opportunity for improvement due to the high energy consumption (Horodytska et al. 2018a; Soto et al. 2020).

The dewatering procedure is, in fact, one of the most important sources for energy consumption, making the recycling process less efficient and, therefore, less green. For dewatering, the centrifugation process is often used, involving large equipment where plastic waste is admitted and centrifuged. Typically, a single centrifugation stage filters out approximately 30% of the water content, requiring additional dewatering steps, which can be additional centrifugation systems or air drying systems (Briassoulis et al. 2013; Horodytska et al. 2018b).

The dewatering solution presented in this work proposes a novel approach to remove the suitable water content following the plastic waste washing procedure. Additionally, this methodology has lower energy costs and a smaller footprint, when compared to currently available solutions.

2 Literature Review

The main steps of the mechanical recycling processes of plastic films are detailed on this topic to contextualize the washing and dewatering procedures within the recycling process workflow.

2.1 Recycling Process

Plastic films have a broad scope of applications across diverse industrial domains. Due not only to their wide range of use, but also considering the volume of plastic films used every day at a global scale, it is key to improve their recyclability, thus providing important contributions towards a more sustainable environment and an improved material circularity. These applications can be grouped into both packaging and non-packaging. Plastic films used on packaging products can be split into consumer and commercial/industrial uses (Briassoulis et al. 2013; Horodytska et al. 2018a; Nowakowski and Mrówczyńska 2018). Thus, solid plastic waste can be classified into post-industrial (PI) waste if it never makes it into consumer, and into post-consumer (PC) waste, when it is disposed of at its end-of-life (Ragaert et al. 2017). Considering the wide use of plastics, and the related waste generated both at PI and PC scale, efficient recyclability solutions are required to value and recirculate solid plastic waste. To such end, mechanical recycling is currently the main solution for post-consumer plastic waste, including the shredding, washing, dewatering and reprocessing procedures. Reprocessing is typically achieved by extrusion to form new raw material or to directly manufacture new products (Soto et al. 2018; Faraca and Astrup 2019). Mechanical recycling of plastic waste is estimated to produce about 1,4 tonne CO_2/tonne plastics, which is less than 30% of the CO_2 emissions associated with the production of new plastic materials (5,1 tonne CO_2/tonne plastics) (Ragaert et al. 2020).

As shown in Fig. 1, the process for recycling waste starts with placing it in a collection point according to its typology, then being transported to separation installations, where the waste is separated according to its characteristics. In the plastic waste line, the waste is separated by different techniques (for example, ballistic separator and optical separators). These techniques allow separation in different types of material, LDPE, HDPE, PP, PET, among others.

After this stage, the waste already separated is sent to the plastic recycling industry. This study applies mainly to plastic film (mainly LDPE), which will be subjected to a shredding process, followed by pre-washing/washing, dewatering and reprocessing. All of these steps are associated with high energy consumption, namely the pre-washing/washing and dewatering.

Fig. 1. Main phases of the recycling process

Figure 1 illustrates the whole process cycle, starting by collecting the plastic solid waste, going through recycling and revaluing it as a new plastic film. When comparing the outputs of a life cycle analysis (LCA) of plastic waste, with different final destinations, like recycling, energy recovery and landfill, it was found out that recycling, in general, has greater environmental benefits (Ferreira et al. 2014; Bernardo et al. 2016).

2.2 Washing and Dewatering

The washing procedure requires a high-intensive use of water due to the presence of dirt or other contaminants that affect the quality of the recycled material at the end. Therefore, an effective dewatering procedure is required for the following recycling steps. Centrifugation systems are often adopted to the high centrifugal acceleration that provides much higher slip velocities (Ragaert et al. 2017; Singh et al. 2017).

Concerning the effectiveness of the dewatering procedure, a single centrifugation system is not enough to achieve the correct moisture content. Typically, a single centrifugation stage filters out approximately 30% of the water content. Thus, are required additional dewatering systems, such as successive centrifugations systems or thermal drying systems, that push energy costs to a much higher level (Briassoulis et al. 2013; Horodytska et al. 2018b; Ügdüler et al. 2020).

The use of centrifugation systems assembled in series, interspersed with fan blowers and cyclones to transport the material to the following centrifugation step, is a commonly used setup to dewater the washed material (Fig. 2).

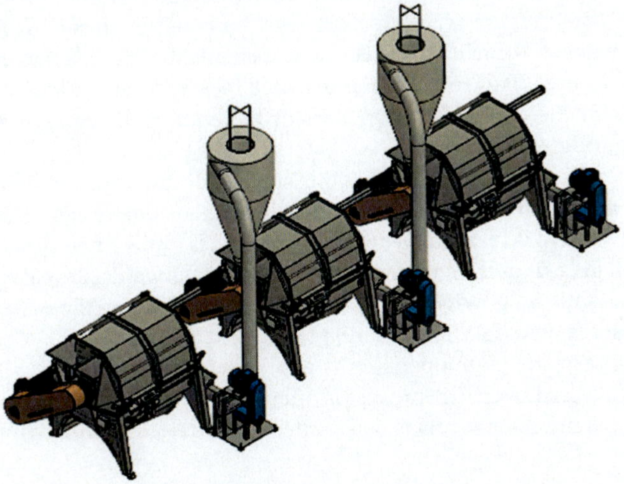

Fig. 2. Centrifugation system setup

3 Novel Dewatering Technique

The goal for this dewatering solution is to remove the excessive water and to achieve the correct moisture content before the material is reprocessed. The equipment works in continuous mode, providing a maximum throughput of 1,5 tonnes/hour.

3.1 Dewatering Concept Based on Compression

The compression approach enables an electromechanical technique to squeeze the excessive water content on a single procedure. Figure 3 depicts the main components of this equipment concept.

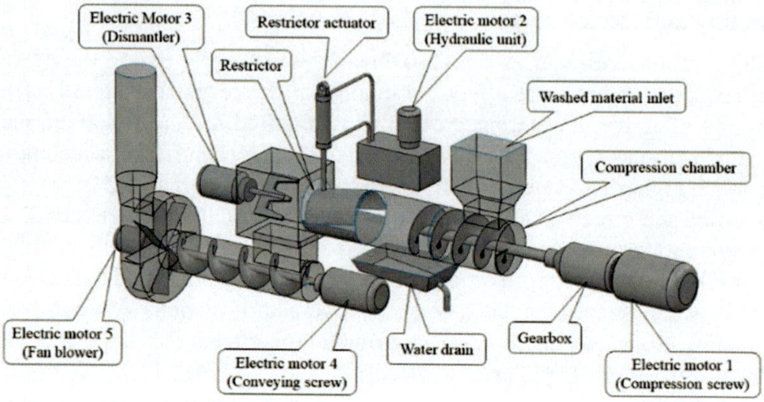

Fig. 3. Main components of the dewatering equipment based on compression

The washed shredded material enters the compression chamber of the system and it is fed into a compression screw that conveys the material ahead. At the end of the compression chamber, the exit cross-section is dynamically restricted to ensure the correct compression level. This restriction is actuated by a hydraulic cylinder and its action is correlated with the operating parameters of the equipment through a specific and proprietary algorithm.

Water content is seriously reduced at this stage and water is filtered out through the drilled holes on the wall of the water drain section of the compression chamber.

The compressed material is then pushed along and it forms a nearly solid cake that is pushed through the exit of the compression chamber. Afterwards, the cake is dismantled and the material falls by gravity into a conveying screw, leading the material into a fan blower that ensures material transportation to the next step.

The dewatering procedure operation is dependent on the photocell signal installed on the washed material inlet. The process is often subject to material supply fluctuations. Therefore, the compression screw is deactivated to save energy, although it is critical to

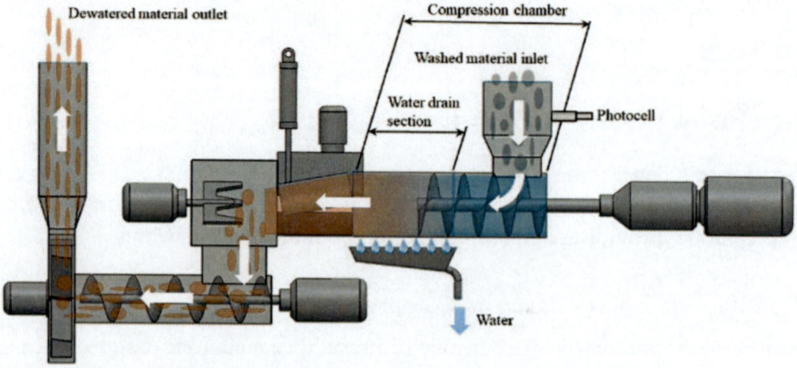

Fig. 4. Schematics of the material flow on the dewatering procedure

keep the remaining actuators active to ensure that the dewatered material already on the conveying screw is duly sent to the following step (Fig. 4).

3.2 Water Content

The washing procedure introduces high water content into the shredded material. After the washing process, the material is forwarded to a compression chamber inlet and it is compressed until most of the water has been removed. The removed water is forwarded to the waste water treatment station. The compression chamber has the capacity to remove different types of retained water within the cake. According to Horodytska *et al.* (2018b), there are four different types of retained water after washing:

- Free water: most of the water within the cake can be represented by water in the pores and voids of the plastic cake;
- Two types of water retained by capillary: the plastic cake is composed mostly by shredded plastic film, flakes, so (1) it is expected a considerable amount of superficial water between the flakes superficies where these capillary forces will keep the flakes stuck between each other. (2) There is also some water in the contact points of the different flakes, pendular water.
- Water trapped: the flakes are material very flexible or with some flexibility, producing folds and curvatures of the film. Water can be trapped within these folds and curvatures due to the tortuosity of the material.

It is estimated that most of the free water is removed in the water drain section due to the compression of the material. At the end of the compression chamber there is a rotating dismantler. In this section the plastic cake is now dismantled, exposing the water retained by capillary and trapped in folds and curvatures to air, removing an additional water. The plastic goes to the conveying screw, being slightly compressed and forwarded to the transportation of material on an air duct, pushed at high speed by the fan blower. A residual water content is required for the reprocessing step to fluidise the dewatered material in order to optimize the extrusion process. Typical values of remaining moisture are kept between 1 and 12%, depending on the extruder characteristics (Horodytska et al. 2018b).

3.3 Energy Spent and Dewatering Process Control

Energy is a critical issue on material recycling. Considering the typical throughput, the dewatering system requires a total installed power of slightly less than 70 kW. Table 1 shows the installed power breakdown for each actuator installed on the compression-based dewatering system.

In which concerns to the process control, the dewatering procedure depends on the dynamic restriction at the end of the compression chamber. The process relies on a previous definition of the compression setpoint for compression screw. Compression setpoint (Cset) is defined according to the residual water content desired at the end of the dewatering process. A Knowledge Database (KDB) is available to retrieve both previously used data as well as to store actual processing data.

Table 1. Installed power breakdown for each actuator.

Actuator	Function	Maximum power
Electric motor 1	Compression screw actuator	37 kW
Hydraulic Cylinder (2)	Compression section restrictor	1,5 kW (hydraulic unit)
Electric motor 3	Cake dismantler	5,5 kW
Electric motor 4	Conveying screw actuator	1,5 kW
Electric motor 5	Fan blower actuator	22 kW

At start-up, all actuators are activated and the restrictor will be on the closed position to ensure initial material compression. The process is monitored continuously in parallel to, either close or release the restrictor according to Cset. The photocell located at the material inlet provides information about material presence at the inlet and such information is used to control the operation of the compression screw. If no material is present on the material inlet for a 30 s delay, the compression screw is deactivated. This delay is required to ensure that the remaining material being dewatered on the compression chamber can be dismantled again to be conveyed to the fan blower. The operation mode established for the dewatering system is briefly described on the flowchart depicted on Fig. 5.

4 Dewatering Techniques Comparison

Considering the same throughput of 1,5 tonnes/hour, centrifugation equipment, air drying equipment and compression equipment are compared in terms of energy consumption, floor area required to install each solution, operation noise and economic impact for the recycling companies.

4.1 Energy

The dewatering of plastic films is a highly energy consuming recycling operation (Horodytska et al. 2018a) whereby the energy consumption of the proposed compression study needs to be analysed and compared to conventional dewatering systems with similar performance. The conventional dewatering system is a centrifugation equipment with a similar throughput and similar dewatering performance. It is composed of three centrifugation systems with a peak power of 90 kW each and three ventilators of 7,5 kW each to transport the plastic cake (Plasmaq 2020). Two fan blowers between each centrifugation unit and the third fan blower to transport from the last centrifugation unit to the storage of the dewatered plastic. The total power of the system is 292,5 kW. It is assumed that the centrifugation systems and fan blowers work at 90% of the power peak, estimating the energy efficiency of 175,5 kWh/tonne of dewatered plastic.

The compression screw actuator of the compression system described works on average of 50% of the power peak, depending on the plastic load at the inlet and it is assumed about 90% for the remaining electric motors and hydraulic cylinder. Based on

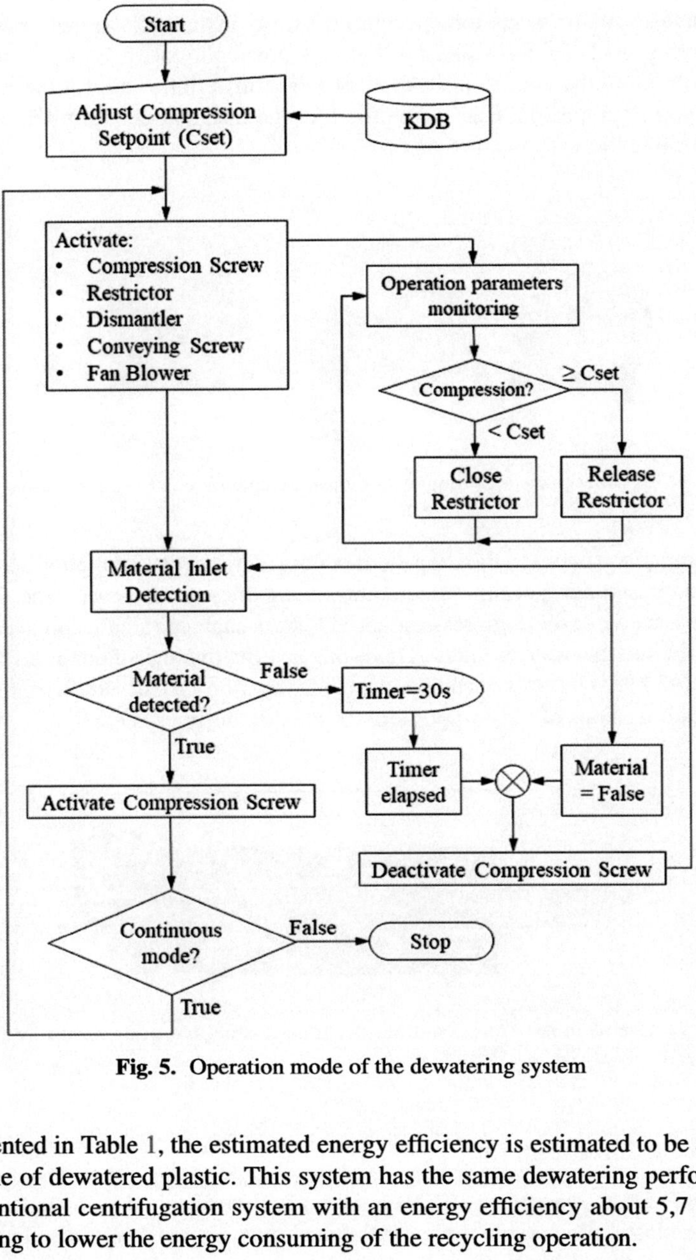

Fig. 5. Operation mode of the dewatering system

data presented in Table 1, the estimated energy efficiency is estimated to be about 30,6 kWh/tonne of dewatered plastic. This system has the same dewatering performance as the conventional centrifugation system with an energy efficiency about 5,7 times less, contributing to lower the energy consuming of the recycling operation.

4.2 Footprint

The installation of dewatering systems may require a large area, considering the shop-floor area effectively occupied by the equipment and additional area required either for safety reasons as well as for maintenance purposes. The use of centrifugation units, as mentioned in the previous topic, requires 3 units installed in series to accomplish the dewatering goal. Additionally, fan blowers are required to supply material from unit to unit and, at the end, to push the dewatered material to the following step. An effective shop-floor area of 30 m^2 is required for the installation of a full assembly of 3 centrifugation units, as shown in Fig. 6.

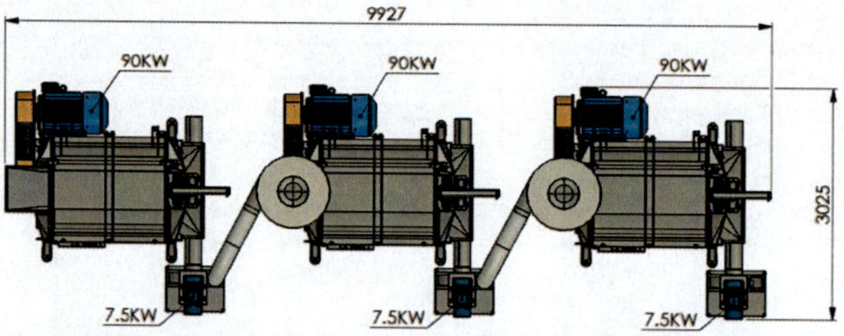

Fig. 6. Overall dimensions in mm for the centrifugation dewatering system setup

The comparison between the shop-floor occupied by the centrifugation setup and the compression-based dewatering system shows a significant difference. The dewatering effectiveness shown of the compression-based system enables a single unit to be installed to achieve the suitable water content. Therefore, an effective shop-floor area of 13,15 m^2 is required (43,8% of the area occupied by a centrifugation system setup). Figure 7 shows the overall dimensions of the compression-based dewatering system.

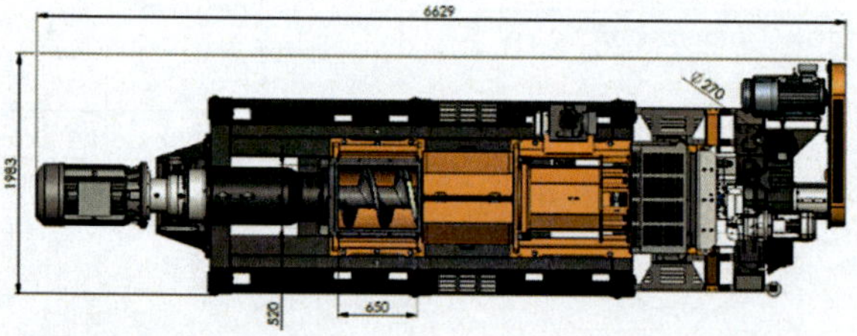

Fig. 7. Overall dimensions in mm for the compression-based dewatering system

4.3 Return of Investment

This brief Return of Investment (ROI) analysis is performed only for the dewatering of shredded plastic films, since these are the materials where the dewatering procedure is more difficult to accomplish and the compression-based solution proved to be more effective as well as sustainable.

When comparing the shop-floor footprint associated with the acquisition costs for each dewatering solution, it allows predicting that the initial costs for the whole centrifugation system setup more than doubles the costs of the compression-based dewatering system. Figure 8 summarizes the estimated values for operational costs per month and acquisition and installation costs as well as the energy required to process a tonne of material and installed power for each dewatering solution.

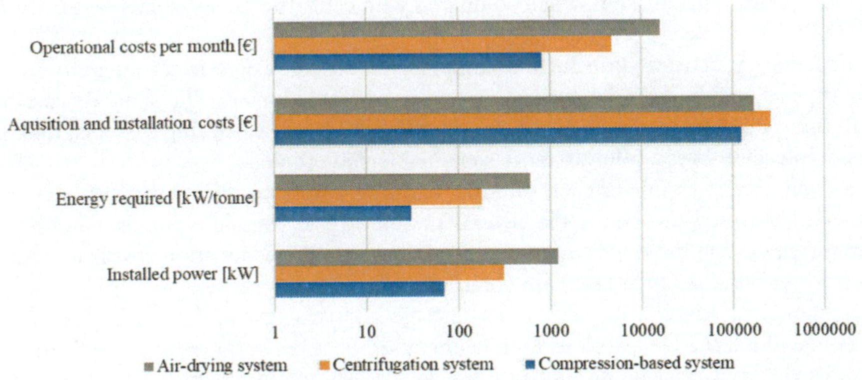

Fig. 8. Estimated costs and energy efficiency

Concerning to estimated operational costs, based on the energy efficiency estimations and predicted maintenance operations, it allows foreseeing significantly higher energy costs for the centrifugation system setup and even higher for the air-drying system.

Acquisition and installations costs are also considered, showing that the centrifugation system setup has the higher value. Although the air-drying system may not be as expensive, however, in terms of operational costs may not be as feasible as the other alternatives. Typically, air-drying systems are combined the centrifugation systems to mitigate the efficiency losses from both dewatering solutions.

The energy required by the compression-based system to process each tonne of plastic material is the lowest from all the compared dewatering solutions.

Considering that centrifugation dewatering efficiency relies on higher rotation, higher maintenance costs are expected during the equipment's life. Thus, the break-even is achieved in just a few months for the compression-based dewatering system, while a centrifugation system setup requires some years to achieve the break-even point.

5 Conclusions and Remarks

The dewatering system presented in this work proved that this is a highly efficient way to remove water from the washed material. Although it is only applicable on shredded plastic films, these are the materials where the dewatering procedure is more difficult to accomplish. Shredded plastic waste from solid packaging or other solid plastic parts do not retain a high content of water. Therefore, the conventional centrifugation systems are still more efficient for such particular use.

Considering the high volume of plastic films used on primary packaging and secondary packaging at a global scale, it is acknowledged that the resulting solid plastic waste must be properly and efficiently recirculated.

From shredding to reprocessing, the solid plastic waste recirculation cycle includes processes that intensively use precious resources, such as water and energy. The dewatering system based on compression is a more efficient way to remove water on a much smaller footprint installation, when compared with centrifugation systems or air-drying systems.

The energy consumption for the compression-based system is significantly lower than the one required by the alternative dewatering solutions for the same dewatering goal. Installed power is much higher for centrifugation and air-drying systems, making the compression-based solution more sustainable and efficient.

Concerning the ROI for each solution, and also assessing the investment related to the space required to operate the several dewatering options, the compression-based dewatering system provides a much higher ROI, since the acquisition cost is up to two times lower and the operation/maintenance costs are also lower.

Although it was not possible to collect the operating noise data, it is possible to state that all alternative solutions to the compression-based option rely on high rotation systems that surely generate higher noise levels. The compression-based dewatering system here presented relies on a 40 rpm screw to compress and to dewater the shredded material, thus with significantly lower acoustic impacts.

Acknowledgements. The authors wish to acknowledge to the Plasmaq technical staff involved in this R&D project, headed by Mr. Jacinto Abrunhosa.

References

Bernardo, C.A., Simões, C.L., Pinto, L.M.C.: Environmental and economic life cycle analysis of plastic waste management options. A review. AIP Conf. Proc. **1779** (2016). https://doi.org/10.1063/1.4965581

Briassoulis, D., Hiskakis, M., Babou, E.: Technical specifications for mechanical recycling of agricultural plastic waste. Waste Manag. **33**, 1516–1530 (2013). https://doi.org/10.1016/j.wasman.2013.03.004

D'Amato, A., Paleari, S., Pohjakallio, M., et al.: Plastics Waste Trade and the Environment. Boeretang (2019)

Faraca, G., Astrup, T.: Plastic waste from recycling centres: characterisation and evaluation of plastic recyclability. Waste Manag. **95**, 388–398 (2019). https://doi.org/10.1016/j.wasman.2019.06.038

Ferreira, S., Cabral, M., da Cruz, N.F., et al.: Life cycle assessment of a packaging waste recycling system in Portugal. Waste Manag. **34**, 1725–1735 (2014). https://doi.org/10.1016/j.wasman.2014.05.007

Horodytska, O., Valdés, F.J., Fullana, A.: Plastic flexible films waste management – a state of art review. Waste Manag. **77**, 413–425 (2018a). https://doi.org/10.1016/j.wasman.2018.04.023

Horodytska, O., Valdés, F.J., Fullana, A.: Centrifugal dewatering performance in plastic films recycling. Waste Manag. **80**, 211–217 (2018b). https://doi.org/10.1016/j.wasman.2018.09.015

Nowakowski, P., Mrówczyńska, B.: Towards sustainable WEEE collection and transportation methods in circular economy - comparative study for rural and urban settlements. Resour. Conserv. Recycl. **135**, 93–107 (2018). https://doi.org/10.1016/j.resconrec.2017.12.016

Ragaert, K., Delva, L., Van Geem, K.: Mechanical and chemical recycling of solid plastic waste. Waste Manag. **69**, 24–58 (2017). https://doi.org/10.1016/j.wasman.2017.07.044

Ragaert, K., Huysveld, S., Vyncke, G., et al.: Design from recycling: a complex mixed plastic waste case study. Resour. Conserv. Recycl. **155**, 104646 (2020). https://doi.org/10.1016/j.resconrec.2019.104646

Singh, N., Hui, D., Singh, R., et al.: Recycling of plastic solid waste: a state of art review and future applications. Compos. Part B Eng. **115**, 409–422 (2017). https://doi.org/10.1016/j.compositesb.2016.09.013

Soto, J.M., Blázquez, G., Calero, M., et al.: A real case study of mechanical recycling as an alternative for managing of polyethylene plastic film presented in mixed municipal solid waste. J. Clean Prod. **203**, 777–787 (2018). https://doi.org/10.1016/j.jclepro.2018.08.302

Soto, J.M., Martín-Lara, M.A., Blázquez, G., et al.: Novel pre-treatment of dirty post-consumer polyethylene film for its mechanical recycling. Process Saf. Environ. Prot. **139**, 315–324 (2020). https://doi.org/10.1016/j.psep.2020.04.044

Ügdüler, S., Van Geem, K.M., Roosen, M., et al.: Challenges and opportunities of solvent-based additive extraction methods for plastic recycling. Waste Manag. **104**, 148–182 (2020). https://doi.org/10.1016/j.wasman.2020.01.003

Performance of Thermal Break Strips in Lightweight Steel Framed Walls

Telmo Ribeiro(✉), Paulo Santos, and Diogo Mateus

ISISE, Department of Civil Engineering, University of Coimbra, Pólo II, Rua Luís Reis Santos, 3030-788 Coimbra, Portugal
telmo.ribeiro@dec.uc.pt

Abstract. An accurate thermal characterization of the envelope components is essential to achieve a reliable evaluation of thermal behaviour and energy efficiency of buildings. In lightweight steel-framed (LSF) building components, the major thermal performance concern is related to the unwanted significant thermal bridge effects originated by the high thermal conductivity of steel. The application of thermal break (TB) strips in the steel stud flanges is one of the most currently used thermal bridge mitigation strategies. In this paper the thermal performance of ten interior LSF walls configurations are measured, using the heat flow meter (HFM) method under laboratory-controlled conditions. Three TB strips materials and three TB locations (inner, outer and both sides of steel stud) are assessed and a comparison with the thermal performance of a reference wall without TB strips is made. Regarding the TB strips materials, it was found that the best thermal performance is achieved by aerogel, which is the material that presents the lowest thermal conductivity. Considering the TB strips location, the application on both sides of steel stud shows a relative significant thermal performance increase comparatively to the application on inner or outer side, presenting these last two configurations very similar performances.

Keywords: Lightweight steel frame · LSF walls · Partition walls · Thermal resistance · Thermal break strips · Experimental measurements

1 Introduction

The thermal comfort and energy efficiency in buildings are strongly influenced by the characteristics of the envelope. In the specific case of LSF walls, the high thermal conductivity of steel frames can lead to significant thermal bridges, that should be predicted and treated appropriately. One of the most used strategy to mitigate steel studs thermal bridge effect is the application of thermal break (TB) strips along stud flanges, being this the main focus of the research project Tyre4BuildIns – "*Recycled tyre rubber resin-bonded for building insulation systems towards energy efficiency*" [1]. The TB strips, usually made of thermal insulating materials, allow to increase the thermal resistance of the LSF walls, by reducing the heat losses due to steel stud thermal bridges [2].

J. R. da Costa Sanches Galvão et al. (Eds.): *Proceedings of the 1st International Conference on Water Energy Food and Sustainability (ICoWEFS 2021)*, pp. 382–389, 2021.
https://doi.org/10.1007/978-3-030-75315-3_43

Nowadays, there are available in the market several TB strips materials, which were specifically developed for this purpose, or could easily be adopted for this use.

In this work, with the aim of evaluating the thermal break (TB) strips performance for the mitigation of thermal bridges originated by the steel studs, the overall surface-to-surface thermal resistance (*R*-value) of ten different configurations of interior partition LSF walls were measured in controlled laboratory conditions. The laboratorial tests were performed using a mini hot box apparatus with a set of two climatic chambers, being the thermal performance of the LSF walls measured using the heat flux meter (HFM) method [3]. For each wall, three tests were performed, applying the sensors at different high positions (top, middle and bottom) within the LSF wall test-sample surfaces, totalizing thirty lab tests. The TB strip materials tested were recycled rubber (MS-R1), extruded polystyrene (XPS) and aerogel (AG), and three different configurations for the localization of the TB strips were considered, along the: inner; outer, and; both steel stud flanges. Furthermore, in order to perform a verification of the experimental values, all the LSF wall measurements results (overall conductive *R*-values) were compared with bi-dimensional finite element numerical simulations.

2 Materials and Methods

2.1 Characterization of LSF Walls

In this section the characterization regarding materials, geometry, dimensions and thermal properties of the LSF reference wall and the thermal break (TB) strips are performed.

The cross-section of the reference LSF interior wall is illustrated in Fig. 1. The vertical steel studs (C90 × 37 × 15 × 0.6 mm) are spaced 400 mm apart and the steel sheet is 0,6 mm thick. The outer and inner sheathing surfaces are constituted by two gypsum plasterboards (GPB) on each side (2 × 12.5 mm thick). The cavity, with a total thickness of 90 mm, is totally filled with mineral wool (MW) batt insulation.

The thickness of each layer and the thermal conductivities of the materials are presented in Table 1.

The thermal break (TB) strips tested are 50 mm wide and 10 mm thick, and the materials used were recycled rubber (MS-R1), extruded polystyrene (XPS), and CBS aerogel (AG) (Fig. 2), with thermal conductivities ranging from 0.122 W/(m·K) to 0.015 W/(m·K), as presented in Table 2. The three configurations considered for the localization of the TB strips, along the: inner, outer, and on both steel stud flanges, are illustrated in Fig. 3.

2.2 Experimental Lab Tests

The laboratorial tests were performed using a mini hot box apparatus, in which the wall sample is placed between two climatic chambers (hot box and cold box), as illustrated in Fig. 4a. The LSF wall test samples used in the measurements have 1030 mm height and 1060 mm width, and are composed by three vertical steel studs spaced 400 mm, being the middle one centered. The measurement of the thermal performance of the

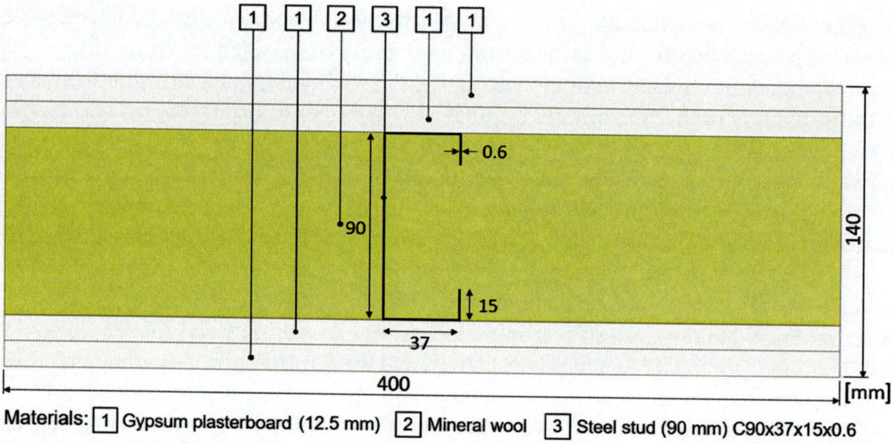

Fig. 1. Horizontal cross-section of the reference interior LSF wall: geometry, dimensions and materials.

Table 1. Thickness (*d*) and thermal conductivities (λ) values of the LSF interior wall constituent materials [4].

Material	*d* [mm]	λ [W/(m · K)]
GPB[a] (2 × 12.5 mm)	25	0.175
MW[b]	90	0.035
Steel Stud (C90 × 37 × 15 × 0.6 mm)	–	50.000
GPB[a] (2 × 12.5 mm)	25	0.175
Total Thickness	140	–

[a]GPB - Gypsum Plaster Board (Standard A: GyptecIberica);
[b]MW - Mineral Wool (AlphaRolo: Volcalis®)

LSF walls was obtained using the heat flux meter (HFM) method [3], adapted to have two HFM sensors [5]. Four heat flux meters were used to measure the heat flux through the LSF wall, being two of them on the hot surface and the other two on the cold wall surface. In order to measure the two distinct thermal behavior zones (steel stud zone and cavity zone) within the LSF wall sample, in both wall surfaces, one HFM was placed in the zone of the central vertical stud, and the another one in the middle of the insulation cavity. Temperature measurements were performed using 12 Type K PFA insulated thermocouples (TCs), being half of them in the cold side, and another half in the hot side. In each side (cold and hot), two of the six TCs measured the environment air temperature inside the chamber, another two measured the wall temperature near the wall surface, and the remaining two measured the wall surface temperatures.

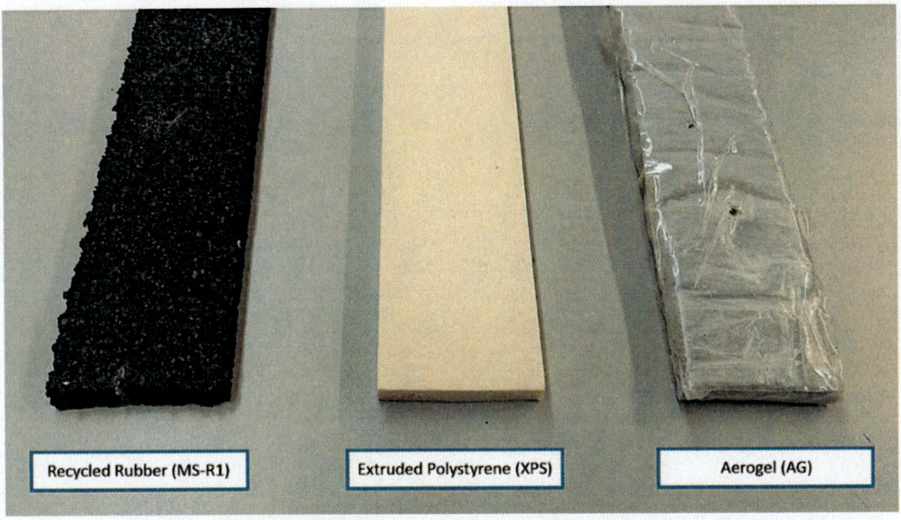

Fig. 2. Thermal break strips materials used on the experimental tests.

Table 2. Materials and thermal conductivity (λ) of the thermal break strips [2].

Material (abreviation)	λ [W/(m · K)]
Recycled Rubber (MS-R1)	0.122
Extruded Polystyrene (XPS)	0.034
CBS[a] Aerogel (AG)	0.015

[a]CBS - Cold Break Strip

The hot and cold boxes were programmed to maintain a temperature of 40 °C and 5 °C, respectively, being the measurements performed in a quasi-steady-state heat transfer condition. Furthermore, in order to ensure the repeatability of the experimental measurements, for each wall, three tests were performed corresponding to three high locations: top, middle and bottom (Fig. 4b). The considered measured overall conductive *R*-value of the LSF walls was the average of these three tests.

2.3 Numerical Simulations

The 2D numerical simulations of the LSF walls were performed using the finite element method (FEM) software THERM® (version 7.6.1). Being a bi-dimensional FEM numerical simulation, the models created consider only a 2D representative part of the walls cross-section (400 mm width), as previously illustrated in Fig. 1 for the reference LSF wall. Regarding the thermal properties of the materials, the values used in these simulations were previously presented in Sect. 2.1 (Table 1 and 2).

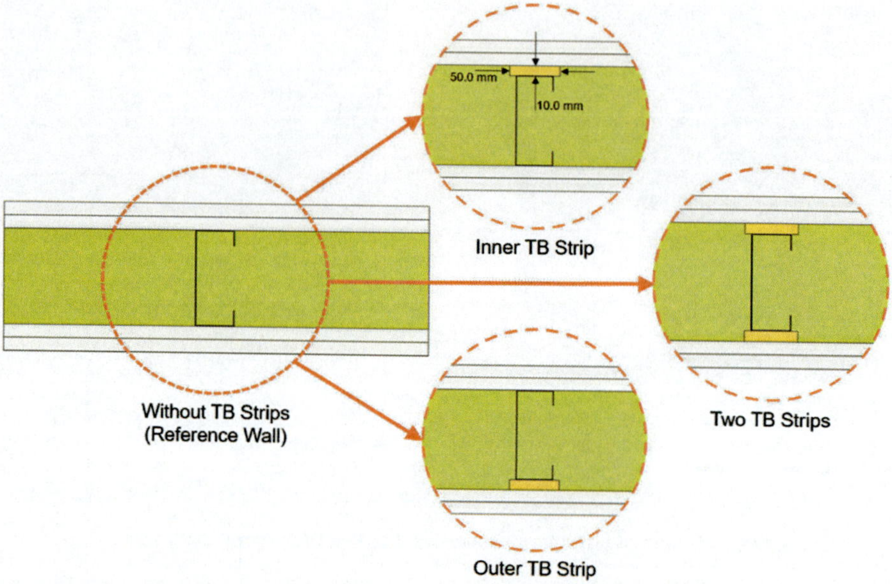

Fig. 3. Geometry and location of the thermal break (TB) strips.

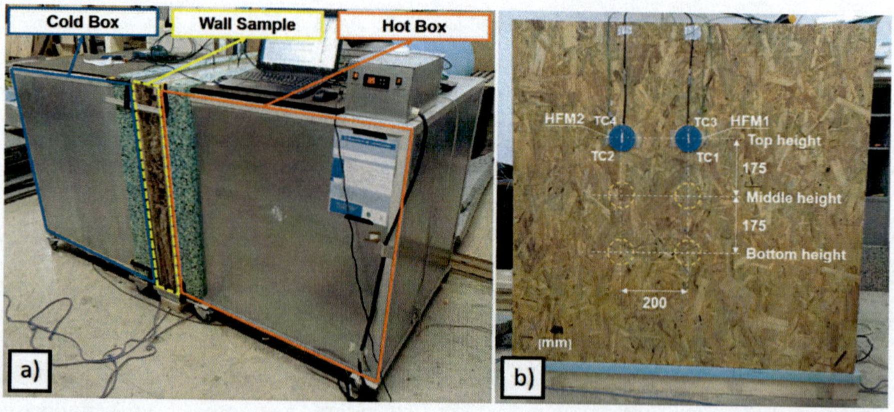

Fig. 4. Laboratorial tests: a) mini hot box apparatus; b) tested LSF wall sample and sensors localization.

In order to apply the boundary conditions to the models, the environment air temperatures and surface thermal resistances were defined. The air temperatures of the warm and cold environments were set equal to the temperature values defined for hot and cold climatic boxes in lab measurements, i.e., 40 °C and 5 °C, respectively. The modelling of the surface thermal resistances was performed using the average values measured for each test and for each LSF wall surface, considering the difference between the air and surface temperatures.

3 Results and Discussion

The values measured in laboratory and the values predicted by THERM software 2D FEM models for the conductive thermal resistances of the LSF walls assessed, as well as the absolute and percentage differences between them, are displayed in Table 3. The results presented in that table are organized into four parts: the first composed by the reference LSF wall (W_{ref}); the second, by the LSF walls with an inner TB strip (W_{int}); the third, by the LSF walls with an outer TB strip (W_{out}), and; the fourth, by the LSF walls with two TB strips (W_{x2}), inner and outer.

Table 3. Predicted (THERM) and measured thermal resistances (conductive *R*-values).

Wall Code Layer Description (mm)	*R*-value		Difference	
	THERM	Measured	Absolute	Percentage
	[(m² · K)/W]	[(m² · K)/W]	[(m² · K)/W]	[%]
W_{ref} 2GPB(12.5) + [C90 + MW(90)] + 2GPB(12.5)	1.719	1.752	+0.033	+2%
$W_{MS\text{-}R1in}$ 2GPB(12.5) + [C90 + MW(90) + **R1(10)**] + 2GPB(12.5)	1.932	1.964	+0.032	+2%
W_{XPSin} 2GPB(12.5) + [C90 + MW(90) + **XPS(10)**] + 2GPB(12.5)	2.162	2.195	+0.033	+2%
W_{AGin} 2GPB(12.5) + [C90 + MW(90) + **AG(10)**] + 2GPB(12.5)	2.359	2.404	+0.045	+2%
$W_{MS\text{-}R1out}$ 2GPB(12.5) + [**R1(10)** + C90 + MW(90)] + 2GPB(12.5)	1.932	1.931	−0.001	0%
W_{XPSout} 2GPB(12.5) + [**XPS(10)** + C90 + MW(90)] + 2GPB(12.5)	2.162	2.211	+0.050	+2%
W_{AGout} 2GPB(12.5) + [**AG(10)** + C90 + MW(90)] + 2GPB(12.5)	2.359	2.414	+0.055	+2%
WMS-R1x2 2GPB(12.5) + [**R1(10)** + C90 + MW(90) + **R1(10)**] + 2GPB(12.5)	2.147	2.142	−0.005	0%
WXPSx2 2GPB(12.5) + [**XPS(10)** + C90 + MW(90) + **XPS(10)**] + 2GPB(12.5)	2.574	2.627	+0.053	+2%
W_{AGx2} 2GPB(12.5) + [**AG(10)** + C90 + MW(90) + **AG(10)**] + 2GPB(12.5)	2.892	2.885	−0.007	0%

GPB – Gypsum plasterboard; C90 – Steel stud type and web dimension in mm; MW – Mineral wool; OSB – Oriented strand board; MS-R1 – Recycled rubber thermal break strip; AG – Aerogel thermal break strip.

Analyzing the results obtained, it is possible to verify that the measured and predicted *R*-values are very similar (percentage differences between 0% and +2%), thus ensuring the reliability of these values. Furthermore, the results presented show that the *R*-value increase depends mainly on two factors: (1) the number of thermal break (TB) strips (single TB strip on inner or outer flange, or TB strips on both flanges), and; (2) the thermal conductivity of the TB strips materials. For a better visualization and comparison, in

Fig. 5, the measured *R*-values are graphically displayed, being indicated, additionally, the *R*-value obtained by THERM for the reference wall without steel studs.

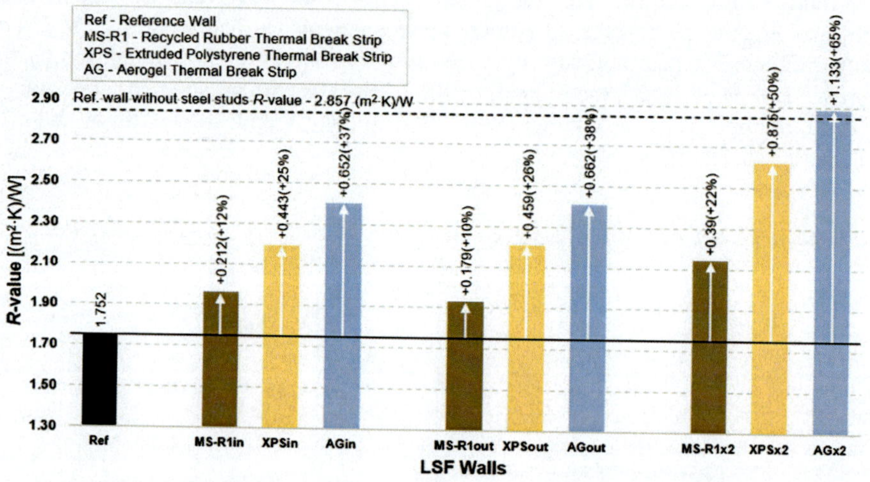

Fig. 5. Measured thermal resistances (conductive *R*-values).

As expected, the heat loss reduction due to mitigation of the steel thermal bridges provided by the application of TB strips allows to increase the *R*-value of the LSF walls. The application of TB strips on the steel studs has led to a thermal resistance increase ranging from 10% (for outer recycled rubber TB strip) up to 65% (for two aerogel TB strips).

Regarding TB strip materials, aerogel exhibited the largest thermal performance improvement: +37% and +38% for inner and outer TB strips, respectively, and; 65% for TB strips at both flanges. The lower *R*-value increase was measured in the recycled rubber TB strips (material with the highest thermal conductivity), while the extruded polystyrene TB strips registered intermediate values. Taking in account the localization of the TB strips on the steel studs, the results demonstrate that, for each material, the application at both flanges achieves the highest thermal performance improvements, increasing significantly the *R*-value comparatively with a single TB strip. Furthermore, the comparison between a single inner and outer TB strip shows that the *R*-values obtained are very similar for each one of the tested materials, as expected given the wall's symmetry.

Among the tested LSF walls configurations, only the two aerogel TB strips solution is able to reach the *R*-value provided by the reference wall without steel studs (2.857 $m^2 \cdot K/W$), fully mitigating the steel frame thermal bridge effect.

4 Conclusions

In this work the thermal performance of lightweight steel frame (LSF) partition walls with thermal break strips was assessed experimentally in laboratory-controlled conditions.

Three TB strips materials (recycled rubber, extruded polystyrene and aerogel) were tested, and three configurations for the localization of the TB strips were considered: inner, outer and on both steel stud flanges.

The main conclusions of this study are presented as follows:

(1) The thermal performance achieved when a single TB strip is applied on inner or outer steel stud flanges is very similar, as expected.
(2) The application of TB strips on both steel stud flanges provides a significant thermal resistance increase, compared to the application of single TB strips.
(3) Aerogel was the TB strip material with the best thermal performance, while the recycled rubber exhibited the worst results. The extruded polystyrene presented an *R*-value increase between the other two materials.
(4) The application of aerogel TB strips on both steel stud flanges was the only configuration able to reach the *R*-value provided for the reference wall without steel studs, fully mitigating the steel frame thermal bridge effect.

Acknowledgments. The authors also want to thank the support provided by the following companies: Pertecno, Gyptec Ibéria, Volcalis, Sotinco, Kronospan, Hulkseflux, Hilti and Metabo.

Funding. This research was funded by FEDER funds through the Competitivity Factors Operational Programme – COMPETE and by national funds through FCT – Foundation for Science and Technology, within the scope of the project POCI-01-0145-FEDER 032061.

Cofinanciado por: POCI-01-0145-FEDER-032061

References

1. Tyre4BuildIns, Research project Tyre4BuildIns - 'Recycled tyre rubber resin-bonded for building insulation systems towards energy efficiency', University of Coimbra, PT. www.tyre4buildins.dec.uc.pt. Accessed 29 Jan 2021
2. Santos, P., Mateus, D.: Experimental assessment of thermal break strips performance in load-bearing and non-load-bearing LSF walls. J. Build. Eng. **32**, 101693 (2020)
3. ISO 9869-1: Thermal insulation – Building elements - In-situ measurement of thermal resistance and thermal transmittance. Part 1: Heat flow meter method. ISO - International Organization for Standardization, Geneva, Switzerland (2014)
4. Santos, P., Lemes, G., Mateus, D.: Thermal transmittance of internal partition and external facade LSF walls: a parametric study. Energies **12**(14), 2671–2690 (2019)
5. Soares, N., Martins, C., Gonçalves, M., Santos, P., Simões da Silva, L., Costa, J.J.: Laboratory and in-situ non-destructive methods to evaluate the thermal transmittance and behaviour of walls, windows, and construction elements with innovative materials: a review. Energy Build. **182**, 88–110 (2019)

Thermal Balance Analysis in a Light Steel Frame Residence with Different Insulation Material Thickness

Isabel Oberderfer Consoli[1](✉), Ney Lyzandro Tabalipa[1], and Carlos Alberto Rodrigues Andrade[2]

[1] Universidade Tecnológica Federal do Paraná, Pato Branco, PR 85503-390, Brazil
[2] ESTIG, Instituto Politécnico de Bragança, 5300-252 Bragança, Portugal

Abstract. Sustainability and efficiency are concepts that have been spreading in civil construction. Among all building types, residential buildings have great potential for energy savings. It is where building strategies need to be carefully planned, as they seek to meet the needs of residents, and also provide security and comfort. Well-designed houses will also provide resource saving over time, as they add climate analysis, design strategy studies and appropriate materials to their design. In order to verify a Light Steel Frame house behavior regarding energy consumption for heating and cooling, this paper aims to evaluate walls and roofs with layer different thickness of rock wool and expanded polystyrene as insulation materials. The applied methodology to obtain energy efficiency will be through analysis in the software Design Builder®. The constructions in Portugal are mainly made of masonry and the search for alternative construction solutions aimed at complying with Directive (EU) 2018/844 is relevant. A drastic change in collective thinking will be necessary. This article aims to show as result the energy efficiency of Light Steel Framing construction solutions, presenting the materials which have the greatest benefits regarding energy consumption.

Keywords: Insulation materials · Energy consumption · Light steel frame

1 Introduction

Application of sustainable strategies in buildings has become mandatory in many countries around the world, especially in the last few years. One of the major challenges faced by sustainable buildings is the achievement of satisfactory levels in efficiency terms, without negative impacts in costs. Therefore, in-depth and reality-driven studies should be developed in building designs, seeking a comparison between solutions based on environmental principles, which perform efficiently, meet user needs, promotes financial savings and reduces material consumption.

One factor that has great influence on thermal and energy performance in a building is the choice of building materials. Elements that separate the interior and exterior environments, such as walls, roof, glazing and others, must have suitable thermal conductivity characteristics to prevent unwanted energy gains or losses. Often the envelope

J. R. da Costa Sanches Galvão et al. (Eds.): *Proceedings of the 1st International Conference on Water Energy Food and Sustainability (ICoWEFS 2021)*, pp. 390–399, 2021.
https://doi.org/10.1007/978-3-030-75315-3_44

can be designed with numerous layers, with structural and insulation materials, further improving its thermal properties. Likewise, other sustainable strategies can be ensured, such as natural lighting, the use of natural surroundings to create an appropriate microclimate, along with energy-generating devices from renewable sources, such as solar panels [1].

Previous studies developed both in computer simulations and applied in real situations show that EPS and rock wool are efficient materials when related to thermal transmission, also showing environmental benefits [2]. They are also more cost-effective when compared to others available for building application. Other studies indicate that the use of local data with statistical basis is essential for the simulation results to be in agreement with the future situation of the building [3].

In order to verify a residential building envelope behavior regarding energy consumption for heating and cooling, different thicknesses of rock wool were analyzed inside walls and roofs. The study case is a Light Steel Frame (LSF) house that will be placed in Bragança, Portugal. The simulations were performed with Design Builder™ software.

2 Thermal Insulation of Building Envelope and Energy Consumption

Elements that directly control heat gains and losses in a building, such as the roof, external walls, floors and windows, are considered envelope elements. The theory of heat transfer says that heat flows from warmer to cooler zones until a thermal balance is achieved. Applying this concept in building designs, there are different situations for winter and summer. During the hottest season, heat transfer occurs from the outer spaces to the interior of the building. In winter, heat is transferred from indoor heated spaces to those at a different and lower temperature (outside or unheated environments) through the envelope [3].

In order to decrease thermal gains and losses, along with reducing energy consumption with heating and cooling devices, walls, roofs, windows and floors must be good thermal insulators (in most situations using layered material compositions to further improve insulating features). Well-designed buildings with suitable materials must provide comfort and low energy costs all year round [3, 4].

In LSF buildings, the steel structure that supports the walls has a sealing layer in both sides (inner and outer) and an insulating layer between them. The appropriate insulating characteristics can be obtained in different ways and using different materials and compositions. Same is true for the roof.

3 Case Study House: Light Steel Frame Residence

Bragança is a city located in northeast Portugal. As stated by IPMA (Portuguese Institute of Sea and Atmosphere), the city has a Csb climate, according to Koppen Geiger. Figure 1 exemplifies the temperature profile presented by the city throughout the year [5]. Figure 1 also shows that summer and winter temperatures can reach extreme values, with a large thermal amplitude between them. Consequently, in order to achieve indoor thermal comfort, passive constructive strategies need to be proposed for the residence. It is also of great importance that the internal temperature remains at a satisfactory level for users all over the seasons.

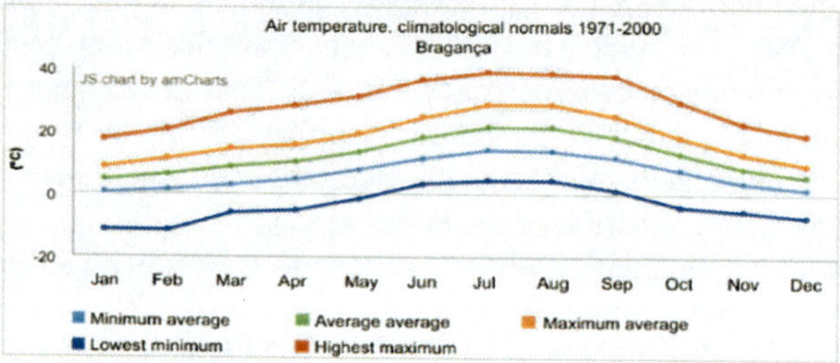

Fig. 1. Annual temperature profile in Bragança.

The case study house is a single-family residence, which has three bedrooms, closet, toilets, home office, an integrated kitchen and dining area, laundry, garage and an outdoor recreation space, totaling 180,00 square meters [m^2]. The residence schematic floor plan is shown in Fig. 2. It is important to highlight that this building was designed in accordance with the Portuguese law-decree 118/2013 (transposition of the European Directive 2010/31/UE), which concerns the energy performance requirements of buildings.

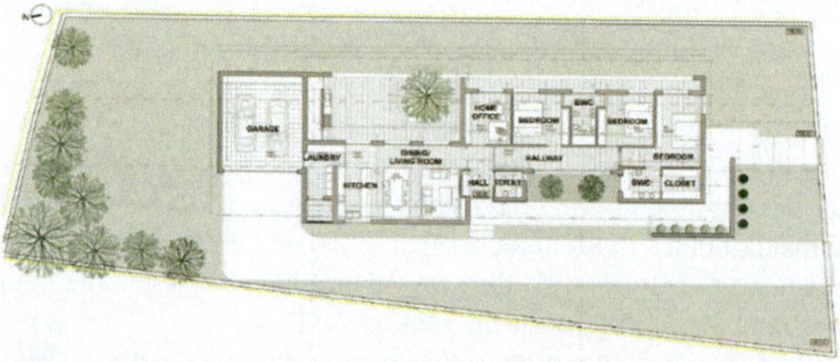

Fig. 2. Residence floor plan.

Some additional information about the case study house that will impact directly on model efficiency are:

- LED lamps will be used for lighting, to complement natural light when necessary;
- Artificial heating will be attained by electric underfloor heating and a pellet fireplace (biomass) located in the dining/living room;
- Domestic Hot Water (DHW) will be heated by a heat pump (mostly renewable energy);
- Natural ventilation provided by window openings will be the main cooling device;

- Garage and the laundry room were purposely positioned in North direction, since they represent less significant spaces regarding the direct sunlight.
- Intimate environments were oriented towards South and East in order to obtain solar gains and adequate natural lighting;
- Shading elements such as marquees and vegetation were proposed on the West façade.

All the windows have been positioned to maximizing natural lighting and heat gains from the sun. Following this principle, the house has no openings directed to the North. Although they represent a large percentage in the building energy efficiency, this paper will not discuss architectural elements, position of the rooms or passive construction strategies applied. Only envelope materials were analyzed, considering the floor plan presented in the Fig. 2.

3.1 Envelope Elements

After the architectural floor plan had been modeled in the software, some settings for indoor spaces use were adjusted. Then, building materials could also be detailed. For the case study house, the constructive elements were configured in DesignBuilder™ so that the simulations could be as close as possible to the real situation.

Layers designed for the floor are (from the innermost layer) of 15 mm of vinyl finishing, 40 mm of concrete floating block (smoothing mortar), 5 mm of electric underfloor heating system, 100 mm of expanded polystyrene (EPS) with equivalent thickness in the surroundings, 300 mm concrete slab. Floor will be built on compacted soil, as in Fig. 3.

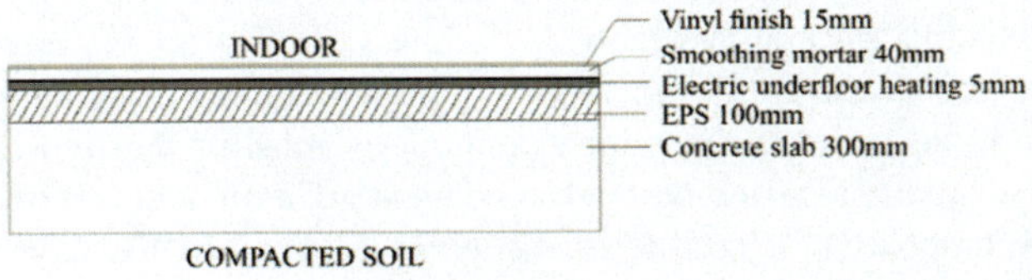

Fig. 3. Floor layers.

External walls layers are, from innermost one, a 13 mm gypsum plasterboard, a variable layer of rock wool as the first insulation material, a variable air gap, a 12 mm oriented strand board, a variable layer of extruded polystyrene and 6 mm of outer cladding, according to Fig. 4.

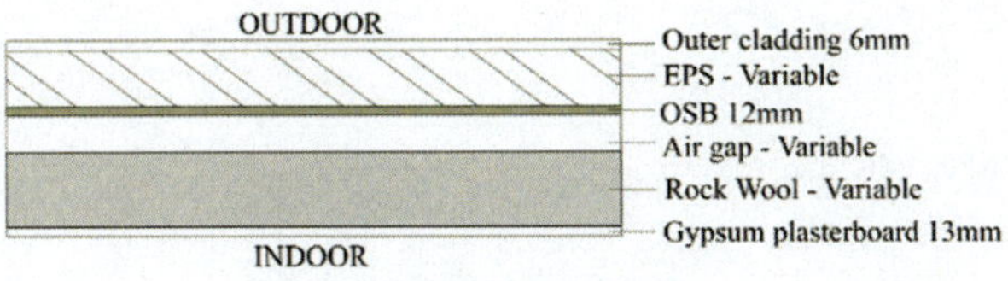

Fig. 4. External wall layers.

As the city has great thermal variation throughout the year, there is a need to design a composite insulation system: the ETICS system from the outside, with the EPS layer, and rock wool inside the walls.

In order to establish a relationship between energy consumption and efficiency of insulation materials, five different arrangements of external walls were simulated.

Thicknesses for insulation materials in each simulation are described in Table 1.

Table 1. Thicknesses of insulation materials in each simulation.

Simulation	EPS [mm]	Air gap [mm]	Rock wool [mm]	Total insulation [mm]
Reference wall	100	50	100	200
Wall 01	100	30	120	220
Wall 02	100	0	150	250
Wall 03	120	50	100	220
Wall 04	150	50	100	250

Internal partition walls are designed with two gypsum plasterboard with 12 mm each, filled with 90 mm of rock wool. Although rock wool does not show significant performance related to heat exchanges in indoor environments, it has great acoustic performance, therefore its use is justified inside these partitions. Windows and glazed elements are made of double glazing with a 16 mm air layer between, with PVC frames. Glazed elements are designed with a Heat Transfer Coefficient (U) of 1,5 [W/m^2 °C]. In the simulations the box shutters (above windows) were externally covered with the ETICS system, in order to minimize thermal bridges. For air exchanges through the openings, it was considered the minimum amount established by the Portuguese law. External doors are made of wood with a Heat Transfer Coefficient (U) of 2,995 [W/m^2 °C]. The roof is composed of a light-colored polyvinyl chloride (PVC) layer (outermost), followed by 100 mm of EPS, an 18 mm oriented strand board (OSB) fixed on the LSF slab, which has 400 mm. Below the slab there is an air gap (variable due to the roof slope), 200 mm of rock wool layer and the 13 mm gypsum plasterboard. Figure 5 details the roof layers, and also the covering of the box shutters in exterior walls.

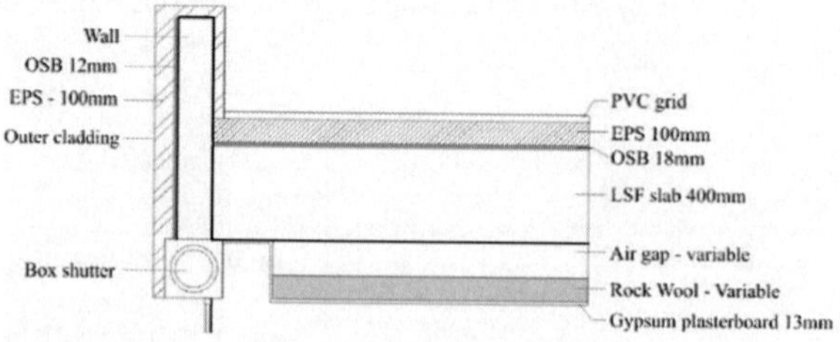

Fig. 5. Roof layers.

The housing area that has the greatest exposure to the sun heat is the roof. The walls also receive direct sunlight, but they may have greater or lesser intensity throughout the day, depending on the solar path. This makes the thermal gains through the roofing elements greater than gains from walls during summer. On the other hand, during the winter situation, the indoor environments are heated using the underfloor heating or the pellet fireplace. Due to the phenomenon of convection, the hot air, which has lower density, tends to rise, causing the heat to be transmitted to the outside through the roof in greater quantity than through the walls.

For these two reasons, it is essential that roof is designed with greater efficiency. Thus, seeking to increase the efficiency of the model, the simulations performed consider only configurations in which the insulation thickness of the roof is greater than the one in the walls. Higher wall thicknesses would end up in unnecessary material consumption, without greater benefits in terms of thermal and energy efficiency.

Specific physical properties that interfere with the thermal performance of materials are density, thickness and respective thermal conductivity. The values corresponding to each of the properties, obtained through the Portuguese manual ITE 50 (see Table 2), are relevant to obtain the thermal resistance of the building elements. In turn, thermal resistance together with indoor and outdoor air temperatures are determining values for obtaining the heat transfer coefficient and verify the behavior of the material when exposed to different thermal situations [2].

Table 2. Properties of construction materials listed in ITE 50 [2].

Material/layer	Building element	Density ρ [kg/m^3]	Thermal resistance R [m^2 °C/W]
Air gap – horizontal flux	Walls	1,23	0,18
Air gap – vertical flux	Roof	1,23	0,16
Material/Layer	Building element	Density ρ [kg/m^3]	Thermal conductivity λ [W/m°C]
Concrete smoothing mortar	Floor	1800–2000	1,300
Rock Wool	Walls and Roof	20–35	0,045
Concrete slab	Floor	2000–2300	1,650
Gypsum plasterboard	Walls and Roof	1000	0,400
Expanded polystyrene (EPS)	Walls, Roof, Floor	20	0,037
Polyvinyl Chloride (PVC)	Roof	1200	0,140
Oriented Strandboard (OSB)	Walls and Roof	650	0,130
Outer cladding	Walls	1800–2000	1,300

All values of thermal properties were entered in DesignBuilderTM during the modeling of the residence.

3.2 Thermal Balance Simulations

By default settings, Design Builder™ performs simulations that consist of summer analysis (or cooling design), winter analysis (or heating design) and simulations for specific time intervals. The designer can specify criteria or perform the simulation with information from an entire year. The developed analyzes were made with the annual interval, in order to find the thermal gains and losses for the case study house throughout the seasons. Thus, it is also possible to compare climatic variations and the characteristics of the building itself.

Within the annual simulation, in the heating season (winter) the software considers that indoor spaces are constantly heated until they reach a set-back temperature. In this residence, the defined value was 18 [°C]. For the summer situation, the residence will not have an artificial cooling device (only natural ventilation).

To obtain thermal balance throughout the year, solar gains through windows, natural ventilation, internal gains through occupation (lighting, equipment, user activity) will be included, in addition to the conduction and heat convection between surfaces. These values will be compared with outside air temperature, making it possible to identify energy consumption profile and heating needs inside the building. For statistical and comparison purposes, the simulations consider 8760 h (365 days).

The first simulation considered the reference wall configuration, with has the lowest thickness of rock wool and EPS (100 mm thickness of each). The following simulations were performed first increasing the thickness of rock wool (keeping the EPS fixed in 100 mm). Then the value of the rock wool remained fixed in 100 mm, while the increase occurred in the EPS layer. Consequently, the results obtained were related to five different situations, where it was possible to make comparisons between the characteristics of the elements and its consequent efficiency.

4 Results and Discussion

At first it was possible to calculate the heat transfer coefficient of each element by entering material data in Design Builder™. These values are shown in Table 2. Regarding the heat transfer coefficient, the higher is the value, the greater is the heat flow through the surface, in other words, lower is the insulating capacity of the material [7] (Table 3).

According to Portuguese law Portaria 379-A/2015, the maximum values allowed for heat transfer coefficient in each element are [7]:

External walls: $U_{max} = 0,35\,[\mathrm{W/m^2\,°C}]$;
Roof: $U_{max} = 0,30\,[\mathrm{W/m^2\,°C}]$;
Floor: $U_{max} = 0,50\,[\mathrm{W/m^2\,°C}]$;
Glazed elements: $U_{max} = 2,20\,[\mathrm{Wm^2/°C}]$.

Therefore, all proposed combinations are within the allowed range.

Regarding energy consumption, heating needs were obtained for each of the simulations. As the residence will have two different heating devices, energy consumption was separated between rooms heated by the electric underfloor heating, and rooms heated by the pellet fireplace (dining/living room and kitchen). In Fig. 6, the red portion of the

Table 3. Heat transfer coefficient of construction elements.

Envelope element	Heat transfer coefficient U [W/m^2 °C]
Floor	0,50
Glazed elements (windows)	1,50
Doors	2,90
Roof	0,133
Reference wall	0,191
Wall 01	0,176
Wall 02	0,162
Wall 03	0,173
Wall 04	0,152

graph is related to the energy needs supplied by the fireplace, while the blue portion represents the electrical energy consumed by the underfloor heating device.

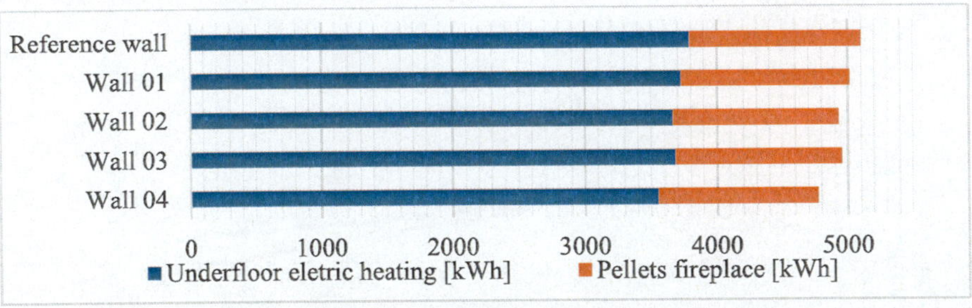

Fig. 6. Energy consumption with heating device.

It is observed that approximately 25% of the energy used for heating inside the residence can be used with the pellet fireplace, which in addition to being an efficient equipment, also uses biomass as a principle for its operation, a renewable and sustainable energy source.

The total energy consumption for heating obtained in each of the performed simulations can be analysed in Table 4. In this table, which presents the absolute values, it is also possible to relate energy consumption with total thickness of insulation material (adding EPS and rock wool).

These results show that lower thicknesses of insulation material will generate higher energy consumption with heating (the reference wall has the highest consumption value). This fact is expected and somehow, intuitive. It is also noticeable that the energy consumption is lower when using greater thicknesses of EPS rather than rock wool. This phenomenon occurs due to thermal properties of each material: EPS has slightly lower thermal conductivity than rock wool, which will result in greater thermal resistance with the increase in the material thickness.

Table 4. Total insulation thickness and total energy consumption with heating in simulations.

	Ref. wall	Wall 01	Wall 02	Wall 03	Wall 04
EPS thickness	100 mm	100 mm	100 mm	120 mm	150 mm
Rock wool thickness	100 mm	120 mm	150 mm	100 mm	100 mm
Total insulation thickness	200 mm	220 mm	250 mm	220 mm	250 mm
Total energy consumption [kWh]	5098,7	5009,63	4926,07	4950,37	4769,44

To establish a relationship between the increase in energy efficiency and the application of thicker layers of insulating material to the wall, the graph shown in Fig. 7 present some interesting analysis. The reference value used corresponds to the simulation with a 100 mm rock wool and 100 mm of EPS layers (reference wall). Initially, a comparison could be made between the reference and the two simulations that resulted in 220 mm of insulation (total). Both simulations had an increase of 20 mm in the insulation layer. It is noticeable that when adding 20 mm on the EPS layer, an increase in 2,91% efficiency is achieved, which corresponds to 0,145% per millimetre. When the increase is in rock wool layer, the percentage improved with the addition of 20 mm is 1,75%, corresponding to 0,087% per millimetre.

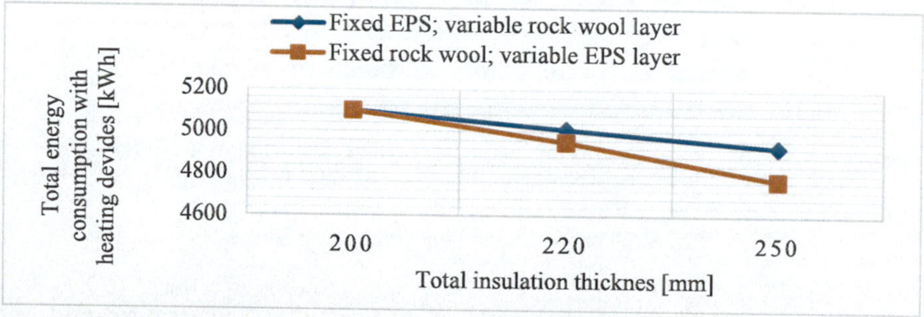

Fig. 7. Relation between insulation thickness and energy consumption.

Now analyzing the increase of 50 mm in both materials, the following behavior was observed: with the 50 mm increase in EPS, the efficiency gain was 6,46%, or 0,129% per millimeter. With rock wool layer, the gain was 3,39%, 0,068% per millimeter, when compared to the reference wall. These results lead to the recognition that the situation with best relationship between energy efficiency and the increase in insulation material thickness is Wall 03 (120 mm EPS + 100 mm rock wool). Performing a projection of consumption with very high insulation thicknesses, there is a tendency for asymptotic behaviour, or at any given time, regardless of the thickness of the insulation material, the reduction in energy consumption will be minimal.

5 Conclusion

Performing the analyses with the insulation layers configuration presented in this paper, it can be concluded that it is not always suitable to use large thicknesses in insulation material layers for envelope elements. Ratio between the material thickness and the gains in energy efficiency do not follow a linear pattern. The simulations performed are important to ensure that nearly real data are obtained. So, it avoids waste of materials and prevents building owners from wasting unnecessary financial resources. In the presented LSF residence, the configuration with better efficiency was simulation with Wall 03, with 100 mm rock wool inside walls and 120 mm EPS layers in the exterior ETICS system, the simulation that showed the best ratio between thickness and energy efficiency.

Further analysis should be performed using different insulation materials, varying their thickness and comparing each other to verify their respective efficiencies and cost-effectiveness.

In this paper, the LSF residence was studied because it is a constructive system that has low thermal inertia. Therefore, it is essential that insulating materials be applied to these buildings, after a careful analysis on their efficiency and benefits in the medium and long term. Furthermore, LSF is considered a sustainable, flexible, versatile system, and with great potential for recycling at the end of the product's life (building), reinforcing the existing interconnection between energy efficiency, resource saving and sustainability.

It is important to notice that in typical masonry constructions in Portugal it is not common to find insulation materials with this thickness (higher than the usual), due to the constraints of the traditional building system itself. Even references such as ITE 50 have maximum thicknesses of 80 mm for calculations of thermal transmission coefficients (U) (Santos & Matias, 2006). As a result, conventional masonry buildings are considered less energy efficient than the solution presented in Light Steel Frame.

References

1. Lamberts, R., Dutra, L., Pereira, F.O.R.: Eficiência Energética na Arquitetura, 3 edn., Ed., Rio de Janeiro, RJ: PW Editores (2014)
2. dos Santos, C.A.P., Matias, L.: ITE 50 - Coeficientes de trasnmissão térmica dos elementos da envolvente dos edifícios. Laboratório Nacional de Engenharia Civil (LNEC), Lisboa (2006)
3. Hu, J., Wu, J.: Analysis on the influence of building envelope to public buildings energy consumption based on DeST simulation. In: 9th International Symposium on Heating, Ventilation and Air Conditioning (ISHVAC) and the 3rd International Conference on Building Energy and Environment (COBEE) (2015)
4. Santos, P., et al.: Influence of climate change on the energy efficiency of light-weight steel residential buildings. Civ. Eng. Environ. Syst. **28**(4), 325–352 (2011)
5. IPMA, "Normais Climatológicas - 1971–2000 - Temperatura do ar Bragança, 25 September 2019. https://www.ipma.pt/pt/oclima/normais.clima/1971-2000/#575
6. Design Builder Software, Design Builder, 6 January 2009, Stroud, UK (2019)
7. Portuguese Republic: Decreto-Lei n.º 118, Lisboa (2013)

Infrasound and Low-Frequency Noise Produced by Wind Turbines in Housings - Case Studies

João Almeida[1](✉), Andreia Videira[1], João Paulo Figueiredo[2], and Ana Ferreira[1]

[1] Coimbra Health School (ESTeSC), Department of Environmental Health, Polytechnic of Coimbra, Coimbra, Portugal
joaoalmeida@estescoimbra.pt

[2] Coimbra Health School (ESTeSC), Department of Complementary Sciences, Polytechnic of Coimbra, Coimbra, Portugal

Abstract. The wind power is considered as a clean energy. However, the noise produced by wind turbines, namely infrasound and low frequency noise (below 500 Hz) are responsible for several pathologies in the humans. The aim of this study was to assess the sound pressure level of infrasound and low frequency noise, produced by wind turbines, in housings nearby wind farms. The measurements were carried out in 3 houses near the "S. João 2" wind farm (Penela, Portugal). It was considered two measuring points for each house (inside and outside). For the measurements it was used a sound level meter CESVA SC420 and the Q-TRAK Plus IAQ Monitor with Probe Model 8592 for measure air velocity. For statistical analysis it was used IBM SPSS Statistics v.26. It was found that, despite the increase in distance (between the different houses and the wind farm), there was no decrease in sound pressure levels. The houses nearby "low vegetation" present higher sound pressure level for infrasound and low frequency noise. Infrasound and low frequency noise levels are higher "outside" the houses. For lower air velocities, the values of infrasound and low frequency noise are higher. Regarding the behavior of the variation of noise frequencies, it was found that the sound pressure levels were higher in the houses furthest from the source. It was concluded that there are several factors that influence the propagation of noise. Therefore, it is important to consider measurements to minimize the impacts on the population and invest in prevention policies concerned with public health and which define safe setbacks.

Keywords: Wind turbines · Infrasound · Low frequency noise · Housing

1 Introduction

In recent years there has been a rapid growth in the installed power of wind energy in Portugal. This fact, combined with technological development, led to an exponential growth in the number of wind turbines, neglecting the negative impacts they cause in their area of influence [1]. This type of energy reduces our dependence on fossil fuels

J. R. da Costa Sanches Galvão et al. (Eds.): *Proceedings of the 1st International Conference on Water Energy Food and Sustainability (ICoWEFS 2021)*, pp. 400–410, 2021.
https://doi.org/10.1007/978-3-030-75315-3_45

and contributes to the reduction of greenhouse gases and other adverse pollution [2]. However, one of the major disadvantages of this structures is the noise emitted by wind turbines, being considered a new source of noise, contributing to the increase in existing noise pollution [3].

Sound is caused by a series of pressure pulsations, or more broadly, by changes in air pressure [4]. We can find three frequency ranges: for the human ear the range of audible frequencies is between 20 Hz and 20000 Hz; the sounds of frequencies below 20 Hz are called infrasound; the sounds of frequencies higher than 20000 Hz is called ultrasound [2].

Measurable infrasound is produced worldwide daily by a variety of natural and artificial sources [5]. Infrasound and Low Frequency Noise (ILFN) are present in all urban areas, in many suburban areas and in some rural areas. In the case of ILFN produced by wind turbines, there is an alternative that could combine, in a balanced way, the existence of facilities or places that produce ILFN and human populations: effective planning laws [6].

This type of noise is considered the frequency superpower and is characterized by having less attenuation by walls and other structures, for traveling long distances with little loss of energy due to atmospheric attenuation and the ground and because the protective devices are much less effective and the ability to produce resonance in the human body [7].

Wind turbines generate two types of noise, mechanical noise and aerodynamic noise. Mechanical noise results from the interaction of mechanical components and the dynamic response between them. This type of noise has been practically eliminated through insulation materials [8]. Aerodynamic noise is mainly produced by the airflow of the blades, meaning that noise generally increases with the velocity of the tip. This type of noise is directly linked to energy production and is therefore inevitable [9]. During the night, wind turbines can produce higher sound pressure levels due the weather conditions (as relative humidity and wind speed). The frequencies of this noise are characterized by their ease of propagation through infrastructure and means, such as walls and windows [10, 11].

This study aimed to determine the noise levels in the houses close to the Wind Farm (WF) and to understand if the surroundings and the distance to the WF would influence the noise levels.

2 Material and Methods

This study was an observational (analytical) type of level II, of a transversal nature. The measurements were carried out in houses close to the “São João 2” Wind Farm, located in Penela, district of Coimbra (Portugal). The sample design was defined as for the non-probabilistic type and for the convenience technique.

For the case study, 3 houses near “São João 2” Wind Farm were selected. For the definition of the measuring points, the gap of 1.5 km to 2 km was considered. Google Earth Pro software version 7.3.3.7699 was used for house selection. In Fig. 1 it is possible to see the location of the target houses. These houses are located on the East (E) and Southeast (SE) of the WF under analysis.

In House A, 6 measurements were made, in Houses B and C, 5 measurements in each, making a total of 16 measurements. All measurements were subject to average air temperature conditions of 12.1 °C and North wind direction.

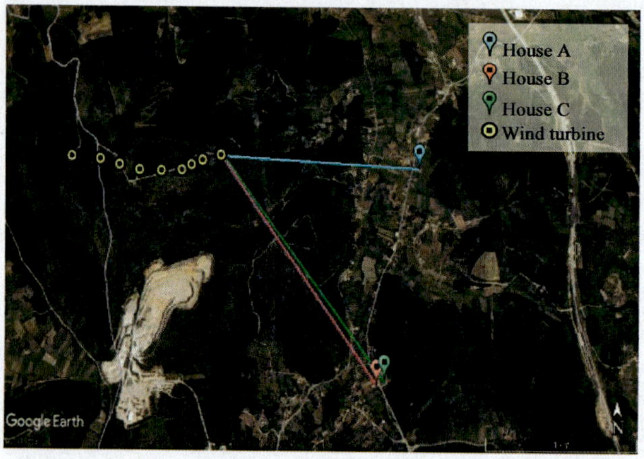

Fig. 1. Location of the WF S. João 2 wind turbines and the houses under study. **Source**: *Google Earth Pro*, version 7.3.3.7699.

In the cross-sections for each WF-House vector, it is possible to verify the absence of obstacles between the houses and the WF (Fig. 2).

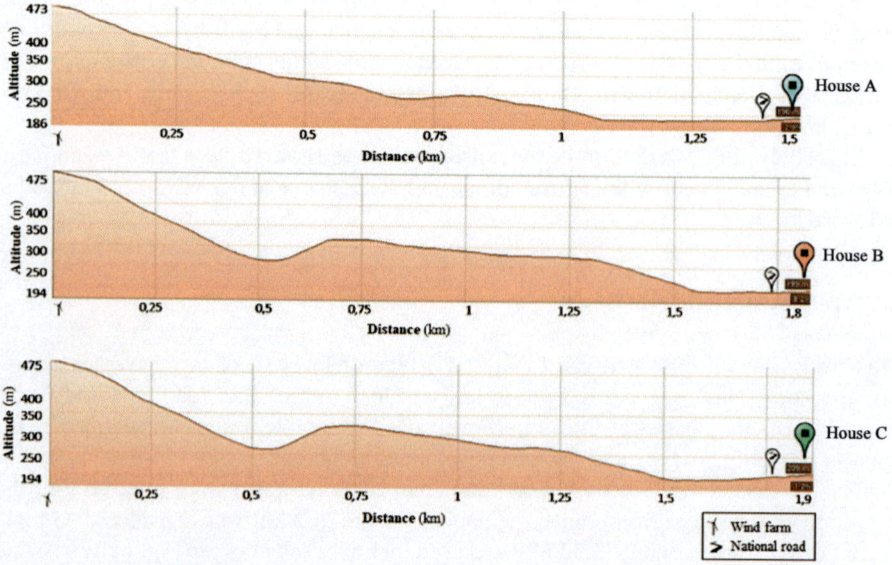

Fig. 2. Transverse profiles between the wind turbine (closest to the Houses) and the Houses. **Source:** *Google Earth Pro*, version 7.3.3.7699.

Data collection took place in February 2020. For data collection related to the sound pressure levels, the CESVA SC420 was used. The sound level meter was calibrated before and after the measurements. The following rest conditions were considered: sound level meter positioned 90 cm from the floor; distance of 2 m from walls, windows and infrastructures; sampling time of 1 min; two measurement points per House; three measurements were performed at each point. Simultaneously, air velocity measurements were made using the Q-TRAK Plus IAQ Monitor with Air Velocity Probe Model 8592. Were considered the following test conditions: equipment placed 90 cm from the floor; stabilization period of 5 min; measurements outside houses.

Data collection was based on the divisions where the residents of each dwelling spend the most time. After collecting the sound pressure levels, the different types of noise were calculated (based on the different frequencies): Infrasound (2 Hz to 20 Hz); Low Frequency Noises (LFN) (20 Hz to 500 Hz); ILFN (2 Hz to 500 Hz) and Environmental Noise (EN) (according to the General Noise Regulation approved by Decree Law n.° 9/2007, of 17 January [12]).

For statistical analysis, IBM SPSS Statistics version 26 software was used. A descriptive analysis was performed as mean values and standard deviation of the different noise levels (Infrasound, LFN, ILFN and EN). Subsequently they were submitted to an inferential analysis, taking into account the validation of assumptions for the correct application of statistical models. The following statistical tests were applied: Kruskal-Wallis non-parametric test and Wilcoxon-Mann-Whitney non-parametric test for two independent samples. To estimate the statistical inference, a 95% confidence level, and a random error less than or equal to 5% were taken into account.

3 Results

It was assessed noise levels in houses as a function of different distances to wind turbines (Table 1). According to the statistic test, no significant differences were observed with respect to Infrasound (p-value > 0.05). Regarding the type of noise LFN, ILFN and EN, there were statistically significant differences (p-value < 0.05). It can be verify that, in the Houses B and C (more distant to the source), the sound pressure levels of LFN and EN are higher.

Using the Bonferroni corrected multiple comparison test, the LFN values were much higher, between House C compared to House A ($Z = -3.180$; $p = 0.004$) and House B ($Z = -2.354$; $p = 0.056$). As for the level of ILFN, the differences occurred between House C compared to House A ($Z = 8.224$; $p = 0.012$). Regarding the type of EN noise, there were statistically significant differences between Houses B and C ($Z = -2.395$; $p = 0.05$), which allowed us to verify that the values were significantly higher in this last house compared to House B.

It was assessed the noise levels according to the type of surroundings (presence or absence of natural vegetation) and it was found that the noise variation was not relatively homogeneous (Table 2). The noise evaluated in the presence of low vegetation having slightly higher mean values of infrasound and ILFN compared to places with high vegetation. The opposite situation was found in the types of noise LFN and EN.

When verifying the relationship between the number of floors in the house and the noise levels (Table 3), it was found that there were no significant differences because

Table 1. Variation of mean noise values between different distances from the emission source.

House	Distance (km)	N		Infrasound(a) [dB]	LFN(a) [dB]	ILFN(b) [dB]	EN (a) [dB(A)]
A	1.5	13	x̄	68.553	55.133	68.785	52.535
			σ	13.151	11.043	13.068	9.770
B	1.8	11	x̄	71.144	55.902	71.903	51.619
			σ	8.227	8.625	7.502	10.095
C	1.9	11	x̄	69.811	63.865	71.674	60.199
			σ	6.358	4.246	5.642	6.319
			p-value	0.170	0.005*	0.016*	0.028*

Tests: (a) Test *Kruskal-Wallis*; (b) Test for Median Independence | N – number of cases; x̄ – mean; σ – standard deviation | * *p-value* < 0.05

Table 2. Comparison of mean noise values regarding the natural vegetation.

Natural vegetation		Infrasound(a) [dB]	LFN (a) [dB]	ILFN (b) [dB]	EN (a) [dB(A)]
High > 1m	x̄	68.083	59.103	69.148	57.119
	σ	10.040	9.916	9.979	8.675
Low < 1m	x̄	69.840	55.186	70.673	51.725
	σ	7.378	8.739	6.636	10.635
	p-value	0.254	0.370	0.464	0.122

Test: Wilcoxon-Mann-Whitney | σx̄ – mean; σ – standard deviation

there was a p-value > 0.05 for all types of noise. However, the Infrasound and ILFN noises showed slightly higher values on the 1st Floor compared to the ground floor.

It was also intended to understand the variation of noise as a function of air velocity (Table 4). The results revealed, at lower air velocities, the infrasound, RBF, IRBF noise values were significantly higher, compared to higher air velocities. However, although there are no significant differences between the different air velocities with respect to EN type noise, it was found that the lower the air velocity, the greater the expression of this type of noise.

It was also tried to describe the behavior of the Infrasound variation at different noise frequencies as a function of the distance to the emission source (Fig. 3). As can be seen, at any of the frequencies, the sound pressure levels were always higher, mostly, the greater the distance from the source. In addition, at higher frequencies, the average noise values tend to be lower.

It was proposed to apply the same analysis strategy as before, but increasing the number of frequencies to low frequency noise, on the X-axis (Fig. 4). It was found that,

Table 3. Comparison of mean noise values according to the floor of the house.

House floor	N		Infrasound[(a)] [dB]	LFN[(a)] [dB]	ILFN[(b)] [dB]	EN[(a)] [dB(A)]
Ground floor	26	x̄	67.594	57.736	68.775	56.103
		σ	5.569	7.862	5.345	8.542
1st floor	6	x̄	73.128	58.499	73.305	52.534
		σ	18.458	16.183	18.382	13.505
		p-value	0.999	0.923	0.923	0.498

Test: Wilcoxon-Mann-Whitney | N – number of cases; x̄ – mean; σ – standard

Table 4. Comparison of mean noise values regarding air velocity classes.

Air velocity (m/s)	N		Infrasound[(a)] [dB]	LFN[(a)] [dB]	ILFN[(b)] [dB]	EN[(a)] [dB(A)]
[0 - 1.000]	20	x̄	68.962	59.542	70.403	56.640
		σ	5.534	8.101	4.892	9.454
[1.000 - 2.000]	12	x̄	68.081	55.108	68.328	53.423
		σ	13.620	11.533	13.540	9.641
		p-value	0.035*	0.043*	0.008*	0.212

Test: Wilcoxon-Mann-Whitney | N – number of cases; x̄ – mean; σ – standard deviation | * p-value < 0.05

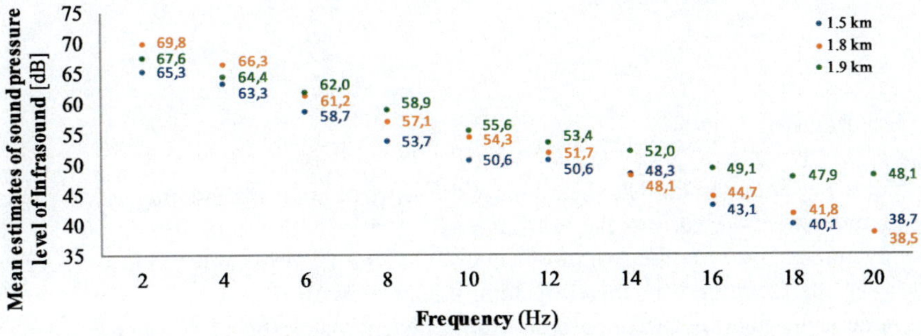

Fig. 3. Variation of sound pressure levels of Infrasound, by frequency.

for lower frequencies, the average values of sound pressure level are higher, ending up stabilizing at higher frequencies (around 120 Hz).

Finally, it was tried to assess if the noise levels of the type Infrasound and ILFN differed according to the measurement location, inside and outside the House (Table 5). It can be seen that the variation in noise levels Infrasound and ILFN is always higher

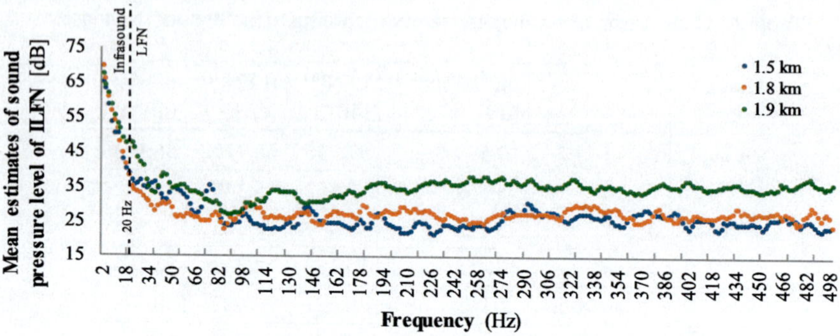

Fig. 4. Variation of sound pressure levels of ILFN, by frequency.

"Outside" the house. It can be also seen that the noise variation showed to have significant differences regarding this factor.

Table 5. Comparison of Infrasound and ILFN noises, attending the measurement location point.

Noise type	N	Measurement location	$\bar{x}$	σ	*p-value*
Infrasound [dB]	16	Inside	68.632	9.360	0.018
	3	Outside	81.822	6,746	
ILFN [dB]	16	Inside	69.625	9.135	0.018
	3	Outside	81.852	6.725	

Test: Wilcoxon-Mann-Whitney | N – number of cases; $\bar{x}$ – mean; σ – standard deviation

4 Discussion

In order to better understand the behavior of the types of noise under study, it was important to understand the various factors that may interfere with sound pressure levels. Air velocity and distance are the two most important factors. Infrasound and low-frequency noises are characterized by traveling long distances with little loss of energy due to atmospheric attenuation and the ground [13]. This characteristic was observed when the distance was related to Infrasound noise (2 Hz to 20 Hz). When crossing frequencies up to 500 Hz, it was found that even at a greater distance from the wind farm (1.9 km) the sound pressure level still high. The houses under study are between 1.5 km and 1.9 km, but even in houses located several kilometers away from wind turbine installations, high sound pressure level are registered [14].

For the types of noise Infrasound and ILFN, it appears that, in House C, which is the most distant from the wind farm, the sound pressure levels tend to decrease slightly, in comparison to Houses A and B. This little decrease, in particular of the low frequency noise, due to a very particular characteristic of this type of noise, its wavelength is extremely long [7]. The amount of ILFN depends simultaneously on several factors, such as the distance to the building, the size of the blade, the velocity and direction of the wind, the geology, the type of soil and the existence and type of vegetation [15]. For types of noise LFN and EN, it appears that, in House C, the sound pressure levels increase significantly, but the existence of high traffic nearby the housing may justify the values of the audible noise.

As for the surroundings, it was found that the presence of trees and plants may have a mitigating effect on the propagation of sound. Attenuation caused by tree trunks and branches depends on the wavelength of the sound emitted, that is, sounds with long wavelengths will not be affected and will be transmitted through tree trunks [15] which was verified for the noises Infrasound and ILFN type, that is, small obstacles and structures in the propagation path of these types of noise do not have a significant interference [7]. The propagation of noise does not depend on vegetation, but it does depend on the "soil" effect, mainly at the level of sound impedance [15].

The velocity of sound is affected by several external factors. Outside, the main influencing factors are wind velocity and direction and atmospheric temperature at the location [15]. Inside, the values can be explained by the type of construction of the house, the place inside the house where the measurements were made, but also the wind velocity and direction at the time of the measurement. When analyzing the "inside" and "outside" data of the houses, we observe that inside the houses, the noise levels are always lower, which can be explained by the direction of sound propagation, since it depends on the position of the source and the receiver of sound, as well as the wind direction [13, 15], as House A is to the east and Houses B and C to the southeast of the wind farm and the wind direction was from the north. The same can be seen in the results obtained for "air velocity classes" in which, for a higher air velocity, lower noise levels were obtained.

Environmental noise is mainly measured using the A-weighted sound pressure level (L_{Aeq}) as the only parameter to determine the perceived disturbance. However, if the noise is produced by sources characterized by producing low frequency noise, the L_{Aeq} measure underestimates the real disturbance [16, 17], reason why the measurements of this noise should be in dB.

Despite the lack of studies in humans, there have already been seen some examples in which a dose-response relationship is established in relation to exposure to infrasound and low frequency noise [18]. Noises that have a frequency less than or equal to 500 Hz are responsible for a set of disturbances in the human organism [6]. Studies present these consequences classified into two distinct categories: Vibroacoustic Disease [18, 19] and Wind Turbine Syndrome [20]. Exposure to this noise can cause a wide range of disorders, such as loss of sleep quality, stress, cardiovascular disorders, gastrointestinal complications, endocrine disorders, changes in respiratory rate and central nervous system, tiredness, harmful effects due to the frequency of resonance internal organs, among others [6, 14, 21].

Balancing the existence of ILFN producing facilities or locations and human populations through effective planning laws [6, 14] will create locations for wind turbines at adequate distances without having negative effects on human health [22]. However, in

Portugal, the legislation does not refer to distance or exposure limits, it only covers limit of exposure to audible noise, that is, environmental noise and occupational noise [14].

5 Conclusion

The conclusions reached with this study allow us to understand the behavior of the types of noise Infrasound, ILFN, LFN and EN in relation to the variables under study.

As for the distance, we can conclude that the values of noise type Infrasound and ILFN tend to be lower in House C, which is the most distant to WF (when compare to house A and B). The audible noises LFN and EN are significantly higher in house C. However, the existence of a national road with high traffic may have influenced these noises. When analyzing the noise frequencies, it was found that these are always higher in the house more distant to the WF (1.9 km). It can be concluded that, despite the distance increases, the propagation of infrasound not significantly decreased as is the case for the audible noise.

Regarding the air velocity, it was observed that, in this case, the wind direction (North) and the measurement point can be a factor of influence of noise propagation noise. Despise the literature, in this study the values of sound pressure level decreased when the air velocity increase.

It is concluded that there are factors that influence, the noise levels such as the air velocity, the distance to the wind farm and the measurement point. Also the surroundings, in this case the natural vegetation is a factor that can attenuate the propagation of the noise and in the present study it was verified that there is an attenuation of the noise levels, although of little significance.

In order to prevent public health consequences, it is important to define strict policies such as delimiting safety zones and distances for the installation of wind turbines in residential areas. It is also important the inclusion of the infrasound and low frequency noise in the environmental impact studies and not just environmental noise.

The present study had limitations in terms of data collection outside the houses under study, it was not possible to take into account the type of construction of each house and the fact that the car traffic was not controlled, this may have influenced the values for the types of audible noise (LFN and EN). If more measurements were made and in other house divisions, as well as outside, it would make the study more representative. It would also be relevant to take into account the air humidity, since the climatic variables air velocity, and temperature were considered.

Therefore, it would be interesting for a future study to focus only on housing, taking into account the type of construction and its materials, the age of the housing and making measurements in various divisions, and whether or not these variables influence noise levels within buildings.

References

1. Saramago, J., Ferreira, A., Almeida, J., Figueiredo, J.: Ruído infravermelho e baixa frequência de parques eólicos - Estudo de caso comparativo em relação à variação sazonal (inverno e verão). Revista Europeia de Saúde Pública (2019)

2. Maia, D.: Ruído de parques eólicos - Análise e caracterização. Dissertação de Mestrado em Engenharia Civil, Faculdade de Engenharia da Universidade do Porto (2010)
3. Costa, A.: O ruído ambiental de aerogeradores de pequena dimensão. Dissertação de Mestrado em Engenharia Civil, Faculdade de Engenharia da Universidade do Porto (2014)
4. Evans, A.: Environmental noise pollution: Has public health become to utilitarian? Open J. Soc. Sci. **05**(05), 80–107 (2017)
5. Drob, D., Picone, J., Garcés, M.: Global morphology of infrasound propagation. J. Geophys. Res. D: Atmos. **108**(21), 1–12 (2003)
6. Alves-Pereira, M., Branco, N.: Sobre o impacto de infrassons e ruído de baixa frequência na saúde pública - Dois casos de exposição residencial. Revista Lusófona de Ciências e Tecnologias da Saúde **2**(4), 186–200 (2007)
7. Berglund, B., Hassmén, P., Job, R.: Sources and effects of low-frequency noise. J. Acoust. Soc. Am. **99**(5), 2985–3002 (1996)
8. Filho, W., Azevedo, A.: Impactos ambientais em usinas eólicas. In: AGRENER GD 2013, Itajubá, Brasil, pp. 1–17 (2013)
9. Pedersen, E., Halmastad, H.: Noise annoyance from wind turbines - a review. Swedish Environmental Protection Agency, 1–26 August 2003 (2003)
10. Seltenrich, N.: Wind turbines: a different breed of noise. Environ. Health Perspect. **122**(1), 20–25 (2014)
11. Møller, H., Pedersen, C.: Low-frequency noise from large wind turbines. J. Acoust. Soc. Am. **129**(6), 3727–3744 (2011)
12. Decreto Lei n.° 9/2007. Diário Da República, n.o 12 - 1.a Série-A. Ministério do Ambiente, do Ordenamento do Território e do Desenvolvimento Regional. Lisboa, 17 de Janeiro 2017
13. Almeida, J.: Análise geográfica do risco de exposição a infrassons e ruído de baixa frequência, com origem em turbinas eólicas. Tese de Doutoramento em Geografia Humana, Faculdade de Letras da Universidade de Coimbra (2018)
14. Boretti, A., Ordys, A., Al Zubaidy, S.: Infrasound cancellation and mitigation in wind turbines. NASA/ADS **7**(1), 73–81 (2018)
15. Gronberg, A.: Sound propagation through Australian forest land with special regards to noise generated by wind turbines. Master thesis in Technology and Engineering. Faculty of Engineering LTH of Lund University (2015)
16. Simões, B., Almeida, J., Simões, H., Ferreira, A., Figueiredo, J.: Ruído Ambiental versus Infrassom e Ruído Baixa Frequência - Caso de estudo de população próxima de turbinas eólicas. Millenium - J. Educ. Technol. Health **2**(1), 397–405 (2016)
17. Caniato, M., Bettarello, F., Fausti, P., Marsich, L., Ferluga, A., Schmid, C.: Low frequency noise and disturbance assessment methods: a brief literature overview and a new proposal. In: 22nd International Congress on Acoustic, Buenos Aires, Argentina, vol. 28, no. 032001, pp. 1–9 (2016)
18. Ferreira, J., Sousa, J., Foreid, P., Antunes, M., Cardoso, S., Alves-Pereira, M., Branco, N.: Abnormal respiratory drive in vibroacoustic disease. Revista Portuguesa de Pneumologia **12**(4), 369–374 (2006)
19. Alves-Pereira, M., Branco, N.: Vibroacoustic disease: biological effects of infrasound and low-frequency noise explained by mechanotransduction cellular signalling. Progr. Biophys. Mol. Biol. **93**(1–3), 256–279 (2007)
20. Pardal, T.: Ruído Ocupacional: Baixa Frequência: Doença vibroacústica vs. Síndrome da turbina eólica. Tese de Mestrado em Segurança e Higiene no Trabalho, Escola Superior de Ciências Empresariais e Escola Superior de Tecnologia do Instituto Politécnico de Setúbal (2013)

21. Bento, E.: A incomodidade do ruído: Relações com o nível de exposição sonoro e identidade de lugar. Mestrado Integrado em Psicologia. Faculdade de Psicologia da Universidade de Lisboa (2011)
22. Mroczek, B., Banaś, J., Machowska-Szewczyk, M., Kurpas, D.: Evaluation of quality of life of those living near a wind farm. Int. J. Environ. Res. Public Health **12**(6), 6066–6083 (2015)

Variation of Infrasound and Low Frequency Noise Produced by Wind Turbines, Regarding Distance and Vegetation - Case Study

João Almeida[1(✉)], Márcia Santos[1], Roksolana Stefuryn[1], João Paulo Figueiredo[2], and Ana Ferreira[1]

[1] Coimbra Health School (ESTeSC), Department of Environmental Health, Polytechnic of Coimbra, Coimbra, Portugal
joaoalmeida@estescoimbra.pt

[2] Coimbra Health School (ESTeSC), Department of Complementary Sciences, Polytechnic of Coimbra, Coimbra, Portugal

Abstract. With the gradual depletion of traditional fossil fuels, it is essential to find sustainable alternatives, such as wind energy. Wind turbines, despite the many advantages, generate mechanical and aerodynamic noise, causing an impact on the health of the nearby population. This study aimed to verify whether the vegetation could be an acoustic barrier. Also aimed to verify the influence of the distance between wind turbines and housings. The measurements were made in the surroundings of 'Lousã I' wind farm (Lousã, Portugal), in areas with and without vegetation and close to housings. Measurement points were defined at 100 m, 500 m, 1000 m, 1300 m and 2000 m, for a total of 80 measurements in order to calculate the levels of infrasound and low frequency noise. For measurements, it was used the sound level meter CESVA SC420, the Q-TRAK Plus IAQ Monitor with Air Velocity Probe Model 8592 for measure air velocity and the software IBM SPSS Statistics version 26, for statistical analysis. It was possible to verify that, in areas with low vegetation, the noise level was higher, *i.e.*, in areas where there are trees, the noise tends to be lower than in areas with low vegetation. It was also found that the highest levels of infrasound and low frequency noise were located closer to the wind turbines. Nevertheless, the levels still high at 2000 m. It is concluded that the characteristics of the vegetation may influence attenuation. However, the attenuation is not enough to prevent it from reaching residential areas. Thus, it is importance to consider this source and this noise as a potential public health issue.

Keywords: Wind turbines · Infrasound · Low frequency noise · Vegetation

1 Introduction

The gradual depletion of fossil fuels such as coal, oil, and natural gas [1], the progressive increase in energy demand by emerging countries, and the environmental concerns,

J. R. da Costa Sanches Galvão et al. (Eds.): *Proceedings of the 1st International Conference on Water Energy Food and Sustainability (ICoWEFS 2021)*, pp. 411–420, 2021.
https://doi.org/10.1007/978-3-030-75315-3_46

contributes to indispensable search for sustainable alternatives [2]. These alternatives should be able to oppose the foreseeable shortage of energy resources in the medium term [2], as is the case with wind energy. Its relevance as a clean source has shown continuous growth over time in several countries. This development entails certain implications for the existing energy system, as well as for human well-being and the environment [3].

The sound, in general, is defined as a variation of atmospheric pressure that the human ear can detect, in any means of propagation [4]. Noise can be defined as any acoustic phenomenon that produces an unpleasant or disturbing auditory sensation, capable of altering the welfare of people [5, 6].

The sound consists of three distinct frequency sets: infrasound (below 20 Hz), sound composed by audible frequencies, and ultrasound (above 20 kHz) [7, 8]. Within the audible sound there is a range considered the low frequencies (below 500 Hz) [9].

Noise has always been an important environmental issue for humans [10]. However, there has always been a greater emphasis on audible noise and its consequences for health, and therefore less emphasis was given to infrasound and low frequency noise, because it was considered that as it was not audible, it would have no consequences. Low-frequency noise is felt by individuals exposed as a pressure instead of sound or through the experience of a sensation or vibration in the chest or throat area [11].

The noise has several harmful effects on human health, depending on the time of exposure, the sound pressure level, the fact that the noise is temporary or permanent or the sensitivity of the individual [7]. The implications, for the Human being, of the noise produced by wind turbines in particular, can be grouped into two large groups of symptoms and pathologies: Wind Turbine Syndrome (WTS) caused by the noise emitted by wind turbines in the low frequency range (between 20 Hz and 500 Hz) and is characterized by the various symptoms related to the organs of the vestibular system (sleep disturbance, headaches, tinnitus, feeling of tremor or vibration, nervousness, cardiac arrhythmia, nausea, difficulty of concentrating and memory and irritability) [12]; and Vibroacoustic Disease (VAD), a pathology developed as a consequence of excessive and prolonged exposure to Infrasound and Low Frequency Noise (ILFN) [9] and involving the whole organism, characterized by the abnormal proliferation of the extracellular matrix in the absence of the inflammatory process [13].

The respective effects, observed in different receptors, namely humans, have determined the increasing and systematic demand for measurement and analysis methodologies, since the presence of low-frequency noise has become a problem for the Society from different countries [14].

Unlike high frequency noise, ILFN is much less disturbed by obstacles and acoustic absorbent materials, which allows it to propagate through the atmosphere with less attenuation [11]. The attenuation caused by vegetation is low for low-frequency sound, while higher frequencies are more affected by interaction with forest components, such as foliage and tree trunks [15].

In order to consider the ILFN as public health issue, it is necessary to recognize that the acoustic phenomena do not affect only the hearing aid. A segmentation of the acoustic spectrum should be made and measuring ranges adapted to each environment to be assessed [16].

It is necessary to understand how the forest and its characteristics may influence, or not, the attenuation of Low Frequency Noise (LFN) and Infrasound. Therefore, the aim of this study was to ascertain whether the forest can serve as an acoustic barrier in the protection of populations, or whether its effect is reduced or null.

2 Material and Methods

This study was observational, level II, descriptive-correlational and cross-sectional, and the measurements were carried out at the Wind Farm of 'Lousã I', in municipality of Lousã, District of Coimbra (Portugal). The study took place in winter of 2019.

The sample design was defined as a non-probabilistic type and the sampling technique used was accidental or convenience. Four wind turbines (of the 14 wind turbines of this wind farm) were selected, around which, measurements points were selected. These points were select at 100 m, 500 m, 1000 m, 1300 m and 2000 m at north and south of the wind farm, considering areas with and without vegetation (or with low vegetation) between the wind farm and the measuring point, near a river, near a forest access and near houses, performing a total of 80 measurements. The area considered with trees, was constituted by young pines, although some were dry, the tree trunks of some of the pines were thin and others of greater size. The observed forest stain was not constant, and in some places a higher density was verified. To mark the measuring points, it was used the Google Earth Pro software (7.1.8.3036 version). Some measurement points were adjusted on site, due safety and access issues.

The study area was the wind farm of 'Serra da Lousã I'. Near this location there are other wind farms, namely the wind farm of 'Coentral-Safra' (at East) and 'Lousã II' (at Southwest). There was no industry and no traffic nearby the measuring points.

After the selection of measurement points, analytical data were collected. The CESVA SC420 sound level meter was used to collect noise levels. The sound level meter was calibrated before and after measurements. To carry out the measurements, the sound level meter was placed on a tripod 1.5 m from the ground and 2 m away from walls, windows and infrastructures. It was selected the Fast Fourier Transform (FFT) module for the analysis of 10000 lines from 2 to 20000 Hz (2 Hz/line) during one minute each measurement. Three measurements were made at each point. Based on the frequencies collected, the following noise types were calculated: Infrasound (2 to 20 Hz, in dB), LFN (20 to 500 Hz, in dB), ILFN (2 to 500 Hz, in dB) and Environmental Noise (EN) (in dB(A), according to the General Noise Regulation [17]).

Wind direction and air velocity were taken into account during the data collection, so that the measurements were carried out under similar conditions. For the purpose of monitoring, the measurement of temperature and relative humidity was also performed, using the Q-TRAK Plus IAQ Monitor with Air Velocity Probe Model 8592.

The study was developed in the winter and 80 measurements were carried out with the wind turbines in operation. The measurements were subjected to air temperature conditions between 4.3 °C and 21 °C) and air velocity ranged from 0.5 m/s to 12.3 m/s.

Statistical analysis of the data was performed in the software IBM Statistical Package for the Social Science, version 26. Data were subjected to a descriptive analysis of the general sample. After that, were subjected to an inferential analysis in which the following statistical tests were used: parametric test T-Student, non-parametric test of Friedman for independent samples and Kruskal-Wallis. About terms of correlation was applied the efficient Coordinal correlation of Spearman.

For the estimation of statistical inference, a 95% confidence level and a random error of less than or equal to 5% were considered.

3 Results

In order to understand if the air velocity influences the noise propagation, after the Spearman's linear correlation coefficient applied, it was observed a correlation pattern with air velocity and Infrasound (rho = 0.007; p = 0.965), LFN (rho = 0.021, p = 0.96), ILFN (rho = 0.012; p = 0.942) EN (rho = 0.278, p = 0.075). In this way, it is possible to verify that, with the increase of the air velocity, the sound pressure levels also increase.

Regarding a correlation between air velocity and distance to the source, after applied the Spearman's linear correlation coefficient, it was found significant correlation for the noise values of the IRBF that were higher in the presence of lower air velocity at a distance of 500 m from the noise source (rho = 0.036, p = 0.036). However, at a distance of 1000 m from the wind farm it was found that at the highest wind speed the noise levels of both Infrasound (rho = 0.421, p = 0.040) and the IRBF (rho = 0.406, p = 0.049) also presented higher values and significantly. Significant and positive correlations were recorded at a distance of 1300m between air velocity and ILFN (rho = 0.572, p = 0.05) and EN (rho = 0.647, p = 0.023). A similar profile occurred between air velocity and EN at a distance of 2000 m (rho = 0.591; p = 0.043). There was no pattern of variation of the noise type with the wind velocity at a distance from the source to 100 m.

The several types of noise were related to the surrounding characteristics (Table 1). There were statistically correlation differences in noise regarding the characterization and Infrasound and ILFN parameters (p-value < 0.05). Regarding Infrasound and ILFN it was found that the measure points with low vegetation had higher sound pressure levels compared to the places with the presence of trees, near the road, rivers with trees near the site and near houses. However, there were no significant variations of LFN and EN regarding measurement point characterization.

It was intended to understand the noise variation according to distance to wind farm (Table 2). According to the results, it was verified that there are statistically significant differences in mean variation of the type of noise according to the different distances in which they were evaluated ($p < 0.05$). It was observed that all types of noise had higher sound pressure levels at a distance of 100 m (except EN). At this distance, Infrasound and ILFN presented higher sound pressure level compared to the others types of noise. At distances of 500 m and 1000 m, all noises recorded lower sound pressure levels. EN registered the higher value at 2000 m.

Table 1. Relation between type of noise ant the characterization of the measuring point.

Type of noise	Characterization	N	x̄	σ	*p-value*
Infrasound [dB]	Forest	27	64.46	14.45	0.031*
	Small vegetation	23	68.11	18.01	
	Near to forest access	12	57.60	10.34	
	River with trees near	9	55.27	4.28	
	Near houses	9	64.88	5.74	
LFN [dB]	Forest	27	53.19	10.26	0.219
	Small vegetation	23	54.68	12.60	
	Near to forest access	12	53.09	9.70	
	River with trees near	9	50.22	8.64	
	Near houses	9	63.19	3.39	
ILFN [dB]	Forest	27	65.20	13.88	0.033*
	Small vegetation	23	68.82	17.31	
	Near to forest access	12	59.67	9.33	
	River with trees near	9	56.88	4.99	
	Near houses	9	67.28	4.64	
EN [dB(A)]	Forest	27	43.44	10.67	0.191
	Small vegetation	23	39.88	11.93	
	Near to forest access	12	41.59	9.47	
	River with trees near	9	44.71	12.79	
	Near houses	9	56.64	5.14	

Test: *Kruskal-Wallis* I N – number of cases; x̄ – mean; σ – Standard deviation I * *p-value* < 0.05

It was also considered pertinent, the study of the frequency behavior in function of characterization of the measurement points (Fig. 1). Therefore, the mean frequency values from 2 Hz to 500 Hz, *i.e.*, from Infrasound (<20 Hz) and ILFN (<500 Hz), were obtained considering the vegetation (forest and small vegetation), the proximity to river and houses.

As observed, at a 100 m from wind turbine, there was a higher sound pressure level in the points without forest or small vegetation. However, the behavior along the frequencies in places with and without trees was similar up to 258 Hz, with a change in behavior thereafter. It was found that the highest values were found in the infrasound range, with a decrease in sound pressure levels from 42 Hz.

Regarding the distance of 500 m from the source, there was a similar behavior in all variables, showing that the vegetation is not a barrier to these types of noise. However, it still can be seen some peak values with bigger expression on the point 'River with trees near'. It was also possible to observe that at this distance, the sound pressure levels are lower compared to the distance of 100 m.

When related to the type of vegetation in the distance of 1000 m, it was verified a smooth variation across frequencies, *i.e.* there were no significant variations in sound level pressure, especially above 82 Hz.

Table 2. Relation between type of noise ant the distance to source.

Type of noise	Distance (m)	N	x̄	σ	*p-value*
Infrasound [dB]	100	23	81,71	8,42	0,039*
	500	15	53,50	6,63	
	1000	30	54,64	7,59	
	1300	6	63,04	6,39	
	2000	6	63,36	5,85	
LFN [dB]	100	23	64,72	5,84	0,011*
	500	15	42,83	3,16	
	1000	30	49,60	8,46	
	1300	6	61,83	4,53	
	2000	6	60,24	4,90	
ILFN [dB]	100	23	81,85	8,30	0,017*
	500	15	54,35	5,95	
	1000	30	56,50	7,27	
	1300	6	65,58	5,48	
	2000	6	65,08	5,52	
EN [dB(A)]	100	23	52,16	5,49	0,046*
	500	15	31,16	3,74	
	1000	30	40,05	11,53	
	1300	6	54,36	6,36	
	2000	6	51,12	5,75	

Test: *Kruskal-Wallis* | N – number of cases; x̄ – mean; σ – Standard deviation | * *p-value* < 0.05

4 Discussion

Despite the fact of impossibility to correlate noise propagation and air velocity (due to different air velocity), studies indicate that the influence of air velocity on sound propagation is representative [18] showing that, the higher the air velocity, the more tendency to higher sound pressure levels [19]. Noise propagation can also be affected by other factors as relative humidity [20].

The existence of a heterogeneity vegetation (tree trunk thin/big, dry trees, forest stain not constant) does not guarantee the same conditions at different distances, as also, between the measurement points and the noise source.

With regard to the behavior of noise in relation to the characteristics of the measurement point, it was observed a lower noise values in forest area when compared to 'small vegetation'. Fact that can be explained by attenuation power of the vegetation. This attenuation is partially caused by the reflection or dispersion of the sound by the forest components, such as foliage and tree trunks, and partly by absorbing the sound energy of these elements [15].

Measurements made 'near houses' (933 m is the distance between the wind turbine and the closest house), revealed high values of infrasound and low frequency noise. this fact may be due to the wavelengths associated with infrasound and low frequency noise,

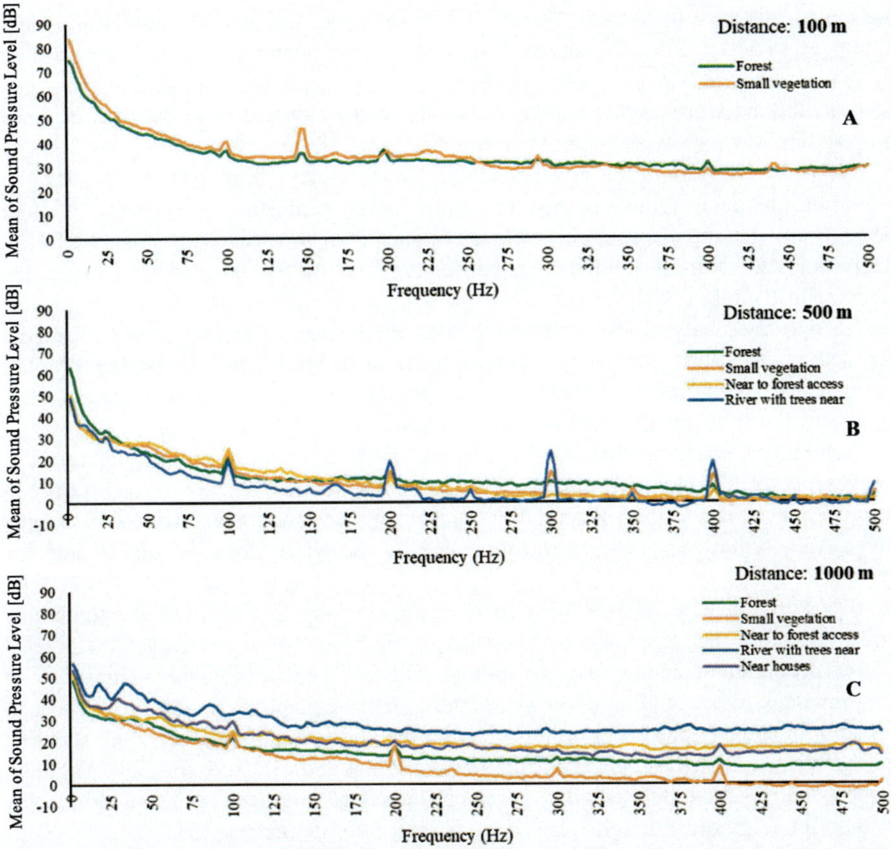

Fig. 1. Variation of the mean of sound pressure level, by frequency, regarding the characteristics of the measuring point. ***Friedman Test*:** **A** - Forest: $X^2 = 914.317$; df = 249; < 0.001; Small vegetation: $X^2 = 2257.094$; df = 249; $p < 0.001$; **B** - Forest: $X^2 = 2383.648$; df = 249; $p < 0.001$; Small vegetation: $X^2 = 1290.007$ df = 249; $p < 0.001$; **C** - Forest: $X^2 = 1863.988$; df = 249; $p < 0.001$; Small vegetation: $X^2 = 703.688$; df = 249; $p < 0.001$; Near to forest access: $X^2 = 621.458$; df = 249; $p < 0.001$; River with trees near: $X^2 = 1499.663$; df = 249; $p < 0.001$; Near houses: $X^2 = 645.032$; df = 249; *p-value* < 0.001.

producing a resonance effect in the most common interior spaces (houses, buildings, etc.) increasing noise levels [21].

Among all the noise types, it was observed that the highest sound pressure levels were recorded in Infrasound and ILFN, since that similar factors influence the propagation of low frequency noise [22]. The significant increase in noise levels was also evident in the area with ‘small vegetation’. It was also observed that in the area ‘near to forest access’ there is an increase in environmental noise, which can be explained by a road nearby, where sporadically vehicles circulate.

With regard to the relation between the distance and the types of noise, all types of noise recorded the highest sound pressure level at 100 m, since the measurement was the closest distance to source. Above 1000 m there was an increase of EN and LFN compared to 500 m from the source. The value of environmental noise can be justified by the presence of a road nearby. In the case of low frequency noise, this result can be due the long wavelengths making the obstacles and structures in the path of noise propagation having low significant interference [23, 24].

From the analysis to the noise behavior, by frequency (from 2 Hz to 500 Hz), in relation to the measurement points, it was possible to verify that, in the areas 'without vegetation', the sound pressure levels are higher than in areas with vegetation. The degree of attenuation provided by a barrier depends on the frequency of noise, the increase in distance and the effect on the line of sight (from the source from the receiver) [25]. It was also analyzed the relationship between the behavior of sound waves and the characterization, and it can be observed that the frequency tends to decrease in the higher frequencies. The literature confirms that, the wind turbine noise emission spectrum has an almost linear shape, decreasing along the frequency [26].

It was also observed that the 1300 m and 2000 m environmental noise tends to increase, since it is closer to the dwellings and consequently, subject to road traffic. In these distances it was also observed an increase of the infrasound and low frequency noise, values that may have been influenced by the wind farms 'Lousã II' and the 'Coentral-Safra'.

The Portuguese legislation, only focuses and characterizes the exposure to audible noise, therefore the infrasound and low frequency noise are neglected [27]. The legal framework should include setback for the installation of wind turbines taking into account the growing number of wind farms across the country. This problem has been a focus of research in some countries in order to protect the public health. However, no setbacks have yet been established between residential buildings and wind turbines [28]. Although some countries have defined some rules on the safety distances for the wind turbine installation (e.g.: tower height; size of the blades; rotor diameter) [19].

5 Conclusion

This study was based on a concern with the lack of Portuguese legislation against infrasound and low frequency noise produced be wind turbines. The existence of few studies related to the attenuation of noise by the forest and the expansion or construction of new wind farms were also a concern on this study.

The vegetation in the vicinity of the wind farm has a noise attenuator effect compared to areas without any vegetation, as is the case of near to forest access, which leads to the conclusion that no matter how low the vegetation will always have some mitigating effect. It can also be concluded that the density of the forest or vegetation may be determinant to absorb more noise, *i.e.*, it will be expected that in dense areas (as a forest) the values of sound level pressure be lower.

It can also be concluded that, despite the distance to the wind turbine increase, sound pressure levels, do not tend to decrease significantly. Exception to environmental noise that have a different behave due a short wavelength. On the contrary, infrasound and low frequency noise, due to the long wavelength, do not find barriers on the path, allowing to reach distances further from the source.

Considering the results found and their analysis, it was considered that high sound pressure levels of infrasound and low frequency noise are a public health issue. Thus, it is extremely important that some measures be taken to minimize the impact of these disease agents. In this regard, it was necessary to devote more attention to the responsible entities, in particular in terms of spatial planning, engineering, environment and health, to reduce the consequences of these impacts. Some measures could include defining standards to delimit areas of the territory best suited to the installation of wind power equipment, establishing a minimum safety distance from wind turbines to dwellings, but also by including these types of noise in environmental impact studies.

The present study presented limitations regarding air velocity, since if there were more measurements the noise values would be more representative and due to the existence of other wind farms nearby, increasing the noise levels.

References

1. Leite, G., Araújo, A., Rosas, P.: Prognostic techniques applied to maintenance of wind turbines: a concise and specific review. Renew. Sustain. Energy Rev. **81**(2), 1917–1925 (2018)
2. Correia, F.: Caraterização do potencial de energia eólica na beira interior. Dissertação de Mestrado em Engenharia Electromecânica, Universidade da Beira Interior (2010)
3. Beckera, R., Thräna, D.: Completion of wind turbine data sets for wind integration studies applying random forests and k-nearest neighbors. Appl. Energy **208**, 252–262 (2017)
4. Goelzer, B., Hansen, C., Sehrndt, G.: Occupational exposure to noise: evaluation, prevention and control. Federal Institute for Occupational Safety and Health. World Health Organization, Dortmund (2001)
5. Coelho, J., Ferreira, A.: Critérios para a análise de relações exposição-impacte do ruído de infra-estruturas de transporte. Instituto Superior Técnico, Lisboa (2009)
6. Freitas, L., Cordeiro, T.: Segurança e saúde do trabalho - Guia para micro, pequenas e médias empresas. Autoridade para as Condições do Trabalho, Lisboa (2013)
7. IA: Ruído e a cidade. Instituto do Ambiente, Lisboa (2004)
8. Silva, M.: Medição e avaliação do ruído ambiente - Estudo dos fatores que condicionam a determinação dos níveis de pressão sonora. Mestrado em Tecnologias Ambientais, Escola Superior de Tecnologia e Gestão do Instituto Politécnico de Viseu (2012)
9. Alves-Pereira, M., Branco, N.: Contribuição para o conceito dose-resposta em exposições a infrassons e ruído de baixa frequência. Revista Lusófona de Ciências e Tecnologias da Saúde **6**(1), 31–44 (2009)
10. Berglund, B., Lindvall, T., Schwella, D.: Guidelines for Community Noise. World Health Organization, Genebra (1999)
11. Pardal, T.: Ruído Ocupacional: Baixa Frequência: Doença vibroacústica vs. Síndrome da turbina eólica. Dissertação de Mestrado em Segurança e Higiene no Trabalho, Escola Superior de Ciências Empresariais e Escola Superior de Tecnologia do Instituto Politécnico de Setúbal (2013)
12. Pierpont, N.: Wind turbine syndrome: A report on a natural experiment. K-Selected Books. Santa Fé (2009)
13. Branco, N., Ferreira, J., Alves-Pereira, M.: Respiratory pathology in vibroacoustic disease: 25 years of research. Revista Portuguesa de Pneumologia XII **I**(1), 129–135 (2007)
14. Aroeira, C., Bicker, A., Leão, C., Conde, L., Rodrigues, C.: Avaliação de ruído de baixa frequência - Situação em Portugal. 48° Congreso Español de Acústica - TECNIACÚSTICA® / Encuentro Ibérico de Acústica. Corunha, Espanha (2017)

15. Gronberg, A.: Sound propagation through Australian forest land with special regards to noise generated by wind turbines. Master thesis in Technology and Engineering. Faculty of Engineering LTH of Lund University (2015)
16. Magalhães, A.: Doença vibroacústica: Haverá ‘ruído’ além do ‘ouvido’?. Dissertação de Mestrado Integrado em Medicina. Faculdade de Medicina da Universidade de Coimbra (2010)
17. Decreto-Lei n.º 9/2007. Diário Da República, n.o 12 - 1.a Série-A. Ministério do Ambiente, do Ordenamento do Território e do Desenvolvimento Regional. Lisboa, 17 de Janeiro 2017
18. Yamamoto, M.-Y., Ishihara, Y., Kanao, M.: Infrasonic waves in Antarctica: a new proxy for monitoring polar environment. Int. J. Geosci. **4**(4), 797–802 (2013)
19. Almeida, J.: Análise geográfica do risco de exposição a infrassons e ruído de baixa frequência, com origem em turbinas eólicas. Tese de Doutoramento em Geografia Humana, Faculdade de Letras da Universidade de Coimbra (2018)
20. Lazzarini, V.: Elementos de Acústica (1998). https://www.fisica.net/ondulatoria/elementos_de_acustica.pdf. Acedido a 15 June 2020
21. Seltenrich, N.: Wind turbines: a different breed of noise. Environ. Health Perspect. **122**(1), 20–25 (2014)
22. Leventhall, G., Pelmear, P., Benton, S.: A review of published research on low frequency noise and its effects. Defra, London, UK (2003)
23. Berglund, B., Hassmén, P., Job, R.: Sources and effects of low-frequency noise. J. Acoust. Soc. Am. **99**, 2985 (1996)
24. Rogers, A., Manwell, J., Wright, S.: Wind Turbine Acoustic Noise University of Massachusetts at Amherst, Massachusetts, USA (2006)
25. Belo, H.: O ruído e o ordenamento do território – caso de estudo: o concelho de Cascais (2013)
26. Lepoutrel, P., Avan, P., Cadene, A., Ecotière, D., Evrard, A., Moati, F., Topilla, E.: Health effects of low frequency noise and infrasound from wind farms: results from an independent collective expertise in France (2017)
27. Pardal, T.: SafeMed - Consequências da exposição ao ruído de baixa frequência (RBF) (2015)
28. Alves-Pereira, M., Branco, N.: In-Home Wind turbine noise is conducive to vibroacoustic disease. In: 2nd International Meet Wind Turbine Noise, pp. 32–42 (2007)

Smart Energy in Sustainable Services Building

João Galvão[1,2](✉), Rita Ascenso[1,3], Sofia Pereira[1], Licínio Moreira[1], and Filipe Neves[1]

[1] Polytechnic of Leiria, ESTG, Leiria, Portugal
jrgalvao@ipleiria.pt
[2] R&D Unit, Institute for Systems Engineering and Computers - INESC Coimbra, Coimbra, Portugal
[3] CIIC - Computer Science and Communication Research Centre, Polytechnic of Leiria, Leiria, Portugal

Abstract. One of the goals of future cities in many regions around the world is that they become smart cities. One of the major contributions that implement the smart cities concept is the use of renewable energy sources, which, in its turn, leads to the desired decarbonized society. In order to make it possible, it is necessary to ensure sustainability policies to take it into account either by using natural resources (*e.g.*, replacing the so far used energy sources by renewable ones) or by saving resources (*e.g.*, energy, water…), clearly resulting in financial and ecological savings.

One of the main objectives of these cities is they can be energetically autonomous, sustainable, and digitally more and more efficient. But for this to happen, it is necessary to depart from the existing services and respective buildings and implement therein efficient energy policies.

Based on these assumptions, this paper presents the research work results derived from a case study focused on a kindergarten infrastructure. In order to implement the smart energy and smart city concepts, endogenous resources were used, such as the Sun, garden facades, a heat pump and remote digital management mode.

Concurrent with the migration to renewable energy sources, other efficiency actions were taken as part of the research. They encompassed the development of consumers' digital capabilities, the control and remote management of energy systems and the decrease of consumption and associated costs. Finally, increase the mobility of employees with the installation of an electric charging station, whose energy source will be generated from the installed system.

Keywords: Consumption · Efficiency · Renewable energy sources · Digitalization

1 Introduction

Governments and organizations are now responsible for implementing international treaties such as the United Nations Agenda 2030 for building a future society based on a model of sustainable development [1].

J. R. da Costa Sanches Galvão et al. (Eds.): *Proceedings of the 1st International Conference on Water Energy Food and Sustainability (ICoWEFS 2021)*, pp. 421–430, 2021.
https://doi.org/10.1007/978-3-030-75315-3_47

The current global trend is that more and more people are living in cities, causing, on the one hand, an increasing need for access to energy and environmental resources, including food, but on the other hand, this growth must also trend to be associated with the reduction of Greenhouse Gas Emissions (GHG), in these cities and in the built area.

In this context, where the planet sustainability is the guiding principle, the intention is to optimize the energy consumption resulting from the human activity, implement and use local energy production systems, while mitigating the leakage and other causes of waste of energy. This guiding principle must, also, include, whenever possible, the production of food, as far as possible, from consumer centres, in a double perspective of reducing the transportation energy of these goods whilst improving the air quality.

One way to cities become more efficient and sustainable is to monitor energy and water consumptions in real time, for quick decision making and jointly proceed to the widespread use of renewable energy sources than current infrastructure. This model and others similar must have the support of integrated energy management systems, which by automation and use of sensors network, for collecting energy variables should be remotely managed. This will increase the digitization levels of whole process and will be a contribution to the implementation of smart city and smart energy concepts [2, 3].

2 Motivation, Objectives and Methodology

The current need to carry out the energy transition of societies to energy sources increasingly free of greenhouse gas emissions, leads to the use of endogenous energies accompanied by a renovation of cities and their buildings to become more energy efficient and sustainable. Based on this assumption, in this work we intend to contribute to this transition, based on a case study building, occupied by more than 100 users daily. This will result in an improvement of the energy performance of the building under study, also in a reduction in the annual bill for energy costs and emissions, using renewable energies and a contribution to the digital society by promoting the use of digitalization of the management energy system. These all results contribute for the increase of electric mobility of the company's employees and better sustainability, in the use of the building [4–6].

The methodology initially followed in our study was based on an energy audit in accordance with National Law [7], which led to a proposal presentation with multiple actions to be implemented. These included the design of new energy systems for an optimization of energy consumption and consequent energy rationalization of the case study building. It is intended that the workers of this service building have a contribution to electric mobility and start to make their transport in electric vehicles, because they will have charging stations, after the installation of the local electricity production system and according to the World Economic Forum, road transport (vehicles, freight and passenger) is responsible for 74.5% emissions [8].

As a starting point, several field works were carried out, including visits to the building facilities to get knowledge about the different spaces' utilization and geographic location. This permitted to collect the needed information within the scope of an energy audit process, which regulates energy efficiency matters and results from the transposition of European Directive No. 2012/2 /EU [9]. In this energy audit, consumption data (gas, electricity, water) were gathered. The drawings of the building's architecture were, also, collected, as well as respective descriptive document relative to the employed materials, energy billing and total working hours of equipment.

After analysing the energy consumption data (electricity, gas, water) of the last three years, it was possible to know the type of consumption and define the load diagrams. A survey of all existing electric equipment was also carried out. Since some of them no longer have the desired information, an approximate calculation of the nominal power was made. All this information provided the possibility to set the energy variables and, so, understand the type of pattern of consumption in the building. For more details on the decomposition of electric energy variables analysers were placed on the main electrical board of services of the building, which collect all types of electrical consumption existing in previously defined time intervals, to make the daily and hourly consumptions load diagrams.

2.1 Analysis of Energy Consumption

In a wide context of analysis to the energy efficiency, four types of consummation were considered: electricity, gas, water and building energy leaks. Supported by applicative software the load diagrams were calculated.

Electricity Energy Particularities

From the collected data, it was possible to characterize a typical consumption day, whose load diagram is presented in Fig. 1. As it can be seen, during the night there is a residual power consumption and from 7 AM onwards, opening time, a rise begins to be noticed that goes on until the end of the day.

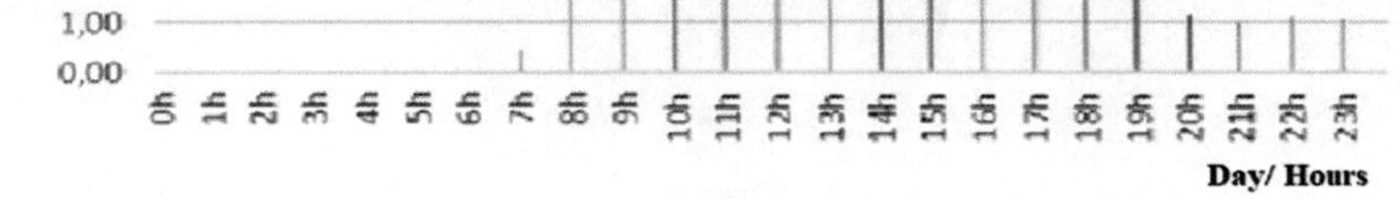

Fig. 1. Daily electricity load diagram of the case study.

There are electric consumption peaks along the day: before noon and after lunch, whereas during the night there is no relevant consumption.

In this audit, special attention was paid to the consumption of the laundry equipment, since both their age and intensive use caused to suspect there were an opportunity cost that was possible to reduce.

As a result, the diagram of Fig. 2 shows the daily laundry load for a period of two weeks in December. The days not representing consummation refer to the weekend days, in which the building and its activity are closed.

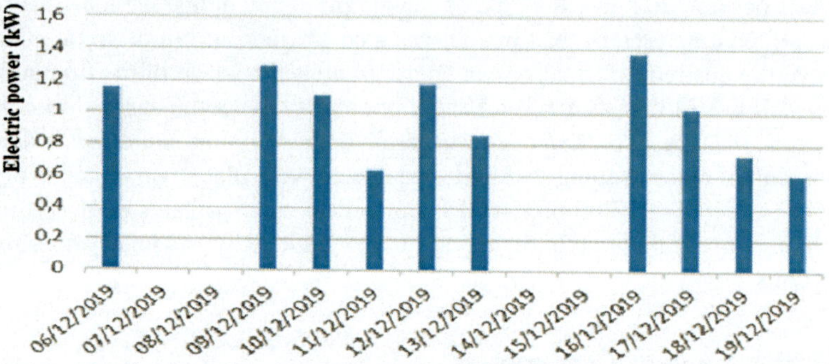

Fig. 2. Daily laundry load diagram in the sampling period.

As it can be seen, on some days on the load diagram present a nil (or almost nil) consumption.

Gas Consumption

The Fig. 3 represents the total natural gas consumption for a year of 11 359 m^3 breakdown corresponding to domestic hot water, meals and for heating at a total annually cost of 8 600€, which was considered relatively significant portion of the annual kindergarten budget, besides the sustainability it compromises due to gas emissions. As it can be seen, the most part of the consummation is due to the heating process (77.86%) and the remaining (22.14%) is due to domestic hot water.

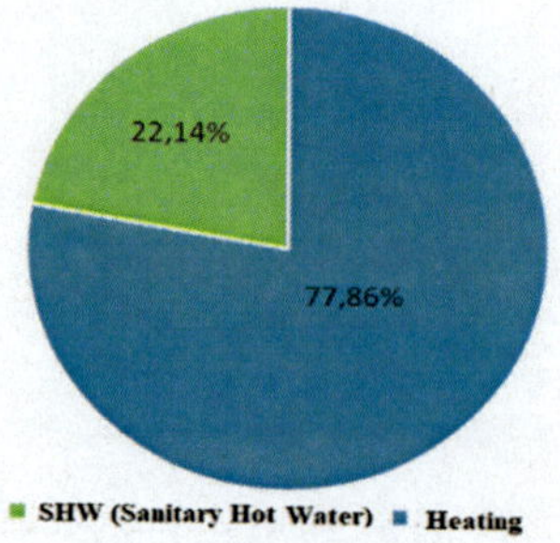

Fig. 3. Distribution of natural gas consumption

Water Consumption

Efficiency in water management is a relevant factor and, therefore, an analysis of water consumption and cost over the period were carried out from 2017 to 2019 (3 years), which results are shown in Fig. 4.

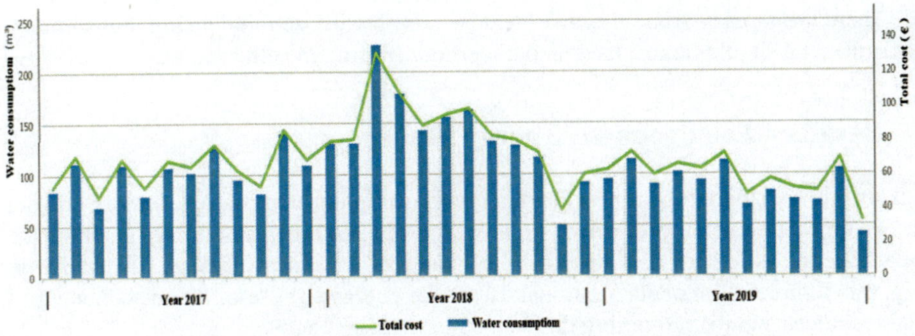

Fig. 4. Water consumption for 3 years from 2017 to 2019

As it can be observed, the values of water consumption do not denote significant discrepancy from an average value of about 100 cubic meters per month, except in March and November of 2018. Furthermore, these values may derive from statistical estimated values made by the water supplier company. As it can be derived, the amount of water that was consumed without reutilization is ~100 m^3/month (~350 L/day), which is considered significant to not be reutilized. The existence of only two peaks in the consumption that occurred in 36 occurrences are not considered as a factor of disturb, for the solution to come from our study, but dual flush toilets and water flow controllers are recommended on all taps.

Building Energy Leakages

To characterise the building energy leaks, a thermographic analysis was done to identify the places with the highest thermal energy leaks are located, and so, to overcome this problem. Figure 5 shows the results. In this figure, cold colours, as blue, represent the leakages. As it can be seen, outside pillars and beams, that are built by concrete, are the worst materials contributing to such leakages [10].

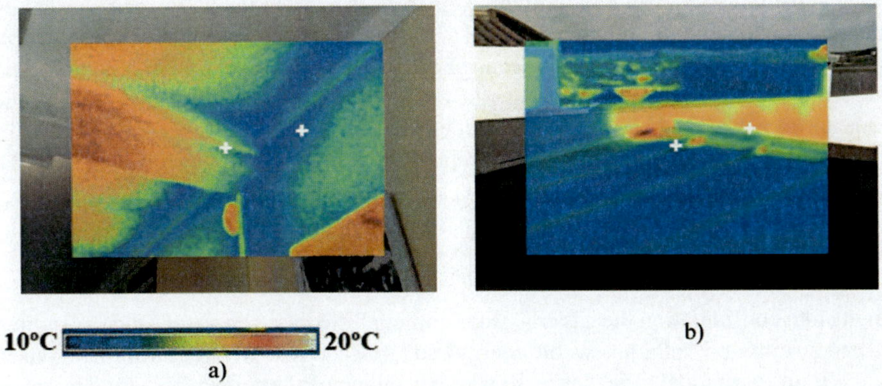

Fig. 5. Thermal leaks and temperature scale: a) The pillars and beams to the outside of the building are areas of thermal leaks; b) Roof with insulation needs.

There are already some thermal break windows with double glazing, but in many areas the roof should be insulated and underfloor heating in some rooms.

3 Results - The Proposed Energy Model

Taking into consideration the previous energy consumption analysis, a new energy model integrating different areas of intervention was developed for this building, in which the smart energy, efficiency and sustainability concepts were incorporated, as detailed in Fig. 6 in a threefold model: i) rational use of the energy; ii) use of renewable energies and resources and iii) promoting digital energy management.

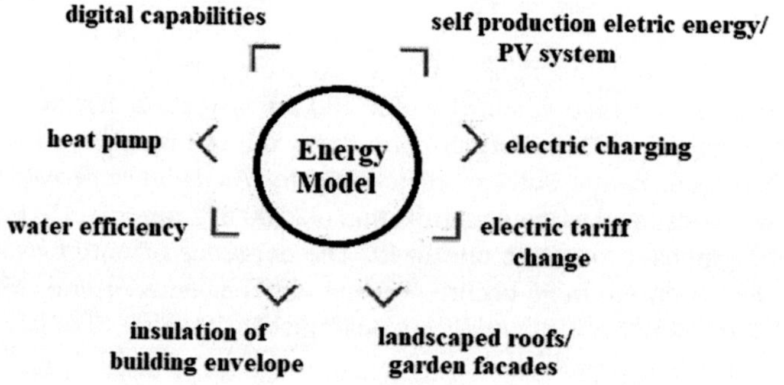

Fig. 6. The more efficiency energy model for the case study building.

3.1 Rational Energy Use

The first aspect that stands out from the electricity consumption analysis is the need to flat the electric consumption curve of a typical day, by open the possibility of deviating power consumptions from day to night. This implies to install a network of sensors by the main consumer equipment, as well as an information and communication platform that permits the energy manager to control each equipment, to increase efficiency and balance their use. In addition, the use of a cell phone or tablet app that permits the manager to consult and take appropriate actions according to the time of the day is part of the model. This permits the manager reprogram equipment to take advantage of the period in which the electric energy tariff is less expensive [11].

3.2 Renewable Energies and Resources

Taking into consideration the gas emissions coming from heat generation and the audited water consumption, solar renewable energy and rainwater resource capturing are advised by the proposed model. Two forms to take advantage of these resources may be taken into consideration: active and passive. Observing the former one, the model proposes electric local production by installing Photovoltaic Panels (PV) aiming to supply heat pomp. Since electric vehicles are expected to need charge, PV will, in a short period, be useful on promoting electrical mobility to the employees of the institution [12].

Still with the thermal consumption mitigation in view, the model proposes the use of garden facades, since the architectural permits.

Another factor that contributes to the decrease in thermal consumption is the external insulation of the walls surrounding the building associated with the placement of vegetation suitable for this climatic zone, which will cause one indoor thermal gradient more constant along the time, in certain parts of the construction and thus, avoiding high heating and cooling needs [13–15].

Next PVSyst program was used to design the PV - photovoltaic system, having self-consumption in mind [16]. Then the 3D model of the installation of solutions for landscaped roof and facade coverings were projected using the Sketchup 2020 [17] and the Fig. 7 illustrates these results.

a) b)

Fig. 7. The local PV production: a) In building roof and the green color are the green/landscaped rooftop; b) Three electric charge stations (bottom side of the image), electrically powered by PV.

3.3 Thermal Energy - Heat Pump

An alternative for production of thermal energy will be the exchange of existing boiler, which consumes natural gas, for a heat pump. The consumption of this energy is very high in the winter months and the existing boiler needs to be replaced in the short term. There is some consumption of electrical energy by the heat pump, but there is local production of the same. Although the kitchen still needs to continue to consume gas in the meals production, that energy is will be greatly reduced.

New consumption was calculated with the replacement of the current equipment for the proposed alternative. Also, the monthly useful daily energy consumption was calculated. The choice of the pump to be used was based on the average hourly power value of daily consumption. The heat pump is for production of heating and domestic hot water (DHW), see Fig. 8. Only the value of the heating consumption was considered, as it would be enough to meet the needs, since the domestic hot water will be stored and produced in hourly periods where the space heating power is lower than the nominal value of the heat pump and during the night.

Fig. 8. Heat pump installation model with design support on Sketchup 2020.

3.4 Water Efficiency

Regarding the efficient use of water, the model proposed here suggests the capture and storage of rainwater collected from the roof of the building. This water can have industrial uses, as well as in cooling towers, fire networks, HVAC systems and for irrigation.

In addition to these general uses, this process can have domestic proposes, such as for bathrooms and car washes. Rainwater carries dirt such as leaves, soot, among others, so first the water must pass through a thick filter that separates the larger dirt, then the water with impurities is discarded. Thereafter, the water that has not been discarded goes through a finest filter again, which retains the minor impurities. Finally, it can be disinfected with chlorine, ozone or ultraviolet and kept in a differentiated reservoir with different links according to the various uses [18, 19].

3.5 Digital Capabilities Energy Management

The implementation of the proposed model lay on the utilization of digital technologies. Here, the sensor networks installed together with the consumer equipment will send, through information and communication platform, the collected data to tablets or cell phones, so that it can be managed in a precise way, avoiding the waste of resources.

The model proposes and suppose the use of digital capabilities from the energy manager of the organization so that, such skills permit to supervise equipment consumptions and the electric charging station in the car park. These skills are useful for controlling the equipment remotely and the dedicated program allows to better rationalize energy consumption.

4 Conclusions

The objectives of this research this research were to identify a maximum number of solution actions applicable in the present case study building that can be replicated, in the future in similar constructions having in mind to promote self-sufficiency in the energy resources, to satisfy its energy needs. The digital technologies have been integrated and support the management and supervision of the energy model actions, while applying the principles of the smart city concept, where energy efficiency enhancement and digitalization are key factors for greater sustainability and low-cost solution.

In this context, despite the model being intended for the just aforementioned building, it is, without doubt, a part of a whole - the smart cities concept implementation solution. A possible contribution of this work to society is that the achieved results should be applied and disseminated in buildings such as centres for the elderly, SPA centers and gyms promoting the energy efficiency, digitization and sustainability.

A possible contribution of this work to society is that the achieved results should be applied and disseminated in buildings such as centres for the elderly, SPA centers and gyms promoting the energy efficiency, digitization and sustainability.

Future work includes the renovation of sanitary facilities and install dry toilets, which generate organic compost and do not consume water, thus increasing water use efficiency. As well as the painting of interior spaces with mineral paints, which improve air quality. Since this type of paint does not release harmful particles to the environment, this case study will be a real benefit for the health of users, in a context of ecological environment protection.

Acknowledgment. This research was supported by the UIDB/00308/2020 project of the Science and Technology Foundation (FCT) of Portugal.

References

1. ONU: The Sustainable Development Agenda. https://www.un.org/sustainabledevelopment/development-agenda/. Accessed 12 Aug 2020
2. Power Technology: Smart cities - smart energy projects revolutionize future energy use. https://www.power-technology.com/features/featuresmart-grid-cities-future-energy/. Accessed 30 Oct 2020
3. Yigitcanlar, T., et al.: Can cities become smart without being sustainable? A systematic review of the literature. Sustain.Cities Soc. **45** (2018). https://doi.org/10.1016/j.scs.2018.11.033
4. Ahvenniemi, H., Huovila, A.: How do cities promote urban sustainability and smartness? An evaluation of the city strategies of six largest Finnish cities. Environ. Dev. Sustain. (2020). https://doi.org/10.1007/s10668-020-00765-3
5. Dameri, R.: Smart city definition, goals and performance. In: Smart City Implementation, vol. 850, pp. 1–22. Springer (2017)
6. Bisello, A.: Multiple benefits of housing regeneration and smart city development: the European Project SINFONIA. Sustainability **12**(19), 8038 (2020). https://doi.org/10.3390/su12198038
7. National Law No. 68-A /2015. https://dre.pt/pesquisa/-/search/67123272/details/. Accessed 05 Feb 2020
8. World Economic Forum - Road transport emissions. https://www.weforum.org/agenda/2020/10/cars-planes-trains-aviation-co2-emissions-transport. Accessed 05 Oct 2020
9. Energy Efficiency Directive 2012/27/EU. https://ec.europa.eu/energy/topics/energy-efficiency/targets-directive-and-rules/energy-efficiency-directive_en. Accessed 02 Aug 2020
10. ICS - Infrared Consulting Services Inc.; Infrared Roof Moisture Survey & Scanning. https://irtest.com. Accessed 03 Jan 2020
11. Ampere Energy. https://ampere-energy.com/smartengineering. Accessed 02 Jan 2020

12. Nuttall, W., Gibson, D., Trzmielak, D., Ibarra-Yunez, A.: Energy and Mobility in Smart Cities. ICE Publisher (2019). ISBN 9780727764256
13. Yang, J., Kumar, D., Pyrgou, A., Chong, A., Santamouris, M., Kolokotsa, D., Lee, S.: Green and cool roofs' urban heat island mitigation potential in tropical climate. Solar Energy **173**, 597–609 (2018). https://doi.org/10.1016/j.solener.2018.08.006
14. Cook-Patton, S., et al.: Potential benefits of plant diversity on vegetated roofs: a literature review. https://doi.org/10.1016/j.jenvman.2012.04.003
15. Coma, J., et al.: Thermal assessment of extensive green roofs as passive tool for energy savings in buildings. Renew. Energy (2016). https://doi.org/10.1016/j.renene.2015.07.074
16. PVSyst - Photovoltaic-software.com photovoltaic software.com › pvsyst. Accessed 03 July 2020
17. SketchUp Pro 2020.1.1 for Windows; google-sketchup.br.uptodown.com › windows. Accessed 10 June 2020
18. Liu, J., et al.: Water scarcity assessments in the past, present, and future. Earths Future **5**, 545–559 (2017). https://doi.org/10.1002/2016EF000518
19. Liu, Y., Engel, B.A., Flanagan, D.C., Gitau, M.W., McMillan, S.K., Chaubey, I.: A review on effectiveness of best management practices in improving hydrology and water quality: needs and opportunities. Sci. Total Environ. **601–602**, 580–593 (2017). https://doi.org/10.1016/jsc iotenv.2017.05.22

Sustainable and Affordable Building Technologies in Northern Kenya

Zeltia Blanco[1(✉)], Eija Laitinen[2], and Vincent Kitio[1]

[1] United Nations Human Settlements Programme, P.O. Box 67553, Nairobi, Kenya
[2] HAMK University, Hämeenlinna, Finland

Abstract. Affordable and sustainable housing is among the critical challenges in most countries worldwide, and specially in developing countries where big proportion of the population live under the poverty line. In the arid territories of Kenya, more than half of the population in urban areas live in informal settlements (UN-Habitat 2017) without provision of basic services such as water, electricity, sanitation or waste management, and housing units made of iron sheets can hardly protect people from the extreme climatic conditions. Housing is hardly affordable for majority of the population, however, the wide availability of appropriate building materials such as soil and stone quarries can greatly reduce the cost of housing without compromising the quality of the construction while contributing reduce its carbon footprint. This article is based on the experience of the project *Climate Resilient Low Cost Buildings In Marsabit County*, Kenya, funded by the Nordic Development Fund, which highlights how the sustainable building design and the use of appropriate building technologies with local materials can facilitate access to housing while contributing to socio-economic development of women, youth and local economies.

Keywords: Energy efficiency in building · Affordable housing · Local building materials · Sustainability · Low-carbon · Hot-arid climate

1 Introduction: Need for Energy and Resource Efficient Affordable Housing

Energy consumption in the building sector constitutes nearly 40% the total direct and indirect CO_2 emissions worldwide (IEA 2020), and 56% in Africa (UN-Habitat 2015). Addressing energy and resource efficiency in the building chain has been raised as a critical aspect to reduce the global greenhouse gas (GHG) emissions.

On the other hand, need of affordable housing for low income people is a reality in most countries in the world. This is specially pressing in Sub-Saharan Africa (SSA), where 41% of the population live with less than 2 dollars per day (World Bank 2015) and where there is a housing shortage of around 56 million units and 90% in the affordable housing bracket (Shelter Afrique 2020).

In the framework of the project "Climate Resilient Low Cost Buildings In Marsabit County, Kenya", funded by the Nordic Development Facility, the United Nations Human

J. R. da Costa Sanches Galvão et al. (Eds.): *Proceedings of the 1st International Conference on Water Energy Food and Sustainability (ICoWEFS 2021)*, pp. 431–444, 2021.
https://doi.org/10.1007/978-3-030-75315-3_48

Settlements Programme (UN-Habitat), Strathmore Energy Research Center (SERC, Kenya) and Häme University of Applied Sciences (HAMK, Finland) worked together to "create models for sustainable low-cost buildings and energy efficient/renewable energy technologies in hot and arid climates in order to address poverty, lack of housing and climate change adaptation/mitigation whilst providing affordable and quality housing options" for low income people. A sustainable and affordable housing development pilot project of 100 units was planned and sustainable building design and renewable energy trainings developed.

The objective of the project was to reduce the carbon footprint of housing in Marsabit County through an overall approach to housing and access to basic services, including energy and resource efficient building design and sustainable neighborhood planning, use of low carbon local materials, renewable energies for energy supply, water efficient technologies and clean sanitation systems. This was initially supported by the development of a legal framework on energy and building that contribute to keep the trend towards sustainable construction and energy efficiency in the long term in the County.

In Kenya there is a total housing deficit of 2,000,000 units, which increases at 200,000 housing units per year (Africa Housing Finance Yearbook 2019). It is estimated that, 75% of the buildings that will be standing in Kenya by 2050 are yet to be built. Considering that most of the modern buildings are built using materials with carbon intensive products like steel, cement and aluminon roof, there is an urgent need to promote green building pilot projects to guide the construction boom forward and to ensure the transition towards environmentally friendly ways of construction. A lot of research has been developed on sustainable building in arid and semi-arid lands (ASAL) in SSA, however, few examples have been implemented. In Kenya, as in many other African countries at the moment, the National Government has given priority to the implementation of affordable housing through the Affordable Housing Programme (AHP) as one of the key pillars of their Big Four Agenda, which aims to construct 500,000 housing units across the country by 2022. The Government of Kenya (GoK) has allocated funds for low-cost housing in all 48 counties, and Marsabit County itself is to build 2000 housing units as part of AHP. The GoK has also committed to a minimum compliance for green building standard.

In this context the Marsabit Housing Project described in this paper was hoped to prevent the "business-as-usual" in the massive construction to happen in Kenya in the coming years. The pilot housing units were planned to serve as sustainable construction demos for the wider community. The project has been aligned with the AHP and the pilot housing units are expected to facilitate integration of green technology with carbon saving objectives to the AHP.

The need for pilot projects on energy efficient housing is obvious not only in Kenya but in the larger East Africa Community, where the low and growing middle-class is looking for decent housing options.

Currently, housing developments in Marsabit county, located in Northern Kenya, are highly dependent on materials provided from at least 400 km away, which greatly increases the cost of building and its carbon footprint. In addition, water supply is not enough to cover the needs of the population over the year, and energy supply usually depends on diesel generation plants, which keeps prices of basic services unaffordable for majority of the population. However, despite scarcity of water and energy provision, there is no use of energy and water efficient technologies.

This article discusses how green building design, the use of local materials such as interlocked soil blocks and thin tile vaulting, as well as the integration of efficient water, sanitation and hygiene (WASH) technologies and renewable energies in housing can contribute in addressing the urgent need of affordable housing development under sustainable building parameters in arid and semi-arid climates in Kenya and other East African countries with similar climatic conditions.

Fig. 1. Left to right: traditional building © HAMK; current ways of construction © UN-Habitat; thin tile vaulting with local materials © MCSPA Turkana.

2 Climate Resilient Architecture: Sustainable Low-Cost Housing Design

2.1 Sustainable Housing Design

An environmentally sound building could reduce the demand for energy and water by 70% and 60% respectively (UN-Habitat 2017), while still responding to the needs of the occupants. Therefore, strategies for green building design should address each of the following aspects according to the local climate: climatic data; building orientation and shape; appropriate allocation of in-door spaces; adequate proportion of openings; natural lighting; natural ventilation; natural cooling and heating strategies; solar protection; building envelope and appropriate materials; external finishes; integration of renewable energies and water efficiency technologies; appropriate drainage and sanitation systems; landscaping; energy efficient appliances.

In Marsabit sustainable building must be designed according to its hot and arid climate and the geographic location, thus taking into consideration temperature, relative humidity, solar radiation, and wind throughout the year as well as its latitude: 03° 27′ North; 38° 57′ East.

High daily temperature variation, high solar radiation, and low minimum relative humidity along most of the year as described in Figs. 1, 2, 3 and 4, in addition to a geographic location close to latitude 0, leads to the following building design strategies (Figs. 5 and 6):

- Orientation and main windows towards the North-South to protect from solar radiation and reduce heat gains;

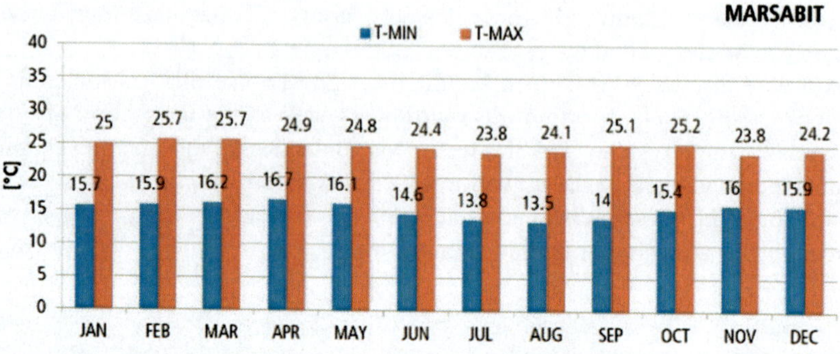

Fig. 2. Average daily temperature variation for Marsabit (30-year period) (WMO 2017)

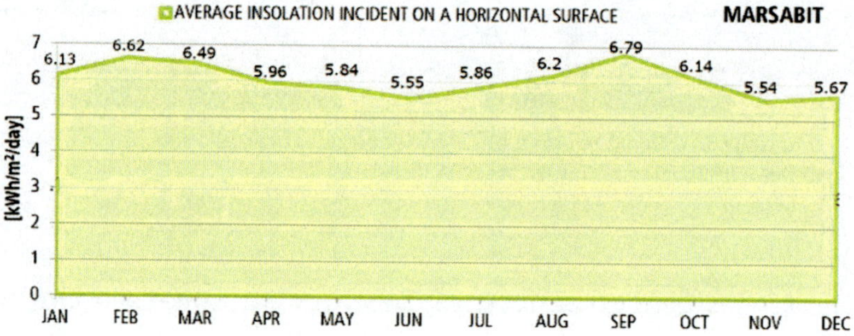

Fig. 3. Average insolation incidence on a horizontal surface in Marsabit (Weather Online 2017) solar radiation

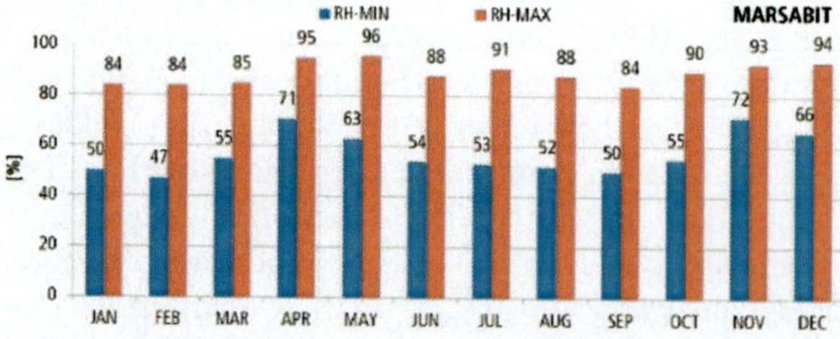

Fig. 4. Average daily relative humidity fluctuations in Marsabit

- Configuration around a courtyard with garden in order to protect from solar radiation and to provide natural evaporative cooling;
- Enough natural lighting as to minimize the use of energy for artificial lighting,
- Natural ventilation and ventilated roof to prevent overheating;
- Protection from solar radiation: solar protections in windows, verandas, garden with local trees that provide a fresher environment in the dry season;
- Natural cooling. Maximizing the vegetation cover to provide evaporative cooling;
- Maximizing the use of natural and local building materials with high thermal mass that helps to keep the housing cool and contribute to reduce the building carbon footprint:

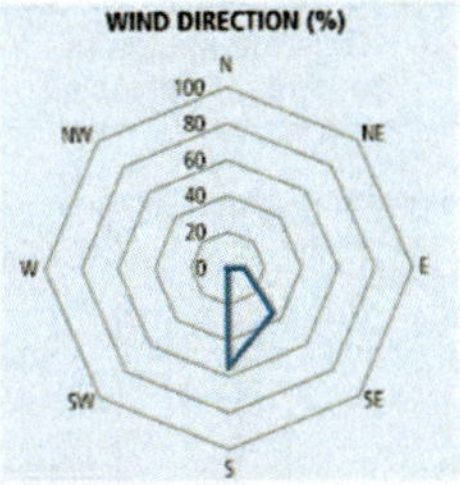

Fig. 5. Windrose diagram for Marsabit showing the prevailing wind direction between January 1996 and December 2016 (Weather Online 2017)

Interlocked Stabilized Soil Blocks (ISSB) for walls and lime/gypsum tiles for thin tile vaulting in slabs, roofs and stairs.

The technical drawings of the project *Climate Resilient Low-Cost Buildings in Marsabit County* for the sustainable low-cost housing units were developed according to the bioclimatic design principles described above in order to reduce energy consumption and improve indoors comfort in building. It included environmentally friendly sanitation models (composting toilet), main concepts of water efficient technologies (rainwater harvesting and greywater recycling systems), and PV solar systems for energy supply in the housing units.

The combination of the sustainable building design with the energy and resource efficient technologies planed can save at least 60% of the energy and resource consumption compared to the average permanent housing unit in urban Kenya, thus contributing to climate change adaptation and mitigation (Figs. 7 and 8).

2.2 Low Cost Building Technologies

The housing prototypes propose the use of local low cost materials such as Interlocked Stabilized Soil Blocks and clay and lime tiles, as well as building technologies such as the "Landie Slabs" or the thin tile vaulting for slabs, roofs and staircases that greatly reduce the cost of construction. These technologies and building materials, besides making use of local available materials and mostly natural materials, can greatly reduce the cost of the housing units without compromising the quality of the construction. Therefore, the proposed models can contribute to provide affordable housing options for the middle-low income people.

Interlocked Stabilized Soil Blocks (ISSB) were proposed for walls, since they make use of local materials (soil and sand) and just need a small proportion of cement, thus bringing down the cost of construction while having a minimal environmental impact. They are considered an appropriate material for construction in this hot and arid climate since the soil keeps the housing units fresh during the daytime. Despite being an unknown building material in Marsabit this is a well-known technology in Kenya.

Use of thin tile vaulting for slabs and roofs as structural and low-cost building technology. This technology makes use of low carbon building materials such as burnt

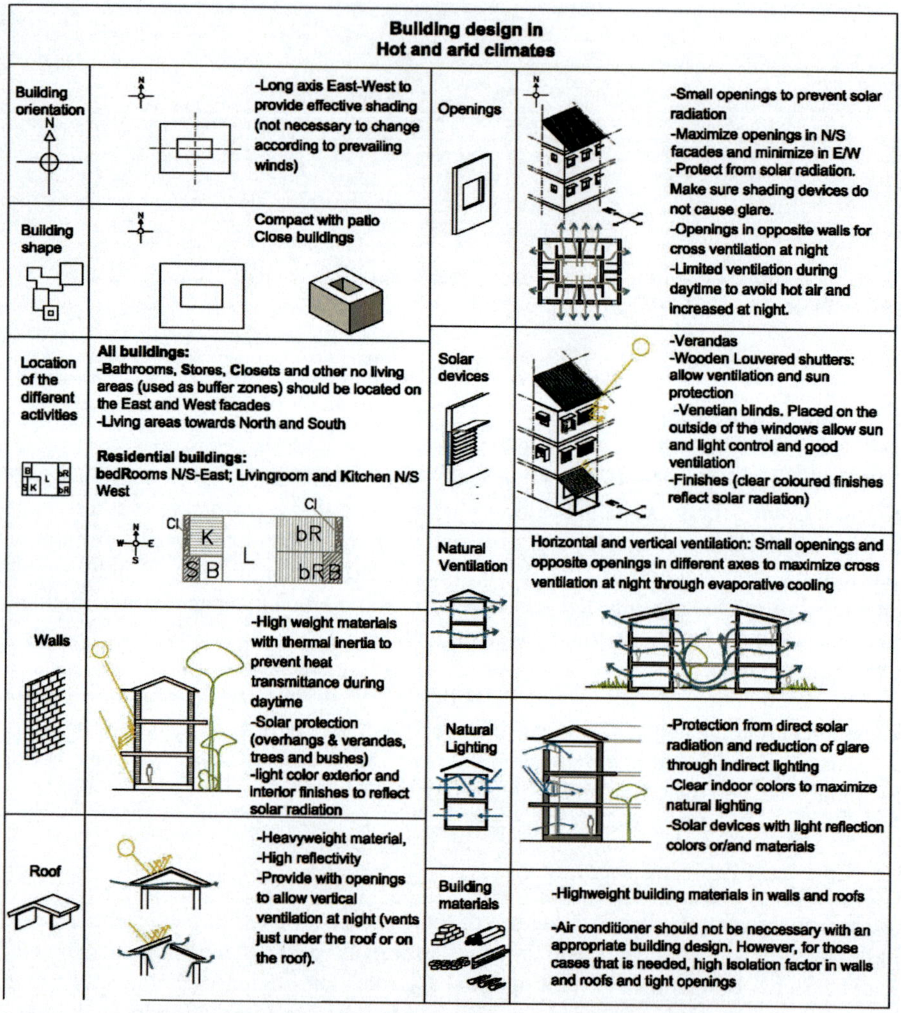

Fig. 6. Sustainable building strategies for hot and arid areas

soil tiles/clay and gypsum tiles, while greatly reducing the price of construction, since it does not need cement or steel (Fig. 9).

The project has taken into consideration the local capacities available to implement the trainings on ISSB and thin tile vaulting in the region. The thin tile vaulting – knowns also as Guastavino vaulting- is a traditional building technique developed in Spain at the beginning of the twentieth century to reduce the cost of the building material and to expand the span of the vaulting. This technique was widely used in the USA for civil work. The municipality owns two diesel powered machines for ISSB production and has two experts on the topic. In the Neighboring County of Turkana there is a project

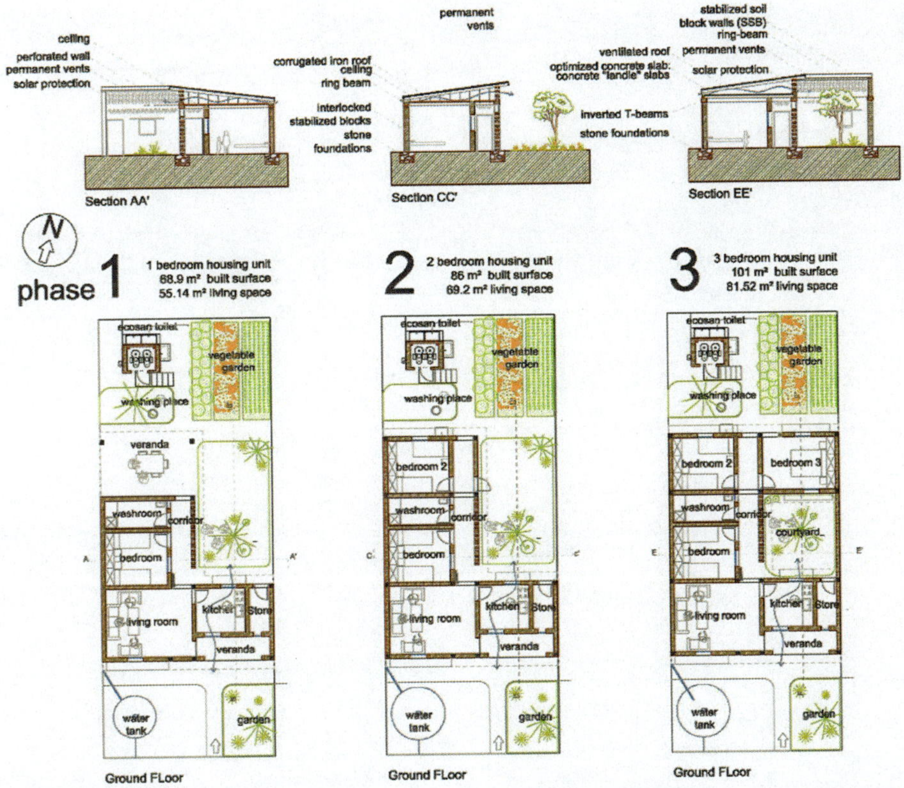

Fig. 7. Sustainable low-cost Housing units proposal. Incremental housing. 1- to 3-bedroom units

that produces clay and lime tiles in a sustainable way and provides thin tile vaulting workshops.

There have been found two examples of buildings done with soil blocks in the region: two buildings of the local secondary school and a private housing unit, both in Marsabit town. According to the local trainers on ISSB the soil is clay rich in most of the area, so a proportion of sand, available locally, should be added. Based on previous experiences it has been estimated that around 5% of cement or lime should be added to stabilize the soil, which keeps the cost of construction of this building technology low. Anyway, a soil test on the site should be done to confirm the quality of the soil and the proportions of stabilizer and sand required.

Despite being a labor intense building technology, with the two hydraulic machines owned by the County Government it is possible to produce the soil blocks for the 100 units in 2 months. This would be an income generation activity for the women groups trained.

The project was planned to be a training field for constructors, students from the Technical School and women and youth groups. A labor-based model implemented by the Government of Kenya in other projects was chosen for the construction process,

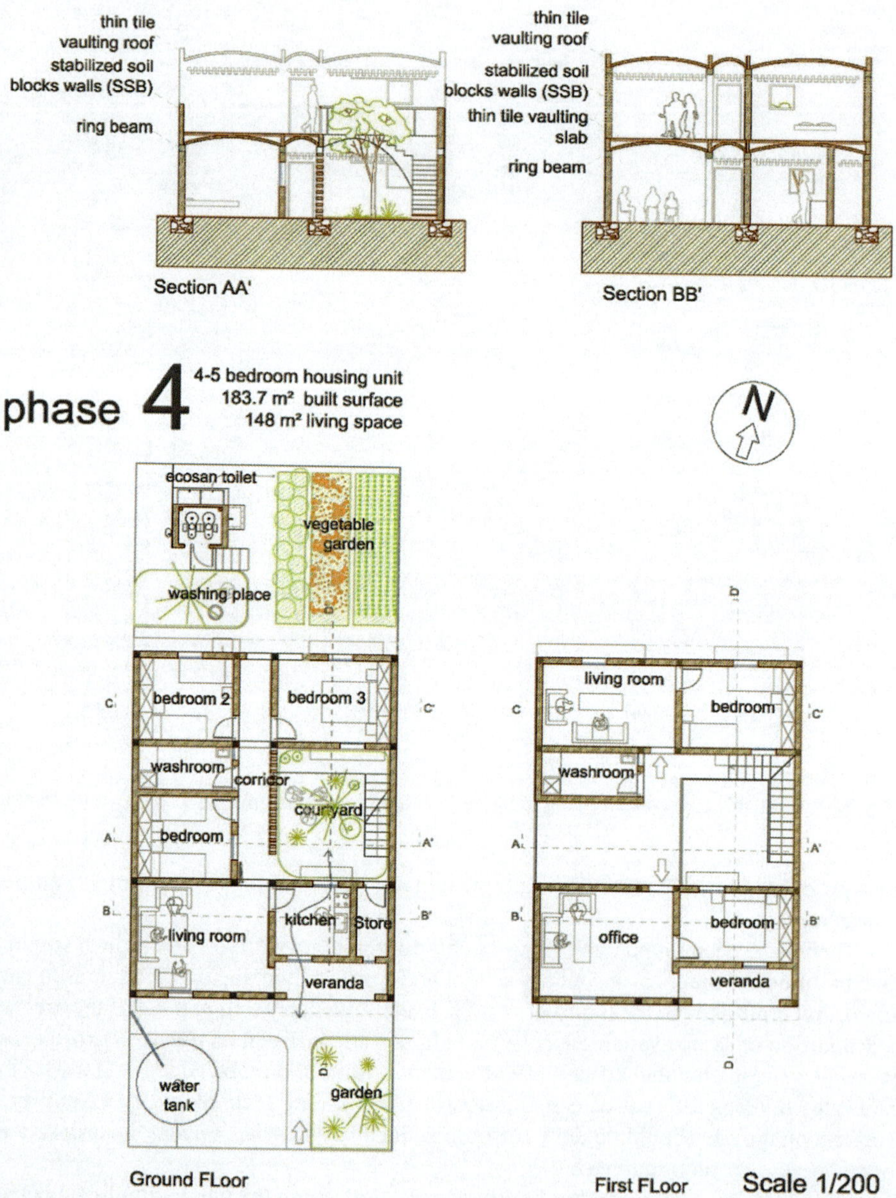

Fig. 8. Sustainable low-cost Housing units proposal. Incremental housing. 4- to 6-bedroom unit or 2 to 3 independent units

which reduces the cost of construction compared to the common tendering process launched for constructors.

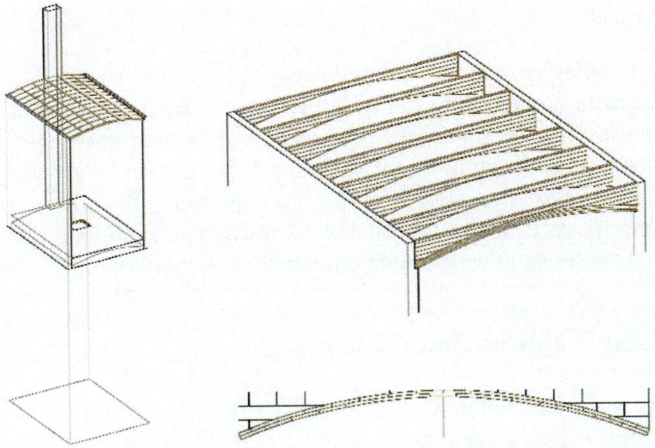

Fig. 9. Thin tile vaulting for slabs and roofs

2.3 Energy and Resource Efficient Technologies

Provision of self-sufficient basic services in Marsabit becomes an important sector to address. The following actions were planned to ensure water and electricity supply as well as clean sanitation systems (Fig. 10):

Maximizing the roof area for rainwater harvesting. Grey water recycling at home level and Fog Water harvesting at the Cluster level have been planned to minimize the dependence on external water sources;
Efficient cooking technologies to reduce the consumption of charcoal and wood thus the environmental impact;
Provision of clean energy from solar PV;
Sanitation solution that decomposes the human waste. The Ecosan latrine, known as well as composting toilet, was chosen in consideration with the local community as the sanitation option.

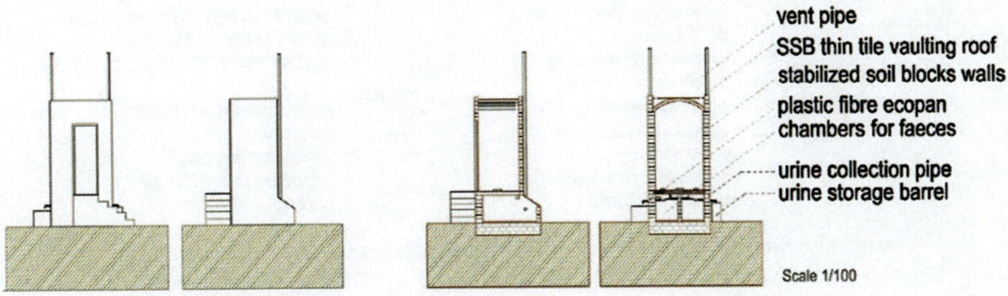

Fig. 10. Composting toilet with ISSB and thin tile vaulting

2.4 Social Aspects

In order to fit the different local and cultural needs (14 tribes in Marsabit County) raised in the meetings with community groups, the units have been designed to provide for: private and semiprivate spaces; separate washroom from latrine; latrine with water tap for cleaning; outdoors space for cooking and washing cloths; garden for growing vegetables; space for parking in the front plot; master room bigger than the other rooms; fencing with vegetation; courtyard with perforated block walls to keep privacy; individual water tanks to prevent conflicts on water; and communal public spaces.

3 Sustainable Neighborhood Planning

The settlement layout and external space are also to be adapted for hot and arid areas. This includes optimal protection against direct or indirect solar radiation by mutual shading resulting to compact settlements with narrow streets. It advocates for compact planning to minimize solar exposure of individual houses thus reduces solar heat gains (Fig. 11).

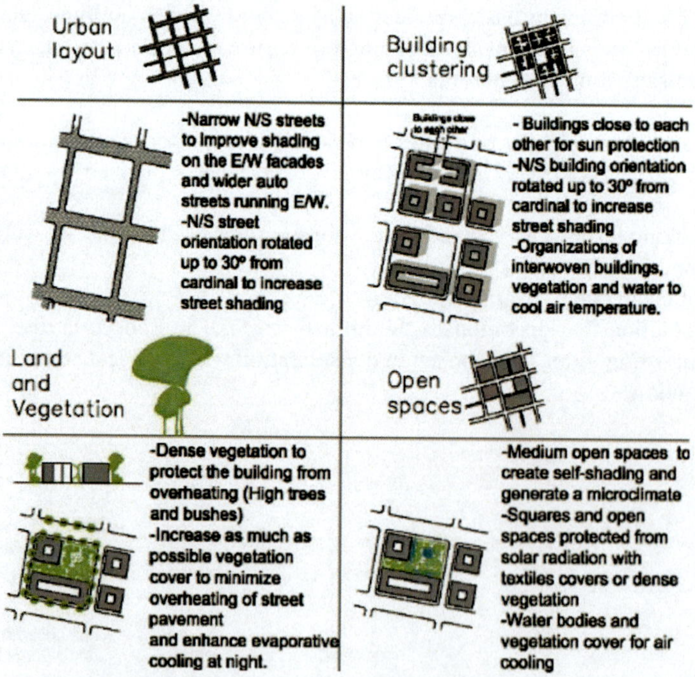

Fig. 11. Site Planning Climate resilient strategies for hot and arid areas

In addition, the neighborhood was planned following sustainability criteria, including economic, social and urban parameters such as: well-balanced public spaces: 50% of spaces allocated to streets, roads, public spaces, gardens and parks; Mixed land use: the development has been planned to allocate housing and commercial uses together, as well as administrative uses and offices in the near future planned by the County Government; adequate density and compact design; connectivity; walkability and active mobility: urban planning has given priority to pedestrian and cycling mobility, which already are the main ways of mobility for low income people in the region, and provided space for public transports and motorbikes as other important ways of transport.

3.1 Climate Resilient Site Planning

The site planning has been organized in clusters of 25 housing units and sub-clusters of 5–7 units. This will make it possible to (Figs. 12 and 13):

- Provide additional basic services per cluster;
- Organize the housing development in a flexible way;
- Organization around public spaces to fit the communal way of living;
- Each cluster is provided with an area located towards the southern border that will supply the cluster with the following services:
  - A barrier of dense vegetation to slow down the strong winds and dust
  - Fog catchers and rainwater harvesting to canalize additional water;
  - A section to locate the solar photovoltaic net. (This meaning: several PV panels set up together that provide electricity to a housing cluster. In order to provide electricity to a large housing area, this system is more efficient than the individual PV panels. This has been planned in case CGM wants to provide electricity through PV panels to the entire housing development.)

3.2 Social Aspects and Capacity Building

The neighborhood has been planned to follow the local live style as agreed with the different community groups consulted. Clusters and sub-clusters have been organized around a public space which serves as open spaces for social live and economic activities. Housing units have been designed as to be prepared to develop small business inside and combine residential and commercial uses.

At the moment the local construction sector is very dependent on non-local personnel. The different phases of the construction were planned to be developed through hands on training packages, where local masons and community groups linked to construction and contacted on previous missions of the project are trained on the different building technologies at the time that the construction is implemented.

The benefit of using local actors during the design and building process contributes to the empowerment of community groups. With this approach, the capacity building impacts and benefits would remain in Marsabit and the building process would contribute to o create business activities locally. For this purpose, Vocational Centres (technical schools) where different construction and energy jobs are taught, were invited to be involved in the trainings and construction works, so the project serves as a hands on

Fig. 12. Centrality of 25 incremental housing units and Cluster of 5 housing pilot units 1. Main roads with recommended directions for hot and arid climates; 2. Vegetal wind barriers with endemic vegetation; 3. Wind Catchers for water harvesting for each centrality; 4. Centrality of 25 units; 5. Public open spaces; 6. Sub-cluster 5–10 housing units; 7. Common space.

Fig. 13. General site planning for 100 housing units

training field and the knowledge remains in the Centres for future generations. Both theoretical and practical trainings are part of the capacity building plan, where local officers in charge of energy, urban planning and building works, masons, architects, engineers, construction companies, women groups and youth groups are involved, so the knowledge on sustainable building and energy/resource efficiency is sprawled at all levels.

3.3 Business Opportunities for Women in Construction

Gender balance was anticipated to be challenging and discussed with County Government of Marsabit (CGM) which committed to mobilize the community to obtain 50% of women participation on trainings. Women in Marsabit are in charge of building the family housing, in charge cooking including sourcing of cooking fuel (firewood and charcoal), water collection and saving measures.

Women groups in Marsabit are very active and a broad interest in participation to trainings on ISSB production, water saving technologies and efficient cooking stoves was confirmed among women groups, businesswomen and the women working at the CGM.

In order to encourage business development through the project, a green business module targeting mainly women and youth was included in the training planning. Women participation on the construction process was focused on the production and selling of ISSB, with which technology they were already acquainted.

4 Conclusions

The need for climate-smart housing options for the low-income sector and the growing middle class is huge. In Kenya only 200,000 homes are needed every year. Achieving this target in a sustainable way remains a challenge, since there is lack of understanding of environmentally friendly and effective construction. There are no examples of housing developments in the area with a holistic approach to sustainable site planning and building where neighborhood planning, architectural design, energy, and resource efficient technologies are integrated and addressed from the sustainability perspective.

During the design process social aspects were carefully considered. The community was consulted at all stages to make sure that their views and experience were taken into considerations so that a new building typology was imposed on them. Special attention was put on the issue of affordability to ensure that the housing will be accessible to the community. A well balanced private and public space were also considered. Local culture and customs were also taken into consideration, for example the community insisted that they should have a small space for keeping their animals as this was also a source of income. Other traditional technic and method of construction were also considered, making sure that they contribute to lowering the both the cost and the carbon emission.

In addition, gender and human rights-based approach became very important in this context. In Marsabit, the fact that women are responsible for construction facilitated the way to plan business trainings and income generation with the use of local materials.

Cooperation with local educational institutions was also important when enhancing local capacity building.

This project demonstrated that is possible to design social houses that are affordable, sustainable, and socially friendly and acceptable.

The experience gained and the information collected during the implementation of this project have contributed to the production of a guidebook on sustainable building design for hot and arid climate of Kenya. This document will be used to guide building practitioners in the social housing sector and decision-makers.

References

Africa Housing Finance Yearbook (2019). https://housingfinanceafrica.org/app/uploads/CAHF-Housing-Finance-Yearbook_East-African-Community-2019-Profile.pdf. Accessed 2 Feb 2021

Gut, P., Ackerknecht, D.: Climate Responsive Building: Appropriate Building Construction in Tropical and Subtropical Regions. SKAT, St. Gallen, Switzerland (1993)

IEA Buildings, International Energy Agency. https://www.iea.org/topics/buildings. Accessed 2 Feb 2021

Konya, A.: Design Primer for Hot Climates. Architectural Press, London (1980)

Shelter Afrique Newsroom. https://www.shelterafrique.org/en/newsroom/view/shelter-afrique-partners-with-world-bank-to-address-affordable-housing-crisis?title=0. Accessed 2 Feb 2021

Traditional Building. https://www.traditionalbuilding.com/features/book-review-guastavino. Accessed 2 Feb 2021

The World Bank Blog. https://blogs.worldbank.org/opendata/85-africans-live-less-550-day. Accessed 2 Feb 2021

UN-Habitat: Sustainable Building for Hot and Arid Areas. Marsabit County. UN-Habitat, Nairobi, Kenya (2017)

UN-Habitat Darfur Early Recovery: Stabilized Soil Blocks for Sustainable Urban Growth Nairobi. UN-Habitat, Kenya (2010)

UN-Habitat: Sustainable Building Design for Tropical Climates: Principles and Applications for Eastern Africa. UN-Habitat, Nairobi, Kenya (2015)

UN-Habitat: Hands-on Training on Energy Efficiency and Renewable Energy for Youth Empowerment. UN-Habitat, Nairobi, Kenya (2016)

UN-Habitat: Energy and Resource Efficiency Urban Neighborhood Design Principles for Tropical Countries. UN-Habitat, Nairobi, Kenya (2018)

Weather Online. https://goo.gl/CLWzLd. Accessed 20 Sept 2017. Wind Direction - Marsabit - Climate Robot Kenya

WMO, World Weather Information Service. https://worldweather.wmo.int/en/home.html. Accessed 20 Sept 2021

The Challenge of Turning Galicia into an Interconnected Distributed Electrical Grid Using CLEW Tools

Antía Míguez Fariña(✉) and Ángeles López Agüera

Department of Physics, Santiago de Compostela University, Rúa Xosé María Suárez Núñez, s/n, (Campus Vida), 15782 Santiago de Compostela, Spain
antia.miguez@rai.usc.es

Abstract. Nowadays, Galicia is one of the main producers of electricity in Spain, mainly from non-renewable sources (thermal plants). However, compliance with current EU environmental policies raises doubts about the continuity of one of the largest thermoelectric plants in the region (As Pontes, 1700 MW). The loss of energy production capacity associated with a possible closure of the plant even if will induce worrying consequences both at the energy and the economic levels in the region, can be seen as an opportunity to the economic recovery of the region.

This article explores the possibility of making this situation the challenge of designing a Distributed Galician electric network interconnected with the Spanish one. The Galician reference energy system (RES) has been defined and implemented in the Open Source Energy Modeling System (OSeMOSYS), studying different long-term scenarios (until 2030). The results show how local renewable resources will be sufficient to fulfil the electrical needs of the region, resulting in a huge reduction in CO_2 emissions.

Keywords: NEXUS · CLEWs · Sustainable energetic transition · Decarbonization

1 Introduction

The energy sector is one of the main topics of discussion at a political, scientific, environmental, and social level. The fossil fuel crisis, its consequences on the environment and the energy poverty in some regions of the world are of concern to the international community. Modelling tools can be useful approaches to collaborate on the design of solutions. These tools allow to think about future scenarios that meet our expectations and wishes in various dimensions -economic, energy, environmental, political…

Within this framework we find the CLEWs (Climate - Earth - Energy - Water) tools, which study the interconnections between the different resource systems, paying special attention to how changes in one of the systems can affect the others. CLEWs tools can help long-term policy planning to achieve the Sustainable Development Goals (SDGs) set by the 2030 Agenda [1]. The European Commission is promoting several research

J. R. da Costa Sanches Galvão et al. (Eds.): *Proceedings of the 1st International Conference on Water Energy Food and Sustainability (ICoWEFS 2021)*, pp. 445–455, 2021.
https://doi.org/10.1007/978-3-030-75315-3_49

projects to carry out scenarios related to the energy sector. The proper use of these models can help the development of sustainable growth in national and regional economies: an efficient energy mix not only helps to face current environmental challenges, but also reduces the cost of electricity [2]. Therefore, the use of these tools is considered an option with high potential when it comes to helping develop a sustainable energy transition for Galicia, one of the largest electricity producers in Spain.

Galician's main sources of production are thermoelectric plants and wind energy. But due to current environmental regulations, it will be necessary to abandon the use of coal as a primary resource in the next years, with its well-known energy and economic consequences. This raises the need to analyze the strengths and weaknesses of the sector to study the possible future scenarios that allow us to draw a road map. These evaluations are intended to help make decisions aimed at a sustainable and fair energy transition, which support progress in the fulfilment of the SDGs.

In this paper, a study of the Galician electricity sector has been carried out following the CLEWs philosophy. The article begins with a description of its main characteristics, then continues with the elaboration of five possible future scenarios (Business-as-usual -BAU-; Conservative with controlled emissions; Decentralized energetic network; Strategic promotion of biogas and biomass), considering the needs and opportunities of the region, using the energy modelling tool OSeMOSYS [3].

2 Galician Framework

Located northwest of the Iberian Peninsula, Galicia is a region with great energy production capacity, both from renewable and fossil sources. The Galician energy sector is strategic for the local economic vector generating 9% of the Spanish electricity needs, being the 34% of the total amount exported or sold abroad. Table 1 resume the details about both generated electricity by and installed power capacity in the actual Galician scenario. As remark, the biodegradable energy waste is included in the category *solar and others* and *cogeneration* covers cogeneration from oil producers, natural gas, and residual energies.

Resuming, the Galician generated energy comes from thermal power plants (both, coal thermal and the combined cycle), 26% from wind power, 11% from cogeneration and other 11% from large hydropower plants. Solar and biogas only provide less than 1% of the total production. Regarding the installed capacity, thermoelectric plants and cogeneration plants represent 34% while hydropower accounts the 31% of the total, followed closely by wind energy with 29% of the total installed capacity [4, 5].

Galicia's present energy situation has advantages and disadvantages. As we said, the region is a large energy producer, therefore this sector plays an important role in the local economy. Although it has a high potential, it must face great challenges in the near future.

As disadvantages we can see that 48% of the energy produced comes from fossil sources. Galicia does not have this resource and must import coal. In addition, thermal power plants must close before 2021 due to the tightening of European regulations regarding decarbonization paradigm.

On the other hand, there are several advantages. Galician installed capacity, especially hydroelectric and wind power, is much higher than that currently generated. Even the

Table 1. Installed power (GW) and generated energy (PJ) for each technology considered.

Type of plant	Installed power	Generated energy
	GW	PJ
Coal thermoelectric plants	1.96	39
Combined cycle thermoelectric plants	1.247	6.38
Cogeneration	0.563	10.36
Waste (non-biodegradable)	0.067	0.63
Hydropower	3.741	12.15
Wind	3.334	24.94
Biomass	0.038	0.79
Biogas	0.011	0.11
Solar and others	0.028	0.12

closure of the electro-intensive power plants can be seen as an opportunity to reformulate the Galician energy system in a more sustainable way. Moreover, it would imply a considerable reduction in the Galician demand.

Based on this situation it is necessary to think of new possible scenarios that facilitate a sustainable energy transition. One option, inspired by the "grids of grids", is creating a distributed Galician electric network interconnected with the Spanish one. Its main advantages son a sharp decrease in transport losses and the facilitation of the integration of renewable sources, among others.

Another idea to maintain the potential of the energy sector, despite the closure of hydroelectric plants, is to make better use of installed power in renewable energy in terms of renewables, and to promote energy fed by local resources, betting specifically on biomass and biogas.

3 Methodology

At a methodological level, the protocol established by the OSeMOSYS tool, -an Open Source modeling system for long-run integrated assessment and energy planning (CITA)-, was followed [6, 7]. The necessary data on the characteristics of the electrical system to be modeled have been obtained from both the *Instituto Enerxético de Galicia* (INEGA) [5] and *Red Eléctrica de España* (REE) [4]. The historical data used goes from 2015 to 2018. The data belonging to the period of the COVID-19 pandemic were not considered as they constitute an exception and are not considered representative.

The crucial point of the modelling process is the definition of scenarios. The scenarios are descriptions of potential futures that allow us to investigate the possible implications of changes that deviate from trends. When defining the scenario, the focus was put on evaluating some of the most widespread proposals within the current socio-political panorama. This means, on the one side, considering the scheduled shutting down of the Galician thermal power plants and the closure of ALCOA (the largest electro intensive plants in the territory). On the other side, it was necessary to include the need for sustainable management of livestock industry waste (biogas) and biomass waste [3]. Table 2 summarizes the main characteristics of the proposed scenarios.

Table 2. Limits and constrains imposed for each scenario.

Scenarios	Description
Business-as-usual (BAU)	– It is the reference scenario. Its input data assumes that Galician's techno-economic conditions do not change. The only restriction imposed is the closure of all thermoelectric plants in 2021 as planned
Scenario 1: Conservative with controlled emissions	– This is the most likely scenario – Due to climate change, hydraulic capacity is reduced by 5% each year – The deadline to close combined cycle plants is extended until 2030 – Limits are imposed on renewable energies, taking into account the maximum potentials already calculated – The program is required to reduce emissions by 30% by 2030 following European regulations [8]
Scenario 2: Decentralized energetic network	– The restrictions imposed on Scenario 2 are maintained – Alcoa electrointensive is closed in 2021 – A decentralized energetic network is proposed (end of exports)
Scenario 3: Strategic promotion of biogas and biomass	– The restrictions imposed on Scenario 3 are maintained – The use of biomass and biogas is promoted. For this, the simulation is carried out reducing the cost of these energies by 80%

The BAU scenario is performed in order to validate the model. On this basis, different possible transition scenarios have been evaluated, the characteristics of which are detailed in Table 2. These scenarios range from the most conservative to some more creative, always depending on Galician needs and opportunities.

The Scenario 1 was designed focusing on the recommendations proposed by the European Union for emission control. The second scenario is based on the construction of a decentralized energetic network, that is on the energy independence of the region. Finally, in the last scenario, the possibility of exploiting biomass and biogas is evaluated. This scenario is raised because it has been considered by the Galician government since there are a great forest wealth and enormous livestock potential in the region, in addition to the fact that slurry is a social problem.

4 Results and Discussions

In all cases, the economic cost per kWh produced has been the parameter used to optimize the proposed energy mixing. From there, the results obtained have focused on the type of new energy source to be installed, the investments necessary to carry them out, as well as the emission levels of each proposal, that is, the level of decarbonisation. The perspective of 2030 has been chosen as a time interval.

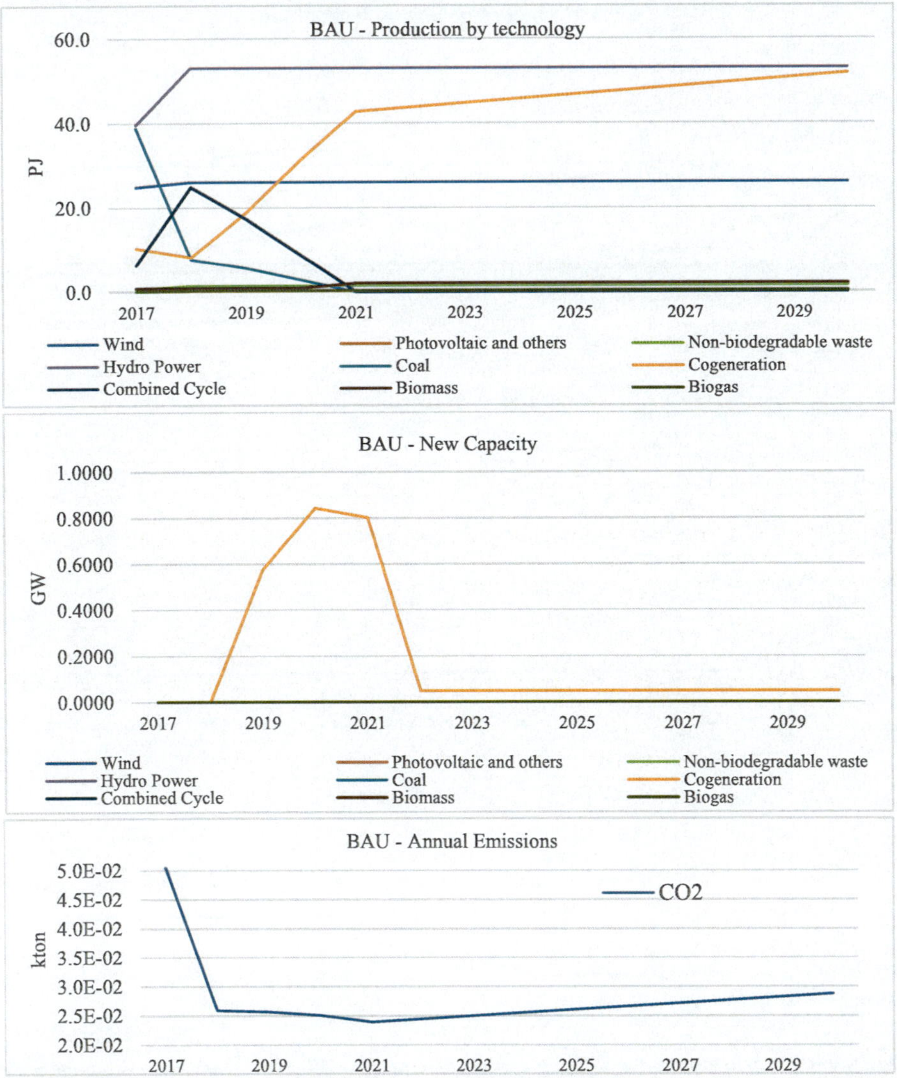

Fig. 1. From top to bottom the following graphs are presented for the BAU scenario: production by technology, need for new production capacity, annual emissions.

4.1 Business-as-Usual (BAU)

This scenario, in which current conditions are maintained, is the most conservative. Figure 1 shows the main results. It is important to note how, the optimization of the system in economic terms, already proposes to advance the closure of the thermos in favor of the combined cycle (2.62 GW new until 2030). The currently installed underutilized capacity (hydro, wind, and others) is sufficient to meet energy needs, without installing

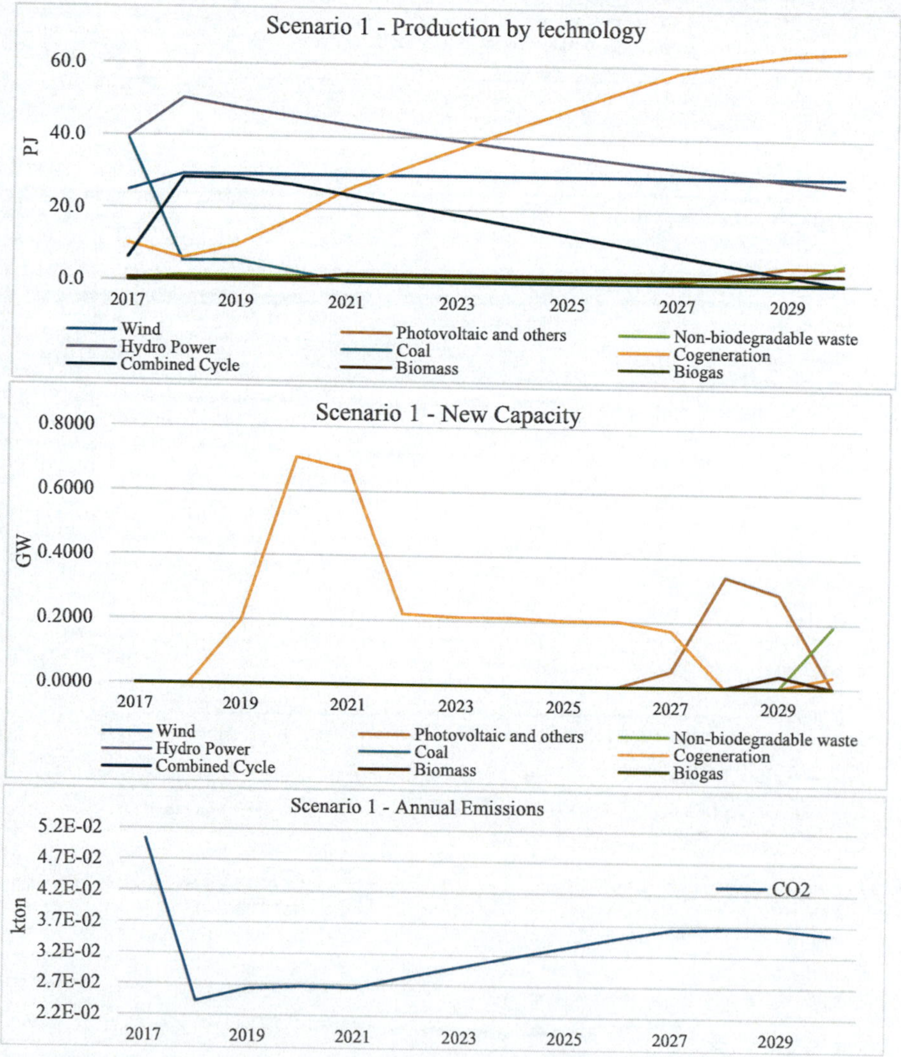

Fig. 2. From top to bottom the following graphs are presented for the Scenario 1: production by technology, need for new production capacity, annual emissions.

new facilities. At the emission level, while drastically reduced, the objectives proposed by the eu are not achieved.

4.2 Scenario 1: Conservative with Controlled Emissions

The European Green Pact imposes a 30% reduction in polluting emissions on EU countries by 2030. In this scenario, the electrical production capacity of fossil fuel, i.e.,

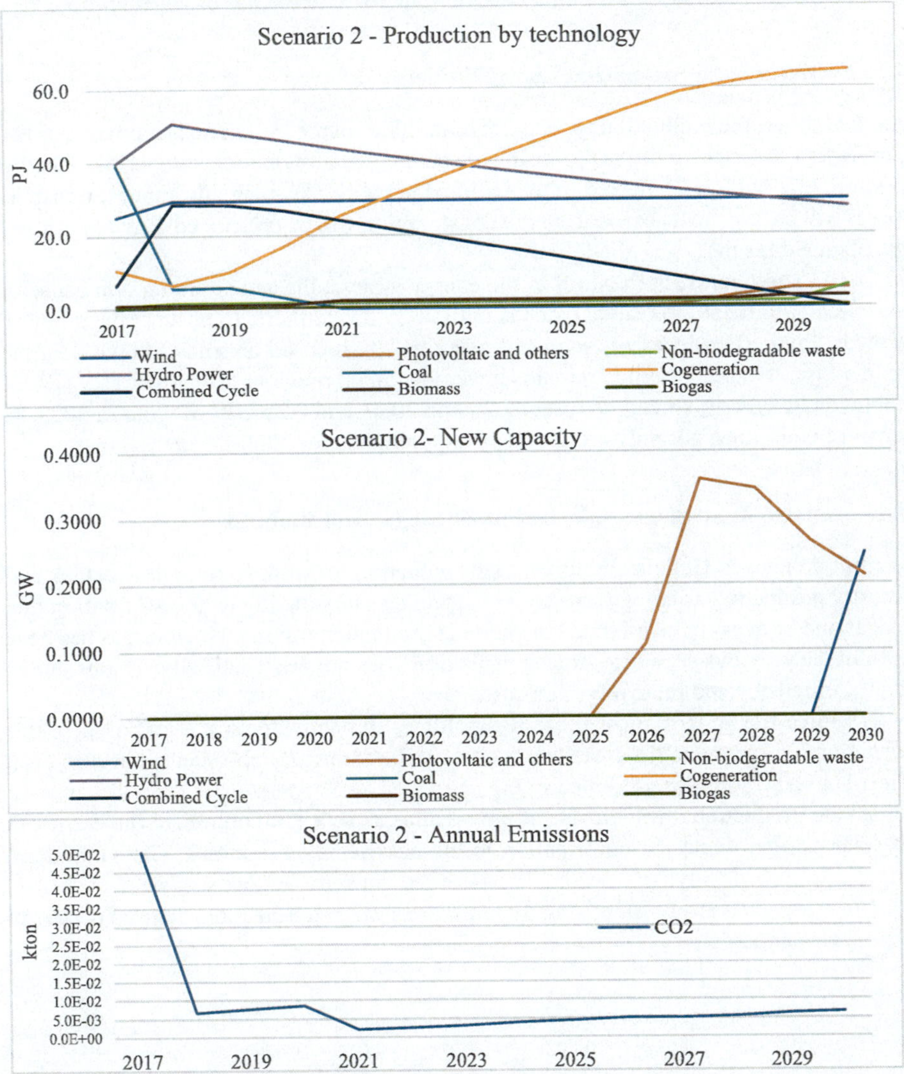

Fig. 3. From top to bottom the following graphs are presented for the Scenario 2: production by technology, need for new production capacity, annual emissions.

cogeneration, is limited to ensure compliance with the requirements. Following the nearby countries tendencies, combined cycle plants are accepted in energetic mix.

As can be seen in Fig. 2, in this scenario, to supply fossil production, to install new power of photovoltaic energy, biomass and non-biodegradable waste is need (0.68 GW, 0.04 GW, 0.19 GW respectively). Even, to ensure demand, cogeneration need be slightly increased.

In these conditions, the proposed emissions limit is effectively met, with the lower limit in 2018. This is associated with the closure of carbon thermoelectric plants. Subsequent increase is linked to the necessary increase in the use of cogeneration.

4.3 Scenario 2: Decentralized Energetic Network

As seen above, the production capacity of renewable source in Galicia is more than 60%. This favour the opportunity of proposing a scenario that turns Galicia into its own electric network that will be connected to the national network. As main advantages, transport losses will be drastically as local energy consumption will be prioritized. Figure 3, shows the obtained results.

As a main remark, at the level of the energy system, the actual production capacity is sufficient to meet the Galician needs until 2025. From there, to ensure consumption growth, it would be necessary to install new capacity focused on solar and wind farms. In this case, the total installed remains lower than in all previous scenarios: 1.29 GW of photovoltaic energy, spread between 2025 and 2030, and 0.25 GW of wind in 2029. In terms of emissions, not only are the EU's requirements for 2030 (55% less) met.

4.4 Scenario 3: Strategic Valorization of Biogas and Biomass

Current policies in Galicia are aimed at the valuation of waste from both livestock and forestry production as energy source. This involves encouraging its transformation into biogas and biomass to be included in Galician production mixing. However, as has been seen in the previous scenarios, this contribution does not arise naturally. This is due to its low efficiency and relatively high cost.

To force the apparition of those technologies, proactive policy is imposed in this scenario: 80% of the initial installation cost can be externally co-financed at zero cost. The obtained results are presented in Fig. 4.

At the production level, the use of biogas appears as a valid option, to the detriment of solar energy, while the contribution to biomass remains minimal. The same trend appears in new deployment needs. The installed capacity is now sufficient, with new facilities expected from 2025. Finally, emissions are reduced more than expected to European standards.

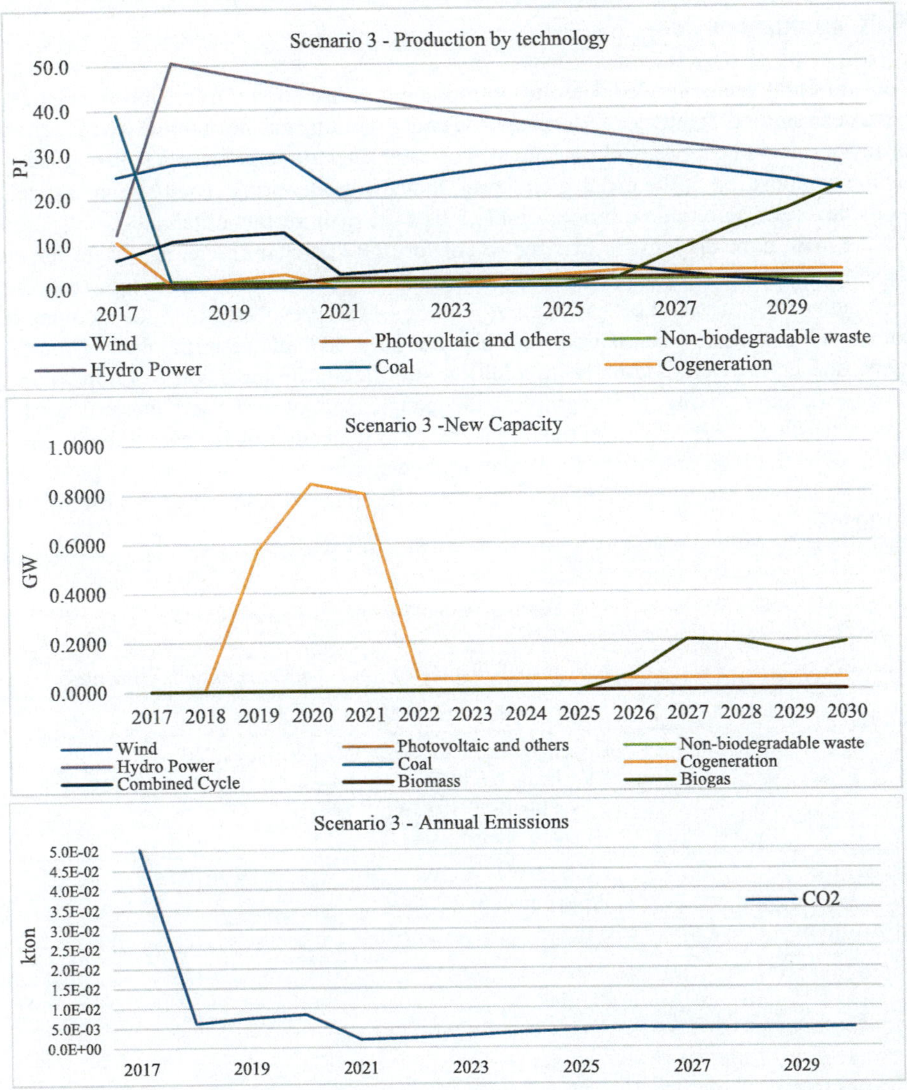

Fig. 4. From top to bottom the following graphs are presented for the Scenario 3: production by technology, need for new production capacity, annual emissions.

5 Conclusions

Four different scenarios dedicated to optimizing a future energy transition for Galicia have been studied. In all cases, the proposed energy mixing has been optimized in terms of lower energy cost per KW consumed. For each scenario, emphasis has been placed on the composition of the energetic mixing, the new production capacity needs as well as the levels of polluting emissions. Table 3 lists the main results obtained.

As main conclusion, it is feasible to convert the Galician energetic system into a decentralized network based on local renewable sources. In addition, promoting the inclusion of biomass and biogas in energy mixing, while resolving the management of forest and livestock waste, is only feasible if at least 80% of the initial cost of implementation is financed externally. In addition, promoting the inclusion of biomass and biogas in energy mixing, while resolving the management of forest and livestock waste, is only feasible if at least 80% of the initial cost of implementation is financed externally. Because of the high potential of Galicia in renewable resources, none of the proposed scenarios should take specific measures to reach the emission level requested for the EU by 2030.

Table 3. Conclusions

Scenario	New installation (GW)	CO_2 emissions reductions
Business-as-usual (BAU)	Cogeneration: 2.62 GW	50% by 2030
Scenario 1	Cogeneration: 2.82 GW PV plants: 0.68 GW Waste plants: 0.19 GW Biomass plants: 0.04 GW	30% by 2030
Scenario 2	PV plants: 1.28 GW Wind power: 0.267 GW	95% by 2030
Scenario 3	Biogas plants: 0.807 GW	95% reduction by 2030

References

1. López González, S.P.: Transformar nuestro mundo. Derecho Glob Estud sobre Derecho y Justicia **16301**(6), 7–10 (1969). https://doi.org/10.32870/dgedj.v0i6.106
2. Howells, M., Hermann, S., Welsch, M., et al.: Integrated analysis of climate change, land-use, energy and water strategies. Nat. Clim. Change **3**(7), 621–626 (2013). https://doi.org/10.1038/nclimate1789
3. Míguez Fariña, A., López Agüera, Á.: Estudio de diferentes escenarios para la transición energética de Galicia mediante el uso de herramientas. CLEWs (2020)
4. Red Eléctrica de España. https://www.ree.es/es. Accessed 20 Jan 2020
5. Inega: Instituto Enerxético de Galicia. https://www.inega.gal/. Accessed 24 Nov 2019

6. Howells, M., Rogner, H., Strachan, N., et al.: OSeMOSYS: the Open Source Energy Modeling System. An introduction to its ethos, structure and development. Energy Policy **39**(10), 5850–5870 (2011). https://doi.org/10.1016/j.enpol.2011.06.033
7. CLEWS - Home. https://www.osimosys.org/. Accessed 3 Jan 2021
8. Un Pacto Verde Europeo | Comisión Europea. https://ec.europa.eu/info/strategy/priorities-2019-2024/european-green-deal_es. Accessed 6 July 2020

Case Studies of Latent Heat Storage with Phase-Change Materials in the Agroindustry Value Chain

Carla Silva[1], Carlos Simão[2], João Murta-Pina[1,3](✉), Fernando Reboredo[1,4], Henrique Paula[5], and Alfredo Martins[6]

[1] NOVA School of Science and Technology, NOVA University of Lisbon, Caparica, Portugal
jmmp@fct.unl.pt
[2] CQS Unipessoal Lda., Lisbon, Portugal
[3] Centre of Technology and Systems, Uninova, Caparica, Portugal
[4] Geobiosciences, Geotechnologies, and Geoengineering (GeoBioTec), Caparica, Portugal
[5] Gráfica Ideal de Águeda, Indústrias Gráficas, SA, Águeda, Portugal
[6] Zêzerovo, Produção Agrícola e Avícola Do Zêzere SA, Ferreira do Zêzere, Portugal

Abstract. Phase-change materials (PCM) can store thermal energy as latent heat, by means of a phase change, thus absorbing or releasing energy at a constant temperature. Distinct stages of the agroindustry value chain require thermal energy, e.g., for food processing (heat) or for its distribution (cold). In all these stages the quality of food needs to be preserved, for which temperature must be kept within tight limits, defined by standards. Additionally, these specific stages may represent an average of nearly 30% (processing) and 10% (logistics) of the agroindustry value chain energy consumption in the European Union. The use of PCMs allows foreseeing the improvement of energy efficiency in the sector while maintaining food quality.

In this work, two case studies of the agroindustry value chain, where the integration of PCMs is expected to meet the previous desiderata, are analyzed. First, PCMs are used in the processing of chocolate, integrated into production equipment. Then, its performance is assessed in the distribution of fresh food, embedded in a transportation box. In both cases, experimental and/or simulation results show that improvements in the energy efficiency of the processes are obtained, while preserving food quality, and, in the processing stage, they may even support the integration of renewables.

Keywords: Agroindustry · Latent heat · Phase-Change Materials (PCM) · Thermal energy storage

1 Introduction

Agriculture and agroindustry systems are responsible for 30% of the energy consumed worldwide, as well as for 22% of greenhouse gas emissions, according to the Food and

J. R. da Costa Sanches Galvão et al. (Eds.): *Proceedings of the 1st International Conference on Water Energy Food and Sustainability (ICoWEFS 2021)*, pp. 456–465, 2021.
https://doi.org/10.1007/978-3-030-75315-3_50

Agriculture Organization (FAO) of the United Nations [1]. Therefore, the research and development of methods for improving energy efficiency are required.

Agroindustry uses heat in several processes, and thermal energy storage and the management of stored energy are thus preponderant aspects in its value chain. Thermal energy storage is usually related to sensible heat, where the latter is stored in a material by increasing its temperature.

Nevertheless, storage of latent heat, in the so-called Phase-Change Materials (PCM) is becoming an attractive solution in several sectors, due to the improved efficiency of storage, where besides heat is stored and released at a constant temperature. PCMs have been applied in the construction sector or transportation of medications, among other areas [2, 3], but agroindustry may have plenty of opportunities for these materials. There are no major regulatory barriers to its application, but the cost and technical challenges are still the key limitations.

In this work, the benefits of using PCMs in two distinct links (each analyzed by a specific case study) of the agroindustry value chain are demonstrated through experiments. These links are related to the processing and distribution of food products.

This paper is organized as follows: in the next section, the agroindustry value chain is described. Section 3 introduces phase-change materials, and in Sect. 4 the two case studies (chocolate processing and distribution of fresh products) are described. Results are analyzed in Sect. 5, and conclusions are drawn in Sect. 6.

2 Agroindustry Value Chain

FAO defines a generical value chain as the set of interrelated economic activities that start with the production of a primary commodity and end with its delivery to a consumer. In the specific case of the agroindustry sector, this value chain includes inputs, production, transport, storage and handling, marketing and distribution, and the end-user, see Fig. 1. In all the links, FAO identifies opportunities for implementing clean energy technologies [4].

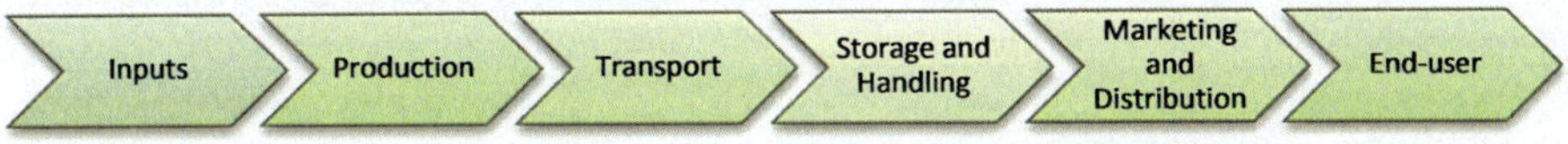

Fig. 1. The agroindustry value chain, according to FAO [4].

Agroindustry includes a multitude of enterprises and supply chains, where several of these links may benefit from thermal energy storage. In the current work, the production and transport links are evaluated, using two case studies, as already mentioned.

3 Phase-Change Materials

Latent heat may be stored or released through the phase change of materials, as between solid and liquid phases, in the context of the present work. Nevertheless, another type of phase change may be used, as solid to solid, or liquid to vapor phase changes. During the phase change, the temperature of the material remains constant, and that phenomenon is exploited here. This process is illustrated in Fig. 2.

The materials that are used to exploit this phase change process are simply called phase-change materials (PCM). Water is probably the most common type of PCM, due to its low cost, availability, and relatively high volumetric thermal capacity, (4.17 MJ/m^3·K) [5]. Water has a solid-to-liquid transition at 0 °C and a liquid to vapor transition at 100 °C, both at 1 atm. Despite the limited temperatures, this is still useful in some agroindustry processes.

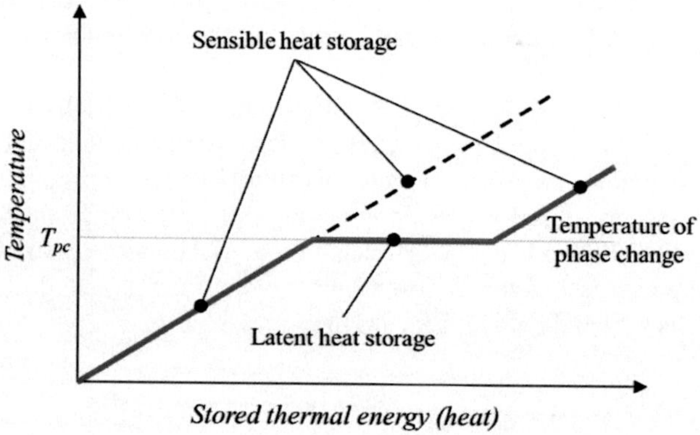

Fig. 2. Storage of sensible and latent heat. The dashed line represents sensible heat storage without the phase change of a material, while the solid line shows sensible and latent heat storage when the phase change occurs. The temperature at which the phase change arises is T_{pc}.

The total (sensible) heat stored in a material is given by (1), while the total (sensible and latent) heat stored under a phase change is given by (2) [3]. In the latter, the latent heat adds to the stored energy, making it a more efficient process than sensible heat storage. In those equations, m is the mass of the medium where heat is stored; $\overline{C}_p$ is the average specific heat between the initial-, T_i, and final temperatures, T_f, of the medium, where, in the current work, the former corresponds to the solid phase and the latter to the liquid phase; T_{pc} is the phase change temperature; $\overline{C}_{sp}$ is the average specific heat between T_i and T_{pc}; $\overline{C}_{lp}$ is the average specific heat between T_{pc} and T_f; a_m is the fraction of the total phase of the medium that is melted; and ΔH_m is the enthalpy of the phase change.

$$Q = m\overline{C}_p(T_f - T_i) \tag{1}$$

$$Q = m\left[\overline{C}_{sp}(T_{pc} - T_i) + a_m \Delta H_m + \overline{C}_{lp}(T_f - T_{pc})\right] \tag{2}$$

PCMs are classified into three major groups, namely organic, inorganic, and eutectic. The detailed description of these groups is beyond the scope of this paper, but it can be found, e.g., in reference [6]. Other references including details and applications are [7] for organic, [2, 8] for inorganic, and [3, 9] for eutectic PCM.

The PCM used in this work is of the organic type, which can be found as paraffin and non-paraffin types. Among other factors, its selection was related to their wide range of phase-change temperatures, which provides flexibility for addressing distinct applications. Nevertheless, they show a relatively low thermal conductivity, they are inflammable, and they are incompatible with several plastic materials [7]. This may pose a technical challenge, as the materials need to be encapsulated to avoid leakage and corrosion [9].

4 Case Studies in the Agroindustry Value Chain

As mentioned, two case studies will be analyzed, to assess the opportunity of using PCMs in the agroindustry, with the envisaged benefits of increasing its energy efficiency.

The first case study is related to chocolate processing, where PCMs are used to maintain the temperature in a mixing machine, allowing longer processing times at constant (hot) temperature, without or with less energy consumption. The main motivation for this case study is also to assess the benefits of directly integrating PCMs in production equipment, upgrading it, and avoiding the need to build new devices.

The second case study is developed in the context of the distribution of fresh food, and PCMs are used to keep the (cold) temperature of the products, also decreasing energy needs. In this case, the motivation was to assess the viability of developing a cardboard-based distribution box for fresh products where the PCMs could be integrated.

These cases are described in the following sections.

4.1 Production: Chocolate Processing

A chocolate tempering water bath-based machine was used in the first case study. Chocolate may be melted in distinct mediums, like milk. Water, placed in an outer jacket, is heated by a 3 kW resistor situated at the bottom of the machine, and the temperature is maintained by on-off control, see Fig. 3 for details of the device. Temperature setpoints are adjustable, and the volume of water is 11.5 L.

Although this machine had originally agitating blades to homogenize the chocolate mixture, these were removed as a proof-of-concept was desired. Water-proof temperature sensors (model DS18B20 from Maxim Integrated and an energy meter (model ACS71020 from Allegro MicroSystems), also depicted in Fig. 3, were assembled in the machine. A programmable system-on-chip (model PSoC 5LP from Cypress) was used to collect data and send it to the Azure platform for processing.

An organic PCM from Phase Change Material Products Limited, model S34 was chosen, see also Fig. 3. Its phase-change temperature is 34 °C, and the material represents around 30% of the available heat storage volume (11.5 L), while the remaining is water.

The chocolate pellets used in the experiments are shown in Fig. 4. It is commercial chocolate made of 55% of cacao with a density of 1.261 kg/m^3. Several mixtures, with different chocolate to milk ratios, were used in the experiments.

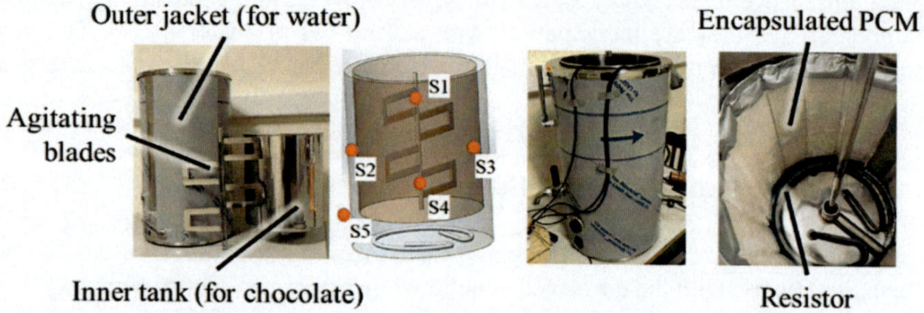

Fig. 3. Chocolate tempering machine used in the experiments. From left to right: the machine disassembled, with the outer jacket where water is usually placed and heated, the inner tank that contains the chocolate mixture, and the agitating blades, which were removed for this case study; location of the sensors, where S1 and S4 measure the chocolate temperature near the axis, S2 measures the PCM/water temperatures, S3 measures the chocolate temperature near the tank wall, and S5 represents the energy meter; the machine with the sensor wires visible; and the interior of the outer jacket, with the heating resistor at the bottom and the encapsulated PCM.

Fig. 4. Chocolate pellets used in the experiments. From left to right: the chocolate pellets; the pellets inside the machine; and the chocolate already melted.

In the tests reported in this paper, 15 kg of chocolate were melted in 2 L of milk, with a total volume of 13.9 L, nearly 60% of the inner tank volume. Temperatures and energy were measured in two distinct cases, namely a water bath used with and without (only water) PCM, heated up to 38 °C, four degrees above the phase change temperature. The following methodology was used:

Water Bath with PCM. The water bath was heated for several hours, to charge the PCM embedded in it. Power was then turned off, to assess the discharge time. Results are shown in Fig. 5.

Water Bath Without PCM. The water bath is heated and then power is turned off, to assess the discharge time. Results are shown in Fig. 6.

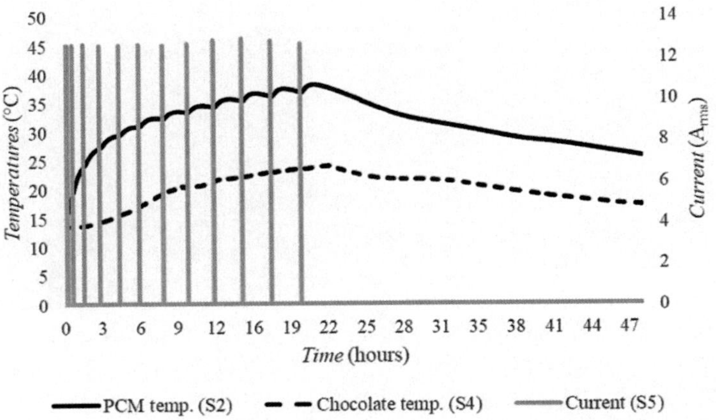

Fig. 5. Evolution of temperatures and current (an image of power) in the experiment with PCM.

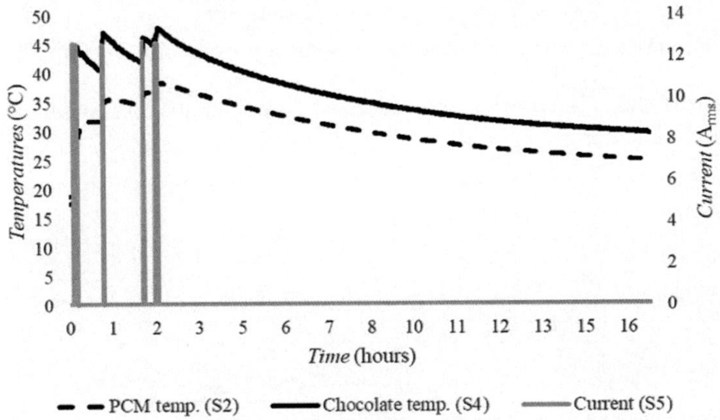

Fig. 6. Evolution of temperatures and current (an image of power) in the experiment without PCM.

4.2 Transport: Distribution of Fresh Food

As mentioned, a cardboard distribution box was developed, integrating PCM for the conservation of fresh food, when in situations of increased temperature, as during distribution processes. The box is shown in Fig. 7 and has a total volume of 24 L. The cardboard is 4 mm thick and the box should be able to transport up to 4 kg of products.

To measure both inner and outer box temperature and humidity, DHT11 sensors from Sunrom Technologies were used, and data is collected by Arduino Uno microcontrollers.

Again, an organic PCM is used, namely the model RT5HC from Rubitherm, see Fig. 7, whose phase-change temperature is around 5 °C. The encapsulated PCM was placed in the bottom of the box, and 2 kg of diverse fruits were used in the experiments.

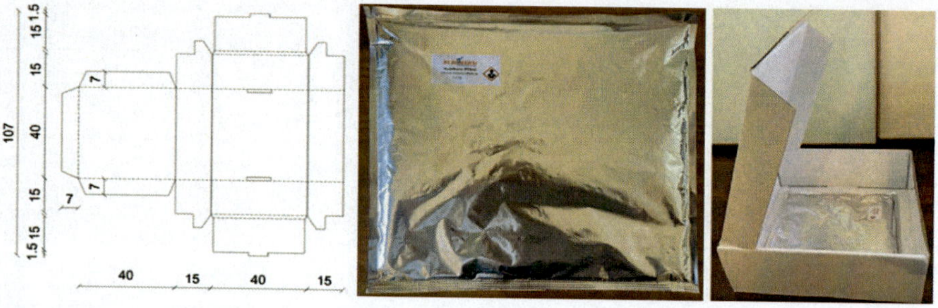

Fig. 7. Cardboard distribution box developed for the case study. From left to right: box dimensions, in centimeters; encapsulated PCM to integrate into the box; box with the encapsulated PCM in the bottom.

The following methodology was used:

Box with PCM. The box was kept with the fruit at 10 °C and then was removed from that environment to assess the temperature evolution. Results are shown in Fig. 8.

Box Without PCM. The previous procedure was used, but the box contained no PCM. Results are shown in Fig. 9.

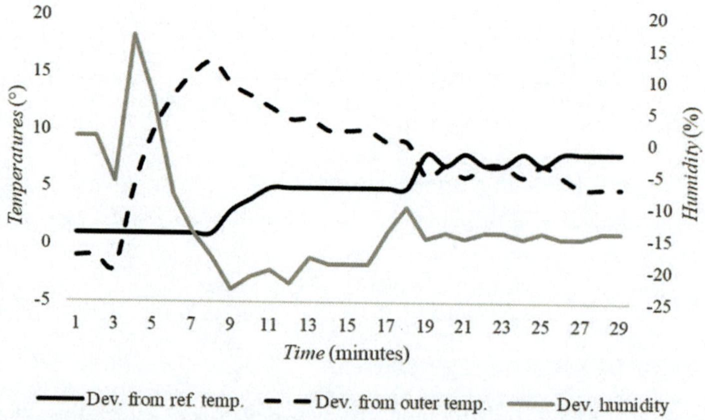

Fig. 8. Evolution of temperatures and humidity in the box with PCM. The deviation of the inner temperature from the reference (10 °C) is shown, as well as the deviation of the inner temperature from the outer one, and the deviation of the inner humidity from the outer humidity.

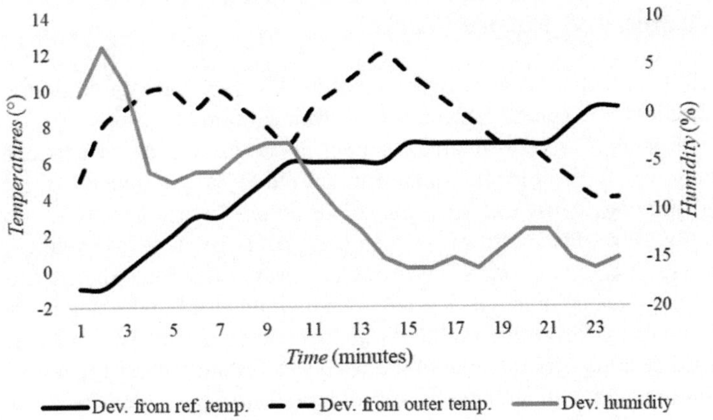

Fig. 9. Evolution of temperatures and humidity in the box without PCM. The deviation of the inner temperature from the reference (10 °C) is shown, as well as the deviation of the inner temperature from the outer one, and the deviation of the inner humidity from the outer humidity.

5 Discussion

In this section, the results from the experiments are discussed.

5.1 Production: Chocolate Processing

According to the observations, the chocolate temperature decreased between 38 and 30 °C at an average rate of −0.53 °C/min when PCM was used. Without PCM this value increased (in absolute value) to −1.20 °C/min. Therefore, when only sensible heat was stored, the cooling rate more than doubled.

Nevertheless, the stored latent heat was provided from the electrical source. Therefore, to be viable, it should come from a renewable source, such as solar thermal heated water.

5.2 Transport: Distribution of Fresh Food

After being removed from the cold storage medium, at 10 °C, the deviation from the inner and outer temperatures increase abruptly and then is slowly decreased, in the box with PCM. After nearly half an hour, this deviation is around 5 °C, and the inner temperature is around 8 °C above the reference one. The deviation between inner and outer humidity is relatively stable along the process.

In the box without PCM, the deviation from reference temperature is faster, and the deviation between inner and outer temperatures has strong variations, meaning that the outer temperature strongly affects the inner one. Humidity deviation has also more variations.

The PCM, even in a very simple integration, thus show good prospects for improving temperature and humidity stabilization.

6 Conclusion and Future Work

In this work, PCM materials were integrated into two links of the agroindustry chain, to assess its viability and energy efficiency improvement opportunity.

Two case studies were evaluated, namely in the processing of chocolate and in the distribution of fresh products. In the former, the PCM was integrated into existing equipment, proving that it can keep chocolate hotter for a longer time without any energy supply. Nevertheless, the energy required to charge the PCM should come from a renewable source, to be economically viable. It was verified that these materials may bring benefits if the equipment is upgraded, without the need to develop new ones.

Yet, if a temperature profile should be followed, as is the case of chocolate tempering, PCMs would become inviable, due to the need to constantly charge and discharge the material. Future work will include the assessment of storing residual heat in equipment like furnaces, to allow reducing the energy required to heat, e.g., the combustion chamber.

In the second case, the PCM was used to maintain the temperature of a fresh products distribution box. Even in a simple integration of the material, just placed at the bottom of the box, it was observed that its inner temperature was kept for a longer time than without the PCM. The same was observed for humidity.

Future work will include more sophisticated integration of the material, as micro-encapsulation in the cardboard, to improve its performance, as well as assessing the organoleptic properties of the food products, and evaluating methodologies for recycling the material.

Acknowledgement. This work was supported by FCT (Fundação para a Ciência e a Tecnologia), within the scope of Research Unit CTS, Centre of Technology and Systems, with the reference UIDB/00066/2020.

References

1. Food and Agriculture Organization: Policy Brief: The Case for Energy-Smart Food Systems (2014). https://www.fao.org/3/i2456e/i2456e00.pdf. Accessed 25 Feb 2021
2. Ge, H., Li, H., Mei, S., Liu, J.: Low melting point liquid metal as a new class of phase change material: an emerging frontier in energy area. Renew. Sustain. Energy Rev. **21**, 331–346 (2013). https://doi.org/10.1016/j.rser.2013.01.008
3. Sharma, A., Tyagi, V.V., Chen, C.R., Buddhi, D.: Review on thermal energy storage with phase change materials and applications. Renew. Sustain. Energy Rev. **13**, 318–345 (2009). https://doi.org/10.1016/j.rser.2007.10.005
4. Flammini, A., Bracco, S., Sims, R., Cooke, J., Elia, A.: Costs and benefits of clean energy technologies in the milk, vegetable and rice value chains: intervention level. Food and Agriculture Organization of the United Nations (FAO) (2018). ISBN 978-92-5-109992-6 (FAO)
5. Cabeza, L.F. (ed.): Advances in Thermal Energy Storage Systems: Methods and Applications. Woodhead Publishing, Sawston (2015)
6. Sarbu, I., Sebarchievici, C.: A comprehensive review of thermal energy storage. Sustainability **10**, 191 (2018). https://doi.org/10.3390/su10010191
7. Oró, E., de Gracia, A., Castell, A., Farid, M.M., Cabeza, L.F.: Review on phase change materials (PCMs) for cold thermal energy storage applications. Appl. Energy **99**, 513–533 (2012). https://doi.org/10.1016/j.apenergy.2012.03.058

8. Pielichowska, K., Pielichowski, K.: Phase change materials for thermal energy storage. Prog. Mater. Sci. **65**, 67–123 (2014). https://doi.org/10.1016/j.pmatsci.2014.03.005
9. Su, W., Darkwa, J., Kokogiannakis, G.: Review of solid–liquid phase change materials and their encapsulation technologies. Renew. Sustain. Energy Rev. **48**, 373–391 (2015). https://doi.org/10.1016/j.rser.2015.04.044

Energy, Thermal Comfort and Air Quality in a Ventilation System Based on Confluent Jets

Eusébio Conceição[1](✉), Mª Inês Conceição[2], Mª Manuela Lúcio[1], João Gomes[3], André Ramos[1], and Hazim Awbi[4]

[1] FCT - University of Algarve, Campus de Gambelas, 8005-139 Faro, Portugal
{econcei,aframos}@ualg.pt
[2] Instituto Superior Técnico, Av. Rovisco Pais, 1049-001 Lisboa, Portugal
ines.conceicao@tecnico.ulisboa.pt
[3] CINTAL, Campus de Gambelas, 8005-139 Faro, Portugal
jgomes@ualg.pt
[4] School of Built Environment, University of Reading, Reading RG6 6AW, UK
h.b.awbi@reading.ac.uk

Abstract. In this study a ventilation system based on vertical confluent jets is developed and applied. The energy consumption, thermal comfort level, indoor air quality level and Air Distribution Index are evaluated and discussed. The study presented in this work is performed in a virtual chamber with dimensions of $4.50 \times 2.55 \times 2.50\ m^3$. The chamber, occupied with twelve virtual occupants, is equipped with six tables, twelve chairs, one outlet system and one inlet system, based in confluents jets system. This simulation considers the descendent inlet airflow, near the lateral walls. In the outlet system are considered six air ducts, located above the head level, connected to the ceiling area. The inlet system, based in two horizontal ducts, with 0.15 m diameter, located 1.8 m above the floor level, are equipped with consecutive holes, that promote vertical descendent jets near the wall. The predicted percentage of dissatisfied index results show that the comfort level is acceptable. The dioxide carbon concentration show that the indoor air quality, in general, is acceptable. When the airflow rate increases the air quality number increases and the thermal comfort number decreases. Thus, the Air Distribution Index increases slightly.

Keywords: Energy · Thermal comfort · Air quality · Confluent jets system

1 Introduction

Several studies have been developed, in the last years, on the area of confluent jets ventilation systems: see as example Arghand et al. [1], Karimipanah et al. [2] and Cho et al. [3]. The confluent jets are associated to a multiple jets system that confluent in a single airflow in the exit ventilation system. Different techniques and methodologies are used. However, in general, in this ventilation system, the inlet is made through holes lines and the outlet is made through an outlet system.

J. R. da Costa Sanches Galvão et al. (Eds.): *Proceedings of the 1st International Conference on Water Energy Food and Sustainability (ICoWEFS 2021)*, pp. 466–475, 2021.
https://doi.org/10.1007/978-3-030-75315-3_51

To evaluate the occupant comfort is used the thermal comfort level and indoor air quality level, to evaluate the occupant local discomfort is used the Draught Risk (DR) and to evaluate the Heating Ventilating and Air-Conditioning (HVAC) system performance is used the Air Distribution Index (ADI).

The Predicted Mean Vote (PMV) index and the Predicted Percentage of Dissatisfied (PPD) index are used to evaluate the thermal comfort level. These indexes, developed by Fanger [4], are applied to determine the thermal comfort conditions in spaces equipped with HVAC system (see ISO 7730 [5]).

The carbon dioxide concentration is used to evaluate the indoor air quality level (Conceição et al. [6]). In this study the carbon dioxide concentration, release by the occupants, is used as indicator of the indoor air quality in occupied spaces (ASHRAE standard [7]).

The ADI is used to evaluate simultaneously the air quality level, thermal comfort level, contaminants removal efficiencies and heat removal efficiencies. ADI was presented and detailed in the studies of Awbi [8], for uniform environments, and Conceição et al. [9], for non-uniform environments.

DR is used to evaluate the local thermal discomfort level that the occupants are subjected. DR, which depends on temperature, velocity and turbulence intensity of the air, was developed by Fanger et al. [10].

The numerical model used in this work is based on a coupling between Computer Fluid Dynamics (CFD), see Conceição and Lúcio [11], and Human Thermal Response (HTR) numerical models. This methodology can be seen in the studies of Conceição and Lúcio [12], Conceição [13], Conceição et al. [14], and Conceição et al. [15].

The coupling numerical model need inputs obtained from Buildings Dynamics Response (BDR) software. Examples of this kind of software can be analyzed in Yang et al. [16], Sailor [17] and Balagi et al. [18].

In this work, the output data from BDR is used as input data in the coupling software. Application examples can be seen in Conceição et al. [19] and Conceição and Lúcio [20], in the evaluation of air temperature distribution, surfaces temperature distribution and energy consumption. Others examples are presented in Conceição et al. [21], Conceição and Lúcio [22] and Conceição and Lúcio [23]. This software considers the evaluation of thermal comfort through the PMV/PPD indexes (Conceição et al. [24]), the adaptive thermal comfort (Conceição et al. [25]), and the temperature preferred control model (Conceição et al. [26]). This software was validated in summer conditions and in winter conditions. An application of summer conditions can be analyzed, as example, in Conceição and Lúcio [27].

In order to evaluate the thermal comfort level is necessary to evaluate the air temperature and air velocity around the occupants. As example, Conceição et al. [28] show an application of this methodology in vehicles using experimental techniques, and Conceição and Lúcio [11] show an application of this methodology in buildings using numerical techniques.

The purpose of this numerical study is to develop a vertical confluent jet system to improve the thermal comfort and air quality levels to the occupants, with low DR levels, and the HVAC performance. To evaluate the thermal comfort, indoor air quality and DR is used three software: a coupling of the CFD and HTR and a third software that simulates the BDR. The study considers three Case studies, with different airflow rates, in winter conditions.

2 Materials and Methods

2.1 Numerical Model

This study considers a numerical model, developed by the authors in the last years, divided in two parts: a coupling of a differential numerical model, CFD (see Fig. 1a), and an integral numerical model, HTR (see Fig. 1b), and an integral numerical model, that simulates BDR.

The integral HTR model, that simulates the human thermal and thermoregulatory and clothing thermal response, evaluates the tissue, blood and clothing temperatures distribution. This numerical model divides the human body in 24 cylindrical elements and 1 spherical element and each one is divided into several concentric layers. Each element can also be protected by several slices. More details can be seen in the works of Conceição et al. [9], Conceição and Lúcio [11], Conceição and Lúcio [12], Conceição [13], Conceição et al. [14], and Conceição et al. [16].

The differential CFD model evaluates the air velocity, air temperature, air turbulence intensity and CO_2 concentration. In this study the DR around the occupants, the thermal comfort, the air quality and the ADI are evaluated. This numerical model, that works in steady-state conditions and non-isothermal thermal conditions, uses the RNG turbulence model and simulates the three-dimensional airflow in Cartesian coordinates. The numerical model, that simulates the presence of higher occupation level, uses partial differential equations (mass continuity, moment equations, energy equation, turbulence kinetic energy and turbulence energy dissipation rate and contaminant concentration equations), resolved by the finite volume method. This numerical model divides the human body in 25 boxes. More details can be seen in Conceição et al. [9] and Conceição and Lúcio [11].

The integral numerical model, that simulates building thermal response, evaluates the air temperature distribution, surfaces temperature distribution and energy consumption. This numerical model divides the building in opaque (wall, ceiling, floor and interior bodies) and transparent (windows bodies). The opaque bodies are divided in several slices. More details can be seen in Conceição et al. [9], Conceição and Lúcio [11], Conceição et al. [19], Conceição and Lúcio [20], Conceição et al. [21], Conceição and Lúcio [22], Conceição and Lúcio [23], Conceição et al. [24], Conceição et al. [25] and Conceição and Lúcio [27].

In the numerical validation, see Conceição and Lúcio [11] and Conceição and Lúcio [12], for the HTR, Conceição and Lúcio [11] for the CFD and Conceição and Lúcio [29] and Conceição et al. [30] for the BDR.

The ADI, that uses simultaneously the occupant thermal comfort, occupant air quality and ventilation effectiveness for heat and contaminants removal levels, evaluates the HVAC system performance.

2.2 Numerical Methodology

This numerical study is performed in a virtual chamber with dimensions of 4.50 × 2.55 × 2.50 m^3 (see Fig. 1):

- occupied with twelve virtual occupants;
- equipped with six tables and with twelve chairs;
- equipped with one exhaust system built with six vertical air ducts, located above the head level in the central of the chamber;
- equipped with one inlet system, based in confluents jets system, based in two horizontal ducts, with 0.15 m diameter, located 1.8 m above the floor level, equipped with consecutive holes with 1.65 m length, located in the central area, that promotes vertical jets near the wall.

The location of the occupants and the number of each occupant, can be seen in Fig. 2a). The inlet (green arrows) and outlet (blue arrows) airflow location used in the CFD numerical simulation can be seen in Fig. 2b).

The numerical simulations were performed for a winter typical day conditions: with indoor mean air temperatures of 20 °C; with indoor air relative humidity of 50% and with outdoor air temperatures of 0 °C.

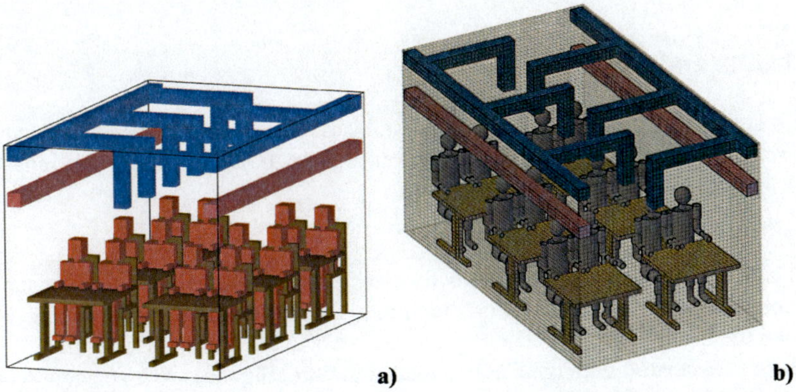

Fig. 1. Scheme of the virtual chamber, equipped with confluents jet systems: a) used in the CFD; b) used in the HTR.

It was considered a typical clothing level of 1 clo for winter conditions, with a typical level of activity of 1.2 met for winter conditions.

In the numerical simulation three case studied are made:

- Case A – with an airflow rate for 12 occupants, an inlet air velocity of 0.7 m/s, an inlet air temperature of 11.6 °C and a thermal power of 1447.3 W;
- Case B – with an airflow rate for 24 occupants, an inlet air velocity of 1.4 m/s, an inlet air temperature of 15.8 °C and a thermal power of 3942.6 W;
- Case C – with an airflow rate for 36 occupants, an inlet air velocity of 2.1 m/s, an inlet air temperature of 17.2 °C and a thermal power of 6438.2 W.

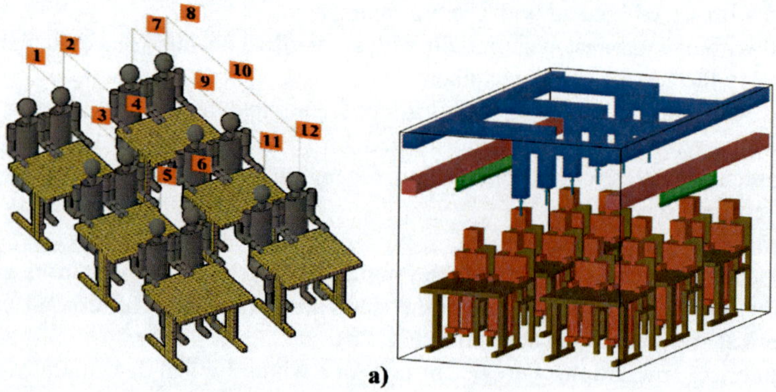

Fig. 2. Location of the occupants and identification of the number of the occupants a) and inlet and outlet airflow location used in the CFD numerical simulation b).

3 Results and Discussion

In this section the distribution of environmental variables around the occupants, namely, the air velocity distribution, the DR distribution, and the ADI, are presented and analyzed.

3.1 Environmental Variables

In Fig. 3, they are presented the air velocity distribution and DR distribution, around the occupants, for winter conditions. Fig. I identified the air velocity distribution and Fig. II identified the DR distribution. The Fig. 3a) are associated with the Case studied A, the Fig. 3b) are associated with the Case studied B and the Fig. 3c) are associated with the Case studied C.

In according to ISO 7730 [5] the Category A of the DR, are associated with (10%), the Category B of the DR are associated with (20%) and the Category C of the DR are associated with (30%).

The results obtained show that:

- The air velocity around the occupants' sections are relatively uniform, being the air velocity in upper sections similar than in lower sections;

- The air velocity around the occupants' sections increases when the inlet air velocity increases;
- The air velocity in the upper member of the two occupants located near the confluents jets system is highest;
- The air temperature around the occupants' sections is relatively uniform;
- The variations of the air temperature around the occupants' sections decreases slightly when the airflow rate increases;
- The DR around the occupants' sections increases when the airflow rate increases;
- The DR in the upper members of the two occupants located near the confluent jets system is not acceptable;
- The DR, except the two people located near the vertical confluents jets, in the Cases A and B is acceptable in accordance with the Category B, while in the Case C is acceptable in accordance with the Category C;
- The DR around the occupants' sections, in general, is relatively uniform.

In accordance with the obtained results is recommended that the occupants should be not located near the confluents jets system and the length of the confluent jet nozzle should be increase. The implementation of these recommendation should reduce the air velocity and DR in the upper bodies sections.

3.2 Air Distribution Index

Table 1 presents the results of the effectiveness for heat removal, the PPD, CO_2 concentration in the breathing area, the effectiveness for contaminants removal, the percentage of dissatisfied (PD) people with indoor air quality and the ADI index, winter conditions, for Case A.

The numerical simulation obtained results demonstrate that:

- The mean body temperature decreases slightly when the airflow rate increases;
- The effectiveness of heat removal decreases when the airflow rate increases;
- The PPD index increases when the airflow rate increases;
- The thermal comfort level is acceptable, in accordance within Category B;

Table 1. ADI obtained for Cases A, B and C.

Occupant number (Mean)	Case A	Case B	Case C
Body Mean Temperature (°C)	22.92	22.83	22.54
Effectiveness for Heat Removal (%)	71.75	60.98	52.56
PPD (%)	6.45	7.63	9.35
Thermal Comfort Number	11.68	8.35	6.07
CO_2 in the Respiration Area (mg/m^3)	2660.51	1585.22	1250.97
Effectiveness for Contaminant Removal (%)	33.88	35.28	37.50
PD with Indoor Air Quality (%)	14.83	7.97	5.26
Air Quality Number	2.28	4.43	7.14
Air Distribution Index (ADI)	4.89	5.72	5.95

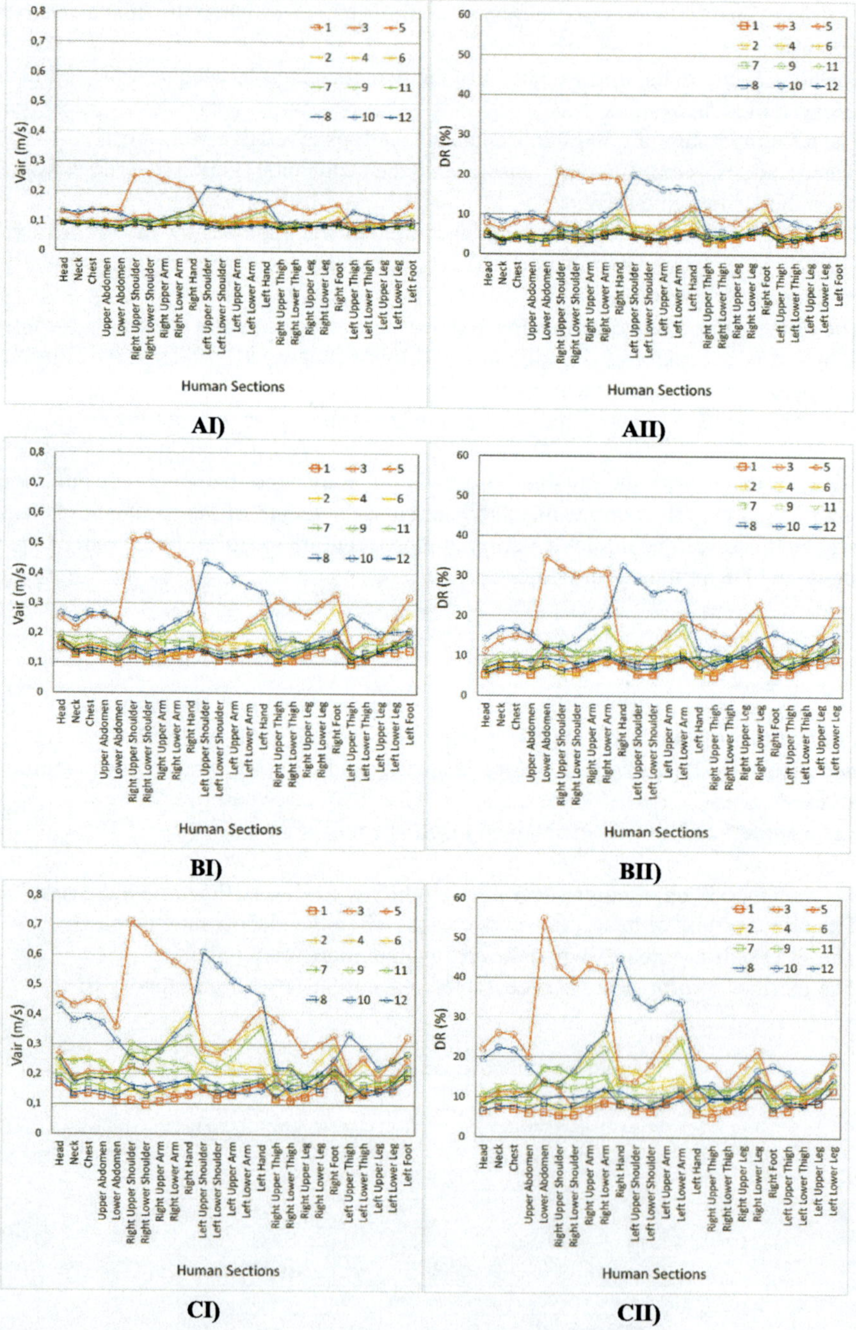

Fig. 3. Air velocity (V_{air}) (I) and Draught Risk (DR) (II) distribution around the occupants for A) case A, B) case B, C) case C).

- The thermal comfort number decreases when the airflow rate increases;
- The carbon dioxide concentration decreases when the airflow rate increases;
- The indoor air quality level is not acceptable in the Case A, however is acceptable in the Cases B and C;
- The effectiveness for contaminant removal increases slightly when the airflow rate increases;
- The air quality number increases when the airflow rate increases;
- The ADI increases slightly when the airflow rate increases.

In accordance with the obtained results, when the airflow increases the thermal comfort number decreases, the air quality number increases and the ADI index increases slightly. Thus, an optimal point, when the two values are similar, should be used as working point.

4 Conclusion

In this work the application of a vertical confluents jets system, in a school virtual chamber, was analyzed. The study, made in winter conditions, considers a virtual chamber occupied with twelve occupants and equipped with six tables, twelve chairs, one outlet system and one inlet system, based on confluents jets system.

In this system the descendent airflow is distributed on the ground floor level. The ascendant airflow transports the contaminants concentration and heat release by the occupants to the outlet system located above the head level.

The air velocity around the occupants are relatively uniform and increases when the airflow rate increases. The air velocity in the upper members of the two occupants located near the confluent jets system presents high values.

The DR in the upper members of the two occupants located near the confluent jets system is not acceptable. However, the DR for the occupants' sections is acceptable.

The thermal comfort number decreases when the airflow rate increases and the air quality number increases when the airflow rate increases. Thus, in accordance with the obtained results, the ADI increases slightly when the airflow rate increases.

In future works, the influence of different vertical confluents jets in the thermal comfort and local thermal discomfort will be analyzed. The length of the nozzle and the location of the vertical confluents jets will be analyzed in detail.

Acknowledgement. The authors would like to acknowledge to the project (SAICT-ALG/39586/2018) from Algarve Regional Operational Program (CRESC Algarve 2020), under the PORTUGAL 2020 Partnership Agreement, through the European Regional Development Fund (ERDF) and the National Science and Technology Foundation (FCT).

References

1. Arghand, T., Karimipanah, T., Awbi, H., Cehlin, M., Larsson, U., Linden, E.: An experimental investigation of the flow and comfort parameters forunder-floor, confluent jets and mixing ventilation systems in an open-plan office. Build. Environ. **92**, 48–60 (2015)

2. Karimipanah, T., Sandberg, M., Awbi, H.: A comparative study of different air distribution systems in a classroom. In: Proceedings of 7th International Conference on Air Distribution in Rooms, ROOMVENT 2000, pp. 1013–1018. Elsevier, Amsterdam (2000)
3. Cho, Y., Awbi, H., Karimipanah, T.: Theoretical and experimental investigation of wall confluent jets ventilation and comparison with wall displacement ventilation. Build. Environ. **43**(6), 1091–1100 (2008)
4. Fanger, P.: Thermal Comfort: Analysis and Applications in Environmental Engineering. Danish Technical Press, Copenhagen (1970)
5. ISO 7730: Ergonomics of the Thermal Environments—Analytical Determination and Interpretation of Thermal Comfort Using Calculation of the PMV and PPD Indices and Local Thermal Comfort Criteria. International Standard Organization, Geneva, Switzerland (2005)
6. Conceição, E., Lúcio, M., Vicente, V., Rosão, V.: Evaluation of local thermal discomfort in a classroom equipped with crossed ventilation. Int. J. Vent. **7**(3), 267–277 (2008)
7. ANSI/ASHRAE Standard 62-1: Ventilation for Acceptable Indoor Air Quality, American Society of Heating, Refrigerating and Air-Conditioning Engineers: Atlanta, GA, USA (2016)
8. Awbi, H.: Ventilation of Buildings. Taylor & Francis, London (2003)
9. Conceição, E., Lúcio, M., Awbi, H.: Comfort and airflow evaluation in spaces equipped with mixing ventilation and cold radiant floor. Build. Simul. **6**(1), 51–67 (2013)
10. Fanger, P., Melikov, A., Hanzawa, H., Ring, J.: Air turbulence and sensation of draught. Energy Build. **12**(1), 21–39 (1988)
11. Conceição, E., Lúcio, M.: Numerical simulation of the application of solar radiant systems, internal airflow and occupants' presence in the improvement of comfort in winter conditions. Buildings **6**(3), 38 (2016)
12. Conceição, E., Lúcio, M.: Numerical and subjective responses of human thermal sensation. In: Proceedings of the 6th Portuguese Conference on Biomedical Engineering, Faro, Portugal, 11–12 June 2001
13. Conceição, E.: Evaluation of thermal comfort and local discomfort conditions using the numerical modelling of the human and clothing thermal system. In: Proceedings of 7th International Conference on Air Distribution in Rooms, ROOMVENT 2000, pp. 131–136. Elsevier, Amsterdam (2000)
14. Conceição, E., Lúcio, M., Farinho, P.: Experimental and numerical study of personalized of ventilation in classrooms desks. In: Proceedings of 10th International Conference on Air Distribution in Rooms, ROOMVENT 2007, vol. 3, pp. 519–526. Elsevier, Amsterdam (2007)
15. Conceição, E., Rosa, S., Custódio, A., Andrade, R., Meira, M., Lúcio, M.: Study of airflow around occupants seated in desks equipped with upper and lower air terminal devices for slightly warm environments. HVAC&R Res. **16**(4), 401–412 (2010)
16. Yang, L., Yan, H., Lam, J.: Thermal comfort and building energy consumption implications - A review. Appl. Energy **115**, 164–173 (2014)
17. Sailor, D.: A green roof model for building energy simulation programs. Energy Build. **40**(8), 1466–1478 (2008)
18. Balaji, N., Mani, M., Venkatarama, B.: Thermal performance of the building walls. In: Proceedings of Building Simulation Applications BSA 2013, pp. 151–159. BU Press, Bozen-Bolzano (2013)
19. Conceição, E., Silva, M., André, J., Viegas, D.: Thermal behaviour simulation of the passenger compartment of vehicles. Int. J. Veh. Des. **24**(4), 372–387 (2000)
20. Conceição, E., Lúcio, M.: Numerical simulation of passive and active solar strategies in building with complex topology. Build. Simul. **3**(3), 245–261 (2010)
21. Conceição, E., Lúcio, M., Lopes, M.: Application of an indoor greenhouse in the energy and thermal comfort performance in a kindergarten school building in the south of Portugal in winter conditions. WSEAS Trans. Environ. Dev. **4**, 644–654 (2008)

22. Conceição, E., Lúcio, M.: Numerical study of the thermal efficiency of a school building with complex topology for different orientations. Indoor Built Environ. **18**(1), 41–51 (2009)
23. Conceição, E., Lúcio, M.: Numerical study of the influence of opaque external trees with pyramidal shape on the thermal behaviour of a school building in summer conditions. Indoor Built Environ. **19**(6), 657–667 (2010)
24. Conceição, E., Gomes, J., Ruano, A.: Application of HVAC systems with control based on PMV index in university buildings with complex topology. IFAC PapersOnLine **51**(10), 20–25 (2018)
25. Conceição, E., Nunes, A., Gomes, J., Lúcio, M.: Application of a school building thermal response numerical model in the evolution of the adaptive thermal comfort level in the Mediterranean environment. Int. J. Vent. **9**(3), 287–304 (2010)
26. Conceição, E., Lúcio, M., Ruano, A., Crispim, E.: Development of a temperature control model used in HVAC systems in school spaces in Mediterranean climate. Build. Environ. **44**(5), 871–877 (2009)
27. Conceição, E., Lúcio, M.: Evaluation of thermal comfort conditions in a classroom equipped with radiant systems. Appl. Math. Model. **35**(3), 1292–1305 (2011)
28. Conceição, E., Silva, M., Viegas, D.: Airflow around a passenger seated in a bus. HVAC&R Res. **3**(4), 311–323 (1997)
29. Conceição, E., Lúcio, M.: Numerical study of thermal response of school buildings in summer conditions. In: Proceedings of HB 2006 - Healthy Buildings: Creating a Healthy Indoor Environment for People, vol. 3, pp 195–200. Curran Associates, New York (2006)
30. Conceição, E., Silva, A., Lúcio, M.: Numerical study of thermal response of school buildings in winter conditions. In: Proceedings of 9th International Conference on Air Distribution in Rooms, ROOMVENT 2004, p. 79. Universidade de Coimbra, Coimbra (2004)

Sustainability

Regional Bioeconomy: From Conceptual Frameworks to Management Approach

Remco Kranendonk(✉) and Raymond Schrijver

Wageningen Research, Wageningen, The Netherlands
remco.kranendonk@wur.nl

Abstract. The growth strategy for circular bioeconomy has a strong regional dimension, and is based on mission driven innovation policies. Regions are challenged to find a balance between economy and ecology, to find new solutions for specific regional problems and opportunities: bio-based value chains, regional carbon balance, new business opportunities for primary sector to avoid emissions, store carbon in bio-based products. Regions differ from each other and develop regional smart specialization strategies, value added economic activities based on unique natural and industrial characteristics and qualities, in balance with the planetary boundaries. The paper will focus in the management concepts and techniques which are necessary for setting conditions for innovations, collaborations and investments as well as for steering, managing and coordinating the deployment of the bioeconomy in European regions. This framework identifies overarching and interacting concepts that need to be addressed during the implementation of the bioeconomy and offer instruments to support the transition to well performing regional innovation systems. The paper will describe these concepts and will design a management tool for approach.

Keywords: Regional innovation strategy · Bioeconomy · Transition management · Regional governance

1 Managing Regional Bioeconomy Innovation – The Problem

European and National frameworks have set clear objectives for agriculture regarding climate change, for circular bioeconomy and for regional innovation. Large budgets are reserved for finding solutions for the climate problem at the different levels of governance. Nowadays also provinces and regions have set their missions and objectives. Following the European Partnership on 'Circular bio-based Europe' [1], the bio-based sector can contribute strongly to the EU's agenda for greener and more inclusive growth and jobs, particularly in coastal and rural areas, through greater participation of primary producers in local bioeconomy's and through the high and fast-growing number of start-ups in the sector. Over 18 million jobs already exist in today's bio-economy sector, with the potential of at least one million new green jobs by 2030. New future-oriented

J. R. da Costa Sanches Galvão et al. (Eds.): *Proceedings of the 1st International Conference on Water Energy Food and Sustainability (ICoWEFS 2021)*, pp. 479–488, 2021.
https://doi.org/10.1007/978-3-030-75315-3_52

jobs could be created for Europe's younger generations of employees, strengthening the European Union's international competitiveness. Currently, the European Green Deal [2] is the EU new growth strategy. ''It will help to cut emissions while creating jobs." The underlying Action Plan on Circularity emphasizes the importance of education and skill development [3].

The bio-based economy is the part of the broader bioeconomy focusing on the transition from the use of fossil fuel to renewable biomass in non-food applications. On their website, the European Union defines the bioeconomy as:

> *"The bio-economy encompasses the production of renewable biological resources and the conversion of these resources, residues, by-products and side streams into value added products, such as food, feed, bio-based products, services and bio-energy."* [4]

The European Green Deal is the EU new growth strategy. "It will help to cut emissions while creating jobs." The growth strategy has a regional dimension and is based on Mission driven innovation policies [5], which focus on problem-specific societal challenges, with many different sectors and domain interactions needed to solve the problems, challenging partners to develop smart and innovative solutions for regional problems as climate change. Regions are searching for a balance between economy and ecology by greening the economy and making it circular. The concept of RIS3 [6] emerged due to insights from mission-driven innovation moving towards RIS4 (smart sustainable specialization strategies). Many regions have mentioned the circular bioeconomy as one of the main drivers for their regional innovation and growth. The volume of the biomass, the maturity of the value chains and the applied technologies influence the future directions of the regional smart specialization strategies emphasized by the European Union. They vary region by region. Economy, production and processing and regional growth, should all take place within planetary boundaries. The focus is on greening the economy, developing bio-based value chains and improving the regional carbon balance, developing new business opportunities for primary sector to avoid emissions and storing carbon in bio-based products. To develop such specific regional solutions and opportunities an integrated approach is needed leading to changes at system level. Approaches focusing on just one of the aspects or sectors will not lead to transformative changes. This will be the research question of this paper: how to manage the transformation to a regional smart and sustainable circular bioeconomy?

In order to find an answer, the following activities will be described. First, relevant concepts and methodologies will be explored, from the bioeconomy perspective as well as from a perspective of regional innovation. The second part will be the design of a management tool based on the structure of the Business Canvas Model.

2 From Conceptual Framework to Framework of Concepts

In order to steer and manage the transition regional partners should be aware of the various underlying concepts relevant for the deployment of the regional bioeconomy. In this paper a distinction is made between the technological and physical concepts of biomass conversion and valorization, the bioeconomy concepts, the socio-economic concepts relevant for regional innovation and the framework of concepts of regional innovation. As regional partners need to deal with various different concepts, we decided to describe the different underlying concepts and operationalize them in order to make them useful for the different actors in regional innovation processes.

2.1 Bioeconomy Concepts

The crux for the bioeconomy is to contribute to environmental sustainability by making the best possible use of available biomass – in terms of societal value and efficiency. An overview of conversion and valorization routes [7] is presented in Fig. 1 which captures all conversion routes in one picture. It all starts with the sun providing the energy for biomass to grow. The biomass can then either be used directly as, for example, food and feed or be further processed and converted into other materials, products and (business) models. In bio-refineries the biomass is split mechanically and/or chemically into components that can be used in various applications, as biochemicals, materials or energy.

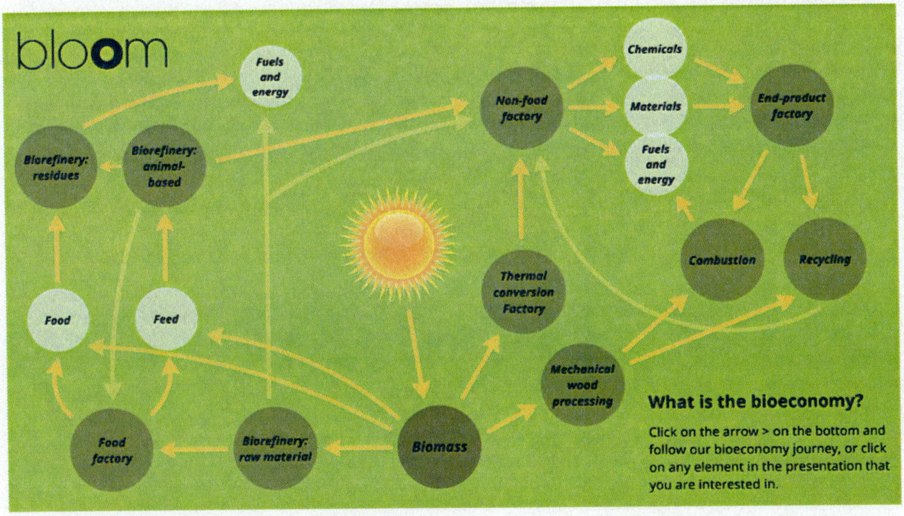

Fig. 1. Overview of biomass conversion and valorization routes within the bioeconomy. (www.bit.ly/bloom-bioeconomy *for a dynamic version)*

By doing so optimal use is made of the biomass. In a circular economy the biomass is not wasted, but reused and recycled by a cascading principle. Higher value use is always preferred over lower value use. Many European regions are in search for the most optimal valorization routes for their biomass and try to discover which kind of processing facilities investments should take place.

In the execution of new strategies and policies, gaps between current practices and bio-based solutions are being identified which hamper the uptake of the bioeconomy in Europe. The main gap is between the technological development and the awareness of the potential value thereof, from the value chain partners to the general public. The current situation is characterised by small-scale pilots and demonstrations. Transitions face lock-in problems such as resistance, current business practices, conflicting interests, (institutional) learning effects, economies of scale, economies of scope, network externalities, informational increasing returns, technological interrelatedness, collective action and the differentiation of power [8].

2.2 Regional Innovation Concepts

To bridge these gaps and to deal with such lock-in situations various overarching concepts have been identified, Fig. 2. These concepts offer instruments to support the transition to well performing regional innovation systems and need to be addressed during the implementation of the bioeconomy. Furthermore, these concepts offer information about patterns of and phases in the transitional process and offer organizational basics about conditions for innovation. They have to be made operational in order to bridge and manage the complex organizational, economic and societal challenges that come with this transition.

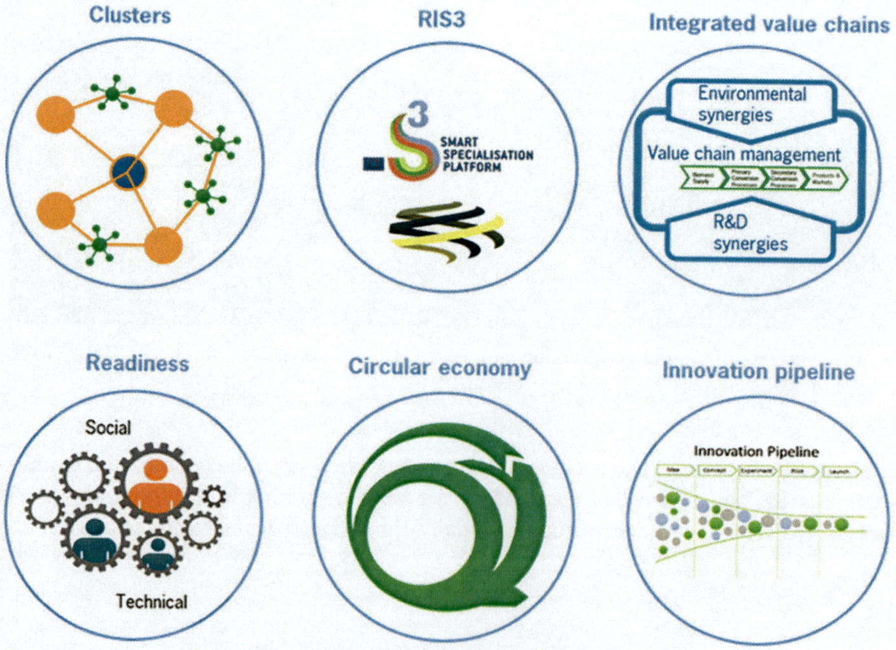

Fig. 2. Framework of regional innovation concepts

Sustainable Smart Specialization Strategies - from RIS3 to RIS4. From European regional policies and funds (EFRO), European regions are challenged to make strategic choices to invest in specific sectors or challenges that fit with the characteristics and qualities of their economies and environment – also compared to other regions. The concept of RIS3 emerged due to insights from mission-driven innovation moving towards RIS4. Many regions have mentioned the bioeconomy as one of the main drivers for regional innovation and growth.

Within the bioeconomy different routes are possible, depending on the combination of availability and components of the regional biomass and the existing industries for valorization. Quadruple helix partners are challenged to discover the unique regional profile and collaboratively come to smart strategic specializations as well as make these operational in developing pathways. Quadruple helix partners commit themselves to these strategies and align their own objectives and resources with these strategies. By doing so, they create broadly supported conditions for innovation and regional growth.

Clusters. Innovation ecosystems and innovation systems are well known concepts [9]. Innovation systems on the regional scale often called clusters [10]. "Cluster initiatives are organized efforts to increase the growth and competitiveness of the regional economy, involving cluster firms, government, the research community." Recently this triple helix partner network has been extended to a quadruple helix with the involvement of the civil society. Among clusters, there is a growing attention for sustainability and green growth [11].

Quadruple Helix Cooperation. The idea behind transitioning towards a regional bioeconomy is the necessity of interplay between the domains of the private sector, knowledge and research, government and the civil society. The perspectives and resources from all domains are relevant in order to go through the phases of innovation and to arrive to implementation, alignment and acceptance that is needed in order for an innovation to be successful. The triple helix model of innovation was firstly suggested by Etzkowitz and Leydesdorff [12] and emphasizes trilateral networks and hybrid organizations of university-industry-government relations to provide the necessary infrastructure for innovation and economic development. The quadruple helix adds as a fourth helix the general public and civil society emphasizing sociological concepts.

Innovation Pipeline. The innovation pipeline represents a logical order in which innovations usually evolve: from fundamental knowledge to applied knowledge, piloting and demonstrations as well as investments, business case development and upscaling. Transitions from one phase to the next do not always run smoothly. The innovation process can be supported by management, facilities and instruments.

Innovations start with (1) ideation. Promising ideas need to be (2) explored and (3) tested or demonstrated. The so-called valley of death often occurs when innovations have successfully reached the demonstration phase. Then, the most difficult part is for an innovation to reach maturity (4) by implementation and full investment in a new business case. It is important to recognize these phases, characteristics and potential hurdles in advance and be prepared in order to support the innovation process as optimal as possible.

Readiness Levels. Technological readiness levels (TRL's) were developed at NASA during the 1970's as a means for estimating the maturity of a technology. Nine TRL's are recognized and adapted by the European Commission in the Horizon 2020 program [13]:

TRL 1. Basic principles observed;
TRL 2. Technology concept formulated;
TRL 3. Experimental proof of concept;
TRL 4. Technology validated in lab;
TRL 5. Technology validated in relevant environment
TRL 6. Technology demonstrated in relevant environment
TRL 7. System prototype demonstration in operational environment;
TRL 8. System complete and qualified;
TRL 9. Actual system proven in operational environment

These nine levels primarily focus on the development of technological aspects – or artifacts – within one innovation and on the status regarding implementation. The uptake of new technology in society, however, also depends on social aspects and acceptance, i.e. social readiness. Often there are innovations hampering due to lack of social conditions or acceptance. Smooth uptake involves topics and issues such as awareness raising, knowledge transfer, coalition building, addressing and coping with resistance – by compensating for real losses instead of just enforcing new realities, trust building and institutionalizing new practices. Vested interest often impose lock-ins that can only be overcome with changes in the institutional settings [14]. Van der Steen et al. [15] conclude that the governance style to steer society best suited to underpin innovations and transitional change is actually a mixed set of styles, oriented from outside in and in a more interactive form of co-creating policies in which civil society is involved. Since there are public goods in place, principles of Ostrom [16] govern the commons. This involves a collective governance approach for appropriation, provision, strategic decision making, monitoring, enforcement and conflict resolution to create collaborative (nested) spaces and organizations.

Integrated Value Chains. All regions have specific biomass from their natural resources, from their specific agricultural production, from arable farming or from bio-waste streams. From this regional biomass, specific components such as sugars, fibres and proteins can be extracted or be processed by refineries in order to produce the building blocks for industries. Within regions the bio-chemical building blocks can be processed into regional specific applications that can subsequently be found in textiles, packaging, paper, bio-plastics, construction and different other sectors. Another option is to connect regions with biomass availability to regions with processing facilities, industries, knowledge centres, capacities and investments. In other words, the primary sector – covering biomass producers – can be linked with industries. The value chain can thus be formed across regions.

3 Design of Management Approach for Regional Bioeconomy Innovations

3.1 Need for Regional Transition Management

It is difficult to develop the regional bioeconomy as it is to be seen as a transformation at all levels: technological, as well as environmental, economic and social. In this paper the organizational challenges are described, focussing on making the underlying concepts explicit and operational in order to develop an overview of relevant conditions and processes in order to optimize the transition management.

There is a need for transition management, which is a responsibility of the partners from different domains. Often, an intermediate organization is being shaped, steered by the quadruple helix partners, as a regional service organization, like a cluster organization, to support the implementation of new policies and strategies of circular bio economy, carbon neutrality, regional innovations and specializations.

Transition management should focus on strategy development, collaboration among quadruple helix partners, innovation processes, strengthening the regional profiling of smart specialization, bringing together and attracting resources such as knowledge, human capital (capacities and capabilities) and financial means (subsidies, funding and investments). The services which are needed should be based on different aspects of change (economic, technological, environmental and social), on the phases of innovation (ideation, exploration, demonstration and implementation), and on the means for action (financial, knowledge, capacities and capabilities).

3.2 Cluster Management Canvas

In the private sector the Business Model Canvas (BMC) developed by Strategyzer,/Osterwalder [17] is broadly used. This canvas is based on an ontological model to express the business logic. The BMC is constituted by four pillars (product, costumer interface, infrastructure and finance) and nine building blocks of the canvas template inspired by management science and business- and process modelling. Both the organizational and process perspectives are strongly business level oriented. At the level of the transition area an analogue ontology can be identified aiming for classification of the processes as identified within the framework of concepts: the Cluster Management Canvas (CMC).

At the *organizational level* most important questions arise regarding the design of the cluster itself. Which organizations are required to be connected and involved, and what are their roles? In general a bio-based value chain is made up of biomass producers, extraction or processing intermediate business where building blocks for different industries are being produced, for new and sustainable applications, each adding more value to end user products, supported by knowledge centres, governments, private sector finance and consultancy, civil society organizations, each performing specific roles within the innovation ecosystem.

The Cluster Model Canvas, Fig. 3, is based on all relevant concepts and aspects of regional bioeconomy innovation, and can be used as a regional innovation management tool. By using this tool in a coherent and structured manner insight in the current situation and the next steps can be obtained, and regions can assure that all conditions can be set. The Model can be used by cluster managers in a collaborative decision making process among quadruple helix partners; exactly the way the BMC is often used in workshop settings. The BMC is a scheme with 9 building blocks for a development and implementation of a strategy (or a business investment). Just like the BMC, the CMC for regional bioeconomy innovation, also distinguishes 9 building blocks and processes behind. These are shortly described beneath and inserted in the CMC scorecard (Fig. 3):

1. Mission (explore possible bioeconomy pathways, find shared interest, gather the stakes of the partners, establish common objectives)
2. Existing conditions (map natural resources/biomass availability, economic structure, Technological readiness level, Social readiness level)
3. Program of requirements (define desired output and outcome, circularity, social inclusion)
4. Innovation pipeline; (innovation pipeline phases: ideation, exploration, demonstration and implementation, and the facilities needed)
5. Regional Specialization (determination of regional strengths, opportunities and comparative advantages; strategic choices for regional innovation and sustainable specialization
6. Learning cycle (knowledge infrastructure, valorization of knowledge, transdisciplinarity, co-creation, monitoring and evaluation)
7. Capacity and capabilities for bioeconomy transitions; mobilization of the quadruple helix partners/network management (assess capacity gaps, knowledge and skills and invest in human capital,)
8. Playing field – Regulations (define boundaries, access and operational rules for cooperation and competition, calls and funding principles, monitoring and accountability)
9. Finance (collect and attract financial resources, funding and investments)

The building blocks are integrated in a template, which can be used by the cluster to manage the regional bioeconomy transition. All building blocks are relevant and practitioners and partners can start at all different building blocks.

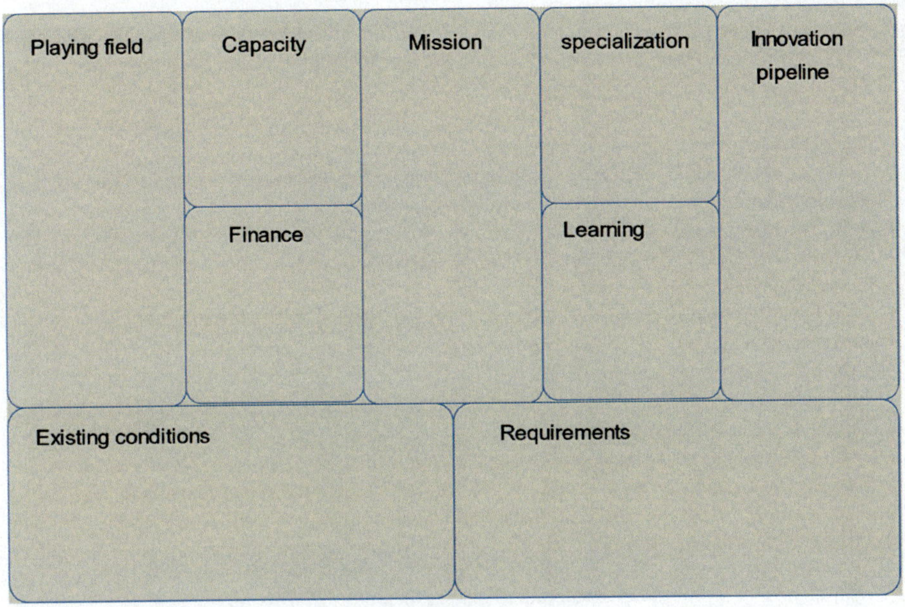

Fig. 3. Cluster Model Canvas Template

4 Conclusion

Many regions have ambitions and growth- and innovation strategies on bioeconomies. But many regions also struggle with the processes, due to the complexity of the different aspects and concepts behind. The Cluster Model Canvas is to be seen as a new methodology and management tool for taking into account all relevant concepts, aspects, conditions and relevant processes. The CMC methodology and the developed template can support regions and their partners in developing and implementation of a regional bioeconomy innovation strategy. The CMC as presented in this paper is a first exercise and elaboration, still at conceptual and strategic level, which needs improvement, testing and to be made operational. Practical and real life application will help to reveal the scope and coherence of the present classification. Is this indeed covering all regional processes within actor networks in a universal and meaningful manner? Meaningful in this sense would be helpful in addressing and resolving issues which hamper so many transition processes today.

References

1. https://ec.europa.eu/info/law/better-regulation/have-your-say/initiatives/11903-European-Partnership-for-a-Circular-bio-based-Europe
2. European Commission: Green Deal Announcement, Brussels (2019)
3. European Commission. For a cleaner and more competitive Europe. EU Green Deal, EU (2020)

4. European Commission: Communication from the Commission to the European Parliament, the Council, the European Economic and Social Committee of the Regions: Sustainable Growth: A Bioeconomy for Europe (2012)
5. Mazzucato, M.: Mission-oriented innovation policies: challenges and opportunities. Ind. Corp. Change **27**(5), 803–815 (2018)
6. Foray, D., et al.: Guide to research and innovation strategies for smart specialisation (RIS 3). European Commission, Regional Policy, Brussels (2012)
7. Bos, H., Kranendonk, R.: Bioeconomy mapping report: an overview of the bioeconomy. Deliverable D1.1 of the European Union H2020 Bloom project, Wageningen, Wageningen Research (2018)
8. Unruh, G.C.: Understanding carbon lock-in. Energy Policy **28**, 817–830 (2000)
9. Granstrand, O., Holgersson, M.: Innovation ecosystems: a conceptual review and a new definition. Technovation **90**, 102098 (2020)
10. Lindqvist, G., Ketels, C., Sölvell, O.: The Cluster Initiative Greenbook 2.0. Ivory Tower Publishers, Stockholm (2013)
11. Derlukiewicz, N., et al.: How do clusters foster sustainable development? An analysis of EU policies. Sustainability **12**(4), 1297 (2020). https://doi.org/10.3390/su12041297
12. Etzkowitz, H., Leydesdorff, L.: The dynamics of innovation: from National Systems and "Mode 2" to a Triple Helix of university–industry–government relations. Res. Policy **29**(2), 109–123 (2000). https://doi.org/10.1016/S0048-7333(99)00055-4
13. European Commission: Technology readiness levels (TRL), HORIZON 2020 – WORK PROGRAMME 2014–2015 General Annexes, Extract from Part 19 - Commission Decision C(2014)4995 (2014b)
14. Hekkert, M.P., Suurs, R.A.A., Negro, S.O., Kuhlmann, S., Smits, R.E.H.M.: Functions of innovation systems: A new approach for analysing technological change. Technol. Forecast. Soc. Change **74**, 413–432 (2007)
15. Van der Steen, M.A., Loorbach, D.: The Governance of Transitions and the Transitions of Governance. In: Van der Steen, M.A., Burger, J., Burger, W. (eds.) The Governance of Transitions - The Transitions of Governance: Cases from Southern Africa. African Sun Media under Sun Press imprint (2018). https://doi.org/10.18820/9781928480051
16. Ostrom, E.: Governing the Commons: The Evolution of Institutions for Collective Action. Cambridge University Press, New York (1990)
17. Osterwalder, A., Pigneur, Y., Tucci, C.L.: Clarifying Business Models: Origins, Present, and Future of the Concept. Commun. AIS **16** (2010). https://doi.org/10.17705/1CAIS.01601

3D Printing Filaments from Kenaf, Poplar and Agricultural Residues

M. Acevedo[1], L. Royano[1(✉)], A. I. Parralejo[1], J. Cabanillas[1], J. F. González[2], and J. González[1]

[1] Centre for Scientific and Technological Research of Extremadura (CICYTEX), Consejería of Economía, Ciencia and Agenda Digital, Junta de Extremadura, 06187 Guadajira, Badajoz, Spain
luis.royano@juntaex.es

[2] Escuela de Ingenierías Industriales, Universidad de Extremadura, Campus Universitario, Av. de Elvas, s/n, 06006 Badajoz, Spain

Abstract. Biocomposite materials manufacturing from vegetal raw substances or vegetal waste is an increasingly fact. Biocomposite materials were made from a mixture of polylactic acid (PLA) and vegetal material fibre, the latter is the reinforcement component. This research tries to get the mixture that provides the best mechanical properties to the biocomposite material. The different mixtures behaviour has been studied in a 3D printer. Currently, bast fibre is used as a reinforcing agent in polymers, substituting those with a bigger ecological footprint. Previous studies determined that kenaf fibre improve mechanical properties of matrix polymer. On the other hand, one way of bringing the circular economy closer to agricultural activity is by reusing its waste. Parts of the rice plant, thistle or paulownia bark are discarded because of their high mineral content. In this work, filaments have been manufactured for use in fused deposition modelling (FDM) from kenaf fibres, cork residues, rice, vine shoots, paulownia, poplar and thistle, and bars of PLA mixed with kenaf fibre through extrusion process.

Keywords: Circular economy · PLA · Natural fibre · Biomaterial · Biocomposite

1 Introduction

The pressing need to replace plastics with other materials less aggressive with the environment has led to the development of biopolymers such as PLA, made from biological sources and also easily compostable. PLA is one of most used 3D printing materials and FDM is a booming technique for the manufacture of objects and as a hobby for many people. Also, distributed manufacturing as well as distributed power generation is one key factor that can contribute to keep people in rural areas [1].

In agriculture and biomass treatment, products as kenaf or poplar and residues such as vine shoots or cork powder are generated. Some are difficult to value due to their high ash content such as rice husk, rice straw, thistle stalks or paulownia bark. The mineral

J. R. da Costa Sanches Galvão et al. (Eds.): *Proceedings of the 1st International Conference on Water Energy Food and Sustainability (ICoWEFS 2021)*, pp. 489–494, 2021.
https://doi.org/10.1007/978-3-030-75315-3_53

content of some of those materials can reach up to 20% [2]. Kenaf is a plant whose fibres have been used on many applications where resistance is an important characteristic such as ropes or bags. Currently, bast fibre is used as a reinforcing agent in polymers, substituting those with a bigger ecological footprint. Previous studies determined that kenaf fibre improve mechanical properties of matrix polymer.[3, 4] In addition, kenaf shows a high porosity so that it could be used as a sound absorber. On this matter is probable that kenaf and PLA mixtures have similar acoustic properties [5] (Fig. 1).

Fig. 1. Milled kenaf fibre (a), Vine shoot (b), Rice husk (c).

Likewise, vegetal residues are capable of being used also as fillers in polymers modifying their initial properties. In this manner, the incorporation of vegetable fibres or residues to a polymeric matrix in order to generate a composite with better mechanical properties is a good way of bringing the circular economy closer to agricultural activity by reusing its waste.

In this work, filaments have been manufactured for use in 3D printing from kenaf fibres, cork residues, rice, vine shoots, paulownia, poplar and thistle, and bars were made of PLA mixed with kenaf fibre using extrusion technique.

The objective of this research is to obtain a simple procedure to manufacture filaments for 3D printing with these residues, creating a catalogue of products with greater added value that expands those already on the market. It is a circular economy proposal based on certain waste that could otherwise end up being disposed of in the environment.

2 Materials and Methods

2.1 Raw Material

The blade mill used for kenaf fibre grinding was a Retsch SM 2000 (Retsch GmbH, Germany). A Noztek Touch extruder (Noztek, UK) was used to make the initial compound coupled with 5.0 mm nozzle. For solid bars, an isolated tube was used instead of a nozzle. Filaments were made in a 3DEVO Composer 350 filament maker (3DEVO Netherlands). Those filaments were tested by printing two pieces on a BQ Hephestos (BQ, Spain). The drying oven used for desiccating all the mixtures was Raypa model DO-20 (Raypa, Spain).

All pieces for testing were designed using the software FreeCAD 0.17 (www.freecadweb.org), exported in stl format and layered in CURA V. 4.1.0 (Ultimaker B.V.).

2.2 Experimental Procedures

Composite Manufacturing

The kenaf fibre is crushed and sieved with a 1.5 mm light sieve. Subsequently, it is re-screened with a 0.1 mm sieve. The sieved sample is dried together with PLA in an oven for 24 h at 80 °C [6]. Afterwards, the dry quantities necessary to make mixtures of 10% by natural fibre mass were weighed. The mixture is introduced into the NOZTEK extruder with the following parameters to reach a high degree of homogeneity: 60 RPM, 180 °C in the compression zone and 220 °C at the die. This achieves the maximum shear stress possible. The 5 mm diameter filament obtained is chopped in a knife mill to a particle size of about 5 mm.

Filament

The composite granulate is introduced into the filament machine with a temperature profile: Feeding 175 °C, melting 187 °C, transitioning 180 °C and metering 175 °C at 5 rpm. The cooler was set to 75%. Filaments of up to biomass 25% were also made in the same way (Fig. 2).

Fig. 2. From left to right kenaf, rice husk, thistle, poplar biomass, paulownia, paulownia bark, cork and rice straw 10% filaments.

Manufacture of Bars By Extrusion

A composite granulate with 40% kenaf was made following the previously described procedure. The NOZTEK TOUCH extruder was modified with a tubular mold as shown in Fig. 3. Several bars of about 18 cm length were made.

In this case the temperatures were for NOZTEK 200 °C for the compression zone and 250 °C in the die. The latter was necessary to be able the mold to heat up to a suitable temperature. The extruder was set to 60 RPM to reach the maximum degree of shear stress.

a b c

Fig. 3. Accessory for molding (a) Resulting pieces (b) detail of the bar (c)

3 Results and Discussion

3.1 Printer

Filament

The 3DEVO filament machine uses a mechanism (puller) that pulls the semi-molten material until the selected thickness is achieved. In some cases this reveals small bumps that are attributed to inhomogeneity. This effect is exacerbated in materials such as cork powder in the form of fine particles. The tendency to form aggregates on the part of all the materials tested may be due to the surface potential of these materials once they have dried. On the other hand, it has been observed that the feed requires a perfectly homogeneous granulometry because if this is not the case, fluctuations in the inlet mass flow result in fluctuations in the outlet velocity and therefore in the diameter. This is why the thickness of the filament shows small fluctuations, remaining between 1.69 mm and 1.80 mm. Also, in general, a darkening of the material is observed, probably due to the strong oxidation process at these temperatures.

Molding

The pieces obtained present a solid appearance, although it has not been possible to obtain bars of greater length. There are some zones with visible fibre, might be necessary additional attention to homogeneity.

Printer

In the first place, it is observed that the more mineralized residues such as rice husk present a better finish, a more homogeneous surface, probably due to a smaller particle size. It has been possible to print the selected pieces with filaments of all proportions. For mixtures of 10%, parts have been printed with 0.4 mm nozzles, although with imperfections. This is consistent with the general recommendation of the industry to use nozzles equal to or greater than 0.5 mm for this type of filament with vegetable fibres. In some cases, obstructions have occurred due to the protrusions that some filaments present. These do not always occur in the mouthpiece, but rather in the final zone of the barrel due to internal inhomogeneities. The most suitable printing parameters were: layer height: 0.3 mm, line thickness 0.4 mm, speed 55 mm/s Printing temperature 205 °C, hot bed at 55 °C.

3.2 Future Lines of Action

The main characteristic to be improved is homogeneity. It has been found that this is a crucial factor to take into account in order to get the best performance when printing. Also, under the same printing and extrusion parameters, the filaments showed a slightly different behaviour. It is necessary to study the most suitable extrusion and printing parameters for each type of filament. In addition, it is important to standardize the composite chopping process so that the grain size is more homogeneous and avoid variations in the thickness of the filament. It is also necessary to study possible additives improving mechanical properties such as polybutylene adipate terephthalate (PBAT).

It is necessary to carry out characterization studies of the materials: mechanical properties and porosity since they are parameters required by the food industry.

Molding. After verifying the possibility of manufacturing parts using this technique, a greater control of the process is necessary, specifically of the temperature of the mold and its cooling. In this way, larger objects can be obtained, as well as better mechanical properties.

4 Conclusions

The accomplishment of the proposed objective seems achievable, extrusion and printing PLA mixed with vegetal fibre, both by overcoming the problems encountered so far, and by the good finish achieved in some of the pieces, whose appearance largely resembles any object printed with a commercial filament.

Acknowledgements. The authors thank the financial support of the MESOCEX project (CICY-TEX - Junta de Extremadura – co-financed with FEDER funds) and IDERCEXA POCTEP Project (Operational Program for Cross-Border Cooperation Spain-Portugal).

References

1. DeVor, R.E., Kapoor, S.G., Cao, J., Ehmann, K.F.: Transforming the landscape of manufacturing: distributed manufacturing based on desktop manufacturing (DM)2. J. Manuf. Sci. Eng. **134**(4), 041004 (2012). https://doi.org/10.1115/1.4006095
2. N.P., Har, Irzaman, Irmansyah: Crystallinity and electrical properties of silicon dioxide (SiO2) from rice straw. In: AIP Conference Proceedings, vol. 2202, no. December (2019). https://doi.org/10.1063/1.5141641
3. Nishino, T., Hirao, K., Kotera, M., Nakamae, K., Inagaki, H.: Kenaf reinforced biodegradable composite. Compos. Sci. Technol. **63**(9), 1281–1286 (2003). https://doi.org/10.1016/S0266-3538(03)00099-X

4. Ku, H., Wang, H., Pattarachaiyakoop, N., Trada, M.: A review on the tensile properties of natural fiber reinforced polymer composites. Compos. Part B Eng. **42**(4), 856–873 (2011). https://doi.org/10.1016/j.compositesb.2011.01.010
5. Chin, D.D.V.S., Yahya, M.N.B., Din, N.B.C., Ong, P.: Acoustic properties of biodegradable composite micro-perforated panel (BC-MPP) made from kenaf fibre and polylactic acid (PLA). Appl. Acoust. **138**(December 2017), 179–187 (2018). https://doi.org/10.1016/j.apacoust.2018.04.009
6. Sawpan, M.A., Pickering, K.L., Fernyhough, A., Zealand, N., Engineering, B., Zealand, N.: Hemp fibre reinforced poly(lactic acid) composites moyeenuddin a. sawpan 1. Adv. Mater. Res. **29–30**, 1–4 (2007). https://doi.org/10.4028/www.scientific.net/AMR.29-30.337

Development of a Pine Tree Thermal Model Used in Forest Fire Environments

Eusébio Conceição[1(✉)], João Gomes[2], Mª Manuela Lúcio[1], Jorge Raposo[3], Domingos Xavier[3], and Mª Teresa Viegas[3]

[1] FCT - University of Algarve, Campus de Gambelas, 8005-139 Faro, Portugal
econcei@ualg.pt
[2] CINTAL, Campus de Gambelas, 8005-139 Faro, Portugal
jgomes@ualg.pt
[3] Universty of Coimbra, Pólo II, Pinhal de Marrocos, 3030-290 Coimbra, Portugal
{xavier.viegas,maria.viegas}@dem.uc.pt

Abstract. A numerical model of the pine tree thermal behavior, nearby a forest fire environment, is developed and applied. The pine tree thermal behavior numerical model is based on energy equations for the tree elements. The energy balance integral and differential equations consider the heat exchange by conduction inside the tree, the heat exchange by convection between the external tree surfaces and the environment and the heat exchanges by radiation between the external tree surfaces and the environment and fire front. In this work the geometry of a virtual pine tree trunk was developed using adaptive grid generation. This geometry is used to evaluate the view factors applied in calculating the heat exchanges by radiation between the virtual pine tree trunk and a fire front. The virtual pine tree trunk geometry is divided into concentric layers. The numerical simulation is characterized by the propagation of a fire front at a constant fire spread rate from a distance of 5 m upstream of the tree trunk to a distance of 25 m downstream of the tree trunk. The tree trunk has a height of 2 m and a diameter of 0.3 m. The fire front has a tilt angle of 45°, 10 m wide and 1.2 m high. The temperature distribution in the pine tree trunk was obtained for two different wind speeds, 0.1 m/s, and 1.0 m/s, and for two flame temperatures, 500 °C and 750 °C. The results obtained from the temperature distribution allowed to define the areas of the tree's tissues that reached the lethal threshold and thus evaluate its sustainability before a forest fire.

Keywords: Forest fire · Numerical model · Pine tree · Sustainability

1 Introduction

Forests cover about 39% of the area of mainland Portugal, being mainly composed of pine trees, eucalyptus, cork, holm oaks, oaks and chestnuts [1]. In particular, pine forests occupy an area of about one million hectares [1]. Official reports prepared in previous years usually refer to more than 20,000 occurrences of forest fires per year [2],

J. R. da Costa Sanches Galvão et al. (Eds.): *Proceedings of the 1st International Conference on Water Energy Food and Sustainability (ICoWEFS 2021)*, pp. 495–504, 2021.
https://doi.org/10.1007/978-3-030-75315-3_54

which makes Portugal one of the countries in all of Europe with the highest incidence rate of forest fires [3, 4]. The high number of forest fires with extensive burnt areas has been causing more and more negative impacts on "carbon storage, biodiversity conservation, hydrologic processes, and economic and social services" [5]. Thereafter, studies on the behavior of forest fires and their impacts on impairment and tree mortality are increasingly relevant. The mechanism of direct tree death from fire is the cambium necrosis via heat transfer to the crown, stem and root tissue [6]. The heat transfer occurs by convection, conduction, and radiation and all of these processes can cause either damage or death to the tree. The lethal threshold for trees is obtain for temperatures above or equal to 60 °C, although longer exposure at lower temperatures can also cause tissue death [7].

The numerical model of the pine tree trunk developed in this study is based on the numerical model that emulates the human body geometry. Like the numerical model of the human body, the numerical model of the pine tree trunk also uses energy balance equations. The numerical model of the human body was applied in the development of the model that simulates the human body thermal response. For example, the human body thermal response model was used in the works presented in articles by Conceição [8, 9], Conceição and Lúcio [10, 11], and Conceição et al. [12–14].

The numerical model uses the generalized mesh to establish the differential energy equations that describe the thermal behavior existing inside the pine tree trunk. In boundary conditions, the numerical model uses energy balance equations. The thermal flow processes defined by these equations are as follows: by conduction with the interior of the pine tree trunk; by convection (forced, natural or mixed) between the surface (bark) of the pine tree trunk and the surrounding environment; by radiation between the surface (bark) of the pine tree trunk and the surrounding environment; and by radiation between the surface (bark) of the pine tree trunk and the fire front. In the calculation of radiative exchanges between the pine tree trunk and the surrounding environment (including the fire front), a concept similar to heat exchanges between surfaces within building compartments is used. The modeling of this heat exchange process between surfaces was implemented in the numerical model of the thermal response of buildings with complex topology or in passenger car cabinets. Its application was effective in the studies developed by Conceição and Lúcio [15–18], and Conceição et al. [19–23].

In buildings, the temperature of the opaque surfaces is calculated layer by layer. In occupants, the temperature is calculated in different concentric layers. In the case of the pine tree trunk, an adapted 2D mesh is used. The temperature distribution inside the pine tree trunk is calculated using a finite differences implicit model.

The main objective of this numerical work, still in an initial phase, is to allow the definition of a methodology that allows assessing the sustainability of a pine forest in the presence of forest fires with different levels of intensity. In this context, this work begins by studying the temperature distribution in a pine tree trunk in the presence of a progressive fire front. Four scenarios were defined for two flame temperatures, 500 °C and 750 °C, and two wind speeds, 0.1 m/s and 1.0 m/s.

2 Numerical Model

The following thermal phenomena are considered for the pine tree trunk: heat conduction inside the pine tree, heat convection between the pine tree surface and the air environment and heat exchange by radiation between the pine tree trunk surface and the surrounding body surfaces, namely, the fire front, the fuel bed and the sky.

The hypothesis used to establish the energy balance equations are as follows:

- The heat flux is two-dimensional;
- The air temperature around the trunk, that is uniform and equal to the environment temperature;
- Use of heat transfer coefficients by convection developed for isothermal surfaces;
- The pine tree trunk is composed by bark and cambium;
- The fire effects around the trunk are not considered.

The type of grid used in the numerical simulation influences the results. In this study a numerical grid generation where the mesh is adapted to the body surface contours was developed using the finite difference method. The mesh used in the pine tree trunk has 30×20 points. In this adaptive grid generation, a physical space and a computational space were considered. The idea of this method consists of transforming the physical domain into the computational plan. This grid transformation is done by two elliptic partial differential equations, of Poisson's type. The adaptive grid generation used in this work can be seen in Fig. 1. The data input of this model are the wind speed, the fire front conditions (dimensions, tilt angle, flame temperature, fire rate spread), the tree dimensions, initial distance of the fire front from the tree and other initial conditions. This model was calibrated and validated in the work of Conceição et al. [24]. In this model, no moisture content was considered in the pine tree trunk. In this model, the following simplifications were also considered: constant dimensions, flame temperature and spread velocity of the fire front; homogeneous trunk constitution.

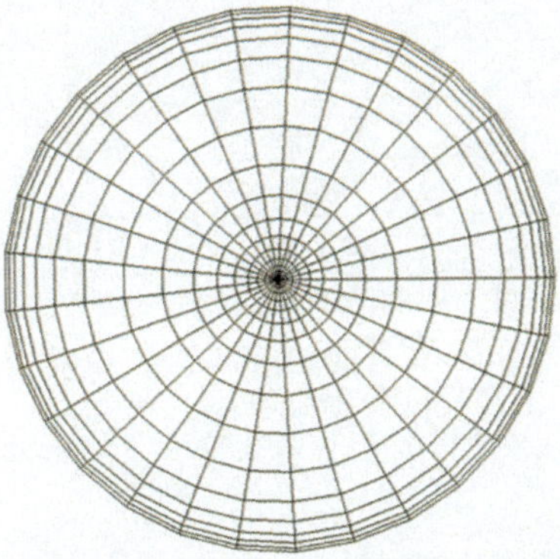

Fig. 1. Adaptive grid generation used in the pine tree trunk (30×20 grid points).

3 Methodology

The scheme of the forest fire scenery used in the numerical simulation is presented in Fig. 2. This scheme is constituted by an inclined fire front, a pine tree trunk and a fuel bed. Figure 2 also shows the symbology used in the representation of the wind speed (v_{air}), the fire spread rate (R), the dimensions of the fire front and the pine tree trunk. The fuel bed is considered to have finite dimensions $a \times b$.

The simulation analyzes the situation in which the fire front moves at a constant fire spread rate from a distance of 5 m upstream of the pine tree trunk to a distance of 25 m downstream of the tree trunk. The input data of the simulation are present in Table 1. The output data of the simulation are the temperature distribution obtained in a plane that cuts the pine tree at a height of 2 m at 30 points (P) equidistant distributed along the bark of the pine tree trunk, Fig. 3a, and at 20 points (Q) distributed along the radius of the pine tree trunk, Fig. 3b. In Fig. 3b, the line of points chosen is in a plane perpendicular to the direction of propagation of the fire front. The temperature distribution is obtained for the following Cases: a flame temperature of 500 °C and a wind speed of 0.1 m/s (Case A) or 1.0 m/s (Case B); a flame temperature of 750 °C and a wind speed of 0.1 m/s (Case C) or 1.0 m/s (Case D).

Table 1. Input data of the numerical simulation.

Case	A	B	C	D
Fire Front				
Flame temperature (T_f)	500 °C		750 °C	
Height (z_f)	1.2 m			
Width (b)	10.0 m			
Tilt angle (ϕ)	45°			
Fire spread rate (R)	0.01 m/s	0.1 m/s	0.01 m/s	0.1 m/s
Tree Trunk				
Height (z_t)	2.0 m			
Diameter (d)	0.3 m			
Others				
Wind speed (v_{air})	0.1 m/s	1.0 m/s	0.1 m/s	1.0 m/s
Environmental air temperature (T_{air})	20 °C			
Fuel bed length (a)	30 m			

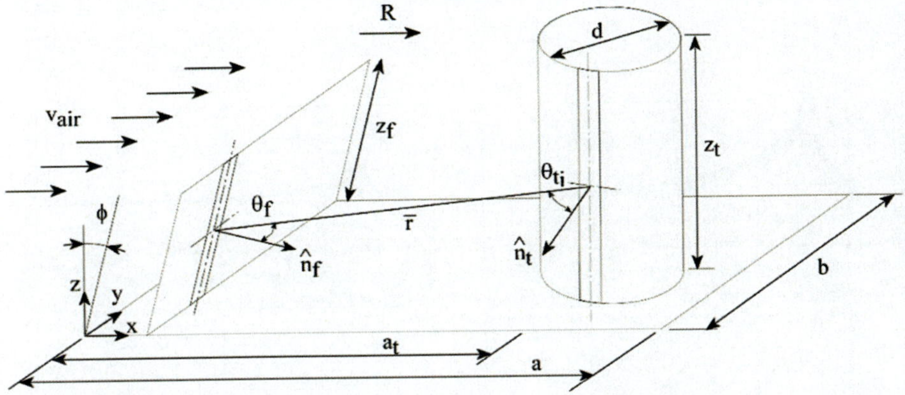

Fig. 2. Scheme of the forest fire scenery used in the numerical simulation: on the left is the fire front and on the right is the tree trunk. The dimensions of the trunk are represented by the diameter d and height z_t and those of the fire front by height z_f, width b and slope ϕ.

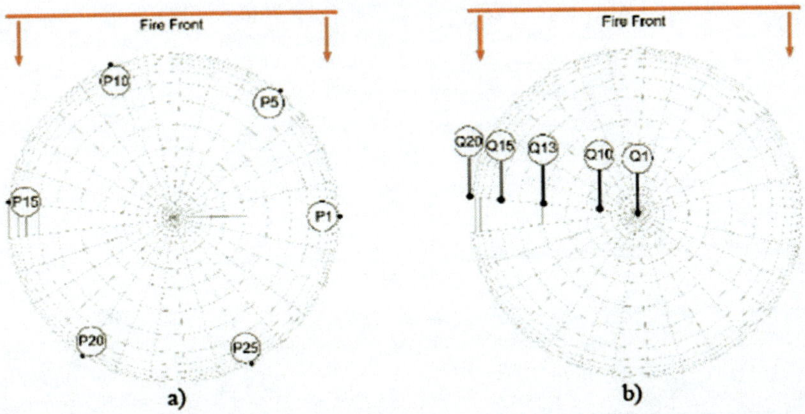

Fig. 3. Location of the points distributed along the pine tree trunk, where the temperature is numerically calculated, and the fire front: a) P-points equidistant distributed along the bark; b) Q-points distributed along the radius.

4 Results and Discussion

The evolution of the temperature values obtained at selected points of the pine tree trunk for the wind speeds of 0.1 m/s and 1.0 m/s points is presented for a flame temperature of 500 °C and 750 °C in, respectively, Fig. 4 and Fig. 5. In these Figures, the P1-P15 points are located on the bark of the pine tree trunk facing upstream of the fire front, the points P16-P30 are located on the bark of the pine tree trunk facing downstream of the fire front, and the points Q1-Q20 are located on the line of the tree trunk radius. In Fig. 4 and Fig. 5, the dashed line represents the pine tree trunk lethal threshold ($T_{trunk} \geq 60$ °C) [7].

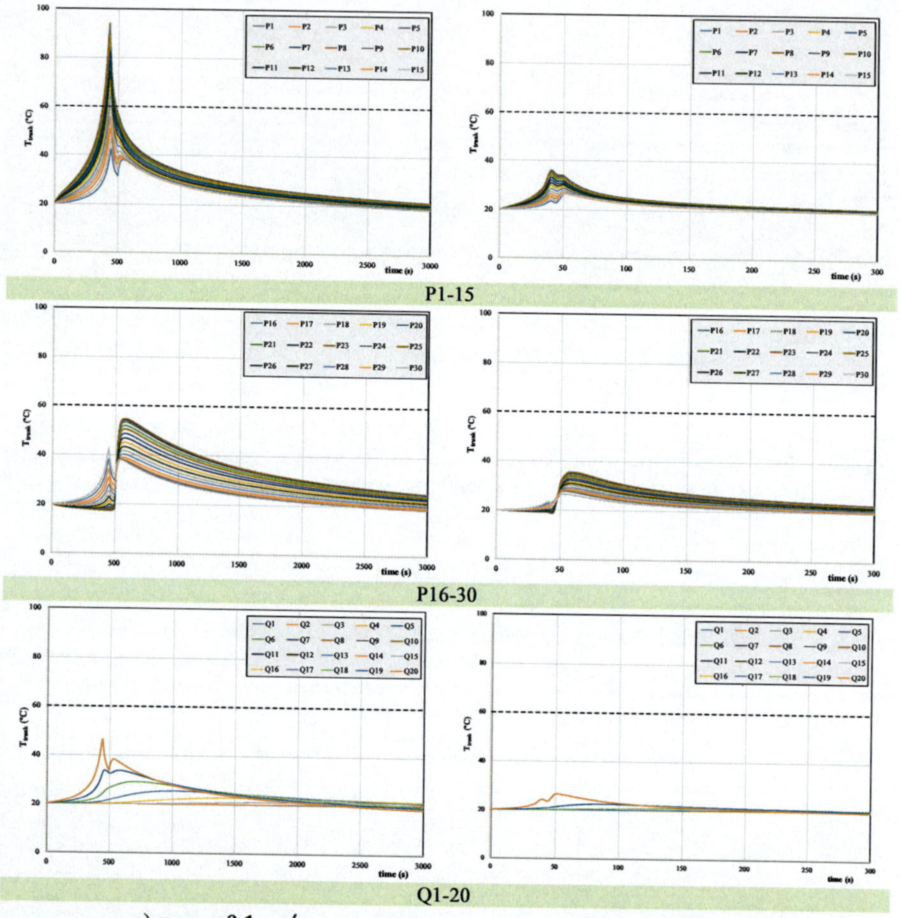

a) $v_{air} = 0.1\ m/s$ b) $v_{air} = 1.0\ m/s$

Fig. 4. Evolution of the temperature values obtained at selected points of the pine tree trunk for a flame temperature of 500 °C when the wind speed is: a) 0.1 m/s; b) 1.0 m/s. The dashed line represents the pine tree trunk lethal threshold ($T_{trunk} \geq 60$ °C).

Regarding to the results, the temperature distribution presents higher values when the flame temperature is 750 °C than when it is 500 °C and, for these flame temperatures the temperature distribution presents higher values when the wind speed is 0.1 m/s than when it is 1.0 m/s. For both flame temperatures, the temperature reaches higher values in the pine tree trunk bark upstream of the fire front than downstream of the fire front. Inside the pine tree trunk, the temperature reaches higher values at its outermost points. The highest temperature is reached at a point in the bark of the pine tree trunk facing upstream of the fire front, this value being about 130 °C for the flame temperature of 500 °C and about 245 °C for the flame temperature of 750 °C. Inside the pine tree trunk, the highest temperature is reached at the point located on the outermost ring, this value being about 46 °C for the flame temperature of 500 °C and about 106 °C for the flame temperature of 750 °C.

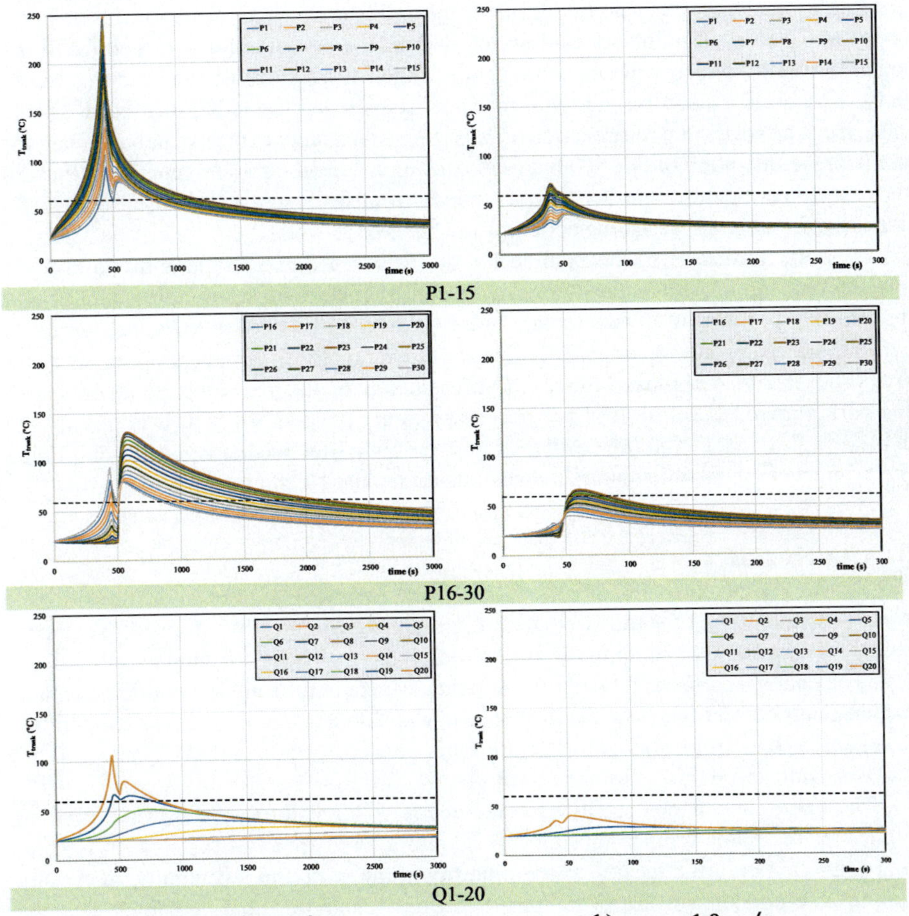

a) $v_{air} = 0.1\ m/s$ b) $v_{air} = 1.0\ m/s$

Fig. 5. Evolution of the temperature values obtained at selected points of the pine tree trunk for a flame temperature of 750 °C when the wind speed is: a) 0.1 m/s; b) 1.0 m/s. The dashed line represents the pine tree trunk lethal threshold ($T_{trunk} \geq 60$ °C).

For the flame temperature of 500 °C, the temperatures in the tree remain below the lethal threshold for a wind speed of 1 m/s, which means that, in this case, the pine tree trunk withstands the passage of the fire front. For a speed of 0.1 m/s, only 11 points of the bark of the pine tree trunk turned upstream of the fire front reach the lethal threshold, which will cause the death of their tissues; in the other points of the bark of the pine tree trunk turned downstream of the fire front and inside the pine tree trunk, the temperature is below the lethal threshold. In this case, it can be stated that the pine tree trunk suffers only minor damage to its bark and that it will also resist fire well.

For the flame temperature of 750 °C, the temperatures in the tree remain below the lethal threshold in the interior of the pine tree trunk for a wind speed of 1 m/s; only for a few points located on the bark of the pine tree trunk reach the lethal threshold. In general, in this case, the pine tree trunk withstands the passage of the fire front. For a speed of 0.1 m/s, all points located on the bark of the pine tree trunk turned reach the lethal threshold, which will cause the death of their tissues; inside the pine tree trunk, only two points located on the two outermost rings reach the lethal threshold. In this case, it can be stated that the pine tree trunk suffers some more extensive damage than for the flame temperature of 500 °C, but only causing the death of more superficial located tissues. However, these injuries are not enough to cause the tree to collapse, so it can be considered that it will resist the action of the fire front.

In short, according to the results, for the assessed flame temperatures and wind speeds considered, the existence of a good cleaning of the land surrounding the tree and an effective pruning up to two meters in height of the tree trunk allows the tree resist a forest fire with only damage to its bark and eventually in some points of its outer ring. Thus, for medium-sized forest fires presented here, the tree can survive, only with the bark burned, therefore it is not necessary to cut it down. In fact, it is possible to achieve a sustainable and fire-resilient forest through effective management that limits the available combustible matter and ensures its regular cleaning.

5 Conclusion

A numerical model of the pine tree trunk thermal behavior, nearby a forest fire environment, was developed and applied. The pine tree thermal behavior numerical model is based on energy equations for the tree elements, which, in turn, are substantiated on heat exchange processes. To assess the resilience of a pine tree to a forest fire, the temperature distribution in its trunk was evaluated for two flame temperatures (500 °C and 750 °C) and two wind speeds (0.1 m/s and 1.0 m/s).

The main conclusions are that the temperature distribution values in the pine are higher for the flame temperature of 750 °C and a wind speed of 0.1 m/s. In general, it appears that only the tissues of the pine tree trunk bark and, eventually, of the first ring located near the surface of the pine tree trunk reach the lethal threshold. The pine tree trunk withstands a flame temperature of 500 °C, for both wind speeds, and a flame temperature of 750 °C, for a wind speed of 1.0 m/s.

Therefore, the results make it evident that a good cleaning of the forest land and that an effective pruning of the trees make them more resilient to low and medium intensity fires.

As future work, it is proposed to study the influence of environmental variables on the thermal behavior of the tree in the presence of a forest fire, to introduce in the thermal behavior tree model the equations that translate the effect of the tree transpiration and, finally, to establish a forest fire prevention index.

Acknowledgement. The authors would like to acknowledge the support of the project reference PCIF/MPG/0108/2017, funded by the Portuguese Foundation of Science and Technology (FCT).

References

1. ICNF – Instituto de Conservação da Natureza e das Florestas. 6º Inventário Florestal Nacional (IFN6). https://www2.icnf.pt/portal/florestas/ifn/ifn6. Accessed 20 Dec 2020
2. Radovanovic, M., Vyklyuk, Y., Stevancevic, M., Milenkovic, M., Jakovljevic, D., Petrovic, M., Milicevic, S., Vukovic, N., Vujko, A., Yamashkin, A., Sydor, P., Vukovic, D., Skoda, M.: Forest fires in Portugal – case study. Thermal Sci. **23**(1), 7–86 (2019)
3. Carvalho, A., Flannigan, M., Logan, K., Johnston, L., Miranda, A., Borrego, C.: The impact of spatial resolution on area burned and fire occurrence projections in Portugal under climate change. Clim. Change **98**, 177–197 (2010). https://doi.org/10.1007/s10584-009-9667-2
4. Botequim, B., Arias-Rodil, M., Garcia-Gonzalo, J., Silva, A., Marques, S., Borges, J., Oliveira, M., Tomé, M.: Modeling post-fire mortality in pure and mixed forest stands in Portugal – a forest planning-oriented model. Sustainability **9**, 390 (2017)
5. Bowman, D., et al.: Fire in the Earth system. Science **324**, 481–484 (2009)
6. Sharon, M., Varner, J., van Mantgem, P., Cansler, C.: Fire and tree death: understanding and improving modeling of fire-induced tree mortality. Environ. Res. Lett. **13**, 113004 (2018)
7. Kelsey, R., Westlind, D.: Physiological stress and ethanol accumulation in tree stems and woody tissues at sublethal temperatures from fire. Bioscience **67**, 443–451 (2017)
8. Conceição, E.: Avaliação de condições de conforto térmico: simulação numérica do sistema térmico do corpo humano e do vestuário. In: V Iberia and Inter-American Air Conditioning and Refrigeration Congress, CIAR 1999, Lisbon, Portugal, 14–16 October 1999. (in Portuguese)
9. Conceição, E.: Evaluation of thermal comfort and local discomfort conditions using the numerical modelling of the human and clothing thermal system. In: Proceedings of the 7th International Conference on Air Distribution in Rooms, RoomVent 2000, Reading, UK, 9–12 July 2000
10. Conceição, E., Lúcio, M.: Numerical and subjective responses of human thermal sensation. In: Proceedings of the 6th Portuguese Conference on Biomedical Engineering, Faro, Portugal, 11–12 June 2001
11. Conceição, E., Lúcio, M.: Numerical simulation of the application of solar radiant systems, internal airflow and occupants' presence in the improvement of comfort in winter conditions. Buildings **6**(3), 38 (2016)
12. Conceição, E., Lúcio, M., Farinho, J.: Experimental and numerical study of personalized of ventilation in classrooms desks. In: Proceedings of the 10th International Conference on Air Distribution in Rooms, Helsinki, Finland, 13–15 June 2007
13. Conceição, E., Rosa, S., Custódio, A., Andrade, R., Meira, M., Lúcio, M.: Study of airflow around occupants seated in desks equipped with upper and lower air terminal devices for slightly warm environments. HVAC&R Res. **16**(4), 401–412 (2010)
14. Conceição, E., Lúcio, M., Awbi, H.: Comfort and airflow evaluation in spaces equipped with mixing ventilation and cold radiant floor. Build. Simul. **6**, 51–67 (2013)
15. Conceição, E., Lúcio, M.: Numerical study of thermal response of school buildings in summer conditions. In: Proceedings of the 8th International Conference and Exhibition on Healthy Buildings (HB 2006), Lisbon, Portugal, June 4–8 2006
16. Conceição, E., Lúcio, M.: Numerical study of the thermal efficiency of a school building with complex topology for different orientations. Indoor Built Environ. **18**, 41–51 (2009)
17. Conceição, E., Lúcio, M.: Numerical simulation of passive and active solar strategies in building with complex topology. Build. Simul. **3**, 245–261 (2010)
18. Conceição, E., Lúcio, M.: Numerical study of the influence of opaque external trees with pyramidal shape in the thermal behaviour of a school building in summer conditions. Indoor Built Environ. **19**, 657–667 (2010)

19. Conceição, E., Silva, M., André, J., Viegas, D.: Thermal behaviour simulation of the passenger compartment of vehicles. Int. J. Veh. Des. **24**(4), 372–387 (2000)
20. Conceição, E., Lúcio, M., Lopes, M.: Application of an indoor greenhouse in the energy and thermal comfort performance in a kindergarten school building in the south of Portugal in winter conditions. WSEAS Trans. Environ. Dev. **4**, 644–654 (2008)
21. Conceição, E., Lúcio, M., Ruano, A., Crispim, E.: Development of a temperature control model used in HVAC systems in school spaces in mediterranean climate. Build. Environ. **44**, 871–877 (2009)
22. Conceição, E., Nunes, A., Gomes, J., Lúcio, M.: Application of a school building thermal response numerical model in the evolution of the adaptive thermal comfort level in the mediterranean environment. Int. J. Vent. **9**, 287–304 (2010)
23. Conceição, E., Gomes, J., Ruano, A.: Application of HVAC systems with control based on PMV index in university buildings with complex topology. IFAC PapersOnLine **51**(10), 20–25 (2018)
24. Conceição, E., Viegas, D., André, J.: Thermal behaviour simulation of a tree trunk nearby a fire front. In: III International Conference on Forest Fire Research, 14th Conference on Fire and Forest Meteorology, vol. 1, pp. 831–842 (1998)

Comparative Analysis of Soil Slope Stability, Using Dynamic and Pseudo-static Methods on the Garrapata - Santa Maria Road, Manabi Province, Ecuador

Lucía Macías[1], Digna Loor[1], Eduardo Ortiz-Hernández[1,2], Gema Casanova[1], and Daniel Delgado[1,3](✉)

[1] Departamento de Construcciones Civiles, Facultad de Ciencias Matemáticas, Físicas y Químicas, Universidad Técnica de Manabí, Portoviejo 130108, Ecuador
daniel.delgado@utm.edu.ec

[2] Department of Civil Engineering, University of Alicante, 03690 Alicante, Spain

[3] Red de Desarrollo Urbano Sostenible de Manabí, Portoviejo, Ecuador

Abstract. Ecuador is exposed to seismic events, which cause a reduction in soil resistance due to vibration loads and their effect is the landslide. This work analyzes the stability of the soil slopes that are located on the abscissa 25 + 280 and 63 + 040 of the Garrapata - Santa María road located between the Chone and El Carmen cantons of the Manabí province, through the determination of the safety factor using the GeoSlope software, performing a dynamic and pseudo-static analysis, to observe the variations that exist when applying horizontal forces (Kh). In addition, the comparison of the Fellenius, Bishop, Janbu and Morgenstern-Price limit equilibrium methods was carried out based on the SPT results, which allowed classifying the soil strata. The results of the triaxial test determined the geotechnical parameters such as cohesion and friction angle.

Keywords: Stability · Safety factor · Seismicity · Soil

1 Introduction

In Ecuador, studies related to the determination of the risk of instability are frequently carried out to avoid large mass movements such as those caused by natural phenomena and catastrophic events in the twentieth century, because it caused a high destructive potential in the socioeconomic and environmental [1].

Slope stability analysis should focus on understanding the process that originates, develops, and triggers instability. In this analysis, three conceptual models can be applied, such as the Limit Equilibrium Method, where different techniques have been developed, and they differ between them basically by the number and type of equilibrium equations they use and, in the assumptions, included in interslice forces. A second group is the Limit Analysis Methods where the Lower-Limit and Upper-Limit theorems are employed. The

J. R. da Costa Sanches Galvão et al. (Eds.): *Proceedings of the 1st International Conference on Water Energy Food and Sustainability (ICoWEFS 2021)*, pp. 505–515, 2021.
https://doi.org/10.1007/978-3-030-75315-3_55

third group and of growing interest are the Numerical Methods, among which the most widespread are the Finite Element Method and the Finite Difference Method. These numerical methods meet the theoretical requirements to obtain the complete solution; that is, that obtained based on the system of equilibrium equation, compatibility, constitutive laws, and boundary conditions [2].

During a seismic event, the stability of a slope can decrease in different ways [3]. This research has the purpose of evaluating the instability of the specific slopes on the Garrapata - Santa María road, with the acceleration records projected in the study area, of the 2016 Pedernales earthquake (Mw 7.8), which indicate values between 0.26 and 0.45 g [4].

When studying the phenomena of slope instability, it is necessary to identify and characterize the different types of behavior and classify them appropriately. A precise analysis of the type of failure allows optimization of slope containment and stabilization measures, since it considers the acting mechanism, its speed and dimensions. On the contrary, an incorrect definition of the failure mechanism can lead to incorrect or ineffective solutions, and in some cases counterproductive [5].

During the rainy season, landslides and collapses of natural or artificial slopes frequently occur. There are two main causes that cause them: the increase in shear stresses on the slopes and the reduction of shear resistance of the constituent materials (soils and rocks). The agent that contributes the most to generating and developing slope failures is water (90% of associated failures).

In general, the slope stability is determined by geometric factors (height and slope), geological, hydrogeological and geotechnical factors. In this way, the combination of these 4 factors can determine the condition of the soil breaking on some type of surface, allowing the movement of its contained mass [6].

This research proposes the analysis of slopes stability in the abscissa 25 + 280 and 63 + 040 of the Garrapata - Santa María road (Chone canton of the Manabí province), based on geomechanical and geotechnical parameters, to mitigate the effects risk of instability on the road.

2 Methodology

Once the results of the SPT site test were obtained, adding the Atterberg Limits tests, natural humidity, granulometry and the geotechnical characterization in conjunction with the Triaxial UU tests (resistance parameters), the friction angle and cohesion of the materials that make up the slopes.

The study points are in the Bourbon geological formation (Fig. 1), which is made up of medium to coarse-grained bluish-gray sandstones in compacted banks with abundant mega fossils in irregular water reserves, intercalations of hardened lama and tuff gray volcanic, conglomerate lenses and generally a basal conglomerate resting discordantly on the Onzole and Playa Grande formations. In the Chone area, the contact is concordant and transitional. According to Bristow [7] and Reyes & Michaud [8], this formation is considered from the Upper Miocene to Pliocene age.

The safety factors of the soil slopes were calculated using the limit equilibrium methods, which consist of dividing the mass of potentially unstable ground into vertical slices,

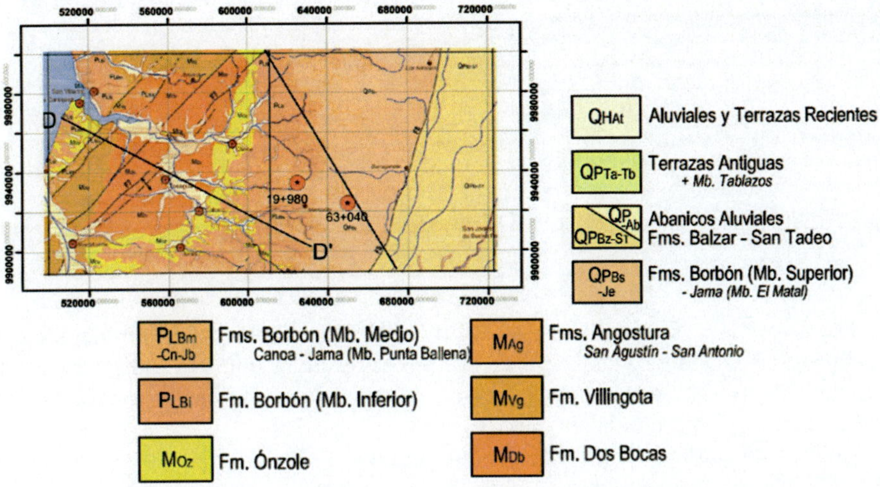

Fig. 1. Geological chart of the central coast of the province of Manabí. The road project refers to the PLB (Borbón) and QPBZ (alluvial fans of Balzar and San Tadeo) formations. Units in yellow is Onzole formation, and orange unit is Fm Dos Bocas [8]

calculating the equilibrium of each one of them and analyzing the global equilibrium, to obtain a factor defined as the relationship between resistant forces/moments and unbalanced forces/moments. Once the value of FS has been calculated for a certain potential failure curve, the process is repeated for a different one, a procedure that is repeated until a minimum value of FS is obtained [5]. The safety factor was calculated using GeoSlope's SLOPE/W software, making a comparison in the static and pseudo-static analyzes.

The limit equilibrium methods applied to determine the stability of the slopes and at the same time evaluate differences between them are detailed below:

- Ordinary or Fellenius method.
- Bishop's method.

Ordinary or Fellenius Method: The precursor methods of limit equilibrium consider the potential failure mass as a free body and apply the equations of global equilibrium. This free body theory has been implemented in some methods, such as the Ordinary Fellenius Method (Fig. 2) [2].

The forces acting on the slope (Fig. 2) are:

Wn: Resulting segment weight.
R: Force acting as a reaction to the weight of the segment.
Nn and N (n + 1): Normal forces acting on each side of the segment.
Tn and T (n + 1): Tangential forces acting on each side of the segment.
Nr: Normal component of reaction R.
Tr: Tangential component of reaction R.

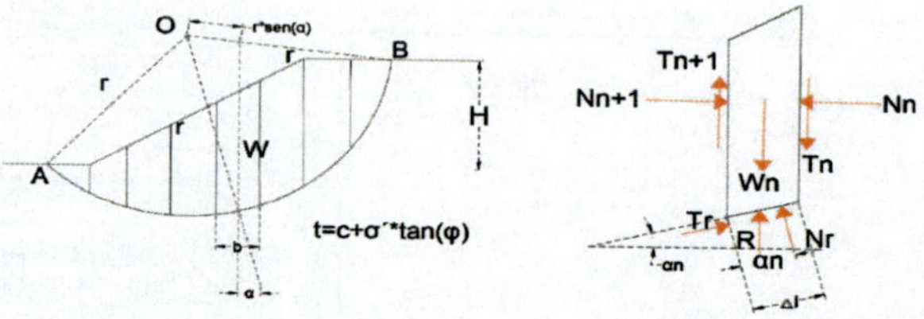

Fig. 2. (Left) Slope stability analysis and (right) forces acting on the segment using the Fellenius method [6]

The forces Nn, N (n + 1), Tn and T (n + 1) are complex to determine, so it is possible to take certain considerations: The forces resulting from Nn and Tn are equal in magnitude to those resulting from N (n + 1) and T (n + 1). Finally, the formula to determine the safety factor is obtained [6].

Taking into consideration the limit equilibrium:

$$N_r = W_n * \cos(\propto_n) \quad (1)$$

However, the shear resistance offered by the floor is considered:

$$T_r = \tau * (\Delta l) = \frac{\tau * (\Delta l)}{FS} = \frac{1}{FS} * \left[c + \sigma' * tg(\varnothing)\right] * \Delta l \quad (2)$$

In the above equation the normal stress is equal:

$$\frac{N_r}{\Delta l} = \frac{W_n * sin(\alpha_n)}{\Delta l} \quad (3)$$

One of the main characteristics of the Fellenius method is to satisfy the balance of moments, considering that the moment of the mobilized forces is equal to the moment of resistant forces [6].

$$\sum\nolimits_{n=1}^{n=p} W_n * sin(\alpha_n) = \sum\nolimits_{n=1}^{n=p} \frac{1}{FS} * \left[c + \frac{W_n * cos(\alpha_n)}{\Delta l} * tg(\varnothing)\right] * \Delta l * r \quad (4)$$

So:

$$FS = \frac{\sum\nolimits_{n=1}^{n=p}[c * \Delta l + W_n * cos(\alpha_n) * tg)(\varnothing)]}{\sum\nolimits_{n=1}^{n=p} W_n * sin(\alpha_n)} \quad (5)$$

FS = Safety factor.
c = Cohesion.
Δl = Length of the segment.
W = Total weight of each segment.

Bishop's Method: Consider the following hypotheses and development [9]:

- The failure surface is assumed to be circular.
- The sliding mass is divided into "n" vertical strips.

- The equilibrium of moments in each slice in relation to the center of the circle is established.
- This method assumes that the contact forces between the slices do not influence, because they are balanced.

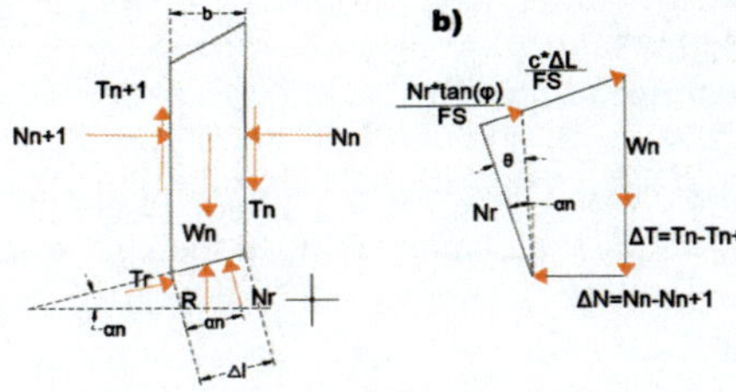

Fig. 3. Slope stability analysis by the Bishop method. a) Diagram of the forces acting on the segment. b) Diagram of forces in equilibrium of the "N" segment [6]

From Fig. 3, the following equations are determined:

$$\boldsymbol{T_r = N_r * tg\,(\varnothing) + c + \Delta l = N_r * \left[\frac{tg\varnothing}{FS}\right] + \frac{c * \Delta l}{FS}} \tag{6}$$

$$\boldsymbol{N_r = \frac{W_n + \Delta T - \left[\frac{c * \Delta l}{FS}\right] * sin\,(\alpha_n)}{cos(\alpha_n) + \frac{tg(\varnothing) * sin(\alpha_n)}{FS}}} \tag{7}$$

It is possible to calculate the equilibrium equations of moments and forces, starting from the free-body diagrams for each segment, finding the expression to calculate the FS [6]:

$$\sum\nolimits_{n=1}^{n=p} W_n * r * sin(\alpha_n) = \sum\nolimits_{n=1}^{n=p} T_r * r \tag{8}$$

Where:

$$T_r = \frac{1}{FS} * \left[c + \sigma' * tg(\varnothing)\right] * \Delta l = \frac{1}{FS} * \left[c * \Delta l + N_r * tg(\varnothing)\right] \tag{9}$$

The value of forces N can be estimated, once the balance of vertical forces for each segment is analyzed [6], allowing to replace and find the following equation:

$$FS = \frac{\sum_{n=1}^{n=p}\left[c * b + W_n * tan(\varnothing) + \Delta T * tg(\varnothing)\right] * \frac{1}{m_{\alpha(n)}}}{\sum_{n=1}^{n=p}\left[W_n * sin(\alpha_n)\right]} \tag{10}$$

Where:

$$m_{\alpha(n)} = cos(\alpha_n) + \frac{tg(\varnothing) * sin(\alpha_n)}{FS} \tag{11}$$

Simplifying, we have:

$$FS = \frac{\sum_{n=1}^{n=p}\left[c * b + Wn * tg(\varnothing)\right] * \frac{1}{m_{\alpha(n)}}}{\sum_{n=1}^{n=p}[W_n * sin(\alpha_n)]} \tag{12}$$

FS = Safety factor.
c = Cohesion.
b = Width of the segment.
Wn = Total weight of each segment.

Pseudo-static Slope Analysis: The seismic stability of structures such as slopes have been analyzed through pseudo-static analysis since the 1920s, in which the effects of an earthquake are represented by horizontal accelerations [10].

The pseudo-static analysis consists of applying a horizontal force on all the elements subject to analysis on the slope, called the k coefficient, which is multiplied by the weight of each element. The point where the force must be located is of great importance in the analysis. For this, Terzaghi [11] recommended that this force should be applied at the center of gravity of each segment [12].

This method is generally based on the same procedure that any limit equilibrium method uses, the only variation is that both horizontal and vertical pseudo-static forces are added because of the seismic event. Said seismic forces must be proportional to the weight of the potential sliding mass and in addition to the seismic coefficients Kh and Kv, they are valued at n times the acceleration of gravity caused by the earthquake. It can be indicated that generally in the analyzes the horizontal seismic force is used and it is assumed that the vertical seismic force is zero. However, in the epicentral area that produces landslides, the Kv would not be significant [13].

The seismic coefficient varies depending on the acceleration or intensity of the earthquake, frequency, and duration of the movement. Usually in a conservative analysis, it is assumed that the horizontal seismic coefficient is equal to the maximum acceleration of a seismic movement [10].

The Chilean Highway Manual [14] establishes that the Kh is determined by the following equation:

$$Kh = 0.5A' \tag{13}$$

"A" represents the maximum effective acceleration coefficient, which depends on the seismic zone of the country. This method is applied when there is no seismic hazard study available to estimate the maximum acceleration value [10].

The slopes analyzed are located between the cantons Chone and El Carmen, coastal zone (red area). Its z factor corresponds to 0.50 g in rock, with a very high seismic hazard characterization (Table 1).

Table 1. Z factor values depending on the seismic zone adopted.

Seismic Zone	I	II	III	IV	V	VI
Z factor	0,15	0,25	0,3	0,35	0,4	≥0,50
Seismic hazard characterization	Intermediate	High	High	High	High	Very high

3 Results

At station 25 + 280, the results of the standard penetration test (Table 2) are observed, determining a homogeneous stratum formed by silty sand. It is established that the compactness in the first 2 m is relatively low and in depths between 2 to 5 m it improves its compactness.

Table 2. Mechanical properties in abscissa 25 + 280.

Depth (m)	Granulometry% through				Limits			W%	Classification	N° of hits SPT	Unit weight (t/m^3)
	4	10	40	200	LL	LP	IP		SUCS		
0.20	98	95	92	71	–	–	NP	32.32	SM	1,2,2	–
1.00	100	98	95	80	–	–	NP	35.24	SM	1,1,2	1.96
2.00	100	99	97	80	–	–	NP	26.15	SM	1,2,3	–
3.00	100	96	91	71	–	–	NP	38.43	SM	4,6,6	–
4.00	100	97	78	52	–	–	NP	44.37	SM	3,2,5	–
5.50	100	99	75	56	–	–	NP	51.32	SM	12,23,41	–

At station 63 + 040 the results of the standard penetration test (SPT) are observed with a depth of 6 m. Table 3 determines that the stratum is homogeneous and corresponds to a low plasticity clay. It is established that the compactness in the first 2 m is relatively low.

The mechanical parameters used to calculate the safety factor for slopes 25 + 280 and 63 + 040 are: unit weight, cohesion, friction angle, seismic conditions, and slope geometry. In addition, the comparison of the static and pseudo-static analysis was carried out using the limit equilibrium methods (Fellenius, Bishop), as shown in Tables 4 and 5, Fig. 4, 5, 6, 7.

As can be seen in Table 6, there are variations in the results of the safety factors. As methods evolved, the safety factor and precision increased. The most conservative method was Fellenius in providing the lowest FS.

The FS produced in the pseudo-static analysis are lower, since the seismic factors were taken according to the NEC-2015 [15] standard, which is related to the most important earthquakes that occurred in recent decades in Ecuador.

Table 3. Input parameters for GeoSlope software, abscissa 25 + 280.

Depth (m)	Granulometry% through				Limits			W%	Classification	N° of hits SPT	Unit weight (t/m^3)
	4	10	40	200	LL	LP	IP		SUCS		
0.20	100	100	99	97	49.30	22.57	26.73	21.38	CL	2,3,4	0.20
1.00	100	100	96	84	36.60	17.70	18.90	24.05	CL	3,4,5	1.00
2.00	100	100	95	70	46.38	29.00	17.39	51.92	CL	2,3,4	2.00
3.00	100	100	99	95	39.78	20.08	19.70	47.81	CL	1,4,7	3.00
4.00	100	100	97	76	34.68	12.37	22.31	62.72	CL	2,3,4	4.00
0.20	100	100	99	97	49.30	22.57	26.73	21.38	CL	2,3,4	0.20

Table 4. Input parameters for GeoSlope software, abscissa 25 + 280.

Description	Value	Unit
Cohesion	100.4	Kpa
Friction angle	11.21	degrees
Slope height	22.16	m
Specific weight	19.21	kN/m^3
Seismic factor [14]	0.25	–

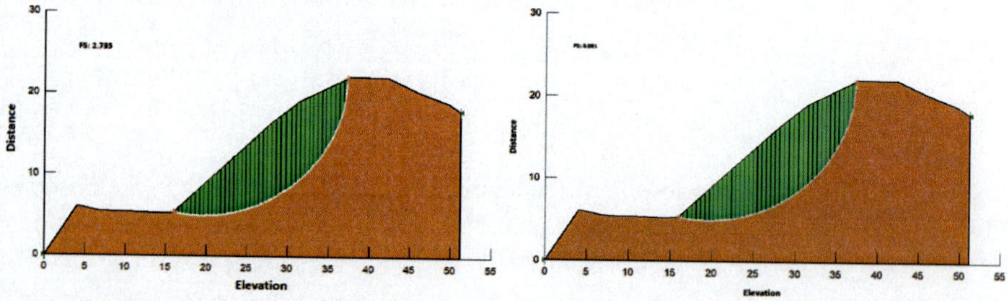

Fig. 4. Calculation of slope stability corresponding to the abscissa 25 + 280 Fellenius method a) static analysis b) pseudo-static analysis.

According to the results, it is determined that Bishop's method is more precise, because it considers the circular failure surface and establishes the moment equilibrium in each segment.

The FS values present high levels of stability because the soil strata of the study area have excellent geomechanical characteristics, favoring the absence of a water table.

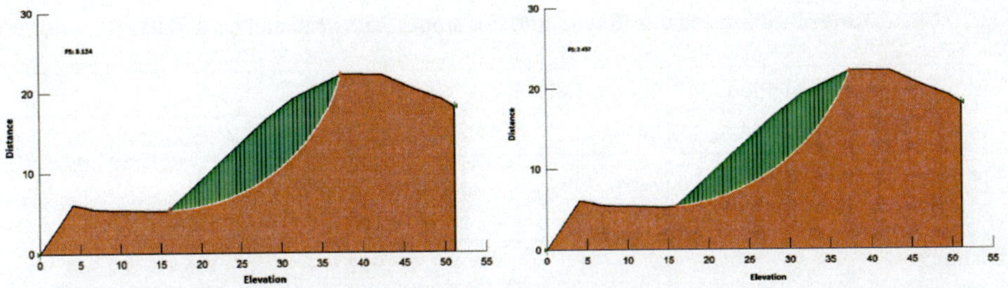

Fig. 5. Calculation of slope stability corresponding to the abscissa 25 + 280 Bishop method a) static analysis b) pseudo-static analysis.

Table 5. Input parameters for GeoSlope software, abscissa 63 + 040.

Description	Value	Unit
Cohesion	85.12	Kpa
Friction angle	25.59	degrees
Slope height	23.23	m
Specific weight	19.55	kN/m^3
Seismic factor [14]	0.25	-

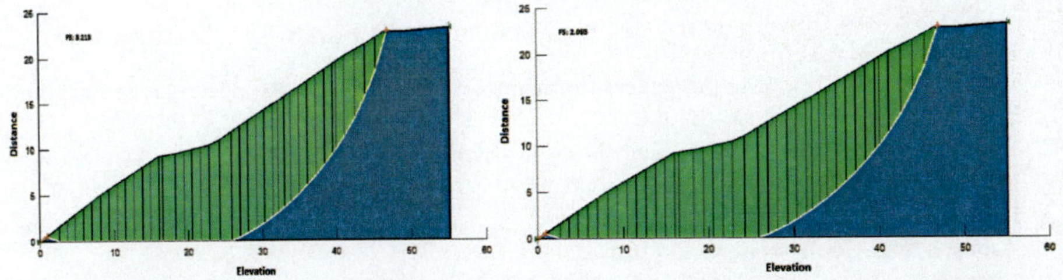

Fig. 6. Calculation of slope stability corresponding to the abscissa 63 + 040 Fellenius method a) static analysis b) pseudo-static analysis.

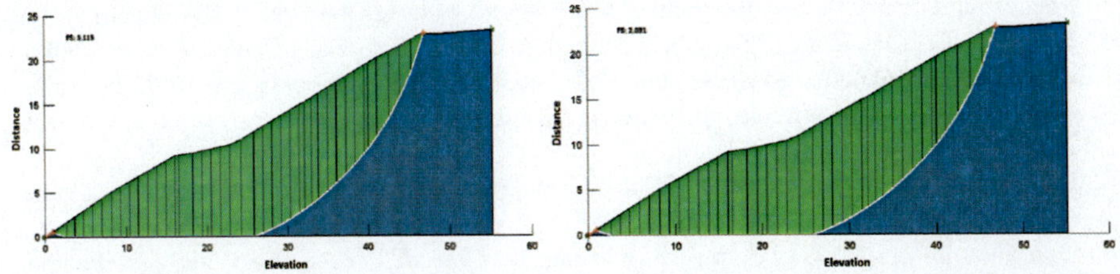

Fig. 7. Calculation of slope stability corresponding to the abscissa 63 + 040 Bishop method a) static analysis b) pseudo-static analysis.

Table 6. Summary of safety factors for slopes 25 + 280 and 63 + 040.

Limit Equilibrium Method	Slopes (Abscissa)	Analysis	
		Static FS	Pseudo-static FS
Fellenius	25 + 280	2,735	2,091
	63 + 040	3,123	2,085
Bishop	25 + 280	2,972	2,406
	63 + 040	3,392	2,201

4 Conclusions

The slope stability studies considered in the present investigation, for the static and pseudo-static cases, show that the envelope of the fault presents a greater decrease in the Bishop method in relation to the Fellenius method.

The variations obtained through the static and pseudo-static analysis determined that the differences between the two cases are drastic and could generate a great destructive potential if the seismic factor is not considered as analysis parameters.

In Ecuador, the consideration of the seismic factor is very important, because the country has frequent earthquakes.

References

1. D'Ercole, R., Trujillo, M.: Amenazas, vulnerabilidad, capacidades y riesgo en el Ecuador. EKSEPTION Editorial, Quito (2003)
2. Bojorque, J.: Métodos para el análisis de la estabilidad de pendientes. Revista científica MASKANA **2**(2), 1–16 (2011)
3. Hack, R., Alkema, D., Kruse, G., Leenders, N., Luzi, L.: Influence of earthquakes on the stability. Eng. Geol. **91**(1), 4–15 (2007)
4. Chunga, K., Livio, F.A., Martillo, C., Lara, H., Ferrario, M.F., Zevallos, I., Michetti, A.M.: Landslides triggered by the 2016 Mw 7.8 Pedernales, Ecuador earthquake: Correlations with ESI-07 intensity, lithology, slope and PGA-h. Geosciences **9**(9), 371 (2019)
5. Valiente Sanz, R., Sobrecases Martí, S., Díaz Orrego, A.: Estabilidad taludes, conceptos básicos, parámetros de diseño y métodos de cálculo. Revista Civilizate **7**, 50–54 (2015)
6. Sanhueza, C., Rodríguez, L.: Análisis Comparativo de métodos de cálculo de estabilidad de taludes finitos aplicados a laderas naturales. Revista de la construcción **12**(1), 17–29 (2013)
7. Bristow, C.R.: The Daule group, Ecuador (Middle Miocene-Pliocene). Newsl. Stratigr. **5**, 190–200 (1976)
8. Reyes, P., Michaud, F.: Mapa Geológico de la Margen Costera Ecuatoriana (1500000), Quito (2012)
9. González de Vallejo, L.I., Ferrer, M., Ortuño, L., Oteo, C.: Ingeniería Geológica. Pearson educación, Madrid (2002)
10. Ramírez, R., Barrera, S., Gómez, P.: El método seudoestático de estabilidad en presas: Un análisis crítico. Obras y Proyectos **9**, 30–37 (2011)
11. Terzaghi, K.: Mechanism of landslides. Application of geology to engineering practice, pp. 83–123. Geological Society of America (1950)

12. Duncan, J.M., Wright, S.G.: Soil Strength and Slope Stability. Wiley & Sons, Hoboken (2005)
13. Suarez Diaz, J.: Deslizamiento: Análisis Geotécnico, vol. 1. Editorial Universidad Industrial de Santander, Bucaramanga (2009)
14. Ministerio de Obras Públicas del Gobierno de Chile: Manual de Carreteras vol. 3, Instrucciones y Criterios de Diseño. Dirección de Vialidad, Santiago (2002)
15. Norma Ecuatoriana de la Construcción: Peligro Sísmico/Diseño Sismo Resistente, Código: NEC-SE-DS, Quito (2015)

Implementation of the Eco-Schools Program in Higher Education - A Path to Environmental Sustainability at IPC

Álvaro Leal[1], Hélder Simões[1](✉), João Figueiredo[2], António Loureiro[3], Sílvia Seco[3], and Ana Ferreira[1]

[1] ESTeSC - Coimbra Health School, Department of Environmental Health, Polytechnic Institute of Coimbra, Coimbra, Portugal
{heldersimoes,anaferreira}@estescoimbra.pt
[2] ESTeSC - Coimbra Health School, Department of Complementary Sciences, Polytechnic Institute of Coimbra, Coimbra, Portugal
[3] Occupational and Environmental Health Service, Polytechnic Institute of Coimbra, Coimbra, Portugal

Abstract. The Rio Conference (1992) was an important moment for the Goals of Sustainable Development through the document Agenda 21. This document constitutes an opportunity to inscribe the theme in the Eco-Schools Program. The Ecological Footprint arises as a concept determined to estimate the quantity of natural resources we consume to support our lifestyle - in 2002 at the Johannesburg Summit it was recognized as a reliable sustainability gauge. The present study is therefore aimed at assessing the students' knowledge about the Eco-Schools/Sustainability program and the Environmental Footprint of the Polytechnic Institute of Coimbra (IPC). A questionnaire divided into three parts was made available, totaling thirty-six questions in digital format on the platform Google Forms, and was disseminated through e-mail and social networks (Facebook and Messenger). The responses to the questionnaire were automatically downloaded to a Microsoft Excel spreadsheet and later coded for statistical treatment with the aid of the statistical software Statistical Package for the Social Sciences (SPSS) version 27.0 for Windows. After the statistical treatment of the responses of the 253 participants we verified, by the results obtained in the calculation of the Ecological Footprint, that the respondents put into practice the knowledge disseminated by the Eco-Schools Program, highlighting its importance for environmental sustainability. The Eco-Schools Program and Sustainability are values already in-stilled in our student community, but even so we must continue to bet on a greater community participation in actions in favor of the Program. In view of the results related to the calculation of the Ecological Footprint, the need to improve this indicator in the student community is evident, as is promoting its performance in this area. It will be therefore important to reinforce the awareness sessions to reduce the Ecological Footprint in favor of the investment made by the IPC.

Keywords: Eco-Schools · Environmental sustainability · Ecological footprint

J. R. da Costa Sanches Galvão et al. (Eds.): *Proceedings of the 1st International Conference on Water Energy Food and Sustainability (ICoWEFS 2021)*, pp. 516–525, 2021.
https://doi.org/10.1007/978-3-030-75315-3_56

1 Introduction

The Eco-Schools program started in 1994 in an international initiative by the Foundation for Environmental Education in the United Kingdom, Germany, Greece, and Denmark, having been implemented in Portugal during the school year of 1996/1997 by the European Blue Flag Association (ABAE) called, in Portuguese, Associação Bandeira Azul da Europa [1]. The Eco-Schools program aims to encourage the development of activities, aiming to improve the environmental performance of schools, contributing to the change of behaviors and the impact of environmental concerns on different generations, recognizing and rewarding the work developed with the attribution of the Green Flag Eco-Schools award [2].

The environmental performance of the Coimbra Health School was recognized in 2009, being this the first High Education Institution in Portugal to get such an award, attributed by ABAE. Currently all six Schools and Institutes of the Polytechnic Institute of Coimbra (SI IPC) hold the award. There were no recent studies for this type of education at the level of higher education, which led us to develop the first part of this study using a questionnaire adapted to the knowledge of the Environmental Sustainability (ES) community. In the second part of the present study, we observe the performance of the students from IPC, with regard to practices for ES, using the Ecological Footprint (EF). The EF in a concept proposed in 1996 by the researchers William Rees and Mathis Wackernagel, designed to estimate the amount of natural resources we consume to support our lifestyle. This concept ended up being recognized as a reliable sustainability gauge in 2002 at the World Summit on Sustainable Development (Johannesburg) [3]. The EF was created so that it was possible to understand how much natural resources are used to sustain a lifestyle. The EF can also be understood as a tool for reading and interpreting reality, through which it is possible to visualize known problems and propose new ways to solve them through the sustainable measures of each individual [4].

2 Material and Methods

We begin the present study with the aim of perceiving the involvement of students about the ES. In the second phase we use an adapted questionnaire, proposed by earthday.net, to verify the EF of the several SI IPC. The study was carried out/applied to the students of each of the six SI IPC, Coimbra Agriculture School (ESAC), Coimbra Education School (ESEC), Coimbra Health School (ESTeSC), Oliveira do Hospital Technology and Management School (ESTGOH), Coimbra Business School (ISCAC) and Coimbra Engineering School (ISEC). The questionnaire was made available in digital format on the platform Google Forms, having been disseminated through e-mail and social networks, thus considerably decreasing the environmental impact with saving paper regarding printing. The structure of this instrument was divided into three parts: the first proposed to identify (the source) and characterize the target population, the second - with adaptation of a form from the University of Passo Fundo of Brasil - aimed to obtain results about the knowledge of the Eco-Schools Program and Environmental Sustainability, which included the questions 6th to 18th inclusively, and lastly, the adaptation of the form from myfoot-print.org, allowed to obtain the EF of the IPC, questions 19th to 36th inclusively. For the calculation of the EF, each reply to the questions from 19th to 36th has an impact on the environment that varies with the greater or lesser aggressiveness we cause to the environment. To calculate the real EF there is the need to convert the results obtained from the sum of the aforementioned questions to hectares (ha), according to Table 1.

Table 1. Interpretation of results of the Ecological Footprint.

Total obtained	Ecological Footprint
Less than 150	>than 4 ha
Between 150 and 400	[4 e 6[ha
Between 400 and 600	[6 e 8[ha
Between 600 and 800	[8 e 10] ha
More than 800	<than 10 ha

For example: if at the table 600 corresponds to 8 ha and the sum has been 550 this means that it will effectively correspond to X, so X = (550 × 8)/600 being the result of 7,3 ha. For the conversion of the results, one should always resort the value closest to the interval considered.

In view of the exposed in Table 2 and to obtain results that could be the subject of the discussion and having no history of the EF of the different bachelor's degrees of the others SI IPC, we chose to process only the responses of the ESTeSC, dividing the school by the bachelor's degrees of Audiology, Biomedical Laboratory Sciences, Dietetics and Nutrition, Pharmacy, Clinical Physiology, Physical Therapy, Medical Imaging and Radiotherapy and Environmental Health. We are using the statistical software Statistical Package for the Social Sciences (IBM SPSS) version 27.0 for Windows. With regard to the applied statistical analyses, these were divided into: descriptive statistics (analysis of absolute and relative simple/accumulated frequencies) and inference (Chi-square test of Independence). For statistical inference we considered a 95% confidence level for a random error <5%.

The responses collected were anonymous and with no relation possible to identify the respondent, thus complying with the rules applied in the General Data Protection Regulation and mentioning that this research has a purely curricular and academic interest, with no commercial or financial gain resulting from this study.

Table 2. Number of answers of the study sample.

Schools and Institutes IPC	N
ESAC	2
ESEC	2
ESTeSC	211
ESTGOH	34
ISCAC	4
ISEC	2
Total	253

3 Material and Methods

The present study has two main goals: the knowledge of the Eco-Schools Program and the contribution to the Ecological Footprint by the student community, also presented in two separate groups. In the 253 responses obtained there is no surprise that the vast majority are female because said gender is the most representative of students and the age group is also within the current parameters of Higher Education students. Regarding the year of attendance, there is a balance between the first, second and third years, being unsurprising the low adhesion of the fourth year due to it being a year of curricular internship (Table 3).

Table 3. Description of responses according to respondents' socio-biographical characteristics.

	Schools and Institutes IPC	Female		Male		Total	
		N	%	n	%	n	%
Age	18–24	176	91.2	50	83.3	226	89.3
	25–30	7	3.6	4	6.7	11	4.3
	31–40	7	3.6	5	8.3	12	4.7
	>41	3	1.6	1	1.7	4	1.6
	Total	193	100.0	60	100.0	253	100.0
Year of Enrollment	Year 0	2	1.0	0	0	2	0.8
	1st Year	80	41.5	18	30.0	98	38.7
	2nd Year	53	27.5	17	28.3	70	27.7
	3rd Year	37	19.2	13	21.7	50	19.8
	4th Year	21	10.9	12	20.0	33	13.0
	Total	193	100.0	60	100.0	253	100.0

Addressing the first large group under study, the Eco-Schools Program, and questioning whether they had already heard of the program, a large majority of respondents (214) replied "Yes" which corresponds to 86.4%. When questioning when they heard there is a balance in the percentages at different levels of education, however, we cannot ignore that there were 39 respondents who replied not knowing.

When speaking about the results obtained on Sustainability, all participants replied affirmatively to the importance of promoting it, however only 61.7% (156) said they were well informed about the topic.

The respondents also mentioned that their SI IPC already acts in a positive manner 84.2% (213), also 89.9% (225) affirmed that their SI IPC promotes sustainability behaviors, but according to the 178 respondents, environmental standards are welcomed by the students in a positive fashion (Table 4).

Table 4. Description of responses regarding the SI IPC adopting behaviors that promote sustainable development, Environmental Standards and Curriculum Programs.

		N	%	Accumulated (%)
Behaviors that promote sustainable development	Yes	225	88.9	88.9
	No	28	11.1	100.0
	Total	253	100.0	
Environmental standards welcomed in the student community	Disagree	19	7.5	7.5
	Don't agree nor disagree	56	22.1	29.6
	Partly agree	133	52.6	82.2
	Agree	45	17.8	100.0
	Total	253	100.0	
Curriculum programs containing aspects of sustainable development	Disagree	46	18.2	18.2
	Don't agree nor disagree	81	32.0	50.2
	Partly agree	82	32.4	82.6
	Agree	44	17.4	100.0
	Total	253	100.0	

With regard to the curricular content of licentiate degrees, there is an equivalent division over Sustainable Development (Table 4).

When asked about the existence of sustainability experts, either in the student community or in the faculty, there are opposite values. The participants, with almost equal percentages of answers, refer to a greater number of students than to faculty in terms of specialization. Let's look at Table 5.

When asked if at the end of the degree they will be better able to act in favor of Sustainability 198 of the respondents say that they "partly agree" and "agree" and this is reflected when 65 say they use public transport because it is more economical and 56

Table 5. Description of responses regarding the presence of specialists both in the student community and in the faculty.

		N	%	Accumulated (%)
Within the student community there are specialists in sustainable development	Disagree	11	4.3	4.3
	Don't agree nor disagree	94	37.2	41.5
	Partly agree	78	30.8	72.3
	Agree	70	27.7	100.0
	Total	253	100.0	
Your teachers are experts in sustainable development	Disagree	41	16.2	16.2
	Don't agree nor disagree	105	41.5	57.7
	Partly agree	54	21.3	79.1
	Agree	53	20.9	100.0
	Total	253	100.0	

because of environmental issues, representing 60.5%, we also have to mention that 53 live close to their SI IPC (Table 6).

Table 6. Usage of Public Transports.

	N	%	Valid %	Accumulated (%)
Larger frequency	30	11.9	15.0	15.0
Larger frequency during rush hours	6	2.4	3.0	18.0
Short travel time	13	5.1	6.5	24.5
It is more economical	65	25.7	32.5	57.0
It is more comfortable	3	1.2	1.5	58.5
Compliance with punctuality	11	4.3	5.5	64.0
Quality of transports	3	1.2	1.5	65.5
Greater environmental efficiency	56	22.1	28.0	93.5
I don't want to take public transports	13	5.1	6.5	100.0
Total	200	79.1	100.0	
Not applicable/Live close to the SI IPC	53	20.9		
Total	253	100.0		

We then proposed to study the relationship between the Ecological Footprint index with certain characteristics of the students who participated in the study (Table 7).

Table 7. Relationship between the Ecological Footprint Index and socio-biographical characteristics of all participants

		[4 - 6[hectares		[6 - 8[hectares		Total		p-value
		N	%	N	%	N	%	
Gender	Female	128	78.5	65	72.2	193	76.3	0,456
	Male	35	21.5	25	27.8	60	23.7	
	Total	163	100.0	90	100.0	253	100.0	
Age	18–24	145	89.0	81	90.0	226	89.3	0,970
	25–30	8	4.9	3	3.3	11	4.3	
	31–40	7	4.3	5	5.6	12	4.7	
	≥41	3	1.8	1	1.1	4	1.6	
	Total	163	100.0	90	100.0	253	100.0	
Year of Enrollment	Year 0	1	0.6	1	1.1	2	0.8	(a)
	1st Year	70	42.9	28	31.1	98	38.7	
	2nd Year	42	25.8	28	31.1	70	27.7	
	3rd Year	29	17.8	21	23.3	50	19.8	
	4th Year	21	12.9	12	13.3	33	13.0	
	Total	163	100.0	90	100.0	253	100.0	

Chi-square test of Independence; (a) – not achievable

According to previous results, there were no differences in the Ecological Footprint index either by gender, age ($p > 0.05$) and year of enrollment. We also proposed the same analysis strategy according to the Licentiate Course.

Table 8. Relation between the Ecological Footprint Index and the bachelor's degrees at ESTeSC.

	[4 - 6[hectares	[6 - 8[hectares	Total
Dietetics and Nutrition	11 (8.0%)	11 (14.9%)	22 (10.4%)
Environmental Health	47 (**34.3%**)	26 (**35.1%**)	73 (34.6%)
Clinical Physiology	11 (8.0%)	3 (4.1%)	14 (6.6%)
Pharmacy	10 (7.3%)	3 (4.1%)	13 (6.2%)
Medical Imaging and Radiotherapy	17 (12.4%)	5 (6.8%)	22 (10.4%)
Audiology	11 (8.0%)	6 (8.1%)	17 (8.1%)
Physical Therapy	16 (11.7%)	8 (10.8%)	24 (11.4%)
Biomedical Laboratory Sciences	14 (10.2%)	12 (16.2%)	26 (12.3%)
Total	137 (64.9%)	74 (35.1%)	211 (100%)

Chi-square test: $x^2 = 7{,}035$; $gl = 7$; p-value = 0,425

According to the results presented above (Table 8), we can see that there were no significant differences in the frequency of responses observed per course regarding the different classes of the Ecological Footprint index ($p > 0.05$). However, we can see that the prevalence of participations (responses from students) with an index of [6 - 8[hectares was 35.1% ($n = 74$). The largest share of contribution of responses came from the Environmental Health course. We also tried to describe the average estimates of the respective Ecological Footprint index per course and for two organic units.

Table 9. Relation between the Ecological Footprint Index and bachelor's degrees at ESTeSC and SI IPC.

	Average	Hectares (ha)	Standard deviation	N
Dietetics and Nutrition	**374.3182**	5.6	69.44157	22
Environmental Health	367.2603	5.5	80.26450	73
Clinical Physiology	**327.8571**	4.9	92.10601	14
Pharmacy	**328.0769**	4.9	74.34543	13
Medical Imaging and Radiotherapy	345.2273	5.2	60.18775	22
Audiology	366.1765	5.5	84.06711	17
Physical Therapy	356.8750	5.4	91.80334	24
Biomedical Laboratory Sciences	**372.4074**	5.6	77.07851	27
ESTeSC	360.3081	5.4	79.37419	211

According to the results presented in Table 9, we verified that the degrees with the highest average Ecological Footprint index were "Dietetics and Nutrition" and "Biomedical Laboratory Sciences". In the opposite sense, the lowest was observed in the degrees of "Clinical Physiology" and "Pharmacy".

4 Discussion

From the results obtained on the themes of Eco-Schools/Sustainability from our respondents, it was easy to see that this is a current and important topic for their day-to-day lives, and even so we still verified a small percentage of students who have not heard about the Program Eco-Schools. It was also noticed that some of them have heard about this on primary school, which is normal as it was created for that same level of education, which luckily has been evolving towards the Higher Education. If the Eco-Schools program is intended to be a methodological contribution to a participatory and enlightened environmental education in schools where educating is to create conscious and active citizens for the environment it may be the fact that sustainable actions are only applied on commemorative school dates that is a common mistake. Nonetheless, the growing concern with environmental issues, both internationally and nationally and its influence on the adoption of environmentally correct behaviors and attitudes has not been significant. Despite recommendations for the school to be involved in promoting

"Education for Sustainable Development", the results of several recent studies "Program Eco-Schools: a contribution to its' evaluation" show that projects aimed at this type of education are concentrated in the first years of education, reducing significantly in secondary education and becoming practically non-existent in Higher Education [5]. The school is, by excellence, the privileged way to materialize a set of behaviors and attitudes, which constitute a contribution to the discovery of solutions and new paths to problems related to the environment, development and "Sustainability" [6]. Being this theme transversal to the entire academic community, it was found that our participants also express some negativity in relation to the curricular programs and specialists in the subject. The teacher, whatever his area of specialization, must have broad contact with environmental issues, which will allow reflection on the subject and the possibility of the emergence of proposals for pedagogical action at school. Educators, from basic education to university, would need to be aware that the in-depth debate on environmental issues can contribute a quite lot to building citizenship and better the quality of life on the planet, but for that, it will be necessary to carry out an interdisciplinary work in which each one learns to communicate in other languages, making interfaces with other areas of knowledge [7].

Analyzing the results obtained in the calculation of the EF of the student community of ESTeSC, by bachelor's degrees, we noticed a variation between the 4.9 and 5.6ha, as well as the EF of the ESTeSC that has reached the value of 5.4ha. Previous studies conducted at ESTeSC, "Environmental Footprint of the Coimbra Health School: A Sustainability Indicator" and "Environmental Footprint of a Portuguese Higher Education Institution" showed values below those we obtained, this fact will not be unrelated to the investment and awareness around sustainability made over the past few years [8, 9]. When we analyze the results now obtained with the national 4,1ha average and the world 2,75ha average in 2016 there is a very significant difference, it is crucial to promote the preparation of higher education institutions for the challenges of sustainable development [10]. The Academic Community certainly plays a decisive role in promoting better overall performance in the indicators used in this study.

5 Conclusion

The Ecological Footprint is, as we all know, an indicator of the environmental performance of populations, and in this study, it is evident the need to continue to promote knowledge aiming at the adoption of more sustainable behaviors and attitudes, such as greater use of public transport that are increasingly becoming more environmental friendly, knowing that sometimes there is some difficulty in using it, sharing car trips can be an excellent alternative. In terms of eating habits, and considering our findings, the number of meals in which meat and fish are present should be reduced, opting for another type of healthier food, without necessarily having to become vegetarian or vegan, promoting the number of meals as well made at home. Recycling should not be an option, be it in the correct separation of waste or reuse, it must be an obligation as responsible and concerned citizens about the Planet on which we live.

It was in a pandemic scenario that we conducted this study, and its development was certainly influenced by this unique context, the number of respondents is characteristic of this - we obtained 253 responses from a universe of more than 7000 individuals. Since it was not a representative sample, we redirected the initial objective of the study, which initially had the scope of all SI IPC to ESTeSC with regard to PE. It will be important for the IPC to continue with its awareness actions and implementation of new environmental measures given the great responsibility it has being one of the largest Portuguese educational institutions with six SI IPC and having all of them the hoisted flag of the Green Flag of the Eco-Schools Program. It is also suggested that in the reception of new students from the six SI IPC study cycles, a moment of information/awareness is created to make the IPC Environmental Policy known.

References

1. EB1/PE Ribeiro Domingos Dias. Acedido a s.p.d. https://escolas.madeira-edu.pt/eb1perddias/ProjetosEventos/ECOESCOLAS/tabid/4494/Default.aspx
2. DGE – Direção-Geral da Educação. https://www.dge.mec.pt/projeto-eco-escolas
3. Câmara Municipal de Anadia. https://www.cm-anadia.pt/pages/551
4. Pegada Ecológica, Um novo conceito no universo da sustentabilidade. Acedido a s.p.d. https://betaeq.com.br/index.php/2019/08/13/pegada-ecologica-um-novo-conceito-no-universo-da-sustentabilidade/
5. Serra, C.J., Moura, G.: Programa Eco-Escolas: um contributo para a sua avaliação. Universidade Aberta (2009)
6. Painço, Cidália, P.D.M.: Mestrado em Cidadania Ambiental e Participação. Universidade Aberta (2009). https://repositorioaberto.uab.pt/handle/10400.2/1498
7. Simone, G.M.S., Maria, T.C.G.: A formação universitária para o ambiente: educação para a sustentabilidade (2003). https://repositorio.furg.br/bitstream/handle/1/7707/898-1855-1-PB.pdf?sequence=1
8. Ferreira, A., Sá, N., Costa, A.C., Rocha, A.T., Vidal, A., Martins, F.L., Delgado, S.M.: Pegada Ecológica da Escola Superior de Tecnologia da Saúde de Coimbra Um Indicador de Sustentabilidade (2008). https://saudeambiental.net/wp-content/uploads/PEGADA_ECOLOGICA_ESTESC.pdf
9. Paixão, S., Sá, N., Simões, J., Gaminha, I.: Pegada ecológica de uma instituição do ensino superior portuguesa. Revista de Geografia e Ordenamento do Território **1**(1), 165-a (2012)
10. National Footprint Accounts 2019 edition (Data Year 2016 com base em Indicadores de Desenvolvimento Mundial, Banco Mundial (2019); Organização das Nações Unidas para Alimentação e Agricultura

Indoor Air Quality, Ultrafine Particles in Offices and Services of Coimbra Health School

Fernando Moreira(✉), Ana Ferreira, João Paulo Figueiredo, and Inês Caseiro

Instituto Politécnico de Coimbra, ESTeSC, Coimbra, Portugal
fernando.moreira@estescoimbra.pt

Abstract. Air quality is related to the pollution in the air quality degree, which is caused by a mixture of chemicals, released into the air or resulting from chemicals reactions, that changes the natural composition of the atmosphere. People working in offices waste a large part of their day in this place. In this way, the indoor air quality of the offices affects the total exposure of humans and therefore, they are considered microenvironments of great importance. This study aimed to evaluate expose the occupants to indoor air pollutants in offices and services of Coimbra Health School, check if pollutants complied with the limits of legislation and evaluate the exposure to ultrafine particles in different environments. Data collection consisted of assessing air quality in three offices and four services, using air pollutants and atmospheric variables. We checked that the average concentration of PM_{10} in some places exceeded the protection threshold and that the average concentrations of ultrafine particles were higher when combustion processes existed. In this study it was possible conclude that some measures are necessary to improve indoor air quality in offices and services of Coimbra Health School, once that the average concentration of PM_{10} in some places exceeded the protection threshold. So the institution needs to do frequent monitoring to avoid exposing occupants to risky situations.

Keywords: Indoor air quality · Offices · Services · Atmospheric pollutants · Ultrafine particles

1 Introduction

In the last two decades, several studies show that air pollution, especially in the urban environment, has become a major concern. According to health guidelines, as the population grows with 54% of the current world population in urban areas, the majority of residents live in cities with inadequate air quality [1].

Air quality is related to the degree of pollution in the air, which is caused by a mixture of chemicals, released into the air or resulting from chemicals reactions, that changes the natural composition of the atmosphere. Air pollution is seen as a public health problem responsible for increasing human mortality and morbidity [2].

Carbon monoxide (CO), carbon dioxide (CO_2) and particulate matter (PM2, 5 and PM10) ate the main pollutants of indoor and outdoor air [3].

J. R. da Costa Sanches Galvão et al. (Eds.): *Proceedings of the 1st International Conference on Water Energy Food and Sustainability (ICoWEFS 2021)*, pp. 526–536, 2021.
https://doi.org/10.1007/978-3-030-75315-3_57

Indoor air pollution can be characterized by presence of chemical, physical or biological pollutants in the indoor air of buildings, for example, in schools, offices, homes and commercial and service buildings and inside the means of transport. Indoor air quality (IAQ) is considered to be one of the greatest threats to human health, once that society stays approximately 90% of time in indoor environments, where the levels of pollutants can be to 100 times higher than those of outdoor air [4].

According to the Public Health Institute, the sick building syndrome (SBS) can be defined as a situation in which the occupants of a building feel occupational discomfort and/or acute health problems. However, is not always possible to identify the causes of the discomfort/health problem, nor establish a specific diagnosis [5].

SBS is due to poor air quality in the various rooms of some buildings, where a poor IAQ can lead to a variety of effects/symptoms on human health, such as lethargy, headache, tiredness, irritation of the nose, eyes and throat, lack of concentration and also breathing problems and lung cancer. In addition to the direct effects, a poor IAQ can also affect the productivity and performance of individuals [4].

In residential buildings, IAQ and effects on human health were determined by the lifestyle choices of the occupants (smoking, pets) and the structure and quality of the building. Building management, such as maintenance operations, are important determinants of occupant's health, because they can contribute to IAQ problems [6].

Other sources that contribute to the increase of indoor air pollution are outdoor air, smoke, cleaning activities, humidity and proximity to main roads and industrial activities [7].

Regarding people working in offices, they spend about a third of their day in this place. In this way, the IAQ of the offices affects the total exposure of the human and, therefore, offices are considered microenvironments of great important [8]. Recently, a study demonstrated that air pollution in offices can present a risk to worker's health [9].

Ultrafine particles are particles with a diameter less than 100 nm, that have a high deposition fraction [10], being able to enter in the human organism by inhalation, through the respiratory tract; by absorption, through the skin; by ingestion, through the mouth or by a combination of these. Inhalation is probably the most important route of human exposure to these particles. This way, and due to their size and other characteristics, ultrafine particles can reach the alveolar region, giving rise to inflammatory processes in lung and the subsequent cardiovascular morbidity and [11].

It has been shown that ultrafine particles are the ones that most affect human health [10], once that they're not limited to the respiratory and cardiovascular systems, as they were found in organs such as the brain and heart [12].

Taking into account that society spends a lot of time in indoor environment, it's important to monitor the IAQ, as well as to implement preventive measures, in order to preserve the health of teachers, workers and students.

This study aimed to evaluate expose the occupants to indoor air pollutants in offices and services of Coimbra Health School, check if pollutants complied with the limits of legislation and evaluate the exposure to ultrafine particles in different environments.

2 Material and Methods

Study design - The present study was an observational analytical type, transversal in nature and level of knowledge II (correlational description).
Population and Sample - The target population were teachers, workers and students who were part of the 3 offices and 4 services evaluated in ESTeSC. The type of sampling used was non-probabilistic, where the sampling technique for convenience was applied.
Instruments and Data Collection - The study made in 2019 and 2020 and the data collection period between the January and March 2020.

Data collection consisted of assessing air quality, through the assessment of air pollutants (CO_2, CO, PM2.5, PM10 and ultrafine particles) and meteorological variables temperature (T°) and relative humidity (Rh).

Measurements were made over a period of 15 min, with samples every minute The measurements should be performed according to the new guideline from APA based in Port. 353-A/2013, 4 of December. The measurements of IAQ of each location were performed on 3 different days, and each day they took place at 3 different times, which relate to before, to the middle and to the final phase of the occupation, placing the equipment in the most central possible position of each evaluated space and, approximately, at the height of the occupants airways in the sitting position. At this moment of investigation, the characterization of each space was also done, taking into account the volume, the existence of windows, doors and ventilation systems, as well as potential sources of contamination. The measurements of outdoor air quality happened in the outside of ESTeSC, in the place that was considered most critical, but at least 1 m away from the walls and at the same height at which the measurements of IAQ were.

Data collection was made using three portable equipment's for reading in real time, namely the Q-Track Plus, brand TSI, model 8552/8554 for the assessment of the concentration of CO, CO_2, T° e Rh; the Particles Counters, brand Lighthouse Worldwide Solutions, model 3016 LAQ for the assessment of the concentrations of PM2.5 and PM10 and the P-Track Ultrafine Particle Counter, model 8525, which has a reagent with 99,5% solution of isopropyl alcohol for counting ultrafine particles. In order to transfer the data from the measuring equipment's to the computer, was used the specific software for each equipment.

It was considered as a reference for the maximum concentration of CO, 10,0 mg/m^3 (9 ppm), for CO_2, 2250 mg/m^3 (1250 ppm), for PM2.5, 0,025 $mg/m^3$and for PM10, 0,05 mg/m^3, as related in Ordinance 353-A/2013, of 4 december (Port n. ° 353-A/2013, 2013). Due to the lack of reference values for the concentration os ultrafine particles, the average analytical values recorded in the measurements of outdoor air were taken as a reference point. According to Decree-Law n. ° 243/86, of 20 august, the environmental comfort conditions of reference for T° must oscillate between 18 and 22° C and for Rh must oscillate between 50 and 70% (DL 243/86 de 20 de Agosto, 1986).

- **Statistical analysis**
  The statistical treatment of data collected in the study was performed using the IBM SPSS Statistics, version 25.0 software. Descriptive statistics techniques were used in the study, such as measures of central tendency (arithmetic mean), dispersion measures (standard deviation) and frequency tables. At the level of statistical inference, was

used a previous assessment of the assumptions of the metric variables (parametric or non-parametric). As for hypothesis testing, the following were applied: Kruskal Wallis, t-Student for 1 sample, Chi-square of Independence, t-Student for Independent Samples, Mann-Whitney and Fisher's Exact Test. In the interpretation of the statistical tests, a 95% confidence level was taken into account for a maximum random error up to 5%.

3 Results

- General sample characterization
  The sample of this study was confined to three offices and four services of ESTeSC and also the outside of this educational institution.
- Inferential analysis
  After the statistical treatment of the data, the values of the concentrations of PM2.5, PM10, CO_2 and CO of the indoor air of the various offices and services were compared with the legally established values, that is, the protection threshold. Let's see Table 1 below.

Table 1. Average concentrations of PM2.5, PM10, CO_2 and CO inside the different offices and services evaluated.

Evaluated places	$PM_{2.5}$			PM_{10}			CO_2			CO		
	Mean	Standard deviation	PL	Mean	Standard deviation	PL	Mean	Standard deviation	LP	Mean	Standard deviation	PL
Office 1	0,006	0,003	<	0,022	0,008	<	846	227	<	1,8	0,1	<
Office 2	0,007	0,002	<	0,024	0,009	<	823	338	<	1,8	0,1	<
Office 3	0,005	0,002	<	0,019	0,004	<	1196	646	<	1,8	0,0	<
Service 1	0,006	0,002	<	0,020	0,009	<	775	283	<	1,8	0,0	<
Service 2	0,010	0,004	<	0,043	0,019	<	385	84	<	1,8	0,1	<
Service 3	0,006	0,003	<	0,018	0,005	<	571	171	<	1,8	0,0	<
Service 4	0,021	0,008	<	0,051	0,016	>	296	61	<	1,8	0,0	<

PL- Protection Limit; PL PM2.5 = 0,025 mg/m^3; PL PM10 = 0,05 mg/m^3; PL CO_2 = 1250 ppm; PL CO = 9 ppm; N = 9: Test: t-Student for 1 sample

There were statistically significant differences between the average concentration value of PM2.5 and the protection threshold in all offices and services evaluated, except for service 4 (p-value = 0.133). In all places, the average values were below the protection threshold. It was service 4 that registered the highest average values, while office 3 had the lowest average values. Regarding the average PM10 concentration values, it was found that there were no statistically significant differences in services 2 and 4 (p-value = 0.280; p-value = 0.861, respectively). The remaining locations showed statistically significant differences. Service 4 showed the highest average PM10 values and, consequently, was the only place that registered PM10 concentration values above the protection threshold. On the other hand, service 3 was the one with the lowest average values. With regard to the average values of CO_2 concentration, it was found that there were statistically significant differences between the average values and the protection threshold in all evaluated locations, except for office 3 (p-value = 0.808). In all places, the average values were below the protection threshold. Differences were recorded in the average CO_2 concentration, with office 3 having the highest average values, while service 4 had the lowest average values. As for the average concentrations of CO, the presence of statistically significant differences between the mean values and the protection threshold was observed in all evaluated places (p-value < 0.05), so the mean values were below the threshold of protection. The average CO values were very similar.

Then we proposed to verify, in general, between offices and services, the occurrence of exceedances to the legislated value for the different pollutants (PM2.5, PM10, CO_2, CO) in indoor air. Let's look at the Table 2.

Table 2. Pollutants (PM2.5, PM10, CO_2, CO) in the indoor air of offices and services.

		PM2.5		PM10		CO_2		CO	
		[0–0,025] mg/m^3	> 0,025 mg/m^3	[0–0,050] mg/m^3	> 0,050 mg/m^3	[0–1250] ppm	>1250 ppm	[0–9] ppm	>9 ppm
Offices	n (%)	27 (100%)	0 (0%)	27 (100%)	0 (0%)	22 (81,5%)	5 (18,5%)	27 (100%)	0 (0%)
Services	n (%)	34 (94,4%)	2 (5,6%)	27 (75%)	9 (25%)	36 (100%)	0 (0%)	36 (100%)	0 (0%)
Total	n (%)	61 (96,8%)	2 (3,2%)	54 (85,7%)	9 (14,3%)	58 (92,1%)	5 (7,9%)	63 (100%)	0 (0%)

Test: x2 of Independence

Evaluating the PM2.5 data, it was possible to verify that of the 63 measurements made in the different offices and services, 3.2% (n = 2) were outside the legally established range [0–0.025 mg/m^3], with this excess was registered in the services. As for PM10, 14.3% (n = 9) of the measurements made showed values that exceeded the reference range [0–0.050 mg/m^3], with all the exceedances being verified in the services. With regard to CO_2, 7.9% (n = 5) of the measurements exceeded the protection threshold (1250 ppm) in the offices. As for the CO, of the 63 measurements made, no results were recorded outside the reference value [0–9 ppm] (Table 3) .

It was found that there were statistically significant differences between the average values of the concentration of ultrafine particles in the different places in study. Service 4 had the highest average values, while service 1 had the lowest average values.

Next, we tried to confirm the occurrence of exceedances to the legally established value for the meteorological variables (T° and Rh) in the indoor air (Table 4).

Taking into account the data referring to T°, we found that of the 63 measurements (n = 21) carried out in offices and services, 33.3% had temperatures outside the legally established range [18°-22 °C], with the majority of these altered temperatures being registered in the offices. Regarding the Rh, of the 63 measurements (n = 21), 33.3% also had values outside the reference range [50–70%]. In this case, it was the services that recorded the most measurements with altered humidity values.

Then, it was intended to evaluate the evolution of the average concentrations of PM2.5, PM10, CO_2, CO and ultrafine particles of the evaluated places, between the different measurement phases. Let's see Table 5.

Table 3. Average concentration of ultrafine particles inside the evaluated offices and services.

Evaluated places	Ultrafine particles	
	Mean	Standard deviation
Office 1	3678	1486
Office 2	3819	1866
Office 3	12005	22457
Service 1	2599	1255
Service 2	7416	3056
Service 3	4702	1682
Service 4	22435	25449

p-value < 0,0001; N = 9; Test: Kruskall Wallis

Table 4. Meteorological variables T° and Rh in the evaluated offices and services.

Evaluated places		Temperature			Relative humidity		
		<18 °C	[18 °C–22 °C]	>22 °C	<50%	[50%–70%]	>70%
Offices	n (%)	1 (3,7%)	14 (51,9%)	12 (44,4%)	8 (29,6%)	19 (70,4%)	0 (0%)
Services	n (%)	1 (2,8%)	28 (77,8%)	7 (19,4%)	13 (36,1%)	23 (63,9%)	0 (0%)
Total	n (%)	2 (3,2%)	42 (66,7%)	19 (30,1%)	21 (33,3%)	42 (66,7%)	0 (0%)

Test: x2 of Independence

It was found that there were no statistically significant differences between the mean values of PM2.5 and the different phases of measurement (p-value = 0.251). The measurement phase with the highest values was the "middle of the occupation", while the "final phase of the occupation" was the one with the lowest values. As for the average concentrations of PM10, it was observed that there were no statistically significant differences between the average values and the measurement phases (p-value = 0.187). In the "middle of occupation" the highest values were presented, while in the "final phase of occupation" the lowest values were presented. Regarding the average CO_2 values, there were statistically significant differences between the average values and the measurement phases (p-value = 0.001). It was in the "final phase of occupation" that the highest average values were registered, unlike "before occupation" that the lowest values were found. With regard to the average values of CO concentration, it was found that there were no statistically significant differences (p-value = 0.057), and the values presented were similar. Regarding the average concentration values of the ultrafine particles, it was

Table 5. Average concentrations of PM2.5, PM10, CO_2, CO and ultrafine particles in the different measurement phases.

		Medication period		
		Before occupation	Middle of occupation	Final phase of occupation
$PM_{2.5}$	Mean	0,009	0,010	0,007
	Standard deviation	0,006	0,008	0,004
	n	21	21	21
PM_{10}	Mean	0,027	0,032	0,025
	Standard deviation	0,016	0,019	0,015
	n	21	21	21
CO_2	Mean	444	756	896
	Standard deviation	157	314	549
	n	21	21	21
CO	Mean	1,8	1,8	1,8
	Standard deviation	0,1	0,0	0,0
	n	21	21	21
Ultrafine particles	Mean	5280	13383	5618
	Standard deviation	3893	22979	4081
	n	21	21	21

Test: Kruskall Wallis

observed that there were no statistically significant differences in the different phases of measurement (p-value = 0.686), with the "middle of occupation" having the highest values and "before occupation" the lowest values.

Table 6 shows the average concentrations of the different pollutants in two situations, one in which there is at least one open ventilation source (window, door and artificial ventilation) and another situation in which all ventilation sources are closed/off.

Analysing the results, regarding PM2.5 and PM10, it was found that there were statistically significant differences between the average concentrations and the different types of assessment (p-value = 0.005; p-value = 0.018, respectively). In both cases, higher mean values were found in the presence of at least one source of ventilation. Regarding CO_2, it was found that there are no statistically significant differences between the average concentrations and ventilation (p-value = 0.840), with the highest values being recorded when there is no ventilation. With regard to CO, it was found that there were no statistically significant differences (p-value = 0.805) and that the mean values

Table 6. Average concentrations of pollutants and the presence or absence of ventilation.

Ventilation source	n	$PM_{2.5}$		PM_{10}		CO_2		CO		Ultrafine particles	
		M	SD	M	SD	M	SD	M	SD	M	SD
Open/On	40	0,010	0,007	0,031	0,019	690	457	1,8	0,1	10943	16905
Closed/Off	23	0,006	0,002	0,023	0,009	713	343	1,8	0,1	3138	1418

M-Mean; SD-Standard Deviation; Open/On: At least one open/connected ventilation source; Closed/Off: All ventilation sources closed/off; Tests: t-Student for 1 sample, x2 of Independence, Fisher's Exact

are similar in the presence or absence of ventilation. As for ultrafine particles, it was observed that there were statistically significant differences (p-value = 0.006), with higher values when there was at least one source of ventilation.

Then we tried to establish a relationship between the average values of the concentrations of air pollutants evaluated inside and outside. Let's look at the following table. As seen in Table 7, it was found that the average values of PM2.5 in service 4 were higher than the average values from outdoor. With regard to PM10 and CO, their average values outside were higher than the average values recorded inside and despite some average variation of PM10 and CO between the evaluated spaces, the estimates proved to be relatively homogeneous. As for CO_2, all the average values recorded inside were higher than the average values outside, with the highest value recorded in office 3. With regard to ultrafine particles, only office 3 and service 4 had higher average values than those from outside, and service 4 showed the highest values.

Table 7. Relation between the average values of the concentrations of air pollutants evaluated

Evaluated places	n	$PM_{2.5}$		PM_{10}		CO_2		CO		Partículas Ultrafinas	
		M	SD	M	SD	M	SD	M	SD	M	SD
Office 1	9	0,006	0,003	0,022	0,008	846	227	1,8	0,1	3678	1486
Office 2	9	0,007	0,002	0,024	0,009	823	338	1,8	0,1	3819	1866
Office 3	9	0,005	0,002	0,019	0,004	1196	646	1,8	0,0	12005	22457
Service 1	9	0,006	0,002	0,020	0,009	775	283	1,8	0,0	2599	1255
Service 2	9	0,010	0,004	0,043	0,019	385	84	1,8	0,1	7416	3056
Service 3	9	0,006	0,003	0,018	0,005	571	171	1,8	0,0	4702	1682
Service 4	9	0,021	0,008	0,051	0,016	296	61	1,8	0,0	22435	25449
Outdoor	8	0,018	0,009	0,057	0,023	233	57	1,9	0,1	9976	6726

M-Mean; SD-Standard Deviation; Tests: t-Student for independent samples, Mann Whitney

4 Discussion

After analysing the results obtained, it was found that, on average, PM2.5 concentrations in offices and services were below the protection threshold (0.025 mg/m^3). When there is at least one source of ventilation, higher values of PM2.5 were found, which indicates that the outside air may be influencing the indoor air, taking into account that one of the origins of the particulate matter is the outside air [4]. When comparing the concentration of PM2.5 from the outside with the inside, it was found that service 4, which concerns the refectory, is the only one with higher average values than the outside. Since the measurements were carried out in the normal period of operation of the refectory and as the particles result from human activities, as is the case of cooking, this may be related to the value of PM2.5 recorded in this place [9].

Regarding the concentration of PM10, it was found that service 4 presented average concentration values above the protection threshold. Bearing in mind that this service concerns the refectory, this higher value may be due to the resuspension of dust due to the activities and movements of the occupants [8], but also to the preparation of food, as previously mentioned in PM2.5. When there is at least one source of ventilation, higher values of PM10 were found and, in addition, when comparing the outside with the inside, higher values were observed outside, which generally indicates that the outside air is the main source of indoor air contamination [9].

Regarding the CO_2 concentration, it was found that all average values were below the protection threshold, however office 3 was the one with the highest average values. Since in some measurements in this office a candle was lit and CO_2 is produced by incomplete combustion [13], so the burning of the candle may be related to the values found in the office. When analysing by measurement phase, it was observed that the average values of CO_2 increased throughout the day, which can be justified with the metabolic activity of the occupants, since this is one of the main sources of CO_2 [13]. Low CO_2 levels indicate that indoor spaces are well ventilated [4], thus, it was found that when there was a ventilation source, lower average CO_2 values were recorded.

Regarding the average values of CO concentration in the evaluated spaces, it was observed that they were below the limit value imposed in the legislation. This pollutant results from human activities, as is the case with incomplete combustion processes of organic materials [14].

With regard to ultrafine particles, it was found that the highest average concentrations were recorded in office 3 and service 4. In office 3, as discussed above, a candle was lit at some measurement times and as the ultrafine particles can be emitted by combustion processes, such as candles [15], this may be associated with the high value of these particles in office 3. As regards the high values recorded in service 4 (refectory), these can be related not only to the combustion processes, since in this space the cooking takes place food, but also with other occupant activities [1]. In addition, this space was always with doors open and outside were often occupants smoking. It was also observed that when there is a source of active ventilation, the concentrations of ultrafine particles inside the spaces evaluated are higher, which was also verified in other studies [1].

T° and Rh in indoor spaces are important aspects for the spaces themselves and for the health of the occupants and are directly related to the structure and quality of buildings and their comfort [16]. From the analysis of the average values of T° found inside the evaluated spaces, it was found that 3.2% had temperatures below the reference value [18–22 °C] and 30.1% had temperatures above this same value. Temperatures below 18 °C may be related to the fact that the measurements were made in the winter and the heating systems are turned off, when temperatures above 22 °C were found, these can be explained by the heating systems in operation. With regard to Rh in the indoor, it was found that 33.3% had Rh levels below the reference range [50–70%]. On the days when measurements were made, there was some precipitation and thus, the Rh in the outside air exceeded 70%, however, due to the low levels of Rh obtained inside, it is believed that the outside air did not influence the indoor air. Several studies report that low Rh values can cause dryness, irritation of the skin and mucous membranes and respiratory tract infections [16].

5 Conclusion

IAQ in homes, schools, companies or other public or private buildings is an essential and determinant factor for the comfort and well-being of the users of these spaces. This was identified by Environmental Protection Agency (EPA) as one of the top five environmental risks to public health. It is estimated that people spend about 90% of their time indoors and therefore some experts believe that people may be more affected by indoor air pollution than outside air [17].

Through the results obtained in this study, it can be concluded that it is necessary to take measures in order to improve the IAQ in the offices and services of ESTeSC, since, on average, the concentrations of PM10 in some evaluated places, were higher than the values legally established. Therefore, continuous monitoring must be carried out in order to prevent the occupants of the establishment from being exposed to risky situations.

It also highlights the importance of improving air renovation systems, opting whenever possible for natural ventilation, using for example the frequent opening of windows.

During this study, some limitations were found, such as the lack of specific legislation regarding the concentration of air pollutants in outdoor and ultrafine particles.

IAQ is a highly important environmental risk factor, since it can affect not only productivity, but also the health and cognitive functions of humans, who spend a large part of their time in indoor, so it is essential to manage and monitor IAQ [18].

References

1. Slezakova, K., de Oliveira Fernandes, E., do Carmo Pereira, M.: Assessment of ultrafine particles in primary schools: emphasis on different indoor microenvironments. Environ. Pollut. **246**, 885–895 (2019).https://doi.org/10.1016/j.envpol.2018.12.073
2. da Conceição Ferreira, A.M.: Qualidade do Ar Interior em Escolas e Saúde das Crianças, pp. 1–393 (2014). https://hdl.handle.net/10316/26262
3. Fernandes, A.R.: Análise da qualidade do ar e preocupações com a saúde, pp. 1–80 (2017). https://repositorio-aberto.up.pt/bitstream/10216/102619/2/180885.pdf
4. Nunes, D.P.: Avaliação da Qualidade do Ar em Ginásios e a Exposição Humana a Poluentes Atmosféricos, p. 228 (2018)

5. Silva, Â.E.S.: Síndrome do Edifício Doente (2017)
6. Thomas, N.M., Calderón, L., Senick, J., Sorensen-Allacci, M., Plotnik, D., Guo, M., Yu, Y., Gong, J., Andrews, C.J., Mainelis, G.: Investigation of indoor air quality determinants in a field study using three different data streams. Build. Environ. **154**(March), 281–295 (2019). https://doi.org/10.1016/j.buildenv.2019.03.022
7. Majd, E., McCormack, M., Davis, M., Curriero, F., Berman, J., Connolly, F., Leaf, P., Rule, A., Green, T., Clemons-Erby, D., Gummerson, C., Koehler, K.: Indoor air quality in inner-city schools and its associations with building characteristics and environmental factors. Environ. Res. **170**(December 2018), 83–91 (2019). https://doi.org/10.1016/j.envres.2018.12.012
8. Kalimeri, K.K., Bartzis, J.G., Sakellaris, I.A., de Oliveira Fernandes, E.: Investigation of the PM2.5, NO2 and O3 I/O ratios for office and school microenvironments. Environ. Res. **179**(2), 108791 (2019). https://doi.org/10.1016/j.envres.2019.108791
9. Morawska, L., Ayoko, G.A., Bae, G.N., Buonanno, G., Chao, C.Y.H., Clifford, S., Fu, S.C., Hänninen O., He, C., Isaxon, C., Mazaheri, M., Salthammer, T., Waring, M.S., Wierzbicka, A.: Airborne particles in indoor environment of homes, schools, offices and aged care facilities: the main routes of exposure. Environ. Int. **108**(April), 75–83 (2017). https://doi.org/10.1016/j.envint.2017.07.025
10. Fonseca, C.C.: Ultrafine particles in children's home environments, July 2015
11. Esteves, H., Gomes, J.F., Miranda, R.M., Albuquerque, P.: Exposição ocupacional a nanopartículas na indústria cerâmica, October 2018
12. Loxham, M., Davies, D.E., Holgate, S.T.: The health effects of fine particulate air pollution. BMJ **367**(November), l6609 (2019). https://doi.org/10.1136/bmj.l6609
13. Coentro, S.: Qualidade do Ar Interior em habitações : Fontes emissoras de poluentes. Dissertação de Mestrado Em Engenharia Do Ambiente Da Universidade Nova de Lisboa (2015)
14. Loureiro, A.: Qualidade Do Ar Interior Em Lares De Idosos E a Exposição Ocupacional Aos Poluentes Atmosféricos, pp. 1–9 (2015)
15. Fonseca, J., Slezakova, K., Morais, S., Pereira, M.C.: Assessment of ultrafine particles in Portuguese preschools: levels and exposure doses. Indoor Air **24**(6), 618–628 (2014). https://doi.org/10.1111/ina.12114
16. Abreu, C.F.C.: Capítulo 3: Qualidade do Ar Interior. O Ambiente Interior e a Saúde Dos Ocupantes de Edifícios de Habitação, pp. 23–68 (2010). https://hdl.handle.net/10400.6/3579
17. OSHA: Indoor Air Quality in Commercial and Institutional Buildings (2011). https://www.osha.gov/Publications/3430indoor-air-quality-sm.pdf
18. Fernandes, S., Igrejas, G., Feliciano, M.: Monitorização e controlo de qualidade do ar interior em gabinetes de edifícios escolares. Revista de Ciências Agrárias **40**(SP), S274–S281 (2017). https://doi.org/10.19084/rca16192

Indoor Air Quality, Ultrafine Particles in Laboratories and Classrooms of Coimbra Health School

Fernando Moreira(✉), Ana Ferreira, João Paulo Figueiredo, and Ruben Ferreira

Instituto Politécnico de Coimbra, ESTeSC, Coimbra, Portugal
fernando.moreira@estescoimbra.pt

Abstract. Indoor air quality refers to the air quality inside buildings and structures. The causes of indoor air pollution are a combination of physical, chemical and biological factors, and are also related to the ventilation. In teaching places, an inadequate indoor air quality can lead to health problems and can also reduce learning performance, attendance and comfort. The present study aimed to evaluate expose students and teachers to indoor air pollutants in laboratories and classrooms of Coimbra Health School, check if pollutants complied with the limits of legislation and evaluate the exposure to ultrafine particles in different environments. The data collection was based on the assessment of air quality in three laboratories and four classrooms, using air pollutants and atmospheric variables. We checked that the average concentrations of PM10 and CO_2 in some laboratories and classrooms exceeded the protection threshold and that the average values of ultrafine particles were higher when combustion processes existed. With this study, it was possible to conclude that some measures are necessary to improve indoor air quality, once that some atmospheric pollutants exceeded the values imposed by legislation. So the institution needs to do frequent monitoring to avoid exposing occupants to risky situations.

Keywords: Indoor air quality · Laboratories · Classrooms · Atmospheric pollutants · Ultrafine particles

1 Introduction

One of the top five environmental risks to public health is indoor air pollution, inside the buildings the levels of pollutants can be two to five times higher than outside and in extreme cases can be a hundred times higher. Nowadays society spend a large part of the day indoors, making these levels of contamination of great importance [1].

Air quality can be translated by the level of air pollution, which is caused by the mixture of chemicals that are released into the air or resulting from chemical reactions and that changes the natural composition of the atmosphere, causing negative consequences for public health and well-being of populations, fauna, flora and buildings [2].

The main indoor and outdoor air pollutants are carbon monoxide (CO), carbon dioxide (CO_2) and particulate matter (PM2.5 and PM10) [3]. The increase of mortality and

J. R. da Costa Sanches Galvão et al. (Eds.): *Proceedings of the 1st International Conference on Water Energy Food and Sustainability (ICoWEFS 2021)*, pp. 537–547, 2021.
https://doi.org/10.1007/978-3-030-75315-3_58

morbidity rates has been associated with exposure to inadequate air, leading to respiratory and cardiovascular diseases, and currently 6,5 million people are estimated to die from air pollution [4].

Indoor air quality (IAQ) refers to the air quality inside buildings and structures. The causes of indoor air pollution are a combination of physical, chemical and biological factors, and are also related to the ventilation [1].

The World Health Organization define the "sick building syndrome" as the set of diseases and discomfort suffered by people who live in a building. This is caused by inadequate ventilation, temperature and relative humidity, lighting condition and others [5].

Poor IAQ can lead to several symptoms, the most common being headaches, fatigue, nasal congestion, coughing, sneezing and mucous irritations. In addition to this symptoms, it may cause even more serious consequences [6].

The Portuguese Environment Agency report that several dangerous compounds are released inside building from various sources, such as construction materials, cleaning products and equipment, office supplies [7] and occupant activities (smoking, pets) [8].

Regarding teaching places, an inadequate IAQ in classrooms can lead to health problems for occupants, and can also reduce learning performance, attendance and comfort [9]. Occupation is a factor associated with the levels of particulate matter, especially PM10, due to the resuspension of deposited particles [10]. Thus, in this context we can verify the existence of two particularly critical parameters, namely, the density of occupation and the duration of teaching periods and breaks [9].

Ultrafine particles are generally defined as the fraction of particles with a diameter less than 100nm and due to their size, they can enter in the bloodstream causing various adverse effects on the health of the population, using inhalation as the major exposure route [4]. Several epidemiological studies associate the presence of ultrafine particles in the atmosphere with total mortality, respiratory and cardiovascular diseases, acute airway inflammation, impaired lung function and increased asthma symptoms. These particles in outdoor environment are formed through natural and anthropogenic sources that can be introduced into the indoor environment through ventilation [11].

So it is essential monitor the IAQ, as well as to implement preventive measures in order to preserve the health of the occupants. The present study aimed to evaluate expose students and teachers to indoor air pollutants in laboratories and classrooms of Coimbra Health School, check if pollutants complied with the limits of legislation and understanding how the ultrafine particles behaved in different environments.

2 Material and Methods

- **Study design -** The present study was an observational analytical type, transversal in nature and level of knowledge II (correlational description).
- **Population and Sample** - The target population were teachers and students who were part of the 3 laboratories and 4 classrooms evaluated in ESTeSC. The type of sampling used was non-probabilistic, where the sampling technique for convenience was applied.
- **Instruments and Data Collection** - The study was carried out in the academic year 2019/2020 and the data collection period took place in March 2020.

Data collection was based on the assessment of air quality, through the assessment of air pollutants (CO_2, CO, PM2.5, PM10 and ultrafine particles) and meteorological variables temperature (T°) and relative humidity (Rh).

Measurements were made over a period of 15 min, with samples every minute APA based in Port. 353-A/2013, 4 of December. The measurements of each indoor location were performed on 3 different days, and each day they took place at 3 different times, which relate to before, to the middle and to the final phase of the occupation of each evaluated class, taking into account the measurement hours, so that the classes taught in a specific laboratory or classroom were evaluated at an approximate time. The measuring equipment was placed in the most central possible position of each assessed space and approximately at the height of the occupant's airways, in the sitting position. In this phase, the characteristics of the spaces were also considered, such as volume, windows, doors, artificial ventilation and potential sources of contamination. The measurements of outdoor air quality happened in the outside of ESTeSC, in the place that was considered most critical, but at least 1 m away from the walls and at the same height at which the measurements of IAQ were.

Data collection was made using three portable equipment's for reading in real time, namely the Q-Track Plus, brand TSI, model 8552/8554 for the assessment of the concentration of CO, CO_2, T° e Rh; the Particles Counters, brand Lighthouse Worldwide Solutions, model 3016 LAQ for the assessment of the concentrations of PM2.5 and PM10 and the P-Track Ultrafine Particle Counter, model 8525, which has a reagent with 99,5% solution of isopropyl alcohol for counting ultrafine particles. In order to transfer the data from the measuring equipment's to the computer, was used the specific software for each equipment.

It was considered as a reference for the maximum concentration of CO, 10,0 mg/m^3 (9 ppm), for CO_2, 2250 mg/m^3 (1250 ppm), for PM2.5, 0,025 mg/m^3 and for PM10, 0,05 mg/m^3, as related in Ordinance 353-A/2013, of 4 December (Port° 353-A/2013, 2013). Due to the lack of reference values for the concentration of ultrafine particles, the average analytical values recorded in the measurements of outdoor air were taken as a reference point. According to Decree-Law n. ° 243/86, of 20 august, the environmental comfort conditions of reference for T° must oscillate between 18 and 22° C and for Rh must oscillate between 50 and 70% (DL no 243/86 de 20 de Agosto, 1986).

- **Statistical analysis**

  The statistical treatment of data collected in the study was performed using the IBM SPSS Statistics version 25.0 software. Descriptive statistics techniques were used in the study, such as measures of central tendency (arithmetic mean), dispersion measures (standard deviation) and frequency tables. At the level of statistical inference, was used a previous assessment of the assumptions of the metric variables (parametric or non-parametric). Regarding hypothesis tests, Kruskal Wallis, t-Student for 1 sample, Chi-square of Independence, t-Student for Independent Samples, Fisher's Exact and Mann Whitney tests were used. Data interpretation was based on a significance level of $p = 0.05$ with a 95% confidence interval.

3 Results

- General sample characterization
  The sample of this study was confined to three laboratories and four classrooms of ESTeSC and also in outsider of the same.
- Inferential analysis
  Following the statistical treatment of the collected data, some tables were prepared. Tus, we started by analysing and comparing the values of the concentrations of PM2.5, PM10, CO_2 and CO of the indoor air of the various laboratories and classrooms with the legally established values (protection threshold). Let's see Table 1 below.

Table 1. Average concentration of PM2.5, PM10, CO_2 e CO inside the various laboratories and classrooms evaluated.

Evaluated places	$PM_{2.5}$			PM_{10}			CO_2			CO		
	Mean	Standard deviation	PL	Mean	Standard deviation	PL	Mean	Standard deviation	PL	Mean	Standard deviation	PL
Classroom 1	0,010	0,001	<	0,041	0,015	<	982	736	<	1,9	0,1	<
Classroom 2	0,012	0,005	<	0,034	0,016	<	865	478	<	1,9	0,1	<
Classroom 3	0,014	0,005	<	0,066	0,040	>	1552	556	>	1,9	0,1	<
Classroom 4	0,009	0,002	<	0,031	0,011	<	1008	510	<	1,8	0,0	<
Laboratory 1	0,005	0,002	<	0,015	0,006	<	671	162	<	1,8	0,0	<
Laboratory 2	0,013	0,005	<	0,046	0,011	<	1214	633	<	1,8	0,0	<
Laboratory 3	0,013	0,006	<	0,045	0,022	<	406	109	<	1,9	0,0	<

PL-Protection Limit; PL PM2.5 = 0,025 mg/m^3; PL PM10 = 0,05 mg/m^3; PL CO_2 = 1250 ppm; PL CO = 9 ppm; N = 9; Test: t-Student for 1 sample

It was found that there were statistically significant differences between the average PM2.5 concentration values and the protection threshold in all classrooms and laboratories (p-value < 0.05), so the average values were below the protection threshold. Regarding the divergences recorded, we can see that classroom 3 presented the highest average values, in contrast to laboratory 1 which presented the lowest average values. Regarding the mean PM10 concentration values, it was found that there were statistically significant differences between the mean values and the protection threshold in classrooms 2 and 4 and in laboratory 1 (p-value $= 0.014$; p-value $= 0.001$; p-value < 0.0001, respectively). The remaining locations did not present significant differences (p-value > 0.05). Classroom 3 had the highest average PM10 values and, in this case, it was the only place that registered values above the protection threshold. On the other hand, laboratory 1 was the one with the lowest average values.

As for the average values of CO_2 concentration, it was found that there were statistically significant differences between the average values and the protection threshold in classroom 2 and in laboratories 1 and 3 (p-value $= 0.042$; p-value < 0.0001; p-value < 0.0001, respectively). The remaining locations didn't show significant differences (p-value > 0.05). Regarding the average CO_2 values, it was also classroom 3 that registered higher average values and the only one to exceed the protection threshold. However, laboratory 3 showed the lowest average values. It was also observed that there were statistically significant differences between the average values of CO concentration and the protection threshold in all classrooms and laboratories (p-value < 0.05), so the average values were below the protection threshold. The average CO values were all very similar.

Next, it was sought, between classrooms and laboratories in general, to assess in which group there were more exceedances compared to the reference value imposed in the legislation for pollutants (PM2.5, PM10, CO_2 and CO) in indoor air. Let's see Table 2.

Table 2. Pollutants (PM2.5, PM10, CO_2 e CO) in the indoor air of labs and classrooms.

		$PM_{2.5}$		PM_{10}		CO_2		CO	
		[0 – 0,025] mg/m^3	>0,025 mg/m^3	[0 – 0,050] mg/m^3	>0,050 mg/m^3	[0 – 1250] ppm	>1250 ppm	[0 – 9] ppm	>9 ppm
Classrooms	n	36	0	25	11	22	14	36	0
	(%)	(100%)	(0%)	(69,4%)	(30,6%)	(61,1%)	(38,9%)	(100%)	(0%)
Laboratories	n	26	1	20	7	22	5	27	0
	(%)	(96,3%)	(3,7%)	(74,1%)	(25,9%)	(81,5%)	(18,5%)	(100%)	(0%)
Total	n	62	1	45	18	44	19	63	0
	(%)	(98,4%)	(1,6%)	(71,4%)	(28,6%)	(69,8%)	(30,2%)	(100%)	(0%)

Test: x2 of Independence

Analysing the results for PM2.5, we found that of the 63 measurements, only 1.6% (n = 1) of the measurements were outside the legally established range [0–0.025 mg/ m^3]. It was in the laboratories that a measurement was found that exceeded the protection threshold.

Regarding the PM10 results, 28.6% (n = 18) presented values outside the reference range [0–0.050 mg/ m^3]. In this case, the exceedances occurred mainly in the classrooms (n = 11). As for CO_2, 30.2% (n = 9) had values outside the reference range [0–1250 ppm], with the majority of cases also occurring in classrooms (n = 14). Regarding the CO, we found that all estimates of 63 measurements in laboratories and classrooms, showed values within the reference range [0–9 ppm].

The following table shows the average results of the ultrafine particles in the different evaluated places (Table 3).

It was found that there were statistically significant differences between the average values of the ultrafine particles in the different evaluated places ($p < 0.0001$). Laboratory 2 was the one with the highest average values, unlike laboratory 1, which had the lowest average values.

Then we tried to check if there was an exceedance of the legislated value for the meteorological variables (T° and Rh) in the indoor air. Let's see Table 4.

We verified the results regarding T° and it was found that 63.5% (n = 40) presented values outside the reference range [18°C-22°C]. It was in the laboratories that we verified the majority of cases outsider of the proper value. Regarding Rh, 25.4% (n = 16) presented values outside the legally established range [50%–70%], with the majority of cases also occurring in laboratories (n = 12).

Table 3. Average concentration of ultrafine particles inside the various laboratories and classrooms evaluated.

Evaluated places	Ultrafine particles	
	Mean	Standard deviation
Classroom 1	4626	893
Classroom 2	5667	2743
Classroom 3	9223	2793
Classroom 4	3877	499
Laboratory 1	3518	3080
Laboratory 2	56020	36184
Laboratory 3	25244	7191

p-value < 0,0001; N = 9; Test: Kruskall Wallis

Table 4. Meteorological variables T° and Rh in the laboratories and classrooms evaluated.

Evaluated places		Temperature Classes			Relative Humidity Classes		
		<18°C	[18 °C–22 °C]	>22 °C	<50%	[50%–70%]	>70%
Classrooms	n (%)	0 (0%)	17 (47,2%)	19 (52,8%)	4 (11,1%)	32 (88,9%)	0 (0%)
Laboratories	n (%)	0 (0%)	6 (22,2%)	21 (77,8%)	12 (44,4%)	15 (55,6%)	0 (0%)
Total	n (%)	0 (0%)	23 (36,5%)	40 (63,5%)	16 (25,4%)	47 (74,6%)	0 (0%)

Test: x2 of Independence

Then we set out to assess the evolution of concentrations of PM2.5, PM10, CO_2, CO and ultrafine particles over the different measurement phases ("before occupation", "middle of occupation" and in the "final phase" occupation" of laboratories and classrooms). Let's look at the following Table 5.

It was observed that there were no statistically significant differences between the mean values of PM2.5 and the different phases of measurement (p-value = 0.198). It was in the "final phase of occupation" that higher average values were registered, in contrast to the phase "before occupation" that lower average values were recorded. Regarding PM10, it was found that there were statistically significant differences between the average values and the different measurement phases (p-value = 0.005). In this case, it was the "middle of occupation" phase that presented the highest average values and the "before occupation" phase with the lowest average values. There were also statistically significant differences between the average CO_2 values and the different measurement phases (p-value < 0.0001). It was in the "before occupation" phase that lower average values were registered, in contrast to the "final phase of occupation" that higher average values were observed. With regard to CO, there were no statistically significant differences (p-value = 0.068) and the highest average value was in the "before occupation" phase. In the case of ultrafine particles, there were no statistically significant differences

Table 5. Evolution of concentrations of $PM_{2.5}$, PM_{10}, CO_2, CO and ultrafine particles in different phases of measurement.

		Measurement phase		
		Before occupation	Middle of occupation	Final phase of occupation
$PM_{2.5}$	Mean	0,010	0,011	0,012
	Standard deviation	0,005	0,005	0,006
	n	21	21	21
PM_{10}	Mean	0,032	0,044	0,043
	Standard deviation	0,032	0,019	0,018
	n	21	21	21
CO_2	Mean	500	1122	1248
	Standard deviation	281	531	618
	n	21	21	21
CO	Mean	1,9	1,8	1,8
	Standard deviation	0,1	0,0	0,1
	n	21	21	21
Ultrafine particles	Mean	9531	20282	16547
	Standard deviation	7925	29438	24169
	n	21	21	21

(p-value = 0.815) and it was found that the "middle of occupation" phase showed the highest average values, while the "before occupation" phase showed the highest lower average values (Table 6).

Table 6. Average concentration of pollutants and the presence or absence of ventilation.

Ventilation source	n	$PM_{2.5}$		PM_{10}		CO_2		CO		Ultrafine Particles	
		M	SD	M	SD	M	SD	M	SD	M	SD
Open/On	51	0,011	0,006	0,040	0,026	862	496	1,8	0,1	16721	24566
Closed/Off	12	0,009	0,002	0,040	0,014	1359	790	1,8	0,1	10066	9140

M-Mean; SD-Standard Deviation; Open/On: At least one source of ventilation; Closed/off: No ventilation source; Tests: t-Student for 1 sample, x2 of Independence, Fisher's Exact

Then, we tried to compare the average concentrations of the different pollutants between two situations, when they are without any type of ventilation and when there is at least one source of ventilation (windows, door and artificial ventilation).

There were statistically significant differences between the mean values of PM2.5 and the presence or absence of ventilation. It was in the presence of at least one ventilation source that the highest mean values were recorded. In the case of PM10 and CO, there were no statistically significant differences (p-value = 0.995; p-value = 0.794, respectively) and it was found that the average values with the presence or absence of ventilation are similar. In the case of CO_2, there were statistically significant differences, in which the highest average values were recorded in the absence of any source of ventilation. Regarding ultrafine particles, there were no statistically significant differences (p-value = 0.362). The highest mean values were recorded in the presence of at least one source of ventilation.

Finally, an attempt was made to establish a relation between the estimated values of air pollutants in the different indoor places evaluated and the average values outside (Table 7).

Table 7. Relation between the average values of the pollutants in the different indoor places evaluated and the values outside

Evaluated places	n	$PM_{2.5}$		PM_{10}		CO_2		CO		Ultrafine Particles	
		Mean	Standard Deviation	Mean	Standard Deviation	Mean	Standard Deviation	Mean	Standard Deviation	Mean	Standard Deviation
Classroom 1	9	0,010	0,001	0,041	0,015	982	736	1,9	0,1	4626	893
Classroom 2	9	0,012	0,005	0,034	0,016	865	478	1,9	0,1	5667	2743
Classroom 3	9	0,014	0,005	0,066	0,040	1552	556	1,9	0,1	9223	2793
Classroom 4	9	0,009	0,002	0,031	0,011	1008	510	1,8	0,0	3877	499
Laboratory 1	9	0,005	0,002	0,015	0,006	671	162	1,8	0,0	3518	3080
Laboratory 2	9	0,013	0,005	0,046	0,011	1214	633	1,8	0,0	56020	36184
Laboratory 3	9	0,013	0,006	0,045	0,022	406	109	1,9	0,0	25244	7191
Outdoor	8	0,019	0,008	0,075	0,034	227	62	1,9	0,1	5379	3449

Tests: t-Student for independent samples, Mann Whitney.

Analysing the previous table, it was found that the average values of PM2.5 and PM10 outside were higher than the average values recorded in the various indoor places. With regard to CO_2, all the average values recorded in the indoor places were higher than the average values obtained outside. The highest average CO_2 values were in classroom 3. As for CO, the average values were very similar inside and outside. In the case of ultrafine particles, classrooms 1 and 4 and laboratory 1 were below the average values from the outside, unlike the other places that had higher average values, with the highest values recorded in laboratory 2.

4 Discussion

After analyzing the mean values of PM2.5, it was observed that all laboratories and classrooms evaluated didn't exceed, on average, the protection threshold. With this, only one measurement was recorded in laboratory 3 that exceeded the threshold. As for the PM10 values, it was found that the measurements that exceeded the protection threshold (0.050 mg/m^3) were mostly registered in the classrooms, however only classroom 3 showed average values that exceeded the threshold. In the indoor, the particulate matter may have origin in the materials present in buildings, but also in the occupants, either by the agitation caused in the air, by the release of small particles of clothing or even by the flaking of the skin [1], hence it is verified that the average values of PM2.5 and PM10 are lower in the phase before occupation. It was also found that the average values of these pollutants existed in greater quantities on the outside, in comparison with the inside, and the outside air may be the main source of contamination of the interior spaces [12].

Regarding the concentration of CO_2, it was found that most of the exceedances to protection threshold were registered in the classrooms; however only the average values in classroom 3 exceeded the value of 1250 ppm. It was observed that classroom 3 was also the one with the highest number of occupants at the time of assessment, which caused a greater release of CO_2 as a result of metabolic activity [13]. Dense occupation, as well as poor ventilation, are conditions that lead to increased CO_2 concentrations [1], thus justifying the higher average values in the absence of ventilation. A study also states that the high concentration of CO_2 in classrooms is a good indicator for the need to open windows, as it impairs students' attention and productivity [14].

From the analysis of the average values of CO concentration, it was found that they were below the protection threshold (9 ppm). CO is mainly associated with combustion processes, garages and infiltrations from outside [15].

With regard to ultrafine particles, the average values recorded in laboratories 2 and 3 and in classrooms 2 and 3 were higher than the average values in outdoor. It was found that laboratories 2 and 3 were the ones that stood out the most due to their high values, which may be related to combustion processes [4] that took place in the normal development of laboratory activities, such as the use of the Bunsen burner. It was also found that the laboratories were the only evaluated places that were on the ground floor, and this context may lead to significantly higher values [4]. As previously mentioned, classroom 3 was the one with the highest occupation density, which may be related to the levels of ultrafine particles, since it was found that there is an association between the levels of these particles and the occupation of the place (area per student) due to activities performed by occupants [4]. In addition, outdoor air pollution is also considered to be one of the biggest sources of indoor air contamination, which may be associated with higher average values in the presence of ventilation [3].

T° and Rh are also relevant factors, as thermal comfort enables greater productivity of man in his leisure and work activities [16]. From the analysis of the average T° values, it was found that 63.5% had temperatures higher than the legislated value, with the majority of the exceedances recorded in the laboratories. Temperatures above the reference value may be related to insufficient ventilation. With regard to the average Rh values, it was found that 25.4% had levels lower than the legislated value, with the majority being registered in the laboratories. Thus, from the values recorded in the laboratories, there was a possible association between high temperature and low relative humidity. In the data collection, there were days with some precipitation and with Rh values in outdoor above 70%, however there was no case in laboratories and classrooms where Rh was higher than the legislated value. It is important to note that low humidity values increase the risks of dryness and irritation of the skin and eyes, respiratory infections and allergic rhinitis, as well as the increase in the levels of bacteria and viruses [16].

5 Conclusion

The IAQ of building is an essential determinant of people's lives and well-being, once that most spend an average of 87% of their time indoors and 5% in vehicles. Indoor air pollution can increase the probability of developing health problems in the short and long term [17].

With this study, it was possible to conclude that some measures are necessary to improve the IAQ, once that some atmospheric pollutants exceeded the values imposed by legislation in some laboratories and classrooms. So the institution needs to do frequent monitoring to avoid exposing occupants to risky situations.

It is important to have a good air renovation system, opting whenever possible for natural ventilation, such as, opening windows frequently.

During the preparation of the study, some limitations were found, such as the lack of specific legislation for the reference values, with regard to the concentrations of air pollutants in outdoor and ultrafine particles.

Indoor air pollution in buildings is directly reflected in the health and consequently in the well-being and productivity of occupants, and the effects of air quality should be seen as a public health issue. In the case of educational establishments, IAQ can affect students and teachers, harming the teaching quality or having effects on attendance. Therefore it is important that there is concern in controlling air quality, using the main tools, such as ventilation [18].

References

1. Nunes, D.P.: Avaliação da Qualidade do Ar em Ginásios e a Exposição Humana a Poluentes Atmosféricos. 228 (2018)
2. Loureiro, A.: Qualidade Do Ar Interior Em Lares De Idosos E a Exposição Ocupacional Aos Poluentes Atmosféricos, pp. 1–9 (2015)
3. Fernandes, A.R.: Análise da qualidade do ar e preocupações com a saúde, pp. 1–80 (2017)
4. Sllezakova, K., de Oliveira Fernandes, E., do Carmo Pereira, M.: Assessment of ultrafine particles in primary schools: Emphasis on different indoor microenvironments. Environ. Pollut. **246**, 885–895 (2019)
5. Longhi, F.T.: Arquitetura Residencial Multifamiliar mais saudável: Proposição de um Guia Orientativo Composto por diretrizes projetuais saudáveis. Вестник Росздравнадзора **6**, 5–9 (2017)
6. Esteves, H., Gomes, J.F., Miranda, R.M., Albuquerque, P.: Exposição ocupacional a nanopartículas na indústria cerâmica, October 2018
7. Agência Portuguesa do Ambiente: Qualidade do Ar Interior (2016). Accessed 18 March 2020

8. Thomas, N.M., Calderón, L., Senick, J., Sorensen-Allacci, M., Plotnik, D., Guo, M., Yu, Y., Gong, J., Andrews, C.J., Mainelis, G.: Investigation of indoor air quality determinants in a field study using three different data streams. Build. Environ. **154**(November 2018), 281–295 (2019)
9. Madureira, J., Paciência, I., Pereira, C., Teixeira, J.P., de O. Fernandes, E.: Indoor air quality in Portuguese schools: levels and sources of pollutants. Indoor Air **26**(4), 526–537 (2016)
10. Majd, E., McCormack, M., Davis, M., Curriero, F., Berman, J., Connolly, F., Leaf, P., Rule, A., Green, T., Clemons-Erby, D., Gummerson, C., Koehler, K.: Indoor air quality in inner-city schools and its associations with building characteristics and environmental factors. Environ. Res. 170(November 2018), 83–91 (2019)
11. Viitanen, A.-K., Uuksulainen, S., Koivisto, A.J., Hämeri, K., Kauppinen, T.: Workplace measurements of ultrafine particles—a literature review. Ann. Work Exposures Health **61**(7), 749–758 (2017)
12. Morawska, L., Ayoko, G.A., Bae, G.N., Buonanno, G., Chao, C.Y.H., Clifford, S., Fu, S.C., Hänninen, O., He, C., Isaxon, C., Mazaheri, M. Wierzbicka, A.: Airborne particles in indoor environment of homes, schools, offices and aged care facilities: the main routes of exposure. Environ. Int. **108**(August), 75–83 (2017)
13. Coentro, S.: Qualidade do Ar Interior em habitações : Fontes emissoras de poluentes. Dissertação de Mestrado Em Engenharia Do Ambiente Da Universidade Nova de Lisboa (2015)
14. Stabile, L., Dell'Isola, M., Russi, A., Massimo, A., Buonanno, G.: The effect of natural ventilation strategy on indoor air quality in schools. Sci. Total Environ. **595**, 894–902 (2017)
15. Ramos, C.A., Wolterbeek, H.T., Almeida, S.M.: Exposure to indoor air pollutants during physical activity in fitness centers. Build. Environ. **82**, 349–360 (2014)
16. Amorim, D.: Otimização das condições de ventilação natural em edifícios de serviço para assegurar a qualidade do ar interior sem consumos excessivos de energia, pp. 1–10 (2012)
17. Yang, J., Nam, I., Yun, H., Kim, J., Oh, H.-J., Lee, D., Jeon, S.M., Yoo, S.H., Sohn, J.-R.: Characteristics of indoor air quality at urban elementary schools in Seoul, Korea: assessment of effect of surrounding environments. Atmos. Pollut. Res. **6**(6), 1113–1122 (2015)
18. Sanguessuga, M.: Síndroma dos Edifícios Doentes. Portal Da Saúde Pública (2012). https://www.saudepublica.web.pt/05-promocaosaude/054-SOcupacional/SED_analise.htm

Towards the Concept of a Digital Green Twin for a Sustainable Product Lifecycle

Jan Michael[1(✉)], Eva-Maria Grote[1], Stefan Achilles Pfeifer[1], Rik Rasor[1], Christian Henke[1], Ansgar Trächtler[2], and Lydia Kaiser[3]

[1] Fraunhofer-Institute-for Mechatronic Systems Design IEM, Zukunftsmeile 1, 33102 Paderborn, Germany
jan.michael@iem.fraunhofer.de

[2] Research Group Control Engineering and Mechatronics, Heinz Nixdorf Institute, University of Paderborn, Fürstenallee 11, 33102 Paderborn, Germany

[3] Technical University of Berlin, Straße des 17. Juni 135, 10623 Berlin, Germany

Abstract. Sustainability these days is required more than before. Resources are becoming rare goods, the pollution of air and water rises. Modern mechatronic systems consist of a high variety of components and materials and achieved high market penetration in recent years. As a result, the percentage of such systems of the total waste per year rises. The worldwide e-waste in 2019 made 53,6 million tons, an increase of 9,2 million tons, compared to 2014 (Global E-waste Monitor). That suggests that the problem of recycling gets more complicated and the risk of an increase of pollution will be higher in future. To handle this challenge, innovative concepts are required. The need of an efficient recycling must be considered much earlier than today. Lifecycle phases (Conceive, Design, Realize, Service, Reuse) show high efficiency in their procedures and tools, used to ensure quality and shorter cycle times. The aspect of sustainability today is not considered over the entire lifecycle. This paper highlights a concept, where different facets of a digital twin within the product lifecycle could create added value for the sustainability. A decomposition-simulation e.g., coming from development, supporting the recycling procedure, or a traceable data-model, including material-data of the product helping to identify the needed recycling steps. Therefore, the associated efforts of further facets in different steps of the lifecycle are compared to the benefit, given by a traceable application of a digital twin. This concept defines the Digital Green Twin, which is shown using consumer electronics industry as an example.

Keywords: Digital twin · Sustainability · Digital Green Twin

1 Introduction

Today the environmental and climate protection gets increasingly interesting for population [1, 2]. For two-thirds of humanity the climate and environmental protection is one basic challenge in future. The rising awareness for these topics is also reflected in

J. R. da Costa Sanches Galvão et al. (Eds.): *Proceedings of the 1st International Conference on Water Energy Food and Sustainability (ICoWEFS 2021)*, pp. 548–557, 2021.
https://doi.org/10.1007/978-3-030-75315-3_59

legislation. One example is the directive for waste of the European Union from 2008 (directive 2008/98/EG) where the basics of product responsibility was defined. Accordingly, manufacturers are obliged to take back and dispose returned products after the use phase [3, 4]. Therefore, the need to design sustainable products over the entire product lifecycle arises in many companies to meet the increasing requirements of society and legislation.

At the same time, modern systems and products become more complex. Mechatronic and intelligent systems consist of different domain specific components like software, hardware or electronic. All components are integrated in these systems, so that the components of one single system already need to be recycled in different ways. Different plastics and metals e.g. must be recognized and separated. Different requirements, coming from user and market lead to a higher variety in modern products so that even one product has to be recycled in many ways regarding the varieties.

It is challenging to decompose the products and dissolve the materials in an efficient way. The recycled raw materials must be competitive with new materials regarding cost and quality. Recycling technologies need to be economically efficient to be attractive for companies [5]. The design of products and material compounds determine the complexity. Optimizing the recycling process is important to regain a large amount of the recyclable materials, such as metals and rare earths [6]. Due to the scarcity of resources, the raw materials become more valuable and the prices will increase. That makes it more important to close the loop for a sustainable circular economy [5]. However, the price development for recyclable materials is uncertain, which makes it difficult to calculate the recycling costs [7]. There is a huge potential for improvement in the development and optimization of recycling technologies [5, 7]. Beside the complexity of recycling processes, the amount of waste has to be taken into account. There are 2,01 billion tons of waste generated worldwide per year [8]. Innovative systems are characterized by mechatronic components. These systems belong to the electronic waste. The global E-waste monitor summarizes the total amount of waste each year [9]. For 2014, the monitor shows an amount of 44,4 million tons of E-waste, which made an amount of 6,4 kg per criteria. In comparison the amount of 2019 made 53,6 million tons of E-waste, which was 7,3 kg per criteria. It becomes clear, that the share of E-waste has increased by nearly 24% in the last years [9]. Furthermore, the monitor points out, that only 17,4% of this waste is recycled, what means, that 82,6% of the waste flows are not recycled or documented properly. The forecast shows an increase of the global waste amount with 74,7 million tons per year in 2030 [9]. Today's approaches and processes are overloaded by regarding these developments.

Summarized, there are different challenges to be faced. The rising complexity of systems, which goes along with variant materials and components, the rising amount of waste, and the actual recycling approaches should be reconsidered and thought in a sustainable way. As a first step, there is a need for transparency over the entire Product Lifecycle to identify sustainability potentials. This transparency is additionally demanded by the German Federal Ministry for the Environment, Nature Conservation and Nuclear Safety [10]. Transparent product data should serve to optimize production, use and recycling of products in form of a digital product passport, which could represent a basis for a digital twin. The publication underlines additionally the need of innovative approaches for a holistic consideration of transparency in different phases of the product lifecycle. Therefore, this paper highlights the concept of the Digital Green Twin. Different facets of digital artifacts collectively form the twin, which creates added value for modern recycling processes. Based on the example of a refrigerator, these benefits are highlighted within a canvas for the Digital Green Twin. Different phases of the product lifecycle are considered, especially regarding the digital artifacts used within the phases.

2 Research Methodology and State of the Art

As part of the research methodology, the current state of the art was examined. In addition to a general literature review, first findings were identified using the literature database Scopus [11]. The first analysis shows that the Digital Green Twin is not yet a popular publishing field. In the last 20 years, no publication was found including «Digital Green Twin» in the search string (title, abstract, keywords). Only one publication from Fraunhofer IPK on a Sustainable Twin is to be mentioned. It deals with a «Life Cycle Sustainability Assessment by Means of Digital Lifecycle Twins» [12].

For this publication, a three-step approach was selected here. These results are shown in Fig. 1. In the first step, the results based on the search string «Digital Twin» were analyzed. This named term has to be included in the title, abstract or keywords. All publications entered in Scopus in the subject areas «Engineering», «Computer Sciences», «Energy», «Materials Science» and «Environmental Science» were examined. In total, more than 2700 publications on the topic «Digital twins» were published between 2000 and 2020. The number of publications has increased significantly, especially in the last three years (321, 912, 1324). 70% of these publications were in the subject area «Engineering».

In the second step, the analysis was carried out for the subject area «Circular Economy». More than twice number of publications (6765) were identified in the last 20 years. As for Digital Twin, there is also an increase but a smaller one in the last three years [1016 (2018), 1628 (2019), 2411 (2020)]. Only 37% of the publications were written in the subject area of «Engineering».

Of importance are publications that focus on both the «Digital Twin» and «Circular Economy» topics. The following word combinations were analyzed: «Digital Twin && Circular Economy» and «Digital Twin && Sustainability». The first publication on this word combination is found in 2017. Between 2017 and 2020, a total of 58 publications were published. 60% of these publications were placed in «Engineering».

The two essential aspects used in the analysis are the Digital Twin and Circular Economy. Therefore, the state of the art is examined in the following section.

Digital Twin: There is no unified understanding of the digital twin. Basically, the concept of a twin was first used by NASA, which used a twin-mockup of the rocket that flew to space. The twin was used to perform tests, based on the surrounding conditions of the rocket in space. The digital twin was defined by Grieves in 2003 at University of Michigan course on Product Lifecycle Management and published in various papers [13, 14]. Grieves defined the digital twin as a digital representation of tangible or intangible objects or processes, which are independent from the real object. This approach

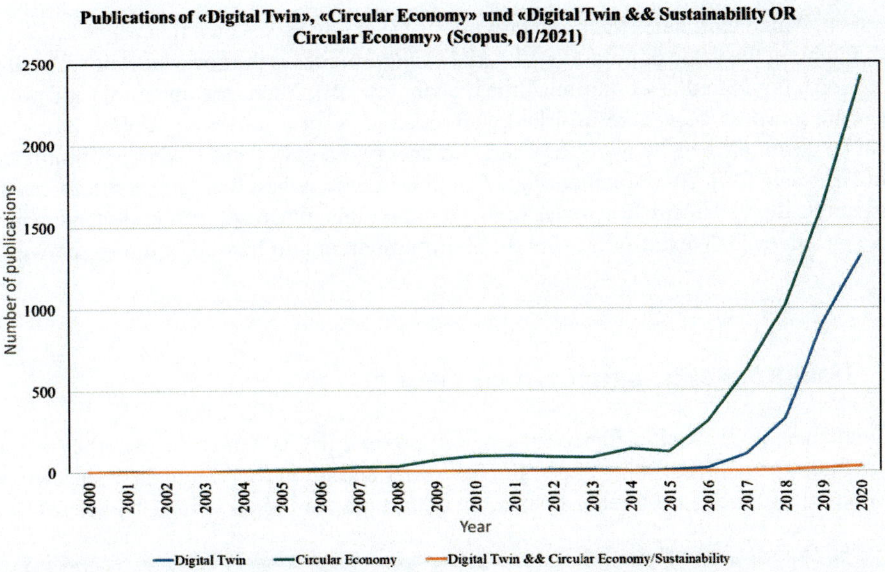

Fig. 1. Publications of «Digital Twin», «Circular Economy» und «Digital Twin && Sustainability OR Circular Economy» (Scopus, 01/2021)

distinguishes between the real and the virtual space. The digital twin consists of interconnected data and models. Data could be represented as system parameters, test data or other data that refers to the system. Models could be software- or simulation models or in case of approaches like Model Based Systems Engineering any others. In case of model-based development of mechatronic systems there are several examples, e.g. [15]. The virtual space is linked to the real space by extracting data, but the presence of the real space is not mandatory. In case of a development process e.g., the conception of the later product is often performed by using simulation models, before a real prototype is built. The vision of a digital twin is, that all the models and data are continuously and consistent linked over a products lifecycle, so that transparency and traceability are given. Chatti et al. provide another definition of a digital twin, that represents a digital image of a product and contains virtual models and data from various product lifecycle phases [16]. Applications of a digital twin could be the simulation or, as given by Uhlemann et al., the optimization of a production process. The different steps of the production are simulated and forecasted, so that logistic actions could be planned more precisely [17]. Further applications of digital twins in design are shown by Tao et al. [18], like real time data-acquisition during manufacturing based on the digital twin. Summarized, there are lots of applications of digital twins to optimize different phases of a product lifecycle. Common to all is the digital link between the artifacts, which create transparency and traceability.

Circular Economy: The concept of a Circular Economy is based on the idea of finite and limited resources and was first mentioned in 1989 by Pearce and Turner [19]. The intention of a Circular Economy is to set up a restorative or regenerative industrial system, which eliminates the production of waste and focusses on a renewable concept. Maintaining the value of products, materials and resources is the key aspect of a Circular Economy compared to a traditional linear economy [20]. Therefore, materials and products are recycled, reused, refurbished, and repaired as far as it is possible [19, 20]. This can be accomplished by using regenerative energy resources and leads to a minimization of waste [20]. By implementing a Circular Economy based on the cradle-to-cradle principle, the transformation towards an overall sustainable, resource efficient economy can be achieved. This also benefits the environment, due to less pollution of the waste [20].

3 Definition of a Digital Green Twin

Today there is no specific approach or definition for a Digital Green Twin in the state of the art. In the context of this paper, the following definition for a Digital Green Twin is proposed and used in reference to existing definitions for digital twins [13, 14, 16]:

> *The Digital Green Twin (DGT) is a digital image of a product or service. It contains and connects virtual models and data that include all important environmental and sustainability aspects from various product lifecycle phases. The goal of the DGT is to transparently present sustainability potentials and to optimize recovery processes.*

Considering the initial situation presented above, the overall objective of the concept is to design a DGT canvas, that faces the challenges shown. These can be derived from the fields of action digital twin and sustainability. Basically, there are three fields and associated goals:

Subgoal 1 - Use Cases of a DGT over the Product Life Cycle: For the expansion of future market services, the diverse use cases of a DGT must be identified and structured based on existing applications of a digital twin. Legal requirements and guidelines must also be considered. Knowledge about the technical possibilities of a DGT enables the derivation of feasible use cases. The analysis should cover all phases of the Product Lifecycle and consider the essential, current regulations and guidelines. The individual use cases are to be evaluated in terms of their requirements for corporate structures.

Subgoal 2 - Design of a Canvas: Based on the structured use cases, an architecture for an overall solution will be developed, which corresponds to a reference architecture for a DGT. Requirements of and synergies between the use cases will be derived for the technical realization. Components of the reference architecture are e.g. a data model as well as the necessary IT infrastructure.

Subgoal 3 – Procedure Model: Based on the use cases and the canvas, a process model is to be made available to interested stakeholders. This process model enables systematic guidance on the integration and implementation of the DGT in e.g. corporate structures.

In summary, this will lead to a more efficient, sustainable product creation process as well as a green Product Lifecycle. In particular, the benefits for recycling should be emphasized here. The concrete experience gained in implementing and integrating the canvas of the DGT forms the basis for further integration of sustainability potentials in the Product Lifecycle Management landscape.

4 Canvas and Procedure Model of Digital Green Twin

Based on the previously presented definition of the DGT, the following chapter shows a canvas of the DGT. For this canvas, existing approaches from the field of the digital twin, e.g. [21], are combined with sustainability aspects. Presented in Fig. 2, the individual processes or stages of a typical product life cycle form the basis of the canvas. The processes were selected based on ISO 24748-1 [22]. Existing IT systems and tools that are typically used in the individual Product Lifecycle stages in engineering are listed under these processes [23]. Thus, e.g., authoring tools such as a specific requirements management tool are used in the product design stage. The IT systems and tools described are used to generate results or outputs in the form of artifacts during the activities in the individual Lifecycle processes. For the requirements management example, a requirement list in Excel could be an artifact. In the final part of the DGT canvas, sustainability potentials are presented that extend the classic canvas of a digital twin. The sustainability potentials result from the consistent use of the virtual models and data of the individual product lifecycle stages. The specific potentials are documented in the form of fact sheets, in which single use cases are also described. A sustainability potential for the example of requirements management is, for example, the use of substance compliance data during the creation of requirements. Already in this early development stage, it can be ensured that only materials and substances are used that fulfill substance compliance guidelines.

Physical simulation, as part of the model-based development, is an effective approach for development processes. Within this approach there are physical models built from the system of interest. In case of this paper, the simulation of the refrigerator is performed by modeling e.g. the compressor of the heat pump, the thermal simulation of the cooling behavior or the logical simulation of the control software. All these sub-models are simulated within a system-simulation, to analyze the interactions of the different components of the refrigerator. These physical models, as part of the digital twin already contain system parameters and material properties of the refrigerator, which could create added value for the recycling process and shorten the recycling steps The refrigerant e.g. does not have to be identified before recycling, when adding the simulation models to the Digital Green Twin. Some of these models must be enriched with further information. This could evoke additional effort in the development phase, but simultaneously optimize the recycling. The effort of adding information to simulation models is slight, compared to the benefit for recycling, and most of the information is already included. In this example, the simulation models from the development process already contain material data and fluid data, which could be used without additional effort. Using this information e.g. saves the time for the recycler to find out which material used.

Further aspects of the DGT are artifacts of the integration phase. Verification and validation deliver a lot of test data, which could enhance the DGT. The refrigerator is

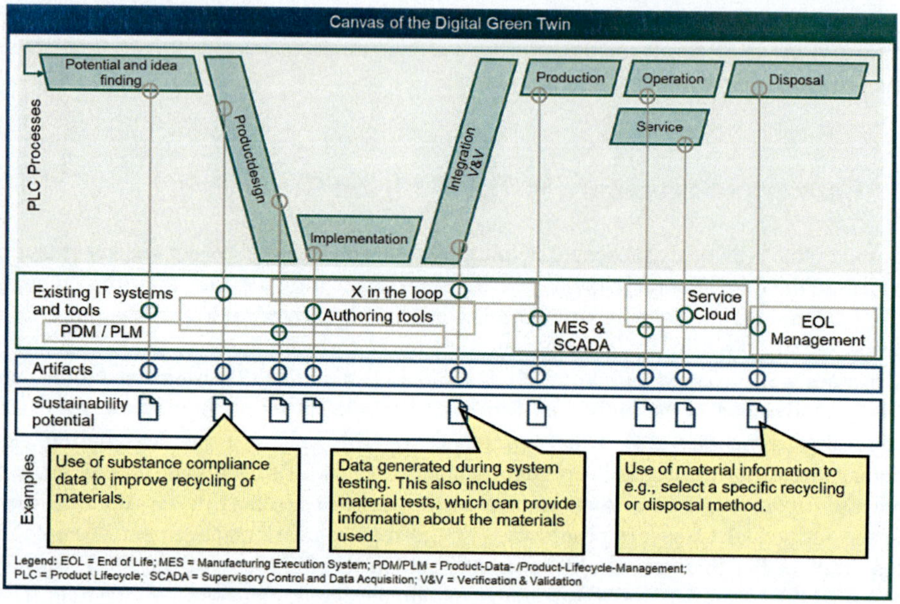

Fig. 2. Canvas of Digital Green Twin based on [20]

subjected to different tests. Mechanical tests, where the mechanic parts like the movement of the door and the handle are tested. Mechatronic tests also deliver a lot of data, like a Hardware-in-the-Loop test [24], which could be used. Furthermore, the materials are tested, often in form of long-term tests und various conditions. These tests and results provide information, which could be integrated in the DGT. Material data is, as said, the most beneficial information for recycling and simplifies the process. When producing modern products, the bill of materials (BOM, mBOM, xBOM) often is used to describe the parts, which have to be assembled. These lists and the assembly instructions of the products are further artifacts, which bring time-benefit for recycling by adding them to the DGT. Today there are assembly simulations, used to simplify the assembly. These simulations often originate from design phase, based on the product structure. In case of the refrigerator the mounting of each component could be simulated and visualized. These simulations could also be used beneficially in recycling. The model enables to perform disassembly-steps, so that the refrigerator could easily be disassembled into the different components, the compressor, the door, the insulation etc.

These examples given already show, how different artifacts could offer added value for recycling. To reach a holistic approach for integrating a DGT, the canvas for the DGT offers a structure to gather the necessary information and elements. Furthermore, to implement the DGT, a procedure model is required. This procedure model supports the detection of potentials and the implementation of a DGT.

The procedure model for the development and integration of a DGT is shown in Fig. 3. It consists of four phases in a sequential order with specific tasks and a defined result in every phase. This procedure leads to a DGT, which can be approached step by step.

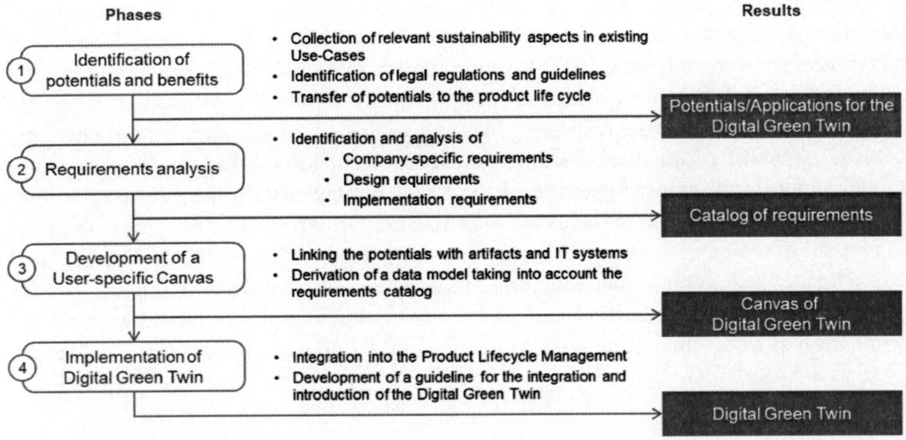

Fig. 3. Procedure model for development and integration of Digital Green Twin

Phase 1 – Identification of Potentials and Benefits: The process starts with the identification of potentials in existing use cases and benefits, where future applications, operations and values of the DGT are considered. The relevant sustainability aspects are collected, and all legal regulations and guidelines must be identified in the first phase. The potentials are transferred to the Product Lifecycle and as a result the applications of a DGT are defined.

Phase 2 – Requirements Analysis: In the next step a requirement analysis is done by identifying and analyzing the company-specific requirements, the design requirements and the implementation requirements, which result in a catalog of several requirements.

Phase 3 – Development of a Canvas: Afterwards a canvas of the DGT is developed by linking the applications with artifacts and IT Systems. A data model is derived with consideration of the requirements catalog.

Phase 4 – Implementation of Digital Green Twin: The DGT is implemented in the final phase, where it is integrated into the Product Lifecycle Management. Besides a guideline for the integration and introduction of the DGT is developed. By going through the whole process, the DGT can be obtained, considering all relevant features in all phases.

5 Summary and Conclusion

The paper introduces the Digital Green Twin as a concept to optimize future processes with. The main idea of the DGT is to regard the entire lifecycle of products and especially the digital artifacts being used within the different phases. These artifacts are analyzed regarding their added value for recycling. Furthermore, the paper delivers a guideline in form of a canvas for the DGT and a procedure model for the development and integration of the DGT in existing processes. With the help of the canvas, data and information from the existing artifacts can be identified and used for sustainability purposes. They could be represented as parameters, test data or simulation models. These artifacts enhance the recycling process, without generating a lot of further effort. The canvas could be used to analyze and visualize the different potentials in the actual Lifecycle and to develop a company-specific canvas, which shows the DGT for a specific Lifecycle. The procedure model supports the implementation of the DGT, defined within the canvas. All these aspects provide a fundamental basis to optimize modern Product Lifecycles with regard to recycling, which gets more important in today's products and systems. Further work could be focused on the implementation of single use cases, where the aspects of effort and benefit are each quantified in detail. This could exemplarily be shown within a simulation of a decomposition process, where simulation models from the development process are used.

References

1. Umweltbundesamt: Environmental awareness in Germany. https://www.umweltbundesamt.de/en/topics/sustainability-strategies-international/environmental-awareness-in-germany. Accessed 17 Jan 2021
2. The Federal Ministry for the Environment, Nature Conservation and Nuclear Safety (BMU): Umweltbewusstsein in Deutschland 2018 - Ergebnisse einer repräsentativen Bevölkerungsumfrage, Berlin, May 2019
3. European Union: REGULATION (EC) No 1907/2006 OF THE EUROPEAN PARLIAMENT AND OF THE COUNCIL of 18 December 2006 concerning the Registration, Evaluation, Authorisation and Restriction of Chemicals (REACH), establishing a European Chemicals Agency, amending Directive 1999/45/EC and repealing Council Regulation (EEC) No 793/93 and Commission Regulation (EC) No 1488/94 as well as Council Directive 76/769/EEC and Commission Directives 91/155/EEC, 93/67/EEC, 93/105/EC and 2000/21/EC, European Union (2006)
4. European Union: DIRECTIVE (EU) 2018/851 OF THE EUROPEAN PARLIAMENT AND OF THE COUNCIL of 30 May 2018 amending Directive 2008/98/EC on waste, European Union (2018)
5. Stolle, T.: Zukünftiges Abfallaufkommen von Photovoltaikmodulen. Materialfluss- und Reifegradanalyse für das Recycling, pp. 80–83. Diplomica Verlag, Hamburg (2016)
6. Kausch, P., Bertau, M., Gutzmer, J., Matschullat, J.: Strategische Rohstoffe — Risikovorsorge, pp. 164–168. Springer Spektrum (2014)
7. Hoyer, C.: Strategische Planung des Recyclings von Lithium-Ionen-Batterien aus Elektrofahrzeugen in Deutschland, pp. 71f. Springer Gabler (2015)
8. Kaza, S., Yao, L.C., Bhada-Tata, P., Van Woerden, F.: What a Waste 2.0: A Global Snapshot of Solid Waste Management to 2050. Urban Development. World Bank, Washington, DC © World Bank. License: CC BY 3.0 IGO (2018). https://openknowledge.worldbank.org/handle/10986/30317

9. Forti, V., Baldé, C.P., Kuehr R., Bel G.: The Global E-waste Monitor 2020: Quantities, flows and the circular economy potential. United Nations University (UNU)/United Nations Institute for Training and Research (UNITAR) – co-hosted SCYCLE Programme, International Telecommunication Union (ITU) & International Solid Waste Association (ISWA), Bonn/Geneva/Rotterdam (2020)
10. Kammerer, F., Schäfstoß, N., Röw, M.: Digital Policy Agenda for the Environment. The Federal Ministry for the Environment, Nature Conservation and Nuclear Safety (BMU) Division for Public Relations, Online Communication and Social Media, February 2020
11. Scopus abstract and citation database of peer-reviewed literature: scientific journals, books and conference proceedings. https://www.scopus.com/search/form.uri?display=basic&zone=header&origin=#basic. Accessed 15 2021
12. Riedelsheimer, T., Neugebauer, S., Lindow, K.: Progress for life cycle sustainability assessment by means of digital lifecycle twins – a taxonomy. In: Kishita, Y., Matsumoto, M., Inoue, M., Fukushige, S. (eds.) EcoDesign and Sustainability II, pp. 329–346. Springer, Singapore (2021)
13. Grieves, M.: PLM-Beyond lean manufacturing. In: Manufacturing Engineering, vol. 130, p. 23. SOC MANUFACTURING ENGINEERS (2003)
14. Grieves, M.: Digital Twin – Manufacturing Excellence through Virtual Factory Replication (Whitepaper). LLC (2014)
15. Holtkötter, J., Michael, J., Henke, C., Trächtler, A., Bockholt, M., Möhlenkamp, A., Katter, M.: Rapid-control-prototyping as part of model-based development of heat pump dryers. In: 4th International Conference on System-Integrated Intelligence (2018)
16. Chatti, S., Tolio, T.: CIRP Encyclopaedia of Production Engineering. The International Academy for Production Engineering, pp. 5–18. Springer, Berlin (2020)
17. Uhlemann, T.H.-J., Lehmann, C., Steinhilper, R.: The digital twin: realizing the cyber-physical production system for industry 4.0. In: The 24th CIRP Conference on Life Cycle Engineering. Procedia CIRP **61**, 335–340 (2017)
18. Tao, F., Liu, A., Hu, T., Nee, A.Y.C.: Digital Twin Driven Smart Design. Academic Press, Elsevier (2018)
19. Delchet-Cochet, K.: Circular Economy: From Waste Reduction to Value Creation, pp. 3–5. Wiley (2020)
20. Mavropoulos, A., Nilse, A.W.: Industry 4.0 and Circular Economy: Towards a Wasteless Future or a Wasteful Planet?, pp. 75–78. Wiley (2020)
21. Rasor, R.: Der digitale zwilling: anwendungsbereiche, architektur und technologien. In: OWL VIPROSIM E.V., 14. Fachtagung Virtual Prototyping and Simulation, Paderborn (2020). https://www.owl-viprosim.de/veranstaltungen-viprosim/fachtagungen/fachvortraege-der-14-fachtagung-virtual-prototyping-simulation/. Accessed 26 Jan 2020
22. ISO/IEC/IEEE 24748-1:2018: Systems and software engineering - Life cycle management - Part 1 - Guidelines for life cycle management (2018)
23. Eigner, M., Koch, W., Muggeo, C.: Modellbasierter Entwicklungsprozess cybertronischer Systeme. Springer, Heidelberg (2017)
24. Millitzer, J., Mayer, D., Henke, C., Jersch, T., Tamm, C., Michael, J., Ramisch, C.: Recent developments in hardware-in-the-loop testing. In: Conference & Exposition on Structural Dynamics (2018)

![Check for updates]

Planning for Urban Sustainability: A Complex Adaptive System Analyses of the Lisbon Green Infrastructure

Ticiana Ponte(✉) and Margarida Queirós

Centre for Geographical Studies, Institute of Geography and Spatial Planning, ULisboa, R. Branca Edmée Marques, 1600-276 Lisbon, Portugal
ticianaponte@edu.ulisboa.pt

Abstract. The 2030 Agenda for Sustainable Development and the Paris Agreement call for a transformation in societies, with interdependent goals, and a complex coupling between social, technical, and ecological systems. Cities are essential drivers for development, innovation, experimentation, and change. They can be obstacles and/or drivers for achieving sustainability. Nevertheless, current models of urban governance may not be able to address 21st century challenges. They are highly dependent on predictability, address problems linearly, and assume that viable solutions can be prescribed to meet established goals when, in reality, society is complex, dynamic, and constantly evolving. An interconnected approach between complex system thinking and planning, can present new perspectives for a sustainable transition. Through adequate governance of green infrastructure in cities, urban systems can benefit through the improvement of citizens health, well-being, conservation of biodiversity, supply of ecosystem services, and strength of urban capacity for resilience and adaptation. This research explores, through the lens of complexity, how governance arrangements and institutions influenced the effective implementation of the Lisbon Green Infrastructure System (LGIS) and the role of actors involved. The analysis shows that the process of construction and implementation of the visions for the future (planning) of the actual LGIS can be explained as a complex adaptive system, shaped by negative and positive feedback loops over time. Through mapping where transformative action was originated, the actors involved, and the effects on the implementation process, can help to embrace uncertainty in urban governance.

Keywords: Complex system · Urban planning · Urban sustainable development · Urban transition · Lisbon · SDGs

1 Introduction

Collective urban life impact is unsustainable, creating negative ecological impacts, social and economic tensions [1], making cities vulnerable to ecological, socioeconomic and political crises. "New" socioeconomic and political problems are pushing issues of ecological security – e.g., climate change, growth of new diseases, water and energy

J. R. da Costa Sanches Galvão et al. (Eds.): *Proceedings of the 1st International Conference on Water Energy Food and Sustainability (ICoWEFS 2021)*, pp. 558–567, 2021.
https://doi.org/10.1007/978-3-030-75315-3_60

security - up the agenda of national governments, once they need to ensure secure and continued access to resources needed to provide their economic and social reproduction [2].

The transformations that cities, societies and environment, together with global threats of resource limits and climate change, induced by human activities, put sustainability at critical importance for urban development. The particular role of cities as obstacles and/or levers for achieving sustainability has been underlined widely [1, 3]. Furthermore, there is a recent trend of decentralization from national to local spheres of governance [3], supported by a global sustainable development agenda [4], with cities getting a specific sustainable development goal (i.e., SDG#11), together with the New Urban Agenda [5] for sustainable urbanization and planning. Rethinking how to foster urban sustainability is an urgent necessity, cities response to these challenges will decide the future configurations of their social-ecological-technical (complex) systems. There is an understanding that urban planning has a major impact on shaping more socially integrated cities and regions and that it can be an instrument for promoting sustainable development [1]. In contrast, inadequate spatial planning can contribute to more inequalities and urban conflicts [3].

Cities are, at the same time, artifacts and dual complex systems. The city, and each of its parts (i.e., the urban agents), are a complex self-organizing system [6]. Cities, as complex systems: a) are open, self-organizing, adaptive, and constantly evolving systems [6–8]; b) comprise multiple actors with varying responsibilities, capabilities, priorities and interests, as well as processes that transcend the institutional compartmentalization of city administration [1]; c) are constituted of many sub-systems of multifarious manifestations, such as, geophysical spaces, physical infrastructures, economic relationships, social entities and cultural practices [1]; d) are embedded in broader ecological, economic, technical, institutional, legal and governance structures that often constrain their systemic function, which cannot be separated from wider power relations [1]; e) are an emerging phenomena on a local scale [8] resulting from processes (i.e. physical, social, ecological, cultural and economic) that are causally interlinked with interactions and feedback loops.

An interlinked approach between complex adaptive system thinking and urban planning can introduce new perspectives for a transition pathway to sustainable development [1, 9, 10]. Although complex systems theory was not developed specifically to explain social-cultural systems behavior, but primarily biological and environmental systems [11], it provides a possible framework for spatial planning [10]. Thereby, this article examines the transitions occurred in the green infrastructure of Lisbon, in Portugal, from late 19th century to 2012 (inauguration of the 1st Lisbon Green Corridor), exploring through the lens of complexity, how governance arrangements and institutions influence the effective implementation of the Lisbon Green Infrastructure System (LGIS) and the role of actors involved.

2 Lisbon Green Structure System

Urban green infrastructure (UGI) is understood as both multifunctional networks of green and blue spaces and the process that govern its development [12, 13]. Various global development organizations have been underlining the determinant importance of UGI [14], once it is believed to offer a pathway to sustainable development [13], with the potential to make contributions to meet policy objectives for improving public health and wellbeing, protecting biodiversity, delivering a wide range of ecosystem services (food, clean water and air, climate regulation, flood prevention, pollination, recreation, among others), [15] enhancing social cohesion and connectivity, supporting the local economy and strength of urban capacity for resilience and adaptation [13].

Connectivity and multi-functionality became central components of plans for the management of green areas in cities and their metropolitan regions. Historically the focus moved from individual green areas to UGI [14]. The idea of continuity was already embedded in the concepts defined by Frederick Law Olmsted and Ebenezer Howard in the end of 19th century. In the last decades of the 20th century those ideas matured through the adoption of the "connectivity" concept [12, 14]. Connectivity and multi-functionality of UGIs are directly related to historical development processes of each city, as urban dynamics (planned and unplanned) change constantly the infrastructure, the shape and the functions of built and non-built urban fabric [14] as well as the social relations they influence.

Until 18th century, Lisbon grew without planning; the development of green areas begins with the construction of the Public Promenade (Passeio Público). The first plan containing the idea of a continuous and integrated UGI for Lisbon was only proposed in 1938, by De Gröer, and finally incorporated on the Lisbon Master Plan in 1948 [16]. It envisaged a green belt around the city that included the Monsanto Forest Park, the area surrounding the airport and protective bands along the main ring roads [16].

The base concept embedded in today's LGIS is the "continuum naturale" [16, 17], introduced and disseminated in Portugal by Francisco Caldeira Cabral in the 1940s. After finishing his studies in Berlin, he starts to teach in the Lisbon Scholl of Agronomy (Instituto Superior de Agronomia – ISA) where he founds the Landscape Architecture course in 1942, and a new profession in Portugal [16–18], being the first to idealize a landscape planning based in ecological corridors, a visionary idea. The School of Landscape Architecture represented an evolution in relation to the scientific knowledge existing in Portugal at that time [19].

After Portugal joined the European Economic Community (today European Union) a shift occurred in planning and environmental legislation in Portugal, affecting national, regional and local spheres. Soon after, started the elaboration of a new Master Plan for the city of Lisbon which was approved in 1994. One of its goals was to improve the city's environmental quality. For that, it incorporated the Lisbon Green Plan (Plano Verde de Lisboa) coordinated by Gonçalo Ribeiro Telles, a landscape architect and agronomist graduated by ISA, in 1952. He was part of the first generation of landscape architects trained in Portugal. Planner and designer, professor, member of the government, legislator, deputy and councilor, he played a significant role in decisive moments in Portuguese contemporary history. The Green Plan - basis of LGIS - was developed from three basic concepts: the "continuum naturale", a continuous system based on ecological principles and on the utility landscape that supports wildlife; the cultural continuity, integrating built and open spaces; and the genius loci, including geology, topography, climate, the flow of water and the relation with the built environment. Ribeiro Telles work anticipated the UGI concept and identified the importance of green corridors, where resources are concentrated, to link natural and cultural landscapes [12, 20].

3 Methods

The research employs the case study of the construction and the visions for the future (planning) of the LGIS. To understand and evaluate the role and the interplay between actors and institutional settings, in this research first stage, we aim to map the LGIS adaptive dynamic change over time in order to identify the tipping points and the feedback loops (positive and negative) that originated then, and the characteristics of system change. To construct the system timeline and to explore the "uprise" of the first ideas that constitutes today's LGIS, and its evolution in the city planning process and legislation, until its effective implementation, we conducted a comprehensive qualitative literature review. Through a keyword search in most relevant scientific literature databases (Scopus, Web of Science, Google Scholar) we defined the corpus for the review, using a Boolean set of terms. References have been included by backward analysis, identifying relevant actors and documentation mentioned on primary documents. The sample was scrutinized through a qualitative content analysis to gather the data to trace the system timeline and map the historical evolution of the LGIS. The outcome was a historical matrix of national, regional and local planning context and legislative initiatives, and most relevant agency interventions and actors involved. From the matrix data we draw a system analysis to visualize the tipping points in time and the system dynamics, using the Multi-phase concept [21] developed by Jan Rotman, René Kemp and Marjolein van Asselt. This concept was originally observed and used in natural systems but proved efficient to social systems as well [10, 11, 22]. It is a simplified model of theory of change for complex systems, but much used in transition research to describe the dynamics of social systems with a sequence of alternating phases over time [11]. It identifies the nonlinear pattern of change, from one dynamic state of equilibrium to another. The Multi-phase graphic at the conceptual level is, commonly, divided in four stages [21], showing the aggregation of the subjacent curves.

Figure 1 presents the possible phases of system dynamic in which: 'Predevelopment', represents a state of dynamic equilibrium where a smaller system is changing slowly in the background [11] but changes are not visible[21]; 'Take-off', is the beginning of the process of change [21]; 'Acceleration', structural changes become visible [11]; 'Stabilization', where the speed of social change decreases [21] and the system adjusts itself while achieving a higher order of organization and complexity (Grin et al., 2010; Rotmans et al., 2001). Other possible outcomes can be: 'Lock-in', in which an increasing path dependence emerges; 'Backlash', represents a block in the system's innovation; and 'Breakdown', a reverse transition and system collapse [22].

The sequence of phases does not lead to a fixed pathway, there is great uncertainty and complexity in which the degree of predictability is relatively small. The framework purpose is not to predict the course of transition over time, but to identify the various phases and provide guidelines and levers to influence the direction of change. It does not describe change, or transition, in terms of different scale dynamics. The usage of the Multi-phase concept is to function a) as a descriptive framework to identify the direction,

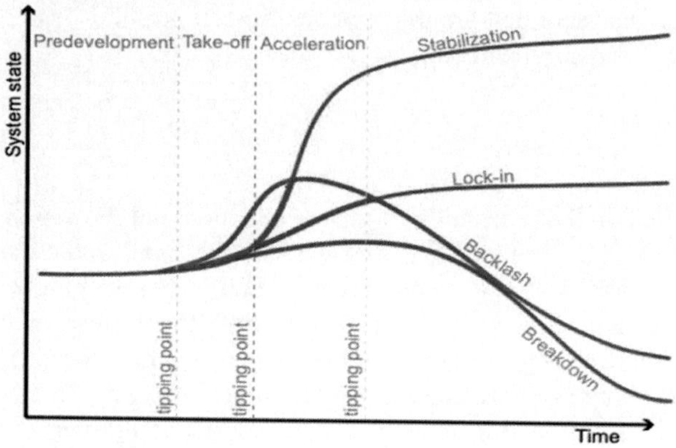

Fig. 1. Multi-phase concept: phases of systems transition (Source: elaborated by Frederico B. Dias based in the original graphic by Jan Rotman, Johan Schot and John Grin [22]).

velocity and magnitude of a transition, describing the changes in phases; and b) as an explanatory framework for the driving forces and mechanisms behind the phases and their changes. Other limitation of the Multi-phase concept is the absence of indicators to measure social development or for systems change and the time period may vary considerably, as well as the tipping points, that are hard to determinate [22].

4 Results

Throughout the analysis of the historical matrix, we were able to map the phases of the process of construction and implementation of the LGIS. Hereafter, we will present the analysis of the institutional context and actors and agency relevance, in a sequential phase, in order to identify the tipping points and the conditions and structures that enabled LGIS change.

The birth of the idea for the creation of the Forest Park of Monsanto may have started in late 1860s, by a report underlining the importance of this green lung for the city and the need for afforestation to supply the city with wood, soften the climate and contribute to public health [23]. From 1847 to 1911, Frederico Ressano Garcia, a political leadership and Head of the Technical Department of the City Council, conducted the Haussmannian Reform in Lisbon. In 1888, he presents the Plans for new Parks (Campo Grande Gardens and the neighborhoods of tree-lined avenues), inspired by international trends from the USA and major urbanization plans in Paris. In 1904, still under monarchy, he approved a Master Plan for the city of Lisbon (Plano Geral de Melhoramentos – PGM), in which one of the most relevant interventions was the opening of Liberdade Avenue over the old Public Promenade. A strongly contested decision, object of a petition by more than two thousand citizens [17]. These events may have preceded the Predevelopment phase. This trend is confirmed when, in the 1920s, after the 1st World War, two projects, including afforestation, were proposed for the Monsanto Park (an arid area of approximated 1.000ha in the city periphery). First, in 1925, by the medical doctors and brothers Mac Bride, a forest park in the culmination of a greenway based on aesthetic and hygienist principles, through a green ring from Campo Grande to Monsanto Park, transforming Eduardo VII Park into a large tree-lined avenue. Later, in 1927, Jean-Claude Nicolas Forestier, a landscape engineer greatly influenced by Olmsted's concepts, proposes the extension of the Avenue and the entire park that extended to the North, including gardens and playgrounds [18, 20]. The proposals were not implemented and, in the political scenario a "coup d'état" (1926) establishes a military dictatorship in Portugal. The system evolves to a Lock-in, in which negative feedback loops dampens the system responses and not much change is perceived.

In 1938, a tipping point occurred. The system initiates a Take-off as changes turn visible. The engineer and politician Duarte Pacheco promoted important changes in planning policies. His strong political and administrative power was determinant: he was the Minister of Government with the portfolio of Public Works and Communications accumulating the post of Lisbon Mayor, in 1938 [16, 17]. In his administration, new legislations were approved to ease land expropriation and to facilitate the city expansion and rehabilitation relying in urban planning as an instrument to city ordering and urban policies [24]. He took the initiative in the process that led to the implementation of Monsanto Park through a Decree-Law, in 1934 [17]. The park construction started in 1938, when the forestry engineer Joaquim Rodrigo and the architect Keil do Amaral were hired. The former responsible for the afforestation of the area and the latter for the project layout [20] and public equipment [17]. Regarded the 'nationalist' regime, Portugal 'imported' several foreign planners to work in different municipalities across the country. They greatly marked and influenced Portuguese urban planning. The Russian architect and urbanist Etienne de Gröer - graduated in the Institut d'Université de Paris and with strong planning links to authors such as Ebenezer Howard, the idealizer of the Garden City planning model [24] - was hired by Duarte Pacheco to coordinate the elaboration of the Lisbon's Master Plan (Plano Director de Urbanização de Lisboa – PDMU). The process started in 1938 with the publication of a preliminary analysis program to support the elaboration of the PDMU, in which he proposes the concept of a continuous and integrated UGI for Lisbon [17]. The PDMU was only approved in 1948, ten years after the beginning of its elaboration process. Despite the structural change in the LGIS, that culminated with the implementation of the Forest Park of Monsanto and the approval of the concept of a UGI for Lisbon in a legally binding plan, the system apparently had evolved to a Lock-in around the 1948 PMDU approval. The Park's limits were not legally institutionalized, although defined in the Gröer Plan, which allowed the consequences of urban pressure to become a real threat to the park. Also, the green belt proposed around the city, foreseeing the expansion of Monsanto Park, never came to be implemented. Furthermore, the following Lisbon Master Plan, approved in 1958, only maintained the concepts of the Gröer Plan, but with some reductions in the proposed green areas [17].

Nevertheless, external events and agency starts slowly to build up in the background, and soon begins to pressure the system to change. Caldeira Cabral starts to disseminate the "continuum naturalle" concept in Portugal [16–18] at the Lisbon School of Agronomy (ISA), turned into a niche for new sustainable ideas and innovation. In this period, we can mention few significant examples of urban interventions that impacted in the LGIS. Between 1955 and 1962, plans were elaborated by the Technical Housing Office of the

City Hall for the neighborhoods of Olivais Sul and Chelas, following the functionalist urbanism preconized in the Athens Charter. Nevertheless, the plans lack morphological rules and did not considered the importance of the ecological structure for a balanced urban environment [17]. A significant green area was the Calouste Gulbenkian Foundation garden, one of the most emblematic of the modern movement in Portugal, a project of Ribeiro Telles and António Viana Barreto. The construction took place between 1966 and 1969, a private closed space, which was opened only to the public after 1974. In the 1970s, public nature conservation policy began in Portugal. The milestone was the creation of the National Park of Peneda-Gerês, at the time with 60 thousand ha of natural and integral reserves and pre-park area [25, 26]. In this period, several proposals for Lisbon illustrating and defending the ecological balance and the "continuous naturalle" concept have been presented by Ribeiro Telles. In 1974, a democratic regime is established in Portugal. New legislation is necessary thus, in 1976, the current Portuguese Constitution is approved, and in the same year, a new Master Plan for Lisbon (Plano Director da Cidade de Lisboa - PDCL). The latter maintained the concept of the previous plan, although it shows a concern with the consequence of urbanization in the environment, presenting an analysis of the city's green spaces and, in relation to the Monsanto Park, draws attention to existing and projected areas corresponding to uses outside the park function and adds that about 29% of the park's area was "in less desirable conditions" [17].

In 1979, finally, the delimitation of the Monsanto Forest Park was published in a law-decree [20]. The relevance of this event indicates a 2nd Take-off in LGIS. Portugal's accession to the European Economic Community, in 1986, may have been determinant, together with the rise of global ecological awareness [20]. Thus, in 1987, the first environmental act (Lei de Bases do Ambiente) institutionalized the "continuum naturale" concept, envisioned by Caldeira Cabral in the 1940s [16, 17]. A positive feedback loop seems to be in course and the system is evolving to a higher stage of dynamic equilibrium, embedding a sustainable concept envisioned more than 40 years before. In this period, a great number of laws regarding more developed concepts in planning system and environment protection and governance where enacted in Portugal, with effects in national, regional and local levels.

The LGIS finally enters in an Acceleration phase. In 1993, a study for an ecological infrastructure for Lisbon (Plano Verde - Green Plan) – conducted by a multidisciplinary team of teachers and practitioners of the Landscape Architecture Course of ISA, coordinated by Ribeiro Telles – is presented. The Green Plan represents the continuation of a sustainable philosophy for the city UGI, which has been developed in a number of proposals and submitted to the City Council in the past decades [16]. The Green Plan was a component of a larger plan – the new Lisbon Master Plan – that was approved in 1994 [17] therefore, making the Green Plan legally binding. One of the objectives of the 1994 Master Plan, was to improve the city's environmental quality, it proposes the connection of the major green areas of the city through greenways, defined as an environmental component of the Master Plan by the Green Plan [20]. The historically most important of the greenways - linking the central city to Monsanto Park - started to be implemented in the 1990s [20]. In 1994, an international tender was launched for the project of Tagus (Tejo) and Tracão Park, in addition to afforestation and a leisure space for the population. The project included the rehabilitation of one of the most polluted areas in Portugal. The intervention in this new "urban area" was driven by the EXPO'98 – 1998 Lisbon World Exposition. The city of Lisbon, then, experiences a total urban change in the Eastern zone, driven by its realization [18].

The 2008 economic crisis, and the subsequent austerity measures, affected strongly the Lisbon Metropolitan Area, with two main socio-political change restructuring parameters: financially driven central state policies, especially as a consequence of the 2011 bailout; and a more socially and territorially driven urban political strategies [27]. In this context, the effective implementation of the Monsanto Green Corridor takes place in 2012, bringing the city and the Monsanto Forest Park closer to each other physically and conceptually. This integration brought a new importance for the area and to respond to the growing demand for additional leisure spaces, new ones were inaugurated, and several improvements were carried out within the Park [20]. In this same year, the revision of the previous 1994 Master Plan is concluded. The new Master Plan reinforces the integration of nature within the urban fabric, through the revalidation of the Green Plan and a Stabilization phase takes place.

5 Discussion and Conclusion

From a complex adaptive view, and using the Multi-phase concept, it is possible to outline a macro-analysis of LGIS. The Predevelopment phase started in the 1920s with the first projects for the area of the Monsanto Forest Park. Here, pioneer ideas were building up very slowly in the background, in a post first world war macro scenario, but at local level, a "coup d'état" (negative feedback loop) forced a Lock-in and the system took more than 10 years until a Take-off occurred. Again, innovative ideas and the external macro-scenario pressed the local regime. Thus, in 1938, changes in the system starts to build-up, not by integrative or participatory agency, but by leadership and "illuminated figures" responsible for importing, planning and implementing those new ideas in the capital of the nation. Despite the undeniable achievements and structural changes in LGIS, the system dynamics changes to a new Lock-in. The LGIS shows a high path dependency, functioning in a dynamic alignment supported by ongoing negative feedback loops, or slightly adaptive activities that maintain stability. For the system to evolve to a higher level of change increasing "irritations" were needed to foster a more radical change. Those irritations came from the build-up of innovative ideas and knowledge until finally could be embedded in the 1994 Lisbon Master Plan through the Lisbon Green Plan.

The 2nd Take-off phase was influenced by the forementioned innovative ideas and concepts and the change in the national political regime. New legislation framework provided the needed conditions for the system to evolve to an Acceleration phase. In this phase, structural changes were effectively implemented through an accumulation of socio-cultural, economic, ecological, and institutional changes. Around 2012, LGIS seems to enter a Stabilization phase as a new dynamic system setting emerged in a transformed stage in which a more sustainable UGI concept was embedded [22]. Although, possibly, the system is still in a slower Acceleration phase considering that new sustainable integrated policies, supported by the global sustainable development agenda [4, 5] and the European Green Deal, are being implemented not only in Lisbon, but also in its metropolitan area, evolving to a network of green areas connected to a sustainable mobility structure. The LGIS tipping points seem highly related to the political context, when the feasibility, support and legitimacy of the old regime ceases, struggles about what should replace it are intense and the result can be a transformed (Stabilization) or a re-formed one that accommodate the "irritations" but stay on a similar development trajectory (Lock-in) [11].

Through the presented macro-analysis, it was possible to map the system dynamics, not to foreseen system change, but to understand how we may intervene in the system, learn how to embed sustainable innovation, and mitigate and adapt to uncertainties. It was possible, also, to identify tipping points, in which levers should be deeper analyzed. Deeper understanding of the conditions that fostered the tipping points and respective feedback loops is necessary. Therefore, further research is needed into the role of politics, power and innovative ideas and knowledge [2] in planning process in order to address urban challenges for reaching a sustainable development and to embrace uncertainty in urban governance [1, 9, 10].

References

1. Bai, X., Surveyer, A., Elmqvist, T., Gatzweiler, F.W., Güneralp, B., Parnell, S., Prieur-Richard, A.H., Shrivastava, P., Siri, J.G., Stafford-Smith, M., Toussaint, J.P., Webb, R.: Defining and advancing a systems approach for sustainable cities. Curr. Opin. Environ. Sustain. **23**, 69–78 (2016). https://doi.org/10.1016/j.cosust.2016.11.010
2. Hodson, M., Marvin, S.: Can cities shape socio-technical transitions and how would we know if they were? Res. Policy **39**, 477–485 (2010). https://doi.org/10.1016/j.respol.2010.01.020
3. UN-Habitat: World Cities Report 2016: Urbanization and Development, Emerging Futures (2016). https://unhabitat.org/world-cities-report. Accessed 02 Feb 2021
4. United Nations: Transforming Our World: The 2030 Agenda for Sustainable Development (2015). https://www.un.org/ga/search/view_doc.asp?symbol=A/RES/70/1&Lang=E. Accessed 02 Feb 2021
5. United Nations-Habitat III: New Urban Agenda (2017). https://uploads.habitat3.org/hb3/NUA-English.pdf. Accessed 02 Feb 2021
6. Portugali, J.: Complexity, Cognition and the City. Springer (2011)
7. Batty, M.: Complexity in city systems: Understanding, evolution, and design. UCL Centre for Advanced Spatial Analysis (2007)
8. Johnson, S.: Emergence: The Connected Lives of Ants, Brains, Cities, and Software. Scribner (2001)
9. Healey, P.: Relational complexity and the imaginative power of strategic spatial planning. Eur. Plan. Stud. **14**, 525–546 (2006). https://doi.org/10.1080/09654310500421196
10. Queirós, M.: Cities and Regions as Complex Systems. Impact on Spatial Planning. In: Farinós, J., Peiró, E. (eds.) Territorio Y Estados, pp. 49–70. Tirant Humanidades, Valencia (2018)
11. Göpel, M.: The Great Mindshift: How a New Economic Paradigm and Sustainability Transformations go Hand in Hand. Springer Open (2016)
12. Ahern, J.: Greenways in the USA: theory, trends and prospects. In: Pungetti, G., Jongman, R.H.G. (eds.) Ecological Networks and Greenways, pp. 34–55. Cambridge University Press (2004)

13. Pauleit, S., Ambrose-Oji, B., Andersson, E., Anton, B., Buijs, A., Haase, D., Elands, B., Hansen, R., Kowarik, I., Kronenberg, J., Mattijssen, T., Stahl Olafsson, A., Rall, E., van der Jagt, A.P.N., Konijnendijk van den Bosch, C.: Advancing urban green infrastructure in Europe: outcomes and reflections from the GREEN SURGE project. Urban For. Urban Green **40**, 4–16 (2019). https://doi.org/10.1016/j.ufug.2018.10.006
14. Madureira, H., Andresen, T., Monteiro, A.: Green structure and planning evolution in Porto. Urban For. Urban Green **10**, 141–149 (2011). https://doi.org/10.1016/j.ufug.2010.12.004
15. European Environment Agency: Green infrastructure and territorial cohesion: The concept of green infrastructure and its integration into policies using monitoring systems, Copenhagen (2011). https://www.eea.europa.eu/publications/green-infrastructure-and-territorial-cohesion. Accessed 02 Feb 2021
16. Magalhães, M.R.: Ecological structure of Lisbon. Ekistics **60**, 159–166 (1993)
17. Telles, G.R.: Plano Verde de Lisboa. Edições Colibri, Lisbon (1997)
18. Soares, A.L., Castel-Branco, C.: As árvores da cidade de Lisboa. In: Árvores e Florestas de Portugal - No7—Floresta e Sociedade - Uma História em comum, pp. 289–334. Fundação Luso-Americana, Lisboa (2007)
19. Magalhães, M.R.: Paisagem-perspectiva da arquitectura paisagista. Philosophica **29**, 103–113 (2007)
20. Matos, M.J.: The green corridor: a vision for lisbon. In: Gee, G.N., Vogelaar, A. (eds.) Changing Representations of Nature and the City: The 1960s–1070s and their Legacies, pp. 165–179. Routledge, New York (2018)
21. Rotmans, J., Kemp, R., Van Asselt, M.: More evolution than revolution: transition management in public policy. J. Foresight **3**, 15–31 (2001). https://doi.org/10.1108/14636680110803003
22. Grin, J., Rotmans, J., Schot, J.: Transitions to Sustainable Development: New Directions in the Study of Long Term Transformative Change. Taylor & Francis, New York (2010)
23. Tostões, A.: Monsanto, Parque Eduardo VII, Campo Grande, Keil do Amaral. Arquitecto dos Espaços Verdes de Lisboa, Salamandra (1992)
24. Marat-Mendes, T., de Sampayo, M.T.: Étienne de Groër: The different scales of the urban intervention in Lisbon territory. In: Proceedings of the 1st International Conference of the EAHN, Guimarães, pp. 32–43 (2010)
25. Telles, G.R.: A propósito do início da política de ambiente em Portugal. Rev. das Autarquias **2**, 73–80 (1995)
26. Queirós, M.: Os Parques Naturais e o Ordenamento do Território em Portugal. Inforgeo. **02**, 141–150 (2001)
27. Seixas, J., Tulumello, S., Corvelo, S., Drago, A.: Potentials and restrictions on the changing dynamics of the political spaces in the Lisbon Metropolitan Area. In: Knieling, J., Othengrafen, F. (eds.) Cities in Crisis: Socio-Spatial Impacts of the Economic Crisis in Southern European Cities, pp. 217–238. Routledge (2016)

Time of Day and Workdays vs. Weekend Differences in the Use of Cultural Ecosystem Services in Urban Parks (Coimbra, Portugal)

Luís Valença Pinto[1,2](✉), Carla S. S. Ferreira[1,3,4], and Paulo Pereira[2]

[1] Research Centre for Natural Resources, Environment and Society (CERNAS), Polytechnic Institute of Coimbra, Coimbra Agrarian Technical School, Coimbra, Portugal
lmpinto@esac.pt

[2] Environmental Management Laboratory, Mykolas Romeris University, 08303 Vilnius, Lithuania

[3] Department of Physical Geography and Bolin Centre for Climate Research, Stockholm University, 10691 Stockholm, Sweden

[4] Navarino Environmental Observatory, Costa Navarino, 24001 Messinina, Greece

Abstract. Urban green spaces (UGS) are a key component to achieve United Nations sustainable development goals (SDGs) such as good health and wellbeing (Goal 3) and sustainable cities and communities (Goal 11). Among other areas, UGS comprise urban parks, which are key to supply a variety of ecosystem services (ES) to humans. However, urban dwellers visit parks at a different time of the day and the week. This study investigates the possible relevance of different times of the day and different week periods (workdays and weekends) for the visitation and use of urban parks. The data was collected through personal interviews in 5 urban parks (n = 1000) in Coimbra, Portugal. Data were collected during August 2020, considering six-time slots from 7h30 to 22h00 (7h30 to 9h30; 9h30 to 12h30; 12h30 to 14h30; 14h30 to 17h30; 17h30 to 20h30; 20h30 to 22h00), also differentiating working days and weekend days. Activity performed by respondents was aggregated into three groups of cultural ES: Physical interactions; Experiential and aesthetical interactions; and Social interactions. Our results showed differences in park use, in timeslots and weekdays. Timeslots were especially relevant in the largest park (Choupal national forest), where users looked for sport activities (e.g., running, biking), particularly in the morning period. Differences in weekday use were relevant in all five parks. In some parks we observed a lower diversity of CES in use during working days and high levels of social interactions during this period. Results also suggested that park usage along the day and between weekdays and weekends depending on the activities performed and park characteristics.

Keywords: Cultural ecosystem services · Urban green spaces · Urban parks · Time of day · Period of the week · Workday vs. weekend

1 Introduction

Urban green spaces (UGS) are key to achieve the United Nations (UN) sustainable development goals (SDGs), especially Goal 3 - good health and wellbeing, and Goal

J. R. da Costa Sanches Galvão et al. (Eds.): *Proceedings of the 1st International Conference on Water Energy Food and Sustainability (ICoWEFS 2021)*, pp. 568–575, 2021.
https://doi.org/10.1007/978-3-030-75315-3_61

11 - sustainable cities and communities [1]. They are also essential under the UN New Urban Agenda [2] and the European Commission Biodiversity Strategy for 2030 [3]. UGS are fundamental components in cities and include all vegetated areas in urban environment [4]. They provide a large set of ecosystem services (ES), including provisioning, regulating and cultural services [5]. Urban parks are a crucial component of UGS, being the place where urban dwellers interact with nature [6]. However, urban dwellers visit parks at different times of the day and at different periods of the week. This study investigates the possible relevance of different daily timeslots and week periods (workdays vs. weekends) in park visitation in 5 different parks located in Coimbra and analyses the different cultural ES (CES) supplied by each park.

2 Materials and Methods

2.1 Study Areas

Coimbra is the largest city in the Portuguese center region (Fig. 1a). With 91,775 inhabitants [7], the city extends over an area of 51 km^2. The consolidated urban area, the area comprising stabilized urban soils and infrastructures (Fig. 1b), covers 25% of the city, settling 31% of its inhabitants [8]. The selected UGS include Manuel Braga park, Botanical garden, Choupal national forest, Mondego Green park and Vale das Flores park (Fig. 1c).

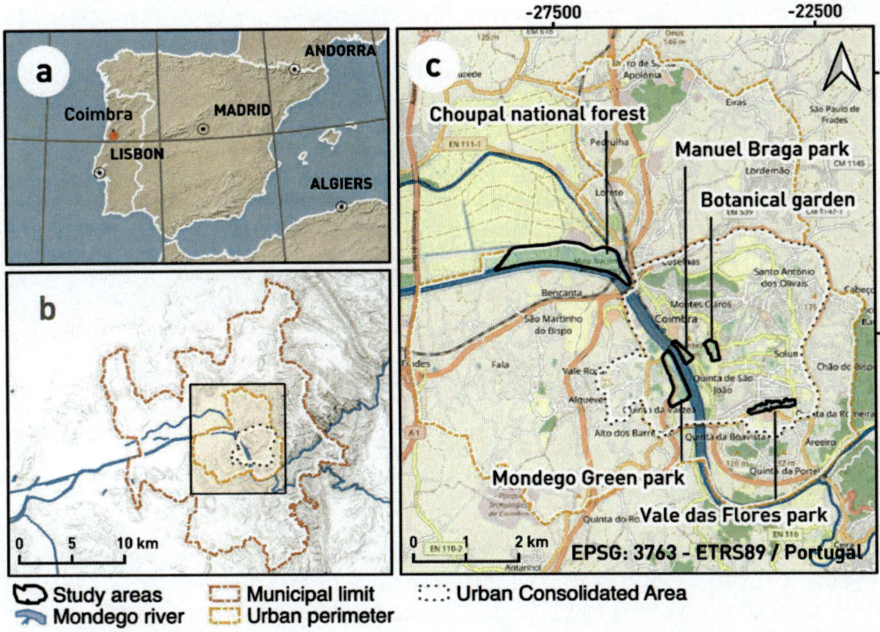

Fig. 1. Location of Coimbra in the center mainland Portugal (a), extent of the consolidated urban area in relation to municipal boundaries (b), and Studied urban parks (c)

Manuel Braga park is a medium-sized park (2.9 ha), located near the Mondego river, designed in 1927. The park has a tree coverage of 53.4%. The Botanical garden was designed in the 18th century, currently covers 4.0 ha and has a tree coverage of 49.0%. The Choupal national forest is located near the Mondego river, in the northwest limit of the city. This park is larger than the other parks (71.9 ha) and has the highest tree coverage (85.3%) compared to the other parks. The Mondego green park has an area of 18.1 ha and is located near the Mondego river. This park offers several services such as water-related sports equipment. However, has the lowest tree coverage (20.8%). Finally, Vale das Flores park covers an area of 3.2 ha and has a similar tree coverage of 20.9%. It is a typical urban park, with sports and leisure equipment (Fig. 1c).

2.2 Data Collection

2.2.1 The Questionnaire and Survey Process

A face-to-face survey was conducted, and different types of information were collected (socio-demographic data, including gender, age group, postal code – to calculate distance to UGS from home, means of transportation to the UGS, education level, frequency of visit to the investigated UGS, average monthly income, timeslot of the registered visit, and workdays vs. weekend). The activity performed at the time of the survey was also collected and aggregated into three groups of CES: 1) Physical interactions; 2) Experiential and aesthetical interactions; and 3) social interactions. In total, 200 users were interviewed per park. The survey was directed exclusively to Coimbra inhabitants. The survey was developed during both workdays and weekends, covering different timeslots (morning – 7h00-9h30; late morning – 9h30–12h30; noon – 12h30–14h30; afternoon – 14h30-17h30; evening – 17h30–20h30; night – 20h30–22h00), during August 2020.

2.3 Data Analysis

Statistical differences between timeslots and week period were assessed applying a non-parametric Kruskal-Wallis test. Significant differences were considered at a $p < 0.05$. A Principal Component analysis (PCA) was conducted to assess the association between the different variables. All statistical analysis was performed in R software (macOS version 1.3.1093), with plots produced with ggplot2. Spatial analysis was performed in QGIS (v3.16).

3 Results

3.1 Descriptive Statistics

Figure 2a shows the users percentage of the studied parks in the week and weekend. Among all the parks, only Vale das Flores registered a high percentage of weekday visitors (circa 55%). Manuel Braga park, Choupal national forest and Mondego green park, registered 58%, while in the Botanical garden 67% of the park users visited it in workdays. The Fig. 2b shows the distribution of users according to the different CES activities and park. Physical interactions are the most common in all parks, especially in Choupal national forest (over 82%). The lowest values of physical interactions were observed in the Botanical garden (circa 40%). Social interactions had similar percentages in all parks (between 20% and 25%), with the exception of Choupal national forest, which registered 14% of users engaging in social CES. Experiential and aesthetical interactions had the highest differences between parks. Choupal national forest and Vale das Flores park had percentages under 5%, while the Botanical garden had the highest percentage of users engaged in experiential and aesthetical interactions (over 32%).

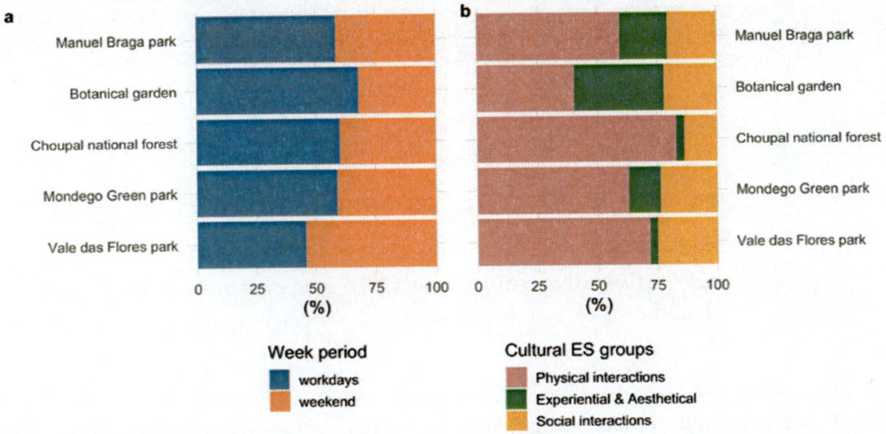

Fig. 2. Percentage of users (on a weekly basis) by week period and park (a) and percentage of users by CES groups by park (b).

Regarding the percentage of visitors of the different parks in the different timeslots in workdays (Fig. 3a) and weekend (Fig. 3b), except for Choupal national forest, all the other parks have at least 60% of the visits on the afternoon-to-night period in workdays. The weekend period has a more balanced visitors' distribution, with all parks registering over 45% of the visits until the end of the noon period (14h30). Over 55% of Choupal national forest users visit the park during the morning and late morning period in workdays. The Vale das Flores park registered the highest percentage of visitors at night during workdays (25%), while the Choupal national forest had the highest percentage of visitors during the same period in the weekend (over 13%). During workdays, the Botanical garden had almost 75% of its users after 14h30, this percentage decreasing to circa 60% on weekends.

Significant statistical differences were observed in gender between 12h30–14h30. In relation to age group, statistically significant differences were identified in 09h30–12h30 and 17h30–20h30 timeslots. With the exception of the timeslots between 07h30–09h30 and 14h30–17h30, statistically significant differences were observed among education levels. The average monthly income level did not differ statistically in the timeslots between 07h30–09h30, 09h30–12h30 and 20h30–22h00. Transport type to UGS and frequency of visit with exception of 12h30–14h30 and 20h30–22h00, differed statistically in all the other timeslots. Finally, the distance to UGS had statistically significant differences in all timeslots, except for 12h30–14h30 and 17h30–20h30. The differences

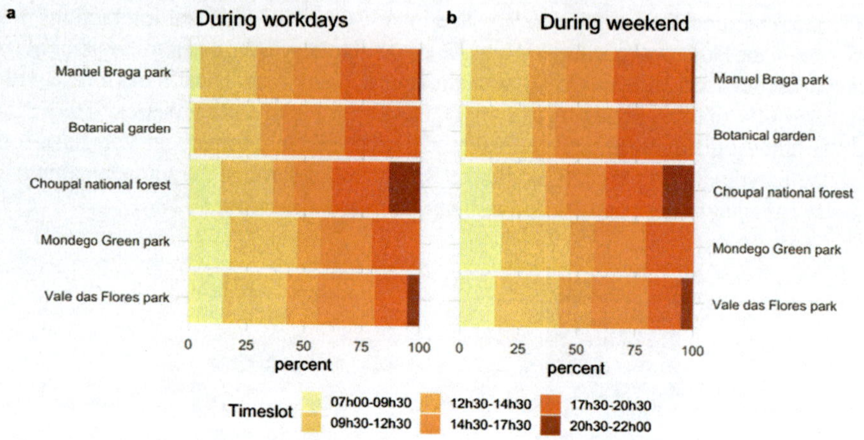

Fig. 3. Percentage of users by park and different timeslots, for workdays (a) and weekend (b)

in working days were statistically significant in all the studied variables. Finally, except for frequency of visit, the other variables showed statistically significant differences in weekend (Table 1).

Table 1. Kruskal-Wallis results for differences between parks between socio-demographic variables and both Timeslots and Week periods (ns: not significant at $p < 0.05$)

Socio-demographic variables	Timeslots						Week period	
	07h30 09h30	09h30 12h30	12h30 14h30	14h30 17h30	17h30 20h30	20h30 22h00	Workdays	Weekend
[1]	[2]	[3]	[4]	[5]	[6]	[7]	[8]	[9]
Gender	ns	ns	<0.05	ns	ns	ns	<0.05	<0.05
Age group	ns	<0.05	ns	ns	<0.05	ns	<0.05	<0.05
Education level	ns	<0.05	<0.05	ns	<0.05	<0.05	<0.05	<0.05
Average monthly income	ns	ns	<0.05	<0.05	<0.05	ns	<0.05	<0.05
Transport type to UGS	<0.05	<0.05	ns	<0.05	<0.05	ns	<0.05	<0.05
Distance to UGS	<0.05	<0.05	ns	<0.05	ns	<0.05	<0.05	<0.05
Frequency of visit	<0.05	<0.05	ns	<0.05	<0.05	ns	<0.05	ns
Workdays vs. weekend	<0.05	<0.05	<0.05	<0.05	ns	<0.05	–	–
Timeslots	–	–	–	–	–	–	<0.05	ns

3.2 Multivariate Analysis

Figure 4 shows the PCA results. The dimension 1 explained 22.2% of the variance, while the dimension 2 explained 19.6%. Distance to park and transport type were positively associated between them and negatively related to frequency of visit. Age group, income and workday vs weekend were also positively related between them and negatively associated with timeslot. Finally gender and education level have an inverse relation. The visits to Botanical garden are influenced by the distance and transport type, while Choupal national forest visitors are influenced by the education level and transport type. Manuel Braga park is mainly influenced by the gender and Vale Flores park by the frequency of visit.

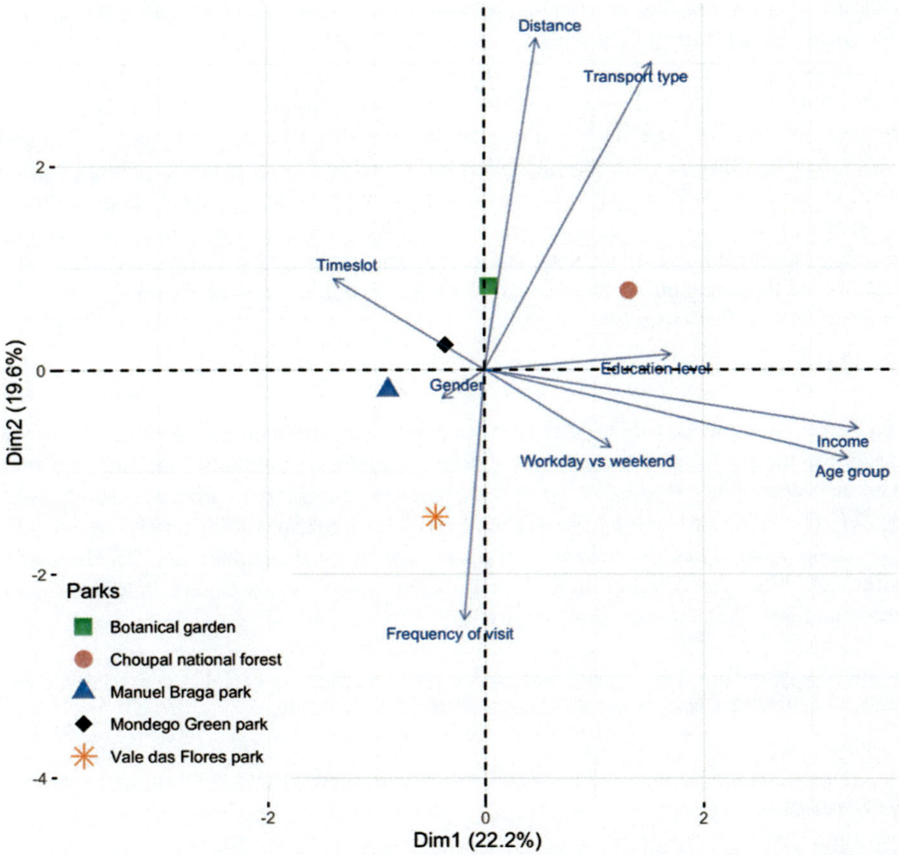

Fig. 4. Relation between PCA Dim1 and Dim 2.

4 Discussion

Physical interactions (e.g. walking, group sports, running, biking or walking the dog) are the most dominant CES identified as the main activities in all parks, similar to previous works (e.g. [9–11]). Timeslots and workdays vs. weekend visits vary across the different parks. Workdays vs. weekend differences showed to be relevant in all the five parks. Significant differences among all the socio-demographic variables were observed as well. Timeslots showed significant differences among parks concerning all other socio-demographic variables, with all timeslots except evening (17h30-20h30) showing significant differences regarding the two week-periods under analysis. Differences in timeslots were especially relevant among the Choupal national forest and the other parks, for the workdays period. This can be related to Choupal park users being mainly looking for sport activities (e.g., running, biking), especially early in the morning during workdays. Except for the Choupal national forest, all parks had most of the visits in the afternoon and evening periods during workdays, while weekend visits were more evenly spread along the different timeslots. The results suggested that park usage along the day and between weekdays and weekends depends on the activities performed and park characteristics. A common characteristic found in all parks is the strong positive association between transport type and distance to UGS, with both of them showing a negative relation with frequency of visit to UGS. Finally, age group and average monthly income are related as observed in previous works. [12–14]. Overall, the results highlighted that the studied variables need to be considered in park planning.

5 Conclusions

UGS are a fundamental tool to achieve sustainability in cities and communities, bringing relevant ecological, social and economic benefits. Socio-demographic variables are fundamental when analyzing UGS use. Among socio-demographic variables, users' usage of UGS can be constrained by time-related variables like time of day and week periods. This study showed that the relevance of these variables is dependent on UGS characteristics and CES offer. The finding of this work can help UGS managers understand user needs and time-related visitation constrains.

Acknowledgements. This research was funded by the Portuguese Foundation for Science and Technology (FCT), through the PhD grant SFRH/BD/149710/2019.

References

1. United Nations education: Transforming Our World: The 2030 Agenda for Sustainable Development (2015). https://sdgs.un.org/sites/default/files/publications/21252030%20Agenda%20for%20Sustainable%20Development%20web.pdf
2. United Nations: New Urban Agenda (2017). https://habitat3.org/wp-content/uploads/NUA-English.pdf
3. European Commission: EU Biodiversity Strategy for 2030 - Bringing nature back into our lives, Brussels, Belgium (2020)

4. World Health Organization: Urban Green Spaces and Health: A Review of Evidence. WHO Regional Office for Europe, Copenhagen (2016)
5. Haase, A., Rink, D., Grossmann, K., Bernt, M., Mykhnenko, V.: Conceptualizing urban shrinkage. Environ Plan A. **46**, 1519–1534 (2014). https://doi.org/10.1068/a46269
6. Andersson, E., Tengö, M., McPhearson, T., Kremer, P.: Cultural ecosystem services as a gateway for improving urban sustainability. Ecosyst. Serv. **12**, 165–168 (2015). https://doi.org/10.1016/j.ecoser.2014.08.002
7. INE: Resident population per NUTS (estimates) (2020). https://www.ine.pt/xportal/xmain?xpid=INE&xpgid=ine_publicacoes
8. INE: General Population Census per statistical unit (2011). https://www.ine.pt/xportal/xmain?xpid=INE&xpgid=ine_publicacoes
9. Bijker, R.A., Sijtsma, F.J.: A portfolio of natural places: using a participatory GIS tool to compare the appreciation and use of green spaces inside and outside urban areas by urban residents. Landscape Urban Planning **158**, 155–165 (2017). https://doi.org/10.1016/j.landurbplan.2016.10.004
10. Palliwoda, J., Banzhaf, E., Priess, J.A.: How do the green components of urban green infrastructure influence the use of ecosystem services? Examples from Leipzig. Germany Landscape Ecol. **35**, 1127–1142 (2020). https://doi.org/10.1007/s10980-020-01004-w
11. Rall, E., Bieling, C., Zytynska, S., Haase, D.: Exploring city-wide patterns of cultural ecosystem service perceptions and use. Ecol. Ind. **77**, 80–95 (2017). https://doi.org/10.1016/j.ecolind.2017.02.001
12. Lo, A.Y.H., Jim, C.Y.: Citizen attitude and expectation towards greenspace provision in compact urban milieu. Land Use Policy **29**, 577–586 (2012). https://doi.org/10.1016/j.landusepol.2011.09.011
13. Riechers, M., Barkmann, J., Tscharntke, T.: Diverging perceptions by social groups on cultural ecosystem services provided by urban green. Landscape Urban Planning **175**, 161–168 (2018). https://doi.org/10.1016/j.landurbplan.2018.03.017
14. Sun, R., Li, F., Chen, L.: A demand index for recreational ecosystem services associated with urban parks in Beijing China. J. Environ. Manage. **251**, 109612 (2019). https://doi.org/10.1016/j.jenvman.2019.109612

Soil Health in Urban Protected Areas and Pathways for Sustainable Development

Snežana Štrbac[1(✉)], Carla S. S. Ferreira[2,3,4], Lucian C. Dincă[5], Nevena Antić[1], Gorica Veselinović[1], Zahra Kalantari[2,3,6], and Milica Kašanin-Grubin[1]

[1] Institute of Chemistry, Technology and Metallurgy, University of Belgrade, 11000 Belgrade, Serbia
snezana.strbac@ihtm.bg.ac.rs

[2] Department of Physical Geography, Stockholm University, Bolin Centre for Climate Research, 10691 Stockholm, Sweden

[3] Navarino Environmental Observatory, 24001 Messinia, Greece

[4] Research Centre for Natural Resources, Environment and Society (CERNAS), College of Agriculture, Polytechnic Institute of Coimbra, Coimbra, Portugal

[5] National Institute for Research and Development in Forestry Marin Dracea, 13 Closca Str., 500040 Brasov, Romania

[6] Department of Sustainable Development, Environmental Science and Engineering, Sustainability Assessment and Management, KTH Royal Institute of Technology, 100 44, Stockholm, Sweden

Abstract. Soils are a natural capital which support life on Earth and are responsible for many ecosystem services that are the foundation for human wellbeing. However, soils have been intensively used and subject to degradation, particularly in urban areas. The main objective of this study is: i) to give an overview of soil conditions in urban protected areas (UPAs), and ii) to provide suggestions how the nature-based solution (NBS) can enhance the soil health in three UPAs located near or in the city of Belgrade, Serbia (Byford's and Zvezdara forests, and Avala mountain). In this study pH values and soil organic matter content were analyzed as main parameters for the soil characterization. The Byford's and Zvezdara forests have similar slightly alkaline soil pH, while the Mt. Avala soils are slightly acidic as a consequence of geological settings. The lower content of soil organic matter (1.26%–5.73%) in the Byford's and Zvezdara forests might be improved with appropriate management practices. Introducing the principles of the NBS (e.g., incorporation of compost and biochar) in soil management practices in the protected areas, combined with conservation and restoration of forests, can provide enhanced soil health and supporting new green infrastructure (GI). The citizens and the society would highly benefit from a network of small green areas connected with UPAs in a number of ways, such as cleaner air, mitigation of noise stress, maintaining everyday wellbeing, recreation opportunities, contribution to employment and supporting social cohesion.

Keywords: Soil health · Urban protected areas · Nature-based solutions

J. R. da Costa Sanches Galvão et al. (Eds.): *Proceedings of the 1st International Conference on Water Energy Food and Sustainability (ICoWEFS 2021)*, pp. 576–584, 2021.
https://doi.org/10.1007/978-3-030-75315-3_62

1 Framework

Soil, as a critically important component of the Earth's biosphere [1], is a natural capital which supports life on Earth, and provides many ecosystem services important for human wellbeing. Soil health has been defined as "the capacity of soil to function as a vital living system, within ecosystem and land-use boundaries, to sustain plant and animal productivity, maintain or enhance water and air quality, and promote plant and animal health" [2]. Criteria for indicators of soil quality and its health relate mainly to their (i) utility in defining ecosystem processes and integrating physical, chemical, and biological properties; (ii) sensitivity to management and climatic variations; and (iii) accessibility and utility to agricultural specialists, producers, conservationists, and policy makers [3]. Due to natural events or human impacts, soil health can be changed over time. Particularly in urban areas soils have been intensively used and subjected to pollution.

Soil compaction, low organic matter content, and high levels of contamination, usually from air deposition or from historical uses, are common attributes of urban soils [4, 5]. As a result, "urban soils" have been viewed as drastically disturbed material of low fertility [6]. The characteristics of soil can vary greatly across the urban landscape, including not only highly disturbed but also relatively undisturbed soils, modified by management and urban environmental factors [7]. Urban disturbance can be caused by direct or indirect effects on soil properties [8]. Direct effects include physical disturbances, incorporation of anthropogenic materials, and burial or coverage of soil by filling materials and impervious surfaces [9], and soil management practices (e.g., fertilization and irrigation). Indirect effects include for example the introduction of exotic plant and animal species [10], atmospheric deposition of pollutants [11], particularly heavy metals [12, 13], and potentially toxic organic chemicals [14]. The poor conditions of soils in the urban areas, can often be a significant impediment to successfully implementing the green infrastructure (GI).

Urban protected areas (UPAs) are those located in, or at, the edge of larger population centers [15], thus suffering from great human pressure. Urbanization is considered a major threat to UPAs [15]. This process changes the landscape and affects biodiversity in the city areas, and may have both positive and negative effects on protected areas [15]. The negative side of urbanization is generally much more evident, linked with, for example, depletion of water resources and forests, air and water contamination, fragmentation of habitat and edge effects, and introduction of invasive alien species [15].

Nature-based solutions (NBS) have become an interesting approach to mitigate the impacts of urbanization on the environment. The NBS are defined as "actions inspired by, supported by, or copied from nature" [16, 17], and can provide affordable, sustainable and feasible benefits that will contribute to public health, social wellbeing and urban resilience against global changes [18, 19]. Considering climate change adaptation and mitigation issues as well as risk management, NBS serve as a valuable tool for reaching sustainability objectives, such as the sustainable development and the restoration of degraded ecosystems. Urban renaturing may be a useful approach to radically transform the landscape.

2 Objectives

The main objective of this study is: i) to give an overview of soil conditions in UPAs, and ii) to provide suggestions how NBS can enhance the soil health in the three UPAs located near or in the city of Belgrade, Serbia.

3 Methodology

3.1 Study Area

The study was developed in Belgrade, Serbia, in three UPAs under different protection regimes: Byford's and Zvezdara forest, and Avala mountain (Fig. 1).

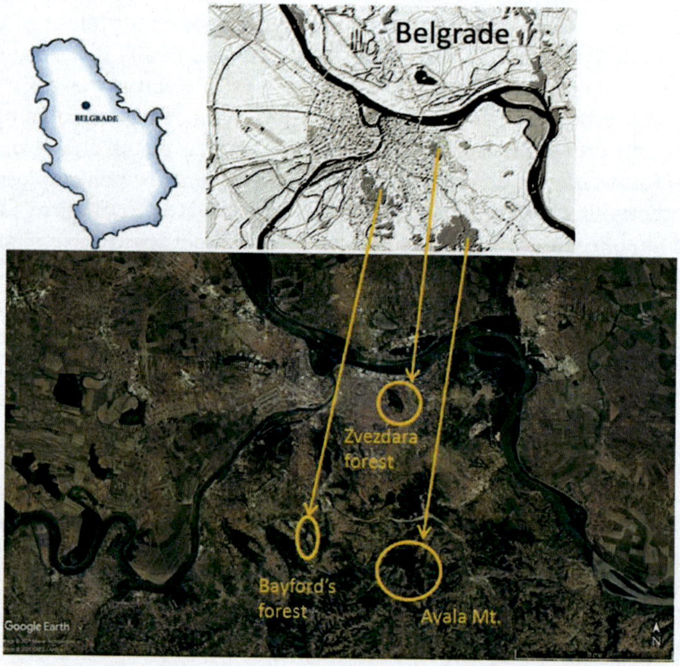

Fig. 1. Study area and location of the three UPAs investigated in Belgrade, Serbia.

The Byford's forest is located in the Belgrade's municipality of Voždovac (44°45′58″N and 20°28′35″E), with a total area of 40.1 ha. It is an artificially raised forest but gives the impression of an autochthonous forest area, including the remains of natural climatogenic and oroclimatogenic mixed deciduous forests from the central and eastern parts of the Balkan Peninsula. The Byford's forest is protected as a Nature Monument, of local significance Category III [20]. According to the categorization of the International Union for Conservation of Nature (IUCN), it is classified in Category III: Natural monument or feature. Management of the Byford's forest Nature Monument may also include guidelines and objectives listed for IUCN Category IV: Habitat/species management [21].

The Zvezdara forest is also an artificially raised forest in the city of Belgrade (44.804°N and 20.510°E), with a total area of 80.6 ha. This area provides a favorable microclimate, binding of air pollutants, enrichment of the soil with humus, protection from erosion and landslides, the existence and spread of forest species [22]. As the largest forest in the urban zone of Belgrade, it represents the most important source of oxygen for the city [22]. It is a Nature Monument, which is a protected area of local significance Category III ("Off. Gazette of RS", No. 36/2009, 88/2010 and 91/2010). According to the categorization of the IUCN, it is classified in Category III: Natural monument or feature [21].

The Mt. Avala is located 15 km south-east of downtown Belgrade (44°41′25″N and 20°30′51″E). The mountain area extends over 489.1 ha, and belongs to the wider Belgrade City area. The Mt. Avala is protected as a Landscape of Outstanding Features, with local significance Category III [23], while according to the categorization of the IUCN it is classified in Category V: Protected landscape/seascape [21].

The geological settings of the Byford's and Zvezdara forests are similar, comprising Miocene clays, sands and limestone. The Mt. Avala consists of rocks of different age and composition, such as Mesozoic serpentinites overlain with limestones, sands, marls and chert, Upper Cretaceous flisch sediments, and dacite rocks. Subsequent paragraphs, however, are indented.

3.2 Soil Sampling

Soil samples were collected at three depths (0–10 cm, 10–20 cm, and 20–40 cm) in nine sites from the Mt. Avala, and in three sites from both Byford's and Zvezdara forest. The number of sampling sites was based on the size of the UPA. A total of 45 soil samples were collected from the three UPAs. Soil samples were collected randomly over each UPA, and taken vertically with a soil tube, and then placed in a clean bucket and mixed thoroughly.

3.3 Chemical Analysis

The soil samples were air dried, milled, sieved through a 2 mm stainless steel sieve, and then analyzed using standard procedures and implemented laboratorial quality controls. Soil pH was measured using a Thermo Orion 3 pH-meter, based on soil-water solutions. Soil organic matter content was determined through the humid oxidation method and the titrimetric dosage, following the Walkley Black method with the Gogoasa modification. Samples were treated with $K_2Cr_2O_7$ and concentrated H_2SO_4, heated for 20 min at 80 °C in a water bath, diluted and titrated with Mohr salt.

4 Results

4.1 Soil Chemical Composition

The pH values and soil organic matter content obtained for all 45 soil samples are shown in Fig. 2.

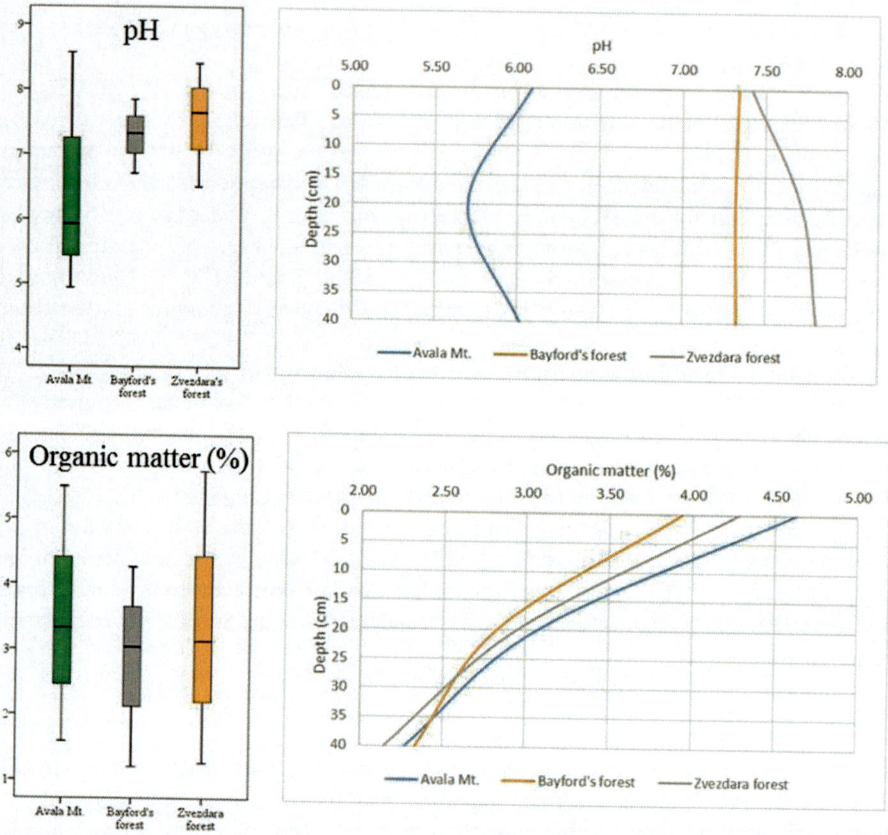

Fig. 2. Changes in average soil pH and organic matter content over the soil profiles sampled in Mt. Avala, Byford's and Zvezdara forest.

Soil pH ranges from moderately acid to slightly alkaline, with average values of 5.93 in Mt. Avala, 7.33 in the Byford's forest, and 7.65 in the Zvezdara forest. These results indicate improved conditions in Byford's and Zvezdara forest soils, since the optimal pH values for plant nutrient availability range between 6.0 and 7.5 [4]. The slightly acidic soil conditions in Mt. Avala can be a consequence of the geological settings.

Soil organic matter content ranged between 1.59% and 5.49% in the Mt. Avala, 1.20% and 4.26% in the Byford's forest, and 1.26% and 5.73% in the Zvezdara forest, although the average values are similar in the three UPAs (3.04%–3.32%) (Fig. 2). In typical well-drained mineral soils, such as those from UPAs in Belgrade, the organic matter content is relatively low, varying from 2% to 6% [24], indicating a moderate soil fertility. As expected, the highest soil organic matter contents (2.61–5.73%) are in the surface layer, due to high accumulation of organic matter driven by litter decomposition [25].

The slightly lower soil organic matter content in Byford's and Zvezdara forests might be a consequence of the lower biomass/vegetation density. In order to enhance the soil health and the overall productivity of the ecosystem, improved soil management practices may be implemented. Since organic matter affects many soil functions, it is important to maintain or even increase its content by incorporating organic waste. Improved soil quality can better support the establishment of new GI, which must be connected with current UPAs to maximize the supply of ecosystem services.

4.2 Recommendations for Improving Soil Management

Soil natural capital can be preserved through protection status of some areas. Byford's and Zvezdara forests are under protection level III, according to the Law on Nature Protection [20] and the Regulation on protection regimes [26]. Protection regimes of I, II and III degree have been established in the area of the Mt. Avala. The I and II protection degree prohibits the performance of any action (except the necessary intervention) that can endanger the protected area. The protection level III, implemented in all three UPAs, establishes that management interventions can be performed for restoration, revitalization and overall improvement of the protected area. It also restricts the use of chemicals (e.g., pesticides, fertilizers, pest control) and other activities that may have a significant adverse effect on the values of the protected area (e.g., natural value).

The Mt. Avala, Byford's and Zvezdara forests have an important role, for example, in providing a favorable microclimate and improving air quality in the city. The implementation of NBS principles in the management of UPAs, combined with conservation activities and restoration of forests, can provide cost-effective solutions to mitigate the impacts of global changes [27]. The Intergovernmental Panel on Climate Change (IPCC) performed several suggestions to support implementation of the United Nations Framework Convention on Climate Change (UNFCCC) to keep global warming, such as reforestation and afforestation [28]. Given the limited use of fertilizers in all three protected areas, improving soil health may be only performed by applying the NBS principles.

In the investigated UPAs, possible alternatives for improving soil health can be incorporation of compost and biochar. Biochar can provide positive effects on nutrient retention [29], cation exchange capacity [30], water-holding capacity [31], soil microbial and mycorrhizal activity [32], soil acidity [33] and soil electrical conductivity [34], thus improving soil fertility and thereby plant growth [35]. Biochar may also improve remediation and biophysical conditions of soils, which could benefit soil biodiversity [27].

5 Conclusions

The study was developed in the city of Belgrade, Serbia, in three UPAs under different protection regimes: Byford's and Zvezdara forest, and Mt. Avala. The Byford's and Zvezdara forests have similar slightly alkaline soil pH, while the Mt. Avala soils show slightly acidic conditions, which is a consequence of geological settings. The relatively low soil organic matter content in the three UPAs, but particularly in Baford's and Zvezdara forests, might be improved with appropriate soil management practices. Implementing the NBS principles (e.g., incorporation of compost and biochar) in soil management practices, combined with conservation and restoration of forests, can provide enhanced soil health and support the establishment of new GI, primarily connecting UPAs with smaller green areas and in that way establishing green networks.

6 Contribution of the Work to Society

The citizens and the society would highly benefit in a number of ways: cleaner air, reducing exposure to pollutants, mitigating noise stress, improving wellbeing, promoting recreation, contributing to employment and supporting social cohesion.

Acknowledgments. The authors would like to thank the Ministry of Education, Science and Technological Development of Republic of Serbia (Grant No: 451–03-9/2021–14/200026) for financial support.

References

1. Glanz, J.T.: Saving Our Soil: Solutions for Sustaining Earth's Vital Resource. Johnson Books, Boulder (1995)
2. Doran, J.T.H.W., Liebig, M.A., Santana, D.P.: Soil health and global sustainability. In: 16th Proceedings of the World Congress of Soil Science, Montepellier, France (1998)
3. Doran, J.W.: Soil health and global sustainability: translating science into practice. Publications from USDA-ARS/UNL Faculty (2002)
4. U.S. EPA (U.S. Environmental Protection Agency): Evaluation of urban soils: suitability for green infrastructure or urban agriculture. EPA 905R11o3. U.S. Environmental Protection Agency, Washington, D.C. (2011)
5. Ferreira, C.S.S., Pereira, P., Kalantari, Z.: Human impacts on soil. Sci. Total Environ. **644**, 830–834 (2018)
6. Craul, P.J.: Urban Soils: Applications and Practices. Wiley, New York (1999)
7. Pouyat, R.V., Russell-Anelli, J., Yesilonis, I.D., Groffman. P.M.: Soil carbon in urban forest ecosystems. In: Kimble, J.M., et al. (eds.) The Potential of U.S. Forest Soils to Sequester Carbon and Mitigate the Greenhouse Effect. CRC Press, Boca Raton (2003)
8. Pouyat, R.V., Effland. W.R.: The investigation and classification of humanly modified soils in the baltimore ecosystem study. In: Kimble, J.M., et al. (eds.) Classification, Correlation, and Management of Anthropogenic Soils. Proccidings Nevada and California Workshop, pp. 141–154. USDA-NRCS, National Soil Survey Center, Lincoln (1999)
9. Schleuß, U., Wu, Q., Blume, H.-P.: Variability of soils in urban and peri urban areas in Northern Germany. Catena **33**, 255–270 (1998)
10. Ehrenfeld, J.G., Kourtev, P., Huang, W.Z.: Changes in soil functions following invasions of exotic understory plants in deciduous forests. Ecol. Appl. **11**, 1287–1300 (2001)
11. Lovett, G.M., Weathers, K.C., Arthur, M.A.: Control of N loss from forested watersheds by soil C: N ratio and tree species composition. Ecosystems **5**, 712–718 (2002)
12. Mielke, H.W.: Lead in the inner cities. Am. Sci. **87**, 62–69 (1999)
13. Ferreira, C.S.S., Walsh, R.P.D., Ferreira, A.J.D.: Degradation in urban areas. Curr. Opinion Environ. Sci. Health **5**, 19–25 (2018)
14. Wong, S.L., Zhang, L.V., Tong, A.H., Li, Z., Goldberg, D.S., King, O.D., Lesage, G., Vidal, M., Andrews, B., Bussey, H., Boone, C., Roth, F.P.: Combining biological networks to predict genetic interactions. Proc. Natl. Acad. Sci. U. S. Am. **101**(44), 15682–15687 (2004)

15. Trzyna, T.: Urban Protected Areas: Profiles and Best Practice Guidelines. International Union for Conservation of Nature, Gland, Switzerland (2014)
16. Cohen-Shacham, E., Walters, G., Janzen, C., Maginnis, S.: Nature-Based Solutions to Address Societal Challenges. International Union for Conservation of Nature, Gland, Switzerland (2016)
17. EC (European Commission): Nature-Based Solutions & Re-Naturing Cities Towards an EU Research and Innovation policy agenda for Final Report of the Horizon 2020 Expert Group on 'Nature-Based Solutions and Re-Naturing Cities'. Publications Office of the European Union, Luxembourg (2015)
18. Fink, H.S.: Human-nature for climate action: Nature-based solutions for urban sustainability. Sustainability **8**(3), 254 (2016)
19. Kalantari, Z., Ferreira, C.S.S., Page, J., Goldenberg, R., Olsson, J., Destouni, G.: Meeting sustainable development challenges in growing cities: coupled social ecological systems modeling of land use and water changes. J. Environ. Manage. **249**, 471–480 (2019)
20. Law on Nature Protection. Official Gazette of the Republic of Serbia, No. 36/2009, 88/2010 and 91/2010
21. Dudley, N.: Guidelines for applying protected area management categories, International Union for Conservation of Nature, Gland, Switzerland (2008)
22. Stavretović, N., Trikić, M., Ostojić, D., Medarević, M., Đorđević, Z., Stojanović, V., Lukić, D., Andonović, J., Panič, N., Nikolić, M., Vuković, A., Sekulić, G., Ajtić, R.: Studija zaštite spomenik prirode Zvezdarska šuma. Zavod za zaštitu prirode Srbije, Beograd (2010)
23. Rulebook on categorization of protected natural goods. Official Gazette of the Republic of Serbia, No. 30/92
24. Brady, N.C.: The Nature and Properties of Soils, 9th edn. Macmillan Publishing Company, New York (1984)
25. Romell, L.G., Heiberg, S.O.: Types of Humus layer in the forests of North-Eastern united states. Ecology **12**(3), 567–608 (1931)
26. Regulation on protection regimes. Official Gazette of the Republic of Serbia, No. 31/12
27. Griscom, B.W., Adams, J. Ellis, P.W., Houghton, RA., Lomax, G., Miteva, D.A., Schlesinger, W.H., Shoch, D., Siikamäki, JV., Smith, P., Woodbury, P., Zganjar, C., Blackman, A., Campari, J., Conant, R.T., Delgado, C., Elias, P., Gopalakrishna, T., Hamsik, M.R., Herrero, M., Kiesecker, J., Landis, E., Laestadius, L., Leavitt, S.M., Minnemeyer, S., Polasky, S., Potapov, P., Putz, F.E., Sanderman, J., Silvius, M., Wollenberg, E., Fargione J.: Natural climate solutions. Proc. Natl. Acad. Sci. **114**(44), 11645–11650 (2014)
28. Healthy Soils to Cool the Planet: A Philanthropic Action Guide. Breakthrough Strategies and Solutions (2019)
29. Major, J., Lehmann, J., Rondon, M., Goodale, C.: Fate of soil applied black carbon: downward migration, leaching and soil respiration. Global Change Biol. **16**, 1366–1379 (2010)
30. Topoliantz, S., Ponge, J.F., Arrouays, D., Ballof, S., Lavelle, P.: Effect of organic manure and the endogeic earthworm Pontoscolex corethrurus (Oligochaeta: Glossoscolecidae) on soil fertility and bean production. Biol. Fertil. Soils **36**, 313–319 (2002)
31. Glaser, B., Lehmann, J., Zech, W.: Ameliorating physical and chemical properties of highly weathered soils in the tropics with charcoal–a review. Biol. Fertil. Soils **35**, 219–230 (2002)
32. Thies, J.E., Rillig, M.C.: Characteristics of biochar: biological properties. In: Lehmann, J., Joseph, S. (eds.) Biochar for Environmental Management–Science and Technology. Earthscan, London (2009)
33. Chan, K.Y., van Zwieten, L., Meszaros, I., Downie, A., Joseph, S.: Using poultry litter biochar as soil amendments. Aust. J. Soil Res. **46**, 437–444 (2008)

34. Asai, H., Samson, B.K., Stephan, H.M., Songyikhangsuthor, K., Homma, K., Kiyono, Y., Inoue, Y., Shiraiwa, T., Horie, T.: Biochar amendment techniques for upland rice production in Northern Laos 1. Soil physical properties, leaf SPAD and grain yield. Field Crops Res. **111**, 81–84 (2009)
35. Novotny, E.H., Hayes, M.H.B., Madari, B.E., Bonagamba, T.J., Azevedo, E.R., Souza, A.A.: Lessons from the Terra Preta de Indios of the Amazon region for the utilisation of charcoal for soil amendment. J. Braz. Chem. Soc. **20**, 1003–1010 (2009)

Asian Wasp as an Environmental Health Problem- Case of Study in the Municipality of Coimbra

Susana Paixao[1,2](✉), Ana Ferreira[1], Paulo Figueiredo[3], and Adrina Felizardo[1]

[1] Polytechnic Institute of Coimbra – Portugal Coimbra Health School Department of Environmental Health, Coimbra, Portugal
supaixao@estescoimbra.pt
[2] CEGOT, Coimbra, Portugal
[3] Polytechnic Institute of Coimbra – Portugal Coimbra Health School Department of Complementary Sciences, Coimbra, Portugal

Abstract. With the increase in the urbanization and population mobility , the biodiversity has been greatly affected. Globalization has facilitated the dispersion of many species that become a threat, reaching the status of invasive species with an impact on the ecosystems. The Asian wasp was spotted in 2011 in Portugal. The biological cycle of Asian wasp is anual, although the climatic conditions attend to its development. The goal of this study is to observe the growth over the years of sightings of nests, relating the life cycle of this species with the time of the year when it's more propitious to its incidence, taking into account the impacts on the environment and the health of the population. The nature of this study was observational zoological and of retrospective cohort. Data and records were provided by the Instituto da Conservação da Natureza e das Florestas (ICNF). The data were evaluated in IBM SPSS Version 25 program. It was found that in the municipality of Coimbra, 2019 was the year with the highest numbers of observation records, the definitive nests were those that had the most prevalent nests located in the trees while the primary nests were found in housing spaces. The parishes that had a higher frequency of registrations were Santo António dos Olivais, União de Freguesias de São Martinho do Bispo/Ribeira de Frades, União de Freguesias de Sé Nova/Santa Cruz/Almedina/São Bartolomeu. This study allowed us to understand which factors are favorable to the appearance of invasive species in Portugal, their reproduction and propagation, as well as the impacts they have on the environment and population, which can lead to the death if a person that has been bitten has an allergic reaction.

Keywords: Wasp · Environment · Health · Nests

1 Introduction

In the last decades, with the increase in the urbanization and population mobility, biodiversity has been greatly affected. This globalization has facilitated the dispersion of

J. R. da Costa Sanches Galvão et al. (Eds.): *Proceedings of the 1st International Conference on Water Energy Food and Sustainability (ICoWEFS 2021)*, pp. 585–593, 2021.
https://doi.org/10.1007/978-3-030-75315-3_63

many species that have become a threat, reaching the status of invasive species with an impact on ecosystems (Álvaro and Castro-Díez 2014).

The concept of pest is defined by Food and Agriculture Organization of The United Nations (FAO) as "organism that damages economic, environmental, socio-political or cultural values" (FAO 2009).

With the growth of cities, natural spaces were the most affected and consequently the species that used them as habitat, leading to an environmental imbalance. With the lack of urban planning, land use and waste treatment policies, all conditions were met for the emergence of pests in the urban environment, since man himself provided these species with water, shelter and food necessary for their survival (Zuben et al. 2006).

The wasps belong to the family Vespidae and to the order Hymenoptera. These insects have a very important ecological role, since they are predators of other insects (pests), but when they play an invasive insect role, they alter the functioning of food chains (Patrício 2018).

The Asian wasp nigrithorax, the focus of this study, is also called Asian wasp, originating in China, Afghanistan, Indochina and Indonesia where 12 varieties of this species have been identified (Maia and Silva 2012).

Although its distribution was apparently restricted to the Asian continent, it quickly spread. With the ease of communication and export between countries, it managed to reach Europe and was detected for the first time in France, in 2004. It is believed that its unintended introduction took place via a container from a boat from the Chinese horticultural trade bound for the port of Bordeaux. When it reached French territory, it quickly spread, becoming the first wasp to successfully invade Europe. Its expansion continues on this continent and in 2010, it was sighted for the first time in Spain, arriving in portuguese territory in 2011, more specifically in the north, to the district of Viana do Castelo (Rome et al. 2011).

This species has the peculiarity of having yellow paws, and the fourth abdominal segment is orange. However, the remaining patterns such as thorax differ from region to region, however, the colors are mostly brown for the chest and black and yellow for the rest of the body (David and Moreira 2017).

The projections of the climate scenarios for 2100 relate the risk of invasion to climate change, since an increase in the climatic adaptation of species in the Northern hemisphere is expected, especially near areas of Europe already invaded (Barbet-Massin et al. 2013; ICNF 2018).

The biological cycle of Asian wasp is annual, although climatic conditions may condition its development. Its cycle can be divided into two phases: the first phase for the founding queen, from February to April; the second phase for colony growth, from April to November (Maia and Silva 2012).

It is important to locate, as well as to know how to identify the nests to control the invasion in Portugal (Maia and Silva 2012).

The detection or suspicion of the existence of a nest or specimens of Asian wasp must be communicated through online platforms or government services (ICNF 2018).

According to the data from the STOPvespa platform provided by the ICNF, in the total balance over the period of 2014 to 2020, in Portugal around 52677 were observed and 48495 nests were exterminated. The districts with the highest number of observations were Porto, Aveiro, Braga, Viseu and Coimbra. From 2014 to June 2018, 12245 observed nests were registered, with possibly an increase in registrations in the following months. In 2019, 23 579 nests were observed. In 2020 until June 4, 1245 nests were sighted and 1124 nests were destroyed in Portugal. According to the ICNF, the location of Asian wasp nests is distributed among trees above 10 m in height (73%), buildings (10%), hedges (3%) and others (14%) (Luz 2020).

The presence of the Asian wasp represents a risk from different points of view: for beekeeping, the effect on the bee population is a direct effect due to the losses produced by the direct predation by Asian wasp, and indirectly, by the decrease in the activities of the bees in the face of the presence of Asian wasp, presenting two direct consequences, in one hand, a lower production of honey and related products and, on the other hand, a decrease in pollination given the importance of honey bees in this important biological function. Although not individually more aggressive to humans than the European wasp, it reacts quite aggressively to threats to its nest; when faced with a threat or vibration at 5 m, a group response is produced that can chase the source of the threat for about 500 m. In addition, the large size that the nests can reach and sometimes their location in urban or peri-urban areas, can result in greater risk for citizens (ICNF 2018) .

The hymenopteran bite allergy can be fatal, but once diagnosed it can be prevented and treated, and these patients should be referred to the Immunoallergology Services in the country. It is important that there are protocols for the therapeutic action of allergic reactions, in the Emergency Services, both in Hospitals and Health Centers, so that these situations are quickly diagnosed and treated.

For the environment, it is a non-indigenous species, a natural predator of bees and other insects, which may eventually have significant impacts on biodiversity in the medium term, in particular on native wasp species and populations of other insects. As side effects of the decrease in the native entomofauna, less pollination of species of natural or cultivated vegetation can occur (ICNF 2018).

This study objective was to observe the growth over the years of sightings of Asian wasp nests, relating the life cycle of this species with the time of year that is more propitious to its incidence. As specific objectives it was proposed to seek to know the prevalence of the typology of registered nests such as their location in terms of support and which parishes where a higher frequency was found and the factors for this to be verified, relating to the impacts on the environment and health of the population.

2 Material and Methods

This study was performed between September 2019 and July 2020, the data collection was carried out between April 2020 and June 2020. This study was presented as an observational zoological and retrospective cohort type. The target population of this study was composed by Asian wasp nests observations in the municipality of Coimbra. The type of sampling was non-probabilistic. The sample consisted of 517 registrations. The data collection was carried out through the provision of official data by the ICNF, where the dates of registration of the nests were presented, the support in which they were found, the date of their extermination, the respective coordinates accompanied by the district and municipality where observations were registered. The STOPvespa platform was used to validate the parishes where the nests were found. Subsequent to the data processing, it was asked to DGAV to provide the number of apiaries and hives in the parishes, to later compare the prevalence of wasps nests with the existence of apiaries in the same areas.

The statistical analysis of the collected data was carried out using the IBM SPSS Statistic software version 25 for Windows 7. With regard to data processing, we used a descriptive statistics consisting of frequency analyzes (absolute and relative) either presented in tables or graphs (Bar and Sector Diagrams). The test used was the Independence Chi-square for independent samples (non-parametric test), that is, for a 95% confidence level, p-value ≤ 0.05.

3 Results

The sample consisted of 517 observations of nests located in the municipality of Coimbra. The variables that were analyzed, such as, nest typology (primary level or definitive level), year of observation, location of the nest, observation in the 18 parishes of the municipality of Coimbra.

It was possible to verify that the year of 2017, obtained a smaller number of observation records of nests both at "primary level" and at "definitive level", which in the year of 2018 regarding its observation there was a slight increase in the identification of nests in the "definitive level". However it can be seen that the year 2019 was marked by a sudden growth in terms of "definitive level" nests, exceeding 350 records.

Applying the Chi-Square Test of Independence, the year of observation and the nest typology were compared and it was verified the presence of statistically significant differences between the years of observation of the nests compared to the typology ($p < 0.05$). We can say that in the year 2019 the number of records of the type of nest differed significantly, that is, with greater expression of the type "definitive level" (79.2%) compared to the "primary level" (67.6%). However, in 2018 the opposite was found regarding the number of registrations by typology. Finally, in 2017 the prevalence of the number of registration of nests of the type "definitive level" was slightly higher compared to the type "primary level".

We then sought to assess the distribution of nest records as to typology over the months of the year, studying the distribution of the number of the registrations by the years under evaluation and the months, we also sought to assess the distribution of the number of the records of Asian Wasp nests according to the locations that were discovered or reported by the region under study.

It was verified that, regarding the support, that 66.34% of the nests were located in "trees", 23.21% in "others" (which includes, for example, buildings, annexes, air conditioning), 3.48% in "roofs", 2.71% on the "soil", 2.71% on the "interior of buildings" and 1.55% were on "walls".

We tried to evaluate the pattern of distribution of the number of registered nests according to the location of detection and the degree of their development. We found, significantly, that the registration of nests in the "primary level" condition was more prevalent in "housing spaces" (64.9%) compared to the condition of the "definitive level" nest. However, it was more prevalent when detected at the "tree" location compared to the "primary level" nest development.

We then proposed to understand the distribution of the Asian wasp nest register on the level of development at the "primary level" according to the parishes in the municipality of Coimbra.

In União de Freguesias de São Martinho do Bispo/Ribeira de Frades from the 4 nests observed, 2 (50,0%) were located on "tree" and 2 (50,0%) in "housing space"; in Torres do Mondego it was observed 1 (100%) nest in "housing space"; in União de Freguesias de Santa Clara/Castelo Viegas, from the observation of 3, from "primary level", 1 (33,3%) was located on "tree" and 2 (66,7%) in "housing space"; in Santo António dos Olivais from the total of 6 nests, 3 (50,0%) were verified on "trees" and 3 (50,0%) in "housing spaces"; in Ceira it was observed 1 in "housing spaces"; in União de Freguesias de Assafarge/Antanhol it was observed 1 in "housing spaces"; in União de Freguesias Eiras/São Paulo de Frades it was observed 1 in "housing spaces"; in Cernache it was verified 1 in "housing spaces"; in União de Freguesias de Taveiro/Ameal/Arzila it was observed 2 in "housing spaces" and in União de Freguesias de Trouxemil/Torre de Vilela it was found 1 on "trees".

At the "primary level" the parishes where it was verified a larger number of registers were, União de Freguesias de Sé Nova/Santa Cruz/Almedina/São Bartolomeu, Santo António dos Olivais and União de Freguesias de Antuzede/Vil de Matos.

We then proposed, the same strategy of analysis as to the degree of nest developement "definitive level".

União de Freguesias de Antuzede/Vil de Matos with a total of 24 nests observed, 23 (95,8%) were observed on the "tree" support, while 1 (4,2%) was observed in "housing spaces"; in União de Freguesias de São Martinho do Bispo/Ribeira de Frades from the 64 nests observed, 43 (67,2%) were located on "tree" and 2 (3,1%) in the "soil", 18 (28,1%) in "housing spaces" and 1 (1,6%) in "wall"; in Torres do Mondego from the 7 nests observed, 3 (42,9%) were found on "tree" and 4 (57,1%) in "housing spaces"; in União de Freguesias de Santa Clara/Castelo Viegas, from the observation of 38 nests, from the "definitive level" type, 21 (55,3%) were located on "tree", 1 (2,6%) in the "soil", 13 (34,2%) in "housing spaces" and 3 (7,9%) in "wall"; in Santo António dos Olivais from the total of 67 nests, 39 (58,2%) were verified on "trees", 3 (4,5%) in the "soil", 24 (35,8%) in "housing spaces" and 1 (1,5%) in "wall"; in the parish of Almalaguês, from the 20 nests observed, 12 (60,0%) were located on "tree" and 8 (40,0%) in "housing spaces"; in Ceira from the 17 nests observed, 13 (76,5%) were found on "tree", 1 (5,9%) in the "soil" and 3 (17,6%) in "housing spaces"; in União de Freguesias de Assafarge/Antanhol from the 21 nests registered, 16 were found on "tree" and 5 (23,8%) in "housing spaces"; in União de Freguesias de São Martinho de Árvore/Lamarosa from the 15 nests observed, 12 (80,0%) were located on "tree" and 3 (20,0%) in "housing spaces"; in União de Freguesias de Eiras/São Paulo de Frades from the 28 nests, were observed 21 (75,0%) on "tree", 2 (7,1%) in the "soil" and 5 (17,9%) in "housing spaces"; in Cernache from the observation of 15 nests at the "definitive level", it was verified that 8 (53,3%) were located on "tree" and 7 (46,7%) in "housing spaces"; in Brasfemes, from the 8 nests observed, 6 (75,0%) were located on "tree" and 2 (25,0%) in "housing spaces"; in União de Freguesias de Taveiro/Ameal/Arzila from the 22 nests observed, 17 (77,3%) were verified on "tree" and 5 (22,7%) in "housing spaces" and in União de Freguesias de Trouxemil/Torre de Vilela from the 17 nests registered, 12

(70,6%) were observed on "tree", 1 (5,9%) in the "soil", 3 (17,6%) in "housing spaces" and 1 (5,9%) in the "wall".

At the "definitive level" the parishes where it was verified a larger number of observations of nests were, União de Freguesias de Sé Nova/Santa Cruz/Almedina/São Bartolomeu, Santo António dos Olivais, União de Freguesias de São Martinho do Bispo/Ribeira de Frades and União de Freguesias de Santa Clara/Castelo Viegas.

4 Discussion

According to (Deutsch et al. 2008) the anthropogenic climate change has the potential to significantly influence the biology of all organisms. Environmental factors (such as temperature) directly affect the survival, development, reproduction and dispersion of insects and pests and, therefore, their biogeographic and biotic potential. It is known that temperature fluctuations are the main factors that affect the biology of insects, the activity and the distribution of natural enemies in agroecosystems (Chidawanyika et al. 2012). In other words, the higher the temperature, their metabolism will be accelerated and their proliferation will be higher (Lozano-menezes et al. 2019).

The months of July, August, September and October were the ones where there it was verified an increase in the registers, comparing these results with the life cycle of the Asian wasp it was found that in this period of time predation and the construction of definitive nests begins, being confirmed that in the referred months there is a greater expression of these nests. It was also found that the month of April was the only one that had only "primary level" nest records, according to the life cycle of the Asian wasp, this is the month in which primary nests will be created (ICNF 2018).

Since the sighting of wasps and observation of nests are dependent on the population regarding the filling out of forms, as well as their georeferencing, accompanied with photographs of the nest, which are later sent to municipalities and parishes, this may be late. Therefore, the reproduction of the species before the nest and its characteristics are confirmed and, subsequently, its extermination.

The summer and autumn of the year 2017 was marked by the occurrence of numerous wildfires, which is why there are difficulties in the surveying of the records, which may be one of the factors for which it was verified that there was a smaller number of observations compared to the other years under study.

The destruction of an Asian wasp nest is not simply the destruction of the cellulose structure built and inhabited by a wasp colony, but the whole set of actions that lead to the death of all individuals in the colony, whatever the method or medium used. If a significant number of worker wasps remain alive, they will give origin to one or more secondary nests which, despite creating only males, will maintain their predatory pressure. If the founding wasp stays survives with some surviving workers, the colony survives the attempted destruction and will build another active nest nearby. The location and destruction of embryonic or definitive nests is particularly effective until the end of July/August, when the founding wasps started to nest in the nests, since it definitively interrupts the reproductive cycle of the Asian wasp in that nest.

Knowing that, from September to the end of October, the founders leave the nests to mate, the sooner the destruction of all previously located nests, the greater the effectiveness and prevention (Marques et al. 2018).

From 2019 it was made available a financial support for the destruction of Asian wasp nests, where the intention was to support municipalities in the task of detecting and destroying colonies. In the years before 2019, there was a small portion of records where the nests were not exterminated or, when the extermination occurred, it was observed that the period of time between observation and extermination was prolonged, in that way, allowing the reproduction of the wasp and thereby existing less control and prevention. Late observations explain the rise of numbers between 2018 and 2019.

According to the ICNF, the location of Asian wasp nests is distributed mainly in trees (73%) and buildings (10%). It was verified in the data collected and studied that in fact the supports where there was a greater number of nests were on the "trees" and the "housing spaces".

The detection of nests is a very difficult task since they were usually located in places of difficult visibility, namely in treetops, sometimes at more than 30 m in height. The existing foliage makes it difficult to identify the nests, especially between June and September. After this period, in the autumn, if the trees are deciduous, the difficulty in identifying and locating nests decreases, as they become visible from the soil (Marques et al. 2018).

According to the study, the "definitive level" nests were those where there was a greater predominance in contrast to the "primary level" nests, these compared to the "definitive level" nests are nests where there is an identification by the observant more difficult and where the access is more restricted.

In total, the parishes with the highest incidence of nests over the 3 years were: Santo António dos Olivais (73), União de Freguesias de São Martinho do Bispo/Ribeira de Frades (68), União de Freguesias de Sé Nova/Santa Cruz/Almedina/São Bartolomeu (66) and União de Freguesias de Santa Clara/Castelo Viegas (41).

A rudimentary method of locating nests results from observation of the movement of wasps that transport bees captured by them in an apiary, and from observing their movement from a bait placed elsewhere. The crossing of both wasps trajectories on their return to the nest indicates their location (Marques et al. 2018).

Through data provided by DGAV in the period of time from 1 of September to 30 of September of 2019 in the parishes where there was a greater number of observations, these being União de Freguesias de Santa Clara/Castelo Viegas, Santo António dos Olivais, União de Freguesias de São Martinho do Bispo/Ribeira de Frades and União de Freguesias de Sé Nova/Santa Cruz/Almedina/São Bartolomeu, it was found that in these parishes there were 3 apiaries and 15 colonies, 11 apiaries and 155 colonies, 4 apiaries and 47 colonies and 1 apiary and 6 bee colonies, respectively.

Knowing that bees are targets for predation of the Asian wasp, the existence of bee apiaries in a parish will generate a greater probability of sighting and construction of nests by the Asian wasp in the localities, therefore it will exist an increase of the concentration of Asian wasp and consequently the it's sighting by the population in these parishes.

5 Conclusion

By observing the results and the information found to carry out the study, climate change was one of the important factors for the modification of ecosystems over the years, which is conducive to the existence of invasive species and to the increase of their predation, in this case, Asian wasp over time.

For this study, the nest extermination variable was not relevant since most of the observed nests were exterminated.

In conclusion over the three years under study, the number of nests observed in the municipality of Coimbra has increased. Knowing that this species reproduces and multiplies easily and that the extermination by the competent entities is based on the sighting and denunciation of the problem by the population, its late denunciation favors Asian wasp growth/proliferation.

The summer is the most favorable time for the incidence of the Asian wasp, relating to its life cycle in the months of June, July, August and September there is the creation of definitive nests and its predation begins. With the existence of bee apiaries in its vicinity, perfect conditions are found for their development, which was verified in parishes such as União de Freguesias de Santa Clara/Castelo Viegas, Santo António dos Olivais, União de Freguesias de São Martinho do Bispo/Ribeira de Frades and União de Freguesias de Sé Nova/Santa Cruz/Almedina/São Bartolomeu, which showed a higher frequency of records over the years, and they are also areas with great vegetation and located close to water points.

The definitive nests were the ones that presented a greater predominance compared to the primary nests, because they are the ones that are more easily identified and that have a better accessibility compared to the primary ones. Primary nests are built and more easily located in housing spaces, while definitive nests are built and more easily located on trees.

We can also conclude that at an environmental level, this species cause a decrease in plant pollination when feeding on bees, so there will be a larger demand for nutrients in areas located near apiaries and water points, thereby being prone to the occupation of urban areas. Becoming in that way a danger to the population, which can lead to death if the person that is bitten has an allergic reaction, knowing that this species has defensive behaviors of its colony which results in an increase of the aggression when it feels threatened.

In the identification of the nest typology, prevention and monitoring methods are necessary so that they are not late and that there is greater control and accessibility in the registration of primary nests that are not identified, as there is a greater definitive typology that means that surveillance at the level primary is not systematic.

The theme is relatively recent in the country, which is important to be discussed by everyone. At the level of Environmental Health, it is important to create strategies and implement good practices and action plans and formation and awareness in all municipalities. It is of utmost importance to provide the population with formation and awareness of how to proceed after the sighting of a nest and which entities should be addressed so that they proceed to their identification and, subsequently, their extermination. It is also important that countries carry out a control of these invasive species before commercial exchanges occur, since they are the cause of the spread of invaders in ecosystems, thereby compromising the health of the population and the environment.

References

Álvaro, A., Castro-Díez, P.: Las invasiones biológicas y su impacto en los ecosistemas. Ecosistemas **24**(1), 1–3 (2014). https://doi.org/10.7818/ecos.2015.24-1.01

Barbet-Massin, M., Rome, Q., Muller, F., Perrard, A., Villemant, C., Jiguet, F.: Climate change increases the risk of invasion by the Yellow-legged hornet. Biol. Conserv. **157**(July 2018), 4–10 (2013). https://doi.org/10.1016/j.biocon.2012.09.015

Chidawanyika, F., Mudavanhu, P., Nyamukondiwa, C.: Biologically based methods for pest management in agriculture under changing climates: challenges and future directions. Insects **3**(4), 1171–1189 (2012). https://doi.org/10.3390/insects3041171

David, S., Moreira, L.: Estudo de Implementação e Gestão Ambiental de Medidas de Recuperação num troço do Rio Leça (2017)

Deutsch, C.A., Tewksbury, J.J., Huey, R.B., Sheldon, K.S., Ghalambor, C.K., Haak, D.C., Martin, P.R.: Impacts of climate warming on terrestrial ectotherms across latitude. Proc. Natl. Acad. Sci. United States Am. **105**(18), 6668–6672 (2008). https://doi.org/10.1073/pnas.0709472105

Fernandes, D.S.S.: Pragas no município do Porto: monitorização e proposta de gestão de três espécies de insetos, vol. 94 (2016). https://repositorio.ul.pt/bitstream/10451/25300/1/ulfc120675_tm_Diana_Fernandes.pdf

Food and Agriculture Organization: Medidas Fitossanitárias Nimf, no. 5, pp. 1–27 (2009)

Institute for Nature Conservation and Forests: Plano de Ação para a Vigilância e Controlo da Vespa velutina em Portugal, pp. 1–46 (2018)

Lozano-menezes, S., Henrique, Á., Souza, C. De, Seron, C.D.C.: Desenvolvimento de Rhyzopertha dominica sob diferentes temperaturas de armazenamento, pp. 29–31 (2019). https://rdu.unicesumar.edu.br/bitstream/123456789/3998/1/DANIELEDESOUZATERASSI.pdF

Luz, P.: Vespa Asiática. Já foram destruídos mais de 1100 ninhos em Portugal este ano. Diário de Notícias (2020). https://www.dn.pt/edicao-do-dia/08-jun-2020/vespa-asiatica-ja-foram-destruidos-mais-de-1100-ninhos-este-ano-em-portugal-12287295.html

Maia, M., Silva, J.: A Vespa Velutina em Portugal Continental e a Apicultura Nacional. **2009**(1), 2–4 (2012)

Marques, A.F., Moreira, T., Casaca, J.: Manual de Boas Práticas na destruição de ninhos de Vespa velutina, pp. 1–29 (2018)

Paiva, R., Cabral, R., Lopes, C.: Vespa velutina ou Vespa asiática, no. 1, pp. 1–16 (2020)

Patrício, F.S.: Contributo para a identificação dos inimigos das abelhas melíferas Apis mellifera iberiensis na Zona do Douro Superior, pp. 1–61 (2018). https://hdl.handle.net/10198/19153

Pedro, M.E.: Alergia ao veneno de Himenópteros Alergologia e o Clínico Geral. Rev. Port. Imunoalergol. **7**(3), 191–194 (1999). https://www.spaic.pt/client_files/rpia_artigos/alergia-ao-veneno-de-himenopteros.pdf

Rome, Q., Perrard, A., Villemant, C., Muller, F.: Monitoring and control modalities of a honeybee predator, the yellow-legged hornet Vespa velutina nigrithorax (Hymenoptera: Vespidae). Aliens: Invasive Species Bull. **31**, 7–15 (2011). ISSN 1173-(31)

Zuben, A.P.B., de von Almeida, M.G.R., de Lira, E.S.: Manual De Controle Integrado De Pragas, pp. 1–68 (2006). https://www.campinas.sp.gov.br/sa/impressos/adm/FO086.pdf

Urban Mobility Characterization and Its Application in a Mobility Plan. Case Study: Bahía de Caráquez – Ecuador

Daniel Delgado[1,2](✉), Santiago Quiroz[1,3], Gema Casanova[1], Marlon Arturo Cedeño Álava[4], and João Pedro Cruz da Silva[5]

[1] Departamento de Construcciones Civiles, Facultad de Ciencias Matemáticas, Físicas y Químicas, Universidad Técnica de Manabí, Portoviejo 130108, Ecuador
daniel.delgado@utm.edu.ec

[2] Red de Desarrollo Urbano Sostenible de Manabí, Portoviejo, Ecuador

[3] Instituto de Posgrado, Universidad Técnica de Manabí, Portoviejo 130108, Ecuador

[4] Instituto de Lenguas, Universidad Técnica de Manabí, Portoviejo 130108, Ecuador

[5] Departament of Civil Engineering, Polytechnic of Leiria, Leiria, Portugal

Abstract. The main aspects that generally cause problems in urban mobility are caused by improper use of the private vehicle as the preferred transport and by poor planning of urban spaces. The objective of this work is to carry out a diagnosis of the urban mobility and public spaces in the city of Bahía de Caráquez, Ecuador, identifying the main deficiencies of the transportation system and proposing solutions as an input for the elaboration of a Sustainable Urban Mobility Plan. The methodology included the collection of data in situ, the use of available official databases, vehicles counts on different days between the years 2018 and 2019 and the application of a digital survey to understand the mobility habits of the city inhabitants. The results obtained showed that the city, despite having a relatively low traffic flow for most of the year, is particularly affected during holidays and festivities, when the number of tourists and visitors is quite high, generating serious mobility problems.

Keywords: Urban mobility · Mobility characterization · Mobility solutions for Bahia de Caráquez

1 Introduction

The development of a city is determined by a set of individual forces and interests, which leads to the use of different types of transportation modes. Understanding the structure of the city, the distribution of activities and the most influential factors in the mobility of people that lead to individual modal choices is one of the main results of a good urban characterization [1].

Growth without proper urban planning and without considering the sustainable development of many cities in South America, has created a massive use of private transport,

J. R. da Costa Sanches Galvão et al. (Eds.): *Proceedings of the 1st International Conference on Water Energy Food and Sustainability (ICoWEFS 2021)*, pp. 594–604, 2021.
https://doi.org/10.1007/978-3-030-75315-3_64

setting aside the use of public transport, due to the territorial and functional distances that make it difficult for the pedestrian to do daily trips to different points of interest and public services.

According to the Andean Development Corporation [2], the future of Latin America is urban, and the public transport system becomes a defining element of the development dynamics in cities. Most of the main cities have been affected by an accelerated growth that has generated a negative and unsustainable impact on the road systems, vehicular congestion, the state of transport, public services and also in the accident rates [3].

The form of occupation of large urban areas, associated with the unequal process of employment location and public services, generates a chaotic pattern of movement of people and goods. These patterns and mechanisms cause serious problems for the most vulnerable users such as pedestrians and cyclists and for the majority of the population that needs public transport. On the other hand, groups with higher income tend to use private vehicles more extensively collaborating towards urban dispersion and the intensive use of a limited road system that, in addition, needs to, adequately, serve public transport vehicles. This mobility pattern generates serious negative externalities such as air pollution, accidents, and road congestion [4].

Bahía de Caráquez is a coastal city of the Republic of Ecuador and cantonal head of Sucre, belonging to the province of Manabí. The city is made up of two urban parishes, Leonidas Plaza and Bahía de Caráquez, developed on an area of 181.75 m^2 with a population of 26112 inhabitants [5].

In recent years, the city has been affected by large-scale natural disasters (earthquakes and floods) that have seriously affected its urban facilities, generating significant changes in the economic and social behavior of its inhabitants.

The situation of visitors and tourists in the city is singular. Most of the time, the city of Bahía de Caraquez has only its own inhabitants, being calm and with low traffic. However, during weekends and holidays the situation is very different (see Fig. 1 on the left). The high rate of tourists and visitors in the city causes the mobility patterns to which the inhabitants are accustomed to be completely different, triggering serious traffic issues and generating problems for urban mobility that affects all modes of transport.

The objective of this work is to characterize urban mobility and public spaces in the city, identifying the negative components that influence it and presenting some solutions to solve them so they can eventually be incorporated in a Sustainable Urban Mobility Plan (SUMP).

2 Methodology

The research began by delimiting the study area, covering the parishes of Leonidas Plaza and Bahía de Caráquez for further data collection according to the information available in the official databases and to information collected in the city by the research team. Vehicle counts were carried out on motorized vehicles on different days between 2018 and 2019, to understand the dynamics of vehicular traffic. Surveys were conducted on a sample of the inhabitants of Bahía de Caráquez using an online form to determine the modal distribution and characterize the behavior of citizens regarding the use of modes of transport.

Fig. 1. Entrance to "Los Caras" Bridge, Bahía de Caráquez parish. (Left) Holidays, (Right) Non-holiday

Based on the analysis of the data some issues were identified that should be modified with several optimization measures and strategies. These measures and strategies were proposed based on several practical experiences, norms, laws, regulations, and guidelines that are applied in this country and in various parts of the world, trying to identify criteria and information structures that are compatible with the situation in the city of Bahía de Caráquez. Afterwards they were combined and adapted to the local challenges, enabling the development of six main measures that can improve existing urban mobility.

3 Current Situation

3.1 State of the Roads

Because of various events such as natural disasters, and a persistent lack of maintenance (Fig. 2) larges areas of the city are deteriorated.

Fig. 2. (Left) Cincinato Estrada Street, (Right) Muñoz Dávila Street

3.2 Pedestrian Crossing

Despite the fact that the city is considered a tourist area, walking is not encouraged, due to the lack of a proper infrastructure as well as an absence of interconnection with green spaces and tourist attractions.

3.3 Bike Paths

Although Bahía de Caráquez has a flat terrain throughout the city and has amazing tourist spaces, accompanied by a pleasant climate for bicycle use, it does not have a proper bicycle network which is essential to assure an ecological, tourist friendly, efficient and sustainable urban mobility. There is only a small bicycle network on the "Los Caras" bridge, which lacks adequate accesses to promote its widespread use.

3.4 Parking Lots

As shown in several studies [6–9], in many cities in Latin America as well as those on the coast of Ecuador, the scarcity of adequate public parking policies is common. Bahía de Caráquez is no exception and therefore parking is mostly unregulated.

3.5 Public Transport

Bahía de Caráquez has several types of urban transport. It has buses (15), taxis (167) and tricycles (129, but only 67 active); the latter are more widely used because Bahía de Caráquez is considered an ecological city (see Fig. 3).

Fig. 3. Ecological public transport – Tricycles

Regarding taxis and tricycles, their number and distribution is adequate throughout the city. Urban buses do not have information on their rout that crosses the main road and entry point of the city, which is frequently congested.

The city's terminal is located on its periphery, where five companies work (Coactur, Reina del Camino, Turístico Manabí, Aray, Tosagua) offering a total of 167 intercity trips throughout the day.

3.6 Vehicle Fleet

The situation of private vehicle transport in the city is complex, because normally the traffic flow is very low, but it changes considerably on holidays and weekends. To study this variation, the behavior of light vehicles was studied during four different dates using vehicle counts for all types of vehicles (without classifying them), from 10:00 AM to 5:00 PM on the entrance avenue to Los Caras bridge, Alberto Santos Avenue.

Table 1. Vehicle counts on different dates and conditions.

Date	Vehicle count (15 min)	Date	Vehicle count (15 min)	Observations
11/09/2018	79	11/09/2019	87	Weekday without any extraordinary event
22/09/2018	143	21/09/2019	129	Saturday (weekend)
23/09/2018	171	22/09/2019	201	Sunday (weekend)
03/11/2018	478	03/11/2019	529	Holiday

Table 2. Question 1: What is your most frequent means of transportation?

Transportation type	Frequency	%
Walking	41	11
Bike	63	18
Private vehicle	141	39
Public transport (taxis, buses and tricycles)	113	32
Others	0	0

The results are shown in Table 1:

If the count corresponding to the normal day is taken as a reference (average of September 11, 2018 and 2019), an average growth can be observed for Saturdays of 164%, for Sundays 224% while for holidays the vehicle increase reached 607%.

3.7 Modal Distribution

To evaluate this important parameter, a survey was conducted using electronic media and distributed to 358 users in the city of Bahía de Caráquez. Table 2 presents the results of the survey.

The private vehicle is the preferred transport type (39%), although the city has a relatively small area. A positive point is the high use of public transport (32%), due to the diversity of the offer of this type of transport (bus, taxi, tricycle). However, "walking and bike", which are the most environmentally friendly types and should be predominate in a sustainable mobility plan, have a low representation in relation to other modes of transport.

4 Measures and Strategies Proposed

4.1 Stimulate Pedestrian Mobility

Bahía de Caráquez is a city with a low pedestrian movement in relation to other modes of transport. One of the measures to overcome this situation is to propose the construction of a good pedestrian infrastructure connecting the main city landmarks and services with the residential neighborhoods and sectors [10]. For Bahía de Caráquez to become an inclusive city for everyone, it is fundamental to have a proper and integrated offer without the exclusion of people with reduced capacities (implementation of access ramps, vertical and horizontal tactile signs, signs hearing systems, platforms, among others).

A "school path" implementation is recommended, which consists of placing eye-catching graphics and colors on pedestrian footwear that motivate children to use it, promoting mobility on foot and generating good mobility habits in the younger generations [11, 12]. The school path should be built from the closest bus stations to the city's schools. Since there are few schools and due to the short distances, a large investment will not be necessary for its implementation.

The creation of "Pedestrian Islands" will be important, implementing urban spaces within them in which the circulation of pedestrians is exclusively reserved. By creating the pedestrian island, it is expected that residents and visitors will gradually increase the use of this mode of transport, which will be encouraged by the availability of a safe, comfortable, and exclusive recreation area for walking. This initiative should be implemented on the boardwalk "Paseo de Roberto" in the parish of Bahía de Caráquez (see Fig. 4b).

The pedestrian island will offer its services on weekends and holidays, when the influx of tourists is higher. The area will also be available for cyclists, taking into account the priority of pedestrians. The access of motorized vehicles should be restricted by means of retractable bollards (which will be removed on Sundays at 11:00 PM. until Friday at 6:00 PM).

For residents of the area who are affected by this measure, they should be issued a safe-conduct that allows them access from 10:00 PM to 11:50 PM on Fridays and Saturdays as well as for emergency reasons.

4.2 Urban Ecological Way

Within Bahía de Caráquez there are areas that correspond to the current pedestrian network that are not used frequently, becoming sectors that are forgotten by the inhabitants and visitors of the city for walking purposes. The lack of interest in the aforementioned sections is the result of the current conditions of the pedestrian infrastructure, which lacks attractive visual aspects that generate interest.

The implementation of a vegetation system is proposed in the pedestrian path where green spaces and rest and shade areas are less frequent, generating a more pleasant visual appearance, encouraging pedestrian movement, helping health and the environment in general [13]. This ecological route should be implemented in the "El Astillero" sector, an intermediate point between the parish of Bahía de Caráquez and Leonidas Plaza.

The road section of the El Astillero neighborhood, which has an average cross-section of 9 m (secondary road with little use), must be reduced to give rise to an extension of the sidewalk, generating a more pleasant and safe aspect, providing shade, and collaborating with the ecosystem.

4.3 Implementation of the Bicycle in Means of Transport (CicloBÍA)

This measure is aimed at establishing mechanisms for the inclusion and promotion of bicycles in the city of Bahía de Caráquez. The following bicycle route is proposed (see Fig. 4a):

Through the proposal outlined in Fig. 4a, the citizens of Bahía de Caráquez are offered an alternative means of urban and intermodal displacement through which the use of private vehicles can be controlled and rationalized, incorporating safety and quality measures for cyclists in urban areas and gradually increasing the demand for bicycles as a means of transport.

The CicloBÍA (BÍA for Bahía de Caráquez) should be incorporated in various sections of the city, creating a link between the main centers of attraction and the generation of trips [14–16]. The measure will incorporate a series of cycle paths, signposted and safe, connecting different sectors of the city. This network may be extended in the future.

In Fig. 4a the proposal for the implementation of the CicloBÍA is presented, defining it in the following types:

- **Type 1:** This section will be delimited by fixed bollards. The vehicular road will have two lanes in one direction. The dimension of the network for cyclists will be 1.60 m plus the line of bollards. The cycling network will have two lanes in both directions (see Fig. 5).
- **Type 2:** This section will be delimited by signposting, without the presence of fixed bollards. The circulation of the bicycle will take place in shared spaces with motorized vehicles, so the maximum speed allowed for the latter will be 20 to 30 km/h. The dimension of the network for cyclists will be 1.0 m plus the signaling line. The section will be restricted to a single lane in one direction for motorized vehicles (see Fig. 5).

In addition to the CicloBÍA layout, there must be parking for bicycles, adapted to the potential of daily mobility, which encourages safe and accessible parking. At least six parking lots for bicycles should be implemented, located in strategic points of the city as shown in Fig. 4a.

Together with the previous measures to promote bicycle use, it is proposed to develop a “BAHÍA BIKE” BICYCLE RENTAL PLAN, preferably administered by the Municipality of Sucre Canton, and to offer its services on weekends, holidays and rest days. This measure can to generate a fair amount of profits, mainly in the tourism sector, attracting foreign and local visitors. This system will be a pioneer in the province, with the potential to become a tourist icon of Bahía de Caráquez.

It will be necessary to consider that this measure, being outdoors, will be affected by seasonality and climate factors and will have the potential to economically compensate months in which income is considerably reduced.

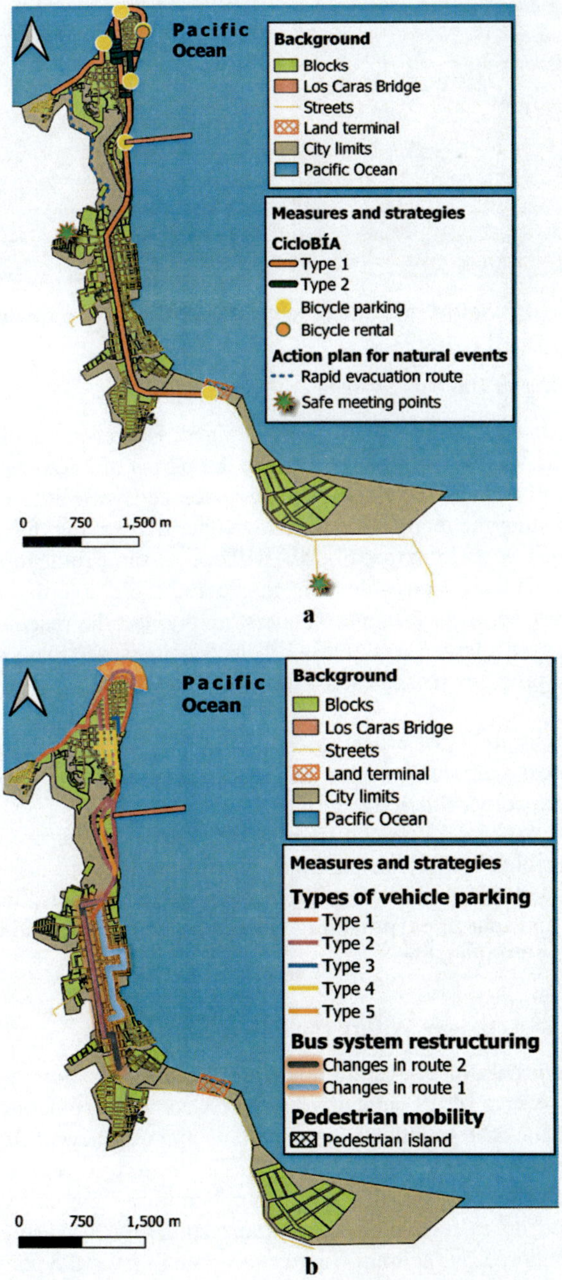

Fig. 4. **a.** Measures and strategies proposed for the SUMP (part 1). **b.** Measures and strategies proposed for the SUMP (part 2).

Fig. 5. (Left) CicloBÍA projection Type 1, (Right) CicloBÍA projection Type 2.

4.4 Restructuring in the Bus System

The number of urban bus lines should be maintained. However, the circulation of both lines will be modified using secondary roads in the parish of Leonidas Plaza (see modifications in Fig. 4b), due to having adequate space and with little use by motorized vehicles, decongesting the main road and bridge entering/leaving the city.

The circulation of both lines (see Fig. 4b) will use the same route for both the outward and return routes, whenever possible. It is considered preferable that the bus stops are connected face to face, to facilitate the understanding and the reach of the users. The new routes will cover a larger area and will help decongest the main artery of the city. Together with the previous strategies, it is proposed to [15, 16]:

- Improve the accessibility of urban buses to users with reduced capacities: the buses must allow access without obstacles and inconveniences for all users, regardless of their physical capacity or their age group;
- Improve modal exchange between buses and bicycles: Buses need to have special racks where bicycles can be transported safely and smoothly;
- Recondition bus stops by implementing physical improvements, including information on routes and schedules, providing better service to users and creating security zones for the most vulnerable.

4.5 Regulation of Activities Within the Public Street

This regulation consists of not authorizing any trade, fair, sporting events, cycling marches or other events of a competitive and non-competitive nature, which implies the occupation of the road or sidewalks, as well as any other event that may affect the fluidity of circulation or road safety, during public holidays or other days of massive influx of tourists, without prior case study and application of temporary protection measures. All the procedures for authorizations corresponding to these events should be regulated and monitored by the municipality, which will have the power to approve or deny them prior to a review of the temporary mobility plan presented by the interested party, which must be viable.

Regulation in the transport of goods is also part of this measure. Goods transport vehicles should be prohibited from circulating during public holidays or other days of massive influx of tourists during the hours between 06:00 AM to 10:00 PM.

4.6 Vehicle Parking Plan

Implementation of parking lots throughout the city is recommended, in physical spaces that cause a minor negative impact on the urban mobility [15, 16]. In certain areas of the city, where the demand for parking is greater, priced parking should be implemented with moderate costs (see Fig. 4b). All these parking lots must be guarded to issue the corresponding sanctions if necessary.

- **Type 1:** Corresponds to areas where the CicloBÍA will be implemented. Traffic signs will be posted to indicate that light vehicle parking is allowed on the opposite road from the cycling network;
- **Type 2:** Corresponds to areas where parking will be permanently prohibited for any type of vehicle (except for force majeure exceptions), due to reduced road widths, priority facilities and exclusive public transport sectors;
- **Type 3:** Corresponds to areas where the parking of any type of vehicles will be prohibited between 6:00 AM to 3:00 PM;
- **Type 4:** Corresponds to tariff zones. Parking will be allowed only to light vehicles and institutional vehicles respectively identified in exchange for a monetary fee that will have to be evaluated by the managing entity;
- **Type 5:** Corresponds to areas where parking will be allowed for all types of vehicles, exclusively on the right bank in relation to the vehicular direction.

It is necessary to take into account that this measure and, especially the tariff parking, will cause, in the beginning phases, a clear social impact, since users may change their behavior. Although a paid parking system seems logical to achieve positive changes within urban mobility, its management can be complex and its social impact highly debatable, even more so in our low-income environment. The authorities must analyze how the rate assignment will be managed and monitored, as it can be done through parking meters or security personnel.

5 Conclusions

Urban mobility problems are common in Latin American cities. The form of occupation of urban areas, the improper use of the private vehicle and the lack of awareness or not use of SUMPs, are some of the underlying causes of the urban mobility problems that are visible in Ecuador.

The dynamics in the traffic flow in the city of Bahía de Caráquez is very heterogeneous. There is an increase of more than 600% vehicles in holidays comparing to a normal day. This causes significant mobility challenges in the city. On the other hand, the lack of attractive and inclusive measures, the absence of pedestrian signage and incentive to use bicycles despite having an ideal geographical space for this type of movement, has caused these two types of transport to be the least used in the city.

The measures proposed should contribute positively to improve the performance of urban mobility throughout the year, promoting the use of more efficient and sustainable modes of transport that will contribute to the identity of the city as an ecological tourist attraction, combining the high quality of its surrounding appellative nature sites with a sustainable urban environment.

The present work can be considered as a reference point for the development of SUMPs in the province of Manabí and the Republic of Ecuador, providing guidelines that can be applied and adapted to the problems of other cities.

References

1. Alcântara de Vasconcellos, E.: Análisis de la movilidad urbana. Espacio, medio ambiente y equidad. Corporación Andina de Fomento (CAF), Bogotá (2010)
2. CAF-scioteca. https://scioteca.caf.com/handle/123456789/642. Accessed 21 Dec 2020
3. Dalkmann, H., Sakamoto, K.: Transport: Investing in Energy and Resource Efficiency. UNEP, Towards a Green Economy: Pathways to Sustainable Development and Poverty Eradication, pp. 374–411 (2011)
4. Corporación Andina de Fomento (CAF): Desarrollo urbano y movilidad en América Latina. CAF, Caracas (2011)
5. Independent National Electoral Commission: VII Censo de Población y VI de Vivienda. Instituto Nacional de Estadísticas y Censos, Quito (2010)
6. Moller, R.: Transporte urbano y desarrollo sostenible en América Latina: El ejemplo de Santiago de Cali. Universidad del Valle, Colombia (2006)
7. Kogan, J., Brennan, P., Ribera, A., Ospina, G.: Colombia: análisis del sector transporte. Sistemas integrados de transporte masivo. Serie informes sectoriales-Infraestructura, Caracas, vol. 3, no. 2 (2005)
8. Hidalgo, D.: Transporte Sostenible para América Latina: Situación actual y perspectivas. Documento de respaldo foro de transportes sostenible FTS de América Latina, pp. 22–24 (2011)
9. Lupano, J., Sánchez, R.: Políticas de movilidad urbana e infraestructura urbana de transporte. Naciones Unidas, Santiago de Chile (2009)
10. Guío-Burgos, F.A.: Recomendaciones de diseño para infraestructura peatonal en Colombia. Facultad de Ingeniería **17**(25), 39–52 (2008)
11. Rivas, M.R.: Camino escolar, más allá de la movilidad. Estudios de Construcción y Transportes **114**, 47 (2011)
12. Rivas, M.R., Canosa, I.S.: Camino escolar: Pasos hacia la autonomía infantil. Federación Española de Municipios y Provincias, Madrid (2011)
13. La Nación. https://www.nacion.com/ciencia/medio-ambiente/capital-inaugura-sendero-ecologico-para-mejorar/YYM2NZ2HWFCNZG6NMK7UHF6IH4/story/. Accessed 05 Jan 2021
14. Muñoz Sotomayor, V.A., Betancourt, D., Jaramillo Sangurima, W.: Diseño de ciclovías para ciudades intermedias, una propuesta para Loja. INNOVA Res. J. **1**(12), 11–22 (2016)
15. Sevilla, A.: Plan de Movilidad Urbana Sostenible del Municipio de Sevilla. Diagnóstico (2020)
16. Cuenca, Municipalidad: Plan de Movilidad y Espacios Públicos. GADM del Cantón Cuenca (2015)

Sustainable Logistics and Operations Design: Sustainable Practices for Emerging Countries - Perspective of the Situation in Bolivia

Agatha Clarice da Silva-Ovando[1,2] and Jocabed Becerra-Soliz[2(✉)]

[1] Universidad de la Sabana, Chía, Colombia
agathadaov@unisabana.edu.co, agathadasilva@upb.edu
[2] Universidad Privada Boliviana UPB, Cochabamba, Bolivia

Abstract. Over the years the concept of sustainable logistics (SL) has been acquiring more importance and has been gaining attention in organizations and in the decisions made in the design of logistics operations. Sustainable logistics is defined by influencing three important factors: economic, environmental and social, which are achieved through good practices that intervene in the strategies developed by companies. These practices have begun to gain much more strength in already established companies, however, small low-budget companies find themselves in difficult situations to be able to develop such practices, not only because of their cost but also because of the ignorance of the importance of sustainability within their organization. For this reason, the purpose of developing this research article is to analyze practical sustainability initiatives within organizations, focusing on existing small businesses in Bolivia. First, a bibliographic review will be carried out on the current best practices related to sustainable logistics, selecting each one of them in a category according to the conditions of the country's companies. Then, a field investigation will be carried out through a survey with companies of different sizes in Bolivia, to understand and characterize how sustainable logistics is understood in this group. Finally, the document concludes with the development of a diagnosis of the state of sustainable operations in Bolivia and recommendations are made to mark a path for the implementation of SL in small companies.

Keywords: Sustainable logistics · Resource efficiency · Sustainable practices.

1 Background

Currently sustainable logistics has become a topic of growing concern for many companies worldwide [1], gaining more strength since the publication of the sustainable development goals by the United Nations (UN) [2], putting pressure on organizations to incorporate into their business system. As a result of this situation, sustainability has become a critical aspect in the supply chain and is also considered a great challenge in terms of the implementation of good practices within the activities in the logistics area.

A clear definition of sustainability that was widely accepted both in industry and academia was proposed by the Brundtland Commission Report. According to this report,

J. R. da Costa Sanches Galvão et al. (Eds.): *Proceedings of the 1st International Conference on Water Energy Food and Sustainability (ICoWEFS 2021)*, pp. 605–615, 2021.
https://doi.org/10.1007/978-3-030-75315-3_65

sustainable development is defined as *"Development that meets the needs of the present without compromising the ability of future generations to meet their own needs"* [3]. Along with this definition, the authors Liu and Yun add an important requirement for organizations, which is the critical evaluation of all activities in relation to environmental, social and economic aspects [4].

Over the years, sustainability has ceased to be considered a cost [5], becoming a very important factor within companies, not only because of the competitive advantage it provides but also because of the impact it has on society and the environment. Therefore, organizations are increasingly required to develop practices that are appropriate to all their activities, corroborating the need to understand the role that sustainability plays in the logistics area. Although this topic has been accelerated worldwide, its development continues to be investigated in different countries in development [6], such is the case of Bolivia in Latin America. The recent growth of its economy and its logistics sector has evolved rapidly and continuously, generating a need to improve its performance in its processes under different perspectives.

Manufacturing companies in countries with emerging or developing economies, such as Bolivia, have recently begun to incorporate sustainability initiatives into their supply chain management [6]; however, achieving this objective within logistics processes is still a challenge for both micro, small, medium or large companies, because implementing different sustainable practices would imply a need of investment and a probable initial reduction of financial benefits. One of the aspects that are taken into account when adopting sustainable initiatives, according to Kannan's article is the economic factor, this being an important point for all companies [7]. Anyway, there are other reasons why such initiatives are implemented such as government rules, corporate image, competitive advantage among others [8].

It is important to mention that different publications that are made in this specific area are focused on reducing the impact on the environment in the operations of the supply chain such as the carbon footprint [9], and it was found that there is a shortage of empirical research related to various sustainability initiatives that can be applied to countries with emerging economies such as Bolivia.

As a result, this article will analyze, develop and evaluate the different sustainable practices in logistics operations in Bolivia, in addition to presenting a situational analysis of different sizes and categories of the existing companies in the region, so this document will focus on two aspects to have an overview: (a) Type of sustainable initiatives that have been implemented and (b) the reasons and factors that influence the application of these practices. To achieve the objective of the article, the document will be divided into different sections, the first being a brief introduction to the topic, then a review of the literature on sustainable logistics systems and the best practices to apply in developing countries. Section three will define the methodology to be applied to be successful in the study, then section four will have the situational diagnostic. Finally, in section five and six there will be a presentation of the results and final conclusions of the investigation.

2 Literature Review

The literature review is divided into three sections. The first part presents the general concepts of logistics systems related to sustainability, where its importance for successful business management in developing countries is analyzed. The second part shows the sustainable initiatives that have existed in general for logistics activities.

2.1 Sustainable Logistics Systems in Countries with Emerging Economies

In this first section we want to show the importance of sustainability within the activities of the logistics system through a synthesis of the research presented in recent years, taking into account social, economic and environmental approaches.

Dey, LaGuardia and Srinivasan, established that the application of sustainable logistics is a relevant factor when it comes to managing resources efficiently and correctly in organizations, avoiding waste, and providing at the same time the opportunity to improve the value of the brand and comply with the regulations established by the government [10]. Corroborating this information, Eroglu et al. argue that organizations recognize that sustainability is a critical factor and should be taken as a competitive advantage and not as a cost [11]. To validate this statement, an evaluation was carried out about the reaction that companies have to awards of this type, reacting positively because it gives them a value before society.

The importance of sustainable logistics is not alien to countries with emerging economies as it has gained a lot of interest in recent years due to not only pressure from different stakeholders [12] but also because it is considered a competitive advantage today. However, there are many studies that have found that there is a deficient amount of research that examined the implementation of logistics systems in developing economies compared to developed economies [13], which increases the difficulty of how sustainability has developed in countries with emerging economies. Despite this, many authors consider that the implementation of sustainable logistics together with its practices can be enhanced in a similar way with countries that have developed economies due to their characteristics [14].

2.2 Sustainable Initiatives in Logistics Activities

The successful implementation of a sustainable system in logistics activities can be achieved through different initiatives, however, there is no specific definition about what a sustainable initiative is within logistics activities [15], but there is a concept made by the author Su-Yol Lee defining what is a sustainable initiative within the supply chain, describing it as *"Programs that allow the dissemination of environmental management"* [16].

According to Ramus, there are three types of sustainable practices or initiatives: (i) Those that reduce the environmental impact on the company (ii) those that solve the environmental problems of companies and (iii) those that develop services that are both efficient and environmentally friendly [17]. Other authors also decide to group these sustainable practices into indicators related to transport and initiatives that go beyond transport [15]. Despite the different ways in which these practices are classified, there is no doubt that their impact on organizations is positive in terms of the sustainable performance of operations [18].

A study carried out by Morgan analyzes [19] the different factors that influence the success of the implementation of sustainable initiatives in logistics activities. A factor considered fundamental in the implementation of sustainable logistics is the commitment to resources since it has been shown that for the good performance of the supply chain it is important to take into account the efficient use of technological resources of transformation, processing of information and handling and transportation operations. However, there are no standard criteria or rules for evaluating such activities in the context of sustainability, which difficults its understanding for companies of all sizes [19], this being a very important topic to deepen in the development of the article.

3 Methodology

The present study is aimed to develop a better understanding of how the nano, micro and small firms perceive the implementation of sustainable logistics practices in their operations, and to identify the best practices to be implemented on those, considering how factors such as budget, motivation, and knowledge are managed. For that, both primary and secondary sources were used through (i) review of the literature related to logistics systems and existing sustainable initiatives for their activities in countries with emerging economies, in addition to understanding the current context of developing countries and verification of the latest practices of efficiency of resources that will be necessary to make a comparison and thus be able to select the best ones, (ii) secondary data gathering from public local and global institutions, to contextualize the data into the Bolivian scenario, (iii) preparation and publication of a preliminary survey to collect information on sustainability in companies of different sectors in Bolivia, to a further analysis of the results of the information collected, and (iv) stakeholder validation through a short interview to have a complete diagnosis about the current situation in Bolivia and finally (v) discussion of results in contrast to the literature reviewed. Finally, the conclusions of the study will be made together with the respective acknowledgments.

3.1 Literature Review Analysis

Sustainable initiatives within logistics activities in countries with emerging economies continue to be an area little explored [19] and therefore the article develops a conceptual analysis about sustainability in logistics systems together with the practices that go hand in hand for its implementation. Given this, an exhaustive review of the literature was carried out in order to support the information collected from the survey and interview that were prepared for the research.

To carry out the literature review, a schematic order was followed to abstract the most relevant information, (i) as a first instance, the search and download of relevant articles in the area was carried out, (ii) later an exhaustive review was carried out according to the topic, category and the author who made the article, (iii) finally the information found was structured to support the investigation.

The bibliographic review is a quite valid and relevant approach for any study [20], since it fulfills two important objectives [21], the first is to develop an objective identifying what patterns, themes and problems have existed, and the second is to identify the conceptual content of the subject being addressed, this being a very important step in every article.

3.2 Data Gathering from Public Local and Global Institutions

The compilation of information about the current characteristics that Bolivia presents was carried out by searching different company websites and documents published both by the government or by different universities when conducting their research. This information would allow us to prepare the survey developed in the following point in order to have a more appropriate context for the current situation in Bolivia. Among the reports, it was possible to obtain characteristics such as the classification of the companies according to their annual turnover, also the general operations in the supply chain that the different companies have and a key factor for the article, which is the current regulations that apply for environmental management.

3.3 Survey Development and Publication

The development and publication of the survey is a very important step for the whole study, this survey was developed in conjunction with the Sustainable Logistics Observatory, of the Technological Institute of Buenos Aires (Argentina), and other universities part of the SCALE LA network of the Massachusetts Institute of Technology (United States), where a semi-structured question guide was used, which was designed based on the information we wanted to collect, so only certain data was changed based on the context of the country. The survey is divided into two sections with a total of 14 questions related to general company information, adoption of sustainability initiatives and the role of sustainability in the company's strategy, and finally the reasons that facilitate the implementation of green practices.

3.4 Stakeholder Validation

Once the survey was carried out, it was necessary to consolidate the results obtained through the validation of the stakeholders, through a short interview. This interview was carried out with different companies with specific characteristics, in order to ask them key questions about the barriers that hinder the adoption of green initiatives.

3.5 Diagnosis and Analysis of Results

Finally, for this section, the information was analyzed and compiled from September and November 2020, identifying the sustainable logistics practices carried out. This information was stored in a database, which would then be structured to prepare the different reports according to the objective of the article. Once this analysis was finished, the validation was carried out with the selected stakeholders.

4 Diagnosis – Survey

This section will present a summary about the findings of the public survey, organizing the information with the most important topics that will allow us to understand how emerging economies manage the implementation of sustainable logistics in logistics

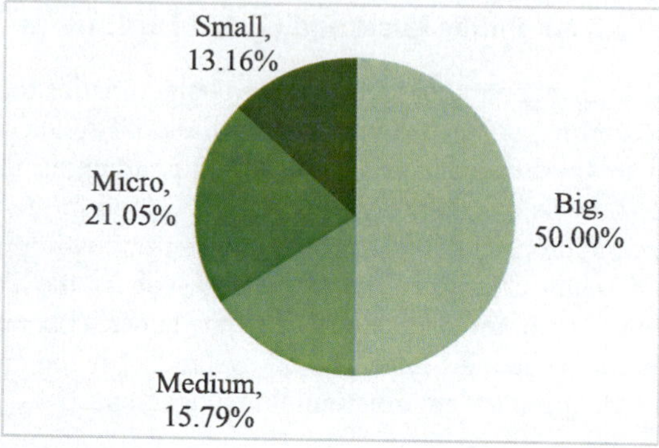

Fig. 1. Graph of percentage of the size of the companies found in the survey carried out according to their invoicing value.

activities. As a result of the survey, 38 companies have collaborated, from micro to large enterprises (see Fig. 1).

The realization of this first stage of research consisted of a series of concise steps that allowed to obtain a general panorama of how the situation in Bolivia is with respect to today's sustainability practices. For that, it began with the elaboration of the questions that would be included in the survey according to the information necessary for the investigation. Following that, the survey was published by different social media to companies in the country, finally the analysis of the information collected was carried out, obtaining results that will be presented below.

Within the analysis of the information obtained, it was found that the largest number of responses correspond to items of the secondary sector or industry, tertiary or services sector and the mass consumption and retail sector of different sizes (see Fig. 2).

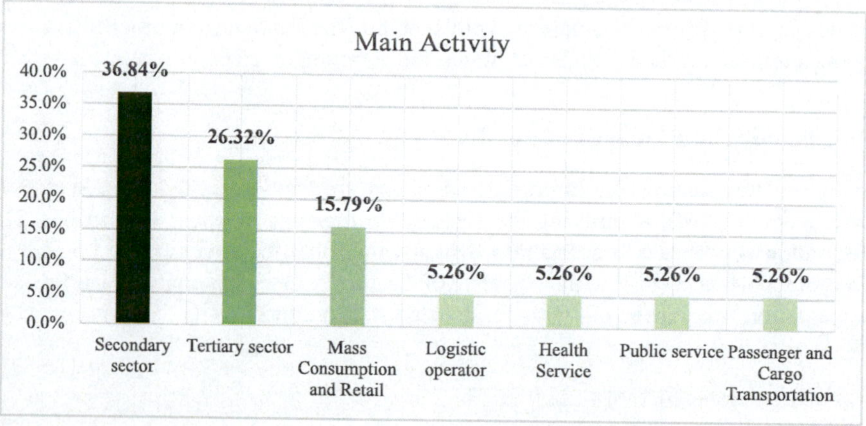

Fig. 2. Graph of percentage of the size of the companies found in the survey carried out according to their invoicing value.

Each sector presents different items, which can be seen below.

- Secondary sector, or Industrial: Agribusiness, Chemical, Petrochemical, Textile, Metallurgy, Energy, Plastic, Cement, Auto parts, Paper, etc.
- Tertiary sector, or Services: Industrial services, Professional services, among others.
- Primary sector or Extractive: Livestock, Agriculture, Oil & gas, Miner.

Each of these organizations gives a certain degree of importance to sustainability practices, where it was identified that large companies related to the secondary sector consider having a fully integrated system with said practices in their processes, whereas the tertiary sector only takes this matter into account without putting it into practice.

As small or medium-sized companies do not usually incorporate these practices or do not know if their application in their field really corresponds, due to different factors and reasons.

To understand how companies choose to implement sustainability practices, one of the fundamental questions of the survey was what motivates organizations to incorporate sustainability aspects into the organization's management, finding that these practices are mostly implemented due to savings in costs, corporate image or resource awareness (see Fig. 3).

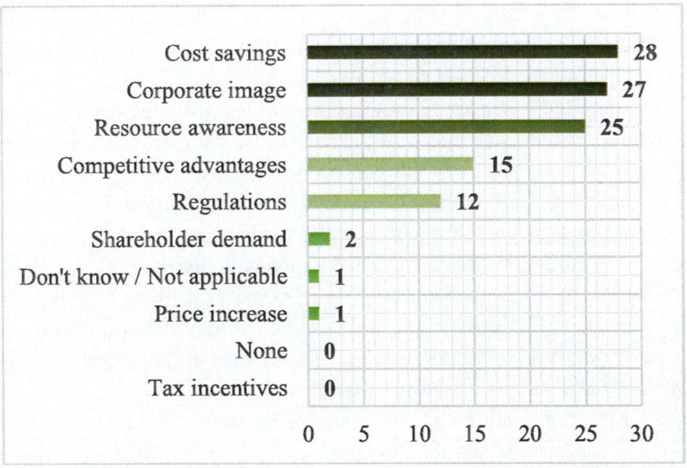

Fig. 3. Aspects that motivate organizations to manage sustainability.

Generally, the most common practices implemented by organizations in these sectors are the EEIA, MAI, PMA, RAI, ISO 14001, RASIM and IAA standards, however, this amount only represents 12.5% of the responses collected. Most companies do not have these standards (Fig. 4).

Based on this preliminary study, the validation is carried out with the interested parties, choosing small and medium-sized companies, where it was witnessed that this type of company has many barriers to implement sustainable management due to lack of knowledge and due to costs.

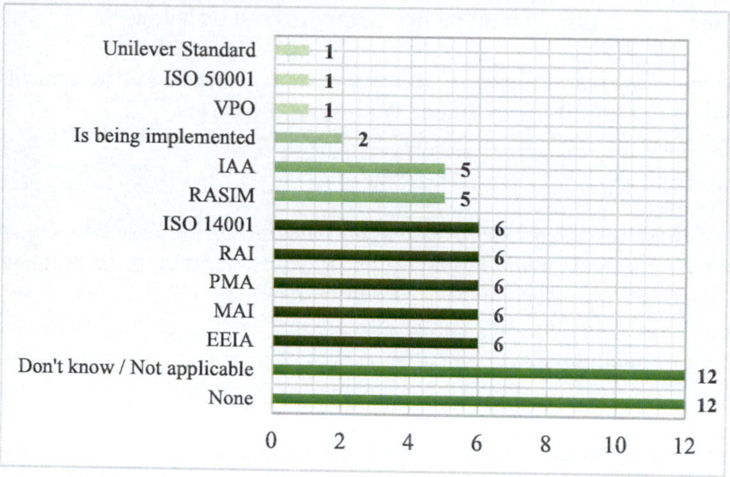

Fig. 4. Standards and initiatives carried out by companies

5 Conclusions

It is clear that Bolivia is still a very young country when it comes to the design of operations motivated by sustainability issues and a focus on the environment.

Like in many other places, very small companies usually do not apply these practices due to their limited budgets, whilst the already established organizations have the opportunity to implement these types of practices, even with some shortcomings.

However, contrasting with the literature reviewed, we can distinguish a series of initiatives that can be applied to emerging economies, breaking it down according to the activities that logistical systems present and the motivation that led them to implement them, the most common practices being the reduction of waste (recycling) and packaging improvement for cost reduction, the reduction of carbon footprint with the improvement of vehicles accompanied by environmental controls along with the improvement of the planning of transport routes, the application of a training related to the sustainability to both employees and customers, among others.

Among the main conclusions, it was possible to notice that there is no clear understanding of what consists in sustainable logistics practices. Many of the respondents believe that the implementation of norms such as ISO 14000 are sufficient to ensure that their practices are sustainable. It is possible to notice that even larger companies do not implement deeper studies to improve the quality of their operations, which leads to understanding and explain why most of them only implement those actions to obtain better brand image, or just to comply with internal or external norms.

This leads to a second level to be explored, which is the better understanding of how sustainable logistics practices may lead to an optimization of the internal operations, reducing costs and benefiting the whole process.

6 Recommendations

From this study, it is recommended that a deeper field work should be developed within companies, through round tables and discussion groups to clarify what is sustainable logistics and develop workshops to generate a better perception of the benefits when implementing sustainable operations (Tables 1 and 2).

In addition, it is expected that companies begin to evaluate the motivations they have with respect to sustainability and thus be able to define the best practices for their company, and then identify these initiatives according to the operational activities that an example can be seen in the following table.

Table 1. Sustainability practices according to activity and motivation

Area	Initiative	Outcome	Motivation
Transport	Clean vehicles/fuel efficiency	Reduce costs by and minimizing waste and increase efficiency	Corporate image
			Regulations
	Reuse of pallets and containers		Cost savings
			Resource awareness

Table 2. Sustainability practices according to activity and motivation

Area	Initiative	Outcome	Motivation
Warehousing	Clean material handling equipment	Increasing the efficiency of the system due to changes in network	Corporate image
			Regulations
	Process optimization		Cost Savings
	Automatic warehousing systems		Resource awareness
	Inventory minimization programs and Just-in-time system		Competitive advantage
	On-site recycling		
	Product disposition		
Waste management	Waste contractor,	Reduce costs by and minimizing waste and increase efficiency	Corporate image
	Trade waste recycling		Regulations

The previous table is an example of the practices that can be implemented according to the activities that they present; it is recommended that companies start doing a planning of activities to companies that are just implementing a sustainable management in the logistics system. The application of sustainable practices will allow organizations to reduce costs and use their resources efficiently, so its implementation is recommended to companies with emerging economies.

Acknowledgements. We want to express our sincere gratitude for the support and contribution of the Sustainable Logistics Observatory, of the Technological Institute of Buenos Aires (Argentina), to our sponsor the Bolivian Private University and to the SCALE LA network of the Massachusetts Institute of Technology (United States). Also thank the efforts of our team that were invaluable for this investigation.

References

1. Caiado, R.G.G., et al.: A literature-based review on potentials and constraints in the implementation of the sustainable development goal. J. Clean. Prod. **198**, 1276–1288 (2018). https://doi.org/10.1016/j.jclepro.2018.07.102
2. Kumar, A.: Green logistics for sustainable development: an analytical review. Int. J. Bus. **1**, 7–13 (2015)
3. Schaefer, A., Crane, A.: Addressing sustainability and consumption. J. Macromarket. **25**(1), 76–92 (2005). https://doi.org/10.1177/0276146705274987
4. Liu, J., Yuan, C., Hafeez, M., Yuan, Q.: The relationship between environment and logistics performance: evidence from Asian countries. J. Clean. Prod. **204**, 0959–6526 (2018)
5. Wichaisri, S., Sopadang, A.: Sustainable logistics system: a framework and case study. In: IEEE, Thailand, vol. 1 (2014). https://doi.org/10.1109/IEEM.2013.6962564
6. Esfahbodi, A., Zhang, Y., Watson, G.: Sustainable supply chain management in emerging economies: trade-offs between environmental and cost performance. Int. J. Prod. Econ. **181**, 350–366 (2016)
7. Brandenburg, M., Govindan, K., Sarkis, J., Seuring, S.: Quantitative models for sustainable supply chain management: Developments and directions. Eur. J. Oper. Res. **233**(2), 299–312 (2014)
8. Malik, M., Abdallah, S.: sustainability initiatives in emerging economies: a socio-cultural perspective. Sustainability **11**(18), 4893 (2019)
9. Wang, D.-F., Dong, Q.-L., Peng, Z.-M., Khan, S.A.R., Tarasov, A.: The green logistics impact on international trade: evidence from developed and developing countries. MDPI **10**(7), 2235 (2018)
10. Dey, A., LaGuardia, P., Srinivasan, M.: Building sustainability in logistics operations: a research agenda. Manage. Res. Rev. **34**(11), 1237–1259 (2011). https://doi.org/10.1108/01409171111178774
11. Eroglu, C., Kurt, A.C., Elwakil, O.S.: Stock market reaction to quality, safety, and sustainability awards in logistics. J. Bus. Logist. **17**(4), 329–345 (2016)
12. Huq, F.A., Chowdhury, I.N., Klassen, R.D.: Social management capabilities of multinational buying firms and their emerging market suppliers: An exploratory study of the clothing industry. J. Oper. Manage. **46**, 19–37 (2016)
13. Rajeev, A., Pati, R.K., Padhi, S.S., Govindan, K.: Evolution of sustainability in supply chain management: a literature review. J. Clean. Prod. **162**, 299–314 (2017)
14. Silvestre, B.S.: Sustainable supply chain management in emerging economies: Environmental turbulence, institutional voids and sustainability trajectories. Int. J. Prod. Econ. **167**, 156–169 (2015)
15. Martinsen, U., Huge-Brodin, M.: Greening the offerings of logistics service providers. In: Proceedings of the 22nd Annual NOFOMA Conference: Logistics and Supply Chain Management in a Globalised Economy (2010)

16. Lee, S.Y.: Drivers for the participation of small and medium-sized suppliers in green supply chain initiatives. Supply Chain Manage.: Int. J. **13**(3) (2008). ISSN 1359–8546
17. Ramus, C.A.: Encouraging innovative environmental actions: what companies and managers must do. J. World Bus. **37**(2), 151–164 (2002)
18. Lu, C.S., Shang, K.C., Lin, C.C.: Examining sustainability performance at ports: port managers' perspectives on developing sustainable supply chains perspectives on developing sustainable supply chains. Pol. Manage. **43**, 909–927 (2016)
19. Watanabe, E.H., Blos, M.F., da Silva, R.M., Junqueira, F., Santos Filho, D.J., Miyagi, P.E.: A framework to evaluate the performance of disperse productive system through sustainability performance indicators. IFAC-PapersOnLine **48**(3), 1664–1669 (2015)
20. Hsu, C.C., Tan, K.C., Zailani, S.H.M., Jayaraman, V.: Supply chain drivers that foster the development of green initiatives in an emerging economy. Int. J. Oper. Prod. Manage. **33**(6) (2013)
21. Esfahbodi, A., Zhang, Y., Watson, G.: Sustainable supply chain management in emerging economies. Int. J. Prod. Econ. **181**, 350–366 (2016)
22. Seuring, S., Müller, M.: Core issues in sustainable supply chain management–a Delphi study. Bus. Strategy Environ. **17**(8), 455–466 (2008)

Sustainable Development and Reverse Logistics

Ricardo Martins[1](✉), Noelle Camello[2], and Carlos Soares[3]

[1] Mestrando da Universidade Federal Fluminense, Niterói, Brasil
ricardoam@id.uff.br
[2] Doutoranda da Universidade de Aveiro, Aveiro, Portugal
ncamello@ua.pt
[3] Professor da Universidade Federal Fluminense, Niterói, Brasil
csoares@id.uff.br

Abstract. Due to the production and consumption patterns currently used by society, a large amount of solid waste is generated, as well as its wrong destination in the environment, which has negative effects, including harm to human health. In this scenario, we can mention the 2030 Agenda for Sustainable Development, an action plan that aims to eradicate poverty, protect the planet and ensure that people achieve peace and prosperity. To this end, companies can contribute by using Reverse Logistics, which aims to efficiently and effectively plan and manage the return of products to their manufacturer (Leite 2017, p. 206). The work aims to understand how Reverse Logistics can be used as a tool for the contribution of Sustainable Development, based on goal number 12 (Ensure sustainable production and consumption patterns), in Brazilian reality. The methodology, in addition to the literature review, is based on a qualitative and exploratory approach. From the research, it is concluded that Reverse Logistics works as a tool that can contribute to the achievement of goal number 12 of Sustainable Development, as both seek to achieve more sustainable behaviors in society.

Keywords: Sustainable development · Reverse logistic · Sustainability

1 Introduction

Any action that comes from man is capable of producing positive or negative effects on the environment. Regarding the negative impacts, society must be attentive to them in order to reduce or eliminate them on the environment. The environmental impacts that result from human activities on nature can cause scarcity and depletion of natural resources as well as degradation, pollution and contamination of nature. To reduce or eliminate these environmental impacts, society must seek to reduce the consumption of non-renewable resources, use more renewable resources and mitigate all forms of environmental degradation and pollution (Fenker et al. 2015).

Since the Industrial Revolution, it has been evident that the demand for natural resources is increasing day by day and this has caused several negative impacts on the environment and human health itself. On the other hand, it is observed that recovery of the environment does not follow the pace of its deterioration (D'Agosto 2018).

J. R. da Costa Sanches Galvão et al. (Eds.): *Proceedings of the 1st International Conference on Water Energy Food and Sustainability (ICoWEFS 2021)*, pp. 616–625, 2021.
https://doi.org/10.1007/978-3-030-75315-3_66

In terms of negative environmental impacts, it is necessary to mention companies, since they consume the most natural resources - especially the factories that are responsible for producing goods – in order to carry out their economic activities. In other words, companies extract raw materials from nature and, most of the time, discard - either during the manufacturing process or after the manufacturing process - waste to the environment, irregularly, without any kind of care or treatment, which further increases the pollution and contamination of soil, air and water (Pereira et al. 2012).

Not only has the production of goods by companies increased, but also consumption by people across the planet, which can lead to environmental problems when products are released into the environment after their useful life without any treatment (Carneiro 2018).

Faced with this scenario, several countries began to think about the way of production and consumption as both require large amounts of resources to meet the needs of people, as well as generate large amounts of waste, as threatening to limit the capacity of the planet itself. The result of all this endangers not only the present, but also the future of the next generations. It is in this context that the concept of Sustainable Development emerges, challenging patterns of production and consumption in face of natural resource depletion and the planet's limited capacity to absorb the waste produced (Barbieri 2016).

For companies interested in operating more sustainably in the marketplace, the main challenge is to find the best way to manage environmental and social dimensions and integrate them with economic interests. In this sense, companies can adopt Environmental Management practices. (Jabbour and Jabbour 2013).

It is worth noting that the International Chamber of Commerce (CCI), in order to help companies from all over the world to improve their environmental performance, created the Business Charter for Sustainable Development, which consists of sixteen principles that must be followed by companies so that Sustainable Development can be achieved (Donaire and Oliveira 2018).

Thus, the growing concern of various social actors for the environment exhorts to reuse and thus reduce waste. And in the corporate context, one of the ways to contribute to the reduction of negative environmental impacts resulting from the disposal of goods in nature is the implementation of reverse logistics (Mattos and Santos 2014).

The work aims to understand how Reverse Logistics can be used as a tool to contribute to Sustainable Development, based on goal number 12 (Ensure sustainable production and consumption patterns), in Brazilian reality.

The research is based on a qualitative and exploratory approach and for its development the bibliographic research method was used, in which we searched for books and scientific articles that deal with reverse logistics and sustainable development.

2 Sustainability and Sustainable Development

The concept of sustainability refers to a systemic process that aims to integrate, relate and balance the economic, environmental and social dimensions in order to maintain the continuity of a given system. Based on this concept, the term sustainable development was coined in 1987 by the Brundtland Report, from the United Nations (UN), which is understood as the need to meet the needs of the present generation without compromising future generations in meeting their needs. Thus, sustainable development must take into account the environmental, economic and social dimensions simultaneously (Pereira 2012).

In 2015, the member countries of UN officially signed the new sustainable development agenda, known as "Transforming Our World: the 2030 Agenda for Sustainable Development", in which it presents 17 goals (see Fig. 1), divided into 169 targets.

Fig. 1. Sustainable development goals.

In this way, Reverse Logistics serves as a tool that companies can use to contribute to the achievement of Sustainable Development, especially in relation to Sustainable Development Objective number 12: ensuring sustainable production and consumption patterns. This is because, in addition to enabling solid waste to reach a correct and safe destination, thus contributing to the environmental issue, it also plays an important economic and social role, as it enables job creation and social inclusion (Carneiro 2018).

3 Reverse Logistics

In view of the environmental problems mainly related to the extraction of raw materials and the disposal of waste in nature, as well as the increase of products with shorter life cycles and the great variety of products available on the market, in the last decades of the last century there has been an increased concern for the return of these products that have been consumed or simply not used at a given time. Thus, the field of reverse logistics in the business sector is beginning to gain importance in the administration field (Leite 2017).

The first studies on reverse logistics took place in the 1970s and 1980s. From the 1990s onwards, the topic began to gain importance in the business world, as companies in a competitive and globalized environment realize that one should not only seek profit from their activities, but that it is also necessary to pay attention to social, environmental and governmental interests in order to maintain business and profitability in the long run (Leite 2017).

Logistics, as one of the areas of management, can be understood as a set of methods and techniques of decision support for the management (planning, organization, operation and control) of the flow of goods and information between the producer of the raw material and the final customer (D'Agosto 2018).

Reverse logistics, in turn, can be understood as "the process of planning, implementing and controlling a flow of materials, products in process, finished products and related information from the point of consumption to the point of origin through reverse distribution channels" (Costa et al. 2014, p. 19). In short, it aims to enable the return of goods or materials to the production cycle after they have been discarded as after-sales or post-consumer products.

In summary, companies can apply two types of production systems: the horizontal production system (see Fig. 2), which is the traditional way and used by most industries, and the production system with the application of reverse logistics (see Fig. 3), which has a cyclical character (Carneiro 2018).

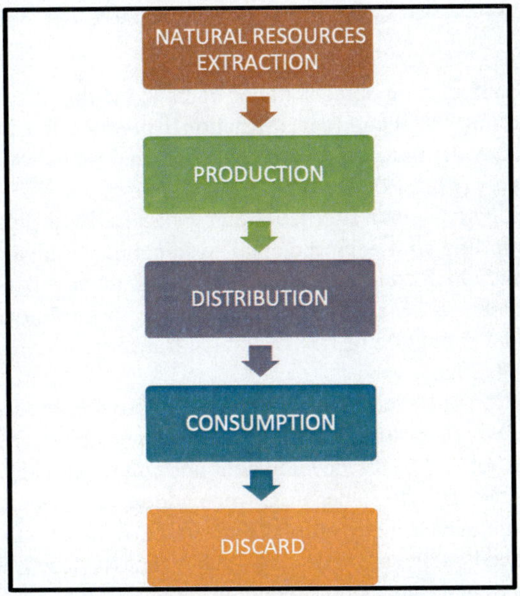

Fig. 2. Horizontal productive system.

In reverse logistics, there are two types: post-sale reverse logistics and post-consumer reverse logistics. The first refers to the return of products that have had little or no use,

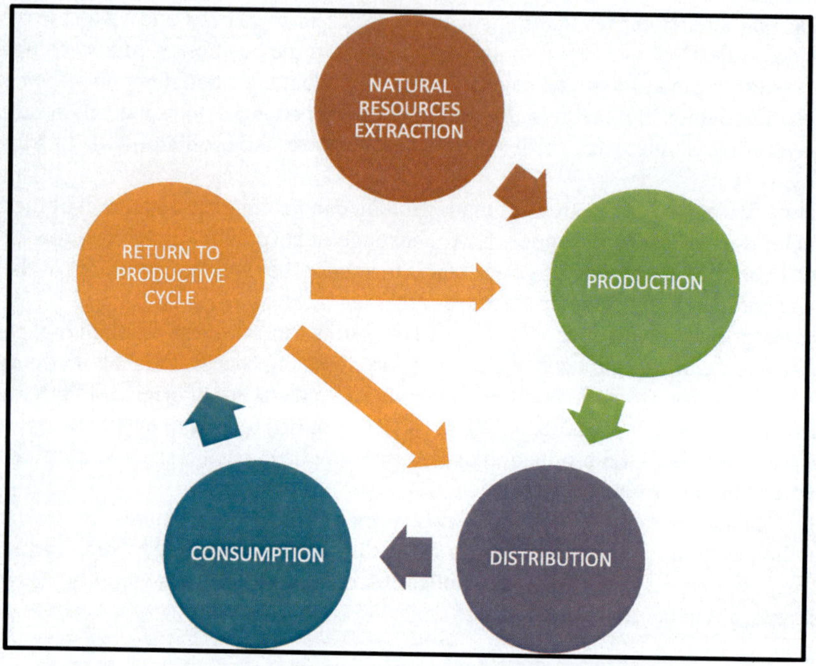

Fig. 3. Productive system with the application of Reverse Logistics

including due to manufacturing defects, goods damaged within the warranty period, or customer dissatisfaction. The second acts in relation to products that have been discarded by society in general, with products that are no longer used for various reasons returning to the production cycle (Leite 2017).

Thus, Reverse Logistics - as an alternative that seeks to solve problems related to solid waste management in line with environmental sustainability - advances the concept of the circular economy (which combines economic development with better use of natural resources), a model that is emerging as an alternative to the linear economic model (which consists of the extraction of raw materials, transformation, use and disposal of waste) (Cosenza et al. 2020).

The implementation of Reverse Logistics is able to provide environmental, financial and social benefits, which in turn, when recognized, are able to stimulate more and more new initiatives related to the development and improvement of reverse logistics processes by organizations. These benefits, when added, allow the company to gain a competitive advantage over its competitors (Faria and Polido 2018).

The environmental benefits are related to the fact that Reverse Logistics enables the reduction of environmental impact, thus reducing the impact that the company generates on the environment. It also ensures that natural resources are used as responsibly as possible, thus avoiding waste. Not to mention that it helps to reduce the extraction of natural resources, such as water, wood and paper (Fernandes et al. 2018; Silva and Pimental, 2019; Oliveira et al 2020).

With regard to the financial issue, Reverse Logistics allows cost reduction in the organizations' operations due to the fact that they can sell waste to secondary markets, as well as cost reduction due to the reuse of waste in their production processes. And the company reusing the materials allows to increase the value of the reusable materials, since it is not necessary to buy a new raw material (Silva and Pimentel 2019; Oliveira et al. 2020).

In terms of social benefits, Reverse Logistics contributes to creating jobs and income for the local community through partnerships with cooperatives and recycling collectors (Faria and Polido 2018).

It is worth noting that in most cases, the implementation of Reverse Logistics is driven by legislation and not by the free, spontaneous will of companies, especially in Brazil. Thus, legislation allows productive companies to review the life cycle of their products and establish reverse logistics structures for themselves (Couto and Lange 2017).

On the other hand, in many companies there are barriers to the implementation of reverse logistics. In this sense, among other things, there is a lack of knowledge about the importance of implementing reverse logistics practices among some managers, in addition to the lack of dissemination of the concept and its applicability within organizations (Faria and Polido 2018).

In addition, there is the fact that post-consumer waste is scattered across large urban centers, which is related to the issue of collecting such waste, as well as the fact that industries are geographically distant from the distribution sites of their products. In the Brazilian case, we can mention, for example, the cities of Rio de Janeiro and São Paulo, where there are outsourced companies that collect ordinary garbage, as well as informal waste collectors, often in a state of neglect. Thus, for managers, this type of collection becomes a costly affair for the company (Mendonça et al. 2017).

Many companies also point out that it is necessary to have a viable amount of solid waste to generate sufficient volume for the implementation of the reverse flow process to make it economically viable, that is, to enable collection while optimizing operating costs (Mendonça et al. 2017).

In turn, the challenges highlighted can cause high costs for the operationalization of the Reverse Logistics systems, making them initially unattractive for managers to implement in companies. This fact reveals that there is still a prevailing notion that environmentally friendly actions are seen as a cost rather than a practice that needs to be carried out to maintain environmental and social balance. Thus, reverse logistics, even if it initially requires resources for investment and innovation in production processes, can initially cost the company, but when implemented, it can bring significant financial and image returns to the company (Faria and Polido 2018).

In this sense, so that there are not many obstacles to the implementation of Reverse Logistics, the importance of the actions of the following actors becomes clear: government (to implement, stimulate and subsidize reverse logistics programs); organizations (with the role of promoting social-environmental responsibility); society (help in the separation of materials so that they can be collected in this way and reused in the production cycle) (Mattos and Santos 2014).

One way to promote Reverse Logistics in Brazilian companies would be, for example, tax incentives that would cause an increasing commitment of companies to Reverse Logistics (Souza et al. 2018).

Another important point is that companies do not have a system that can monitor the performance of the processes that make up Reverse Logistics. In this sense, there are few studies that address this issue, which shows that it is still an under-researched field by researchers (Fernandes et al. 2018).

Finally, it is worth mentioning that there is another concept that is very widespread in the field of logistics, which for the time is being confused with the concept of Reverse Logistics and which also has an ecological orientation: Green Logistics, which plans and mitigates the negative environmental impact of logistics as a whole (D'Agosto 2018).

In this sense, green logistics seeks to integrate a balance between efficiency in the economic and environmental fields, while creating sustainable value for the company, evaluating the financial returns of the company in the short and medium term. Green logistics has a wider scope in the field of logistics, in which it takes into account the consumption of non-renewable natural resources, the emission of contaminated wastewater, the use of rural roads, noise pollution and the dumping of waste into nature (Gallegos et al. 2019).

3.1 Reverse Logistics in the Brazilian Legislative Context

In Brazil, in 2010, a federal law was created that establishes the National Solid Waste Policy. It is possible to verify the existence of reverse logistics as an instrument of economic and social development. According to Brazilian legislation, discussed in Reverse Logistics, there is an obligation in Brazil to implement reverse logistics for manufacturers, importers, distributors and traders of pesticides, batteries, tires, lubricating oils and their residues and packaging, fluorescent lamps, and electrical and electronic products and their components (Brasil 2010).

From the analysis of this law, it appears that there are no drastic penalties for non-compliance with its provisions. In addition, the lack of control measures is visible, so companies feel more pressured to adopt Reverse Logistics practices, which was even highlighted in the studies of Silva and Pimentel 2019.

Another relevant point regarding the law is the lack of a visible definition about the limitation of liability of each responsible agent involved in the production chain, as it is difficult to determine exactly where the responsibility of each responsible agent begins and ends during the production chain (Mendonça et al. 2017).

A critical point regarding the legislation is the fact that it mentions that Reverse Logistics is only for the private sector, where it must be implemented independently from the public urban sanitation (Velazquez and Marcon 2017). However, municipalities must have an interest in developing projects that can include Reverse Logistics actions, as it is possible to reduce the amount of solid waste in landfills, reduce the waste deposited on urban streets and enable the environmental protection of the community (Oliveira et al. 2020).

Although such legislation exists, there is no mandatory inclusion of more materials of other types, such as plastic. Although plastic producing industries are not required to implement Reverse Logistics, some studies indicate that there are those that have already implemented them and others that have the potential to implement them, but there are barriers to their effectiveness.

In this sense, Mendes et al. (2016, p. 44) conducted a survey in which they described how a company in the plastics segment organizes its industrial processes based on the use of reverse logistics. The results showed that the organization had a "thrift in its resources for the development of a new product, reducing the environmental impact in its operations, efficient in reducing operating costs, in obtaining a higher percentage of profit and, at the same time, increase your competitiveness".

In turn, Silva and Neto (2011), p. 250, while conducting a study in plastic factories, found that most of them "can reuse their post-consumer material in the production process itself, reducing their costs in purchasing the raw material". They also point out, in relation to reverse logistics, that the industry has had difficulty in implementing it because there is no consolidated separate collection program in the city where the company is located. In addition, they point to the lack of commitment of companies to label their products with the codes introduced by the Brazilian Association of Technical Standards (ABNT), which favored their return to industrial processes.

In a study by César and César (2019), it was shown that after the enactment of the Solid Waste Law, the number of Brazilian scientific articles dealing with reverse logistics increased.

However, as Chaves et al. (2019) point out, although they acknowledge that Reverse Logistics has gained importance in academic, business and government environments, there is still a greater need for research on this topic in Brazil compared to the United States and European countries. And for this, the authors still mention that it is important that there are more partnerships with other research institutions abroad to promote and enrich the discussions internationally, as well as to seek the best practices that can be implemented in the brazilian reality.

It should also be noted that even after its approval, the law has not been effectively applied in all brazilian regions. Again, it cannot be denied that there is progress in the issue of waste management, but slowly, considering the accelerated way in which the production and consumption of goods, as well as the disposal of waste, takes place into the environment, causing environmental pollution.

4 Final Considerations

The generation of large amounts of solid waste, the scarcity of natural resources and the growing awareness of the population regarding the preservation of the environment are putting pressure on companies to act in favor of more sustainable production.

Taking into account the concept of sustainable development, it can be said that it is necessary for companies to review the way they carry out their production.

Based on the theoretical discussion presented, it can be said that Reverse Logistics is a valid tool for promoting Sustainable Development Goals. Companies that intend to include both post-consumer and post-sale goods in their production cycle have not only environmental but also social and economic benefits.

For example, developing the systematisation of waste streams discarded by companies and their possible reuse both inside and outside the original production chain contributes to the reduction of natural resources as well as to the reduction of environmental impacts. It also allows, for example, the creation of jobs and the reduction of production costs. Furthermore, it allows the company to continue to make its profits while offering quality to meet the demands of the market and, in particular, of its customers, that is, it manages to stay on the market in a sustainable way.

Thus, from the research, it can be seen that Reverse Logistics functions as a tool that can contribute to the achievement of the number 12 objective of Sustainable Development Goals as both seek to achieve more sustainable behaviors in society.

References

Barbieri, J.C.: Gestão ambiental empresarial: conceitos, modelos e instrumentos. 4th edn. Saraiva, São Paulo (2016)

Brasil. https://www.planalto.gov.br/ccivil_03/_ato2007-2010/2010/lei/l12305.htm. Accessed 20 Jan 2021

Carneiro, E.F.: Desenvolvimento sustentável e logística reversa: um passo no caminho das práticas sustentáveis. Rev. Direitos Difusos **2**(70), 213–230 (2018)

César, R.M., César, M.B.: Bibliometric analysis of scientific production on reverse logistics in Brazil. Rev. São Luís Orione **1**(14), 1–14 (2019)

Chaves, G.L.D., Balista, W.C., Comper, I.C.: Logística reversa: o estado da arte e perspectivas futuras. Eng. Sanit. Ambient **24**(4), 821–831 (2019)

Cosenza, J., Mamede, E., de Andrade, G., de Assunção, M.: Economia circular como alternativa para o crescimento sustentável brasileiro: análise da política nacional de resíduos sólidos. Revis. Gestão Ambiental e Sustentabilidade **9**(1), e16147 (2020). https://doi.org/10.5585/geas.v9i1.16147

Costa, L., Mendonça, F.M., Souza, R.G.: O que é logística reversa. In: Valle, R., Souza, R.G. (eds.) Logística Reversa: Processo a Processo. Atlas, São Paulo (2014)

Couto, M.C.L., Lange, L.C.: Analysis of reverse logistics systems in Brazil. Eng. Sanit. Ambient **22**(5), 889–898 (2017)

D'Agosto, M.A.: Logística Sustentável: vencendo o desafio contemporâneo da cadeia de suprimentos. 1st edn. Elsevier, Rio de Janeiro (2018)

Donaire, D., Oliveira, E.C.: Gestão Ambiental na Empresa, 3rd edn. Campos Elíseos, São Paulo (2018)

Faria, H.C.G., Polido, A.F.: Logística Reversa: um interesse em constante crescimento. Simpósio de Tecnologia - Faculdade de Tecnologia de Taquaritinga, pp. 167–176 (2018)

Fenker, E.A., Diehl, C.A., Alves, T.W., Kalinowski, C.: Gestão Ambiental: Incentivos, Riscos e Custos, 1st edn. Atlas, São Paulo (2015)

Fernandes, S.M., et al.: Revisão sistemática da literatura sobre as formas de mensuração do desempenho da logística reversa. Gest. Prod. **25**(1), 175–190 (2018)

Gallegos, G.M.C., Jiménez, M.A.V., Lara, B.N.: Estudio bibliométrico comparativo entre la logística inversa y la logística verde. Revista CIMEXUS **14**(2), 153–169 (2019)

Jabbour, A.B.L.S., Jabbour, C.J.C.: Gestão Ambiental nas Organizações: Fundamentos e tendências, 1st edn. Atlas, São Paulo (2013)

Leite, P.R.: Logística Reversa, 1st edn. Saraiva, São Paulo (2017)

Mattos, W.C., Santos, S.S.: A Logística Reversa como Ferramenta Competitiva e de Sustentabilidade Ambiental. Revista Ensaios & Diálogos **1**(7), 94–104 (2014)

Mendes et al.: Logística Reversa: estudo de caso em uma indústria de artefatos plásticos. Exacta – EP **14**(1), 37–45 (2016)

Mendonça, J.C.A., et al.: Reverse logistics in Brazil: a study on the environmental mechanism, corporate social responsibility and relevant legislation. Capital Cientifico Magazine **15**(2), 130–147 (2017)

Oliveira, E.F., et al.: Logística reversa: importância econômica, social e ambiental. Braz. J. Anim. Environ. Res. **3**(4), 4325–4337 (2020)

Pereira, A.L., Tadeu, H.F.B., Silva, J.T.M., Boechat, C.B., Campos, P.M.S.: Logística Reversa e Sustentabilidade. Cengage Learning, São Paulo (2012)

Silva, E.A., Neto, J.M.M.: Logística reversa nas indústrias de plásticos de Teresina-PI: um estudo de viabilidade. Polímeros **21**(3), 246–251 (2011)

Silva, A.M.G., Pimentel, M.S.: Logística reversa na construção civil: um estudo de caso sobre o gerenciamento dos resíduos de construção civil e sua reintegração na cadeia de valor. Rev. Fatec Zona Sul **6**(2), 18–33 (2019)

Souza, E.D., Hammes, G., Rodriguez, C.M.T.: Barreiras na implementação da logística reversa nas empresas catarinenses. S. Am. Dev. Soc. J. **4**(11), 195–213 (2018)

Velazquez, V.H.T., Marcon, V.T.B.: Aspectos relevantes da logística reversa na política nacional de resíduos sólidos. Rev. Direito Ambiental e Soc. **7**(3), 201–229 (2017)

An Overview of Green Roofs in Urban Areas: Impact on Buildings and Food-Energy-Water Nexus

Ricardo Gomes[1,3](✉), João Galvão[1,3], Pedro Gala[1,3], Luís Prola[1], and Vânia Ribeiro[2,4]

[1] School of Technology and Management, Polytechnic of Leiria, Leiria, Portugal
ricardo.gomes@ipleiria.pt
[2] School of Health Sciences, Polytechnic of Leiria, Leiria, Portugal
[3] INESC Coimbra, DEEC, Rua Sílvio Lima, Polo II, 3030-290 Coimbra, Portugal
[4] CiTechCare, Centre for Innovative Care and Health Technology, School of Health Sciences, Polytechnic of Leiria, 2411-901 Leiria, Portugal

Abstract. Nowadays, one of the applications with most interest in environmental and sustainability of urban areas, concerns green roofs design. A green roof is a vegetative layer grown on a rooftop and consists of a waterproof membrane, growing medium (soil) and vegetation (plants) overlying a traditional roof. The use of green roofs improves the building's energy comfort and reduces the stormwater runoff discharge in urban drainage systems, reducing the risk of flooding in urban areas, if implemented in a large-scale. The use of green roofs has potential to improve carbon retention in urban areas and the water holding capacity can be improved with a proper selection of the soil substrate and drainage layer. At the same time, it can be used for growing edible herbs, berries, vegetables, and flowers or simply for creating a peaceful place to relax at open air environment. In this paper an overview of the use of green roofs in urban areas is presented, highlighting its impact on buildings and food-energy-water nexus.

Keywords: Green roof · Sustainability · Urban areas · Impact on buildings · Food-energy-water nexus

1 Introduction

The global socioeconomic development model that has been followed since the 70's of the 20th century is not sustainable from the point of view of the use of natural resources and of the environmental impact (pollution of water bodies, soil and air). The 2030 Agenda, adopted in 2015 by all the Member States of the United Nations [1], provides guidelines for the present date and into the future society based on a model of Sustainable Development. As example, in December 2019, the European Commission announced its growth strategy based on combating climate change and promoting a new green economy – evidently, this strategy has a strong impact on the construction sector, particularly in

J. R. da Costa Sanches Galvão et al. (Eds.): *Proceedings of the 1st International Conference on Water Energy Food and Sustainability (ICoWEFS 2021)*, pp. 626–635, 2021.
https://doi.org/10.1007/978-3-030-75315-3_67

urban areas. The 'Green Deal' thus brings the environment to the center of Europe's development policy, assuming Europe's ambition to be carbon neutral in 2050.

Urban areas require a significant amount of energy, water, and others natural resources, which have a strong environmental impact. One solution to ensuring the sustainable urban development and the environmental protection is to make buildings more sustainable, using new construction solutions, technologies, and materials. One can then think of achieving smart and progressively self-sustaining buildings. In order to aid this strategy, one can also think of using buildings as a local food production, as well as to save on water and energy demand.

One of the most interesting new features of buildings regarding its environmental and sustainability upgrading, concerns green roofs design [2]. The decision for planting a green roof should be anticipated by a proper assessment of its advantage and disadvantages. Literature provides a number of multifunctional benefits of green roof [3–10]: i) reduction of in the urban heat island effect; ii) reduction or delaying rainwater runoff; iii) cleaning of runoff water and the surrounding air; iv) improvement of the natural biodiversity; v) reducing energy requirements concerning building heating and cooling; vi) amenity value, aesthetics and marketing; vii) improving sound insulation; viii) promoting roof's protection from UV radiation and acidic rain; ix) improvement of carbon retention, and x) providing an area for rest and relaxation. As disadvantages: i) structural overload limitations; ii) possible damages arising from roof leaks with the potential to result in structural damages; iii) limited choice of plants; and iv) installation costs and periodic maintenance costs.

Green roofs are generally classified as extensive or intensive, according to the substrate thickness [9, 11–13]. Extensive Green Roofs typically have shallow soil layers (5–20 cm). They are intended to function with minimal irrigation and maintenance and use a narrow set of plant species that are suited to the water stress created by this environment. These green roofs are easier to install and have a lighter and less expensive soil substrate [14]. On the other hand, Intensive Green Roofs typically have deeper soil layers (20–200 cm) with edible crops or landscaped gardens that require irrigation and relatively high maintenance. Intensive Green Roofs are more often installed on new buildings, because the additional weight can then be taken into account in the building's structural design, in contrast to what happens with Extensive Green Roofs, typically installed on existing buildings.

In this paper, an overview of the use of green roofs in urban areas is presented, including its impact on buildings and food-energy-water nexus.

2 Main Research Core and Results

2.1 Energy Efficiency and Thermal Comfort

According to International Energy Agency, the buildings are responsible for over than 40% of world primary energy demand and 24% of greenhouse gas emissions [15]. Because more and more people are living in cities, more urban areas are emerging with higher temperatures than those of rural areas. Therefore, the urban heat island appears in cities [16, 17] due to the concentration of buildings, with high energy consumption and thermal comfort requirements, particularly in residential and service buildings.

Since the 1960s, the use of green roofs is a growing mitigation technology for lowering the surface temperatures of roofs and thus decreasing the corresponding heat flux released to the atmosphere. Now this green infrastructure is one contribution for a sustainable construction to create living and heathy environment, which include the following main elements: vegetation, organic mixture/soil, water management and insulation screen. This technology of green or living roofs is a good solution to decrease the heat flux released to the atmosphere and to mitigate the effect of high temperatures of the roof's surface. Consequently, the temperature gradient inside buildings will be lower [18].

It is difficult to determine how vegetation grows on green roofs, given that the climatic conditions, the design (shadow areas of the building) and the structure of the building for some extra load from the vegetation beds affect these decisions. It is intended that the vegetation is a shading system and visually more pleasant, in contrast to the brick walls and building exterior color, but this vegetation absorbs part of the thermal energy in the process of growth and photosynthesis.

Quezada-Garcia et al. [19] demonstrated the analytical solutions of the heat flow of the green roofs, but there are features to determine the best solutions in each building, related to the cooling mechanisms according to the type of plant growth, which affects the heat absorption capacity and, consequently, the thermal comfort inside the building. The transpiration of plants, the evaporation of moisture from the soil, have the consequence of lowering the temperature on the roof (uncooling), because there is a loss of latent heat from the plants, resulting from these elements.

The main objective of green roofs' contribution to energy efficiency and sustainability in cities is to reduce the urban heat island effect and consequently the reduction of energy, inherent costs, the increase of energy efficiency, maintaining levels of comfort inside the houses [19]. On the other hand, the heat gain of the buildings will be reduced with the consequence of a decrease in the consumption of air conditioning inside the building. The energy balance for the green roof will be related by the following variables: $Rn = H + \lambda E + G$ (MJ/(m^2*day), where: Rn is net radiation. It represents the gain of energy by the surface from radiation. It is a positive number when it is towards the surface; H is the flow of sensitive heat. It represents the loss of energy by the surface by heat transfer to the atmosphere; λE is the latent heat flow resulting from soil evaporation or plant evapotranspiration. It represents a loss of energy from the surface due to evaporation. (λ is the specific latent heat of evaporation, units J kg^{-1} and E is the evaporation rate, with units kg m^{-2} s^{-1}; G is the heat flow for buildings or heat storage. It is the loss of energy by heat conduction through the lower boundary. It is a positive number when it is directed away from the surface into ground. For instance, if Rn is a constant value, the green roofs increase the latent heat flow compared to the sensitive heat flow, therefore to a lower value of heat stored in the buildings and, consequently, to the reduction of thermal energy consumption (cooling) inside the building [20].

Recent research in several cities with different climates, highlights, the features, and best practices in the use of the green roof technology, such as the following: I) In Italy a quarter in Bolzano city, which participated in the smart cities Europe vegetation SINFONIA project (Smart INitiative of cities Fully cOmmitted to iNvest In Advanced large-scaled energy solutions), showed for the use of green roofs: a) reduces air temperature during hot summers by up to 0.5 °C; b) Improves thermal comfort, with the greatest reduction of temperature amplitude registered during the day; and c) Raises air quality from 21% to 28%, made possible by adding vegetation to horizontal surfaces; II) In Loutraki region (Greece) a study about the thermal properties of green roofs concluded: a) inside the building, with a green roof, the measured temperature is lower than the temperature in a traditional roof building; b) heat insulation performance of green roofs is considerable for buildings with little to no insulation; c) the estimated heating and cooling loads are lower in the building with green roofs regardless of insulation [21] III) In United States of America, in New York, the proprieties of green roofs have lowered the urban heat island effect and in Chicago the green roofs are being encouraged to manage stormwater. These cities had different climatic conditions and the function of a green roof is distinct, as for example: to reduce stormwater runoff, but in hotter climates, in buildings, to provide thermal comfort and energy consumptions reductions [22].

2.2 Rainwater Retention

The urbanization increases the impervious surfaces of buildings, roads, and parking lots and, consequently, produces a change in the natural hydrological cycle, and represent a threaten to the water quality by washing pollutants from impervious surfaces. Green multilayer roofs play a very important role in the development of sustainable urban areas [3]. Several researchers prove that Green Roofs are one of the best sustainable solutions for runoff reduction in urban areas and for the protection of water resources from being contaminated [23–27]. However, these measures only have impact in preventing urban flooding if applied on a large-scale [28–30].

In an urban context, multilayered green roofs are characterized by the combination of conventional green roof technology with a rainwater catchment system for a storage tank. Recall that this type of roofs consists of a vegetative/surface layer, a substrate layer, a geotextile filter membrane, a storage/drainage layer, and waterproof membrane to protect the building. In this context, the vegetation and subtract layers operate as a bio-filtration system, reducing the pollutants of stormwater drainage network [25]. The thickness, type and age of the substrate layer is known to have more impact on water retention, than the type of vegetation used [35]. The main contribution for water retention it related with storage/drainage layer, designed to support plants development. Water retention in green roof's structure reduces the overload to urban stormwater drainage network, while the time spent to cross the green roof's layers increases the lag time to peak flow and, consequently, reduce the risk of urban flooding [30].

A green roof can change the stormwater runoff because of its ability to retain water on its structure. According to Mohapatra et al. [31], the intensive use of green roofs permits an annual runoff reduction of 65–85% of the annual precipitation, while the corresponding value for extensive green roofs is in the range of 27–81%. Additionally, Li and Yeung [32] shown that the retention of semi-intensive vegetated green roof and intensive vegetated green roof depends on the coverage area of the roof. The presented retention percentage ranges from 26% to 88% by green roofs when the rainfall recording was 12 mm in the rain gauge. Also, roof slope plays an important role in the water retention for different rainfall intensities. Green roofs can be classified according to roof slope. Therefore, one has: i) typical flat roof: 0–5° roof slope (without additional support); ii) normal sloped roof: 5–15° roof slope (with optional support depending on project); iii) Steep sloped roof: 15–45° roof slope (with appropriate measures to cope with shear forces and erosion); iv) Green Wall: over 45° slope. An experiment showed that a slope of 2% allow retain up to four times rainwater as compared with slope of 14% [31].

2.3 Substrates, Vegetation, and Edible Plants

As mentioned before, green roofs can be characterized as extensive or intensive, according to the substrate thickness. The thickness of the soil and the type of substrate limit the length of the root and, consequently, the type of vegetation that can be installed [33]. Green roof substrates are mainly inorganic, providing long-term stability, while organic components (<20% v/v) reduce the weight of the substrate and increase water retention, depending on the rate and type. Usually, the substrates used on green roofs have low organic matter levels (5–20%) to discourage excessive plant growth and weed invasion [34]. The selection of organic components depends on availability and organic waste can improve sustainability and reduce costs [3]. A study by Xue and Farrel (in 2020) aimed to study the effect of five organic residues (coarse coconut, fine coconut, composted green residue, almond and pistachio shell) with the addition of slag in different percentages for the creation of substrates for lettuce (*Lactuca sativa*) on green roofs. The results allowed to conclude that rates that varied between 15–20% v/v of coarse coconut, fine coconut and composted green residues reduced the weight of the substrate, allowing a good aeration and improved the plant available water. Substrates with pistachio shell and almond shells showed phytotoxicity. The study confirmed the efficiency of the use of organic waste in the use of a slag-based substrate, where the maximum organic matter limit (20% v/v) was not exceeded, according with the FLL Green Roof Guidelines [14] (FLL – Landscape Development and Landscaping Research Society). The use of recycled or locally available by-products and materials in the formulation of substrates has played a crucial role in soil sustainability.

Various materials have been tested in mixtures and their performance evaluated. It is worth mentioning the study carried out by Eksi (2020) that used on the substrates of green roofs blends composed of: locally available materials such as inorganic (pumice, lava rock, perlite and zeolite) and organic (sheep manure), as well as recycled materials such as garbage compounds municipal, crushed concrete, crushed brick and sawdust with a proportion of 8 inorganic material: 2 organic material per volume. It was found in this 58-week study that the substrates made up of pumice and perlite corrected with municipal compost outperformed the other mixtures in terms of overall performance. However, if this compound was successful in general for most plants, it was found that the substrates that contained pearlite and lava rock were strongly affected by the fluctuations in ambient temperature, which contributed to the death of the Thymus and Helichrysum genus. It is essential that the selection of the substrate considers the edaphoclimatic conditions for the adaptation of the plants. When selecting the substrate, the porosity and particle size distribution of the mixtures should be considered as they influence water retention, infiltration, plant growth and survival in dry periods [35].

In many studies the selection of plants for green roofs is based on the edaphoclimatic conditions of the region where they are implemented (tropical, mediterranean or others). The use of local native plants, in addition to being already properly adapted to the climate, has the potential to improve moisture retention in the soil of green roofs [36]. The selection of native species largely determines the functionality and efficiency of extensive green roofs [37]. A study in Puerto Rico found that of the 26 plant species introduced to a tropical green roof, only 6 survived after 4 years (*Arachis hypogaea, Cymbopogon ambiguus, Passiflora foetida, Sedum stahlii, Talinum paniculatum, Tulbaghia violacea*) [38]. Considering the mediterranean-climate, the careful selection of species able to withstand drought and with an optimized growth, is mandatory [39]. The amount of water retained in the aerial part of the plants is an important component of the green roofs' water retention capability [40]. Edible plants that can to be incorporated in the mediterranean green roofs are thyme (*Thymus vulgaris*), immortelle (*Helichrysum italicum*) [35], rosemary (*Rosmarinus officinalis*) [40], sea fennel (*Crithmum maritimum*) [39], among others. One of the current trends to green roofs utilities is utilizing these spaces for the growing of edible species, decreasing the carbon footprint and increasing this infrastructure functionality, promoting the urban agriculture. The introduction of edible plants for the purpose of domestic vegetable production has been studied to verify the impact on food sustainability, in reducing the import of vegetables from other regions or countries. In a study carried out in Singapore it was found that the adequacy of the roofs of the public housing infrastructures allowed the implementation of edible green roofs and that if a national program was successfully implemented, it would increase the domestic production of vegetables by 700%, which would supply in domestic demand 35.5%, reducing food imports as well as reducing annual carbon emissions by 9,052 metric tons [41]. Another study carried out in Bologna, Italy in order to verify the production capacity of rooftop gardens also showed that they could supply more than 12,000 tons/year of vegetables, satisfying 77% of the population's needs [42]. Variety of vegetables, fruits, flowers, medicinal and ornamental plants have been introduced to green roofs, in a perspective of home economics due to the price of fruit and vegetables and the scarcity of land to cultivate in densely populated cities. Some of the plants grown in this perspective stand out, such as kale, cauliflower, broccoli, chili, lettuce, brinjal, onion, cucumber, tomato, spinach, peppers, even fruit trees such as lemon trees, apple trees, banana trees, among others [43]. A systematic analysis of the literature finds that edible green roofs are one of the structures that can offer greater resilience and quality of life in cities, especially in food production, with social, economic, and environmental contributions to urban sustainability and food security. However, the potential risks to human health from exposure to heavy metals and chemical contaminants must be considered [44].

2.4 Overload Structural Effects

The adoption of green roofs requires performing specific verifications concerning the structural stability of the buildings. This is true for both new and existing buildings, particularly concerning intensive green roofs. Cascone et al. [13] studies the additional loads that are needed to be considered when adopting green roofs. According to these authors, the weight of intensive green roofs can be very significant particularly when they are under a saturated condition. Information concerning green roofs loadings can also be found in FLL Green Roof Guidelines [45]. Peck and Kuhn [46] addressed that if a green roof is installed on an existing building, the design will be limited to the carrying capacity of the existing roof. If not, the owner should upgrade the structure, which can be a significant cost.

For existing buildings, designed according to the European standard regulations (Eurocodes), Cascone [13] estimates that only a few types of intensive green roofs can be adopted without surpassing the load-capacity forecasted by Eurocodes, and therefore requiring structural upgrades. Evidently, these loads must also be considered for the design of new buildings, see also Dunnett and Kingsbury [47].

Green roofs can also generate relevant modifications on wind actions. FLL Green Roof Guidelines [45] presents some indications concerning these loads. This can be particularly relevant for steel buildings, which are particularly prone to the effect of these actions. This is particularly relevant since steel, being a sustainable structural material [47, 48] is thought to be used in new sustainable constructions. Calculations concerning the design of steel structure supporting a green roof were performed by Murphy [49]. In his master's dissertation Raposo [50] indicates the aspects that should be considered when verifying the safety of the structures that support a green roof, regarding their actions, namely: vertical loads, influence the slope of the roof and wind load. The adoption of green roofs has also implications in the dynamic proprieties of existing and new buildings. In fact, the placement of a significant mass in the top of a building can significantly change its dynamic proprieties (natural frequencies) turning it more flexible. Again, this is particularly relevant for steel buildings. Consequently, implications can arise in the effects of seismic actions. In this context, it is also worth mention the ongoing studies on the potential for the green roofs to be used as energy dissipators, see [50, 51].

3 Conclusions

The review presented in this paper has shown that the introduction of green roofs in urban areas has many advantages, the most significant being the comfort of buildings, the rainwater retention, especially in Mediterranean countries, the potential food production aimed at reducing family costs in the purchase of fruit and vegetables, and the increase in green and leisure areas that provide well-being to people without leaving home.

Research work is needed to determine the cooling effect using materials to promote the cooling proprieties inside the buildings, the variable U-value of the roof and its thermal capacitance for different climatic conditions. Also, a sensitivity analyses by computational package software for transient building simulation, with several boundary conditions, should be done. More research is required about the effect of green roofs to be used as energy dissipators of seismic actions, as well as to optimize the substrate structure and the most appropriate vegetation selection to improve the water retention and water quality discharged into the stormwater drainage network. Also, to take full advantage of the opportunities that digitalization can offer, remote control and supervision systems should be developed to manage and enhance the potential of green roofs for controlling energy, water, vegetation, and comfort variables in buildings.

Acknowledgments. Study supported by UIDB/00308/2020 project.

References

1. United Nations. https://www.un.org/sustainabledevelopment/. Acessed 15 Jan 2021
2. Kim, E., et al.: Economic and environmental sustainability and public perceptions of rooftop farm versus extensive garden. Build. Environ. **146**, 206–215 (2018)
3. Cristiano, E., Deidda, R., Viola, F.: The role of green roofs in urban water-energy-food-ecosystem nexus: a review. Sci. Total Environ. **756**, 143876 (2021)
4. Shafique, M., Xue, X., Luo, X.: An overview of carbon sequestration of green roofs in urban areas. Urban Forestry Urban Greening **47**, 126515 (2020)
5. Shafique, M., et al.: An overview of life cycle assessment of green roofs. J. Clean. Prod. **250**, 119471 (2020)
6. Liu, J., et al.: Assessing the mitigation of greenhouse gas emissions from a green infrastructure-based urban drainage system. Appl. Energy **278**, 115686 (2020)
7. Polo-Labarrios, et al.: Comparison of thermal performance between green roofs and conventional roofs. Case Stud. Thermal Eng. **21**, 100697 (2020)
8. Tadeu, A., et al.: Drainage and water storage capacity of insulation cork board applied as a layer on green roofs. Constr. Build. Mater. **209**, 52–65 (2019)
9. Reyes, R., et al.: Effect of substrate depth and roof layers on green roof temperature and water requirements in a semi-arid climate. Ecol. Eng. **97**, 624–632 (2016)
10. Cascone, S., et al.: The evapotranspiration process in green roofs: a review. Build. Environ. **147**, 337–355 (2019)
11. Lu, J., et al.: Effect of substrate depth on initial growth and drought tolerance of Sedum lineare in extensive green roof system. Ecol. Eng. **74**, 408–414 (2015)
12. Eksi, M., et al.: Effect of substrate depth, vegetation type, and season on green roof thermal properties. Energy Build. **145**, 174–187 (2017)
13. Cascone, S., et al.: A comprehensive study on green roof performance for retrofitting existing buildings. Build. Environ. **136**, 227–239 (2018)
14. Xue, M., Farrell, C.: Use of organic wastes to create lightweight green roof substrates with increased plant-available water. Urban Forestry Urban Greening **48**, 126569 (2020)
15. International Energy Agency Solar Heating & Cooling Technology Collaboration Programme. https://iea-shc.org/. Acessed 10 Jan 2021
16. Gago, E.J., et al.: The city and urban heat islands: a review of strategies to mitigate adverse effects. Renew. Sustain. Energy Rev. **25**, 749–758 (2013)
17. Santamouris, M.: Cooling the cities – a review of reflective and green roof mitigation technologies to fight heat island and improve comfort in urban environments. Solar Energy **103**, 682–703 (2014)
18. Yang, J., et al.: Green and cool roofs' urban heat island mitigation potential in tropical climate. Solar Energy **173**, 597–609 (2018)
19. Quezada-García, S., et al.: Heterogeneous model for heat transfer in green roof systems. Energy Build. **139**, 205–213 (2017)
20. Cirkel, D., et al.: Evaporation from (Blue-)green roofs: assessing the benefits of a storage and capillary irrigation system based on measurements and modeling. Water J. **10**(9), 1253 (2018)
21. Bisello, A., Antoniucci, V., Marella, G.: Measuring the price premium of energy efficiency: a two-step analysis in the Italian housing market. Energy Build. **208**, 109670 (2020)
22. Semaan, M., Pearce, A.: Assessment of the gains and benefits of green roofs in different climates. Proc. Eng. **145**, 333–339 (2016)

23. Gong, Y., et al.: Factors affecting the ability of extensive green roofs to reduce nutrient pollutants in rainfall runoff. Sci. Total Environ. **732**, 139248 (2020)
24. Shafique, M., Kim, R., Rafiq, M.: Green roof benefits, opportunities and challenges – a review. Renew. Sustain. Energy Rev. **90**, 757–773 (2018)
25. Loiola, C., Mary, W., Pimentel da Silva, L.: Hydrological performance of modular-tray green roof systems for increasing the resilience of mega-cities to climate change. J. Hydrol. **573**, 1057–1066 (2019)
26. Vijayaraghavan, K., Reddy, D.H.K., Yun, Y.S.: Improving the quality of runoff from green roofs through synergistic biosorption and phytoremediation techniques: a review. Sustain. Cities Soc. **46**, 101381 (2019)
27. Liu, W., et al.: The impacts of substrate and vegetation on stormwater runoff quality from extensive green roofs. J. Hydrol. **576**, 575–582 (2019)
28. Ercolani, G., et al.: Evaluating performances of green roofs for stormwater runoff mitigation in a high flood risk urban catchment. J. Hydrol. **566**, 830–845 (2018)
29. Versini, P.A., et al.: Fractal analysis of green roof spatial implementation in European cities. Urban Forestry Urban Greening **49**, 126629 (2020)
30. Gomes, R., Garcia, R.: Sistemas Urbanos de Drenagem Sustentável: Influência sobre as Infraestruturas Existentes. In: 14.° Simpósio de Hidráulica e Recursos Hídricos dos Países de Língua Portuguesa, pp. 1–4. APRH, Cidade da Praia, Cabo Verde (2019)
31. Mohapatra, S., et al.: A critical appraisal of green vegetated roofs: Energy and environment in focus. Mater. Today: Proc. (2020)
32. Li, W.C., Yeung, K.K.A.: A comprehensive study of green roof performance from environmental perspective. Int. J. Sustain. Built Environ. **3**(1), 127–134 (2014)
33. Skjeldrum, P.M., Kvande, T.: Moisture-resilient upgrading to blue-green roofs. Energy Proc. **132**, 417–422 (2017)
34. Nagase, A., Dunnett, N., Choi, M.S.: Investigation of weed phenology in an establishing semi-extensive green roof. Ecol. Eng. **58**, 156–164 (2013)
35. Eksi, M., et al.: Assessment of recycled or locally available materials as green roof substrates. Ecol. Eng. **156**, 105966 (2020)
36. Galvão, J., et al.: Green roofs for energy efficiency and smart food production to sustainability. In: Environment, Green Technology, and Engineering International Conference (EGTEIC), Caceres, Spain (2018)
37. Lundholm, J., et al.: Plant species and functional group combinations affect green roof ecosystem functions. PLOS ONE **5**(3), e9677 (2010)
38. Grullón – Penkova, I.F., Zimmerman, J.K., González, G.: Green roofs in the tropics: design considerations and vegetation dynamics. Heliyon **6**(8), e04712 (2020)
39. Azeñas, V., et al.: Performance evaluation of five Mediterranean species to optimize ecosystem services of green roofs under water-limited conditions. J. Environ. Manage. **212**, 236–247 (2018)
40. Brandão, C., et al.: Wet season hydrological performance of green roofs using native species under Mediterranean climate. Ecol. Eng. **102**, 596–611 (2017)
41. Astee, L.Y., Kishnani, N.T.: Building integrated agriculture utilising rooftops for sustainable food crop cultivation in Singapore (2010)
42. Orsini, F., et al.: Exploring the production capacity of rooftop gardens (RTGs) in urban agriculture: the potential impact on food and nutrition security, biodiversity and other ecosystem services in the city of Bologna. Food Secur. **6**(6), 781–792 (2014)
43. Safayet, M., Arefin, M.F., Hasan, M.M.U.: Present practice and future prospect of rooftop farming in Dhaka city: a step towards urban sustainability. J. Urban Manage. **6**(2), 56–65 (2017)

44. Russo, A., et al.: Edible green infrastructure: An approach and review of provisioning ecosystem services and disservices in urban environments. Agric. Ecosyst. Environ. **242**, 53–66 (2017)
45. Landscape Development and Landscaping Research Society (FLL), Green Roof Guidelines: Guidelines for the Planning, Construction and Maintenance of Green Roofs (2018)
46. Peck, S., Kuhn, M.: Design Guidelines for Green Roofs. Environment (2000)
47. Dunnet, N., Kingsbury, N.: Planting Green Roofs and Living Walls. London, UK (2008)
48. Zhong, Y., Wu, P.: Economic sustainability, environmental sustainability and constructability indicators related to concrete- and steel-projects. J. Clean. Prod. **108**, 748–756 (2015)
49. Murphy, P.B.: Sustainable Solutions for Energy efficiency and Acoustic Performance, AE Senior Thesis. Galludet University, Washington (2007)
50. Raposo, F.M.F.: Manual de Boas Práticas de Coberturas Verdes, Análise de Casos de Estudo, Instituto Superior Técnico, Universidade de Lisboa, Lisbon (2013)
51. Carmody, M., et al.: Seismic Response of Green Roofs. In: NZSEE Conference (2009)

How Smart is Green? Sustainable and Intelligent Mobility Solutions for Transporting People and Goods

Filipe Ferreira[1], Miguel Ribeiro[1], Luís Serrano[1,3](✉), Marcelo Gaspar[1,3], and Jorge Julião[2,3]

[1] Polytechnic Institute of Leiria, Leiria, Portugal
luis.serrano@ipleiria.pt
[2] Católica Porto Business School, Universidade Católica Portuguesa, Porto, Portugal
[3] Centre for Smart and Sustainable Mobility, Leiria, Portugal

Abstract. Smart mobility relates to the ability of integrating technology into transportation. When considering the transition to a greener mobility paradigm, several key elements of smart and sustainable dimensions have to be addressed. To allow outlining more efficient transportation solutions, current exploratory research presents and discusses a dedicated prospective conceptual framework. In this research, the discussion is supported by empirical data collected via a dedicated exploratory survey. The prospective framework structure and dimensions were approached in order to derive the focus from mere technical features, but also considering the final users' perspectives and needs, so that its acceptability can be more effectively promoted. The main findings of this study are related to the greening effect of smart mobility, which allowed foreseeing increased benefits from integrating smart and sustainable solutions to improve mobility.

Keywords: Smart mobility · Transportation · Sustainable mobility

1 Introduction

Smart and electric mobility are closely intertwined, as they both use digital-based solutions to improve the transportation of people and goods. Nonetheless, it is widely accepted that these alternative mobility solutions can be more sustainable than current internal combustion-based solutions, these are still far from comprising the sustainability goals envisaged by the World Commission on Environment and Development [1].

Smart mobility [2] can be defined as the ability of the movement associated with technology. To foster a greener mobility paradigm [3], several key elements of smart and sustainable dimensions have to be considered and addressed, so that more efficient transportation solutions can be outlined and promoted. To such end, current research uses a qualitative research methodology to better understand the end-users' perceptions and points of view on the subject, thus contributing to an immersion into the specifics

J. R. da Costa Sanches Galvão et al. (Eds.): *Proceedings of the 1st International Conference on Water Energy Food and Sustainability (ICoWEFS 2021)*, pp. 636–645, 2021.
https://doi.org/10.1007/978-3-030-75315-3_68

and identifying important patterns and inter-relationships regarding the sustainability characteristics of smart mobility.

Current study uses a holistic approach to perceive and foster efficiency improvements of a future greener mobility paradigm based on smart and interconnected solutions. This approach focuses not only on environmental protection and the reduction of environmental impacts, but also integrates the environmental aspects with the socio-economic issues with the technology level. The final prospective framework aims at attaining the identification of the key drivers of smart mobility that contribute and enhance the sustainability dimensions of a greener mobility paradigm.

2 Literature Review

Mobility can be rightly considered a basic human need [4]. Among current literature, it is possible to find increased recognition in the sustainability transitions regarding the significance of mobility in the transformation of transportation [5]. This allows for promoting interactions between people and enables an efficient exchange and transportation of goods and materials [6].

Smart mobility is a broad and complex concept. Nonetheless, to allow for identifying its importance and potential for current, and future society, this can be defined as the ability of the movement associated with technology [2]. A holistic overview of the smart mobility ecosystem instead of valuing partially any of its many dimensions may help perceiving, and better understanding, the many interconnections that occur between the digital and physical worlds [6].

2.1 Smart Mobility Ecosystem

The smart mobility ecosystem allows identifying and discussing many of its dimensions and strengths [7]. Considering this approach, it is possible to highlight the increase in efficiency related to smart mobility, the positive effect on safety and security for all those that integrate such an ecosystem, or the flexibility as a key-feature of these solutions, just to mention a few.

The increase in efficiency is usually assigned to smart mobility as a direct consequence of optimal and rational use of material resources [8]. Additionally, it also allows valuing one of the most important resources of humankind: the time savings generated by such efficient mobility solutions.

On what concerns safety and security, combining smart mobility infrastructures with smart systems on vehicles are foreseen as key issues on the enhancement of active and passive safety solutions. These aspects are even more relevant, when considering the envisaged evolution of current vehicles to autonomous intelligent systems integrating different levels of assisted driving [9].

Flexibility can be also considered a key dimension of smart mobility. Thus, it is widely recognized that smart flexible transport systems positively contribute to enhancing the efficiency and performance of mobility [10, 11].

This flexibility allows for individual drivers, public transportation passengers or companies to choose the desired mobility solution among a large variety of alternatives, selecting dedicated navigation routes, times and modes of transportation, service providers, or even desired payment options. This is mainly the result of integrating a wide array of transportation modes and solutions, as well as multiple service providers for the same, or similar ends. Thus, this flexibility translates into more comfortable and cost-effective mobility of people and goods [12].

2.2 Sustainable Mobility

The sustainable mobility notion derives from a broader concept related to sustainable development [1] as its main focus is to "ensure that our transport systems meet society's economic, social and environmental needs whilst minimizing their undesirable impacts on the economy, society and the environment" [13].

As what was mentioned for smart mobility to be addressed in a holistic approach to better understand all of related dimensions and perspectives, the sustainability aspects of mobility also require a wider approach. Thus, even though the aspects of pollution and emissions to the environment associated with transportation are of primary importance, the strategies to pursue the objective of developing sustainable mobility solutions have to take into account the social and economic impacts of mobility [14].

When analyzing the sustainable mobility policies promoted by governance, it is possible to identify its focus merely on environmental protection and the reduction of environmental impacts. Nonetheless, other combined topics have to be considered, thus integrating the environmental aspects with the socio-economic and the technological level. These integrate topics of air pollution, car-sharing, ecodriving, intelligent transportation systems or taxes and incentives, just to mention a few [15].

Considering the need to evolve from the current mobility paradigm on what concerns the integrated sustainability issues, a new model using concepts of transition theory has been developed. Thus, this modelling effort relies on combining agent-based modelling solutions with a system dynamics structure to attain a transitions model for sustainable mobility [5, 16].

Finally, when considering all environmental economic and social dimensions of sustainability, the final goal is to plan and reach a sustainable mobility paradigm based on four types of actions, namely making the best use of technology, regulating transportation, work on planning and focus on the mobility's user [3, 14]. Thus, great effort has to be put towards evolving from technological innovation to societal learning to reach a sustainable mobility paradigm [17].

3 Methodology

Qualitative research is a holistic approach to discover and understand the complex interrelation of facts that occur in reality [18]. Moreover, it uses inductive processes of organizing data into categories to identify the relationships between groups of data according to specific frameworks. Different researchers may develop distinct frameworks to describe the same reality, as the complexity of a given phenomenon may allow developing different views and insights around it. These frameworks also depend on the level of detail of that research and from the active involvement of the researchers in those real experiences [19].

Data Collection. On what refers to the data collection techniques of current research, the resources for the literature review and the state of the art focused on published articles concerning smart and sustainable mobility. Some of the databases accessed for published articles and dissertations were ScienceDirect, Google scholar, ResearchGate and Springer. Complementary to the empirical data collected based on a dedicated exploratory survey, this survey comprises the perceptions and points of view of representatives of companies, associations and public sector organizations that are related to mobility.

Data Analysis. Data analysis techniques for current research was based on coding and processing of information using dedicated software tools. The raw data of current research was systematically analyzed and grouped according to the research objectives using clustering techniques. The data collected from the reference literature and the survey was organized into several main categories, thus allowing for comparing and correlating the data into dedicated categories. Subsequently, the related patterns were identified and discussed by analyzing their inter-relationships using a clustering technique used to process qualitative data.

Data Processing. Based on the results of the analysis techniques, the data collected and analyzed was further processed and discussed using conceptual frameworks. Conceptual frameworks are graphical tools that allow linking abstract concepts to empirical data [20]. Thus, the basic theoretical relationships and definitional issues related to mo-bility were further discussed and systematized using conceptual frameworks. These were used as analytical tools to organize data and highlight fundamental concepts and relations associated with smart mobility and sustainable mobility.

4 Results and Discussion

Current research comprised two complementary steps, namely the exploratory approach and the proposed conceptual framework.

4.1 Exploratory Approach

In the exploratory phase of current research, based on the literature survey and using dedicated data analysis techniques, the most relevant papers for current research were grouped and coded according to the research objectives using clustering techniques. As a result, considering the major dimensions related to the literature findings on smart and sustainable mobility issues, an exploratory framework was developed (Fig. 1). This framework will be further discussed considering the main presented dimensions.

As can be observed in Fig. 1, the data collected from the reference literature and the survey was organized and clustered into several main categories, thus allowing for comparing and correlating the data into dedicated categories. Subsequently, the related patterns were identified and discussed by analyzing their inter-relationships.

Intelligent Cities. Intelligent, or Smart Cities comprise technological advances that allow instrumenting key points of urban locations to deliver data to be used with the aim of providing sustainable prosperity for citizens and businesses. To such end, these cities are increasingly interconnected, thus allowing governance to collect and process data to operationalize in real-time solutions envisaging an increase in efficiency of the overall urban infrastructure [21]. Urban lights and traffic lights are the precursor systems used in urban locations to improve the flow of local traffic and, thus, improve the mobility of that location.

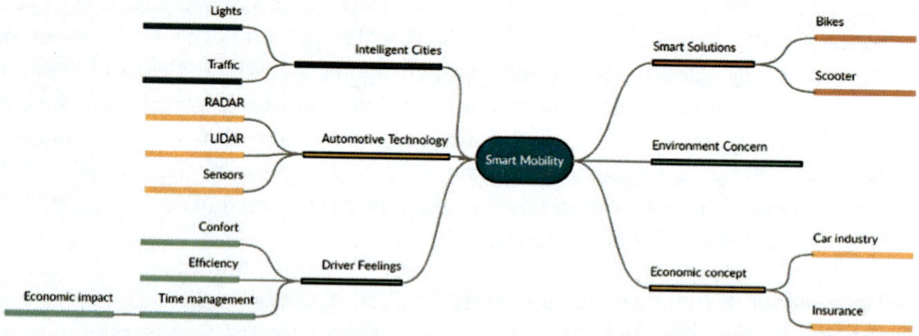

Fig. 1. Exploratory framework resulting from coding literature review data.

Smart Solutions. Smart devices are those that have the ability to carry-out autonomous decisions. When integrated on a smart grid, like that of a smart city, these devices allow for bidirectional communication between the user of that device and the grid they are integrating [22]. This flow of data allows a dynamic control of the smart device to improve and enhance the users experience and safety. Current micromobility solutions integrate lightweight vehicles, such as bicycles or scooters, that improve urban mobility as short-distance transports. These vehicles are used in distances usually shorter than 5 miles. One key feature of urban available e-bikes or e-scooters is that these can usually be monitored and controlled in real-time using dedicated protocols that allow them to interact with a computing and communication core at a distance [23].

Automotive Technology. Considering that micromobility only refers to lightweight vehicles, automotive technology is a main topic that must be addressed when focusing on smart mobility. Current technological advances concerning the smart mobility of automobiles involve complex sensors and cameras. Two of these key technologies are the Radio Detection and Ranging (RADAR) and the Light Detection and Ranging (LIDAR) technologies that can detect the environment involved, thus as drivers from monotonous tasks or by reducing the driver stress [24]. These are intelligent and autonomous features in modern vehicles that improve users' experience as a life-saving automatic technology.

User Experience. When developing new products or services, the designed propositions aim at providing specific solutions for the problems felt by their end-users. Conversely, when considering smart-mobility solutions, these have increased chances of success if they use a consumer-centered approach. Thus, any new and improved technological solutions will only thrive if accepted by their end user. This is why the user experience is so relevant for any mobility solution.

The user's experience relates to what people need and want when using a vehicle. As mentioned above, safety is a key feature for drivers and passengers alike when using a vehicle. To such end, active and passive automotive safety systems focus on improving the usability of those vehicles. Nonetheless, when considering mobility in a holistic approach, the user's experience exceeds the vehicle, as it allows integrating other systems to improve the driver's (and passengers) experience. As an example, the ability to use a GPS (Global Positioning System) to identify the best navigation solution (thus avoiding traffic jams, or by pointing-out a scenic route) effectively improves the user's journey in the vehicle. This allows for improving the vehicle users experience and time-management. As a result, this allows achieving overall efficiency for the end-user.

Economical Concept. The above-mentioned vehicle efficiency is not a concern that solely affects the vehicle users. All of the industry related to automotive manufacturing and servicing, as well as insurance companies depend on vehicles efficiency and safety. Thus, it is expected that smart mobility will promote a positive impact of both safety and efficiency in the near future [25]. Nonetheless, many advantages can be foreseen with the possible evolution of vehicles to include autonomous and assisted driving, these still comprise high uncertainty and many unsolved questions regarding both human interaction and social interaction on the road, with these new forms of driving [26].

Environmental Concerns. Environmental concerns are one of the underlying foundations on vehicle development and, also, as a key driver for improving mobility. Major efforts have been carried-out by academia and industry alike to minimize fuel consumption and to use greener forms of energy. As a consequence, environmental impacts have been significatively reduced and much effort is still being put to improve existing and future alternatives based on smart solutions. These allow for keeping the systems on their best usability by monitoring the vehicle use and controlling dedicated systems to improve overall efficiency, thus minimizing the environmental impacts of vehicle use.

4.2 Complementary Survey

Based on the initial exploratory approach, data collected from the literature review was coded using clustering techniques to derive the exploratory framework of Fig. 1. Considering the cluster nodes of that framework, a dedicated survey was carried-out to further discuss the insights resulting from the state of the art. In the current section, the implications and new insights of the original framework will be discussed based on the dedicated survey comprising the perceptions and points of view of representatives of companies, associations and public sector organizations that are related to mobility.

Smart Mobility. Automotive technology is currently still the keystone for the mobility in urban areas. Even though increased effort has been put to inhibit automotive circulation and to complement automotive vehicles by alternative mobility solutions, there are still no short-term effective alternative solutions on sight.

Nonetheless, micromobility solutions based on the use of small and lightweight vehicles are foreseen as an efficient alternative to automotive vehicles in urban centers of smart cities. These human-powered, or electric vehicles, allow for people to move in short distances (up to 5 miles). These transportation methods are particularly relevant for people's mobility when used combined with public transportation solutions.

Companies rely on automotive transportation to transport goods in urban areas due to its high flexibility of use and to the ability for customizing navigation plans to optimize routes and navigation.

Automotive technology is strongly related to smart cities, as these vehicles will soon be able to communicate and automatically interact with their surroundings. Thus, increased autonomy is envisaged with autonomous or assisted driving. Nonetheless, significant legal and ethical discussions are still needed to overcome major issues related to both human interaction and social interaction on the road.

The future vehicle user's experience is expected to improve significantly with the envisaged solutions related to smart mobility. Different sensors are and will be further used in vehicles to avoid collision, to detect pedestrians and cyclists and to complement the driver vision with camera-sensing systems. These and other combined solutions will contribute to reduce the driver's stress by adding life-saving automatic devices.

Sustainability. In a smart city, global interconnectivity allows for a more efficient use of resources. One example of such connectivity in smart cities is the combined use of public transportation with the micromobility of small lightweight vehicles. This results in less automotive vehicles traveling in cities and a significant decrease in energy consumption. These combined solutions translate into increased saving on both economic and environmental dimensions.

Also, vehicle manufacturers have made an increased effort to change the ongoing internal combustion engine propulsion paradigm to electrified alternatives. The recent widespread use of hybrid and electric vehicles are a direct result of such effort. Companies and researchers share their perceptions towards a significant environmental sustainability increase as a result of this vehicle electrification shift.

The use of smart grids and a holistic approach to more efficient and smart mobility may envisage alternative solutions to the movement of people and goods. As an example, current digital transformation with online and distance learning allows for students and teachers to participate in classes without having to leave their homes. The same applies to remote working or to sales pitches carried-out online and at a distance. Thus, the less people are on the move, the more easily and efficient is the flow of vehicles for the remaining population.

Thus, one can conclude that the greening effect on mobility works both on reducing the need for people to move, combined with the more efficient mobility of people and goods whenever it is effectively necessary for them to move. These resource savings and decreased environmental impacts are related to several of the UN Sustainable Development Goals [27].

5 Summary and Discussion

The mobility of people and goods are amongst the major concerns for a more efficient and sustainable future. Current digital revolution allowed ushering in new and improved outlooks on the issues related to such mobility of people and goods, having recently been coined as smart mobility.

Current research comprised two complementary steps, namely the exploratory approach and the proposed conceptual framework. In the exploratory phase of research, this study considers all major dimensions related to the literature findings combined with the survey on current and future smart mobility. On the latter step of this study, a dedicated framework was developed as presented in Fig. 2. This framework will be further discussed considering the main presented dimensions.

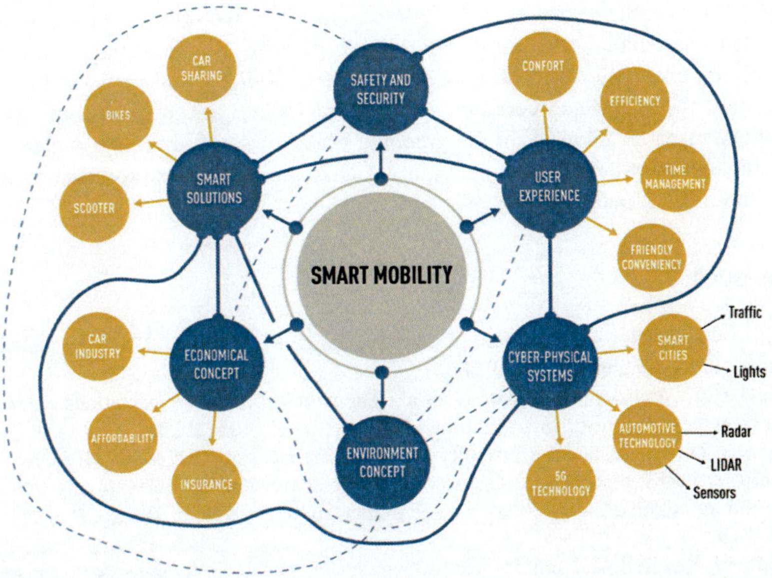

Fig. 2. Conceptual framework comprising the main dimensions related to smart and sustainable mobility.

In the proposed framework of Fig. 2, smart mobility is presented as the focal point due to its integration ability to interact with all of the surrounding dimensions. When observing the main nodes of the framework (identified in a blue color), it can be observed that these are proximally located to the smart mobility focus, as none of them surpasses the other in such holistic approach. Considering the secondary nodes, presented in a yellow color, these refer to the secondary dimensions discussed in the previous section of this text. As discussed above, smart solutions are an omnipresent precondition in every smart mobility solution. User safety and increased benefits for mobility experience contribute to the overall efficiency of smart mobility.

To conclude, mobility efficiency directly affects the economic dimensions with a smarter use of resources and an effective reduction in energy consumption. These combined solutions translate into increased saving on both economic and environmental dimensions. These combined effects can be coined as the greening effect of smart mobility. Finally, in a nutshell, considering all of the above, the greening effect of smart mobility allows foreseeing increased benefits not only from current electrification of transports combined with the widespread use of smart solutions used as a holistic approach to the mobility of people and goods.

6 Conclusions

Current exploratory research allowed developing and discussing a dedicated prospective conceptual framework focusing on the smart and sustainable dimensions related to the mobility. It was realized that even though the aspects of pollution and emissions to the environment associated with transportation are of primary importance, the strategies to pursue the objective of developing sustainable mobility solutions have to take also into account the social and economic impacts of mobility. The main findings of this exploratory research focused on the greening effect of smart mobility, which allowed foreseeing increased benefits from integrating smart and sustainable solutions to improve current mobility of people and goods.

References

1. United Nations: Report of the World Commission on Environment and Development. General Assembly Resolution 42/187 (1987)
2. van Oers, L., de Hoop, E., Jolivet, E., et al.: The politics of smart expectations: interrogating the knowledge claims of smart mobility. Futures **122**, 102604 (2020)
3. Banister, D.: The sustainable mobility paradigm. Transp. Pol. **15**, 73–80 (2008)
4. Sunio, V., Peckson, P., Ugay, J.C.: How urban social movements are leveraging social media to promote dignified mobility as a basic human right. Case Stud. Transp. Pol. **9**(1), 68-79 (2020)
5. Köhler, J., Geels, F.W., Kern, F., et al.: An agenda for sustainability transitions research: state of the art and future directions. Environ. Innov. Soc. Transitions **31**, 1–32 (2019). https://doi.org/10.1016/j.eist.2019.01.004
6. Flügge, B.: Smart mobility - connecting everyone: Trends, concepts and best practices (2017)
7. Benevolo, C., Dameri, R.P., Auria, B.D.: Empowering Organizations: Enabling Platforms and Artefacts, vol. 11, p. 315 (2016). https://doi.org/10.1007/978-3-319-23784-8
8. Ning, Z., Xia, F., Ullah, N., et al.: Vehicular social networks: enabling smart mobility. IEEE Commun. Mag. **55**, 16–55 (2017). https://doi.org/10.1109/MCOM.2017.1600263
9. Tokody, D., Albini, A., Ady, L., et al.: Safety and security through the design of autonomous intelligent vehicle systems and intelligent infrastructure in the smart city. Interdiscip. Descr. Complex Syst. **16**, 384–396 (2018)
10. Porru, S., Misso, F.E., Pani, F.E., Repetto, C.: Smart mobility and public transport: opportunities and challenges in rural and urban areas. J. Traffic Transp. Eng. (Engl. Ed.) **7**, 88–97 (2020)
11. Groth, S.: Multimodal divide: reproduction of transport poverty in smart mobility trends. Transp. Res. Part A Pol. Pract. **125**, 56–71 (2019)

12. Butler, L., Yigitcanlar, T., Paz, A.: How can smart mobility innovations alleviate transportation disadvantage? Assembling a conceptual framework through a systematic review. Appl. Sci. **10** (2020)
13. Council of Union European Union: Review of the EU Sustainable Development Strategy (EU SDS)—Renewed Strategy; 10917/06. Brussels, Belgium (2006)
14. Hickman, R., Hall, P., Banister, D.: Planning more for sustainable mobility. J. Transp. Geogr. **33**, 210–219 (2013). https://doi.org/10.1016/j.jtrangeo.2013.07.004
15. Gallo, M., Marinelli, M.: Sustainable mobility: a review of possible actions and policies. Sustainability **12** (2020). https://doi.org/10.3390/su12187499
16. Köhler, J., Whitmarsh, L., Nykvist, B., et al.: A transtions model for sustainable mobility. Ecol. Econ. **68**, 2985–2995 (2009). https://doi.org/10.1016/j.ecolecon.2009.06.027
17. Vergragt, P.J., Brown, H.S.: Sustainable mobility: from technological innovation to societal learning. J. Clean. Prod **15**, 1104–1115 (2007)
18. Collins, C.S., Stockton, C.M.: The Central role of theory in qualitative research. Int. J. Qual. Methods **17**, 1 (2018). https://doi.org/10.1177/1609406918797475
19. Marvasti, A.: Research methods. In: Treviño, A. (ed.) The Cambridge Handbook of Social Problems, pp 23–38. Cambridge University Press (2018)
20. Leshem, S., Trafford, V.: Overlooking the conceptual framework. Innov. Educ. Teach. Int. **44**, 93–105 (2007). https://doi.org/10.1080/14703290601081407
21. Deakin, M., Al Waer, H.: From intelligent to smart cities. Intell. Build. Int **3**, 140–152 (2011). https://doi.org/10.1080/17508975.2011.586671
22. Koutitas, G.: Control of flexible smart devices in the smart grid. IEEE Trans. Smart Grid **3**, 1333–1343 (2012). https://doi.org/10.1109/TSG.2012.2204410
23. Gnanavelbabu, A., Saravanan, P., Rajkumar, K., Karthikeyan, S.: Experimental investigations on multiple responses in abrasive waterjet machining of Ti-6Al-4V alloy. Mater Today Proc. **5**, 13413–13421 (2018). https://doi.org/10.1016/j.matpr.2018.02.335
24. Al-Hourani, A., Evans, R.J., Farrell, P.M., et al.: Millimeter-Wave Integrated Radar Systems and Techniques. Elsevier Ltd. (2017)
25. Parker, M.: Automotive radar. Digit. Signal Process. **101**, 253–276 (2017). https://doi.org/10.1016/b978-0-12-811453-7.00020-2
26. Brown, B., Laurier, E.: The trouble with autopilots: assisted and autonomous driving on the social road. In: Conference on Humam Factors Computer System- Proceedings, May 2017, pp. 416–429 (2017). https://doi.org/10.1145/3025453.3025462
27. United Nations: General Assembly Resolution A/RES/70/1. Transform our world 2030 Agenda Sustain Dev (2015). https://doi.org/10.1057/978-1-137-45443-0_24

Climate Governance and Sustainable Development: A Sight to the Intermediate and Border Cities of Ecuador

Vanessa Karina Duque-Rengel[1], Marco Heredia-R[2](✉), Ketty Daniela Calva-Cabrera[3], Bolier Torres[4,5], and Theofilos Toulkeridis[6,7]

[1] Universidad Técnica Particular de Loja, Loja 110150, Ecuador
vkduque@utpl.edu.ec
[2] AgSystems, Ceigram, itdUPM, Centro de Innovación en Tecnología Para el Desarrollo, Universidad Politécnica de Madrid (UPM), 28040 Madrid, Spain
[3] Universidad Nacional de Loja, Loja 110150, Ecuador
[4] Universidad Estatal Amazónica (UEA), Pastaza 160101, Ecuador
[5] Universidad Regional Amazónica IKIAM, Km. 7 Vía Muyuna, Tena, Ecuador
[6] Universidad de las Fuerzas Armadas ESPE, Sangolquí, Ecuador
[7] Universidad de Especialidades Turísticas, Quito, Ecuador

Abstract. Although urban areas in Latin America are not major GHG emitters, they play crucial, little-studied roles in the field of climate change. In Ecuador, there are 21 intermediate cities, which include at least 7 border cities. Intermediate cities are key to achieving sustainable urban development, merged with rural networks and environmental systems. The present work aims to conduct a sociodemographic characterization of intermediate and border cities of Ecuador and to develop a mapping of climate change actors. Three intermediate and border cities (Esmeraldas, Loja and Nueva Loja), belonging to the country's regions: Coast, Highlands and Amazonian Lowland, respectively, were analyzed. The methodology was proposed from a qualitative-quantitative approach, with two stages: 1) Identification of the sociodemographic structure and distribution through the survey of secondary sources; and 2) Development of a mapping of actors based on a given methodology, which groups four steps: a) Clarify the objectives of the mapping, b) Define the variables and design the scheme, c) Collect the information, and d) Analyze the mapping of actors. The results contrast the social characteristics and demographic indicators between these cities and the climate governance instruments. Based on these data, a dynamic graphic mapping was developed that identifies the active strategic actors of climate change at the national, provincial and cantonal levels, determining the National Government as the leading actor of climate governance, and the absence of key actors such as media, journalists and private companies.

Keywords: Climate change · Demography · Strategic actors · Intermediate cities · Sustainable development

J. R. da Costa Sanches Galvão et al. (Eds.): *Proceedings of the 1st International Conference on Water Energy Food and Sustainability (ICoWEFS 2021)*, pp. 646–655, 2021.
https://doi.org/10.1007/978-3-030-75315-3_69

1 Introduction

1.1 Climate Change and Intermediate Cities

Cities are considered a complex ecosystem of connected elements or parts, where human activities are linked by communications that interact as mentioned system evolves and is transformed [1]. In this sense, cities are living organisms, composed of a series of elements that interact with each other and that seek to generate a balance in these urban environments, and that also have to attend to at least three basic components, being ecological, economic and social [2].

Nowadays, 54% of the world's population lives in urban areas, a proportion that is expected to increase to 66% by 2050 [3]. In addition, climate change (CC) generates pressures, uncertainties and different perceptions that will pose challenges to society, the economy and the environment. In this case, by focusing on impacts on urban life [4], this can affect basic urban services, such as water or energy supply, making the city's Córdova ability to function continuously crucial for most of the world population. As stated by the United Nations, the management of urban areas has become one of the most important development challenges of the 21st century [3]. Cities face an ever-increasing range of adversities and challenges in the 21st century, and increasing urban resilience is the only way to survive and adapt to the shocks and stresses that lie ahead [5]. CC is one of the main problems that urban ecosystems have to overcome, there are different perceptions at the Ecuadorian level about CC and its implications [6–8]. CC scenarios at the national (Ecuador) and sub-national levels are changing [9, 10], but intermediate cities can be decisive for adaptation and mitigation plans and strategies for the effects of the CC, or the protection of biodiversity in their respective areas of influence [11].

Intermediate cities are those with a population of between 50,000 and one million inhabitants that generally play a fundamental role in connecting important rural and urban areas to basic facilities and services [11]. due to their complexity of analysis intermediate cities are gave a more holistic definition [12, 13]: a) The intermediate city is a complex system that works in networks, b) due to its scale, it has a greater capacity to implement strategies with high added value that allow it to position itself in regional, national and even international settings, using fewer resources than large cities, c) they constitute in themselves a breaking element of the status quo derived from the impact of globalization, since it contributes to questioning the hierarchies of the system urban, opening new horizons of territorial cooperation [14].

1.2 Climate Governance and Strategic Actors

Cities as the main GHG emitters require a governance process focused on planning and decision-making that promote actions in order to address CC. In consequence, climate governance refers to "collaboration and coordination between government and social actors (social, private, academic) to define and implement mechanisms and measures aimed at prevention, mitigation and adaptation to the risks posed by climate change" [15]. Therefore, governance has as its central axis the collaborative and cooperative work of those who constitute key actors in CC management, for decision-making in these urban ecosystems [2]. For cities to reach their CC governance potential, there are several actions, like to integrate mitigation and adaptation, to coordinate risk reduction and climate adaptation; cogenerate risk information, to focus on disadvantaged populations, and, to improve the governance of knowledge and communication networks [16].

It is at this point, where the social actors of the CC acquire greater prominence, since, to aim at the construction of sustainable cities and promote actions that promote this end, their joint work will be required [17]. Currently, in Mesoamerica, the CC adaptation processes have been promoted mainly from the public sector, with the support of regional and international organizations and technical cooperation agencies [18]. However, on the role of other actors CC, such as private companies, the media, schools or universities, social groups, among others, little or nothing is known. Furthermore, if one delves into the level of interference of these strategic actors in climate governance processes, such as the promotion of public policies or the development of instruments to guide the actions and decision-making of local and national governments. It is in this context that this study aimed to identify the actors of the climate change frame of reference in intermediate cities of Ecuador.

2 Methodology

The descriptive study made it possible to obtain an overview of the object of study and presents a mixed approach, since it uses quantitative and qualitative research techniques, which allowed to collect data, analyze results and establish a global perspective of the sociodemographic characterization and strategic actors of the CC in intermediate cities of Ecuador.

Sample Selection and Geographic Location. The study areas were three intermediate and border cities of the country, being a) Esmeraldas (C1), b), Loja (C2) and c) Nueva Loja (C3), distributed in the geographic regions of the Coast, Highland and Amazonian Lowland, respectively. They are also part of the network of intermediate cities in Ecuador [19]. The research development was established in two moments.

Stage 1: Sociodemographic Characterization of the Cities. This moment of the research aimed to identify the population structure and sociodemographic indicators in the intermediate and border cities of Ecuador. Therefore, the following method has been established:

For the sociodemographic characterization, a demographic projection was performed for the year 2020. We used the PRODEX program (version 5.6) developed by CEPAL [20], as well as the inputs were data from the national censuses 2001, 2010, 2015 (local) and information demographic of the Development Plans and Territorial Organization (PDyOT) of each intermediate city. In addition the population structure by intermediate city and its distributions by sex and age were studied and the corresponding indices were calculated, being 1) Proportions of young population (<14 years), 2) adult population (15 to 64 years), 3) percentage of women of childbearing age, 4) child/woman ratio, defined as: the number of children under 5 years of age for each woman of childbearing age reproductive rate, 5) masculinity rate, which consists of the ratio of men to every 100 women in a given population (it is considered the first indicator to analyze the distribution by sex in the population), 6) the juvenile dependency index, which is the relationship between the potentially dependent population (<15 years) and the potentially active population (between 15–64 years) [21, 22].

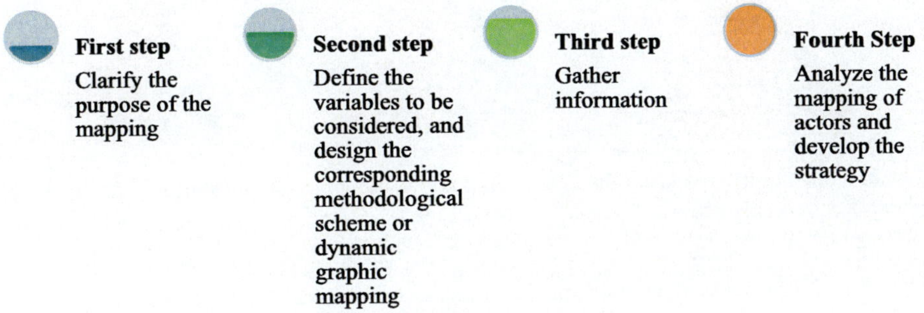

Fig. 1. Steps for mapping actors

Stage 2: Mapping of Actors. This moment of the study was aimed at identifying the climate governance instruments towards sustainable development and developing a mapping of climate change actors. This mapping was based on the methodology proposed by Ortiz et al. [23] that allows a process based on four (Fig. 1).

3 Results and Discussion

3.1 Population Structure and Sociodemographic Indicators

Population pyramids were performed for intermediate cities, demonstrating a structural difference between men and women. In Esmeraldas the population is of about 189,737 inhabitants (92,076 men and 97,661 women), in Loja the population is of about 216,504 inhabitants (903.59 men and 126,145 women) and for Nueva Loja the population is of about 131,605 inhabitants (77,872 men and 53,733 women). The pyramidal profile of the study indicates that cities are of a progressive type (Fig. 2), similar to the population studied by Heredia-R et al. [24], which are typical of underdeveloped countries, having high birth rates, and a very low life expectancy due to high mortality. They are very young populations, so they present a high growth, while their dynamics is a large contingent of young population at the base, which disappears rapidly as age groups advance, at the top very few remain [25].

In reference to the demographic indicators (Table 1) in the city of Loja there is less young population but it has the largest adult population compared to Esmeraldas and Nueva Loja. In the three intermediate cities the fertility age of women exceeds 50%, while the average child-woman ratio in the studied cities is of around 0.3. In Loja there is the lowest percentage of masculinity, with a difference of 7 and 11 points with Esmeralda and Nueva Loja, respectively. Likewise, the youth dependency ratio is lower than the intermediate cities studied.

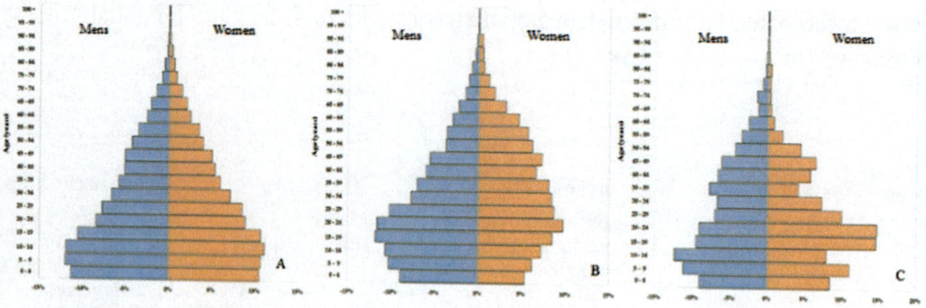

Fig. 2. Population pyramids of the intermediate and border cities of Ecuador; People per 100 of both sexes in A) Esmeraldas. B) Loja, C) Nueva Loja.

Table 1. Demographic indices by intermediate and border city of Ecuador.

Index	Intermediate and border cities		
	Esmeraldas	Loja	Nueva Loja
Population ages, 0–14 (%)	33.72	22,61	30,54
Adult population, 15–64 years (%)	60.79	70,27	66,28
Women of childbearing age (%)	52	57	63,2
Child – woman ratio	0,4	0,2	0,3
Male rate (%)	49	42	59
Youth dependency ratio (%)	55	32	46

3.2 Mapping of Actors

This mapping was established in order to contextualize the governance of the CC in the intermediate and border cities of Esmeraldas, Loja and Nueva Loja and their respective macroenvironments (province) and supraenvironments (country). For this, they were developed in four steps, based on Ortiz et al. [23].

First Step: Object of mapping of CC governance actors.

- Identify the instruments and projects that affect the governance of the CC of the studied cities, developed in the last 10 years.
- Determine the organizations (actors) that lead, sponsor and participate in the projects and instruments of governance and management of climate change.
- Analyze the role of the strategic actors in the governance and management of the CC.

Second Step: Definition of variables and design of dynamic graphic mapping.

The following variables were taken as reference:

a) Jurisdictional: national, provincial and cantonal.
b) Power level: high, medium and low.

Third Step: Gather information.

Based on the objectives and variables proposed in the first and second steps, they are collected in the climate change framework; 19 instruments at the national level, 6 at the provincial level; and 8 at the city level, as listed in Table 2.

Fourth Step: Actor mapping analysis.

Figure 3 establishes the dynamic graphic mapping, which, based on the previous step, allows identifying the strategic actors who have led and participated in the construction of the CC's referential framework in Ecuador, with emphasis on the three studied intermediate and border cities. The actors identified in are grouped into three profiles, being 1) Government or state body; 2) International organizations or NGOs; and, 3) Academy[1].

Government or State Body: This is demonstrated as the actor with the greatest power, due to his decision-making capacity. Those who comprise it are the State and its different portfolios, as well as the local and provincial administrations. The function entrusted to this group is based mainly on the creation of public policies, management bodies, design and development of macro strategies and plans, that is, at the highest level, which must guide the actions of the entire country, province and country. canton respectively. This actor plays a multilevel role in climate governance. As evidenced in Table 2, this actor has led 28 of 33 documents drawn up in the study, that is, he has participated in 84.8% of instruments that make up the reference framework for climate change in Ecuador. **International organizations or NGOs:** This actor is present in the sphere of national analysis, however, at the provincial level his participation is only evident in Loja, and at the city level, it is present in Loja and Esmeraldas. Although, it exercises a multilevel role in climate governance, its level of power is limited to being an actor with influence, but without decision-making capacity, since it mainly exercises roles as sponsor, project leader, proposal designer and plans, and management support. The percentage of participation of in the reference framework of the CC of the country is 42.4%. **Academy:** This strategic actor is present in the three spheres of action, being national, provincial and cantonal or city. It focuses on his participation in dialogue tables, but mainly in the development of research, which generates diagnoses for the

[1] MAE, Ministry of Environment and Water; IDB, Inter-American Development Bank; CAN, Department of Countries of the Andean Group; UNCCD, National Focal Point of the United Nations Convention para la Lucha contra la Desertificación; ONU REDD, Programa Nacional Conjunto ONU REDD Ecuador; FOCAM Ecuador, Fomento de Necesidades para la Mitigación de Cambio Climático en el Ecuador; PNUD, Programa de Naciones Unidas para el Desarrollo; SENPLADES, Secretaría Nacional de Planifi-cación y Desarrollo; GEF, Fondo para el Medio Ambiente Mundial; FAO, Organización de las Naciones Unidas para la Alimentación y la Agricultura; CIIFEN, Centro Internacional para la Investigación del Fenómeno del Niño; USAID, Agencia de los Estados Unidos para el Desarrollo Internacional; USFS, Servicio Forestal de los Estados Unidos; UNL, Universidad Nacional Loja; ONU-HABITAT, Programa de las Naciones Unidas para los Asentamientos Humanos; CAF, Banco de Desarrollo de América Latina; AFD, Agencia Francesa de Desarrollo; LAIF, Facilidad de Inversión en América Latina; UTPL, Univer-sidad Técnica Particular de Loja; CeALCI, Fundación Carolina; USFQ: Universidad San Fran-cisco de Quito; PNUD, Programa de las Naciones Unidas para el Desarrollo; y, MDMQ, Mu-nicipio del Distrito Metropolitano de Quito.

Table 2. Instruments for climate governance

	Document	Author
National	Constitution of the Republic of Ecuador (2008)	National Constituent Assembly
	Organic Code of Planning and Public Finance	SENPLADES
	Second National Communication of Ecuador and the United Nations Framework Convention on Climate Change (2011)	MAE
	Impact of climate change on subsistence agriculture in Ecuador (2012)	CeALCI / USFQ
	Organic Code of Territorial Organization, Autonomy and Decentralization – COOTAD (2012)	National Constituent Assembly
	Ecuador's National Climate Change Strategy (2012)	MAE
	Analysis of the vulnerability of emblematic hydroelectric plants to the effects of climate change in seven hydrographic sub-basins of Ecuador (2012)	MAE
	Ecuador: Mitigation and Adaptation of Climate Change (2013)	BID, Climate Change and Sustainability Division, CAN
	Integrated Management to Combat Desertification, Land Degradation and Adaptation to Climate Change (2014)	MAE / UNCCD
	Agreement No. 248	MAE
	How to incorporate climate change into local planning? (2014)	MAE
	National Climate Change Plan (2015)	MAE / FOCAM / PNUD
	First Biennial Update Report of Ecuador to the United Nations Framework Convention on Climate Change (2016)	MAE / GEF / PNUD
	Organic Law of Territorial Ordering, Use and Land Management (2016)	Superintendency of Territorial Planning, Land Use and Management
	Measures and Actions REDD + Ecuador: Application of the landscape approach and stock flow (2015)	ONU / REDD / MAE
	Climate Change Mitigation Capacity Building Project in Ecuador (FOCAM) (2016)	PNUD / MAE
	National Plan for Good Living 2017 – 2021	SENPLADES
	Organic Code of the Environment (2017)	MAE
	Third National Communication of Ecuador to the United Nations Framework Convention on Climate Change (2017)	MAE /PNUD/ONU/REDD/FAO/GEF
	Tool for the integration of Climate Change criteria in Development Plans and Land Use Planning (2019)	MAE, PNUD

	Document	Author	Esmeraldas	Loja	Nueva Loja
Provincial	Development Plan and Territorial Organization Province of Esmeraldas 2015-2025 (2015)	Prefecture-Esmeraldas	X		
	Manual of the environmental management process of the Decentralized Autonomous Government of the Province of Esmeraldas (2018)	Prefecture-Esmeraldas	X		
	Climate change plan for the provincial decentralized autonomous government of Loja (2015)	CIIFEN, GAD-Loja		X	
	Study of "Vulnerability to climate change in the Southern Region of Ecuador: Potential impacts on ecosystems, biomass production and water production" (2015)	UNL, USAID, USFS		X	
	Land Use Plan for the province of Loja 2015 – 2025 (2015)	Prefecture-Loja		X	
	Sucumbíos land use plan	Prefecture-Sucumbíos			X
Local	Climate Pact	MDMQ			X
	Development plan and territorial ordering of the canton Lago Agrio	GAD- Lago Agrio			X
	Climate Change Plan (2015)	GAD-Esmeraldas		X	
	Climate Change adaptation and mitigation strategy for the Canton Esmeraldas (2011)	ONU-HABITAT	X		
	Development Plan and Territorial Organization of the Canton Esmeraldas 2014-2019 (2014)	GAD-Esmeraldas	X		
	Land use plan for the Municipality of Loja 2014 – 2022 (2014)	GAD-Loja		X	
	Loja canton action plan to reduce carbon footprint and water footprint (2016)	CAF, AFD, GAD-Loja		X	
	Loja Climate Change Vulnerability Index 2019	UTPL, CAF, AFD		X	

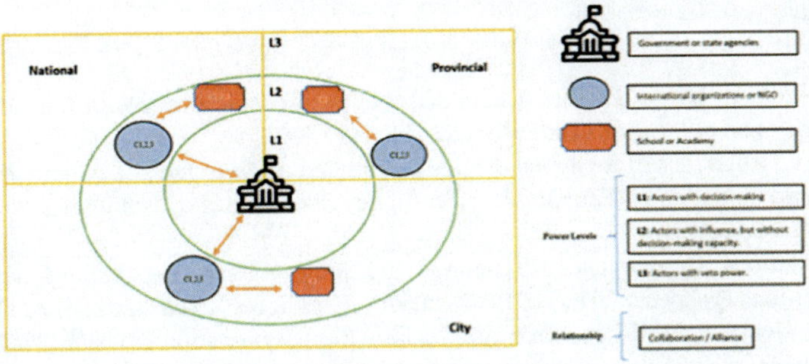

Fig. 3. Graphical actor mapping of intermediate and border cities in Ecuador

decision-making of governments in favor of mitigation, adaptation and resilience to climate change. Table 2 lists their participation at the national level in one of the 19 management instruments raised, as well as at the provincial and local levels. Thus, the academy becomes an actor of a technical character of medium-level power, that is, with influence, but without decision-making capacity.

4 Conclusions

The profiles of the population pyramids and indicators of the intermediate border cities studied represent the dynamics of a developing society, for which the potential existence of human capital for the development of capacities in terms of climate governance could be inferred.

This study makes visible the participation of key actors in the country's climate governance, showing that: the strategic actor that has led the CC's frame of reference in intermediate and border cities of Ecuador is the actor "government or state organisms" with the highest level of power, having the capacity to make decisions and participate in the 84.8%. In second place, there is the actor "international organizations or NGOs" that has a medium level of power, since it has influence without decision-making capacity, however its level of presence is only 42.4%. Thirdly, the actor "Academy" is observed, with a medium power level and with a minimum participation of 9%. Additionally, it is observed that the relationship of these three actors is based on cooperation and/or alliances. Additionally, one of the most important findings of this research is the absence of other actors, within the framework of the country's CC, with great power of influence and decision-making capacity, such as the media, private companies, social groups, citizenship, among others.

Finally, the results presented show that climate governance in intermediate cities of Ecuador needs to be strengthened with the active participation of all actors in discussions, negotiations and decision-making in order to promote effective actions in the face of climate change challenges in urban ecosystems.

References

1. Von Döhren, P., Haase, D.: Ecosystem disservices research: a review of the state of the art with a focus on cities. Ecol. Indicators **52**, 490–497 (2015)
2. Duque, V., Rojas, Y.: El rol de los medios de comunicación frente a la gobernanza y gestión del cambio climático en Ecuador. Revista Ibérica de Sistemas e Tecnologias de Informação **E26**, 150–163 (2020)
3. United Nations, Department of Economic and Social Affairs, Population Division. World Urbanization Prospects: The 2014 Revision, Highlights (ST/ESA/SER.A/352); United Nations, Department of Economic and Social Affairs, Population Division: New York, NY, USA (2014)
4. ARUP International Development and the Rockefeller Foundation: City Resilience and the City Resilience Framework. London, UK, Arup (2015)
5. Rockefeller Foundation, 100 Resilient Cities. https://www.100resilientcities.org/. Accessed 26 Jan 2021
6. León Alvear, V., Torres, B., Luna, M., Torres, A., Ramírez, P., Andrade-Yucailla, V., Muñoz-Rengifo, J.C., Heredia-R, M.: Percepción sobre cambio climático en cuatro comunidades orientadas a la ganadería bovina en la zona central de los Andes Ecuatorianos. Livestock Research for Rural Development, vol. 32, Article #165 (2020)
7. Toulkeridis, T., Tamayo, E., Simón, D., Merizalde, M.J., Reyes, D.F., Viera, M., Heredia, M.: Cambio Climático según los académicos ecuatorianos - Percepciones versus hechos. La Granja: Revista de Ciencias de la Vida, vol. 31, no. 1, pp. 21–46 (2020). https://doi.org/10.17163/lgr.n31.2020.02
8. Heredia-R, M., Barreto, D., Toulkeridis, T.: Percepción de las poblaciones indígenas kichwa sobre el cambio climático y sus implicaciones en Puyo, Ecuador, pp. 209–230. En: Carrión A y Ariza-Montobbio P.(Eds) 2020. La acción climática en las ciudades latinoamericanas: aproximaciones y propuestas . FLACSO. Quito, Ecuador, 254 pp (2020)
9. Manchego, C.E., Hildebrandt, P., Cueva, J., Espinosa, C.I., Stimm, B., Günter, S.: Climate change versus deforestation: Implications for tree species distribution in the dry forests of southern Ecuador. PloS One **12**(12), e0190092 (2017)
10. Carvajal, P.E., Anandarajah, G., Mulugetta, Y., Dessens, O.: Assessing uncertainty of climate change impacts on long-term hydropower generation using the CMIP5 ensemble—the case of Ecuador. Clim. Change **144**(4), 611–624 (2017)
11. UCLG Co-Creando el Futuro Urbano. La Agenda de las Metrópolis, las Ciudades Intermedias y los Territorios. GOLD IV (GOLD - Cuarto Informe Mundial sobre la Descentralización y la Democracia Local (2017)
12. Bellet, C., Llop, J.: Ciudades intermedias: entre territorios concretos y ciudades y espacios globales, en Ciudad y Territorio. Estudios Territoriales, vol. 36, núms. 141–142, pp. 569–582 (2004)
13. Bellet, C., Llop, J.: Miradas a otros espacios urbanos: las ciudades intermedias, en Geo-Crítica Scripta Nova. Revista Electrónica de Geografía y Ciencias Sociales, vol. 8, núm. 165 (2004)
14. Marull, J., Font, C., Boix, R.: Modelling urban networks at mega-regional scale: are increasingly complex urban systems sustainable? Land Use Pol. **43**, 15–27 (2015)
15. Bojórquez, G.C., Aguilar, M.D.L.R. Gobernanza climática: actores sociales en la mitigación y adaptación en el estado de Coahuila, México. Nóesis. Revista de Ciencias Sociales y Humanidades **24**(47–2), 129–146 (2015)
16. Rosenzweig, C., Solecki, W.: Action pathways for transforming cities. Nature Clim. Change **8**, 756–759 (2018). https://doi.org/10.1038/s41558-018-0267-x

17. Duque, V., Rojas, Y.: Análisis del rol de los medios de comunicación como actor clave de la gobernanza y gestión del cambio climático en la ciudad de Loja en el período 2015–2019. Tesis de posgrado, Facultad Latinoamericana de Ciencias Sociales, Ecuador (2020). https://bit.ly/36ieBjn
18. Neff, H., Pearsall, D.M., Jones, J.G., de Pieters, B.A., Freidel, D.E.: Climate change and population history in the Pacific lowlands of southern Mesoamerica. Quat. Res. **65**(3), 390–400 (2006)
19. Llop, J., Vivanco Cruz, L.: El derecho a la ciudad en el contexto de la Agenda Urbana para Ciudades Intermedias en el Ecuador, Cuenca, Universidad de Cuenca (2017)
20. Cavenaghi, S.: Estimaciones y proyecciones de población en América Latina: desafíos de una agenda pendiente. Serie e-Investigaciones N **2**, 108–112 (2012)
21. Rowland, D.: Book Review: Demographic Methods and Co8ncepts, Oxford: Oxford University Press, 2003, xiv + 546 pp. Eur. J. Popul, vol. 20, pp. 291–292 (2004)
22. Wilson, T.: Visualising the demographic factors which shape population age structure. Demogr. Res. **35**, 867–890 (2016)
23. Ortiz, M.D., Matamoro, V., Psathakis, J.: Guía para confeccionar un mapeo de actores. Bases conceptuales y metodológicas, Documento de trabajo (2016)
24. Heredia-R, M., Torres, B., Cayambe, J., Ramos, N., Luna, M., Diaz-Ambrona, C.G.: Sustainability assessment of Smallholder agroforestry indigenous farming in the Amazon: a case study of Ecuadorian Kichwas. Agronomy **10**(12), 1973 (2020)
25. Caniglia, J., Leapard, B.: Analysing the world population: using population pyramids and if the world were a village. Teach. Stat. **32**(2), 41–44 (2010)

Climate Change and Culicidae Vectors - a Systematic Review of the Evolution on Southern European Countries

Susana Paixao[1,2(✉)], Ana Ferreira[1], Paulo Figueiredo[3], and Inês Ferreira[1]

[1] Portugal Coimbra Health School Department of Environmental Health, Polytechnic Institute of Coimbra, Coimbra, Portugal
supaixao@estescoimbra.pt
[2] CEGOT, Coimbra, Portugal
[3] Portugal Coimbra Health School Department of Complementary Sciences, Polytechnic Institute of Coimbra, Coimbra, Portugal

Abstract. Culicidae vectors are insects that have the ability to transmit several diseases such as malaria, zika or dengue. The development of these vectors is mainly influenced by temperature, relative humidity and precipitation of their environment. Therefore, climate change can increase the geographical distribution of diseases transferred by these vectors. The predominant goals of this review is to identify the primary Culicidae vectors in Southern European countries, understand how climate change can influence the geographical distribution of these vectors as well as their impact on the health of the population of the affected areas. The bibliographical research started on the 1st of October 2019 and ended on the 20th of April 2020 and was carried out resorting to data bases suchlike Google Schoolar, Web of Science, BioMed Central and European Centre for Disease Prevention and Control. Whilst writing this review 23 original scientific articles, 1 master's degree dissertation and 10 scientific reports were consulted, mainly in English and in Portuguese. After having analysed the 32 selected articles, we concluded the primary environmental alterations will most likely be the increase of the global temperature, decrease of precipitation, and decrease of the relative humidity's percentage on Southern European countries. We gathered that this vectors will continue to have ideal conditions to thrive in the countries mentioned above and that no country in Southern Europe will have an adequate climate to the development of vectors pertaining to the genus Aedes which can lead to the disappearance of these species of vectors.

Keywords: Culicidae vectors · Climate change · Southern Europe countries · Species · Genus · Malaria · Zika West Nile Virus

J. R. da Costa Sanches Galvão et al. (Eds.): *Proceedings of the 1st International Conference on Water Energy Food and Sustainability (ICoWEFS 2021)*, pp. 656–669, 2021.
https://doi.org/10.1007/978-3-030-75315-3_70

1 Introduction

– Framing of the theme

Culicidae vectors are insects that belong to the family *Culicidae*, one of the most primitive families of the order *Diptera* that is constituted by over 3500 species and subspecies from all over the world. These type of vectors develop primarily in fresh and clean water nurseries exposed to the sun, the water can be slightly brackish much like, for example, water found in coastal swamps, irrigation channels and paddy fields or water from domestic nurseries, such as potted plants. These insects are found all over the world except in Antarctica (Silva 2017; CEVDI 2016).

Despite having significant functions in the ecosystem, Culicidae vectors can be vectors for pathological microorganisms that can endanger wildlife conservation and have harmful effects on human health, in consequence, these vectors are a highlighted threat to public health. Culicidae vectors can transmit arboviruses among vertebrate hosts such as birds, rodents, domestic animals and humans, causing high mortality diseases like malaria, dengue or yellow fever (Alexandra et al. 2014; Mixão et al. 2016; Nasir et al. 2017; Roiz et al. 2014).

The diseases mentioned above are some of the ailments that deeply affect human health. Consequently, it comes as no surprise that about 80% of the worldwide population is at risk of being exposed to the diseases passed on by these vectors. Due to favourable climate conditions, tropical countries have higher transmission rates regarding the maladies transmitted by Culicidae vectors, although, there have always been autochthonous epidemics in Europe (Alexandra et al. 2014; Franklinos et al. 2018).

Some diseases passed on through these vectors are surfacing or reappearing while others are expanding. The most recent data unveil that, in 2018, in Europe, there were reported several imported and native (not imported) cases of the ailments caused by Culicidae vectors. In regards to the native cases: Portugal registered 102 patients with malaria; Spain diagnosed 6 people with dengue, 851 of malaria and 1 case of west nile virus infection; France announced 8 cases of dengue, 2840 cases of malaria, 29 patients infected by the west nile virus and 7 cases of yellow fever; Italy registered 722 cases of malaria and 618 cases of west nile virus infection; Malta flagged 7 cases of malaria; Croatia highlighted 2 cases of malaria and 63 cases of west nile virus infection; Slovenia reported 3 cases of malaria and 5 cases of west nile virus infection and Greece announced 55 cases of malaria and 316 cases of west nile virus infection (Alexandra et al. 2014; ECDC 2018).

The appearance of these diseases is frequently associated with alterations in the ecosystem, modifications of the human behaviour and most importantly climate change. In the Mediterranean, the climate is characterized by the occurrence of hot and dry summers that can perpetually lead to higher temperatures, further evapotranspiration and a decrease in the quantity of soil water (Alexandra et al. 2014; Gravina et al. 2019). The likelihood of survival of the invading species increases if there are prospering ecological conditions that enable the development of their population. According to some studies, the heightening of temperature, humidity and precipitation are excellent for the evolution of a specific mosquito's habitats which are fit for neither too low nor too high temperatures (Alexandra et al. 2014; Williams et al. 2014).

In the last 100 years, climate change has been extremely noteworthy in the continents belonging to the Northern hemisphere, specifically Europe and Asia. These weather alterations focus primarily on the decrease in precipitation and the elevation by 1 °C of the global temperature which led to an increase of the number of heatwaves and droughts as well as the emergence of new infectious diseases. In the future, these alterations will become even more severe (Hoerling et al. 2012; Watts et al. 2019).

Climate change and globalization are both crucial factors that can aid in the geographical distribution of some diseases caused by the vectors. Considering that climate change can influence several aspects of the vector-disease-host relationship, it has been put forward that there might be a bond between the climate and the infection. (Liu-Helmersson et al. 2016; Roiz et al. 2014). On Southern European countries, there is a high likelihood that there will be weather alterations mostly related to the temperature and precipitation. Primarily in the summer, the temperature will most probably increase and this boost will be accompanied by a decrease in precipitation. In consequence, these variations can impact the spread and distribution of the Culicidae vectors (Varga et al. 2019).

To sum up, climate alterations will induce changes not only in the development of Culicidae vectors and their habitat, but will also influence human's vulnerability to the diseases passed on through these vectors, due to the rise of sea levels, extreme temperatures and more difficult access to potable water (Chen et al. 2013; Kolivras 2010).

– Goals

  Clime change has been a well-discussed theme in our current society, seeing as it is incredibly important to understand how this phenomenon is affecting the development of Earth's living beings but specifically of Culicidae vectors. This is because not only these vectors can cause severe ailments but also because the ideal conditions for their development, such as temperature, relative humidity and precipitation, can differ within the various species of vectors.

  This systematic review is based on the analysis of documents that focus on identifying the main Culicidae vectors in Southern European countries, on understanding in what way climate change can influence the geographical distribution of these vectors and their impact on the health of populations. This review also has, as specific goals, the determination of the ideal conditions, regarding temperature, relative humidity and precipitation, necessary for the vectors' expansion as well as the diseases they cause.

2 Material and Methods

– **Elegibility criteria**

  The revision method PRISMA was adopted for this systematic review (Moher et al. 2009).

  To fulfil the criteria of a systematic review, the research performed centred mainly on observational cohort transversal studies, longitudinal, retrospective and prospective studies.

  The selected articles, written between 2010 and 2020, were chosen to complete the goals set out for this review. To collect data 23 original scientific articles, 1 master's

degree dissertation and 10 scientific reports were consulted, mainly in English and in Portuguese due to the availability of information.
Original scientific articles and systematic reviews were the most sought out type of documents whilst researching. The scientific magazines taken into account were all thoroughly identified and classified according to their denomination, impact factor and index H classification.
To write this review primary sources of information (original scientific articles) were analyzed as well as secondary sources of information (systematic reviews). The latter aimed at gathered works, on the subject, that have already been published as articles.

- **Sources of information**
  Platforms such as Scholar Google, ScienceDirect, Web of Science, Elsevier, BioMed Central (BMC), European Centre for Disease Prevention and Control (ECDC), Vector Vigilance Network (REVIVE program), The Intergovernmental Panel on Climate Change (IPCC), Portugal contemporary base data (PORDATA) and World Meteorological Organization (WMO) were resorted to in order to analyse the occurrence of Culicidae vectors in Europe.
  The bibliographical research, regarding the keywords of this investigation, started on the 1st of October 2019 and ended on the 20th of April 2020.

- **Information Selection criteria**
  Scholar Google and Web of Science are databases that withhold summaries and quotes of multidisciplinary articles destined to be published on several academical magazines. The platforms ScienceDirect and Elsevier contain literature revisions, concerning technology and health care, available for consultation. BMC harbours a portfolio that compilates over 300 revision articles on themes related to technology, science, engineering and medicine. ECDC is an agency that belongs to the European Union that strives to protect Europe against infectious diseases whose functions are epidemic surveillance, scientific counselling and the development of the scientific magazine Eurosurveillance. The platforms IPCC, PORDATA and WMO lodge several articles on climate change, temperature variation, humidity and precipitation.
  Throughout the online research, several keywords and expressions, whether in English or in Portuguese, were employed such as "climate change", "vectors", "Culicidae Vectors", "temperature variation", "precipitation variation" and "humidity variation". The definition of Culicidae vectors and climate change put forward by Medical Subject Headings (MeSH) were taken as an inclusion selection criteria during the online research.
  Culicidae vectors are insects that belong to the order Diptera. The larvae of this species develop in aquatic environments and their adults are mainly characterized by the morphology of their wings. These vectors are of extreme medical importance.
  Climate change refers to long term significant changes in the climate and can be caused because of natural or anthropogenic factors. This review will only focus on temperature, humidity and precipitation alterations, this is because these are the elements that influence the development of the vectors mentioned above.
  Regarding the climatic factors, the temperature is defined as the ability to determine the heat flow direction when an object is exposed to direct thermal contact. The amount of

water vapour in the air is also known as humidity and precipitation is the development of water particles that later fall from the atmosphere creating rain as we know it.
The study of other species of vectors (except Culicidae vectors) and the development of vectors in other European countries were defined as an exclusion criteria. Therefore, these terms were not included in the bibliographical research. In addition to the exclusion criteria mentioned above, articles that analysed other factors as either cause or consequence of climate change, such as the rise of seawater levels or the increase of CO_2 concentration in the atmosphere, were also disregarded.
The bibliographical references were retrieved from the program Mendeley Desktop (1.19.4 version). Two databases were created, one on the theme "Climate Change" and other on "Culicidae Vectors", both documents are on a pdf format.
Result wise, some of the analysed studies conveyed contradictory information regarding the climate factors that influence the development of vectors. As far as statistical tests, most of the articles assessed do not reveal the type of tests used (parametric or non-parametric tests) for data treatment. However, it is mentioned that ordinal Spearman's correlation tests were used to assay the association between the abundance of adult Aedes aegypti mosquitos, temperature and humidity. Chi-square tests of independence were employed on articles whose goals were to assort the variables that affect this species' development. To analyze the correlation between the average annual temperature and the average temperature of both warmest and coolest seasons on Southern European countries Pearson's correlation coefficient test was applied. Finally, One-way ANOVA tests were utilized on studies that analyzed the effects of high temperatures on the development of the species Anopheles arabiensis. The articles that assess temperature, relative humidity and precipitation all employed impact statistics and calculated minimum and maximum values, median, mean and standard deviations.

3 Results

- **Articles' Characteristics**
  On the primary stage of the investigation on the topic of interest, "Climate change and Culicidae vectors- A systematic review of the evolution on Southern European countries" we realized that there were several species of vectors, spread out through various countries, that caused different diseases and whose ideal development conditions were distinguishable and could be affected by climate change. Consequently, the review was subdivided in "Culicidae vectors on Southern European countries" and "climate change".

Culicidae Vectors

1. ***Culex* Genus**

According to CEVDI (2017), the genus *Culex* was described by Linnaeus in 1758 and encompasses over 300 species, the most abundant in Southern European countries is *Culex pipiens* that has been found in Portugal, Spain, Italy, Greece, France, Croatia and

Slovenia, the species *Culex theileri* that was detected in Portugal, Italy and France and the species *Culex univitattus* detected in countries like Portugal and Spain. (Capelli et al. 2018; CEVDI 2017).

As attested by Mixão et al. (2016), the species of vectors belonging to the genus culex have a high ability of ecological adaptation. According to CEVDI (2018), the nurseries of these species were gatherings of temporary or permanent water that were extremely polluted and rich in organic matter or, in case of *Culex univitattus* and *Culex theileri,* clear and lacking in organic matter. The three species mentioned above are abundant during summer and Autumn, in the spring adults begin their activity (CEVDI 2018; Mixão et al. 2016).

Even though the growth of these species might be related to factors like precipitation and relative humidity, which was stated by Semenza et al. (2016), Groen et al. (2017) defended that the temperature is the main factor that affects the survival, reproduction, abundance and distribution of the genus Culex (Groen et al. 2017; Semenza et al. 2016).

Relatively defended by the authors Mixão et al. (2016), Roiz (2014) and Bisanzio et al. (2011), the evolution of these species caused multiple diseases such as dirofilaria repens, dirofilaria immitis and west nile virus, in the case of *Culex theileri* vector. The remaining vectors mentioned above are responsible for the spreading of the west nile virus, however the vector *Culex pipiens* was the main source of transmission on Southern Europe which was mentioned by Groen et al. (2017). According to the authors Groen et al. (2017) and Paz (2015), these vectors affected not only rural areas but also on urban sites transmitting diseases to birds, humans, horses and other mammals (Bisanzio et al. 2011; Groen et al. 2017; Mixão et al. 2016; Paz 2015; Roiz et al. 2014).

2. *Anopheles* Genus

Seven of the many species belonging to the genus *Anopheles*, that according to CEVDI (2017), were firstly discovered in 1818, were found in several countries of Southern Europe like Portugal, Spain, Greece, Italy and France. The species *Anopheles macupilennis* s.I., *Anopheles atroparvus, Anopheles labranchiae* e *Anopheles sacharovi*, were the most abundant and widely distributed, which was discovered through the analysis of an article written by Piperaki and Daikos (2016) (CEVDI, 2017; Piperaki and Daikos 2016).

As stated in CEVDI (2017), the larvae of these vectors develop in fresh and clean water nurseries exposed to the sun, the water can be slightly brackish much like, for example, water found in coastal swamps, irrigation channels and paddy fields. Normally these species were associated with domestic animals and breeding stock, ergo they were found in closed shelters, such as rabbit hutches, pigsties and stables (CEVDI 2017).

Piperaki and Daikos (2016) declared that these vectors are the main transmitters of malaria, which still remains one of the biggest current infectious disease causing a higher number of fatalities than any other ailment passed on by these vectors (Piperaki and Daikos 2016).

As it was stated by Piperaki and Daikos (2016) and Hertig (2019), the transmission of these species of vectors across the Southern European countries was seasonal, due to factors such as temperature and humidity, this is because weather conditions, mainly on the summer and winter, were favourable to the evolution of these vectors (Hertig 2019; Piperaki and Daikos 2016).

3. *Aedes* Genus

In accordance with a study carried out by Petrić et al. (2018) the species belonging to the genus *Aedes* were first identified on 1762 by Linnaeus and were widely distributed throughout the countries of the European Union. The species *Aedes aegypti, Aedes albopictus* and *Aedes japonicus* were the most abundant in Portugal, Spain, Italy, Greece, Malt, France, Croatia and Slovenia. (CEVDI 2017; Petrić et al. 2018).

CEVDI (2018) defended that these species were of great medical importance as they were the main vectors for diseases such as dengue, yellow fever, zika and chikungunya and they can also pass on the west nile virus, myxomatosis, aviary plasmodium and canine filariasis. According to Liu-Helmersson et al. (2016), the species *Aedes aegypti* was the most efficient in the transmission of dengue (CEVDI 2018; Liu-Helmersson et al. 2016).

CEVDI (2017) claimed that the development of the larvae takes about 10 days. Their nurseries were usually small water reservoirs, either clean or brackish, found in urban agglomerates (potted flowers, abandoned cans and gutters). An adult mosquito is of small dimensions and has a life cycle of about a month. Females were most active at dusk and stung inside out (CEVDI 2017).

Schaffner et al. (2013) study put forward that the main developing factors of the vectors belonging to the genus *Aedes* are globalization, mainly through emigration and the transportation of goods, and climate factors. The effects that climate factors have over the reproduction and growth of vectors can differ from species to species. Nasir et al. (2017) believes that the vector *Aedes aegypti* develops easier on environments whose average annual temperature is over 20ºC and that have a high relative humidity (over 50%). Fischer et al. (2011) assures that *Aedes albopictus's* reproduction takes place in areas in which the average annual temperatures ranges from 10 ºC to 25 ºC, with an annual precipitation that varies from 500 mm to 2000 mm and high relative humidity. Lastly, Montarsi et al. (2019) attests that the species *Aedes japonicus* is found in mild climate areas, characterized by cold winters and warm summers, an average daily temperature ranging from 17 ºC to 22 ºC in the summer and –2 ºC to –3 ºC in the winter and an average precipitation under 1000 mm (Fischer et al. 2011; Montarsi et al. 2019; Nasir et al. 2017; Schaffner and Mathis 2014).

4. *Culiseta* Genus

In 1838 the gender *Culiseta* was identified by J. Macquart, multiple species within this genus were found all over the world except in Southern America. Becker and Hoffmann (2011) defended that, in Europe, the species with greater distribution were *Culiseta longiareolata* and *Culiseta morsitans* and were detected in Portugal, Spain, Greece, Italy, France and Croatia (Becker and Hoffmann 2011; CEVDI 2017).

As defended by Schaffner et al. (2013) and Becker and Hoffmann (2011), these species were the main transmission vectors for diseases such as the west nile virus and sindbis virus (Becker and Hoffmann 2011; Schaffner et al. 2013).

According to CEVDI (2018), the expansion of these species happens in stagnant water rich in organic matter, its nurseries are especially diverse like paddy fields, irrigation channels and irrigation tanks. The vectors' nurseries can be exposed to solar radiation or be in the shade, they can be made out of sweet or salty water that can be either clear or brackish. (CEVDI 2018).

Even though these vectors can be found year round in their habitats, in the spring and summer maximum density population is reached, in accordance with a study performed by CEVDI (2018). On a survey carried out by Ferreira et al. (2015) stated the these vectors were found in countries with an average annual temperature of over 20 °C, low relative humidity (under 50%) and an average annual precipitation under 500 mm. (CEVDI 2017; Ferreira et al. 2015).

5. *Ochlerotatus* Genus

As maintained by CEVDI (2018), the genus *Ochlerotatus* was first ascertained by Pallas, in 1771, that contains several species widely distributed throughout the palearctic region (a region which includes Europe, Northern Africa and Northern Asia). Because of the analysis perpetrated by Bisanzio et al. (2011), it has come to the conclusion that, in Europe, the primary species identified was *Ochlerotatus caspius*, located in Portugal, Italy and France (Bisanzio et al. 2011; CEVDI 2017).

CEVDI (2018) unveils the vector *Ochlerotatus caspius* is a halophilic mosquito abundant in wet regions such as estuaries, saltpans, swamps and paddy fields. The larvae are found in brackish water nurseries that are often near dense vegetation (Bisanzio et al. 2011; CEVDI 2017; Roiz et al. 2014).

Roiz et al. (2014) claimed that this species' reproduction usually occurs in places where the average temperatures ranges from 20 °C to 25 °C, however they are resistant to higher temperatures (33 °C–36 °C), and places in which precipitation does not rise above 516 mm, there hasn't been identified any correlation between relative humidity and the growth of these mosquitos (Roiz et al. 2014).

As stated by the authors Roiz et al. (2014), CEVDI (2018) and Bisanzio et al. (2011), this species of vectors is considered an anthropophilic mosquito that mainly passes on myxomatosis, tahyna and west nile virus (Bisanzio et al. 2011; CEVDI 2018; Roiz et al. 2014).

Climate Change

On the article written by Alexandra et al. (2014), it was noticeable that the climate alterations happening all over Europe have long term effects that can be prompted by natural causes. Nevertheless some authors believe that the climate changes that have taken place over the last century have been brought up by human activities that led to the increase of greenhouse gases, emission of pollutants and other aerosols, as well as ground changes such as urbanization an deforestation, as it was stated by Collins et al. (2019) (Alexandra et al. 2014; Collins et al. 2019).

In accordance with a study elaborated by Paz (2015), these alterations, that vary from region to region, are currently one of the biggest current global threats, with both social and economic repercussions that affect the planet and humanity. The impacts of these alterations can be seen in the rise of temperature, decrease of precipitation and of relative humidity. On a report executed by Piperaki and Daikos (2016) it became clear that these changes, specifically the elevation of temperatures, will affect the abundance and distribution of arthropods and as a consequence so will the geographical distribution of diseases passed on by these vectors (Paz 2015; Piperaki and Daikos 2016).

Some studies, namely the ones written by Amendola et al. (2019) and Lopes et al. (2013) defended that Southern Mediterranean countries will be the most affected by climate change and the variation of the three factors being analysed: temperature, relative humidity and precipitation (Amendola et al. 2019; Lopes et al. 2013).

Regarding the temperature changes, Semenza et al. (2016) came to the conclusion that the temperature will rise more intensively and for longer periods of time, which will increase the frequency and duration of heatwaves in several regions of Europe. On an investigation carried out by Paz (2015), the author defended that the elevation of temperatures felt in multiple European countries registered a higher augmentation than the global average, especially on Northern Europe during the winter, and higher than both the European average in mountainous areas and during the summer in the Mediterranean, where dry periods tend to intensify in quantity and frequency, a phenomenon named by Hertig (2019). On his study Fischer et al. (2011) was of the opinion that, by the end of the century, Central Europe would have as many warm days as the ones that Southern Europe is currently experiencing (Fischer et al. 2011; Hertig 2019; IPCC 2014; Paz 2015; Semenza et al. 2016).

Due to the research carried out by the authors Collins et al. (2019) and Fischer et al. (2011) they came to the conclusion that, when it comes to the variation of precipitation, by the end of the century, this climate factor will increase in Northern Europe and decrease in Southern Europe mainly during Spring, Autumn and Summer as it was claimed by Franklinos et al. (2018). As a consequence hydrological systems might be affected which can influence the quantity and quality of our water's resources, this is mentioned on the scientific report created by IPCC (2014) (Collins et al. 2019; D'Oria et al. 2017; Fischer et al. 2011; IPCC 2014; Franklinos et al. 2018).

Concerning the factor relative humidity, Silva (2017) assessed that dryer climates are most favourable to wildlife. The authors Quinteiro et al. (2019) and Bador et al. (2017) realized that factors such as temperature and precipitation will likely increase and decrease, respectively, in several areas of Southern Europe, consequently heatwaves will be more frequent and more intense. Therefore, there will be a reduction in water evaporation and there will be less water available in the atmosphere which becomes dryer, this was supported by Cardoso et al. (2019). Thus, all of Europe will experience a decrease in relative humidity (Silva 2017; Bador et al. 2017; Cardoso et al. 2019; Quinteiro et al. 2019).

4 Discussion

After analysing all the selected articles, we concluded that, regarding the distribution of Culicidae vectors on Southern European countries, most of the studies suggest that vectors belonging to the *Culex* genus were the most abundant species in Southern European countries. Other articles suggest that the species within *Aedes* genus are the most abundant in these regions.

Concerning the diseases caused by these vectors, the majority of articles defend that dengue is the ailment that most affects the health of populations. However, other studies put forward that west nile virus is the most prejudicial disease to Man.

Regarding climate change, there is a general agreement that temperature is the main factor that influences the development of the vectors and is the one that will suffer more variations as climate alterations carry on. Other studies suggest that relative humidity and precipitation impact the development of vectors and will also fluctuate as result of climate change. It has come to conclusion that climate change will affect the evolution and growth of Culicidae vectors in Southern Europe, and as a consequence will impact the distribution of diseases passed on by these vectors.

Climate change is thought of as a problem of the future as it is overlooked when compared to other current quagmires. Within half a century, climate alterations will have had an impact on both the environment and the human health, therefore aggravating already existing problems. During the 21st century it is expected that these alterations will lead to concerning health problems in numerous regions of the world. By 2100, the increase of temperature and relative humidity will compromise many activities such as agriculture or outdoors labour (Cardoso et al. 2019; IPCC 2014; Varga et al. 2019).

Culicidae vectors require adequate climate conditions for their development since climate factors have severe and direct impact on their abundance, distribution, longevity, pathogeny (period of incubation, replication and lineage), on their hosts (abundance, distribution and behaviour) and on their interactions. Therefore, the predicted climate change, will, in the future, influence the distribution of vectors as well as the diseases they transmit. (Chen et al. 2013; Roiz 2014; Williams et al. 2016).

Concerning the factors temperature, precipitation and relative humidity, exposure to high temperatures can have several effects on poikilothermic animals, such as insects, precipitation is likely to decrease and there is a high likelihood that there will be a rise in the number of heatwaves which will lead to a decrease in relative humidity. These are some of the climate phenomenon that will probably take place with higher frequency, duration and intensity, due to climate change, in the near future. (Gaitán et al. 2019; Hertig 2019; Oliver and Brooke 2017).

According to Roiz (2014), and having in mind the likely climate change scenarios, it is predictable that the abundance of the vector *Ochlerotatus caspius* will increase in Southern Europe which will result in the spreading of myxomatosis, tahyna arbovirus and west nile virus. Piperaki and Daikos (2016), assure that there will be a growth in the proliferation of the vectors belonging to the genus *Anopheles*, as a consequence the transmission of malaria will be easier. Semenza et al. (2016) defends that, in the future there will be more outbreaks of west nile virus caused by the vectors of the genus *Culex*, that according to Capelli et al. (2018), will be one of the vectors that will develop more frequently over the next few years and that the vectors of the genus *Culiseta* will continue to have favourable conditions for their evolution (Capelli et al. 2018; Piperaki and Daikos 2016; Roiz 2014; Semenza et al. 2016).

On the other hand Liu-Helmersson et al. (2019) states that, by the end of the 21st century, no Southern European country will have a climate adequate for the development of vectors within the genus *Aedes,* the mains carriers of dengue, yellow fever, zika e chikungunya, which will lead to the disappearance of this species on the European area mentioned above (Liu-Helmersson et al. 2019).

Thus, it can be claimed that climate change will have different effects on the reproduction, development and survival of the multiple species of vectors studied. Also, there will be noticeable changes in the distribution and abundance of the vectors as some will have appropriate conditions for their development whereas other will face the risk of disappearing.

Even though there are several articles on the theme "Climate change and Culicidae vectors in Southern European countries" it is important that this topic receives the coverage it needs, in order to, in the near future, further studies be performed with the objective of gaining more knowledge regarding the existence of vectors belonging to the genus *Culiseta* and *Ochlerotatus* and the relationship between precipitation and relative humidity and the development of these species' vectors.

References

Marques, M.: Mosquitos invasores na Europa e importância da sua vigilância em Portugal (2014)

Amendola, S., Maimone, F., Pelino, V., Pasini, A.: New records of monthly temperature extremes as a signal of climate change in Italy. Int. J. Climatol. **39**(4), 2491–2503 (2019). https://doi.org/10.1002/joc.5952

Bador, M., Terray, L., Boé, J., Somot, S., Alias, A., Gibelin, A.L., Dubuisson, B.: Future summer mega-heatwave and record-breaking temperatures in a warmer france climate. Environ. Res. Lett. **12**(7), 074025. (2017). https://doi.org/10.1088/1748-9326/aa751c

Becker, N., Hoffmann, D.: First record of Culiseta longiareolata (Macquart) for Germany. European Mosquito Bulletin Journal of the European Mosquito Control Association, 29 December, pp. 143–150 (2011). https://e-m-b.org/content/first-record-iculiseta-longiareolatai-macquart-germany

Bisanzio, D., Giacobini, M., Bertolotti, L., Mosca, A., Balbo, L., Kitron, U. and Vazquez-Prokopec, G.M.: Spatio-temporal patterns of distribution of West Nile virus vectors in eastern piedmont region, Italy. Parasites Vectors **4**(1), 1–11. https://doi.org/10.1186/1756-3305-4-230

Capelli, G., Genchi, C., Baneth, G., Bourdeau, P., Brianti, E., Cardoso, L., Danesi, P., Fuehrer, H.P., Giannelli, A., Ionică, A.M., Maia, C.: Recent advances on Dirofilaria repens in dogs and humans in Europe. Parasites Vectors **11**(1), 1–21 (2018). https://doi.org/10.1186/s13071-018-3205-x

Cardoso, R.M., Soares, P.M.M., Lima, D.C.A., Miranda, P.M.A.: Mean and extreme temperatures in a warming climate: EURO CORDEX and WRF regional climate high-resolution projections for Portugal. Clim. Dyn. **52**(1–2), 129–157 (2019). https://doi.org/10.1007/s00382-018-4124-4

Centro de Estudos de Vetores e Doenças Infeciosas Doutor Francisco Cambournac. (2016). REVIVE 2016 Culicídeos e Ixodídeos. Retrieved from file:///C:/Users/asus/Downloads/Relatorio-REVIVE_2017.pdf

CEVDI: Revive 2011–2015 - Culicídeos e Ixodídeos: Rede de Vigilância de Vectores (2016)

CEVDI: Relatório REVIVE 2017 - Culicídeos e Ixodídeos: Rede de Vigilância de Vetores (2017). https://repositorio.insa.pt/handle/10400.18/5543

CEVDI: Relatório REVIVE 2018 Culicídeos e Ixodídeos (2018). Retrieved from file:///C:/Users/asus/Downloads/Relatorio-REVIVE_2017.pdf

Chen, C.C., Jenkins, E., Epp, T., Waldner, C., Curry, P.S., Soos, C.: Climate change and West Nile virus in a highly endemic region of North America. Int. J. Environ. Res. Publ. Health **10**(7), 3052–3071. https://doi.org/10.3390/ijerph10073052

Collins, M., Sutherland, M., Bouwer, L., Cheong, S.-M., Frölicher, T., Jacot Des Combes, H., Koll Roxy, M., Losada, I., McInnes, K., Ratter, B., Rivera-Arriaga, E., Susanto, R.D., Swingedouw, D., Tibig, L.: Extremes, Abrupt Changes and Managing Risks, pp. 589–656 (2019)

D'Oria, M., Ferraresi, M., Tanda, M.G.: Historical trends and high-resolution future climate projections in northern Tuscany (Italy). J. Hydrol. **555**, 708–723 (2017). https://doi.org/10.1016/j.jhydrol.2017.10.054

ECDC: Surveillance Atlas of Infectious Diseases (2018 data) (2018). https://atlas.ecdc.europa.eu/public/index.aspx

Ferreira, C.A.C., De Pinho Mixão, V., Novo, M.T.L.M., Calado, M.M.P., Gonçalves, L.A.P., Belo, S.M.D., De Almeida, A.P.G.: First molecular identification of mosquito vectors of Dirofilaria immitis in continental Portugal. Parasites Vectors **8**(1), 1–11 (2015). https://doi.org/10.1186/s13071-015-0760-2

Fischer, D., Thomas, S.M., Niemitz, F., Reineking, B., Beierkuhnlein, C.: Projection of climatic suitability for Aedes albopictus Skuse (Culicidae) in Europe under climate change conditions. Global Planetary Change **78**(1–2), 54–64 (2011). https://doi.org/10.1016/j.gloplacha.2011.05.008

Gaitán, E., Monjo, R., Pórtoles, J., Pino-Otín, M.R.: Projection of temperatures and heat and cold waves for Aragón (Spain) using a two-step statistical downscaling of CMIP5 model outputs. Sci. Total Environ. **650**, 2778–2795 (2019). https://doi.org/10.1016/j.scitotenv.2018.09.352

Gravina, H.D., Suzukawa, A.A., Zanluca, C., Segovia, F.M.C., Tschá, M.K., da Silva, A.M., Faoro, H., da Silva Ribeiro, R., Torres, L.P.M., Rojas, A., Ferrerira, L.: Identification of insect-specific flaviviruses in areas of Brazil and Paraguay experiencing endemic arbovirus transmission and the description of a novel flavivirus infecting Sabethes belisarioi. Virology **527**(September 2018), 98–106 (2019). https://doi.org/10.1016/j.virol.2018.11.008

Groen, T.A., L'ambert, G., Bellini, R., Chaskopoulou, A., Petric, D., Zgomba, M., Marrama, L., Bicout, D.J.: Ecology of West Nile virus across four European countries: empirical modelling of the Culex pipiens abundance dynamics as a function of weather. Parasites Vectors **10**(1), 1–11 (2017). https://doi.org/10.1186/s13071-017-2484-y

Hertig, E.: Distribution of Anopheles vectors and potential malaria transmission stability in Europe and the Mediterranean area under future climate change. Parasites Vectors **12**(1), 1 (2019). https://doi.org/10.1186/s13071-018-3278-6

Hoerling, M., Eischeid, J., Perlwitz, J., Quan, X., Zhang, T., Pegion, P.: On the increased frequency of Mediterranean drought. J. Clim. **25**(6), 2146–2161 (2012). https://doi.org/10.1175/JCLI-D-11-00296.1

IPCC: Climate change 2014: synthesis report. Contribution. In: Contribution of Working Groups I, II and III to the Fifth Assessment Report of the Intergovernmental Panel on Climate Change (2014). https://doi.org/10.1046/j.1365-2559.2002.1340a.x

Kolivras, K.N.: Changes in dengue risk potential in Hawaii, USA, due to climate variability and change. Clim. Res. **42**(1), 1–11 (2010). https://doi.org/10.3354/cr00861

Liu-Helmersson, J., Quam, M., Wilder-Smith, A., Stenlund, H., Ebi, K., Massad, E., Rocklöv, J.: Climate change and Aedes vectors: 21st century projections for dengue transmission in Europe. EBioMedicine **7**, 267–277 (2016). https://doi.org/10.1016/j.ebiom.2016.03.046

Liu-Helmersson, J., Rocklöv, J., Sewe, M., Brännström, Å.: Climate change may enable Aedes aegypti infestation in major European cities by 2100. Environ. Res. **172**(February), 693–699 (2019). https://doi.org/10.1016/j.envres.2019.02.026

Lopes, M., Monteiro, A.C., Ribeiro, I., Sá, E., Martins, H., Coutinho, M., Borrego, C. Alterações Climáticas e Gestão da Água em Portugal (Climate Change and Water Management in Portugal). Revista Brasileira de Geografia Física **5**(6), 1333 (2013). https://doi.org/10.26848/rbgf.v5i6.232926

Franklinos, L.H., Jones, K.E., Redding, D.W., Abubakar, I.: The impact of global change on mosquito-borne disease Lydia. **44**(0), 1–43 (2018)

Mixão, V., Barriga, D.B., Parreira, R., Novo, M.T., Sousa, C.A., Frontera, E., Venter, M., Braack, L., Almeida, A.P.G.: Comparative morphological and molecular analysis confirms the presence of the West Nile virus mosquito vector, Culex univittatus, in the Iberian Peninsula.Parasites Vectors **9**(1), 1–13 (2016). https://doi.org/10.1186/s13071-016-1877-7

Moher, D., et al.: PRISMA 2009 checklist. In: PLoS Medicine, vol. 2 (2009). https://doi.org/10.1371/journal.pmed1000097

Montarsi, F., Martini, S., Michelutti, A., Da Rold, G., Mazzucato, M., Qualizza, D., Di Gennaro, D., Di Fant, M., Dal Pont, M., Palei, M., Capelli, G.: The invasive mosquito Aedes japonicus japonicus is spreading in northeastern Italy. Parasites Vectors **12**(1), 1–8 (2019). https://doi.org/10.1186/s13071-019-3387-x

Nasir, S., Jabeen, F., Abbas, S., Nasir, I., Debboun, M.: Effect of climatic conditions and water bodies on population dynamics of the dengue vector, Aedes aegypti (Diptera: Culicidae). J. Arthropod-Borne Dis. **11**(1), 50–59 (2017)

Oliver, S.V., Brooke, B.D.: The effect of elevated temperatures on the life history and insecticide resistance phenotype of the major malaria vector Anopheles arabiensis (Diptera: Culicidae). Malaria J. **16**(1), 1–13 (2017). https://doi.org/10.1186/s12936-017-1720-4

Paz, S.: Climate change impacts on West Nile virus transmission in a global context. Philos. Trans. Roy. Soc. B: Biol. Sci. **370**(1665), 1–11 (2015). https://doi.org/10.1098/rstb.2013.0561

Petrić, M., Lalić, B., Pajović, I., Micev, S., Durdević, V., Petrić, D.: Expected changes of Montenegrin climate, impact on the establishment and spread of the Asian tiger mosquito (Aedes albopictus), and validation of the model and model-based field sampling. Atmosphere **9**(11), 1–18 (2018). https://doi.org/10.3390/atmos9110453

Piperaki, E.T., Daikos, G.L.: Malaria in Europe: emerging threat or minor nuisance? Clin. Microbiol. Infect. **22**(6), 487–493 (2016). https://doi.org/10.1016/j.cmi.2016.04.023

Quinteiro, P., Rafael, S., Vicente, B., Marta-Almeida, M., Rocha, A., Arroja, L., Dias, A.C.: Mapping green water scarcity under climate change: a case study of Portugal. Sci. Total Environ. **696**, 134024 (2019). https://doi.org/10.1016/j.scitotenv.2019.134024

Roiz, D., et al.: Climatic effects on mosquito abundance in Mediterranean wetlands. Parasites Vectors **7**(1), 1–13 (2014). https://doi.org/10.1186/1756-3305-7-333

Schaffner, F., Bellini, R., Petrić, D., Scholte, E.J., Zeller, H., Marrama Rakotoarivony, L.: Development of guidelines for the surveillance of invasive mosquitoes in Europe. In: Parasites and Vectors, vol. 6 (2013). https://doi.org/10.1186/1756-3305-6-209

Schaffner, F., Mathis, A.: Dengue and dengue vectors in the WHO European region: past, present, and scenarios for the future. Lancet Infect. Dis. **14**(12), 1271–1280 (2014). https://doi.org/10.1016/S1473-3099(14)70834-5

Semenza, J.C., Tran, A., Espinosa, L., Sudre, B., Domanovic, D., Paz, S.: Climate change projections of West Nile virus infections in Europe: implications for blood safety practices. Environ. Health: Global Access Sci. Source **15**(Suppl 1) (2016). https://doi.org/10.1186/s12940-016-0105-4

Silva, M.: R I A S E Mosquitos invasores na Europa e importância da sua vigilância em Portugal Invasive mosquitoes in Europe and importance of their surveillance in Portugal, vol. 3 (2017)

Varga, D., Roigé, M., Pintó, J., Saez, M.: Assessing the spatial distribution of biodiversity in a changing temperature pattern: the case of Catalonia, Spain. Int. J. Environ. Res. Publ. Health **16**(20) (2019). https://doi.org/10.3390/ijerph16204026

Watts, N., Amann, M., Arnell, N., Ayeb-Karlsson, S., Belesova, K., Boykoff, M., Byass, P., Cai, W., Campbell-Lendrum, D., Capstick, S., Chambers, J.: The 2019 report of the Lancet countdown on health and climate change: ensuring that the health of a child born today is not defined by a changing climate. Lancet **394**(10211), 1836–1878 (2019). https://doi.org/10.1016/S0140-6736(19)32596-6

Williams, C.R., Mincham, G., Faddy, H., Viennet, E., Ritchie, S.A., Harley, D.: Projections of increased and decreased dengue incidence under climate change. Epidemiol. Infect. **144**(14), 3091–3100 (2016). https://doi.org/10.1017/S095026881600162X

Williams, C., Mincham, G., Ritchie, S.A., Viennet, E., Harley, D.: Bionomic response of Aedes aegypti to two future climate change scenarios in far north Queensland, Australia: implications for dengue outbreaks. Parasites Vectors **7**(1), 1–8 (2014). https://doi.org/10.1186/1756-3305-7-447

Life Cycle Analysis and Sociocultural Impact for a Sustainable Nexus Design

Mahamat Habib Bechir[1,2], A. Mannelli[2(✉)], and A. López Agüera[2]

[1] Institute of Water and Energy Sciences, Pan African University, Tlemcen, Algeria
[2] Department of Physics, Santiago de Compostela University, Santiago de Compostela, Spain

Abstract. The eradication of multidimensional poverty remains the primary objective of the United Nations Sustainable Development Goals (SDGs). In this context, the Sustainable Community Project (SCP, UNESCO) aims to develop demonstrative pilot projects to identify comprehensive development success models in small isolated communities. This article explores the importance of applying, in addition to CLEWs tools, Lifecycle analysis to the technologies to be implemented. Similarly, special emphasis is placed on the crucial need to incorporate appropriate qualitative social studies, particularly ethnography, at all project steps. The results show the importance of conducting a Life Cycle Analysis (LCA) during the formulation of the development master plan to select the most appropriate technologies in terms of life time with respect to the environmental and economic context. Under the sociocultural impact, the involvement of the local community in each decision-making is unavoidable in order to ensure the success of this project, as well as to guarantee its sustainability.

Keywords: Climate land energy water · Nexus · Life cycle analysis · Ethnographical studies

1 Introduction

The eradication of multidimensional poverty remains the primary objective of the United Nations Sustainable Development Goals (SDGs) [1]. Although due to the global efforts, the obtained results are insufficient to achieve the desired change proposed in the Agenda 2030. The Sustainable Development Goals Report 2020 shows that some areas worsened, and global inequalities continue to grow. Global poverty suffered the first increase in decades, food insecurity continued to increase, and billions of people still lack clean water, sanitation services and electricity [2]. These inequalities are mostly identified in small populations isolated in rural areas thus contributing to increased vulnerability due to precarious access to basic resources that favors social exclusion.

Advancing the 2030 agenda must therefore focus its efforts on rural small communities as working target, making a deep reflection on the reasons for the relative failure of both working methodologies and initiatives applied so far. One of the most common issues is the challenge to correlate the whole deficiencies to offer global solutions which

J. R. da Costa Sanches Galvão et al. (Eds.): *Proceedings of the 1st International Conference on Water Energy Food and Sustainability (ICoWEFS 2021)*, pp. 670–679, 2021.
https://doi.org/10.1007/978-3-030-75315-3_71

respond to the expectations of target communities. In fact, SD11 (Sustainable Cities and Communities) focus the dignified and sustainable development of such human settlements. The fundamental idea behind, is that any given community, in any cultural, climatologic and socio-economic context can satisfy their needs, wishes and expectations without endangering the environment and resources for current and future generations. So, SD11 aims to chart roadmaps that promote fair adaptation and reduce inequalities in the scenario of the impacts created by climate change on every human settlement, creating resilient communities.

In this line of thinking, in recent years the NEXUS concept of Integral Development is being made way where the different deficiencies are strongly interrelated. In the NEXUS frame, Climate, Energy, Water, Food (safe food, agriculture, forestry, and others) and Income Generation approach are considered as an overall.

To clarify the connections between the different actions and their possible consequences arose a modelling framework named CLEWS (Climate, Land, Energy, Water Systems). The CLEWS tool allows to model each of the resource systems separately, identifying the points at which they interact (water needs in land use or energy production, energy needed to supply water, etc.). The output of each model forms the input of the other two; the entire model is resolved sequentially, and a convergent solution is found after a series of iterations [3]. In most cases synergies are greater than trade-offs.

Nexus concept and CLEWs tools although intended to be valid for all types of scenario, regardless of its size, (country, region, village) decreases its reliability levels the smaller the application environment. This effect is associated with the highest level of external dependency when the most local is the operating environment. For the socio-ecological systems that make up rural communities that rely primarily on local resources as a form of livelihood, either by direct food or as a source of income. In this scenario, there is no doubt that NEXUS concept and CLEWS needs to be carefully applied being probably necessary to incorporate additional specific working methodologies.

The search for technological solutions that consider economic, environmental, and social aspects represent the greatest challenge. The last 25 years of experience should serve to know that in the field of technological development, problems grow faster than solutions, and this applies not only to poor countries but also to rich and so-called developed countries. As an example, the complex management of the huge amount of technological waste derived from their programmed obsolescence. Therefore, continuing with the modern technological development model as we know it is not feasible if we seek to alleviate existing multidimensional poverty.

To think about a new approach to achieving sustainability at all levels we start from the lessons we learned from previous experiences that represent an invaluable contribution: the projects carried out from the network of NGOs, universities and institutions called Greening Africa Together (GATo); Humanity in Action, an international non-profit organization that educates and connects young leaders on issues related to human and minority rights; the Sustainable Community Project (SCP), an initiative from the UNESCO Committee of Local & Regional Leaders Appointment (CLRLA) and others. From all of them, some extra constraints which must be considered as the basics pillars to ensure a sustainable development of rural communities have been identified: work with initiatives that arise from the communities with the bottom-up logic, with socially appropriate technologies (SAT) developed to work towards self-sufficiency, within the framework of models of low external dependence model (LEDD). Actually, what is sought is to increase community resilience and foster the ability to face adverse situations without compromising their identity and structure as a group.

Bottom-up initiatives are those that arise from the community itself, where the direct beneficiaries are at the same time the creators and the main actors in the decision-making of the project. Normally these initiatives are proposed by local agents (community organizations, government institutions, entrepreneurs) due to the identification of specific needs that triggers an organizational process for putting them into action. These types of initiatives give rise to the social demands of different groups that make up the communities. Thus, both hegemonic and minority groups can propose projects for the resolution of problems identified as relevant by them.

Low External Dependence Distributed (LEDD) Models, also named zero philosophy models, are community self-managed based models in which a responsible valorization and exploitation of available local resources (human, energetic, cultural, environmental, etc.) is the structural basis for sustainable community development [4]. The application of LEDD will encourage the self-empowering of the communities ensuring the decline of the multi-poverty levels.

But, without any doubt, one of the keys to success in the development of a sustainable community is the identification and implementation of so-called socially appropriate technologies (SAT). SAT are those that are designed with special attention to the social, environmental, cultural, and economic aspects of the community (i.e., the human, material, and cultural resources of the economy) [5]. The SAT concept is complex and polyhedric. Understanding it as the balanced integration of economic execution, social inclusion and environmental resilience for the benefit of current and future generations is crucial.

Usually, the work done with SAT prioritizes one of its dimensions. There is a tendency to focus on the social acceptance of technologies (without a clear definition of what it means and implies) or on the use of local materials or resources, without adequate training of local populations for their use, maintenance and control. Thus, this way of working not only generates greater external dependence - which is the opposite of the objective sought - but also leaves aside the consideration of the sustainability of the production processes of the technologies and their associated materials.

Even if social acceptance is defined in different ways according to the authors, in this paper we understand social acceptance as an active process where the population adopts or not the use of certain innovations, in this case those associated to the technological development. The social acceptance is not a passive approval, it implies an active attitude [6]. From this perspective, the way to assess social acceptance must be approached from a different view, considering sociocultural specificities, fostering in parallel the capacity for learning and transformation.

From the technological point of view, although SATs favor the use of non-captive technologies (of controlled price), they do not necessarily consider aspects to retard the obsolescence (with training for its fix) or the sustainability of the materials and the manufacturing process of the technologies used.

The ideas and concepts are clear, and it is known that there are a large number of initiatives that are working on this path. The procedures to do so vary from team to team, thus generating disjointed practices and without a continuity relationship where they feed each other. Working together on the same line could lead to greater success. We find it necessary to develop a protocol that has sufficient support to become a standard guide capable of adapting to different contexts.

Ensuring that a technological solution is socially and environmentally appropriate requires the development of an unbiased procedure that considers all these variables and capable of being reproducible in different projects. This article will discuss the need to identify and evaluate a set of practices for working with TSA in all its dimensions that we believe should be part of a good practice guide for future projects.

An adequate evaluation process will allow the development of a methodology that serves as a "roadmap" for the design and implementation of sustainable community projects.

2 Assessments of Sociocultural Aspects

Considering socio-cultural aspects within technological innovation programs for the development of sustainable communities is not new. In fact, its importance was recognized in the 70s after analyzing the failures of the projects that had been carried out before. Anyway, many initiatives continue failing by treating technological aspects as if they are not part of the socio-cultural system in which they acquire meaning, as if they were an external and independent addition and not an integral part of the same system.

The most widely used methodology to assess sociocultural factors in this type of project, such as SCP, is based on the identification of relevant variables and the use of indicators for diagnosis and evaluation. The benefits of working with this type of methodology are undeniable. Indicators make it possible to work with a large amount of data, classify them, providing a synthetic representation of reality that can be used to assess the situation of a community.

However, even if all this information is available, it is usual that some obstacles appear when implementing the project. In the last decades, it has been increasingly recognized that there is a factor that can potentially be a barrier to the implementation of this type of project: social acceptance. The general acceptance of the implementation of renewable energy, generally based on opinion polls, is high even in places where there is no active policy to promote it. However, other limiting non-technical factors condition effective support and influence social acceptance [7].

There are different reasons that respond to a complex network of meanings and relationships that develop in the local context and are therefore permeated by sociocultural differences. Although at a general level a community may have a favorable opinion on the use of clean energies, this situation may change due to different conceptions and value of the territory, its landscape, the perceived procedural and distributive justice of the project, trust in the project leaders and their intentions, the recognition of local knowledge, etc. [7].

In addition to the senses and meanings of the communities, the forms of social organization also deserve special attention. The proposed project with a community that has egalitarian organizational structures will not be the same as that of a hierarchical one. It is also important to consider the informal networks that coexist with the formal institutional structure since they play a fundamental role in the life of the community.

Qualitative research methodologies and specifically ethnography continue the best option to access this type of information. Qualitative studies - and at the local level - are necessary to understand the dynamics of the processes in the development of programs for the transition towards sustainability. The specific case studies help us to access these vicissitudes, the complex cultural fabric, the practices of the communities and thus the understanding of the community in a holistic way.

When we think about the design, implementation and evaluation of projects that involve new technologies, we cannot forget that technology is part of society, it is not an external addition [8]. This implies that it is not only necessary to consider its function and technical benefits, but that there is a set of values, interpretations and meanings (socially and culturally determined) that are associated with it. The study of cultural patterns and local practices must be included in all stages if we want the project to be successful [9].

Ethnographic fieldwork can generate this knowledge because as an approach seeks to understand social phenomena from the perspective of its members [10] and presents behaviors as actions within the frameworks of interpretation where they make sense. This means that, when working with a group, what is sought is the interpretation of the agents themselves. Working in a participatory way and in continuous contact with the communities allows understanding the significant intentions involved in the behavior. This is important because it could help to understand the apparent contradictions mentioned before that appear between opinion polls and the implementation. In other words, ethnography helps us frame the survey results in the local context where they make sense.

In addition, one of the characteristics of ethnographic fieldwork is its flexibility. It includes a variety of techniques, such as participant observation, formal and informal talks, interviews that serve to accommodate the heterogeneous nature of work situations and data [11]. Within participant observation there are dialogues, encounters, and events of the everyday and the extraordinary, which are very useful for the researcher. "The meaning of social life is expressed particularly through discourses that constantly emerge in daily life, informally through comments, anecdotes, terms of treatment and conversations". [10, p. 75][1]. Flexibility serves to warn of the unpredictable, the researcher must be open to everything that happens in the field, to be able to recognize, identify notions or circumstances that are relevant to the group and that had not been previously considered. This view, where are heterogeneous elements, networks, representations that mobilize projects, allows analyzing the characteristics of development actions in a holistic way. Observe the daily life, local practices and expectations of the target population to contrast them with the parameters that are handled in the framework theories.

The most important benefit of this approach is that it facilitates the active participation of the community in the entire decision-making process and the joint construction of knowledge. Community participation in the decision-making process of the project is a

[1] Own translation.

key factor for social acceptance [7]. The continuous contact of the participants from all the sectors involved generates communication networks, trust and ownership [12].

Furthermore, through participation under conditions of equality - and the recognition that there are multiple types of knowledge - we seek not to fall into neocolonial practices, where projects are imposed, sometimes naively and sometimes not, for the supposed benefit of the community, without considering local knowledge [13].

3 Assessment of Technological Sustainability

The question about whether technological development responds to our needs without putting resources and environmental conditions at risk has been studied for decades and progress is undeniable. To do not start from scratch, among the number of existing procedures developed to address sustainable technological development, it is necessary to identify and adapt some of them to be able to address the environmental, economic and social sustainability of the SAT in an appropriate way.

Although initially the LCA was born as an analysis of environmental impacts in industrial processes, in recent years the so-called Life Cycle Analysis (LCA) has been making its way as an identifier of the levels of technological sustainability. The LCA makes possible to evaluate the environmental loads associated with a technology, a production system or activity, identifying and quantifying the amounts of matter and energy used, the waste generated and evaluating the environmental impacts derived from these. It is a simplified model that studies the entire life cycle of a product, process or system from its design and development, starting from the extraction of raw materials, transportation, processing, manufacturing, usage to the end-of-life of the activities (reuse, recycling, final disposal, etc.) [14]. This is the so-called "cradle to grave" philosophy.

The application of LCA to assess the level of sustainability in the case of TSA needs some modifications of its original formulation [15, 16] associated with the different estimators that make up the LCA, their definitions and dimensions. Thus, the characteristic equation of LCA sustainability is defined as:

$$LCA = WP(1)\,IRR + WP(2)\,EPBT + WP(3)\,IMP \tag{1}$$

Where IRR represents the economic return time of the initial investment, including maintenance costs, EPBT (energy payback time) and IMP represent the environmental impact. All of them evaluated in the philosophy of "cradle to grave". WP (i) represent the specific weights that can be modified depending on the scope of the technology under study. As an example, we can think that the economic factor does not have the same weight in a technology designed for the industrial sector than one designed to be used in a domestic context, the priorities will be different. Figure 1 shows a simple scheme of the differences.

Another modification to improve the LCA effectiveness in evaluating the levels of sustainability in the development of non-captive technologies is the unification of all the estimators in the same unit of measure. This allows a direct sum and therefore a numerical value for LCA that represents a compromise between the three aspects considered, this is economic, energy and environmental. The measure chosen is the time, specifically in

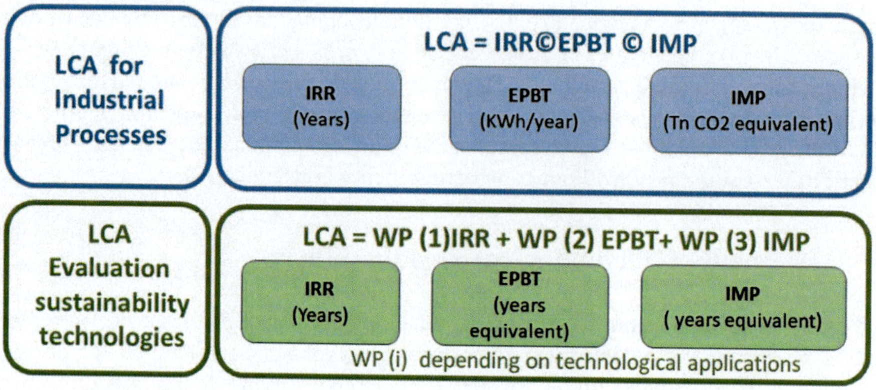

Fig. 1. Comparative scheme of the LCA with the proposed modifications.

years, which are obtained from comparing the estimators with the installations carried out with conventional technology without the use of renewable sources.

This, while slightly complicating the calculation procedure, facilitates the interpretation of LCA. Thus, LCA now offers as a result the number of years of operation of the technology to be absolutely amortized. The direct comparison of LCA with the useful life of the technological solution allows not only to ensure but also to quantify the sustainability always associated with positive values of LCA.

In the case of IRR and IMP, both are normalized to the calculation of the same parameters if the design of the installation was carried out with non-renewable technologies. This allows the evaluation of LCA as well as a direct comparison between different technologies potentially applicable in the case of projects. In addition, in the evaluation of the EPBT, the energy cost of fuel necessary in the useful life of the installation is introduced, which favors the use of own energy sources, in particular those associated with the reuse of waste.

As an example, in Fig. 2, the EPBT values associated with the PV solar panels are shown as function of the distance between the technological installation and the technology production. In this case of study, two different levels of irradiation are considered. As the main remark, the transport effect in the EPBT is more than 50%, being slightly smaller for low irradiated locations [17].

In summary, the following table shows the main modifications and their associated effects. As can be seen, LCA makes it possible to evaluate how a technology favors local production or not, the use of own resources and avoid the effects of programmed obsolescence. In addition, the fact of considering the economic, environmental and energy estimators and weighting them according to the situation, allows suitable solutions depending on the desired objective. The application of the specific weights of each estimator must be studied to adapt to the application characteristics of each technology under study (Table 1).

A simple direct comparison of the standardized LCA with the Useful Life of the installation or technology under study will allow assessing not only the degree

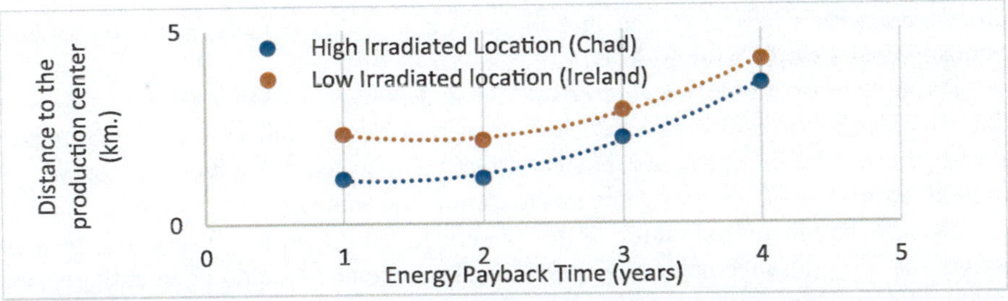

Fig. 2. Energy payback time dependence with the distance between the production point and the corresponding installation.

Table 1. Effects on the assessment of technological sustainability related to proposed modifications in the LCA calculations process.

Estimator	Modifications in the LCA calculation process	Effects on the assessment of technological sustainability	Dimension
IRR	Includes maintenance costs Comparison with conventional technology	Promotes avoid the effects of programmed obsolescence and extend its useful life Favors the independence of the group for the maintenance and repair of technologies	Years
EPBT	The energy cost of transport is included Incorporates the energy cost of fuel Includes energy expenditure associated with demand	Drives local production Stimulates the use of own resources and waste management Promotes energy efficiency	
IMP	Measured compared to non-renewable technology	Favors CO_2 compensation mechanisms and generates economic incoming	

of sustainability but also choose, according to this parameter among competing technologies.

4 Concluding Discussion

Although great efforts are being made to fight multidimensional poverty, the transition to sustainability remains a difficult goal to achieve and the results are not as expected. The small rural communities are the human settlements that are in a more precarious situation and therefore vulnerability.

The focus of projects that seek the development of sustainable communities must change to overcome the evident failures. It is necessary to improve the established procedures with a new approach that includes the LEDD models, based on a nexus philosophy and works with TSA.

Throughout the article, we focused on TSA and on two issues that we consider essential to achieve better results in the transition to sustainability: qualitative and ethnographic studies to evaluate the social acceptance of technologies and life cycle analyzes for the product improvement and to reduce its environmental impacts.

The application and validation of the models in specific projects that allow the analysis of the TSA from the point of view of social acceptance - including quantitative and qualitative variables - as from technological sustainability are necessary. Right now, the Life Cycle Analysis of the technologies and ethnographic fieldwork are presented as the best options. This does not mean that they are the only or the definitive ones. Further work is needed on the proposed modifications to this type of analysis paying special attention to the specific weight (WP) for different applications.

As we have seen, regarding the socio-cultural variables that we must consider when preparing, implementing and evaluating a project, we can access and work with them in two different but complementary ways. Social indicators provide us with valuable information that describes social conditions and situations that should be placed in contexts, through ethnographic research, for their correct interpretation. Access to the network of local meanings, desires, expectations can only be achieved in and with the local community, through participatory processes where initiatives and each part of the decision-making process decisions are shared, and no relevant agent or sector is left out of it.

Regarding the analysis of the environmental impact of technologies, it is essential, but not sufficient, to work with renewable energies and non-captive technologies that respond to local needs. The analysis of the complete life cycle of products and processes must be studied to achieve true sustainability.

A more dedicated study is needed on this type of research in order to have tools that allow us to carry out sustainable community programs. The goal is ambitious: finding a sustainable model at an economic, ecological and social level that serves to improve the living conditions of populations that are in precarious and vulnerable situations. From this moment, a dedicated study to find ways of evaluating that allow the creation of decision tables to compare technologies based on their sustainability in all dimensions. Interdisciplinary work and the active participation of all stakeholders is essential.

References

1. United Nations: Transforming our world: the 2030 Agenda for Sustainable Development (2015)
2. United Nations Department For Economic And Social Affairs. Sustainable development goals report 2020. S.l.: United Nations (2020)
3. Howells, M., et al.: Integrated analysis of climate change, land-use, energy and water strategies. Nat. Clim. Change **3**(7), 621–626 (2013)
4. de Acevedo, J.F.D.: Sustainable communities project: protocol and appropriate technologies for water management. Universidade de Santiago de Compostela, Santiago de Compostela (2016)

5. Hazeltine, B., Bull, C. (eds.): Field Guide to Appropriate Technology. Academic Press, Amsterdam (2003)
6. Mallett, A.: Social acceptance of renewable energy innovations: the role of technology cooperation in urban Mexico. Energy Policy **35**(5), 2790–2798 (2007)
7. Wüstenhagen, R., Wolsink, M., Bürer, M.J.: Social acceptance of renewable energy innovation: an introduction to the concept. Energy Policy **35**(5), 2683–2691 (2007)
8. Rip, A., Kemp, R.: Technological change», en Human choice and climate change. In: Rayner, S., Malone, E.L. (eds.) 2: Resources and Technology, pp. 327–399. Battelle Press, Columbus (1998)
9. Carceles-Breis: Towards a framework of socio-cultural indicators for development process. UNESCO, dic. (1992)
10. Guber, R.: La etnografía: método, campo y reflexividad. Grupo Editorial Norma, Bogotá (2001)
11. Velasco Maíllo, H.M., de Rada, Á.D.: La lógica de la investigación etnográfica: un modelo de trabajo para etnógrafos de escuela. Trotta, Madrid (2018)
12. Walter, J.: Technological adaptation and "learning by cooperation". A case study of a successful onshore technology transfer in Tierra del Fuego. J. Technol. Transf. **25**(1), 13–22 (2000)
13. Lander, E., Castro-Gómez, S. (eds.): La colonialidad del saber: eurocentrismo y ciencias sociales: perspectivas latinoamericanas, 1. ed. Buenos Aires : [Caracas, Venezuela]: Consejo Latinoamericano de Ciencias Sociales-CLACSO ; UNESCO, Unidad Regional de Ciencias Sociales y Humanas para América Latina y el Caribe (2000)
14. Wu, P., Ma, X., Ji, J., Ma, Y.: Review on life cycle assessment of energy payback of solar photovoltaic systems and a case study. Energy Proc. **105**, 68–74 (2017)
15. Kloepffer, W.: Life cycle sustainability assessment of products: (with Comments by Helias A. Udo de Haes, p. 95). Int. J. Life Cycle Assess. **13**(2), 89–95 (2008)
16. Zamagni, A.: Life cycle sustainability assessment. Int. J. Life Cycle Assess **17**(4), 373–376 (2012)
17. Habib Bechir, M.: Life cycle estimation of rural electrification PV based for extreme climatic Sahel region (Algeria, Mauritania, Burkina Faso, Senegal and Chad). Máster thesis, Pan African University, Algeria (2020)

Researching on Sustainable Urban Qualification Scenarios

Pedro Bento(✉) and Margarida Louro

CIAUD, Faculdade de Arquitectura, Universidade de Lisboa, Lisbon, Portugal
{pedrobento,mlouro}@fa.ulisboa.pt

Abstract. The focus of this work is to establish applied urban and architectural research on very particular urban fabrics in the metropolises of Lisbon and Barcelona. Objectively, these are two expectant areas located on the limits of their central cities, considering both their administrative and physical borders, fundamentally marked by their waterfronts, and which open up to different scenarios of critical and project appropriation.

The objectives are the understanding of the phenomenology related to terrain vague and waterfronts and the development of different urban and territorial qualification scenarios for them. The research methodology used consisted of defining the object of study and the hypothesis and, the technique used was "Research by Design". The hypothesis is that the urban qualification scenarios in these areas are open and admit different morphological, density and layout typologies.

In short, it is intended to address both the debates on rehabilitation policies in the inherited and canonical city, and the particular morphological problems with respect to the city-territory with which these areas are adjacent. The reflection ends with a set of designing proposals and intervention strategies that point out possible scenarios of new urbanity, capable of binding the surroundings and highlighting their relationship with water.

Keywords: Sustainability · Designing proposals · Intervention strategies

1 Introduction

The object of study of this research are the expectant and interstitial spaces of two water fronts: Doca de Pedrouços, in Lisbon and the coastal front of the three chimneys, in Sant Adrià de Besòs, Barcelona. The objectives are: on the one hand, the understanding of the phenomenology related to the "terrain vague" and the waterfronts and, on the other hand, the development of different scenarios of urban and territorial qualification of the same. The research methodology used consisted of defining the object of study and the hypothesis and, the most used technique, the "Research by Design". The hypothesis that has been worked with is that the urban qualification scenarios in these areas are open and admit different morphological, density and layout typologies. Thus, the article begins by making a critical review, as a theoretical framework, of some of the most important reflections carried out on the object of study; a brief description of the physical framework is then presented, stating the problem; to then present the appropriate project responses. Finally, the work ends with two conclusive notes (Fig. 1).

J. R. da Costa Sanches Galvão et al. (Eds.): *Proceedings of the 1st International Conference on Water Energy Food and Sustainability (ICoWEFS 2021)*, pp. 680–690, 2021.
https://doi.org/10.1007/978-3-030-75315-3_72

Fig. 1. Sant Adrià de Besòs. Bento, 2020

2 Theoretical Framework

The present work can be inserted in different fields of contemporary urban and morphological reflection, but there are at least two that we cannot escape referring to: that of expectant, empty or interstitial spaces and that of water fronts.

Regarding the first field, we must say that this has been and continues to be the subject of extensive reflection. Solà-Morales (1996, 2002) defined this concept as *terrain vague* and applied it to the void, vacant, free of defined limits and affirmed that it constituted an important category in the metropolitan city. Fausto and Rábago (2001), Machín (2009) and Messen (2005) approached these spaces as *urban voids*. The first applied the concept to land or even intraurban buildings that have no defined use or are not inhabited; the second, he applied it to free intra-urban lands, preferably private and, the third, to residual, interstitial, obsolete spaces or on vague landscapes. The anthropologist Marc Augé (2000) and the sociologist Amendola (1997) described it as *non-places*. The first referring to empty spaces, where people transit, and referring to spaces opposed to the concept of *place*; and, the second, to the spaces created by the urban dynamics caused by postmodernity, also being spaces opposed to the place. Bazant (2001), Méndez (2009, 2011), Peimbert (2008) and the anthropologist Cedeño (2007) called it *urban interstices*. The first, he applied it to intermediate wastelands, spaces that are being left by uncontrolled urban expansion in the periphery; the second, to the intermediate space between plateaus, the third to the forced relationship with the built: interstitial landscape and the fourth applied it to the discontinuous space of the city. The architect Clichevsky (2007) defined it as *vacant land* and applied it to the void, vacant, unused private land, within the perimeter of agglomeration. Finally, Águila Flores (2014: 54), in a recent publication, defined interstitial spaces as "those (empty spaces) that accompany the phenomenon called urban expansion".

With regard to the second field - the waterfronts - it should be said that it is also the subject of enormous academic reflection. Numerous academics such as Joan Busquets, Francisco Monclús Fraga and Juan Alemany or María del Carmen Blasco Sánchez and Francisco Martínez Pérez in Spain; Victor Matias Ferreira and Jorge Gaspar, Manuel Salgado or João Costa in Portugal; Dominique Mashini, Daniel Talesnik and Alejandro Gutiérrez in Chile; Hans Meyer in Holland; Regine Keller in Germany; Laurent Vermeersch in France; Rinio Bruttomessi in Italy, and organizations such as Project for Public Spaces and La Città d'Acqua, among many others, have developed important works and research on the subject (Mashini 2010).

3 Physical Framework/Problem

The research carried out during the last two years addresses two interstitial peri-urban territories, which have been the subject of various studies by different administrations and academies. In Lisbon, this area is formalized by the area of the Pedrouços docks, and specifically by the action area defined by the Doca de Pedrouços Strategic Document, approved in 2011. An area of the river front, formerly occupied by the activities of the Port of Lisbon, located in the western part of the city, formalizing the border with the municipality of Oeiras and which is assumed as a potential space for multiple transformations. In Barcelona, the view falls on part of its promenade, on the land occupied by the building of the old Thermal of Sant Adrià, located on the left bank of the Besós river, and specifically on the area of action of the PDU d'Ordenació del Front Litoral in l'Àmbit de les Tres Xemeneies, from January 2018 and which is also defined as a strategic area for contemporary reflection (Fig. 2).

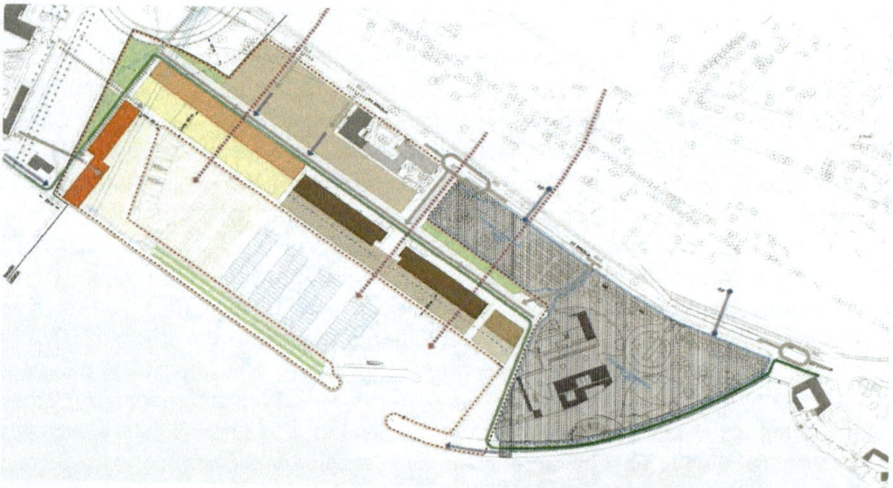

Fig. 2. Doca de Pedrouços Strategic Document - Implementation Proposal, Câmara Municipal de Lisboa and APL -Administração do Porto de Lisboa, S.A. (2011).

The Pedrouços dock area is located in the parish of Santa María de Belém and covers approximately 125,000 m^2. The whole area was occupied until the beginning of 2012 by industrial buildings, largely unused, and without architectural interest, meanwhile demolished, except for the "Edifício da Lota" from the 60 s, which due to its unique relationship with the pier is intended to requalify. The reconversion program focuses in particular on two distinct areas, which complement each other and work in an integrated manner. The first zone constitutes the entire front of the dock and includes the area destined for the future recreational marina of the Pedrouços dock, which will integrate the necessary valences for the realization of the main national and international ocean navigation events. The second zone consists of the strip along Av. Brasilia and the final auctions with the east and west ends. It is configured in an area of accessibilities, parking, green spaces and new construction. The objectives proposed in the Doca de Pedrouços Strategic Document approved and published in 2011 for the area in question aim to contribute to:

- requalify an area that is currently disqualified, giving it characteristics that last over time;
- equip Lisbon with a recreational marina and its infrastructures with the conditions to host important national and international ocean navigation events and large recreational vessels;
- urban reorganization of the northern strip of the intervention area in contact with Av. Brasilia, in terms of uses, equipment, accessibility and green spaces;
- dignify the landscape and urban integration of the Torre de Belém, eliminating the metal viaducts that connect Av. De Brasilia and Av. Da Índia, and
- establish a formal and functional articulation with the reconverted public outdoor spaces within the scope of contiguous operations, in particular, the areas surrounding the Champalimaud Foundation (Fig. 3).

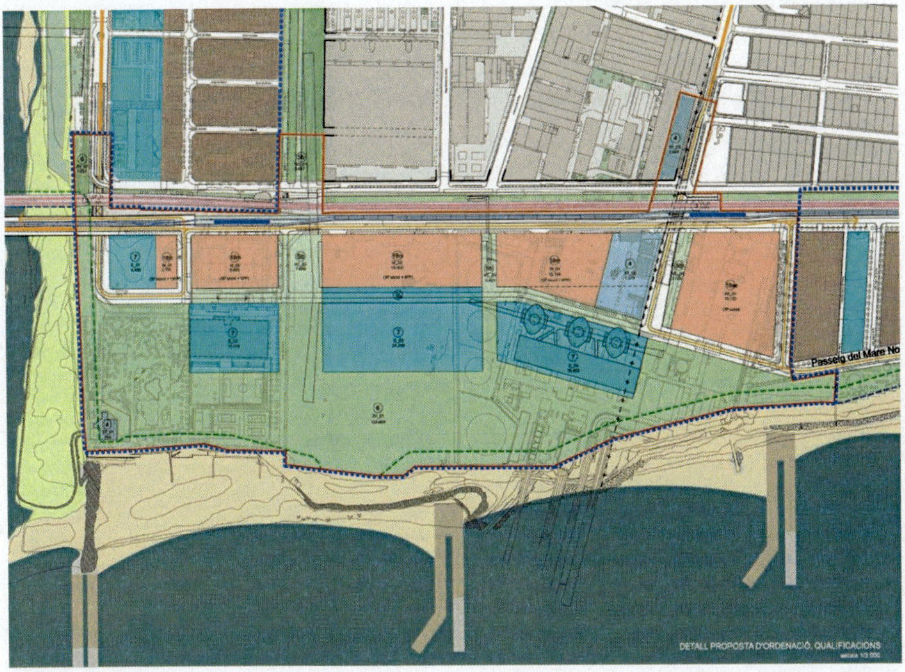

Fig. 3. Detail of the Front Litoral PDU Proposal, Generalitat de Catalunya and Estudi Ramón Folch, January 2018.

In Barcelona, although the area of the PDU will be specified throughout the drafting work, it starts from a land that occupies a total area of 124.9 Ha, corresponding to the area of the sector that delimits and orders the PDU, of about 32, 6 Ha and an area of 92.3 Ha corresponding to the urban fabrics adjacent to it, which are located in a more interior position, in order to promote their transformation and to guarantee the continuity of these fabrics of Sant Adrià and Badalona with the façade coast. The sector that delimits the PDU, called the Three Chimneys, and object of new planning, is located on the coastline, on the left bank of the Besòs river. It ranges from the mouth of the river Besòs to the new urban development of the Port of Badalona, between the municipalities of Badalona and Sant Adrià del Besòs. The sector covers an area with an area of 326,712 m^2, (32.67 Ha), which is delimited as described below:

- To the Northwest, along Avenida Eduard Maristany and the land reserve linked to the railway system of the Barcelona Mataró Line, together with incisions at the level of Calle Olímpic, Calle de la Playa and Avenida del Maresme.
- To the Northeast, by the boundary of the residential sector made up of the blocks between Paseo del Mare Nostrum and Avenida de Eduard Maristany, south of the Badalona marina, called Barrio de la Mora.
- To the south-east, by the delimitation of the Terrestrial Maritime Public Domain (DPMT) towards the sea.
- South-West, along the same delimitation of the DPMT corresponding to the Besòs river up to Avenida de Eduard Maristany.

4 Project Responses

The project strategies are developed based on two premises: on the one hand, through a critical review of the planning instruments presented for the two case studies and, on the other hand, through the response to a functional program of utopian and exploratory nature, alternative to the aforementioned plans and which is detailed below. Finally, it should be noted that this project research "Research by Design" is carried out in the curricular units of Laboratory Project IV and V of the 4th year of the Master in Architecture developed during the academic years 2018/2019 and 2019/2020 by the group of students who make up the aforementioned curricular units and their respective teaching team.

Lisbon, and especially its waterfront, constitutes a potential space for multiple transformations and appropriations. The existing dysfunction in the Port of Lisbon, caused by the deactivation of many of its areas, promotes new uses that allow the connection between the built network and the river. Recent large-scale interventions, such as the New Cruise Terminal, the expansion of the Museum of Art, Architecture and Technology, the construction of the Champalimaud Foundation or even the restructuring of public spaces, such as the Ribeira das Naus and the Plaza del Comercio, are configured as incursions that effectively open a new stage in the permanent and qualified experience of the river front.

Associated with these uses (perhaps more stable and assumed as new polarities and emerging urbanities) coexist other phenomena of an especially temporary nature that activate, during certain moments, attraction actions in strategic areas of the city. Some examples would be events such as the global Web Summit or the summer concerts in Parque das Nações or in the Passeio Ribeirinho de Algés. Thus, in this context, the Lisbon Volvo Ocean Race event stands out as a field for project reflection. One of the best known sailing regattas, started in 1973, which already has thirteen editions completed, and where the last three were established with a connection to Lisbon, settling precisely in the western area of the Pedrouços dock, which we are now dealing with.

In this way, assuming the pretext of the temporary qualification of this area of the city, combined with the opportunity for a permanent regeneration of the dock and surrounding areas, this object is chosen as the privileged space for project reflection.

Therefore, an urban program is established for which project strategies must be developed that qualify the space in question and its surroundings. On the one hand, it must articulate an ephemeral part of the urban space, related to the implementation of the Lisbon Ocean Race Village program with its own functional areas defined by: exhibition area, activity area, team area and others; and, on the other hand, the permanent part, qualified with new uses, such as: hotel, marina and commercial areas with adjacent restaurants and a nautical sports complex.

The ever-expanding Barcelona, and in particular its promenade, continues to be territories with great urban potential for carrying out numerous transformations and appropriations. In fact, the dysfunctionality of some sectors, among which, that delimited by the land occupied by the old Sant Adrià Thermal Power Plant - located on the left bank of the Besós, municipal limit of Sant Adrià de Besós, province of Barcelona - constitute magnificent opportunities and are defined as strategic areas of contemporary reflection. This particular and extensive area of the promenade - where there is still part of the impressive built structure of what had once constituted the Three Chimneys Thermal Power Plant, operating for approximately 38 years (between 1973 and 2011) and which is widely known for its emblematic chimneys of approximately 200 m high (recently classified as a cultural asset of local interest), - remains a controversial focus of discussion on the directions to take in terms of future strategies, and how to operate or maintain the existing architectural heritage legacy.

Thus, the exercise that has been attempted consists in taking up some of those discussions of a more architectural scope and, simultaneously, outlining different urban qualification strategies that reconstruct the city's relations with water and with its different nearby urban environments. The only programmatic requirement at the level of urban proposals is that they contemplate the creation of a new urban facility - a museological complex and research center that is related to the issue of energy and its paradigms - that can be developed below.

Also in this case, nine urban occupation scenarios have been deployed, very different from each other, but which as a whole have wanted to offer urban improvement solutions capable of qualifying the aforementioned interstitial space and its surroundings.

4.1 Morphological Typology, Density and Tracing

After the detailed analysis of the 18 proposals studied (9 in Lisbon and 9 in Barcelona), it is very curious to observe how the urban qualification scenarios for these areas can be so open and admit morphological, density and layout typologies so different from each other. With regard to morphology, at least four types of action have been read: the orthogonal type proposals, the organic type, the radial type and, a few, that by mixing elements of at least two of the morphological types listed above, make up some hybrid type proposals.

Orthogonal Morphology Proposals

From the reading carried out, it is concluded that the most explored morphological types are still orthogonal, adding up to seven proposals in total. However, although they may keep certain elements in common with each other, such as wide, straight and perpendicular streets, creating a checkerboard shape, the spatial diversity achieved and the urban elements put in value are very rich and heterogeneous (Fig. 4).

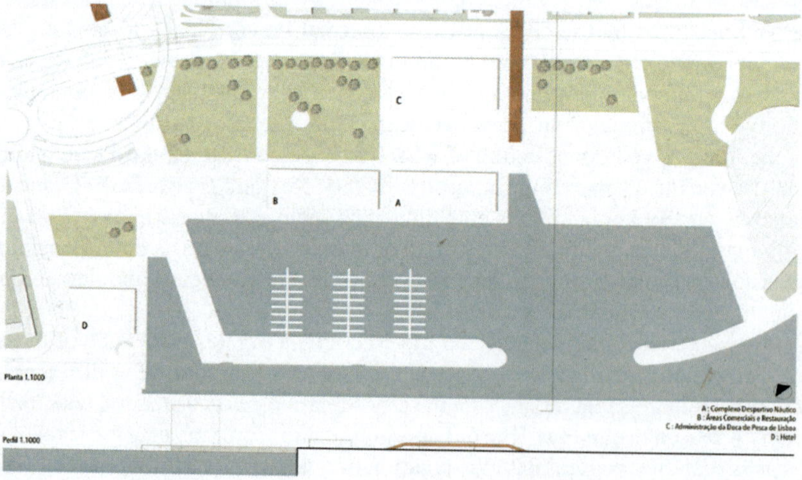

Fig. 4. Group 1Lx Proposal, Correia, Reis and Gordalina, 2019

Organic Morphology Proposals

There are fewer proposals that develop an organic urban morphology (2 in Lisbon and 2 in Barcelona), but the results are no less interesting and the proposed urban public spaces are very diverse, creative and innovative (Fig. 5).

Fig. 5. Group 4Lx Proposal, Junior, Tumpej, Alves and Reis, 2019.

Radial Morphology Proposals

The urban proposals that are framed in a radial type morphology have been the least numerous (only 2) and have been developed only for Lisbon (Fig. 6).

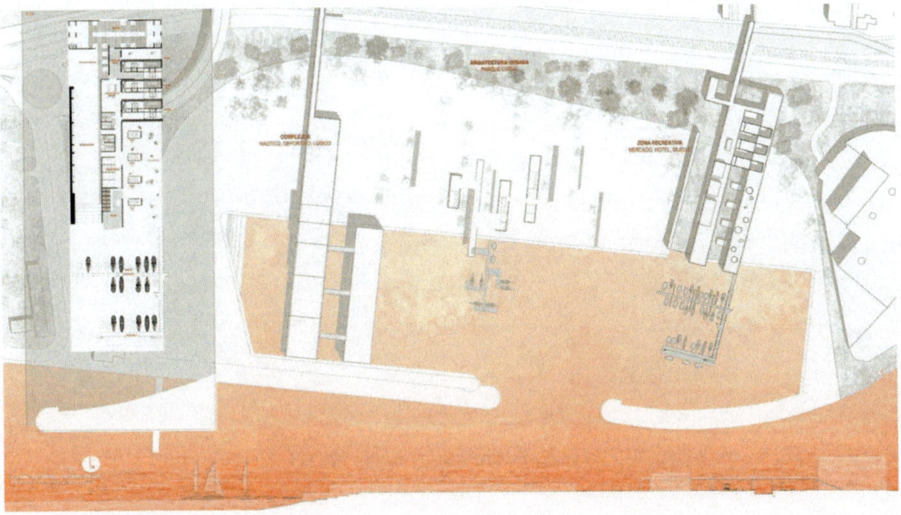

Fig. 6. Group 9Lx Proposal, Alvarenga, Atristain, Redondo and Rivera, 2019.

Hybrid Morphology Proposals

Therefore, we can highlight five proposals that configure a type of hybrid morphology (2 in Lisbon and 3 in Barcelona) that also present a great diversity of urban solutions for the terrain vague studied (Figs. 7 and 8).

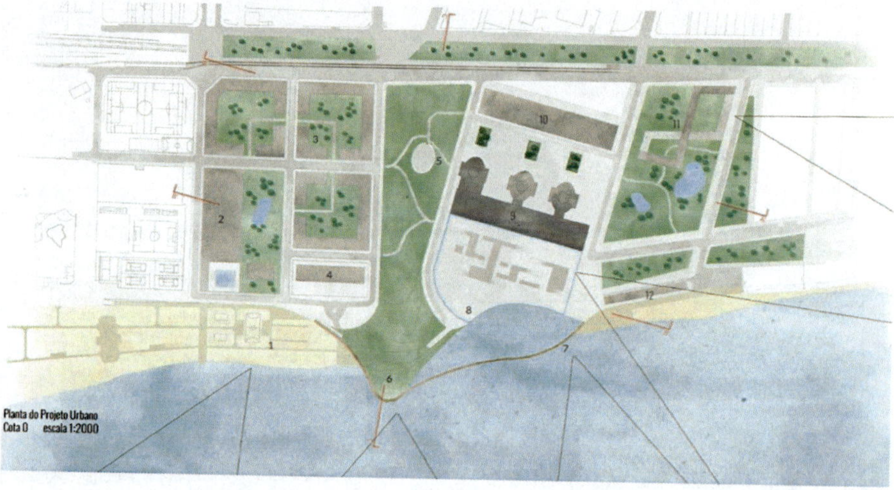

Fig. 7. Group 2Bcn Proposal, Barbosa, Ramos, Rocha and Lourenço, 2020.

Fig. 8. Pedrouços, Louro, 2020

5 Concluding Notes

From urban planning and architecture there are many different answers that we can offer to areas that present the same type of problems and opportunities. In other words, both territories - Doca de Pedrouços and Front Marítim de les tres Chimeneas - suffer from similar problems: territorial isolation; physical barriers raised by heavy mobility infrastructures that are difficult to overcome; social stigma due to previously existing industrial activities; abandonment and obsolescence of the built heritage, etc. and opportunities with a certain resemblance: a location that was previously peripheral but currently of extraordinary centrality; well communicated areas; and, given the scarcity of land in both Lisbon and Barcelona, these plots have a high economic interest, reinforced because the land is located next to the coastline. But the actions to be carried out can be all that we want them to be (preferably consensual and shared with society) and the research presented here is nothing more than a set of examples of different scenarios to choose from: if for more organic morphologies, orthogonal, radial or hybrid; if by a high, low or intermediate density and if by paths that break with the existing ones or otherwise they are drawn by continuity.

Finally, the relationship with preexistence, that is, the existing heritage legacies in the two contexts (cold stores in Pedrouços and chimneys and turbine room in Sant Adrià de Besòs) and the way in which these have more or less influenced the various proposals. Ultimately, all the proposals presented here not only try to integrate in one way or another the historical memory of these preexistence, but the preexistence constitutes the starting point for the project of a new urbanity and that attitude marks a position very clear about the strategy of projecting with what has been built. It is evident that this is an open process… and we believe that there is no better laboratory than the academic one to carry out all the inquiries and project explorations that allow us to qualify the contemporary urban.

References

Águila, J.L.: Espacio intersticial. Surgimiento y transformación. Caso: Tonalá, Jalisco en México. Ediciones Universidad Internacional de Andalucía, Sevilla (2014)

Amendola, G.: La ciudad postmoderna. Celeste Ediciones, Roma (1997)

Augé, M.: Los No lugares, espacios del anonimato. Una antropología de la sobremodernidad. Editorial Gedisa, Barcelona (2000)

Bazant, J.: Interpretación teórica de los procesos de expansión y consolidación urbana de las periferias. Rev. Estud. Demogr. Urbanos **16**(2) (47). El Colegio de México, pp. 351–374 (2001a). - Periferias urbanas. Expansión urbana incontrolada de bajos ingresos y su impacto en el medio ambiente. Ed. Trillas, México (2001b)

Busquets, J.: Los proyectos de renovación de los “frentes de agua”. In: Nueva arquitectura urbana en la ciudad europea. VIII Curso Intern. Rehab. Urbana: Las experiencias europeas de los 80; Edición Univ. Intern. Menéndez Pelayo, Cuenca (1990) - Spanish Waterfronts/Los frentes de agua españoles. In: Portus, nº1, Marsilio Editori, Venice (2001). - A Reabilitação das “frentes de agua”. In: Estuário, nº1 y nº2, AML, Lisboa (2003)

Cedeño, M.C.: Homepage (Espacios y periferias). https://espaciosyperiferias.blogspot.com/2007/10/intersticios-urbanos.html. Accessed 12 Apr 2020

Clichevsky, N.: La tierra vacante, revisitada. Elementos explicativos y potencialidades en su utilización. Cuaderno Urbano **6**, 195–220 (2007)

Documento Estratégico Doca de Pedrouços, Memória Descritiva/Proposta de Implantação. Câmara Municipal de Lisboa. APL-Administração do Porto de Lisboa, S.A., Lisboa (2011)

Fausto, A., Rábago, J.: ¿Vacíos urbanos o vacíos de poder metropolitano? In: Ciudades, revista trimestral Red Nacional de Invest. Urbana. Núm. 49, México D.F., pp. 33–39 (2001)

Machin, H.: Homepage (Espacios urbanos olvidados. De-DEDedos). https://dedededos.wordpress.com/2009/07/25hueco-urbano-1/. Accessed 08 May 2020

Mashini, D.: Homepage (Waterfronts: éxitos y fracasos de íconos de regeneración urbana). https://www.plataformaurbana.cl/archive/2010/07/27/waterfronts-exitos-y-fracasos-de-regeneracion-urbana/. Accessed 08 Apr 2020

Méndez, E.: Arquitectura simulacro. Posgrados PNP CONACYT, UdeG, CUAAD, CUC, Guadalajara, (2009). - Imaginarios urbanos del turismo en la morfología de Puerto Peñasco. In: Méndez, E.C., Isabel, R., Jesús, E. (eds.) Imaginarios y paisajes del turismo. Ciudades y relatos frente al mar. Ed. Acad. Española, Saarbrucken (2011). - Imagen de ciudad: de Al Capone a los gobernadores en Puerto Peñasco. In: 53er Cong. Intern. Americanistas (2009)

Messen, R.: Homepage (Exploración y puesta en valor de vacíos urbanos, como estrategias de desarrollo para barrios periféricos marginales de Santiago). https://es.scribd.com/document/343997689/Exploracion-y-Puesta-en-Valor-de-Vacios-Urbanos-Como-Estrategia-de-Desarrollo-Para-Barrios-Perifericos-Marginales-de-Santiago. Accessed 5 May 2020

Peimbert, A.J.: Paisaje intersticial. Mexicali: de la entropía al vacío. Tesis para obtener el grado de Maestro en Arquitectura (not published). Universidad Autónoma de México (2008)

Plano Geral de Intervenções da Frente Ribeirinha de Lisboa – Documento de Enquadramento. Câmara Municipal de Lisboa, Lisboa (2008)

Pla Director Urbanístic d'Ordenació del Front Litoral en l'Ambit de les Tes Ximenees. Doc. d'objectius i propòsits generals: Gen. Catalunya. DTS-DGOTU, Barcelona (2018)

Solá-Morales, I.: Presente y futuros. La arquitectura en las ciudades. In: A.A.W., Presente y Futuros. Arquitectura de las grandes ciudades, pp. 10–23, Col.legi Oficial de Arquitectes de Catalunya, Centre de Cultura Contemporánea, Barcelona (1996); - Territorios. Ed GG, Barcelona (2002)

Energy-Water-Food Nexus: The Case of the Sahel Region

Moses S. Bass[1,2(✉)], A. Lopez-Agüera[2], and Míguez Farina Antía[2]

[1] Pan African University, Institute for Water and Energy Sciences-PAUWES, c/o, Tlemcen University, Pôle Chetouane, B.P. 119, 13000 Tlemcen, Algeria

[2] Department of Physics, Santiago de Compostela University, Rúa Xosé María Suárez Núñez, s/n. (Campus Vida), 15782 Santiago de Compostela, Spain

Abstract. The NEXUS philosophy is one of the most significant aspects of the United Nations Agenda 2030. One of the cardinal principles of the Agenda 2030 is "leaving no one behind", and as such, the provision of universal access to energy, water and food to all and sundry is a key aspiration. In that regard, this paper gives an overview of the progress registered using the Sahel region of Africa as a case study. The article Proposes a method for assessing the current situation of countries in the Sahel based on energy, water and food data. The proposed method is based on 9 indicators which have been identified as sensible for the analysis. Ten countries including Burkina Faso, Chad, Mauritania, Niger, Cameroon, The Gambia, Mali, Senegal, Guinea, and Nigeria were studied, and the results shows that only eight countries have attained fifty percent progress in the overall nexus index and two are below fifty percent. The findings of this study are important for policy makers, international donor agencies as well as philanthropies.

Keywords: Energy-water-food nexus · Sustainable development goals · Composite index · Sahel region

1 Introduction

The availability of energy, water and food in any given society is quite significant for the sustenance of life. Therefore, ensuring the adequate, affordable, accessible and sustainable supply of these three is hugely important. Energy, water and food are seen to be in a nexus (Rasul 2016). This is based on the fact that while water treatment, supply and extraction requires energy, energy production requires water and in the same vein, food production requires both energy and water (Allan et al. 2015; Bazilian et al. 2011). This nexus philosophy has been highlighted by the former United Nations Secretary General, Ban-ki Moon during the world water day in 2011 as one of the challenges facing mankind (Simpson and Jewitt 2019). It however became widespread and drew attention among international development agencies, policy makers, academics among others during the Bonn2011 conference on Water Energy and Food Security Nexus (Hoff 2011).

According to (NIC 2012), the world shall witness a substantial increase in demand for energy (50%), water (40%) and food (35%) by 2030 and this will be as a result of

J. R. da Costa Sanches Galvão et al. (Eds.): *Proceedings of the 1st International Conference on Water Energy Food and Sustainability (ICoWEFS 2021)*, pp. 691–698, 2021.
https://doi.org/10.1007/978-3-030-75315-3_73

unsustainable economic and population growth, rapid urbanization as well as climate change (Hoff 2011). With such projections in mind, the challenge of providing universal access to energy, water and food is undoubtedly enormous and what makes matters more unfavorable is that the world is already off-track in achieving universal access by 2030 (FAO, IFAD, UNICEF, WFP and WHO 2019; IEA, IRENA, UNSD, WB, WHO 2019; UN Water 2018). The advent of COVID19 is also likely to hamper already registered progress in terms of achieving the Sustainable Development Goals (Filho et al. 2020). Nevertheless, it is still crucial to monitor progress registered in terms of energy, water and food in an integrated way. A few researchers (Mahlknecht and González-Bravo 2018; Simpson and Berchner 2017) have recently attempted to do so and both proposed a water-energy-food (WEF) Index wherein the former uses three indicators from each sector while the latter used only two indicators from each sector. This paper adopted a similar method to that of (Mahlknecht and González-Bravo 2018) with slight modifications in the indicators in order to suite the purpose of the study but nevertheless, the selected indicators serve the purpose of highlighting the level of sustainability and, country's population vulnerability in terms of energy, water and food access.

2 The Sahel in Context

The Sahel region of Africa lacks a definition that is universally accepted and as a result, there is a plethora of definitions that one can come across when attempting to study the region (Epule et al. 2017; Helly et al. 2015). According to (Walther and Retaillé 2010) The Sahel is located between south of the Sudanese savannah and north of the Sahara. It is characterized by different and complex economic structures, geopolitical forces, ethnic and social relations as well as commercial networks that existed over centuries. Another unique fact about the Sahel is that there is no recognition of it as an international geographical boundary under the administration of a single state or form of government. It however stretches across several African countries to which there is no consensus (Helly et al. 2015). For the purpose of this paper, we have adopted the ten countries included in the United Nations Support Plan for the Sahel (2018–2030) and they are Burkina Faso, Chad, Mauritania, Nigeria, Cameroon, The Gambia, Mali, Senegal, Guinea, and Niger.

The region comprises semi-humid and humid tropical climate of tropical-zone and equatorial rain forest in the south but grass and woodlands in the North (Potts et al. 2013). A unique climatic feature noted about the Sahel is that it is experiencing an increase in temperatures coupled with uncertainties in rainfall patterns (World Bank 2020). The inhabitants of the Sahel are predominantly nomads and pastoralists whose main occupation is farming, fishing or livestock rearing (UN.ECA 2016) but the rapid increase in population remains a matter of great concern. (Potts et al. 2013) warned that conflict and large-scale starvation could escalate if sustainable development plans are not instituted due to the fact that population is gradually exceeding the availability of natural resources which is already undermined by climate change. From a socioeconomic standpoint, almost all countries in the Sahel are classified as least developed countries (UN-OHRLLS 2018) wherein fifty-one percent of the people are affected by mass poverty (UNDP 2015) and forty-one percent live below two dollars per day (World

Bank 2020). According to (UN 2018), a large number of people in the Sahel do not have access to basic services such as energy, clean water and food.

3 Methodology

In order to measure the energy-water-food nexus index, a total of nine parameters were used, three parameters for each sector (Energy-Water-Food). The scale of measurement for each of the parameters is from 0 to 100%. The index values for each country were obtained by averaging all the nine parameters and the more a country's index is closer to the number 1 indicates its closeness to achieving universal access to energy, water and food and high level of sustainability. The data used in this study is quantitative in nature and was obtained from online sources run by international organizations such as World Health Organization & United Nations Children's Fund (WHO/UNICEF), International Renewable Energy Agency (IRENA), Food and Agriculture Organisation (FAO) as well as the World Bank Group (Table 1). The study period covered was only limited to the year 2017.

Table 1. Energy-water-food nexus index dimensions, parameters and sources.

Dimension	Parameters	Definition	Source
Energy	Electricity access	Population share with access to electricity	(IRENA 2020)
	Access to clean cooking	Population share with access to clean fuels for cooking	
	Regulatory Indicators for Sustainable Energy (RISE) score	Country's support and preparedness for SDG 7	(RISE 2020)
Water	Basic water service	Sources of water available for drinking, cooking and other domestic purposes and time of water access not more than 30 min	(WHO/UNICEF 2020)
	Basic sanitation service	Proper management of human waste	
	Basic hygiene service	Conditions for averting the spread of diseases	
Food	Cereal import independency ratio	Country's own production of cereals as opposed to imports	(FAO 2020)
	Non-prevalence of undernourishment	Share of the population that is not undernourished	
	Average dietary energy supply adequacy	Dietary energy supply of calories as a percentage of the average dietary energy requirement estimated by F.A.O. for the population	

4 Results

Pertaining to energy, access to electricity is recorded highest in Senegal (61.7%), Cameroon (61%) and The Gambia (56%) whereas it is lowest in Chad (12%), Niger (20%) and Burkina Faso (25.5%) respectively. There is as well a general low access to clean cooking fuels in the entire region with the exception of Mauritania which achieved close to fifty percent. The achievement of SDG 7 requires a concerted effort from both government and the private sector. On the one hand, the former ensures the creation of a level playing ground by instituting the right policies and regulatory frameworks while on the other hand the former supports with the necessary investments. However, Chad, Mauritania, Mali and Nigeria all scored poorly in terms of their preparedness and support for sustainable energy for all Fig. 1.

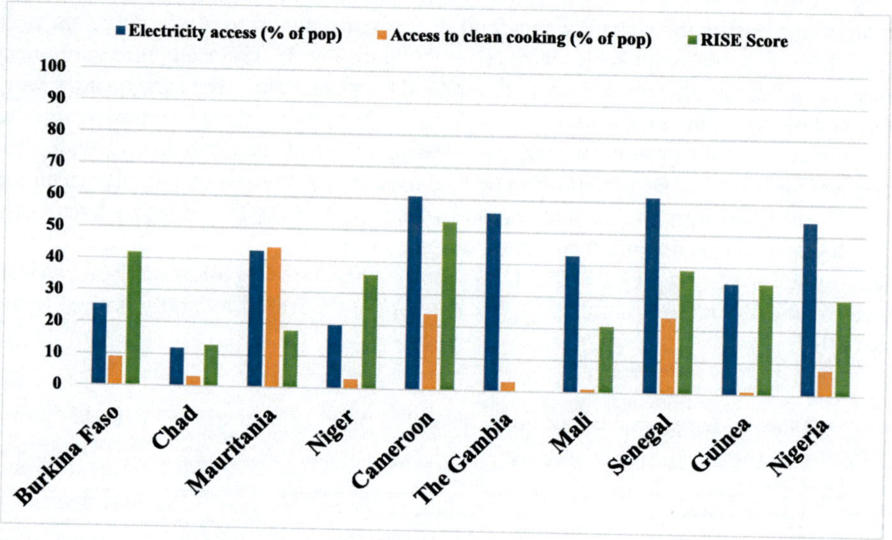

Fig. 1. SDG 7 progress registered in the Sahel (2017). Source: authors elaboration with data obtained from (IRENA 2020) and (RISE 2020)

With regards to Water, more progress is registered in the provision of basic drinking water compared to basic sanitation and hygiene respectively. There are more people in Senegal (81%), Mali (78%) and The Gambia (78%) that have access to basic drinking water in contrast to Chad, Burkina Faso as well as Niger where access is still very low. Access to sanitation is generally higher than hygiene in the region but only Mauritania managed to register up to 40% for both in the entire Sahel. Apart from Chad, the rest of the countries have surpassed their dietary supply adequacy. Self-production of cereals is highest in Chad (92.7) but nevertheless, it is also the country with the highest prevalence of undernourishment (37.5%). Mauritania and Senegal are highly dependent on cereal imports with each failing to meet 50% of its supply (Fig. 3). The WEF index which is a quantitative aggregation of progress registered in the provision of universal access to energy, water and food in consideration of all nine parameters is presented in Fig. 4. Mali (0.60), Cameroon (0.58) and Senegal (0.56) records the highest with Chad (0.37), Niger (0.46) and Burkina Faso (0.50) registering the lowest.

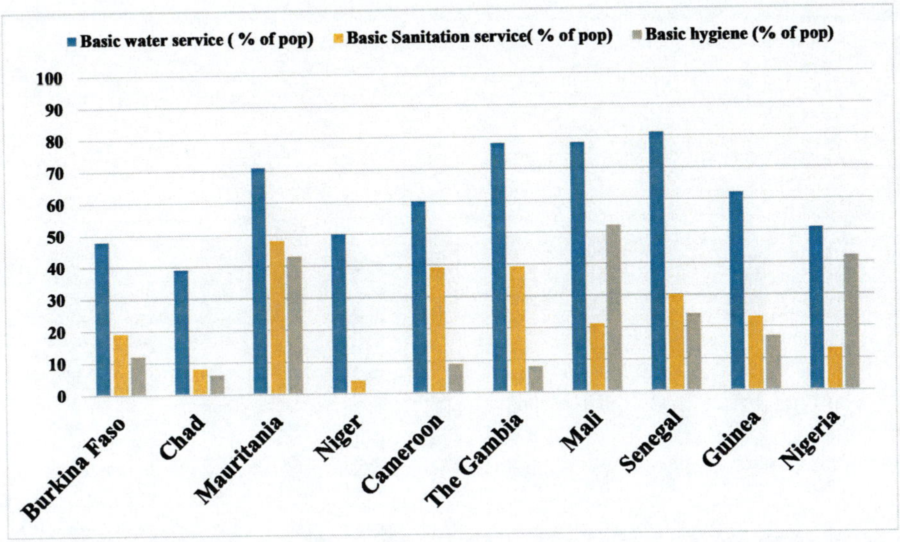

Fig. 2. SDG 6 progress registered in the Sahel (2017). Source: authors elaboration with data from (WHO/UNICEF, 2020)

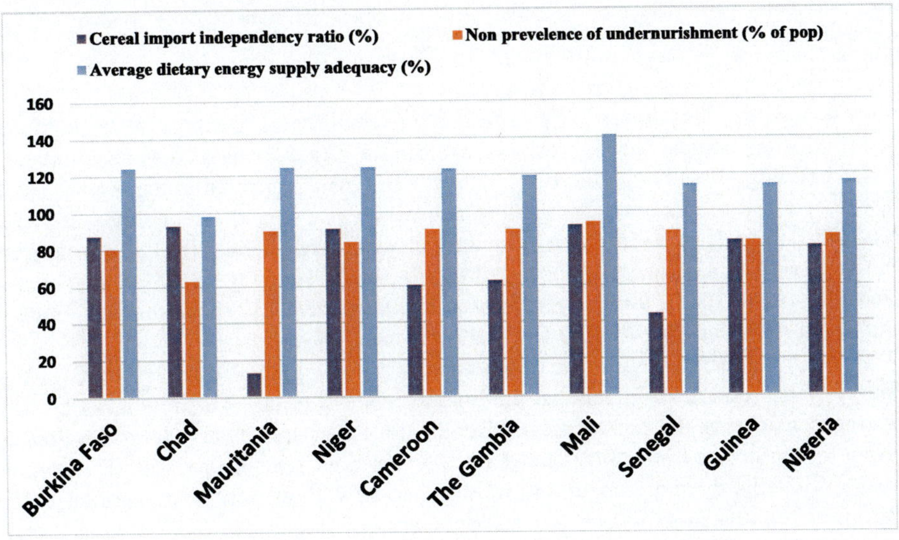

Fig. 3. SDG 2 progress registered in the Sahel (2017). Source: authors elaboration with data from (FAO 2020)

5 Discussion

The Sahel is diverse in many ways and this is also reflected in the various country's performance in terms of meeting universal access to energy, water and food. The results

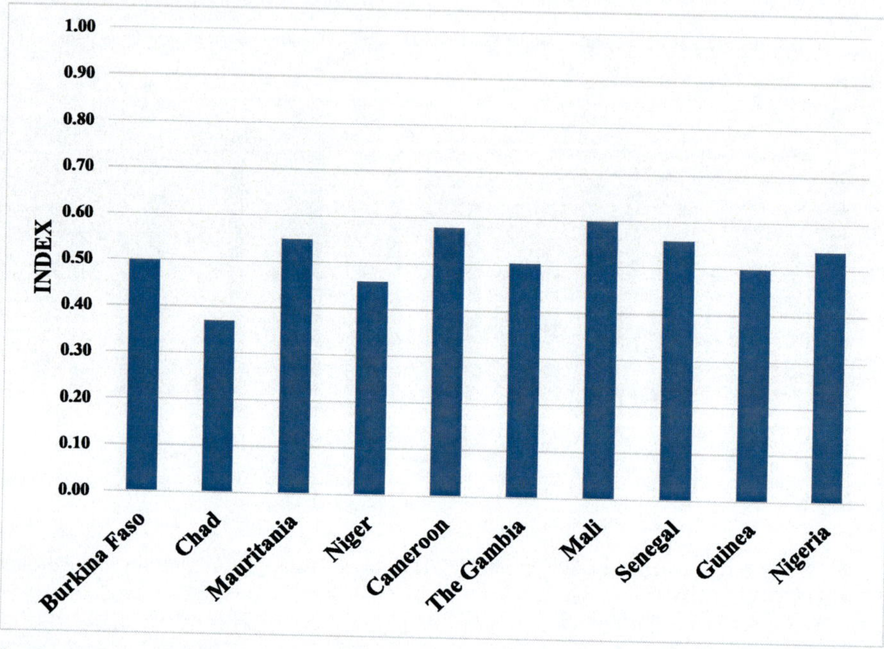

Fig. 4. Country performance in the overall EWF nexus in the Sahel (2017). Source: authors elaboration.

show that none of the countries is yet to attain universal access to energy, water and food and Mali is the country that is most advanced in the region as far as the nexus index is concerned, Fig. 4. Although there is noticeable progress registered in the region, other researchers that used a nexus index approach which also considered energy, water and food albeit slightly different parameters have shown higher indices for Latin American countries such as Uruguay (0.96), Brazil (0.93) and Paraguay (0.92) (Mahlknecht and González-Bravo 2018). Similarly, (Simpson and Berchner 2017) have shown Germany and South Africa scoring (0.877) and (0.728) respectively.

From a comparative view the Sahel appears to be lagging behind in terms of the energy-water-food nexus given that the highest index is (0.60). This might not be surprising considering the fact that countries in the Sahel are majorly least developed, prone to conflict and instability and have also suffered several droughts (Essoungou 2013). Globally, there are about 840 million people without access to electricity, 844 million without access to basic water service and more than 840 million hungry people and Africa and the Sahel in particular have the highest level of vulnerability (IEA, IRENA, UNSD, WB, WHO 2019), (UN Water 2018), (FAO, IFAD, UNICEF, WFP and WHO 2019). Given this current situation, it is clearly manifested that the global target of reaching universal access to energy, water and food by 2030 have been diminished and the persistence of COVID-19 will to a large extent retard progress unless more robust strategies are installed. The nexus approach is therefore crucial in this regard and can immensely contribute to providing universal access to energy, water and food in a sustainable manner.

6 Conclusions

The purpose of this study was to develop an energy-water-food nexus index for the Sahel region of Africa in order to enable the prioritization of countries for the implementation of "Sustainable Communities Projects (SCP)". The project aims at developing sustainable and resilient communities in the Sahel through technological cooperation that takes energy-water-food nexus into consideration. The EWF nexus index proposed in this paper helps to identify which countries are advancing in terms of meeting the sustainable development goals particularly SDG 7, SDG 6 and SDG 2 and can also be used to assess regional performance. Countries underperforming are also identified. The findings of this study are crucial not only for the SCP but by extension the monitoring of progress in the Sahel and can equally help policy makers, international donor agencies as well as philanthropies to get informed about the wellbeing of people living in the region. Though this study is basically concerned with reporting on current registered progress and consideration of countries for SCP, further studies could be undertaken in order to develop more sustainable development scenarios that could see the Sahel emerging as a region that is able to provide universal access to energy, water and food.

References

Allan, T., Keulertz, M., Woertz, E.: The water–food–energy nexus: an introduction to nexus concepts and some conceptual and operational problems. Int. J. Water Resour. Dev. **31**(3), 301–311 (2015). https://doi.org/10.1080/07900627.2015.1029118

Bazilian, M., Rogner, H., Howells, M., Hermann, S., Arent, D., Gielen, D., Steduto, P., Mueller, A., Komor, P., Tol, R.S.J., Yumkella, K.K.: Considering the energy, water and food nexus: towards an integrated modelling approach. Energy Policy **39**(12), 7896–7906 (2011). https://doi.org/10.1016/j.enpol.2011.09.039

Epule, T.E., Ford, J.D., Lwasa, S., Lepage, L.: Climate change adaptation in the Sahel. Environ. Sci. Policy **75**, 121–137 (2017). https://doi.org/10.1016/j.envsci.2017.05.018

Essoungou, A.-M.: The Sahel: One region, many crises. Afr. Renew. **27**(3), 22–23 (2013). https://doi.org/10.18356/982a70b2-en

FAO: FAOSTAT (2020). https://www.fao.org/faostat/en/#country. Accessed 23 Nov 2020

FAO, IFAD, UNICEF, WFP and WHO: The State of Food Security and Nutrition in the World 2019. Safeguarding against economic slowdowns and downturns. Rome, FAO (2019)

Filho, W.L., Brandli, L.L., Salvia, A.L., Rayman-Bacchus, L., Platje, J.: COVID-19 and the UN sustainable development goals: threat to solidarity or an opportunity? 14 (2020)

Helly, D., Théroux-Bénoni, L.-A., Galeazzi, G., Maïga, I., Ouédraogo, F.: Sahel strategies: why coordination is imperative. Inst. Secur. Stud. Policy Brief **12** (2015). https://www.files.ethz.ch/isn/190285/PolBrief76Eng.pdf

Hoff, H.: Understanding the nexus. In: Background Paper for the Bonn2011 Conference: The Water, Energy and Food Security Nexus. Stockholm Environment Institute, Stockholm (2011). https://mediamanager.sei.org/documents/Publications/SEI-Paper-Hoff-UnderstandingTheNexus-2011.pdf

IEA, IRENA, UNSD, WB, WHO: Tracking SDG 7: The Energy Progress Report (2019). https://www.irena.org/-/media/Files/IRENA/Agency/Publication/2019/May/2019-Tracking-SDG7-Report.pdf

IRENA: Statistical Profiles. /Statistics/Statistical-Profiles (2020). https://www.irena.org/Statistics/Statistical-Profiles. Accessed 15 Nov 2020

Mahlknecht, J., González-Bravo, R.: Measuring the water-energy-food nexus: the case of latin america and the Caribbean Region. Energy Proc. **153**, 169–173 (2018). https://doi.org/10.1016/j.egypro.2018.10.065

NIC: National Intelligence Council Global Trends 2030: Alternative Worlds | Public Intelligence (2012). https://publicintelligence.net/global-trends-2030/

Potts, M., Henderson, C., Campbell, M.: The Sahel: a Malthusian challenge? Environ. Resour. Econ. **55**(4), 501–512 (2013). https://doi.org/10.1007/s10640-013-9679-2

Rasul, G.: Managing the food, water, and energy nexus for achieving the sustainable development goals in South Asia. Environ. Dev. **18**, 14–25 (2016). https://doi.org/10.1016/j.envdev.2015.12.001

RISE: Regulatory Indicators for Sustainable Energy (2020). https://rise.esmap.org/. Accessed 17 Nov 2020

Simpson, G.B., Jewitt, G.P.W.: The development of the water-energy-food nexus as a framework for achieving resource security: a review. Front. Environ. Sci. **7**, 8 (2019). https://doi.org/10.3389/fenvs.2019.00008

Simpson, G., Berchner, M.: Measuring integration—towards a water-energy-food nexus index. Water Wheel 22–23 (2017)

UN: UN SUPPORT PLAN FOR THE SAHEL: Working together for a prosperous and peaceful sahel (2018). https://www.un.org/africarenewal/sites/www.un.org.africarenewal/files/English%20Summary%20Report_0.pdf

UN Water: Sustainable Development Goal 6: Synthesis report 2018 on water and sanitation. United Nations (2018)

UNDP: UNDP strategy for the Sahel. UNDP (2015). https://www.undp.org/content/undp/en/home/librarypage/crisis-prevention-and-recovery/undp-strategy-for-the-sahel.html

UN.ECA: ECA Sahel study: Conflict in the Sahel region and the developmental consequences. United Nations. Economic Commission for Africa (2016). https://repository.uneca.org/handle/10855/23474

UN-OHRLLS: List of Least Developed Countries (2018). UN-OHRLLS. https://unohrlls.org/

Walther, O., Retaillé, D.: Sahara or sahel? The fuzzy geography of terrorism in West Africa. SSRN Electron. J. (2010). https://doi.org/10.2139/ssrn.1803996

WHO/UNICEF. (2020). Sub-Saharan Africa. https://washdata.org/data/household#!/, last accessed 2020/11/20

World Bank: Sahel Adaptive Social Protection Program (ASPP) (2020). https://www.worldbank.org/en/programs/sahel-adaptive-social-protection-program-trust-fund. Accessed 24 June 2020

Industry 4.0 for Sustainable Production in Footwear Industry

Tatjana Spahiu[1](✉), Henrique Almeida[2], Athanasios Manavis[3], Panagiotis Kyratsis[3], and Antonio Jimeno-Morenilla[4]

[1] Faculty of Mechanical Engineering, Textile and Fashion Department, Polytechnic University of Tirana, Tirana, Albania
tspahiu@fim.edu.al

[2] School of Technology and Management, Polytechnic Institute of Leiria, Leiria, Portugal

[3] Department of Product and Systems Design Engineering, University of Western Macedonia, Kila, Kozani, Greece

[4] Department of Computer Technology, University of Alicante, Alicante, Spain

Abstract. Technological advancements as part of fourth industrial revolution occurring every day have changed the way or the process that various products are created from concept to the final product. In this way by contributing even to a sustainable production in various industries. Due to the fact that footwear industry is the one classified as the fifth regarding negative impact on environment among various sectors of industry, sustainable production is required. The aim of this paper is to present some evidence related with footwear consumption and its negative impact on the environment. Moreover by focusing on the role of sustainable materials an analysis conducted from other works regarding the implementation of sustainable production in footwear industry. As Industry 4.0 is well known to support sustainable production, the presented work will reveal through cases of its implementation, the contribution of Industry 4.0 within this scope, and the role of additive manufacturing as an enabler to support the sustainable production in footwear industry.

Keywords: Industry 4.0 · Sustainable production · Footwear · 3D printing

1 Introduction

Footwear has a long history since their use for the first time. They have evolved over time and come in different shapes and materials. At first their role was to protect the foot and now the other one is linked with aesthetic part. Moreover, the last one has become an important factor in buying decision of consumers. Usually, the shoes production is related with physical and chemical processes ordered to merge the upper and lower part of the shoes. Earlier their production was done mainly as a manual work but with the inventions of various equipment's used for shoes manufacturing, they have shifted to a mechanical production but still footwear industry is considered as a labor intensive work.

J. R. da Costa Sanches Galvão et al. (Eds.): *Proceedings of the 1st International Conference on Water Energy Food and Sustainability (ICoWEFS 2021)*, pp. 699–707, 2021.
https://doi.org/10.1007/978-3-030-75315-3_74

Further developments as advanced manufacturing systems part of Industry 4.0 or forth industrial revolution has changed the way of production with application in every industrial area. In a globalized word, they make possible to offer products for every consumer. These technologies are able to produce in a more cost-effective way and to support sustainable production, which is required in order to reduce waste and pollution or negative impact on environment caused by various sectors of industry. Initiatives taken by governments worldwide to encourage or support the implementation of Industry 4.0 in order to be aware and shift their production to the new paradigm of Industry 4.0 will help them to raise productivity, quality, reducing waste, in other words to be competitive [1], and moreover in a sustainable development [2]. The traditional manufacturing sector as footwear, textiles and clothing can benefit from the implementation of Industry 4.0 [3] where simulation contribute to process optimization to increase productivity [4, 5]. Introducing technologies that link the use of CAD/CAM with old machinery is of a great interest for footwear sector [6] and to outsource intensive aspects of CAD/CAE functions with minimum costs by managing CAD/CAE as a Service [7].

Among various sectors of industrial production, footwear industry is the one classified as the fifth regarding negative impact on environment. As we are faced everyday with pollution caused by various industries, in order to maintain and help to lower it, solutions are required from various players to change their traditional practices of manufacturing in a more sustainable way. These solutions can be projected since at the designing stage and development of innovative materials. Product design is evaluated as the biggest contributor to sustainable production and design for sustainability should be integrated in the strategy and process of organizations [8, 9]. Collaborations between companies and other stakeholders are seen as opportunities in order to reduce emissions within apparel and footwear sector [10]. According to Footwear Distributors & Retailers of America (FDRA) shoe sustainability is the shoe design, development, manufacturing, distribution, and selling processes that minimize negative environmental impacts, conserve energy and natural resources, are safe for employees, communities and consumers, and are economically sound [11].

Several technologies even called as pillars of Industry 4.0 or Smart revolution have an important role to enhance sustainability in Textile & Apparel Industry by contributing to whole system of production, merchandizing, advertising, retailing, etc. [12]. Despite the fact that industrial revolution is now in the middle, the integration of digital manufacturing is still not spread at a high level. Measuring the integration of Industry 4.0 by companies would be of a great interest in order to push them to this new digital paradigm. As companies requires specific competences in the field of Industry 4.0, changing and adapting the curricula will increase the number of students interested in technical areas [13] and focusing on sustainability [14]. Products like clothing and shoes have faster changes compared to other products and buyers seek for personalized products which is of a great interest for modern consumer [15]. The EU package includes the use of economic instruments and other measures to provide incentives for the application of the waste hierarchy, such as the extended producer responsibility, and appears to introduce a change in the concept of waste that is increasingly treated as a resource, with a significant impact on the fashion industry [16]. Figure 1 presents a simplified workflow of the main process used for footwear production.

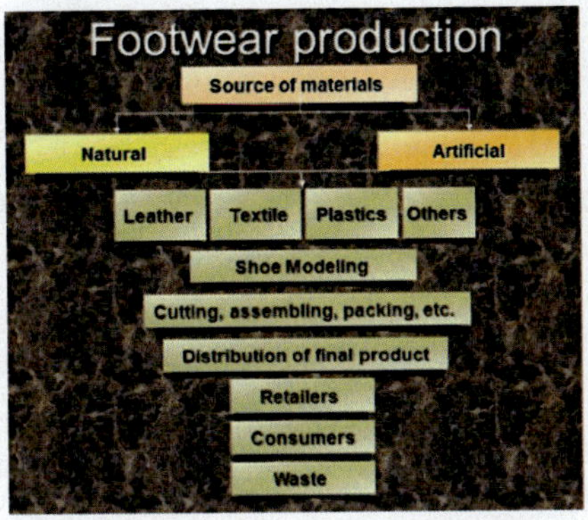

Fig. 1. Footwear production.

The aim of this research is to analyze current sustainable trends in footwear production. The structure of this paper is as follows: Sect. 2 presents sustainable trend that are occurring in footwear manufacturing. Third section includes the role and use of additive technology for footwear products, as one of the main pillars of Industry 4.0. Last section deals with the conclusions of this study.

2 Sustainable Trends in Footwear Manufacturing

2.1 Emerging Needs to Shift into Sustainable Production

Materials used for footwear production vary from leather, plastic, textiles, etc. These materials have a direct impact on comfort sensation experienced by users, especially the lining and insole which are in direct contact with the foot. In order to achieve the optimal conditions of temperature and humidity for users during physical effort, material combinations present great results [17]. Due to the negative effect on environment that footwear production contributes, nowadays sustainable footwear is a strategic drive for every company operating in this sector. As a result, footwear manufactures should be continuously aware to find innovative materials and new ways of production, merchandizing, retailing, where each one of them contributes to sustainability. If we came to values, it is evaluated that footwear industry contributes upwards of 700 million metric tons (MT) of CO_2. There is an increase of companies that highlights sustainability [18, 19]. The global sustainable footwear market size was valued at USD 7.5 billion in 2019 and is expected to grow at a compound annual growth rate (CAGR) of 5.8% from 2020 to 2027 [20]. However, there is a lot of work to be done in order to reduce the negative impact. If we take into account the role of consumer is evident that his purchasing behavior raises this negative impact. Despite the known fact of the negative impact that fashion industry contributes to the environment, the fast fashion has a fundamental role to these higher values. As a result companies should shift their production to slow fashion [21].The carbon footprint caused by footwear products is not only during the shoe production but even during its use, in other words using life cycle assessment for a pair of running shoes made of synthetic material the value is 14 ± 2.7 kg CO_2-equivalent

[22]. As a result manufactures should reduce materials consumption and select materials produced under sustainable criteria [23, 24]. Despite the fact that leather products which are natural materials that offer various advantages regarding fit and comfort they have the most detrimental environmental impacts across their life cycle [25]. Moreover due to the known fact that leather and footwear industries produce a major solid waste, waste recycling will help in waste free environment [26].

2.2 Eco Friendly Solution as an Alternative for Footwear Products

Among various types of the footwear products, athletic footwear presents a market with a tremendous growth. According to report presented by Grand View Research is expected to reach 95.14 billion (USD) by 2025 [27]. But the problem faced with sneaker production is the pollution. It is evaluated with a value of 1.4% of the global greenhouse gas emissions [28]. According to a study for lifecycle assessment conducted by MIT, pair of running shoes is evaluated to generate 30 lb of carbon dioxide emissions [29]. All these facts bring raise awareness where well-known companies as Timberland, Adidas, Nike, New Balance, etc., are continuously working to more sustainable production. Adidas a sportive shoes brand by developing sustainable strategies has started with shoe production using plastic waste and by the end of 2050; it aims to achieve climate neutrality [30]. Even Nike products present sustainability in production through the use of recycled plastic, natural dyes, or recycled leather [31]. New Balance in its campaign for sustainable production includes recycling in order to reduce carbon footprints by 30% on 2030 [32]. The uses of biodegradable materials in textile and footwear industry have attracted a great attention due to the ability that these materials have to biologically degrade by reducing carbon dioxide. Eco leather materials are a good alternative to replace leather materials. They are environmentally friendly and fulfill customers' requirements. There is a great number of vegan leather produced from waste of fruit as apple, banana, grape, pineapple, green tea, kombucha tea, mushrooms, cork, etc. Several products as shoes, hand bags, and accessories are produced. Studies presented by various authors depict the possibilities to produce shoes using sustainable materials [33] and products costs evaluation [34]. Bacterial cellulose as one of bio-based materials offers several advantages as a leather substitute [35]. The multi layered cellulosic material developed by bonding cellulosic fiber mat, denim and hemp fabric presents no difference or higher properties of heat loss, break force and permeability by promising an alternative of leather substitutes [36]. The case of mycelium composite developed by researchers presents an application for sole by providing a great possibility for sustainable materials used for footwear production [37].

Apart from the materials used for footwear production, adhesives used to assemble various components of the shoes play an important role to ensure the quality of the final product [38]. They are the most widely used technology attaching the sole with the upper part of the shoe. In addition adhesives are one of the most harmful elements used in footwear production. It is well-known the fact that without them, the shoe wouldn't have the shape and the structure. Usually the type of adhesive used is related with the type of materials that will join together. According to a study by INNESCOP 15% of carbon footprint of footwear is part of assembly process, by presenting a bio based adhesive as an innovative and sustainable solution together with the low-pressure plasma treatment as an alternative and sustainable technology [39, 40]. Reducing environmental impact caused by CO_2 emissions; the last one is used as a way to produce polyols, the main component of polyurethane adhesives [41]. But yet eco-friendly adhesive are used in small quantities and more efforts have to be taken in order to replace petroleum based adhesives.

Sustainable solutions for footwear products include even the packaging, which usually are boxes but with an essential impact on the buying experience of consumers. Its role it is not only about brands image but include the functional part of logistics in order to protect them. But recently due to the pandemic situation there has been an increase in online shopping, which has worsened the situation where an extra packing is required for individual sales. Known brands, despite in a small number are trying to shift to sustainable packing. Figure 2 presents a circular chain of production that should be a priority in footwear production, and further.

Fig. 2. Sustainable production in footwear industry.

3 Additive Manufacturing for Footwear Products

Additive manufacturing is a term known even as rapid prototyping or 3D printing. It produces objects directly from CAD data, which are 3D models generated by digitalization or 3D modeled on CAD software. It offers great advantages compared with standard manufacturing [42]. The widespread use of 3D printing technologies is well known even in fashion industry. Thinking about these applications, cases presented by various companies, researchers, or individual users mention again here its role in development of several products. Besides various applications of 3D printing in fashion industry, the case of footwear products has undergone an immediate increase. Especially these applications are mainly focused for sole production. This due to the fact that the material used for sole production is more similar to the one used for 3D printing. Moreover, the freedom offered by 3D printing to print complex geometries which can be difficult to be produced with the traditional way of manufacturing. *But a key question that arises here is how 3D printing contributes to sustainable manufacturing?* In fact 3D printing is one of the pillars of Industry 4.0, which is even called as sustainable production. This is since 3D printing reduces materials waste. 3D printing technologies have attracted attention of well-known companies operating in production of sportive shoes. They have experimented and several prototypes are tested and evaluated. Nowadays 3D printed footwear products are available for mass market.

A long-term strategic planning should be considered in order that companies shift to the new model of manufacturing, by integrating digital technologies part of forth Industrial revolution. Digitalization in products development offers several benefits, as sustainable production. Cutting off industrial waste produced by several attempts for product prototyping is one of these advantages. The environmental impacts caused by industrial manufacturing with the focus on footwear industry should be considered by all actors operating in the whole chain of production. As a result, these data must be used in order to raise awareness and efforts should be performed by all players. Industry 4.0 through digitalization implemented on the entire chain of production is able to fulfill consumer requirements for personalized products. Opportunities offered by Industry 4.0 are not only at manufacturing level, but these technologies have impact even on social aspects of human life. According to a report by SmarTech, footwear 3D printing revenues currently represent roughly 0.3% of global footwear market revenues by making footwear arguably the single most relevant consumer vertical for 3D printing [43] and projected to generate over US $4.2 billion in annual revenues in 2025. For the next decade is predicted that among the 3D printing technologies, Vat Polymerization will continue to be the main technology used for footwear production followed by Powder- Based Fusion (PBF) [44]. Integration of digital design and advanced manufacturing technologies part of the Industry 4.0 aims to support the personalized production of footwear [45]. Sole production using 3D printing has gained a wide range of applications. Cases presented as personalized soles in order to secure a topological optimization according pressure maps and reducing material waste. Obviously additive manufacturing brings several opportunities as the case of footprint developed and patented in Spain, which present a solution for healthcare environments, due to the antimicrobial properties of the material used [46]. Some of the main steps related with digital manufacturing in footwear industry are depicted in Fig. 3.

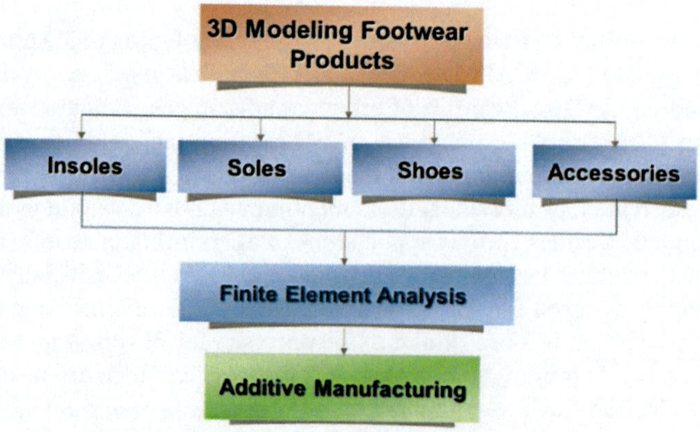

Fig. 3. Digital manufacturing of footwear products.

4 Conclusions

Integrating digital design and advanced manufacturing technologies help companies to raise productivity, quality and reduce time to market in order to be competitive. But from the other side the negative impact that footwear industry has on the environment should be taken under consideration. As a result, initiative taken by governments and organizations try to push companies to implement sustainable production. Parts of the chain of sustainable production are sustainable materials. As a whole process is material waste. Nowadays sustainable footwear is a strategic drive for every company operating in this sector and manufactures should be continuously aware to find innovative materials and new ways of production, merchandizing, retailing, where each one of them contributes to sustainability. Although the role of the manufacturing, consumers itself should be able through their purchasing behavior to lower the number of footwear products. Smart consumption through increase in mass customization, e-commerce, and new business models, such as clothing as a service will be a key driver to sustain sustainable economy. As a result we all benefit from all these actions, by reducing carbon footprint. The feasibility offered by 3D printing as an alternative of traditional manufacturing due to the advantages as customization, waste reduction regarding material and stock minimization and the ability to produce complex geometry compared to traditional manufacturing are important elements in the predictions that project the raise of 3D printing technology in footwear industry.

References

1. Ulusoy, T., Yasar, E., Aktan, M.: Critical Issues Impacting Science, Technology, Society (STS), and Our Future, 1st edn. IGI Global, USA (2019)
2. Stock, T., Obenaus, M., Kunz, S., Kohl, H.: Industry 4.0 as Enabler for a sustainable development: a qualitative assessment of its ecological and social potential. Process Saf. Environ. Protect. **118**, 254–267 (2018)
3. Jimeno-Morenilla, A., Azariadis,P., Molina-Carmona, R., Kyratzi, S., Moulianitis, V.: Technology enablers for the implementation of Industry 4.0 to traditional manufacturing sectors: a review. Comput. Ind. **125** (2021)
4. Calderón-Andrade, R., Hernández-Gress, E.S., Montufar Benítez, M.A.: Productivity improvement through reengineering and simulation: a case study in a footwear-industry. Appl. Sci. **10**, 5590 (2020)
5. Román-Ibáñez, V., Jimeno-Morenilla, A., Pujol-López, F.A.: Distributed monitoring of heterogeneous robotic cells. A proposal for the footwear industry 4.0. Int. J. Comput. Integr. Manuf. **31**(12), 1205–1219 (2018)
6. Gómez-Hernández, J.F., Davia-Aracil, M., Sanchez-Romero, J.L., Jimeno-Morenilla, A.: An approach to implement STEP-NC in the footwear industry. Comput. Ind. **125**, 103384 (2021)
7. Azariadis, P., Papagianis, P., Koutkalaki, Z.: Integrating modern virtual engineering tools in footwear design and development. In: IOP Conference Series: Material Science and Engineering, Lesvos, Greece, vol. 459 (2018)
8. Ceschin, F., Gaziulusoy, I.: Evolution of design for sustainability: from product design to design for system innovations and transitions. Design Stud. **47**, 118–163 (2016)
9. Ashton, E.G.: Analysis of footwear development from the design. Strategic Design Res. J. **11**(1), 2–8 (2018)

10. Sadowski, M., Yan, C., Cummis, C., Aden, N.: Apparel and footwear sector science-based targets guidance. World Resources Institute (WRI) (2019)
11. Shoe Sustainability Homepage. https://www.shoesustainability.com. Accessed 26 Feb 2021
12. Ahmad, S., Miskon, S., Alabdan, R., Tlili, I.: Towards sustainable textile and apparel industry: exploring the role of business intelligence systems in the era of industry 4.0. Sustainability **12**(7), 2632 (2020)
13. Earsmus+: Study on Industry 4.0 applied to the footwear industry in Europe, September 2020. FEET IN 4.0. Accessed January 2021
14. Argento, D., Einarson, D., Mårtensson, L., Persson, C., Karin, W., Westergren, A.: Integrating sustainability in higher education: a Swedish case. Int. J. Sustain. High. Educ. **21**(6), 1467–6370 (2020)
15. Saniuk, S., Grabowska, S., Gajdzik, B.: Social expectations and market changes in the context of developing the industry 4.0 concept. Sustainability **12**, 1362 (2020)
16. Jacometti, V.: Circular economy and waste in the fashion industry. Laws **8**, 27 (2019)
17. Serweta, W., Olejniczak, Z., Matusiak, M.: Improve of footwear comfort sensation with material packages and knitted fabrics. Fibres Text. Eastern Eur. **27**(3), 85–90 (2019)
18. Footwear Distributers and Retailers of America: FDRA Homepage. https://fdra.org. Accessed 26 Feb 2020
19. Azambuja, P., Marques, A., Broega, A.: Sustainability in the Portuguese fashion accessory &footwear industry (case studies). Leather Footwear J. **1**(1) (2019)
20. Sustainable Footwear Market Size, Share & Trends Analysis Repor t By Type (Athletic, Non-athletic), By End User (Men, Women, Children), By Region, And Segment Forecasts, 2020–2027 (2020)
21. Niinimäki, K., Peters, G., Dahlbo, H., Perry, P., Rissanen, T., Gwilt, A.: The environmental price of fast fashion. Nat. Rev. Earth Environ. **1**, 189–200 (2020)
22. Cheah, L., Ciceri, N.D., Olivetti, E., Matsumura, S., Forterre, D., Roth, R., Kirchain, R.: Manufacturing-focused emissions reductions in footwear production. J. Clean. Prod. **44**, 18–29 (2013)
23. Herva, M., Álvarez, A., Roca, H.: Sustainable and safe design of footwear integrating ecological footprint and risk criteria. J. Hazard. Mater. **192**(3), 1876–1881 (2011)
24. Ramchandani, M., Coste-Maniere, I.: Leather in the age of sustainability: a norm or merely a cherry on top? In: Leather and Footwear Sustainability. Springer, Singapore (2020)
25. Van Rensburg, M.L., S'phumelele, L., Ntandoyenkosi M.: Life cycle and end-of-life management options in the footwear industry: a review. Waste Manag. Res.: J. Sustain. Circ. Econ. **38**(6), 599–613 (2020)
26. Jadhav, N.C., Jadhav, A.C.: Waste and 3R's in footwear and leather sectors. In: Leather and Footwear Sustainability. Springer, Singapore (2020)
27. Grand View Research, April 2018. https://www.grandviewresearch.com/industry-analysis/athletic-footwear-market. Accessed Feb 2021
28. Paleviciute, R.: Make Fashion Better, May 2020. https://www.makefashionbetter.com/blog/environmental-impact-of-sneaker-production. Accessed Feb 2021
29. Chu, J.: Massachusetts Institute of Technology, May 2013. https://news.mit.edu/2013/footwear-carbon-footprint-0522. Accessed Feb 2021
30. Adidas Homepage. https://www.adidas-group.com/en/sustainability/managing-sustainability/general-approach/. Accessed 20 Feb 2021
31. Nike Homepage. https://purpose.nike.com/innovating-sustainably/. Accessed 20 Feb 2021
32. New Balance Homepage. https://www.newbalance.com/responsible-leadership/environment.html. Accessed 20 Feb 2021
33. Fernandes, A.M.: The Importance of ergonomics in the development of sustainable materials applied to footwear design. In: Advances in Ergonomics in Design AHFE. Advances in Intelligent Systems and Computing, vol. 955 (2019)

34. Zavodna,L.S., Trejtnarova, L., Pospisil, J.Z.: A sustainable materials for footwear industry: designing biodegradable shoes. Appl. Res. Tech. Technol. Educ. **8**(1) (2020)
35. Garcia, C., Auxiliadora Prieto, M.: Bacterial cellulose as a potential bioleather substitutefor the footwear industry. Microbial Biotechnol. **12**(4), 582–585 (2018)
36. Changhyun, N., Young-A, L.: Multilayered cellulosic material as a leather alternative in the footwear industry. Cloth. Text. Res. J. **37**(1), 20–34 (2018)
37. Silverman, J., Cao, H., Cobb, K.: Development of mushroom mycelium composites for footwear products. Cloth. Text. Res. J. **38**(2), 119–133 (2020)
38. Paiva, R.M.M., Marques, E.A.S., da Silva, L.F.M., Antonio, C.A.C., Aran-Ais, F.: Adhesives in the footwear industry. Proc. Inst. Mech. Eng. Part L: J. Mater.: Design Appl. **230**(2), 357–374 (2015)
39. Aran-Ais, F., Ruzafa-Silvestre, C., Carbonell-Blasco, M., Perez-Limi~nana, M., Orgiles-Calpena, E.: Sustainable adhesives and adhesion processes for the footwear industry. Proc. IMechE Part C: J. Mech. Eng. Sci. **235**(3) (2020)
40. Ruzafa-Silvestre, C.-B.P., Orgilés-Calpena, E., Arán Aís, F.: Low-pressure plasma treatment applied to polimeric materials for a sustainable footwear industry. In: 8th International Conference on Advanced Materials and Systems, pp. 455–460 (2020)
41. Orgilés-Calpena, E., Arán-Aís, F., Torró-Palau, A.M., Orgilés-Barceló, C.: Sustainable polyurethane adhesives derived from carbon dioxide. Polym. Renew. Resour. **7**(1), 1–12 (2016)
42. Pereira, T., Kennedy, J.V., Potgieter, J.: A comparison of traditional manufacturing vs additive manufacturing, the best method for the job. Proc. Manuf. **30**, 11–18 (2019)
43. 3D Printing in Footwear, SmarTech Analysis (2019)
44. Intrado GlobeNewswire Homepage. https://www.globenewswire.com/news-release/2020/08/20/2081371/0/en/Additive-Manufacturing-in-the-Footwear-Industry-to-Generate-US-4-2-Billion.html. Accessed 26 Feb 2021
45. Di Roma, A.: Footwear design. The paradox of "tailored shoe" in the contemporary digital manufacturing systems. Design J. **20** (2017)
46. Davia-Aracil, M., Hinojo-Pérez, J.J., Bertazzo, M., Orgilés-Calpena, E.: Design and functionalisation of shoe outsoles with antimicrobial properties using additive manufacturing technologies: industrial applications. Comput. Ind. **121**, 103246 (2020)

Sustainable Materials for Additive Manufacturing: Earth-Based Concrete

Yeşim Tarhan[1](✉), İsmail Hakkı Tarhan[2], Flávio Craveiro[3,4], and Helena Bártolo[3,4]

[1] Engineering Faculty, Ardahan University, Ardahan, Turkey
yesim.tosun@atauni.edu.tr
[2] Engineering Faculty, Atatürk University, Erzurum, Turkey
[3] School of Technology and Management, Polytechnic of Leiria, Leiria, Portugal
[4] CIAUD, Lisbon School of Architecture, Universidade de Lisboa, Lisbon, Portugal

Abstract. Earth-based materials have been used as a construction material since ancient times, as it is an affordable and sustainable material, widely available and recyclable. Additive manufacturing has become a promising production technology for the construction industry, allowing to produce complex structures with optimum material use, zero waste and a great design freedom. A brief review is provided on the prospects and challenges introduced by the extrusion of earth-based materials through additive manufacturing technologies. A numerical simulation was carried out replacing a 3D printed concrete sample by an earthen based one, and its behavior compared with the behavior of the concrete study. Results suggests that extruding earth-based materials can reach approximately half of the number of layers of the ones using concrete materials.

Keywords: Additive manufacturing · Earth-based materials · 3D printing · Finite element analysis

1 Introduction

Over the most recent years, earth was one of the most long-established structural materials, has been used as a construction material in the developed countries [1], as it is a low cost and easily accessible material. Earthen buildings use 100% natural local resources, have a very low carbon footprint, present great possibilities for indoor climate and provide climatic control because of absorbed solar energy. Earth elements are entirely breathable and have the ability to absorb toxins and smells from the indoor air environment. Earth is fully recyclable, so creates zero waste and is very agreeable to work with, offering a huge potential for personal creativity [2].

Earth construction is currently under strong development, likely due to its sustainable properties, though conventional earth construction techniques are time-consuming and labor-intensive [3], which is a limitation. It also suffers from a poor image, does not meet modern productivity standards, and has difficulty in overcoming some strength and durability tests designed for construction [4]. Nevertheless, earthen materials have very

J. R. da Costa Sanches Galvão et al. (Eds.): *Proceedings of the 1st International Conference on Water Energy Food and Sustainability (ICoWEFS 2021)*, pp. 708–716, 2021.
https://doi.org/10.1007/978-3-030-75315-3_75

important advantages in terms of sustainable development, so they must be integrated with new construction techniques.

Additive manufacturing, a digitally fabrication technique, also called 3D printing, is a new technology for the construction sector, enabling a fast production rate, improved safety labor and providing architectural design freedom. Additive manufacturing allows the design and fabrication of complex 3D objects, layer-by-layer, with a wide range of materials. Recent developments in digitally based construction methods have been introducing earth-based materials, mix-design, and processing methods for 3D printing technology. However, there are very few 3D printing studies aiming at either earth materials' printability or searching for printed earthen real properties.

Studies on 3D printed structural elements are increasing, though more effort, time and resources are required until it reaches sufficient maturity. On the other hand, studies on advancing knowledge on the subject through numerical simulations are very limited. However, to make realistic numerical simulations on material characterization, there is a need either to adapt existing numerical methods or to develop new methods. This work was conducted as a brief review on the sustainability of buildings produced through earth-based materials and the capabilities and shortcomings of the extrusion of earth-based materials. A numerical simulation was used to investigate and compare the mechanical behavior of a 3D printed element model using earth-based material instead of concrete.

2 Historical Earth Building and New Technologies

2.1 A Sustainable Material: Earth

Using earth as a construction material has a very low environmental impact on top of being fully recyclable. Many historical buildings built using soil are still enduring, which is the most important evidence that earth-based materials are sustainable. Earthen structure designs date back to ancient times and this structural material is still utilized in many environments (see Fig. 1) and societies [5]. It is estimated that more than two billion people are living in earthen buildings and about 10% of the world heritage properties incorporate earthen structures [6]. The earth materials are local, so with no transportation costs. Raw earth is generally nontoxic and nonpolluting with a large adsorption capacity for volatile compounds, thanks to its clay content. Earth structures have a good hygrothermal performance due to its relatively low thermal conductivity, large thermal mass and strong breathing capability (easy water/vapor phase change and transport, important for passive air conditioning) [7, 8]. When used for indoor surfaces, earth plasters may show important advantages, in particular hygrothermal comfort, as earth plasters can easily absorb water vapor in excess or release it when it is scarce [9–11]. Whenever environmental impacts of embodied energy need to be considered, there is a need to include non-renewable resources [12]. The awareness that construction materials have a substantial role in environmental sustainability is gradually increasing, especially when considering zero-energy buildings [13] or passive houses [14] where energy consumption during the operational phase is reduced to negligible amounts [15]. The most conventional construction materials, like cement, steel, aluminum, and concrete, all need a fairly large number of technological operations to be produced in an industrial scale. This requires large amounts of energy [16] and a non-negligible consumption of fuel and non-renewable resources [17]. For this reason, the concrete industry worldwide is facing growing challenges in conserving material and energy resources, as well as reducing its CO2 emissions [18]. Considering all this, it should be emphasized that, environmental effects of earthen materials are significantly better than cementitious materials.

Fig. 1. a) Shibam Hadramawt is a mudbrick-made high-rise buildings town in Yemen [19], b) Musgum mud huts units are traditional domestic structures built of mud, in Cameroon [20], c) The Ksar of Ait- Ben-Haddou is a perfect synthesis of earthen architecture of the pre-Saharan regions of Morocco [21], d) Arg-e Bam is the most representative example of a fortified medieval town built in vernacular technique using mud layers [22].

2.2 Towards 3D Printing Earth Concrete

There are different traditional earth building techniques as follows: cob [23–25], masonry with adobe (mud bricks) [26], and rammed earth [27, 28]. Masonry with compressed earth blocks was introduced in the middle of the 20th century. In this case, the blocks are manufactured by pressing sieved soil into block molds with a manual, mechanical, or hydraulic press. Different stabilizers have been used, like clay, tree resins, natural bitumen, gum, agave juice, cactus juice, cowpats, casein, to improve strength, resistance to disaggregation and erosion, or to correct the mixture, to obtain a self-compacting earth concrete, adding various additives used in ordinary Portland cement-based concrete [5, 29–31]. The French laboratory CRATerre and the University of Mokpo have made experimental tests with cast earth concrete by adding cement plasticizers to reduce the yield stress, including around 8 to 10 wt.% of cement to allow the setting [32]. Similarly, the mixture produced at ETH Zurich using the addition of 5% calcium sulfoaluminate cement and polycarboxylate ether was called self-compacting clay concrete. Moevus et al. [33] formulated a cement-free castable self-compacting clay concrete by increasing the population of fine particles and using a clay dispersant. The cement setting has been simulated by using biopolymers such as alginate or using a combination of hydraulic binders. Different research groups are investigating the production of a mixture design considering both fast casting and sufficient strength in the dry state. Currently, novel processing methods like extrusion [1, 34–39] are being introduced adding concrete to raw earth and accelerating the building rate. Figure 2 shows a small building sample produced by San Fratellon and Rael using 3D printing techniques and earth-based materials [38], and a 3D earth-based concrete mixture produced by Perrot et al. [1].

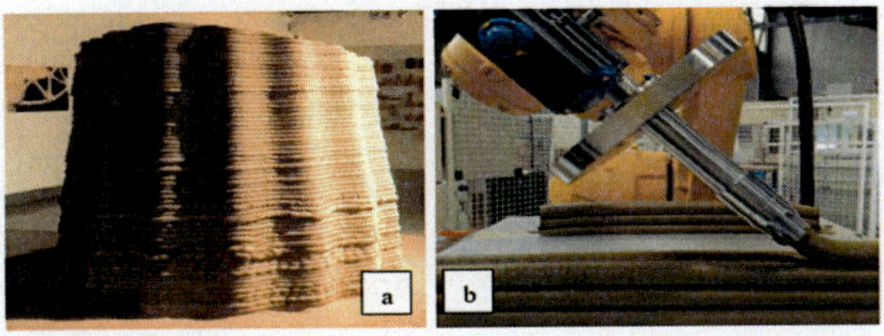

Fig. 2. 3D printed earth concrete examples: a) The MUD Frontier large-scale structure and vessels at the Rubin Center Gallery as part of the New Cities, Future Ruins Exhibit [38], b) 3D printing of earth concrete [1].

It is an important challenge trying to combine the world oldest construction material with the newest construction processing techniques [1]. Distinct research groups have been investigating 3D printing construction techniques for cement-based materials [40, 41]. 3D printing technologies allow obtaining lightweight products and the design of multifunction components, as well low material usage, reduced costs, increased functionality, and more sustainability [42, 43], thus it is a promising technique that can make huge changes in the construction industry. The success of this printing process depends on the strength of the deposited layers, as the printed material must be hard enough to support the increasing load induced by subsequently deposited layers. A faster development of the so-called green strength of the material is required to ensure fast production and structural stability [1].

2.3 Brief Literature Review

3D printing construction elements, quite new in the construction sector, are generally focused on concrete. There are only a few 3D printing earth-based concrete studies in the literature. Fratello and Real [38], used soil, salt and grape skins materials for 3D printing, low-cost materials many in opposition to expensive patented powders and resins, which makes 3D printing more accessible, less expensive, more ecological and sustainable. Arrieta-Escobar et al. [39], made a literature review of recent applications (2000 to 2019) of 3D printing with earth-based materials, identifying main challenges and opportunities presented by using 3D printing with earth-based materials. According to Arrieta-Escobar et al. [39], despite the remarkable and rapid progress made in the development of 3D printing in recent years, this technology is still in its infancy. Perrot et al. [1] showed that it is possible to print a structure with an earth-based material, adding alginate seaweed biopolymers to raw earth to develop earth concrete's green strength. Bajpayee et al. [36] discussed the development of a new palette of natural construction materials, based on natural soils, by additive manufacturing, gathering knowledge from soil chemistry and composite science, such as the modelling of mesoscale phenomena, rheological studies of extrudable soil "inks", generative design.

On the other hand, numerical simulation studies on material characterization of 3D printed concretes are still limited. In the research conducted by Wolfs et al. [44], a numerical model based on a time dependent Mohr-Coulomb fracture criterion was developed to analyze the mechanical behavior of 3D-printed concrete in its fresh state. The model developed in the Abaqus program, based on the finite element method (FEM), has been validated by comparison with 3D printing experiment results. The model was then used to simulate structural behavior during a real printing process. In the study conducted by Wolfs et al. [45], the effect of concrete age on failure behavior was experimentally tested and validated by numerical analysis of flat using the FEM. The printing process was simulated by applying the model change function, in which layers are added on top of each other step by step. After activation of a new layer, the interface with the previous layer was defined by the constraints. In numerical simulations using the properties obtained with a customized triaxial test procedure, each layer was subjected to gravitational load. Wall samples were printed until failure, during the printing trials, and the results were compared with numerical estimates. According to this work, triaxial compression tests are a suitable method to characterize the structural fresh properties of 3D printed concrete. The comparison between the numerical model and the printing trials reinforces the need to characterize the material properties to be printed as a function of additional state parameters such as temperature.

Along the printing process, layers are deposited on top of each other with no formwork to stabilize the build. Thus, the fresh concrete mixture should be sufficiently strong, stiff, and stable to carry its self-weight and the weight of the other layers above it and limit deformations [44]. In this work, it was also numerically simulated the printability of fresh earth concrete based on the development of mechanical properties of fresh earth concrete over time. According to Wolfs et al. [44], the circular geometry model produced with concrete was redesigned using 3D printing earth-based materials experimental data produced by Perrot et al. [1]. Thus, the load-bearing capacities of 3D printable building elements composed by earth-based materials were evaluated.

3 Method

In this work, the buildability properties of earth-based materials for 3D printing were investigated through a numerical analysis. A Grasshopper component library called VoxelPrint developed by Vantyghem et al. [46] in C# using RhinoCommon SDK was used for the numerical simulations. VoxelPrint, which uses a voxel-based numerical approach, was used to simulate failure caused by elastic buckling and material yield.

Wolfs et al. [44] is the first work proposing a FEM approach to accurately predict printed product quality (during and after printing) and avoid failures. In this study, it was performed an experimentally modelling method of the 3D concrete printing process to evaluate the printability of the time-dependent development of fresh concrete during printing. Numerically simulated by Wolfs et al. [44], this model is used to simulate the structural behavior during an actual printing process, as well to study the failure-deformation mode. Vantyghem et al. [46] has numerically tested an axisymmetric cylindrical model used by Wolfs et al. [44] to validate the VoxelPrint.

In this work, we used the same model used by Wolfs et al. [44] and validated by Vantyghem et al. [46], to make a meaningful comparison between concrete and earth-based materials. The model under consideration is a cylinder with 250 mm radius, 40 mm bed width and 10 mm sheet height. The printing speed was taken as 83.33 mm/s, which is the same defined by Vantyghem et al. [46]. The earth material properties were obtained from the work of Saleh et al. [47] regarding hardening soil model parameters of marine clay. According to this work, the Poisson ratio is 0.2, the cohesion is 33 kPa, the internal friction angle is 17.51° and the dilatation angle 1.6°. Perrot et al. [1] showed that the earth material elasticity module does not show a time-dependent behavior in its fresh state. Thus, the modulus of elasticity was considered as 0.025 N/mm^2 and the material density 1750 kg/m^3, based on the work of Perrot et al. [1].

4 Results and Discussion

The resulting failure mechanism for the printing simulation is illustrated in Fig. 3. According to Wolfs et al. (2019) and Vantygem et al. (2021), the plastic yielding of the bottom layers and elastic buckling are two of the most frequent failure mechanisms reported in 3D concrete printing. In these two simulation works, both based on the assumption that the printing process was perfect, the failure was triggered by a plastic yielding in the bottom layers.

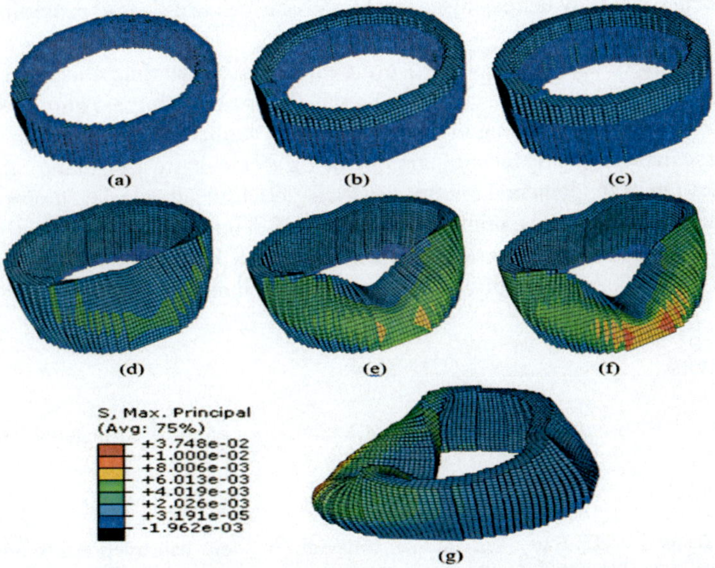

Fig. 3. Results from the numerical analysis and different steps during the printing simulation: (a) 80 mm, (b) 130 mm, (c) 180 mm, (d) 220 mm buckling started (e-f-g) 220 mm buckling failure.

Conversely, in the present study, the model produced by 3D printing of the earth-based materials was examined and it was defined that the failure occurred due to the elastic buckling of the upper layers. In addition, the plastic deformation did not occur as the yield limit of the material was not exceeded.

This study suggests that it might be possible to add more layers if sufficient stability conditions are fulfilled. The model using concrete material properties failed at the 46th layer, while the model using earth-based material properties was only able to arrive at the 22th layer. As previously mentioned, the number of layers might be higher because the problem is not resulting from the yielding of the material. Moreover, a comparison cannot be made between concrete and earth-based materials without any strengthening.

On the other hand, material characterization studies requiring an accurate simulation of the 3D printed concrete and earth-based materials are scarce, which makes it difficult to perform an effective numerical simulation. It is also important to state that the properties of the earth-based 3D printable material simulated within the scope of this study could only be partially obtained.

5 Conclusions

This study used a numerical analysis simulation to predict the buildability properties of earth-based materials through 3D printing technologies. This simulation work was carried out by using other studies and methods available in the literature. Buckling failure can be minimized if the stability in earth-based materials is provided. For example, it is possible to make 3D printed structures with earth-based materials making some improvements, either by increasing adhesion with various additives or providing support during printing.

The time-dependent parameters to print earth-based materials must be accurately determined and presented, especially to simulate its fresh properties. Further work will include the experimental testing of these parameters and its numerical simulation.

Additive manufacturing technologies are being increasingly used in the construction sector. They are of fundamental importance for a global sustainable development as they enable saving raw materials, minimizing waste, and reducing costs. The application of these technologies using highly sustainable materials, such as earth-based materials, will constitute a remarkable step forward the construction industry.

References

1. Perrot, A., Rangeard, D., Courteille, E.: 3D printing of earth-based materials: processing aspects. Constr. Build. Mater. **172**, 670–676 (2018)
2. Solid Earth Adobe Buildings Homepage. https://www.solidearth.co.nz/. Accessed 18 Feb 2021
3. Parra-Saldivar, M.L., Batty, W.: Thermal behaviour of adobe constructions. Build. Environ. **41**(12), 1892–1904 (2006)
4. Van Damme, H., Houben, H.: Earth concrete stabilization revisited. Cem. Concr. Res. **114**, 90–102 (2018)
5. Landrou, G., Brumaud, C., Winnefeld, F., Flatt, R.J., Habert, G.: Lime as an anti-plasticizer for self-compacting clay concrete. Materials **9**(5), 330 (2016)

6. Eloundou, L., Joffroy, T.: Earthen architecture in today's world. In: Proceedings of the UNESCO International Colloquium on the Conservation of World Heritage Earthen Architecture. UNESCO Publishing, Paris (2013). www.whc.unesco.org/en/series/36
7. Melià, P., Ruggieri, G., Sabbadini, S., Dotelli, G.: Environmental impacts of natural and conventional building materials: a case study on earth plasters. J. Clean. Prod. **80**, 179–186 (2014)
8. Hall, M., Allinson, D.: Analysis of the hygrothermal functional properties of stabilised rammed earth materials. Build. Environ. **44**(9), 1935–1942 (2009)
9. Reddi, S., Jain, A.K., Yun, H.B., Reddi, L.N.: Biomimetics of stabilized earth construction: challenges and opportunities. Energy Build. **55**, 452–458 (2012)
10. Liuzzi, S., Hall, M.R., Stefanizzi, P., Casey, S.P.: Hygrothermal behaviour and relative humidity buffering of unfired and hydrated lime-stabilised clay composites in a Mediterranean climate. Build. Environ. **61**, 82–92 (2013)
11. Minke, G.: Building with Earth: Design and Technology of a Sustainable Architecture. Walter de Gruyter (2012)
12. Shukla, A., Tiwari, G.N., Sodha, M.S.: Embodied energy analysis of adobe house. Renew. Energy **34**(3), 755–761 (2009)
13. Thiel, C.L., Campion, N., Landis, A.E., Jones, A.K., Schaefer, L.A., Bilec, M.M.: A materials life cycle assessment of a net-zero energy building. Energies **6**(2), 1125–1141 (2013)
14. Dahlstrøm, O., Sørnes, K., Eriksen, S.T., Hertwich, E.G.: Life cycle assessment of a single-family residence built to either conventional-or passive house standard. Energy Build. **54**, 470–479 (2012)
15. Himpe, E., Trappers, L., Debacker, W., Delghust, M., Laverge, J., Janssens, A., Van Holm, M.: Life cycle energy analysis of a zero-energy house. Build. Res. Inf. **41**(4), 435–449 (2013)
16. Wang, E., Shen, Z.: A hybrid Data Quality Indicator and statistical method for improving uncertainty analysis in LCA of complex system–application to the whole-building embodied energy analysis. J. Clean. Prod. **43**, 166–173 (2013)
17. Cabeza, L.F., Barreneche, C., Miró, L., Morera, J.M., Bartolí, E., Fernández, A.I.: Low carbon and low embodied energy materials in buildings: a review. Renew. Sustain. Energy Rev. **23**, 536–542 (2013)
18. Schneider, M., Romer, M., Tschudin, M., Bolio, H.: Sustainable cement production—present and future. Cem. Concr. Res. **41**(7), 642–650 (2011)
19. Shibam Hadramawt is a town in Yemen. https://en.wikipedia.org/wiki/Shibam_Hadramawt. Accessed 18 Feb 2021
20. Musgum mud huts are traditional domestic structures. https://en.wikipedia.org/wiki/Musgum_mud_huts. Accessed 18 Feb 2021
21. The ksar, a group of earthen buildings. https://whc.unesco.org/en/list/444/. Accessed 18 Feb 2021
22. Arg-e Bam is the most representative example vernacular using mud. https://whc.unesco.org/en/list/1208. Accessed 18 Feb 2021
23. Castrillo, M.C., Philokyprou, M., Ioannou, I.: Comparison of adobes from pre-history to-date. J. Archaeol. Sci.: Rep. **12**, 437–448 (2017)
24. Hamard, E., Cazacliu, B., Razakamanantsoa, A., Morel, J.C.: Cob, a vernacular earth construction process in the context of modern sustainable building. Build. Environ. **106**, 103–119 (2016)
25. Phung, T.A., Le Guern, M., Boutouil, M., Louahlia, H.: Mechanical and thermal performance of cob materials. Acad. J. Civ. Eng. **35**(2), 166–172 (2017)
26. Illampas, R., Ioannou, I., Charmpis, D.C.: Adobe bricks under compression: experimental investigation and derivation of stress–strain equation. Constr. Build. Mater. **53**, 83–90 (2014)
27. Easton, D., Easton, T.: Modern rammed earth construction techniques. In: Modern Earth Buildings, pp. 364–384. Woodhead Publishing (2012)

28. Kebao, R., Kagi, D.: Integral admixtures and surface treatments for modern earth buildings. In: Modern Earth Buildings, pp. 256–281. Woodhead Publishing (2012)
29. Ma, C., Chen, B., Chen, L.: Effect of organic matter on strength development of self-compacting earth-based construction stabilized with cement-based composites. Constr. Build. Mater. **123**, 414–423 (2016)
30. Ma, C., Chen, L., Chen, B.: Experimental study of effect of fly ash on self-compacting rammed earth construction stabilized with cement-based composites. J. Mater. Civ. Eng. **28**(7), 04016022 (2016)
31. Ouellet-Plamondon, C., Habert, G.: Self-compacted clay concrete: a concrete with CSA and higher clay content. In: 8th International RILEM Symposium on Self-compacting Concrete (SCC 2016), pp. 37–47. RILEM Publications SARL (2016)
32. Kang, N., Hwang, H.: Study on high strengthening of an earth wall using earth and high performance lime. In Proceedings of the International Conference on Earth Architecture in Asia, Mokpo, Korea, vol. 1114, October 2011
33. Moevus, M., Jorand, Y., Olagnon, C., Maximilien, S., Anger, R., Fontaine, L., Arnaud, L.: Earthen construction: an increase of the mechanical strength by optimizing the dispersion of the binder phase. Mater. Struct. **49**(4), 1555–1568 (2016)
34. Khelifi, H., Perrot, A., Lecompte, T., Ausias, G.: Design of clay/cement mixtures for extruded building products. Mater. Struct. **46**(6), 999–1010 (2013)
35. Maskell, D., Heath, A., Walker, P.: Laboratory scale testing of extruded earth masonry units. Mater. Design **45**, 359–364 (2013)
36. Bajpayee, A., Farahbakhsh, M., Zakira, U., Pandey, A., Ennab, L.A., Rybkowski, Z., Banerjee, S.: In situ resource utilization and reconfiguration of soils into construction materials for the additive manufacturing of buildings. Front. Mater. **7**, 52 (2020)
37. Craveiro, F., Nazarian, S., Bartolo, H., Bartolo, P.J., Duarte, J.P.: An automated system for 3D printing functionally graded concrete-based materials. Addit. Manuf. **33**, 101146 (2020)
38. San Fratello, V., Rael, R.: Innovating materials for large scale additive manufacturing: salt, soil, cement and chardonnay. Cem. Concr. Res. **134**, 106097 (2020)
39. Arrieta-Escobar, J.A., Derrien, D., Ouvrard, S., Asadollahi-Yazdi, E., Hassan, A., Boly, V., Dignac, M.F.: 3D printing: an emerging opportunity for soil science. Geoderma **378**, 114588 (2020)
40. Craveiro, F., Duarte, J.P., Bártolo, H., Bártolo, P.J.: Additive manufacturing as an enabling technology for digital construction: a perspective on construction 4.0. Autom. Constr. **103**, 251–267 (2019)
41. Van Damme, H.: Challenges and opportunities for concrete in the digital era. In: Gulf Conference on Sustainable Built Environment, pp. 27–56. Springer, Cham (2020)
42. Tarhan, Y., Remzi, S.: Developments of 3D concrete printing process. In: International Civil Engineering and Architecture Conference (2019)
43. Wangler, T., Roussel, N., Bos, F.P., Salet, T.A., Flatt, R.J.: Digital concrete: a review. Cem. Concr. Res. **123**, 105780 (2019)
44. Wolfs, R.J., Bos, F.P., Salet, T.A.M.: Early age mechanical behaviour of 3D printed concrete: numerical modelling and experimental testing. Cem. Concr. Res. **106**, 103–116 (2018)
45. Wolfs, R.J.M., Bos, F.P., Salet, T.A.M.: Triaxial compression testing on early age concrete for numerical analysis of 3D concrete printing. Cem. Concr. Compos. **104**, 103344 (2019)
46. Vantyghem, G., Ooms, T., De Corte, W.: VoxelPrint: a grasshopper plug-in for voxel-based numerical simulation of concrete printing. Autom. Constr. **122**, 103469 (2021)
47. Saleh, S., Yunus, N.Z.M., Ahmad, K., Said, K.N.M.: Numerical simulation with hardening soil model parameters of marine clay obtained from conventional tests. SN Appl. Sci. **3**(2), 1–13 (2021)

A Review on the Application of Palm Oil Fuel Ash in Concrete as a Cementitious Material for Construction

Arpan Joshi[1](✉), Flávio Craveiro[1,2], and Helena Bártolo[1,2]

[1] ESTG – Politécnico de Leiria, Leiria, Portugal
[2] CIAUD – Universidade de Lisboa, Lisbon, Portugal

Abstract. Sustainable construction is a rapidly increasing research area. For concrete production, cement is the most important and expensive raw material that binds together the solids. However, during the production of concrete, a large amount of carbon dioxide is produced, causing environmental concerns resulting in a growing interest in the field of supplemental cementitious material. One of the potential materials is palm oil fuel ash (POFA), the by-product of palm oil industries. These by-products, if dumped in the landfill sites, create an environmental problem and health hazard. Several research works have been carried out to investigate the potential use of palm oil fuel ash as a supplementary cementing material to develop high strength and aerated concrete. The experiments conducted for the use of POFA as a pozzolanic material to improve the strength and durability of concrete have obtained satisfactory results. POFA can be used as an environmentally friendly replacement for the cement to produce concrete. This study a state-of-the-art of possibilities for the sustainable construction of concrete through the partial replacement of concrete with palm oil fuel ash.

Keywords: POFA · Palm oil fuel ssh · Concrete · Cement replacement

1 Introduction

At Present, the demand for conventional building materials such as cement, bricks, steel, aluminum, wood, cladding, and partitioning materials is increasing with the high pace of urbanization and growth in population. To meet the housing demand for the growing population, there is an exponential need to produce these building materials (Meyer 2009). The production of conventional building materials not only consumes a large amount of thermal and electrical energy but is also polluting the environment and emptying our natural resources. On the other hand, the management of wastes generated from agriculture and agro-based industries is a serious challenge, and there is a need to pay attention to save our environment and resources (Tambichik 2018). The major wastes generated from agro-based industries are sugarcane bagasse, rice husks ash, palm oil fuel ash, cotton stalks, and others that are creating environmental pollution and increasing the expenditure of the industries in disposing of these wastes (Hussin et al. 2009). The

J. R. da Costa Sanches Galvão et al. (Eds.): *Proceedings of the 1st International Conference on Water Energy Food and Sustainability (ICoWEFS 2021)*, pp. 717–726, 2021.
https://doi.org/10.1007/978-3-030-75315-3_76

reuse of such agricultural wastes in building materials not only reduces the high cost of production but also promotes green construction and provides a viable solution to environmental problems (Raut et al. 2011).

Concrete is considered one of the most destructive and expensive building materials, with more than 10 billion tons of production annually (Meyer 2009). It is expected to rise to approximately 18 billion tons a year by 2050 (Mehta et al. 2006). Subsequently, the concrete industries are going to invade a large number of natural resources to meet this demand. The current status of concrete industries is not sustainable; however, the incorporation of agricultural and agro-based mineral additives as a cementitious material in concrete production could be a breakthrough to make the industry more sustainable and environmentally friendly (Shafigh et al. 2014), resulting in increased concrete´s durability and reduced cost of production (Buczkowski 2017; Ma et al. 2016).

To reduce the environmental impact, some measures were already introduced, such as incorporating agricultural wastes as an alternative way to produce more sustainable building materials. However, before establishing agricultural wastes as a substitute for building construction materials, it is essential to verify the potential benefits, costs, and potential risks. Only the wise understanding and systematic application of agro-based wastes into building construction materials will maximize its potential, concurrently, higher strength of materials, and low environmental risks.

2 Integrating POFA in Green Concrete Development

2.1 Origin of Palm Oil Fuel Ash

The palm oil industry is one of the most important agro-industries in certain countries like Malaysia, Thailand, Indonesia, and Nigeria. These industries produce many by-products in the form of empty fruit brunches, fibers, and kennels (Johari et al. 2011; Yusoff 2006). These enormous volumes of by-products are then used as fuel at a temperature of about 800–1000 °C to heat the boilers to produce steam for electricity generation and palm oil extraction (Shafigh et al. 2014). The ash produced from this heating process has been known as palm oil fuel ash (POFA) (Chavalpariti 2006). Since a large volume of POFA is produced after burning, which, if not utilized, will make a massive impact on the environment. So, an investigation was performed in Malaysia to examine the potential of POFA to be used as fertilizer (Hussin 1955). However, no significant amount of nutrients was observed in POFA, consequently was dumped in open fields without any profitable return, causing environmental pollution and human hazards (Awal and Shehu 2013). To counter this, several researchers initiated experiments on exploring the feasibility to use POFA as a sustainable binding material in concrete (Hussin et al. 2009). It was found that POFA possesses pozzolanic properties and can be used as a supplementary cementitious material in concrete (Tangchirapat et al. 2007; Munir et al. 2015). Though the use of POFA as an additive mineral in concrete had already started in 1990. Tay and Show (1995) and Awal and Hussin (1997) showed that the POFA has the pozzolanic property, and the replacement of cement up to 10% by weight will add to the strength of a material.

2.2 Pozzolanic Reaction

Pozzolan in itself possesses little or no cementitious properties, and the pozzolanic activity of POFA can be increased when they are finely divided (Khalid et al. 2018). In the presence of moisture, it chemically reacts with calcium hydroxide, which then reacts with the silica in POFA to produce a secondary calcium silicate hydrate (C-S-H) that enhances the concrete strength by filling up the voids between cement and aggregate and generate a stronger bond (Tambichik et al. 2018).

Hence, the binding properties of POFA depend on the particle's fineness, particle size and amorphousness of its structure, and the SiO, Al_2O_3, and Fe_2O_3 content (Kroehong et al. 2011) (Table 1).

Table 1. POFA Pozzolanic reaction (Tambichik et al. 2018).

Cement + Water	→	Calcium Silicate Hydrate (C-S-H) + Calcium Hydroxide ($Ca(OH)_2$
Silicon + Oxygen	→	Silica Oxide (Si(OH))
Silica Oxide (Si(OH)) + Calcium Hydroxide (Ca(OH)2)	→	Secondary Calcium Silicate Hydrate

2.3 Effect of Particle Size, Mix Proportion, and Fineness on Pozzolanic Activity

Several researchers investigated the effect of the POFA particle size within cementitious behavior. Tangchirapat et al. (2007) showed that palm oil fuel ash's original particle size is 183.0 μm, rounded with porous texture and Wi et al. (2018) confirmed 70% amorphous content in POFA. The grounding of POFA into medium (15.9 μm) and small particle size (7.4 μm) shows irregular and angular shape, which increases not only the fineness but also the specific gravity and the decrease in porosity (Sofri et al. 2015). Thus, the grounded POFA enhances and increases the interfacial bonding (Alsubari et al. 2018), improves the compressive strength of concrete when supplemented with finer POFA particles due to its filling effect and reactivity between the particles compared to the rounded large size particles (Usman 2015). Hussin et al. (2009) showed that POFA with a fineness of 10 μm achieved a higher strength than that of concrete produced with a fineness of 45 μm. Similarly, an experiment conducted by Sofri et al. (2015) with POFA of particle size 50 μm observed the compressive strength lower than the standard concrete (48.379 N/mm^2); however, the replacement of cement by 10% POFA showed higher strength (47.732 N/mm^2) compared to the substitution of 30% (37.065 N/mm^2) and 50% (20.644 N/mm^2) at 28 days and suggested that the fineness of POFA determines its pozzolanic activity. The replacement of POFA higher than 30% makes the concrete dry and porous but adding up to 30% stimulates the workability (Alsubari et al. 2016). The substitution up to 20% in the concrete will give a better strength performance in terms of compressive, splitting tensile and flexural strength than ungrounded POFA. However, Nano POFA enhances the hydration of cement and acts as fillers to refine the microstructure of pastes compared to micro POFA particles (Johari et al. 2011).

3 Chemical Composition

The chemical composition of POFA contains a significant amount of silica, but the lime content is relatively low compared to ordinary Portland cement, as shown in Table 2. The palm oil fuel ash can be grouped in between class C and class F as specified in ASTMC618-92a. However, the chemical composition of POFA can be varied according to the processing process in the oil plant (Altwair 2010).

The chemical composition of POFA used by various researchers is shown in Table 2. From the table, it can be noticed that the chemical composition of POFA used by various researchers varies, which may be due to the factors like burning temperature, parts of the palm oil tree burned, and fineness of particle (Tangchirapat and Jaturapitakkul 2010). The higher silica content can be obtained by burning the palm tree's kernel, and the burning of brunches brings more carbon to the palm oil fuel ash (Zulkefli et al. 2018). The strength and durability of concrete are directly proportional to the amount of silica present in the POFA. From the Table 2, we can see that the POFA contains a high percentage of silica (43% to 70%) that can react with the calcium hydroxide to produce a secondary C-S-H gel. The other pozzolanic components are AL_2O_3 and Fe_2O_3. A lower percentage of lime content in POFA was observed compared to OPC that will help in the prevention of thermal cracking (Tambichik 2018). In most cases, the loss of ignition (LOI) was much higher than the limit specified in ASTM C618 (Chavalparit 2006). The amounts of LOI and SO_3 for pozzolanic are within limits of 10% and 4% respectively as specified by ASTM C618. The POFA contains a higher amount of K_2O as the trees consume K_2O from the soil during its growing period. In all the cases (see Table 2), the sum of silica, alumina, and iron oxide is 50% to 80% of the total components of POFA. According to ASTM C618, POFA can be classified as class C Pozzolan when the sum of SiO2 + AL2O3 + Fe2O3 is above 50% while the sum is greater than 70% then the range of class Pozzolan is F (Abdullah et al. 2006; Skariah et al. 2017). The result was justified by the calcium oxide content in the POFA (4.12%) and the pozzolanic activity index of 65.12%. Although POFA is not a natural Pozzolan, Tangchirapat and Jaturapitakkul (2010) classified POFA as class N (natural) Pozzolan according to ASTM C618.

Depending upon the results of the chemical composition obtained by several works, there is an argument regarding the classification of POFA to meet ASTM C18. Hence, more research is required to address this contradiction.

Table 2. Physical properties and chemical composition of cement and POFA.

Chemical analysis %	OPC	POFA					
		(Mohamad 2016)	(Tabichik 2018)	(Rajak 2015)	(Awal et al. 2013)	(Wi et al. 2018)	(Tangchirapat et al. 2010)
Aluminium Oxide (Al_2O_3)	5.7	11.4	4.65	7.40	2.54	5.63	2.5
Calcium Oxide (CaO)	62.5	8.4	5.70	14.00	4.92	3.78	6.4
Ferric Oxide (Fe_2O_3)	3.0	4.7	8.12	4.47	5.02	3.51	1.9
Loss on Ignition (LOI)	2.7	18.0		9.3	-	10.1	10.0
Magnesium Oxide (MgO)	2.6	4.8	3.52	4.14	4.52	3.61	3.0
P_205	0.28		-	-	3.58		-
Potassium Oxide (K_2O)	0.87	3.5	9.05	-	7.52	5.66	5.7
Pozzolanic Activity Index with OPC	28.9	59.7	75.37	66.67	67.18	80.04	69.7
Silicon Dioxide (SiO_2)	20.2	43.6	62.60	54.80	59.62	70.9	65.3
Sodium Oxide (Na_2O)	0.16	0.39	-	-	0.76	0.39	0.3
Sulphur Trioxide (SO_3)	1.8	2.8	1.16	0.71	1.28	-	0.4

4 Mechanical Properties

4.1 Tensile Strength

Deepak et al. (2016) made high strength concrete using POFA with substitution ratio of 0%, 5%, 15%, 25%, 35%, and 45% and found that the tensile strength at 28 days of curing of the concrete was comparatively higher than the 7 days concrete. With 15% replacement POFA, there is a slight increase in the tensile strength (OPC:3.7 MPa, 15% POFA replacement: 4.31 MPa).

Sooraj (2013) performed experiments with 10%, 20%, 30%, and 40% replacement levels of POFA in concrete and found that the tensile splitting strength increases as the percentage of the POFA increase from 0% to 10%. However, the tensile strength developed was the same as the control sample obtained when the replacement was increased to 20%.

Furthermore, the strength decreased with an increasing percentage of POFA in the concrete. Similarly, Musa and Alkali (2014) evaluated the mixtures of 5% POFA passing through 75 μm and asphaltic concrete (AC14) and showed better indirect tensile strength values, good ability to resist fatigue, rutting deformations, and moisture susceptibility compared to control AC14.

4.2 Compressive Strength

Ismail et al. (2010) observed an increasing compressive strength when OPC was replaced with POFA 45 and POFA10 at a replacement level of 20%. The strength with POFA 45 and POFA 10 increased approximately by 0.91 and 8.21%. The POFA10 performed better because of higher fineness, resulting in more pozzolanic reaction, densification of microstructure, compared to the coarser POFA or OPC. Similarly, Sata et al. (2004) and Tangchirapat et al. (2007) confirmed the satisfactory performance of concrete with cement replacement with POFA up to 20%. The replacement above 20% may reveal lower concrete strength. Hussin et al. (2009) observed maximum compressive strength of concrete (85.9 MPa or about 111% of the control concrete) at the age of 28 days of curing when replaced with 20% POFA. The POFA with a high fineness of 10 μm can be used to produce high strength concrete than that of control concrete and POFA with 45 μm particle size concrete. Likewise, Pone and Ash (2018) showed greater early compressive strength with the replacement of POFA up to 5%. Sanawung et al. (2017) explained the influence of water to binder ratio on compressive strength. They revealed that the POFA concrete with W/B ratio of 0.40 had the compressive strengths varying from 45.8 to 55.9 MPa or 82–94% of OPC concrete at 90 days.

4.3 The Durability of Concrete Containing POFA

Chloride Penetration

There are few studies to demonstrate the resistance of POFA supplemented concrete towards chloride penetration. Hussin et al. (2009) determined the application of 20% of fine palm oil fuel ash of particle size 10 μm and 45 μm, towards the durability of high strength POFA concrete subjected to sulfate, acid, and chloride attacks. Irrespective of the degree of fineness, the POFA concrete displayed better in decreasing chloride diffusion than OPC concrete. The application of fine POFA speeds up the pozzolanic reaction, resulting in C-S-H gel formation, making the concrete denser. Consequently, POFA decreases concrete permeability and enables it to exhibit better resistance to chloride attack (Ismail et al. 2010). Hence, chloride penetration depends on the ash fineness. Similar results have been reported by other researchers (Zhang et al. 2000). Further, Sanawung et al. (2017)) suggested replacing ground POFA from 15 to 35% for increasing the chloride ion penetration resistance of concrete than for increasing its compressive strength.

Sulfate Attack
The soluble sulfate of calcium, sodium, magnesium, and ammonium present in soil and groundwater are harmful to concrete as these elements can lead to swelling and consequently cracking of concrete (Rusli 2010). The durability of the concrete prepared with POFA reveals better resistance to sulfate attack in size, number of cracks formed, and loss of integrity (Hussin et al. 1998). The finer POFA in the specimen speeds up the pozzolanic reaction. Having the low-lime content shows better resistance to sulfate attack (Hussin et al. 2009) due to gypsum and ettringite's low degree formation (Ismail 2008). Therefore, it can be concluded that the application of finer POFA as a partial cement replacement enhances the resistance of concrete towards sulfate attack (Hussin et al. 2009; Philip et al. 2019).

Resistance Towards Acid Attack
The acid resistance of high-strength concrete is better with the use of POFA (Ofuyatan and Edeki 2018); however, the resistance increases with the increase in fineness of POFA (Philip et al. 2019). The concrete prepared with fine ash POFA10 withstands acid attack better than POFA45 due to the increase in concrete densification due to pozzolanic reaction, enhancing the pore refinement compared to POFA45 (Hussin et al. 2009). The leaching of $Ca(OH)_2$ causes weight loss in concrete resulting in more disintegration as the immersion period increases. The POFA test samples have less CaO; as a result, less $Ca(OH)_2$ resulting in better performance than the OPC sample (Awal 1998).

Acid attacks mainly the calcium hydroxide. The pozzolanic reaction in finer POFA enables to reduce the hydrated lime that is vulnerable to acid attack, thus making it more resistant to acid attack (Ismail et al. 2010). Comparing the compressive strength of the experimental samples after immersing into the hydrochloric solution for 1800 h with the samples stored in water was performed by Ismail et al. (2010), and a significant reduction in the strength was observed in the samples immersed in acid solution. The control sample's compressive strength at 1800 h reduced by about 24%, and that of POFA10 reduced by 15.9%.

5 Conclusions and Recommendations

From literature revision, it can be concluded that the POFA exhibits excellent pozzolanic property. The performance of POFA regarding compressive strength, durability, resistance to chemicals increases with the increase in fineness of the POFA. Optimal replacement rates of 20% of cement with POFA maximize the utility as a supplementary cementitious material. However, several studies revealed that there would be no significant reduction in the strength of concrete up to about 30% substitution of POFA. Hence, proper utilization of POFA in cement production will be a considerable step to saving energy and reducing the cost of concrete. The effective management and application of POFA as supplementary cementitious material in concrete production will protect ecology and the environment and solve the problem of landfilling areas required for the disposal of POFA. Palm oil fuel ash should be considered as an important ingredient of concrete and not merely for the substitution of cement or reduction of cost.

To conclude, there is a need for concerted effort to promote the technically sound, environmentally safe, and economically justified utilization of POFA. The performance of high-strength and high-performance concrete containing POFA needs to be studied in-depth as only a few studies were noticed. The potential of POFA in partial replacement of material in aerated concrete to produce high strength concrete is needed.

References

Alsubari, B., Sha, P., Zamin, M.: Utilization of high-volume treated palm oil fuel ash to produce sustainable self-compacting concrete. J. Clean. Prod. **137**(7), 982–996 (2016)

Altwair, N.: Palm oil fuel ash (POFA): An environmentally-friendly supplemental cementitious material for concrete production. In: International Conference on Material Science, pp. 234–247 (2010)

Awal, A.S., Hussin, M.W.: The effectiveness of palm oil fuel ash in preventing expansion due to alkali-silica reaction. Cem. Concr. Compos. **19**(4), 367–372 (1997)

Awal, A.S., Shehu, I.A.: Evaluation of heat of hydration of concrete containing high volume palm oil fuel ash. Fuel **105**(10), 728–731 (2013)

Rajak, M., Zaiton, A., Ismail, M.: Morphological characteristics of hardened cement pastes incorporating nano-palm oil fuel ash. Procedia Manufact. **2**(2), 512–518 (2015)

Buczkowski, R.: Development of cement industry in Poland - history, current state, ecological aspects. A review. J. Clean. Prod. **141**(6), 702–720 (2017)

Chavalparit, O.: Clean technology for the crude palm oil industry in Thailand. Dissertation. Wageningen University (2006).

Donza, H., Cabrera, O., Irassar, E.F.: High-strength concrete with different fine aggregate. Cem. Concr. Res. **32**(5), 1755–1761 (2002)

Hussin, M.W., Ismail, M.A., Budiea, A., Muthusamy, K.: Durability of high strength concrete containing palm oil fuel ash of different finess. Malays. J. Civ. Eng. **21**(2), 180–194 (2009)

Ismail, M.: High performance blended cement concrete in Malaysia. In: 8th International Symposium on Utilization of High-Strength and High-Performance Concrete, pp. 639–646 (2008)

Ismail, M.A., Budiea, A.M.A., Hussin, M.W., Muthusamy, K.B.: Effect of POFA fineness on durability of high strength concrete. IndIan Concr. J. **84**(11), 21–28 (2010)

Johari, M.A.M., Brooks, J.J., Kabir, S., Rivard, P.: Influence of supplementary cementitious materials on engineering properties of high strength concrete. Constr. Build. Mater. **25**(5), 2639–2648 (2011)

Abdullah, K., Hussin, M.W., Zakaria, F. Muhamad, R., Abdul Hamid, Z.: POFA: a potential partial cement replacement material in aerated concrete. In: Proceedings of the 6th Asia-Pacific Structural Engineering and Construction Conference, pp. B132–B140 (2006)

Khalid, N.H. Rasid, A.N.N.A., Mohdsam, A.R., Lim, N.H.A.S., Ismail, M., Zardasti, L., Mohamed, A., Majid, Z.A., Ariffin, N.F.: Characterization of palm oil fuel ash and eggshell powder as partial cement replacement in concrete. In: IOP Conference Series: Materials Science and Engineering, vol. 431, issue 3 (2018)

Ma, K., Feng, J., Long, G., Xie, Y.: Effects of mineral admixtures on shear thickening of cement paste. Constr. Build. Mater. **126**(5), 609–616 (2016). https://doi.org/10.1016/j.conbuildmat.2016.09.075

Meyer, C.: The greening of the concrete industry. Cem. Concr. Compos. **31**(8), 601–605 (2009). https://doi.org/10.1016/j.cemconcomp.2008.12.010

Mohamad, M.E.: A review of the mechanical properties of concrete containing biofillers. In: IOP Conference Series: Materials Science and Engineering, vol. 160, issue 12, p. 012064 (2016)

Rusli, M.E.: Performance of concrete by using palm oil fuel ash as a cement replacement material. B.Sc. dissertation. Universiti Malaysia Pahang (2010)

Munir, A.: Utilization of palm oil fuel ash (POFA) in producing lightweight foamed concrete for non-structural building material. In: Procedia Engineering, vol. 125, pp. 739–746. Elsevier B.V. (2015)

Musa, A., Alkali, I.A.: The indirect tensile strength of palm oil fuel ash (POFA) modified asphaltic modified asphaltic concrete. Appl. Mech. Mater. **587–589**(7), 1270–1275 (2014)

Ofuyatan, O.M., Edeki, S.O.: Dataset on the durability behavior of palm oil fuel ash self compacting concrete. Data Brief **19**, 853–58 (2018)

Philip, J., Ismail, M.A.K., Al-Subari, B.: Strength and sulphate resistance of high performance concrete containing different fineness of palm oil fuel ash. In: IOP Conference Series: Materials Science and Engineering, vol. 495, issue 1 (2019)

Pone, J., Ash, A.: Palm oil fuel ash as a cement replacement in concrete. Mod. Approaches Mater. Sci **1**(1), 4–8 (2018)

Sanawung, W., Cheewaket, T., Tangchirapat, W., Jaturapitakkul, C.: Influence of palm oil fuel ash and W/B ratios on compressive strength, water permeability, and chloride resistance of concrete. In: 2017Advances in Materials Science and Engineering (2017)

Sata, V., Jaturapitakkul, C., Kiattikomol, K.: Utilization of palm oil fuel ash in high-strength concrete. J. Mater. Civ. Eng. ASCE **0899–1561**(12), 623–628 (2004)

Shafigh, P., Bin, H., Zamin, M., Zargar, M.: Agricultural wastes as aggregate in concrete mixtures – a review. Constr. Build. Mat. **53**(10), 110–117 (2014). https://doi.org/10.1016/j.conbuildmat.2013.11.074

Skariah, B., Kumar, S., Sahan, H.: Sustainable concrete containing palm oil fuel ash as a supplementary cementitious material – a review sustainable concrete containing palm oil fuel ash as a supplementary cementitious material – a review. Renew. Sustain. Energy Rev. **80**, 550–561 (2017)

Sofri, L. A., Zulham, M., Mohd, A., Isa, N. F., Azizan, M. A., Ahmad, M. M., Ahmran, A.: Performance of concrete by using palm oil fuel ash (POFA) as a cement performance of concrete by using palm oil fuel ash (POFA) as a cement replacement material. In: ICONET 2015, Applied Mechanics and Materials (2015)

Sooraj, V.M.: Effect of palm oil fuel ash (POFA) on strength properties of concrete. Int. J. Sci. Res. Publ. **3**(6), 691–697 (2013)

Tambichik. M.A.: Utilization of construction and agricultural waste in Malaysia for development of green concrete: a review. In: IOP Conference Series: Earth and Environmental Science p. 9 (2018)

Tangchirapat, W., Jaturapitakkul, C.: Cement & concrete composites strength, drying shrinkage, and water permeability of concrete incorporating ground palm oil fuel ash. Cem. Concr. Compos. **32**(10), 767–774 (2010)

Tangchirapat, W., Saeting, T., Jaturapitakkul, C., Kiattikomol, K., Siripanichgorn, A.: Use of waste ash from palm oil industry in concrete. Waste Manage. **27**(2), 81–88 (2007). https://doi.org/10.1016/j.wasman.2005.12.014

Tay, J., Show, K.: Use of ash derived from oil-palm waste incineration as a cement replacement material. Resour. Conserv. Recycl. **13**(1), 27–36 (1995)

Usman Jamo, H.: Characterization of a treated palm oil fuel ash. Sci. World J. **10**(1), 27–31 (2015)

Wi, K., Lee, H., Lim, S., Ismail, M.A., Hussin, M.W.: Effect of using micropalm oil fuel ash as partial replacement of cement on the properties of cement mortar. Adv. Mater. Sci. Eng. **2018**, 1–8 (2018)

Yusoff, S.: Renewable energy from palm oil e innovation on effective utilization of waste. J. Clean. Prod. **14**(9), 87–93 (2006)
Zhang, M.H., Malhotra, V.M., Bouzoubaa, N.: Laboratory-produced high-volume fly ash blended cements compressive strength and resistance to the chloride-ion penetration of concrete. Cem. Concr. Res. **30**, 1037–1046 (2000)
Zulkefli, R.N.A.R., Yaacob, H., Putra, R., Aris, S.M., Muthusamy, K.: Properties of fresh and hardened sustainable concrete due to the use of palm oil fuel ash as cement replacement. In: IOP Conference Series: Materials Science and Engineering, vol. 342, p. 012035 (2018)

Water

eSmartWatering - A More Sustainable Solution for Irrigation

João Ascensão[(✉)], Rui Alves, and Paulo Matos

Instituto Politécnico de Bragança, Campos de Santa Apolónia, 5300-253 Bragança, Portugal
a34505@alunos.ipb.pt, {rui.alves,pmatos}@ipb.pt

Abstract. The management of natural resources, through the conscious control and use of energy sources, is essential and necessary, so implementing mechanisms that enable better management is today an increasingly frequent effort. Due to its unsustainable consumption, water use has long one become of the main problems of modern times. Much of the current efforts use the concept of the Internet of Things (IoT), present in vast modern sectors, transforming high water consumption into a greener and more sustainable consumption. In areas often affected by severe droughts, the management of this consumption is critical, because water it's an important resource for humanity, since it affects countless sectors that depend on it, such as the development of various agricultural crops. The use of manual processes to irrigate these crops remains frequent, leading to much higher water consumption than necessary for optimal crop growth. Focused mainly on reducing water consumption, and ensuring minimum conditions for crop growth, the solution presented in this article, through technologies such as NB-IoT, a small set of sensors and based on the analysis of the humidity variation of the soil throughout the day, time and possibility of precipitation, allows to set efficient and sustainable rule times. Additionally, given the characteristics of the NB-IoT network, the solution allows not only remote consultation of the humidity levels of the soil, as the scheduling of manual watering cycles.

Keywords: Water · Irrigation · Sustainable · Agriculture

1 Introduction

Water management, as a natural resource, must be in a clear balance between its availability and demand, avoiding conflicts that may result from its use. The existence of water is what distinguishes our planet from all the others we know. Although the available global freshwater supply is more than adequate to meet current demand, it's spatial and temporal distributions aren't [1]. It's estimated that about 1.8 billion people in seventeen countries [2] may be approaching a severe water shortage crisis in the coming years. With this negative trend, it's predictable that water resources will not have a very encouraging future. Although no one can accurately predict the future of water, based on current data [3, 4], the world population will face several wars related to its consumption. It will certainly be one of the main challenges of this century, to reverse this negative

J. R. da Costa Sanches Galvão et al. (Eds.): *Proceedings of the 1st International Conference on Water Energy Food and Sustainability (ICoWEFS 2021)*, pp. 729–735, 2021.
https://doi.org/10.1007/978-3-030-75315-3_77

trend, leading to the idea of a new planning for the management of water consumption, transforming the current one into something more sustainable. A more sustainable consumption will certainly go through several transformations in the management of water consumption in the different areas of activity. The urban, industrial, agricultural and energy sectors are where the use of water is considered more unsustainable. Of the total water consumption in Europe [5], it's estimated that around 44% is used in the agricultural sector for activities such as irrigation, 40% is consumed in industry and energy production and 15% is used in the public water supply. Analyzing the previous numbers, it's possible to understand the importance of water in agricultural activities. Of the total value of cultivated land [6], around 20% are associated with irrigated agriculture, contributing 40% of the total food produced worldwide. On average and per unit of land, irrigation agriculture is at least twice as productive as non-irrigated agriculture, thus allowing greater intensification of production and diversification of crops. With the current population growth, it's estimated that agricultural production will need to expand by approximately 70% by 2050 [6], which will necessarily increase water consumption. The water level of the soil and the atmospheric humidity have a profound influence on the development of the plants [7]. Given the influence of water consumption on plant growth and given the forecast of expansion of the agricultural sector, it's important to find mechanisms that allow the correct growth of production, but, in parallel, enable the sector to have a sustainable consumption of water. The eSmartWatering solution, illustrated in this paper, will certainly help to achieve a more sustainable balance in the management of agricultural water resources. Using a set of sensors and chips and a mobile application, the solution presented allows the automatic and optimized management of soil moisture, based on meteorological factors (humidity of the atmosphere, temperature, probability of rain timetable, type of crop and daily history of soil moisture variation throughout the day. The remainder of the paper is organized in the following sections: Sect. 2 illustrates a brief analysis between the described solution and similar solutions; Sect. 3 describes the architecture designed for the problem presented; Sect. 4 illustrates the conclusions of the article and the objectives for the future work.

2 Related Work

There are some solutions with similar functionalities as presented in this paper. This section analyses the most relevant aspects, including a comparison with the features presented by eSmartWatering. The irrigation automation system of Hortau [8] allows producers to have more control in the irrigation process. One of the features offered by the system is to allow producers to start watering cycles using their mobile phone or PC. This cycle start can be done manually or automatically, depending on the soil water readings or temperature. Compared to what is presented in eSmartWatering, the activation of the system offered by Hortau is not influenced by climatic factors, that is, this system does not consider the probability of precipitation at the time of activation of a watering cycle. In addition, the history of variation of the water level in the land is also not included, contrary to what happens in eSmartWatering. Arable Mark [9] offers a complete irrigation management tool. Synthesizes climate, plant and soil data to produce useful information for plant growth. To achieve the capture of various climatic parameters, the offered solution incorporates many sensors for reading the values. Although both solutions can do treatment of climate parameters, eSmartWatering does not need so many sensors, since by using NB-IoT it can read various climate parameters of the terrain where it's configured. Finally, the solution presented by Arable Mark does not analyze the historic of the variation in the percentage of water in the land. The approach used by Tule [10] for better water management is somewhat different from the approach used in eSmartWatering. The solution presented by the company Tule is able to quantify the actual evapotranspiration - the process of evaporating the surfaces of the plants and the soil and from within the tissues of the plants, which is done by putting hardware over each plant. This approach is close to the analysis of the historic of variation of the percentage of water in the soil performed by eSmartWatering, but it achieves more accurate values that can allow a more precise performance at the moment of activating the watering cycle. However, eSmartWatering only needs a sensor to measure water in the soil, while the solution presented by Tule requires specific hardware. Through many sensors for measuring different parameters, the company Cropx [11] can offer a solution that makes it possible to collect and send data to the cloud for later consultation. However, contrary to eSmartWatering, this solution requires multiple sensors to collect the various weather data, while in the solution presented in this paper, integration with the OpenWeather API is sufficient to access the climate data of the terrain region where eSmartWatering is configured.

3 Proposed Architecture

The presence of the IoT concept in the various sectors of activity of today's society is increasingly frequent and the agricultural sector is no exception. Building mechanisms for the management of irrigation cycles in agriculture is no easy task, since numerous factors can influence the activation/deactivation of these cycles. When it comes to automatic irrigation management, it is important to understand what factors can lead to the need to activate a normal watering cycle. Based on the solutions presented in Sect. 2, it was understood to consider the following factors: precipitation probability, time of day (time) and daily historic of percentage of water present in the soil for the land in question. One of the key points of the solution presented is the analysis of the probability of precipitation. The presence of excess of water [12] in the soil negatively affects a large number of plants and causes other environmental problems (e.g., land displacement, etc.). At the time when it is verified whether a cycle should be activated or not it is important to understand what the probability of rain in the following hours. Being high, the watering cycle should not start. This small analysis makes it possible, not only to reduce water consumption, but also to mitigate other environmental problems that arise as a result of excess of water in the soil. Another important point of this solution is the analysis of the historic of the percentage of water in the soil at the time when the watering cycle should be started. Performing the analysis of the water variation in the soil along a time window, allows to understand the absorption/evaporation rate of the water. It's expected that there will be a variation in the presence of water in the soil, which can be higher or lower depending on the climatic conditions and the time at which the measurement is carried out. Understanding the absorption/evaporation rate allows adjusting the percentage of water that is introduced into the soil at each watering cycle. The eSmartWatering settings define the percentage value of water that must be kept in the soil and its safety margin. Knowing the absorption/evaporation rate and the percentage value of water that should be kept in the soil, it is possible to manage the amount of water used in each irrigation cycle. For example, the evaporation rate of water tends to be lower as the atmospheric temperature varies [13]. Thus, by keeping the percentage of water in the soil between the defined percentage value and its withdrawal interval, based on the evaporation/absorption rate calculated at the beginning of each cycle, a reduction in water consumption between the different watering cycles is possible. The eSmartWatering architecture, represented in Fig. 1, consists in the following components: a mobile application; an Arduino MKR 1500 [14]; a small set of Restfull services with the OpenWeather API integration [15]; multiple pairs composed of a chip ESP32 [16] with several Soil Moisture Sensors [17] and a Solenoid Valve [18]. It's based on the area of the terrain where the eSmartWatering will be configured that it is calculated the number of chips ESP32 and Soil Moisture Sensors required. Each ESP32 chip will be responsible for reading the values of the sensors it controls. Using BLE 5, which supports communication up to 200 m distance [19], all ESP32 chips, which function as Slaves in this solution, receive a provident parameter of the Arduino MKR 1500 that have the role of master. The received value corresponds to the query frequency of the value of the various Soil Moisture sensor associated with each chip. This value is transmitted to the Arduino board via the mobile application shown in Fig. 1. In addition to this functionality, it is through the mobile application that the identifiers of all ESP32 chips will be transmitted to Arduino. Finally, this application allows a manual watering cycle to be started at any time throughout the land where the solution is configured or only in a section of it. When the area exceeds the possible distance limit for communication with the BLE resource, a mesh network [20] was conceived for data transmission. With this solution, it is possible to cover large areas using a single master device. For each read cycle, all ESP32 chips send the read values of the sensors they control to the master. Using the NB-IoT [21] network, the master sends to the cloud all the received data adding the identifier of the respective ESP32 chip that performed the reading. When data is received in the cloud, it is initially recorded in the database. After recording the data, they are analyzed to verify if there is any section of land that requires the activation of a watering cycle. This verification shall be carried out based on the reference value and its defined safety margin for the land concerned, the precipitation probability read from the OpenWeather API integration and the evaporation/absorption velocity value that was calculated based on the historic of readings already performed for that instant. After the completion of this check is sent as a response to the connection established by the Arduino the sections of the terrain that need the activation of the watering cycle.

When the Arduino board receives the response with a request to activate the watering cycle of certain areas of the land, it sends requests to the ESP32 chips that control the targeted areas, to start the watering cycle, opening the associated solenoid valves.

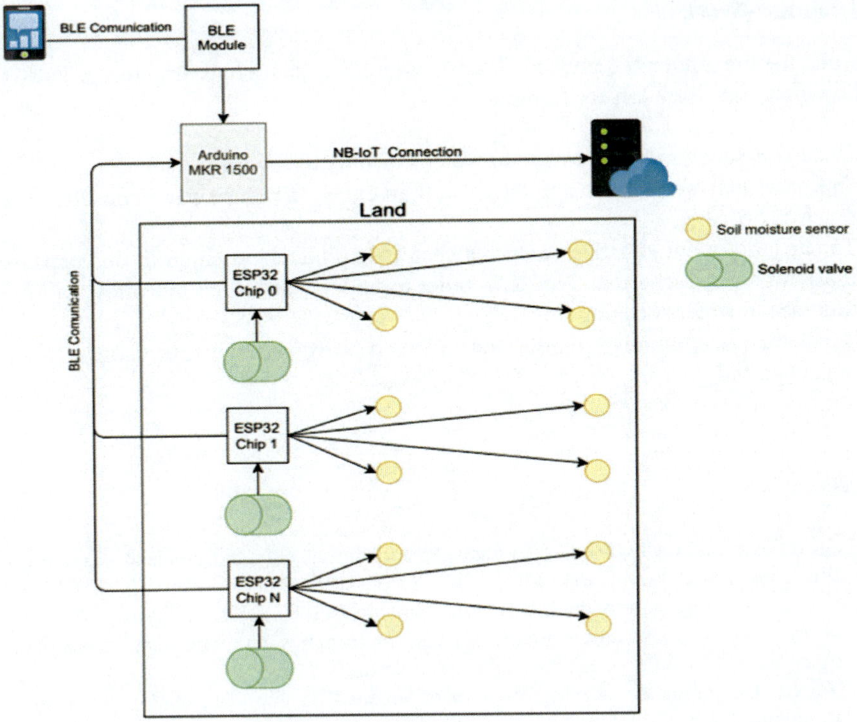

Fig. 1. eSmartWatering architecture.

4 Conclusion

This paper introduces a new tool that can contribute to improve water management in current agricultural activities. Through highly optimized watering cycles, this tool makes it possible to use only the volume of water strictly necessary for each cycle. Although data capture already started and solenoid valves actuation is also possible, Winter does not promote the climate conditions to assess the solution.

One of the many applications of this project is to understand the impact of water levels variation on the soil in the evolution of crop growth. Carrying out this analysis not only allows a more sustainable management of water, but also promotes new opportunities at the level of the business model of the agricultural sector. The availability of freshwater [22, 23] limits the growth of plants on the earth's surface. For terrestrial plants, freshwater is a basic requirement for life, where its soil absorption facilitates inorganic mineral nutrition. In addition, water retention drives the expansion of plant cells contributing to their shape and size. Controlling this dependency between water level of the soil and growth of the crops allows, even if, indirectly, controlling the demand for certain productions enabling new business models in the agricultural sector.

4.1 Future Work

Despite the great potential presented, this solution still has a long way to go. Thus, the following points were left for future work:

- Understand the exact percentage of water saved by the presented solution.
- Optimize analysis processes to the maximum extent to reduce water consumption as much as possible.
- To carry out a comparative study between the solution presented in this paper and the solutions presented in Sect. 2 in order to understand which one allows a further decrease in water consumption.
- To idealize an alternative solution for regions where communication using NB-IoT is not supported.

References

1. Cosgrove, W.J., Loucks, D.P.: Water management: current and future challenges and research directions. Water Resour. Res. **51**(6), 4823–4839 (2015). https://doi.org/10.1002/2014WR016869. https://agupubs.onlineli-brary.wiley.com/doi/abs/10.1002/2014WR016869
2. Dormido, H.: These countries are the most at risk from a water crisis. https://www.bloomberg.com/graphics/2019-countries-facing-water-crisis/
3. Orlović, M., Krajnovic, A.: In: Water management - an important challenge for modern economics, October 2015
4. Gómez-Llanos, E., Durán-Barroso, P., Robina-Ramírez, R.: Analysis of consumer awareness of sustainable water consumption by the water footprint concept. Sci. Total Environ. **721**, 137743 (2020). https://doi.org/10.1016/j.scitotenv.2020.137743. http://www.sciencedirect.com/science/article/pii/S0048969720312547
5. Agency, E.E.: Water use by sectors. https://www.eea.europa.eu/archived/archived-content-water-topic/water-resources/water-use-by-sectors
6. Bank, T.W.: Water in agriculture. https://www.worldbank.org/en/topic/water-in-agriculture
7. Wangermann, E.: The effect of water supply and humidity on growth and development. In: Melchers, G. (ed.) External Factors Affecting Growth and Development Aussenfaktoren in Wachstum und Entwicklung, pp. 618–633. Springer, Heidelberg (1961)
8. hortau.com: Hortau's irrigation automation technology is a complete wireless system that integrates flawlessly with existing operations. https://hortau.com/irrigation-automation/. Accessed 29 Dec 2020
9. arable.com: A Complete Water-Budgeting Solution. https://www.arable.com/solutions_irrigation/. Accessed 29 Dec 2020
10. tuletechnologies.com: What is ET?. https://www.tuletechnologies.com/tule-faq. Accessed 29 Dec 2020
11. cropx.com. https://www.cropx.com/.Accessed 29 Dec 2020
12. Macie, E.: How to correct excessive soil moisture and drainage problems. https://trees-energyconservation.extension.org/how-to-correct-excessive-soil-moisture-and-drainage-problems/
13. Brutsaert, W.: Daily evaporation from drying soil: universal parameterization with similarity. Water Resour. Res **50**(4), 3206–3215 (2014). https://doi.org/10.1002/2013WR014872
14. Arduino.CC: ARDUINO MKR NB 1500. https://store.arduino.cc/arduino-mkr-nb-1500-1413. Accessed 29 Dec 2020

15. OpenWeather: OpenWeather API. https://openweathermap.org/api. Accessed 29 Dec 2020
16. espressif.com: ESP32. https://www.espressif.com/en/products/socs/esp32. Accessed 29 Dec 2020
17. Environment, M.: Which soil sensor is perfect for you? https://www.metergroup.comenvironment/articles/which-soil-sensor-is-perfect-for-you/. Accessed 29 Dec 2020
18. tameson.com: Solenoid Valve - How They Work? https://tameson.com/solenoid-valve-types.html. Accessed 29 Dec 2020
19. vinter.se: What's New with BLE5? And how does it compare to BLE4? https://vinnter.se/whats-new-with-ble5-and-how-does-it-compare-to-ble4. Accessed 29 Dec 2020
20. www.bluetooth.com: Mesh networking is blue. https://www.bluetooth.com/learn-aboutbluetooth/recent-enhancements/mesh/. Accessed 29 Dec 2020
21. GSMA: NarrowBand-Internet of Things (NB-IoT). https://www.gsma.com/iot/narrow-baninternet-of-things-nb-iot/. Accessed 29 Dec 2020
22. Robbins, N.E., Dinneny, J.R.: Growth is required for perception of water availability to pattern root branches in plants. Proc. Natl. Acad. Sci. **115**(4), E822–E831 (2018). https://doi.org/10.1073/pnas1710709115, https://www.pnas.org/content/115/4/E822
23. Blatt, M.R., Chaumont, F., Farquhar, G.: Focus on water. Plant Physiol. **164**(4), 1553–1555 (2014)

Impact of Permeable Pavement on Stormwater Runoff in Urban Areas: A Case Study of Leiria

Nahum Santos[1](✉), Ricardo Gomes[1,2], and Anabela Veiga[1,3]

[1] School of Technology and Management, Polytechnic of Leiria, 2411-901 Leiria, Portugal
nahum.hijaz@gmail.com, ricardo.gomes@ipleiria.pt
[2] INESC Coimbra, DEEC, Rua Sílvio Lima, Polo II, 3030-290 Coimbra, Portugal
[3] University of Coimbra, Geoscience Center, 3004-531 Coimbra, Portugal

Abstract. The human occupation in high demographics density areas causes a series of environmental impacts. Among these impacts, special attention should be given to the alteration of hydrological cycle caused by the impermeabilization of the surface, which creates an increase in the peak flow stormwater runoff and a decrease in lag time, which are involved in urban flooding. This paper evaluates the impacts of permeable pavement as a sustainable solution to reduce stormwater runoff in Capuchos Urbanization, Leiria city. This area was chosen because it is a consolidated urban environment with information about water catchment and drainage network. A Storm Water Management Model (SWMM5.1) was developed to evaluate the performance of drainage network, for different rainfall intensity and with/without permeable pavement. The results show that permeable pavement preset to be a good solution to reduce the peak flow of stormwater drainage network and, consequently, the mitigation of the risk of flooding in urban areas.

Keywords: Urban area · Permeable pavement · Stormwater drainage network · SWMM

1 Introduction

It is undeniable that human presence has a great impact over the changes in the natural environment, mainly in regions with a great agglomeration of people, such as large cities. According to Araújo et al. [1], urban occupation through impermeable areas, such as roofs, sidewalks, pavements, parking lots and others, changes the hydrological cycle in urban areas, resulting in an increase of urban floods and degradation of the quality of water bodies in the receiving environment. It is also noteworthy that intense convective rains can be aggravated by the phenomena of heat islands, which are formed over the more urbanized surfaces [2].

According to Broekhuizen et al. [3], the stormwater drainage network traditionally adopted in urban areas is based on the use of pipes for a quick removal of local runoff, ignoring the possible consequences of floods and water quality downstream,

J. R. da Costa Sanches Galvão et al. (Eds.): *Proceedings of the 1st International Conference on Water Energy Food and Sustainability (ICoWEFS 2021)*, pp. 736–743, 2021.
https://doi.org/10.1007/978-3-030-75315-3_78

being also financially expensive to construct and sustain networks. In this way, a sustainable stormwater drainage network tries to correct these problems by carrying out runoff management in a more natural way [4].

As defined by the Construction Industry Research and Information Association [5, 6], sustainable stormwater drainage network is a set of sustainable rainwater control and management techniques and has emerged as an alternative to the traditional stormwater drainage network in urban areas. In this study, the permeable pavements were conceived to manage the stormwater runoff management, to reduce the risk of flooding in Leiria's downtown [7]. The permeable pavements provide a paving system suitable for both pedestrians and vehicles, allowing rainfall to infiltrate superficially towards the lower layers [5, 6].

2 Study Area Description

The study was focused on the area of the Capuchos Urbanization, close to downtown Leiria (Portugal). This is a consolidated urban area where the characteristics of water catchment and drainage network are known (Fig. 1).

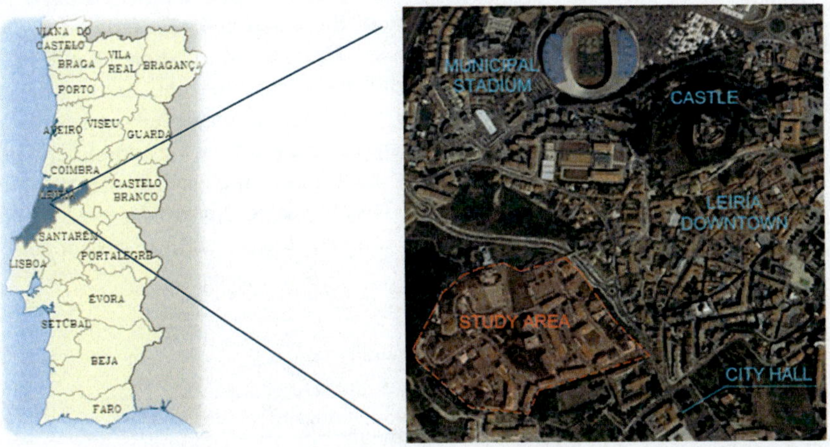

Fig. 1. Study area identification (Adapted from Google Earth, 2018)

The case study has a total area of 10.82 hectares, of which 62% are occupied by buildings, 31% are covered with pavements and sidewalks and 7% are green or permeable areas. The ground elevation is quite irregular, with land plots with slopes of 20%, and others flat. The occupation is mainly residential, with a single-family and multi-family buildings. Two important infrastructures that stand out are a school unit and a parking lot. Some commerce activities can also be found, but without great relevance.

The stormwater drainage network includes three outlet points: E1, E2 and E3. The E2 outlet is connected to Leiria downtown's stormwater drainage network, where the risk of flooding is real. Each outlet point defines a small water catchment of the stormwater drainage network for study area (Fig. 2): i) Network E1 has 218 m and the pipe diameter is 300 mm, ii) Network E2 has 409 m and pipe diameters of 300 and 400 mm, and iii) Network E3 has a 411 m and pipe diameter of 300 mm, and 400 mm closer to the outlet point. The pipe material is concrete [8].

Fig. 2. Stormwater drainage network.

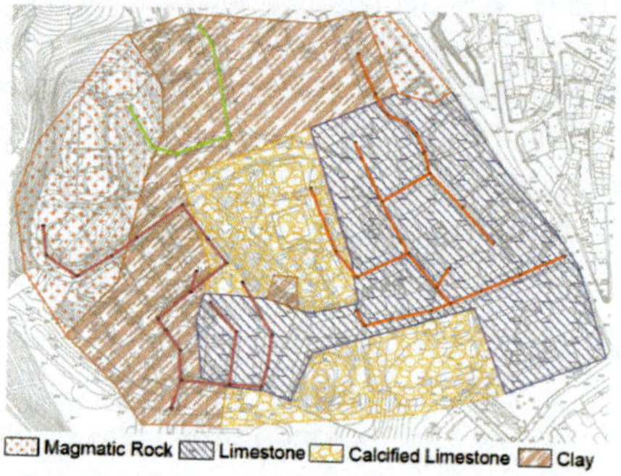

Fig. 3. Definition of the geological units.

Regarding the soil permeability, it was assessed based on the Geological Map of Portugal provided by the National Energy and Geology Laboratory - LNEG, namely "Folha 23-C – Leiria" [9], and the geological nature of the outcropping materials (magmatic rock, clays and limestone). The magmatic rock, which was much altered and fractured, was considered permeable, the clays were considered impermeable and the limestone, which was found in thin layers and sometimes calcified, was considered semi-permeable.

Based on a visual inspection carried out in the study area, it was possible to establish more precisely the geological units that have more influence on the stormwater drainage networks in the study area (Fig. 3). Also, it was possible to describe the correlation between the types of soil found and the respective permeability rates adopted (see Table 1).

Table 1. Permeability rates.

Soil type	Permeability rates (m/s)
Magmatic rock, altered and very fractured	10^{-2} m/s (high permeability)
Limestone rock	10^{-5} m/s (average permeability)
Calcified limestone rock	10^{-7} m/s (low permeability)
Clays	10^{-9} m/s (very low permeability)

3 Hydrological and Hydraulic Modelling

To evaluate the advantage of permeable pavement, different scenarios were performed for the case study – with different rainfall intensity. A Storm Water Management Model

(SWMM5.1) was developed to evaluate the performance of stormwater drainage network. The Infrastructural Department of Leiria Council provided the set data of the drainage network.

For each small water catchments shown in Fig. 2 (E1, E2 and E3) three categories for land cover were defined: i) buildings; ii) pavements and sidewalks; and iii) green areas. The land cover categories were established considering the different level of rainwater infiltration. The categories "buildings" was carried out to respect the configuration of the existing land plot subdivisions. The Municipal Planning Directive of Leiria (2018) establishes 80% for construction area. It is important to highlight that the plots without occupation were considered built, thus admitting possible future construction. All plots were considered earthworks to simulate construction conditions, so a slope of 1% was considered. The categories "pavements and sidewalks" were defined as the area currently intended for sidewalks and street public roads. In this case, it was decided to consider the value of 100% impermeabilization, since the infiltration capacity in this type of covering material can be neglected. The categories "green areas" include the areas defined by the Municipal Planning Directive of Leiria as leisure place, where the maximum construction area is 20%.

The soil's infiltration capacity is another parameter that will have influence in the modelling since it determines the rainwater runoff. The Soil Conservation Service (SCS) method was chosen to obtain its numerical results. According to the Environmental Protection Agency (EPA, 2015), this method is an approximation for estimating the runoff, as this value is obtained from the difference between the soil infiltration capacity, determined through the Curve Number (CN), and the accumulated precipitation volume. The CN varies according to the land cover, the soil type and the soil drying time. According to Environmental Protection Agency (EPA) [10], the soil drying time varies between 2 and 14 days, and the soil types are divided into groups A, B, C and D, as established by the United States Department of Agriculture [11]. The parameters used are summarized in Table 2.

Table 2. Parameters for establishing SCS-CN.

Soil type	Soil group	Drying time (days)
Magmatic rock, altered and very fractured	A	7
Limestone rock	B	7
Calcified limestone rock	C	14
Clays	D	14

According to EPA [10], the layers of permeable pavement system are, generally, surface layer, soil layer and storage layer. The surface layer indicates the portion that directly receives the rainfall and runoff from boundaries areas, stores excess inflow in depressions, and generates runoff into the stormwater drainage network or for downstream areas. It is in this layer that the paving blocks and filling material is used, in the case of modular systems, or porous concrete/asphalt. The soil layer consists of a sand layer with the purpose of providing filtration and accommodation base for a paving blocks layer. The storage layer represents the base and sub-base of the pavement, created by a bed of pebble or gravel used for rainwater retention. For the support soils with low permeability or low resistance drains must be installed in the storage layer.

For the case study, the permeable pavements with two layers were specified, one composed of the pavement, the layer that will receive the runoff, and the other composed of the storage layer, which will be responsible for rainwater retention. A drain has been established to empty the storage layers. The pavement layer includes a thickness of 100 mm, void ratio of 0.2 and permeability of 150 mm/h. The storage layer has thickness of 300 mm, void ratio of 0.4 and permeability according to the soil type. Finally, a drain network with 100 mm diameter was used at the bottom of the storage layer, whose contribution is measured from the flow exponent. The value recommended by EPA [10] is 0.5 and the flow coefficient that is obtained through Eq. 1.

$$C = \frac{2 \times D^{\frac{1}{2}}}{T} \tag{1}$$

where: C is the flow coefficient; D is the distance from the drain network to the surface (mm); and T is the time in which the volume rainwater retained must be drained (hours). It was established a 24-h period for emptying the storage layer.

Figure 4 shows the precipitation for the return period of 10 and 20 years (time step: $\Delta t = 10$ min), used to evaluate the performance of stormwater drainage network with/without permeable pavement.

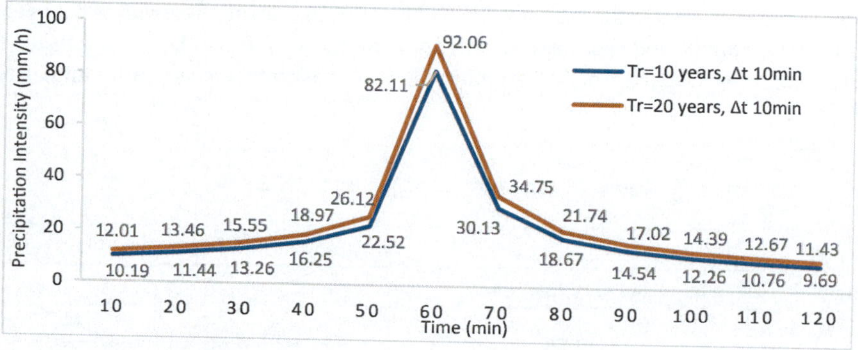

Fig. 4. Precipitation for the return period of 10 and 20 years, $\Delta t = 10$ min.

4 Results

4.1 Lag Time

For the lag time analysis of the stormwater drainage networks, a modelling of precipitation with uniform intensity was performed for the return periods of 10 and 20 years (time step: 10, 15 and 20 min). The lag time is defined as a time taken to reach the maximum flow at reference section (E1, E2 and E3). The use of three different precipitation intensities is justified by increasing the database for estimating thc value of lag time. Table 3 presents the results of lag time for all networks and return periods. It was noticed that all networks had some reduction in the lag time recorded. This happens because the permeable pavement temporarily retains the runoff.

Table 3. Results of lag time.

Permeable pavement		Return period 10 years	Return period 20 years
Network E1	Without	30 min	30 min
	With	60 mins	60 min
Network E2	Without	35 min	30 min
	With	115 min	150 min
Network E3	Without	25 min	25 min
	With	80 min	75 min

4.2 Peak Flow

For the peak flow analysis of the stormwater drainage networks, a modelling of precipitation for the return periods of 10 and 20 years was performed (Fig. 4). The peak flow is defined as a maximum flow at reference section (E1, E2 and E3). Table 4 presents the results of peak flow for all networks and return periods.

Table 4. Results of peak flow.

Permeable pavement		Return period 10 years	Return period 20 years
Network E1	Without	412.91 l/s	465.82 l/s
	With	280.33 l/s	331.60 l/s
Network E2	Without	735.14 l/s	756.78 l/s
	With	338.99 l/s	448.88 l/s
Network E3	Without	638.13 l/s	669.04 l/s
	With	411.13 l/s	522.91 l/s

It was noticed that all networks had some reduction in the peak flow recorded, and the greatest reduction was observed in network E2. This happens because network E2 is over soil with average permeability (Fig. 3), which increases the infiltration of rainwater. For the return period of 20 years, it is possible to conclude that the registered peak flow is higher, because this return period presents rains of greater intensity.

4.3 Stormwater Runoff

Here a comparative study of the stormwater runoff on water catchments will be carried out, before and after the implementation of the permeable pavement. Return periods of 10 and 20 years were used (Fig. 4). Table 5 presents the runoff results for all networks and return periods.

Table 5. Results for stormwater runoff.

Permeable pavement		Return period 10 years	Return period 20 years
Network E1	Without	362.69 l/s	420.13 l/s
	With	241.45 l/s	288.54 l/s
Network E2	Without	227.35 l/s	263.10 l/s
	With	82.02 l/s	97.88 l/s
Network E3	Without	201.07 l/s	234.00 l/s
	With	91.77 l/s	110.98 l/s

For the return period of 10 and 20 years, the stormwater runoff presents reduction values in the same order of magnitude, which indicates similar behavior. However, a larger stormwater runoff is observed for the return period of 20 years, because of the higher rain intensity. The best results were obtained from network E2, because it is over soil with average permeability (Fig. 3), which increases the infiltration of rainwater.

5 Conclusions

With the increase in the world's population, the influence of human beings on the environment becomes increasingly clear, such as the alteration of the urban hydrological cycle. The results of a case study show that permeable pavements can be used as a sustainable solution to reduce the stormwater runoff in urban water catchments, as well as to reduce the peak flow and increase the lag time in stormwater drainage network, reducing the risk of flooding. It was possible to observe lag time increments greater than 30 min (reaching the maximum value of 1 h and 55 min). The reduction of the peak flow was also significant, ranging from 32% to 54%. And the reduction of the stormwater runoff reduced between 10% to 20%, on soil with low permeability, and 100% on soils with high permeability.

It is important to note that network E2 offers the best performance with permeable pavements, which shows how much this kind of technology can have its potential increased based on the soil infiltration capacity: network E2 is on area of predominantly calcareous soil (average permeability), while networks E1 and E3 there are predominantly installed over clayey soil (very low permeability). It was observed that the soil in water catchments of the network E2 is 10,000 times more permeable than the soil found in water catchment of the networks E1 and E3.

As a future work, more research is required about the cost benefits analysis of the project proposed.

Acknowledgments. This study was supported by UIDB/00308/2020 project. The authors would like to thank the support of Infrastructural Department of Leiria Council and the Water Company of Leiria, for providing the information necessary to achieve the study. The authors also would like the thank of Jean Noèl Fernand Mercereau, lecturer in the Department of Language Sciences at the Polytechnic of Leiria, for revising the English text.

References

1. Araújo, P.R., Tucci, C.E.M., Goldenfum, J.A.: Avaliação da eficiência dos pavimentos permeáveis na redução do escoamento superficial. Revista Brasileira dos Recursos Hídricos **5**(3), 21–29 (2000)
2. Miguez, M.G., Veról, A.P., Rezende, O.M.: Drenagem Urbana - Do Projeto Tradicional à Sustentabilidade. Elsevier, Rio de Janeiro (2015)
3. Broekhuizen, I., Muthanna, T.M., Leonhardt, G., Viklander, M.: Urban drainage models for green areas: Structural differences and their effects on simulated runoff. J. Hydrol. X **5**, 1–13 (2019)
4. Eckart, K., McPhee, Z., Bolisetti, T.: Performance and implementation of low impact development – a review. Sci. Total Environ. **607**(608), 413–432 (2017)
5. Construction Industry Research and Information Association (CIRIA): Site handbook for the construction of SUDS. CIRIA C698. Londres (2007)
6. Construction Industry Research and Information Association (CIRIA): The SUDS Manual. CIRIA C753. Londres (2015)
7. Gomes, R., Garcia, R.: Sistemas Urbanos de Drenagem Sustentável: Influência sobre as Infraestruturas Existentes. In: 14.º Simpósio de Hidráulica e Recursos Hídricos dos Países de Língua Portuguesa, APRH, Cidade da Praia, Cabo Verde (2019)
8. Santos, N.: Utilização de pavimentos permeáveis para dar resposta à redução do escoamento superficial em áreas urbanas - caso de estudo. Dissertação de Mestrado em Engenharia Civil – Construções Civis, Instituto Politécnico de Leiria (2020)
9. Teixeira, C., Zbyszewski, G., Torre de Assunção, C., Manupella, G.: Carta Geológica de Portugal, na escala 1/50 000. Notícia explicativa da folha 23-C (Leiria), Serviços Geológicos de Portugal (1968)
10. Environmental Protection Agency: Storm Water Management Model User's Manual Version 5.0. National Risk Management Research Laboratory Office of Research and Development, Cincinnati, OH: EPA (2015)
11. United States Department of Agriculture (USDA): SCS Urban Hydrology for Small Watersheds. 2nd edn., (TR-55), Washington, DC (1986)

Potential for the Use of Rainwater in Higher Education Institutions: A Case Study of Building D of Campus 2 of the Polytechnic of Leiria

Sara Bona[1](✉) and Ricardo Gomes[1,2]

[1] Polytechnic of Leiria, School of Technology and Management, 2411-901 Leiria, Portugal
ricardo.gomes@ipleiria.pt
[2] INESC Coimbra, DEEC, Rua Sílvio Lima, Polo II, 3030-290 Coimbra, Portugal

Abstract. As water scarcity increases, the search for alternative water sources, the development of new technologies for water efficient use and the need for planning and integrated water resources management have been shown to be relevant in several fields. The rainwater harvesting system appears in this context as one of the most promising alternative of water sources for buildings, as the system can be easily installed, and rainwater can be used without significant treatment for non-potable purposes. The study presented here, focuses the potential for the use of rainwater harvesting system in higher education institutions, taking as case study the Building D of Campus 2 of the Polytechnic of Leiria. The results show that economic viability, reliability, and resilience of the system depends on factors related to precipitation, the area of rainwater catchment, the storage capacity, and the demand/value/tax for drinking water from public services. Additionally, the use of rainwater may reduce the drinking water demand between 46% to 89% and enabling the building taken as in case study to become more sustainable.

Keywords: Higher education institutions · Rainwater harvesting system · Water efficiency

1 Introduction

Currently, with the increase in global discussions involving efficiency in the use of water resources, water is at the center of sustainable development and concerns the central promise of 6th Goal of the 2030 Agenda [1]. World water consumption has increased six times in the past 100 years and its demand is expected to remain constant until 2050 [2]. This increase is due to a combination of population growth, economic development, and changes in water demand patterns. Associated with an increasingly irregular and uncertain water supply, climate change will intensify the situation in regions that already have water scarcity and will cause water stress in regions where this resource is still abundant [3]. Recently, in 2017, but without the phenomenon being isolated, the Mediterranean region, in which Portugal is inserted, faced a situation of severe and extreme drought [4, 5].

J. R. da Costa Sanches Galvão et al. (Eds.): *Proceedings of the 1st International Conference on Water Energy Food and Sustainability (ICoWEFS 2021)*, pp. 744–753, 2021.
https://doi.org/10.1007/978-3-030-75315-3_79

In this sense, solutions for the use of water resources that present alternatives with less environment impact, aligned to the protection of water resources by the efficiency in its use, must be fundamental in the development of public strategies and a measure of economic interest for citizens and companies. As far as buildings are concerned, there is a great potential for the implementation of water efficiency practices, without affecting the effectiveness of its use, maintaining comfort, safety, and water quality and, thus, generating benefits associated with the reduction of drinking water demand and the associated energy [4, 6, 7]. Currently, many countries are motivated and committed to the use of rainwater and to the development of technologies that facilitate and guarantee the safe use of this alternative source. In Portugal, this practice is still incipient, but several good examples of its application are already known [8–10].

In this context, the present study aims to analyze the potential for the use of rainwater in the Building D of Campus 2 of the Polytechnic of Leiria, using the methodology suggested by the Portuguese ANQIP association (ANQIP – *Associação Nacional para a Qualidade nas Instalações Prediais*). Additionally, this study intends to identify the advantage of water efficient use in buildings and sensitize the academic community and the population in general for this issue.

2 Materials and Methods

A methodology was developed to identify solutions for the case study, that allows the evaluation of savings potentials and implement appropriate measures to reduce drinking water demand, through the adoption of rainwater harvesting systems. The flowchart in Fig. 1 shows the methodology proposed.

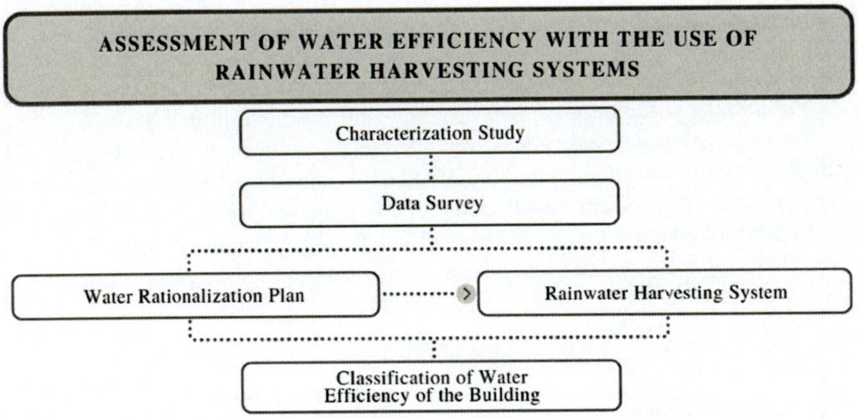

Fig. 1. Flowchart of the water efficiency assessment and classification method [11].

2.1 Case Study Description

The water-sanitary projects analysis refers to the initial stage of the study, with the aim to identify and characterizing the existing devices toilets in the building. Building D of the Campus 2 of the Polytechnic of Leiria, was used as case study. The building started operating in 2003 and has approximately 14.225 m^2, spread over five floors plus the roof, used as a technical area. The building includes several amphitheaters, classrooms, laboratories, teachers' offices, and computer services. There are 9 bathrooms, including female and male, and the water demand allowed are, as a rule, for flushing toilets, urinals, and washbasins (see Fig. 2). So far, the building has no certification or evaluation of water efficiency and there is no mechanism for real-time control of drinking water demand. The public water supply is the only source of drinking water, where it meets demand across the Campus. Regarding sewer, it is transported until the public drainage network and, finally, launched at the Wastewater Treatment Station in Coimbrão. The rainwater, collected from impermeable surfaces and peripheral drains in the building envelope, is directed to the public stormwater drainage network and discharged into the Lena River.

Fig. 2. Sanitary devices in Building D of the Polytechnic of Leiria (Campus 2).

2.2 Data Survey

Through a building audit, it was possible to validate and update the records of technical water and sewage installations, made available by the institution's Technical Office. Also, it was possible to identify and characterize the devices of water demand in the building, with emphasis on the functionality and its state of conservation, and the recent changes in terms of technical installations. The volume and individual demand of each water device took place through analysis of manufacturer's catalogs. Table 1 shows the characteristics of the water demand devices, according to the analysis performed.

Table 1. Characterization of building devices - Building D.

Device	Number	Consumption
Toilet	47	4.27 L/washing
Urinals	26	1.75 L/washing
Bathroom faucets	46	9.45 L/min
Kitchen/Laboratory faucets	17	15.09 L/min
Showers	6	11.00 L/min

The overall of the building water demand was also analyzed, where there was a consumption trend related to the period of school activities, temperature variations and the number of students. Through a flowmeter installed at the entry point of the building it was possible to analyze the drinking water demand in the last five years – the annual average water demand corresponds to 206.5 m^3/month. The building's occupation over the day was established according to data provided by the Head of the School of Technology and Management of Polytechnic of Leiria and the Pedagogical Organization Office, considering the maximum capacity of the classrooms, laboratories, teachers' offices, and computer services. The estimated number of users of Building D corresponds to 618 users/day, that corresponds a water demand unit per user to 15 L/(user.day), considering 22 workdays per month. Due to the pandemic control measures implemented in 2020 (COVID-19), resulting in the replacement of presential classes for online classes, the percentages affect for each water use was based on similar higher education institution [12].

2.3 Water Rationalization Plan

The water rationalization plan provides for the possibility of leading improvements to the facilities in terms of water saving measures. Commonly, there are several areas of a building with potential for waste reduction and, under a diagnosis of the situation, it is possible to make improvements to existing installations in terms of direct savings measures (for example: replacement of devices or adjustment of volume with flow reducers) and indirect measures in terms of user behavior (for example: awareness-raising or training actions) [4, 5].

In the rationalization plan implementation, it is important to study the limitations so that the adoption of economy measures is carried out in a balanced way, ensuring the proper functioning of the facilities. In this sense, according to the institution's Technical Office [13], due to actions carried out in recent years with the objective of implementing water saving solutions (flushing toilet replacement, awareness signs, device tuning), results about water demand reduction were obtained, however associated difficulties were identified, especially at the level of the sanitation network (for example: blockage of urinals, malfunctioning of pumping systems and obstruction of sewer drainage network). Thus, for the present case study, water demand in existing devices has been controlled, so they cannot be reduced from the point of view of potentiating deficiencies and, thus, financially exceeding the savings obtained in reducing consumption. With this knowledge, it is assumed that to increase water efficiency, one of the solutions is the implementation and adequate management of an alternative source of water for non-potable uses.

2.4 Rainwater Harvesting System

In Portugal, the rainwater harvesting systems design must meet the requirements imposed by legislation, regulation, and standardization in force at national and European level. ANQIP established two Technical Specifications for rainwater harvesting system (ETA): ETA 0701 – establishes technical and calculation criteria for design, implantation, and operation [14]; and ETA 0702 – refers to the conditions for their certification [15].

The methodology for rainwater harvesting systems design starts from a reservoir with a predefined volume to simulate the behavior of the water stored in the reservoir from a rainfall historical series. This calculates the months in which the selected reservoir meets or does not meet the water demand for rainwater. Therefore, it is essential to know the characteristics of the building, the precipitation of the region and the volume of water that can be used in the building [5, 15]. Initially, for the hydrological study, precipitation series provided by the Portuguese SNIRH system (SNIRH – *Sistema Nacional de Informação de Recursos Hídricos*) were used (Leiria, rain gage: 15E/01UG, average monthly rainfall data from 1987 to 2017). The identification of area for rainwater harvesting took place through architectural projects, where the surface of the roof represents 1,962 m^2 and the balconies correspond 978 m^2. In Building D, the non-potable water demand, that is, the volume that can be supplied through the rainwater harvesting system, refers to the flushing toilet and urinals.

2.5 Classification of Water Efficiency of the Building

In Portugal, ANQIP has developed a voluntary water efficiency certification and labeling system for water devices marketed at the national level, where A^{++} represents the highest degree of efficiency and E, the lowest (see Fig. 3a). The requirements for the implementation and use of the labels are established in Technical Specifications ETA 0802 [16] and ETA 0803 [17].

Currently, the ADENE agency (ADENE – *Agência para Energia*) is promoting the $AQUA^{+}$ system, a voluntary instrument to assessment and classification of water efficiency in buildings (see Fig. 3b) [18]. The certification system allows to diagnose the water level use and performance, on a rating scale from less efficient (F) to more efficient (A^{+}).

Fig. 3. a) Water efficiency labels in Portugal [6]; b) $AQUA^{+}$ water efficiency logo [18].

In this study, the ANQIP methodology proposed for the evaluation of the water efficiency of non-residential buildings was applied, according to its Manual [6]. In summary, the method consists of adding the flows of the water devices that are part of the sanitary facilities, relating them to their usage factors, in this way it is possible to determine the daily water demand per user. Usage factors are determined for each sanitary device and establish parameters of frequency use. As shown in Table 2, the classification categories range from A^{++} to E depending on water demand patterns for administrative buildings [6]. In a similar way, similarities are admitted between the consumption of these buildings and university centers, consequently, the same classification was established.

Table 2. Classification of university buildings according to the consumption pattern.

Category	Public network drinking water consumption [liters/(user.day)] *(c)*
A^{++}	$0 < c \leq 20$
A^{+}	$20 < c \leq 25$
A	$25 < c \leq 30$
B	$30 < c \leq 40$
C	$40 < c \leq 55$
D	$55 < c \leq 75$
E	$c > 75$

3 Results and Discussions

From the ANQIP methodology, it is possible to evaluate and classify the water efficiency in Building D of the Campus 2 of the Polytechnic of Leiria. It is important to note that there are no washing machines (correction factor = 1.0), and all showers are disabled.

Table 3 indicates the results of calculating the water demand patterns and the final classification of water efficiency in the building. As expected, it is noted that the building presents a positive result, with a total drinking water demand equal to 16.37 L/(user.day) and, therefore, to a water efficiency rating corresponding to Category A^{++}.

Table 3. Water demand calculator.

Device	Units	Volume/Flow (Average)	Usage factor	Total liters/(user.day)
Toilet	Liters	4.27	0.54	2.31
Urinals	liters	1.75	4.32	7.56
Bathroom faucets	Liters/minutes	9.45	0.64	6.05
Kitchen/Laboratory faucets	Liters/minutes	15.0s9	0.03	0.45
Showers	Liters/minutes	11.00	0.00	0.00
Total calculated consumption	Liters/(user.day)			16.37
Correction factor				1.00
Total consumption	Liters/(user.day)			**16.37**
Water efficiency categorization of the building				A^{++}

In the study of rainwater harvesting systems for Building D, two different scenarios were considered, where the rainwater catchment area, in the first scenario (see Table

4) is equivalent only to roof area (1,966 m^2) and, in the second includes the roof and balconies areas (2,940 m^2).

Table 4. Calculation of rainwater harvesting system (scenario 1).

(C) = 0.9			(η) = 0.9			(A) [m^2] = 1962.9			
Month	Monthly average rainfal	Monthly consumption	Usable volume of monthly rain	Difference	Calculation differences	Volume adopted for the reserve	Volume of water in the reservoir (m^3)		Supply
	(mm)	(m^3)	(m^3)	(m^3)	(m^3)	(m^3)	Start	End	(m^3)
	(1)	(2)	(3)	(4)	(5)	(6)	(7)	(8)	(9)
Jan	112.5	134	178.9	44.9	44.9	15,00	15.0	15.0	0.0
Feb	78.1	134	124.2	−9.8	0.0		15.0	5.2	0.0
Mar	49.4	134	78.5	−55.5	0,0		5.2	0.0	50.3
Apr	66.0	134	104.9	−29.1	0.0		0.0	0.0	29.1
May	47.1	134	74.9	−59.1	0.0		0.0	0.0	59.1
Jun	21.6	134	34.3	−99.7	0.0		0.0	0.0	99.7
Jul	7.4	67	11.8	−55.2	0.0		0.0	0.0	55.2
Aug	13.6	13	21.6	8.6	8.6		0.0	8.6	0.0
Sep	41.3	134	65.7	−68.3	0.0		8.6	0.0	59.7
Out	125.9	134	200.2	66.2	66.2		0.0	15.0	0.0
Nov	98.1	134	156.0	22.0	22.0		15.0	15.0	0.0
Dec	106.1	134	168.7	34.7	34.7		15.0	15.0	0.0
Total	767.1	1420							353.1
Use of Total Rainwater [m^3] = 1066.9 (10)						→	75%		(11)

On the other hand, by simulating the system for scenario 2, and with a reservoir with the same capacity, a total of 89% of non-potable water demand in the Building D can be met with rainwater (1,259 m^3).

From the analysis of the results, for both scenarios, the following points can be highlighted: i) In general, the available volume of average monthly precipitation allows to cover the monthly demand for non-potable uses, with the negative values in column (4) being the lack of rainwater to meet this monthly demand (non-potable water); ii) A reserve of 15 m^3 it was considered as ideal capacity of the system. The rainwater harvesting system can support the non-potable demand for six months in the first scenario, and seven months in the second scenario. This means that in months where there are deficits greater than the capacity of the reservoir, other sources of water are needed to supply the demand. iii) The water in the reservoir can be replaced by the public water supply, whenever the usable volume of rainwater is lower than the demand of non-potable water in the building. When there is a volume above the maximum capacity of the reservoir, this surplus must be properly discharge to the public stormwater drainage network.

By inserting the average contribution of rainwater use in the ANQIP methodology, it is possible to evaluate the gains in water efficiency in Building D. Therefore, for scenario 1, when considering this contribution, the total drinking water demand reduces to 8.97 L/(user.day) (see Table 5). Following the same calculation principle for scenario 2, the total drinking water demand reduces to 7.59 L/(user.day).

Table 5. Water consumptions calculator (scenario 1).

Device	Units	Volume/Flow (Average)	Usage factor	Total liters/(user.day)
Toilet	Liters	4.27	0.54	2.31
Urinals	Liters	1.75	4.32	7.56
Bathroom faucets	Liters/minutes	9.45	0.64	6.05
Kitchen/Laboratory faucets	Liters/minutes	15.09	0.03	0.45
Showers	Liters/minutes	11.00	0.00	0.00
Total calculated consumption	Liters/(user.day)			16.37
Correction factor				1.00
Contribution of use of rainwater Liters/(user.day)				7.40
Total consumption	Liters/(user.day)			**16.37**
Water efficiency categorization of the building				A^{++}

Due to the irregularity of the precipitation, it is important to evaluate the performance of the rainwater harvesting system to meet the water demand in dry seasons. In this context, to simulate the potential of the system in drier years, average monthly rainfall corresponding to 2005 was considered for Scenario 1, the year with lower annual average rainfall in the region (see Table 6).

The performance of the rainwater harvesting system in a critical context reaches an efficiency of 46%. Therefore, the remainder of the demand must be provided by the public water supply network (approximately 763.3 m^3 of water). Regarding the water efficiency rating of the building, for the most unfavorable context, Category A^{++} remains with a total water demand equivalent to 11.83 L/(user.day).

It is worth mentioning that there are gains in terms of water efficiency, even if the scale of categorization of installations (A^{++}) is maintained before and after the inclusion of the contribution of the use of rainwater harvesting system in the building, as there was a significant reduction in the consumption of drinking water in the different scenarios and simulated contexts.

Briefly and objectively, the feasibility of implementing rainwater harvesting system for the case study was evaluated. When considering the cost of the potable water supply in Leiria (1.575 €/m^3), an annual savings of 1,680 € was obtained for the first scenario and 1,983 € for the second scenario. In dry years, it is expected to obtain an annual savings of 1,034 €/year for the first scenario. The cost of rainwater harvesting system includes the implementation cost (building networks and reservoir), and the maintenance cost. The results show the potential for the use of rainwater in Building D of Campus 2 of the Polytechnic of Leiria, however, as future work, a project net present value must be assessed to evaluate the return on investment.

Table 6. Calculation of rainwater harvesting system in 2005 (Scenario 1).

(C) = 0.9			*(η) = 0.9*			*(A) [m²] = 1962.9*			
Month	Monthly average rainfal	Monthly consumption	Usable volume of monthly rain	Difference	Calculation differences	Volume adopted for the reserve	Volume of water in the reservoir (m^3)		Supply
	(mm)	(m^3)	(m^3)	(m^3)	(m^3)	(m^3)	Start	End	(m^3)
	(1)	*(2)*	*(3)*	*(4)*	*(5)*	*(6)*	*(7)*	*(8)*	*(9)*
Jan	8.8	134	14.0	−120.0	0.0	15,00	15.0	0.0	105.0
Feb	23.8	134	37.8	−96.2	0.0		0.0	0.0	96.2
Mar	40.8	134	64.9	−69.1	0.0		0.0	0.0	69.1
Apr	33.3	134	52.9	−81.1	0.0		0.0	0.0	81.1
May	23.3	134	37.0	−97.0	0.0		0.0	0.0	97.0
Jun	2.6	134	4.1	−129.9	0.0		0.0	0.0	129.9
Jul	10.7	67	17.0	−50.0	0.0		0.0	0.0	50.0
Aug	11.7	13	18.6	5.6	5.6		0.0	5.6	0.0
Sep	11.2	134	17.8	−116.2	0.0		5.6	0.0	110.6
Out	107.7	134	171.2	37.2	37.2		0.0	15.0	0.0
Nov	81.9	134	130.2	−3.8	0.0		15.0	11.2	0.0
Dec	61.9	134	98.4	−35.6	0.0		11.2	0.0	24.4
Total	417.7	1420							763.3
Use of Total Rainwater [m^3] = 656.7 *(10)*						→	**46%**		*(11)*

4 Conclusions

Water resources management is, and will be gradually more, a decisive factor for environmental and economic sustainability development. From the analysis of the results, the advantages of potential use of rainwater in higher education institutions in Portugal are evident, since it is possible to achieve: i) an increase in the water efficiency of buildings and, thus, promote environmental sustainability; ii) a reduction of rainwater for the drainage network and, minimizing the impacts of urban floods; iii) a reduction of the cost with drinking water; iv) generating benefits (economic, environmental and strategic), and opportunities for social awareness. Additionally, we are living in a period which can become remarkable, regarding the rehabilitation of buildings, a time that constitutes an opportunity to establish radical changes that motivate the application of sustainable projects.

Next step, research work is needed to determine a net present value of the project, regarding the real implementation of this Rainwater Harvesting System.

Acknowledgments. This study was supported by UIDB/00308/2020 project. The authors would like to thank the Technical Office of the Polytechnic of Leiria, and the Head of the School of Technology and Management of Leiria, for providing the information necessary to achieve the study. The authors also would like the thank of Jean Noèl Fernand Mercereau, lecturer in the Department of Language Sciences at the Polytechnic of Leiria, for revising the English text.

References

1. United Nations. https://www.un.org/sustainabledevelopment/. Accessed 15 Jan 2021
2. UNESCO WWAP: Relatório Mundial das Nações Unidas sobre Desenvolvimento dos Recursos Hídricos 2019: Res. Executivo (2019)
3. UNESCO WWAP: Relatório Mundial das Nações Unidas sobre Desenvolvimento dos Recursos Hídricos 2020: Res. Executivo (2020)
4. Aqua eXperience. https://www.aquaexperience.pt/. Accessed 30 Feb 2020
5. ANP/WWF, Vulnerabilidade de Portugal à Seca e Escassez de Água: Relatório Outubro 2019. https://www.natureza-portugal.org/. Accessed 22 Oct 2020
6. Silva, A., Pimentel, C.: Manual de Eficiência Hídrica em Edifícios. ANQIP - Associação Nacional para a Qualidade nas Instalações Prediais, Portugal (2017)
7. Valenzuela, C., Muñoz, F., Gomes, R.: Diseño de un sistema de aprovechamiento de agua lluvia bajo criterios de eficiencia hídrica en edificios. Caso de estudio: edificio de clases y Laboratorio de Hidráulica de la Facultad de Ingeniería, Ciencias Físicas y Matemática de la Universidad Central del Ecuador. INGENIO, 2(2), 25–37 (2019)
8. ANA Aeroportos de Portugal - VINCI Airports. https://www.ana.pt/pt/system/files/documents/ana_rda2018_04.pdf. Accessed 28 Apr 2020
9. Santos, C., Imteaz, M.A., Ghisi, E., Matos, C.: The effect of climate change on domestic rainwater harvesting. Sci. Total Environ. **729**, 138967 (2020)
10. Silva, C., Sousa, V., Carvalho, N.: Evaluation of rainwater harvesting in Portugal: application to single-family residences. Resour. Conserv. Recycl. **94**, 21–34 (2015)
11. Bona, S.: Aproveitamento da água da chuva para usos não potáveis no Edifício D do Campus 2 do Politécnico de Leiria. Dissertação de Mestrado em Engenharia Civil – Construções Civis, Instituto Politécnico de Leiria (2020)
12. Portilla, C.: Water Efficiency Analysis in the Classroom Building and Hydraulics Laboratory of the Faculty of Engineering, Physical Sciences and Mathematic – Central University of Ecuador. Polytechnic Institute of Leiria and Central University of Ecuador (2018)
13. Armindo, D.: Medidas de Poupança de Água em Instalações Sanitárias de uma IES e sua Consequência no funcionamento da Rede de Saneamento. Webinar organizado pelo Grupo de Eficiência Hídrica, Rede Campus Sustentável (2020)
14. ANQIP-ETA 0701: Sistemas de Aproveitamento de Águas Pluviais em edifícios (2015)
15. ANQIP-ETA 0702: Certificação de Sistemas de Aproveitamento de Águas Pluviais (2015)
16. ANQIP-ETA 0802: Regulamento do sistema voluntário ANQIP de certificação e rotulagem de eficiência hídrica de produtos (2015)
17. ANQIP-ETA 0803: Rótulos de eficiência hídrica de produtos. Características e condições de utilização (2015)
18. Sistema AQUA+ Homepage. https://www.aquamais.pt/. Accessed 30 Aug 2020

Check for updates

Real and Virtual Water in Civil Construction

M. D. N. Silva[1(✉)], Duarte André Montenegro[2], and M. V. A Vinagre[1]

[1] University of Amazonia (UNAMA), Belém, PA, Brasil
md.silva@uepa.br

[2] Federal University of Pará (UFPA), Belém, PA, Brasil
amonte@ufpa.br

Abstract. Water is an abundant natural resource in some regions such as the Amazon, but is finite and limited. The growing demand for various social and economic actors, there may be situations of water scarcity. This study evaluates the importance of water in the construction industry, as a real or virtual element. We intend to quantify the water in monetary or financial calculated based on rational principles, specifically in construction sites, introducing the budgeting of water as raw material in construction. The case study methodology is real, the value calculated of water was slightly more than 0.5% of the total cost of the work. This percentage is apparently little, but is not covering all elements inherent to the principles of sustainability (economic, social and environmental), which could not be worked out in this article, especially the fact that studies on this subject and approach are new, incipient, with little theoretical.

Keywords: Water resources · Virtual water · Construction · Sustainability · Water budgeting

1 Introduction

Brazil has one of the largest surface water reservoirs. However, due to the inadequate use of water resources, associated with an increased demand for several social and economic aspects, some Brazilian areas face situations of water scarcity.

For mankind, water has always been essential and irreplaceable, since there are no other inputs that can be used in its replacement, such as what happens to oil and rice, for example, which can be replaced by solar energy and wheat, respectively. However, the fact that it is essential and irreplaceable, even for the life of all beings, makes the right of access to water transcend the individual issue, becoming diffuse and collective, as already stated by Petrella (2004).

In building construction, water is an important element for the composition of materials and also for the development of the services required by the construction, but its connection with sustainability at the construction site is still a highly discussed issue. Alternatives have been discovered to make materials and components more sustainable, such as the use and reuse of water in construction sites and in the operation of the project.

J. R. da Costa Sanches Galvão et al. (Eds.): *Proceedings of the 1st International Conference on Water Energy Food and Sustainability (ICoWEFS 2021)*, pp. 754–762, 2021.
https://doi.org/10.1007/978-3-030-75315-3_80

Professor Vahan Agopyan from USP (2016), in his academic research, states that: Concrete itself loses only to water in the table of most consumed products. In addition, the production chain needed to obtain cement consumes one third of the planet's natural resources. Besides, according to Agopyan, 40% to 75% of the consumption of materials in society happen in the construction industry.

According to Gomes (2015), the amount of real and virtual water present in the concrete varies according to its strength. For a material with 25 Mpa (Mega Pascal), used for conventional or small structures, 12,664 m^3 of water (12,474 virtual and 0,190 real) are incorporated, while for a concrete with 45 Mpa (large structures), 13,860 m^3 of water (13,670 virtual and 0,190 real). It appears that, according to this author, the variation in the amount of water is in the virtual context.

According to Barbosa et al. (2008), the other way of seeing water is to consider it as a natural, essential and liable to adding economic, political, social and ecological value. This vision is increasingly present considering, for example, its unequal distribution, which makes it an extremely valuable resource in regions of scarcity, which has been discussed in national and international meetings, such as World Water Forums and the Social Forum of Waters.

The introduction of the term "virtual water" was made by Tony Allan1[1]. This idea was brought to the public by this author for almost a decade until he got recognition for the theme, which involves issues related to the environment, food engineering, agricultural production engineering, international trade and so many other areas that relate to water (RISOMAS 2021).

As part of the composition of civil construction inputs, water is present in almost all materials used, whether in the condition of "virtual water" or as part of the construction site production process, however, economic value is not attributed, unlike other materials or inputs, which are considered in the construction budget, such as concrete and steel. This means that water is not accounted for, financially, economically or environmentally, in the composition of unit costs, which contains the inputs of services with their respective indexes and values.

When used to produce a good or service, water, as an input, is often not visible or tangible, being incorporated into the final product. In this condition, the concept of "virtual water" emerges.

Water, a natural resource with multiple purposes, including the maintenance of life, has become scarce due to problems of increased demand and climate change, which has already become a global consensus.

Although water is a very important element and used on a large scale in civil construction, it is not budgeted, that is, its importance in economic and financial terms is not quantified.

It is possible to study water as an input in civil construction, incorporating it into the budget, as a component of the composition of construction services, so that the budget will incorporate sustainability principles. Such principles and approaches related to sustainability require that this natural resource has its appropriate economic value, that is, that it be valued.

[1] https://www.eni.com/en-IT/low-carbon/concept-of-virtual-water.html.

The Sustainable Development Goals (SDGs 2021) have already been working on the implementation of indicators, plans and targets that can make cities and communities more sustainable, as well as providing responsible consumption and production. Within this scenario, the sustainable use of water becomes present.

Water should be considered as one of the most important resources, among the many existing, holding an enormous economic value, especially in the construction stages where there is a variety of water-dependent services.

2 Materials and Methods

In this research, a case study was carried out in 4,420 popular houses built in structural concrete and masonry walls in ceramic brick.

For the purpose of presenting and understanding the proposed budget mechanisms, a popular residential design project was used, which was financed by *Caixa Econômica Federal* (CEF 2010). This project contains buildings with two floors, each comprising units of 35 m^2 of built area, with the following internal division: two bedrooms, living room, kitchen, circulation, toilet, bathroom and yard. Figure 1 and 2 demonstrate the layout of the residences.

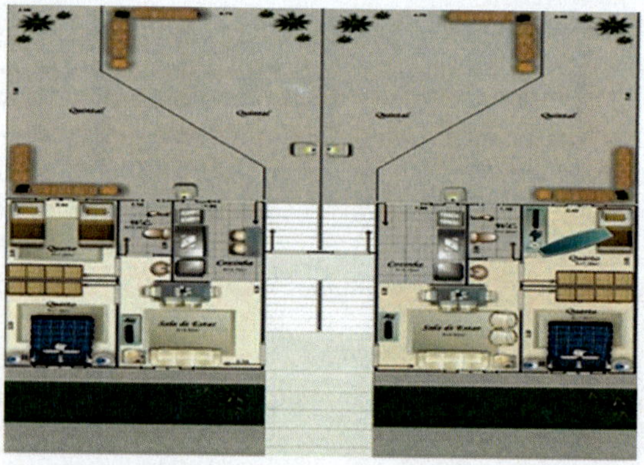

Fig. 1. Floor plan of the models of the houses

Figure 1 shows the layout of the compartments, the suggestion of furniture for the building and, at the background, the location of the yard with a division for four apartments in each block.

Fig. 2. Perspective plan of the houses

In Fig. 2, there is a part of the number of houses that will be constructed, with paving around the building, access to each block and a landscaping that makes up the front of the buildings.

Each building reflects a different reality. However, when choosing materials, there is an indispensable input in construction, which is water. Such input can be analyzed in any building standard, sometimes with a higher or lower consumption, depending on how the construction was planned to be carried out.

After the budgetary and financial analysis, the most used materials in civil construction were selected, as a way to facilitate the data investigation. A comparative study was carried out with the data collected in the technical literature, by unit of measure, with the data from the budget spreadsheet (CEF). A comparative analysis was performed with the inputs of the budget spreadsheet, in which the consumption of both real and virtual water is present, however, this consumption is not inserted directly in the composition of the materials.

The data collection used for the development of the research occurred taking as a reference the composition of the inputs of two manuals: *Mãos à Obra* (ABCP 2010) and Fiorito (1994), and the budget spreadsheet practiced by the Secretariat of public works of the state of Pará (SEDOP 2019). All of these references provide information with technical understanding within the standards adopted in construction sites for civil construction.

Figure 3 presents the Methodology Workflow of the developed study:

Research question: What is the value (or cost) of the water, whether real or virtual, used in a civil construction?

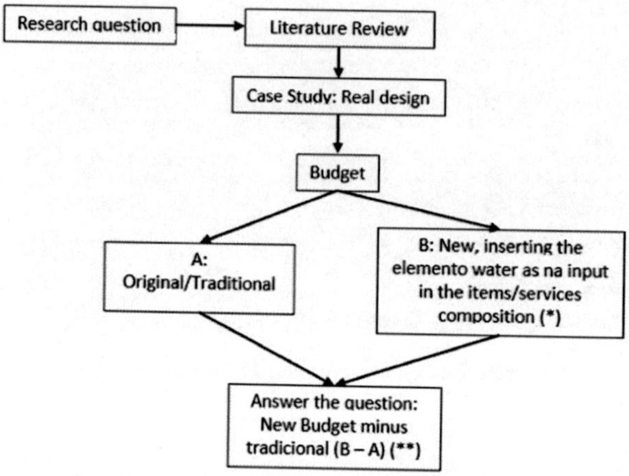

(*) Quantifications of water required in each item / service, whether real or virtual. They were defined based on the technical literature, without measurements on site, and the unit prices were based on reliable institutions in the civil construction segment and in the water supply concessionaire.
(**) Only the "total service cost" was considered, which is approximately equivalent to the "values for direct water users and net benefits return flow", that is, the "total cost" or "total value", which contemplate the environmental, economic and social externalities and the opportunity cost, were not taken into account.

Fig. 3. Methodology workflow

3 Results

Of the 4,420 popular houses built in structural concrete, which cost a total of 150,830,266.82 (one hundred and fifty million, eight hundred and thirty thousand, two hundred and sixty-six reais and eighty-two cents), it was elaborated a spreadsheet of budgetary values resulting from the research developed with the main inputs used in this work, with a total cost of R$ 49,547,227.16 (forty-nine million, five hundred and forty-seven thousand, two hundred and twenty-seven reais and sixteen cents), and with water tariff values of R$ 14,40 applied during the budget period, in the Belém-Pará market.

Table 1 shows the quantities and the amount of water used in each service, the total consumption in liters and m^3, the price charged by the water supply concessionaire and the total price for services with and without the water input.

4 Discussion

Based on the results obtained, as shown in Table 1, there is a comparison of the cost of water as an input and without valorization as an input. Water as an input has a value of R $ 49,820,598.92 (forty-nine million, eight hundred and twenty thousand, five hundred and ninety-eight reais and ninety-two cents) and without accounting for inputs the value is R $ 49,547,227.16 (Forty-nine million, five hundred and forty-seven thousand two hundred and twenty-seven reais and sixteen cents). In the budgeted amount for water within the budget spreadsheet in Table 1, the total water used in cubic meters

Table 1. Budget spreadsheet with data and research results

	Services	Un	Service quantities	Water amount (liter)	Total consumption liter (un)	Total consumption (m^3)	Water price per m^3 (Cosanpa)	Water consumption's total price (R$)	Service's total price (R$) With water	Service's total price (R$) Without water
FOUNDATION	RUNNING SHOE AND BALDRAME	m^2	7,867.60	168.8	1,327,657.50	1,327.65	14.04	18,640.21	3,800,202.15	3,781,561.94
	CONCRETE BALLAST	m^3	2,055.30	144	295,963.20	295.96	14.04	4,155.32	402,369.70	398,214.38
	LEVELING LEVEL	m^3	10.574,85	144	1.522.778,40	1.522,77	14,04	21,379.69	3,679,114.56	3,657,734.87
STRUCTURE	STRUCTURAL CONCRETE	m^3	5,768.10	160.7	926,991.35	926.99	14.04	13,014.94	1,676,477.30	1,663,462.36
	PRE-MOLDED SLAB WITH CONCRETE	m^3	6,313.09	160,7	1.014.576,69	1,014.57	14.04	14,244.56	7,040,709.28	7,026,464.72
WALLS AND PANELS	MASONRY SETTLEMENT	m^2	468,210.60	2.23	1,044,109.64	1,044.10	14.04	14,659.16	13,105,827.54	13,091,168.38
	CONCRETE ROD	m^3	481.78	160.7	77,426.86	77.42	14.04	1,086.98	291,494.33	290,407.35
COATINGS	ROUGHCAST	m^2	178,081.80	0.67	119,314.80	119.31	14.04	1,675.11	701,536.58	699,861.47
	PLASTER	m^2	33,879.30	6.68	226,313.72	226.31	14.04	3,177.39	403,630.72	400,453.33
	PLASTERING	m^2	1,006,478.20	6.68	6,723,274.38	6,723.27	14.04	94,394.71	13,523,582.15	13,429,187.44
PAVING	SMOOTH FINISH	m^2	156,711.10	2.67	418,418.63	418.41	14.04	5,874.48	3,478,592.46	3,472,717.98
	PROTECTION SIDEWALK	m^2	10,806.90	2.67	28,854.42	28.85	14.04	405.05	526,376.87	525,971.82
	TINY FINISH	m^2	6,364.80	2.67	16,994.02	16.99	14.04	238.54	177,689.16	177,450.62
	TOTAL				19,453,999.59	19,453.93			49,820,598.92	49,547,227.16

Source: Authors

is 19,453.93. Considering the value adopted by the concessionaire (COSANPA 2019) is R $ 14.04/m^3. Table 2 below shows the relationship between the value of water as an input, the percentage in relation to the work and the value adopted for the cost of water.

Table 2. Budgeted amount for water

Budgeted amount for water	% in relation to construction	Unit value adopted for water (R$/m^3)
R$ 273.371	**0,55%**	**14.04**

Source: Authors

The amount of R$ 14,04/m^3 used to budget water as an input in construction is not, in principle, very significant, increasing the cost by just over 0.50% of the total amount surveyed in the worksheet. This magnitude, which is insignificant in relative terms, may be due to the fact that the real value of water was not considered.

The volume of water calculated for budget purposes in this case study is restricted to the construction steps shown in Table 1. Therefore, it is worth mentioning that other construction steps, such as cleaning the work and hydraulic tests, require a significant amount of water, but in the present research it was not possible to identify or even estimate this quantity. According to Roges et al. (1998) apud Abad (2007), the real value or the value in sustainable use is the one shown at drawing 01.

Drawing 01– **General Principles for Water Cost and Value**

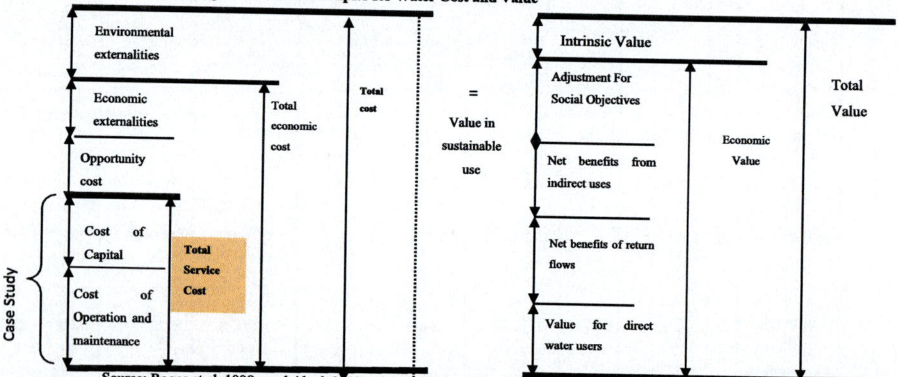

Source: Roges et al. 1998 apud Abad, 2007, adapted by the authors

From drawing 01, it can be inferred that the case study presented in this work was restricted to the **total cost of service**. However, the sustainable value of water in civil construction is not limited to what was calculated in this study. Therefore, it would also be necessary to quantify the environmental, economic, social externalities, which should constitute the collection cost (including the onerous grant) and the opportunity cost.

5 Conclusions

The world is in an environmental dilemma, including with regard to climate change, and the scientific community must seek to contribute so that natural resources are used correctly and sustainably.

This research seeks to contribute to the segments of civil construction and water resources management in the proposal to introduce water input into the worksheet of the construction.

Considering that water is an indispensable and irreplaceable natural resource in construction processes and that, so far, it is not considered as an input with a price in a construction's budget, it is proposed, in this study, that its value should be budgeted.

In times of global warming, scarcity of water resources and society's awareness of the importance of natural resources, the principles of sustainability and the consequent need to attribute these resources, such as water, economic values compatible with their importance, it can be considered that the value of water is much higher than what is currently budgeted, which can be observed when this value is compared to other natural inputs, such as sand and pebble, for example, which have costs of R $ 59.00/m^3 respectively and R $ 123.08.00/m^3 (SEDOP 2019). This is probably due to the fact that not all stages of the life cycle of a building have been worked on, but only the implantation (construction site), with water not being priced at the project design and planning stage, nor in the later stages of implantation. Therefore, virtual water is not being budgeted as a whole, only virtual water or even real/physical water that is used at the construction site as an input. A natural resource, such as water, has real value that includes all sustainability requirements, that is, economic, social and environmental aspects.

In this research, only the economic aspect was considered, this being the budget for the water supply service by the concessionaire, with a unit value of R$ 14.04/m^3 which, in the case study, has represented 0.55% of the total cost of construction. Certainly, this value, even when calculated only for the moment of implementation (execution of construction services), as in the present study, should be much higher if the other phases of a civil construction are considered (project, manufacture of materials such as steel, cement and masonry) and other aspects (externalities and opportunity cost), which can be analyzed in future works, which improve and expand the real and sustainable quantification of water, considering all the principles of sustainability, as well as in other stages or phases of the building life cycle that, in this research, was restricted to the implementation/execution.

Referências

BRAZILIAN CEMENT PORTLAN ASSOCIATION - ABCP - Concrete Sustainability. https://www.abcp.org.br/conteudo/sustentabilidade/sustentabilidade-do-concreto. Accessed Sept 2010

Barbosa, F., Barreto, F., César de Sá, F.: Angle of Water: Integration challenges. In: Barbosa, F. (Org.). Different Views of Water. Belo Horizonte: Federal University of Minas Gerais (UFMG) (2008)

Eugenia Patrícia Garcés, A.B.A.D.: Proposed water pricing for irrigation in agriculture, using the methodology of positive mathematical programming. Dissertation (Doctoral degree in Water Resources and Environmental Sanitation) - Hydraulic Research Institute, Federal University of Rio Grande do Sul. Porto Alegre (2007)

CEF - CAIXA ECONÕMICA FEDERAL. Discriminative Budget Worksheet of a Residential of popular houses. https://www.caixa.gov.br/Paginas/home-caixa.aspx . Accessed Oct 2020

PARÁ SANITATION COMPANY - COSANPA. Service tariff table. https://www.cosanpa.pa.gov.br/wp-content/uploads/2018/05/Tarifas-Cosanpa-atual_dez2019-2.pdf. Accessed 8 Dec 2020

Fiorito, A.J.S.I.: Mortar and coating manual: studies and execution procedures. PINI, São Paulo (1994)

Gomes, L.P.L.B.: The study of real and virtual water in machined concrete. 51f. Master's Dissertation in Civil Engineering. Postgraduate Program in Civil Engineering. Federal University of Pará, Belém, Pará (2015)

Petrella, R.: The water. The challenge of the common good. Translation from Fabiola kochenborger. NEUTZLING, Inácio (Org.). Water: universal public good. Publishing company Unisinos (2004)

Education and culture portal - RISOMAS. Rodrigo Travitzki (2008). https://rizomas.net/quem-somos/o-que-e-um-rizoma.html. Accessed 9 Jan 2019

ODS - Sustainable Development Goal - 2030 Agenda Platform. Available: https://www.agenda2030.org.br/. Accessed 26 Jan 2021

Secretary of State for Urban Development and Public Works - SEDOP - Price Composition Worksheet. https://www.sedop.pa.gov.br/. Accessed 7 Dec 2020

UNIVERSITY OF SÃO PAULO - USP - Civil construction is the branch that consumes the most materials in the world. Polytechnic School. Edition nº110 - Year 45 – Science and Technology (2012). https://www.usp.br/aun/antigo/exibir?id=4848&ed=853&f=2. Accessed 7 Jan 2021

A Contribution to Overcome Water Management Challenges in Brazil

Lisiane Ilha LIbrelotto[1](✉), Paulo Cesar Machado Ferroli[1], Flávio Craveiro[2,3], and Helena Bártolo[2,3]

[1] Federal University of Santa Catarina, Florianópolis, SC, Brazil
[2] School of Technology and Management, Polytechnic of Leiria, Leiria, Portugal
[3] CIAUD, Lisbon School of Architecture, Universidade de Lisboa, Lisboa, Portugal

Abstract. Water is a vital resource for life on earth. The importance of an effective and efficient management of water in buildings and cities is highlighted, as well the need to preserve soil and ecosystems to avoid water waste and degradation. The Brazilian context regarding water conservation and management is explained, as despite owning the biggest water resources worldwide, there are still people suffering from water supply and sanitation problems. Some technologies used in Brazil, promoting better water management use in buildings and cities, are presented. The need to progress towards net zero water buildings is emphasized, as well the urgency in finding new strategies for net zero water and waste requirements.

Keywords: Water management · Technologies for buildings · Strategies for Net Zero Water (NZW)

1 Introduction

Water is a vital resource for life on earth, though it is a scarce resource at risk due to increased urbanization and surface degradation, as well to climate change. In 2015, the UN adopted the SDGs and urged countries to take action and protect the planet (United Nations 2015a, b), and improve and improve water conservation and management. Presently, there is a wide international awareness and commitment towards the importance of an effective and efficient management of water for the environment (Aither 2018).

In Brazil the water is abundant, the climate and rainfall patterns are decisive for Brazil's water resources, though unequally distributed across regions, as there are water shortages in some regions, as well water shortages, declining water quality, and difficulty in access to water in poorer economic areas (OECD 2015), so there was a need to create the National Water Resources Council (CNRH) in 1998 and the National Water Agency (ANA) in 2000, in order to implement quite a few water reforms and a water management plan, though the water management has not been effective. According to

J. R. da Costa Sanches Galvão et al. (Eds.): *Proceedings of the 1st International Conference on Water Energy Food and Sustainability (ICoWEFS 2021)*, pp. 763–771, 2021.
https://doi.org/10.1007/978-3-030-75315-3_81

Johnson (2014), there is a need to reform hydraulic infrastructure and to improve water supply in metropolitan areas.

Brazil is a very rich country in water resources, holding the largest share of the world's surface fresh water (12% of the world's freshwater), surpassing 7% of the European continent and 10% of African surface reserves. According to conservative estimates by the ANA – "Agência Nacional de Águas e Saneamento"/National Water and Sanitation Agency (2020), Brazil has 112000 km^3 in reserved volume of water that occupies about 48% of the area of the national territory. Brazil has 27 aquifers. In addition, its subsoil contains the two largest world aquifers: the Guarani Aquifer System (SAG - with 1200000 km^2 in the countries of Brazil, Argentina, Uruguay and Paraguay) - and the Greater Amazon Aquifer System (SAGA – "Sistema Aquífero da Grande Amazônia"- or "Alter do Chão") (see Fig. 1). Besides this, other aquifers such "Cabeças", "Urucuia-Aredo", "Bauru", "Pré-Cambriano", "São Paulo", "Serra Geral", "Taubaté", "Tubarão", besides the "Guarani", are constantly monitored by CETESB – "Companhia Ambiental do Estado de São Paulo"/Environmental Company of São Paulo (CETESB 2020). Amongst superficial and deep waters, it is the world's largest water reserve, though the lack of access to safe water for consumption is still a reality in the poorest areas of some Brazilian States.

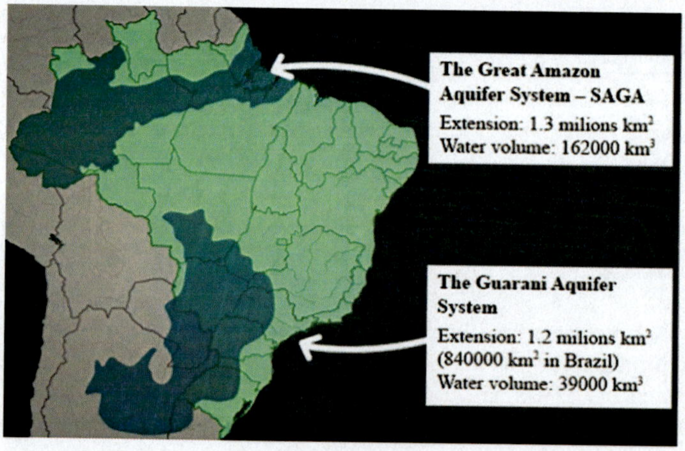

Fig. 1. Main Brazilian aquifers (adapted from Globo 2018).

Most Brazil's population has migrated from rural areas to cities, quite often to steep slopes areas called "favelas" in metropolitan areas with no infrastructure for water supply, sanitary disposal, or protection against urban flooding and landslides, where both domestic and industrial waste accumulates producing pollution, as well sediment and solid waste negatively affecting nature and ecosystems (Tucci 2013). Actually, a set of activities and a complex network of interconnections has contributed to the lack of water for consumption in Brazil. Human activities divert water from its natural cycle, in applications such as agriculture, industry, and human activities.

Conversely, in cities, soils compacted by urban development are highly impervious, so water retention and undersized drainage networks (sizing criteria based on average rainfall) can eventually cause flooding in urban centers. The pipes and sidewalks can also cause an increase in the surface flow velocity, promoting urban floods, which associated with the absence of water retention devices, can generate major disasters in cities (Fig. 2). In addition, the compaction and densification of the soil are causing soil degradation, severely affecting water supplies and its quality. On the other hand, cultivation near Tshorelines and agrarian movements and deforestation are causing erosion, occupying the areas of riparian vegetation in watercourses, modifying natural flows, contributing to water contamination the devastation of watercourses, mangroves and streams and maximizing the effect of tidal floods on urbanized coastlines. Furthermore, there is poor sanitation (16% of the population has no treated water and 51% has no access to sewage system), 39,2% of wastes in drinking water distribution networks (IBGE 2017; SNIS 2019). These factors indicate an inefficient planning and water management. There is a tendency to transform drainage urban surface waters into large open sewage systems.

Fig. 2. Floods in Brazil (Agência Brasil 2020).

Approximately 26% of the water collected is destined to the built environment, that is, to the cities. About 10% of this water is destined for consumption, which in Brazil is about 153.9 L/inhabitant/day, excluding industry and agribusiness (SNIS 2019). More than one-third of the water removed from nature by water utilities is lost during the distribution process. Moreover, water use is unequal in the states, municipalities, and neighborhoods of Brazilian cities, with higher-income regions showing high water usage.

This work intends to discuss how buildings and cities can contribute to water management in the Brazilian reality by implementing technologies and measures to optimize the use of this precious resource.

2 Sustainable Water Management

The sixth Sustainable Development Goal (SDG) of the 2030 agenda aims to establish the guarantee of water supply and availability, and a sustainable water management and sanitation for all: (i) universal access to safe water, (ii) sanitation and hygiene should be assured by 2030; (iii) water quality should be improved by reducing pollution, chemical discharges and hazardous materials; (iv) the proportion of untreated wastewater halved; (v) increased recycling and reuse at a global scale; (vi) increased efficiency in water use with sustainable extractions and freshwater supply to address water scarcity; (vii) integrated water resources management should be implemented; (viii) international cooperation and capacity building support for developing countries in activities and programs related to water and sanitation; (iX) support and strengthen the participation of local communities (UN 2015a, b).

The CIB (International Council for Research and Innovation in Building Construction) Agenda for Sustainable Construction in Developing Countries, established the efficiency of energy and water as one of the main objectives for sustainability. Water management is also an important criteria in several certification and labeling to assess building sustainability, as shown in Fig. 3, where the blue color represents the water management weight for some Green Building Rating Systems.

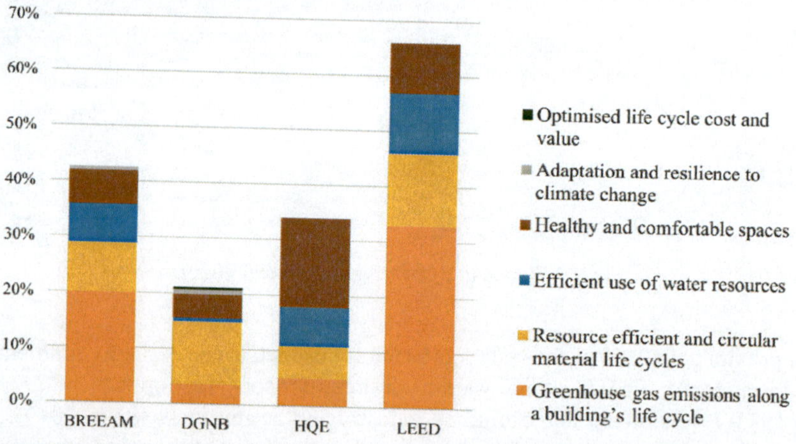

Fig. 3. User operation item weight breakdown per Green Building Rating Systems (BREEAM - Building Research Establishment Assessment Method, DGNB - Deutsche Gesell-schaft für Nachhaltiges Bauen, HQE - Haute Qualité Environnementale, and LEED - Leadership in Energy & Environmental Design) (Sánchez Cordero et al. 2019).

Water management incorporates a complex web of correlations that includes the re-education of the population (with a more distant horizon), the redesign of domestic equipment for lower consumption, positive legislation to encourage the reuse of wastewater and rainwater, planning measures, and monitoring of solutions implemented in buildings and cities or even the critical discussion of solutions already established.

There are direct ways of saving water and indirect strategies, which can interfere with the quality and availability of water in urban environments.

In buildings, there are many ways to contribute to water management and create Net ZeroWater (NZW). The NZW concept, similarly to the concept of Net Zero Energy in buildings, means a sustainable balance between production and consumption, where the amount of water used is equal to the water returned to the original water source. The goal is to support the quality of natural water resources with minimal depletion. It can be achieved by legislation to optimize the reuse of water, efficient technologies avoiding water contamination, as well designing the building considering the characteristics of the context (urban environment). A NZW Building must use strategies to: minimize total water consumption; maximize the use of alternative water sources; minimize the building's wastewater discharge; and return water to its original source.

In the USA, a Federal Energy Management Program (FEMP) published in 2017 guidelines for planning, designing, and constructing net zero energy, water, and waste building facilities and campuses (Fowler et al. 2017). According to EERE (2020), to compose a closed system for water consumption and use, it is necessary: ensure the correct use of potable water; think about possible ways of collecting non-potable water (from freshwater sources) and protecting water sources; analyze the potential of using reused rainwater; ensure that treated wastewater on-site returns to the original water source; and, provide for the return of infiltrated rainwater to the original water source through green infrastructure.

2.1 Strategies for Water Management

A sustainable building design comprises reducing the environmental impacts of buildings by sustainable planning, water and energy efficiency, materials and resource preservation, and quality of the indoor environment (Chau et al. 2010). There are simple measures for saving water in buildings include re-educating the population to consume fewer resources and save drinking water, directly: turning off the tap while brushing teeth, shaving, soaping dishes, among other domestic uses; not using the hose to wash floors, sidewalks, cars, etc. Replacing flow valves on toilet flushes with close coupled toilets with volume limiter(s) per flush and per type of waste; using dry toilets or with water reuse, reducing bath time, and adjusting the water flow; using the washing machine and dishwasher only at the maximum volume of the machine and reusing the water from the machine rinsing for new washing cycles or floor cleaning (Ribeiro et al. 2018).

There are other indirect ways for a more efficient water management, such as planting trees, efficient urban drainage solutions, avoiding contaminating springs and groundwater by discarding effluents. Even when water is used for livestock or agricultural activities, responsible for a large part of water consumption, even if returning to the cycle after use (the water cycle is closed), there is a risk of contamination by pesticides, bacteria, and heavy metals, which can reduce its quality and lead to health problems. Several technologies can be used for water management in buildings, for direct consumption or to support other sustainable building requirements, as illustrated in Fig. 4. This figure presents several technologies available/developed in Brazil to ensure a most efficient use of water: (i) aerator devices for faucets reduce water consumption by 80%, (ii) dry toilets that do not need water for flushing, including models with tanks for depositing the dry content, and (iii) the addition of sawdust or chemical components for the dry decomposition of organic material. Similarly, special toilets, quite common in Austria and Germany for some commercial establishments, which separate solids from liquids, an important way to reduce water consumption when flushing. Other equipment, also shown in Fig. 4, allows the direct reuse of greywater by reusing water from sinks and showers. These technologies can significantly reduce the consumption of treated water for non-potable uses, as well reducing the amount of effluent discharged into the sewage system. Figure 4 also presents some measures to improve water quality, preventing the waste that is not collected, or incorrectly dumped in the streets, from reaching the waterways.

Many other indirect measures can also be used in buildings for better water management, as follows:

- Using roof gardens, which increases the water retention time, promotes the thermal insulation in buildings and reduces the phenomenon of heat island formation. They are also important as a space for growing plants for food, contributing to the reconstitution of the landscape, regulating the humidity and temperature of the environment. A limitation is the need of both an effective maintenance and an efficient waterproofing system;
- Using drainage wells and infiltration ditches, which can be used to collect water from the ground, to use geothermal energy for heating or cooling water, or just to collect groundwater;
- Using geothermal energy for heating and cooling, which in addition to using collected or reused water, also contributes to energy efficiency;
- maximizing green areas and permeable sidewalks, which provides water infiltration into the soil and controlling local climate;
- rainwater harvesting - NBR 15.527/2007 deals with the harvesting of rainwater from the buildings roofs in urban areas for non-potable purposes, presenting six calculation methods for water reservoir systems. In Brazil, there are several of these commercial systems, for example, the AcquaSave (2020) brand collectors.

Other water management strategies can include effluent treatment solutions, such as root tanks, biodigesters, and individual effluent treatment strategies to reduce overloads in the urban collection network.

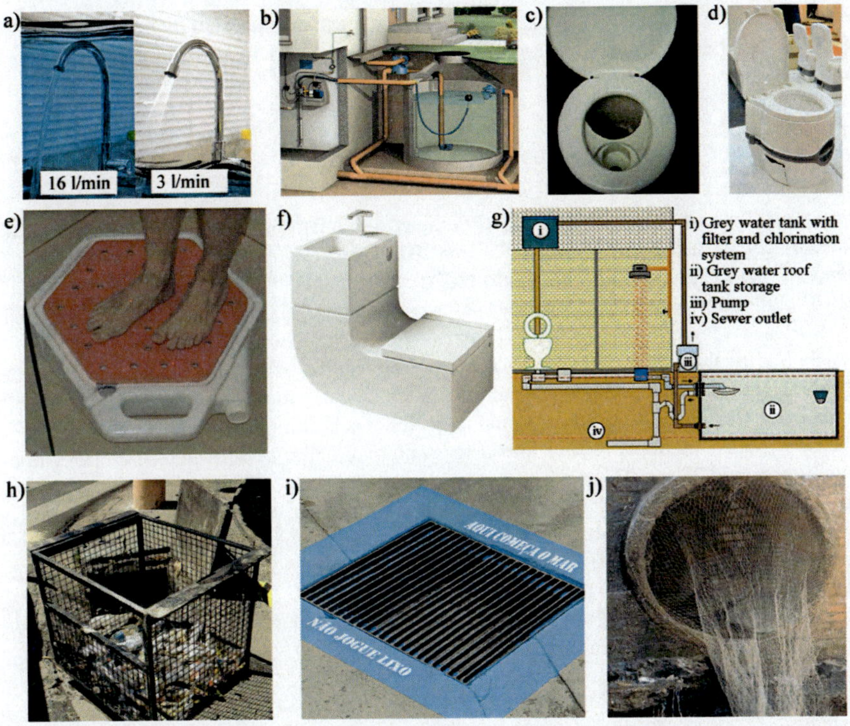

Fig. 4. Strategies for water management in buildings: a) tap aerators (Blukit 2021), b) rainwater harvesting system (Acquasave 2020), c) solid-liquid separation toilet bowl (Paulo et al. 2019), d) dry toilet (Diniz 2020), e) greywater recycling system for shower (Sustentarqui 2021), i) greywater reuse from showers (Sustentarqui 2021), g) Manhole filter (Câmara Municipal de Dourados 2018), h) toilet using recycled grey water from wash basin (Roca 2019), i) solutions for drainage systems: the sea starts here (Colatusso 2020), j) filter for drainage network outlet (Prefeitura de Itaí 2019).

3 Conclusions

The importance of an effective and efficient management of water in buildings and cities is highlighted, as well the need to preserve soil and ecosystems to avoid water waste and degradation, supporting this way the quantity and quality of natural water resources with minimal depletion. In this way, we tried to show the importance of water for sustainable development, which can only be obtained by adopting water management strategies in buildings and cities.

Some technologies developed and/or implemented in Brazil promoting better water management use were presented, involving either the reducing of water consumption or changing habits and more technological efficient solutions, which can be achieved by the re-education of the population and better city planning. The need to progress towards Net Zero Water conceptual buildings is emphasized, as well to find new strategies for net zero water and waste requirements.

Most technologies developed in the Brazilian context can also be employed in other countries for better water management in buildings and in the context of the urban environment.

References

Acquasave: Sistema para aproveitamento de água de chuva – AcquaSave/3P Technik (2020). https://acquasave.com.br/. Accessed 21 Feb 2021

Agência Brasil: Enchentes em São Paulo (2020). https://agenciabrasil.ebc.com.br/geral/noticia/2020-02/desenvolvimento-desordenado-e-uma-das-causas-de-enchentes-em-sp. Accessed 23 Feb 2021

Agência Nacional de Águas (ANA) & Conselho Empresarial Brasileiro para o Desenvolvimento Sustentável (CEBEDS) (2009). Água: fatos e tendências. https://arquivos.ana.gov.br/institucional/sge/CEDOC/Catalogo/2006/AguaFatosETendencias.pdf. Accessed 05 Feb 2021

Blukit: Arejador Tipo Spray (2021). https://www.blukit.com.br/produto/detalhe/arejador-tipo-spray-1. Accessed 23 Jan 2021

Câmara Municipal de Dourados: Projeto de Marçal propõe «bueiros inteligentes» em Dourados. Projeto de Marçal propõe «bueiros inteligentes» em Dourados (2018). https://camaradourados.ms.gov.br/noticia/projeto-de-marcal-propoe-bueiros-inteligentes-em-dourados. Accessed 29 Jan 2021

Companhia Ambiental do Estado de São Paulo (CETESB) (2020). Aquíferos monitorados. https://cetesb.sp.gov.br/aguas-subterraneas/programa-de-monitoramento/consulta-por-aquiferos-monitorados/aquifero-guarani. Accessed 16 Dec 2020

Colatusso, P.: Aqui começa o mar (2020). https://imaculada.redesagradosul.com.br/unidade/noticias/aqui-comeca-o-mar-colegio-imaculada-adere-a-projeto-de-conscientizacao-ambiental-que-pretende-chegar-as-ruas-de-santa-felicidade. Accessed 23 Jan 2021

Diniz, P.: Vaso sanitário portátil: Como escolher o melhor em 2021? (2020). https://www.reviewbox.com.br/vaso-sanitario-portatil/. Accessed 23 Jan 2021

Department of Energy's Office of Energy Efficiency and Renewable Energy (EERE). Officce of Energy Efficiency and Renewable Energy. https://www.energy.gov/eere/femp/net-zero-water-building-strategies. Accessed 10 Jan 2021

Globo: Aquífero Alter do Chão é a maior reserva de água descoberta até o momento (2018). https://g1.globo.com/pa/santarem-regiao/bom-dia-santarem/videos/v/aquifero-alter-do-chao-e-a-maior-reserva-de-agua-descoberta-ate-o-momento/6599709/. Accessed 02 Jan 2021

Instituto Brasileiro de Geografia e Estatística (IBGE). Pesquisa Nacional de Saneamento Básico (PNSB) (2017). https://www.ibge.gov.br/estatisticas/multidominio/meio-ambiente/9073-pesquisa-nacional-de-saneamento-basico.html?=&t=destaques. Accessed 10 Jan 2021

United Nations (UN): Transforming our world: the 2030 Agenda for Sustainable Development (2015a). https://sdgs.un.org/2030agenda. Accessed 10 Jan 2021

Sustentarqui: Equipamento brasileiro para reaproveitar a água do chuveiro (2021). https://sustentarqui.com.br/equipamento-brasileiro-para-reaproveitar-a-agua-do-chuveiro/. Accessed 20 Feb 2021

Sistema Nacional de Informações sobre Saneamento (SNIS). País desperdiça 39,3% de água potável (2019). https://www.snis.gov.br/. Accessed 10 Jan 2021

Paulo, J., Prim, M., Rubin, J.: Banheiro Seco: um exemplo de ecotécnica (estudo de caso). Maiêutica - Ciência, Tecnologia e Meio Ambiente **2**(1), 29–34 (2019)

Prefeitura de Itaí: Prefeitura de Itaí instala a primeira Eco Rede do Brasil (2019). https://www.itai.sp.gov.br/noticia/972/prefeitura-de-itai-instala-a-primeira-eco-rede-do-brasil/. Accessed 10 Jan 2021

Roca. Bacia suspensa com lavatório integrado (2019). https://www.br.roca.com/servlet/getProductPdf?productId=932&designlineId=263. Accessed 10 Feb 2021

Sánchez Cordero, A., Gómez Melgar, S., Andújar Márquez, J.M.: Green building rating systems and the new framework level(s): a critical review of sustainability certification within Europe. Energies **13**(1), 66 (2019)

Aither: A guide to managing water for the environment. The Australian Water Partnership. Canberra, Australia (2018)

United Nations (UN): Transforming our world: the 2030 Agenda for Sustainable Development (2015b). https://www.un.org/sustainabledevelopment/sustainable-development-goals/. Accessed 12 Jan 2021

Fowler, K.M., Demirkanli, I., Hostick, D.J., McMordie-Stoughton, K.L., Solana, A.E., Sullivan, R.S.: Federal Existing Buildings Handbook for Net Zero Energy, Water, and Waste Prepared for the U.S. Department of Energy Federal Energy Management Program, USA (2017)

OECD: Water Resources Allocation: Sharing Risks and Opportunities. OECD Studies on Water. OECD Publishing (2015)

Ribeiro, J.M., Bocasanta, S.L., Avila, B.O., Magtoto, M., Jonck, A.V., Gabriel, G.M., Guerra, J.B.: The adoption of strategies for sustainable cities: a comparative study between Seattle and Florianopolis legislation for energy and water efficiency in buildings. J. Cleaner Prod. **197**, 366–378 (2018)

Chau, C.K., Tse, M.S., Chung, K.Y.: A choice experiment to estimate the effect of green experience on preferences and willingness-to-pay for green building attributes. Build. Environ. **45**(11), 2553–2561 (2010)

Johnson, R.M.: Water Resources Management in Brazil: Challenges and New Perspectives. The World Bank (2014). https://www.worldbank.org/content/dam/Worldbank/Feature%20Story/SDN/Water/events/Rosa_Formiga_Johnson_Presentacion_Ingles-3.pdf. Accessed 10 Feb 2021

Tucci, C.E.: Brazil: Progress towards the integration of water resources management. Global Water Partnership (2013). https://www.gwp.org/globalassets/global/toolbox/case-studies/americas-and-caribbean/brazil.-progress-towards-the-integration-of-water-resources-management-289.pdf. Accessed 10 Feb 2021

Agência Brasil: IBGE: enchentes deixaram 1,4 milhão de desabrigados entre 2008 e (2014). https://agenciabrasil.ebc.com.br/geral/noticia/2014-04/ibge-27-dos-municipios-brasileiros-foram-atingidos-poe-enchentes-afetando-14#. Accessed 20 Feb 2021

Urban Development and Water Supply System: A Case Study on Comilla City Corporation

G. M. Sarwar Badhan[1], Reetika Ali[2](✉), and Sabkat Kamal[3]

[1] Engineering and Planning Consultants Ltd., 7/4, Block-A, Lalmatia, Dhaka 1207, Bangladesh
badhangunn@gmail.com
[2] Onushandhani Creeds Ltd., House #135, Road #5, New DOHS, Mohakhali, Dhaka 1206, Bangladesh
[3] Department of Urban and Regional Planning, Jahangirnagar University, Savar, Dhaka 1342, Bangladesh

Abstract. Water is another name of life. The standard of quality of life in an urban area is much dependent on the extent of the availability of drinking water. Water supply is the basic responsibility of any city or town authority. Getting standard quality water is the basic need of the people. In Bangladesh, people who are living in urban areas receive piped water supply. A planned water supply system is compulsory for serving urban people with an adequate quantity of supply water of good quality. Comilla Municipality started supplying water in 1925 with few distribution mains. After forming a new Comilla City Corporation (CoCC) in 2011, the water supply system was newly maintained and operated jointly by DPHE (Department of Public Health and Engineering) and CoCC. But the planning has drawbacks and the amount of pure water supply is inadequate compared to the need. Roughly 30% of families in slums suffer from acute water shortages for daily requirements inside CoCC area. This unhygienic condition causes severe health problems for these people. Urban expansion is prevalent with the constant increase in population but water supply provision is not increasing at the same rate. Even though the urban area is expanding constantly, the political boundary is an area of 53.04 km^2 which was 37.50 km^2 in 1982. The growth of built-up areas expands over the political area. The expansion of residential areas to the northern part of the city is the main problem for decreased agricultural land.

Keywords: Water supply · Comilla City Corporation · Water shortage · Urban expansion

1 Introduction

Water is one of the most important components to survive on earth. People use water for multiple purposes starting from drinking to household activities, in the construction and industrial sectors. Safe, adequate, and accessible supplies of water, combined with proper sanitation, are basic needs and essential components of primary health care. Water supply is one of the basic services of any municipality. In Dhaka City water supply coverage

J. R. da Costa Sanches Galvão et al. (Eds.): *Proceedings of the 1st International Conference on Water Energy Food and Sustainability (ICoWEFS 2021)*, pp. 772–784, 2021.
https://doi.org/10.1007/978-3-030-75315-3_82

is only 65% and in other urban areas, water supply coverage is around 50% [1]. Most of the municipal areas are promote insufficient supply and fail to co-opt with growing trends what is a threat to development [2].

As the urbanization trend in Bangladesh is very high, at present, according to United Nation's data, about 35 million people, or approximately 25% of Bangladesh's population, currently live in urban areas (World Bank Report, 2000). With urbanization, urban facilities must be provided to the people and rapid urbanization also creates pressure on the basic utility services. So, for the effective management of basic utility services like water supply in the present study is chosen for the study purpose. The municipalities with their limited resources are trying to increase this facility to its inhabitants, but due to many constraints such as scarcity of water sources, inadequate planning, and technical assistance, limited financial capability, etc. cannot provide the services to the larger mass [3].

Comilla City Corporation is selected as the study area for this study because it is growing at a rapid rate and within the City Corporation jurisdiction area, it is very important to provide an adequate amount of water. The water supply system requires to follow a development trend. But the present scenario of supply is disappointing. Therefore, the study is important to give some perspective planning proposals to improve the water supply system of the study area i.e., Comilla City Corporation [4]. This study focuses on water supply operation techniques that are related to urban development, such as legislation, policy, and plan, institutional arrangement, local action plan, etc. And in this study, a social component like population will be included as a sustaining aspect of the study.

1.1 Study Area

The precise location of the study area in Comilla City Corporation, a self-governed municipal administration in Bangladesh that administers and supervises development and maintenance works in the city. The corporation covers an area of over 53.04 square kilometers in the Comilla city corporation where over 5 lac people live as permanent residents [5] (Figs. 1).

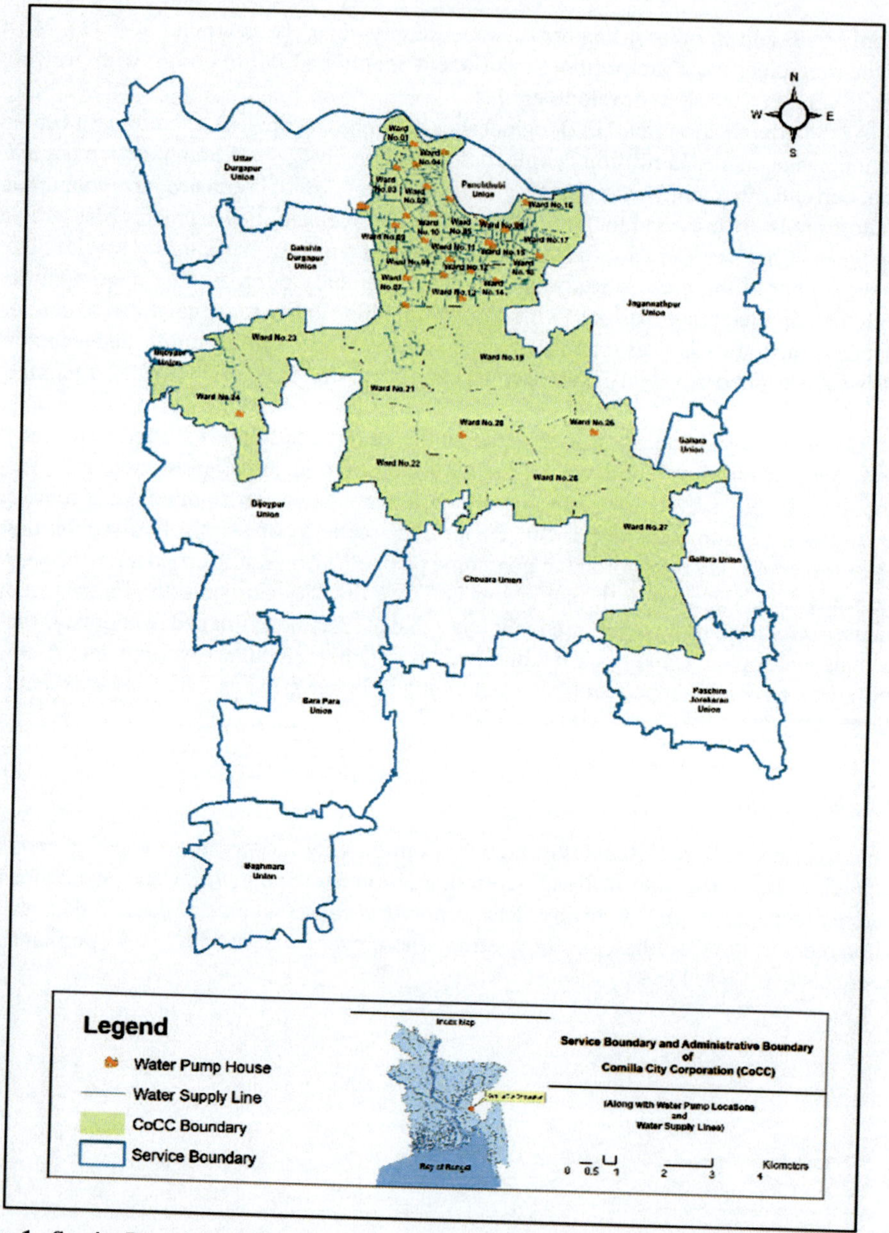

Fig. 1. Service Boundary of Comilla City Corporation (Source: Comilla City Corporation, 2017)

2 Framework

(See Fig. 2).

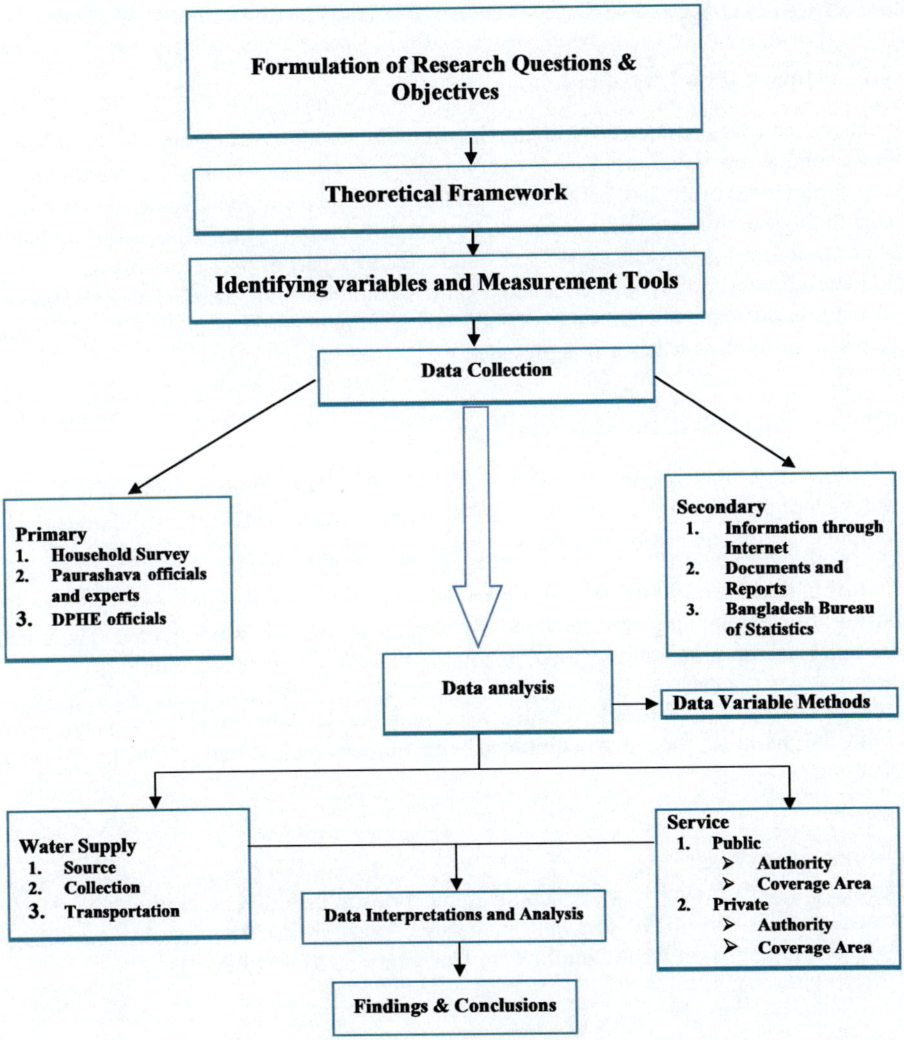

Fig. 2. Framework of the Study

3 Objectives of the Study

- To access the existing condition of the water supply system.
- To explore urban expansion and land use trends complying with the water supply system.
- To identify gaps of water supply system based on urban expansion scenario.

4 Methodology

4.1 Primary Data Investigation

Primary data was collected by contacting and discussing with experts in the relevant field, conducting a household survey of two categories of users namely: household users and commercial users and, contacting The Project Director of the Department of Public Health Engineering (DPHE) of Comilla Project. In addition to this, observation and field study, and focus group discussion was conducted as a part of primary analysis.

The primary survey was conducted on the community of the City Corporation to investigate the present water supply conditions, quality of services, and their willingness to take the supplied water if it is provided.

4.2 Secondary Data Investigation

To know about the theoretical side of the utility services planning, especially for the water supply, the secondary data investigation was executed with different related books, journals, reports, and websites. The secondary investigation was divided as:

Planning and Development: To gain knowledge about the planning procedure of the utility services planning, various books, newspapers, journals were studied. There is a relation between urban and regional planning and utility services planning.

Internet Browsing: Related websites were searched to collect the latest information about the planning procedure combined with modern technology for utility services planning.

Collection of Water

There are 220 Community water supply connections in the city corporation area. There are also the numbers of 110 tube wells in the area. There are 4570 household connections in the city corporation area (Comilla City Corporation, 2017) (Table 1).

Table 1. List of the water pump stations

Pump station no	Pump station name	Recent condition	Diameter	Depth
01	Sujanagar	Open	**150 mm**	**450 mm**
02	Godhirpukurpar	Open	**150 mm**	**450 mm**
03	Chawk bazaar shapla market	Open	**150 mm**	**450 mm**
04	Chawk bazaar Bus Stand	Open	**150 mm**	**450 mm**
05	Gangchar	Open	**150 mm**	**450 mm**
06	Nawab Bari Chaowmuhoni	Open	**150 mm**	**450 mm**
07	Dakshin Chartha	**Off**	**150 mm**	**450 mm**
08	Mohila College	Opening Soon	**150 mm**	**450 mm**
09	Gobindapur	Open	**150 mm**	**450 mm**
10	Thakurpara	Open	**150 mm**	**450 mm**
11	Badurtola	Open	**150 mm**	**450 mm**
12	Bagicha gaon	Open	**150 mm**	**450 mm**
13	Ranir Bazar	Opening Soon	**150 mm**	**450 mm**
14	Kaliajuri	Open	**150 mm**	**450 mm**
15	Vatpara	**Off**	**150 mm**	**450 mm**
16	Bishnu pur	Open	**150 mm**	**450 mm**
17	Kaptan Bazar	Open	**150 mm**	**450 mm**
18	Dharmasagarpar	**Off**	**150 mm**	**450 mm**
19	Old Chawdhuri para	Open	**150 mm**	**450 mm**
20	Badsha Miar Bazar	Opening Soon	**150 mm**	**450 mm**
21	Ranir Dighirpar	Open	**150 mm**	**450 mm**
22	Jangalia	**Off**	**150 mm**	**450 mm**
23	Rampur	**Off**	**150 mm**	**450 mm**
24	Muradpur	Open	**150 mm**	**450 mm**
25	Jangalia DOHS	Open	**150 mm**	**450 mm**

Transportation of Water

Water is mainly transported by 2 types of pipes i.e., Pressure pipes and Gravity pipes (Table 2).

Table 2. List of Water supply Pipeline

Sl. no	Form	Up to	Diameter	Meter
1	Racecourse, old passport office to police line, Fosdari Chomoni, Rajgong to trank road	Gudir pukarpar	200 mm	2591.46
2	Policeline Diabatic Hospital form jawtala	Kandirpar Pubali bank Chomoni	200 mm	609.72
3	Kandirpar mosque market	Rajgong Chomoni	200 mm	457.31
4	Tomsom Bridge Chomoni to Eastern Road	Kusaitali Adonik Hospital	200 mm	1829.26
5	Tomsom Bridge Chomoni to Western Road	Gobindapur Railgate	200 mm	914.63
6	Gobindapur Chomoni to Southern	Rab Office	150 mm	91.46
7	Gobindapur Chomoni	Majar	150 mm	289.63
8	Gobindapur Khanka sarif	Gobindapur Kazi bari	150 mm	365.85
9	Gobindapur Khalifa bari	Gobindapur Chomoni	150 mm	426.82
10	Gobindapur Khalifa bari	Mosjid south west road	150 mm	121.95
11	Gobindapur Chomoni	western road Railgate	150 mm	609.75
12	Dolipara Chomoni,Rajapara Chomoni, Laxmipur, Disabon,Kazipara, Unaisar	Khuchu Chomoni	200 mm	5487.80
13	South chorta Sonar bagla Hospital,Sahara nurse line	Woman Collage	150 mm	914.63
14	Chortha Chomoni	Boro pukarpar	150 mm	182.92
15	Kusaitali, Houseging	Norpur Chomoni	150 mm	609.75
16	Godirpukarpar	Barapara moynamoti hospital	150 mm	457.31
17	Laksam road salauddin hotal	Tomsom Bridge	200 mm	609.75
18	Kasium uolom Madrasa behind		150 mm	457.31
19	Zangalia to	Tomsom Bridge	200 mm	1677.43
20	Godir pukarpar	Houseging	150 mm	182.92
21	Norpur matri sadan	Insaf houseging	150 mm	304.87
22	Talichona chomoni	Sujanagar chomoni	150 mm	365.85
23	Kaliajuri Shahid minar	Morapara Railgate	150 mm	457.31

(*continued*)

Table 2. (*continued*)

Sl. no	Form	Up to	Diameter	Meter
24	Kaliajuri Shahid minar	Bodurpur Chomoni	150 mm	609.75
25	Bisnopur Raja store Bari	English Grave	150 mm	140.24
26	Mogbari Chomoni western	Chotora Katalia pool	150 mm	274.39
27	Bisnopur Grave Road		150 mm	121.95
28	Old Chowdaripara	Delni Road Asif Bari	150 mm	182.92
29	Barapara Moynamoti Hospital	Kusiatali Adunik Hospital	150 mm	457.31
30	Kaptan Bazar Shajan Komisnar bari	Late minto mia bari	150 mm	121.95
				=21924.2

4.3 Data Analysis, Interpretation, and Presentation

After collecting all the data from the primary and secondary sources, the data have been analyzed very carefully and presented in tabular forms to show the existing condition and demand for quality and quantity of water by the consumers of Comilla City Corporation. The descriptive and unstructured data, documents, and other records have been summarized and arranged in a sequential order to attain the objective of the study by the software SPSS and Microsoft Excel. Some maps were produced by the GIS software.

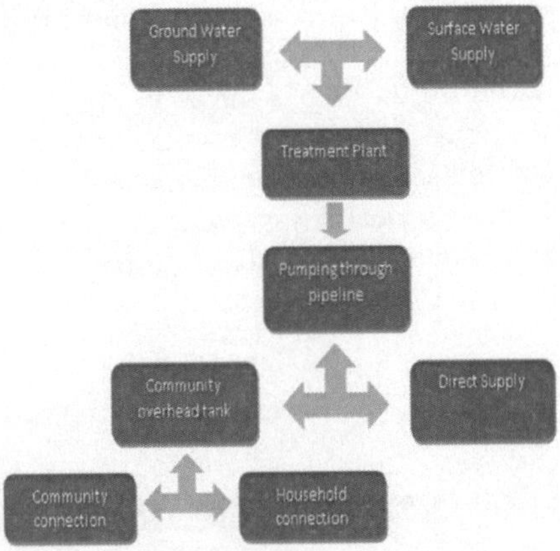

Fig. 3. Distribution Process of water in Comilla City Corporation

Distribution of Water

There are three types of distribution systems of water supply is available in Comilla City Corporation.

- Gravity System
- System with direct pumping
- System with pumping and storage

5 Local Policy Model

National Policy for Safe Water Supply & Sanitation, Bangladesh, (1998) gives some idea about both rural and urban water supply and sanitation and point out some strategies to gain 100% water supply and total sanitation. This policy states that the authority must provide safe drinking water within its jurisdiction area. The need for building the institutional capacity of local governments and communities and promoting sustainable water and sanitation services is recognized. The policy also acknowledges ensuring proper storage, management, and use of surface water. **Pourashava Ordinance, (1977)** gives some specific provisions of water supply to the Municipality and also gives the provisions of the private source of water supply.

But there are not enough connections according to the need. Additionally, the pipelines and other equipment are not maintained properly. Due to lack of transparency and strict execution of policies from the authority the water supply and maintenance is not up to the mark. Use of modern equipment, as well as an inventory of water supply, should be made to execute the policies regarding water supply and maintenance. CoCC has already taken the initiative to treat iron contamination in the supplied water.

6 Analysis of Urban Expansion and Land Use Trend Complying with Water Supply System

A survey of land use in 2016 found that this city had 52% of residential areas, 17% of commercial areas, and 9% of agricultural areas. The change of land use in Comilla City during 2001–2016 is shown in Fig. 3.

Urban growth in Comilla City has been rapid and disorganized, resulting in acute environmental deterioration. The rapidly increasing urban settlements continue to invade what previously was protected land, and land use has shifted from green areas to agricultural areas, to finally, urban.

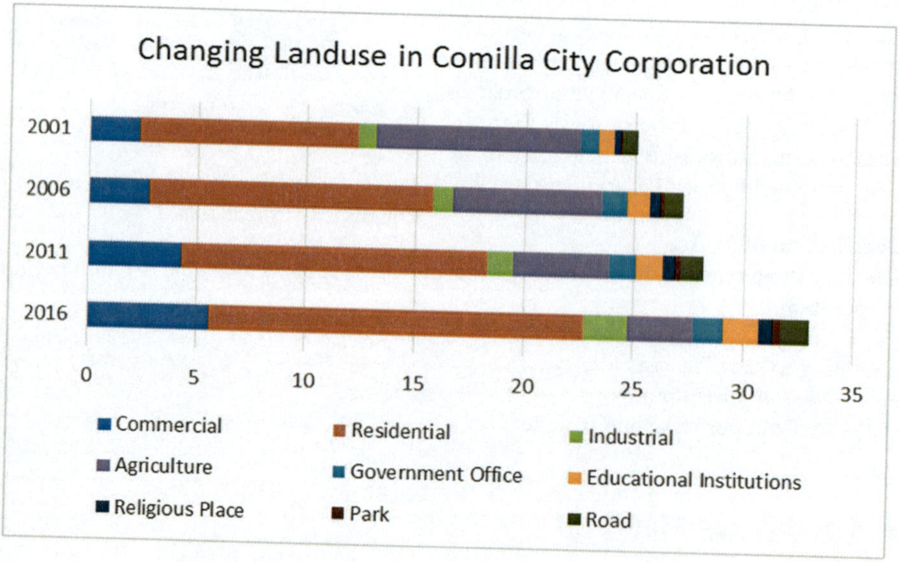

Fig. 4. Changing Land use in Comilla City Corporation

An analysis of the change of land use found that residential areas are growing rapidly and the growth of built-up areas destroys the precious remaining agricultural areas in the northern part of the city. Agricultural areas decreased around 6.36 km^2/yr. Community settlements were confined within the old city limits. Although the urban area expands continuously, the area boundary of Comilla City Corporation still has the same area size as 53.04 km^2 since 2010. So, the water demand is increasing rapidly (Fig. 4).

7 Results

The daily demand for water in Comilla City Corporation is 1.5 crore liter. The number of the pumps is 25, but from only 18 pumps the supplied water is being pumped up daily. But City Corporation does not meet the demand, because they can supply only 10840000 L of water per day. So, there is a deficit of 27.73% water in the Comilla City Corporation area.

7.1 Population and Water Demand

There is a considerable gap between the present and future population of CoCC and water demand. The demand exceeds the supply (Figs. 5 and 6).

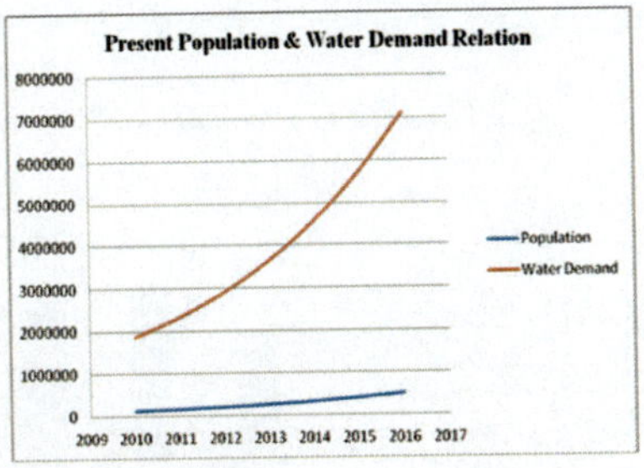

Fig. 5. Present Population and Water Demand

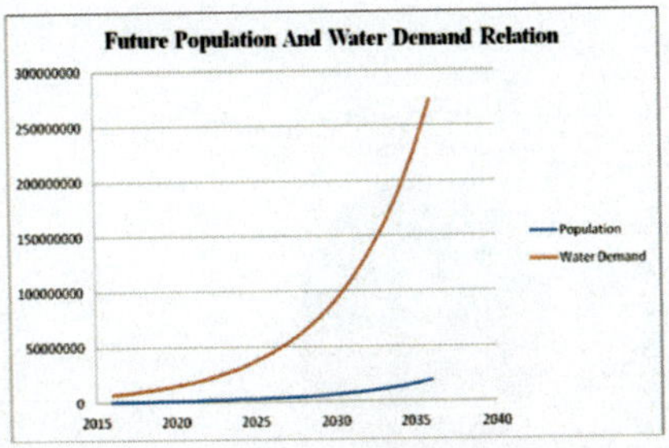

Fig. 6. Future Population and Water Demand

7.2 The Time Duration of Water Supply and Water Provision System

Comilla City Corporation supplies water a minimum of 2 h to 4 h per day, which is not enough considering the consumption rate.

Moreover, if the population in Comilla City Corporation still has the water consumption pattern as it was, this city will certainly face the water problem that will affect urban development, human security. Among them 56% direct supply, 21% temporary reservoir, 16% overhead tank, and 7% hand tube well (Fig. 7).

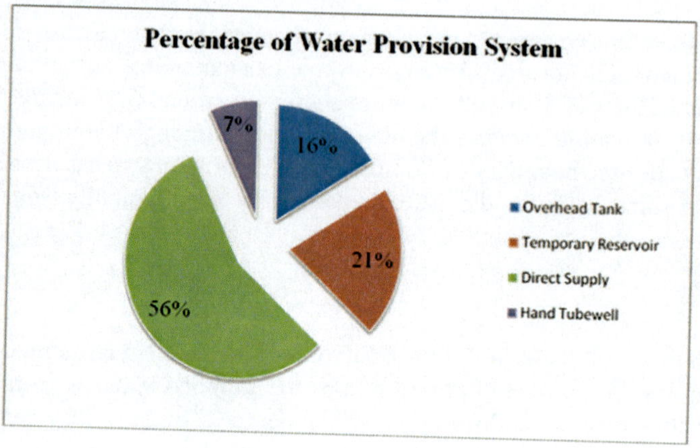

Fig. 7. Water Provision System

7.3 Reasons for not Using Supplied Water and Users Perception of Water Supply

There are many reasons for not using the supplied water in the study area.

Due to insufficient supply, dirty water, low pressure of water the people of Comilla City Corporation are not fully satisfied. The satisfaction level of the user of water supply from 100 household surveys is shown in the chart (Figs. 8 and 9).

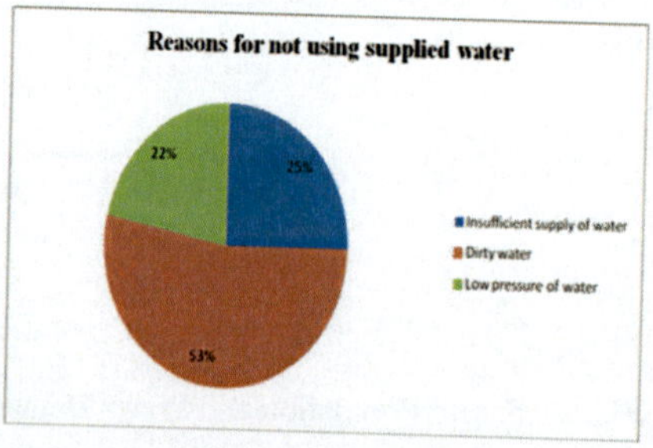

Fig. 8. Reasons for not using supplied water

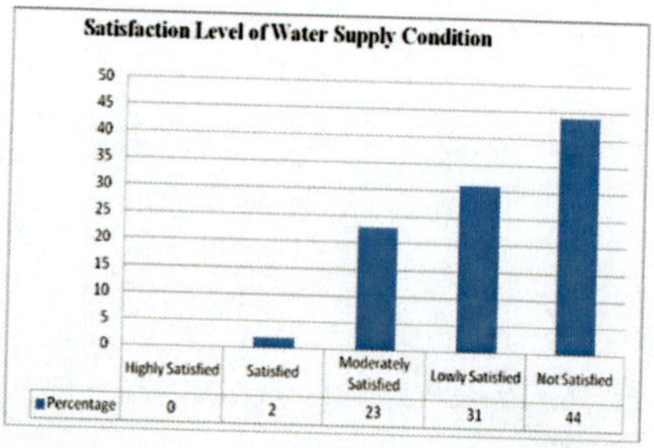

Fig. 9. Satisfaction level of water supply condition

7.4 Management of Pipes and Pumps

Water supply management is a daily routine work of the Comilla City Corporation. But they don't maintain the system and don't monitor them on regular basis due to insufficient manpower and technology. The result is dirty water, low pressure of water supply in Comilla City Corporation. The user perception conducted from 100 household surveys about water supply management and monitoring condition is shown in the chart (Fig. 10).

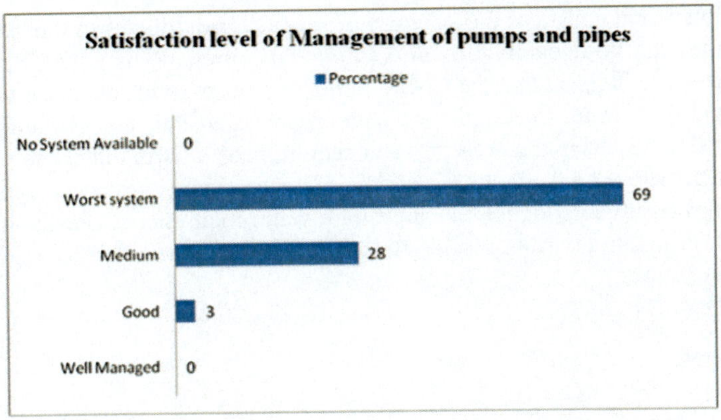

Fig. 10. Satisfaction level of Management of pumps and pipes

8 Conclusions

- Water supply is provided by Comilla City Corporation two times per day in maximum areas. Comilla City Corporation supplies water a minimum of 2 h to 4 h per day. It is not enough for the dwellers in those areas.
- For fulfilling community needs they should provide the water supply continuously.
- Due to insufficient supply, dirty water, low pressure of water the people of Comilla City Corporation are not fully satisfied.
- 300 taka is the monthly bill for the water supply. The monthly bill should be less than the present rate.
- As the existing water supply system is not satisfactory to the people, providing new water supply lines to support the growth of water consumption demand can be considered as the solution to the water shortage problem.
- The pumps and pipes are not well maintained. The majority of the household samples are in the worst condition and are not managed properly.
- Low water prices are a known offender within increased water consumption. Appropriate pricing of water should be used for controlling water demand.
- When municipal water users will pay realistic water prices, utilities will be able to maintain these systems, minimize losses and maximize the quality and the level of service.

9 Contribution of the Work to the Society

Urban expansion and water supply are interrelated. As urban expansion occurs, the quality of life of the people also improves. This study found out that the water quality of Comilla City Corporation is getting worse day by day because of iron transmission. Despite this iron factor the water supply system of CoCC is quite reasonable and maintains the need of people which directly contributes to society because water is the basic need of humans. In addition to this, the water of good quality ensures a disease-free healthy society and enhances the standard of life. This contributes to the overall social and economic development by providing improved community facilities. And the findings of this study indicate that the CoCC authority is also giving attention to providing a better quality of water in compliance with urban expansion. The planning proposals are prepared considering the present water requirement, existing land use pattern, and population growth. The population, after the design period, is projected using the Geometric Progression method and thus the future water requirement is estimated. This is significant for planning utility services to supply adequate water to the residents of the City in near future.

References

1. Sustainability in Water Supply | IWA Publishing. https://www.iwapublishing.com/news/sustainability-water-supply. Accessed 29 Jan 2021
2. Bangladesh: Improving Water Supply and Sanitation. https://www.worldbank.org/en/results/2016/10/07/bangladesh-improving-water-supply-and-sanitation. Accessed 29 Jan 2021
3. Rahman, M., Islam, M.: Scarcity of safe drinking water in the south-west coastal Bangladesh. J. Environ. Sci. Nat. Resour. **11**, 17–25 (2019). https://doi.org/10.3329/jesnr.v11i1-2.43361
4. Master Plan for Comilla City and its Influence Area (2014–2034) (2015)
5. কুমিল্লা সিটি কর্পোরেশন. https://cocc.gov.bd/

Building a Digital Twin for the Management of Pressurised Collective Irrigation Systems

Madalena Moreira[1,2], Sandra Mourato[1,3](✉), Carlos Rodrigues[1,4], Sandra Silva[1], Rita Guimarães[1,4], and Carlos Chibeles[5]

[1] MED - Mediterranean Institute for Agriculture, Environment and Development, Pólo da Mitra, Ap. 94, 7006-554 Évora, Portugal

[2] Department of Architecture, University of Évora, Escola dos Leões, Estrada dos Leões, 7000-208 Évora, Portugal
mmvmv@uevora.pt

[3] ESTG – Polytechnic Institute of Leiria, Leiria, Portugal
Sandra.mourato@ipleiria.pt

[4] Department of Rural Engineering, University of Évora, Pólo da Mitra, Ap. 94, 7006-554 Évora, Portugal
{camr,rcg}@uevora.pt

[5] Odivelas Associations of Irrigation Water Users, 7900-562 Ferreira do Alentejo, Portugal
cchibeles@aboro.pt

Abstract. Digital Twins (DT) can be exploited to improve Collective Irrigation Systems (CIS) performance and be fundamental for decision-making. Developing a DT is an enormous challenge with continuous adjustments and learning from the physical system. Agriculture represents one of the most water demanding sectors, and changing from gravity to pressurised water distribution turning the efficient use of water and energy as one of the main purposes of managing pressurised CIS. The DT is created based on field and laboratory tests, a systematic application of hydraulic modelling, water balance, energy balance and performance indicators. The data connection between the physical system and the DT and data advanced analytics provides a means to the operational setting, optimising, studying scenarios, and adapting to external changes of the pressurised CIS. In this paper, the authors propose a methodology to build a DT of a primary and secondary pressurised CIS network to evaluate water use efficiency, global energy efficiency, pumping facilities efficiency and energy inefficiency associated with water losses. A preliminary application to Block 4, a pressurised sub-system of Odivelas CIS, is presented in Southern Portugal. It was possible to ascertain the great potential of the physical twin – digital twin set to understand system processes, the corrective measures, the priority of maintenance and management decision of the pressurised collective irrigation system. This formulation is a good way to move towards smart water management in CIS.

Keywords: Collective Irrigation Systems · Digital Twin · Water balance · Energy balance · Performance indicators · EPANET

J. R. da Costa Sanches Galvão et al. (Eds.): *Proceedings of the 1st International Conference on Water Energy Food and Sustainability (ICoWEFS 2021)*, pp. 785–795, 2021.
https://doi.org/10.1007/978-3-030-75315-3_83

1 Introduction

Agriculture represents one of the most water demanding sectors with consumption of about 70% of freshwater [1], and its role is central to defining measures to get optimal water use efficiency, meeting farmers demands. Irrigation systems with significant water volumes supplied and important energy consumption have to be innovative in a changing world, especially in the Mediterranean region, where irrigation areas have increased, and water scarcity is already happening.

Digital Twin (DT) is at the core of the entire Industry 4.0 development, enabling automation, data exchange, and manufacturing processes, producing endless opportunities for industries to grow [2]. This concept can be adapted to all sectors, including the water sector and Collective Irrigation Systems (CIS), to enhance water and energy efficiency.

Improving water efficiency in CIS reduces physical water losses, reduces overflow, and ensures the reduction of non-revenue water. The reduction of losses and discharges can be guaranteed by the maintenance, conservation and rehabilitation of infrastructures, for which the association of irrigation water users has to be economically sustainable, i.e., it has to reduce unbilled water to a minimum. Non-revenue water also depends on apparent losses due to measurement errors in the water delivered to farmers, usually measured by default. On the other hand, transitions from gravity to pressurised water distribution systems that have been implemented in CIS worldwide [3] require a simultaneous analysis to get optimal energy efficiency. Energy efficiency may be improved by increasing pumps' efficiency, reducing physical water losses, controlling the required energy at the farmer's water delivery section, and promote energy recovery, when available.

The technological advances that have been introduced at CIS, in the framework of the flow, pressures, water levels in reservoirs and canals monitoring, and the application of IoT to transmit and store recorded data have significantly increased the complexity of the physical system. Incorporating the DT of the physical system into a single system with information connection between them, it is possible to mitigate system complexity by providing more and better information about the Physical Twin (PT) [4, 5] and use that information to implement improvements in the management of the physical system.

This paper applies a DT concept as a virtual representation of infrastructures and processes in pressurised CIS, from abstraction to water delivery to farmers. It absorbs the PT data obtained from georeferenced information, infrastructure data, laboratory fieldwork results using component operation calibration, or obtaining error curves in measuring equipment. The supply of equipment and construction works must be monitored, and the components' proper functioning must be guaranteed to assess measurement errors. Most construction companies or equipment suppliers consider that their work is over after the equipment has been delivered and assembled. The DT's performance assessment allows transmitting information to the PT on the necessary interventions or measures to augment water and energy use efficiency in the CIS.

This global system built by the PT, DT and connections (data and information) can be useful throughout the CIS lifecycle since the design, build, exploration, conservation and rehabilitation, ensuring continuous useful life. Systems and services must be seen within a dual nature perspective: physical and virtual. This approach adapts what was presented by [6].

The DT concept model in water sectors contains three main parts, adapted from [7]: a) system infrastructures and managers processes in real space, b) Geographic Information System (GIS), hydraulic models, water balance, energy balance, performance indicators in virtual space, and c) data and information connections between real and virtual systems and processes.

The data related to the physical system and processes have progressed rapidly in the last decades. Also, the simulation and exploration of the virtual system have progressed rapidly. The two-way connection between real and virtual space has been lagging. The connection between the data about the processes in water conveyance and distribution systems and the virtual model's information must be synchronised.

In the past, information was manually collected, mostly paper-based, which was a challenge for process managers, and there was a small quantity of information. Nowadays, this collection has progressed to being digital and being collected by various physical non-destructive sensing technologies, including sensors and gauges [7]. On the virtual side, the amount of available information has been increasing. With all these advances, there is a danger that system managers can consider that all the data collected is correct and represent the process. It is essential to realise that the more progress we have in data acquisition, the greater the need to check the quality of data and calibration and validation. An organisation's ability to innovate depends not only on the overarching system and information technology (IT) infrastructure but also on how humans employ it to maximise innovation.

SCADA (Supervisory Control and Data Acquisition) has been implemented in many water conveyance and distribution systems, mainly in urban water supply systems [8]. These systems require a high first investment and operation and maintenance costs and also specialised human resources. CIS managers can apply for European funds for the first investment, but their operation and maintenance are part of the water users' current costs. The price that the irrigators pay for the water must cover the expenses associated with their service. If the water prices rise, the crops' economic sustainability may be put at risk, jeopardising their production. Therefore, CIS managers must find solutions that allow them to guarantee the highest water and energy use efficiency at the lowest cost. Metrics for performance systems create a baseline for our research on better irrigation water systems' performance, diagnose the problems, and identify priorities for intervention. For this system of performance indicators, it is necessary to apply instruments such as Water Balance (WB) and Energy Balance (EB) and develop a CIS hydrodynamic model and all the processes inherent to its application.

This paper presents the methodology for creating a pressurised CIS DT, the methods, data and information required. It also presents the preliminary assessment of the methodology applied to a practical case, Block 4 (Odivelas CIS).

The paper outline is as follows: Sect. 2, the methodological approach and methods applied are presented. Section 3 presents the study area and describes data acquisition. The results are presented in Sect. 4, and finally, some preliminary conclusions are drawn in Sect. 5.

2 Methodology to Build a Digital Twin for Pressurised Collective Irrigation Systems

The methodology to build a DT for evaluating the water and energy efficiency in pressurised CIS, proposed by the authors, is based on a PDCA (Plan- Do- Check-Act) cycle for continuous improvement.

The first step (Plan) is the definition of the system boundaries and the period for which it will be applied. It is followed by identifying the necessary data and the data's origin. The application of Performance Indicators allows diagnosing water and energy use efficiency. To determine the variables' values in the indicators, it is necessary to apply the water balance and the energy balance. The water balance allows the calculation of physical water losses, leaks and interruptions in the network, discharges to the reservoirs, and unbilled water. The energy balance allows calculating the energy supplied (gravity and pumping) upstream of the system and the minimum energy required (energy needed to deliver water to the watering reservoirs). Applying the water balance and the energy balance is necessary to record the water input into the system, water delivery to the irrigation system, level variation in the reservoirs, energy consumed in electric pumps and recovered energy (if any). The registration and data transfer use IoT and data storage in the cloud to store a large amount of data. The representation of the water conveyance and distribution system through a hydrodynamic model, from the capture to the water delivery in the irrigation system, is essential to the system's evaluation of the energy flows. It is also essential in projections and simulations for the future or for system operation scenarios to simulate and analyse improvement or correction measures. The data quality and the numerical tools used are essential to obtain the proper diagnosis of the system's situation concerning water and energy use efficiency. It is also essential to evaluate the flow measurement equipment's error curve.

The second step (Do) corresponds to the Digital Model's creation requires high investment in evaluating the quality of the data transferred from the physical model (Fig. 1). It is necessary to create the system's hydrodynamic model that needs to be correctly calibrated and validated based on pressure and flow measurements in the physical model. The calibration and validation criterion corresponds to the minimisation of the pressure and flows diagram adjustment indicators over time and in defined sections, with several best-fit indicators. The water balance applied to the system and the sub-systems allow the data's coherence recorded in the physical system to be evaluated [9]. The quality of the water balance data influences the energy balance data quality. It is an example that the gravitational energy available in the system depends on the volume of water entering the system, and the minimum energy required in the delivery of the water to the irrigation system is a function of the volume of water delivered.

The calculation of performance indicators is based on the variables obtained by water balance and energy balance results (Table 1). This step requires an in-depth and lengthy comparison between the data recorded in the physical system and the information taken from the DT, which may require changes to the physical system and thus successively. It corresponds to a cycle within the cycle of continuous improvement.

The third step (Check) allows assessing if the DT represents the system. With the evolution of the systems, amendments are necessary by increasing the area irrigated

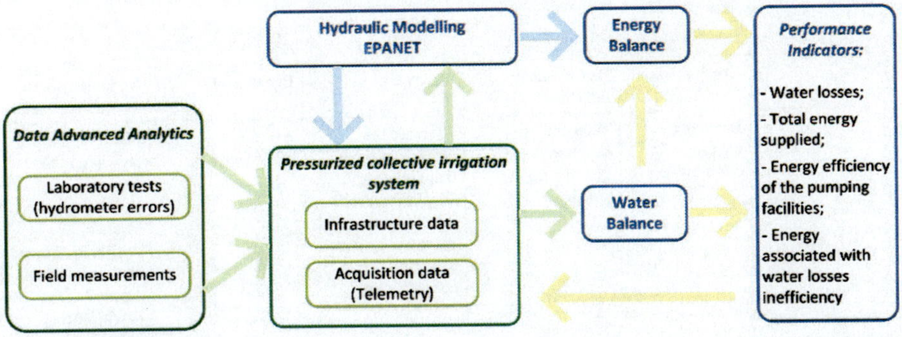

Fig. 1. The Digital Twin methodology framework for a CIS (Green boxes – Physical Twin; Blue boxes – Digital twin).

or changing crops' allocation. Consequently, the volume of water delivered to irrigation may require updating the DT, so the fourth step (Act) implements the necessary improvements. Depending on the result, it may be necessary to change the DT plan.

Table 1. Performance indicators: goals, criteria, performance indicators, calculation and variables [10].

Goals	Criteria	Performance indicator	Description	Equation	Variables/Source
Sustainability of irrigation water services	Economic and financial sustainability	Non- revenue water (%) (AH04)	It assesses the economic losses that correspond to water that, despite entering the system, is convoyed, stored and distributed but not invoiced to users	AH04 = dBH06/dBH05 × 100	dBH05 - Water entering the conveyance and distribution system (m^3)/water balance dBH06 - Unbilled water (m^3)/water balance
Water use Sustainability	Water use efficiency	Water efficiency (%) (AH17)	Evaluates the fraction of incoming water that is authorised or lost by natural processes	AH17 = dBH12/dBH05	dBH12 authorised consumption + evaporated volume (m^3)/water balance dBH05 Input water to conveyance and distribution system (m^3)/water balance
Energy use sustainability	Energy efficiency	Energy efficiency of the pumping facilities (%) (AH15)	Evaluates the overall efficiency of pumping stations	AH15 = dBE04/dBE02 × 100	dBE02 Pumping power consumption (kWh)/energy balance dBE04 Useful energy from pumping stations (kWh)/energy balance

(continued)

Table 1. (*continued*)

Goals	Criteria	Performance indicator	Description	Equation	Variables/Source
	Reducing energy consumption	Total energy supplied to the system (%) (AH16)	Assesses the number of times that the energy supplied exceeds the minimum energy to guarantee irrigation consumption	AH16 = dBE05/dBE06	dBE05 Energy supplied to the system (kWh)/energy balance dBE06 Minimum energy for water supply (kWh)/energy balance

3 Case Study

The Odivelas CIS is located in the Alentejo region, in the Sado and Mira hydrographic region (Fig. 2a). The exploitation began in 1974, with abstraction at the Odivelas reservoir. This exploitation has a beneficiated area of 12 698 ha and is composed of 117.58 km of water conveyance and distribution open channel network; 298.10 km of water conveyance and distribution network in pressurised pipes; 825 pressure watering outlets and 1311 gravity watering outlets, 5 pumping stations with collection in 5 intermediate reservoirs (Fig. 2b). The average volume of water supplied between 2016 and 2018 was 5 917 411 m^3. Energy consumption at the 5 pumping stations was 5 628 705 kWh in 2016, with an increase of 30% in 2017 and a reduction of 15% in 2018. The average consumption between 2016 and 2018 was 5 917 411 kWh. With the installation of a photovoltaic panel park, energy production in 2018 was 769 268 kWh, about 16% of the total energy consumed.

The methodology presented in this paper was applied in Irrigation Block 4, one of the 12 irrigation blocks that constitute the CIS (Fig. 2c).

Block 4 was chosen due to several aspects, namely: is the most recent sub-system; availability of a large amount of information either from infrastructures or from data acquired through telemetry and stored on local servers; in February 2015, Block 4 underwent modernisation, replacing the gravity distribution network by a pressure distribution network; has a relatively small size that allows the processes to be identified; the existence of equipment for recording flow rates, from the Parshall flume at the inlet to flow meters in the pressure pipes and hydrometers in the outlet; and continuous records of water level at the Parshall flume and at the reservoir. For all these reasons, the sub-system has the potential for integrated management within the Industry 4.0 framework applied to the water sector taking advantage of new technologies, IoT and decision support tools.

Block 4 benefits an area of approximately 1,685 ha (13% of the total CIS) in a set of 32 plots (mainly olive groves and corn), located in the parish of Figueira de Cavaleiros, in the municipality of Ferreira do Alentejo.

Irrigation Block 4 with abstraction from the main channel *Canal Conductor Geral* (CCG), consists of a reservoir, the Parshall flume to measure the flow into the regulator

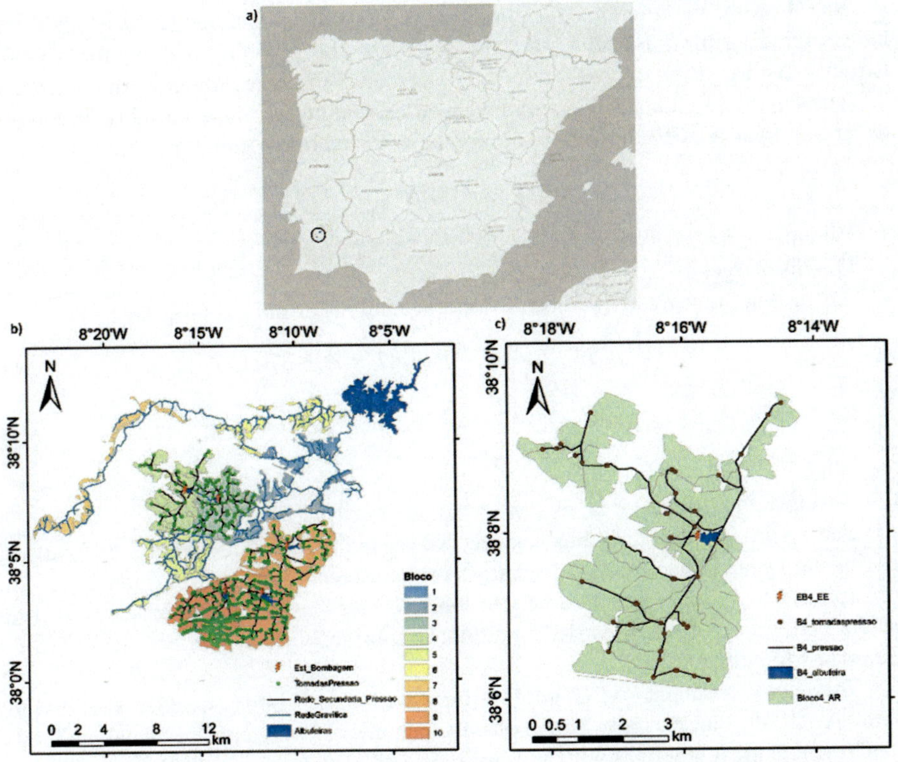

Fig. 2. a) Location of the Odivelas CIS; b) Odivelas CIS; c) Irrigation Block 4 infrastructures.

tank and a pumping station to deliver water under pressure to the watering systems of pipeline 41 (north) and pipeline 42 (south). The irrigation system uses the water under pressure, and hydrometers control the flow. Block 4 has 50 hydrants (6% of the total CIS), and the total length of pressurised pipes is 25.07 km (7% of the total CIS). The system's topographic elevation varies between 67 and 89 m, with an average elevation of 80 m. The pipes are in Ductile iron with diameters between 600 and 1000 mm and HDPE with diameters between 200 mm and 600 mm.

The annual volume measured at the Parshall flume is 4 499 618 m^3. The annual volume of water delivered (pumped) is 4 334 115 m^3 (12% of the total CIS). The energy consumption of Block 4 is 896 142 kWh (15% of the total CIS).

4 Results

This application aimed to show the importance of the connection between the physical model and the digital model. The equipment for records, storage of these records, verification of the quality of records and their mathematical treatment has to be appropriately treated to build a global Physical Twin - Digital Twin system that meets the objectives, which in this case refer to the efficient use of water and energy in pressurised CIS.

Based on the georeferenced information (GIS), the detailed model of Block 4 was built, including the Parshall flume, the reservoir, the pumping station, the pipeline network, the irrigation hydrant outlet and accessories that affect the flow in the system.

The Parshall flume's calibration has been carried out with the portable flow meter River Surveyor ADCP, allowing to obtain the adjusted rating equation law:

$$Q = 0.5904(3.28 \times \mathrm{h})^{1.4097} \tag{1}$$

Where:

Q – discharge, m^3/s; h – water height (m)

According to the design project, the storage curves of the reservoir are:

$$V = 5.812 \times Z^2 - 927.82 \times Z + 37033 \tag{2}$$

$$A = 1059.2 \times Z^2 - 161978 \times Z + 6.185 \times 10^6 \tag{3}$$

Where:

Z – elevation, m; A – área, m^2; V – volume, dam^3

The reservoir water level has been recorded, and it does not overpass the spillway during the 3 years. Therefore, no discharges from the reservoir have been considered. The precipitated and evaporated volume into the reservoir was calculated based on records at the Ferreira do Alentejo weather station; evapotranspiration was obtained with the Penman-Monteith equation.

The pumping station has 12 pump groups with a total power of 2528 kW, divided into the 2 sub-blocks. Two equal electromagnetic flowmeters are installed in the electromechanical pipelines with an error rating of approximately ±0.8%.

Field tests were carried out to evaluate the error of the hydrometers installed in the water intakes, compared with an ultrasonic flow meter, and the value of ±2% was obtained. It was found that the flow delivered to the hydrometers are in the flow rate class where the error is lower, i.e., between the transitional and the permanent flow rate.

The hydrodynamic model was created on the EPANET software (Fig. 3), a freeware software run for 2016–2018. The time scale is different according to the objectives: Hourly - in the calibration of the hydrodynamic model; Daily - in the analysis of losses, consumption patterns and evaluation of measurement errors; Monthly – in the water balance for evaluation of components that are not measured; Annual – in the water balance, energy balance and performance assessment system.

Field data are used to calibrate and validate the EPANET model. Consumption patterns are represented by the records' mathematical advanced analytics from 2016 to 2018 in each of the water intakes.

The water balance has been applied to the two sectors of Block 4: sector 1 - Parshall flume, reservoir, pumping station; Sector 2 - Pumping station distribution network and water intakes.

The energy balance application, based on the water balance and the hydrodynamic modelling of Block 4, made it possible to obtain the necessary variables to calculate performance indicators. The results are presented in Table 2.

The system's diagnosis based on these results allows the conclusion that there is the capacity to recover the value that is not invoiced by registering more reliable consumption

Fig. 3. The hydrodynamic model implemented in EPANET.

Table 2. Results of the performance indicators.

Performance indicators	Equation	Variables/Source
Non-revenue water (%) (AH04)	AH04 = 11,9%	dBH05 = 4020181 m^3 dBH06 = 4020 181 − 3539920 m^3
Water efficiency (%) (AH17)	AH17 = 89,3%	dBH12 = 3539 920 + 49291 m^3 dBH05 = 4020181 m^3
Energy efficiency of the pumping facilities (%) (AH15)	AH15 = 77,8%	dBE02 = 822642 kWh dBE04 = 822642 − 183008 kWh
Total energy supplied to the system (%) (AH16)	AH16 = 1,6	dBE05 = 1727495 kWh dBE06 = 1106969 kWh

and controlling physical water losses. The energy efficiency of the electric pump groups must be analysed group by group to identify where the efficiency loss is, and there is a very high value of energy available in relation to what is required by the farmers.

The collected data at the physical model has been transferred to the digital model, allowing the following information:

- Sector 1: Was detected a water balance closing error in the reservoir storage curve or the monthly water level record.
- Sector 2: the correlation between the flow pumped and the flow delivered to the farmers allowed to visualise that there are errors by a defect in the records measured and transferred by telemetry in the irrigators' water intakes. The error due to physical water losses in the pipes and errors by telemetry was evaluated through mathematical analysis. This information sent to the physical model allowed an intervention in the telemetry system to reduce this error.

Concisely, the physical model application - virtual model and its connections had the following contribution to improve the water use and energy efficiency in Block 4, Table 3.

Table 3. Preliminary assessment of the Digital Twin in the CIS.

Identified problem	Where it was identified	How	Corrective measure adopted
Doubt about the volume of water measured at the Parshall flume	Virtual Model	Mathematical correlation between input flow and output flow in water intakes	- Parshall flume calibration and validation - Identification and validation of telemetry and transmitted values
The fragility of the record of variation in the volume of water inside the reservoir	Virtual model	Water balance in Sector 1	- Reservoir bathymetry - Identification and validation of water level values in the reservoir
Guarantee of a good record of the volume of water pumped	Physical Model	Measured value uncertainty curve	- Design of histograms of the volume of water pumped
Occurrence or non-occurrence of breakages in the system	Physical Model	No breakout events were identified during the period under review	- Records sheet of faults or interventions due to pipe breakage or detectable leaks
Guarantee of good registration and transfer of the volume of water delivered in each hydrometer	Physical Model	Uncertainty curve of the hydrometers	- Parshall flume calibration and validation - Identification and validation of telemetry and transmitted values

5 Conclusions

The development of a DT of a pressurised CIS is an essential step in developing a tool that enables integrated system management and decision-making. Developing a DT is an enormous challenge with continuous adjustments and learning from the physical system.

The DT should be created based on the infrastructure design and corrected based on what has been built. It should also be calibrated based on the field measurements. Performance indicators allow diagnostic water and energy use efficiency over time, based on water balance, energy balance and hydrodynamical modelling. The improvements must be tested through the performance of the virtual model with different scenarios simulated. Specifically, by merging how the process runs in the physical system, virtual processes performance with the measures implemented is possible to get a better result. More assessments must be repeated to identify the hydrometers errors tendency and preferably to test the hydrometers in the laboratory.

The goal is to have this complex system occurring in real-time. The connection between the physical and virtual models enables managers to compare the physical system and the virtual system's performance, collaborate with other managers to benchmark the results, and finally conceptualise measures to implement.

Focusing on this connection between the physical and virtual systems will improve water and energy use efficiencies. The DT model's success will rest on its value for both the managers and irrigation water users. There will also be new issues, security among the most important, that need to be addressed.

This paper presented a preliminary application of the methodology, implementing some fieldwork and having awareness about what more fieldwork is needed. It was possible to define some corrective measures to adopt.

References

1. FAO: Water for Sustainable Food and Agriculture. A report produced for the G20 Presidency of Germany, Rome (2017)
2. Alabi, M., Telukdarie, A., Van Rensburg, N.: Industry 4.0: innovative solutions for the water industry. In: Proceedings of the International Annual Conference of the American Society for Engineering Management, Huntsville, Alabama, USA (2019)
3. Corominas, J.: Agua y energía en el riego, en la época de la sostenibilidad. Ing. del Agua **17**, 219–233 (2010)
4. Grieves, M.: Origins of the Digital Twin Concept, August 2016
5. Grieves, M.: Virtually intelligent product systems: digital and physical twins. In: Flumerfelt, S., et al. (eds.) Complex Systems Engineering: Theory and Practice, pp. 175–200. American Institute of Aeronautics and Astronautics (2019)
6. Grieves, M.: Virtually Perfect: Driving Innovative and Lean Products through Product Lifecycle Management. Space Coast Press, Cocoa Beach (2011)
7. Grieves, M.: Digital Twin: Manufacturing Excellence through Virtual Factory Replication (2015). https://www.researchgate.net/publication/275211047. Accessed 14 Jan 2021
8. Conejos, P., Martinez Alzamora, F., Hervas Carot, M., Alonso Campos, J.C.: Building and exploiting a digital twin for the management of drinking water distribution networks. Urban Water J. **17**, 704–713 (2020)
9. Cunha, H., Loureiro, D., Sousa, G., Covas, D., Alegre. H.: A comprehensive water balance methodology for collective irrigation systems. Agric. Water Manag. **223**(C), 1 (2019)
10. Loureiro, D., Alegre, H., Moreira, M., Chibeles, C., Sousa, G., Matos, M., Cordeiro, D., Arranja, C.: Proposta de sistema de avaliação do desempenho no uso da água e de energia em aproveitamentos hidroagrícolas. VIII Congresso Nacional de Rega e Drenagem (2020)

Evaluation of Chemical Impacts of Rice Irrigated with Urban Treated Wastewater

Isabel Maria Duarte[1,2](✉), Ana Rita Cardoso[1], Helena Marques[1], Filipe Melo[1], José Borralho[1], Paula Amador[1,2], Kiril Bahcevandziev[1,2], Teresa Vasconcelos[1,2], José M. Gonçalves[1], and Mário Pina Barreto[3]

[1] Instituto Politécnico de Coimbra, Escola Superior Agrária, Coimbra, Portugal
iduarte@esac.pt

[2] IIA – Institute of Applied Research, CERNAS - Research Centre for Natural Resources, Environment and Society, Coimbra, Portugal

[3] Águas do Centro Litoral AdCL, Grupo Águas de Portugal, Coimbra, Portugal

Abstract. Improving practices of water saving in irrigation is a priority facing global climate changes. To cope with water scarcity, the irrigation with treated wastewater (TWW) is becoming a solution, posing however several health and environmental risks on agricultural fields. This preliminary study aimed to assess the physicochemical impacts of rice irrigated with TWW applied with a subsurface drip irrigation (SDI) system, focusing the rice food, soil and environmental safety. Normal water and flooding was used as a reference for comparison. The experimental scheme considered three treatments with five repetitions, namely the TWW irrigation with SDI, the normal water with SDI, and the flooded irrigation with normal water. The pots filled with 15 L of soil, sowed with a local traditional rice variety, were kept outdoors. The fertilization scheme followed the usual one under field conditions. The irrigation frequency varied from three to five times a week. Measurements of physicochemical properties of irrigation water, drainage and rice grain were carried out, according to the analytical reference methods. Results showed that the irrigation with TWW rises the electric conductivity of drainage water, but that the rice grain does not present increased risks to public health due to the low content of arsenic, cadmium, lead and mercury. However, the irrigation method must be adapted to a SDI to avoid human and animal contact with putative contaminants present in this water and thus safeguarding Environmental and Food Safety. Long term soil effects of TWW, including winter percolation, will be assessed in the future.

Keywords: Rice irrigation · Chemical impacts · Subsurface drip irrigation · Treated wastewater

1 Introduction

Rice is the world most important food crop, and the staple food for more than half the human population. Its demand in the world market has increased, and it is foreseeable a

J. R. da Costa Sanches Galvão et al. (Eds.): *Proceedings of the 1st International Conference on Water Energy Food and Sustainability (ICoWEFS 2021)*, pp. 796–805, 2021.
https://doi.org/10.1007/978-3-030-75315-3_84

continuos rise around 24% over the next 20 years [1]. The growth of African and Asian populations, as well as the changes in diet, explain the increase of human rice consumption. In the Mediterranean countries it is cultivated on about 1.30 million hectares [2], mainly in flat alluvial regions, with a relative abundance of water. Traditionally, it is grown under continuous flooding.

To respond to the growing market demand, rice production can be increased enlarging its cultivation area; however, there are limitations hard to overcome. The main constrainment is the water supply, due to its unavailability, high costs in many regions, and for countering a demand from society to restrict its use in irrigation. Moreover, flooding is effective on flat, deep and poorly permeable soils, which are precisely the soils where rice is traditionally cultivated; conditions that are scarce in other areas to allow the expansion of rice paddies. Many studies had explored the sustainability of innovative irrigation options, in order to reduce water consumption for rice production and its negative environmental impacts, and to extend rice cultivation outside traditional rice areas [3].

Nowadays, improving irrigation practices to save water is a priority. Therefore, the use of drip irrigation for rice cultivation has been one of the recommended solutions [3, 4], since it solves some of the most critical issues of rice continuous flooding, namely, it allows: i) reducing water consumption, especially the fraction of deep percolation; ii) the use of lower quality water, in terms of salinity and microbiology (high microbiological load in subsurfce to do pose direct risk to human and animal healty); iii) facilitating the automation of irrigation, reducing labor; and iv) to reduce the residual cadmium and arsenic contents in the seeds.

In this context, the use of treated wastewater (TWW) for irrigation is a solution to consider to deal with water scarcity. In certain places, TWW it is a significant source of water for irrigation with guaranteed supply during the summer, despite posing several health and the environmental risks [5].

The aim of this study was to evaluate the physicochemical impact of the rice irrigation with urban TWW on drainage water from soil percolation (environmental safety) and on rice grain quality (food safety). For this purpose some physicochemical quality parameters were monitored in the irrigation and drainage water, as well as the effect of the irrigation method on the rice grain quality.

2 Material and Methods

2.1 Experimental Layout

To assess the impacts of reusing municipal TWW for rice irrigation in Lis valley, the environmental exposition of drainage water to chemical coumpounds of TWW was determined through the physicochemical analysis of soil water samples; wheareas the rice consumer exposition to heavy metals was assessed through rice grain analysis.

Two sources of irrigation water (Treated WasteWater, TWW and Normal Water, NW, used for comparison) and two irrigation methods (Subsurface Drip Irrigation, SDI and Continuous Flooding, F) were used for this evaluation in a pot tests.

The experimental layout considered three treatments with five repetitions, namely: i) irrigation by SDI with TWW, ii) irrigation by SDI with NW, and iii) irrigation by continuous flooding with NW. The NW was collected from a well, whereas the Urban TWW was supplied by the Coimbrões WWTP, AdCL SA, Leiria. The pots with 15L of soil, seeded with a traditional variety of local rice (*Oryza sativa* L. cv Ariete), were kept outdoors. The fertilization scheme followed the usual field conditions. The irrigation frequency varied between three to five times a week for pots with SDI and flooding, respectively. The SDI with dripper at a depth of 15 cm, supplied through a small reservoir. A rice field owned and managed by a farmer, irrigated with NW, was used for comparison. Measurements of soil texture, irrigation water volume were recorded. Drainage water samples were collected by free percolation on two different dates, after harvest and after the Winter period.

The physicochemical parameters of water samples and of heavy metals in the rice grain were determined according to the analytical reference methods referred in Table 1.

3 Results and Discussion

3.1 Effect of Irrigation Water on the Quality of Soil Drainage Water

Physicochemical Parameters

The rice crop cycle was completed in 147 days. The mean values of pH, electrical conductivity (EC) and chlorides of the irrigation and drainage water of the pots (Tables 1 and 2) reveal significant differences between the treatments associated with the type of irrigation water. The significantly higher values of conductivity and chloride content of the water drained from pots irrigated with TWW are clearly associated with the origin of the irrigation water, which also has higher values than NW.

The pots irrigated with NW had a significant impact on the pH increase of the drainage water (from 6.7 to 7.2). In contrast, in the pots irrigated with same water (NW) by SDI conditions, the pH dropped from 6.7 to 6.1.

The total dissolved solids (TDS) were lower in the irrigation water compared to drainage, with the highest value in the TWW treatment. In turn, the total suspended solids (TSS) were higher in irrigation water, NW compared to TWW, possibly because TWW are filtered during the treatment process, with NW inversion of the order relationship in the drainage water (SDI-TWW> F-NW or SDI-NW) (Table 1).

There was an increase between the Electrical Conductivity (EC) values of TWW irrigation samples and the respective drainage, which was higher than those with NW. Regarding the NW, an increase was observed about EC values between the irrigation water and that of the respective drainage (1610μS/cm with the NW and 2100μS/cm with the TWW). Instead, the sodium adsorption ratio (SAR), being lower in irrigation water (NW and TWW) than in drainage water (SDI-TWW and F-NW+SDI-NW), and also increasing between irrigation water and the respective drainage, now has a higher increase in the case of NW (1.7) than in TWW (1.5) (Table 1). The salinity of the drainage water, deriving from that existing in the soil solution, increased due to the contribution of TWW high saline content in sodium and chlorides (sodium chloride), although other calcium and magnesium salts were also present (Table 1).

Table 1. Average values of the physical-chemical analysis of the water used for irrigation and the drainage water from the pots

Parameter (units)	Method	RV	MAV	Irrigation		Drainage	
				NW	TWW	SDI-TWW	(SDI,F)-NW
pH (Sorensen scale)	NP411:1966	6,5–8,4	4,5–9,0	6.7	7.2	7	6.6
Nitrates (mg/L)	ASTM D 4327.2017	50	—	2.9	30	66	10
Sulfates (mg/L)	ASTM D 4327.2017	575	—	20	54	230	130
Chlorides (mg/L)	ASTM D 4327.2017	70	—	43	150	730	470
Fluorides (mg/L)	ASTM D 4327.2017	1,0	15	<0.30	<0.30	<0.30	<0.30
Total Suspended Solids (TSS) (mg/L)	SMEWW 2540 D, 23ªEd	60	—	7.6	1.9	8	3.6
Salinity: Electrical Conductivity (μS/cm)	MI n°013 (03.05.2011)	1	—	290	1100	3200	1900
Sodium Adsorption Ratio (SAR)	Calculation	8	—	1.5	8.5	10	3.2
Total Dissolved Solids (TDS) (mg/L)	Electrometry	—	—	150	650	1700	1000
Free residual chlorine (mg/L)	SMEWW 4500 G, 22.Ed	—	—	<0.1	<0.1	<0.1	<0.1
Aluminum (mg/L)	DIN EN ISO 11885	5	20	0.04	0.31	0.09	0.09
Arsenic (mg/L)	DIN EN ISO 11885	0, 1	10	<0.005	<0.005	<0.005	<0.005
Barium (mg/L)	DIN EN ISO 11885	1.0	—	0.042	0.055	0.15	0.15
Beryllium (mg/L)	DIN EN ISO 11885	0,5	1	<0.002	<0.002	<0.002	<0.002
Boron mg/L	DIN EN ISO 11885	0,3	3,75	<0.05	0.09	<0.05	<0.05

(continued)

Table 1. (*continued*)

Parameter (units)	Method	RV	MAV	Irrigation		Drainage	
				NW	TWW	SDI-TWW	(SDI,F)-NW
Cadmium (mg/L)	DIN EN ISO 11885	0,01	0,05	<0.001	<0.001	<0.001	<0.001
Calcium (mg/L)	DIN EN ISO 11885	—	—	12.7	52.1	102	133
Chromium (mg/L)	DIN EN ISO 11885	0,1	20	<0.005	<0.005	<0.005	<0.005
Cobalt (mg/L)	DIN EN ISO 11885	0,05	10	<0.005	<0.005	<0.005	<0.005
Copper (mg/L)	DIN EN ISO 11885	0,2	5	0.027	0.004	0.009	0.007
Iron (mg/L)	DIN EN ISO 11885	5.0	—	0.03	1.3	0.11	0.02
Lead (mg/L)	DIN EN ISO 11885	5	20	<0.005	<0.005	<0.005	<0.005
Lithium (mg/L)	DIN EN ISO 11885	2,5	5,8	0.01	<0.01	<0.01	<0.01
Magnesium (mg/L)	DIN EN ISO 11885	—	—	5.33	21.5	26.9	33.2
Manganese (mg/L)	DIN EN ISO 11885	0,2	10	<0.005	0.15	0.022	0.02
Molybdenum (mg/L)	DIN EN ISO 11885	0,005	0,05	0.009	0.009	0.009	0.009
Nickel (mg/L)	DIN EN ISO 11885	0,5	2	<0.005	<0.005	0.01	<0.005
Potassium (mg/L)	DIN EN ISO 11885	—	—	4.3	14.4	73.8	23.7
Selenium (mg/L)	DIN EN ISO 11885	0,02	0,05	<0.01	<0.01	<0.01	<0.01
Sodium (mg/L)	DIN EN ISO 11885	—	—	24.7	290	458	155
Vanadium (mg/L)	DIN EN ISO 11885	0,1	1,0	<0.005	<0.005	<0.005	<0.005
Zinc (mg/L)	DIN EN ISO 11885	2	10	0.16	<0.01	0.01	0.02

RV, Recommended Value; MAV, Maximum Allowable Value,
Irrigation water: NW, normal water; TWW, treated wastewater.
Drainage water: SDI-TWW, underground drip irrigation with TWW; (SDI,F)-NW, average values of underground drip irrigation and continuous flooding with normal water.

The SAR indicates the ratio of exchange sodium to other non-toxic cations, such as calcium and magnesium. As sodium degrades the soil structure and raises the pH, it increases the risk of immobilization and the consequent deficiency of some micro-elements, especially zinc.

The quality of drainage water with respect to EC and Chlorine (Table 2), improved after the Autumn/Winter period, without cultivation, during which the rainfall occurred allowed the leaching of the soil, with a very relevant reduction in salinity.

Table 2. Average values of physical-chemical parameters of irrigation and drainage water from the pots after harvest and at the end of the Autumn/Winter period.

Parameter	Irrigation		Drainage					
	2019*		At harvest/2019**			After Winter/2020***		
	NW	TWW	SDI-TWW	SDI-NW	F-NW	SDI-TWW	SDI-NW	F-NW
pH	6.70	7.20	6.60	6.10	7.20	5.97	5.74	5.92
ECondutivity(μS/cm)	290.0	1100.0	2758.0	1064.0	1128.0	202.2	278.0	324.6
Chlorine (mg/L)	43.0	150.0	541.0	165.0	184.0	19.2	41.0	56.4
Nitrates (mgNO$_3$/L)	<3.0	30.0	66.0	10.0	10.0	0.5	0.8	0.4
TDS (mg/L)	150.0	650.0	1000.0	1000.0	1700.0	145.6	195.6	229.4

Irrigation water: NW, normal water; TWW, treated wastewater;
Drainage water: SDI-TWW, underground drip irrigation with TWW; SDI-NW, underground drip irrigation with normal-water; F-NW, irrigation by continuous flooding with normal-water.
*samples of irrigation water, during irrigation season. Determinations made at the *Tomaz Laboratory*
**samples taken from 10/Oct to 10/Dec/2019. Determinations made at the ESAC Chemistry Laboratory
***sample taken at the end of the autumn/winter period. Determinations by the ESAC Chemistry Laboratory

The values of arsenic, cadmium, cobalt, lead, chromium, lithium, nickel, selenium and vanadium from TWW were below the detection threshold of the method, which leads us to conclude, in the first approach, that in relation to these parameters, this water has no restrictions for irrigation use. The only sample where boron was detected was in TWW (90 μg/L) used for irrigation, which was not in the drainage water of the pots irrigated with this water. It should be noted that the boron content of TWW did not exceed one part per million (1 ppm = 1 mg/L), a limit that should not be reached given its toxic nature for rice. The values of arsenic, chromium, lead found in irrigation and drainage water are below 0.005 mg/L, and those for cadmium below 0.001 mg/L. These values are below the maximum limits stipulated by the legislation [6, 7] (Table 3).

3.2 Effect of Irrigation Water Quality on Rice Production. Heavy Metals in the Grain

Total arsenic, cadmium, lead and mercury concentrations found in rice samples are shown in Table 4. Portuguese legislation sets for limits on the content of heavy metals in water for human consumption and irrigation water; however, limits for food products

Table 3. Maximum levels of heavy metals in the water, according to the legislation in force

Heavy metal	Irrigation		Potable for human consumption			Residual	Surface
	MRV (mg/L)	MAV (mg/L)	MRV (mg/L)	MAV (mg/L)	VP (mg/L)	VLE (mg/L)	MAV (mg/L)
Arsenic (As)	0.10	10	0.010	0.050	0.010	1	0.1
Cadmium (Cd)	0.01	0.05	0.001	0.005	0.005	1	0.01
Lead (Pb)	5	20		0.050	0.010	0.20	0.05
Mercury (Hg)	—	—	0.0005	0.001	0.001	0.05	0.001
Nickel (Ni)	0.5	2	—	—	0.020	2	0.05

VLE, emission limit value (concentration that must not be exceeded) DL-236/98
MAV, maximum allowable value (value that must not be exceeded) DL-236/98
MRV, maximum recommended value (value that must be respected or not exceeded) [6].
VP, parametric value (maximum residual concentration of the polymer in contact with water) [7].

are not yet defined. Considering the maximum levels of 0.2 mg/kg, fixed for inorganic arsenic, total cadmium and lead in rice by the Regulation of the European Communities [8, 9], it appears that lead is below this threshold in all samples.

Table 4. Concentration of total arsenic, cadmium, lead and mercury in the grain of brown rice produced in the three treatments and under traditional field conditions

Heavy metal (mg/kg)	SDI-TWW	SDI-NW	F-NW	FC
Arsenic	<0.1	<0.1	0.4	0.9
Cadmium	0.1	0.3	0.2	<0.01
Lead	0.1	<0.02	0.2	<0.02
Mercury	<0.00	<0.00	<0.00	<0.00

Treatments: SDI-TWW, underground drip irrigation with TWW; SDI-NW, underground drip irrigation with normal water; F-NW, irrigation by continuous flooding with normal water; FC, field conditions, produced by a rice farmer field, with flood irrigation under field conditions.

Mercury was not detected in any of the samples. In any case, there are no legislated maximum values for this metal for rice. The cadmium content of rice irrigated with NW (0.3 mg/kg) was above the limit for this heavy metal.

The soil of the SDI pots was kept in oxic conditions, which would have increased the solubility of this element, due to the favorable conditions for the desorption of aluminum, iron and manganese oxy-hydroxides or solubilization of CdS or $CdCO_3$. In this drained soil, it is procuced cadmium sulfate ($CdSO_4$), which, being soluble in water, allows its greater absorption by rice roots. Cadmium is strongly retained in flooded soils (reducing condition), accumulating less cadmium than that developed under oxidizing conditions, which is related to the formation of cadmium sulfide in anoxic conditions [10]. TWW had higher values of pH (7.2), aluminum (0.31 mg/L), iron (1.3 mg/L) and manganese (0.15 mg/L) compared to normal water, which had values pH 6.7, aluminum 0.04 mg/L, iron 0.03 mg/L and manganese less than 0.05 mg/L.

The arsenic content found in rice shows a direct relationship with growing conditions (aerobic or anaerobic). The rice produced in flooding contains the highest levels in pot (F-NW, 0.4 mg/kg) and in field conditions (0.9 mg/kg), well above the maximum limit of 0.2 mg/kg regulated for arsenic for rice [9]. Similar results had already been described by Simões [11] in traditional carolino rice (japonica variety) produced in Lower-Mondego Valley, Portugal, with average values of 0.2426 mg/kg.

The accumulation of arsenic in plants depends essentially on its bioavailability (arsenic speciation) and on the levels of arsenic present in the soil [12]. Given that in this trial the same soil (composition and type) was used in all treatments and that all irrigation water contained trace levels of arsenic (<0.005 mg/L) (Table 4), the differences found are due to the effect other factors that resulted in different arsenic bioavailability in treatments.

Under conditions of almost permanent flooding, in anaerobiosis, arsenic exists mainly dissolved in the form of arsenite, its most toxic and most bioavailable form for the plant [13], which justifies the capture of arsenic by cultivated rice in flooding with NW order of magnitude greater than that of the drip irrigated with TWW. Arsenic is carried by the phloem to the seeds where it is stored in vacuoles and other tissues of the edible parts of the grain. Their storage in the grain is done mainly in the form of arsenite and dimethylarsinic acid [14].

4 Final Considerations

The use of TWW in rice irrigation is a challenging issue. The experiment carried out in pots during the 2019 season, based on physicochemical analysis of the water samples, confirmed the soil salinity risk associated with TWW. The high EC of drainage water, with 2758μS/cm, when TWW was used to irrigate (corresponding 96μS/cm for NW), with a predominance of the elements sodium and chlorine. This issue is very important for productivity and soil conservation. It should be noted that after the autumn-winter period, in which the pots were subject to precipitation and free drainage, the EC values dropped to values similar to the initial ones. The results on heavy metals in irrigation and drainage water, lead to the conclusion that they are at levels below the maximum stipulated by legislation.

Regarding the levels of heavy metals detected in rice grains, it was found that the grain produced in the SDI-NW treatment has a cadmium content above the legal limit. The greater solubility of this element is explained by the redox potential and the pH of the soil in these test conditions, although no progress has been made in this evaluation. The arsenic level was higher in grains produced in the F treatment, whereas that value in the SDI treatment is insignificant. Results showed that the irrigation with TWW rises the EC of drainage water, but that the rice grain does not present increased risks to public health due to the low content of arsenic, cadmium, lead and mercury. However, the irrigation method must be adapted to a SDI to avoid human and animal contact with putative contaminants present in this water and thus safeguarding Environmental and Food Safety. The use of TWW must imply a careful plan to monitor the soil salinity throughout the irrigation campaign, to prevent critical situations of reduced production, or soil conservation. Long term soil effects of TWW, including winter percolation, will be assessed in the future.

Acknowledgments. Project MEDWATERICE: Towards a sustainable water use in Mediterranean rice-based agro-ecosystems. The project is financed in the context of the PRIMA Programme (PRIMA-Section-2018; Topic: 1.1.3: Irrigation technologies and practices). (MEDWATERICE-PRIMA-0005-2018, www.medwaterice.org).

References

1. Nguyen, N.V., Ferrero, A.: Meeting the challenges of global rice production. Paddy Water Environ. **4**, 1–9 (2006). https://doi.org/10.1007/s10333-005-0031-5
2. FAOSTAT Online Database. Food and Agriculture Organization, Rome (2016). https://faostat3.fao.org
3. Adekoya, M.A., Liu, Z., Vered, E.: Agronomic and ecological evaluation on growing water-saving and drought-resistant rice (Oryza sativa L.) through drip irrigation. J. Agric. Sci. **6**(5), 110–119 (2014). https://doi.org/10.5539/JAS.V6N5P110
4. Parthasarathi, T., Vanitha, K., Mohandass, S., Vered, E., Meenakshi, V.: Variation in rice root traits assessed by phenotyping under drip irrigation. F1000 Res. **6**(125), 1–19 (2017). https://doi.org/10.12688/f1000research.9938.2
5. Asgari, K., Cornelis, W.M.: Heavy metal accumulation in soils and grains, and health risks associated with use of treated municipal wastewater in subsurface drip irrigation. Environ. Monit. Assess. **187**(7), 1–3 (2015). https://doi.org/10.1007/s10661-015-4565-8
6. Decree-Law 236/98, of 1998-08-01. Establishes Water Quality Standards. Republic Diary, 76/1998, Series I-A, 3676–3716, Ministry of the Environment, Portuguese Presidency of the Council of Ministers, Lisbon, Portugal (1998). https://dre.pt/pesquisa/-/search/430457/details/maximized
7. Decree-Law 306/2007, of 2007-08-27. Establishes Water Quality Standards for human consumption. Republic Diary, 164/2007, Series I, 5747-5765, Ministry of the Environment, Territorial Planning and Regional Development, Portuguese Presidency of the Council of Ministers, Lisbon, Portugal (2007). https://data.dre.pt/eli/dec-lei/306/2007/08/27/p/dre/pt/html
8. EC 2006. European Communities, Commission Regulation N. 1881/2006, of 19 December 2006. Sets the maximum levels for certain contaminants in foodstuffs (Text with EEA relevance) Official Journal of the European Union, L 364, 5–24 (2006). https://data.europa.eu/eli/reg/2006/1881/oj
9. EC 2015. European Communities, Commission Regulation N. 2015/1006, of 25 June 2015. Amends Regulation (EC) No 1881/2006, as regards maximum levels of inorganic arsenic in foodstuffs. Official Journal of the European Union, L 161, 14–16 (2015). https://data.europa.eu/eli/reg/2015/1006/oj
10. Alloway, B.J., Jackson, A.P., Morgan, H.: The accumulation of cadmium by vegetables grown on soils contaminated from a variety of sources. Sci. Total Environ. **91**, 223–236 (1990). https://doi.org/10.1016/0048-9697(90)90300-J

11. Simões, A.C.P.: Avaliação da presença de arsénio em arroz e produtos derivados de arroz (Doctoral dissertation, ISA, Lisbon) (2014). https://www.repository.utl.pt/handle/10400.5/6783
12. Bundschuh, J., Nath, B., Bhattacharya, P., Liu, C.W., Armienta, M.A., López, M.V.M., Lopez, D.L., Jean, J.S., Cornejo, L., Macedo, L.F.L., Tenuta Filho, A.: Arsenic in the human food chain: the Latin American perspective. Sci. Total Environ. **429**, 92–106 (2012). https://doi.org/10.1016/j.scitotenv.2011.09.069
13. Rahman, M.A., Hasegawa, H.: High levels of inorganic arsenic in rice in areas where arsenic-contamined water is used for irrigation and cooking. Sci. Total Environ. **409**(22), 4645–4655 (2011). https://doi.org/10.1016/j.scitotenv.2011.07.068
14. Smith, E., Juhasz, A.L., Weber, J., Naidu, R.: Arsenic uptake and speciation in rice plants grown under greenhouse conditions with arsenic contaminated irrigation water. Sci. Total Environ. **392**, 277–283 (2008). https://doi.org/10.1016/j.scitotenv.2007.11.023

Check for updates

Exploring the Potential of Remote Sensing in Irrigation Management at District Scale. Study on Lis Valley, Portugal

Susana Ferreira[1(✉)], Juan M. Sanchéz[2], and J. M. Gonçalves[1]

[1] IPC Instituto Politécnico de Coimbra, Escola Superior Agrária de Coimbra, Coimbra, Portugal
{susana.ferreira,jmmg}@esac.pt

[2] UCLM Universidade de Castilla-La Mancha, Inst. de Desarrollo Regional, Albacete, Spain
juanmanuel.sanchez@uclm.es

Abstract. Irrigation plays an important role in agriculture, with impact on land productivity and economic prosperity of farmers. Water management is complex but can benefit from remote sensing satellite images, which provide a monitoring of soil, crops and water, in harmony with the requirements of precision agriculture. The surface data from the combined constellation Sentinel-2A, 2B and Landsat 8 offer unique opportunities with this aim. The SpiderwebGIS tool, developed by the Instituto de Desarrollo Regional, from the University of Castilla La-Mancha (UCLM), allows a friendly management of satellite images, providing information to the community in a simple and accessible way. This communication explores the potential applications of remote sensing in irrigation management at district scale, aiming to improve water management in the plots, under the responsibility of farmers, and in collective distribution, by water users' associations. This study focuses on the Lis Valley Irrigation District. The methodology considers ground observations at a local scale of climate, soil and plant parameters, for the development and calibration of models based on satellite images. The ultimate objective is the operational use of temporal series of spatially distributed information to support agricultural management and optimize the use of water, energy and economic resources. This study is supported by Instituto Politécnico de Coimbra and UCLM and is integrated into the activities of the Lis Valley Water Management Operational Group.

Keywords: Remote sensing · Sentinel-2A and 2-B · Landsat 8 · Rational management of water · Lis Valley Irrigation District

1 Introduction

New risks to global water security are emerging due to global changes [1], urging water saving in irrigated agriculture and irrigation intensification to increase the food production for a growing population [2]. Collective irrigation schemes play a major role in irrigated agriculture around the world, impacting land productivity and peasants welfare [3], and their governance is complex, pressed to improve the effectiveness of scheme

J. R. da Costa Sanches Galvão et al. (Eds.): *Proceedings of the 1st International Conference on Water Energy Food and Sustainability (ICoWEFS 2021)*, pp. 806–811, 2021.
https://doi.org/10.1007/978-3-030-75315-3_85

systems, demanding knowledge, funds and the participation of farmers to adapt to each particular water saving context [4]. Irrigation management demands timely and frequent information on soil, crop and water, with a spatial resolution meeting the requirements of the precision agriculture (PA) paradigm [5, 6]. Remote sensing (RS) is largely used in the context of irrigated agriculture [7, 8], exploring the quantitative relations between remotely sensed spectral information and crop characteristics. RS used in irrigated agriculture is accepted by end-users and has become a valuable tool to support water management in irrigation schemes, complementing ground level and conventional monitoring [9]. As a first approach, this study analyses the RS tools to support irrigation management in Lis Valley, Portugal [10], This paper is a preliminary step of an ambitious research project aiming to survey and systematize the potential of RS in irrigation management at district scale.

2 Remote Sensing Tools and Techniques

Since the 1970s, satellite images have been extensively used in PA, namely to monitor the changes of physical and physiological traits and characteristics of crops over time. Sensors used for RS differ based on the spatial, spectral, radiometric and temporal resolution that they offer. In the past years, also aerial platform, including aircraft and unmanned aerial vehicles (UAVs) are being used to collect data about crop growth and provide this information at fine temporal and spatial resolutions [11].

The combined use of Sentinel 2 (VNIR, spatial resolution of 10-20m with a revisit time of 2–3 days) and Landsat 8 (spatial resolution of 30 m in the VNIR and 100 m for TIR, with a revisit time of 8–15 days) imagery, provides the suitable resolution to capture surface variability. Surface reflectance and radiance data are used for the calculation of the classic vegetation indices (VI), such as NDVI and SAVI, Vegetation development, crop water requirements, water consumption or field water productivity can be explored from biophysical models using these index data as inputs. In addition, land surface temperature (LST) can be an attractive variable in water stress scenarios. In addition to satellite images.

For the management of satellite image data sets, a webGIS platform (SpiderwebGIS©) developed by the Universidad de Castilla La-Mancha – Instituto de Desarrollo Regional (UCLM-IDR) was used (https://maps.spiderwebgis.org/webgis/). This platform makes the information available to the community in an easy and accessible way.

UCLM-IDR has also developed HidroMORE©. This is a software that applies FAO 56 methodology, assisted by RS images, for the determination of water availability on the plant surface. This is based on the retrieval of the basal crop coefficients (K_{cb}) through the dependence NDVI-K_{cb} [12]. This model has been applied in several studies [13, 14], including water accounting on the Water Users' Association (WUA) management scales [15, 16].

3 Lis Valley Case Study: The Challenges of RS Application

The study area is located in the Irrigation District of Vale do Lis (LVID), which is a public irrigation district, located in the Coastal Centre of Portugal (coordinates 39°51′22.1″ N

8°50′56.1″ W), belonging to the Administrative District of Leiria and managed by WUA. The total area is about 2000 ha, mainly with modern alluvial soils of high agricultural quality, although some have poor drainage conditions, and the main crops cultivated include maize, fodder, horticulture, orchards and rice [17].

RS can give a valuable contribution to monitor irrigation, due to the synoptic nature of the data and easily available image archives [18]. Due to its operationally and constant progress, RS operative applications are being offered to different water managements, as it introduces the advantages of monitoring large areas at plot spatial scale [15].

The issues related with water management in LVID that could be effectively supported by RS are presented below:

- **Mapping of irrigated areas** – The crop distribution on several fields is a determinant information for irrigation management at district scale. It allows to predict the irrigation scheduling and the water requirements per secondary or tertiary branches, during the irrigation season. Mapping crop distribution of irrigated areas is required to carry out an optimal planning of water allocation, particularly during the peak period, safeguarding spatial equity in distribution and its management in water scarcity periods. This mapping, if assessed retrospectively, also allows the analysis of crop cultivation and its economic and environmental effects [19].
- **Crop water requirements and water accounting** – The calculation of crop water requirements could be favored by satellite RS data, processed by informatic platforms. It complements the agrometeorological data observed on ground and provides a more detailed spatial information. Note that the great strength of the reflectance-based models from the point of view of crop irrigation management is the capability to estimate the potential crop transpiration, based on the temporal evolution of the RS-based K_{cb} and the actual ET_o values [9]. This information, integrated with the mapping of irrigated areas, and the evaluation of crop development through vegetation indexes (e.g., NDVI), allows to calculate the crop water consumption, and the water used at on-farm and off-farm levels [20].
- **Management of sector water distribution** – Further applications are the monitoring of the irrigation system, the water accounting at several levels, and the performance analysis of collective distribution, including the equity of water distribution to several fields within the same sector [15]. This information, regularly updated by satellite data, allows the optimal planning of water supply, with a more rational pumping and more efficient distribution during the scarcity periods.
- **Development of precision irrigation** – PA systems applicable to on-farm systems are highly dependent of crop and soil data gathered by in situ sensors. The RS data obtained at irrigation district level could support spatially-distributed crop information to farmers. RS allows monitoring crop development and evapotranspiration (ET) in order to support crop growth models [21] and irrigation advisory services [9], as well as analysis of crop uniformity within the field [22]. Therefore, it can help to create a knowledge base to make precision irrigation control systems more robust and efficient.
- **Management of the drainage riparian vegetation** – The management of riparian vegetation on the drainage system is very important in many irrigation districts, becoming a costly and complex process. This vegetation decreases the ditches flow capacity and obstructs gates and hydrants. The management operation represents an

added difficulty for workers, because it is often located in areas of difficult access. The RS monitoring of this vegetation is a very useful task, since it provides information on its development and density, which is required to adjust the management plans, namely, to program each operation, avoids previous actions by workers, saving efforts and resources and allowing prevention of occupational risks [23, 24].

As an example, to illustrate the use of the SpiderwebGIS tool, Fig. 1 shows the temporal evolution of the NDVI, for the period 2019–2020, in a selected point of a chosen experimental parcel. An NDVI subset from a Sentinel-2 image, corresponding to 9th September 2020, is also shown in Fig. 1. The field chosen is cultivated with maize, in spring/summer and with forage crops, in autumn/winter. This type of information will allow to determine indicators related with crop and water management, namely, the spacial crop development and irrigation uniformity, the relationship of NDVI with crop productivity and soil fertility, and the identification of subareas with specific problems, such as drainage, or crop protection.

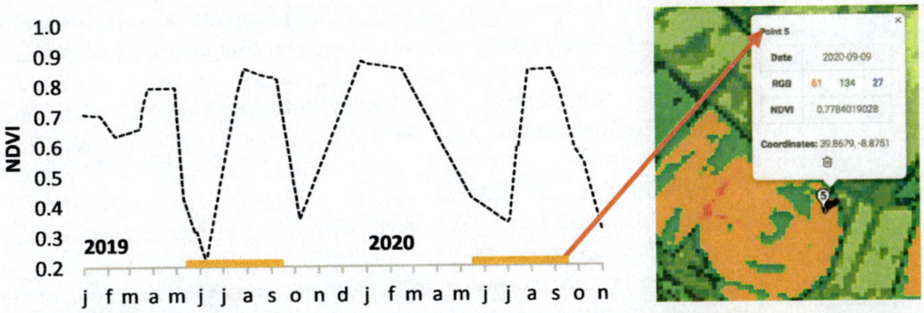

Fig. 1. Example of NDVI evolution of an Experimental Parcel, between 2019 and 2020, based on satellite remote sensing (note: orange bar represents the maize crop season). At right, details of the NDVI chart obtained with Sentinel-2, image of 9th September 2020 (source: SpiderwebGIS: https://spiderearth.appspot.com).

4 Final Considerations

The enrichment of RS applications over large and diverse irrigation surface areas offers both public and private water managers a complementary tool to obtain information. The combination of RS imagery and ancillary ground data (soil types, crop characteristics, meteorological), provides temporal and spatial information about the net irrigation requirements at on-farm level, and about the water distributed by conveyance system, as managed by WUA. The effective application of RS spatial information to support agricultural management and optimize the use of water, and linked resources, becomes a challenge, but the task will be embraced by this project, in collaboration with farmers and WUA.

Acknowledgments. Project Grupo Operacional para a Gestão da Água no Vale do Lis (PDR2020-1.0.1-FEADER- 030911), funded by PDR2020, cofunded by FEDER, on framework of Acordo de Parceria Portugal 2020, Programa PDR2020, Medida Inovação.

References

1. Thenkabail, P.S., Knox, J.W., Ozdogan, M., Gumma, M.K., Congalton, R.G., Wu, Z.T., Milesi, C., Finkral, A., Marshall, M., Mariotto, I., You, S.: Assessing future risks to agricultural productivity, water resources and food security: how can remote sensing help? Photogram. Eng. Remote Sens. **78**(8), 773–782 (2012)
2. Harmel, R.D., Chaubey, I., Ale, S., Nejadhashemi, A., Irmak, S., DeJonge, K.C., Evett, S.R., Barnes, E.M., Catley-Carlson, M., Hunt, S., Mani, I.: Perspectives on global water security. Trans. ASABE **63**(1), 69–80 (2020)
3. Molden, D.J., Burton, M., Bos, M.G.: Performance assessment, irrigation service delivery and poverty reduction: benefits of improved system management. Irrig. Drainage **56**, 307–320 (2007)
4. Sordo-Ward, A., Granados, A., Iglesias, A., Garrote, L., Bejarano, M.D.: Adaptation effort and performance of water management strategies to face climate change. Impacts in six representative basins of Southern Europe. Water **11**, 1078 (2019)
5. Evett, S.R., O'Shaughnessy, S.A., Andrade, M.A., Kustas, W.P., Anderson, M.C., Schomberg, H.S., Thompson, A.: Precision agriculture and irrigation: current U.S. perspectives. Trans. ASABE **63**(1), 57–67 (2020)
6. Soto-Garcia, M., Del-Amor-Saavedra, P., Martin-Gorriz, B., Martínez-Alvarez, V.: The role of information and communication technologies in the modernization of water user associations' management. Comput. Electron. Agric. **98**, 121–130 (2013)
7. Atzberger, C.: Advances in remote sensing of agriculture: context description, existing operational monitoring systems and major information needs. Remote Sens. **5**, 949–981 (2013)
8. Bégué, A., Arvor, D., Bellon, B., Betbeder, J., de Alelleyra, D., Ferraz, R.P., Lebourgeois, V., Lelong, C., Simões, M., Véron, S.R.: Remote sensing and cropping practices: a review. Remote Sens. **10**, 99 (2018)
9. Calera, A., Campos, I., Osann, A., D'Urso, G., Menenti, M.: Remote sensing for crop water management: from ET modelling to services for the end users. Sensors **17**, 1104 (2017)
10. Gonçalves, J.M., Ferreira, S., Nunes, M., Eugénio, R., Amador, P., Filipe, O., Duarte, I.M., Teixeira, M., Vasconcelos, T., Oliveira, F., Gonçalves, M.: Developing irrigation management at district scale based on water monitoring: study on Lis Valley, Portugal. AgriEngineering **2**, 78–95 (2020)
11. Sishodia, R.P., Ray, R.L., Singh, S.K.: Applications of remote sensing in precision agriculture: a review. Remote Sens. **12**, 3136 (2020)
12. Moreno, R., Arias, E., Sánchez, J.L., Cazorla, D., Garrido, J., Gonzalez-Piqueras, J.: Hidro-MORE 2: an optimized and parallel version of HidroMORE. In: 2017 8th International Conference on Information and Communication Systems (ICICS), Irbid, pp. 1–6 (2017)
13. Sánchez, N., Martínez-Fernández, J., Calera, A., Torres, E., Pérez-Gutiérrez, C.: Combining remote sensing and in situ soil moisture data for the application and validation of a distributed water balance model (HIDROMORE). Agric. Water Manag. **98**, 69–78 (2010)
14. Garrido-Rubio, J., Sanz, D., González-Piqueras, J.: Application of a remote sensing-based soil water balance for the accounting of groundwater abstractions in large irrigation areas. Irrig. Sci. **37**, 709–724 (2019)

15. Garrido-Rubio, J., González-Piqueras, J., Campos, I., Osann, A., González-Gómez, L., Calera, A.: Remote sensing–based soil water balance for irrigation water accounting at plot and water user association management scale. Agric. Water Manag. **238** (2020)
16. Garrido-Rubio, J., Calera, A., Arellano, I., Belmonte, M., Fraile, L., Ortega, T., Bravo, R., González-Piqueras, J.: Evaluation of remote sensing-based irrigation water accounting at River Basin District management scale. Remote Sens. **12**, 3187 (2020)
17. Ferreira, S., Oliveira, F., Silva, F.G., Teixeira, M., Gonçalves, M., Eugénio, R., Damásio, H., Gonçalves, J.M.: Assessment of factors constraining organic farming expansion in Lis Valley Portugal. AgriEngineering **3**, 111–127 (2020)
18. Ozdogan, M., Yang, Y., Allez, G., Cervantes, C.: Remote sensing of irrigated agriculture: opportunities and challenges. Remote Sens. **2**, 2274–2304 (2010)
19. Ranghetti, L., Cardarelli, E., Boschetti, M., Busetto, L., Fasola, M.: Assessment of water management changes in the Italian rice paddies from 2000 to 2016 using satellite data: a contribution to agro-ecological studies. Remote Sens. **10**, 416 (2018)
20. Sánchez, J.M., López-Urrea, R., Valentín, F., Caselles, V., Galve, J.M.: Lysimeter assessment of the Simplified Two-Source Energy Balance model and eddy covariance system to estimate vineyard evapotranspiration. Agric. Forest Meteorol. **274**, 172–183 (2019)
21. Balbontín, C., Campos, I., Odi-Lara, M., Ibacache, A., Calera, A.: Irrigation performance assessment in table grape using the reflectance-based crop coefficient. Remote Sens. **9**, 1276 (2017)
22. Campos, I., Neale, C.M.U., Arkebauer, T.J., Suyker, A.E., Gonçalves, I.Z.: Water productivity and crop yield: a simplified remote sensing driven operational approach. Agric. Forest Meterol. **249**, 501–511 (2018)
23. Kang, B., Kim, D.: Applications of unmanned aerial vehicles in fluvial remote sensing: an overview of recent achievements. KSCE J. Civil Eng. **22**, 588–602 (2018)
24. Eninger, R., Johnson, R.L.: Unmanned aerial systems in occupational hygiene – learning from allied disciplines. Ann. Occup. Hyg. **59**(8), 949–958 (2015)

Microbiological Monitoring of Superficial Water of Lis Valley Irrigation District. Impacts of Agricultural Activities

Isabel M. Duarte[1,2](✉), Helena Marques[1], Maria Corte-Real Araújo[1], Paula Amador[1,2], Teresa M. Vasconcelos[1,2], Susana Ferreira[1], and José M. Gonçalves[1,2]

[1] Instituto Politécnico de Coimbra, Escola Superior Agrária, Coimbra, Portugal
iduarte@esac.pt

[2] IIA – Institute of Applied Research, CERNAS - Research Centre for Natural Resources, Environment and Society, Coimbra, Portugal

Abstract. Society's demand for the preservation of water resources has led to major technological advances in irrigation management, simultaneously promoting socio-economic development and natural resources conservation. This study aims to assess the impact of agricultural activities at the Lis Valley Irrigation District (LVID) on the quality of superficial water. This information will support the improvement and the minimization of risks regarding the conservation of groundwater and surface water bodies in the valley. The physicochemical and microbiological quality of water samples collected at the inlets and outlet of the LVID was determined according to the analytical reference methods. Results showed that the water at the inlets of the LVID contains microbiological contamination, due to external sources. In general, the concentrations of total coliforms (TC) in the inlet water were 1.4 times lower than in the outflow water; however, an inverse situation was found with faecal coliforms (FC), faecal enterococci (FE), and ampicillin-resistant (Amp^r) *Enterobacteriaceae*. These data reveal that the LVID's agricultural activities have not aggravated (and have even improved) those microbiological indicators of the outlet water when compared to those of the water which reaches the entrance of the district, available for agriculture. Concluding, the monitoring in a continuum of a few years will allow the assessment of risk situations of microbiological contamination to support the improvement of regional planning and agricultural management, being recommended to increase the sampling representativeness.

Keywords: Microbiological indicators · Coliforms · Enterococci · *Enterobacteriaceae* · Irrigation · Conservation of water bodies

1 Introduction

The development, sustainability, and entrepreneurial competitiveness of Portuguese agriculture depend heavily on irrigation [1], namely on the improvement of irrigation water

J. R. da Costa Sanches Galvão et al. (Eds.): *Proceedings of the 1st International Conference on Water Energy Food and Sustainability (ICoWEFS 2021)*, pp. 812–820, 2021.
https://doi.org/10.1007/978-3-030-75315-3_86

management. Also, in the climate change scenarios that indicate a decrease in available water resources [2], water-saving is a strategic priority to irrigation development. The management of irrigation districts with water running systems has specific operative and water distribution characteristics; therefore, the water reuse from the drainage ditches allows mitigating the water shortage, with higher distribution equity and water productivity. This rises however some concerns regarding the health safety of farmers, food and consumer safety, as well as soil salinization. Thereafter, it is important to control the physicochemical and microbiological water quality, to supply the information system with data required for the rational management of the collective irrigation. In this context, water monitoring is a basic process for improving the management of the irrigation district, providing information about irrigation and drainage water quality, both spatially and temporally, identifying health, and environmental risks, and the required farm practices adjustments to cope with putative water scarcity and quality problems. These aspects of the irrigation systems governance should support farmers in adopting more environmentally friendly practices, thus contributing to irrigation sustainability [3].

This paper describes a study carried out on the framework of Lis Valley Water Management Operational Group, integrated on the agricultural European Innovation Partnership (EIP-AGRI) [4]. It aimed to assess the impact of agricultural activities at the Lis Valley Irrigation District (LVID) on the microbiological quality of superficial water, focusing on the results of the 2018 to 2020 seasons. This information will support the improvement and the minimization of risks regarding the conservation of groundwater and surface water bodies.

2 Materials and Methods

2.1 Study Area

This study was conducted in the LVID with about 2000 ha, mainly with modern alluvial soils of high agricultural quality, although some have poor drainage conditions. The main crops grown include forage maize, forage grass, horticultural, orchards and rice. The climate in the Lis River watershed has a Mediterranean influence, with hot and dry summers, as well as an Atlantic influence, with predominant frontal winter surfaces, moving from west to east, which are liable for most of the precipitation [5]. This is a public irrigation district managed by a Water Users' Association (WUA, located in the counties of Leiria and Marinha Grande, in the Coastal Center of Portugal (Fig. 1)). The irrigation water is supplied through several small temporary weirs from the Lis River and its tributaries, and the conveyance system comprises, for each sector, the main canal, gravity-fed by a river diversion from those weirs. During the irrigation peak period, the downstream irrigated areas are recharged by pumping from the river or drainage ditches [5].

2.2 Water Sampling

The methodology for monitoring the water quality of the irrigation and drainage networks followed published guidelines [8]. The sampling aimed to assess the representativeness

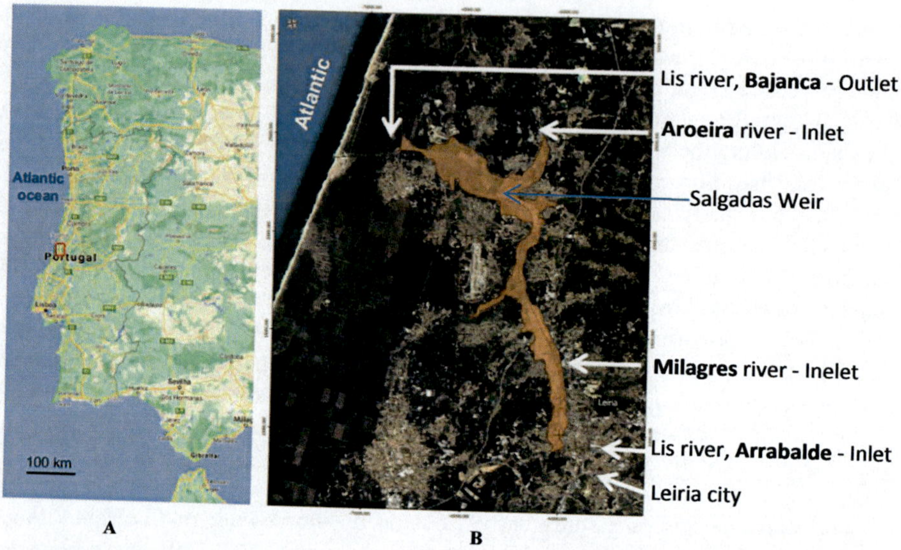

Fig. 1. Location of Lis valley irrigation district in Portugal (red rectangle) [6] (A) and the sampling sites in the LVID [7] (B).

of the agricultural actives at the LVID in the water quality released downstream, into the environment. The sites studied are located at the main inlets: Lis River at Leiria, Arrabalde [39°45′8,29″; 8°48′28,49″], Milagres River [39°47′04,542″; 8°49′29,750″], Aroeira River [39°53′34,644″; 8°50′54,316″], and at the outlet of the perimeter: Lis River at Bajanca [39°53′00,295″; 8°56′05,295″]. The sampling at the downstream Lis river section (Bajanca bridge site) was carried out at low tide, to avoid the influence of brackish seawater (Fig. 1). Within the scope of this article, these sites are hereinafter referred to as, Arrabalde, Milagres, Aroeira, and Bajanca, respectively.

Sampling for microbiological quality was conducted during the irrigation periods to guarantee seasonal representativeness. It was carried out in three periods: i) at the beginning of the irrigation season when there is plenty of water in the water lines, ii) at the end of the summer, before the autumn rains, when there is a shortage of water in the water lines and iii) in the middle of those periods, during the peak of the irrigation period. Water samples were collected in sterile plastic bottles and maintained at 4 °C until the microbiological processing, which was always carried out within four hours after sampling.

2.3 Physicochemical Analysis

The analysis of the physicochemical quality of the water samples was performed with a precalibrated *in-situ* portable multiparametric probe (SmarTROLL RDO Handheld, Fort Collins, CO 80524, USA) measuring, pH, Electrical Conductivity (EC, μS/cm), Saturation of Dissolved Oxygen (SDO, %), Temperature (T, °C), Total Dissolved Solids (TDS, ppm). Nitrates (NO_3^-, mg/L) were evaluated with ion chromatography [9]. The results obtained were compared to the maximum recommended values (MRV) of the Portuguese Irrigation Water Quality Standards [10].

2.4 Microbiological Analysis

Microbiological analysis of water samples included the enumeration of Total Coliforms (TC), Fecal Coliforms (FC), and Fecal Enterococci (FE) by the dilution method with multiple fermentation tube technique, the results being expressed as the Most Probable Number (MPN/100 mL). The mesophilic microorganism counts at 37 °C and 22 °C were carried out by the plate count incorporation method, the results being expressed as Colony Forming Units (CFU/mL), following the analytical reference methods [10].

The ampicillin-resistant bacteria in water samples were enumerated with the membrane filtration technique. Decimal dilutions of each sample were prepared in 0.9% sterile saline solution, from which 100 mL were filtered through cellulose membranes of 0.45 μm in pore size (Millipore, Bedford, MA, USA) under a vacuum system, and the filters aseptically placed on the surface of a selective medium for *Enterobacteriaceae* VRBG (Violet Red Bile Glucose) agar (Oxoid, Hampshire, England) supplemented with 20 μg/mL of sodium ampicillin (AppliChem, Darmstadt, Germany). Plates were aerobically incubated at 37 °C for 24 h to 48 h. All the *Enterobacteriaceae* isolates resistant to ampicillin (Amp^r) were counted [11].

The salinity of inlet water samples is significantly lower than those of the valley outlet samples. This is explained by the hydrogeological characteristics of saline springs on the downstream area and by the discharge of the LVID drainage ditches, as well as by the high tide of Atlantic water. This significant increase of water salinity of Lis river towards the mouth prevails, although the Aroeira contributes with much less saline water. The river water downstream Salgadas weir, located in the middle of the irrigation district (Fig. 1), is not feasible for irrigation.

3 Results and Discussion

3.1 Physicochemical Parameters of Water Quality

The average results of the physicochemical analysis obtained from 2018 to 2020 are presented in Table 1. The pH values of water samples from the irrigation and drainage are within an acceptable range [10], in general slightly above 7.0, except for the Aroeira river, a Lis tributary.

The saturation dissolved oxygen values of inlet and outlet water samples are similar and above 86% (Table 1). Although there are no references to irrigation water [12], it is recommended a value higher than 50% for drinking water, allowing to conclude that dissolved oxygen in these water bodies is good or acceptable.

The average nitrate content of inlet water samples (14.27 ± 3.7 mg/L) is lower than at the outlet (16.8 ± 6.8). Aroeira samples had fewer nitrates than other sites. Countering this trend, the content of upstream water samples evidence the intensive production systems, outside of the LVID and also the practice of soil manure amendment. Although the data is still scarce and does not include groundwater, it indicates that irrigation in the Lis Valley has no significant negative effect on the nitrate load on water resources [10].

Table 1. Physicochemical parameters of water samples, per site for Lis valley inlet and outlet water (average ± standard deviation).

Site	Role	Water quality parameters				
		pH (6.5–8.4)*	EC, μS/cm (1000)*	SDO, %	TSD, ppm (640)*	Nitrates, mg/L (50)*
Arrabalde	Inlets	7.80 ± 0.37	591.50 ± 18.9	87.26 ± 11.94	370.00 ± 30.01	17.0 ± 3.0
Milagres		7.73 ± 0.31	493.80 ± 14.8	88.09 ± 10.87	298.33 ± 32.48	16.8 ± 5.7
Aroeira		6.84 ± 1.17	286.97 ± 38.37	91.07 ± 10.87	298.33 ± 102.92	9.0 ± 2.5
Bajanca	Outlet	7.39 ± 0.29	1272.00 ± 850	86.33 ± 8.37	803.44 ± 571.90	16.8 ± 6.8

* Maximum recommended values, according to the Portuguese Irrigation Water Quality Legislation [10]. Legend: EC - Electrical Conductivity; SDO - Saturation of Dissolved Oxygen; TDS - Total Dissolved Solids

3.2 Microbiological Parameters of Water Quality

The analysis of water microbiological quality, assessed through the enumeration of indicator groups at the main inlet and outlet of the perimeter (Fig. 2A, B), shows an effect of the agricultural activities within the Valley on the water quality.

The TCs are widespread in the basin with average counts above 700 MPN/100 mL in all sites, being the counts in drainage water 1.4 times higher than those in irrigation water. Nevertheless, those enumerations, both from irrigation and drainage water samples are in between the maximum recommended value (500 MPN/100 mL) and the maximum admissible value (10,000 MPN/100 mL) regarding aquaculture and bathing purposes [10].

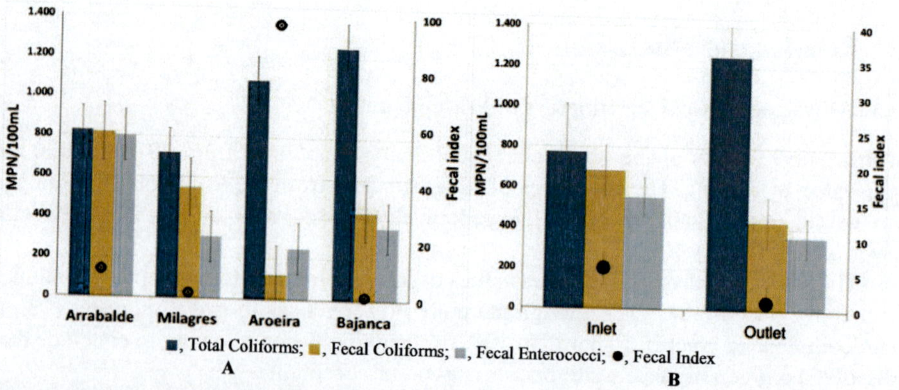

Fig. 2. Total Coliforms, Fecal Coliforms, Fecal Enterococci and Fecal Index (average counts and standard deviation) per sampling site (A) and at inlets and outlets of the LVID (B).

The FC counts in all samples are above 100MPN/100 mL, therefore not fulfilling the quality environmental objectives established for irrigation water [10]. The higher average

FC counts were registered in the inlet samples, namely in those collected at Arrabalde (816.5 MPN/100 mL) and Milagres (547.7 MPN/100 mL). These high FC values of irrigation water (Lis river water at Arrabalde weir) is explained by contamination from external sewage sources from irrigated area [13]. Nevertheless, the FC counts in the outlet water were 1.14 times lower than that in the average inlet water. This might be explained by the soil effect on the reduction of *Enterobacteriaceae* numbers [14] and also due to the type of crops grown at the outflow of the valley (pastures), which soils are not manure-amended, contrary to those with other crops in upstream areas of the valley. Considering aquaculture and bathing uses, the FC counts of LVID effluent water are above the maximum recommended value (100 MPN/100 mL), but below the maximum admissible value (2,000 MPN/100 mL) [10].

The quality standard set for irrigation water [15] refers to quality requirements for the reuse of urban wastewater treated in irrigation of agricultural, forestry, ornamental, nursery, lawn, and other green spaces. Relative to the microbiological quality requirements, that standard [15] has taken into consideration the use of irrigated crops (e.g. human or animal consumption, or industrial crops) and the respective irrigation method applied, less restrictive and more flexible than it is set in the Portuguese legislation. [10]. According to Monte and Albuquerque [16], the Portuguese legislation [10] is very demanding, since the water in most rivers does not have an FC content below 100 MPN/100 mL.

Regarding the FE (Fig. 1), the higher average counts were recorded at the valley inlet, specifically at Arrabalde weir (800.9 MPN/100 mL), with a similar explanation given to FC [13]. Although, the FE counts in the outlet water were 1.26 times lower than those at the inlets, they are above the maximum recommended value (100 MPN/100 mL), considering aquaculture and bathing uses of LVID effluent water.

The analysis of the Fecal Index proposed by Pepper and co-authors [17] (Table 2) reveals that at the initial section of Lis River (Arrabalde) and Milagres there is evidence of fecal contamination of human origin, thus entering the LVID. Differently, the water sampled at the final sections of Lis river evidence fecal contamination of animal waste in mixed pollution, possibly related to the use of animal manure as a soil fertilizer.

Table 2. Source of pollution according to the Fecal Index (FC/FE).

FC/FE	Source of pollution
>4.0	Human pollution
2.0–4.0	Human waste in mixed pollution
0.7–2.0	Animal waste in mixed pollution
<0.7	Wild animal pollution

FC, Fecal Coliforms; FE, Fecal Enterococci;
Source: [17].

3.3 Enumeration of Ampicillin-Resistant Bacteria

The mean counts of *Enterobacteriaceae* isolates resistant to ampicillin showed high variability throughout the irrigation season in Milagres and Bajanca sites (Fig. 3A), with particularly high enumerations in 2019 samples. The water samples with higher mean counts were collected in Milagres river (seven orders of magnitude higher than those from Aroeira and Arrabalde134). The average counts of the inputs are two orders of magnitude higher than the output (Fig. 3B).

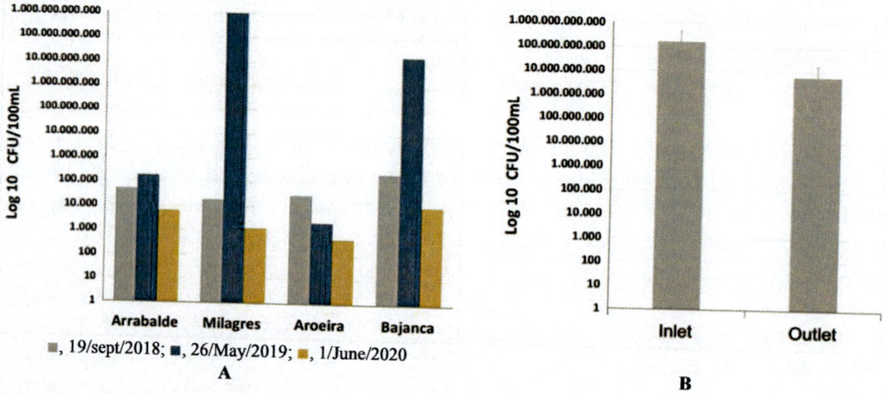

Fig. 3. Average Amp^r *Enterobacteriaceae* per sampling site and sampling dates (A) and at the LVID inlet and outlet (B).

The frequent resistance to antibiotics found in the aquatic systems, namely to the β-lactams (ex. ampicillin), is an additional critical point associated with faecal contaminants, placing the water quality and public health in a critical situation. The identification of situations of microbiological contamination risk in irrigation water of LVID, whose main responsibility is external to the Lis valley, requires special precautionary measures, regarding the safety of farmers and consumers. These risks will also need to be assessed for the real influence of agricultural activity within the irrigation district, particularly at the drainage network level.

4 Final Considerations

The analysis performed to the physicochemical and microbiological parameters of water samples reveals that a significant source of the problems identified in the water is external to the irrigation district.

The agricultural activities in the LVID seems to have a positive impact in the quality of the outlet water, despite the poor quality of the water resources reaching the district and used for irrigation (the counts of faecal coliforms and ampicillin resistant bacteria were reduced 1.14 times and two orders of magnitude, respectively). The monitoring of soil and water matrixes play an important role supporting decisions in the most severely affected sectors.

The WUA and farmers should adopt the operating mode and technologies to minimize the negative impacts of the low-quality irrigation water, which monitoring within the district is crucial to allow effective adjustments [8, 19, 20]. The organic load of livestock applied to the soil, in certain areas of the district, makes the microbiological risk possibly an emerging issue, due to the presence of antibiotic resistance genes [21]. The downstream part of the district has hydrogeological characteristics of significant salinity which is a risk factor for soil due to capillary rise or drainage water reuse [22].

The poor quality of the fresh water resources available for irrigation leads to multiple risk problems, requiring mitigation measures, namely: i) Prudent use of manure as soil fertilizer, which should be appropriately composted [23], mainly when applied to horticultural crops, to protect environmental and public health problems due to microorganisms and other contaminants, like antibiotics [21]; and ii) Choice of irrigation technology, which should be adjusted according to the risk assessment for the farmer and the consumer of the product. For fresh crops, preference should be given to drip irrigation; surface furrow irrigation has no special restrictions for other crops and sprinkler irrigation is preferentially dedicated to foraging crops or crops with post-harvest processing. Farmers should be informed about special hygiene and safety precautions where microbiological contamination is most at risk [24].

Acknowledgments. Project "Lis Valley Water Management Operational Group" ("PDR2020–1.0.1-FEADER-030911 – Grupo Operacional para a gestão da água no Vale do Lis"), funded by IFAP/PDR2020, Measure Innovation, Intervention Operational Groups.

References

1. Silva, F.G.: The future of irrigation in an integrated logic of water resources management. AGROTEC, vol. 27, pp. 46–49 (2018). (Portuguese)
2. Sordo-Ward, A., Granados, A., Iglesias, A., Garrote, L., Bejarano, M.D.: Adaptation effort and performance of water management strategies to face climate change. Impacts in six representative basins of Southern Europe. Water **11**(5), 1078 (2019). https://doi.org/10.3390/w11051078.
3. Lenton, R.: Irrigation in the twenty-first century: Reflections on science, policy, and society. Irrig. Drainage **63**(2), 154–157 (2014). https://doi.org/10.1002/ird.1844
4. RRN (Rede Rural Nacional): Grupo Operacional para a Gestão da Água no Vale do Lis, https://inovacao.rederural.gov.pt/2/81-grupo-operacional-para-a-gestao-da-agua-no-vale-do-lis. Accessed 01 Feb 2021
5. Gonçalves, J.M., Ferreira, S., Nunes, M., Eugénio, R., Amador, P., Filipe, O., Duarte, I.M., Teixeira, M., Vasconcelos, T., Oliveira, F., Gonçalves, M.: Developing irrigation management at district scale based on water monitoring: study on Lis Valley, Portugal. AgriEng. **2**(1), 78–95 (2020). https://doi.org/10.3390/agriengineering2010006.
6. DGADR. https://www.dgadr.gov.pt/. Accessed 01 Feb 2021
7. Google Maps. https://www.google.com/maps/. Accessed 01 Feb 2021
8. Lothrop, N., Bright, K.R., Sexton, J., Pearce-Walker, J., Reynolds, K.A., Verhougstraete, M.P.: Optimal strategies for monitoring irrigation water quality. Agric. Water Manage. **199**, 86–92 (2018). https://doi.org/10.1016/j.agwat.2017.12.018
9. ASTM D4327–17. Standard Test Method for Anions in Water by Chemically Suppressed Ion Chromatography. West Conshohocken, PA: ASTM (2017)

10. Decree-Law 306/2007, of 2007–08–27. Establishes Water Quality Standards for human consumption. *Republic Diary*, 164/2007, Series I, 5747–5765, Ministry of the Environment, Territorial Planning and Regional Development, Portuguese Presidency of the Council of Ministers, Lisbon, Portugal (2007). https://data.dre.pt/eli/dec-lei/306/2007/08/27/p/dre/pt/html. (In Portuguese)
11. Amador, P., Ruben, M.F., Prudêncio, M.C., Barreto, M.P., Duarte, I.M.: Antibiotic resistance in wastewater: occurrence and fate of *Enterobacteriaceae* producers of class A and class C β-lactamases. J. Environ. Sci. Health, Part A: Toxic/Hazard. Subst. Environ. Eng. **50**(1), 26–39 (2015). https://doi.org/10.1080/10934529.2015.964602
12. Shukla, A., Shukla, S., Hodges, A.W.: Recovering nitrogen from farm-scale drainage: Mechanism and economics. Trans. ASABE **61**(1), 197–206 (2018). https://doi.org/10.13031/trans.12277
13. Vieira, J., Fonseca, A., Vilar, V.J., Boaventura, R.A., Botelho, C.M.: Water quality in Lis river, Portugal. Environ. Monitor. Assess. **184**(12), 7125–7140 (2012). https://doi.org/10.1007/s10661-011-2485-9
14. Tapias, J.C.: Estratégia de gestion de los recursos hídricos en los campos de golf. Tesis Doctoral. Universitat de Barcelona, Spain, 432 p. (1997). https://dialnet.unirioja.es/servlet/tesis?codigo=238410
15. IPQ (Instituto Português da Qualidade). Portuguese Standard on Urban Wastewater Reuse Treated in Irrigation; NP 4434; IPQ, Lisbon, Portugal (2006). (In Portuguese)
16. Monte, H., Albuquerque, A.: Wastewater Reuse; Guias Técnicos Series; ISEL-ERSAR: Lisbon, Portugal (2010). (Portuguese). https://hdl.handle.net/10400.6/1144
17. Pepper, I.L., Gerba, C.P., Gentry, T.J., Maier, R.M. (eds.): Environmental Microbiology. Academic Press (2011)
18. Böger, B., Surek, M., Vilhena, R.O., Fachi, M.M., Junkert, A.M., Santos, J.M.M.F., Domingos, E.L., Cobre, A.F., Momade, D.R., Pontarolo, R.: Occurrence of antibiotics and antibiotic-resistant bacteria in subtropical urban rivers in Brazil. J. Hazard. Mater. **402**, 123448 (2021). https://doi.org/10.1016/j.jhazmat.2020.123448
19. Stine, S.W., Song, I., Choi, C.Y., Gerba, C.P.: Application of microbial risk assessment to the development of standards for enteric pathogens in water used to irrigate fresh produce. J. Food Prot. **68**(5), 913–918 (2005). https://doi.org/10.4315/0362-028X-68.5.913
20. Winfield, M.D., Groisman, E.A.: Role of nonhost environments in the lifestyles of *Salmonella* and *Escherichia coli*. Appl. Environ. Microbiol. **69**(7), 3687–3694 (2003). https://doi.org/10.1128/AEM.69.7.3687
21. Amador, P., Fernandes, R., Prudêncio, C., Duarte, I.: Prevalence of antibiotic resistance genes in multidrug-resistant *Enterobacteriaceae* on Portuguese livestock manure. Antibiotics **8**(1), 23 (2019). https://doi.org/10.3390/antibiotics8010023
22. Hoffman, G.J., Shalhevet, J.: Controlling salinity. In: Design and Operation of Farm Irrigation Systems, 2nd edn., American Society of Agricultural and Biological Engineers, pp. 160–207 (2007)
23. Cameira, M.R., Li, R., Fangueiro, D.: Integrated modeling to assess N pollution swapping in slurry amended soils. Sci. Total Environ. **713**, 136596 (2020). https://doi.org/10.1016/j.scitotenv.2020.136596
24. Ricart, S., Rico, A.M., Ribas, A.: Risk-yuck factor nexus in reclaimed wastewater for irrigation: Comparing farmers' attitudes and public perception. Water **11**(2), 187 (2019). https://doi.org/10.3390/w11020187

Monitoring a Gravity-Flow Conveyance System to Improve Water Management. Lis Valley Irrigation District Case Study

Manuel Nunes[1], Susana Ferreira[1], Rui Eugénio[2], Henrique Damásio[2], and José M. Gonçalves[1](✉)

[1] Instituto Politécnico de Coimbra, Escola Superior Agrária de Coimbra, Coimbra, Portugal
mnunes@esac.pt

[2] Associação de Regantes e Beneficiários do Vale do Lis, Leiria, Portugal

Abstract. This paper presents a study of the monitorization of the conveyance system of the collective water network in the framework of the Lis Valley Water Management Operational Group, aiming the improvement of water management by the Water Users' Association. The methodology considered the measurement of water allocation to main canals, its distribution throughout the conveyance system, and the on-farm irrigation demand. This paper presents the results of Canal 2, for 2019 and 2020 irrigation seasons. This canal supplies an area of 410 ha, is gravity-fed from Lis river, on the Arrabalde weir, in Leiria, and has a pumping recharge during the irrigation peak period. A water balance of supply and irrigation demand was carried out at 10-day period, to evaluate the performance of the distribution water management. The analyzed system reveals the specificity of the drainage water reuse, which leads to a more flexible management of water allocated by gravity-fed, and rational of pumping supply, in order to adjust it to irrigation demand, through energy consumption. Results point to actions that prioritize consolidating the improvement management of water resources, like the implementation of optimal operational plans to adjust the water demand with distribution, accompanied with the improvement of the on-farm irrigation systems.

Keywords: Hydraulic monitoring · Water demand · Canal distribution · Water Users' Association · Lis Valley

1 Introduction

Conveyance canals distribution in irrigation districts have specific operative characteristics. According to the farmers' calendars of irrigation, which aim to achieve equitable water partition, the management becomes complex when water flow is manually controlled. The water management in irrigation districts without upstream reservoirs is based in water running availability, according to inlet flows, which requires high distribution flexibility on the collective conveyance. In this case, water reuse from the drainage ditches allows mitigating the water shortage. Under such conditions, these

J. R. da Costa Sanches Galvão et al. (Eds.): *Proceedings of the 1st International Conference on Water Energy Food and Sustainability (ICoWEFS 2021)*, pp. 821–831, 2021.
https://doi.org/10.1007/978-3-030-75315-3_87

systems require a highly interactive management approach between the water users association and the farmer-beneficiaries themselves [1]. The rational management of collective irrigation should be based on an information system, with data from field monitoring of irrigation water volumes demanded and supplied, energy consumed on pumping, and costs of infrastructure maintenance. Several examples of models and procedures have been developed and applied: Playán et al. [2] developed a database tool for enhancing irrigation district management; and Mateos et al. [3] presented a scheme irrigation management information system.

The irrigation system of Lis Valley does not have an upstream reservoir, guaranteeing the water supply during the irrigation peak period. The water available in the summer is from river basic flow runoff, being its water managed by weirs, which allow a temporary reservoir and level control for gravity-fed and pumping irrigation. The water reuse is an endemic process, where the drainage water returns back to the river, recharging its flow downstream, which can be used again for irrigation. This water reuse through the irrigation pumping, from drainage ditches, allows mitigating the water shortage. In this context, the usual meaning of "water losses" is not valid, since the main criteria for managing water distribution is not to minimise the water losses, but instead, to optimise the equity of spatial water distribution, guaranteeing that all farmers have the conditions to avoid water stress at plot level.

This paper presents a monitoring study of "Canal 2" supply of the Lis Valley Irrigation District, on 2019 and 2020 irrigation seasons, carried out on the framework of an Operational Group of EIP-AGRI [4]. This study considered the measurement of gravity-fed and pumping continuous discharges, the off-farm distribution, the on-farm irrigation demand, and the assessment of demand-supply balance.

2 Materials and Methods

2.1 Study Area

Lis Valley Irrigation District (LVID) is a public irrigation district managed by a Water Users' Association (WUA), and located in the Coastal Center of Portugal [5]. The total area is about 2000 ha, and the main crops are forage corn, forage grass, horticultural, orchards, and rice. The hydraulic infrastructures have the objectives of perimeter drainage defense through slope collectors and valley ditches, the irrigation water supply through a canal conveyance system, and the field drainage based on a ditch network. Water is supplied by an open-channel conveyance network from weirs installed along the Lis river and tributaries, having the primary network with a length of about 44.5 km, and by pumping from drainage ditches.

The water conveyance and distribution for the irrigation district is subdivided into supply sectors, comprised by a main canal, gravity-fed by a river diversion from a weir. During the irrigation peak period, the downstream areas of some supply sectors are not fully supplied, so recharging water from the river or drainage ditches is pumped to reinforce canals flows (Fig. 1). A main canal and a set of small lined or earthen secondary channels to distribute the water to the field hydrants compose each sector. The supply of secondary channels is ensured by duckbill structures to control water level in the main canal (Fig. 1a).

Despite the monitoring work has been applied to the entire irrigation district, Canal 2 was selected to be presented, corresponding to the sectors C2A and C2B (Fig. 2). The Canal 2 is gravity-fed, upstream, on the Arrabalde weir of Lis river, in Leiria city, a section with a limited flow. Downstream Arrabalde there are two tributaries, Lena and Milagres rivers. Canal 2 flow is recharged at the Necessidades Pumping Station (here referred by Necessidades), located 7 km downstream the Arrabalde weir, corresponding to the supply of C2B sector (Fig. 2b,c and Table 1).

Fig. 1. Conveyance system infrastructures: (a) duckbill water level controller, of Canal 2; (b) river weir on Necessidades pumping station; (c) canal reinforcement by pumping at Necessidades.

Table 1. Canal 2 and sectors C2A and C2B characteristics.

Sector	Area (ha)	Canal length (km)	Max. canal heigth (cm)	Max. canal discharge (L/s)	Max. pumping discharge (L/s)
C2A	195.5	4.8	0.75	380	–
C2B	214.7	4.4	0.70	350	150

2.2 Water Supply

The methodology for monitoring the water supply system for irrigation was based on a network of observations to assess the flow rates at strategic points. Flow measurements in canals and pumping stations followed methodology previously published [6], including the observations of operative daily management practices.

For the evaluation of the canals' flows, the canal section velocity method was used. The measurements were performed with the TELEDYNE ISCO 2150 Area Velocity Flow Module and Sensor, which obtains the flow velocity based on the emission of ultrasonic waves together with the Doppler effect (Fig. 3a). In order to obtain a continuous flow measurement, water level sensors were installed in the studied sections of the canals. These sensors record the total pressure, obtaining the respective height of the water by the difference with the value of the atmospheric pressure, measured simultaneously. Based on flow curves calibrated for each section (with the flowmeter), continuous hydrographs were obtained corresponding to water levels measurement, allowing to know the daily variation of the flow level in the canal. These field measurements followed the procedures presented by Replogle et al. [7]. The discharge measurement on the upstream section of Canal 2 was made continuously by the ISCO 2150 sensor, located on a buried pipe component that supplies the open main canal (Fig. 3b).

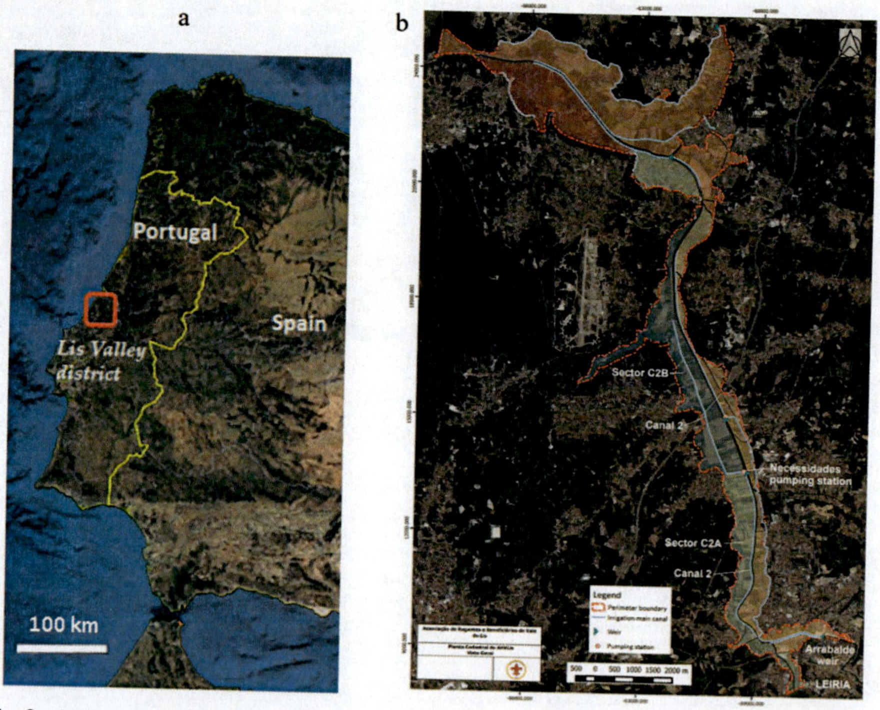

Fig. 2. (a) Location of the Lis valley irrigation district in Portugal (red square) (source: Google Maps, https://maps.google.pt); (b) Location of Canal 2 and Sectors C2A and C2B, in Lis Valley Irrigation District (source: Lis Valley Irrigation District Water Users' Association).

2.3 Irrigation Water Demand

Water demand during the irrigation season was obtained based in meteorological data (reference evapotranspiration, ET_0, and precipitation, P) from three stations positioned along the valley. The analysis was applied between May and October, with a 10-day (decade) time step. The crop evapotranspiration (ET_c) of each sector, corresponding to the water used by the crop consumptively, was calculated applying the crop coefficient (Kc) to the reference evapotranspiration ($ET_c = ET_0 \times Kc$). To obtain a global crop coefficient to each sector, the different crop occupation was recognized based on WUA records, allowing a weighted crop coefficient. The evolution of Kc along crop growth stages was considered according to the usual sowing dates practiced [8]. The net irrigation demand (NID) was calculated by the difference between ET_c and precipitation ($ET_c - P$), assuming that the initial soil water and the groundwater capillary rise contributions were not significant.

(a) (b)

Fig. 3. Flow measurements in canals and pumping stations with ultrasonic flow meter: (a) used as portable device on an open canal section; (b) used fixed on a buried pipe.

3 Results

3.1 Water Supply

The Canal 2 water supply to C2A and C2B sectors in the mid-season months (July and August, of 2019 and 2020) is presented in Fig. 4. Hydrographs of upstream section (C2A inflow) evidence a responsive operation of the water uptake in the weir with the meteorological conditions.

A day/night oscillation of the inflow was observed throughout the whole period, being the night records about 30 to 50% of daytime ones. The main gate of Canal 2 in the Arrabalde weir was programmed to take into account this reduction in the use of on-farm irrigation water at night. Day of higher evaporative demand, corresponded to a higher distribution to on-farm irrigation, and, in consequence, an higher difference between upstream and downstream discharges (e.g. 11st July 2019, or 17th July 2020). On the other hand, on rainy days, this difference was reduced (e.g. 21st July 2019, or 20th August 2020). During rainy days, the response of allocation reduction is unclear; the summer rains are usually occasional, not spatially uniform and of small depth. Therefore, for precaution, it is normal to maintain the collective distribution, and the irrigation adjustments carried out by farmers, according to specific on-farm requirements.

The supply discharge data of C2B sector, including the Canal 2 discharge upstream Necessidades pumping station and the reinforcement of this pumping, are presented in Fig. 5. The pumping period is comprised from early July to mid-September. Necessidades station is composed by two pumps, each one with 75 L/s of discharge capacity. Usually pumping period starts at 8:00 am and ends at 6:00 pm. At days with higher demand, both pumps may work together for a few hours, and when the demand is reduced, the daily pumping period decreases up to zero. The monitoring of the distribution of the water to each one of the irrigated parcels is very complex, since it involves the commitment of a high number of local farmers that do not have records of irrigation time and discharges. Even those who choose to pump from drainage ditches to facilitate pressurized irrigation systems do not have those record.

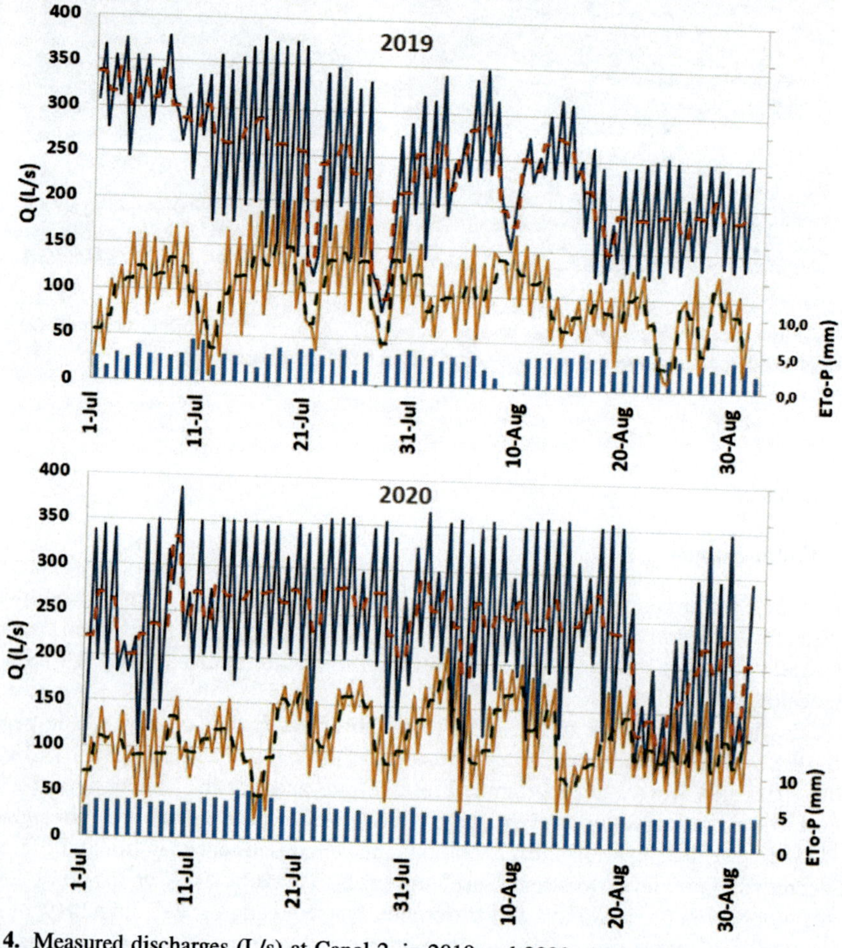

Fig. 4. Measured discharges (L/s) at Canal 2, in 2019 and 2020: C2A supply: —continuous, —daily average; C2B supply: —continuous, —daily average; —ET_0 - P (mm).

The real-time operation of the collective distribution system at sector level, depends of on-farm irrigation demand, related with the crop development, soil and climatic conditions, and the field irrigation system. This operation is based on a strong cooperation between the WUA and the farmers, to get the best water distribution to the several field parcels of the same branch canal. In practice, adjustments are made in real time, with both dynamic adjustments between supply and demand and distribution of water being based on the empirical knowledge of off- and on-farm water losses, and the cooperation between farmers on the same canal branch to share the water available.

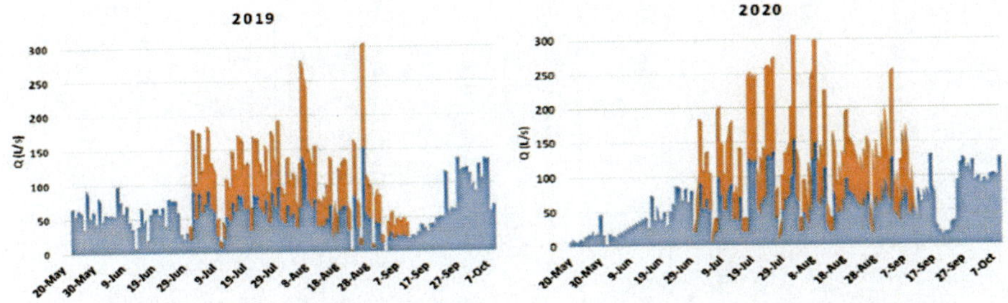

Fig. 5. Supply discharge of C2B sector, on 2019 and 2020: ■gravity-fed discharge from upstream component of Canal 2 (not used on C2A sector); ■pumping by Necessidades.

3.2 Irrigation Water Demand

The average C2A and C2B 10-days crop coefficients are present in Table 2. The crop pattern followed the data presented by Ferreira et al. [9]. Sowing and harvesting schedule of annual crops was assigned considering the common practices in the area. Crop coefficients were estimated considering maximum production, without water stress. Table 3 presents the 10-days meteorological data and crop evapotranspiration for 2019 and 2020. There is a contrast in the meteorological conditions of these two years, leading to a maximum water demand period in August 2019, whereas in 2020 it occured in July. Crop water demand was assessed by the Net Irrigation Demand NID = ET_C–P, a simplified approach, neglecting the capillary rise contribution and the initial soil water storage use. When NID had negative values, it means that in the 10-days period, precipitation exceeded the ET_C. However, a daily analysis in the same period, might reveal the need to irrigate some crops, because the precipitation occours occasionally.

3.3 Sector Water Balance

Comparing the supply and the demand data on the C2A sector, for 10-days periods, of the irrigation seasons of 2019 and 2020 (Fig. 6), it is concluded that the supply significantly exceeded the demand, in many periods. There are several factors that explain this over-supplying practice. On the one hand, being a gravity-fed allocation, this does not entail any additional cost for WUA. On the other hand, being the conservation status of the canal in certain places deficient, leading to uncontrollable water losses to the drainage network. This water that apparently is lost to the drainage network, could be reused during the periods of greatest water shortage, especially when using pressure irrigation systems, pumping from the drainage ditches. Therefore, this recharge of the drainage network is largely a beneficial use of water. In addition, these drainage ditches flow into

Table 2. Crop coefficients of C2A and C2B sectors (10-days period)

Crop	Area (%)		May			June			July			August			September		
	C2A	C2B	1	2	3	1	2	3	1	2	3	1	2	3	1	2	3
Corn	43	33	0.30	0.30	0.40	0.50	0.60	0.70	0.90	1.00	1.00	1.00	1.00	1.00	0.90	0.80	0.60
Pasture	7	24	0.90	0.90	0.90	0.90	0.90	0.90	0.90	0.90	0.90	0.90	0.90	0.90	0.90	0.90	0.90
Vegetab	14	11	0.80	0.80	0.80	0.90	0.90	0.90	0.90	0.90	0.90	0.90	0.90	0.90	0.90	0.90	0.90
Rice	0	15	1.00	1.00	1.00	1.00	1.00	1.00	1.00	1.00	1.00	1.00	1.00	1.00	1.00	1.00	1.00
Fruits	36	17	0.40	0.50	0.60	0.70	0.80	0.80	0.80	0.80	0.80	0.80	0.80	0.80	0.80	0.80	0.70
Kc (C2A)			0.43	0.47	0.58	0.69	0.75	0.79	0.88	0.92	0.92	0.92	0.92	0.92	0.86	0.81	0.68
Kc (C2B)			0.61	0.63	0.71	0.78	0.81	0.85	0.92	0.95	0.95	0.95	0.95	0.95	0.90	0.85	0.77

Table 3. Meteorological data and net irrigation demand per sector C2A and C2B, for the 10-days period of 2019 and 2020

		May			June			July			August			September		
		1	2	3	1	2	3	1	2	3	1	2	3	1	2	3
		2019														
ET_0		35.2	45.1	54.2	41.6	38.8	35.9	36.9	36.9	45.4	36.6	42.0	42.1	39.2	32.7	25.8
P		12.0	0.6	1.0	12.2	3.4	6.0	0.0	0.6	6.2	11.6	0.0	0.0	0.0	7.0	21.8
NID	C2A	3.1	20.6	30.5	16.5	25.7	22.4	32.5	33.4	35.5	22.1	38.7	38.8	33.7	19.5	−4.3
	C2B	9.5	27.8	37.5	20.2	28.0	24.5	33.9	34.5	36.9	23.2	39.9	40.0	35.3	20.8	−1.9
2020																
ET_0		18.6	22.3	22.0	40.2	41.4	40.4	49.3	56.7	51.0	39.1	35.7	47.0	42.3	31.5	25.6
P		21.8	12.8	15.2	0.2	4.2	0	0	0	0.4	0	0.2	20.4	0	43.6	2.8
NID	C2A	−13.8	−2.3	−2.4	27.5	26.9	31.9	43.4	52.2	46.5	36.0	32.7	22.8	36.4	−18.1	14.6
	C2B	−10.5	1.3	0.5	31.2	29.4	34.3	45.4	53.9	48.1	37.2	33.7	24.2	38.1	−16.8	16.9

ET_0, Reference evapotranspiration (mm); P, Precipitation (mm); NID, Net irrigation demand (mm).

the River Lis upstream from the Salgadas weir, being reused for irrigation. These factors explain the differences observed between supply and demand.

Comparing the corresponding supply and the demand data on the C2B sector (Fig. 7), it is concluded that the supply matches relatively well with the on-farm demand. Contrarely to the sector C2A, the pumping supply is operationally well adjusted to the irrigation demand, because this allocation has a high cost charge for WUA. Effectively, the pumping has an important role in water supply, particularly at daytime.

4 Final Considerations

The good practice of water management in an irrigation district should be based on the knowledge of water supply, its off-farm distribution and use at on-farm level. Therefore, monitoring plays a decisive role in improving water management, as it allows the identification of problems in the infrastructure, in the operational practices, and, especially, in the critical situations during the irrigation peak period.

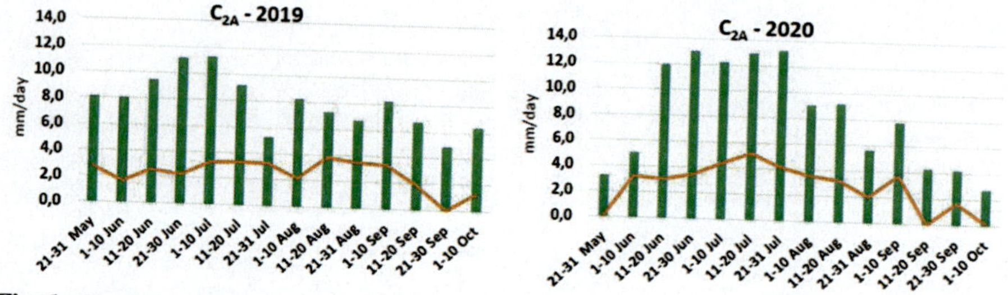

Fig. 6. Gravity-fed supply (▬, mm/day) and net irrigation demand (—, mm/day) of sector C2A, for 2019 and 2020 irrigation seasons.

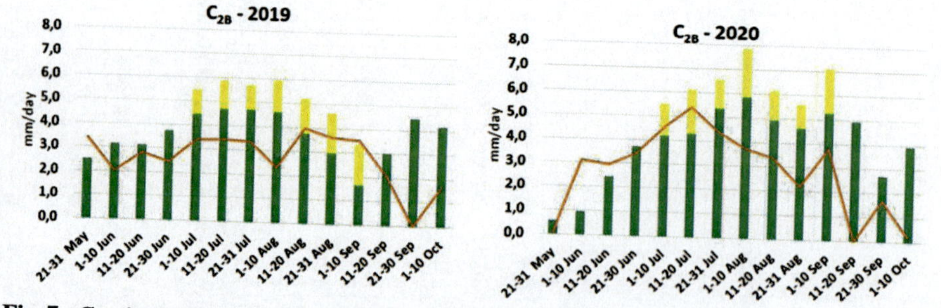

Fig. 7. Gravity-fed supply (▬, mm/day), pumping recharge (▬, mm/day) and net irrigation demand (—, mm/day) of sector C2B, for 2019 and 2020 irrigation seasons.

Lis river has no regularization by reservoirs, and the flow suffers a significant decrease in the summer period. Given this reality, irrigation water management is based on quasi-real-time supply adjustment in the very short time of a few hours or a few days. It takes into account the relative independence of the supply of several sectors, allowing higher flexibility and cooperation among the group of users at secondary canal level, facilitating the management and favor the equity of water distribution. The supply water scarcity risk and the deficiency of the conveyance system lead to the need of recharge pump stations from drainage water. So, the drainage system also works complementary with the irrigation one, like a off-field water reservoir, allowing the pumping directly from the ditches to the farmers' private pressurized systems. The monitoring of this system provides information and knowledge leading to a more flexible management of water allocated by gravity-fed and rational of pumping supply, to adjust it to irrigation demand through energy consumption. These results point to guidelines to consolidate improved water management: better maintenance and conservation of hydraulic infrastructures to reduce water losses and better flow control; implementation of optimal operational plans, to adjust the water demand with distribution; improvement of the on-farm systems with better water application control and maintenance procedures, reducing labor and increasing the distribution uniformity, applying irrigation scheduling plans based on monitoring systems; and improvement of water quality control on the water reuse from drainage ditches.

Acknowledgments. Project "Lis Valley Water Management Operational Group" ("PDR2020-1.0.1-FEADER-030911 – Grupo Operacional para a gestão da água no Vale do Lis"), funded by IFAP/PDR2020, Measure Innovation, Intervention Operational Groups.

References

1. Playán, E., Sagardoy, J.A., Castillo, R.: Irrigation governance in developing countries: current problems and solutions. Water **10**, 1118 (2018)
2. Playán, E., Cavero, J., Mantero, I., Salvador, R., Lecina, S., et al.: A database program for enhancing irrigation district management in the Ebro valley (Spain). Agric. Water Manage. **87**, 209–216 (2007)
3. Mateos, L., López-Cortijo, I., Sagardoy, J.A.: SIMIS: the FAO decision support system for irrigation scheme management. Agric. Water Manage. **56**, 193–206 (2002)
4. RRN: Grupo Operacional para a gestão da água no Vale do Lis. https://inovacao.rederural.gov.pt/2/81-grupo-operacional-para-a-gestao-da-agua-no-vale-do-lis. Accessed 01 Feb 2021
5. Gonçalves, J.M., Ferreira, S., Nunes, M., Eugénio, R., Amador, P., Filipe, O., Duarte, I.M., Teixeira, M., Vasconcelos, T., Oliveira, F., Gonçalves, M.: Developing irrigation management at district scale based on water monitoring: study on Lis valley, Portugal. AgriEngineering **2**(1), 78–95 (2020)
6. Replogle, J.A., Kruse, E.G.: Delivery and distribution systems. In: Hoffman, G.J., Evans, R.G., Jensen, M.E., Martin, D.L., Elliot, R.L. (eds.) Design and Operation of Farm Irrigation Systems, ASABE: St. Joseph, MI, USA, 2nd edn., pp. 347–391 (2007)
7. Replogle, J.A., Howell, T.A., Solomon, K.H.: Measuring irrigation WATER. In Hoffman, J., Howell, T.A., Solomon, K.H., (eds.) Management of Farm Irrigation Systems, ASABE: St. Joseph, MI, USA, pp. 315–370 (1990)
8. Allen, R.G., Pereira, L.S., Raes, D., Smith, M.: Crop Evapotranspiration. Guidelines for Computing Crop Water Requirements; FAO Irrig. and Drain. Paper 56, Rome (1998)
9. Ferreira, S., Oliveira, F., Silva, F.G., Teixeira, M., Gonçalves, M., Eugénio, R., Damásio, H., Gonçalves, J.M.: Assessment of factors constraining organic farming expansion in Lis valley. AgriEng. **3**, 111–127 (2020)

The Application of Multiple Correspondence Analysis Method to Irrigation Water Quality in Tejo Alluvial River Plain

D. Ferreira[1](✉), M. Simões[1], F. Reboredo[1], F. Pessoa[1], A. Almeida[2], M. R. Carvalho[3], A. Coelho[1], C. Pessoa[1], A. Marques[1], I. Luís[1], D. Daccak[1], M. P. Mendes[4], and F. Lidon[1]

[1] Department of Earth Sciences and GeoBioTec, FCT NOVA School of Science and Technology, University NOVA de Lisboa, Lisbon, Portugal
djo.ferreira@campus.fct.unl.pt

[2] National Institute for Agrarian and Veterinary Research (INIAV), Elvas, Portugal

[3] Instituto D. Luiz, Faculdade de Ciências, Universidade de Lisboa, Ed. C1, Campo Grande, 1749-014 Lisbon, Portugal

[4] CERIS, Civil Engineering Research and Innovation for Sustainability, Instituto Superior Técnico, Universidade de Lisboa, Lisbon, Portugal

Abstract. The lower Tejo river basin is an important crop production area in Portugal. The mild climate, diverse water sources and availability along with flat and fertile land areas provide the condition to a highly productive region. Although the historically rainfed cultures installation since the eighteen century, irrigated cultures, such as rice, started to gradually occupy the region in the beginning of the twentieth century. Nowadays, rice cultivation prevails in the area, which requires reliable sources to fulfil water demand for this culture. Paddy-rice fields installation in Tejo tributaries alluvium, such as Sorraia and Almansor rivers, directly supply those parcels in the vicinity. Nevertheless, the largest production area is installed in Lezírias de Vila Franca de Xira, located in Tejo alluvium plains. Such area requires large amounts of water during crop season, from April to October, mainly withdrawn from Tejo river. The Tejo estuary proximity and its salinity influence along with demand peak during summer compromise water quality which, in a long term, can cause salinization and alkalinization of soils. Rice is a light salinity tolerant culture, which is adapted with the current water quality, but the region faces serious sustainability challenges if a shift for some other crop type occurs, as happened in the past. Multiple Correspondence Analysis was used as a statistical method to assess water quality in the region. Water management, driven with water quality allocation by crop and soils requirements can be an answer to minimize permanent soil damages achieving natural resources sustainability in the long term.

Keywords: Tejo river basin · Irrigation water · Water quality · Multiple Correspondence Analysis

J. R. da Costa Sanches Galvão et al. (Eds.): *Proceedings of the 1st International Conference on Water Energy Food and Sustainability (ICoWEFS 2021)*, pp. 832–840, 2021.
https://doi.org/10.1007/978-3-030-75315-3_88

1 Introduction

Agriculture is an important activity located in the Lower Tejo river basin. The highly productive soil along with extensive plain areas were virtues that attracted crops establishment. This area is inserted in National Agricultural Reserve, implemented in Ordinance 113/91 from February 7, 1991, and has the scope of protection of soils and environmental resources to achieve sustainability in the agricultural practices. Rainfed cereal cultures implantation began in the eighteenth century, with wheat, barley, and maize production, essentially in Tejo alluvial plains, known as Lezírias. The irrigation water availability along with favorable climate were key factors to introducing paddy-rice production in the beginning of the twentieth century in Tejo and its tributaries, Sorraia and Almansor rivers alluvium. Paddy-rice production spread out reaching 17 455 hectares in 2018, gathering 44,8% from the total Portuguese paddy-rice cultivation area [1] (Fig. 1). In Portugal, rice cultivation is a spring-summer crop. Tejo runoff is the main source of irrigation waters, especially in the Lezírias, conveyed by ditches across cultivation parcels. Nevertheless, other water sources are used, such as Sorraia and Almansor runoff, agricultural supply systems, canals, and reservoir. Since the sources are directly or indirectly surface water dependent, they are very sensitive for climate factors, such as rainfall. Precipitation variations, driven by climate instability [2], can lead to less predictive seasons, with serious harmful implications in water quantity and quality [3]. Due to the estuary proximity, beside with runoff reduction during summer months, irrigation waters from Tejo suffer often from considerable salinity, which influence irrigation water quality. Rice crops, as being a summer culture, requires large amounts of water as a thermoregulator. Paddy-rice culture tolerates light salinity in its irrigation but, the continuous salt input in soils through watering can lead to soil salinization and alkalinization phenomenon. This fact can be reflected in serious influences in land productivity and capability to accommodate other cultures aside from rice.

Recently, in the summer of 2005, 2012 and 2019, the supply of freshwater to irrigate crops was seriously compromised. The reduction of Tejo runoff along with the intense swell in Atlantic Ocean started to severely affect the quality of Tejo water, with the saline intrusion reaching unprecedent upstream areas, affecting the withdraw for irrigation. The solution to save the crops was to build a dike in Sorraia river to retain water, a very discussed decision with heavy social and environmental impact. These occurrences, which trends to be more frequently, denotes the fragility and sensitivity of this agricultural system and the need of preservation procedures. Serious impacts with soil salinization could arise in this fragile area, as already noticed [4].

The region extent as well as the amount of water quality results requires an unified solution to merge quality data and locations, in order to assess the best sources providing irrigation water according to crops or soils requirements.

Multiple Correspondence Analysis statistical method is presented in this work as an important tool to relate large amounts of data to help selecting the appropriate decisions to stablish an accurate water management, preserving the future of this area.

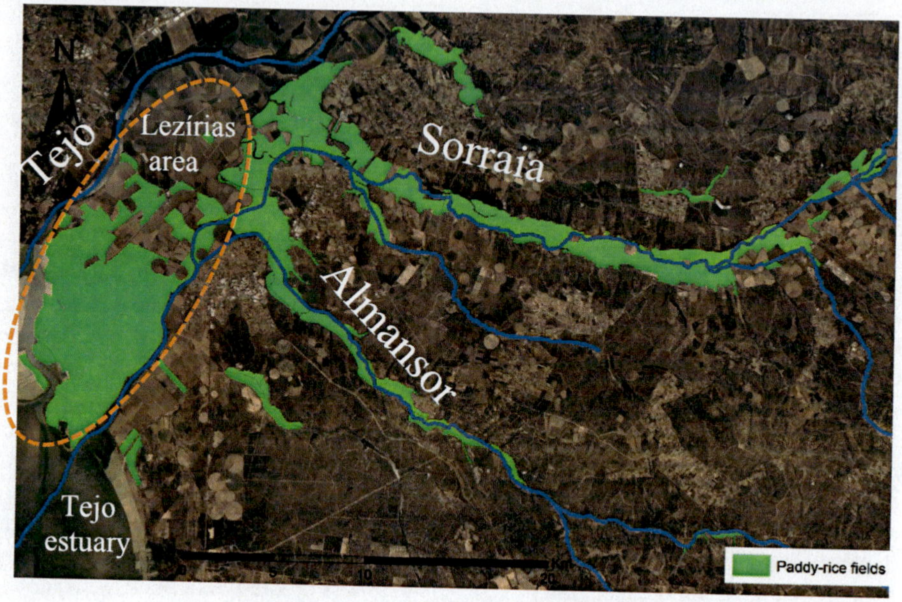

Fig. 1. Study area and paddy-rice fields.

2 Sampling and Methods

In April 2017, fifty-six water samples were collected in the area from different sources: rivers runoff (6), canal (3), reservoir (1), supply systems (2) and ditches (44) (Fig. 2). Electric conductivity (EC) and pH were measured with multiparametric electrode. Bicarbonate was determined with titration method. Sulfate (SO_4), chloride (Cl) and nitrate (NO_3) were determined by photometry. Sodium (Na), potassium (K), calcium (Ca) and magnesium (Mg) were determined by Ion chromatography. All analysis were performed in hydrogeochemical laboratory in Earth Sciences Department in School of Science and Technology, NOVA University of Lisbon. Descriptive statistics of the samples are represented in Table 1. Inside the domain of multivariate statistics, factorial analysis methods allow to summarize big sets of data to reveal relations between variables. The most common multivariate statistical method is Principal Component Analysis (PCA). PCA operates with continuous and quantitative variables, which revealed to be insufficient in gathering all the types of data in this study. Multiple correspondence analysis (MCA) is a multivariate descriptive statistical analysis technique applied to categorical variables, where the idea is to define how strongly and in which way they are interrelated. For using MCA, the continuous variables were coded into categories. The distance in MCA is measured using a "chisquare" (χ^2) distance, which is used to estimate the total inertia (i.e. the measure of the variability contained in the data matrix) [5]. Descriptive statistics and MCA was performed with ANDAD software [6]. The quantitative variables are categorized with indexes from 1 to 3, depending on if they represent the lowest (1), intermedium (2) or highest (3) values. The categories were stablished considering equal frequency discretization.

Table 1. Samples descriptive statistics.

Variable		Mean	Minimum	Maximum	Std. Deviation
pH		7.52	6.22	8.80	0.57
EC	μS cm^{-1}	2146.47	170.00	22400.00	4931.15
HCO_3^-	mg L^{-1}	146.42	52.46	554.50	93.21
Cl^-	mg L^{-1}	765.46	19.60	9000.00	1963.23
SO_4^{2-}	mg L^{-1}	283.57	19.70	2715.00	574.80
Na^+	mg L^{-1}	366.37	3.00	4137.05	887.61
K^+	mg L^{-1}	11.64	0.25	133.22	21.34
Ca^{2+}	mg L^{-1}	49.05	9.37	315.62	51.40
Mg^{2+}	mg L^{-1}	49.34	1.61	545.59	116.04
NO_3^-	mg L^{-1}	5.18	2.20	17.50	6.05

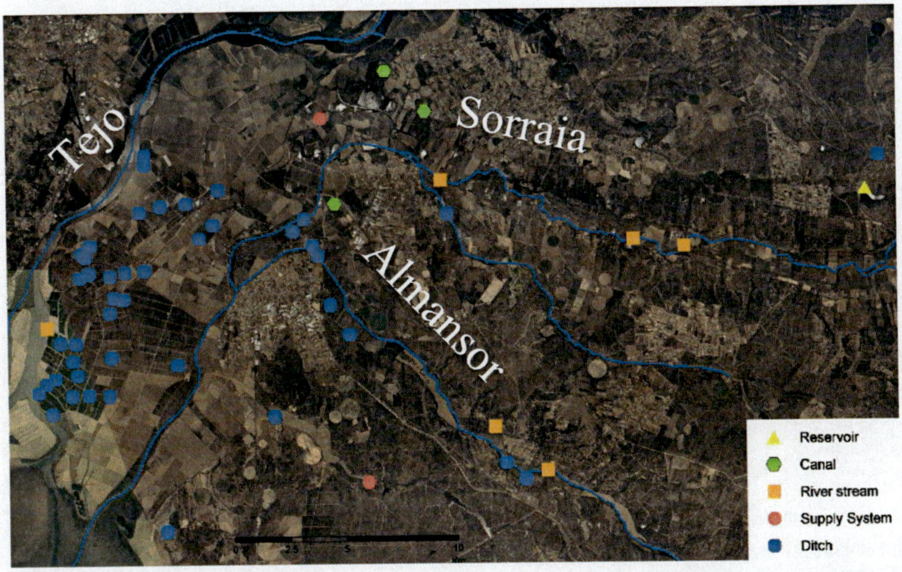

Fig. 2. Water samples location in Tejo river basin and Sorraia and Almansor tributaries.

The labels for the qualitative variables used in this study are expressed in Table 2.

Table 2. Qualitative variables indexes correspondence.

Irrigation water source	MCA label
Rivers runoff	Rive
Canal	Cnal
Reservoir	Rese
Supply Systems	SSys
Ditches	Dtch

3 Results and Discussion

The results from the application of MCA to the samples are addressed in Table 3.

Table 3. Multiple Correspondence Analysis axis eigenvalues results.

	Eigenvalue	Explanation %	Cumulative %
1	**0.36**	**58.43**	**58.43**
2	**0.16**	**25.84**	**84.27**
3	0.02	3.76	88.03
4	0.02	2.75	90.78
5	0.01	2.31	93.08

For the present study, were retained two factors, which explain altogether 84,27% of the samples variability. The factorial plan projection with F1 and F2 factorial axis are represented in Fig. 3.

The variables categories are distributed across F1/F2 factorial plan F2 (percentage inertia of 84.3%). Most same index categories are grouped. The lower index categories are essentially located in the positive F1 semi-axis, corresponding to the lower ions concentration. As opposite, in the negative F1 semi-axis, are located the highest categories of the variables, with the highest indexes, which represents the highest ions concentration. The F2 negative semi-axis describe essentially the variability of the intermedium categories. The projection of the qualitative variables, that represents the sources of irrigation

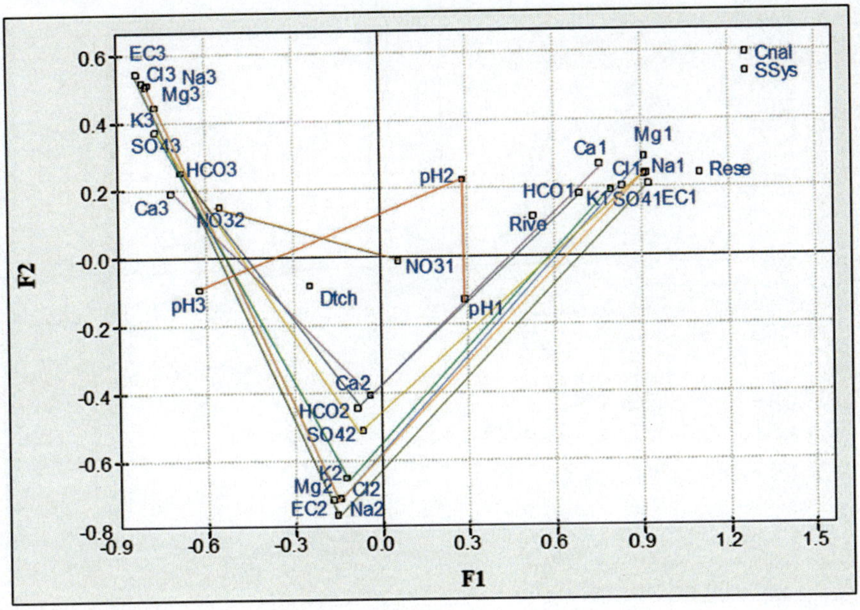

Fig. 3. Factorial plan F1/F2 with variables projections categories (percentage inertia in map: 84.3%).

waters in the region, show a clear separation concerning its quality. Focusing the positive F1 semi-axis, the variables "Cnal", "SSys", "Rese" and "Rive", means canal, supply systems, reservoir, and river streams as irrigation water provenience. Its projections indicate that are sources with low mineralization because of its vicinity with the lower variable categories, as well its position in the positive component of F1 axis. Conversely, in the F1 negative semi-axis, the variable "Dtch" is represented. This projection represents the irrigation waters that have origin in irrigation ditches and considering the proximity in the factorial plan with the clusters of the intermediate and highest categories, can be described as irrigation waters with significant mineralization, with implications in water quality.

Regarding the samples, Fig. 4 express the projection of the factorial plan F1/F2.

Samples from irrigation ditches have the highest representation. There is a clear division between samples clusters, where in the positive F1 semi-axis are represented those samples that have lower mineralization, marked as a blue circle. In the other end of F1 axis are represented the samples that have the highest mineralization, marked as an orange circle. A third cluster is assumed in F2 negative semi-axis and corresponds to the samples with intermediate mineralization.

Geographically, the clusters have expression in their locations, as can be seen in Fig. 5.

The cluster that is characterized by lower mineralization samples, located in the positive F1 and F2 quadrant, have preferential locations in the North region of Lezírias de Vila Franca de Xira and in the East of the area, essentially related with streams apart from the estuarine areas. The samples that are integrated in the cluster projected in F1

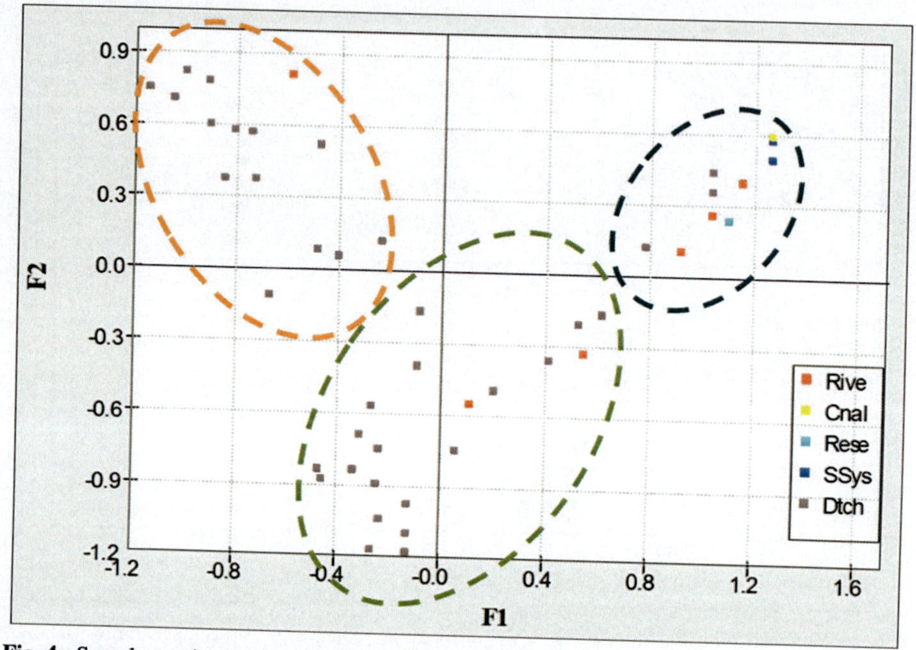

Fig. 4. Samples projections in Multiple Correspondence Analysis factorial plan F1/F2 (percentage inertia in map: 84.3%) with marked clusters as low mineralization samples (blue), intermediate mineralization samples (green) and high mineralization samples (orange).

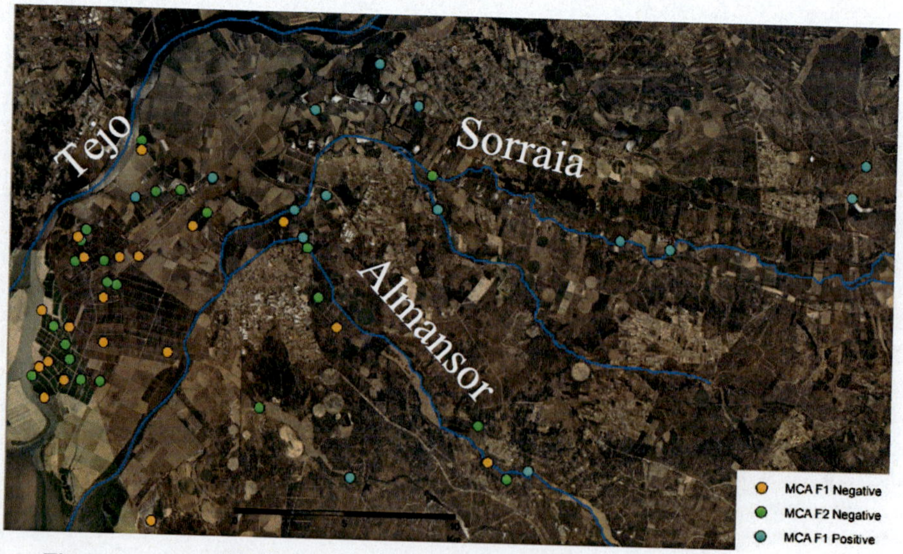

Fig. 5. Location of the samples according to their projections in factorial plan F1/F2.

negative and F2 positive quadrant, which have an intimate relation with the highest categories of the variables, which represents samples with a higher mineralization, have special distribution across West and South areas of Lezíria de Vila Franca de Xira. The projections that belong to the negative F2 semi-axis, which represents the samples with intermediate mineralization, are spread across the region in a general way, and cannot be related to some location.

4 Conclusion

Agriculture is a dynamic and essential economic activity which is highly dependent of water supply, that its availability is very sensitive to climate factors. As seen, irrigated cultures, such as rice, have a strong expression across the studied region. Rice production is, nowadays, adapted and well established to current environmental conditions.

Nevertheless, the application of factorial statistical methods to irrigation water samples was important to arise some sustainability concerns. The main production areas, Lezírias, are essentially supplied by Tejo runoff, where its estuary has a reasonable influence. The application of MCA shown that the worst quality of irrigation waters was mostly located in Lezírias area, with particular incidence in West and South region, along with ditches provenience. Conversely, the lower mineralization was observed in irrigation waters from supply systems, canal, rivers runoff and reservoir, with a strong expression in the North of the region and Tejo tributaries. Samples with intermediate mineralization, which does not pose quality concerns for now, are found in a homogeneous dispersion in the region.

The application of MCA method to irrigation waters revealed being a remarkable technique, appliable in vast areas, to help managing water sources, according to crops or environmental requirements. The water quality poses a serious concern in soil preservation, specially in the locations where high ions concentration are most likely to be found.

In the future, the installation of other crops in the area could be seriously conditioned regarding the health of the soils and irrigation capabilities from diverse sources. If in the past paddy-rice established as an irrigated culture, taking the place of rainfed cultures, climate alterations could force a shift in the type of agriculture that is practiced nowadays in the region. Agricultural decisions have environmental impact, and all the choices must be taken considering sustainability, especially in this important, highly productive, and extremely sensitive region.

Acknowledgments. This work had the support of the research grant SFRH/BD/121892/2016 from National Funds through FCT - Portuguese Foundation for Science and Technology and the financial support of GeoBioTec (UIDB/04035/2020).

References

1. Direção-Geral do Território, Carta de uso e ocupação do solo de Portugal Continental para 2018. Lisbon (2019)

2. Mendes, M.P., Cherubini, P., Plieninger, T., Ribeiro, L., Costa, A.: Climate effects on stem radial growth of Quercus suber L.: Does tree size matter? For. An Int. J. For. Res. **92**(1), 73–84 (2019)
3. Ferreira, D. et al.: Effects of climate change (precipitation variations), on rice crop yields in alluvial plains of the Tejo and Sado rivers. Emirates J. Food Agric. **30**(6) (2018)
4. Tóth, G., Adhikari, K., Varallyay, G.Y., Toth, T., Bódis, K., Stolbovoy, V.: Updated map of salt affected soils in the European Union, Luxembourg (2008)
5. Greenacre, M.: Correspondence Analysis in Practice, Second edn. Chapman and Hall/CRC, Boca Raton (2007)
6. Sousa, P., Sousa, A.J.: Sistema ANDAD. CVRM - Centro de Geosistemas. IST, Lisbon (2000)

Big Data in Smart Cities –Water Demand Prototypes Using Clustering Techniques

Pedro Miguel[1] (✉), José Gonçalves[2], and Luís Neves[3]

[1] INESC Coimbra, DEEC, Rua Sílvio Lima, Pólo II, 3030-290 Coimbra, Portugal
[2] E-REDES - Distribuição de Eletricidade, Lisboa, Portugal
[3] School of Technology and Management, Polytechnic Institute of Leiria, Leiria, Portugal

Abstract. Considering the path that electricity smart grids have taken and the management, control and optimization involved, one may think that is possible to transpose such experience and results to other infrastructures that perform poorly, e.g. the water supply infrastructure. Also, according to Pardo-Bosch & Aguado [1], the infrastructure construction, one of the biggest driving forces of the economy, requires a huge analysis and clear transparency to decide what projects may be executed with the few resources available. Additionally, losses in the water sector are still generally relevant and need to be addressed [2]. In fact, addressing the efficient use of resources is a key goal for the United Nations 2030 Agenda for Sustainable Development.

The purpose of this work is to make use of big data provided by one-year operation of a water utility company and perform an analysis, similar to the one already established for electricity smart grids, contributing to a more effective decision-making process.

The proposed methodology consists of applying clustering algorithms to historical data, namely, the hierarchical and self-organizing neural network methods, from a harvesting station to a reservoir. Such methodology allows obtaining cluster diagrams trends, which represent typical daily profiles of water flow rate and electric power consumption.

The obtained prototypes represent the most common profiles in water capture, for the selected facility. Such analysis enables the definition of a lower losses working profile (optimized profiles) and allow assessing the outcome of their combined existence.

Keywords: Smart grid · Water supply systems · Load profiling · Data clustering

1 Smart Grids Applied to Water Supply Systems

It is recognized by the scientific community that smart grids have an increasing role in the development of the global power sector.

According to [3], a possible definition for a smart grid is an advanced digital two-way power flow power system capable of self-healing, adaptive, resilient and sustainable with foresight for prediction under different uncertainties. A systematic review of the

J. R. da Costa Sanches Galvão et al. (Eds.): *Proceedings of the 1st International Conference on Water Energy Food and Sustainability (ICoWEFS 2021)*, pp. 841–850, 2021.
https://doi.org/10.1007/978-3-030-75315-3_89

environmental and economic impacts of smart grids was conducted by Moretti [4]. This author provided a cost-benefit analyses and environmental impact assessment which are used to support policy on the deployment of smart grid systems and technologies from nine articles and eight reports published between 2000 and 2015. The results show that no standardized method currently exists for assessing the economic and environmental impacts of smart grid systems. The costs varied between 0.03 and 1143 M€/yr, while the benefits ranged from 0.04 to 804 M€/year. The primary energy savings ranged from 0.03 to 0.95 MJ/kWh, whereas the GHG emission reduction ranged from 10 to 180 gCO_2/kWh, depending on the country grid mix and the system boundary of the SG system considered. The study also [4] demonstrates that several factors influence the outcome of smart grids, but it also shows its versatility and robustness.

The concept of Smart Grid unites several technologies, consumer solutions and addresses several policy and regulatory drivers, worldwide. As a research concept, this allowed for the "Smart Grid" not to have one single definition. Different definitions aroused from different institutions, such as the European Technology Platform; Smarter Grids; U.S. Department of Energy; IEC and IEEE.

The electricity sector always had a concern in controlling power losses which are an inevitable consequence of transporting electricity across the power grid [5]. There are even historical data of losses in the electricity grid since 1960 [6], with European Union electricity losses averaging 7,06% [5].

The technical component of losses, i.e., the energy converted to heat and power, is the most prominent one, but non-technical components, in other words, the energy delivered but not metered or billed, are very influential in certain Council of European Energy Regulators (CEER) Member countries and often depend on socio-economic aspects. To the power sector reducing losses, or at least maintaining them at a low level, plays an important role financially, environmentally and technically. Reducing power losses helps with Europe's energy efficiency and security of supply objectives. Reducing power losses is also an important part of CEER's mission of putting consumers first, as the costs of power losses are currently passed on to consumers. CEER advocates implementing incentives to reduce power losses.

The example of the management and practices of the electricity grid and its technological know-how may provide a good basis to bring efficiency to other less efficient sectors, e.g. the water sector. These infrastructures are rather similar in terms of population connected to its service, with the electricity grid having a 100% coverage and the water infrastructure having also that goal in all EU countries.

According to The European Federation of National Water Services [2] the mean values for losses are 23% and 2171 m^3/km/year in EurEau member countries. Water losses can be appointed to theft, broken ducts and upstream branches, under counting of distribution counters, over-counting of input counters, duct cleaning, reservoir cleaning, losses in treatment and pumping, especially when starting up capture stations, losses in reservoirs (pipes in poor condition, overflow).

It should be noted that in the EurEau report, although one of the most complete reports found, it does not discuss the fact that there are countries that do not provide information regarding their country, nor does it address a vision of the future of investment. The lack of information in the water sector is a barrier, that is not as common as in the electricity sector.

The water problem is felt worldwide. The United Nations 2030 Agenda for Sustainable Development Preamble [7] presented a plan of action for people, planet and prosperity. The agenda presents 17 Sustainable Development Goals and 169 targets which demonstrate the scale and ambition of this new universal Agenda. For example, goal 11 and goal 15 intend to, respectfully, make cities and human settlements inclusive, safe, resilient and sustainable and protect, restore and promote sustainable use of terrestrial ecosystems, of which water has a central role.

Figure 1 presents water and electricity losses in European countries.

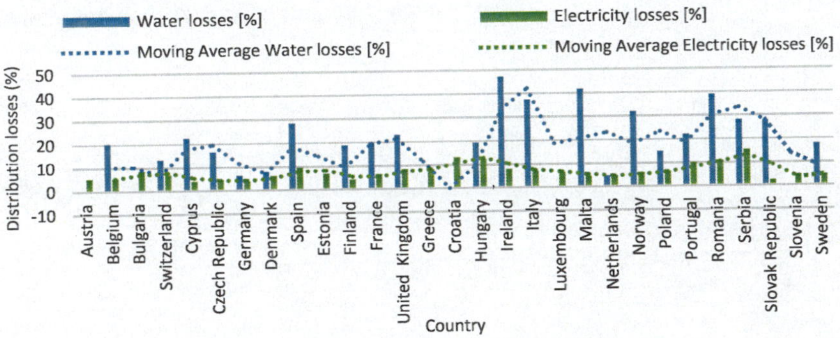

Fig. 1. Water [2] and electricity losses [6].

According to Fig. 1 the electricity losses in European countries range from 2,46% in Slovak Republic to 15,44% in Serbia. Water losses vary from 4,77% in the Netherlands to 47,06% in Ireland. As stated before, there are seven countries in Europe that do not disclose their water losses percentage.

2 Case Study

This case study uses one year of data of a public water distribution system of a utility company, focusing in one capture or harvesting station (input 1), as represented in Fig. 2.

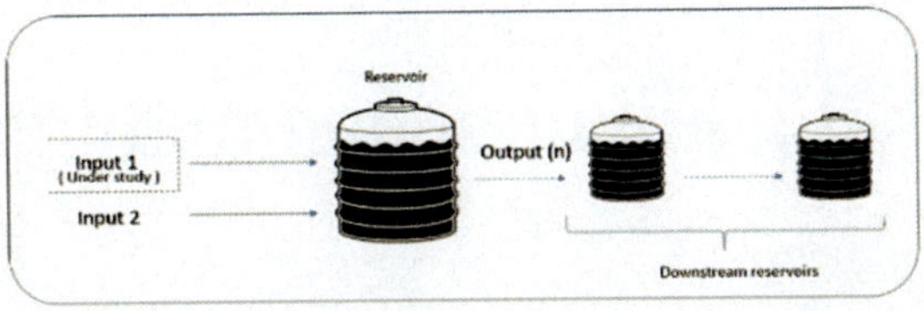

Fig. 2. Schematic representation of the case study, being input 1 the harvesting/capture station.

Each installation or station of the water distribution system is comprised of a Remote Terminal Unit (RTU) which includes the programming of the desired functionalities and also a Global System for Mobile Communications (GSM) and a General Packet Radio Service (GPRS) modem. For the purposes of the automated control only GPRS data is used. Each station is continuously connected to the internet, enabling its communication to the server and to stations that will be required to provide a certain type of service to the requiring station. The data is segmented in 96 quarter hour blocks.

When the reservoir sensors reach certain water heights it activates the pumps of the harvesting stations. which are called upon to supply water, as shown in Fig. 3. Input 1 is the harvesting station under study, and called by the reservoir to supply water, between 3,20 m and 3,60 m. Input 2 which is also an harvesting station, supplies water to the reservoir between 3,30 m and 3,60 m. Considering this programming, input 1 is also the last station to supply water to the reservoir. Input 1 was selected to be studied due the fact, that is called only in last instance, being interesting the analysis of such programming.

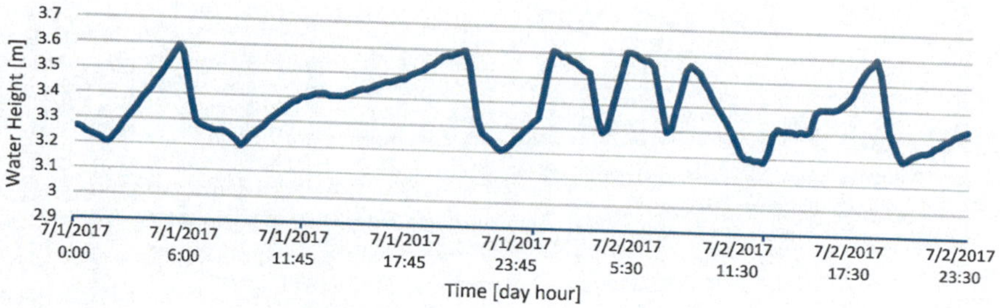

Fig. 3. Water height in the reservoir.

The time period under study is comprised between 01-07-2017 and 01-07-2018. The available information at the station is output water pressure in the station, voltage per phase, power, energy, and instantaneous flow.

Figure 4 represents two months of pressure measuring in the harvesting station. It is noticeable that in normal conditions, pressure goes from five bar to never above seven bar. The pump is driven by a soft-starter. In the graph it is perceivable that there are five events that exhibit unusual pressure in the duct. These events represent either, rupture in ducts (zero and near zero bar events) and grid maneuvers, since this duct has the possibility to be bypassed to supply directly the distribution system.

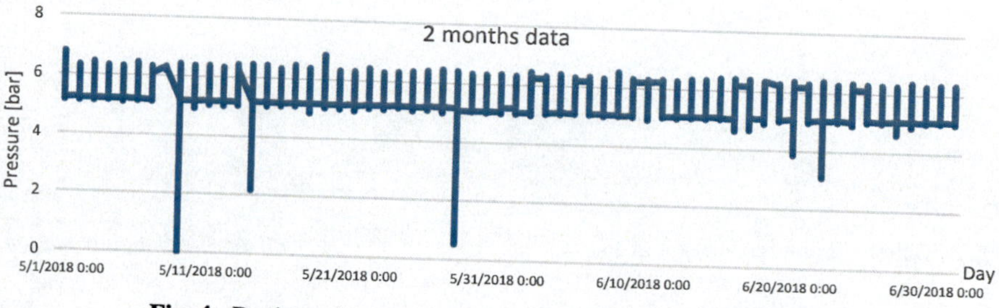

Fig. 4. Registered pressure between 01/05/2018 and 30/06/2018.

Table 1 exhibits the summary of the electrical performance of the station. Two indicators were developed, [m^3/kW] and [kW/m^3], to help determine which station will operate more time. It was determined that there are differences in the performance of the stations, and this type of indicators are helpful to optimize service and decrease pumping time and energy consumption.

Table 1. Harvesting station, operational results from 01/07/2017 to 01/07/2018.

Pump Power [hp]	Energy consumption [kWh]	Extracted water [m^3]	[m^3/kW]	[kW/m^3]	Number of pumping hours	% pumping time
30	83.563,80	114197,288	1,37	0,73	3055	34,77%

3 Methodology to Obtain Data Clusters

3.1 Clustering Methods as in the Electricity Grid

Research in smart grids present an opportunity for data clustering, considering the available amount of information. Big data is ideal for the dynamic analysis of operational parameters.

Clustering methods may provide a response for this need by performing an automatic classification of daily diagrams over a long period of time, such as: the station's power usage and flow rate. Cluster analysis is a convenient method for identifying homogenous groups of objects or clusters [8, 9], aiming to group objects by similarity, but keeping a significant difference between groups. Clustering techniques can be divided in two main groups, Partitional clustering methods and Hierarchical clustering methods [10].

Partitional algorithms use an interactive optimization paradigm. They start with an initial partition and, in each iteration, objects are swapped if the quality of the clustering is improved, by applying an interactive control strategy. The definition of the clusters will strongly depend of the initially selected partition. Another relevant issue is the representation of clusters. In the K-mean algorithms a cluster is represented by its center of gravity while the K-medoid algorithms use the closest object from the center as the center representation. However, according to [11] the limitations of this method are the dependence on the prototypes initially assigned to clusters and also the lack of a clear assignment of objects to clusters, creating difficulties to the definition of prototypes.

The Hierarchical Methods (HM) create sequenced partitions, each one containing the previous one in a hierarchical way. There are two types of HM [12], the agglomerative clustering algorithms and the divisive clustering algorithms, the former performing a bottom-up approach and the latter a top-down approach, using a similar clustering rule based on similarity or distance [12]. The agglomerative clustering algorithms are one of the most used hierarchical methods, and therefore one of the choices made for the current work was based on such methods.

An alternative procedure introduced by Teuvo Kohonen in the early 1980's, the Self-Organizing Maps (SOM), is a type of artificial neural network (ANN) with an unsupervised learning process and unknown output partitions [13].

3.2 Clustering Electric Power and Flow Rate Profiles

To achieve that goal, different clustering techniques were applied to the historical data sets of hourly power and flow rate of a complete year.

The proposed methodology based in [14] is represented in its major steps in Fig. 5.

In order to define the appropriate use of gathered data, an experimental setup was performed using two approaches for the data clustering techniques, one using the input data without any treatment and a second one where the input data were subjected to a normalization procedure such that the sum of all the elements of each vector equals 1, in order to cross-check their validity [11].

In electric smart grids, this experiment showed that in the case of the electricity profiles, the shape of the diagram is the relevant attribute while in the case of price diagrams, the amplitude similarity between diagrams should be used. Considering this knowledge, despite both shape and amplitude were obtained in this exercise, only the shape [p.u] profiles will be presented, since it is what the authors want to analyze. The diagram shape comparison is performed by using the Euclidian distance between individual normalized vectors or/and intermediate vectors that represent the data clusters. In both cases, it was necessary to initially adjust the data to the daylight savings clock changes of the daylight savings scheme, namely on the last Sunday of March and October.

At the clustering stage, the HM helps in choosing the appropriate number of clusters, graphically, using the tree dendrogram and using the inflection point of the curve that relates the distances between clusters with the number of clusters [15]. A comparison was made of the output of both methods regarding the assignment of day types (weekday, weekend + holidays, summer day, winter day) to clusters.

The ANN application was performed using the available MATLAB toolboxes. The Initial set-up block includes the annual input data (normalized) gathered in [96;366] matrixes, which represent the total days under study on 1/4 h samples.

The clustering block is where data will be clustered based on similarity and topology, with a preference for assigning the same number of instances to each class. This step considered a total of 1000 training steps input space coverage, an initial neighborhood size of two, and a layer topology function that calculates neuron positions for layers whose neurons are arranged in an N dimensional grid (e.g. gridtop function). The N-dimensional neuron layer definition was guided by the hierarchical method results and using a layer distance function to find the distances between the layer's neurons given their positions. This step uses both the input data and the neuron layer network to cyclically adjust the weights associated with the input values. During the training phase, the network learns by adjusting the weights to be able to predict accurately the class label of input samples. Finally, the input data is actually allocated/clustered according to the obtained weights in the neuron layer network. For this work the authors chose to implement the HM and SOM/ANN methods due to their adequacy.

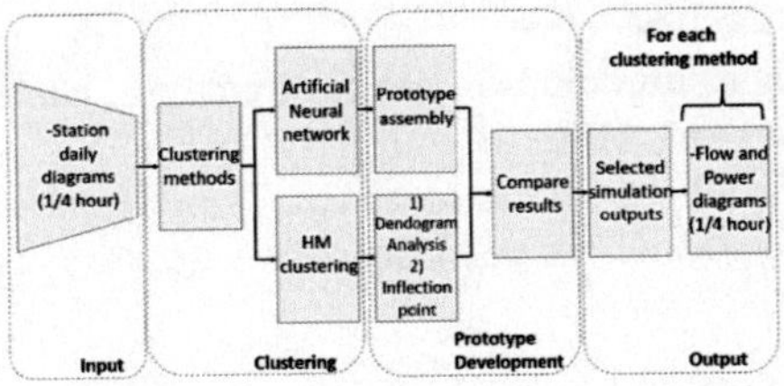

Fig. 5. Methodology to obtain cluster prototypes of flow rate and power, based in [14].

4 Results

Three clusters of prototypes were selected to flow rate and electric power.

4.1 Flow Rate [p.u.]

Table 2 and Table 3 display the clusters analysis for the HM and ANN, regarding flow rate in [p.u.]. It is perceivable from the results shown in Table 2 and Table 3, that cluster 1 from the HM and cluster 1 from the ANN method exhibit the highest number of hits. Cluster 2, from both methods, present the most approximate number of hits.

Table 2. Flow rate results for the HM.

	Total hits	Weekdays	Weekend + holidays	Summer day	Winter day
Cluster 1	188	135	53	119	69
Cluster 2	109	66	43	59	50
Cluster 3	69	47	22	41	28
	366	248	118	219	147

Cluster 1 from the HM method displays a higher number of weekdays occurrences, than the other two clusters. Cluster 2 from the HM has an approximate number between weekdays and weekend + holidays, and winter and summer days. Cluster 3 is more representative of week and summer days.

Table 3. Flow rate results for the ANN.

	Total hits	Weekdays	Weekend + holidays	Summer day	Winter day
Cluster 1	144	98	46	87	57
Cluster 2	110	71	39	64	46
Cluster 3	112	79	33	68	44
	366	248	118	219	147

Figure 6 presents the developed prototypes for the HM and ANN methods. It is possible to observe shape similarities between prototypes, namely: HM #p1 with ANN #p1, HM #p2 with ANN #p2, HM #p3 with ANN #p3.

The difference in the number of hits for specific prototypes is probably due to outliers (frontier values) that could be assigned differently in the clustering process of each method [16].

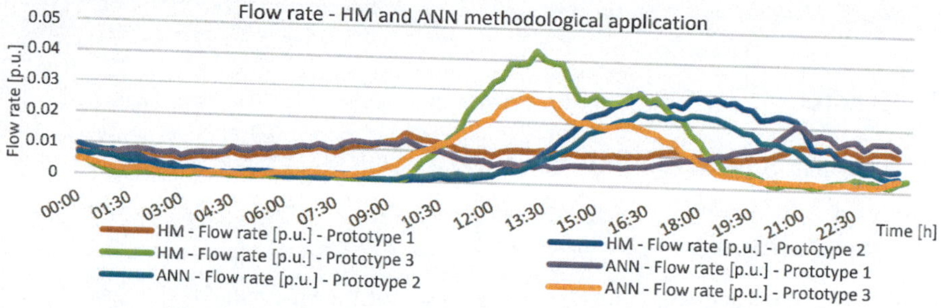

Fig. 6. Flow rate prototypes for the HM and ANN methods.

From Fig. 6 it is perceivable that HM #p1/ANN #p1 represent the prototype with the constant reduced flow, between 0,01 and 0,02. Prototype HM #p2 and ANN #p2 increase flow rate from 12:00 to 18:00 and a decrease from 18:00 onwards. HM #p3 with ANN #p3 exhibit a flow rate increase trend from 8:00/9:00 till 13:00, flow rate is stable from around 14:00 till 17:00, the decreasing flow rate trend occurs after 17:00.

4.2 Electric Power [p.u.]

Table 4 and Table 5 display clusters for the HM and ANN for electric power [p.u.].

Table 4. Electric power results for the HM.

	Total hits	Weekdays	Weekend + holidays	Summer day	Winter day
Cluster 1	62	36	26	28	34
Cluster 2	220	152	68	142	78
Cluster 3	84	60	24	49	35
	366	248	118	219	147

From Table 4 and Table 5 it is possible to observe that clusters #2 from the HM and cluster #1 from the ANN are the most representative clusters, both with predominance to week. Cluster 2 from the HM and cluster 1 from the ANN display the highest number of hits in winter days. Clusters #3 from both methods present similar results in terms of total hits and its distribution (week and summer day).

Table 5. Electric power results for the ANN.

	Total hits	Weekdays	Weekend + holidays	Summer day	Winter day
Cluster 2	126	79	47	74	52
Cluster 1	169	117	52	101	68
Cluster 3	71	52	19	44	27
	366	248	118	219	147

Figure 7 presents the developed prototypes for the HM and ANN methods. It is possible to observe shape similarities between prototypes, namely: HM #p1 with ANN #p2, HM #p2 with ANN #p1, HM #p3 with ANN #p3.

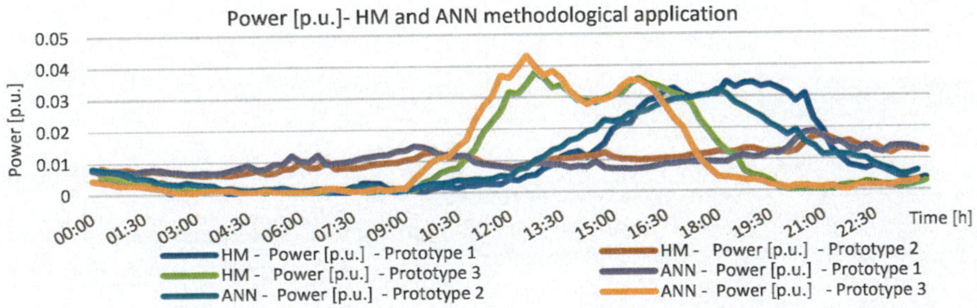

Fig. 7. Electric power prototypes for the HM and ANN methods.

Prototypes HM #p1/ANN #p2 show a growth in power consumption from 9:45 till around 20:15 and a decrease after that. HM #p2/ANN #p1 exhibit prototypes with reduced and constant power consumption, between 0,005 and 0,02 p.u.. HM #p3/ANN #p3 exhibit a flow rate increase trend from 8:00/9:00 till 12:45, power consumption decreases from around 12:45 till 14:15, power consumption increases from 14:15 till 15:45, the decreasing flow rate trend occurs after 15:45 until 19:45 reaching zero.

Finally, it is graphically recognizable cases of combined existence between flow rate and electric power prototypes, as expected.

5 Conclusions

The authors highlight the importance that big data and sensing bring to this sector.

This work presented a possible approach to help implementing technological measures or taking decisions, e.g., the indication of pressure loss, allows faster human intervention, as an alternative to consumers indicating that they are without service. Avoiding overflow losses in reservoirs is another easy to implement technique. However, many losses continue to occur, namely in network accessories, e.g., valves.

The selected methodology can be used in the larger smart grid concept, including water distribution systems. Studying water demand is crucial to ensure an optimized water supply system, as well as helping to fight situations of inefficiency.

This works also intends to increase awareness to the need of good water management. The optimization of the water sector contributes to the environmental and economic results of companies and benefit us all.

The case study results show that in the case of filling a reservoir profiles are not very susceptible to winter, summer and weekdays and weekends. Results also show similarities between obtained prototypes in the HM and ANN methods, for both flow rate and electric power.

Future work should be carried out for installation that directly supply consumers, to assess if the number of workable prototypes is changed.

Acknowledgments. This work has been supported by the Portuguese Foundation for Science and Technology (FCT) under project grant UIDB/00308/2020.

References

1. Pardo-Bosch, F., Aguado, A.: Sustainability as the key to prioritize investments in public infrastructures. Environ. Impact Assess. Rev. **60**, 40–51 (2016)
2. The European Federation of National Water Services, Europe's water in figures, p. 22 (2017)
3. Dileep, G.: A survey on smart grid technologies and applications. Renew. Energy **146**, 2589–2625 (2020)
4. Moretti, M., et al.: A systematic review of environmental and economic impacts of smart grids. Renew. Sustain. Energy Rev. **68**, 888–898 (2017)
5. CEER: CEER Report on Power Losses (2017)
6. IEA: Api_Eg. (2019)
7. UN General Assembly: Transforming our World: The 2030 Agenda for Sustainable Development, *A/RES/70/1*. UN General Assembly (2015)
8. Mahmoudi, N., Eghbal, M., Saha, T.K.: Employing demand response in energy procurement plans of electricity retailers. Int. J. Electr. Power Energy Syst. **63**, 455–460 (2014)
9. Mooi, E., Sarstedt, M.: A concise guide to market research. Harv. Bus. Rev. **68**(1), 105–116 (2011)
10. Pujari, A.K., Rajesh, K., Reddy, D.S.: Clustering techniques in data mining—a survey. IETE J. Res. **47**(1–2), 19–28 (2001)
11. Sousa, J.M.C.: Metodologia para determinação de perfis de consumidores de Baixa Tensão, Faculdade de Ciências e Tecnologia da Universidade de Coimbra (2006)
12. Maimon, L., Rokach, O.: Data Mining and Knowledge Discovery Handbook. Springer, New York (2005)
13. Kohonen, T.: The self-organizing map. Neurocomputing **21**(1–3), 1–6 (1998)
14. Miguel, P., Gonçalves, J., Neves, L., Martins, A.G.: Using clustering techniques to provide simulation scenarios for the smart grid. Sustain. Cities Soc. **26**, 447-455 (2016)
15. Everitt, D., Landau, B.S., Leese, S., Stalh, M.: Cluster Analysis (5th edn.). A John Wiley and Sons, Ltd. (2011)
16. Salvador, S., Chan, P.: Determining the number of clusters/segments in hierarchical clustering/segmentation algorithms. In: 16th IEEE International Conferenceon Tools with Artificial Intelligence, 2004, ICTAI 2004, pp. 576–584 (2004)

Comparison of Reconstruction Methods for Water Supply Systems Flow Rate Time Series

Carlos Ascensão(✉), Bruno Ferreira, Raquel Barreira, and Nelson Carriço

INCITE, Barreiro School of Technology, Polytechnic Institute of Setúbal, Setúbal, Portugal

Abstract. The purpose of this paper is to compare the performance of five univariate models for the reconstruction of flow rate time series. Errors in the measurements may occur due to problems in the sensor or in the communication system with data logger, thus generating missing data in the flow rate time series. The presence of missing values in flow rate data restricts its use in network operation processes. The performance of seasonal ARIMA, Standard and double seasonality Holt-Winters, and original and improved Quevedo approach are assessed. The analysis is made considering a real Portuguese case study and 1-month of flow rate data at 1-h and 10-min period. The holidays compared to the weekdays show great differences in consumption patterns. For this reason, the effect of forecasting holidays is assessed. Obtained results evidence that the improved Quevedo model can cope with different time step intervals and type of day being forecasted, with a reduced computation time.

Keywords: Flow rate · Forecasting · Reconstruction methods · Time series · Water supply systems

1 Introduction

Flow rate monitoring is an increasingly recurring practice in water utilities, due to the larger accessibility and availability of telemetry equipment and remote management systems. The stored time series data can be used for many tasks in operating and monitoring systems (e.g., demand forecasting, burst detection). The measured data are acquired by sensors and stored in the data logger, which communicates remotely to the management system [1]. Errors in the measurements can happen and may be caused by problems in the sensor or in the communication system with the data logger due to power failures, storage limitations or working outside the operational range generate the missing data in time series [2].

The treatment of flow rate time series is a challenging task for water utilities. The validation processes are based on simple heuristics. Usually non-validated data is replaced using reconstruction methods by predicting the measures with multivariate or univariate statistical models for flow rate time series [3]. More advanced techniques such as machine learning may be applied to forecast water demand in water supply systems which consider air temperature, precipitation, and flow rate. However, multivariate models that

J. R. da Costa Sanches Galvão et al. (Eds.): *Proceedings of the 1st International Conference on Water Energy Food and Sustainability (ICoWEFS 2021)*, pp. 851–858, 2021.
https://doi.org/10.1007/978-3-030-75315-3_90

require many variables represent a greater challenge in their application and data acquisition, thus making the operationalization of the models a hard task. Also, the application of advanced techniques, in terms of system monitoring, requires real-time operation and multivariate models do not provide good results when applied in real-time [1]. For these reasons water utilities search for simple forecasting models with low difficulty and complexity in required data, model application and operationalization.

Flow rate time series may show great evidence of daily and weekly cycles that must be considered by forecasting models [4]. Literature shows that autoregressive models and exponential smoothing models with components and parameters that express seasonality are able to obtain reasonable results [1, 9]. Caiado [6] assessed the performance of three different univariate models for water demand forecasting, namely, Holt-Winters, ARIMA and GARCH model and results suggest that all the univariate time series models can be quite useful for short-term forecasting. Quevedo *et al.* [2] developed a short-term forecasting methodology with the purpose of reconstructing missing data in water supply systems. This methodology considers an aggregate daily flow model based on ARIMA models and a 10-min model based on distributing the daily flow using a 10-min demand pattern. Cugueró-Escofet *et al.* [7] applied a methodology for reconstruction of missing flow rate data using a double seasonal Holt-Winters.

This paper compares the performance of five forecasting models, namely, a seasonal ARIMA, a seasonal and double seasonal Holt-Winters, a method proposed in Quevedo *et al.* [2] and our improvement of Quevedo approach. The performance assessment is carried out for a complete forecasted day using the root-mean-squared-error (RMSE) and was applied to a real Portuguese case study.

2 Reconstruction Methods

2.1 Seasonal ARIMA

The ARIMA models are derived from the family of Auto Regressive Moving Average (ARMA) models. Their difference is the integration component that allows differentiating the series to be possible to apply to non-stationary time series. In order to forecast, ARIMA models use a polynomial of the previous values together with the previous prediction errors. Seasonal ARIMA models considers an additional polynomial for the seasonal component [1].

The function of the ARIMA models can be represented by the degrees of the model (p, d, q), where p represents the number of autoregressive terms, d represents the number of differentiations and q the number of lagged forecast errors in the prediction equation. The polynomial function dedicated to the seasonal component works only with a periodicity. Similar to the polynomial of the regular component it can also be represented by (P, D, Q)s where P, D and Q represent the degrees of the model and s represents the number of seasonal periods.

The degrees of the seasonal ARIMA model (p, d, q)(P, D, Q)s can be selected based on the Bayesian information criterion and assessing the fitted values. For the current study, and since 1-h and 10-min time intervals are considered, we used the seasonal ARIMA parameters $(2, 0, 0)(2, 0, 0)_{24}$ and $(1, 0, 0)(1, 0, 1)_{144}$, respectively.

2.2 Quevedo Approach

The present approach shows the implementation and improvements to the model for the reconstruction of time series data from water supply systems presented in Quevedo *et al.* [8].

The procedure for reconstructing missing data consists of two modules. The first module gives the prediction of aggregated daily flow based on the seasonal ARIMA models, denominated as the aggregated daily flow model. This model requires a historic record of daily aggregate flow to be able to predict the daily volume taking advantage of the main components of the ARIMA model. For the selection of the polynomial degrees (p, d, q)(P, D, Q)s it was based on the Bayesian information criterion [9] evaluating the set of models generated by $0 \leq p \leq 3, 0 \leq P \leq 1, 0 \leq q \leq 3$ e $0 \leq Q \leq 1$.

The second module determines a set of flow distribution patterns at 10-min intervals, consisting of 144 average flow rate values for each pattern. Distribution patterns consider the variation in measurements between weekdays and weekends. For this reason, patterns must be determined for the days of the week (Monday to Friday), and of the weekend (Saturdays and Sundays). Holidays should also be considered for the impact they have on the analysis. Quevedo *et al.* [8] determined that consumption habits for holidays are the same as on Sundays. However, by considering ARIMA as the aggregate daily flow forecast model it is not possible to take into account the effect of a holiday during a weekday.

Improvements to the Quevedo approach are proposed to estimate the daily aggregate flow of a holiday. When initializing the model, it is verified that the date to be forecasted is within the subset with holiday date. If the date does not coincide, the model runs according to the initial approach and estimates the aggregate daily flow with the seasonal ARIMA model. If the date coincides, the aggregate daily flow estimation starts with a simple exponential smoothing model, for which the input is a subset with the aggregate daily flow for Sundays. Estimation for the aggregate daily flow of a holiday is carried as a new Sunday.

To reconstruct flow rate time series, Quevedo *et al.* [8] distributes the aggregate daily flow estimate by the distribution pattern of the day to be reconstructed.

2.3 Exponential Smoothing

Holt Winters method is an exponential smoothing method considering trend and seasonal components [10]. In the urban water sector, exponential smoothing methods are well known and have been used in automatic forecasting models [1]. The main characteristic is its simplicity, considering it can be optimized only with the least squares.

Holt-Winters method is based on level, trend and seasonality [11] and models can be divided into two versions based on seasonality patterns, namely, additive or multiplicative seasonality. Depending on the type of seasonal pattern presented in the data, one of the reference versions can be chosen [12]. In additive seasonality, the difference in seasonal fluctuation between successive is constant, while in multiplicative seasonality the variation is a percentage [12]. In this article, only the Holt Winters models with multiplicative seasonality are considered.

Standard Holt-Winters
Initial values of the components (i.e., level, trend and a seasonal index) are required to start the Standard Holt-Winters multiplicative seasonality model. According to [10] initial level is obtained by averaging the observations from the first seasonal period. The estimated initial trend uses a moving average of the first seasonal period and the seasonal indices are estimated using the average of first seasonality period.

The model components are based on three smoothing parameters: α, β and δ. These parameters represent the level parameter, trend parameter, the seasonal parameter for the seasonal cycle (daily in our case), respectively. These parameters can be estimated by minimizing the RMSE and are usually restricted to lie between 0 and 1.

Double Seasonal Holt-Winters
The double seasonal Holt-Winters accommodates two seasonal periods, in this case daily and weekly. To consider the effect of weekly seasonality, the double seasonal Holt-Winters model requires one more seasonality component than the standard model.

The double model requires initial values for initializing – trend, level, daily seasonality, and weekly seasonality index. Taylor [13] formulation was used to represent the initial values, using a s1-period cycle for the daily seasonality and s2-period cycle for weekly seasonality. According to Taylor [13] the initial trend, was chosen as the average of (1) 1/s2 of the difference between the mean of the first s2 and second s2 observations and (2) the average of the first differences for the first s2 observations. The initial level was chosen as the mean of the first two s2 observations minus s2 and half times the initial trend. The initial values for the daily seasonal index are defined by the average of the ratios of actual observation to s1-point centered moving average, taken from the corresponding half-hour period in each of the first 7 days of the time series. The initial values for the weekly seasonal index were set as the average of the ratios of actual observation to s2-point centered moving average, taken from the corresponding half-hour period on the same day of the week in each of the first 2 weeks of the demand series, divided by the initial value of the smoothed within-day seasonal index.

The model components are based on four smoothing parameters α, β, Υ and δ. The first three parameters are similar to the ones previously presented for standard Holt Winters. In addition, a seasonal parameter for bigger seasonal cycle (weekly in our case) is considered. Similarly, these parameters can be estimated by minimizing the RMSE and are usually restricted to lie between 0 and 1.

2.4 Performance Assessments

A complete day of missing data (i.e., long gap duration) was assumed to assess the performance of the presented models. For this reason, models will be trained with historical data with a duration no smaller than a month. The last day will be forecasted using the trained models to assess its performance.

The different forecasting models' parameters were calibrated by minimizing the RMSE of the fitted model. The performance of the models' predictions was assessed using the RMSE between real and predicted measurement.

3 Applications and Discussion

In this section, an analysis to the performance of the five flow rate time series reconstruction models is carried.

The flow rate time series of the case study is collected using an impulse flow rate meter installed at the inlet of the water distribution network of a residential area with up to 3,000 inhabitants and was provided by a water utility located in Lisbon metropolitan area. This case study considers a 1-month of historical flow rate data recorded in intervals of 1-h and 10-min. As such, the model with 1-h intervals will predict 24-steps and the model with 10-min intervals the model will predict 144-steps.

Initially, forecasts were carried considering a weekday and 1-h intervals. Figure 1 presents the obtained results for each model as well as the real flow rate data.

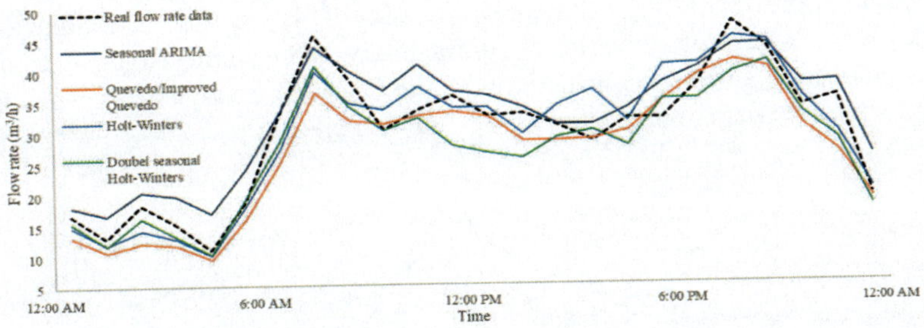

Fig. 1. Comparison for five forecasting techniques considering weekday with hourly intervals.

All models performed reasonably well predicting a day in the series at 1-h intervals for all models. The seasonal ARIMA model presented the best values (RMSE = 3,78), followed by the Holt-Winters (RMSE = 3,89), double seasonal Holt-Winters (RMSE = 4,09), and lastly the Quevedo (RMSE = 4.41). The computation time of each model was assessed, and it was concluded that all models were relatively fast (i.e., less than 40 s). Table 1 shows a summary of computational time (in seconds) and the RMSE obtained for all models.

Time series with 1-h intervals have a very limited use in real-time water supply systems operation. Application of advanced techniques, such as machine learning requires time series with shorter time intervals to operate water supply system in real-time. As such, the same day of the week is predicted with the series of 10-min intervals. The obtained results for each model are presented in Fig. 2 and Table 1.

All models predicted reasonably well, with the Quevedo presenting the best overall results (RMSE = 4,69), followed by Holt-Winters (RMSE = 5,97), seasonal ARIMA (RMSE = 6,10) and lastly double seasonal Holt-Winters (RMSE = 6.76). Nonetheless, the difference in computation time amongst methods is quite significant, with the Quevedo (t = 1 s), Holt-Winters (t = 108 s), Double seasonal Holt-Winters (t = 286 s) and lastly seasonal ARIMA (t = 591 s).

Problems may arise when forecasting holidays during weekdays since the expected behavior of the water demand of a holiday is usually related to the behavior of the

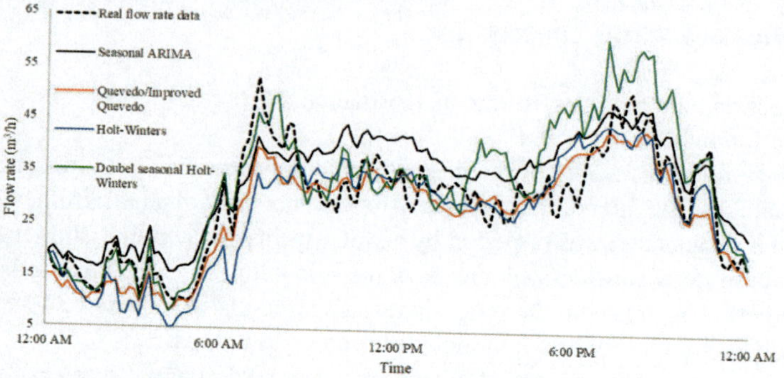

Fig. 2. Comparison for five forecasting techniques considering weekday (10-min intervals).

Sundays (in opposition to the behavior of weekday) [14]. As such, a holiday during a weekday is predicted with the series of 10-min intervals. In addition to the four methods already compared, a fifth is considered with the Quevedo improvement. The obtained results for each model are presented in Fig. 3 and Table 1.

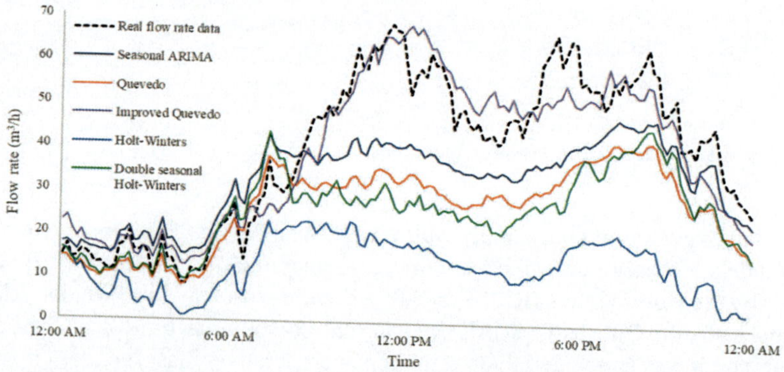

Fig. 3. Comparison for five forecasting techniques considering holiday with 10-min intervals.

Based on Fig. 3 and Table 1 it is possible to conclude that, overall, the improved Quevedo performed reasonably well (RMSE = 5,84). The remaining models failed to capture the holiday variations. The autoregressive and exponential smoothing models make predictions based on seasonality periods and cannot "look" outside the seasonality periods. Similarly, the original Quevedo approach forecast the aggregate daily flow using an autoregressive model (i.e., seasonal ARIMA).

4 Conclusions

Missing data from flow rate time series resulting from a validation process must be reconstructed to apply advanced techniques that requires validated data. Usually, the reconstruction of the flow rate time series is performed by forecasting models. This

Table 1. Comparison of model's performance and computational time

Models	Weekday (1-hour)		Weekday (10-minute)		Holiday (10-minute)	
	RMSE	Computation time (s)	RMSE	Computation time (s)	RMSE	Computation time (s)
Seasonal ARIMA	3.78	12	6.10	591	12.50	624
Quevedo	4.41	1	4.69	1	16.91	1
Holt-Winters	3.89	38	5.97	108	30.53	101
Double seasonal Holt-Winters	4.09	12	6.76	286	19.21	284
Improved Quevedo	4.41	1	4.69	1	5.84	1

paper presents a comparison of five methods to the reconstruction of flow rate time series, namely, the seasonal ARIMA, the standard and double seasonal Holt-Winters, the original and improved Quevedo approach. The comparison is based on 1-month of historical flow rate time series at 1-h and 10-min intervals of a real Portuguese case study. A complete day was forecasted and analyzed for weekdays and holidays. In weekdays, forecasts with 1-h intervals obtained a reasonable RMSE result for all methods. Similarly, reasonable RMSE results were obtained for all models considering weekdays with the 10-min intervals. Nonetheless, great difference in computation time were observed amongst methods. On the other hand, and when forecasting a holiday, only the improved Quevedo approach produced reliable results.

Future research may include the forecast of flow rate data when in absence of reliable historical data, for instance, due to changes in patterns motived by recent lockdowns.

Acknowledgement. The authors want to acknowledge Fundação para a Ciência e a Tecnologia, (grant number DSAIPA/DS/0089/2018) through the Data Science and Artificial Intelligence in Public Administration Programme for supporting WISDom project. The authors also acknowledge the participating water utilities for their contribution.

References

1. Puig, V., Ocampo-Martínez, C., Pérez, R., Cembrano, G., Quevedo, J., Escobet, T.: Real-time Monitoring and Operational Control of Drinking-Water Systems. Springer International Publishing, Cham (2017)
2. Quevedo, J., Puig, V., Cembrano, G., Blanch, J., Aguilar, J., Saporta, D., Benito, G., Hedo, M., Molina, A.: Validation and reconstruction of flow meter data in the Barcelona water distribution network. Control Eng. Pract. **6**(18), 640–651 (2010)
3. Adamowski, J., Fung Chan, H., Prasher, S.O., Ozga-Zielinski, B., Sliusarieva, A.: Comparison of multiple linear and nonlinear regression, autoregressive integrated moving average, artificial neural network, and wavelet artificial neural network methods for urban water demand forecasting in Montreal, Canada. Water Resour. Res. **48**, 1–14 (2012)
4. De Marinis, G., Gargano, R., Tricarico, C.: Water demand models for a small number of users. In: Water Distribution Systems Analysis Symposium 2006 (2007)
5. Taylor, J.W., de Menezes, L.M., McSharry, P.E.: A comparison of univariate methods for forecasting electricity demand up to a day ahead. Int. J. Forecast. **22**, 1–16 (2006)
6. Caiado, J.: Performance of combined double seasonal univariate time series models for forecasting water demand. J. Hydrol. Eng. **15**, 215–222 (2010)

7. Cugueró-Escofet, M., García, D., Quevedo, J., Puig, V., Espin, S., Roquet, J.: A methodology and a software tool for sensor data validation/reconstruction: application to the Catalonia regional water network. Control Eng. Pract. **49**, 159–172 (2016)
8. Quevedo, J., Puig, V., Cembrano, G., Blanch, J., Aguilar, J., Saporta, D., Benito, G., Hedo, M., Molina, A.: Validation and reconstruction of flow meter data in the Barcelona water distribution network. Control Eng. Pract. **18**, 640–651 (2010)
9. Schwarz, G.: Estimating the dimension of a model. Ann. Stat. **6**, 461–464 (1978)
10. Spyros, M., Steven C.W., Rob, J.H.: Forecasting: Methods and Applications. Wiley (1997)
11. Hyndman, R.J., Koehler, A.B., Ord, J.K., Snyder, R.D.: Springer Series in Statistics Forecasting with Exponential Smoothing (2008)
12. Galvas, G.: Time Series Forecasting used for Real-Time Anomaly Detection on Websites. Vrije Universiteit, Amsterdam (2016)
13. Taylor, J.W.: Short-term electricity demand forecasting using double seasonal exponential smoothing. J. Oper. Res. Soc. **54**, 799–805 (2003)

Sustainable Water Package: Technical Analysis Versus Consumer Perception

Henrique Almeida[1,2], Liliana Vitorino[1](✉), and Raquel Antunes[1,3]

[1] School of Technology and Management, Polytechnic of Leiria, Leiria, Portugal
liliana.vitorino@ipleiria.pt

[2] Computer Science and Communication Research Centre, Polytechnic of Leiria, Leiria, Portugal

[3] Faculty of Architecture and Arts of Lusíada University, Porto, Portugal

Abstract. Every year, an average of 1000000 L of mineral water are sold in Portugal. An amount of plastic waste from the bottles, plastic and labels could arise ecological problems. Portuguese are now paying more attention to environmental issues and most consumers consider sustainable issues in their purchase intentions. This study analyses how sustainable Portuguese water bottles considering two important functions of packages: technical analysis (material usage for plastic bottle, plastic cap, label and complete amount of water) and communication role (how consumers perceived the bottle in terms of sustainability). Therefore, two studies are carried out. The first measures the weight variations of bottle components and the second inquiries the consumers to rank the bottle pictures in terms of sustainable design attributes.

Keywords: Plastic water bottles · Costumer perception · Sustainability · Labelling · Package design

1 Introduction

The majority of Portuguese bottled water (natural and spring minerals) are hyposaline, that corresponds to more than 75% of the total consumed water, reflecting the Portuguese taste for bottled water [1]. According to the Portuguese Association of Natural and Spring Mineral Water Industrialists (APIAMNN) and its latest statistical data, the Portuguese consume more still water than sprinkle water [2]. From January to September 2019, the sales statistics by brands indicate that the most sold water brands in the national market correspond to these five brands: *Penacova*, *São Martinho*, *Luso*, *Vitalis* and *Serra da Estrela*.

According to Baynast et al. [3], the package comprises two main important functions: technical and communicational functions. The technical functions are related to protection and conservation of the product, ways of utilization (ergonomics), storage and transportation and environmental protection. The communicational role of the package is linked to visual impact, product identification, brand recognition, positioning, consumer information and purchase impulse.

J. R. da Costa Sanches Galvão et al. (Eds.): *Proceedings of the 1st International Conference on Water Energy Food and Sustainability (ICoWEFS 2021)*, pp. 859–873, 2021.
https://doi.org/10.1007/978-3-030-75315-3_91

Sustainability science is a new paradigm that contemplates the importance of understanding the interaction between nature and society [4]. Everyone has a responsibility to contribute to a better world. If sustainability becomes a reality, consumer behaviour and perspectives will have to change [5]. Organizations should change to a "green" management and developing initiatives in accordance with circular economy principles and keep in mind that the organization goals are now dual: social and financial [6]. On the other hand, consumers should pay more attention in their attitudes and concerns with environmental issues.

The consideration of environmental sustainability on package design decisions have mainly been disregarded and the fact that Portuguese are more concerned with sustainable issues [7] draw to this research about sustainable package consumer perception (communication function) and, technically which brand is more sustainable (technical function). In this paper, packages of 1,5l of the most sold water in Portugal brands was studied. On the one hand, one of the main questions is if the package has a great importance to the consumers' purchase intentions, what they consider at the time of purchase: design attributes of sustainability in terms of form, graphics, package information or price. On the other hand, it tries to understand the structure of the form about the material contained in the 1,5l water bottles.

1.1 Water Package Sustainability

Bottle packaging enables to preserve the quality of the product for the final consumer, ensuring food security, as well maintaining its original properties. The quality of bottled water packages is essential to ensure the quality and the purity of the water. In the water sector, the materials used for the primary packaging are either glass or plastic, namely Polyethylene Terephthalate plastic (PET). The PET package is the preferred material of the consumer which is supplied to the traditional food shops and supermarkets. The PET package is a light package making it easier to transport. Plastic bottles vary in capacity ranging from a few centilitres to several litres of water, and its sale is done either individually or in packs [8].

The challenge for this industrial sector is to reduce the weight of packaging, while preserving the quality and characteristics of the water. This challenge can bring significant environmental benefits in terms of production and transportation. Increasingly, consumers and manufacturers want to know how the various types of containers actually stack up against one another, in order to minimize storage space, which also decreases its environmental impacts. For instance, more water can be transported in the same vehicle [8].

But in terms of sustainability, the plastic bottle is not the only issue of importance, the bottle cap and the label are also key issues that need to be addressed in order to obtain a global understanding of the water bottles. Each material, Polyethylene Terephthalate (PET), Polypropylene (PP), Low-Density Polyethylene (LDPE) or Paper have different ecological values for extraction (obtaining the material), processing or transforming and for end-of-life if the material is recycled or if it ends up in a landfill.

1.2 Perception of Sustainability and Package Design

Considering the new paradigm of sustainability, makes sense that sustainable issues are considered in packages [9] and specifically take into account their functions. It could be distinguished two types of packaging's functions: logistical, functional, or technical; and marketing and communication [3]. The first one essentially protects the product during its movement through distribution channels, from production to disposal and the conservation and safe product [10]; the second concerns about packaging as a key vehicle in marketing communication and managing brands, particularly at points of sale [11] and it could be an important tool to differentiate products. In addition, according to Boz et al. [12] "sustainability in the packaging value chain can be improved by facilitating collection and sorting for recycling, composting, reuse, and waste-to-energy processing, and other proper disposal and then processing of sorted packaging as well as more sustainable material sourcing and reducing material and resource use, while preserving essential functions of packaging".

According to several authors, the two main categories of packaging design are: visual elements (graphics and shape) and informational [10, 13, 14]. Some researchers have found packaging features which convey eco-friendliness in consumers' opinions based on three attributes of the packaging: form/structure, graphics and on-package information [15, 16].

The form/structure concerns about material used, material type (recycled/recyclable) and packaging (reusable) [16]. Wansink [17] concluded that larger packages are perceived as less expensive and encourage greater use and Gomes et al. [10] refers that "consumers usually perceive more stretched-out forms as larger, even when they usually buy the product and know its exact volume."

Graphics include layout, colour combination, typography and photography. Gomes et al. [10] explain that colour of a package influences the consumer's perception of the health attributes of the product. The study of Chu and Rahman [18] based on colour and sustainability concluded that some colours are perceived as more 'sustainable' than others" like green, brown, blue and natural.

On package information, the written information on packaging can help consumers make their decisions based on product characteristics, specific in this case about environmental claims and logos, and labels from environmental organizations [16]. This is important because consumers now pay more attention to information on labels; since they are more concerned with their health and nutrition [19].

Design and Consumer Purchase Intentions

Several authors have described the increasing importance of packaging in the current context of greater competitiveness in the food sector, not only as a fundamental marketing tool at the point of sale but also as a major determinant of purchasing behaviour and food consumption [10, 13, 14, 11].

The packaging plays a major role in attracting consumer attention and influencing consumer purchase decisions [20]. Thus, all packaging elements have to be combined to attract the consumer when purchasing the product [21]. Several studies concluded that people are aware of the impact of packaging on the environment and the main reasons for purchasing are environmental protection, recycle and the feeling of being responsible

[22, 23, 24]. On the same way Steenis et al. [26] concluded that sustainable design strategy has impact in behaviour intentions, attitude and perceived sustainable.

The sustainable packaging design is a challenge to designers because they need to develop what is acceptable to consumers, while considering sustainability levels for consumers aligned to actual sustainability of the materials [27]. The understanding of sustainability included in the message and the signals [15, 27] are another key factor to designer.

2 Methodology and Methods

2.1 Experimental Quantification of Plastic and Paper in 1,5l Water Bottles

To determine the amount of existing material in the 1,5l bottles, the following procedure was adopted for each of the six bottles of each brand:

1. Each bottle (bottle, cap, label and water) was weighed with a precision scale (Fig. 1.a).
2. After performing the weight of each bottle, the water temperature was determined with a precision thermometer (Fig. 1.b).
3. The water was then removed, and the bottles were left to dry.
4. The bottle (bottle, cap and label) was again weighed with a precision scale (Fig. 1.c).
5. The cap and the cap's fixation ring were removed and weighed with a precision scale (Fig. 1.d).
6. The label was then removed and weighed with a precision scale (Fig. 1.e).
7. And finally, the remaining plastic bottle was then weighed with a precision scale (Fig. 1.f).

The weighing process and temperature reading was performed three times in each measurement were the average value was used in the study. Due to the weights and object dimensions, two types of precision scales were used. One with two decimal units of precision and another with four decimal units of precision and a closed chamber. The two decimal precision scale was used for steps 1, 4 and 7. The four decimal precision scale was used for steps 5 and 6. In step 2, the water temperature was determined so that an accurate water density is used in order to determine the existing water volume in each bottle. In step 6, the labels were removed as gentle as possible in order not to damage the labels. The glue used in binding the label to the plastic bottle was also removed from both the label and the bottle. This process was performed as carefully as possible in order not to damage either the labels or the plastic bottles.

2.2 Sustainable Package Consumer Perception

To understand the sustainability perception from a consumer's point of view, an online survey (February 2021) was developed. The survey was applied in several consumers that belong to the researchers' network.

Images of the five main water brands were presented for consumers rank with their perception about material/form, graphics and package information, in a 5-point Likert

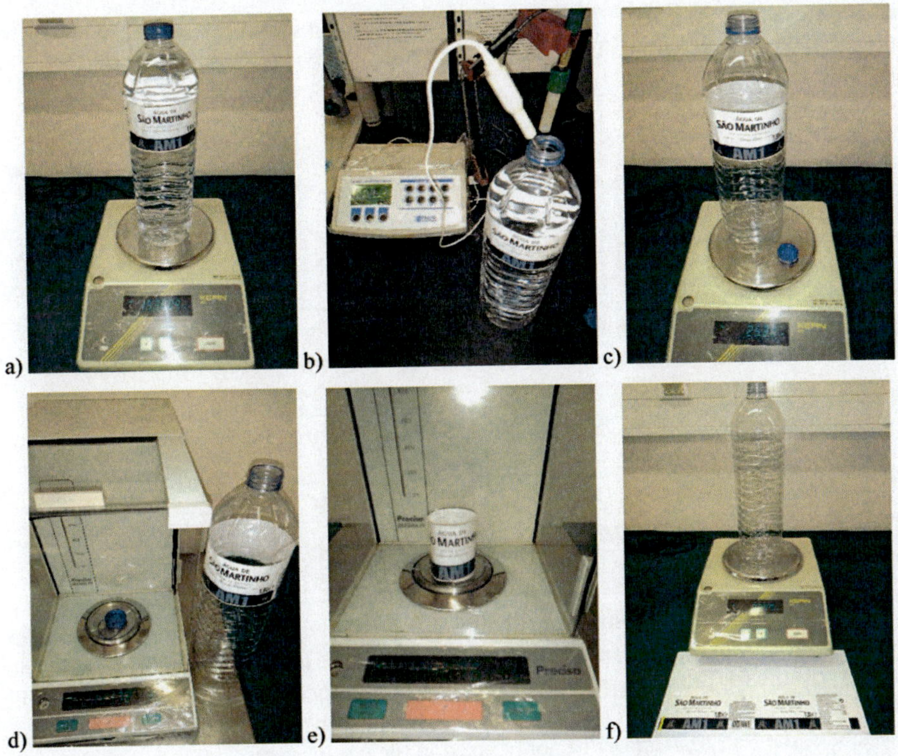

Fig. 1. Images illustrating the weighing process and the temperature reading defined in the procedure.

Scale (1- Nothing Sustainable, 5 – Very Sustainable). Consumers also rank the brands with an image containing all five water bottles from the 1st more sustainable to 5th more sustainable and a question about purchase intention was also questioned with another image with the five bottles and their price. Data statistics were performed using IBM SPSS version 27 software.

3 Discussion and Results

3.1 Experimental Quantification of Used Plastic and Paper in 1,5l Water Bottles

Considering the experimental setup, Table 1 presents the results for step 1. Based on the results, it is possible to verify that the brand Luso presents the lowest standard deviation meaning that this brand presents higher control over its production process, namely in the production of the plastic bottles, caps and labels, binding the labels and the plastic bottles and the water filling process. The brand with the highest standard deviation value is São Martinho, meaning that this brand presents a higher variability in the production process. Considering the average values, Serra da Estrela presents the lowest weight and Penacova presents the highest weight. The average temperature reading of the water inside the bottle was 16,3 °C. The water volume was determined with the aid of the Relative Density equation. The considered water density at a temperature of 16,3 °C is 0,99886 g/ml [28].

Table 1. Weight of the complete package before opening (plastic bottle, plastic cap, label and complete amount of water) for each one of the six bottles.

Brand	Weight [g] (bottle, cap, label and water)						Average	Standard Deviation
	#1	#2	#3	#4	#5	#6		
Luso	1559,65	1564,09	1565,28	1564,51	1562,26	1563,03	1563,14	**1,84**
São Martinho	1537,54	1577,92	1545,49	1557,34	1548,15	1538,09	1550,76	***13,85***
Serra da Estrela	1542,48	1554,28	1544,00	1538,86	1547,10	1543,71	**1545,07**	4,78
Penacova	1572,45	1589,43	1575,85	1579,99	1574,44	1574,78	***1577,82***	5,67
Vitalis	1545,82	1537,92	1553,61	1544,51	1545,02	1549,06	1545,99	4,76

After separating each component of the bottle (Table 2), a simple material observation was performed and the material used for the plastic bottles is Polyethylene Terephthalate, Polypropylene for the caps, paper for the labels except for São Martinho that uses a thin polymer printed film (Low-Density Polyethylene - LDPE). Therefore, when comparing the labels, the São Martinho will not be used in the comparison.

Table 2. Average weight of all components separately and the Average Volume.

Average Weight [g]				Brand	Average Volume [ml]
Bottles	Caps	Labels	Water		Water
23,62	1,3691	**1,3003**	1564,51	Luso	1536,86
23,73	**1,2251**	**0,4101**	1557,34	São Martinho	1525,40
26,56	***1,4025***	***1,6474***	**1538,86**	Serra da Estrela	**1515,46**
23,34	1,3706	1,3108	***1579,99***	Penacova	***1551,80***
23,54	1,3695	1,3491	1544,51	Vitalis	1519,75

Considering the weight of the components separately, it is possible to observe that the brand Penacova presents the lowest weight and Serra da Estrela the highest weight for the water bottles. For the caps, the brand with the lowest weight is São Martinho and Serra da Estrela again is the highest. Regarding the labels, the lowest weight is presented by Luso and the highest is presented by Serra da Estrela. And finally, the average water volume, Penacova presents the highest value while Serra da Estrela presents the lowest.

With a detailed analysis of the labels, the Serra da Estrela brand states, "bottle with 25% of recycled plastic – rPET". This might be a possible reason why this brand presents an average of 3 g of more plastic when compared to the other bottles in order to maintain structural robustness of the bottle. In this case, the Serra da Estrela brand uses 19,92g of new plastic and 6,64 g of recycled plastic. The amount of recycled plastic is not available in the other brands, but also, the focus of this study is the whole amount of used plastic and not difference between new and recycled plastic.

These values seem insignificant, but by transforming these results into the number of sold bottles according to each brand, the numbers become very significant. Table 3 presents the number of sold 1,5l water bottles for the year 2019 according to the given brands. The water volume and the weight of the separate components is also present according to the number of sold bottles (APIAMNN, 2020). The totals of each brand are also presented.

Table 3. Number of sales per brand and the volume and weight of each component according to the number of sold bottles.

Number of Sold Bottles	Brand	Volume [l]	Weight [kg]		
		Water	Bottles	Caps	Labels
30729000	Luso	47226273,37	725665,34	42071,07	39955,38
62917000	São Martinho	95973277,22	1493125,27	77078,57	25804,36
14440000	Serra da Estrela	21883266,47	383478,27	20252,34	23788,46
39217000	Penacova	60856875,24	915324,78	53752,13	51406,30
16273000	Vitalis	24730945,99	383093,54	22285,60	21953,63
163576000	Total	250670638,28	3900687,20	215439,71	162908,13

To perform an adequate comparison of the brands regarding the water volume and the separated components, the previous table was converted to an average number of sales. The average number of sold bottles for the listed brands was 32715200 bottles. Table 4 presents the adjusted data according to this average number of sold bottles. The presented data illustrates that by comparison, the brand Serra da Estrela would have the lowest amount of available water while Penacova would have the highest amount. For the bottles, the brand Penacova presents the lowest amount of used plastic for the bottles while Serra da Estrela presents the highest. For the caps, the brand São Martinho presents the lowest amount of used plastic for the caps while Serra da Estrela presents the highest. Regarding the paper printed labels, the brand Luso presents the lowest amount of used paper for the labels while Serra da Estrela again presents the highest amount of used paper for the labels.

For a better understanding of the amount of material in comparison, Table 5 considers a calculated difference between all brands with the brand that presented the lowest value for a given component. Considering the data in Table 5, it is possible to observe that the other brands present in average more 588574 l of water compared to Serra da Estrela, the other brands present in average of 33411 kg of additional bottle plastic compared to Penacova, the other brands present in average of 5001 kg of additional cap

Table 4. Volume and weight of each component according to the average number of sold bottles.

Brand	Average Volume [l]	Average Weight [kg]		
	Water	Bottles	Caps	Labels
Luso	50278791,32	772569,45	44790,38	**42537,94**
São Martinho	49903602,50	776386,22	**40078,85**	**13417,59**
Serra da Estrela	**49578631,52**	***868806,66***	***45883,61***	***53895,02***
Penacova	***50767392,83***	**763572,77**	44840,54	42883,63
Vitalis	49719034,25	770170,33	44802,92	44135,53

plastic compared to São Martinho and the other brands present in average of 4434 kg of additional label paper compared to Luso.

Table 5. Volume, weight and average of each component according to the lowest value of each component.

Brand	Average Volume [l]	Average Weight [kg]		
	Water	Bottles	Caps	Labels
Luso	700159,81	8996,68	4711,53	**0,00**
São Martinho	324970,99	12813,45	**0,00**	**13417,59**
Serra da Estrela	**0,00**	***105233,89***	***5804,77***	***11357,08***
Penacova	***1188761,32***	**0,00**	4761,70	345,69
Vitalis	140402,73	6597,57	4724,07	1597,59
Average	588573,71	33410,40	5000,52	4433,45

3.2 Sustainable Package Consumer Perception

The survey was answered by 201 subjects but some of them do not conclude it. The completed survey were answered by 113 females, 44 males and 1 that preferred not to answer this question. In average, the respondents are 33,10 years old (standard deviation = 12,112 years old) and 47,1% of them have graduation degree qualifications.

Table 6 shows the main results classified by package design attributes for sustainable packages. In terms of Material/Form, respondents consider Serra da Estrela more sustainable (higher average), followed by Penacova. Apparently, all the water bottles are from plastic material with some form/design variations. The consumer perception could be focused in term of form/material quantity (indeed are very similar) or by seeing the label details about the material. The Serra da Estrela label has the following information written: "bottle with 25% of recycled plastic – rPET" and the Penacova label has a circular economy diagram graphic with the planet at its centre (Fig. 2).

a)

b)

Fig. 2. The five water brands illustrating a) the material/form attributes and b) the graphics and c) package information of the labels.

All the plastic bottles are of colourless plastic, so the Graphics were analysed according to the label contents. The colour combination was analysed in terms of hue, the different types of saturation or brightness was not considered, and the white colour was seen as a background (not as a colour). All the sample bottles have a white background label, the São Martinho has one additional colour (dark blue), the Luso, Vitalis and Penacova have two additional colours on the label (red and blue) and Serra da Estrela has also two, blue and a hint of yellow. Almost all brands include graphics that communicate the water message; however, São Martinho does not have any pictograms to communicate the visual concept.

Penacova was considered the most sustainable brand with a higher average but also with the highest standard deviation. Serra da Estrela is the second brand with a higher average value and with the lowest standard deviation. This choice might not be because of the colour used, but in terms of the image composition that explains the whole brand – yellow star with a mountain.

In terms of Package Information, this attribute was also analysed considering the label. The written information about environmental claims and logos of environmental organizations was analysed. Some of the brands have more information about this topic and other less, and the respondent's perception attended this factor, so they consider the Penavova the more sustainable (higher average) and the Serra da Estrela with lower standard deviation. The reasons discussed here could be used to explain at Material/Form attribute.

The five brands were showed to respondents to decide which brand they considered more sustainable Table 7 reflects the main results. Serra da Estrela brand was considered the more sustainable (55 answers) but Penacova brand had 54 answers. The answers are consistent with the results of individual attributes of Table 6 because Serra da Estrela has the highest values in terms of perception of Material/Form and the second of Graphics and Package Information. On the other hand, Penacova is stated as the second brand with highest value attributes, namely second choice in Material/Form but first in terms of Graphics and Package Information. The less sustainable brand perception considered was São Martinho.

Table 7 shows the consumer choice considering the retail prices. When the prices are showed to respondents, the choice the brand that will buy, they answered Penacova (in Portugal is considered as a retail brand). This sample has an aggregated income lower than 1500€ (54,8%), which can justify this choice. Indeed, the results aren't extremely different from the others considering that the Penacova was the second most sustainable bottle and the contextual environment attributes are of higher level.

The second consumer choice is Serra da Estrela, but doesn't have so many variations between Luso ou São Martinho. Vitalis is the less frequent, but in terms of price is similar to Luso, considering that they are both the more expensive waters (Table 8).

3.3 Experimental Quantification vs Sustainable Package Consumer Perception

From the data presented in Sect. 3.1, Penacova is the brand that provides more water to its consumers while still presenting the lowest amount of used plastic for the bottles. Regarding the caps, São Martinho presents the lowest amount of used plastic. For the labels, the most sustainable option is São Martinho because it uses a thin polymer film

Table 6. Attributes with attributes material/form, graphics and package information related with study brands.

Attributes	Brand	N	Average	Standard deviation
Material/form	Luso	201	2.98	0.943
	Penacova	201	3.39	1.025
	Serra da Estrela	201	3.40	1.015
	Vitalis	201	2.74	1.006
	S.Martinho	200	2.69	0.938
Graphics	Luso	176	3,03	0.988
	Penacova	176	3,47	1.077
	Serra da Estrela	176	3,36	0.94
	Vitalis	176	2.94	1.007
	S.Martinho	176	2.81	1.066
Package information	Luso	163	3.01	1.077
	Penacova	163	3.82	0.968
	Serra da Estrela	163	3.69	0.872
	Vitalis	163	2.72	1.056
	S.Martinho	163	2.61	1.051

Table 7. Sustainable consumer choice ranking.

	Luso count	Penacova count	Serra da Estrela count	São Martinho count	Vitalis count
1st Brand More sustainable	34	54	**55**	9	6
2nd Brand More sustainable	39	**51**	42	10	16
3rd Brand More sustainable	**45**	29	36	14	34
4th Brand More sustainable	23	17	21	41	**56**
5th Brand More sustainable	20	8	12	**83**	35

for the labels that doesn't need separation from the bottle when being placed in the recycle bin. Issue that isn't considered by the general public, the separation of the paper label from the plastic bottle when placed in the recycle bin. Regarding the paper labels, it is the brand Luso that presents the lowest amount of used paper in its labelling.

Table 8. Consumer choice with price known.

Brand	Absolute frequency	Relative frequency (%)
Luso	21	13,3
Penacova	86	54,4
Serra da Estrela	26	16,5
São Martinho	21	13,3
Vitalis	4	2,5
Total	158	100,0

In other hand, when consumers evaluate the bottles separately, in terms of material/form, graphics and package information attributes, Serra da Estrela has a higher value in the first attribute. Penacova presents more graphics and package information attributes for sustainability. In global, consumers consider Serra da Estrela the most sustainable brand, but when the prices are showed, the choice changes. In this case, consumers prefer to buy Penacova, what reveals that people demonstrate their concern about sustainability, but in practice don't apply it.

Globally, Penacova is the most sustainable brand from the experimental quantification (Table 5) and from a consumer's perception is the second one (Table 7). A deeper analysis of the data also shows a possible trend, namely if the size of the population of the questionnaire increases, the results might be different and Penacova may be considered as the first choice in the ranking.

4 Conclusions

Every year in average, 1000000 L of mineral water are sold in Portugal. As a result, several ecological problems arise from this massive market, namely the amount of plastic waste from the bottles and plastic caps and additionally the labels. At same time, costumers are now paying more attention to environmental issues and most of them take into account sustainable issues in the moment of purchase. The consideration of environmental sustainability on package design decisions have been disregarded, so this study shows novelty because it crosses different types of analysis for the same product: an experimental quantification versus consumer sustainability perception and purchase intentions.

The study was carried out with the top five Portuguese sold water brands. First, the technical part of study was divided in several items (water, plastic bottle, cap and label) and in terms of weight variations of the six 1,5l water bottles of each brand. The heaviest unopened and complete bottle is the Penacova brand and the Serra da Estrela the lightest. But when the bottle components are analysed separately (water, plastic bottle, cap and label) the results are different, i.e. the more sustainable brand is Penacova. The weight components individually reflects that the most lightest bottle is Penacova, the cap is São

Martinho, and the label paper is Luso and the water weight is Serra da Estrela. The results are not unanimous.

Second, the consumer perception about sustainable form attribute is Serra da Estrela. In the same way the label graphics to this brand has the lowest deviation. For package information, the Penacova brand was considered the most sustainable. The results in this study are ambiguous, the material form belongs to Serra da Estrela, but regarding the label information, the higher value package information attribute doesn't correspond to that brand.

Thirdly, the discussion about these two different analyses types concludes that Penacova is the retail brand 1,5l water bottle that the inquiries prefer to buy, and that the sustainable issues aren't considered by consumers. In the other hand, the brand that uses the lowest amount of plastic is Penacova as the technical study shows. So Penacova is bought not because of the sustainability concept, but about the price, but curiously it‘s the most sustainable in materials terms. A possible suggestion would be to provide this information to the Penacova brand, so that they may advertise this information as a brand that cares about sustainability and increases the sympathy and commitment between their consumers and attract new ones. As Penacova is only sold by a retail brand (Lidl), this retail can use this information as well.

The analysis also shows that the Serra da Estrela communication in terms of plastic information (rPET – 25% of recycled plastic) could influence the consumer choice.

This study has some survey limitations because of the sample size. The results could be different in terms of more sustainability consumer choices ranking: Penacova instead of Serra da Estrela. Also, in terms of experimental quantification, the analysis of only 6 bottles compared to the amount of sold bottles, is very insignificant. Future research could be done considering a higher number of bottles obtained from packages of 6 bottles of different productions. The amount of recycled plastic of the other brands, the second plastic package (the plastic around the six bottles) and the wood pallet containing the 6pack packages, could also be considered in the study to determine the global ecological footprint. The production method of each brand could be also considered in future analyses.

References

1. Lourenço, C., Ribeiro, L.: Classificação Das Águas Minerais Naturais E De Nascente De Portugal Segundo As Suas Características Físico-Químicas. 7o Congresso Da Água, 11 (2004)
2. APIAMNN Associação Portuguesa dos Industriais de Águas Minerais Naturais e de Nascente (2020). Dados estatísticos e principais resultados 3o trimestre Jan-Set 2019–2020
3. Baynast, A., Lendrevie, J., Lévy, J.: Mercator 25 anos: o marketing na era digital (17ª ed.). Alfragide: Dom Quixote (2018). ISBN 978-972-20-6591-7
4. Kates, R.W., Clark, W.C., Corell, R., Hall, J.M., Jaeger, C.C., Lowe, I., Mccarthy, J.J., Schellnhuber, H.J., Bolin, B., Dickson, N.M., Faucheux, S., Gallopin, G.C., Grubler, A., Huntley, B., Jager, J., Jodha, N.S., Kasperson, R.E., Mabogunje, A., Matson, P., … Svedin, U.: Sustainability Sci. **292**(5517), 641–642 (2001)
5. Antunes, R., Vitorino, L., Lisboa, A.: Young Consumers' Intention to Purchase Products with Sustainable Packages intentions (2019). https://www.csrconferences.org/wp-content/uploads/2019/10/Working-PapersSeries-ICSR2019.pdf. Accessed 23 Sept 2020

6. Harris, C., Tregidga, H.: Hr managers and environmental sustainability : strategic leaders or passive observers ? Int. J. Hum. Resour. Manag. **23**(2), 236–254 (2012)
7. Amaro, C.: Segundo Grande Inquérito sobre Sustentabicapade em Portugal. Imagens de Marca (2019). https://www.imagensdemarca.pt/artigo/portugueses-cada-vez-mais-preocupados-com-a-sustentabicapade/. Accessed 12 Feb 2021
8. Almeida, H.A., Ramos, C.A., Bártolo, H., Bártolo, P.J.: Evaluating the impact of glass and PET packaging for bottled water. In: Bártolo, H., et al. (eds.) Green Design, Materials and Manufacturing Processes. CRC Press (ISBN: 978–1–138–00046–9), Chapter 61, pp. 345–350 (2013). https://doi.org/10.1201/b15002-67
9. Magnier, L., Schoormans, J.: Consumer reactions to sustainable packaging: the interplay of visual appearance, verbal claim and environmental concern. J. Environ. Psychol. **44**(Suppl. C), 53–62 (2015). https://doi.org/10.1016/j.jenvp.2015.09.005
10. Gomes, S., Nogueira, M., Ferreira, M., Gregorio, M.J.: Portuguese Consumers' Attitudes Towards Food Labelling. World Health Organization Europe, October 2017, 58 (2017)
11. Estiri, M., Hasangholipour, T., Yazdani, H., Nejad, H.J., Rayej, H.: Food products consumer behaviors: the role of packaging elements. J. Appl. Sci. **10**(7), 535–43 (2010). https://doi.org/10.3923/jas.2010.535.543
12. Boz, Z., Korhonen, V., Sand, C.K.: Consumer considerations for the implementation of sustainable packaging: a review. Sustainability (Switzerland) **12**(6), 1–34 (2020). https://doi.org/10.3390/su12062192
13. Underwood, R.L., Klein, N.M., Burke, R.R.: Packaging communication: attentional effects of product imagery. J. Product Brand Manag. **10**, 403–422 (2001). https://doi.org/10.1108/10610420110410531
14. Silayoi, P., Speece, M.: Packaging and purchase decisions: an exploratory study on the impact of involvement level and time pressure. Br. Food J. **106**(8), 607–28 (2004). https://doi.org/10.1108/00070700410553602
15. Magnier, L., Crie, D.: Communicating packaging eco-friendliness an exploration of consumers' perceptions of eco-designed packaging. Int. J. Retail Distrib. Manag. **43**(4–5), 350-+ (2015). https://doi.org/10.1108/IJRDM-04-2014-0048
16. Borgman, I., Mulder-Nijkamp, M., De Koeijer, B.: The influence of packaging design features on consumers' purchasing & recycling behaviour. In: 21st IAPRI World Conference on Packaging 2018 - Packaging: Driving a Sustainable Future, pp. 276–284 (2019). https://doi.org/10.12783/iapri2018/24397
17. Wansink, B.: Can package size accelerate usage volume? J. Market. **60**(3), 1–14 (1996). https://doi.org/10.1177/002224299606000301
18. Chu, A., Rahman, O.: What color is sustainable? Examining the eco-friendliness of color. Res. Gate 36–49 (2010)
19. Coulson, N.S.: An application of the stages of change model to consumer use of food labels. Br. Food J. **102**(9), 661–8 (2000). https://doi.org/10.1108/00070700010362031
20. Tarango, R.J.A.: Introducción a la ingeniería en envase y embalaje. In J.A, Rodríguez (2003)
21. Ji, M.F., McNeal, J.U.: Children's visual memory of packaging. J. Consumer Market. **20**(5), 400–427 (2003). https://doi.org/10.1108/07363760310489652
22. Orzan, G., Cruceru, A.F., Balaceanu, C.T., Chivu, R.G.: Consumers' behavior concerning sustainable packaging: an exploratory study on Romanian consumers. Sustainability (Switzerland) **10**(6), 1–11 (2018). https://doi.org/10.3390/su10061787
23. Khan, M.M., Waheed, S., Ahmad, N.: Product packaging and consumer purchase intentions. Market Forces Coll. Manag. Sci. **13**(2), 97–114 (2018). https://www.researchgate.net/publication/329736165_Product_Packaging_and_Consumer_Purchase_Intentions.
24. CGS. Retail and sustainability survey (2019). https://www.cgsinc.com/en/news-events/CGS-Survey-Reveals-Sustainability-Is-Driving-Demand-and-Customer-Loyalty#_edn1/. Accessed 24 Jan 2020

25. Steenis, N.D., Van der Lans, I.A., Van Herpen, E., Van Trijp, H.C.M.: Effects of sustainable design strategies on consumer preferences for redesigned packaging. J. Cleaner Prod. **205**, 854–865 (2018). https://doi.org/10.1016/j.jclepro.2018.09.137
26. Steenis, N.D., Van Herpen, E., Van der Lans, I.A., Ligthart, T.N., van Trijp, H.C.M.: Consumer response to packaging design: the role of packaging materials and graphics in sustainability perceptions and product evaluations. J. Cleaner Prod. **162**(Suppl. C), 286–298 (2017). https://doi.org/10.1016/j.jclepro.2017.06.036
27. Jerzyk, E.: Design and communication of ecological content on sustainable packaging in young consumers' opinions. J. Food Products Market. **22**(6), 707–716 (2016). https://doi.org/10.1080/10454446.2015.1121435
28. https://www.engineeringtoolbox.com/water-density-specific-weight-d_595.html

Water Use and Its Perception in the Hotel Industry in the Iberian Peninsula

Elena Terradillos and Aldina Soares(✉)

Instituto Politécnico de Setúbal – Escola Superior de Tecnologia de Setúbal, Campus do IPS, 2914-508 Estefanilha, Portugal
aldina.soares@estsetubal.ips.pt

Abstract. Tourism industry is one of the fastest growing in recent years, before the pandemic, and is a well-known industry with excessive consumption of water. Climate change together with the abusive use of water resources change the hydrological cycle, thus causing water supply problems. Consequently, tourism companies must undertake actions to change the direction based on a more structured knowledge and vision.

This work is based on the study of water consumption in the various activities of tourist accommodations in the Iberian Peninsula. It aims to compare the perception of their consumption against the published reference values, as well as to apply measures to reduce water consumption.

The applied methodology is based on an initial bibliographic research of the state of the art, reference values for consumption, and the influencing factors in tourist accommodation. Following, the perception of consumption is assessed through an online survey targeting Portuguese and Spanish hoteliers. The survey was conducted in the first months of 2020 and 174 responses were obtained, 101 from Spain and 73 from Portugal.

The results show that the surveyed hoteliers do not have the perception of water consumption, since in general they do not make measurements, are unaware of the consumption of each sector within the hotel and mostly refer to consumption below the reference values. Regarding the measures that have already been implemented to reduce water consumption, they represent measures of lower cost and easier application and that do not require major changes.

Keywords: Tourist accommodation · Water use · Consumption perception

1 Introduction

Tourism is an industry that induces transversal development and the evolution of society. Until the pandemic COVID, tourism was one of the sectors that had grown the most in the world. It offers the most investment opportunities also in the Iberian Peninsula, which is the territory where this work is applied. According to the Turismo de Portugal Homepage it has had a great evolution in this area with an increase till 2019, for example, in Portugal the revenues increase from 2018 to 2019 has grown 8.1% [1]. Spain and Portugal together

J. R. da Costa Sanches Galvão et al. (Eds.): *Proceedings of the 1st International Conference on Water Energy Food and Sustainability (ICoWEFS 2021)*, pp. 874–880, 2021.
https://doi.org/10.1007/978-3-030-75315-3_92

have had a great growth in tourism, it was exponential between 2014 and 2017 (21%) and from 2017 to 2019 it grew more moderately increasing about 2% [2, 3].

Tourism is a modifier for the economic, social, cultural, and environmental structure of each region, country. The modifying element can be positive, causing the areas close to the touristic development to develop, and in a negative way, inflation of the cost of living in the area, economy dependent on tourism, and the most negative side is the impact it has on the consumption of natural resources on the zone [4].

The excessive consumption of natural resources, such as fresh water, causes a decrease in natural water reserves and consequently has an impact on the local flora and fauna. The negative effect, together with climate change, makes it increasingly problematic to obtain water for human consumption. Long periods of the year with extreme scarcity together with the irregular and reduced rainfall: that will be the climatic situation in the Iberian Peninsula. Thus, it is increasingly important to have water resource management policies to adapt tourism enterprises to climate change [4].

Climatic data show that the climate in the Iberian Peninsula in recent years has been very unstable [5–7]. In this territory, precipitation in recent years has been quite irregular, and the asymmetry that exists between North and South in relation to precipitation shows two very different realities in relation to the impact that climate change is making today. The Iberian Peninsula has a very humid North and an increasingly dry South [4]. Climate change models anticipate heat waves and worsening drought in the south. According to a study carried out by the Portuguese Environment Agency, between 1931 and 2000 the climate in Portugal is predominantly colder than in the remaining years, from 2000 to 2019 there is a generalized increase in temperatures. In Spain, according to AEMET between 1971 and 2018, it is observed that the temperature and precipitation in this country is very fickle, highlighting the problem of climate change that is felt in the Iberian Peninsula. Regarding the Spanish territory, the temperature has increased approximately 1.5 °C over the last 53 years, in relation to the precipitation which has been very uneven and inconsistent [7].

2 Water Use in Hotels

As already mentioned, tourism is a modifying element of the area surrounding it. With high tourist demand, the surrounding area will also have a great responsibility for accommodation. And to satisfy the needs of each guest, there is a problem: excessive water consumption. The excessive use of water resources, namely fresh water, can have negative effects on time and space beyond the place and the moment in which the action takes place, however, climate changes such as the increase in water stress in the area where the tourism enterprise operates will, sooner or later, present consequences.

There are many reasons why the hotel industry should take a different approach to water use. This is because water consumption in hotels is generally carried in large volume to cover the need for a wide variety of facilities and leisure services/attractions. Water consumption is influenced by a set of basic consumptions that are associated with: location, climate, category/luxury, hotel size, number of beds, year of construction/renovation, swimming pools, SPA, water parks, golf, among others. Another reason why water consumption in hotels is excessive is due to the behaviour of guests. This increase is caused by a change in the behaviour when compared with what is common practice at home, since it is adopted a "pleasure approach" in daily routines, such as bath time [8–10].

2.1 Reference Values

For the evaluation of the performance of a hotel unit, reference values must be considered, which can be based on an evaluation system, with objectives, criteria, and metrics. In order to be able to compare water consumption in the various areas of the tourist accommodation, water consumption must be presented per guest and per night or per room. Additionally, it must be taken into account that the use of water varies throughout the year, due to occupancy rate and climate factors, and basic consumption. Although trying to have reference values, water consumption in a hotel can vary a lot depending also on the type and category of the hotel and its location. In a study carried out in 21 countries by Almeida, Loureiro, Poças & David [11], it was found that water consumption varies between 40 and 2500 L per overnight guest. In European countries water consumption can vary between 200 to 1000 L for each guest depending on the type, size and category of the tourist resort, in Asian countries the consumption is about 900 L for each guest [11–14].

For the case study, analysis has considered the listed values as well as the values shown in Fig. 1 and chosen the lowest value of 439 L/guest/night, as the reference indicator.

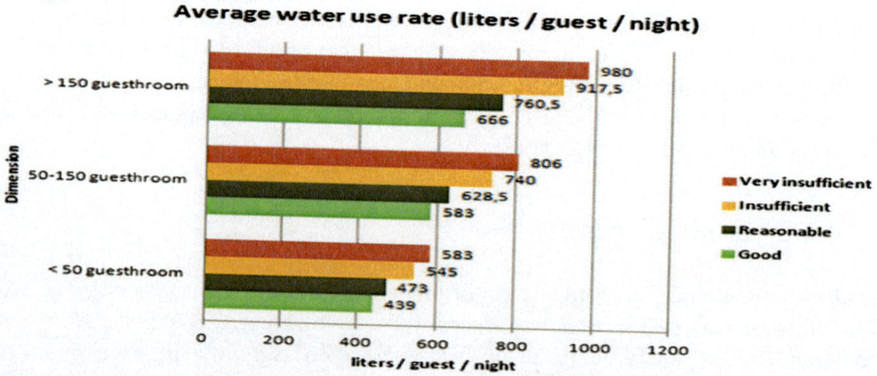

Fig. 1. Efficiency in the use of water for accommodation of different dimensions. The scale ranges from very low water use efficiency (Very Insufficient) to good efficiency (Good), measured by the indicator Average Water Use Rate in liters/guest/night (adapted from [11]).

In order to have water consumption per guest, a performance indicator must be used. The most appropriate indicator for the efficient use of water in tourist developments is the consumption of water that each guest spends per night. Thus, consumption must be calculated using the following formula, used by several researchers [11–14]:

$$C_{guest\,per\,night} = \frac{(C_{Total} * 1000)}{N_{Nights\,for\,each\,guest}} \quad (1.)$$

$C_{\text{guest per night}}$ - Consumption in liters per guest per night,
C_{Total} - total consumption in m^3 for the calculation period (water meter),
$N_{\text{nights for each guest}}$ –Total amount of overnight stays, in the same calculation period.

3 Developments in Methods and Techniques to Reduce Water Consumption in Hotels

The answer to climate change involves adapting the accommodation units to climate change, that is, by adapting the accessories/equipment of the water distribution system with more appropriate elements and implementing more appropriate measures to increase efficiency as well as its management. In this way, an action is defined that has the direct result of a more efficient use of water, by reducing both losses and the amount of water for each use.

The most relevant measures for the adequacy of tourist accommodations are as follows [11–14]:

- Measures at the level of supply systems that are aimed at reducing water consumption in supply systems (example: by reducing pressures on the supply system and using treated urban wastewater). However, this application is the responsibility of the entity that manages the public supply system with coordination with consumers. Flow measurement at various points of use can be a way to identify and evaluate potential zones that can be readjusted.
- Measures at the level of building systems and collective installations within these measures include the reduction of pressure in the building supply system, the thermal insulation of the hot water distribution system, the use or reuse of rainwater and gray water and the reduction of water losses in the system with the help of maintenance and monitoring actions of the supply system.
- Measures at the level of uses like residential ones and specific to collective use facilities are understood to be the uses associated with the personal hygiene of guests, sanitary facilities, washing of clothes and dishes, as well as heating and air-cooling systems. It aims to promote proper use by users; reduce water losses and waste. The most relevant measures are the following: adaptation or replacement: cisterns, showers, taps, washing machine, dishwasher, urinals, air conditioning system.
- Measures at the level of outdoor uses include watering gardens, green spaces in general, washing yards, accesses, filling swimming pools, lakes and water uses for ornamental purposes and washing vehicles. The most relevant measures are the following: adaptation or replacement irrigation system, outdoor washing system, pool supply system. Introduction of systems such as the capture of rainwater for the use of tasks that do not require drinking water, use of grey water.
- Awareness and information measures, these must be developed internally for the organization's employees and for the users of the accommodation unit. In relation to employees, an action plan for the efficient use of water must be defined, through communication and training. For the remaining users, it is necessary to inform and reinforce this information through notices, signs, and awareness actions, both by employees and by the environment where they are.

4 Case Study

4.1 Methodology

To achieve the research objective, an online survey was conducted to Portuguese and Spanish hoteliers. The survey included three parts: first part is about accommodation unit general questions, such as the type and category, occupancy rate, if it has a garden, and if the hotel has golf. In the second part of the survey, the questions are about water consumption total and in some areas, if it was measured or there is a perception of the consumption, the origin of the water used, and especial questions for gardens and maintain green spaces. The third part refers to questions about the measures already implemented and those that intend to be implemented in the accommodation in the future to reduce water consumption.

The survey took place in the first months of 2020. The GoogleForms platform was used to carry out the survey, in which 850 inquiries were sent, 500 to tourism enterprises in Spain and 350 in Portugal. The data were processed individually and subsequently compared between the two countries.

4.2 Survey Results

A total of 174 inquiries answers were obtained, that is, a response rate of about 21%, 101 were Spanish and 73 Portuguese.

Regarding the type and category of tourist accommodations (hotel stars or LA): 1-star (9%), 2-star (14%), 3-star (22%), 4-star (26%) and 5-star (6%) and local accommodation-LA (7%). In both countries regarding the tourist resorts that responded to the survey, about 68% have less than 50 rooms, 20% have between 50 and 150 rooms and 13% have more than 150 rooms.

4.3 Results Analysis

In relation to annual water consumption, in Portugal around 30% consume less than 5000 m^3 per year and 25% have no knowledge about the total amount of annual consumption. On the other hand, in Spain, the same is observed, about 36% of tourist developments consume less than 5000 m^3 per year, and 23% cannot estimate annual water consumption. Regarding the perception of annual total consumption, it is noted that most of the hoteliers, of both countries, have difficulty in knowing or even estimating them.

Relating to the indicator of consumption per guest/night, comparing to the reference value about 59% consume less than 439 L/guest/night and about 23% do not know how to estimate this consumption. When analyzing consumption data per guest per stay, accounting for all consumptions related to your stay (restaurant, pool, golf, among other activities/amenities), it appears that the answer is similar when not counting all water consumption.

For green spaces (garden, golf), the 174 tourist accommodations that participated in the survey, about 51% have green spaces. In relation to water consumption in this sector, about 33% do not know how much water is necessary to maintain these areas. Regarding the origin of the water for irrigation, there are differences between the tourism projects in both countries. In Portugal, about 41% use water from an own borehole, and 35% from the mains water supply, while in Spain it was noted that, on the contrary, about 56% of tourist developments use water from the mains water supply for irrigation and only 20% from an own borehole. This difference can originate from several causes, such as the price of mains water, or from legal difficulties in opening own boreholes, or even the access to water reserves in the region, or other causes; but no data were collected in the survey to conclude why this difference between countries.

Regarding the actions to reduce water consumption and through the literature it is possible to find several technical measures. One of these actions is to measure the water consumption by zones. In view of this, only about 30% of respondents have counters by zones and around 70% do not have. Concerning the measures implemented to reduce water consumption, there is consensus on the responses in both countries. The most used measures are signs, warnings (save water, etc.), use of washing machines always at maximum capacity, use of the right dosage both for washing clothes and other washing/cleaning tasks, double flush toilet discharge. In relation to the future measures to be implemented to reduce water consumption, there were some responses that indicate "*they have no intention*", and the remaining responses were those previously mentioned, signs, warnings and among others.

5 Conclusions

The purpose of this research was to study water consumption in the various activities most common in Iberian Peninsula tourist accommodations, comparing these values with published reference values and also identify whether the techniques and good practices for saving water are known and whether they are applied. In order to obtain the data, a survey was carried out based on interviews with reference tourist resorts, and an analysis of the cases in the literature.

About water consumption, it appears that about 59% of hotels in the studied region say they consume relatively little water, noting that the consumption to accommodate one guest per night is less than the chosen reference value of 439 L/guest/night, and about 23% cannot estimate water consumption. This leads us to deduce that with great probability the values that answered are not the reality, or the greatest part does not know and wanted to give a good environmental image.

Regarding the techniques and measures they already have or intend to use to combat the excessive use of water, it appears that the hoteliers both in Portugal and Spain use the most economical and easy measures to implement. The most used measures are: "Double flush toilet", "Warnings (*save water, don't scrub your teeth with running water, use the flush for each function, reuse towels and bedding when possible, give preference to the shower*, etc.)", "Washing machines are always used to their maximum capacity", % "Use of correct dosage of detergents for washing clothes", "Low flow regulators", "Use of correct dosage of detergents for cleaning". Still analyzing this important issue for the reduction of water consumption, it is noted that when questioned by the measures they wish to implement in the future there is very little desire to implement more expensive measures that could help to overcome this problem of excessive water consumption as they are more effective to place merely a sign or a warning.

References

1. Turismo de Portugal Homepage. https://business.turismodeportugal.pt/pt/Conhecer/Apresentacao/Desempenho_Turistico/Paginas/default.aspx. Accessed 01 June 2020
2. INE Instituto Nacional de Estadística – Hostelería y Turismo (2019). https://www.ine.es/dyngs/INEbase/es/categoria.htm?c=Estadistica_P&cid=1254735576863. Accessed 15 June 2020
3. Figuerola, M.: Evolución del turismo en España durante el año 2018, España (2019)
4. Ferreira, P.: Alterações climáticas e desenvolvimento, Projeto: coerência.pt – O eixo do Desenvolvimento, Lisboa, novembro (2017)
5. Alfonso, C., Estévez, R.: Emergencia climática en España, Observatorio Sostenibilidad, diciembre (2019)
6. Guerra, M., Ribeiro, R., Rodrigues, S.: Relatório do estado do ambiente, Agência Portuguesa do Ambiente, junho (2019)
7. AEMET: Meteorología - Informe anual 2018, AEMET, Gobierno de España – Ministerio para la Transición Ecológica Madrid (2018)
8. Torres-Bagur, M., Ribas, A., Vila-Subirós, J.: Incentives and barriers to water-saving measures in hotels in the mediterranean: A case study of the Muga River Basin. University of Girona, (Girona, Spain) (2019)
9. Rodrigues, G.: Análise técnico-econômica de um sistema de aproveitamento de água pluvial num empreendimento do setor hoteleiro em Delmiro Gouveia, Relatório final – Bacharel em Engenharia Civil. Universidade Federal de Alagoas, Engenharia Civil, Delmiro Gouveia-AL (2018)
10. Tróia Resort: Declaração ambiental 2018, EMAS & SGS, 2ª Atualização. https://emas.apambiente.pt/sites/default/files/files/emas/declaracoes/Declara%C3%A7%C3%A3oAmbiental_2018_vf.pdf. Accessed 21 May 2020
11. Almeida, M., Loureiro, D., Poças, A., David, L.: Água em empreendimentos turísticos – Estado da arte, AdaPTAC:T: LNEC. IPMA, Junho (2015)
12. Pinto, A.: Estado da arte: Aspetos gerais e identificação de indicadores, LNEC e IPMA (2015)
13. Amazona, I.: A solução que cai do céu: revisão das normas para a captação de água pluviais e proposição para o setor hoteleiro no Brasil, Revista Direito Ambiental e sociedade, 6, 239–263. Brasil, Julho (2016)
14. Styles, D., Schonberger, H., Martos, J.: Best Environmental Management Practice in the Tourism Sector, Institute for Prospective Technological Studies – European Commission, ISBN: 978-92-79-30895-6 (pdf) (2013)

Evaluation of the Sediment Rate of the Water Reservoir of a Hydroelectric Plant Using the ^{210}Pb Dating Technique: Case Study of the Vigário Reservoir, Piraí, RJ/Brazil

C. D. Vieira[1], J. M. Godoy[1], M. N. Frota[2(✉)], L. Q. Viana[3], C. A. S. Farias[3], and D. F. C. Oliveira[1]

[1] Chemistry Department, Pontifical Catholic University of Rio de Janeiro, Rio de Janeiro, RJ 22951-900, Brazil

[2] Metrology Programme, Pontifical Catholic University of Rio de Janeiro, Rio de Janeiro, RJ 22951-900, Brazil

mfrota@puc-rio.br

[3] Maintenance Management of Power Plants, Light Energia S.A., Rio de Janeiro, RJ 22951-900, Brazil

Abstract. The Vigário reservoir is part of the Paraíba do Sul basin hydrological complex, operated by Light Energia S/A., which supplies water to three integrated hydroelectric plants, with an installed capacity of 612 MW, put into operation in the 1950s. Considering the increase in the supply and reliability of the electricity provision, the operation of a hydroelectric plant generates changes in the profile of land occupation, with a consequent impact on erosion rates. As a consequence, the original parameters related to the reservoir silting are no longer applicable, resulting in a useful life shorter than that specified in the original design of the reservoir. Among the alternatives for the recovery of historical data related to the sedimentation speed, which allow reconstructing the speed variation on an annual scale, the study highlights the use of naturally occurring radioisotopes, such as ^{210}Pb (Gu, Chen and Salem 2011; Palinkas and Russ 2019; Zhang et al. 2019). This paper discusses the study of samples taken from four sediment cores acquired from different locations in the Vigário reservoir, using a 1meter long gravity sampler. The moisture profile obtained qualified three of the cores to be analysed using the ^{210}Pb dating technique. Considering that the validation of dating with ^{137}Cs in the Southern Hemisphere is not always applicable, metal determination was carried out along the sediment's cores, while the observed variation was compared with the industrial development of the Paraíba do Sul basin. The results obtained demonstrate an acceleration in the silting process over the past decades.

Keywords: Sediment rate in water reservoir · Sediment core · ^{210}PB dating technique

J. R. da Costa Sanches Galvão et al. (Eds.): *Proceedings of the 1st International Conference on Water Energy Food and Sustainability (ICoWEFS 2021)*, pp. 881–891, 2021.
https://doi.org/10.1007/978-3-030-75315-3_93

1 Introduction

The damming of watercourses is a common practice throughout the world and has been present since ancient times, moving watermills. Its construction for the purpose of producing electricity started in the late 19th century, with the Fox River plant, Wisconsin (USA) in 1882. Not long after, in 1889, the first hydroelectric plant in Latin America was built, the *Marmelos* power plant, in the Paraibuna river (MG/Brazil). With the growth in demand for electricity, there was an increase in the construction of small plants for local use. The world and national scenarios changed drastically after the Second World War, especially after 1950, with the acceleration of economic and industrial development [1]. The use of direct current to transport electricity allowed the extension of the distance between the generating plant and the consumer centres, favouring the construction of large hydroelectric plants, such as the *Itaipu* and *Belo Monte*. Despite their benefits, large plants cause changes in the profile of land occupation and, consequently, in the erosion rates.

One of the ways of evaluating the sedimentation rates of a reservoir is made by the ecobatimetric survey, a technique that, however, requires an interval of years between two surveys, in order to allow the calculation of the sedimentation speed (cm/year). In the case of medium-sized reservoirs, 10 m^3–50 x 10^6 m^3, the recommendation of the Brazilian regulatory body (ANEEL) is to consider an interval of 5 years between two rounds of surveys [2, 3]. However, considering that ecobatimetric surveys are costly, the time interval usually employed extrapolates endorsed recommendations. Recovery of sedimentation speed historical data allows reconstructing the variation in speed on an almost annual scale, which is feasible to be achieved by the use of naturally occurring radioisotopes, as is the case with ^{210}Pb [4, 5, 6].

The object of study of this work is the Vigário reservoir, located in the municipality of Piraí, RJ/Brazil. The Vigário reservoir is fed by the Santana reservoir (reservoir of passage and accumulation of water), positioned on the Piraí River and fed by the waters of the Paraíba do Sul River, as a result of the diversion of its water course in 1945, with the purpose of increasing the water capacity for electricity generation of the Light Energia S/A system. The implementation of the Santana reservoir (whose waters has a short detention time as they are pumped to the Vigário reservoir 35.8 m above) causes the accumulation of suspended solids. The water in the reservoirs suffers chemical pollution from industrial activities along the Paraíba do Sul river basin, especially those in the metallurgical and steel sectors, in addition to pollution by domestic sewage from surrounding cities [7].

Bathymetric studies carried out by the concessionaire Light in 1989 (only 37 years after its construction) indicated that the Santana reservoir had already lost 34% of its original volume, even though it had been dredged more than once [7]. A recent study [8] of bathymetric surveys carried out by Light in the years 2001 and 2017 confirm a volume of silted material of 1,280,620 m^3, in 16 years of observation, as a result of the reduction in the volume of water in the reservoir between 2001 (22831611 m^3) and 2017 (21550990 m^3). Based on these two surveys, an average sedimentation speed of 2.7 cm/year was estimated. Through determination of metals, granulometry in sedimentary cores and ^{210}Pb dating of the sediment cores studied, this paper discusses how the useful life of the reservoirs can be affected by the history of the variation in the sedimentation rate.

2 Materials and Methods

Based on the criteria of water column depth and low influence of the river current, four sedimentary cores were collected (Feb/2020) for analysis in different locations of the Vigário reservoir. Figure 1 identifies the depth and the georeferenced coordinates of the collection sites; image Fig. 1a illustrates the slicing procedure, while image Fig. 1b illustrates the preparation/packaging process of the samples produced.

Sample location	Latitude Longitude	Sample depth (m)
Vigário-1	22°38'19.10" S 43°53'46.99" O	6,0
Vigário-2	22°38'49.43" S 43°53'26.21" O	6,5
Vigário-3	22°38'42.10" S 43°52'37.62" O	8,0
Vigário-4	22°40'05.43" S 43°53'04.72" O	6,0

Fig. 1. Sediment core: location; slicing (Fig. 1a) and packaging (Fig. 1b)

Sampling was performed with the help of a gravitational sampler, whose cores were subsequently subdivided into 2 cm thick layers, producing samples that were weighed and cooled to a temperature of 4 °C for further laboratory treatment.

The moisture content of the core samples was determined by the individual insertion of each sample into an oven, maintained at 60 °C, for a period of about 60 h, until the sample mass remained constant, thus indicating absence of water in the sample. The moisture content was then determined on an analytical balance (Sartorius, resolution: 0.0001 g) by the difference between the wet and dry masses of each sample.

The methodology used for the determination of metals in total sediment was adapted from the EPA-3050B method, for the digestion of samples and, by EPA-6010, for the determination by Induced Coupled Plasma - Optical Emission Spectrometry ICP OES. About 0.5 g of previously dried sediments were weighed in duplicate and sprayed in an agate mortar mill. Then, the following procedures were performed: (i) a volume of 10 mL of concentrated HNO_3 was added to the samples that, after addition, were left to stand for 12 h; (ii) the samples were heated to 90 °C, in a digester block for 5 h; (iii) 1 mL of H2O2 was added to the samples, which remained at rest for 1 h; (iv) exposed to an ultrasound bath and heated again for the same time and left to rest for another 12 h and, after this period, (v) the samples were centrifuged for 3 min, at 3,000 rpm and washed. Only then, the extract was separated and added to 30 mL with ultrapure water for later quantification of the metals by ICP OES. All samples were attended by two blank solutions and SRM 2704 certified reference material (Buffalo River Sediment), warranting the quality of the results and verification of the methodology.

The determination of Hg was performed separately, using wet samples. The method used was EPA-7471, for the determination of Hg in solid samples by the cold vapor technique. Thus, as for the other metals, the analysis was performed on a sample for every three, with each slice being analysed in duplicate, accompanied by two "white" solutions and MESS-3 certified reference material (Marine Sediment Reference Materials for Trace Metals and other Constituents). In the sequence, the ^{210}Pb dating was performed. For this, a mass of three grams of each selected layer was double leached, conform the following sequential procedures: (i) with 40 mL HBr 0.50 mol/L and (ii) with 40 mL HBr 0.50 mol/L, with 1.0 g of hydroxylamine chloride, for two hours, stirred at 80° C. After the addition of 1 mL Pb carrier (20 mg/mL), it is separated by ion chromatography, in hydrobromic medium. The feed solution and the two washing solutions (20 mL HBr 0.50 mol/L) are then collected for later determination of ^{226}Ra. Pb elution is carried out with 100 mL 1.0 mol/L HNO_3. The eluate is evaporated to dryness, the solution is taken up with water and the pH adjusted ≈ 5.0 with sodium acetate. Lead is precipitated with the addition of Na_2CrO_4 ($PbCrO_4$), filtered and weighed to determine the chemical yield. ^{210}Pb activity is then determined by beta counting, using a low background proportional detector, for 400 min, after a minimum growth period of 10 days.

The measured ^{210}Pb represents the so called $^{210}Pb_{total}$, as described in Eq. (1).

$$^{210}Pb_{total} = ^{210}Pb_{matrix} + ^{210}Pb_{atmospheric} \quad (1)$$

Taking into account the ^{238}U radioactive decay series, it is possible to consider that the $^{210}Pb_{matrix}$ is in radioactive balance with the ^{226}Ra and, for this specific reason, the dating of sediments with ^{210}Pb also implies the determination of the ^{226}Ra [9].

The granulometry was performed using the particle analyser Cilas 1190 Particle Size Analyzer, which allows measurements of particles with sizes ranging from 0.04 to 2500 μm at once. The wet samples, after defrosting, were used without the need for straining, as they appeared muddy, without gravels. In his study, Jordão (2017) characterized the suspended solids of the Piraí River and the waters of the Vigário Elevatory Plant as silty, with samples of medium size (d_{50}) of 7 μm [10].

3 Discussion and Results

3.1 Moisture Content

Figure 2 presents measurement results of the moisture profiles for the four sediment cores studied, used as an auxiliary tool in the decision process for qualifying sediment cores that have the potential to provide a "datable" ^{210}Pb profile. In the absence of a mixture of sedimentary layers, an exponential decrease in humidity with depth is observed. The core Vigário-1 was discarded due to disturbances explained by the location of its collection, near the dam, where the movement of the current resulting from the pumping is still high.

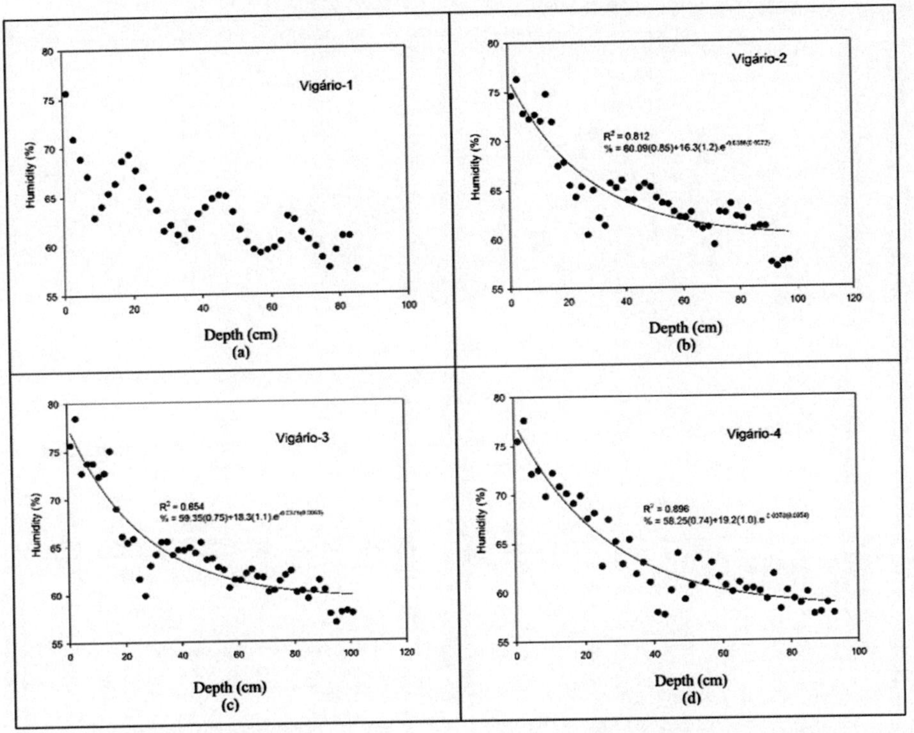

Fig. 2. Moisture profiles in the cores collected in the Vigário reservoir

3.2 Datation with ^{210}Pb

The concentration of ^{210}Pb was performed in one of each three layers, while the concentration of ^{226}Ra, in one of every five layers. Figure 3 shows the results of the ^{210}Pb concentration in excess (Bq/kg) of the samples, not exhibiting the classic exponential, indicating the existence of variations in the sedimentation speed.

The age of each layer was calculated using the Constant of Rate Supply (CRS) model, taking into account the reference date of the last layer of 1954, related to the closure of the Vigário dam. Figure 4 illustrates the graphical representation of the obtained results. Figure 4a shows the adjustment of the linear curve, segmented in three stages, of the sediment age dated by the ^{210}Pb technique, between 1955 and 2020, for the different depths measured along the sediment core. It is observed that the sedimentation speed calculated for the last years is consistent with the estimate obtained by processing the data associated with bathymetric surveys carried out in 2001 and 2017 [9]. Analysis of this graph indicates a gradual increase in sedimentation speed, more specifically in the early 1980s and about 10 years later. This second change is very evident when working with the sedimentation mass velocity (MAR), whose results are shown in Fig. 4b, illustrating detail of the study of the mass depth of the sedimentary layer versus the age calculated by the CRS model.

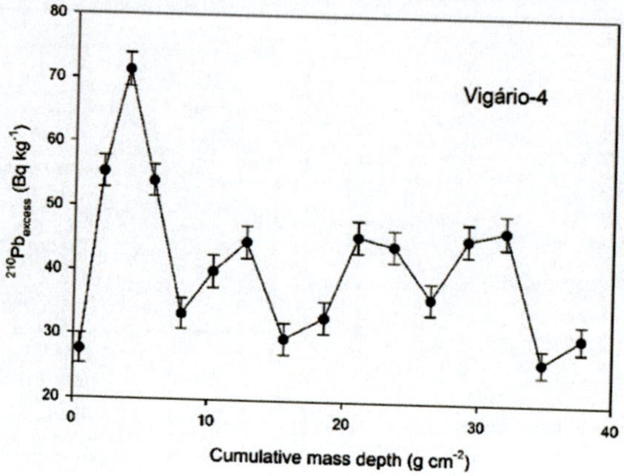

Fig. 3. Results of ^{210}Pb concentration in excess (Bq/kg) in the Vigário-4 sediment core

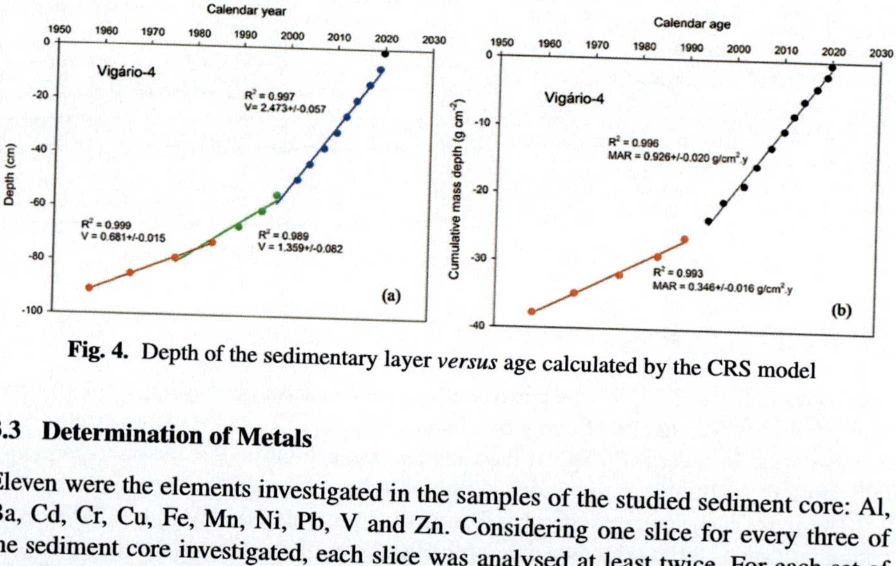

Fig. 4. Depth of the sedimentary layer *versus* age calculated by the CRS model

3.3 Determination of Metals

Eleven were the elements investigated in the samples of the studied sediment core: Al, Ba, Cd, Cr, Cu, Fe, Mn, Ni, Pb, V and Zn. Considering one slice for every three of the sediment core investigated, each slice was analysed at least twice. For each set of samples, two aliquots of the certified reference material were analysed, two blanks and a standard of 1.0 mg/L. The descriptive statistics are presented in Table 1, confirming that all samples have a value of the Coefficient of Determination below the limit of quantification of 0.10 mg/kg.

An analysis of the statistical significance of the data was performed to assess an acceptable statistical difference between the averages obtained for the different elements in the two Vigário-4 cores collected at the same collection site. It was found that the elements Mn, Al and Fe have higher concentrations since it was observed that the Vigário sediment is very muddy. Cu and Ni exhibited the opposite behaviour. As the Brazilian environmental legislation still does not establish standards for organic and inorganic chemical substances found in sediments, limits of references established by Organs international environmental protection agencies were adopted, the same ones used by

Table 1. Descriptive statistics of the analysis results (values expressed in mg/kg).

Statistics component	*Ba*	*Pb*	*Zn*	*Ni*	*Mn*	*Cr*	*V*	*Cu*	*Al*	*Fe*
Average	179,8	31,00	161	12,21	1169	88	81,77	25,48	38624	61031
Standard error	2,0	0,88	12	0,21	69	16	0,89	0,80	539	751
Median	180,1	31,29	163	12,05	1090	55	81,75	25,73	38462	61807
Standard deviation	8,1	3,51	47	0,82	277	63	3,56	3,20	2157	3003
Relative stand deviation (%)	4,5	11	29	6,7	24	72	4,4	13	5,6	4,9
Mínimum value	162,8	25,91	91	11,01	874	48	76,24	21,47	35408	55506
Maximum value	195,9	37,56	238	13,70	1802	272	88,25	30,66	42802	64701
Score	16	16	16	16	16	16	16	16	16	16

the State Environmental Institute of the region (TEC, Threshold Effects Concentrations) [11]. Table 2 shows data related to the assessment of the concentration of pollutants in sediments according to these limits.

Table 2. Metal concentrations: results of this work versus literature.

	This work (Vigário)	Medium High RPS [12]	Medium Low RPS [13]	Low RPS [14]	Santana Reservoir [15]	(TEC/PEC) [16]
Cr ($mg.kg^{-1}$)	38.8 - 47.8	35 - 46	11 - 57	85 - 97	90 - 610	43 - 110
Mn ($mg.kg^{-1}$)	1444 - 1485	-	33 - 794	729 - 1178	460 - 940	460 - 1100
Zn ($mg.kg^{-1}$)	80.9 - 105.1	52 - 109	79 140	195 - 244	220 - 550	120 - 460
Cu ($mg.kg^{-1}$)	20.2 - 21.6	17 - 29	4.7 - 18.2	64 - 91	63 - 156	32 - 150
Pb ($mg.kg^{-1}$)	25.6 - 26.6	25 - 38	11 - 80	-	50 - 135	36 - 130
Hg ($mg.kg^{-1}$)	0.04 - 0.046	0.02 - 0.28	-	-	-	0.18 - 1.10
Fe (%)	5.5 - 6.5	2.3 - 4.2	2.0 - 4.2	10.4 - 11.6	4.7 - 7.6	2 - 4

The TEC value indicates that from this reference value, adverse effects can be frequent in the studied organisms, while the PEC (Probable Effect Concentration) represents the concentration above which adverse effects are commonly expected in these organisms. Table 2 also compares data acquired in this study with results from other studies carried out in the Paraíba do Sul river basin. From this data it is possible to observe that the variation in concentration (maximum and minimum values) for Cr, Zn, Cu, Pb and Hg are below the range that indicates a certain probability of causing toxic effects on organisms (TEC).

Dosages of Mn and Fe, whose results have fallen above the range, suggest that benthic organisms may be suffering some pathogenic effect. The decrease in the ichthyological fauna of the Paraíba do Sul River, as a result of the changes that the river has undergone over the years was reported by Araújo (1996) [17], most likely can be explained by the misuse of the land in the riverside areas and by the introduction of effluents from chemical and steel industries in the region.

Figure 5 presents the results of the historical tracing studied for the different elements.

The graphs detail each element surveyed over the years of observation of the Vigário reservoir: Pb, Ni, V, Cu, documented in Fig. 5a; Ba, Cr, V, Zn, in Fig. 5b; Al, Fe, Mn, in Fig. 5c and Hg, in Fig. 5d.

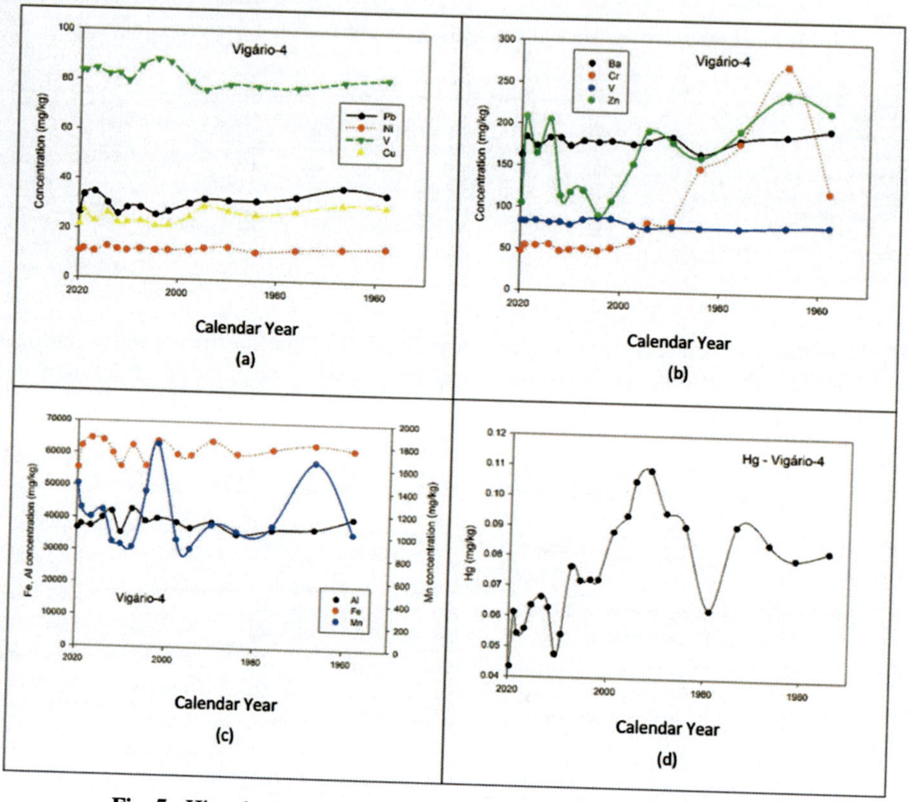

Fig. 5. Historical distribution of metals in the Vigário-4 sediment core

It is interesting to note that, with the exception of Mn and Fe, the dosages of metals present in the samples of the core collected in the Vigário reservoir present either lower values or within the range of the values reported by Pereira et al. (2006) [13], with the exception of Cu, according to analyses performed on samples collected downstream of the studied region [14, 15]. The analyses suggest that the presence of these metals is not correlated with the carrying of waters from the Paraíba do Sul river to the Piraí river, but that the occupation of land by the population along rivers and reservoirs may be the cause of this contamination.

Based on the dating results, it is possible to convert the depth of each sediment layer into chronological age and build the profiles of the temporal variation of the concentration of the various elements analysed in the samples of the studied sediment core. The observation of the results above highlights some points of attention. With regard to Cr, it reached a peak in 1969 (272 mg/kg) and a significant reduction in its concentration until the mid-2000s, remaining practically constant at the level of 50 mg/kg. As far as Zn is concerned, the results also show a higher concentration in 1969, with a decrease in this value between 1995 and 2012, increasing its concentration again in 2014. The main sources of zinc pollution are associated with metallurgy and mining, activities that have been present in the Paraíba do Sul basin since the beginning of the region's industrialization process [18]. As can be seen from the results shown in the images of Fig. 6, Ni, Cu, V, Pb and Ba remained practically constant over the period of time studied. Regarding the elements present in greater concentration, Fe, Al and Mn, are generally associated with the clay fraction; Fe and Al show the same behaviour, while Mn shows two peaks, also, in periods that correspond to the observed anomalies, in other elements, between the beginning of the 2000s and the end of the 1960s. The data in Fig. 5d confirm that the Hg concentration in the analysed sediments is below the established standard limit. In the mid-1970s, the concentration of Hg was high when compared to results of more recent analyses, reflecting the industrialization process that was established in the Paraíba do Sul region [19]. In the 1980s, activities that used to apply organo-mercurial pesticides and the cultivation of sugar cane and its processing, were taking place in the same region, which may explain the peak concentration of mercury observed in that period. It is possible to relate the decay of the Hg concentration in sediments with the efficient control of industrial emissions and/or with the ban implemented later on the use of organo-mercurial pesticides [20].

3.4 Granulometry

The results of determining the particle size were organized in three groups, according to the grain size of the particulate: clay ($Ø \leq 0.02$ mm), silt ($0.02 < Ø \leq 0.06$ mm) and fine sand ($Ø \geq 0.06$ mm). The percentage ratio of these ranges can be seen in Fig. 6.

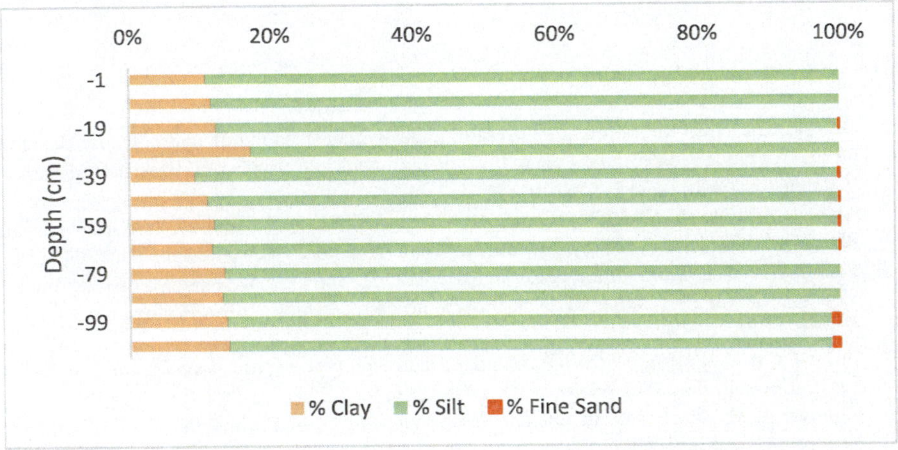

Fig. 6. Particle size percentage of clay, silt and fine sand fractions in the Vigário-4 sediment core

The sedimentation of fine particles identified occurs at the margins and backwaters, regions chosen for the acquisition of the sediment cores [12]. Silt is the predominant portion (>80%) in all depths, corroborating studies by Jordão (2017), who highlights Vigário as a reservoir of suspended solids, exhibiting a median decrease of 15 mg/L in the concentration of suspended solids, between intake and the outflow of the reservoir [11].

4 Conclusions

Results of analyses in sediment cores collected in the Vigário reservoir (i) qualified the moisture content profiles present in the sliced samples as a pre-qualification tool for identifying sediment cores that are feasible to generate datable samples by the ^{210}Pb technique and (ii) allowed to determine the variations in the concentrations of metals present in the sediments. Although only the analyses of the samples from the Vigário-4 core have been completed, the results so far obtained are promising to validate the ^{210}Pb dating technique as an effective strategy for assessing the age of the sediments and estimating the associated sedimentation rate. From this perspective, the work contributes to the understanding of the complex phenomenon related to the silting up of water reservoirs, which so drastically impact the environment and reduce the generation capacity of hydroelectric plants. The study also contributes to the management of the reservoir, guiding interventions and preventive measures to control land use and to contain silting rates.

Acknowledgements. Credits are due to ANEEL (The regulatory body for the Brazilian electric sector) and to Light Energia S/A, for financing the research, through the project of R&D MoVaSC (Flow, Sediment and Climate Modeling), Light/Aneel 5161–0016 / 2019, Contract 4500428146. Acknowledgements are also due to CAPES/Ministry of Education /Brazil (Coordination for the Improvement of Higher Education Personnel), through the Brazilian governmental institutional support (Financing Code 001) to the Graduate Programmes in Chemistry and Metrology of the Pontificia Universidade Católica do Rio de Janeiro.

References

1. DE Oliveira, N.C.C.: The great acceleration and the construction of hydroelectric dams in Brazil (A grande aceleração e a construção de barragens hidrelétricas no Brasil). Portuguese. Varia Historia **34**(65), 315–346 (2018)
2. ANA – AGÊNCIA NACIONAL DE ÁGUAS. ANEEL – AGÊNCIA NACIONAL DE ENERGIA ELÉTRICA (2010). Resolução Conjunta nº 03/2010, 10 de agosto de 2010
3. Carvalho, N.O., Filizola Júnior, N.P., Santos, P.M.C., Lima, J.E.F.W.: Reservoir silting assessment guide. (Guia de avaliação de assoreamento de reservatórios). Iin Portuguese. Brasília: ANEEL, 140p. (2000)
4. Gu, J., Chen, Z., Salem, A.: Post-Aswan dam sedimentation rate of lagoons of the Nile Delta. Egypt. Environ. Earth Sci. **64**(7), 1807–1813 (2011)
5. Palinkas, C.M., Russ, E.: Spatial and temporal patterns of sedimentation in an infilling reservoir. Catena **180**(March 2018), 120–131 (2019)
6. Zhang, Y., Long, Y., Lu, X., Pei, Z., Wu, Z., Zheng, J., Xu, S., Wang, X., Li, W.: Dating reservoir deposits to assess the recently changing sediment yields from a medium-sized agricultural catchment. Land Degradation Dev. **30**(16), 1939–1953 (2019)
7. Coelho, V.M.B.: **Paraíba do Sul**: Um rio estratégico, p. 336 Casa da Palavra, Rio de Janeiro (2012)
8. Frota, M.N., Godoy, J.M., Viana, L.Q.: R&D Project Report, Flow, Sediment and Climate Modeling (MoVaSC), Ref.: Light/Aneel 5161-0016/2019, Contract number: 4500428146

9. Godoy, J.M.: Sediment dating with 210Pb: Rescuing the History of Environmental Pollution in the Anthropocene. [210Pb Sediment Dating: Redeeming the History of Environmental Pollution in the Anthropocene]. (Datação de Sedimentos com 210Pb: Resgate do Histórico da Poluição Ambiental no Antropoceno. [210Pb Sediment Dating: Redeeming the History of Environmental Pollution in the Anthropocene]). In Portuguese. Revista Virtual Química **10**(6), 1733–1757 (2018). Published in the Web: December 18, 2018. Accessed https://rvq.sbq.org.br
10. Jordão, M.D.M.: Influence of Paraíba do Sul-Piraí-Guandu Basin Transposition on Discharge of Suspended Solids to Sepetiba Bay. (Influência da Transposição de Bacias do Paraíba do Sul-Piraí-Guandu na Descarga de Sólidos Suspensos para a Baia de Sepetiba). In Portuguese. Rio de Janeiro: UFRJ/COPPE (2017)
11. INEA. Critérios e procedimentos para o monitoramento da ictiofauna do Rio Paraíba do Sul no trecho Funil – Santa Cecília. Instituto Estadual do Ambiente - INEA, Superintendência Regional do Médio Paraíba. Rio de Janeiro. Caderno Técnico. Revisão 04. 98p. (2010)
12. Neves, L.F.F.: Environmental comparison between sections of the Paraíba do Sul river, with and without river islands. Rio de Janeiro, 2017. (Comparação Ambiental entre Trechos do rio Paraíba do Sul, com e sem Ilhas Fluviais. Rio de Janeiro, 2017). In Portuguese. (Master in Environmental Engineering). Programa de Engenharia Ambiental, Escola Politécnica e Escola de Química, Universidade Federal do Rio de Janeiro, Rio de Janeiro (2017)
13. Pereira, M.O., Calza, C., Anjos, M.J., Lopes, R.T., Araújo, F.G.: Metal concentrations in surface sediments of Paraíba do Sul River. J. Radioanal. Nucl. Chem. **269**, 707–709 (2006)
14. Molisani, M., Salomão, M., Ovalle, A., et al.: Heavy metals in sediments of the lower Paraíba do Sul River and Estuary, R.J. Brazil. Bull. Environ. Contam. Toxicol. **63**, 682–690 (1999)
15. Malm, O., Pfeiffer, W.C., Fiszman, M., Azcue, J.M.P.: Heavy metal concentrations and availability in the bottom sediments of the Paraiba do Sul-Guandu river system. RJ, Brazil Environ. Technol. Lett. **10**, 675–680 (1989)
16. WISCONSIN. Department of Natural Resources, 2003, Consensus-Based Sediment Quality Guidelines. Madson, WI. December, 2003. 40 p. https://dnr.wi.gov/files/PDF/pubs/rr/RR088.pdf. Accessed January 2021
17. Araújo, F.G.: Composition and structure of the fish community in the middle and lower Paraíba do Sul River (in Portuguese). RJ. Revista Brasileira de Biologia **56**(1), 111–126 (1996)
18. de Boffi, S.A.N.O., Ricci, F., Oliveira, E.A.A.Q.: Historical analysis of the Paraíba Valley region as an attraction for the implantation of the base industry (in Portuguese). 193- 1980. Revista UNIVAP **13**, 1 (2006)
19. Lacerda, L.D.: Mercury contamination in Brazil: Industrial sources vs. gold panning (in Portuguese). Química Nova **20**(2), 196–198 (1997)
20. Azevedo, L.S., et al.: Organotropism of methylmercury in fish of the south-eastern of Brazil. Chemosphere **185**, 746–757 (2017)

Interrelations Between Soil Erosion Conditioning Factors in Basins of Ecuador: Contributions to the Spatial Model Construction

Daniel Delgado[1,2](✉), Mahrez Sadaoui[1], Henry Pacheco[3], Williams Méndez[2,4], and Wolfgang Ludwig[1]

[1] Centre de Formation et de Recherche sur les Environnements Méditerranéens, Université de Perpignan, Via Domitia, UMR 5110, 52 Avenue Paul Alduy, 66860 Perpignan CEDEX, France
daniel.delgado@utm.edu.ec

[2] Departamento de Construcciones Civiles, Facultad de Ciencias Matemáticas, Físicas y Químicas, Universidad Técnica de Manabí, Avenida José María Urbina, Portoviejo 130105, Manabí, Ecuador

[3] Departamento de Ingeniería Agrícola, Facultad de Ingeniería Agrícola, Universidad Técnica de Manabí, Lodana, Santa Ana 13132, Manabí, Ecuador

[4] Centro de Investigación Estudios del Medio Físico Venezolano, Instituto Pedagógico de Caracas, Universidad Pedagógica Experimental Libertador, Venezuela University, El Paraíso 1020, Caracas, Distrito Capital, Venezuela

Abstract. Soil erosion due to the effect of surface runoff is one of the main environmental concerns inherent in the management plans of the hydrographic basins in Ecuador. The objective of this work was to analyze the interrelationships between the conditioning factors of soil erosion in hydrographic basins of the Ecuadorian territory, as an input for the construction of a spatial model of soil loss. The methodology considered the delimitation of the hydrographic basins, the calculation of elevations and slopes, the classification of land cover, lithology and R factor in Ecuador. The basins delimitation, calculation of elevations and slopes, were carried out from a DEM of 4 km of resolution; the land cover information was obtained from the Global Land Cover Database; the lithology factor was retrieved from the World Global Lithological Map Database; and the R factor were processed from the Global Rainfall Erosivity Database. The results indicated that the average elevation of the Ecuadorian basins ranges between 57 m and 2200 m. The average of the slopes decreases the greater the distance to the Andes Mountain, showing a greater variation in the basins that drain towards the Pacific. Most of the vegetation cover of the ground corresponds to closed forests and herbaceous vegetation. There is a great variety of lithological outcrops of different natures, which gives them different levels of erodibility. The R factor showed a similar behavior in the areas close to the Andes while, in the sectors close to the Pacific Ocean, the values were more dispersed.

Keywords: Erosion · R factor · Spatial modeling · RUSLE · Satellite images

J. R. da Costa Sanches Galvão et al. (Eds.): *Proceedings of the 1st International Conference on Water Energy Food and Sustainability (ICoWEFS 2021)*, pp. 892–903, 2021.
https://doi.org/10.1007/978-3-030-75315-3_94

1 Introduction

Human activity related to the use and exploitation of the soil resource is the main cause of its accelerated erosion, which has a direct impact on the nutrient cycle, particularly on carbon, as well as on the quality and quantity of water and, in agricultural and economic productivity worldwide [1, 2].

The erosivity of rainfall is the main factor in soil degradation by reducing infiltration rates, water retention capacity, nutrients and organic matter [3].

All regions of the world have been affected by soil erosion. However, this affects developing countries more seriously, due to the lack of adequate technologies and methodologies to assess risks and reduce their effects [4].

Various works [5, 6] have revealed the occurrence of intensive soil erosion processes at a local scale in different parts of the world, through the development of the RUSLE (Revised Universal Soil Loss Equation), which can be processed and applied with the implementation of Geographic Information Technologies (GIT) [7] and represents the erosion model most used to estimate long-term average annual soil loss due to laminar erosion [8].

On a global scale, a mean annual soil erosion is estimated to be around 35.9 Pg yr^{-1} [2]. However, this value is highly variable from one region to another and the results indicate a possible increase in global soil erosion driven by various factors, including the expansion of croplands.

Between 2001 and 2012, the largest increases have been in Sub-Saharan Africa, South America, and Southeast Asia [2].

This research was based on the analysis of elevations and slopes, land cover, lithology and R factor in 32 basins of Ecuador that have areas greater than 500 km^2, covering 80% of the national territory, through the use of recovered and geoprocessed satellite raster information using tools and software of Geographical Information Technologies, allowing a better analysis of conditioning factors of soil loss due to erosion in Ecuadorian hydrographic basins, as well as the interrelationships between them. The objective of the research is to present a methodology that allows building a spatial model of soil erosion in Ecuador, a valuable tool for the management and conservation of natural resources that coexist in its hydrographic basins.

2 Methodology

2.1 Delimitation of Drainage Basins

To study the factors that condition rainfall erosivity in the rivers that drain to the Pacific and tributaries of the Amazon from satellite-based rainfall and measurement stations, it is necessary, first, to delineate the basins considered in this study at a level of defined resolution.

DEM information from the HydroSHED database of 30 arc-sec [9], which is equivalent to 1 km^2 of the surface of a pixel, was used to identify the basins of areas greater than 500 km^2 in the Ecuadorian territory.

We worked with a resolution of 4 km^2, considered intermediate between a fine resolution and a coarse resolution. This resolution was obtained using the R software.

The approximate value of each cell of this resolution (4 × 4 km) represents the average of the 16 cell values corresponding to a fine resolution (1 × 1 km).

From the altimetric values of the DEM, the surface water flow was simulated, starting from the principle that the water flow is distributed to all neighboring cells with lower elevation, using the slope as a proportional distribution factor (multiple flow direction algorithm, MFD) [10].

According to Tarboton [11], the flow takes the exit direction of the line of maximum slope, dividing it proportionally between the neighboring cells of said sector.

Because this model produces a dispersion of the surface runoff flow, the hydrological tool r.watershed of the GRASS GIS software implements an algorithm by estimating the geometry of the surface network that simulates runoff, allowing to obtain accumulation rasters of flow, direction of drainage, location of streams and watersheds that will be useful to use the RUSLE method, based on an existing digital elevation model.

When going from a DEM of a fine resolution to an intermediate resolution, the modification of the mean altimetry can generate depressions that must be eliminated, forming an unreal sink when all the neighboring cells are higher than the treatment cell, leaving the water trapped unable to circulate.

Although these automatic corrections are carried out by the r.watershed module, even more distortions can occur during the change of resolutions, such as a false connection between two sub-basins or between two basins, in addition to the fact that the algorithm does not always allow to remove all sunk pixels.

Manual corrections were made using the SERVAL plugin in QGIS, achieving detailed pixel modification.

Finally, the delimitation of the basins will be defined with their exit points, on the border with the Pacific Ocean or the border with neighboring countries (Peru and Colombia).

2.2 Topography and Slope

For the morphological parameters, the mean values of elevation and slope were extracted by basins using the Software R. The DEM of HydroSHEDS was used, which is equivalent to 1 km^2 of the resolution (surface of a pixel) [9] and it was modified to a resolution of 4 km^2, as in the delimitation of the basins. The slope raster (in degrees) was created from the topography, using the Terrain command in R, adapting by basins to obtain mapping results.

2.3 Land Cover

The land cover characteristics were obtained from the Global Land Cover Database (version 2.0.2) through the global component of the Copernicus Land Monitoring Service, derived from PROBA-V satellite observations [12]. This information, which represents the year 2015, refers to the first global maps, in a 100 m^2 spatial resolution grid. To include the studied study area, it was necessary to download four mosaics of 20 × 20 km of land cover.

For the land cover classification, the base classification that follows the United Nations FAO Land Cover Classification System (LCCS) was used. The global map includes a total of 23 detailed classes and is simplified with 12 main classes, but the land cover for our area study includes only 13 detailed classes and 10 main classes (Fig. 3 up). However, these classes can be reduced to seven main classes: (i) closed forest, (ii) open forest, (iii) scrub and herbaceous, (iv) cropland, (v) urban, (vi) snow and ice and (vii) bodies of water (Fig. 3 down). Subsequently, the area of each class (in percentage) was estimated for each basin (Fig. 3 down).

2.4 Lithology

Information on lithology was obtained from Global Lithological Map v1.1 [13], which has a resolution of 500 m^2. The original GIS data (ArcGis geodatabase) can be downloaded via this link (https://www.dropbox.com/s/9vuowtebp9f1iud/LiMW_GIS%202015.gdb.zip?dl=0).

Once this information valid only for ArcGis had been downloaded, it was taken to QGIS to later be processed and modified using the R Software based on the requirements of this research (geographic space and resolution of 4 km^2).

This map originally contains 16 lithological classes and can be viewed from a 3D viewer using the following link: (https://www.clisap.de/fileadmin/B-Research/IA/IA5/LITHOMAP/dev/lithomap.html). The lithological map classes for our study area were reduced to 13 (Fig. 4 up) and we simplified it to seven main classes: (i) metamorphic rocks, (ii) plutonic rocks, (iii) pyroclastics, (iv) rocks sedimentary, (v) unconsolidated sediments, (vi) volcanic rocks and (vii) bodies of water (Fig. 4 down).

2.5 Rainfall Erosivity (R Factor)

The R factor data was compiled from a global erosivity map that was estimated from the Global Rain Erosivity Database (GloREDa) [1]. This map with a resolution of 30 arc-sec (~1 km) is available from the European Soil Data Center (ESDAC) and provides a complete set of pluvial erosivity data for the whole world based on 3625 precipitation stations and surrounding 30–40 years of temporal coverage (predominant in the last decade: 2000–2010) at high temporal resolution (1 to 60 min).

The global map was adapted to our study area (geographic space and resolution of 4 km^2) (Fig. 5 up) and the mean values of the R factor for each basin were calculated (Fig. 5 down, Table 1).

3 Results and Discussion

3.1 Drainage Basins

According to the basins delimitation detailed above, the entire Ecuadorian territory was divided into 32 sectors, where basins with areas greater than 500 km^2 were considered (Fig. 1). The area of these basins varies from 542 km^2 (ID 11 in Fig. 1, Ayampe basin in Table 1) to 38 10^3 km^2 (ID at 31 in Fig. 1, Napo basin in Table 1). The entire study covers more than 227 10^3 km^2 and represents 80% of the Ecuadorian surface. Of all the selected basins, 24 basins are located on the western slope of the Andes Mountains and were called

Pacific slope basins (CVP); of which 21 discharge directly to the Pacific Ocean within the Ecuadorian territory (basins in yellow, Fig. 1) and the remaining 3 basins discharge to the Pacific Ocean, but outside the Ecuadorian territory (basins in pink, Fig. 1). On the eastern slope of the Andes Mountains, 8 basins have been delimited, corresponding to basins of the Amazon Slope (CVA) (basins in green, Fig. 1).

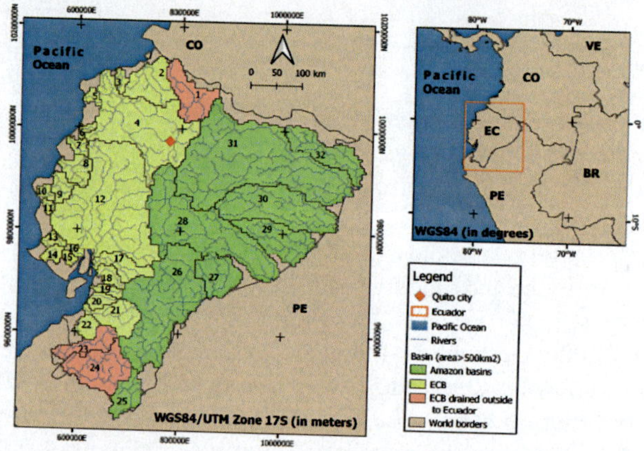

Fig. 1. Location map of the Pacific basins and the Amazon basins

3.2 Topography and Slope

The basins near the Andes Mountains, those that drain to the Pacific as well as the Amazon, begin with altitudes above 2000 m (central strip in Fig. 2 left). Certain basins such as Moyo (ID 25 in Fig. 2 left) and Santiago (ID 26 in Fig. 2 left) are located almost entirely with altitudes greater than 900 m.

Regarding the distribution of average elevations by basin, the Mira river basin (ID 1 in Table 1) is the highest, at 2200 m, originating in the Sierra region of Ecuador and ending in Colombia (Pacific Ocean). On the other hand, the basin with the lowest average elevation belongs to the Daule River (ID 15 in Table 1) with 57 m, which begins at an altitude of 145 m in the Coast region and empties into the Pacific Ocean. 50% of the basins have average elevations above 400 m.

In the Amazon basin, the Moyo River is the one with the highest average elevation, at 2,100 m (ID 25 in Table 1). The maximum slopes in the spatial distribution of the Ecuadorian basins occur in the vicinity of the Andes mountain range (central strip in Fig. 2 right), with slopes >12° (maximum 12.71°, ID 19 in Fig. 2 right) and decrease as move away from the mountain range. Much of the basins of the Southwest of Ecuador have large ranges of slopes, like those identified near the Andes mountain range. According to Peña et al. [15], soil erosion increases as the slope of the land increases, due to the simultaneous increase in the amount and energy of surface runoff. Therefore, a higher percentage of erosivity would be expected in the central zone of Ecuador, specifically in

Table 1. Average environmental characteristics of Ecuadorian basins

ID	River	Type	Area (km^2)	Elevation (m)	Slope (degrees)	R Factor (mm/hahyear)	Erosivity Class [14]
1	Mira	ECB_O	6214	2198	3.910	4164	3-Low
2	Cayapas	ECB	6108	537	1.808	12593	6-Very high
3	Verde	ECB	944	162	0.535	12001	6-Very high
4	Esmeraldas	ECB	21301	1485	2.376	7452	4-Moderate
5	Muisne	ECB	816	147	0.710	15305	7-Severe
6	Coaque	ECB	656	247	0.737	9481	5-High
7	Jama	ECB	1137	238	0.709	6453	4-Moderate
8	Chone	ECB	2898	160	0.737	6487	4-Moderate
9	Portoviejo	ECB	1905	194	0.801	3787	3-Low
10	Caña	ECB	833	222	0.775	1852	2-Very low
11	Ayampe	ECB	592	360	1.155	2988	3-Low
12	Guayas	ECB	32499	709	1.890	6682	4-Moderate
13	Javite	ECB	993	179	0.844	2868	3-Low
14	Zapotal	ECB	1121	89	0.367	2605	3-Low
15	Daule	ECB	640	57	0.265	3139	3-Low
16	Chongon	ECB	656	106	0.599	4190	3-Low
17	Taura	ECB	3472	1369	3.260	6248	4-Moderate
18	Cañar	ECB	1568	928	3.671	6253	4-Moderate
19	Jagua	ECB	1008	1523	5.303	4160	3-Low
20	Gala	ECB	1376	1006	4.935	4024	3-Low
21	Jubones	ECB	4544	2172	4.567	2411	2-Very low
22	Santa Rosa	ECB	1729	405	2.427	4959	3-Low
23	Puyango	ECB_O	3761	1233	3.321	5975	4-Moderate
24	Chira	ECB_O	8258	1595	3.670	4445	3-Low
25	Moyo	AB	3503	2150	4.001	5088	4-Moderate
26	Santiago	AB	23272	2034	3.575	7690	5-High
27	Morona	AB	5457	518	1.245	10514	6-Very high
28	Pastaza	AB	22830	1824	2.129	5437	4-Moderate
29	Tigre	AB	7767	275	0.182	9060	5-High
30	Conocoto_C	AB	14501	296	0.255	9249	5-High
31	Napo	AB	38322	1058	1.586	7490	4-Moderate
32	Putumayo	AB	6987	229	0.102	9837	5-High
Total			227669				

the vicinity of the Andes Mountains, which corresponds to the high zones of the country, while, in the low zones, erosion would be less.

In contrast, flooding and sediment accumulations would predominate in the lower areas. For the average slope parameter, 36% of the basin area has a range between 1.8° and 2.4° (the most prevalent slope in the Ecuadorian basins). The maximum average slope corresponds to the Jagua River (ID 19 in Fig. 2 right), with 5.3° and discharge to

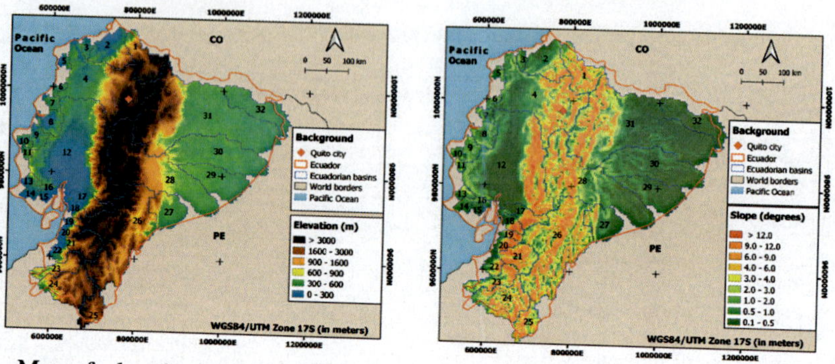

Fig. 2. Map of elevations spatial distribution (left) and slope spatial distribution (right) in Ecuadorian basins

the Pacific Ocean. The minimum average slope belongs to the Putumayo River (ID 32 in Fig. 2 right) with 0.1° located in the Amazon region. The measurement of the average slopes in basins of the Amazon slope and basins of the Pacific slope that flow out of the Ecuadorian territory, are calculated within the limits of Ecuador, clarifying that the real average of the total basin would show variations in the results.

3.3 Land Cover

Based on the previously determined data processing, the results of the land cover for the 32 Ecuadorian basins are shown (Fig. 3).

Through this procedure, it was determined that 84% of the Ecuadorian territory is covered with forests, of which 60% correspond to closed forests and 14% correspond to open forests. 26% of the surface belongs to other types of coverage (Fig. 3). Closed forests correspond to a continuous set of trees that covers more than 40% of an area, while open forests correspond to a discontinuous set of trees with a density between 10% and 40% of an area [16]. The high percentage of forest cover in the Ecuadorian basins would suppose a favorable behavior in the reduction of erosive effects, considering the benefits that this type of cover brings to the soil. In this sense, the basins that would have the greatest erosion problems would be those belonging to the Guayas River and the Taura River (ID 12 and 17 in Fig. 3), due to the large percentage of soil destined for cultivation (cropland).

3.4 Lithology

Once the area of each class was estimated (in percentage) for each basin (Fig. 4), it was determined that 71% of the studied territory is made up of sedimentary soils, of which 46% correspond to siliciclastic sedimentary rocks. 17% from mixed sedimentary rocks and 8% from unconsolidated sediments. 29% of the surface of the study area is made up of the remaining 11 lithological classes.

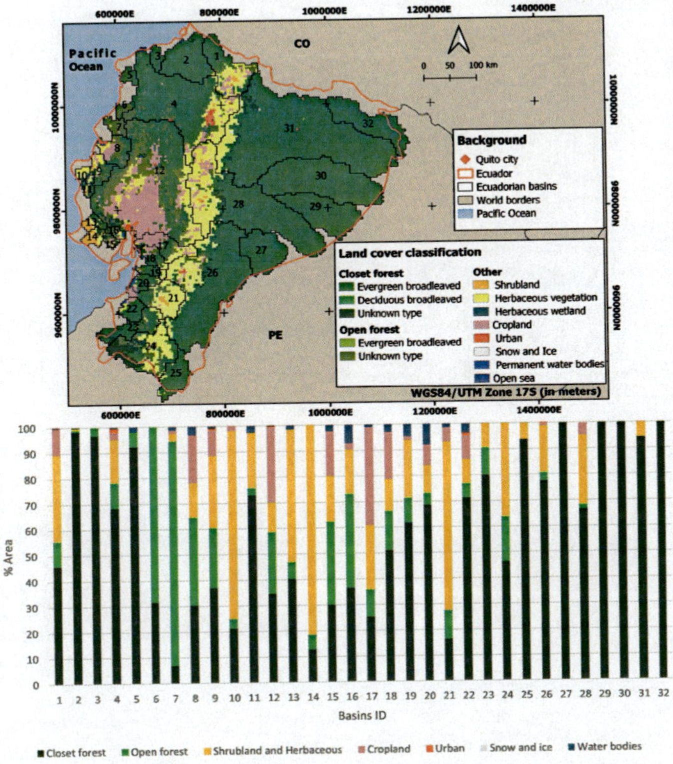

Fig. 3. Land cover in Ecuadorian basins

As there is a great variety of lithological outcrops of different compositions (some in small quantities), the levels of erodability conferred on the Ecuadorian hydrographic basins are very varied. Among these, mixed sedimentary rocks, siliciclastic and unconsolidated, have a level of erodibility that depends on the composition of their sediments (clay, sand, gravel, pebbles, others) and the degree of cementation, so it is difficult to assign a exact erodibility level, which will be between medium and high levels. Intermediate volcanic rocks are the third type of lithological outcrop with the highest concentration in the study area; Its composition is based on the crystallization of materials expelled by volcanoes and these are very susceptible to deterioration through humidification and drying processes, so their level of erodibility is generally high. Therefore, it is expected that the erosion index, product of the lithological classes present in the Ecuadorian basins, is high in most of the Ecuadorian territory.

The basins of the Verde, Muisne, Chone, Portoviejo and Ayampe rivers (ID 3, 5, 8, 9, 11 in Fig. 4) are completely composed of Sedimentary Rocks. Due to the presence of volcanoes in the Andes mountain range (intermediate strip of Ecuador in Fig. 4), a high percentage of Volcanic and Pyroclastic Rocks is observed.

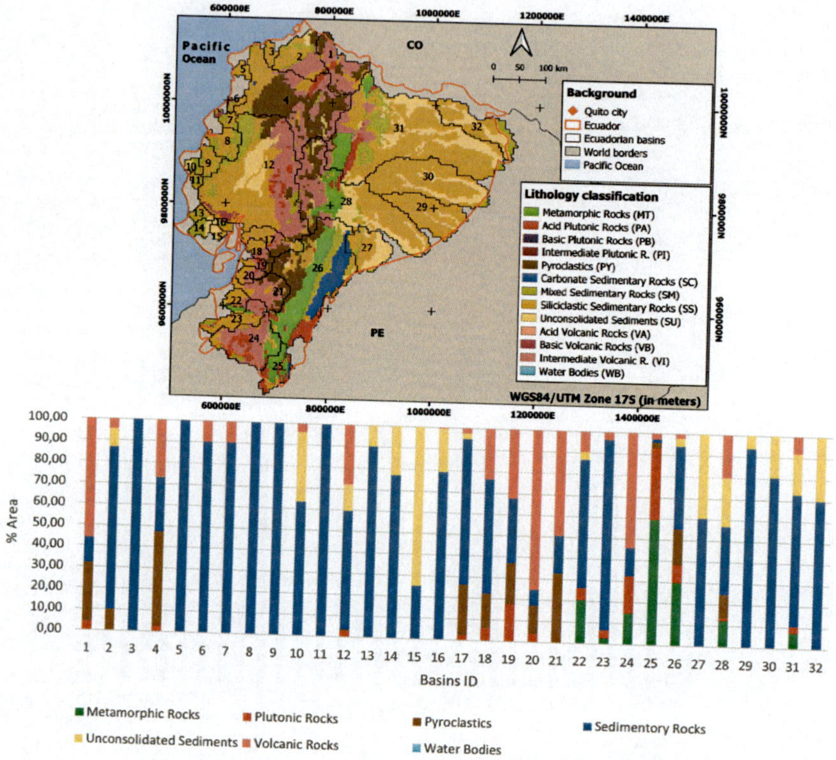

Fig. 4. Lithological classification of the Ecuadorian basins

3.5 R Factor

In Fig. 5 it can be seen how the rainfall erosivity index in the Ecuadorian basins, expressed by the R factor, varies considerably throughout the Ecuadorian territory. The areas close to the Andes Mountains have the lowest values of rainfall erosivity and maintain similar ranges throughout (Fig. 5).

Analyzing the R factor by basin (Table 1), Muisne river basin (ID 5) has the highest R factor, with more than 15,000 MJ mm/ha h year (severe erosion), and the basin with the lowest R factor is Caña River (ID 10), with 1800 MJ mm/ha h year (very low erosion), both on the Ecuadorian coast (Pacific slope) and with similarities in their areas (816 km^2 and 833 km^2 respectively). Regarding the Amazon slope, all basins present moderate to very high erosivity rates (Table 1), expecting considerable percentages of soil loss, due to its correlation with the R factor and the amount of sediments transported by surface runoff. The highest erosivity indices are concentrated in the Northwest of Ecuador (ID 2, 3, 5 in Fig. 5, Table 1), in the province of Esmeraldas.

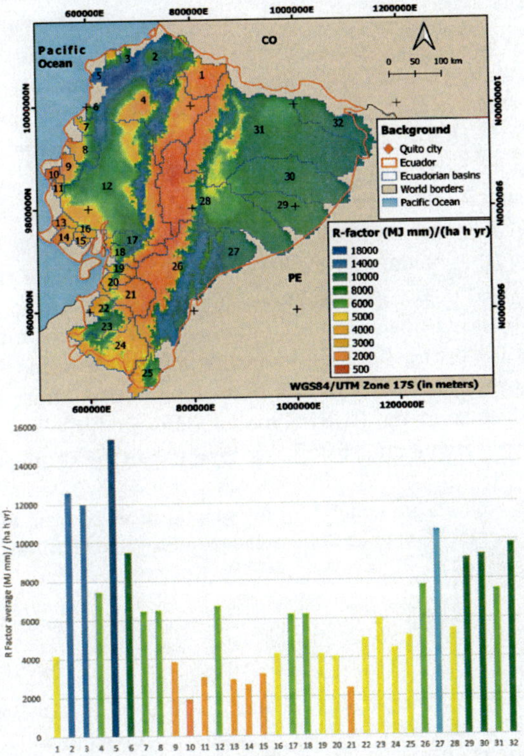

Fig. 5. Spatial distribution of the R factor in Ecuadorian basins

4 Conclusions

The processing of existing DEMs of good resolution and the application of raster information obtained from global databases (Global Land Cover Database, World Global Lithological Map and Global Rainfall Erosivity Database) using Geographic Information Technologies, allowed determining and classifying factors determining factors of soil erosion in Ecuadorian basins.

The average elevation of the Ecuadorian basins ranges between 57 m and 2200 m. A total of 88240 km^2 (39% of the study area, area of 19 basins) present average elevations less than 750 m. 31% (area of 6 basins) presented average elevations between 750 m and 1500 m. The remaining 30% (area of 7 basins), presented average elevations higher than 1500 m.

In spatial distribution of slopes in the Ecuadorian basins, the maximum slopes were concentrated in the vicinity of the Andes mountain range reaching up to 12.71°, and decrease as move away from this sector, both for the left bank (Pacific slope) as well as for the right bank (Amazon basins), until reaching a minimum slope of 0.1°.

84% of the study area is made up of forests, of which 60% correspond to closed forests, conferring a possible reduction in erosive effects if the protection that this biome offers to the soil is considered. Basins with the largest cultivation area correspond to the Taura River (38%) and the Guayas River (30%) in Guayas province.

Geologically, there is a great diversity of lithological outcrops in varied proportions, corresponding to rocks of different natures, conferring different levels of erodivility. A total of 161,645 km^2 (71% of study area), is made up of sedimentary material, presenting higher concentrations in the basins of the Province of Manabí.

The results of the R factor are very changeable throughout the territory studied. The province with the highest risk of erosivity caused by rain is Esmeraldas, because three of the four accounts that contain the highest values of the R factor (>10,000 MJ mm/ha hour year) are geographically located here.

The individual study of each conditioning factor in soil erosion presented different results, demonstrating the need for a joint evaluation to determine the actual soil loss.

The information on the R factor and other factors that affect soil erosion will be useful in the application of the RUSLE model. This analysis will provide a basis for comprehensive management and sustainable land use in the study area.

Acknowledgement. Authors are grateful for the support received from the French Embassy in Ecuador under the Program "Fondo de Solidaridad para Proyectos Innovadores (FSPI)".

References

1. Panagos, P., Borrelli, P., Meusburger, K., Bofu, Y., Klik, A., Lim, K., Yang, J., Ni, J., Miao, C., Chattopadhyay, N., Sadeghi, S., Hazbavi, Z., Zabihi, M., Larionov, G., Krasnov, S., Gorobets, A., Levi, Y., Erpul, G., Birkel, C., Hoyos, N., Naipal, V., Paulo, T., Oliveira, C., Bonilla, M., Nel, W., Dashti, H., Boni, M., Diodato, N., Van Oost, K., Nearing, M., Ballabio, C.: Global rainfall erosivity assessment based on high-temporal resolution rainfall records. Sci. Rep. **7**, 4175 (2017). https://doi.org/10.1038/s41598-017-04282-8
2. Borrelli, P., Robinson, D.A., Fleischer, L.R., Lugato, E., Ballabio, C., Alewell, C., Meusburger, K., Modugno, S., Schutt, B., Ferro, V., Bagarello, V., Van Oost, K., Montanarella, L., Panagos, P.: An assessment of the global impact of 21st century land use change on soil erosion. Nat. Commun. **8** (2017)
3. Panagos, P., Borrelli, P., Poesen, J., Ballabio, C., Lugato, E., Meusburger, K., Montanarella, L., Alewell, C.: The new assessment of soil loss by water erosion in Europe. Environ. Sci. Policy **54**, 438–447 (2015)
4. Eswaran, H., Lal, R., Reich, P.F.: Land degradation: an overview. response to land degradation. In: Bridges, E.M., Hannam, I.D., Oldeman, L.R., Pening de Vries, F.W.T., Scherr, S.J., Sompatpanit, S. (eds.) Proceedings of the 2nd International Conference on Land Degradation and Desertification, Thailand, pp. 25–29 (1999)
5. Rellini, I., Scopesi, C., Olivari, S., Firpo, M., Maerker, M.: Assessment of soil erosion risk in a typical Mediterranean environment using a high-resolution RUSLE approach (Portofino promontory, NW-Italy). J. Maps **15**(2), 356–362 (2019)
6. Teng, H.F., Hu, J., Zhou, Y., Zhou, L.Q., Shi, Z.: Modelling and mapping soil erosion potential in China. J. Integr. Agric. **18**(2), 251–264 (2019)
7. Pacheco, H.A., Mendez, W., Moro, A.: Soil erosion risk zoning in the Ecuadorian coastal region using geo-technological tools. Earth Sci. Res. J. **23**(4), 293–302 (2019)
8. Wischmeier, W.H., Smith, D.: Predicting rainfall erosion losses: a guide to conservation planning. Department of Agriculture, Science and Education Administration, p. 537 (1978)

9. Lehner, B., Verdin, K., Jarvis, A.: New global hydrography derived from spaceborne elevation data. Eos, Trans. Am. Geophys. Union **89**(10), 93–94 (2008)
10. Holmgren, P.: Multiple flow direction algorithms for runoff modelling in grid-based elevation models: an empirical evaluation. Hydrol. Process. **8**(4), 327–334 (1994)
11. Tarboton, D.G.: A new method for the determination of flow directions and upslope areas in grid digital elevation models. Water Resour. Res. **33**(2), 309–319 (1997)
12. Buchhorn, M., Smets, B., Bertels, L., Lesiv, M., Tsendbazar, N.-E., Herold, M., Fritz, S.: Copernicus global land service: land cover 100m, epoch, 2015, Globe (Version V2.0.2), [Dataset] Zenodo (2019)
13. Hartmann, J., Moosdorf, N.: The new global lithological map database GLiM: a representation of rock properties at the Earth surface. Geochem. Geophys. Geosyst. **13** (2012)
14. Rivera, J., Gómez, A.: Erosividad de las lluvias en la zona cafetera central colombiana (Caldas, Quindío y Risaralda). Cenicafé **42**(2), 37–52 (1991)
15. Peña-Cortés, F., Limpert, C., Andrade, E., Hauenstein, E., Tapia, J., Bertrán, C., Vargas-Chacoff, L.: Dinámica geomorfológica de la costa de La Araucanía. Revista de Geografía Norte Grande **58**, 241–260 (2014)
16. Food and Agriculture Organization of the United Nations: Evaluación de Recursos Forestales: Aportes de la información del Inventario Forestal Nacional a las políticas que involucran el sector forestal de Guatemala (2004)

Mobilization Towards an Integrated Research Network for Studying Runoff, Sediment Transport and Climate of the Paraiba do Sul Basin

M. N. Frota[1(✉)], G. B. França[2], E. P. Souza[3], J. M. Godoy[4], A. A. M. Araujo[5], and L. Q. Viana[6]

[1] Metrology Programme, Pontifical Catholic University of Rio de Janeiro, Rio de Janeiro, RJ 22951-900, Brazil
mfrota@puc-rio.br
[2] Meteorology Department, Federal University of Rio de Janeiro, Rio de Janeiro, RJ 21941-916, Brazil
[3] Department Atmospheric Sciences, Federal University of Campina Grande, Campina Grande, PB 58051-900, Brazil
[4] Chemistry Department, Pontifical Catholic University of Rio de Janeiro, Rio de Janeiro, RJ 22951-900, Brazil
[5] POLI/UFRJ, Federal University of Rio de Janeiro, Rio de Janeiro, RJ 21941-916, Brazil
[6] Maintenance Management of Power Plants, Light Energia S.A., Rio de Janeiro, RJ 22951-900, Brazil

Abstract. The article describes the fundamentals associated with the construction of an integrated computational code to model the Paraíba do Sul River basin, considering specificities that are intrinsic to the basin's hydrology and the impact of external aspects related to weather forecasting, climatology and silting. With an area of 60 thousand km^2, the basin is located in the south-eastern region of Brazil, covering the three most industrialized states in the country (Rio de Janeiro, São Paulo and Minas Gerais). The proposal for this hydrometeorological model integrates complementary efforts by research groups from three universities and an energy concessionaire responsible for the operation of the hydroelectric complex powered by waters from the studied basin. The proposed modelling was mainly motivated by the absence an integrated modelling able to carry out a reliable hydro-meteorological forecast. Among the results stand out: (i) the construction of a consistent knowledge base capable of guiding strategic decision making in the operation of the hydrological system and in the simulation of environmental events; (ii) the establishment of a cause-effect relationship capable of explaining the increase in silting up, guiding the water reservoir manager to establish a policy aiming at its mitigation. As results, the project will offer an integrated approach capable of ensuring forecast periods of up to 7 days, with hourly output, in places of interest in the river basin. Unlike the available models, it incorporates the history of variation in the silting up of the reservoirs that integrate the hydrological complex.

Keywords: Coupled atmospheric-hydrological model · Paraiba do Sul hydrological basin · Cooperative research network

J. R. da Costa Sanches Galvão et al. (Eds.): *Proceedings of the 1st International Conference on Water Energy Food and Sustainability (ICoWEFS 2021)*, pp. 904–913, 2021.
https://doi.org/10.1007/978-3-030-75315-3_95

1 Introduction and Motivation

The motivation for studying sedimentation, climate and to develop an integrated hydro-meteorological model for the Paraiba do Sul basin arose the need for the energy concessionaire—Light Energia S/A, which operates two pumping stations (Santa Cecília and Vigário) and six hydroelectric plants (Santa Branca, Nilo Peçanha, Fontes Nova, Pereira Passos, Lajes and Ilha dos Pombos)—to be able to make predictions, seven days in advance, of the water level and flow in strategic locations of the basin hydrographic. The motivation was also due to the fact that the existing computational codes do not consider, in an integrated way, the siltation rate, climate changes and low frequency meteorological phenomena in the seasonal precipitation regime. In this sense, the proposed computational code innovates by proposing a model that considers the silting rate and integrates climatology into the weather forecast to feed the hydrological model in an automatic and interconnected fashion, so that the output of a module is processed as the input of the next along the chain of desired results. The need to provide reliable results on an hourly basis, with a seven-day advance forecast, constitutes the essence of the intended strategy for decision making aimed at guiding the management of the operation of the complex of reservoirs, pumping stations and hydroelectric plants fed by the waters from the Paraíba do Sul basin.

As an illustration, Fig. 1 shows an artistic conception of the hydroelectric complex installed in the Paraíba do Sul hydrographic basin, integrated by several affluent rivers, sub-basins and reservoirs, which, today, feed the six hydroelectric plants in operation in the region.

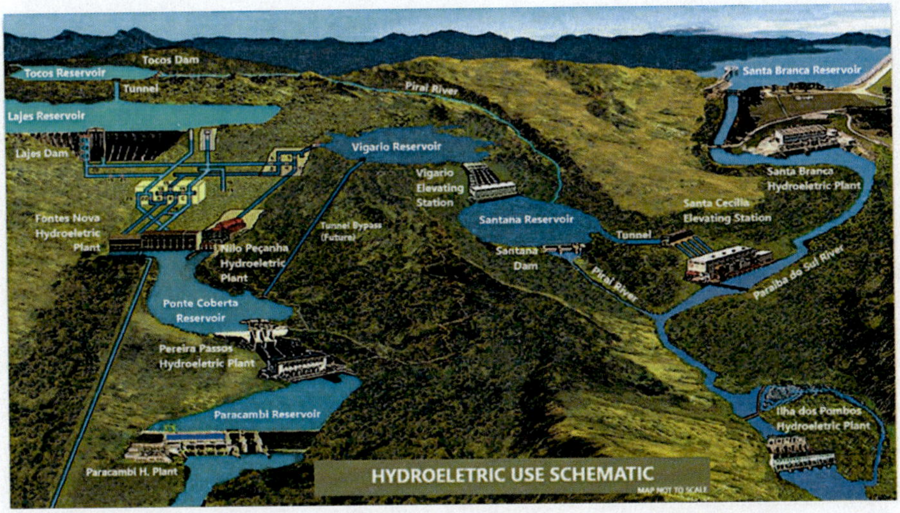

Fig. 1. Hydroelectric complex in operation in the Paraíba do Sul basin.

2 The Integrated Character of the Computational Code

The computational code developed, operationally integrated for rain-flow forecasting, is based on coupled atmospheric-hydrological models, for application in locations considered strategic for the concessionaire Light Energia S/A. With a wide scope, it was designed to produce forecast periods of up to 7 days, with hourly output, integrating the history of variation in the silting of the reservoirs that make up the hydrological complex.

Regarding its atmospheric component, the MoVaSC code (i) allows meteorological forecasts in the basin to assess impacts associated with low frequency phenomena in the precipitation regime of the area of interest to the Light concessionaire and (ii) incorporates the weather forecast for periods of 12 h, 24 h and periods of up to 7 days, in the river basin.

The modeling of sediments by reservoir and the assessment of the real silting level are determined by the ^{210}**Pb** technique (a natural radionuclide). This is a technique that allows to recover the loss of information on the temporal variation of the sedimentation rate over a period of time of up to 100 years, the consequences of which reduce the predictability of the accumulated water volume and the useful life of the reservoir.

In an integrated way, the computational model of *rain–defluvium* simulation, aggregates information related to land use (changes in the *defluvium* coefficient); urban and rural areas, with different types of crops, pastures or forests; whether for human supply, agriculture or any other end or industrial activity. It also incorporates demands for accumulation reservoirs (flood dampening; flow regulation and hydroelectric use), as well as transposition of watersheds.

2.1 Methodology for the Development of Thematic Modules

Four are the thematic modules (weather forecast, climatology, silting and hydrology) that integrate the computational code proposed to model the Paraíba do Sul basin, characterized below:

- **Weather forecast modeling**

The numerical model used to simulate the weather forecast is the Weather Research and Forecasting, WRF [1], computed in the following chronological sequence: (i) it generates the boundary conditions for the model in the chosen domain; (ii) process the data for seven days of forecast and (iii) export the results as data files, to feed the hydrological model. The global data come from the GFS (Global Forecast System) model, operated by the National Centers for Environmental Prediction/National Oceanic and Atmospheric Administration (NCEP/NOAA). The external domain covers an area of approximately 2200 km^2 x 2200 km^2, thus ensuring that large-scale atmospheric systems (e.g.: cold fronts, cyclones, South Atlantic Convergence Zone) can be resolved. The resolution of the external grid is 8 km. To optimize data processing, the rain-producing systems in the Paraíba do Sul river basin use a nested grid of 2.7 km of resolution.

The quality of the distribution and intensity of the expected rain is ensured by the use of appropriate parameterizations, thus feeding the hydrological model continuously, so that it processes the flow forecast for the next days.

- **Climatology**

The main objective of the thematic module Climatology is to correlate the low-frequency oceanic phenomena *El Niño-South Oscillation* (ENOS), *Pacific Decadal Oscillation* (ODP), North Atlantic Oscillation (OAN) and *Atlantic Multidecadal Oscillation* (OMA) with behavior precipitation in the Paraíba do Sul river basin. To fulfill this purpose, it uses precipitation data measured by rainfall stations operated by the concessionaire Light Energia S/A and grid precipitation data generated by NOAA (National Oceanic and Atmospheric Administration), whose information is submitted to a careful quality control, to ensure data consistency and outlier eradication [2].

From these data, indexes of climatic extremes of precipitation and their respective trends will be elaborated, using the software RClimdex, developed by NOAA [3]. The intensity or phase of each oceanic phenomenon will be defined by the time series technique of the descriptive indices of each event, namely: IOS (southern oscillation index), IODP (Pacific decadal oscillation index), IOAN (Atlantic oscillation index) north) and IOMA (Atlantic multidecadal oscillation index). Then, a scenario integrated with all these phenomena in the same time series will be elaborated. Such results will allow the correlation between the precipitation extreme indices and the oceanic indices. Subsequently, the integrated scenario will be updated with the correlated precipitation regime. Fig. 2 illustrates the methodological flowchart for obtaining the integrated scenario on the influence of oceanic oscillations on the behavior of precipitation in the Paraíba do Sul river basin.

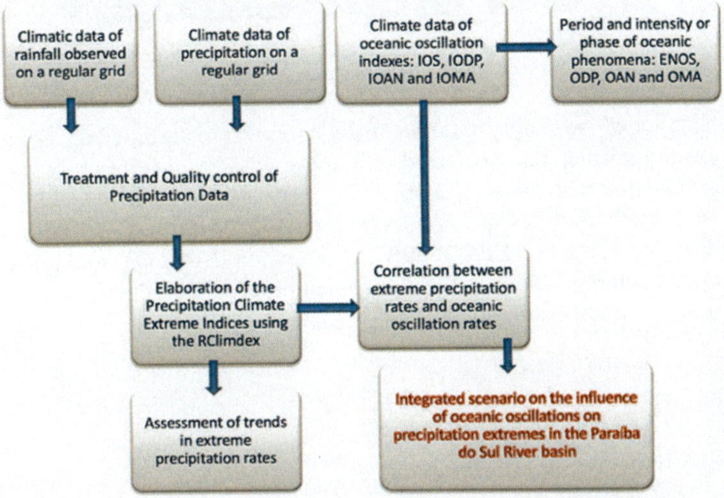

Fig. 2. Methodological flowchart for obtaining the integrated scenario.

- **Sediments**

The assessment of the temporal evolution of the silting rate has been studied by the ^{210}**Pb** technique (a natural radionuclide) in different locations of two of the reservoirs (Santana and Vigário) in the Paraíba do Sul basin. Fig. 3 illustrates one of the sediment cores (100 mx Ø 6 cm) collected (Santana-4), before and after they have been sliced to produce 2-cm samples for analysis of its complex decay chain.

Fig. 3. Sediment core, before and after sliced into samples.

The evaluation of the quality of the sediment cores was made based on measurements of their moisture profile, being selected only those whose moisture profile showed an exponential variation with the core depth.

In the selected cores, metals were determined by ICP-OES (Method EPA-6010D) after acid extraction (Method EPA-3050B), Hg (Method EPA-7474) and granulometry, by the laser diffraction technique. The determination of ^{210}**Pb** concentration, followed the methodology described by [4].

Given the changes in land use and occupation in the region of the Paraíba do Sul River drainage basin, variations in the silting rate in the studied reservoirs are expected. Given this behavior, the scheme adopted for the evaluation of ^{210}Pb data and calculation of sedimentation speed was the CRS, Constant of Rate Supply model [5].

- **Hydrological modeling**

Throughout its existence, supplying water for livestock, pasture, reforestation, agriculture, industrial activity and, since 1908, for electricity generation, the Paraíba do Sul basin has been exposed to intense anthropic activities, which have a great impact on the dynamics of variables of land use and cover, with marked consequences in their hydrological and hydrodynamic regimes. Figure 4 illustrates the methodological formulation for building the desired atmospheric-hydrological model.

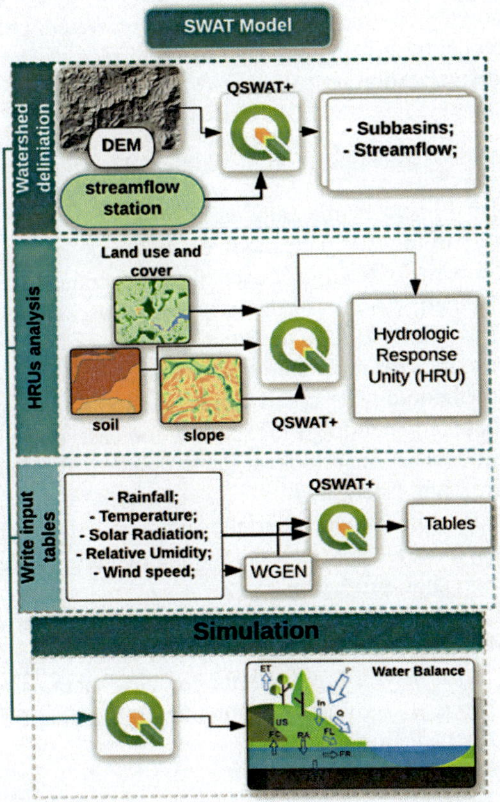

Fig. 4. Methodological flowchart for development of the integrated model.

The maintenance of flow regularization reservoirs for power generation, flow transposition for supplying populations and intense agricultural activity are just some of the mechanisms that define a complex and challenging scenario for the hydrological modeling of this basin. In addition to these adverse effects, the hydrological model also aims to incorporate the phenomena associated with climatology, weather forecasting, and sediment transport, which directly impact the basin's hydrology [6, 7].

The SWAT model implemented converges to the delimitation of the sub-basins, built from (i) topography data of the HydroSHEDS product [8, 9], in the resolution of 90 m; (ii) land use and land cover data from the MapBiomas project (Carlos et al., 2020), with a resolution of 30 m and (iii) a drainage network with sufficient density to reproduce the natural drainage network provided by the National Agency for Waters (ANA). Based on these primary data, the modeling grid will be optimized to include the sensitivity analysis, calibration and model validation processes.

The SWAT model implemented converges to the delimitation of the sub-basins, built from (i) topography data of the HydroSHEDS product [8, 9], in the resolution of 90 m; (ii) land use and land cover data from the MapBiomas project [10], with a resolution of 30 m and (iii) a drainage network with sufficient density to reproduce the natural drainage network provided by the National Agency for Waters (ANA). Based on these primary data, the modeling grid will be optimized to include the sensitivity analysis, calibration and model validation processes.

2.2 Integration of the Atmospheric-Hydrological Model

Schematically, Fig. 5 illustrates the integration process of the silting, climatology, weather forecast and hydrology thematic modules, understood as substrates of the desired coupled model. In the center of the figure, each block illustrates the respective method of processing for each thematic module; on the left, the databases (inputs) that feed the modeling of each thematic module and, on the right, the output of each model, sequentially exported as data files to feed the integrated hydrometeorological model. This is the operationally integrated methodology designed to provide a robust atmospheric-hydrological coupled model, capable of generating the rain-flow forecast, in the Paraíba do Sul river basin, as an integrated management tool for the hydroelectric complex operated by the energy concessionaire Light Energia S/A.

Figure 6 presents a preliminary result of the first step of the methodology for implementing the integrated SWAT code to model the Paraíba do Sul basin, particularly the delimitation of the associated sub-basins.

This network of sub-basins works as a modeling grid, underlying the previously selected points for the model to perform flow predictions. The figure also shows the hydrometeorological data network and the fluviometric stations used to support the model's calibration process. Highlighted in the detail, the image shows the reservoir system and their flow transpositions.

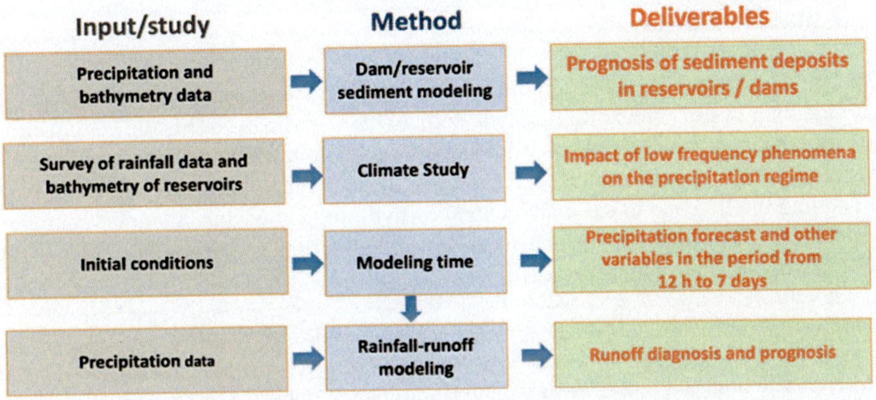

Fig. 5. Scope characterization and interrelationship of thematic modules.

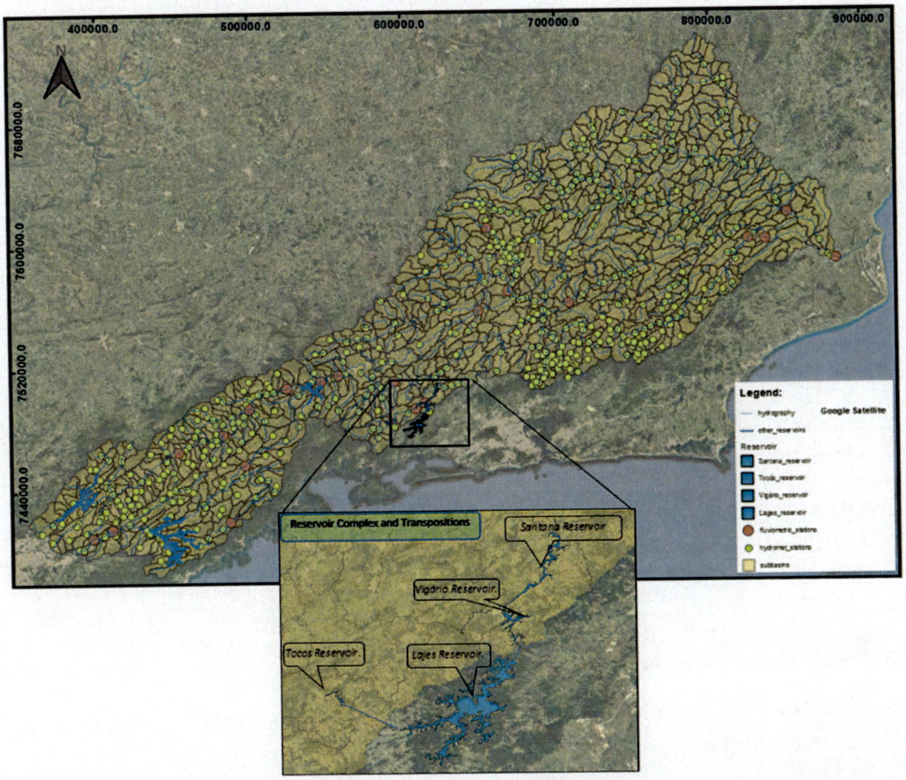

Fig. 6. Output of the integrated model: sub-basins delineation and modeling grid. The highlighted image illustrates the reservoir system and flow transposition.

3 Final Remarks

This paper describes the methodological conception of the proposed hydrometeorological modeling of the Paraíba do Sul hydrological basin and correlated studies that aim to improve the understanding of climatic impacts on precipitation and on the sedimentation evolution of the reservoirs that feed the hydroelectric plants in operation in the basin. By adding intelligence to the management of hydrological basins, integrated computational tools of this nature contribute to the preservation of the environment and to a more effective use of their potential.

Acknowledgments. To the Coordination for the Improvement of Higher Education Personnel (CAPES/Ministry of Education /Brazil), for the financial support (Financing Code 001) to the Research and Graduate Programs of the three universities mobilized to develop the integrated hydrometeorological model, discussed in this publication. Acknowledgments are also due to the regulated Research and Development (R&D) Program Light/Aneel, for funding the research (R&D Light/Aneel 5161–0016 / 2019, Contract # 4500428146, Flow, Sediment and Climate Modeling Project (MoVaSC).

References

Weather Forecast

1. Skamarock W.C., Klemp J.B., Dudhia J., Gill D.O.: A description of the advanced research WRF V. 4. NCAR/TN-556+STR, 145 p. 2019

Climatology

2. Zilli, M.T., Carvalho, L.M.V., Liebmann, B., Silva Dias, M.A.: A comprehensive analysis of trends in extreme precipitation over southeastern coast of Brazil. Int. J. Climatol. **37**(5), 2269–2279 (2017)
3. Marengo, J., Alves, L., Ambrizzi, T., Young, A., Barreto, N., Ramos, A.: Trends in extreme rainfall and hydrogeometeorological disasters in the Metropolitan Area of São Paulo: a review. Ann. New York Acad. Sci. **1472**(1), 5–20 (2020)

Sediments

4. Godoy, J.M., Moreira, I., Bragança, M.J., Wanderley, C., Mendes, L.B.: A study of Guanabara Bay sedimentation rates. J. Radioanal. Nucl. Chem. **227**, 157 (1998)
5. Sanchez-Cabeza, J.A., Ruiz-Fernandes, A.C.: 210Pb sediment radiochronology: an integrated formulation and classification of dating models. Geochimica Cosmochimica Acta **82**, 183 (2012)

Hydrology

6. Arnold, J.G.; Moriasi, D.N.; Gassman, K.C., Abbaspour, K.C., White, M.J., Srinivasan, R., Santhi, C., Harmel, R.D., Van Griensven, A., Van Liew, M.W., Kannan, N., JHA, M.K.: SWAT: model use, calibration and validation. Trans. ASABE. Am. Soc. Agric. Biol. Eng. **55**, 1491–1508 (2012). ISSN 2151-0032
7. Abbaspour, K.C., Rouholahnejad, E., Vaghefi, S., Srinivasan, R., Yang, H., Kløve, B.: A continental-scale hydrology and water quality model for Europe: Calibration and uncertainty of a high-resolution large-scale SWAT model. J. Hydrol. **524**, 733–752 (2015)
8. Lehner, B., Verdin, K., Jarvis, A.: New global hydrography derived from spaceborne elevation data. Eos, Trans. AGU **89**(10), 93–94 (2008)
9. Lehner, B., Grill, G.: Global river hydrography and network routing: baseline data and new approaches to study the world's large river systems. Hydrol. Process. **27**(15), 2171–2186 (2013)
10. Carlos Jr., M.S., Shimbo, J.Z., Rosa, M.R., Parente, L.L., Alencar, A.A., Rudor, B.F.T., Hasenack, H., et al.: Reconstructing three decades of land use and land cover changes in Brazilian biomes with landsat archive and earth engine. Remote Sens. **12**(17), 2735 (2020). https://doi.org/10.3390/rs12172735

Development of a Reactor in a Continuous System Using Advanced Oxidation Processes to Reduce Chemical Oxygen Demand in Industrial Effluent

Cassiano Ricardo Brandt[1], Daniel Kuhn[1], Ytan Andreine Schweizer[1], Sabrina Grando Cordeiro[1], Ani Caroline Weber[1], Gabriela Vettorello[1], Bruna Costa[1], Letícia Hoehne[2], Elisete Maria de Freitas[1], Eduardo Miranda Ethur[1], and Lucélia Hoehne[1](✉)

[1] Universidade do Vale do Taquari – Univates. Programa de Pós-Graduação em Biotecnologia, Rua Avelino Tallini, 171. Bairro Universitário, Lajeado, Rio Grande do Sul 95914-014, Brazil
luceliah@univates.br

[2] Universidade Estadual do Rio Grande do Sul, Unidade Hortências, Rua Santos Dumont, 450. São Francisco de Paula, Rio Grande do Sul 95400-000, Brazil

Abstract. Industrial effluent is the liquid discharged by industrial production. This effluent must be treated and comply with the current environmental legislation before being released in a receiving body. However, in order to remove some specific contaminants or complement the conventional effluent treatment system, advanced oxidation processes (AOPs) are used. Among AOPs, there is the treatment with ultraviolet lamp (UV) and hydrogen peroxide (H_2O_2). Thus, this study develops a continuous reactor in laboratory-scale using UV light and H_2O_2 as treatment to reduce the chemical oxygen demand (COD) of a synthetic dairy effluent. For this purpose, a reactor was built using polyvinyl chloride ducts, a peristaltic pump, and three 54-W UV lamps. Effluent flow rate of 40 L/h, total volume of 10 L, and H_2O_2 concentration of 10 ppm were used. The synthetic effluent was prepared in order to simulate an effluent of the dairy industry. The COD of the effluent was monitored over three hours. It was observed that it is possible to fit the COD parameter to legislation requirements after one hour of treatment. In addition, 67.27% and 72.73% of COD were reduced after two and three hours of treatment, respectively. It is concluded that the reactor elaborated assists in the final polishing of an effluent of the dairy industry, considerably reducing COD.

Keywords: Reactor · AOPs · Treatment effluent · Dairy industry

1 Introduction

1.1 Industrial Residues and Effluents

Water is one of the most used inputs in industries of different fields, being generally contaminated within the production process and discharged into the environment in the

J. R. da Costa Sanches Galvão et al. (Eds.): *Proceedings of the 1st International Conference on Water Energy Food and Sustainability (ICoWEFS 2021)*, pp. 914–921, 2021.
https://doi.org/10.1007/978-3-030-75315-3_96

form of effluent [1]. The food sector is one of the most demanding regarding the use of water. Some authors mentioned that water can have several uses in the production process, being used as an ingredient in some product, as a transport channel and/or used for washing or cooking. Approximately 80 to 95% of the water used becomes effluent and must pass through the effluent treatment plant (ETP) of the company for proper treatment [2].

According to CONAMA [3], all material that may affect the environment is the responsibility of the generating company. To be discharged to receiving bodies, effluents must be treated according to the limits established by legislation, avoiding possible changes in the characteristics of the environment where it is discharged.

1.2 Industrial Effluent Treatment

Effluent treatment is a mechanism that removes important pollutants for the environment. This treatment aims to achieve the desired quality pre-established by the current reference. For this purpose, physical or chemical treatment processes must be used according to the need of the location, which are classified into preliminary, primary, secondary, and tertiary [4].

The primary treatment has the objective of removing sedimentable solids, fats, and much of the organic matter through physical removal [4]. Unlike the primary system, biological treatments predominate in the secondary treatment, which can remove large amounts of organic matter and nutrients (nitrogen and phosphorus). The tertiary or advanced treatment is used when it is required to remove specific pollutants that are not removed by the methods aforementioned. Advanced oxidation processes (AOPs) are among the forms of tertiary treatment.

1.3 Advanced Oxidation Process

AOPs are water and effluent treatments that use the generation of highly reactive hydroxyl radicals, which are capable of degrading highly complex compounds, such as micropollutants [5]. AOPs are based on the formation of highly reactive hydroxyl radicals with oxidative potential in the range of $E^o = 2.80$ eV, which is more reactive than ozone ($E^o = 2.07$), hydrogen peroxide ($E^o = 1.77$), and hypochlorite ($E^o = 1.43$) [6]. For being highly reactive, AOPs do not require a long reaction time, also not depending on the presence of other compounds to be effective [7].

AOPs have attracted the attention of the scientific community for their high oxidative capacity, being considered a solution for the large amount of emerging pollutants that have appeared in contemporary times, as well as for being an advanced treatment to remove residues with physico-chemical characteristics present in industrial effluents [8–10].

When compared to conventional treatments, one of the main characteristics of AOPs is the mineralization capacity of the compounds, which considerably increases the biodegradability of the effluent or liquid waste [11]. In addition, another advantage of the method is the capacity to detoxify the treated sample for degrading highly toxic organic pollutants into less toxic compounds [12].

AOPs encompass a number of methods, with the main methods being the Fenton, photo-Fenton, ozonation, hydrogen peroxide catalytic oxidation (UV/H_2O_2), and heterogeneous photocatalysis. The classification is also extended to the homogeneous or heterogeneous phase, when there is the presence of a solid-phase catalyst [13]. The method is selected by considering some parameters in the effluent, such as the potential of hydrogen (pH), physical and chemical properties, use or not of catalyst, and the temperature of the process, among others [14].

UV/H_2O_2. The UV/H_2O_2 method occurs through the action of UV radiation below 400 nm, which photolysis hydrogen peroxide molecules in highly reactive hydroxyl radicals [15] due to their high oxidation potential (2.80 eV). Another advantage is that once reacted, the radical ●OH exists for a short period, not requiring additional treatment for its removal [16].

On the other hand, H_2O_2 concentration requires more attention for being essential to the success of the technique. If the concentration is very low, the reaction that occurs is insufficient, not removing all the organic matter required [17]. However, if the H_2O_2 concentration is high for the effluent demand, there will be residual hydrogen peroxide, which can be harmful to the environment where the effluent is released, as well as to the microbiota of the ETP [18, 19]. For removal of residual H_2O_2, some studies indicate the use of the enzyme catalase or sodium bisulfate ($NaHSO_4$) [20, 21].

To ensure the efficiency of the method combined with a good cost-benefit ratio, the most suitable UV light source must be selected. Ultraviolet lamps (UV) can be of low pressure (LP), medium pressure (MP), and dielectric. For application to AOPs, low-pressure mercury vapor lamps with wavelength range of 254 nm are the most recommended for generation of UV rays, as H_2O_2 absorbs up to 220 nm [22, 23].

1.4 AOP Reactors

For the best performance and efficiency of AOPs, reactors of different types and forms have been developed, which can be continuous, in batch, and in series, among others. Such projects may vary regarding dimensions, shape, model, number of lamps, power, lamp intensity, having or not the presence of catalysts, and H_2O_2 concentration [23]. In this sense, the authors in [7] used a pilot reactor with a capacity of 0.5 L for the treatment of biodiesel, by means of solar irradiation for 480 min and the addition of 1.25 mL of H_2O_2 every 60 min. Through this treatment, it was possible to remove 92% of the organic load from the effluent.

The authors in [22] developed two types of reactors based on the UV/H_2O_2 method. Both prototypes used the same conditions of effluent flow (mix of drugs), H_2O_2 concentration, and lamp model, with variations only in the dimensions and the number of lamps used. In addition, both were built in a cylindrical shape, with length and diameter of 2300 mm and 300 mm for reactor 1 and 2050 mm and 450 mm for reactor 2, respectively. Mercury vapor lamps with 120-W power and 254-nm wavelength were used.

The authors cited in [24] developed a reactor to treat the effluent of a dairy industry. For this, they used a black light bulb with 8 W of power in a process of heterogeneous photocatalysis. As a catalyst, the authors used 100 mM H_2O_2 and 1.5 g/L of TiO_2. In the end, it was possible to remove approximately 34% of COD in 20 min of treatment, with a cost of 0.05 U$/L.

Other work used the combination of UV/TiO_2/H_2O_2 to treat wastewater from a fish industry. The lamp used was low pressure mercury vapor with 18 W of power, combined with 0.45 g of TiO_2 in an annular reactor with a capacity of 0.9 L of effluent. Adding 2500 ppm H_2O_2, it was possible to remove 72.8% COD and 62.9% TOC [25].

1.5 Chemical Oxygen Demand

COD is an important indicator of organic matter in the medium. COD estimates the oxygen consumed by chemical reactions in the medium [26], and its evaluation includes organic matter of low and high degradability through analysis that requires approximately two hours to obtain results [4].

Then, the aim of this paper was the beginning of development of a continuous reactor in laboratory-scale using UV light and H_2O_2 as treatment to reduce the chemical oxygen demand (COD) of a synthetic dairy effluent.

2 Methodology

The assembly and execution of the continuous reactor for application of AOPs was conducted in the chemistry laboratories at the University of Vale do Taquari - UNIVATES. Similarly, to assist in the development and optimization of the reactor for applications to industrial effluents, a standard effluent containing a COD load above the limit established by the current legislation of the state of Rio Grande do Sul, Brazil (CONSEMA), was developed in the laboratory.

To assist in the treatment and reduction of the parameters aforementioned, the UV/H_2O_2 method was applied using daylight lamps as energy source.

Physico-chemical analysis of COD of the standard sample before and after application of AOPs was conducted in the laboratory bioreactor at UNIVATES.

2.1 Preparation of Synthetic Effluent

To prepare the synthetic effluent, the methodology adapted from [27] was used. It was sought to simulate an effluent of a dairy industry with COD content of approximately 600 mg/L. In the effluent composition, 2.88 g of powdered milk, 4.8 g of glucose, 0.54 g of urea, 1.17 g of ammonium chloride, 1.8 g of monobasic sodium phosphate, 1.2 g of potassium carbonate, 3.12 g of sodium hydrogen carbonate, 1.2 g of magnesium sulphate heptahydrate, 0.048 g of ferrous sulphate heptahydrate, 0.048 g of manganese sulphate monohydrate, and 0.072 g of calcium chloride dihydrate added to 20 L of deionized water were used. Subsequently, the synthetic effluent was frozen to be preserved during analysis.

2.2 Development of the Reactor for Application of Advanced Oxidation Processes

The continuous reactor developed in laboratory-scale used five polyvinyl chloride (PVC) ducts (10 cm diameter and 80 cm length) distributed side by side, forming a serpentine linked by connectors of the same material, resulting in a total length of 4 m. For pumping the effluent, a peristaltic pump (PROINOX, AWG-5000) was used. Over the serpentine system, three UV lamps (OSRAM, LUMILUX T5 HO) with 54-W power and a nominal color temperature of 6500 K, with highest wavelength of 550 nm, were suspended to simulate daylight. The distance between lamps and the effluent was 22 cm. The Fig. 1 shows the reactor.

The effluent flow rate was 40 L/h, H_2O_2 concentration of 10 ppm and effluent volume of 10 L, which continuously recirculated between the gutters and the tank of origin. The natural pH of the synthetic effluent obtained was 7.4.

Fig. 1. AOP reactor developed on a laboratory scale for the treatment of effluent

2.3 Chemical Oxygen Demand

To evaluate the effect of the treatment on COD reduction, the COD of the effluent was checked prior to passage to the reactor and after one, two, and three hours of treatment, enabling to verify the minimum ideal time for the treatment.

All COD tests were performed in accordance with the 5220C closed reflux titrimetric method from Standard Methods for Examination of Water and Wastewater [28].

3 Results and Discussion

3.1 Reactor Modeling

The initial tests showed that there were no changes in flow rate of the effluent. The lamps used were just in the same distance and intensity in all the tests. In each hour a sample was transferred for a vials for the COD analysis.

3.2 Initial Characterization of the Synthetic Effluent

The initial COD concentration obtained for the synthetic effluent was 598 ± 22 mg O_2/L. Thus, it can be verified that the synthetic effluent has a COD reference value similar to the theoretical value (600 mg/L) obtained in the literature [27]. In addition, it is highlighted that the synthetic effluent has an organic load above the maximum limit for COD required by the legislation, which is 330 mg O_2/L for flows of up to 100 m^3/day. Thus, the initial sample proved to be ideal for simulation as an effluent in the final stage of treatment, which is not yet appropriate to be discharged in environmental matrices and requires final polishing to fit the parameters required by CONSEMA.

3.3 Evaluation of COD After UV/H_2O_2 Treatment

The COD showed considerable removal at pH 7.4 and the results obtained after two hours of treatment were equal according to the statistical test ($p < 0.05$). Considering that the initial concentration was 598.64 mg O_2/L, it is highlighted that the COD concentration reduced to 326.56 mg O_2/L, in compliance with the legislation, already in the first hour of treatment. After two and three hours of action, COD reduced to 195.92 mg O_2/L and 163.26 mg O_2/L, respectively.

The results with a pH close to neutral for COD removal corroborate the literature. Authors such as cited in [29] reached a removal of 49.2% for this parameter, considering 60 min of irradiation in the UV/H202 method. Moreover, authors in [30] also reached a COD removal of 53%, considering pH 7 and 60 min of action.

4 Conclusion

It is concluded that it is possible to develop an AOP reactor with simple materials and using the UV/H_2O_2 method. Using three 54 W-UV daylight lamps, H_2O_2 concentration of 10 ppm, effluent flow rate of 40 L/h, and total effluent volume of 10 L, it is possible to fit the COD parameter to the current legislation after one hour of treatment. In addition, after two hours of treatment, there is COD reduction of 67.27% and after three hours of treatment, 72.73%. Therefore, the reactor and the method under study are effective for reducing COD in food industry effluents.

References

1. Hosnar, J., Kralj, A.K.: Reduction of water usage in industry by using the MINLP coordinates technique. J. Loss Prev. Process Ind. **43**, 158–164 (2016)
2. Santos, J.J.N., Sousa, I.C.S., Bezerra, D.C., Coimbra, V.C.S., Chaves, N.P.: Desafios de adequação à questão ambiental em frigoríficos na cidade de São Luís, Maranhão: diagnóstico de situação. Environ. Protect. **81**(4), 315–321 (2014)
3. Brasil, Ministério do Meio Ambiente.: Resolução CONAMA 430, de 13 de maio de 2011 (2011)
4. Jordão, E.P., Pessoa, C.A.: Tratamento de esgotos domésticos, 7th ed. Rio de Janeiro (2014)
5. Wang, J., Zhuan, R.: Degradation of antibiotics by advanced oxidation processes: an overview. Sci. Total Environ. **701**(20), 135023 (2020)
6. Araújo, K.S., Antonelli, R., Gaydeczka, B., Granato, A.C., Malpass, G.R.P.: Processos oxidativos avançados: Uma revisão de fundamentos e aplicações no tratamento de águas residuais urbanas e efluentes industriais. Ambiente & Água **11**(2), 387–401 (2016)
7. Babu, D.S., Srivastava, V., Nidheesh, P.V., Kumar, M.S.: Detoxification of water and wastewater by advanced oxidation processes. Sci. Total Environ. **696**, 133961 (2019)
8. Brito, G.F.S., Oliveira, R., Grisolia, C.K., Guirra, L.S., Weber, I.T., De Almeida, F.V.: Evaluation of advanced oxidative processes in biodiesel wastewater treatment. J. Photochem. Photobiol. A: Chem. **375**, 85–90 (2019)
9. Silva, D.C.V.R., Queiroz, L.G., Alamino, D.A., Fernandes, J.G., Silva, S.C., Paiva, T.C.A., Pompêo, M.L.M.: Avaliação da eficiência de um índice de estado trófico na determinação da qualidade da água de reservatórios para abastecimento público. Engenharia Sanitária Ambiental, **23**(4) (2018)
10. Popat, A., Nidheesh, P.V., Singh, T.S.A., Kumar, M.S.: Mixed industrial wastewater treatment by combined electrochemical advanced oxidation and biological processes. Chemosphere **237**, 124419 (2019)
11. Roshini, P.S., Gandhimathi, R., Ramesh, S.T., Nidheesh, P.V.: Combined electro-fenton and biological processes for the treatment of industrial textile effluent: mineralization and toxicity analysis. J. Hazard. Toxic Radioactive Waste **21**(4), 04017016 (2017)
12. Sharma, A., Ahmad, J., Flora, S.J.S.: Application of advanced oxidation processes and toxicity assessment of transformation products. Environ. Res. **167**, 223–233 (2018)
13. Ribeiro, A.R., Nunes, O.C., Pereira, M.F.R., Silva, A.M.T.: An overview on the advanced oxidation processes applied for the treatment of water pollutants defined in the recently launched Directive 2013/39/EU. Environ. Int. **75**, 33–51 (2015)
14. Asaithambia, P., Govindarajanc, R., Yesufa, M.B., Alemayehu, E.: Removal of color, COD and determination of power consumption from landfill leachate wastewater using an electrochemical advanced oxidation process. Sep. Purif. Technol. **233**, 115935 (2020)
15. Liu, N., Sijak, S., Zheng, M., Tang, L., Xu, G., Wu, M.: Aquatic photolysis of florfenicol and thiamphenicol under direct UV irradiation, UV/H_2O_2 and UV/Fe (II) processes. Chem. Eng. J. **260**, 826–834 (2015)
16. Suzuki, H., Araki, S., Yamamoto, H.: Evaluation of advanced oxidation processes (AOP) using O_3, UV, and TiO_2 for the degradation of phenol in water. J. Water Process Eng. Osaka **7**, 54–60 (2015)
17. Jo, C.H., Dietrich, A.M., Tanko, J.M.: Simultaneous degradation of disinfection byproducts and earthy-musty odorants by the UV/H_2O_2 advanced oxidation process. Water Res. **48**(8), 2507–2516 (2011)
18. Pereira, V.J., Linden, K.G., Weinberg, H.S.: Evaluation of UV irradiation for photolytic and oxidative degradation of pharmaceutical compounds in water. Water Res. **41**(19), 4413–4423 (2007)

19. Lutterbeck, C.A., Baginska, E., Machado, E.L., Kümmerer, K.: Removal of the anti-cancer drug methotrexate from water by advanced oxidation processes: aerobic biodegradation and toxicity studies after treatment. Chemosphere **141**, 290–296 (2015)
20. Yuan, F., Hu, C., Hu, X., Wei, D., Chen, Y., Qu, J.: Photodegradation and toxicity changes of antibiotics in UV and UV/H_2O_2 process. J. Hazard. Mater. **185**, 1256–1263 (2011)
21. Giannakis, S., Vives, F.A.G., Grandjean, D., Magnet, A., De Alencastro, L.F., Pulgarin, C.: Effect of Advanced Oxidation processes on the micropollutants and the effluent organic matter contained in municipal wastewater previously treated by three different secondary methods. Water Res. **84**, 295–306 (2015)
22. Hofman-Caris, R.C.H.M., Harmsen, D.J.H., Beerendonk, E.F., Knol, T.K., Houtman, C.J., Metz, D.H., Wols, B.A.: Prediction of advanced oxidation performance in various pilot UV/H_2O_2 reactor systems with MP- and LP- and DBD-UV lamps. Chem. Eng. J. **210**, 520–528 (2012)
23. Wols, B.A., Harmsen, D.J.H., Van Remmen, T., Beerendonk, E.F., Hofman-Caris, C.H.M.: Design aspects of UV/H_2O_2 reactors. Chem. Eng. Sci. **137**(1), 712–721 (2015)
24. Dias, F.F., Silva, C.V.A., Santos, A.F.M.S., Andrade, J.G.P., Albuquerque, I.L.T.: Tratamento de efluente da indústria de laticínios aplicando processo oxidativo avançado (H_2O_2/TiO_2/UV). Revista GEAMA **4**, 10–15 (2018)
25. Tenório, M.A.R.: Avaliação dos processos oxidativos avançados como tratamento terciário do efluente líquido de uma indústria de pescado (2016)
26. Derisio, J.C.: Introdução ao controle de poluição ambiental, 4th edn. Oficina de Textos, São Paulo (2012)
27. Silva, L.V.C.: Tratamento aeróbio de efluente sintético de laticínio usando reatores inoculados com fungos *Aspergilus niger* AN 400 (2010)
28. Rice, E.W., Baird, R.B., Eaton, A.D., Clesceri, L.S.: Standard Methods for the Examination of Water and Wastewater, 22nd edn. American Public Health Association, Washington (DC) (2012)
29. Ortiz-Marin, A.D., Amabilis-Sosa, L.E., Bandala, E.R., Guillén-Garcés, R.A., Treviño-Quintanilla, L.G., Roé-Sosa, A., Moeller-Chavez, G.E.: Using sequentially coupled UV/H_2O_2-biologic systems to treat industrial wastewater with high carbon and nitrogen contentes. Process Saf. Environ. Protect. **137**, 192–199 (2020)
30. Morais, J.L.M., Zamora, P.P.: Use of advanced oxidation processes to improve the biodegradability of mature landfill leachates. J. Hazard. Mater. **123**, 181–186 (2005)

The Technology of Reagent Sewage Treatment of Industrial Enterprises

Yevgen Zbykovskyy[1], Oksana Turchanina[2], and Iryna Shvets[1](✉)

[1] Donetsk National Technical University, Pokrovsk 85300, Ukraine
[2] CIAUD – Research Centre of Architecture, Urbanism, Design Lisbon School of Architecture, University of Lisbon, 1349-063 Lisbon, Portugal

Abstract. The technology of reagent purification of coke plant waste water after biochemical treatment has been developed. The technology involves the introduction of a natural organic-mineral additive before coagulation. It is proposed to use the sludge formed during the enrichment of coal, or small-scale fractions of the coking coal charge. The positive effect of carbon particles is due to their ability to adsorb a significant amount of hydrolysis products of coagulant and substances present in water. It is proposed to use anionic polyacrylamide as a flocculant.

The technology provides a quick and almost complete afterpurification of water from suspended impurities of organic and inorganic origin and bacteria. Reducing the amount of harmful chemicals in wastewater by 20–50% is expected. The proposed technology allows disposing of sludge, which is formed during waste water afterpurification, by using it as an additive in the charge for coking. Based on the results of technology implementation, technical and operational parameters of reagent waste water afterpurification have been developed.

Ecological and economic assessment of the technology of reagent waste water afterpurification has been carried out. The reduction of pollutant emissions into the atmosphere from the implementation of waste water treatment technology is almost 68%.

Keywords: Waste water treatment · Reagent afterpurification · Coagulation · Organic-mineral additive · Sludge

1 Introduction

Ensuring environmental security is one of the most urgent strategic priorities of the state policy in European countries and a condition for sustainable development. In recent decades, the coal production and its processing including high-temperature coking, has increased significantly. That poses one of the world's acute environmental problems of the present, namely pollution of the biosphere and destruction of its genetic resources. The innovative technologies in coke production aim at not only protecting not the environment but also developing a waste-free technological cycle of production.

In the structure of the modern coke production, cooling hot coke by wet quenching has become the most widespread method. According to this technology, phenolic waters,

J. R. da Costa Sanches Galvão et al. (Eds.): *Proceedings of the 1st International Conference on Water Energy Food and Sustainability (ICoWEFS 2021)*, pp. 922–932, 2021.
https://doi.org/10.1007/978-3-030-75315-3_97

previously treated by biochemical methods, are used at the majority of enterprises. However, treated waste water which is used to coke quenching, contains a large number of contaminants (phenols, rhodanides, volatile and bound ammonia, hydrogen sulfide, tars and oils, cyanides). When they get on hot coke, they go into a vapour state, partially decompose and are removed from the quenching tower. At the same time, they have a detrimental effect on the environment and human health. They contain 3.4-benzpyrene and some polycyclic hydrocarbons (carcinogens) which are extremely dangerous [1, 2]. In addition, the water after being treated at the biochemical treatment plants contains a considerable number of microorganisms (activated sludge). This water is poorly filtered by high-performance industrial filters. Other recommended methods are extremely energy consuming and dangerous. Getting with water on the hot coke, the organic mass of activated sludge burns. At the same time, a strong unpleasant smell is produced.

A range of methods and techniques are used to waste water afterpurification [3]. The simplest and most effective methods of waste water afterpurification include coagulation using inorganic coagulants, coagulation and flocculation, coagulation and flocculation using inorganic opacifier, filtration, sorption and others.

One of the well-known methods of reagent waste water treatment involves adding powdered coagulant into the waste water for further treatment. It contains lime and aluminosilicate minerals in a mass ratio of 1: (0.05–0.3): (0.05–0.7) [4]. The disadvantages of this method include the insufficient efficiency of waste water treatment from suspended impurities of organic origin and microorganisms (bacteria) and the difficulty of sludge disposal. This sludge is formed during waste water treatment, and it additionally contains mineral opacifiers which are used in the coagulation process.

The purification of phenolic waste water from coal tar by biological treatment can be carried out by technology that involves the use of bentonite with the addition of flocculant. The physical and chemical essence of the proposed process is in the combined adsorption of contaminants by the surface of bentonite and subsequent coagulation of the formed aggregates [5]. The process extraction of tars substances from the waste water of coke plants can be carried out by the method of reagent flotation using aluminum oxychloride $Al_2(OH)_5Cl$. This reagent is more effective than other coagulants, and it is the least corrosive and effective one in a wide range of temperatures [6].

Another well-known method of waste water treatment from biological pollution involves the treatment of waste water by adding powdered bentonite (opacifier) and coagulant of aluminum sulfate in an amount of 50 mg/l, then mixing and standing the mixture for 1.5 h [7]. The disadvantages of this method include the following:

- insufficient efficiency of waste water treatment from suspended impurities of organic and inorganic origin, bacteria, as well as from harmful substances dissolved in water;
- insufficient amount of coagulant and, accordingly, incomplete destabilization of contaminant particles;
- low speed of formation and deposition of flakes formed during coagulation;
- the difficulty of the disposal of sludge, which is formed in the process of waste water treatment by coagulation using mineral opacifier (bentonite).

During the coke production, the amount of waste water is more than 38% of the amount of dry coal charge to be coked. The waste water from a coke plant contains a large number of harmful substances [1, 8]:

- phenols up to 2500 mg/l;
- rhodanides up to 500 mg/l;
- cyanides up to 200 mg/l;
- hydrogen sulfide up to 200 mg/l;
- ammonia up to 400 mg/l;
- dissolved salts up to 15000 mg/l;
- other substances.

According to the existing technologies at coke plants, waste water treatment is carried out using physic-chemical treatment processes: extraction, coagulation and flocculation, sorption, biochemical treatment and others [9–11]. At the same time, the water after such treatment is significantly contaminated with suspended impurities of organic, inorganic and mechanical origin, which are dissolved in water with harmful substances. This water also contains a large number of microorganisms: Zoogloearamigera, Opercularia, Epistylis, Pseudomonas, Amoeba. Therefore, the water after biochemical treatment must be purified again. The simplest and most effective way to treat waste water is coagulation and flocculation using inorganic opacifiers and filtration [12, 13].

2 Experimental Part

2.1 Experimental Laboratory Research

In order to determine the most effective coagulants and flocculants, certain chemical reagents for the waste water purification after biochemical treatment at a coke plant were selected [14]. The experimental tests were done at the treatment facilities of PJSC "Makeyevkoks" (Ukraine) using the following reagents:

- inorganic coagulant aluminum hydrochloride "POLVAK 40";
- inorganic coagulant aluminum hydroxychloride "POLVAK 68";
- inorganic coagulant iron chloride;
- inorganic coagulant aluminum sulfate;
- organic coagulant "PolyDADMAX EXTRAFLOCK PD 60";
- organic flocculants - polyacrylamide "EXTRAFLOCK".

The source water for experimental tests was selected only after a biochemical treatment. The experiments were performed at a laboratory flocculator Velp Scientifica (Italy). The coagulants were dosed using Biohit microdispensers. The flocculant was as a 0.1% of working solution prepared by dissolving a portion of 0.1 g of crystalline substance in 100 g of water by mixing it constantly. The preparation time of the flocculant was 60 min. During the tests, the intensity of flake formation, lightening and turbidity reduction, and sludge structure were evaluated.

The sequence of the experiment:

- idle mixing for 2 min at 200 rpm;
- adding of coagulant, mixing for 1 min at 200 rpm;
- adding of flocculant, mixing for 1 min at 200 rpm;
- slow mixing for 2 min at 40 rpm;
- standing.

2.2 Industrial Experiment

In order to confirm the results of the laboratory experiment and conduct research in industrial conditions a plant was designed. It was installed in the department of biochemical water treatment of PJSC "Makeyevkoks". The unit productivity as for water was 1.5 m^3/h, the consumption of coagulant was 6 l/h, the consumption of flocculant was 1 l/h. The volume of the reactor for mixing water and reagents was 2.5 m^3. The unit was additionally provided with two filters with a coke packed bed: at the inlet of water to the reactor and at the outlet of water from the reactor. Mixing of the reagents in the reactor was carried out using barbator. In the industrial experiment, EXSTRAFLOK N 180 in the amount of 4 g/l was used as a flocculant.

In order to improve coagulation, facilitate the formation and deposition of flakes in the purification of water with coagulants and flocculants, some opacifiers were used. Fine coal charge (<0.5 mm) of PJSC "Makeyevkoks" and fine waste of coking coal beneficiation of mine "Pokrovske" were used as opacifiers.

3 Results and Discussion

3.1 The Choice of the Effective Options for Reagent Afterpurification of Waste Water at the Lab

The most effective options for reagent afterpurification of waste water, reagent dosage, which was obtained during experimental laboratory research, are presented in Table 1.

The main experimental laboratory results are as follows:

- the most effective removal of contaminants (discoloration of the solution) is carried out by a combination of reagents: 300 mg/dm^3 of coagulant Pro-AQUA 18 and 5 g/m^3 of flocculant EXTRAFLOCK N 160, sequentially.
- in order to reduce costs, it is advisable to combine the reagent afterpurification: 100 mg/dm^3 with ferric chloride, 200 mg/dm^3 Pro-AQUA 18 and 5 g/m^3 EXTRAFLOCK N 160, sequentially. This option enables reducing the cost of reagent afterpurification by 25%. However, it has a more complex hardware design. At the same time, the side processes related to the oxidation of dissolved components by ferric iron can occur.
- the least expensive option of reagent afterpurification is the following combination: 300 mg/dm^3 of ferric chloride and 5 g/m^3 of flocculant EXTRAFLOCK N 160, sequentially. The disadvantages of this method are related to the presence of residual iron in the treated water and oxidative processes due to the content of ferric ions.

Table 1. The results of laboratory study.

Option of coagulation purification	Option of flocculation purification		Study result
300 mg/dm³ POLVAK 40	5 mg/dm³ EXTRAFLOCK N 160		Flake formation is instantaneous, lighting is high, and discoloration is incomplete.
300 mg/dm³ POLVAK 68	5 mg/dm³ EXTRAFLOCK N 160		Flake formation is slow, lighting is high, and discoloration is weak.
300 mg/dm³ Iron chloride	5 mg/dm³ EXTRAFLOCK N 160		The formation of flakes is rapid, lighting is high, and discoloration is weak.
100 mg/dm³ Iron chloride + 200 mg/dm³ POLVAK 40	5 mg/dm³ EXTRAFLOCK N 160		Flake formation is instantaneous, lighting is high, and discoloration is incomplete.

The laboratory studies proved that the most effective coagulants for waste water afterpurification of a coke plant after biochemical treatment are $Al_2(SO_4)_3 \cdot 18H_2O$ and $FeCl_3$, as well as the flocculant EXSTRAFLOK N 180 (anionic polyacrylamide). These reagents, as well as their mixtures, were used to study the processes of water purification after biochemical treatment, taken at PJSC "Makeyevkoks".

In order to intensify coagulation and deeper water afterpurification, it is proposed to use opacifier in the coke chemical waste water treatment technology after its biochemical treatment. The opacifier is a solid, fine material (<0.5 mm) that acts as a coagulation centre for the formation of the first flakes. When the waste water is turbid, the flakes of the coagulated suspension get bigger. At the same time, their hydraulic size gets bigger, and the speed of sedimentation increases. If the opacifier has ability to sorb harmful dissolved impurities, its use will helps to increase the depth of water afterpurification. That will improve coagulation conditions. The opacifiers remain effective at any pH value. The use of opacifiers does not require the involvement of additional reagents in the technological process of waste water afterpurification. The use of opacifier does not affect the organoleptic characteristics of the water to be treated. The use of opacifier in combination with flocculants and coagulants provides positive results in the water afterpurification from microorganisms. Such benefits of the opacifier contribute to the economical use of coagulants. One of the significant advantages of the carbon-containing opacifier is a possibility of using the sludge formed in coagulation process as an impurity in the coal charge for coking.

The criteria for choosing an opacifier are as follows: very small size of solid particles, organic origin, its availability in sufficient quantities and a low price. Based on these criteria, it is proposed to use a fine coal charge (<0.5 mm) or waste coal beneficiation of coking coal as opacifiers. The organic origin of the proposed opacifiers is its significant advantage. This advantage simplifies the disposal of waste from afterpurification processes and the possibility of further processing. The use of coal beneficiation waste in the process of water purification belongs to the technological methods of solid dangerous waste disposal of the coal industry. Fine coal charge of the coke plant and coal sludge of the coal beneficiation factory of the mine "Pokrovske" (Ukraine) were used in the proposed technology of waste water afterpurification. The results of the studies are shown in Table 2 and Table 3.

It was found that the use of a mixture of coagulants $Al_2(SO_4)_3 \cdot 18H_2O$ and $FeCl_3$ in the ratios of 50–70% $Al_2(SO_4)_3$ and 30–50% $FeCl_3$ carries out the most effective purification of contaminated water from activated sludge and colloidal suspension. The use of mixtures of coagulants in the proposed ratio, as well as the use of the recommended opacifiers enables to completely purify waste water from activated sludge and colloidal suspension. At the same time, there is also a reduction of harmful impurities by 10–25% (phenols, etc.).

Table 2. The results of the use of coagulants and their mixtures in the waste water afterpurification using the coal charge PJSC "Makeyevkoks" as an opacifier.

Impurities	Unit	Waste water	Coagulant				
			$Al_2(SO_4)_3$	$Al_2(SO_4)_3:FeCl_3$			$FeCl_3$
			100%	70:30%	50:50%	30:70%	100%
			Opacifier – coal charge PJSC "Makeyevkoks" Flocculant – EXSTRAFLOK N 180				
Phenols	mg/l	624	502	533	502	624	502
	%*	–	*20*	*16*	*20*	*0*	*20*
Rhodanides	mg/l	106	97	101	89	106	97
	%	–	*8*	*5*	*16*	*0*	*8*
Volatile ammonia	mg/l	85	25	17	25	85	25
	%	–	*70*	*80*	*70*	*0*	*70*
Hydrogen sulfide	mg/l	39	37	34	34	23	37
	%	–	*4*	*10*	*10*	*30*	*4*
Dry residue	mg/l	4580	4490	4270	4480	4580	4490
	%	–	2	*7*	2	*0*	2
Salt content	mg/l	9700	8300	8150	8200	9700	8300
	%	–	14	16	15	0	14
pH		7.92	5.25	5.95	6.08	7.92	5.25

*the percentage reduction of the impurity content with afterpurification by reagents.

3.2 Confirmation of the Experiment Results in Industrial Conditions

During the industrial experiment, the quality of waste water was analyzed at three stages of the process: after biochemical treatment, after purification with reagents and after additional purification in the coke filter (Table 4).

The results of industrial experiment show that the proposed technology is effective. During the experiment, not only the removal of residues of microorganisms, but also a significant reduction of harmful chemicals was proved. Thus, the content of phenols was reduced by more than 5 times, rhodanides – by 9 times, volatile NH_3 – by more than 4 times, H2S – by more than twice, salt residue – by 5.5 times. At the same time, the pH value indicates the neutral state of the after purifying water.

In order to increase the efficiency of waste water afterpurification, it is necessary to use as an opacifier fine (<0.5 mm) coal charge of the coke plant or fine waste of coking coal beneficiation. The recommended amount of opacifier should be 0.5–3.0 kg/m^3 of water. If the opacifier is used in an amount of less than 0.5 kg/m^3, there is no improvement in waste water quality. This amount of opacifier is not enough for complete destabilization of contaminant particles. The use of opacifier in the amount of more than 3.0 kg/m^3 of waste water is technologically and economically inappropriate. In order to implement

Table 3. The results of the use of coagulants and their mixtures in the waste water afterpurification using the coal sludge of the coal beneficiation mine "Pokrovske" as an opacifier.

Impurities	Unit	Waste water	Coagulant				
			$Al_2(SO_4)_3$	$Al_2(SO_4)_3$:$FeCl_3$			$FeCl_3$
			100%	70:30%	50:50%	30:70%	100%
			Opacifier – coal sludge of the coal beneficiation Mine "Pokrovske" Flocculant – EXSTRAFLOK N 180				
Phenols	mg/l	627	549	549	470	627	549
	%*	–	*13*	*13*	*25*	*0*	*13*
Rhodanides	mg/l	103	97	90	101	99	100
	%	–	*6*	*12*	2	*4*	*3*
Volatile ammonia	mg/l	91	31	34	31	34	31
	%	–	*66*	*63*	*66*	*63*	*66*
Hydrogen sulfide	mg/l	44	43	44	44	44	43
	%	–	2	*0*	*0*	*0*	2
Dry residue	mg/l	4450	3910	4100	4170	4180	4200
	%	–	*12*	*8*	*6*	*6*	*5*
Salt content	mg/l	10000	8500	7800	9030	9250	9450
	%	–	*15*	*22*	*10*	*7*	*5*
pH		7.66	6.30	6.12	6.30	6.30	6.24

*the percentage reduction of the impurity content with afterpurification by reagents.

waste-free purification technology, it is recommended to use the sludge formed in the coagulation process as an additive to the coal charge.

The use of opacifier of natural organic-mineral origin in the amount of 0.5–3.0 kg/m^3 is an innovation in the proposed method. Before adding the coagulant, these opacifiers should be added to the waste water which is afterpurificated. The main difference of the method is mixing the water for 2–3 min after adding to water the opacifier, coagulant and anionic flocculant. It is proposed to use the flocculant EXTRAFLOK N 180 as one of the variants of the anionic flocculant. The advantages of the proposed reagent method of waste water afterpurification are as follows:

- quick and almost complete purification of water from suspended impurities of organic, inorganic origin and bacteria;
- reduction of the content of harmful chemicals by 20–50%, which are dissolved in waste water (phenols, rhodanides, hydrogen sulfide and others);
- the possibility of utilization of sludge, which is formed in the process of waste water afterpurification, and can be used in the future as an additive to the coal charge;
- the possibility of using waste of coking coal beneficiation in the case of their use as an opacifier.

Table 4. Waste water afterpurification of a coke plant by coagulation.

Samples	Concentration, mg/l					pH
	Phenols	Rhodanides	Volatile ammonia	Hydrogen sulfide	Salt content	
Waste water after biochemical treatment	0.86	253.8	85.0	49.3	5520	6.99
Waste water after purification by reagents	0.53	241.2	51.0	47.6	1610	6.29
Waste water after additional purification in the coke filter	0.17	28.2	21.0	23.8	1000	6.65

The positive effect of adding an opacifier, coagulants and flocculants to waste water is acceleration of coagulation and flocculation processes and improvement the quality of waste water. This effect is related to the role of small particles of opacifier during mixing, which act as additional centers of coagulation and flocculation. One of the advantages of using carbon particles is their ability to adsorb a significant amount of hydrolysis products of coagulant and harmful substances in water.

The positive effect of opacifiers is enhanced by the presence of oxygen-containing functional groups on the surface of carbon particles and paramagnetic centers due to the presence of unpaired electrons in the coal. This fact can lead to interactions between the particles of the opacifier and chemicals in the water, including bacteria. These processes will be accompanied by the formation during coagulation of large, stronger and denser flakes. The use of the flocculation process significantly influences the formation of flakes and, accordingly, the water purification.

3.3 Development of the Technology

It is proposed to implement the technology of reagent waste water afterpurification as follows. Waste water after biochemical treatment is forwarded into the tank for further purification by coagulation. The opacifier at the rate of 0.5–3.0 kg/m^3 of waste water is added to the tank when it is filled with waste water. Waste water with an opacifier is mixed for 5–10 min. Depending on the level of water pollution, the required amount of coagulant (150–300 g/m^3) is added. The obtained mixture with the coagulant is mixed for 2–3 min. As a coagulant, it is proposed to use aluminum sulfate, which provides the most complete purification of waste water from bacteria. Then, the anionic flocculant is added at the rate of 2–5 g/m^3 of waste water. The mixture is mixed for 1–2 min again. For more convenient and reliable use of coagulant and flocculant, their aqueous solutions of 1.0 and 0.1% should be used. Upon completion of the coagulation process, the water should stand for 30 min to remove sediment. After that, the waste water is passed through a coal or coke filter and can be used to quench hot coke. The use of afterpurified water for quenching coke will significantly reduce emissions of harmful chemicals into the atmosphere. This technology will improve the state of the environment not only of the coke plant, but also of the town where the plant is located. The economic effect of the introduction of the waste water afterpurification results in reduction of the losses of the enterprise by fever emissions of pollutants getting into the atmosphere. The losses can be approximately reduced by 70% and the annual cost savings for the coke plant are over 4.3 million euros.

4 Conclusions

1. The proposed technology of reagent waste water afterpurification involves the use waste of coking coal beneficiation as an opacifier which is relatively efficient, affordable and cheap.
2. It is proposed to use anionic flocculants because they are most effective during waste water afterpurification after biochemical treatment. It was found that their amount in the technological process is insignificant (2–5 mg/l).
3. It was found that the use of a mixture of coagulants $Al_2(SO_4)_3 \cdot 18H_2O$ (50–70%) and $FeCl_3$ (30–50%) carries out the most effective purification of contaminated water from activated sludge and colloidal suspension. The use of these mixtures of coagulants, as well as the use of the recommended opacifiers enables to completely purify waste water from activated sludge and colloidal suspension. There is also a reduction of harmful impurities by 10–25%.

References

1. Yue, J., Zhao, Y., Sheng, Y., Cao, H., Wen, H.: Simultaneous optimization of structure and operation for coking wastewater biological treatment process. Ind. Eng. Chem. Res. **59**(11), 5022–5031 (2020)
2. Crini, G., Lichtfouse, E.: Advantages and disadvantages of techniques used for wastewater treatment. Environ. Chem. Lett. **17**(1), 145–155 (2019)
3. Sueviriyapan, N., Suriyapraphadilok, U., Siemanond, K., Quaglia, A., Gani, R.: Industrial wastewater treatment network based on recycling and rerouting strategies for retrofit design schemes. J. Clean. Prod. **111**, 231–252 (2016)
4. Henze, M., Harremoes, P., Jansen, C., Arwin, E.: Waste Water Treatment: Biological and Chemical Processes, 4th edn. Springer, Heidelberg (2002)
5. Grady, C.P.L., et al.: Biological Wastewater Treatment, 3rd edn. Taylor and Francis Group LLC, Abingdon (2011)
6. Bingxiong, L., Huang, S., Grossmann, I.: Optimal synthesis and operation of wastewater treatment process with dynamic influent. Ind. Eng. Chem. Res. **56**(30), 8663–8676 (2017)
7. Filatov, Y., Ilyshov, M., Tamko, V., Zbykovskyy, Y.: The method of reagent waste water treatment. Patent No. 87177. Bul. No.2, Ukraine (2014)

8. Filatov, Y., Kolomychenko, O., Zolotariov, I., Bazov, S., Ilyshov, M., Tamko, V., Zbykovskyy, Y., Ivashenko, O.: The method of reagent waste water treatment. Patent No. 105878. Bul. No.12, Ukraine (2014)
9. Wang, Y.-P., Smith, R.: Design of distributed effluent treatment systems. Chem. Eng. Sci. **49**, 3127–3145 (1994)
10. Tchobanoglous, G., Burton, F.L.: Wastewater Engineering: Treatment, Disposal, Reuse, 3rd edn. McGraw-Hill Inc., New York (1991)
11. Padrón-Páez, J.I., Almaraz, S.D.-L., Román-Martínez, A.: Sustainable wastewater treatment plants design through multiobjective optimization. Comput. Chem. Eng. **140**, 106850 (2020)
12. Snoeyink, V.L., Jenkins, D.: Water Chemistry, 2nd edn. Wiley, New York (1988)
13. Christ, C.: Production Integrated Environmental Protection and Waste Management in the Chemical Industry. Wiley, Weinheim (1999)
14. Zbykovskyy, Y., Sievriukov, A.: Improvement of reagent purification technology of coke production sewage. Environ. Protect. Ration. Use Nat. Resour. **2**, 86–88 (2014)

Ecotoxicity of Plastic in the Marine Environment - Contributions to an Environmental Impact Model

Iasmin Godinho[1], Lizete Heleno[1], and Nelson Oliveira[2](✉)

[1] School of Technology and Management, Polytechnic of Leiria, Leiria, Portugal
[2] Laboratory of Separation and Reaction Engineering-Laboratory of Catalysis and Materials (LSRE-LCM), School of Technology and Management (ESTG), Polytechnic Institute of Leiria, 2411-901 Leiria, Portugal
nelson.oliveira@ipleiria.pt

Abstract. The pollution caused by plastic materials is a social and environmental challenge whose consequences are yet to be understood and measured. Regarding the pollution of the marine environment, one can notice that the concentration of chemicals in plastic debris affects the general health of the aquatic ecosystem and are not well understood due to the complexity of the microplastic phenomenon. More efforts are needed to design new plastic materials using ecodesign tools, like Life Cycle Assessment (LCA) and it is used to analyze quantitively the life cycle of a product or service. The inventory is classified and characterized on the phase of Life Cycle Impact Assessment (LCIA), using an impact assessment method. It wasn't found in the literature an environmental impact assessment model to analyze and compare the toxicity of plastics in the marine water and it is essential for a proper ecodesign of new plastic materials and products. In this work, it is proposed an adaptation of one of the impact assessment method specific for ecotoxicity, USEtox. This method was adapted to compare two different polymers, Polyethylene (PE) and Polydimethylsiloxane (PDMS). The preliminary results showed that the Potentially Affected Fraction of species integrated throughout the time and the volume by unity of mass of PE emitted is about 6 times superior to the fraction of PDMS.

Keywords: Plastic · Impact assessment method · Life cycle · Toxicity · Marine environment

1 Introduction

Pollution caused by plastic is a social and environmental challenge with consequences that are still not understood and immeasurable. The society consumption choices and their priorities will determine whether this pollution will exceed any limit of the Earth system or if it will be interrupted and mitigated [1]. Thus, the imminence of changes in socioeconomic and environmental standards led to an increase in the discussion about the circularity of plastic, as its functionality in the modern world is a reality.

J. R. da Costa Sanches Galvão et al. (Eds.): *Proceedings of the 1st International Conference on Water Energy Food and Sustainability (ICoWEFS 2021)*, pp. 933–942, 2021.
https://doi.org/10.1007/978-3-030-75315-3_98

In order to replace conventional plastic, new alternatives have been applied, the use of biodegradable or biocompatible compounds, for example. However, the effects that these compounds can have on the environment are not yet clear. Plastic produced from biodegradable compounds has higher impact rates related to acidification and eutrophication, compared to conventional plastic [2]. The difficulty of replacing common plastic is evident.

In addition to the barriers associated with excessive consumption and manufacturing, the inefficient management of plastic waste is identified as an obstacle experienced in cities around the world. The plastic waste enters the ocean via inland waterways, wastewater outflows, and transport by wind or tides [3]. Thus, a local problem becomes an irreversible global impact, since it is impossible to completely remove plastic debris in the marine environment [1].

The use of plastic, therefore, is concentrated on externalities related to three areas: degradation of natural systems as a result of leaks, especially in the ocean; greenhouse gas emissions resulting from post-use production and incineration; and impacts on health and the environment from potentially toxic substances [4].

It is understood that the concentrations of chemicals in plastic waste affect the health of aquatic life and are poorly understood [5, 6] due to the complexity of the microplastics phenomenon. Therefore, understanding how plastic and its variations interact with contaminants, for example, is essential to assess the impact caused to the environment [7], enabling decision making that favors the preservation of aquatic life.

In view of this, it is necessary to determine the current status of the processes, as well as to evaluate the environmental aspects and impacts involved in the ecodesign and production of plastic and its disposal so that one can have reference in adapting and monitoring improvements [8]. For this, there are several tools that allow the in-depth evaluation of a product or service, such as Materials, Energy and Toxicity (MET) matrix, Checklists Ecodesign, Life Cycle Assessment (LCA), Life Cycle Design Strategies, Recycling Checklists, Life Cycle GHG Accounting, Life Cycle Inventory, Life Cycle Costing, Material Flow Analysis, etc.

The LCA is one of the tools most used since it has the particularity of identifying potential trade-offs that occur in the stages of the life cycle and, consequently, offers the possibility of optimizing the overall environmental performance of a product or service [9]. Thus, LCA is a standardized methodology with pre-established steps that make it possible to quantify and analyze environmental impacts along the production, use and end-of-life chain [10]. The result of a LCA is associated with the choice of the impact method that will guide the stage of Life Cycle Impact Assessment (LCIA). This is because each of the methods follows a modeling with different premises, characteristics and environmental database.

In this context, this work has the main goal to contribute to the development of a methodology for LCIA of plastic, considering the environmental aspects and impacts involved mainly in the process of development and disposal of this product. The developed methodology is based in the modification of the USEtox model, a scientific model endorsed by the United Nations Environment Program/Society of Environmental Toxicology and Chemistry (UNEP/SETAC) Life Cycle Initiative for characterizing human and ecotoxicological impacts of chemicals. To validate the methodology, it is applied to two plastic materials: Polyethylene (PE) and Polydimethylsiloxane (PDMS). PE is the most common plastic used today, hydrophobic and it is produced from fossil sources. PDMS has been widely used in the biomedical industry due to its biocompatibility, its biomechanical behavior similar to biological tissues and its hydrophobicity, but it can be treated to transform its surface more hydrophilic.

2 Plastic Pollution

Plastic is associated with environmental issues, mainly because it is produced from a fossil source, a finite natural resource and for being a material with a very long degradation time. It is therefore relevant to identify the paths of plastic waste to assess the potential impacts on the environment.

2.1 Plastics in the Marine Environment

One of the consequences of poor waste management is the directing of plastic waste to the ocean. The proportion of plastic produced that reaches the ocean depends on the effectiveness of the reuse, recycling and waste management chain [11].

Regarding the classification of plastic, despite the variety of sizes in which it can be found in the ocean, the main classifications are macroplastic, microplastic and nanoplastic. Macroplastics are larger than 5 mm, while microplastic is the term used generically to characterize small plastic fragments smaller than 5 mm. Microplastics are susceptible to fragmentation due to the weak interactions of their polymer chains. Thus, small particles are formed in orders of magnitude of the nanometer, micrometer and millimeter, with nanoplastic particles smaller than 100 nm [11].

Macroplastics are large pieces of waste, such as fishing nets, buoys and packaging that sink on the coastlines or swirl around the huge circular current systems of the Pacific, Atlantic and Indian oceans [12]. While aquaculture and the illegal disposal of merchant ships are some examples of sources from aquatic activities, the use of disposable packaging and spilling from industry productions are examples from terrestrial activities.

Regarding the origin of microplastics, it is understood that there are two main sources associated. The first is also related to the aquatic or terrestrial activities, such as fishery, aquaculture, tires, abrasives, paints, antifouling, pellet spillage, wastewater treatment, washing synthetic clothes, illegal landfill, municipal and industrial plastic litter and sewage sludge reuse in agriculture.

The secondary source of formation of microplastics is caused by the degradation of plastics, since plastic has a tendency to degrade and lose its initial properties over time depending on the physical, chemical and biological conditions to which it is exposed. The main types of degradation are [13]: (a) Biodegradation - action of living organisms, usually microbes; (b) Photodegradation - action of light (usually sunlight in outdoor exhibitions); (c) Thermo-oxidative degradation - slow oxidative decomposition at moderate temperatures; (d) Thermal degradation - action at high temperatures (it is not an environmental degradation mechanism); and (e) Hydrolysis - reaction with water. It is noteworthy that the rate of degradation can vary according to the type of polymer, presence of chemical additives, availability of oxygen to the system and environmental temperature [13].

With regard to impacts on the marine environment, it appears that they are both physical, entanglement, suffocation and ingestion [14], and chemical, such as the bioaccumulation of the chemical components of plastic or chemicals adsorbed by plastic from seawater [5]. Plastic fragmentation causes an increase in the surface area exposed to chemical pollutants. Thus, these pollutants are adsorbed mainly to microplastics until they are ingested by animals or accumulated in marine sediment, characterizing the environmental impact that needs further research, since the chemical dynamics of bioaccumulation are not clear [15].

2.2 Marine Plastic Waste and Chemicals

The substances that are present in marine plastic waste can be classified as chemical additives added to plastics in their production and the hydrophobic substances adsorbed by the surfaces of the fragments due to the hydrophobicity of the plastic surfaces [5].

Additives are chemical compounds added to polymers to improve performance during modeling, functionality and/or degradation properties [16]. The main chemical additives are plasticizers, heat stabilizers, UV stabilizers, flame/fire retardants, lubricants, antioxidants, fillers, colorants/pigments, antistatic, cross-linking agents, blowing agents, impact modifiers, processing aids, flavoring and biocides [17].

Regarding the adsorption of hydrophobic chemicals, each polymer, which has different surface structures, has different affinities for pollutants and thus can serve as carriers of toxic products to the environment [18]. As an example of these hydrophobic products, identified as potential environmental impacts, are Organochlorine Pesticides, Polychlorinated Biphenyls (PCBs), Hexachlorocyclohexanes (HCHs) and Polycyclic Aromatic Hydrocarbons (PAHs) [5, 19].

3 Environmental Impact Model for Plastic in Marine Water

The development of an environmental impact model for plastic in water assumes that most of the plastic debris end up in the marine environment and the environmental impact is related with its toxicity in water. There are several impact models to study the toxicity of chemical substances in water, such as USEtox and Recipe. In this work, the developed methodology is based in the main principles of the USEtox since it is encouraged by the UNEP/SETAC Life Cycle Initiative and is the model recommended by the International Reference Life Cycle Data (ILCD) [20]. While the Recipe method is not officially guaranteed by an international authority.

USEtox is an environmental modeling tool, at the midpoint level, capable of determining the characterization factors (CF) that represent the potential damage whose substance under analysis can cause in a compartment (air, soil or water) [20].

Regarding ecotoxicity, which quantifies the factors that characterize a contaminant, a calculation is carried out in three stages: determination of the fate factor (FF), exposure (XF) and the effects (EF) of chemicals emitted to the environment [20]. Thus, the combination of these factors is represented by the CF calculated by the Eq. 1.

$$\mathrm{CF} = \mathrm{FF} \times \mathrm{XF} \times \mathrm{EF} \tag{1}$$

Where **CF**, characterization factor of substance x (Potentially Affected Fraction of species (PAF) integrated over volume (m^3) and the duration of 1 day (d) per kg emission, $PAF.m^3.d/kg$), **FF**, fate factor (days), **XF**, ecosystem exposure factor (dimensionless) and **EF**, freshwater aquatic ecosystem toxicity effect factor ($PAF\ m^3.kg^{-1}$) [20].

By associating the characterization factors with the mass of the emitted contaminant, the Impact Index (IS) is determined, which addresses the entire mixture of chemicals resulting from all chemical emissions related to a product or service, throughout the entire cycle of life [20].

Ecosystem Exposure Factor (XF)

With regard to the calculation of XF, it is based mainly on sorption-diffusion mechanisms, intermolecular interactions of the substance between water, organic phase and dissolved suspended solids. In a sorption mechanism there is adsorption and absorption, and the adsorption to the surface will depend mostly on the relationship between the specific surface area and the particle size [22].

Nanoplastics can have a greater impact than microplastics due to their dimensions and colloidal properties, since they can cross biological barriers and become potentially dangerous for living organisms. In other words, they are much smaller than microplastics, but with a sorption capacity that can be much larger than microplastics [22].

Considering the different sizes in which plastic can be found in the marine environment, it is necessary to take into account the contaminated plastic residues that can function as suspended solids that absorb substances and release others. In addition, they give rise to biofilms by superficial adsorption on plastic, which when ingested by biota, also cause substance transfer, in addition to the problems caused by the plastic itself.

In this context, it is proposed that the calculation of XF be carried out by Eq. 2.

$$XF_x = \frac{1}{1 + K_{susp,x} \times \frac{C_{susp,x}}{1000} + K_{doc,x} \times \frac{C_{doc,x}}{1000} + BAF_{fish,x} \times \frac{C_{biota}}{1000}} \quad (2)$$

Where:

XF_x: ecosystem exposure factor of microplastic x in water; $K_{doc,x}$: partition coefficient of contaminants between nanoplastic (colloidal, up to 1000 nm) and water (L/kg); $K_{susp,x}$: partition coefficient of contaminating substances between suspended microplastics (above 1000 nm and less than 5 mm) in water (L/kg); C_{doc}: concentration of nanoplastic (colloidal) in water (9.03×10^{-8} kg/m^3 for marine water [22, 23]; C_{susp}: concentration of microplastics in water (2.71×10^{-07} kg/m^3 for marine water [22, 23]); C_{biota}: concentration of substance associated with biota (0.001 kg/m^3); BAF_{fish}: plastic bioaccumulation factor in water (L/kg); x: microplastic x.

The variables K_{DOC}, K_{susp} and BAF_{fish} in the USEtox model can be determined from the value of K_{ow} (n-octanol/water partition coefficient), but in order to compare the types of plastic materials with the ability to adsorb hydrophobic organic substances, it is suggested that these variables have taking into account the coefficient of polymer-marine water partition (log $K_{plastic}$). $K_{plastic}$ is justified by considering that plastic represents a potential risk of contamination to the marine environment, since it can carry toxic hydrophobic substances. In addition to being a toxic substance in its own right, it will be evaluated later by calculating the EF factor.

It is suggested that the partition coefficient applied in the context of plastic be calculated by the geometric average of log $K_{plastic,\,n}$ per family of compounds (PAH, HCH and CB).

$$\log K_{plástico} = \sqrt[n]{\prod_{n} \log K_{plástico,n}} \tag{3}$$

Where log $K_{plastic,n}$ per family of compounds is the geometric average of log $K_{plastic,i}$ of each substance i within the family of compounds.

$$\log K_{plástico,n} = \sqrt[i]{\prod_{i} \log K_{plástico,i}} \tag{4}$$

The Table 1 presents the results obtained for the polymer-marine water partition coefficient [22].

Table 1. Polymer-marine water partition coefficient of hydrophobic organic substances [22].

Family of compounds	log K_{PDMSsw}	log K_{PEsw}
PAHs	4.747	6.047
HCHs	2.311	2.209
CBs	4.701	4.916
Geometric mean	**3.722**	**4.035**

The partition coefficients for PDMS and PE were determined for 3 families of hydrophobic organic substances - PAHs, HCHs and CBs, where log K_{PDMSsw} is 3.722 and log K_{PEsw} is 4.035 [22].

It should be noted that in relation to hydrophilic substances, it is assumed that they have low sorption to hydrophobic plastic and the USEtox model is assumed as usually. Considering the hydrophilic plastic, it is considered that the degradation in the marine environment is very high and the metabolites and reactive mechanisms should be considered but out of scope of this study. The plastic additives are also not considered in this study since it is not expected to affect the sorption capacity of the substances in the microplastics [24].

Ecosystem Toxicity Effect Factor (EF)

The EF calculation is essentially based on the effective concentration of a substance that reduces 50% of an exposed population (EC_{50}) [20]. The population considered in the USEtox is at least 3 species from 3 different trophic levels. The average of the EC_{50} logarithm for each species per trophic levels is the avlogEC_{50}, the dangerous affecting concentration (HC_{50}) [21]. However, only experimental data for species of crustaceans (invertebrates) were found [22]. Thus, considering the experimental data, the avlogEC_{50} is calculated with the geometric average of EC_{50} for each species. EF is calculated by Eq. 5 [22]:

$$EF = \frac{0.5}{avlogEC_{50}} \tag{5}$$

Fate Factor (FF)
The determination of the FF is associated with the calculation of the parameters of the matrix K, which mainly takes into account the data of the test substance, degradation processes, mass balances between compartments and the default data for the study scenario, available in USEtox and EPI Suite estimation program interface. The degradation rate in water, sediments and soil ($K_{deg,W}$, $K_{deg,Sd}$, $K_{deg,Sl}$) were estimated using EPI Suite tool, considering standard average molecular weight (<1000 Da) and standard vapor pressure [22]. However, more studies need to be carried out in order to obtain more representative experimental data, since polymers usually have a molecular weight greater than 1000 Da, vapor pressure less than 0.01 bar and Henry's Law Constant below 10^{-8} bar.m^3/mol.

4 Results and Discussions

The Characterization Factors obtained for PE and PDMS, through the application of the proposed modified USEtox model are presented in Table 2.

Table 2. Result of Ecotoxicity for PE and PDMS

Factors	Units	Results	
		PDMS	PE
XF	–	1.00	1.14x10^{-2}
FF	kg.kg^{-1}.day^{-1}	39.5	39.5
EF	PAF.m^3.kg^{-1}	2.71 × 10^{-2}	15.4
CF	**PAF.m^3.day/kg$_{emitted}$**	**1.07**	**6.93**

Although these are preliminary results, it appears that the Potentially Affected Fraction of species integrated over time and volume per unit mass of PE emitted is about 6.5 times higher than the Fraction of PDMS. This result was expected, since it was expected a higher impact for PE due to the biocompatible characteristic of PDMS when compared with PE.

FF results are similar for both polymers. EPI Suite, being a tool that is based on group contribution, wasn't designed for polymers, mainly about characteristics of the polymers such as the average molecular weight and vapor pressure. The result of FF has some limitations due to lack of experimental data or more adequate estimation tool.

In the scope of the polymers, one of the limitations of EPI Suite is the molecular weight. The maximum value suggested by EPI Suite is 478.94 Da for PE and a more realistic polymer molecular weight is about 14 000 Da (1000 backbone carbons), about 29.2 times more. Analyzing the effect of the molecular weight of PE, the FF result increases only 3% in relation to the FF exposed by Table 2. That is, the method is robust in relation to changes in molecular weight, since a high variation in molecular weight has a low effect. Analyzing and adjusting the vapor pressure data to values in the range of 10^{-5} Pa, there is also a very minor change in the FF result, about 2% in the PDMS and without change in the PE.

Regarding the rates of degradation in water, the results show equivalent for both polymers, since in the absence of experimental data, the data estimated from the USEtox default were used. However, there is an uncertainty regarding the data, since each polymer has different characteristics of durability and fragmentation capacity.

As for EF, it was found that experimental EC_{50} data is needed for different trophic levels, since data relating only to invertebrates were used. Despite this, it was found that the toxicity potential of PE was 568 times greater than that of PDMS, a result that needs to be explored through representative experimental data, however, it demonstrates that PE has a higher toxicity potential in relation to PDMS, as expected. However, they have BAF and consequently XF, with close values, representing that the capacity of the two polymers to be absorbed by the organisms, including by ingestion, are correlated, but the potential to be contaminated by PE is much higher (according to EC_{50}).

In this context, the obtained results show that USEtox has the potential to be modified so that it is possible to assess marine ecotoxicity in the context of plastic. It should also be noted that the development of the incorporation of ecotoxicological impacts on seawater into the USEtox structure is already foreseen, but more experimental data are needed for effect and fate factors [22, 25], as well as the development of effect factors that quantify the effect of plastic waste on marine biodiversity [9].

5 Conclusions

The environmental damage associated with the production-consumption-disposal system is the result of a linear economic model that is unsustainable. The importance of the paradigm shift is verified in order to reduce the annual cost of negative externalities associated with the consumption of plastic, while valuing the understanding of the way in which this material and its variations interact with contaminants. Thus, it is possible that the real impact caused to the environment is evaluated, corroborating the decision making that favors the preservation and mitigation of damages.

The present work contributed to the comparison of two types of plastics usually used in the industry due to their characteristics and applicability. The evaluation consisted to modify USEtox method, also considering its methodological guidelines, to the data of the substances PE and PDMS. The results shows that the CF of PE is about 6.5 times higher than the CF of PMDS. In other words, PE is represented by a higher marine ecotoxic potential than PMDS. However, the results show the demand for experimental data mainly regarding exposure and effect factors, in addition to more representative degradation rates due to limitations of the EPI Suite tool.

There was a limitation in the availability of data in the literature. It is understood that in view of the high durability characteristic of plastic, associated with the common fragmentation of this material (in micro and nanoplastic), the importance of considering degradation is verified. Since it is possible to evaluate the influence of the resistance time of the plastic to the marine compartment and to understand the degradation products, in order to reduce or manage potential environmental impact.

References

1. Villarrubia-Gómez, P., Cornell, S.E., Fabres, J.: Marine plastic pollution as a planetary boundary threat – the drifting piece in the sustainability puzzle. Mar. Policy **96**, 213–220 (2018)
2. Gironi, F., Piemonte, V.: Bioplastics and petroleum-based plastics: strengths and weaknesses. Energy Sour. Part A: Recov. Utilization Environ. Eff. **33**(21), 1949–1959 (2011)
3. Jambeck, J.R., Geyer, R., Wilcox, C., Siegler, T.R., Perryman, M., Andrady, A., Law, K.L.: Plastic waste inputs from land into the ocean. Science **347**(6223), 768–771 (2015)
4. Ellen MacArthur Foundation: The New Plastics Economy: Rethinking the future of plastics. Ellen MacArthur Foundation (2016)
5. Teuten, E.L., Saquing, J.M., Knappe, D.R.U., Barlaz, M.A., Jonsson, S., Björn, A., Takada, H.: Transport and release of chemicals from plastics to the environment and to wildlife. Philos. Trans. R. Soc. B: Biol. Sci. **364**(1526), 2027–2045 (2009)
6. Wagner, M., Lambert, S.: Freshwater microplastics: emerging environmental contaminants? Springer (2018)
7. Hüffer, T., Weniger, A.K., Hofmann, T.: Sorption of organic compounds by aged polystyrene microplastic particles. Environ. Pollut. **236**, 218–225 (2018)
8. Saidani, M., Yannou, B., Leroy, Y., Cluzel, F., Kendall, A.: A taxonomy of circular economy indicators. J. Clean. Prod. **207**, 542–559 (2019)
9. Woods, J.S., Veltman, K., Huijbregts, M.A.J., Verones, F., Hertwich, E.G.: Towards a meaningful assessment of marine ecological impacts in life cycle assessment. Environ. Int. **89**, 48–61 (2016)
10. Venkatachalam, V., Spierling, S., Endres, H.-J., Siebert-Raths, A.: Integrating life cycle assessment and eco-design strategies for a sustainable production of bio-based plastics. In: Designing Sustainable Technologies, Products and Policies. Springer (2018)
11. Kershaw, P.: Sources, fate and effects of microplastics in the marine environment: a global assessment. International Maritime Organization (2015)
12. Katsnelson, A.: News feature: microplastics present pollution puzzle. In: Proceedings of the National Academy of Sciences of the United States of America, vol. 112, no. 18, pp. 5547–5549 (2015)
13. Thompson, R.C.: Microplastics in the marine environment: sources, consequences and solutions. In: Marine Anthropogenic Litter. Springer (2015)
14. Casagrande, N.M.: Inclusão dos impactos dos resíduos plásticos no ambiente marinho em avaliação de ciclo de vida. UFSC (2018)
15. Horton, A.A., Walton, A., Spurgeon, D.J., Lahive, E., Svendsen, C.: Microplastics in freshwater and terrestrial environments: evaluating the current understanding to identify the knowledge gaps and future research priorities. Sci. Total Environ. **586**, 127–141 (2017)
16. Hahladakis, J.N., Velis, C.A., Weber, R., Iacovidou, E., Purnell, P.: An overview of chemical overview of chemical additives present in plastics: migration, release, fate and environmental impact during their use, disposal and recycling. J. Hazard. Mater. **344**, 179–199 (2018)
17. Piatti, T.M., Rodrigues, R.A.F.: Plásticos: características, usos, produção e impactos ambientais. Maceió: Edufal, 51 (2005). https://www.usinaciencia.ufal.br/multimidia/livros-digitais-cadernos-tematicos/Plasticos_caracteristicas_usos_producao_e_impactos_ambientais.pdf
18. Mato, Y., Isobe, T., Takada, H., Kanehiro, H., Ohtake, C., Kaminuma, T.: Plastic resin pellets as a transport medium for toxic chemicals in the marine environment. Environ. Sci. Technol. **35**(2), 318–324 (2001)
19. Ziccardi, L.M., Edgington, A., Hentz, K., Kulacki, K.J., Kane Driscoll, S.: Microplastics as vectors for bioaccumulation of hydrophobic organic chemicals in the marine environment: a state-of-the-science review. Environ. Toxicol. Chem. **35**(7), 1667–1676 (2016)

20. Fantke, P., Bijster, M., Guignard, C., Hauschild, M.Z., Huijbregts, M.A.J., Jolliet, O.: USEtox 2.0: Documentation (Version 1) (2017)
21. Payet, J.: Assessing toxic impacts on aquatic ecosystems in LCA. Int. J. Life Cycle Assess. **10**(5), 373 (2005)
22. Godinho, I.F.C.: Ecotoxicidade do plástico em ambiente marinho – contributos para um modelo de impacte ambiental, Politécnico de Leiria (2021)
23. Everaert, G., Van Cauwenberghe, L., De Rijcke, M., Koelmans, A.A., Mees, J., Vandegehuchte, M., Janssen, C.: R: Risk assessment of microplastics in the ocean: modelling approach and first conclusions. Environ. Pollut. **242**, 1930–1938 (2018)
24. Mei, W., Chen, G., Bao, J., Song, M., Li, Y., Luo, C.: Interactions between microplastics and organic compounds in aquatic environments: a mini review. Sci. Total Environ. **736**, 139472 (2020)
25. Fantke, P., Aurisano, N., Bare, J., Backhaus, T., Bulle, C., Chapman, P.M., Hauschild, M.: Toward harmonizing ecotoxicity characterization in life cycle impact assessment. Environ. Toxicol. Chem. **37**(12), 2955–2971 (2018)

Author Index

J. R. da Costa Sanches Galvão et al. (Eds.): *Proceedings of the 1st International Conference on Water Energy Food and Sustainability (ICoWEFS 2021)*, pp. 943–947, 2021.
https://doi.org/10.1007/978-3-030-75315-3